AIR POLLUTION, THE AUTOMOBILE, AND PUBLIC HEALTH

Ann Y. Watson, Sc.D.
Richard R. Bates, M.D.
Donald Kennedy, Ph.D.
Editors

Sponsored by
THE HEALTH EFFECTS INSTITUTE
Cambridge, Massachusetts

NATIONAL ACADEMY PRESS
Washington, D.C.   1988

National Academy Press • 2101 Constitution Avenue, NW • Washington, DC 20418

Air Pollution, the Automobile, and Public Health was sponsored by the Health Effects Institute (HEI), Cambridge, Massachusetts. HEI, established in 1980, is an independent nonprofit corporation structured to define, select, support, and review research that is aimed at investigating the possible health effects of motor vehicle emissions.

HEI's annual operating budget is contributed equally by the U.S. Environmental Protection Agency and domestic and foreign manufacturers of motor vehicles. None of these contributors has any control over the conduct or conclusions of HEI studies. HEI's funds are controlled, within the limits of its charter, solely by its Board of Directors. The board operates independently, and is responsible for all HEI activities, including a scientific mission carried out by two autonomous committees, the Health Research Committee and the Health Review Committee. The board and the committees are supported in carrying out their day-to-day operations by HEI's executive director and a scientific and administrative staff.

HEI makes no recommendations on regulatory and social policy, but seeks rather to gain acceptance by all parties of the data that may be necessary for future regulations.

The Health Effects Institute

Foreword

WILLIAM D. RUCKELSHAUS

The job of protecting Americans from involuntarily imposed environmental risks forms a fundamental challenge for the rest of this century. The public simultaneously fears the health effects of potentially toxic chemicals, while being justly skeptical about our ability to understand and differentiate the true risks of these chemicals. As a society, we have spent billions to control pollution, but we have sometimes done so in a reactive way, responding to the pressures generated by the latest bit of evidence. The result of all this is that our progress has been impressive, but uneven. We have tight controls on risks that may be very small, but remain nearly in the dark about other risks that may constitute much larger public health threats.

I am excited about this volume because it represents, in several ways, how I believe our society should respond to the challenge of tailoring our investments to meet the real environmental threats that remain. Fundamentally, the Health Effects Institute has enlisted some of our best environmental scientists to think hard about how we can quantify the risks as well as narrow the uncertainty about risk in relation to the automobile industry, an industry that is at the heart of our national economy. In so doing, the Institute recognizes that a critical part of our ability to deal credibly with the public about risk resides in improving the scientific base underlying risk assessment. Perhaps as important, the book is not just a set of research recommendations. Its authors recommend a strategic concept within which to view mobile source research. In so doing, they provide a valuable service to decision makers who are constantly faced with deciding how to rationally invest scarce research dollars. The task of deciding how much to invest in science relevant to next year's standard, and how much to buttress our overall scientific base is a daunting one. This book gives concrete and expert guidance about how to make that choice.

My excitement about the science must always be tempered by the knowledge that I am not a scientist myself. I do have considerable experience, however, in understanding how important the institutional context is to believable decisions. And, on that score, I am delighted that the source of this book is the Health Effects Institute. For several years now, the Health Effects Institute has demonstrated that it is possible to combine public and private resources to do credible, objective science that will be accepted by parties that have often fought tooth and nail about environmental policy. We must have more of these enterprises if we are to fashion environmental policy that protects the public health, while not tearing apart the societal fabric we need if we are to restore our economic competitiveness in world markets. That this book was done under the auspices of the Health Effects Institute gives me confidence not only in the wisdom of the work recommended, but in our ability to marshall our society's resources to deal effectively with its environmental problems.

The combination of scientific and institutional integrity represented by this book is unusual. It should be a model for future endeavors to help quantify environmental risk as a basis for good decision making.

Contents

AIR POLLUTION, THE AUTOMOBILE, AND PUBLIC HEALTH

Part I: Overview

Automotive Emissions Research

DONALD KENNEDY
Stanford University

THOMAS P. GRUMBLY
Health Effects Institute

We must not forget the policy objective behind auto emission regulations—to protect the public health. Short-term tactical battles over standards cloud much more fundamental issues—namely, will reduction of auto pollution actually improve ambient air quality significantly, and will better air actually improve public health? These questions have not yet been answered satisfactorily.

Rep. David Stockman
Harvard University
October 19, 1978

Emissions Control: The History of Conflict

The United States has had laws to control air pollution for the past 30 years. Since the passage of amendments to the Clean Air Act in 1970, our society has made an heroic effort to reduce automotive emissions, as one cornerstone of an overall air pollution control policy. And as a result of these laws, we have, in fact, seen major reductions in several pollutants. Some of the

Air Pollution, the Automobile, and Public Health. © 1988 by the Health Effects Institute. National Academy Press, Washington, D.C.

benefits of control are obvious: in parts of the country, the air is noticeably cleaner; and with recent improvements in many inspection and maintenance programs, fewer cars will present gross problems that lead to emissions beyond current standards.

The costs of this control have also been great—and not only in terms of dollars expended by taxpayers, stockholders, and car owners. Throughout the 1970s, our efforts to make meaningful emission reductions and to ensure that these reductions were carried out provided difficult tests for some in our society. For a time, perceived attempts by some car companies to resist regulation brought the industry into disrepute with much of the public. In addition, the Clean Air Act amendments raised expectations that have not yet wholly been achieved.

With the advantage of nearly two decades of hindsight, it seems indisputable that the 1970 law was an important exercise in symbolic politics. The car companies had permitted themselves, one could argue, to be cast as forces of evil. The Congress, responding to the burgeoning environmental movement so well symbolized in the Earth Day of that year, seized the high moral and political ground as a way to transform a policy situation that had be-

3

come increasingly unsatisfactory. The Congress overwhelmingly enacted standards and mandated pollutant reductions that would protect the public and provide, in the words of the statute, an "adequate margin of safety," for all people—including particularly sensitive populations. The Congress further empowered the president, now acting through the new Environmental Protection Agency, to take strong enforcement action against companies and states that did not comply with tough timetables for action. The struggle to meet these new standards in a timely way led, at various times during the 1970s, to threats by major corporations to shut down, and to governmental plans for state compliance that seemed to many as exercises in futility. In the eyes of many citizens, the reputations of government as well as industry suffered during this period.

Benefits of Regulation

Have the dollars, antagonisms, and lost credibility been worth the benefits our society has derived from mobile source regulation? Should we continue to control a major sector of the economy in the same way? Should we continue to explore alternative methods of regulation, such as emission fees, that have always proven politically unacceptable in the past? The answers will only be found by assessing the benefits we have already achieved, and in looking at what might be achieved in the years ahead. We need to see whether we are achieving the fundamental purposes of the Clean Air Act. Are we adequately protecting the public health? In particular, are we meeting the special needs of the old, the young, and those already compromised by disease processes—or by other biological processes that we are only just beginning to understand?

In the early days of program implementation, the answers to these questions seemed easy. Anyone familiar with the famous London fog of 1952 that ultimately killed 4,000 people, or with the "incident" at Donora, Pennsylvania, in which another 48 people died, knew that air pollution at high levels causes disastrous health problems. We also knew that automobiles were significant contributors to the air pollution problem because of seminal work done in the Los Angeles basin. On any hot day in Washington in the late 1960s, the Senate Environment and Public Works Committee knew without asking that automotive emissions just could not be good.

The initial results of control were visible. Air pollution alerts decreased (independent of any changes in how these alerts were defined), and auto emissions came down. Hydrocarbons, carbon monoxide, and oxides of nitrogen have all been significantly reduced. After much struggle, we have also made considerable progress in reducing lead emissions from autos. In noteworthy contrast, we have not sought to control heavy-duty truck and bus emissions until more recently because of energy concerns. Accordingly, many of us still complain bitterly about these emissions, especially if we are behind a vehicle that is not well maintained.

Current Problems for Motor Vehicles

The air is cleaner. Not surprisingly, as progress was made, public pressure about motor vehicle emissions declined. Because of the determined efforts of many scientists and environmentalists, the leaded gasoline issue has retained a high profile, and the government took strong action during 1985. Ozone control is still a vexing problem, from both a health and an environmental perspective, and no strategy for control is without substantial practical problems for industry or state and local governments. Inspection and maintenance, and associated tampering, also appear on the public agenda from time to time, and we have not yet fully explored the health ramifications of the diesel engine.

For the most part, however, other air pollution and environmental issues have displaced mobile sources at the top of our environmental policy agenda. Problems of hazardous waste, radon, acid rain, and pesticides now seem more pressing. This is no sign of failure, but of success, despite the presence of a revisionist movement that castigates the efforts at emissions control as a case of regulatory overkill.

But, to paraphrase the famous question from another context, "What do we know, and when did we know it?" A variety of commentators concerned about control policy, regardless of their political persuasion, quickly dispose of the science of the health effects of auto emissions, citing the inherent problems of dealing with uncertainty in assessing benefits. Science in this, as in some other policy arenas, often seems like the poor cousin—one to be quickly introduced and just as quickly sent away, lest the rest of the family be embarrassed. This treatment is often meted out by lawyers, economists, and various other professionals who somehow assume that the rest of the risk assessment/risk management equation is more certain.

The hard fact still is, of course, that we don't know how successful our air pollution and motor vehicle emission control policies are from a public health perspective. Stockman's question at the opening of this preface is still the relevant one. We have reduced acute health problems through our regulation, but we cannot be sure how much. We know that the policies have had effect, but we cannot assume that biological damage reduction is proportionate to pollution reduction, particularly as we learn more about biological systems. And all of our modes of analysis, whether they involve experimentation with people or animals, epidemiology, or in vitro laboratory work, are still imperfect. Given the comparative expenditures in control technology and relevant environmental health research, however, it is surprising that we know so much about the health benefits of air pollution control. There is obviously nothing we can do about past policy, but a great deal that we can do about policy for the future.

Importance of Continuing Health Research

The question inevitably arises about the value of additional expenditures to find out whether auto emissions and other air pollutants are, in fact, risks worth worrying about. Even if we take it for granted that current emission control regulations have reduced the most serious acute health effects problems, why is research still important? The answers flow directly from the social context as we just described it.

Regulations: Form or Substance?

If it is true that the structure and some of the imperatives of the current Clean Air Act sprang significantly from an exercise in "symbolic politics," that is, a political situation in which positive societal action required the construction of a stage on which good and evil played clearly defined roles, then it is now important to see whether rational inquiry conducted in a less-heated time supports the symbols. This is important not only as an exercise in political science and evaluation, but as a step in rebuilding a base of public trust. We Americans are just now emerging from two decades of internal conflict. Perhaps we now have the luxury of reexamining the choices we made in more turbulent times. And, as a matter of the continuing credibility of the government, the need for intellectual integrity alone justifies a continuing examination of the reasons we have undertaken a cleanup costing billions of dollars. We need to see if our past choices were right—or just expedient.

Safety Margins and Susceptible Groups

The Clean Air Act is quite specific in setting out the need for adequate margins of safety in establishing standards. The debate in the Congress, and subsequent reports from the relevant congressional committees, also makes it clear that these safety margins need to take into account so-called susceptible populations. Neither in the statute nor in the legislative history is the concept of susceptible populations adequately described. The courts and policy makers are left to their own devices in interpreting the term. From a standard-setting process, then, identifying and dealing with groups of people who may be either more susceptible to disease from any given dose of pollutant, or who begin to exhibit "health effects" at levels below those seen in the general population, are critical. Indeed, it can be said to drive the regulatory system with respect to criteria pollutants, and it can only be dealt with through continuing research.

Complicating the problem, of course, is that the current "revolution in biology" was not envisioned by the framers of the amendment. Accordingly, we are developing a range of techniques that can show, even at the subcellular level, the effect of a particular dose of toxic pollutant. We do not, however, yet know how to translate much of this knowledge into terms that can help us quantify risk. But to leave the science at its current stage of development, or simply to leave to chance its capacity to generate information that can help regulation, would be to hinder the regulator in an already difficult task. Accordingly, this commandment to seek out and deal with susceptible populations forms a fundamental scientific driving force in the implementation of the Clean Air Act.

Costs and Benefits

From a practical perspective, it is important to understand whether additional investments by taxpayers and consumers are necessary to further reduce emissions of the so-called criteria pollutants. Many economists (including those with biases toward environmental regulation) now believe that we are in a condition of declining marginal benefits for each dollar invested, and that we have been in this condition for some time. If that is right, then our society should be thinking hard about reallocating motor vehicle emission control dollars to other problems in risk reduction where the return will be greater. The only practical way to approach this issue is through research investments designed explicitly to quantify the risks to human health from pollutants at levels that approximate current use, or that are likely to exist without additional control in the next decade. That this research investment has not already been made is undoubtedly a function of several things. These include the absence of

the scientific tools with which to make major progress, and—to put the matter plainly—the reluctance of government and industry to move the issue away from symbolism and into reality. Fortunately, the tools are being developed, and as we describe below, the attitudes of many former combatants seem to be changing for the better.

We would not argue, of course, that uncertainty about risk can ever be eliminated through science. To say this, however, is different from saying that we cannot reduce uncertainty into the range in which rational policy makers could have more confidence in their control and resource decisions.

Emergence of a New Regulatory Structure in Mobile Source Emission Regulation

The 1970 Clean Air Act amendments directed special attention to a particular group of pollutants, the so-called "criteria pollutants." Given the technology of 1970, this was arguably appropriate. Atmospheric chemists and biologists have known for some time, however, that our approach to understanding and controlling air pollutants, including motor vehicle emissions, is too simple. Not only are many more products actually emitted than are currently controlled, but atmospheric and biological reactions multiply significantly the numbers and the complexity of the ultimate pollutants that hit the "target" organ. We say this not to downplay the need to control "criteria" pollutants, but to emphasize the importance of using those parts of the law that focus on noncriteria or "unregulated" pollutants. We live in a world of complex chemical mixtures, and new technology or fuels further complicate that mixture.

For the Environmental Protection Agency, these new facts may mean that we are about to enter an era in which the EPA begins to look more like its older regulatory cousin, the Food and Drug Administration. Because of what seemed at the time in 1977 to be rather minor amendments to the Clean Air Act, the EPA and the manufacturers now have the responsibility to determine whether new emissions or the technologies that produce them will result in an "unreasonable risk" to the public health. The EPA will find itself more and more in the classic regulatory mode of premarket approval in an era in which quantitative risk assessment provides the standard of evidence. The EPA will only be able to quantify the risk of new mixtures or new technology in the presence of considerable health research information. It will be able to sustain its decisions only if such research is credible within the wider scientific and public policy community.

Research and Comprehensive Risk Management

The assessment of risk through the application of science will never be, nor should it be, the sole criterion for decision making. However, it seems clear that future corporate leaders and regulatory managers will increasingly look at issues of comparative risk in deciding how to deploy limited resources. Just as important, it seems likely that we are also entering an era in which we are beginning to look at the risk of a substance from each of its sources before making control decisions, so that we maximize control in a cost-effective way and do not transfer risks from one media to another. In the motor vehicle arena, this means looking closely at all the sources of nitrogen oxides, carbon monoxide, particulates, formaldehyde, benzene, and other chemicals before implementing additional controls. The techniques to do these analyses are still lacking, and research is required particularly in the critical risk assessment arena.

Regulatory Credibility and the Maturation of Toxicology

In the past several years, we have witnessed a determined attack on what some have called the "new social regulation," including many areas of environmental protection. Indeed, it can be argued that a president of the United States was elected in part because of the success of these arguments.

This attack goes well beyond arguments concerning the costs and benefits of any particular regulation, and to the core of whether much of the government's regulatory behavior was inherently just. Science per se cannot deal with this latter question, but science can narrow the argument. Over time and with effort, science can substantially reduce the uncertainties in extrapolating from animal and laboratory data to human experience. Ironically, then, science can change the debate from a technocratic one to one in which we really do grapple with the fundamental question of "how much control and regulation do we want as a society?"

Our ability to understand and manipulate basic genetic material offers the possibility of radically improving toxicology. It does this by providing ways in which we can compare the results of traditional experiments with results by some of the new methods. That does not mean that traditional toxicology will be less important. Indeed, we will need more chronic testing in order to maximize the utility of some of the newer methods such as "computer modeling." To take advantage of new opportunities will require, however, the long-term effort to bridge historic gaps between the basic and the applied biological sciences. It requires ways to bring together scientists of a variety of disciplines under mission-oriented banners. In short, it will require money as well as the right institutional structures.

The Health Effects Institute and the Social Context

This preface has briefly concentrated on some of the antagonisms, concerns, and possibilities that surround the science and regulation of auto emissions. The concern about credibility and the search for an institutional structure that could marry the basic and applied sciences formed the fundamental bases for establishing the Health Effects Institute, the institutional framework from which this book has developed.

In 1979, it had become apparent that

health research related to mobile source emissions needed to be organized in a different fashion. There was simply too strong an adversarial tone; government-sponsored research, especially if it was done within the regulatory agency, was mistrusted on the grounds that it was aimed at regulatory outcomes, whereas industry-sponsored research was seen as directed toward the exoneration of its own products. Fortunately, realization arrived almost simultaneously at the Environmental Protection Agency and at the automotive industry. A group of motor vehicle and engine manufacturers and Douglas Costle, then Administrator of the EPA, asked Archibald Cox, Donald Kennedy, and William Baker to be the founding Directors of a new organization called the Health Effects Institute (HEI). It came into being in 1981 under the Executive Directorship of Charles Powers, and its research program was launched during 1983.

The basic idea behind the HEI was simple enough. It was to solve two problems: first, public mistrust of the sponsorship of research on the health effects of mobile source emissions; and second, the lack of involvement of the best academic scientists in that kind of research. Clearly the two problems are related. Most scientists want to work on interesting questions, but—all other things being equal—would prefer that the answers yield useful outcomes. But, where controversy and suspicion surround the sponsorship and evaluation of research, the best scientists will exercise their option to busy themselves elsewhere. The HEI sought to change this situation by clearing the sponsorship of suspicion and finding ways to reengage the interest of first-rate people in the hope that others would follow.

The basic concept, then, was that the HEI would draw research resources in equal measure from the regulatory agency and from regulated industry. It would then assemble groups of scientific leaders who would elicit proposals from good research organizations and make the awards. In that way, public confidence in the objectivity of the work could be established and maintained. At the same time, the continuing

availability of resources and credible sponsorship would build interest and confidence within the research community, so that first-rate scientific groups would be prepared to make long-term commitments to work in the area. That would further reinforce public belief that the science could be trusted; and so on.

The history of the HEI has, so far, borne out these expectations. The first step was to recruit a group of outstanding scientists for the two committees that would do the main work of the Institute: the Research Committee, responsible for the development of research objectives and the evaluation of proposals; and the Review Committee, responsible for scientific peer review of the work done. Under the respective chairmanships of Walter Rosenblith of the Massachusetts Institute of Technology and Robert Levy of Columbia University, these committees have been able to develop rosters of outstanding scientists, retain them, and—above all—engage their attention in a sustained way. Everyone who has served on policy committees knows how difficult this last task is. The committee flies in to the meeting site, having (in many cases) done its homework on the airplane; over 36 hours many problems are discussed and tentative conclusions reached; and then everyone goes home before a permanent memory trace has been established. This mode of scientific committee work, once termed BOG-SAT (the acronym for Bunch of Guys Sitting Around a Table), is fraught with hazards. But the HEI committees have largely been free of these, in large part because of the Institute's insistence on purchasing a greater commitment of time and energy from the committee members.

In its equilibrium state, then, HEI seems to be working. It has an annual budget of approximately $6–7 million and over 40 research projects under contract. Only sixteen have been completed, but that is enough to give us confidence in the processes of review and evaluation. The Institute is constantly being asked to do more, or different, things; and as is usual for any research field, priorities are constantly shifting. It is better not to let the setting of priorities fall entirely to external demand, because often the urgent preempts the important and long range, and potentially comprehensive solutions may be missed.

By 1985 then, the Institute had recognized that its own intellectual capital needed replenishment. Accordingly, we brought together a group of people from government, industry, and academia to write and critique papers that would not only review the relevant fields, but that would also identify approaches and a strategy that would most likely lead to improving our risk assessment capability. Only by improving this capability will we ultimately be able to answer Stockman's opening question: What are the health effects of emissions?

The noted political scientist Aaron Wildavsky has written, "There is no point in having good ideas, if they cannot be carried out." We believe that this book, far from being some "academic" exercise in research planning, represents another step in the social policy context of trying to get a good idea, that is, preventing health problems from air pollution, to work. This book is an important link in helping us determine whether further control is necessary, whether symbolic politics can be made real in our society, and whether the public should believe, as we do, that government intervention, so visibly represented by emissions regulation, can ultimately be justified by the facts.

Project History and Organization

ANN Y. WATSON
Health Effects Institute

Rationale

The Health Effects Institute (HEI) is an independent nonprofit corporation that, according to its charter, is "organized and operated . . . specifically to conduct or support the conduct of, and to evaluate, research and testing relating to the health effects of emissions from motor vehicles." Because resources are always limited, funding decisions need to be made in the context of the most important scientific needs. Formulation of a research strategy that takes into account current problems as well as long-range goals provides a mechanism for resource allocation. Accordingly, the HEI undertook a major endeavor to identify specific research problems within automotive air pollution and toxicology. The Institute recognized the need for a comprehensive, integrated research strategy and committed the resources to develop the necessary intellectual input and to make available this information to the scientific and regulatory communities. The focal point of the project was to identify and explain obstacles that prevent the quantification of risk and to propose future research to solve these problems.

Although the Institute, as the name implies, focuses primarily on health effects, the identification of pollutants and quantification of exposure are essential when de-

termining the dose at which effects may occur. Hence, the Institute chose to evaluate topics relevant to exposure analysis and biological effects. The project had two primary goals: (1) to assess the pattern and extent of population exposure to automotive air pollutants; and (2) to assess the toxicity to humans—including unusually susceptible persons—so that societal risk can be estimated.

Organization

Implementation of this project called for the formation of a parent Steering Committee and two Subcommittees (the members of which are listed at the end of this chapter), one for Exposure Analysis and the other for Biological Effects. The Steering Committee, chaired by Donald Kennedy and composed of members from the HEI's Research and Review Committees and staff, established policies and approved of the plans and reports of the two subcommittees. The Exposure Analysis and Biological Effects Subcommittees, chaired by Robert Sawyer and Gerald Wogan, respectively, included members of the Institute's Research and Review Committees and other scientists with expertise in some of the topics studied in the project. The Subcommittees developed plans for evaluating research needs, selected topics for investigation, recruited specialists to write background documents, and guided their efforts.

Air Pollution, the Automobile, and Public Health. © 1988 by the Health Effects Institute. National Academy Press, Washington, D.C.

It was not feasible to commission papers on every aspect of automotive emissions and their potential health effects. One criterion used for selection was the topic's relevance to the goal of the project and the information available. Because inhalation is the primary route of exposure and the lung represents the first line of defense, asthma, cancer, chronic obstructive pulmonary disease, and respiratory infection were chosen for study. Consideration of other physiological systems or health outcomes, however, was not as clear. For example, atherosclerosis was evaluated because of its prevalence in society and its association with cigarette smoke, another air pollutant; however, birth defects or kidney disorders, also very important conditions, were not studied because their connection to automotive emissions has been less clearly established. Available expertise was used as another criterion for determining which topics to include. For example, the investigation of neurotoxicology uses behavioral as well as biological approaches. The identification of a single author equally familiar with both of these approaches was difficult. The decision was made to explore the relationship between behavioral effects and automotive emissions as this relationship may relate to issues of the quality of life. Finally, an effort was also made not to duplicate ongoing Institute studies. For example, carbon monoxide was not chosen as a topic for evaluation because the Institute is currently conducting a multicenter study of this pollutant.

The development of each paper was followed by one member of the appropriate subcommittee and by either Richard Bates or Ann Watson. Early in the project, autumn 1985, a meeting was held in Dearborn, Michigan, at which the authors presented outlines of their chapters to the Steering Committee and Subcommittees, the scientific staff of the Institute, the sponsors of the Institute, and to each other. After brief presentations, comments were invited from the audience. The authors then prepared initial manuscript drafts, which were submitted for external peer review (colleagues serving as peer reviewers are acknowledged at the end of this chapter) and internal review by subcommittee members. Suggestions from reviewers and subcommittee members were transmitted to the authors for their use in the preparation of their final manuscripts.

Messages To and From the Authors

In order to produce a series of focused papers, a set of guidelines was distributed to the authors. They were asked to concentrate on two questions: (1) What do we need to know to be able to quantify levels of risk? and (2) How do we get the necessary information? Papers were to be written as concise and critical reviews, to identify gaps in our knowledge, and to propose research directions necessary to fill those gaps or resolve controversies. These papers were not to be comprehensive reviews, and, much to the concern of some of the authors, extensive bibliographies were discouraged. The authors were encouraged, however, to evaluate creatively the most important and feasible research opportunities in their areas of expertise.

The authors were also asked to write in a style that would be appropriate for a broad readership. It was the Institute's desire to make this information available to an audience that included scientists in fields other than the authors', policymakers, and informed citizens of public interest groups, among others. It was not easy for specialists to step back and realize that what had become his or her first language was often beyond the grasp of many—even fellow scientists in different fields. It was sometimes a struggle to strike a balance between insulting the experts and informing the interested persons.

Within these guidelines, the authors had the freedom to present their own interpretation of the problems and, most importantly, their solutions. "Solutions" are presented in the form of recommendations of specific research projects. Recommendations and their rationale are described in the text and are ranked in priority by the authors at the end of each chapter.

An integrated response to the authors' recommendations is provided in the following chapter, entitled "Motor Vehicle Emissions: A Strategy for Quantifying Risk." This chapter attempts to put together a more overall approach to their recommendations. Surprisingly, even though a wide range of disciplines were represented in this project, several consistent themes emerged and are highlighted in the overview.

The authors have responded to our request to share their knowledge. Now it is up to the scientific and regulatory communities, among the ultimate consumers of this project, to use this information.

Acknowledgments

The realization of such an endeavor from inception to published form required the participation of numerous, too often underthanked, individuals: Richard Bates initiated this project, and then convinced the Institute of the necessity of defining future directions in research in order that the proper scientific data base could be obtained for quantification of risk; Linda Buchin assisted in the massive task of administration; Ellen Williams helped to translate technical expertise into discussions appropriate for the intended audience; Richard Maurer transformed hand-drawn sketches and computer printouts into illustrations and advised on book design; and Virgi Hepner coordinated and proofed the final stages of production. The Executive Director of HEI, Thomas Grumbly, supported the project enthusiastically from beginning to end, and provided the necessary resources. From the National Academy Press, Virginia Martin recognized the potential contribution of this project, and she was instrumental in encouraging the Press to publish a work unrelated to Academy reports. In addition, numerous individuals at the Press worked with and guided us through the intricacies of the publication process. Finally, the Institute is indebted to the authors for their enthusiasm and cooperation.

Steering Committee

Exposure Analysis Subcommittee

Biological Effects Subcommittee

Gerald N. Wogan
(Chairman) Massachusetts Institute of Technology

David V. Bates
University of British Columbia

Richard R. Bates
Health Effects Institute

Earl P. Benditt
University of Washington

Joseph D. Brain
Harvard University School of Public Health

Curtis C. Harris
National Cancer Institute

Donald E. McMillan
University of Arkansas Medical School

Sheldon D. Murphy
University of Washington

Mark J. Utell
University of Rochester School of Medicine

Ann Y. Watson
Health Effects Institute

Peer Reviewers

Zoltan Annau
The Johns Hopkins University

Anne P. Autor
University of British Columbia

John C. Bailar III
Harvard University School of Public Health

Göran Bondjers
University of Göteborg
Göteborg, Sweden

Jack G. Calvert
National Center for Atmospheric Research

Julius S. Chang
National Center for Atmospheric Research

David P. Chock
General Motors Research Laboratories

Steven D. Colome
University of California, Irvine

Ramzi S. Cotran
Harvard Medical School

James D. Crapo
Duke University Medical Center

Jack H. Dean
Chemical Industry Institute of Toxicology

Kenneth L. Demerjian
State University of New York, Albany

Donald L. Dungworth
University of California, Davis

Bruce A. Egan
Environmental Research and Technology, Inc.

Ludwig A. Engel
Westmead Hospital
Sydney, Australia

Hugh L. Evans
New York University Medical Center

James E. Fish
Jefferson Medical College

Inge F. Goldstein
Columbia University School of Public Health

Daniel Grosjean
Daniel Grosjean and Associates, Inc.

Joachim Heyder
Gesellschaft für Strahlen-und Umweltforschung mbH Frankfurt, Federal Republic of Germany

Millicent Higgins
National Institutes of Health

John R. Hoidal
University of Tennessee Center for the Health Sciences

George J. Jakab
The Johns Hopkins University School of Hygiene and Public Health

Fred F. Kadlubar
National Center for Toxicological Research

Samuel S. Lestz
Pennsylvania State University

Richard B. Mailman
University of North Carolina School of Medicine

Kevin T. Morgan
Chemical Industry Institute of Toxicology

Paul Nettesheim
National Institute for Environmental Health Sciences

Günter Oberdörster
University of Rochester

Wayne R. Ott
U.S. Environmental Protection Agency

Robert F. Phalen
University of California, Irvine

Charles G. Plopper
University of California, Davis

Gerald M. Saidel
Case Western Reserve University

Peter W. Scherer
University of Pennsylvania

Dennis Schuetzle
Ford Motor Company

Bernd Seifert
Institute for Water, Soil, and Air Hygiene, Federal Health Office Berlin, Federal Republic of Germany

John H. Seinfeld
California Institute of Technology

Dean Sheppard
University of California, San Francisco

Thomas J. Slaga
The University of Texas System Cancer Center

James A. Swenberg
Chemical Industry Institute of Toxicology

Ira Tager
University of California, San Francisco

Thomas R. Tephly
University of Iowa

Lance Wallace
Harvard University School of Public Health

Peter A. Ward
University of Michigan

David Warshawsky
University of Cincinnati Medical Center

I. Bernard Weinstein
Columbia University

George T. Wolff
General Motors Research

Motor Vehicle Emissions: A Strategy for Quantifying Risk

RICHARD R. BATES
ANN Y. WATSON
Health Effects Institute

Air Pollution, the Automobile, and Public Health. © 1988 by the Health Effects
Institute. National Academy Press, Washington, D.C.

The ultimate objective of a regulatory-oriented research program that focuses on air pollution from mobile sources is to identify and quantify any effects that those emissions may have on human health. But before we invest intellectual and financial resources, we must first understand the limitations of current information and methodologies that preclude accurate estimates of risk to human health. Future research programs should be justified by their promise to overcome these limitations. The goal of this volume, then, is to identify issues and select a research agenda that will be most effective in advancing our ability to quantify the health risks associated with air pollution.

To understand risk, it is essential to understand the relation between the amount—or dose—of pollutants, and the response, in terms of human illness, to those pollutants (National Research Council 1983). Yet the obstacles that challenge our understanding of that relation are formidable. Biologists ask, "To what substances are people exposed, and what dose levels are most relevant for experimental studies?" Engineers and physical chemists inquire, "Which of the hundreds of compounds and chemical reactions are biologically important?" This volume alone cannot answer these questions, but its discussions are critical to beginning a dialogue among the various scientific communities.

The first part of this book addresses research about the exposure of humans to vehicular emissions and their reaction products. The first three chapters examine the nature of automotive emissions (Johnson), their chemical transformation (Atkinson), and the mechanisms of their transport through the atmosphere (Samson). Subsequent chapters explore the use of ambient measurements (Graedel) and mathematical models (Russell) as ways of describing pollutant concentrations, which in turn are key descriptors of human exposure (Sexton and Ryan). The remaining chapters discuss the principles and mechanisms by which, during exposure, particulate (Schlesinger; Sun, Bond, and Dahl) or gaseous (Overton and Miller; Ultman) pollutants are inhaled and distributed throughout the respiratory tract.

In the second half of this book, the chapters examine whether specific biological responses to airborne pollutants constitute a health hazard. The problem is approached from two complementary points of view. One view starts with the chemical constituents of automotive emissions and their transformation products and investigates their effects on human health. In this approach, oxidants (Bresnitz and Rest), polycyclic aromatic hydrocarbons (PAHs) and nitro-PAHs (Hecht), and alcohols and aldehydes (Marnett) are discussed.

In contrast, the second view begins with a set of human diseases and investigates whether automotive emissions or their derivatives play a role in disease development. Attention is given to health effects of obvious relevance to airborne contaminant exposure. Thus, asthma (Bromberg), respiratory infection (Pennington), cancer (Kaufman), fibrosis (Last), and emphysema and small airways disease (Wright) are considered. Coronary heart disease (Clarkson) is examined because cigarette smoke increases mortality, and it is not known if other combustion products affect morbidity and mortality rates. Neurobehavioral effects (Wood) of hazardous substances, which may not be an obvious health effect, also warrant investigation. One's perception of, and response to, the environment may be just as significant for the "health" of society as clinical presentations of organic disease.

Thus, this book examines the existing evidence about risks from automotive emissions and evaluates the methodologies presently available to quantify human risks. However, not all possibilities are explored. Areas for which the evidence was lacking or the rationale for associating automotive emission exposure with health risks was not compelling were omitted; examples include the role of automotive-derived pollutants in the etiology of reproductive disorders or renal dysfunction. By the same token, we did not want to duplicate previous analyses; for example, numerous pollutants are described in the U.S. Environmental Protection Agency (EPA) Criteria Documents. Unfortunately, even though human exposure is not to a single material or even to a uniform mixture, we were unable to properly explore the issue of complex mixtures. Many of the chapters stress the necessity of acquiring better information about atmospheric constituents and quantitative markers of human disorders before a research protocol can be formulated to test, within the scope of limited resources, the bioactivity of such mixtures.

Although the primary focus of this book is automotive emissions—a focus fully justified by the omnipresence of motor vehicles in our society—the scientific principles underlying many of the discussions are applicable to other air pollutants and even to other sources of exposure. First, auto emissions and their derivatives are not the only source, or even the predominant source, of human exposure to many of these chemical substances. Second, air pollutants, whether they originate from automobiles, power plants, or wood-burning stoves, are all governed by similiar principles of transport through the atmosphere and into the body. Finally, the research needed to identify people who are susceptible to emphysema, pulmonary fibrosis, or lung cancer is similar whether the causative agents derive from auto emissions, occupational exposures, or cigarettes.

The primary purpose of this overview chapter is threefold: (1) to describe the issues and difficulties involved in providing reliable scientific data for risk assessment; (2) to highlight and integrate the authors' recommendations into the issues raised; and (3) to propose a research strategy. We start with putting into perspective the automobile as a source of potentially harmful substances. We next describe components that determine risk of injury and discuss the strengths and weaknesses of various research approaches. We then summarize the authors' choices of future research directions. Finally, after having carefully read the authors' suggestions, we conclude with an attempt to translate their individual priorities into a feasible, cross-disciplinary research strategy.

Automotive Emissions: a Brief Perspective

Motor vehicles contribute significantly to emission inventories in certain regions of the United States, particularly in urban areas. Table 1 shows the relative contribution of various emission sources to the total inventory of five pollutants for the country as a whole and for two metropolitan areas. Typically, in urban areas, motor vehicle emissions dominate carbon monoxide inventories and can contribute up to 50 percent or more of nitrogen oxides and hydrocarbons. It is, however, important to remember that these figures reflect outdoor pollutant concentrations. Most people in industrialized societies spend more time indoors than outdoors, and the contribution of motor vehicles to pollution of indoor air has not been adequately characterized.

Two additional points are worth noting. First, volatile organic compounds and total suspended particulates are broad categories that include many individual compounds of divergent toxicologic importance. Thus, table 1 may not adequately portray the relative impact of these emission sources on human health. Second, comparable data are not available for unregulated emissions which include many compounds of possible toxicologic interest. Therefore, even though motor vehicles emit numerous chemical species into the environment, we cannot determine, with current data bases, the relative contribution of mobile sources to the health risks of our society.

Understanding the Components of Risk

Pollutants and their derivatives cause harm by interacting with, and impairing, molecules crucial to the biochemical or physiological processes of the human body. Figure 1 illustrates the pathways from pollutant sources to toxic effects. Clearly, it is difficult to link a specific pollutant to a particular ill effect until the processes in between are understood. The risk of toxic injury from a substance depends on three factors: the chemical and physical properties of the substance, the dose of the material that reaches critical tissue sites, and the responsiveness of these biological sites to the substance. The relationship of each

Table 1. Relative Contribution of Various Sources to Emissions of Regulated Pollutants

Source	Carbon Monoxide	Nitrogen Oxides	Volatile Organic Compounds	Total Suspended Particulates	Sulfur Oxides
United States: 1984[a] 10^6 tons/year (%)					
Transportation[b]	48.5 (69)	8.7 (44)	7.2 (33)	1.3 (19)	0.9 (4)
Fuel combustion[c]	8.3 (12)	10.1 (51)	2.6 (12)	2.0 (29)	17.3 (81)
Industrial process	4.9 (7)	0.6 (3)	8.4 (39)	2.5 (36)	3.1 (15)
Solid waste disposal	1.9 (3)	0.1 (1)	0.6 (3)	0.3 (4)	0.0 (0)
Other	6.3 (9)	0.2 (1)	2.7 (13)	0.9 (13)	0.0 (0)
Metropolitan Los Angeles: 1981[d] 10^3 tons/year (%)					
Transportation[b]	3,859 (92)	399 (62)	406 (44)	431 (59)	57 (26)
Fuel combustion[c]	45 (1)	193 (30)	27 (3)	20 (3)	64 (29)
Industrial process	110 (3)	48 (7)	223 (24)	31 (4)	97 (44)
Solid waste disposal	116 (3)	2 (1)	39 (4)	16 (2)	0 (0)
Other	76 (2)	1 (1)	239 (26)	235 (32)	0 (0)
New Jersey–New York–Connecticut: 1981[d] 10^3 tons/year (%)					
Transportation[b]	4,248 (94)	349 (58)	397 (38)	361 (53)	50 (12)
Fuel combustion[c]	91 (2)	229 (38)	26 (2)	42 (6)	315 (75)
Industrial process	26 (1)	9 (2)	276 (27)	194 (29)	47 (11)
Solid waste disposal	125 (3)	10 (2)	31 (3)	41 (6)	7 (2)
Other	24 (1)	1 (1)	311 (30)	39 (6)	0 (0)

[a] From U.S. Environmental Protection Agency (1986).
[b] Includes motor vehicles, rail, aircraft, vessels.
[c] Includes stationary sources such as residential, electric generation, industrial, commercial-institutional.
[d] From U.S. Environmental Protection Agency (1984).

component to the scheme presented in figure 1 is discussed below.

Chemical and Physical Properties of Pollutants

Exhaust from the tailpipe of motor vehicles is a complex mixture that contains hundreds of chemicals in the form of gases as well as solid and liquid aerosols. The composition of the mixture depends on the fuel, the type and operating conditions of the engine, and the effects of any emission control devices. Upon their release, emitted substances are transformed by complex atmospheric chemical reactions. Airborne pollutants, therefore, consist of primary tailpipe emissions (for example, carbon monoxide, nitric oxide), and new chemical species formed as a result of atmospheric reactions (for example, nitrogen dioxide,

ozone); the opportunity for chemical diversity is immense. In addition, the chemical composition is dynamic; the air we breathe today is different than the air we breathed 10 years ago. Therefore, a changing constituency must be recognized in any evaluation of chronic health effects.

The physical form of airborne contaminants will influence their distribution both in the atmosphere and in biological tissues. As emissions cool, some vapors are adsorbed onto particles or condensed into droplets, and fine particles and droplets coalesce into large ones. Upon inhalation, the regional tissue distribution of gases or aerosols depends on such physical properties as size and solubility. In addition, the deposition, metabolism, and clearance of volatile organic compounds are dramatically altered if they are adsorbed onto respirable particles. Eventual dose, there-

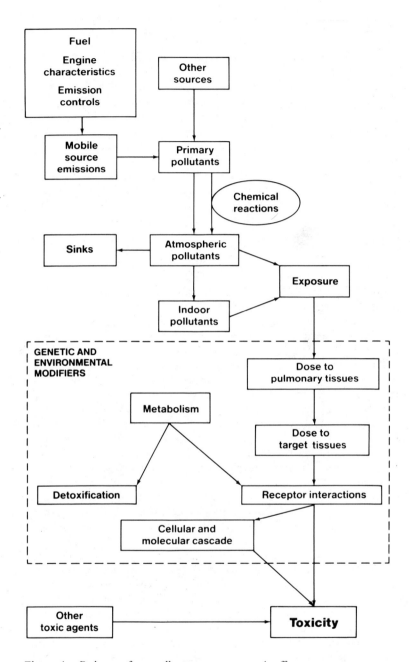

Figure 1. Pathways from pollutant sources to toxic effects.

fore, is intimately associated with the physical properties of the airborne contaminants.

Because chemical structure and physical characteristics are important determinants of toxicity, an improved understanding of these properties is essential. For example,

slight modifications of functional groups on PAHs can markedly alter the mutagenic potential. Nasal, but not bronchial, tumors have been found in rats after prolonged exposure to formaldehyde, which, because of its solubility, is absorbed in the nasal cavity. In addition, some compounds have similar properties but different biological effects. Ozone and nitrogen dioxide are oxidants, but only nitrogen dioxide produces emphysema in laboratory animals. Other as-yet-unidentified factors must influence outcome. Information about structure and toxicity is available for some classes of closely related materials that have been studied extensively. For most substances in the environment, however, data necessary for quantitative prediction of toxicity based on chemical or physical properties are inadequate or unavailable.

Pollutant Exposure and Dose to Biological Tissues

Pollutants derived from vehicular exhaust are transported away from their sites of release by wind and diffusion. Once airborne, they mix both with pollutants from other sources and with materials of natural origin, and they may be chemically transformed into other species. Ultimately, atmospheric pollutants are removed from the air either by rain or by dry deposition to the earth's surface. Highly reactive chemicals may be transformed or removed within a few minutes; stable substances may persist for years. Meteorological conditions and physical structures can also profoundly influence atmospheric concentrations. Thus, individual exposure is determined by the location a person occupies in relation to emission sources, as well as by patterns of atmospheric transport, transformation, and dilution.

Airborne pollutants that are inhaled may deposit onto surfaces of the respiratory tract. Deposited insoluble material is moved (either intra- or extracellularly) toward the pharynx by mucociliary action and then swallowed. Alternatively, particles may be sequestered for long periods within pulmonary tissue or in adjacent lymph nodes. Inhaled chemicals that dissolve in body fluids may pass from the respiratory tract into the bloodstream and circulate throughout the body. As a result, air pollutants may affect extrapulmonary organs.

Pollutants may be chemically transformed within the body by metabolic enzymes. The liver is particularly active in the metabolism of foreign chemicals, but the lung and other organs also have this capacity. In general, metabolism facilitates excretion of pollutants from the body and thus reduces pollutant levels in body tissues. In addition the toxic potential of some parent compounds may be reduced by metabolic conversion. Paradoxically, though, metabolism may also generate products with increased toxicity. The balance between metabolic processes that increase toxicity, decrease toxicity, or favor elimination is an important factor in the sensitivity of an individual to a toxic chemical.

The definition of "dose" may vary widely. It may be based on the concentrations of inhaled pollutants at any point from their deposition on respiratory tract surfaces, to the concentrations of reactive material at "target sites," where damage occurs. The latter is the most valuable definition of dose and the most difficult to determine. Dose-to-target measurements require identification of the sites of damage, understanding the mechanisms by which the toxic material produces damage, and knowledge of the reactive materials responsible for toxicity. Obtaining such information from laboratory animals or tissue cultures is not trivial and is seldom feasible in living human beings.

Instead, for most practical purposes, we must rely on some surrogate measurement of dose-to-target sites. This may be the dose to some other, but more accessible body tissues, such as the blood or respiratory tract. Alternatively, it may be a measure of exposure, or even of atmospheric concentration. The more removed the surrogate measurement is from the target site in humans, the more the approximation is confounded by intervening environmental and biological variables. Nevertheless, surrogate measurements may be fully adequate for some applications. In other cases, mathematical models of the relationships between exposure and dose can extend the applicability of surrogate measures. Dosimetry models that correlate dose between laboratory animals and humans also broaden the use of surrogate measures.

Surrogate measures of dose are commonly used in experimental and epidemiologic studies. Dose/response studies of toxicity in laboratory animals usually measure exposure, not dose. But differences in the respiratory tract anatomy and ventilation of humans and rodents mean that identical exposure may not result in identical dose. Such studies, however, may nevertheless become increasingly important in predicting human health effects as mathematical models capable of supporting interspecies extrapolations are developed; these models aim to relate, in rodents and humans, the physical and chemical properties of pollutants to the sites and amounts of pollutant deposited. Interspecies comparisons of active metabolite dose or extrapulmonary dose-to-target sites will require additional studies of the species-specific pharmacokinetics (absorption, metabolism, distribution, and excretion) of the chemical.

Epidemiologic investigations of the health effects of air pollutants often examine the relationships between the health of people in a community and the concentrations of pollutants at a few stationary monitoring sites within the community. The people studied are not congregated around the monitoring stations but are widely situated, indoors as well as outdoors. Although more localized measurements of subjects' exposure levels would provide better data, such measurements are expensive and, for some pollutants, not technically feasible at this time. Mathematical models of the dispersion and transformation of pollutants from their sources, coupled with a better understanding of time/activity patterns within the community, would provide an alternative approach for making better exposure estimates. Improved accuracy in exposure data for the study population would aid in the application of the conclusions to other populations.

Another area of considerable interest is the development of indices that reflect the actual amount of pollutants that reach target tissues. Although it may not be technically feasible to measure levels of the original substance, quantification of metabolites or reaction products may be possible. The discovery that carcinogenic metabolites covalently bind to macromolecules such as DNA or protein, which in turn can be measured, has stimulated extensive research in this area. It should be noted, however, that the use of such approaches for vehicular emissions is complicated by the fact that many of the pollutants of interest are present in very low concentrations. Therefore, more sensitive analytical techniques for measuring molecular dosimeters are needed than are currently being used.

Biological Responses to Pollutants

The interaction of pollutants with biological molecules (often called receptors) triggers the mechanisms of toxic response. Some responses are direct, the immediate result of the effect on the receptor. For example, some forms of asthma may be precipitated by the direct stimulation of airway irritant receptors by an inhaled pollutant. In contrast, some other manifestations of toxicity are highly indirect and, hence, poorly understood; this is frequently true of chronic or delayed effects of toxicity. In these cases, the initial interaction of the toxic chemical with a target site receptor may trigger a cascade of molecular and cellular events that ultimately damage tissue other than the original target site. Pulmonary fibrosis and emphysema illustrate such indirect manifestations of toxicity. Damage to the pulmonary connective tissue, the hallmark of these diseases, does not result from the direct actions of inhaled toxic pollutants, but rather, it results from a complex, multistep process: pollutant exposure initiates an inflammatory response, which in turn stimulates excessive production of connective tissue or causes the enzymatic digestion of elastic tissue. Through a self-enhancing process, damaged tissue and inflammatory cells release chemical mediators that stimulate the recruitment and proliferation of more inflammatory cells. Products of these cells include highly reactive and toxic oxygen-derived radicals, digestive enzymes, and growth factors.

In addition to the complexity of cellular and molecular events, it is not clear what regulates the balance between normal defense or repair functions and abnormal processes. Components of the inflammatory system, which include macrophages, neutrophils, and complement, are essential factors in the defense against infectious agents and inhaled dusts. These components, in proper balance under biological control mechanisms, kill invading microorganisms and help to prevent or repair tissue damage. Yet under certain pathological circumstances, they produce pulmonary diseases.

Two inferences can be drawn from such complex and poorly understood mechanisms of toxicity. First, a research strategy useful for evaluating exposures—that is, tracing the path of a pollutant from its source to sites of human exposure, and vice versa—cannot be used for many biological problems. We simply know too little about the "middle phase" of most chronic diseases to follow the molecular path from pollutant deposition to aberrant tissue structure or function. The situation is further complicated because a single pollutant may trigger a variety of immediate biochemical and physiological actions as it passes through the body. Many, if not most, are inconsequential, but it is impossible to recognize the few of toxicologic importance without understanding the chronic disease process.

The second implication of the preceding discussion is that a combination of factors, including genetics, nutrition, and other environmental chemicals, influence each of the multiple stages in the development of adverse effects, as well as the interactions of enhancing or inhibiting agents at each stage. These factors undoubtedly determine the sensitivity of any person to the toxic effects of a pollutant. The definition and enumeration of sensitive individuals could be pursued more effectively if the steps between the initial interaction of toxic substances with their target sites and the ultimate manifestations of chronic toxicity were better understood.

Sensitive Individuals

The ability to identify population subgroups particularly vulnerable to health effects induced or exacerbated by hazardous substances is a critical aspect in the assessment of human exposure to automotive emissions. In particular, the young or the elderly may be especially susceptible to deleterious effects; persons with asthma may experience aggravated symptoms upon exposure.

We currently cannot determine whether differential responses in sensitive individuals result from a greater delivered dose or from inherent biological differences. Improved measurements of exposure and dose will aid in the identification of people who exhibit greater responses primarily because they receive more of the pollutant. The challenge, however, is to identify those individuals who are inherently more prone to disease states. Most likely, the genetic background of individuals contributes significantly to their biological response. The multiple cellular and biochemical processes set into motion in response to reactive materials are under complex genetic control. Taken together, such processes determine individual sensitivity and, hence, the outcome from exposure to toxic substances.

Research Approaches

The utilization of multiple approaches will most likely enable us to make reasonable predictions about the risk of toxic injury from automotive emissions or their constituents. An effective research strategy will consist of epidemiologic studies, experimental studies, and/or model systems. The advantages and limitations of these approaches are described below.

Epidemiology

Epidemiology represents the most direct approach to demonstrate in humans that exposure to a particular substance results in an increase in the incidence of a specific disease. Limitations in the methodologies that assess exposures and biological effects, however, prevent better use of epidemiological approaches. Although epidemiologic studies are valuable for detecting ex- acerbations of existing chronic disease or the occurrence of acute respiratory disease, their use for revealing causes of chronic respiratory disease is limited. The time lag between exposure to automotive emission products and the appearance of a chronic disease makes it difficult to prove a causal relationship unless early markers of the disease are known. It is highly desirable, therefore, to determine that adducts or other proposed molecular dosimeters are indeed reliable indices of exposure and to identify early markers of chronic disease. People who should be examined in epidemiologic studies include those with well-defined exposures, specific diseases, or inherent factors of sus- ceptibility. Selection of high-risk groups pro- vides a more sensitive basis for the detection of any increase in disease.

Experimental Methods

Clinical. Two important advantages of human clinical studies are controlled expo- sure and data derived from human subjects. The use of exposure chambers permits the delivery of known quantities of pollutants. Construction of dose/response curves de- mands, though, that careful attention be given to pollutant generation and measure- ment. Numerous methods that cover a range of human response, from physiolog- ical to biochemical effects, are available. Traditionally, pulmonary function tests have been used to assess functional capabil- ities of the lung. Some of these tests form the clinical definition of disease, such as asthma or chronic obstructive pulmonary disease. Because pulmonary function tests reflect end points of respiratory malfunc- tion, they have limited value in detecting the onset of chronic disease. A more recent innovation, the fiber-optic bronchoscope, permits access to human tissues by biopsy and to human cells by lavage. Sampling of human tissues will be of central importance to confirm animal studies. Although bron- choscopy can be performed with minimal risk, it is not ideal for large-scale screening; therefore, the development of noninvasive techniques should also be pursued.

Animal. Several issues must be consid- ered for proper implementation of labora- tory investigations. Animal studies should address three components of experimental design: exposure conditions, disease state, and extrapolation to humans. Ideally, ani- mal exposure should mimic conditions of human exposure. Pollutant concentration should be low and administered by inhala- tion. To establish links between acute and chronic effects, wide ranges of exposure regimes are necessary. Since it is not always feasible or practical to conduct chronic, low-dose, inhalation studies, deviations from these parameters should be validated. Sedation of laboratory animals should be avoided, since anesthesia will alter their ventilatory parameters. An improved un- derstanding of the influence of chemical structure on toxicity is necessary to help guide the choice of appropriate mixtures of compounds that should be tested.

Experimentally induced disease should be similar to the human counterpart. In some cases, it may be necessary to develop more appropriate animal models for some forms of toxicity, as well as for susceptible populations. In addition, if relevant cofac- tors exist in human ailments, these must be considered. For example, research on the effects of air pollutants on morbidity or mortality from atherosclerosis cannot ig- nore the role of diet. Likewise, studies on the susceptibility to infection should use infectious organisms that are appropriate models of human pathogens.

Finally, before accurate dose/response relationships in animal studies can be ap- plied to human risk evaluation, it will be necessary to develop reliable extrapolation or scaling factors. This will require more extensive baseline information on the rele- vant anatomy, histology, physiology, and metabolism in the various species.

Cell Culture. Cell cultures can be used for the isolation of metabolites or mediators, as well as for the elucidation of biochemical mechanisms of toxicity. The use of culture systems allows compounds or complex mixtures to be quickly screened for toxic effects. It is essential, however, that results from in vitro studies be related to in vivo outcomes; for example, the generation of DNA adducts in culture must ultimately be correlated with tumor formation in the whole organism.

Mathematical Models

Mathematical models can be used to estimate or predict emission levels, air quality, human exposure, and respiratory tract distribution of inhaled particles and gases. As alluded to above, the primary asset of these mathematical models is their potential to provide a surrogate measure of dose. In addition, models can be used to distinguish individual contributions from various sources to the total pollutant load. Several models have been developed for these purposes. Such models, though, are only as good as the chemical and physical data upon which they are based. There is a need to improve data bases and to perform model validation studies. New models also need to be developed, as well as incorporating additional concepts into current systems; of primary importance is the inclusion of indoor air quality into exposure assessment and source apportionment models. In addition, predictions of noncriteria pollutant concentrations could be improved if better mathematical descriptions of atmospheric chemical reactions were used. With respect to dosimetry models, reactive gas uptake in the upper respiratory tract and particle deposition in regional areas are inadequately developed.

Physical Models

Some research questions are poorly suited to analysis by mathematical modeling methods. In such cases, it is often easier to construct a small-scale physical analogy to the problem of interest, and then test that scale model to learn how the system behaves. Physical models of atmospheric fluid mechanics problems, such as the dilution of pollutants near buildings, can be built and tested in wind tunnels. Atmospheric chemical reactions can be studied in smog chambers that are constructed to replicate the irradiation of pollutant mixtures by the

sun. Physical models of the geometry of the human respiratory tract can be used to study the patterns of aerosol deposition or gas transport in the lung.

Pursuing Quantitative Toxicology: A Strategy for Research

Armed with the knowledge of what is known about the components of risk, and of what research approaches are available, we can focus our attention on how to go about supplying the missing pieces between the point of pollutant emission and the generation of toxic response. Returning to figure 1, we see that the investigator can link chemical substances to toxic biological responses by starting at either end of the puzzle. In other words, the hypothesis can be posed in either of two ways: (1) Can automotive emissions adversely affect human health under conditions of human exposure? (2) Do human ailments or physiological malfunctions result from exposure to automotive emissions?

Numerous factors complicate the design and interpretation of any experiment. For example, pursuit of the first hypothesis can be confounded by the transformation of primary pollutants and the generation of new chemical species; the "molecule of interest" and its concentration are not always readily identifiable. With respect to the second hypothesis, the contribution of other agents capable of inducing disease makes the assignment of attributable risk to automotive emissions more difficult. Such factors cannot be ignored, and any research strategy must take their presence into account.

Regardless of the number of chemicals that might be tested for toxicity or the number of toxic effects that might be sought, the evaluation of health effects lacks a satisfying sense of unity. No satisfactory paradigm provides a coherent explanation of the relationships between chemical structure, dose, and toxic effects. Perhaps such a paradigm could be said to have existed when the principal concern of toxicology was with immediate and readily observable toxic actions. For these, the maxim of the sixteenth century Swiss physician and alchemist, Paracelsus, applied: "In all substances there is a poison, and there is nothing without a poison. It depends only upon the dose whether a poison is poison or not" (Paracelsus 1958). According to this guidance, chemicals could be categorized into groups on the basis of the magnitude of dose required to cause death or some other toxic end point (Klaassen and Doull 1980). Although this precept served a useful purpose, it is insufficient for evaluating health effects of automotive emissions and their derivatives. For health effects that occur with low frequency or low magnitude, which are the more likely result of these materials, the poison is not only in the dose but also in the genetics, environment, and lifestyle of the exposed individual.

An unfortunate consequence of the lack of a unifying theoretical framework for toxicology is the phenomenological approach used for investigating health risks of pollutants. The battery of tests that should be done to assess each important type of health hazard is large and still growing. The task is magnified immensely when the pollutant is as complex and as variable in its composition as automotive emissions and their transformation products. It is impractical to study the effects of each chemical individually; nor could such an approach adequately assess the synergistic and antagonistic influences of the mixture on the toxicity of each component. On the other hand, it is equally impractical to test all possible variants of the complex mixture that would be produced under differing conditions of fuel, engine operation, and climate. Moreover, no single set of conditions can be assumed that might result in a representative exhaust emission mixture, which, in turn, could serve as a toxicity testing standard for experimental work.

This dilemma will most likely exist for some time into the future. In the short term, we can look toward the scientific developments that provide guidance on the most important applications of phenomenological toxicology. Perhaps the development of batteries of inexpensive, short-term tests for the most important toxic hazards will direct attention to a small number of the components most in need of thorough evaluation. Alternatively, such tests might be applied to varied samples of the complete mixture to define subsets with distinctly different toxicological characteristics. Each subset might then be more fully investigated.

It is possible, however, to speculate on directions of research that may eventually lead to improved quantification of human exposure and risk. Well-defined and well-focused investigation can provide some of the necessary information. But the puzzle will not be solved tomorrow, or all at once. We must take it one piece at a time, and if we choose our pieces wisely, the picture can be visualized sooner.

Highlights of the Authors' Recommendations

Within each chapter, authors provide their interpretations of the most important research directions in their fields of expertise. Their recommendations can be placed within the context of the various components of risk of injury, and the distillation of their suggestions are represented in tables 2–5. However, any comprehensive research agenda will be severely limited by resource constraints; major choices will have to be made. Before outlining a strategy for action, we must first outline our criteria for prioritization.

Criteria for Prioritization

To achieve the primary objective of improving our ability to quantify risks, multiple criteria were used to guide our selection of future research efforts. Consideration must be given to the relative importance of the information sought and the feasibility of obtaining it. Importance may be related to the seriousness of the health risk or to the likelihood that it may result from exposures near ambient levels. Importance is also linked to societal needs. Before industry utilizes a new technology

Table 2. Chemical and Physical Properties

Topic	Author(s)
Formation and Transformation of Unregulated Pollutants	
Vehicle exhaust	Johnson
• Formaldehyde: There is a need for real-time concentration data under various driving conditions in methanol-fueled vehicles operated with and without catalysts.	
• Nitro-polycyclic aromatic hydrocarbons (nitro-PAHs): Kinetics of formation in the exhaust system and dilution tunnel for diesel emissions should be determined.	
• Diesel exhaust: Detailed characterization of particulate and gas-phase hydrocarbons is needed.	
• Diesel particulates: Research on particulate control technology is needed. Data on particle size distribution, metal species, adsorbed hydrocarbons, and effects of additives should be collected.	
Atmospheric reactions	Atkinson
• Oxides of nitrogen: Further investigation of the transformations of oxides of nitrogen under atmospheric conditions is needed. This topic is important for indoor environments as well.	
• PAHs and nitro-PAHs: Atmospheric transformation products of PAHs in gaseous as well as adsorbed phases require study. Quantitative information on reaction pathways leading to nitro-PAHs and on removal processes for nitro-PAHs is needed.	
• Aromatic hydrocarbons: The products arising from hydroxyl radical–initiated reactions of the aromatic hydrocarbons should be identified.	
Aerosol Processes	
Data are needed on aerosol formation, particle size distribution, chemical composition, and chemical transformations.	Graedel; Russell
Instrumentation and Analytical Methods	
Real-time measurement methods should be refined in order to more accurately quantify emission constituents.	Johnson
Analytical techniques for the nondestructive, nonintrusive, in situ study of transformation products of gaseous as well as particle-associated chemical species should be improved.	Atkinson

or government agencies formulate regulatory policies, certain scientific knowledge may be desirable. Finally, some pieces of information must be obtained before other ideas can be pursued. For example, the application of dosimetry models requires that crucial input data be representative.

Some research topics are extremely important, but unfortunately, with current knowledge or methodology, their near-term solutions are not very feasible. For example, it is essential that the science of toxicology evolve to the point where complex mixtures can be analyzed for their toxicity. A prerequisite will be the development of quantitative markers associated with the onset of disease. If sufficient knowledge and adequate methodologies are available, then long-range research goals should be pursued.

A Proposed Research Strategy

If we cross-stitch the recommendations from the various authors with the criteria for prioritization, we can begin to chart a course for future research. Two major themes emerge that appear to be at the heart of our objective of generating an adequate scientific data base for risk assessment: (1) an increased emphasis on the quantification of exposure and dose, and (2) an improved understanding of basic disease processes.

Quantification of Exposure and Dose.
The development of molecular dosimeters should constitute a long-range research objective. Currently, the formation of adducts has been exploited the most for our purposes; however, validation of specific adducts and identification of different do-

Table 3. Exposure Characterization

Topic	Author(s)
Air Quality	
Atmospheric evaluation	
• The chemistry of atmospheric aerosol particles should be monitored in more detail. Chemical differences as a function of particle size should be determined.	Graedel
• Routine monitoring of alcohol and aldehyde levels should be done in areas where alcohol-based fuels are or will be in heavy use.	Graedel; Marnett
• Better mathematical descriptions of appropriate chemical reactions for use in models should be made.	Russell
• Improved descriptions of pollutant transport and dispersion are needed for street canyons and other complex situations in which air movement is restricted. Inclusion of chemical reactions within a street canyon model is also needed.	Russell; Samson
• The potential exposure of passengers in closed and open vehicles along roadways should be examined. Research should continue to focus on the role of vehicular turbulence in initial dispersion of exhaust.	Samson
Indoor evaluation	
• Indoor pollutants should be measured and their sources identified.	Graedel; Sexton and Ryan
• Models should be improved for apportioning specific emission sources to individual exposures and for relating outdoor to indoor air quality.	Russell; Sexton and Ryan
Exposure Assessment	
Studies should be undertaken to provide information on the spatial and temporal distributions of human populations as they relate to exposure. To obtain this information, valid and reliable questionnaires should be developed.	Bresnitz and Rest; Sexton and Ryan
Instrumentation	
• Development and use of more suitable instruments for indoor and personal monitoring are needed.	Bresnitz and Rest; Graedel; Sexton and Ryan
• A reliable, sensitive formaldehyde or aldehyde monitor should be developed.	Graedel
Dosimeters	
• Sensitive methods to detect adducts should be developed.	Hecht; Kaufman; Marnett
• Available biological measurement techniques should be adapted to air pollution monitoring. The relationship among exposure, dose, and health outcome requires better understanding.	Sexton and Ryan
Model Validation	
Detailed and accurate sets of input data should be used to assess the adequacy of current air quality and exposure models.	Russell; Sexton and Ryan

simeters associated with other disease processes are needed. To accomplish these goals, more knowledge about the molecular events related to the onset of disease should be obtained. Furthermore, if molecular dosimeters are to serve as an effective link between pollutant exposure and subsequent health effects, the quantitative relationship among exposure, dosimeter levels, and effects will have to be established.

Therefore, until we are able to identify and validate molecular dosimeters, more emphasis should be placed on refining surrogate measures of dose. For this, better characterizations of exposure, as well as improved models for dose assessment are needed. It is becoming increasingly apparent, however, that exposure estimation is not as straightforward as initially supposed and may have uncertainties as great as, or greater than, those associated with dose/response estimates. We need more information about environmental levels of most unregulated pollutants, indoor exposure levels, and how to integrate indoor and outdoor levels into a comprehensive picture of human exposure. This information is also needed to conduct and interpret experimental studies as well as to validate dosimetry models.

Table 4. Dose Assessment

Topic	Author(s)
Expanded Data Bases	
Comparative analyses in normal species	
• Morphometric measurements at all levels of the respiratory tract should be done in adult humans and laboratory animals. Comparisons among and within species should be made and statistical variability determined.	Schlesinger
• More emphasis should be placed on the upper respiratory tract, including the development of dosimetry models for this region.	Overton and Miller; Schlesinger
• Description of the liquid lining of the lung in humans and laboratory animals is needed.	Overton and Miller; Schlesinger
• Data on deposition and clearance kinetics of particles in different species should be gathered.	Schlesinger
• Scaling factors should be developed based on measurements of total uptake of gases in different animal species.	Ultman
High-risk groups	
• Morphometric measurements are needed for sensitive groups such as the young, old, and diseased. New anatomic models should be developed.	Overton and Miller; Schlesinger
• Airflow patterns and distribution could be better described in sensitive subgroups in laboratory animals and humans.	Schlesinger
Diesel Exhaust and Particle-Associated Organics	
Chronic studies should be undertaken at low concentrations of diesel exhaust, and long-term clearance, translocation, and retention of diesel particulates should be assessed. Coexposures to other pollutants should also be conducted. Healthy animals and models of sensitive populations should be evaluated.	Schlesinger
The effects of carrier particles on the ultimate disposition of adsorbed organics should be determined.	Schlesinger; Sun, Bond, and Dahl
Rates of desorption of adsorbed compounds from inhaled particles should be quantified.	Sun, Bond, and Dahl
Gases	
Chemical reactions of specific pollutants with mucus, blood, and tissues should be quantified.	Overton and Miller; Ultman
Analysis of mass transport through individual diffusion barriers, particularly the mucous layer, the bronchial wall, and the alveolar capillary network is needed.	Ultman
Noninvasive techniques to evaluate the transport of soluble and reactant pollutants should be developed. These techniques coupled with appropriate mathematical models can be used to obtain information on regional inhomogeneities in dose and uptake.	Ultman
Methods should be developed and experimental data should be obtained for dosimetry model validation.	Bromberg; Overton and Miller

The methodology to obtain useful exposure information on the regulated pollutants is now available. In contrast, environmental levels of most unregulated pollutants have not been adequately characterized. Because government standards do not exist, there appears to be little incentive to develop monitoring equipment or programs. But we cannot be lulled into inaction by thinking that trace amounts of a particular substance will be of little consequence. For example, in the atmosphere, the concentration of the hydroxyl radical is only 10^{-6} to 10^{-7} ppm; however, the hydroxyl radical is one of the most reactive atmospheric chemical species known and participates in the majority of atmospheric scavenging reactions. Biologists will have to alert chemists and engineers to potentially toxic pollutants that should be measured.

Data on the concentrations of regulated

Table 5.   Biological Responses

Topic	Author(s)
Noncarcinogenic Pulmonary Effects	
Chronic diseases	
• An epidemiologic cohort study of the effects of long-term exposure to oxidants on respiratory morbidity should be conducted.	Bresnitz and Rest
• Role of injury:	
The pathophysiological significance of alterations in bronchoalveolar epithelial permeability in animals and humans should be determined.	Last
Mediators derived from damaged epithelial cells and inflammatory cells should be identified and characterized by use of cell culture systems.	Last; Wright
The relationships among the various stages of injury and changes in populations of pulmonary macrophages and interstitial cells should be established.	Last; Wright
• Regulation of repair:	
The reversibility of abnormal collagen structure or deposition should be determined.	Last
The relationship among digestive enzymes, connective tissue molecules, and abnormal lung tissue should be examined.	Wright
A better understanding of disease progression is needed. Lung structure and biochemistry should be evaluated during a postexposure recovery period.	Last; Wright
Asthma	
• Panels of individuals with bronchial hyperreactivity should be studied to determine whether exposure to ambient oxidants affects respiratory symptoms or morbidity.	Bresnitz and Rest
• Controlled clinical studies should be conducted with asthmatics exposed to chamber atmospheres similar in composition to ambient atmospheres associated with increased symptomology.	Bromberg
• The effect of the presence of allergens either during or after pollutant exposure should be determined in extrinsic asthmatics.	Bromberg
• The role of airway C-fiber sensory systems in ozone effects in epithelial properties, bronchial reactivity, and airways inflammation should be clarified.	Bromberg
Respiratory infection	
• Epidemiologic surveys should be performed in high-risk populations using serologic and cultural tests to confirm infection.	Bresnitz and Rest; Pennington
• In animal or human studies under controlled laboratory conditions, components of respiratory antiviral defense mechanisms should be evaluated.	Pennington
Carcinogenic Effects	
Critical data on the corrections used in extrapolations should be obtained for quantitative assessments.	Kaufman
Diesel emissions	Kaufman
• Additional studies should be performed on the carcinogenicity of diesel exhaust.	
• Methods should be developed for assessing the carcinogenicity of mixtures.	
PAHs and nitro-PAHs	Hecht
• Structures of the major DNA adducts and protein adducts formed from representative PAHs and nitro-PAHs should be identified.	
• Under conditions of chronic administration of PAHs or nitro-PAHs to experimental animals, the relationship between DNA or protein adduct formation and tumor development should be determined.	
• In order to determine the feasibility of monitoring DNA and protein adducts in humans, pilot studies in individuals potentially exposed to PAHs or nitro-PAHs should be performed.	
• Major pathways of metabolic activation and detoxification should be determined.	
Aldehydes	Marnett
• A chronic inhalation toxicology study of acrolein should be undertaken in rats, with emphasis on carcinogenicity.	
• A chronic inhalation toxicology study of mixtures of formaldehyde and acrolein should be undertaken in rats and hamsters, with emphasis on carcinogencity.	
• Experiments should be undertaken in cells cultured from the upper respiratory tract to determine the mechanisms by which aldehydes exert pathological changes.	

(Table continued next page.)

Table 5. *Continued*

Topic	Author(s)
Other	
Coronary heart disease	Clarkson
• Using ongoing cohort studies or other existing data bases, increased risk of heart disease should be evaluated in persons exposed to varying levels of automotive emissions.	
• The effect of automotive emissions on various components of atherogenesis should be evaluated in animal studies that use cynomolgus monkeys.	
Neurobehavioral effects	Wood
• Tests to detect functional disorders and sensory impairments should be applied in a systematic manner to evaluate the magnitude and prevalence of behavioral alterations due to components of automotive emissions.	

as well as unregulated pollutants are inadequate for indoor environments. Although residences and office buildings are the most logical indoor environments to investigate, the interiors of vehicles should also be considered. Some segments of the population (for example, commuters and truck drivers) spend considerable time, on a regular basis, in vehicles where concentrations of pollutants could be high.

In addition to improved characterization of human microenvironments, more information on the amount of time spent in various locations is needed. Documentation of this sort is inadequate for different sectors of the population, but it is not technically difficult to obtain. Individual activity constitutes an essential component in the assessments of human exposure; the construction worker experiences a markedly different exposure than the businessperson who walks down the same street on his or her way to lunch.

To improve dose assessment, descriptive data bases about respiratory tract anatomy and ventilation must be expanded. Pollutant dose to the lung depends on airway geometry. Most particle deposition and gas transport dosimetry models use the symmetrical lung model developed by Weibel in 1963 (Weibel 1963); this model is based on measurements from one human specimen. Recently, more elaborate models have been proposed that attempt to account for variations in lung geometry. Better anatomic models will improve the accuracy of dose estimates. The development and validation of more realistic models requires representative input data, which are largely unavailable. In addition, measurements of, and a greater emphasis on, the upper airways should be pursued. Large particles and their associated organics, as well as soluble gases, will contact respiratory tissues, and perhaps gain access to the circulation, in the upper airways. Without accurate physical and biochemical descriptions of the liquid lining, the fluid's role as a protective layer for underlying epithelial tissues cannot be determined. Finally, the distribution of inhaled particles and gases is also influenced, but poorly characterized, by ventilation. The effects of increased ventilation have a practical significance for physically active individuals.

Measurements taken in a systematic manner on more specimens within and between species are needed. Experimental data from animals cannot be confidently extrapolated to humans until intra- and interspecies variability has been determined. Data are needed on normal individuals as well as on high-risk subjects. Few descriptive data are available on the young, elderly, or special disease groups. The anatomic and physiological status of these individuals may profoundly influence pollutant dose to their respiratory systems.

Basic Disease Processes. The contribution of any pollutant to the initiation or exacerbation of a disease cannot be accurately determined until we have more insight into the pathogenesis of that disease. By focusing our efforts on early events, we would like to identify the cellular and molecular alterations that may link pollutant exposure to ill health. This information would also aid in the identification of susceptible subpopulations. Thus, real progress in defining human risks is not likely to occur until the cellular and biochemical mechanisms of various diseases are better characterized.

Early events of many respiratory diseases involve damage to epithelial tissues; altered epithelial permeability is implicated in the etiology of an asthmatic attack; damaged epithelial cells release biochemical signals for inflammatory cells, which in turn appear to regulate interstitial cell function; and proliferation of damaged cells is associated with cancers of the respiratory tract. Therefore, it is imperative to focus more attention on the mechanisms by which the epithelium translates insult to disease. Once the integrity of the pulmonary epithelium is compromised by injury, underlying tissues are more susceptible to harm from inhaled substances. In addition to their function as a protective barrier, the extent to which epithelial cells actively contribute to pathological processes is unclear. The recent evidence that interactions between epithelial and interstitial tissue layers affect mutual structure and function suggests a more active role for epithelial cells. In addition to the identification of biochemical mediators in response to injury, it is also unknown how much of the specificity of outcome resides in the signals generated. Whether a pollutant induces the formation of free radicals or macromolecular adducts may influence tissue responses—normal and abnormal.

Another important area for investigation is improved characterization of the inflammatory response. One cellular component, the macrophage, produces growth factors and mediators of activity for several other cell types. These macrophage-derived products probably play a role in the development of fibrosis, emphysema, small airways disease, and atherosclerosis. In addition, evidence suggests that inflammation influences bronchial hyperreactivity. The balance between defense and dysfunction is more than likely regulated, in part, by biochemical signals generated during an inflammatory reaction. In addition to the

characterization of these signals, the circumstances that govern their production require clarification. For example, the timing of insults during repeated exposure may be critical for normal repair. Ideally, we would like to elucidate the conditions that solicit physiological "backfire"—that is, when defense mechanisms no longer operate to protect the organism, but instead, promote disease. Inflammatory events after exposure to initiating agents should be examined in established model systems of specific disease states.

Finally, research efforts should continue in the arena of genetic control mechanisms. The explosion of research in this area provides toxicologists with a potentially important methodology—the use of DNA adducts as dosimeters of carcinogenic interactions with target sites. However, we must still face the mystery—and challenge—of relating genome alterations to human disease. It is not enough to just identify adducts relevant to emissions or emission products. Genetic factors that affect individual susceptibility should be explored further. Tremendous variation exists among human individuals in the metabolic activation of substances to carcinogenic compounds, the molecular basis of which has not been characterized. In addition, factors that influence the capacity of pulmonary tissues to repair genome damage should be investigated. Comparisons of biochemical profiles between sensitive and resistant strains of laboratory animals may aid in the elucidation of these factors. In conjunction with animal studies, increased emphasis should be placed on applying current methodologies to or developing more appropriate analytical techniques for human samples. Finally, if we hope to assess the carcinogenic potential of complex mixtures of airborne pollutants, we need a better understanding of the molecular basis for the actions of promoters and cocarcinogenic substances.

Among these interconnected areas of research lie clues to many fundamental problems in abnormal tissue structure and function that constitute toxicity and disease. Opportunities for more quantitative toxicology should be pursued using the rapid advances in scientific methodology and the recent knowledge of molecular and cell biology.

Summary

Currently, our abilities to obtain data for risk determination directly from humans is limited. Epidemiologic studies that evaluate chronic diseases should be limited until, or unless, adequate estimates of exposure and early biological markers are available. Because of the possibility of correlating exposure with outcome, the incidence of asthmatic attacks or respiratory infection holds the most promise for epidemiologic investigation, assuming that satisfactory methodology for the detection of infection can be applied. Controlled clinical trials are best applied to the analysis of neurobehavioral effects or bronchial hyperreactivity. Even in cases of asthma, data collection is restricted to a subpopulation of mildly affected subjects. We need more information on what determines the severity of the disease.

The example of asthma illustrates another important point; results in clinical studies have not supported epidemiological evidence of an effect of oxidants on airway sensitivity. Whether we have been unable to accurately duplicate ambient atmospheres or identify appropriate susceptible individuals for study is not known. Both factors require additional consideration when conducting further studies.

For most health effects animal experiments must be used as a source of information from which human risks can be approximated. But how can the data be linked more effectively to the human situation? Since risk is related both to dose and to sensitivity, each should be considered a research area that might improve our ability to quantify human risks from animal experiments. Fruitful directions that would improve dosage extrapolation are readily apparent. They lie partly in further development and application of mathematical models for comparing pulmonary deposition in laboratory animals and humans. As the harmful components of automotive emissions are identified, biological markers of their presence in tissues can also be developed and used to compare target site levels. Definitive comparisons of sensitivity of laboratory animals and humans to equivalent target site doses probably awaits better understanding of the molecular determinants in chronic disease development. Elucidation of missing links between acute effects and chronic disease would provide investigators with a powerful tool—early indicators of subsequent injury. Knowing biological markers and factors that determine individual sensitivity will not only provide better extrapolations from animals to humans, but will also improve the feasibility of using epidemiologic approaches.

Ideally, a comprehensive research strategy would link molecular dosimetry to disease in laboratory animals and finally to exposure in humans. This strategy is best illustrated using PAHs—carcinogenic components in vehicle emissions. First, it will be necessary to identify the appropriate DNA and protein adducts in PAH-exposed animals. Second, and probably most crucial, will be to determine the quantitative relationship between adduct levels and tumor formation. Finally, if a predictive correlation can be made between marker and disease, it will then be appropriate to apply these methods to PAH-exposed humans.

Conclusion

Each chapter in this book addresses some aspect of the problem of assessing exposure and risks to humans from automotive emissions and their transformation products. Invariably, and inevitably, the authors found that the research needed to answer questions about automotive emissions is also applicable to a much broader array of issues about human exposure and health. The development of new techniques and a better understanding of atmospheric chemistry and physics, all of which would improve our knowledge of exposure to automotive emissions, are also applicable to other sources of air pollution. Similarly, the methods and the knowledge needed to assess health risks from motor vehicle emissions are also useful to evaluate the effects of other substances in the environment. Thus, rather than there being a sharp boundary of research questions around automotive emissions, a series of concentric circles that overlap other topics of environmental exposure and health exists.

Before proceeding too far toward the more peripheral circles, it is worth returning to the basic question: How important are automotive emissions as a risk to human health? Referring to air pollution in general, a committee of the National Research Council recently concluded that "evidence from controlled human exposures, toxicology, and epidemiology is sufficient to warrant concern that current air pollution still produces substantial adverse health effects in some segments of the U.S. population." Furthermore, "The Committee finds that current air pollution can cause acute and perhaps chronic health effects, particularly respiratory effects, in the population of the United States. Respiratory disease is a major cause of work loss and disability. Even if only a small proportion of very prevalent disease is due to air pollution, the absolute amount of illness that could be prevented by reducing air pollution would be large" (National Research Council 1985). These comments were balanced by another statement: "The impact of ambient air pollutants on the total respiratory disease burden in the United States must be small relative to the impact of cigarette smoking, and occupational exposures might also have greater effects than pollution of ambient air" (National Research Council 1985).

Motor vehicle emissions are responsible for a substantial proportion of atmospheric exposure to carbon monoxide, nitrogen oxides, volatile organic compounds, and, in urban areas, total suspended particulate matter. Atmospheric organic compounds and nitrogen oxides derived from automotive emissions also contribute significantly to the formation of ozone. Thus, to the extent that these substances are a source of the health concerns of the National Research Council Committee, automotive emissions must also be of concern. Questions about this issue have been raised, but not fully answered, as the chapters of this book indicate.

Much less information is available on the hundreds, or perhaps thousands, of individual organic chemicals that contribute to the volatile organic and particulate fractions of atmospheric pollution. Neither the exposure levels, the atmospheric reactions, nor the health effects of many of these have been well characterized. Among these materials are the PAHs, nitro-PAHs, and aldehydes which are discussed in this book, as well as many other chemicals. It seems likely that many of the organic chemicals derived from motor vehicle operation are present at such low atmospheric levels that they do not threaten human health, even in sensitive individuals. On the other hand, highly toxic chemicals can harm health even at very low concentrations, especially when they are mixed with other substances and synergism occurs. Some balance needs to be struck between unnecessary research on trivial problems and insufficient research on possibly important ones. Locating the balance point is not easy. It depends on a continuing dialogue between the experts on atmospheric exposure and the experts on biological effects of toxic substances. The former can contribute information about what is in the atmosphere, the latter on what may be important for health. Thus, each group can help guide the research priorities of the other. We hope that this book will contribute to the dialogue.

References

Klaassen, C. D., and Doull, J. 1980. Evaluation of safety: Toxicologic evaluation, In: *Casarett and Doull's Toxicology: The Basic Science of Poisons* (J. Doull, C. D. Klaassen, and M. O. Amdur, eds.), 2nd ed., pp. 11–27, MacMillan, New York.

National Research Council (Committee on the Institutional Means for Assessment of Risks to Public Health). 1983. *Risk Assessment in the Federal Government: Managing the Process*, p. 191, National Academy Press, Washington, D.C.

National Research Council (Committee on the Epidemiology of Air Pollutants). 1985. *Epidemiology*

Correspondence should be addressed to Ann Y. Watson, Health Effects Institute, 215 First Street, Cambridge, MA 02142.

and Air Pollution, p. 224, National Academy Press, Washington, D.C.

Paracelsus. 1958. *Selected Writings* (J. Jacobi, ed., and N. Guterman, trans.), Bollingen series XXVIII, p. 290, Pantheon Books, New York.

U.S. Environmental Protection Agency. 1984. National Emissions Report, 1981, National Emissions Data System of the Aeromatic and Emissions Re-porting System, Report EPA-450/4-83-022, Research Triangle Park, N.C.

U.S. Environmental Protection Agency. 1986. National Air Quality and Emissions Trends Report, 1984, Report EPA-450/4-84-029, Research Triangle Park, N.C.

Weibel, E. 1963. *Morphometry of the Human Lung*, Springer-Verlag, Berlin.

Part II: Exposure Analysis

Automotive Emissions

JOHN H. JOHNSON
Michigan Technological University

Air Pollution, the Automobile, and Public Health. © 1988 by the Health Effects
Institute. National Academy Press, Washington, D.C.

Pollution from Automobiles—Problems and Solutions

Concern about the automobile as a source of air pollution has been expressed periodically, but national concern was first evidenced in the 1960s when California established the first new car emission standards. The scientific basis of this effort is the pioneering atmospheric chemistry research of A. J. Haägen-Smit, who showed that photochemical reactions among hydrocarbons (HC) and nitrogen oxides (NO_x) produce the many secondary pollutants that reduce visibility and cause eye and nose irritation in the Los Angeles area.

This paper reviews our current knowledge of automotive emissions, including standards, control technology, fuel economy, fuels and additives, in-use emissions, measurement methods for unregulated pollutants, and models for predicting future automotive emissions. Fuel economy is included because achieving high fuel economy and low emissions together makes the engineering effort more difficult. Emissions and fuel economy are interrelated because both are influenced by the engine combustion system design. In practice, the stringency of emission standards determines the importance of this interrelationship. After current knowledge in each area has been reviewed, important gaps in our knowledge are identified and research needed to fill these gaps is described.

Emissions Standards and Control Approaches

Evolving emission standards have resulted in three levels of stringency, and in turn, three types of control technology. Figure 1 describes the technologies applied in each of the three phases and the general time periods in which they were applied to cars, light-duty trucks, and heavy-duty trucks (Ford Motor Co. 1985a). The percent reduction in the HC, carbon monoxide (CO), and NO_x emissions are also shown.

Air/fuel (A/F) ratio, which is controlled by the carburetor or fuel injection system, is the most important variable in determining emissions and in applying catalyst technology. Figure 2 (Heinen 1980) is a plot of NO_x, HC, and CO concentrations in the exhaust versus A/F ratio for a typical gasoline engine. It is impossible to achieve the low emissions demanded by federal standards by A/F ratio control alone since the concentrations of the three pollutants are not minimums at the same A/F ratio. In fact, when CO and HC concentrations are a minimum, at an A/F ratio of around 16:1, NO_x production is close to a maximum. Also shown is the A/F ratio for maximum power (13.5:1) and maximum fuel economy (17:1). The region where A/F ratio exceeds 17.5:1 is the lean burn region where misfires can occur along with slow flame speeds, causing increased HC concentration. The A/F ratio effects are used in all phases of control. The stoichiometric ratio of 14.7:1 is necessary in the Phase III control using three-way catalysts since the A/F ratio must be in a narrow window within ± 0.05 of the stoichiometric ratio to achieve high HC, CO, and NO_x control efficiencies simultaneously. Exploring the lean burn region is an important area of research and development because of the potential of improved fuel economy and adequate emission control with only an oxidation catalyst.

Fuel Economy

Federal regulations also mandate automotive fuel economy. The period from 1968 to 1974 resulted in primary emphasis on emission control with loss of fuel economy from lower compression ratios, changes in spark timing, A/F ratio and axle ratio changes, and exhaust gas recirculation. Figure 3 (Heavenrich et al. 1986) shows the U.S. fleet combined, city, and highway fuel economy data for each model year since 1974. The figure also includes information on average vehicle weight. The fuel economy from 1977 to 1980 improved almost exactly in proportion to the decreasing weight of vehicles. If the data in figure 3 were normalized to the 1978 weight mix, it would show that fuel economy improvements leveled out in 1982. With the intro-

Major Emission Reduction Phases

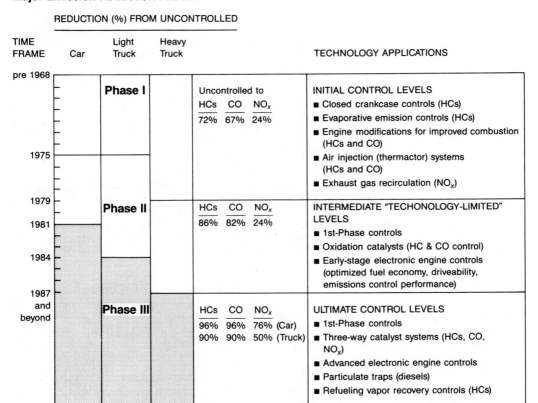

Figure 1. Major phases in the reduction of automotive emissions. (Adapted with permission from Ford Motor Co. 1985a.)

duction of the oxidation catalytic converter in 1975, improved fuel economy and reduced emissions occurred simultaneously. Further emission reduction with simultaneous fuel economy improvement continues through application of new technology, especially computer engine control.

In-Use Passenger Car Emissions

The in-use emissions from passenger cars exceed the new car standards mandated by law. Nonetheless, emissions continue to decrease in spite of high tampering rates and fuel switching (that is, using leaded fuel in engines developed to run on unleaded fuel). From field surveys in 14 cities, Greco (1985) found the overall tampering rates and catalyst tampering rates shown in figure 4.

Figure 5 shows EPA emission factors data as analyzed by General Motors Corp. (1985a). The measured emission concentrations for various model years are compared to the standards that were in effect during those years. The measured NO_x concentrations follow the standards fairly well. Although the measured HC and CO concentrations are higher than the standards, the difference between the actual emissions and the standards appears to be narrowing (although the ratio is not decreasing) as improved technology, more frequent inspection and maintenance, and better training of mechanics has occurred. Even though the overall trend of emissions is down, a few vehicles have high emission levels, as shown in figure 6, probably because of electronic problems rather than catalyst removal or misfueling problems.

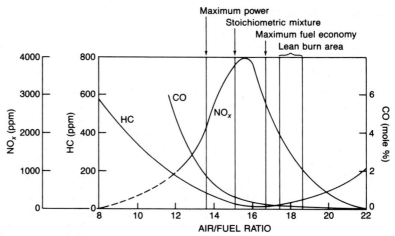

Figure 2. Concentrations of HC, CO, and NO$_x$ emissions as a function of air/fuel ratio in a typical gasoline engine. (Adapted with permission from Heinen 1980, and the Society of Automotive Engineers, Inc.)

Emissions Regulations

In the 1960s, motor vehicles were identified as one of the primary sources of air pollutants in urban areas. Emission standards for passenger cars were first imposed in California in 1965. These were followed by U.S. federal standards in 1968. The 1970 Clean Air Act further imposed stringent HC, CO, and NO$_x$ reductions for 1975 and 1976. These reductions were subsequently delayed and changed by the 1974 Energy and Environmental Coordination Act and the 1977 Clean Air Act Amendments. Recognition of the motor vehicle as a major source of pollutants has spread to other countries, of which many have imposed diverse standards and test procedures reflecting various degrees of stringency. The differences have come about because of different regulatory philosophies and air quality goals, in combination with concerns about the conflicting goal of improved fuel efficiency (Barnes and Donohue 1985).

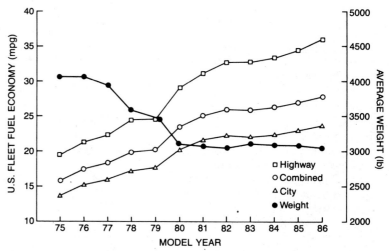

Figure 3. U.S. fleet fuel economy and average vehicle weight by model year. (Adapted with permission from Heavenrich et al. 1986, and the Society of Automotive Engineers, Inc.)

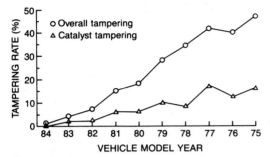

Figure 4. Overall and catalyst tampering rates by vehicle model year, based on 1984 survey. (Adapted from Greco 1985.)

Emission Test Procedures

Passenger Cars. Emissions come principally from three automotive sources: the exhaust, the fuel system (evaporative), and crankcase ventilation gases. To give the standard (maximum allowable level of emission in grams per mile) operational meaning, two major aspects must be defined: the driving cycle and the emissions sampling method. Driving cycles are discussed below and sampling methods will be covered in a later section.

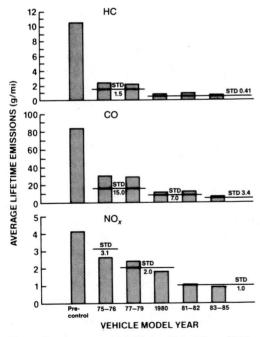

Figure 5. Average vehicle lifetime HC, CO, and NO$_x$ emissions compared with standards (STD), by model year for all industry passenger vehicles. (Adapted with permission from General Motors Corp. 1985a.)

Regulations require exhaust emission measurements during the operation of the vehicle (or engine) on a dynamometer during a driving cycle that simulates vehicle road operation. The approach to driving cycles by various regulatory authorities represent two basic philosophies. According to the first, the driving cycle is made up of a series of repetitions of a composite of various vehicle operating conditions representative of typical driving modes. The European Economic Community and Japanese cycles reflect this philosophy. According to the second, the composite of driving modes is an actual simulation of a road route. The United States, Canada, Australia, Sweden, and Switzerland all use a version of the federal test procedure (FTP). The FTP cycle is divided into a "transient" portion and a "stabilized" portion with a total cycle time of 1,372 sec, a driving distance of 7.5 miles, and an average speed of 19.7 miles per hour (mph). Two such cycles are run: one with the vehicle at an ambient temperature of 16–30°C before start ("cold" cycle), and one with the engine control system hot ("hot" cycle) after a 10-min shutdown after running the cold cycle.

Trucks. Many of the light-duty trucks intended primarily for the carrying of goods are also capable of use as passenger vehicles. The gross vehicle weight for light-duty trucks in the United States is less than 8,500 lb; trucks heavier than 8,500 lb

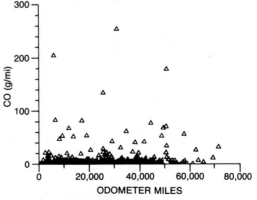

Figure 6. Scatter plots of CO emissions from 703 1981 model-year federal cars. (Adapted with permission from General Motors Corp. 1985a.)

Table 1. Motor Vehicle Emission Standards in the United States

	Emission Rates[a]											
	Federal						California					
Model Year	HC (g/mi)	CO (g/mi)	NO$_x$ (g/mi)	Evap (g/test)	Partic-ulates[b] (g/mi)	HC+ NO$_x$	HC (g/mi)	CO (g/mi)	NO$_x$ (g/mi)	Evap (g/test)	Partic-ulates[b] (g/mi)	HC+ NO$_x$
Passenger Cars												
Pre-control	10.60[c]	84.0	4.1	47			10.60[c]	84.0	4.1	47		
1966							6.30	51.0	(6.0)			
1968	6.30	51.0	(6.0)[d]				6.30	51.0				
1970	4.10	34.0					4.10	34.0		6		
1971	4.10	34.0					4.10	34.0	4.0	6		
1972	3.00	28.0					2.90	34.0	3.0	2		
1973	3.00	28.0	3.0				2.90	34.0	3.0	2		
1974	3.00	28.0	3.0				2.90	34.0	2.0	2		
1975	1.50	15.0	3.1[e]	2			0.90	9.0	2.0	2		
1977	1.50	15.0	2.0	2			0.41	9.0	1.5	2		
1978	1.50	15.0	2.0	6[e]			0.41	9.0	1.5	6[e]		
1980	0.41	7.0	2.0	6			0.39[f]	9.0	1.0	2		
1981	0.41	3.4	1.0	2			0.39	7.0	0.7	2		
1983	0.41	3.4	1.0	2			0.39[g]	7.0	0.4[h]	2		
1984	0.41	3.4	1.0	2			0.39	7.0	0.4	2	0.60	
1985	0.41	3.4	1.0	2			0.39	7.0	0.4	2	0.40	
1986	0.41	3.4	1.0	2	0.60		0.39	7.0	0.4	2	0.20	
1987	0.41	3.4	1.0	2	0.20		0.39	7.0	0.4	2	0.20	
1989	0.41	3.4	1.0	2	0.20		0.39[i]	7.0	0.4	2	0.08	
Light-Duty Trucks												
1975	2.00	20.0	3.1	2					2.0			
1976	2.00	20.0	3.1	2			0.90	17.0	2.0			
1978	2.00	20.0	3.1	6			0.90	17.0	2.0	6		
1979	1.70	18.0	2.3	6			0.50[j]	9.0	2.0[k]	6		
1980	1.70	18.0	2.3	6			0.50[l]	9.0	2.0[k]	2		
1981	1.70	18.0	2.3	2			0.50[m]	9.0	1.5[n]	2		
1983	1.70	18.0	2.3	2			0.50[o]	9.0	1.0[p]	2		
1984	0.80	10.0	2.3	2			0.50	9.0	1.0[p]	2		
1986[q]	0.80	10.0	2.3	2			0.50	9.0	1.0	2		
1988[r]	0.80	10.0	1.7[s]	2			0.50	9.0	1.0	2		

(*Table continued next page.*)

NOTE: Evap = evaporative HC.
[a] Emission rates for HC, CO, NO$_x$ by (or adjusted to equivalent) 1975 Federal Test Procedure.
[b] Diesel passenger cars only.
[c] Crankcase emissions of 4.1 g/mi not included; fully controlled.
[d] NO$_x$ emissions (no standard) increased with control of HCs and CO.
[e] Change in test procedure.
[f] NMHC standard (or 0.41 g/mi for total HC).
[g] 0.7 NO$_x$ optional standard 1983 and later, but requires limited recall authority for 7 yr/70,000 mi.
[h] Optional standard, 0.3 g/mi, requires 7 yr/75,000 mi limited recall authority.
[i] Primary standard = 0.4 g/mi required on 90% of production after 1989.
[j] 0.41 for <4,000 lb.
[k] 1.5 for <4,000 lb.
[l] 0.39 for <4,000 lb (nonmethane) and in following years.
[m] 0.6 for >6,000 lb and in following years.
[n] 2.0 for >6,000 lb.
[o] California: 1.0 NO$_x$ optional standard for 1983 and later, but requires limited recall authority for 7 yr/75,000 mi. Primary standard = 0.4 g/mi.

Table 1. *Continued*

| | Emission Rates | | | | | | | | | | | |
| | Federal | | | | | | California | | | | | |
Model Year	HC (g/bhp-hr)	CO (g/bhp-hr)	NO_x (g/bhp-hr)	Evap (g/test)	Partic-ulates (g/bhp-hr)	HC+NO_x	HC (g/bhp-hr)	CO (g/bhp-hr)	NO_x (g/bhp-hr)	Evap (g/test)	Partic-ulates (g/bhp-hr)	HC+NO_x
					Heavy-Duty Truck and Bus Engines[r]							
1969							u	u				
1972							v	v				
1973								40.0[v]				16.0
1974		40.0				16.0		40.0				16.0
1975		40.0				16.0		30.0				10.0
1977		40.0				16.0	1.0	25.0	7.5			5.0[w]
1978		40.0				16.0	1.0	25.0	7.5	6		5.0[w]
1979		25.0				10.0	1.5	25.0	7.5	6		5.0[w]
1980	1.5	25.0				10.0	1.0	25.0	7.5	2		6.0[w]
1984	1.5	25.0				10.0	0.5	25.0	7.5	2		4.5[w]
1985	1.9[x]	37.1[y]	10.6[z]	3[aa]			0.5	25.0	7.5	2		4.5[w,bb]
1987	1.1	14.4[y]	10.6	3			0.5	25.0	7.5	2		4.5[w]
1988[cc]	1.1	14.4	6.0	3	0.6		0.5	25.0	7.5	2		4.5[w]
1991	1.1	14.4	5.0	3	0.25[dd]		0.5[ee]	25.0[ee]	7.5[ee]	2		4.5[w]

NOTE: Evap = evaporative HC.
[p] 1.5 for >6,000 lb.
[q] Full useful life requirement = 11 yr/120,000 mi (was 5 yr/50,000 mi).
[r] NO_x federal standard = 1.2 g/mi under 3,751-lb loaded vehicle weight (LVW), 1.7 g/mi for ≥3,751 lb LVW, and 2.3 g/mi for ≥6,000 lb LVW.
[s] 1.2 for <3,751 lb.
[t] Various test methods, values are not strictly comparable.
[u] 275 ppm HC, 1.5% CO.
[v] 180 ppm HC, 1.0% CO.
[w] A combined standard is optional in lieu of separate HC and NO_x standards (for example, 1 g HC + 7.5 g NO_x or 5 g [HC+NO_x]).
[x] 1.3 for diesel.
[y] 15.5 for diesel.
[z] 10.7 for diesel.
[aa] 4.0 for >20,000 lb.
[bb] Gasoline only and in following years.
[cc] 1988 federal standards for NO_x have been postponed until 1990.
[dd] Separate standard of 0.1 for all 1991 urban buses and all 1994 engines.
[ee] 1.3 HC, 16.5 CO, 5.1 NO_x for diesel.
SOURCE: Adapted with permission from General Motors Corp. 1986.

are classified as heavy-duty vehicles. The driving-cycle philosophies for the light commercial vehicles follow those for passenger cars. For heavy commercial vehicles, engine dynamometers are used, not chassis dynamometers; that is, the engine rather than the vehicle is certified. The new (effective 1985) U.S. transient test procedure for heavy-duty vehicles combines the two philosophies just described in that the cycle is made up in a random way from actual driving cycle data. The use of this cycle replaces the 13-mode steady-state cycle in use since 1973 in California and since 1974 nationally (U.S. Environmental Protection Agency 1972).

Emission Standards

United States. Emissions standards and test procedures in the United States have changed significantly since the first automobile emission standards were imposed in California in 1966 (see table 1) (General

Motors Corp. 1986). Light-duty truck standards are somewhat higher than the car standards because of the differences in weight.

The U.S. passenger car regulations require that the vehicle comply with the emission standards for five years or 50,000 miles, whichever occurs first. Certification testing of prototype vehicles for 50,000 miles of use is based on the Automobile Manufacturers' Association (AMA) 40.7-mile durability cycle. The cycle consists of numerous stops, acceleration, and high/medium-speed driving (maximum of 55 mph) (U.S. Environmental Protection Agency 1973).

Europe. The European Economic Community, an inter-Europe regulatory body, has announced future model standards for passenger cars based on three engine size (displacement) categories. Large-car (> 2-liter engine displacement) standards are roughly equivalent to current U.S. standards although there is no valid correlation between the distinct U.S. and European emission test cycles. Standards for medium cars (1.4–2.0 liters) are considered to fall in the Phase I/Phase II range shown in figure 1. Requirements for small-car levels (< 1.4 liters) are comparable to Phase I requirements. The standards include diesels; however, large diesel cars are only required to meet medium-car levels.

Japan. Catalyst forcing standards currently in effect for passenger cars are 0.25 HC/2.1 CO/0.25 NO_x g/km for the unique 10-mode hot start and 7.0 HC/60 CO/4.4 NO_x g/test for the 11-mode cold-start test procedures. These standards are generally considered to be equivalent to current U.S. California levels (Ford Motor Co. 1985a).

U.S. Fuel Economy Standards

There have been passenger car and light-truck fuel economy standards since 1978 and 1979, respectively. The manufacturers are required to conduct passenger car fuel economy tests according to the U.S. Environmental Protection Agency (EPA) urban or "city" driving cycle—the FTP for emis-

sion testing described earlier. The EPA also has a suburban or "highway" cycle that includes a significant amount of simulated highway driving. A combined fuel economy number based on these two tests is published by the EPA and the U.S. Department of Energy and used by manufacturers in their sales literature.

Manufacturers each have to meet the Corporate Average Fuel Economy (CAFE) standards for their sales-weighted fleet. Car standards started at 18 miles per gallon (mpg) in 1978, went to 27.5 mpg in 1985, but were reduced to 26 mpg by the U.S. Department of Transportation for 1986–1988.

Vehicle and Emission Control System Technology

The technology used for emission control in cars changed rapidly in the 1970s as the automotive industry spent considerable research and development funds to meet the stringent emission standards originally set by the 1970 and 1977 Clean Air Act Amendments. This technology is now being optimized to reduce the product cost associated with emission controls while improving the in-use durability of the emission control systems. Heavy-duty gasoline-powered vehicles have used this technology as allowable emissions have progressively decreased.

Control technology is being developed to meet proposed standards and anticipated changes in fuels. Proposed 1988, 1991, and 1994 particulate standards require new control systems for heavy-duty diesels. For the United States to become less dependent on imported petroleum fuels, there is interest in using methanol in passenger cars and diesel-fueled buses. There are continued efforts to develop stratified-charge engines for passenger cars because of their potential for better fuel economy at equivalent emissions. There is also a demand for development of direct-injection diesels that give 15 percent better fuel economy than pre-chamber or swirl-chamber engines with equivalent or better emissions. An additional demand exists for an adiabatic diesel engine (more precisely, a low-heat-rejection engine) that would have improved fuel economy and lower emissions with a simpler cooling system, particularly for vehicles in the heavy-duty class.

Spark-Ignition Gasoline-Powered Vehicles

During the past 15 years, emissions have been significantly lowered by improved design of the engine and fuel system while still achieving the high fuel economy demanded by the federal standards and the consumer market. These reductions have come about by A/F ratio control, cylinder-to-cylinder distribution of air and fuel, choke operation, combustion chamber design, fuel injection, exhaust gas recirculation (EGR), ignition systems, spark timing, valve timing, and many additional design details. The computer scheduling of spark timing, EGR, A/F ratio, and transmission gear ratio as a function of engine operating conditions are done very precisely with sensors and actuators. This scheduling is referred to as the engine calibration. With all of this technology, vehicles still do not meet HC/CO/NO$_x$ standards of 0.4/3.4/1 g/mi without aftertreatment devices. The period after 1983 has seen better optimization of systems and removing of components to reduce costs, but nevertheless, catalysts are still necessary.

Catalyst Control Systems. Meeting the 1975 HC/CO standards of 1.5/15 g/mi and at the same time increasing the fuel economy was achieved through the broad introduction of the oxidizing catalytic converter. The catalyst is cold (16–30°C) at the start of the FTP cycle and must warm up to 250–300°C before oxidation of CO and HC occurs. The time required for this is a function of catalyst design and position but can be from 20 to 120 sec. The HC emitted during this period can be one-fourth to three-fourths of the allowable limit (Hilliard and Springer 1984). The amount of NO$_x$ emitted during the cold start is only about 10 percent of the allowable limit.

The time period from 1975 to 1984 saw

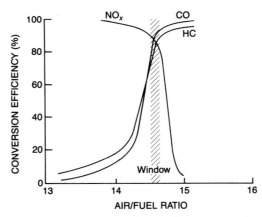

Figure 7. Conversion efficiency characteristics of a three-way catalyst. (Adapted with permission from Amann 1985.)

increased fuel economy and improved emission control through exploitation of the high HC and CO removal efficiency of the oxidizing catalytic converter, so that the engine calibration could be optimized for efficiency. Progress was made by decreasing the cold-start engine-out HC and CO emissions, by achieving faster converter light-off, by reducing heat loss from the exhaust system, and by reducing the deterioration of catalyst performance with cumulative driving distance (Amann 1985).

Reducing combustion temperature by spark retard and/or diluting the incoming mixture with EGR provided NO$_x$ control during the time period from 1973 to 1980. The 1981 standards stipulate no more than 1 g/mi NO$_x$, which could not be achieved either by EGR or engine design and calibration. Two additional catalytic approaches have gained widespread application along with the microprocessor control system, to provide the necessary control: the "three-way" catalyst and the "dual" catalyst.

Three-way catalysts are capable, within a narrow range of exhaust stoichiometry, of simultaneously decreasing NO$_x$, HC, and CO, as shown in figure 7. Within a narrow range of values of the A/F (approximately ± 0.05 from the optimum), all three emissions are decreased with a reasonably high efficiency. An oxygen sensor is used in the exhaust in conjunction with a microprocessor to make this technology feasible.

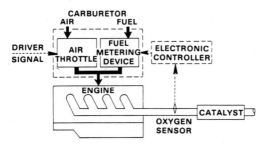

Figure 8. System for closed-loop control of A/F ratio. The oxygen sensor inserted in the exhaust pipe ahead of the catalyst measures oxygen concentration and signals the electronic controller to adjust fuel rate continuously. (Adapted with permission from Amann 1985.)

In a dual catalyst, two catalysts are used in series—a three-way catalyst followed by an oxidizing catalyst. Air is injected into the exhaust gas between the two catalysts to provide the oxygen necessary for the oxidizing catalyst to operate efficiently. Once more, precise A/F ratio control is required to make the three-way catalyst function. During the cold-start portion of the FTP cycle, the air supply to the oxidizing catalyst can be diverted to the exhaust ports to add oxygen to the combustion products of the rich start-up mixture for faster catalyst light-off and to achieve higher HC and CO control efficiencies in the three-way catalyst. The dual-bed converter is more complex than the single-bed three-way catalyst, because it requires an extensive air management system.

To maintain A/F ratio control within the narrow window, closed-loop control (feedback control of fuel delivery on oxygen level in exhaust) was introduced on many cars in 1981. The schematic of a typical system is shown in figure 8 (Amann 1985). The key element in the closed-loop system is the oxygen sensor inserted in the exhaust pipe ahead of the catalyst. It measures exhaust oxygen concentration and signals an electronic controller to adjust fuel rate continuously so that the mixture is maintained at the stoichiometric ratio.

Current Control Approaches. Since 1983 the number of engines with some type of fuel injection has grown drastically, but carburetors are still used on many engines.

No particular trend in emission systems is evident except for the use of heated oxygen sensors to initiate closed-loop operation faster and more predictably and to maintain it during long idling periods. The heated sensors also deteriorate less with extended mileage (Way 1985). Most cars use closed-loop control with a three-way catalyst; many also have an oxidation catalyst that is a dual catalyst and one of three air supply systems (pulse air, air pump, or programmed pump).

Lean-Burn Combustion Systems. An important engine emission control system under development is the lean combustion system. This system uses a closed-loop microprocessor in conjunction with lean mixture sensor and an oxidation catalyst. This alternate emission control approach achieves good fuel economy (potential 10–15 percent improvement) and also meets the emission standards by operating beyond 22:1 A/F where NO_x emission is low enough to meet the 1-g/mi standard. In this lean operating region, the engine needs a different sensor design to provide feedback, and also a highly turbulent fast-burn combustion system so that slow flame speed and misfires do not cause emissions and driveability problems. Toyota has developed and marketed such a system in Japan but not yet in the United States (Kimbara et al. 1985).

It may be possible to introduce this type of system into the U.S. market, but durability and driveability under hot and cold conditions need to be examined further (Kimbara et al. 1985). The other important technological limit might be that lean burn could be restricted to cars under 2,500–3,000 lb because NO_x generally increases with vehicle weight.

Diesel-Powered Passenger Cars: Particulate Control

There has been a major research and development effort during the past seven years to develop aftertreatment devices for diesel passenger cars to meet the federal 0.2 g/mi standard first proposed for 1985 and later put off until 1987. California has a 0.2-g/mi

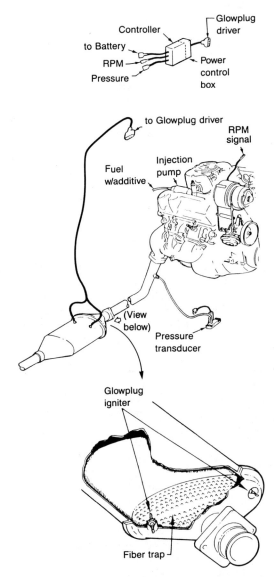

Figure 9. Diesel particulate trapping system utilizing a ceramic fiber trap, a fuel additive, glow plug igniters, and exhaust backpressure regeneration controls. (Adapted with permission from Simon and Stark 1985, and the Society of Automotive Engineers, Inc.)

standard that was initiated in 1986, and will be lowered 0.08 g/mi in 1989. A number of prototype systems have been built and field tested to meet the 0.2 g/mi standard. Mercedes-Benz (Abtoff et al. 1985) introduced a catalytic trap oxidizer in 1985, in conjunction with careful modification of the engine (in particular, the turbocharger). The system meets and is certified to the 1986

California standards and has been sold in the 11 western states. Volkswagen has developed a prototype system that uses a Corning ceramic particulate filter in conjunction with Lubrizol 8220 manganese (Mn) additive. The additive consists of nonstoichiometric Mn fatty acid salts dissolved in naphtha, which is metered from a separate fuel-additive storage tank on the vehicle (lifetime filling) and mixed with the fuel (Wiedemann and Neumann 1985). Emissions of Mn oxide of all valence states, as well as $MnSO_4$, may occur. Data suggest that most of the Mn residue is in the form of sulfate.

General Motors has also tested a system, shown in figure 9, with on-board tank-blending, additive dispensing, and ceramic fiber trap (Simon and Stark 1985). This system uses pressure and engine speed to provide a measure of particulate loading for triggering the glow plug igniters for regeneration. Simon and Stark (1985) investigated three different additives: cerium (0.13 g Ce/liter), manganese (0.07 g Mn/liter), and cerium plus manganese (0.07 g (Ce + Mn)/liter). Their tests showed that vehicles equipped with properly tuned 4.3-liter engines and operated using a fuel additive would not, on a production basis, be able to meet the 1987 federal emissions standards at sea level or at altitude. Equipped with particulate traps, however, the vehicles would probably meet the 1987 federal standards and might, with further engine tailoring, be able to meet the 1989 California standards on a production basis.

Diesel-Powered Heavy-Duty Vehicles

Diesel-powered heavy-duty vehicles use direct-injection turbocharged engines of two-cycle as well as four-cycle design. Diesel engines are designed for a commercial market and hence durability, reliability, and fuel economy drive their development. The approaches enforced to date to meet the standards for particulates, HCs, and NO_x have involved improved turbochargers, intercooling, improved fuel systems and nozzles, and electronic fuel injection control. To reduce NO_x emissions and improve fuel economy, some manufacturers use heat exchangers to lower air inlet temperature. The industry believes that the 1988 standards can be met with advanced electronic fuel systems and possibly with mechanical fuel systems with electronic-governing, air-to-air intercoolers (or low-flow radiators) and improved turbochargers, but that the 1991 particulate standard of 0.25 g/brake horsepower (bhp)-hr will need trap technology. EPA standards have emphasized particulate control rather than NO_x control. There is the feeling that the 1988 standards will result in some loss of fuel economy. GM-Detroit Diesel Allison Division (DDAD) has also decided to remove their 2-cycle engines from the on-highway market because of the disadvantages of this engine under these tight emissions constraints. This is now being reevaluated under the new Detroit Diesel Corp.

In-Use Vehicle and Engine Characteristics

How vehicles and engines perform in the hands of the operator ultimately determines their emissions and, in turn, their impact on air quality. This section examines the emission characteristics of gasoline- and diesel-powered passenger cars and trucks as actually used by owners. The effects of field environmental conditions such as temperature, tampering (removal of and changes in components), or misfueling on emissions are discussed.

Gasoline-Powered Passenger Cars and Trucks

To develop an understanding of the in-use characteristics of gasoline-powered passenger cars, it is important to know whether tampering and misfueling with leaded fuel occur. Misfueling has a twofold impact on the environment: increased lead (Pb) emissions and increased regulated HC, CO, and NO_x emissions due to poisoning of the catalyst. Tampering has a direct effect on the regulated emissions and can also affect the unregulated emissions. In this section,

Table 2. Tampering Prevalence in Light-Duty Vehicles for Critical Emission Control Components, April–October 1984

Component/System	Tampering Rate (%)		
	Trucks	Cars	Overall
Catalytic converter	14	5	7
Filler neck restrictor	14	9	10
Air system	12	7	7
PCV system	3	2	2
Evaporative control system	5	2	3
EGR system	10	10	10
Overall	27	21	22
Fuel switching	19	13	14

SOURCE: Adapted from Greco 1985.

the latest tampering data gathered by the EPA and the Motor Vehicle Manufacturers' Association (MVMA) are examined first. This is followed by data showing the emissions and fuel economy of vehicles in use.

Tampering and Misfueling. The latest EPA report on tampering is based on a survey of 4,426 light-duty vehicles conducted in 14 cities between April and October 1984 (Greco 1985). These inspections were performed with the consent of the vehicle owners and therefore may underestimate tampering rates. Four categories were used to summarize the condition of the inspected vehicles:

1. Tampered—at least one control device removed or rendered inoperative;
2. Arguably tampered—possible but not clear-cut tampering;
3. Malfunctioning;
4. Okay—all control devices present and apparently operating properly.

Greco's overall survey averages of vehicle condition were as follows: tampered, 22 percent; arguably tampered, 29 percent; malfunctioning, 4 percent; okay, 46 percent. The rates for tampering with selected components and the rates of fuel switching are shown in table 2. These results have not been weighted to compensate for inspection and maintenance program representation and probably underestimate the actual

Table 3. Incidence of Misfueling in Large Urban[a] and Nonurban Areas, 1981–1982

Area	Overall Catalyst-Equipped Fleet			Occasional Misfuelers		Persistent Misfuelers	
	Purchase Volume Rate (%)	Vehicle Involvement Rate (%)	Leaded Fuel Used[b] (%)	Vehicle Involvement Rate (%)	Leaded Fuel Used (%)	Vehicle Involvement Rate (%)	Leaded Fuel Used (%)
Large Urban	4.0	14.0	21.6	7.3	3.7	1.6	9.6
Nonurban	11.8	22.1	41.8	5.9	1.9	6.7	28.1
Total U.S.	7.5	18.0	100.0	7.1	8.6	3.5	53.1

[a] Large urban areas are defined as standard metropolitan statistical areas with populations over 1 million. In the National Panel Diary Inc. data base these areas account for about one-half the total urban vehicle population.
[b] Percent of all misfueled leaded fuel purchased by the entire catalyst-equipped fleet.
SOURCE: Adapted with permission from McNutt et al. 1984, and the Society of Automotive Engineers, Inc.

nationwide rates. The tampering rates for catalytic converters and filler inlet restrictors (the insert in the fuel tank neck that prevents insertion of the larger leaded fuel nozzle) have increased steadily since 1978, whereas the rates for other components have fluctuated. The increasing tampering rates for catalytic converters and inlet restrictors may be partly due to the increasing age of the vehicles surveyed. In addition, the presence of inspection and maintenance programs affected tampering rates. The catalyst was removed in 3 percent of the vehicles in areas with mandatory inspection and maintenance programs and in 11 percent of the vehicles in areas having no programs.

Removing the catalytic converter increases HC and CO emissions by an average of 475 percent and 425 percent, respectively (U.S. Environmental Protection Agency 1983). For vehicles equipped with three-way catalysts, substantial increases in NO_x emissions would also be expected to occur. Tampering with the EGR system can increase NO_x emissions by an average of 175 percent (Greco 1985).

Fuel switching, defined as the presence of a tampered fuel filter inlet restrictor, a positive Plumbtesmo tailpipe test, or a gasoline Pb concentration of more than 0.05 g/gal, was found in 14 percent of the unleaded gasoline-powered vehicles in the 1984 survey (see table 2). Regional distribution in the prevalence of misfueling is shown in table 3. The impact of fuel switching on emissions depends upon its duration and certain vehicle characteristics, but emission increases of 475 percent for

HCs and 425 percent for CO can easily occur (Greco 1985).

The tampering rate for light-duty trucks was equal to or higher than that for automobiles in every tampering category, as shown in table 2. The difference in prevalence of catalytic converter tampering is particularly striking—nearly three times as prevalent in light-duty trucks as in passenger cars (14 percent versus 5 percent) (Greco 1985).

To confirm the EPA tampering and misfueling data, the MVMA recently studied catalyst removal and defeat of the fuel filler restrictor. The vehicles used in the MVMA survey were a sample of 1975–84 model year cars and light-duty trucks from scrapyards and impoundment areas in 10 cities (Motor Vehicles Manufacturers' Association 1985; Survey Data Research 1985).

The MVMA study sampled 1,865 vehicles, allowing the following conclusions to be reached to a 95 percent confidence level by Survey Data Research (1985):

1. Nationwide, 8.3% of all the vehicles in the sample were found to have their catalytic converters removed. This removal rate is significantly higher among older (i.e., 1975–1978) model year passenger cars and light-duty trucks.

2. The rate of fuel filler neck restrictor tampering on a national basis (7.3%) is slightly lower than the rate of catalytic converter removal (8.3%). Again, this tampering rate is higher among older (i.e., 1975–1979) model year cars and light-duty trucks.

3. Both catalytic converter removal and fuel filler neck restrictor tampering rates are substantially lower in the sample of Inspection/Mainte-

nance area locations than the sample of Non-Inspection/Maintenance area locations.

As a result of this study, MVMA is now confident that the much more detailed EPA studies, covering in addition such components as the air pump, EGR system, the positive crankcase ventilation (PCV) system, the evaporative emissions control system, and others, are yielding results that are reasonably representative of the in-use fleet (Motor Vehicles Manufacturers' Association 1985).

Effects of Tampering on Emissions. Recently, the Automobile Club of Southern California conducted a test program using its 1981 fleet vehicles (General Motors, Buick, and Pontiac vehicles) in an effort to better understand the effect of system component failures. The primary objective of the program was to determine the degree to which fuel economy, exhaust emissions, horsepower, and driveability are affected by disabling key components of a computer-controlled system; a secondary objective was to establish a method of accurately and efficiently identifying vehicles with disabled components (Jones et al. 1982).

Jones and coworkers (1982) found that disabling the coolant temperature sensor, the throttle position sensor, or the mixture control solenoid has a major effect on vehicle performance. Disconnection of the coolant temperature sensor increased HC emissions an average of 549 percent and CO emissions an average of 1,120 percent over baseline emission levels; disconnection of the throttle position sensor increased HCs by 1,195 percent and CO by 3,113 percent; and disconnection of the mixture control solenoid increased HCs by 1,293 percent and CO by 3,438 percent. Each of these disablings is the disconnection of a single electrical connector (Jones et al. 1982).

■ **Recommendation 1.** Tampering and Misfueling. Tampering and misfueling statistics are fairly well developed but their effect on emissions is not as well known. Therefore, work should be done to better characterize the effect of tampering and

misfueling on emissions from vehicles and to better assess their effect on ambient pollutant concentrations.

Diesel-Powered Passenger Cars

The available in-use data are much more limited for diesel passenger cars than for gasoline-powered cars for two reasons: first, there are far fewer of them, and second, most of the diesel cars that are in use were sold between 1979 and 1983 so only a small proportion of them are more than seven years old.

Hyde and coworkers (1982) drew the following conclusions about the relation between cumulative mileage and rate of emissions from a sample of 20 in-use light-duty diesel vehicles from General Motors, Volkswagen, and Mercedes-Benz.

1. Particulate emissions do not show a mileage-related deterioration (increase) in the Volkswagen group and the Mercedes-Benz group, but show a large deterioration in the General Motors group because of a large increase in extract emissions.

2. Federal Test Procedure HC emissions do not show a mileage-related deterioration in the Volkswagen and Mercedes-Benz groups, but show a deterioration in the General Motors group.

3. FTP CO emissions show a deterioration in the General Motors and Volkswagen groups but not in the Mercedes-Benz group.

4. FTP NO_x emissions show a decrease with accumulated mileage in the General Motors and Mercedes-Benz groups but no change for the Volkswagen group.

Diesel-Powered Trucks

There are limited data on diesel engines in operational use although there are some recent laboratory data obtained by the various manufacturers in an EPA/Engine Manufacturers Association (EMA) in-use emission factor test program for heavy-duty diesels. As part of this emission factor testing effort, the following classifications were included in the sample of 30 engines: tampered engines, poorly maintained engines, and rebuilt engines. The engines

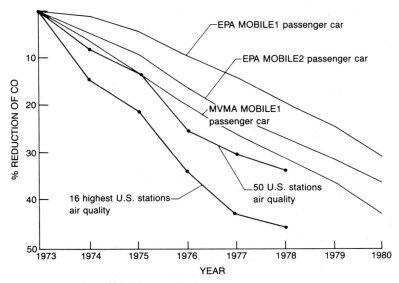

Figure 10. CO air quality and emission factor trend as calculated from three computer models and as measured from the base year 1973. CO measurements were averaged from 50 U.S. stations and from 16 U.S. stations reporting the highest 8-hr yearly concentrations. (Adapted from General Motors Corp. 1985a.)

were tested on FTP 13-mode steady-state and FTP Heavy Duty transient cycle.

The EMA reached the following tentative conclusions from this program (General Motors Corp. 1985b):

1. The in-use control of gaseous emissions from heavy-duty diesel engine from the 1979-80 model year is quite good.

2. Tampering and poor maintenance do not result in excessive gaseous emissions.

3. The lab-to-lab variability of transient emission test results of unburnt HCs as well as particulates needs to be improved.

■ **Recommendation 2.** Diesel Particulate Emissions and Control. There is a need for continued research on particulate control technology, including the regeneration systems, to reduce the cost and complexity of these systems and the associated fuel economy penalties. Work needs to continue with various additives, substrate materials, regeneration systems, and controls to develop optimum systems that are able to decrease the diesel particulate emissions to the levels of 0.1 g/bhp-hr for heavy-duty diesels and 0.08 g/mi for light-duty vehicles required in California. In conjunction with this research there is a need to measure

the metal species and the size distribution of the particles coming from diesel particulate traps.

Models for Predicting Future Emissions

Computer models are used for predicting future emissions from in-use vehicles. The EPA publishes the vehicle emissions model most used at present (U.S. Environmental Protection Agency 1985). The highway source data are based on MOBILE3, a computer program issued by the EPA in June 1984 and recently updated (U.S. Environmental Protection Agency 1985). Figure 10 shows the predicted trends of various models compared to actual air quality data for CO. The curve for percent reduction of CO predicted by MOBILE2 does not correspond to the curves generated by air quality measurements from 50 U.S. stations or the 16 highest U.S. stations (General Motors Corp. 1985a).

There are large differences between General Motors' analysis of the actual emissions data and the EPA MOBILE3 emission factor data beyond 50,000 miles, with

MOBILE3 being the higher (General Motors Corp. 1985a). General Motors attributes the difference to EPA's choosing too high a bhp-hr/mi constant for gasoline-as well as diesel-powered heavy-duty vehicles.

In addition, General Motors is concerned that the evaporative submodel in MOBILE3 uses a value of 11.5 psi for Reid Vapor Pressure (RVP) for gasoline volatility whereas the national average is 10.5 psi. In fact, the RVP varies with the season of the year in different parts of the country as formulations matched to seasonal conditions are refined and delivered to the pumps. It is unlikely that the model will ever give good results if a single RVP number is used to represent evaporation characteristics in all places at all seasons. Instead, the United States should be subdivided into the American Society for Testing Materials (ASTM) class regions to allow for seasonal changes in RVP. The model would then use RVP values that are representative of the season and region of the country. Other problems include estimating the number of trips per day for an average vehicle, identifying an appropriate ambient temperature, and understanding the effect of fuel aging.

Furthermore, a new approach using a proportion of vehicles in each model year with emission rates in each of a number of incremental ranges, that is, a distribution for emission rates within each model year, should be developed for modeling emission rates. The model needs to account for the few high-emission vehicles as well.

The evaporative emissions submodel needs additional work so that it better simulates the actual field fuel and control system effects, because the actual and test fuels have different vapor pressures. Regional and seasonal differences in RVP should be incorporated in the model along with the effects of alcohols.

■ **Recommendation 3.** Evaporative Emission Model. An improved vehicle evaporative emissions model should be developed that is valid over various types of operating conditions for a variety of ambient temperatures. At the same time, changes should be made in the EPA test procedure to obtain the

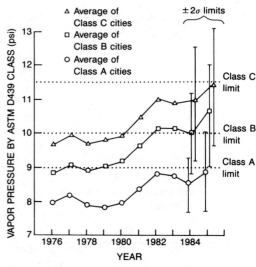

Figure 11. Trends of gasoline RVP averaged by class of cities. Classification of cities by the ASTM D439 is based on weather conditions and geographical location. (Adapted with permission from Ford Motor Co. 1985b.)

data necessary to properly design and size the evaporative system for the high-temperature soak situation, and data should be sought that can be used in EPA's MOBILE3 computer model for other use patterns of cars.

Fuels and Fuel Additives

Trends in Gasoline Fuel Properties

The EPA limited the use of Pb in gasoline to 0.5 g/gal after July 1, 1985, and to 0.1 g/gal after January 1, 1986. This has increased refineries' interest in the use of alternative low-cost octane boosters, particularly light alkanes and methanol and/or ethanol alcohols blended with gasoline.

Figure 11 shows the recent upward trend in the RVP that has resulted from these industry trends (Ford Motor Co. 1985b). Through 1980, the average RVP of the fuels sampled stayed reasonably close to the specification of the certification fuel (9.0 psi RVP). The increase since 1980 results from the petroleum industry's use of more light stock, such as butanes, in the gasoline. Historically, the petroleum industry has favored adding light HCs to gasoline for economic reasons. Large quantities of bu-

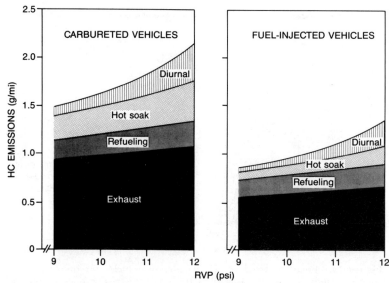

Figure 12. Contribution to total HC emissions from various routes and processes as function of gasoline RVP. Evaporative emissions are derived from nonoperating vehicles parked overnight (diurnal); recently turned off, nonoperating vehicles (hot soak); and vehicles at filling stations receiving fuel (refueling). Exhaust emissions are derived from the tailpipe. (Adapted with permission from Stebar et al. 1985, and the Society of Automotive Engineers, Inc.)

tane, a volatile HC, are produced during the refining of crude oil and natural gas. Butane has a high research octane number (about 94), and it is a good substitute for Pb in gasoline blending. The addition of butane increases the "front end" volatility of a gasoline. High fuel volatility increases automotive evaporative emissions and increases vapor losses from fuel tanks by displacement during refueling (Stebar et al. 1985).

Stebar and coworkers (1985) analyzed a large data base (267 cars from 1978 to 1985 with 141 from 1981) to develop figure 12. The figure illustrates the importance of different HC emission routes and the contributions via individual routes to total vehicle HC emissions for carbureted as well as fuel-injected cars. They observed that:

1. Evaporative emissions (primarily diurnal losses) are the major contributor to the increase in vehicle HC emissions with increase in RVP.

2. Hot soak emissions (particularly with carbureted cars) are a larger contributor to HC emissions than are diurnal losses.

3. Refueling and exhaust HC emissions have low sensitivity to changes in RVP.

4. Exhaust emissions are the largest con- *tributor to total HC emissions and constitute about the same proportion of the total for both carbureted and fuel-injected cars. At 12 RVP, exhaust emissions represent about half of total HC emissions for both types of engines.*

Furey (1985) measured the vapor pressures and distillation characteristics of a large number of gasoline/alcohol and gasoline/ether fuel blends. In that study, the maximum increase in RVP above that of gasoline ranged from 0.2 psi for *tert*-butyl alcohol to 3.4 psi for methanol. As little as 0.25 percent methanol, ethanol, and Oxinol[TM] 50 (a 1:1 mixture of methanol and gasoline-grade *tert*-butyl alcohol) was found to produce measurable increases in RVP.

The EPA estimates that the difference in volatility between the certification fuel and commercial gasoline is responsible for about half of the evaporative emissions from late-model light-duty vehicles and that this trend will continue, as shown in figure 13, if no action is taken to change it (Schwarz 1985).

The Coordinating Research Council–Air Pollution Research Advisory Committee (CRC-APRAC) is also investigating an-

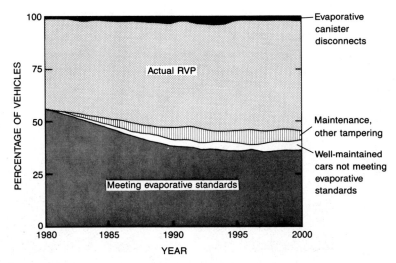

Figure 13. Predicted trend of the fraction of vehicles meeting evaporative emission standards and the reasons why the remaining fraction does not meet the standards. Because the RVP of commercial gasoline is different from the RVP of certification fuel, a significant percentage of vehicles will not meet the standard for this reason alone. (Adapted with permission from Schwarz 1985, and the Air Pollution Control Association.)

other important gasoline fuel issue—benzene emissions. Their preliminary findings from testing specially blended fuels in five late-model cars with three-way catalysts show that the benzene fraction of exhaust HCs increases with increasing benzene content and aromaticity. In refueling and evaporative emissions the benzene fraction increases with benzene content but not with aromaticity (Coordinating Research Council–Air Pollution Research Advisory Committee 1985).

Fuel Usage Trends

Although total gasoline usage has moved slowly upward during the past three years (from 6.5 million barrels per day (MMB/D) to 6.8 MMB/D by 1985), this trend may be temporary. The U.S. Department of Energy (1985) projects that gasoline demand will turn downward in the balance of the 1980s and remain flat in the early 1990s as shown in figure 14. By 1995, total gasoline consumption is projected to be 6.1 MMB/D (8.1 percent below 1983 levels). This number could be somewhat higher if the U.S. Department of Trans-

portation establishes the post-1988 fuel economy standards at 26 mpg.

Total diesel highway fuel demand will continue to grow over the next two decades primarily because of increased use of diesel engines in heavy-duty vehicles. Total highway diesel fuel usage is projected to rise 30 percent from 1.0 MMB/D in 1982 to 1.3 MMB/D in 1995, as shown in figure 14. Figure 15 shows the breakdown of projected fuel usage by application including off-highway usage (U.S. Department of Energy 1985).

Methanol-Fueled Vehicles

From an energy perspective, methanol is one of the most promising long-term alternative fuels for motor vehicles. It can be made from natural gas now and from coal later. One major practical problem is that motor vehicle consumption for a fuel has to reach 10 percent of the present market to create an economically viable free market (Society of Automotive Engineers/U.S. Department of Energy 1985). For the use of methanol to become widespread, it should be competitive in price with gasoline. Gas-

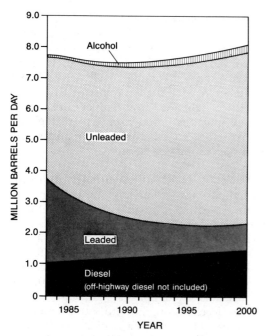

Figure 14. Projected motor fuel consumption by fuel type. (Adapted from the U.S. Department of Energy 1985.)

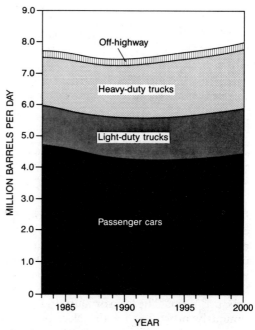

Figure 15. Projected motor fuel consumption by vehicle type. (Adapted from the U.S. Department of Energy 1985.)

oline prices would probably have to exceed $1.50/gal (1985 dollars) for a significant period of time to provide the necessary confidence for investors in methanol processing facilities and car buyers (Sobey 1985). The EPA is, overall, encouraging the use of methanol as outlined by Gray (1985).

Spark-Ignition Engines for Passenger Cars. The technology for methanol-fueled vehicles exists and demonstration fleets have been tested. A summary of the emission results obtained to date for the 1983 Ford Escort fleet is given by Nichols and Norbeck (1985). Overall the vehicles averaged 6,800 miles with a range of 3,100 to 20,300 miles. The average formaldehyde emission rate varied from 54 mg/mi to 79 mg/mi and accounted for 7.0 to 8.8 percent of the reactive HC mass on a mole-of-carbon basis. The formaldehyde as a percent of reactive HC in the exhaust for any individual vehicle ranged between 5.0 and 17.9 percent.

A recent EPA summary of methanol emissions data has been documented by Alson (1985). The HC emissions are largely

methanol, and the aldehydes are nearly all formaldehyde. Figure 16 is a summary of formaldehyde emission data (using the FTP) comparing methanol-, diesel-, and gasoline-powered vehicles, the latter with three-way catalysts, oxidation catalysts, and no catalyst (Alson 1985). Methanol-powered vehicles have higher formaldehyde emissions than diesel-powered or catalyst-controlled gasoline-powered vehicles.

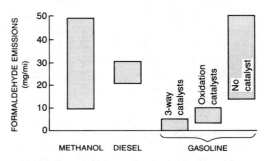

Figure 16. Comparison of formaldehyde emissions from methanol-, diesel-, and gasoline-powered vehicles, the latter with three-way catalysts, oxidation catalysts, and no catalyst. (Adapted from Alson 1985.)

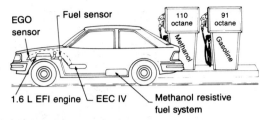

Figure 17. Schematic of Ford Escort modified to use a flexible fuel system installation. Electronic engine controller (EEC); electronic fuel injection (EFI); exhaust gas oxygen (EGO). (Adapted with permission from Wineland 1985, and the Ford Motor Co.)

Ford has recently discussed the concept of a methanol/gasoline flexible fuel system that would accept either methanol or gasoline. An electronic fuel-injected Escort was modified to use an optical fuel sensor for determining the methanol/gasoline mixture ratio. The sensor output is continuously processed by the electronic engine controller which optimizes fuel quantity and spark timing in response to the methanol/gasoline mixture ratio. This system was tested on a 1983 Escort-Lynx having a 1.6-liter electronic fuel injection engine in a production vehicle. The vehicle used the production engine compression ratio of 9.0, while the fuel tank, fuel filter, and the electric fuel pump were replaced with parts that methanol would not corrode. Figure 17 shows a schematic of the system in the vehicle (Wineland 1985).

Spark-Assisted, Compression-Ignition and Stratified-Charge, Spark-Ignition Buses.
Methanol is considered to be a good choice as an alternative fuel for buses for several reasons. First, buses are usually a fleet operation so that methanol fuel distribution should be significantly easier than in the consumer market. Second, the particulate and odor emissions are less than those of the diesel engines it would replace. Third, the use of methanol should improve the reactivity of the exhaust, although the methanol and formaldehyde emissions could be a problem if control systems are not properly developed and maintained.

Lipari and Keski-Hynnila (1985) studied the effect of a catalyst on formaldehyde emissions of a methanol-fueled, two-stroke diesel bus engine and found that even with this catalyst, emissions were still higher than those from conventional diesels in the steady-state 13-mode cycle. Additional data are needed for the transient FTP cycle and for light-load low-temperature operation, since the production of formaldehyde across the catalyst could occur under certain operating conditions.

Areas in Need of Additional Research.
Methods for accurately measuring HC emissions from methanol-fueled vehicles are lacking. The HC unburnt fuel of a neat (100 percent) methanol vehicle is basically methanol. Measurements of HC and formaldehyde concentrations have not been developed yet that can show how high the individual excursions are under acceleration, deceleration, and other transient conditions. There is a particular lack of data taken under light-load or idling conditions, especially of operation at low temperatures. Data show that the NO_x concentration can increase as exhaust from methanol-fueled vehicles passes across a catalyst. Further work needs to be done to understand this effect and to control it properly in the field or determine if it is merely a measurement problem. There is a need to study the worst-case dispersion situations outlined by Harvey et al. (1984) using these new methanol and formaldehyde concentration data. Similar experimental field studies with real-time instrumentation should also be gathered so as to ensure that methanol technology is safe in the hands of the consumer. In these latter studies, misfueling and tampering should be monitored and their effects measured, for we know they occur in gasoline-powered vehicles.

■ **Recommendation 4.** Formaldehyde Measurements. Real-time measurements of formaldehyde concentration should be performed under transient and extreme conditions such as acceleration and deceleration, low temperature, light load, and extended idling with restricted ventilation. This research work should be done with and without catalysts since worst-case conditions in the field will occur with catalysts removed.

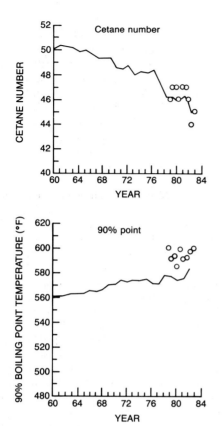

Figure 18. Historical trends of diesel fuel properties. Curve represents data from a DOE survey of type T-T diesel fuel. Open circles represent data from a MVMA survey of No. 2 diesel fuel. (Adapted with permission from Wade and Jones 1984, and the Society of Automotive Engineers, Inc.)

Similar measurements should also be made on bus engines.

Trends in Diesel Fuel Properties

Recent trends in diesel fuel properties have an adverse effect on particulate emissions. They make it harder to meet stringent particulate emission standards for cars and trucks (0.2 g/mi in 1987 for cars and 0.6 g/bhp-hr in 1988, 0.25 g/bhp-hr in 1991 for trucks) because the EPA certification is based on typical in-use fuels. An automotive quality No. 2 diesel fuel with low sulfur and low aromatics is necessary if low particulate emissions are to be achieved (Weaver et al. 1986). Two important fuel characteristics affecting diesel engine emissions have been deteriorating in recent years:

the cetane number has been falling and 90 percent boiling point has been rising, as shown in figure 18 (Wade and Jones 1984).

■ **Recommendation 5.** Automotive Quality No. 2 Diesel Fuel. An automotive quality No. 2 diesel fuel with low sulfur and low aromatics is necessary if low particulate emissions are to be achieved. Research should be undertaken in cooperation with the automotive and petroleum industries to decide on effective and economical cetane number, sulfur and aromatic content, and 90 percent boiling point temperature specification limits for automotive quality No. 2 diesel fuel, and to formulate fuels that meet the specifications. The need for this research becomes more urgent as diesel fuel usage continues to increase. Improved emissions and more control options require a quality low-sulfur diesel fuel.

Refueling Emissions

The basic source of HC emissions associated with the vehicle refueling process is the vapors contained in vehicle fuel tanks that are displaced by gasoline during refueling operations. However, additional emissions are associated with vehicle refueling operations as the result of "breathing losses" from underground storage tanks at gasoline service stations. Stage I (delivery of gasoline to station) vapor recovery is approximately 95 percent efficient (Austin and Rubenstein 1985).

At Stage II (dispensing of gasoline to vehicle fuel tanks) vapor recovery, gasoline vapors are collected at the vehicle fillpipe opening using a nozzle spout. The nozzle is also equipped with a vapor passage in the body of the nozzle that connects the annular space between the spout and the boot to the vapor space in the underground storage tank (Austin and Rubenstein 1985). A Stage II system reduces fillpipe emissions by 85–95 percent. Such systems are being used successfully in California.

The automotive industry, the petroleum industry, and the EPA are debating whether refueling losses should be controlled by on-board vehicle systems or by Stage II systems (Austin and Rubenstein 1985; Schwarz 1985).

Table 4. General Fuel Additive Classification and Typical Bulk Treatment Ranges

Additive Type	Use Classification	Treatment Ranges			
		Gasoline		Diesel	
		PTB[a]	ppm	PTB[a]	ppm
Amine detergent	Performance	3–30	12–120	10–60	33–200
Polymeric dispersant	Performance	5–150	20–600	10–100	33–330
Fluidizer oils	Performance	50–250	200–1,000	NA	
Antiicers	Performance	4–15	16–60	NA	
Combustion modifiers[b]	Performance	(Up to 2.5 g/gal)		400–1,000	1,300–3,300
Corrosion inhibitors	Distribution	1–10	4–40	1–10	3–33
Antioxidant	Quality	3–5	12–20	2–8	7–26
Metal deactivator	Quality	1–4	4–16	1–4	4–16
Demulsifiers	Distribution	0.1–2.5	0.4–10	0.1–2.5	0.3–8
Flow improver	Performance	NA		15–150	50–500

[a] Pounds of fuel additive per 1,000 barrels of fuel where 1 barrel = 42 gal.
[b] Combustion modifiers = octane (gasoline) or cetane (diesel) number improvers.
NOTE: NA = not applicable.
SOURCE: Adapted with permission from Tupa and Doren 1984, and the Society of Automotive Engineers, Inc.

By use of onboard control systems, vapors displaced from the vehicle tank are vented to an enlarged canister where they are absorbed and subsequently purged into the engine (Austin and Rubenstein 1985; Schwarz 1985). A separate canister to control refueling emissions or an enlarged evaporative canister could be used.

In a further detailed analysis of Stage II and onboard control, Austin and Rubenstein (1985) reached the following general conclusion: "the implementation of Stage II controls is a clearly superior alternative to the onboard control concept." Their specific reasons for this conclusion were:

1. Stage II controls have been proven in California and they can achieve about 85 percent control.
2. Stage II controls are the more cost-effective, that is, $0.21/lb HC are reduced versus $0.66 to $2.25/lb HC depending on whether the EPA's or Ford's cost estimate is used for onboard control. The onboard systems can be made more cost-effective with additional evaporative emissions control.
3. Stage II controls give better short-term control because of lead time, and vehicle turnover due to replacement, among others as shown by Austin and Rubenstein (1985).

■ **Recommendation 6.** Evaporative Emissions Control. Some combination of field

RVP control along with a test fuel with typical RVP (or calculation corrections for RVP differences) should be developed. Controlling RVP in motor gasoline, an approach successfully applied in California, is needed generally for controlling field evaporative emissions. The question of whether car manufacturers should be testing with a worst-case RVP test fuel or a typical fuel needs further study.

Additives

Additives are used to improve engine performance and durability and to ensure that fuel specifications and quality are maintained during transport and storage. They are an integral part of today's fuels. Tupa and Doren (1984) discuss in great detail the specific functions and benefits of additives, typical use levels, and test methods for evaluation. Generic types of additives and their uses are shown in table 4 along with general levels of additive treatment for the various types of additives. The variety of chemical compounds used in gasolines today are listed in table 5 and table 6 (Tupa and Doren 1984).

How additives may affect control technologies needs additional research. Knowledge of the size distribution of particles from diesel particulate traps and the metal species they contain. Data on operation

Table 5. Chemicals Typically Used for Gasoline Additives

Additive Type	Chemicals
Amine detergent	Amines
	Alkanol amines
	Amides
	Amido-amines
	Imidazolines
Polymeric dispersants	Alkenyl succinimides
	Hydrocarbyl amines
	Polyether amine
Fluidizer oils	Selected mineral oil
	Thermally stable polyolefin (polypropylene) of moderate molecular weight (800–1,000) and narrow molecular weight distribution
	Ester-type synthetic lubricant
Combustion modifiers	Tetraethyl lead (TEL) (only in leaded fuels, which are less than 35% of gasoline sold today)
	Tetramethyl lead (TML)
	Methylcyclopentadienyl Manganese tricarbonyl[a] (MMT)
Antiicers	Alkenyl succinates and amine salts
	Monocarboxylic acids and amine salts
	Imidazolines and carboxylic acid salts
	Amine alkylorthophosphates
	Ethoxylated alkyl phenols
Corrosion inhibitors	Alkenyl succinic acids, esters, and amine salts
	Dimer acid and other carboxylic acids and amine salts
	Mixed alkyl orthophosphoric acids and amine salts
	Alkyl phosphonic acids and amine salts
	Aryl sulfonic acids and amine salts
	Mannich amines
	Carboxylic acid salts of Mannich amines
Demulsifiers	Long-chain alkylphenols
	Long-chain alcohols
	Long-chain carboxylic acids
	Long-chain amines

(Table continued next column.)

Table 5. *Continued*

Additive Type	Chemicals
Antioxidants	Alkyl- or aryl-substituted phenolenediamines
	Alkyl- or aryl-substituted aminophenols
	Alkyl- or aryl-substituted phenols
Metal deactivators	Fuel treated with N,N'-disalicylidene-1,2-propane
	Diamine which produces copper chelate

[a] These additives are not used in unleaded gasoline, and MMT was specifically banned by the 1977 Clean Air Act Amendments.
SOURCE: Adapted with permission from Tupa and Doren 1984, and the Society of Automotive Engineers, Inc.

with as well as without the working traps are needed since tampering of control devices can occur in the field. The effect of the additive compounds that plug the trap pores also needs further study.

■ **Recommendation 7.** Diesel Fuel Additives. Data should be obtained about the size distribution of particles in diesel exhaust and about the metal species they contain, with and without a particulate trap, with a diesel fuel containing a typical additive under consideration for production use. Data on the HCs bound to the particles and the vapor-phase HCs should also be obtained.

Methods for Measuring the Unregulated Pollutants*

The unregulated pollutants in automotive exhaust have been measured with varying degrees of sophistication for the past 20 years. Interest in a particular pollutant varies as studies of its potential health effects are reported; benzo[a]pyrene represents a good example. Since most of the unregulated pollutants are present only in small amounts in exhaust (in the parts-per-

*This section was written by David Leddy, Michigan Technological University.

million [ppm] range or less) and very small amounts in the ambient air (in the parts-per-billion [ppb] range or less) their amount or concentration is measured only with great difficulty and usually at high expense. There is little doubt that there is a need to balance the degree of difficulty, the cost, and the sensitivity against the real value the procedure produces in assessing health effects or engine performance.

Sampling

Since most of these pollutants are found at low concentrations, nearly all methods of analysis call for collecting a sample over an extended time interval and concentrating it before analysis. Samples are frequently collected by the use of impingers, filters (Evans 1980; Perez et al. 1980; Gorse and Salmeen 1982; Gross et al. 1982; Fox 1985), and solid sorbents (Hampton et al. 1982; Fox 1985).

Sampling of the exhaust may be more important in determining the value of the analysis than the actual measurement itself. For example, there is every reason to believe that during the sampling of particulates on a filter, chemical reactions take place between the organic compounds in the particulates and gaseous or aerosol compounds such as nitric acid (HNO_3), NO_2, and sulfuric acid. These reactions produce the so-called "artifacts of sampling" that are of concern to all who work in this field (Lee et al. 1980; Perez et al. 1980; Pierson et al. 1980; Gorse and Salmeen 1982; Herr et al. 1982; Risby and Lestz 1983). One of the possible sampling artifacts of greatest concern is the formation of the biologically active compounds nitropyrene and nitrobenz[a]pyrene from the reaction of NO_2 with the relatively innocuous compounds pyrene and benzo[a]pyrene, respectively (Gibson et al. 1980; Schuetzle et al. 1980). Other artifacts of concern are the formation of HNO_3 and sulfuric acid on the surface of the sampling material. The effects of artifact formation can be minimized by reducing the length of time a filter is exposed to the exhaust stream to the minimum required to collect

Table 6. Chemicals Typically Used as Diesel Additives

Additive Type	Chemicals
Detergents	Same as for gasoline (table 5)
Polymeric dispersants	Same as for gasoline (table 5)
Combustion modifiers	
Cetane improvers	Alkyl nitrates and nitrites
	Nitro and nitroso compounds
	Peroxides
Combustion catalysts	Organo compounds of barium, calcium, manganese, and iron for changing output particulate emissions
	Organo compounds of manganese, copper, lead, cerium, or combinations of above metals in particulate traps to reduce regeneration temperatures
Deposit modifiers	Barium, calcium, or manganese
Flow improvers	Ethylene vinyl acetate polymers
	Chlorinated hydrocarbons
	Polyolefins
Demulsifiers	Same as for gasoline (table 5)
Corrosion inhibitors	Same as for gasoline (table 5)
Biocides (to inhibit bacteria growth at water/ fuel interface)	Most fuels in United States
	Borate esters
	Quarternary ammonium salts of salicylic acids
	Diamine complexes of nickel
	Organo barium compounds
	Glycol ethers
Stabilizers	Same as for gasoline (table 5)
Antioxidants	Tertiary amines
	Imidazolines
	Tertiary alkyl primary amines
Metal deactivators	Same as for gasoline (table 5)

SOURCE: Adapted with permission from Tupa and Doren 1984, and the Society of Automotive Engineers, Inc.

Table 7. Summary of Analytical Methods for Characterizing Unregulated Emissions from Spark-Ignition and Compression-Ignition Engines

Emission	Sampling	Analytical Method	Reference
Aldehydes	Impinger/DNPH	GC, HPLC, colorimetry	Perez et al. (1980, 1984) Fox (1985)
HCN, cyanogen	Impinger/KOH	GC/ECD	Cadle et al. (1979); Perez et al. (1980)
Organic sulfides	Tenax traps	GC/FID	Cadle et al. (1979); Perez et al. (1980)
Ammonia	Impinger/H_2SO_4	IC	Cadle et al. (1979); Perez et al. (1980)
Organic amines	Impinger/H_2SO_4	GC/NPD	Cadle et al. (1979); Mulawa and Cadle (1979)
NO_x	Tedlar bag	GC/ECD	Cadle et al. (1979)
SO_2	Impinger/H_2O_2	Colorimetry, IC	Perez et al. (1980)
	Real-time	Electrochemical, fluorescence spectroscopy; 2nd derivative spectroscopy	Fox (1985)
HCs	Tedlar bag, trap	GC/FID	Perez et al. (1980); Fox (1985)
Phenols	Impinger/KOH	GC/FID	Perez et al. (1980)
N-nitrosodimethylamine	Impinger/NaOH	GC/MS	Krost et al. (1982)
Benzo[a]pyrene	Filter, trap, polyurethane foam	GC/MS, HPLC	Schuetzle et al. (1980); Eisenberg (1983); Eisenberg et al. (1984)
Sulfates	Filter	Colorimetry, IC	Perez et al. (1980); Schuetzle et al. (1980)
Nitric acid	Filter	IC	Okamoto et al. (1983)
Metals	Filter	Atomic absorption, x-ray fluorescence, emission spectroscopy	Perez et al. (1980)

NOTE: DNPH = 2,4-dinitrophenylhydrazine; GC = gas chromatography; GC/ECD = GC with electron-capture detection; GC/FID = GC with flame ionization detection; GC/NPD = GC with nitrogen-phosphorus detection; GC/MS = GC/mass spectrometry; HCN = hydrogen cyanide; H_2O_2 = hydrogen peroxide; HPLC = high performance liquid chromatography; H_2SO_4 = sulfuric acid; IC = ion chromatography; KOH = potassium hydroxide; and NaOH = sodium hydroxide.

a suitable sample, by using inert materials for filter construction, and by cooling and diluting the exhaust stream prior to sample collection.

Analytical Methods

Not every analytical method used for characterizing emissions from spark-ignition engines is applicable for analysis of emissions from compression-ignition engines, because of interference from combustion products found in compression-ignition engines. Diesel engines produce higher levels of particulates, NO_x, sulfur oxides (SO_x) and certain HCs, all of which can interfere with one or more of the analyses that are commonly used on spark-ignition engine emissions. Some real-time monitoring techniques based on the absorption of light fail when applied to diesel exhaust analysis, either because of scatter-ing of light by suspended particulates or absorption of light by aromatic HCs present in the gaseous phase. Electrochemical methods are affected because particulates foul the membranes and electrode surfaces used in the measuring cells. Applying some of the methods used for continuous monitoring of chemical species in ambient air is even more difficult when one considers that spark-ignition as well as compression-ignition engines generate interfering species that affect the sensitivity, accuracy, and repeatability of the analyses.

The analytical methods used to measure concentration or amount of unregulated pollutants are summarized in table 7. It should be emphasized that these are the analytical methods presently used in laboratories where measurements are being made on a regular basis for judging engine performance. They meet current requirements but will not necessarily meet the requirements of the future.

We need to find out which of the unregulated pollutants must be measured to evaluate advanced technology for the control of emissions from gasoline as well as diesel engines. It would be folly to measure the concentrations of pollutants just because they are there. Present methods of analysis are so tedious, expensive, and unreproducible that unnecessary analyses are to be avoided whenever possible.

Areas in Need of Additional Research. Gaps in unregulated emission measurement methods center on the lack of real-time measurement methods that have the sensitivity required for producing results at moderate costs. If advanced emission control technology is to be studied with transient-cycle test protocols, these real-time techniques are necessary.

Real-time measurements based upon piezoelectric devices, tunable diode laser systems, thermal lens spectroscopy, long-path differential optical absorption spectroscopy, ultraviolet fluorescence spectroscopy, and differential absorption lidar have been reported by Fox (1985). The studies reported in these cases are normally of ambient air with no concern for interferences that may be present in gasoline and diesel exhaust. Pitts et al. (1984) reported the measurement of gaseous HNO_3, NO_2, formaldehyde, SO_2, and benzaldehyde in the exhaust of light-duty vehicles, using an instrument that coupled a multiple reflection cell to a differential optical absorbtion spectrometer. The techniques hold promise for the future and should be explored in more extensive studies.

■ **Recommendation 8.** Emissions Measurement Methods. Studies should begin immediately with an evaluation of the best available emissions data on engines operating with and without emission control devices, to determine which of the unregulated pollutants really pose a potential threat to human health. Other unregulated pollutants might be added to this list if their concentrations reflect engine or emission control device performance. Next, every

effort should be made to improve the analytical procedures presently used to measure the concentrations of those pollutants, to the point where they can be readily carried out by technicians. This may require that packaged sets of reagents and equipment be marketed for a specific analysis. For example, prepacked traps might be available for collecting gaseous HC prior to thermal desorption onto a gas chromatograph with a specified capillary column for the analysis of specific HCs at predetermined conditions.

Current Regulated and Unregulated Emissions

The main focus of this section is unregulated organic emissions, for significant data on regulated emissions and non-organic unregulated emissions from in-use vehicles have been presented already. MOBILE3, a computer model discussed earlier, is the best source of data about regulated emissions since the EPA analyzes all manufacturers' data and develops sales-weighted emission factors (U.S. Environmental Protection Agency 1985). Emission factors for regulated pollutants, based on California Air Resources Board (1980) data, are also available.

Regulated Emissions

Table 8 from the National Research Council (1983b) shows a summary of the regulated emissions from light-duty vehicles. Imposing the HC, CO, and NO_x standards has resulted in 84, 79, and 56 percent reductions, respectively, in 50,000-mi emissions from gasoline-powered, spark-ignition vehicles.

Unregulated Emissions

Gas-Phase Hydrocarbons. The components detected as gas-phase HCs are listed in table 9 (National Research Council 1983b). In another report, the National Research Council (1983a, appendix A) prepared an extensive list of vapor-phase compounds

Table 8. Exhaust Emission Rates[a] for Light-Duty Gasoline-Powered Vehicles

Emission Component	Model Year	Zero-Mile Emission Rate (g/mi)	50,000-Mile Emission Rate[b] (g/mi)
HCs	Pre-1968	7.25	8.15
	1968–1969	4.43	5.68
	1970–1971	3.00	4.85
	1972–1974	3.36	4.21
	1975–1979	1.29	2.74
	1980	0.29	1.74
	1981	0.39	1.34
	1982+	0.39	1.34
CO	Pre-1968	78.27	89.52
	1968–1969	56.34	69.09
	1970–1971	42.17	57.82
	1972–1974	40.78	52.98
	1975–1979	20.16	34.46
	1980	6.14	20.44
	1981	5.60	19.35
	1982	5.21	19.01
	1983+	5.00	18.80
NO_x	Pre-1968	3.44	3.44
	1968–1972	4.35	4.35
	1973–1974	2.87	3.07
	1975–1976	2.43	2.63
	1977–1979	1.69	2.19
	1980	1.56	2.06
	1981+	0.75	1.50

[a] Emission rates are for low-altitude 49-state vehicles. High-altitude and California emission rates are different.

[b] The 50,000-mile emission rates are calculated from zero-mile rate by addition of term that takes account of EPA-projected deterioration rate of vehicle combustion and emission-control systems.

SOURCE: Adapted with permission from the National Research Council 1983b.

in both diesel- and gasoline-powered vehicles by reviewing 250 papers in the literature.

Diesel Exhaust Particulate.

Diesel exhaust particulate material has been the subject of extensive study in the past five years. It is typically about 25 percent extractable into organic solvents, although different vehicles may have extractable fractions of 5–90 percent, depending to some extent on operating conditions. More than half the extractable material is aliphatic HC of 14–35 carbon atoms, alkyl-substituted ben-

Table 9. Unregulated Gaseous Hydrocarbons Emitted from Vehicles

All n-alkanes from n-butane through n-hexacosane
Four methyl-substituted butanes
Ten methyl- and ethyl-substituted pentanes and 11 cyclopentanes
Eleven methyl- and ethyl-substituted hexanes and 35 cyclohexanes
Fifteen methyl- and ethyl-substituted heptanes
Five methyl-substituted octanes
One methyl-substituted nonane
One methyl-substituted decane
One methyl-substituted undecane
Decalin and two methyl-substituted decalins
Two C_{10} alkanes
Eleven C_{11} alkanes
Nine C_{12} alkanes
Thirteen C_{13} alkanes
Eleven C_{14} alkanes
Eight C_{15} alkanes
Eight C_{16} alkanes
Five C_{17} alkanes
Three C_{18} alkanes
Seven methyl-substituted butenes and two methyl butadienes
Eighteen pentenes and pentadiene
Fourteen hexenes
Six heptenes
Four octenes
Decene and dodecene through heneicosene
Seven cyclic olefins
Seventy-one alkyl-substituted benzenes
Eight styrenes and the three xylenes
Fourteen indans and three indenes
Twenty-eight alkyl-substituted naphthalenes
Three alkylthiophenes and two benzothiophenes
Two alkylsulfides and one alkylamine
Six nonaromatic alcohols and eight aromatic alcohols
Eighteen aliphatic and aromatic aldehydes
Six furans, 17 ketones, and six organic acids

SOURCE: Adapted with permission from the National Research Council 1983b.

zenes, and derivatives of the polycyclic aromatic hydrocarbons (PAH) such as ketones, carboxaldehydes, acid anhydrides, hydroxy compounds, quinones, nitrates, and carboxylic acids. There are also heterocyclic compounds containing sulfur, nitrogen, and oxygen atoms within the aromatic ring. The alkyl-substituted PAHs and PAH derivatives tend to be more abundant than the parent PAH compound (National Research Council 1983b).

Table 10. Qualitative Analysis of Nonpolar and Moderately Polar Fractions of Diesel Particulate Extract

Compound	Approximate Concentration in Oldsmobile Extract (ppm)
Nonpolar fractions	
Phenanthrenes and anthracene	600
Methylphenanthrenes and methylanthracenes	1,400
Dimethylphenanthrenes and dimethylanthracenes	3,000
Pyrene	1,700
Fluoranthene	1,400
Methylpyrenes and methylfluoranthenes	800
Chrysene	100
Cyclopenta[cd]pyrene	20
Benzo[ghi]fluoranthene	100
Benz[a]anthracene	500
Benzo[a]pyrene	40
Other PAHs, heterocyclics	30,000
HCs and alkylbenzenes	500,000
Total nonpolar fractions	539,700
Moderately polar fractions	
PAH ketones	
Fluorenones	4,000
Methylfluorenones	400
Dimethylfluorenones	200
Anthrones and phenanthrones	1,600
Methylanthrones and methylphenanthrones	1,600
Dimethylanthrones and dimethylphenanthrones	1,300
Fluoranthones and pyrones	1,200
Benzanthrones	200
Xanthones	300
Methylxanthones	200
Thioxanthones	1,600
Methylthioxanthones	900
Total	13,500
PAH carboxaldehydes	
Fluorene carboxaldehydes	1,600
Methyl fluorene carboxaldehydes	400
Phenanthrene and anthracene carboxaldehydes	2,600
Methylanthracene and methylphenanthrene carboxaldehydes	1,600
Dimethylanthracene and dimethylphenanthrene carboxaldehydes	400
Benz[a]anthracene, chrysene, and triphenylene carboxaldehydes	400
Naphthalene dicarboxaldehydes	300
Dimethylnaphthalene carboxaldehydes	300
Trimethylnaphthalene carboxaldehydes	1,000
Pyrene and fluoranthene carboxaldehydes	1,600
Xanthene carboxaldehydes	600
Dibenzofuran carboxaldehydes	400
Total	11,200
PAH acid anhydrides	
Naphthalene dicarboxylic acid anhydrides	3,000
Methylnaphthalene dicarboxylic acid anhydrides	1,000
Dimethylnaphthalene dicarboxylic acid anhydrides	500
Anthracene and phenanthrene dicarboxylic acid anhydrides	600
Total	5,100

(Table continued next page.)

Table 10.   *Continued*

Compound	Approximate Concentration in Oldsmobile Extract (ppm)
Hydroxy PAHs	
Hydroxyfluorene	1,400
Methylhydroxyfluorene	400
Dimethylhydroxyfluorene	1,500
Hydroxyanthracenes and hydroxyphenanthrenes	600
Hydroxymethylanthracenes and hydroxymethylphenanthrenes	900
Hydroxydimethylanthracenes and hydroxydimethylphenanthrenes	1,300
Hydroxyfluorenone	2,000
Hydroxyxanthone	1,300
Hydroxyxanthene	1,000
Total	10,400
PAH quinones	
Fluorene quinones	700
Methylfluorene quinones	600
Dimethylfluorene quinones	500
Anthracene and phenanthrene quinones	1,900
Methylanthracene and methylphenanthrene quinones	2,000
Fluoranthene and pyrene quinones	200
Naphtho[1,8-*cd*]pyrene 1,3-dione	600
Total	6,500
Nitro-PAHs	
Nitrofluorenes	30
Nitroanthracenes and nitrophenanthrenes	70
Nitrofluoranthenes	10
Nitropyrenes	150
Methylnitropyrenes and methylnitrofluoranthenes	20
Total	280
Other oxygenated PAHs	8,000
PAH carryover from nonpolar fraction	6,000
Phthalates, HC contaminants	30,000
Total, moderately polar fractions	91,000

SOURCE: Adapted with permission from the National Research Council 1983b.

The particulate-extract in the high-performance liquid chromatograph (HPLC) eluent can be separated into nonpolar, moderately polar, and highly polar fractions. The fractions can then be further analyzed by gas chromatography/mass spectrometry (GC/MS). Table 10 lists the results of such an analysis of the nonpolar and moderately polar fractions of a particulate extract from an Oldsmobile diesel vehicle, including the approximate extract concentrations for this particular vehicle. The highly polar fraction has not been fully characterized. It contains the PAH carboxylic acids, acid anhydrides, and probably sulfonates and other highly polar species (National Research Council 1983b).

Most (75 percent) of the direct bacterial mutagenicity resides in the moderately polar fraction. The remaining direct mutagenicity is in the highly polar fraction. These aspects are discussed further in the National Research Council's report (1983b).

Over 50 chromatographic peaks of nitro-PAH compounds have been identified in diesel particulate extracts, as listed in table 11. The most abundant of the nitro-PAHs is 1-nitropyrene, ranging from 25 to 2,000 ppm in the vehicle extracts studies. The other nitro-PAHs are present at concentrations from below the ppm range to a few ppm. The nitropyrenes have been studied in greater detail than other PAH compounds. They are released in diesel and

Table 11. Nitroarenes in Diesel Exhaust Particulate Extracts

Mononitroarenes	Polynitroarenes
Nitroindene	Dinitromethylnaphthalene
Nitroacenaphthylene	Dinitrofluorene
Nitroacenaphthene	Dinitromethylbiphenyl
Nitrobiphenyl	Dinitrophenanthrene
Nitrofluorene	Dinitropyrene
Nitromethylacenaphthylene	Trinitropyrene
Nitromethylacenaphthene	Trinitro(C_5-alkyl)fluorene
Nitromethylbiphenyl	Dinitro(C_6-alkyl)fluorene
Nitroanthracene	Dinitro(C_4-alkyl)pyrene
Nitrophenanthrene	
Nitromethylfluorene	Nitro-oxyarenes
Nitromethylanthracene	Nitronaphthaquinone
Nitromethylphenanthrene	Nitrodihydroxynaphthalene
Nitrotrimethylnaphthalene	Nitronaphthalic acid
Nitrofluoranthene	Nitrofluorenone
Nitropyrene	Nitroanthrone
Nitro(C_2-alkyl)anthracene	Nitrophenanthrone
Nitro(C_2-alkyl)phenanthrene	Nitroanthraquinone
Nitrobenzofluorene	Nitrohydroxymethylfluorene
Nitromethylfluoranthene	Nitrofluoranthone
Nitromethylpyrene	Nitropyrone
Nitro(C_3-alkyl)anthracene	Nitrofluoranthenequinone
Nitro(C_3-alkyl)phenanthrene	Nitropyrenequinone
Nitrochrysene	Nitrodimethylanthracene carboxaldehyde
Nitrobenzoanthracene	Nitrodimethylphenanthrene carboxaldehyde
Nitronaphthacene	Other nitrogen compounds
Nitrotriphenylene	Benzocinnoline
Nitromethylnaphthacene or	Methylbenzocinnoline
nitromethylchrysene	Phenylnaphthylamine
Nitromethylbenzanthracene	(C_2-alkyl)phenylnaphthylamine
Nitromethyltriphenylene	
Nitrobenzopyrene	
Nitroperylene	
Nitrobenzofluoranthene	

SOURCE: Adapted with permission from the National Research Council 1983b.

gasoline exhaust (according to particulate extracts) at rates of approximately 8.0 (diesel fuel), 0.30 (leaded gasoline), and 0.20 mg/mi (unleaded gasoline) (National Research Council 1983b).

1-Nitropyrene has been the only nitro-PAH detected in spark-ignition particulate extracts. Very low 1-nitropyrene particulate extract concentrations have been found recently in on-road heavy-duty diesel and light-duty spark-ignition vehicles (National Research Council 1983b).

Gasoline-Powered Vehicle Refueling Hydrocarbons. Williams (1985) has reported the concentration of HCs in the breathing zone of individuals during vehicle refueling. Gas chromatographic data for gasoline and the refueling vapor indicate that only the lower molecular weight, more volatile compounds are emitted. Williams concluded that:

1. Vapor composition does not equal gasoline composition;

2. Range of total HC concentrations varied widely with the environmental conditions, resulting in exposures from 8 to 3,000 ppmC; and

3. Propane, butane, and pentane provide more than 80 percent of total exposure.

Areas in Need of Additional Research. To do efficient particulate control development work and to better understand emission characteristics, there is a need for a fast-response real-time particulate mass measurement instrument. The tapered element oscillating microbalance (TEOM) holds the most promise, but there is a gap between what is known about its principle of operation and the reality of making its use practical for measuring particulates.

The other instrument gap involves measuring methanol accurately. Formaldehyde has also been identified as a potentially important unregulated pollutant that needs careful real-time measurement and control because it is generally considered to be a carcinogen.

Measurement of particulate emissions from heavy-duty diesel engines using the EPA test procedures with dilution tunnels is inadequate. The current repeatability of measurements is poor. Barsic (1984) showed from a round-robin test that the root mean square of the 2-σ standard deviations were 76 percent of a 0.25 g/bhp-hr standard for six heavy-duty diesel engines tested in seven laboratories. For measurements intended to implement the 0.25 g/bhp-hr or the 0.1 g/bhp-hr standard, this variation is unacceptable.

It is uncertain whether particulate emission standards should be based on amount of total particulate matter, on which current standards are based, or amount of soluble organic component extracted from the particulates. The soluble organic component is the portion of the particulate that has been shown to be mutagenic and possibly carcinogenic (Claxton 1983), suggesting that future health-related regulations should be based on this fraction. Basing standards on the soluble organic component poses the problem of separating and quantifying the specific toxic components by one of the present methods—solvent extraction, vacuum sublimation, or thermogravimetric analysis. Variability associated with the separation methods and sampling condition affects the mass of the soluble fraction collected, compounding the previously stated measurement variability problem for the total particulate matter.

Present measurement methods for the collection of vapor-phase HCs from diesel engines do not collect all the compounds. Characterization of potentially toxic HCs is not possible if they cannot all be collected.

There is need to continue the development and use of advanced HPLC and GC/MS techniques in conjunction with separation methods to more accurately measure the amounts of key biologically active HCs in the particulate as well as the vapor phases. The nitroaromatics are important compounds whose concentrations in diesel exhaust with and without particulate traps should be measured more accurately.

There is a need to investigate and develop measurement methods that quantify diesel odor. Pioneering work was carried out in the late 1960s and early 1970s, but was dropped around 1978 because of the potential health effects of diesel particulate emissions. Diesel odor, along with particulates, is still the typical person's perception of the diesel pollutants that are of concern. There is a need to apply odor measurement methods to new engines used in light-duty and heavy-duty vehicles and advanced engines that use particulate traps or incorporate advanced high-temperature materials.

Refined organic compound measurement is particularly important to advance the development of low-heat-rejection (or commonly called adiabatic, as an ideal goal) diesel engines because their combustion chamber wall and gas temperatures will be higher. This elevated temperature will increase the amount of lubricating oil appearing as particulate emissions and has the potential of producing reactions between the HCs and oxygen/HNO_3/NO_x and other such gaseous mixtures to form toxic and biologically active species.

A particular need in unregulated pollutant characterization data for gasoline engines is additional nitrous acid (HNO_2) data as an extension to Pitts et al. (1984). That paper showed higher levels of HNO_2 from older (1974 and earlier) light-duty vehicles than from 1982 and newer cars that use three-way catalyst systems. The data show that even though the number of older cars is small, their HNO_2 emission levels are so high that they may be the major source of all gaseous HNO_2 from automotive emissions. HNO_2 is a key precursor to photochemical air pollution and is also an inhalable nitrite.

There is little known about how and when the nitro-PAHs are formed in the exhaust system (or the dilution tunnel) of diesel engines. Flow reactor studies with the basic species—NO, NO_2, CO, CO_2, N_2, O_2, SO_2, HCs—present in the exhaust, along with detailed engine studies that include the effects of the particles in the reactions, could help resolve this issue.

■ **Recommendation 9.** PAH Measurements. There is a need for a program of comparative measurements of PAHs in partial-exhaust sampling systems and in full-flow dilution tunnel systems, with measurements made in the atmosphere downwind from the plume, for the purpose of determining how well laboratory data reflect the true composition of emissions into the atmosphere.

■ **Recommendation 10.** Kinetics of Nitro-PAH Formation. Research is recommended to discover how and when nitro-PAHs are formed in the diesel engine

exhaust system and dilution tunnel. This work can best be done by flow reactor studies of the basic gases in conjunction with detailed engine studies that include the actual HCs and particles.

■ **Recommendation 11.** Particulate Measurement Variability. Research is required to reduce the variability in heavy-duty diesel particulate measurements. Work needs to be undertaken to determine how to better control the parameters that influence this variability.

■ **Recommendation 12.** HC Characterization. There is a need for research on the complete characterization of particulate-phase and gas-phase HCs in diesel exhaust.

■ **Recommendation 13.** Diesel Odor. There is a need to investigate and develop analytical methods that quantify diesel odor. This research should take advantage of the knowledge gained in the past eight years about measuring particulate-bound and vapor-phase HCs.

■ **Recommendation 14.** Nitrous Acid. Additional data should be obtained about HNO_2 emissions from older gasoline-powered vehicles. The literature shows high levels of HNO_2 from older cars that may be contributing significantly to increased photochemical smog and direct effects.

Summary of Research Recommendations

HIGH PRIORITY

Based upon current information, the following research studies are most likely to yield useful data.

Recommendation 1 *Tampering and Misfueling*	Tampering and misfueling statistics are fairly well developed but their effect on emissions is not as well known. Therefore, work should be done to better characterize the effect of tampering and misfueling on emissions from vehicles and to better assess their effect on ambient pollutant concentrations.

Recommendation 2
Diesel Particulate Emissions and Control

There is a need for continued research on particulate control technology, including the regeneration systems, to reduce the cost and complexity of these systems and the associated fuel economy penalties. Work needs to continue with various additives, substrate materials, regeneration systems, and controls to develop optimum systems that are able to decrease the diesel particulate emissions to the levels of 0.1 g/bhp-hr for heavy-duty diesels and 0.08 g/mi for light-duty vehicles required in California. In conjunction with this research there is a need to measure the metal species and the size distribution of the particles coming from diesel particulate traps.

Recommendation 4
Formaldehyde Measurements

Real-time measurements of formaldehyde concentration should be performed under transient and extreme conditions such as acceleration and deceleration, low temperature, light load, and extended idling with restricted ventilation. This research work should be done with and without catalysts since worst-case conditions in the field will occur with catalysts removed. Similar measurements should be made on bus engines.

Recommendation 7
Diesel Fuel Additives

Data should be obtained about the size distribution of particles in diesel exhaust and about the metal species they contain with and without a particulate trap, with a diesel fuel containing a typical additive under consideration for production use. Data on the HCs bound to the particles and the vapor-phase HCs should also be obtained.

Recommendation 10
Kinetics of Nitro-PAH Formation

Research is recommended to discover how and when nitro-PAHs are formed in the diesel engine exhaust system and dilution tunnel. This work can best be done by flow reactor studies of the basic gases in conjunction with detailed engine studies that include the actual HCs and particulates.

Recommendation 11
Particulate Measurement Variability

Research is required to reduce the variability in heavy-duty diesel particulate measurements. Work needs to be undertaken to determine how to better control the parameters that influence this variability.

Recommendation 12
HC Characterization

There is a need for research on the complete characterization of particulate-phase and gas-phase HCs in diesel exhaust.

MEDIUM PRIORITY

Recommendation 3
Evaporative Emission Model

An improved vehicle evaporative emissions model should be developed that is valid over various types of operating conditions for a variety of ambient temperatures. At the same time, changes should be made in the EPA test procedure to obtain the data necessary to properly design and size the evaporative system for the high-temperature soak situation, and data should be sought that can be used in EPA's MOBILE3 computer model for other use patterns of cars.

Recommendation 5
*Automotive Quality
No. 2 Diesel Fuel*

An automotive quality No. 2 diesel fuel with low sulfur and low aromatics is necessary if low particulate emissions are to be achieved. Research should be undertaken in cooperation with the automotive and petroleum industries to decide on effective and economical cetane number, sulfur and aromatic content, and 90 percent boiling point temperature specification limits for automotive quality No. 2 diesel fuel, and to formulate fuels that meet the specifications. The need for this research becomes more urgent as diesel fuel usage continues to increase. Improved emissions and more control options require low-sulfur diesel fuel.

Recommendation 6
*Evaporative
Emissions Control*

Some combination of field RVP control along with a test fuel with typical RVP (or calculation corrections for RVP differences) should be developed. Controlling RVP in motor gasoline, an approach successfully applied in California, is needed generally for controlling field evaporative emissions. The question of whether car manufacturers should be testing with a worst-case RVP test fuel or a typical fuel needs further study.

Recommendation 8
*Emissions
Measurement
Methods*

Studies should begin immediately with an evaluation of the best available emissions data on engines operating with and without emission control devices, to determine which of the unregulated pollutants really pose a potential threat to human health. Other unregulated pollutants might be added to this list if their concentrations reflect engine or emission control device performance.

Next, every effort should be made to improve the analytical procedures presently used to measure the concentrations of those pollutants, to the point where they can be readily carried out by technicians. This may require that packaged sets of reagents and equipment be marketed for a specific analysis. For example, prepacked traps might be available for collecting gaseous HCs prior to thermal desorption onto a gas chromatograph with a specified capillary column for the analysis of specific HCs at predetermined conditions.

Recommendation 9
PAH measurements

There is a need for a program of comparative measurements of PAHs in partial-exhaust sampling systems and in full-flow dilution tunnel systems, with measurements made in the atmosphere downwind from the plume, for the purpose of determining how well laboratory data reflect the true composition of emissions into the atmosphere.

Recommendation 13
Diesel Odor

There is a need to investigate and develop analytical methods that quantify diesel odor. This research should take advantage of the knowledge gained in the past eight years about measuring particulate-bound and vapor-phase HCs.

Recommendation 14
Nitrous Acid

Additional data should be obtained about HNO_2 emissions from older gasoline-powered vehicles. The literature shows high levels of HNO_2 from older cars that may be contributing significantly to increased photochemical smog and direct effects.

Acknowledgment

I would like to thank Peter V. Woon for all of his assistance in the preparation of this paper.

References

Abtoff, J., Schuster H. C., and Langer, J. 1985. The trap oxidizers—an emission control technology for diesel engines, Society of Automotive Engineers Paper 850015, Warrendale, Pa.

Alson, J. 1985. EPA methanol vehicle emissions test programs, talk at EPA Region VI Methanol Workshop, May 1985, Dallas, Tex.

Amann, C. A. 1985. The powertrain, fuel economy and the environment, *General Motors Corp. Research Publication* 4949, Warren, Mich.

Austin, T. C., and Rubenstein, G. S. 1985. A comparison of refueling emissions control with onboard and Stage II systems, Society of Automotive Engineers Paper 851204, Warrendale, Pa.

Barnes, G. J., and Donohue, R. J. 1985. A manufacturer's view of world emissions regulations and the need for harmonization of procedures, Society of Automotive Engineers Paper 850391, Warrendale, Pa.

Barsic, N. J. 1984. Variability of heavy-duty diesel engine emissions for transient and 13-mode steady-state test methods, Society of Automotive Engineers Paper 840346, Warrendale, Pa.

Cadle, S. H., Nebel, G. J., and Williams, R. L. 1979. Measurements of unregulated emissions from General Motors' light-duty vehicles, Society of Automotive Engineers Paper 70694, Warrendale, Pa.

California Air Resources Board. 1980. Procedure for estimating on-road motor vehicle emissions, Sacramento, Calif.

Claxton, L. D. 1983. Characterization of automotive emissions by bacterial mutagenesis bioassay: a review, *Environ. Mutat.* 5:609–631.

Coordinating Research Council–Air Pollution Research Advisory Committee. 1985. Automotive benzene emissions, CRC-APRAC Project No. CAPE-35-83, Preliminary Report, October 1985, Atlanta, Ga.

Eisenberg, W. C. 1983. Polycyclic aromatic hydrocarbon round robin study, *Final Report to the Coordinating Research Council*, August 1983.

Eisenberg, W. C., Schuetzle, D., and Williams, R. L. 1984. Cooperative evaluation of methods for the analysis of PAH in extracts from diesel particulate emissions, Society of Automotive Engineers Technical Paper 840414, presented at the International

Correspondence should be addressed to John H. Johnson, Michigan Technological University, Department of Mechanical Engineering and Engineering Mechanics, Houghton, MI 49931.

Congress & Exposition, Detroit, Mich., February 27–March 4, 1984.

Evans, S. B. 1980. Information report: 1979 progress of the chemical characterization panel of the composition of diesel exhaust project and results of particulate extraction roundrobin, *Coordinating Research Council Report* No. 516, Atlanta, Ga.

Ford Motor Co. 1985a. Regulatory requirements and forecast tables, Dearborn, Mich.

Ford Motor Co. 1985b. MVMA Vapor pressure survey data, Dearborn, Mich.

Fox, D. L. 1985. Air pollution, *Anal. Chem. Appl. Rev.* 57:223–238.

Furey, R. L. 1985. Volatility characteristics of gasoline-alcohol and gasoline ether fuel blends, Society of Automotive Engineers Paper 852116, Warrendale, Pa.

General Motors Corp. 1985a. Vehicle In-Use Emissions and Vehicle Emissions Modeling, Warren, Mich.

General Motors Corp. 1985b. EPA/EMA in-use emission factor test program, information available from Detroit Diesel Allison Division, Detroit, Mich.

General Motors Corp. 1986. *Pocket Reference.* General Motors Corporation, Environmental Activities, April 1986, Warren, Mich.

Gibson, T. L., Ricci, A. I., and Williams, R. L. 1980. Measurement of polynuclear aromatic hydrocarbons, their derivatives, and their reactivity in diesel automobile exhaust, General Motors Corporation Research Publication 3478, ENV #5. Warren, Mich.

Gorse, R. A., Jr., and Salmeen, I. T. 1982. Effects of filter loading and filter type on the mutagenicity and composition of diesel exhaust particulate extracts, *Atmos. Environ.* 16:1523–1528.

Gray, C. 1985. The EPA perspective on methanol as a motor vehicle fuel, Paper presented at EPA Region VI Methanol Workshop, May 20, 1985, Dallas, Tex.

Greco, R. 1985. Motor Vehicle Tampering Survey—1984, U.S. Environmental Protection Agency, Office of Air and Radiation, July 1984, Washington, D.C.

Gross, G. P., MacDonald, J. S., and Shahed, S. M. 1982. Informational report on the measurement of diesel particulate emissions, Coordinating Research Council Report No. 522, Atlanta, Ga.

Hampton, C. V., Pierson, W. R., Harvey, M. T., Updegrove, W. S., and Marano, R. S. 1982. Hydrocarbon gases emitted from vehicles on the road. A qualitative gas chromatography/mass spectrometry survey, *Environ. Sci. Technol.* 16:287–298.

Harvey, C. A., Carey, P. M., Somers, J. H., and Garbe, R. J. 1984. Toxicologically acceptable levels of methanol and formaldehyde emissions from methanol fueled vehicles, Society of Automotive Engineers Paper 841357, Warrendale, Pa.

Heavenrich, R. M., Murrell, J. D., and Cheng, J. P. 1986. Light-duty automotive trends through 1986, Society of Automotive Engineers Paper 860366, Warrendale, Pa.

Heinen, C. M. 1980. We've done the job—What's it worth? Society of Automotive Engineers Paper 801357, Warrendale, Pa.

Herr, J. D., Dukovich, M., Lestz, S. S., Yergey,

J. A., Risby, T. H., and Tejada, S. B. 1982. The role of nitrogen in the observed direct microbial mutagenic activity for diesel engine combustion in a single-cylinder DI engine, Society of Automotive Engineers Paper 820467, Warrendale, Pa.

Hilliard, J. C., and Springer, G. S. 1984. *Fuel Economy in Road Vehicles Powered by Spark Ignition Engines*, Plenum Press, London.

Hyde, J. D., Gibbs, R. E. Whitby, R. A., Byer, S. M., Hill, B. J., Hoffman, T. D., Johnson, R. E., and Wenner, P. C. 1982. Analysis of particulate and gaseous emissions data from in-use diesel passenger cars, Society of Automotive Engineers Paper 820772, Warrendale, Pa.

Jones, A. D., Appleby, M. R., and Bintz, L. J. 1982. Effects and detection of implanted defects on new technology emission control systems, Society of Automotive Engineers Paper 820977, Warrendale, Pa.

Kimbara, Y., Shinody, K., Koide, H., and Kobayashi, N. 1985. NO_x reduction is compatible with fuel economy through Toyota's lean combustion system, Society of Automotive Engineers Paper 851210, Warrendale, Pa.

Krost, K. J., Pellizzari, E. D., Walburn, S. E., and Hubbard, S. A. 1982. Collection and analysis of hazardous organic emissions, *Anal. Chem.* 54:810–817.

Lee, F. S. C., Pierson, W. R., and Ezike, J. 1980. The problem of PAH degradation during filter collection of airborne particulates—an evaluation of several commonly used filter media. In: Proceedings of the Fourth International Symposium on Polyaromatic Hydrocarbons: Chemistry and Biological Effects (A. Bjoerseth and A. J. Bennis, eds.), pp. 543–563, Columbus, Ohio: Battelle Press.

Lipari, F., and Keski-Hynnila, D. 1985. Aldehyde and unburned fuel emissions from methanol-fueled heavy-duty diesel engines, General Motors Research Publication 5173, July 15, 1985, Warren, Mich.

McNutt, B. D., Elliot, D., and Dulla, R. 1984. Patterns of vehicle misfueling in 1981 and 1982: where, when, what vehicles, and how often? Society of Automotive Engineers Paper 84135, Warrendale, Pa.

Motor Vehicle Manufacturers' Association. 1985. MVMA newsletter—vehicle tampering, *J. Air Pollut. Control Assoc.* 35:1–8.

Mulawa, P. A., and Cadle, S. H. 1979. Low-molecular weight aliphatic amines in exhaust from catalyst-equipped cars, General Motors Research Publication 2946 ENV-57, Warren, Mich.

National Research Council. 1983a. *Feasibility and Assessment of Health Risks from Vapor Phase Organic Chemicals in Gasoline and Diesel Exhaust,* National Academy Press, Washington, D.C.

National Research Council. 1983b. Polycyclic Aromatic Hydrocarbons: Evaluation of Sources and Effects, National Academy Press, Washington, D.C.

Nichols, R. J., and Norbeck, J. M. 1985. Assessment of emissions from methanol-fueled vehicles: implications for ozone air quality, Ford Motor Co., Dearborn, Mich.

Okamoto, W. K., Gorse, R. A., and Pierson, W. R. 1983. Nitric acid in diesel exhaust, *J. Air Pollut. Control Assoc.* 33:1098–1100.

Perez, J. M., Hills, F. J., Schuetzle, D., and Williams, R. L. 1980. Informational report on the measurement and characterization of diesel exhaust emissions, Coordinating Research Council–Air Pollution Research Advisory Committee Project No. CAPI-1-64, CRC Report No. 517, Atlanta, Ga.

Perez, J. M., Lipari, F., and Seizinger, D. E. 1984. Cooperative development of analytical methods for diesel emissions and particulates—solvent extractables, aldehydes and sulfate methods, Society of Automotive Engineers Paper 840413, Warrendale, Pa.

Pierson, W. R., Brachaczek, W. W., Korniski, T. J., Truex, T. J., and Butler, J. W. 1980. Artifact formation of sulfate, nitrate, and hydrogen ion on backup filters: Allegheny Mountain experiment, *J. Air Pollut. Control Assoc.* 30:30–34.

Pitts, J. N., Jr., Biermann, H. W., Winer, A. M., and Tuazon, E. C. 1984. Spectroscopic identification and measurement of gaseous nitrous acid in dilute auto exhaust, *Atmos. Environ.* 18:847–854.

Risby, T. H., and Lestz, S. S. 1983. Is the direct mutagenic activity of diesel particulate matter a sampling artifact? *Environ. Sci. Technol.* 17:621–624.

Schuetzle, D., Lee, F. S-C., Prater, T. J., and Tejada, S. B. 1980. The identification of polynuclear aromatic hydrocarbon derivatives in mutagenic fractions of diesel particulate extracts, Paper presented at the 10th Annual Symposium on the Analytical Chemistry of Pollutants, Dortmund, Germany, May 28–30, 1980.

Schwarz, M. J. 1985. Onboard refueling control—not a simple task, Paper presented at 78th Annual Meeting of Air Pollution Control Association, July 1985.

Simon, G. M., and Stark, T. L. 1985. Diesel particulate trap regeneration using ceramic wall flow traps, fuel additives, and supplemental electrical igniters, Society of Automotive Engineers Paper 850016, Warrendale, Pa.

Sobey, A. J. 1985. GM's viewpoint on alternative fuels, presentation at joint U.S./Canadian Meeting on Alternative Fuel, Windsor, Canada, June 25–26, 1985.

Society of Automotive Engineers and U. S. Department of Energy. 1985. Conference and workshop on transportation fuel alternatives for North America into the 21st century, Washington, D.C., November 13–15, 1985, Society of Automotive Engineers, Paper 190, Warrendale, Pa.

Stebar, R. F., Benson, J. D., Sapre, A. R., Brinkman, N. D., Dunker, A. M., Schwing, R. C., and Martens, S. W. 1985. Gasoline vapor pressure reduction—an option for cleaner air, Society of Automotive Engineers, Paper 852132, Warrendale, Pa.

Survey Data Research. 1985. MVMA Vehicle Tampering Study. Final Report, July 1985.

Tupa, R. C., and Doren, C. J. 1984. Gasoline and diesel fuel additives for performance, distribution, quality, Society of Automotive Engineers Paper 841211, Warrendale, Pa.

U.S. Department of Energy. 1985. The Motor Fuel

Consumption Model, 12th Periodic Report, U.S. DOE PE/77000-1, Washington, D.C.

U.S. Environmental Protection Agency. 1972. Control of air pollution from new motor vehicles and new motor vehicle engines. *Fed. Regist.* 37:24250–24337.

U.S. Environmental Protection Agency. 1983. Anti-tampering and anti-misfueling programs to reduce in-use emissions from motor vehicles, EPA-AA-TTS-83-10, Research Triangle Park, N.C.

U.S. Environmental Protection Agency. 1985. *Compilation of air pollution emission factors, Vol. II, Mobile Sources*, Washington, D.C.

Wade, W. R., and Jones, C. M. 1984. Current and future light duty diesel engines and their fuels, Society of Automotive Engineers Paper 840105, Warrendale, Pa.

Way, G. 1985. 1986 Engines and Emission Systems Tables (unpublished; Tables can be obtained by correspondence to Gilbert Way, 2371 Dorchester North, Troy, MI 48084).

Weaver, C. S., Miller, C., Johnson, W. A., and Higgings, T. S. 1986. Reducing the sulfur and aromatic content of diesel fuel: costs, benefits, and effectiveness for emissions control, Society of Automotive Engineers Paper 860622, Warrendale, Pa.

Wiedemann, B., and Neumann, K. H. 1985. Vehicular experience with additives for regeneration of ceramic diesel filters, Society of Automotive Engineers Paper 850017, Warrendale, Pa.

Williams, R. 1985. Emissions measurement and unregulated emissions, Ford Motor Co., Dearborn, Mich.

Wineland, R. J. 1985. The Ford flexible fuel vehicle. Ford Motor Co., Dearborn, Mich.

Atmospheric Transport and Dispersion of Air Pollutants Associated with Vehicular Emissions

PERRY J. SAMSON
University of Michigan

Air Pollution, the Automobile, and Public Health. © 1988 by the Health Effects
Institute. National Academy Press, Washington, D.C.

Definitions of Transport and Dispersion

The movement of pollutants in the atmosphere is caused by transport, dispersion, and deposition. Transport is movement caused by a time-averaged wind flow. Dispersion results from local turbulence, that is, motions that last less than the time used to average the transport. Deposition processes, including precipitation, scavenging, and sedimentation, cause downward movement of pollutants in the atmosphere, which ultimately remove the pollutants to the ground surface. This chapter deals only with transport and dispersion.

During the past decade, the complexities of transport and dispersion of airborne pollutants associated with vehicular emissions have been studied with elaborate field and modeling experiments. In the first part of this article, the terms used in the study of transport and dispersion of pollutants and the scales of motion (time and distance) over which vehicular emissions may affect air quality, precipitation quality, or both, are defined. Since pollutants can travel distances from meters to hundreds of kilometers, the relative scales of motion involved in distinguishing transport phenomena from dispersion phenomena may vary from problem to problem.

The second part of the chapter, Transport and Dispersion: Theory and Applications, outlines the observational information on transport and dispersion, describes the theoretical tools that have been used to simulate the transport of vehicular emissions, identifies the limitations of these tools, and recommends specific areas for further research. Mathematical formulations of transport and dispersion are developed only as needed to identify the parameters of interest.

Meteorological Parameters

The concentration of pollutants associated with moving vehicles is determined by several factors: the emission rate of pollutants from the vehicle, mixing induced by vehicle motion, wind speed and direction relative to the axis of the highway, inten-

sity of ambient atmospheric turbulence, reactions to or from other chemical species, and rate of removal to the ground surface (deposition). Concentrations associated with nonmoving vehicles—as might be encountered in traffic queues, parking structures, and street canyons—are determined by emission rates and the wind flows and turbulence produced by the interaction of the local wind with complex structures such as buildings and roadside sound barriers. Weather plays a role in most of these components, generally causing higher emission rates at lower temperatures (Chang and Norbeck 1983b), diluting pollutants at higher wind speeds, mixing pollutants vertically during unstable thermal conditions, and influencing the rates of homogeneous and heterogeneous chemical reactions and the rate at which pollutants are scavenged from the atmosphere by moisture or dry deposited.

Regardless of the distance over which a pollutant is transported, the change in its concentration can be described by the conservation-of-mass equation. The temporal change of the concentration of a pollutant is expressed mathematically as

$$\frac{\partial \chi}{\partial t} + \frac{\partial (u\chi)}{\partial x} + \frac{\partial (v\chi)}{\partial y} + \frac{\partial (w\chi)}{\partial z} =$$

$$Q + R + S \qquad (1)$$

where χ is the concentration in $\mu g/m^3$; u, v, and w are the east-west, north-south, and vertical components of the wind, respectively, in m/sec; t is the time in seconds; Q is the emission rate in $\mu g/m^3/sec$; R is the rate of increase or decrease in concentration due to chemical reaction, in $\mu g/m^3/sec$; and S is the rate of removal by deposition, in $\mu g/m^3/sec$.

The continuous calculation of χ would be untenable and would not allow for extrapolation of the results to other sites. Instead, motions are divided into mean and turbulence components to simplify the calculation and allow generalized parameterization. The turbulence component (dispersion) is defined as the deviation of the actual wind from a mean wind vector. The averaging time used to define the mean varies depending on the scales of motion associated with the problem. For example, dispersive components of wind motions considered in the long-range transport of pollutants include motions that are also considered to be mean components in the dispersion of pollutants within 100 m of a highway.

The variables u, v, w, and χ can each be described as the sum of a mean and a turbulent component as follows: $u = \overline{u} + u'$, $v = \overline{v} + v'$; $w = \overline{w} + w'$, and $\chi = \overline{\chi} + \chi'$, where, for example, u is the instantaneous measurement of the east-west component of the wind, $\overline{u}$ is the mean component, and u' is the deviation of u from the mean. Substituting these equations into equation 1 and assuming an incompressible, nondivergent atmosphere (reasonable assumptions for most scales of motion affecting vehicular emissions) produces

$$(\partial\overline{\chi}/\partial t) + \overline{u}(\partial\overline{\chi}/\partial x) + \overline{v}(\partial\overline{\chi}/\partial y) + \overline{w}(\partial\overline{\chi}/\partial z) =$$
$$- [\partial\overline{(u'\chi')}/\partial x] - [\partial\overline{(v'\chi')}/\partial y]$$
$$- [\partial\overline{(w'\chi')}/\partial z] + Q + R + S \qquad (2)$$

The first term in equation 2 describes the change of concentration with time, the second through fourth terms on the left side describe the changes due to mean motions (transport), and the first three terms on the right side describe the changes due to turbulence (dispersion). Many of the unsolved problems associated with the transport and dispersion of vehicular emissions arise from the need to characterize the turbulence components of this equation in some universal manner.

The turbulence fluxes ($\overline{u'\chi'}$, $\overline{v'\chi'}$, $\overline{w'\chi'}$— defined as the mass of pollutant deposited per unit area per unit time due to turbulence) are difficult to measure directly. It is common therefore to assume that the turbulence flux is proportional to the gradient of the mean concentration

$$\overline{w'\chi'} = - K_z(\partial\overline{\chi}/\partial z) \qquad (3)$$

where the proportionality, K_z, is called the vertical eddy diffusivity. Hence equation 2 becomes equation 4:

$$(\partial\overline{\chi}/\partial t) + \overline{u}(\partial\overline{\chi}/\partial x) + \overline{v}(\partial\overline{\chi}/\partial y) + \overline{w}(\partial\overline{\chi}/\partial z) =$$
$$(\partial/\partial x)\,[K_x(\partial\overline{\chi}/\partial x)] + (\partial/\partial y)\,[K_y(\partial\overline{\chi}/\partial y)]$$
$$+ (\partial/\partial z)[K_z(\partial\overline{\chi}/\partial z)] + Q + R + S \qquad (4)$$

Furthermore, assuming that the eddy diffusivity values, K_x, K_y, and K_z, are invariant along their respective axes (Fickian diffusion, an assumption often made to simplify the calculation, but not necessarily physically realistic), this expression can be simplified to the parabolic form

$$(\partial\overline{\chi}/\partial t) + \overline{u}(\partial\overline{\chi}/\partial x) + \overline{v}(\partial\overline{\chi}/\partial y) + \overline{w}(\partial\overline{\chi}/\partial z) =$$
$$K_x(\partial^2\overline{\chi}/\partial x^2) + K_y\,(\partial^2\overline{\chi}/\partial y^2) + K_z\,(\partial^2\overline{\chi}/\partial z^2)$$
$$+ Q + R + S \qquad (5)$$

Generally, when the flow is parallel to the x-axis, the diffusion term $K_x(\partial^2\overline{\chi}/\partial x^2)$ is small compared to the transport term $\overline{u}(\partial\overline{\chi}/\partial x)$ and is ignored. Likewise, for a continuous infinite line source (for example, a long, straight section of highway) oriented along the y-axis, both the K_x and K_y terms are usually ignored.

Although values of K_z have been determined for relatively simple surface characteristics, values for complex surfaces are nonunique and difficult to establish. The problems associated with parameterizing dispersion over a variety of scales are presented in the Scales of Motion section.

Apart from natural transport and dispersion processes, moving vehicles themselves cause considerable mixing that influences pollutant concentrations within about 100 m of a highway. The aerodynamic drag of a moving vehicle causes a turbulent wake in which pollutants initially mix. This mixing is influenced by the shape and speed of the vehicle, and its influence on concentration is most pronounced when the wind is nearly parallel to the axis of the highway. Figure 1 shows a schematic representation of the wake area behind a moving vehicle.

Vehicular emissions that are transported beyond about 100 m from a highway are generally at levels well below any National Ambient Air Quality Standard (NAAQS). However, in some urban areas and under conditions of limited atmospheric ventila-

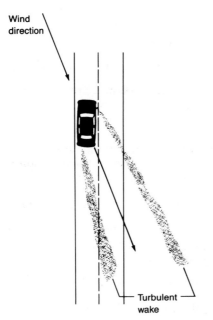

Wind direction

Turbulent wake

Figure 1. Schematic representation of the transport of a turbulent wake away from a highway with an oblique wind angle. (Adapted with permission from Eskridge and Hunt 1979, and from the American Meteorological Society.)

tion (that is, low wind speeds and topographical barriers), potential problems still exist. For example, Los Angeles experiences periods of limited ventilation because of topographic constraints. Schultz and Warner (1982) have demonstrated the recirculation of pollutants in the Los Angeles basin. Other researchers have shown that the recirculation of urban air can lead to high concentrations of primary pollutants (those emitted directly into the atmosphere) and secondary pollutants (those resulting from chemical reaction). Anchorage and Fairbanks, Alaska, although small compared to other U.S. cities, frequently have days on which the carbon monoxide (CO) concentrations exceed the NAAQS. These days occur during the coldest winter months when air stagnates in the valleys (Zimmerman and McKenzie 1974; Bowling 1984; Fernau and Samson 1985). Also, there are potential problems in urban street canyons, where the complexities of wind flow can lead to high pollutant concentrations under certain conditions. The introduction of methanol in fuel, for example, could lead to in-

creases in formaldehyde in urban street canyons.

Although problems still exist, the incidence of elevated concentrations of primary pollutants in urban areas has been declining. This is not true of secondary pollutants such as ozone (O_3). High O_3 concentrations occurring in and downwind of urban areas have been documented for many cities in the United States and elsewhere (Spicer et al. 1982; Clark and Clarke 1984). High O_3 concentrations generally occur during the summer months under sunny, but not necessarily stagnant, atmospheric conditions. To diagnose the cause of high O_3 concentrations in urban areas, better measurement of wind flow is needed. Determining the transport of pollutants in urban areas is often difficult, if not impossible, due to the lack of relevant measurements of wind flow above the surface. Hydrodynamic simulations of wind flow over urban areas have been performed but were constrained by a lack of relevant meteorological data for boundary and initial conditions.

During the past 20 years, transport and dispersion of pollutants over long distances have been implicated in the degradation of air and precipitation quality in remote areas. Concentrations of O_3 in excess of 80 ppb were found in rural areas by Coffey and Stasiuk (1975) and Quickert et al. (1976), and the transport of O_3 over long distances has been investigated by a number of researchers (Husar et al. 1976; Lyons and Cole 1976; Chung 1977; Samson and Ragland 1977; Wolff et al. 1977; Spicer et al. 1979).

Pollutants other than O_3 and its precursors are believed to be transported over long distances and to influence precipitation quality in nonurban areas. The direct contribution of vehicular emissions to acidic deposition is thought to be small given the magnitude and emissions of nitrogen oxides ($NO_x = NO + NO_2$) relative to other sources and the relatively lower mean height of emissions. However, if sulfur dioxide (SO_2) emissions are decreased as part of an emissions control program for acidic deposition, NO_x emissions may contribute a larger percentage of total acidic deposition.

An assessment of the NO_x contributions to the production of O_3 and atmospheric acids requires an estimation of the amount of pollutants moved to the surface near the source by dry deposition. (Dry deposition is the removal of pollutants from the atmosphere in the absence of precipitation. See Atkinson, this volume.) The rate of dry deposition also affects the vertical distribution of pollutants, and hence their residence time in the atmosphere, and is dependent upon the rate of pollutant mixing to a surface as well as the type and condition of the surface. The study of the rate of dry deposition of vehicle-related pollutants is of special interest since pollutants released near the surface should be deposited closer to the source area than those emitted from elevated sources. Presumably, a unit of pollutant emitted close to the surface is available to the surface in higher concentration than is a similar amount emitted from a higher point and thus is more likely to be dry deposited in the near field. This implies that vehicular emissions of NO_x would be less likely to travel long distances than NO_x emitted from elevated sources.

Scales of Motion

Human exposure to vehicular-related pollutants is considered in terms of three scales of distance: near field (0.0–0.2 km), urban (0.2–20 km), and regional (20–2,000 km). Direct exposure to primary vehicular emissions may occur inside vehicles as they idle, or from entrainment of air from other vehicles. Generally, vehicular emissions are produced by moving vehicles, but it is also possible for idling vehicles to contribute to high pollutant concentrations, for example, in a parking structure or lot, a street canyon, or a highway queue at rush hour. Consequently, the dispersions associated both with moving vehicles (line sources) and with aggregates of parked or queued vehicles (area sources) are considered. Elevated (nonsurface) emissions from sources such as smokestacks (point sources) are of little importance in the field of vehicular

emissions except perhaps when comparing the relative human exposure resulting from surface versus nonsurface emissions on urban and regional scales.

The exposure of humans to vehicular emissions has been studied principally in the near-field environment (0–0.2 km) as will be shown. Exposures in this range are experienced by bicyclists, motorcyclists, pedestrians, people in nearby buildings, and vehicle passengers. This exposure has been fairly well documented for simple configurations in open countryside, but the exposure obtained in complex settings such as street canyons is poorly understood. Knowledge of concentrations on highways is also sketchy. In terms of human exposure, however, this information is critical. Air entering a vehicle or breathed by a cyclist is a direct sample of initially mixed emissions. Hence, understanding the initial mixing process is very important in estimating exposure.

Exposures obtained on an urban scale (0.2–20 km) due to the mixture of vehicular and nonvehicular emissions are also fairly well documented. Urban-area exposure to primary pollutants has generally declined due to improvements in emission control technology. However, some urban areas of high exposure still exist, as mentioned previously, and these problems are generally exacerbated by a combination of adverse meteorological conditions and topographic constraints. Secondary pollutants formed from vehicular emissions, such as O_3, can become fairly high in urban areas.

Estimating the contribution of vehicular emissions to regional-scale (20–2,000 km) air quality problems is even more challenging. The NO_x and reactive hydrocarbon emissions of vehicles are thought to contribute to regional-scale O_3 levels and the levels of other oxidizing compounds such as hydrogen peroxide (H_2O_2) and hydroxyl radical (OH^-). These oxidizing agents are known to influence the rate of conversion of SO_2 to sulfate (SO_4^{2-}) and NO_x to nitric acid (HNO_3) and nitrate (NO_3^-). Thus, while the direct influence of emitted NO_x on acidic deposition is not thought to be large relative to other sources, the indirect influence of vehicular emissions on the production of acidic species is probable but poorly understood.

Transport and Dispersion: Theory and Applications

This section describes the theories of transport and dispersion of vehicular emissions for the various scales of motion outlined above and discusses their application to available observational data. Mathematical descriptions are presented only to the point necessary to identify areas of understanding and ignorance.

Near Field

Open Highway.   Most automotive emissions presumably result from moving vehicles. The dispersion from a continuous line source (highways) assuming Fickian diffusion can be described using the Gaussian equation

$$\chi(x, y) = \frac{q}{\sqrt{2\pi}\sigma_z(x)U}$$

$$\left\{ \exp\left[-1/2\left(\frac{z + H}{\sigma_z(x)}\right)^2 \right] \right.$$

$$\left. + \exp\left[-1/2\left(\frac{z - H}{\sigma_z(x)}\right)^2 \right] \right\} \qquad (6)$$

where q is the emission rate in g/m/sec; $\sigma_z(x)$ is the standard deviation of concentration expected in the vertical direction as a function of travel distance, x, downwind; H is the elevation of the source above the surrounding terrain; the $z + H$ term on the right side of the equation describes the effect of surface reflection on the vertical distribution of the plume; and U is the mean wind speed near the highway. (The choice of a representative wind speed is not straightforward. Wind speed and direction can change substantially near the surface, especially near the highway where they are influenced by the motion of the vehicles.)

The theoretical description of transport and dispersion processes has been developed for a continuously emitting line source by Calder (1973) and for an instantaneous emission from a line source by Shair (1974). A variety of line-source models using this basic approach are available (Zimmerman and Thompson 1975; Benson, 1979). Such models have been shown to reproduce concentrations with some accuracy when winds are perpendicular to the highway with near neutral stability (that is, temperature decreases with height at a rate of approximately 1° C/100 m). However, under conditions with winds nearly parallel to the axis of the highway, these models tend to overpredict concentrations of pollutants (Noll et al. 1978; Rao et al. 1979a; Sistla et al. 1979).

One reason for the observed overprediction in these models has been the lack of description of the effect of mixing by vehicles themselves. Figure 2 shows the relationship of the concentration of the tracer gas sulfur hexafluoride (SF_6) versus vehicle speed for unstable, neutral, and stable atmospheric conditions 2 m downwind of a roadway. This plot shows that concentrations generally decrease with increasing vehicle speed regardless of ambient atmospheric stability, suggesting that vehicle-

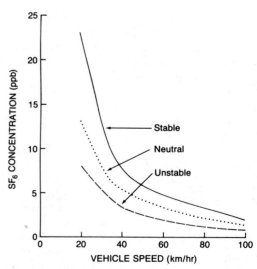

Figure 2.   Concentration of SF_6 tracer gas versus vehicle speed for unstable, neutral, and stable atmospheric conditions at 2 m downwind of a roadway. (Adapted with permission from Petersen et al. 1984.)

induced turbulence dominates ambient turbulence. Green et al. (1979) also noted that the ground-level pollutant concentrations near a highway did not increase as rapidly with decreasing wind speed as predicted by most Gaussian line-source models. Furthermore, using a Gaussian model for a city in Czechoslovakia, Hesek (1981 p. 144) also found that "the diffusion . . . caused by moving cars must also be considered in the calculation."

Most modeling of pollutant transport and dispersion in the vicinity of highways has ignored the turbulence generated by the moving vehicles themselves. Since a theory to predict the wind velocity and turbulence in the wakes of vehicles did not exist, models either ignored the changes in velocity and the mechanical mixing caused by the vehicles, or parameterized them in a simple manner, such as assigning a large eddy-diffusion coefficient to the roadway.

Rao et al. (1979b), using data from a study of pollutant dispersion along the Long Island Expressway (Rao et al. 1978), found enhancement of turbulence energy in the range of 0.1–1.0 Hz downwind of the highway during moderate-to-heavy traffic

conditions. This corresponded to turbulent elements of a few meters in diameter, which is roughly the height of the automobile and truck traffic on the nearby highway. Sedefian et al. (1981) found similar "traffic turbulence" enhancement in wind data collected during the General Motors (GM) dispersion experiment (Cadle et al. 1977). The calculated values of $\sigma_z(x)$ downwind of the GM line source are plotted in figure 3 and compared with the suggested curves for open, flat country from Briggs (1975). The lack of variation in $\sigma_z(x)$ values despite the changing ambient stability illustrates the dominance of traffic-induced turbulence over ambient turbulence in the near field of a line source.

To compensate for the poor predictions of simple Gaussian line-source models under certain conditions, a number of modifications to the basic Gaussian approach have been developed. Bullin and Maldonado (1977) developed a model that used an empirical equation near the highway and the Gaussian dispersion equation downwind. Benson (1982) suggested that the initial mixing of pollutants by vehicles could be described mathematically by redefining $\sigma_z(x)$ values. Chock (1977b) and Rao and Keenan (1980) suggested improvements to the Gaussian line-source model through redefinition of $\sigma_z(x)$ to account for traffic-induced mixing near the highway. The former suggestions were used to improve the predictive capability of the EPA HIWAY model (Petersen 1980). These modifications improved the performance of the model, but they were tuned to vehicle speeds of about 80 km/hr and a vehicle distribution that was not necessarily typical of other traffic situations.

Others have used numerical simulation of the conservation-of-mass equation to predict pollutant transport near highways. Malin (1972) reported on a two-dimensional numerical model that included the likely role of topography on the wind field. Danard (1972) allowed for horizontal variation of vertical eddy diffusion (K_z) over the highway. It was found that even numerical models incorporating variable eddy diffusivity (see, for example, Ragland and Peirce 1975) failed to improve significantly

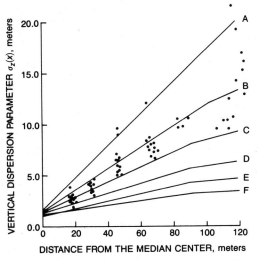

Figure 3. Measured values of $\sigma_z(x)$ as reported by Rao and Keenan (1980) compared with $\sigma_z(x)$ curves suggested by Briggs (1975) for open, flat country point sources under various ambient stability categories, A–F (A represents least stable atmospheric conditions; F, most stable).

Table 1. Suggested Values of $K_x{}^t$ and $K_z{}^t$ to Account for Traffic-Induced Turbulence as a Function of Wind Speed, U (m/sec)

Distance from Road (m)	Height (m)	$K_x{}^t$ (m²-sec)			$K_z{}^t$ (m²-sec)		
		$U > 2$	$1 < U \leq 2$	$U \leq 1$	$U > 2$	$1 < U \leq 2$	$U \leq 1$
1.7 u	4.5	0	0	0.45	0	0	0.30
1.7 u	1.5	0.45	0.45	0.45	0.30	0.30	0.30
Median	4.5	0.45	0.45	0.45	0.30	0.30	0.30
Median	1.5	0.45	0.45	1.05	0.30	0.30	0.45
1.7 d	4.5	0.60	0.60	0.60	0.45	0.45	0.45
1.7 d	1.5	1.05	1.05	1.05	0.45	0.45	0.45
15 d	4.5	0	0.60	0.60	0	0.45	0.45
15 d	1.5	0	0.75	0.75	0	0.45	0.45
30 d	4.5	0	0	0.45	0	0	0.15
30 d	1.5	0	0	0.45	0	0	0.15

NOTE: u = upwind; d = downwind.
SOURCE: Adapted with permission from Chock 1978a, and from the American Meteorological Society.

upon the simulated concentrations using Gaussian line-source models.

Fay (1975) and Lane (1976) made some earlier estimates of the dispersion of pollutants in automobile wakes, but recent work has had the advantage of new data on near-highway turbulence and tracer gas concentrations. Eskridge and Hunt (1979) and Eskridge et al. (1979) developed a finite-difference model for calculating pollutant concentrations on and near a highway that incorporates a vehicle wake theory. The wake theory was modified and verified in wind tunnel experiments by Eskridge and Thompson (1982). They caution that their results were restricted to conditions where vehicle speed was much greater than wind speed. Eskridge et al. (1979) compared the predictions of a line-source model that included wake theory with observed SF$_6$ tracer gas concentrations in a controlled field experiment conducted by the General Motors Research Laboratory (Cadle et al. 1976; Chock 1977a). The results showed that predictions made using wake effect were closer to the observations than those of the EPA HIWAY model (Zimmerman and Thompson 1975). Eskridge et al. (1979) found that ignoring wake turbulence led to overpredictions in concentration when the wind was nearly parallel to the axis of the highway. An increase in mechanical mixing led to a decrease in

predicted concentrations under these conditions. Eskridge and Rao (1986) used vertical and lateral profiles of tracer gas concentrations obtained in the wake of a simulated moving vehicle (Eskridge and Thompson 1982) to determine the best turbulence scale lengths (a measure of the average size of turbulent elements). This information was used to improve upon the earlier concentration predictions, which included wake effects.

Chock (1978a) developed a model for pollutant dispersion near roadways that approximated the influence of traffic by increasing the eddy diffusivity near the highway. This approximation implied that turbulence generated by traffic did not interact strongly with ambient turbulence. For an automobile moving at 80 km/hr, Chock suggested the use of $K_x{}^t$ and $K_z{}^t$, the downwind and vertical eddy diffusivities due to traffic, respectively, as shown in table 1.

Chock (1977a, 1978b) showed that under light wind conditions a measurable heat flux can be discerned from temperature and wind data collected at the GM dispersion experiment and at a field experiment conducted by SRI International (Dabberdt 1977). The exhaust heat convection could play an important role in the dispersion of pollutants near highways under light wind conditions with heavy traffic by lifting the exhaust plume (plume rise).

The transport and dispersion processes described above were designed to predict concentrations of inert gaseous pollutants. Some modeling of the transport and dispersion of particles from highways has also been recently attempted. Katen (1977) used the method of modified Gaussian solutions, developed through analogy with finite-differencing techniques, to simulate the transport, dispersion, and deposition of lead particles from a highway. Sheih (1980) developed a finite-difference model to evaluate the importance of particulate coagulation during dispersion of emissions from a highway.

In summary, there have been several recent advances in the description of transport and dispersion on the open highway. The advances have included the observations that traffic enhances dispersion near highways, traffic-wake-induced turbulence can dominate ambient turbulence, and a measurable plume may rise over highways having high traffic density under light wind conditions.

These advances have stemmed from the availability of high-quality measurements of turbulence and tracer gases. New investigations of vehicle wake turbulence and its effect on near-field pollutant concentrations hold considerable promise for improved description of transport and dispersion of pollutants near highways. But these new studies are constrained by the fact that the data, and hence the models, represent a somewhat biased subset of meteorological conditions and generally represent straight-line sources in open, flat terrain.

Research on transport and dispersion from line sources has greatly benefited from field studies of tracer gas dispersion near straight highways. The recent incorporation of vehicle-wake theory into line-source models offers more realistic descriptions of the vehicle-induced turbulence and its effect on near-field pollutant concentrations.

■ **Recommendation 1.** Future research should continue to focus on the role of vehicular turbulence in the initial dispersion of pollutants. Since high-quality field data have been collected in the past decade, this research would be best served by the use of physical modeling facilities to simulate highway situations. This research should focus on air quality on the highway to which vehicle passengers and open-vehicle (bicycle, motorcycle, moped, and so on) passengers and pedestrians would be exposed.

■ **Recommendation 2.** The importance of plume buoyancy on near-highway and/or street-canyon pollutant concentrations needs to be clarified. The amount of heat being generated on both the highway and vehicles should be quantified. The dispersion of that heat should be modeled to identify the atmospheric conditions most likely to produce significant thermal plume rise. These calculations should be verified against wind tunnel simulations.

Urban Street Canyons and Parking Structures. The transport and dispersion of pollutants from urban street canyons, bus or truck terminals, and parking structures and lots is less well understood than the transport and dispersion of pollutants around open highways. Human exposure to pollutants emitted in these environments could be high in such congested areas.

Simulating the transport of pollutants within a street canyon is very difficult and virtually impossible to extrapolate to other sites or times. Application of line- or point-source equations to vehicular emissions in complex situations is generally not warranted. The airflow in street canyons is neither steady nor homogeneous, and street segments do not approximate infinite line sources. Likewise, exhaust from parking structures is rarely vented through an isolated, elevated point source. Parking structure emissions tend to be relatively close to the ground in areas of complex wind flow because of nearby structures. The initial rise (if any) of the nonbuoyant plume is difficult to estimate and often difficult to measure because of local air circulation variation (Hosker 1984). Transport through open parking lots, such as are found at most truck stops, is also difficult to describe because of the inhomogeneities produced by the interaction of the wind with the configuration of trucks.

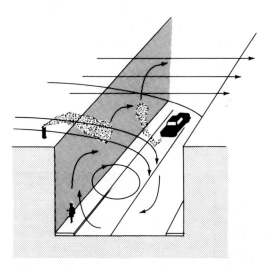

Figure 4.   Schematic representation of the complexity of airflow in a two-dimensional street canyon with wind perpendicular to the axis of the street.

Simple models have been developed by Benesh (1978) and Messina et al. (1983) to simulate pollutant concentrations at urban intersections. These models incorporate modified Gaussian plume equations to try to reproduce concentration fields in complex settings. The errors in simulated short-term concentrations can be large (Rao 1984). This is not surprising since the meteorological conditions at an intersection, and especially in a street canyon, are not homogeneous; consequently, it is difficult to obtain representative values of wind direction, wind speed, and stability. Moreover, the large concentration gradients that could occur in a street canyon because of the complexity of airflow make extrapolation of results to estimates of "typical" human exposure difficult (Brice and Roesler 1966; Cortese and Spengler 1976; Petersen and Allen 1982).

Very few computer simulation models have been developed for street canyons because of the complexity of building geometry and wind flow. A vortex of airflow is expected when the flow above building height is perpendicular to the axis of the canyon, as shown schematically in figure 4. Air moves down the windward side of the canyon, returning as it approaches the street in a direction opposite to the direction of the airflow above the roof, and rises

on the leeward side. For situations where the wind flow problem can be approximated two-dimensionally (such as in the midsection of a long uniform street canyon), some numerical approaches have been attempted. Johnson et al. (1973) developed an empirical model based on data from San Jose, California. The model predicted decreasing concentration away from the line source on the leeward side of the canyon but a linear decrease in concentration with height on the windward side. Sobottha and Leisen (1980) modified the approach of Johnson et al. (1973) to allow concentrations to increase away from the line source on the windward side. Chock (1983) suggested improvements to the Johnson model by allowing the leeward-side concentration to be not only a function of the path lengths from the point closest to the line source along the flow trajectory to the receptor but also a function of the proximity of the trajectory to the source. Yamartino and Wiegand (1986) described the flow and turbulence fields within a street canyon using a simple Gaussian line-source model following the flow field with time-dependent coefficients.

Georgii et al. (1967) conducted one of the first field experiments of street-canyon pollutant concentrations. They found that concentrations were higher on the leeward side of the canyon than the windward side because of the presence of a vortex. Johnson et al. (1973) and Leisen and Sobottha (1980) found that curbside concentrations decreased with height faster on the leeward side than on the windward side. Leisen and Sobottha also observed that the dispersion due to turbulent flows generated around buildings and by moving vehicular traffic was greater than the dispersion generated in the atmosphere through natural means.

Nicholson (1975) developed a micrometeorological approach to the airflow in a two-dimensional street canyon. She compared her results to observed data from Frankfurt, West Germany, and Madison, Wisconsin. Additionally, Hotchkiss and Harlow (1973) attempted numerical simulations of street canyon flow. Although these studies were all able to reproduce the qualitative features of airflow in the street canyon, no attempt has been made to extrapolate these techniques to other situations.

The air quality associated with complex-area sources is not well represented by simple modeling techniques. The best, albeit limited, method for assessing impact from such diffuse sources may be physical modeling. Skinner and Ludwig (1976) conducted experimental studies of CO dispersion from a highway model in an atmospheric-simulation wind tunnel. Their street canyon model included scaled, moving "automobiles" that were able to emit various tracers for visual or ambient study. Kennedy and Kent (1977) used two lengths of nichrome wire to simulate two lanes of traffic. The transport and dispersion of heat from the resistance wires were measured as a surrogate for automobile emissions. They found that the addition of two large buildings to their domain could significantly reduce ground-level concentrations.

Hoydysh and Chiu (1971) examined flow in street canyons using a tracer gas released in a street-level line source. They concluded that the flow is wake or convection dominated, depending on the crosswind component in the street. Hoydysh et al. (1974) concluded from their experiments that a high-density configuration of uniform-size buildings generally increased concentrations and caused pockets of high concentration near building corners, whereas an intermittent high-rise configuration allowed pollutants to escape. Wedding et al. (1977) used a ^{85}Kr–air mixture as a tracer in their wind tunnel to study the effects of building geometry on street-level tracer concentrations. They found that pollutant dilution in a street canyon was controlled by the mean airflow through the canyon rather than by turbulence diffusion. Kitabayashi et al. (1977) used both neutral and stably stratified air in a wind tunnel to simulate a street canyon situation in Tokyo.

The review of the exposure potential for indirect sources requires techniques for estimating transport, and dispersion associated with, for example, truck stops, shopping center parking lots, and intersections. Generally this analysis requires estimates of the number of queued vehicles, idling traffic emission rates, and description of emission density geometry. Transport and dispersion are then estimated using Gaussian line-source approximations (Hanisch et al. 1978; Rao 1984). Although there is potential for short-term exposure in such situations, there are few comprehensive data with which to evaluate the transport of these pollutants. For example, given the possibility that truck drivers, while sleeping in their cabs, could be exposed to long-term (overnight) exposures of pollutants from diesel exhaust, the study of transport and dispersion from idling vehicles must also be explored.

In summary, studies of the transport and dispersion of pollutants from complex urban situations suffer from a lack of reliable data and the uniqueness of each situation. Street canyons in particular pose special problems because of the complicated nature of the wind flows relative to specific geometries of buildings. Computer simulation of airflow in urban street canyons would require elaborate descriptions of the equations of motion and would be very expensive. Tracer studies can provide much useful information about a particular location but are very labor intensive and awkward to use in many situations. Physical modeling (such as wind tunnel simulations) can be adapted to the study area of interest and modified to represent particular atmospheric thermal structures. Although this approach is also quite expensive, it is a reasonable way to evaluate potential or existing pollution hot spots at urban intersections.

■ **Recommendation 3.** Transport and dispersion of pollutants within street canyons and parking structures is so variable that it is unlikely that generalizations will be useful. Nonetheless, better definition of the transport and dispersion of pollutants in such complex situations has a high priority. It is recommended that this problem be studied through use of tracers and that environmentally safe and easily measured tracers be developed. Ideally, the tracer sampling system would be portable and make use of remote sensing. Release of a tracer that could be tracked through the canyon under a variety of conditions would generate new, useful information with which to evaluate the reasonableness of flow generalizations. A portable system could be used at several sites to assess site-to-site deviations from flow generalizations. The sites for this program should include urban street canyons, parking structures (including truck stops), and other complex environments, for example, where sound-reducing roadway barriers and covered highway cuts influence air movement.

■ **Recommendation 4.** Hand in hand with the development of tracer technology, emphasis must be placed on improving wind tunnel facilities to meet the demands of inexpensively simulating many different complex settings with a variety of atmospheric conditions. Given the uniqueness of each urban setting, physical modeling may hold the best hope for identifying potential pollutant hot spots due to complex surface flows. The advantage of physical modeling is that it may be possible to identify physical alterations to the setting that would decrease human exposure to the pollutants.

Urban-Scale Transport and Dispersion

Modeling of urban-scale (0.2–20 km) transport and dispersion of vehicle-related pollutants is limited by the lack of wind observations other than hourly measurements at the surface and twice-daily measurements in the upper atmosphere. Generally, airflow over an urban area is poorly described by surface wind measurements, and upper air observations are made at stations roughly 400 km apart.

The transport of pollutants over an urban area depends, in part, on regional-scale (20–2,000 km) wind flows. Even when regional winds are strong, the flow at the surface is modified (usually slowed and turned) by contact with the surface. When the regional flow is light, the flow at the surface is generally altered by inhomogeneous surface characteristics, both physical (for example, differential friction of forests versus fields) and thermal (differential heating of water versus land). The inhomogeneities in wind flow that have been of

interest to urban-scale transport are those that occur above building height. Goodin et al. (1979) used a dense network of meteorological measurements to describe objectively the three-dimensional wind flow in the Los Angeles basin. Their estimates included a local terrain adjustment technique (from Anderson 1971, 1973) and successive solutions of the divergence equation to reduce anomalous divergence in the complete field.

Transport of pollutants in urban areas has also been described with trajectory analysis. Lurmann (1979) used surface-based observations of wind direction and speed to calculate a transport path through the Los Angeles basin. Lack of observed winds above the surface can lead to significant errors in the estimated path of the air. Liu and Seinfeld (1975) assessed the uncertainty imposed on urban-scale trajectory calculations by the effects of wind shear and horizontal dispersion. Chang and Norbeck (1983a) found that inclusion of wind shear corrections in trajectory estimations of source input significantly improved photochemical model predictions, since pollutant emission sources were not uniform.

Numerous tracer and tetroon (constant-pressure balloon) experiments have shown that under some conditions the air parcel trajectories can vary significantly as a function of height. Angell et al. (1975) observed three-dimensional air trajectories from tetroon flights in the Los Angeles basin. Drivas and Shair (1974) tracked tracer concentrations across the Los Angeles basin to estimate pollutant transport and dispersion. Bornstein (1975) used a hydrodynamic model to describe the expected wind field over the New York metropolitan area. Although their results were not applied to the urban air quality problem, they do represent a reasonable alternative for data-sparse regions where objective analysis techniques are not adequate.

Keen et al. (1979) diagnosed the three-dimensional flow fields associated with lake breeze circulations in the Chicago area. They were able to demonstrate the potential for return, via lake breeze circulation, of pollutants transported over the lake. Vertical circulations such as those associated with a lake or sea breeze will generally not be diagnosed by static spatial interpolation techniques. The development of hydrodynamic models has aided the understanding of complex three-dimensional wind fields over urban areas.

Many transport and dispersion models over urban areas ultimately rely on some estimate of the mixing height (the height to which the atmosphere is uniformly mixed). Holzworth (1964) defined the mixing height as the intersection of two temperature profiles: the observed temperature profile and a profile constructed from the observed maximum surface temperature and adiabatic lapse rate (the theoretical rate at which temperature decreases with height in the absence of any heat input). The method has been tested extensively, using visual observations of clouds, analyses of data collected with acoustic sounders, and comparisons with temperature measurements made through well-mixed layers. In many cases all methods give similar results, but other studies have shown that this method overestimates the overall mixing height by about 50 percent (Pendergast 1974; Schubert 1976).

The estimation of morning mixing heights, presumably important during the morning rush hour, has also been highly generalized. Holzworth (1967) assumed that the urban morning mixing height could be approximated from the intersection of the morning temperature profile and a profile constructed from the observed morning minimum temperature plus some constant value and the adiabatic lapse rate. Although useful for assessment of the general air pollution potential for an urban area, this morning mixing height may be of limited value for diagnosing specific pollution episodes.

The mixing height is considered useful for evaluating the pollutant potential of an area, but its use in problems related to vehicular emissions has been limited. Aron (1983) showed that variations in mixing height were inconsistent in explaining day-to-day variations in O_3 and CO in several urban areas. The concentrations of CO were not consistently related to the morning mixing height, possibly because there was insufficient time for the pollutant to mix to the mixing height, and possibly

because of the unreliability of the method for calculating the morning mixing height. The concentrations of O_3 were not related to the maximum mixing height, possibly because of the dominance of other factors such as solar intensity and upwind pollutant concentrations.

In summary, diagnosing the airflow over an urban area remains a challenge. With only a couple of exceptions (for example, Los Angeles), most cities have inadequate networks of meteorological stations for diagnosing wind flow. Although hydrodynamic models hold promise for interpreting transport, they still require adequate initial and boundary conditions for reliable calculations. The data needed to estimate mixing heights are also generally lacking for most cities. However, better mixing height data may still not yield a useful measure of dispersion potential for vehicular emissions.

■ **Recommendation 5.** As noted earlier, few measurements have been made of the transport and dispersion of air from an urban area. There are several important reasons why this is true and will probably remain true for some time: (a) the logistics of surface measurements of tracers can be very difficult in an urban setting; (b) the possibility of aircraft measurement can be severely limited by airspace restrictions; and (c) the measured concentrations can be influenced by local air circulations caused by the physical setting. This research could be significantly aided by the development of automated, remote-sensing, wind-profiling instruments. The technology for remote sensing of wind fields is currently being developed for measurement of winds above about 2 km from the earth's surface, but development of a sensor for lower levels is still in its infancy. Therefore it is recommended that research into the development of methods for sensing of near-surface winds remotely be conducted.

Regional-Scale Transport

Modeling of regional-scale pollutant transport (20–2,000 km) has so far focused mainly on the transport of sulfur oxides (SO_x), which are emitted principally from elevated point sources (such as smokestacks) rather than vehicles. In this discussion the transport and dispersion of pollutants from elevated point sources is presented because debate continues on the relative residence time of pollutants released at the surface versus those released from an elevated stack.

The relative contribution of elevated point-source emissions versus surface-based emissions could be simulated using a numerical or Gaussian solution to estimate ground concentrations from each type of source. The rate of dry deposition can be approximated by using the boundary condition

$$-\overline{(w'\chi')}_o = v_d\overline{\chi} \qquad (7)$$

where the turbulent flux of pollutant to the surface, $-\overline{(w'\chi')}_o$, is proportional to the mean surface concentration. The proportionality is called the deposition velocity and varies considerably depending on the type of pollutant (gas or particle), its solubility, and the ambient atmospheric stability, among other influences (Sehmel 1984).

Observations of deposition velocities for NO, NO_2, and hydrocarbons have shown considerable variability. Much of the variability in deposition velocities may be related to variability in meteorological conditions. The description of dry deposition has often been addressed in a manner similar to an electrical resistance series (Wesley and Hicks 1977; Fowler 1978). The deposition velocity (v_d) is assumed to be inversely proportional to the sum of three resistance terms

$$v_d = (r_a + r_b + r_s)^{-1} \qquad (8)$$

where r_a is the aerodynamic resistance to pollutant transport through the lowest atmospheric layer, r_b is determined by the rate of molecular and diffusive transport processes over a surface, and r_s is any additional resistance encountered at the surface itself.

Various methods have been proposed for evaluating each of these resistance terms, but few data have been compiled with which to generalize deposition rates for vehicle-related pollutants over a variety of surfaces and meteorological conditions. For

HNO_3 the surface resistance is thought to be near zero (Huebert and Robert 1985), but other pollutants may have significant values of r_s and those may vary significantly with surface type and conditions. Methods for estimating r_a and r_b are described by Hicks and Liss (1976).

The diagnosis of regional-scale transport of vehicle-related pollutants also suffers from the lack of information about the three-dimensional wind fields in the lower troposphere. Horizontal wind measurements are conducted routinely by the National Weather Service (NWS) rawinsonde network, but these may not be adequate for diagnosing pollutant transport or for parameterizing vertical mixing. These measurements have a spatial and temporal resolution of about 400 km and 12 hr, respectively. The NWS rawinsonde network, although capable of resolving features with length scales of about 1,000 km, is too coarse to identify accelerations in the wind field associated with frontal passages and diurnal phenomena such as the low-level nocturnal jet—a layer of high-speed winds occurring at night at an altitude of approximately 300–700 m. It is most common over the Great Plains and the Midwest (Bonner 1968) during the summer months and is characterized by generally southerly flow. It has been speculated (Lyons and Cole 1976; Smith et al. 1978) that this jet could transport pollutants rapidly northward during the night. Thus the use of current 12-hr upper-air data may produce significant biases in atmospheric transport calculations.

Models that utilize the hydrodynamic equations of motion are capable of resolving smaller-scale features (Anthes and Warner 1978), but the computational expense of using these models is often prohibitive for long-term pollution studies. Smaller-scale wind fields may also be obtained by any number of objective analytical techniques (Goodin et al. 1979; Haagenson 1982). Objective analysis offers a less expensive solution to the problem of inadequate data resolution, but the errors involved in spatial and temporal interpolation may distort the "true" meteorological representation. Unfortunately, few high-resolution upper-air wind data are available to test the reliability of interpolation techniques.

Kuo et al. (1985) quantified the effects of low-resolution meteorological data on transport calculations. They conducted several "observing system simulation experiments" in which trajectories were computed using meteorological information derived from a hydrodynamic model. The model output, considered to represent the "true" state of the atmosphere, was systematically degraded to simulate the effects of low-resolution rawinsonde measurements. Isentropic trajectory calculations based on the degraded output suggest that, after 72 hr of travel, horizontal displacement errors greater than 400 km may result. Vertical displacement errors were also shown to increase with decreasing data resolution. The results indicate that increasing the temporal resolution of rawinsonde measurements improves the accuracy of transport calculations to a greater extent than increasing the spatial resolution.

The quantitative estimates of trajectory errors provided by Kuo et al. (1985) were based on the analysis of a single weather system. Long-term air pollution studies such as those required for the estimation of source-receptor relationships for acid precipitation require analysis of a wide range of meteorological situations. Verification of these results over a larger sample of meteorological conditions is difficult due to the lack of high-resolution wind measurements.

The Cross Appalachian Tracer Experiment (CAPTEX), conducted in the northeastern United States and eastern Canada during September and October 1983, reduces this deficiency (Ferber 1984). The CAPTEX data base contains fine-scale meteorological measurements (both spatial and temporal) and offers an excellent opportunity for analyzing the errors in transport calculation resulting from spatial and temporal interpolation of low-resolution wind data. Kahl and Samson (1986) evaluated several spatial and temporal interpolation techniques against the high-resolution measurements available from CAPTEX. They concluded that an increase in spatial

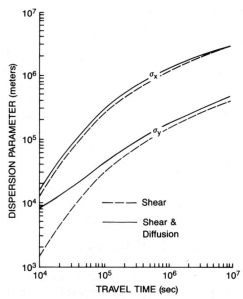

Figure 5. Along-wind (σ_x) and crosswind (σ_y) dispersion parameters from Taylor's (1982) solution of dispersion without (dashed line) and with (solid line) turbulence diffusion. This assumes a 10 m/sec wind speed at about 40° latitude. (Adapted with permission from Draxler and Taylor 1982, and from the American Meteorological Society.)

resolution influenced the position of calculated trajectories more than did an increase in temporal resolution for the CAPTEX period. This dichotomy between Kuo et al. and Kahl and Samson may indicate that there are different observational needs associated with the storm period analyzed by the former and the relatively quiescent period analyzed by the latter.

Over regional scales, the estimation of dispersion rates is intimately connected to the intensity of vertical mixing and vertical wind velocity shear. Saffman (1962) concluded from his analytical investigation that wind-shear-generated dispersion dominates turbulence dispersion after several hours of travel. Likewise, Tyldesley and Wallington (1965), using numerical solutions of Saffman's approach for different wind and diffusivity profiles, reached the same conclusion. Csanady (1969) derived an analytical solution for a hypothetical wind spiral profile to estimate the crosswind spread of pollutants as a function of wind speed. Taylor (1982) expanded this approach to estimate the along-wind spread. Figure 5 shows the alongwind

(σ_x) and crosswind (σ_y) dispersion parameters derived from Taylor (1982) without (dashed line) and with (solid line) turbulence diffusion. This graph illustrates that the effects of wind shear generally dominate the dispersion produced by ambient turbulence after a few hours of travel time. Calculations made by Draxler and Taylor (1982) suggest that the rate of horizontal dispersion of a "puff" of tracer over regional scales is roughly proportional to the wind speed. Draxler and Taylor also conclude that for a given wind speed the rate of growth of a pollutant puff is proportional to travel time, consistent with the empirical rate suggested by Heffter (1965).

In summary, the study of transport and dispersion of vehicle-related pollutants over regional scales suffers from many of the same uncertainties outlined for urban transport. Although the present observational networks are thought to be inadequate, there is still debate over how to most efficiently increase the resolution. The rates of dry deposition of vehicle-related pollutants are not well known and need to be quantified if the relative contributions of surface-based vehicular emissions and elevated emissions are to be estimated.

■ **Recommendation 6.** In view of the possible need to develop efficient strategies for reducing tropospheric O_3 and acidic deposition, the relative contribution of urban, ground-level emissions versus elevated point-source emissions should be assessed. The first step should be a feasibility study using transport, dispersion, and deposition modeling. If this study suggests that the difference in potential impact is relatively large, then a second stage should consist of a dual tracer field experiment to track pollutants over a period of 12 to 24 hr. This experiment would demonstrate the sensitivity of transport to height of release and quantify ground-level concentrations from the two source heights for use in estimating rates of dry deposition.

■ **Recommendation 7.** Research on the transport of pollutants over regional scales should focus on defining the most efficient method for augmenting the current NWS rawinsonde measurement network. It is not now clear whether, on the scale of eastern North America, it would be more advantageous to measure the winds at the existing network sites at a higher frequency or to select additional sites for measurements at the standard frequency. With the development of remote-sensing technology for the measurement of winds, it is essential that the needs of the transport and dispersion researchers be considered in the deployment of the newer equipment. Some recent research has begun to explore this issue, but more work is needed to optimize the information gained from system augmentation. This research should encompass numerical studies of hypothetical atmospheric situations and analysis of data from previous network augmentation programs.

Summary

The level of human exposure to vehicular pollutants depends upon the emission rate of the pollutant from the vehicle, the direction of transport, the rate of dispersion, and the location of the population relative to the ensemble of sources. During the past decade, the complexities of the transport and dispersion of air pollutants associated with vehicular emissions have been made more evident through elaborate field and modeling experiments.

The exposure of humans to vehicular exhaust is considered in terms of three scales of distance: near field (0.0–0.2 km), urban (0.2–20 km), and regional (20–2,000 km). The direct exposure of humans to vehicular emissions in the near-field environment has been the focus of most of the research. It is known that the concentrations of pollutants associated with moving vehicles are determined by several factors, including their emission rate from the vehicle, mixing induced by vehicle motion, wind speed and direction relative to axis of the highway, intensity of ambient atmospheric turbulence, reactions to or from other chemical species, and the rate of removal to the surface (deposition).

The concentrations of pollutants that would be expected in the absence of moving vehicles are determined by emissions rates and the complex wind flow and tur- bulence produced by the interaction of the local wind with complex structures (for example, buildings, sound barriers, other vehicles). Weather plays a role in most of these components, generally causing higher emission rates at lower temperatures, dilut- ing pollutants at higher wind speeds, mix- ing pollutants vertically during unstable thermal conditions, and influencing the rates of homogeneous and heterogeneous chemical reactions and the rate of scaveng- ing of pollutants from the atmosphere by precipitation and dry deposition.

Apart from the natural transport and dis- persion processes, moving vehicles exhibit a considerable mixing potential that influences pollutant concentrations within about 100 m of a highway. The aerodynamic drag of a moving vehicle causes a turbulent wake in which pollutants initially mix. This mixing is influenced by the shape and speed of the vehicle. It is recommended that new studies of the effects of vehicular turbulence on pol- lutant concentrations be initiated.

Simulating the transport of pollutants within street canyons and parking struc- tures is very difficult and virtually impos- sible to extrapolate to other sites or times. The flow of air in street canyons is neither steady nor homogeneous, and represent- ative wind flow measurements are difficult to obtain. It is suggested that additional effort be put into developing artificial tracer systems that can easily be deployed in a variety of complex situations.

Exposures due to urban-scale transport and dispersion of direct emissions and their chemical products are less understood than are the near-field exposures. Urban area air pollution problems still exist, but the inci- dence of elevated concentrations of primary pollutants has been declining. Nonetheless, concentrations of O_3 that exceed the Na- tional Ambient Air Quality Standards still exist and require special consideration. Iden- tifying transport and dispersion in urban ar- eas is often very difficult, if not impossible, because of the lack of relevant measurements of wind flow above the surface. Future re- search should focus on better methods for determining the flow fields over urban areas.

The estimation of the contribution of vehicular emissions to regional-scale air and precipitation contamination is even more challenging than that encountered on an urban scale. It is generally believed that pollutants transported over long distances influence precipitation quality in nonurban areas. An assessment of the vehicle-induced NO_x contributions requires an estimation of the amount of dry deposition to the surface near the source. The rate of dry deposition of pollutants depends upon the rate of pollutant mixing to a surface, as well as the type and condition of the surface. Faster deposition will result in more depo- sition in the vicinity of the emission and, consequently, fewer pollutants will be transported longer distances. Future studies should be aimed at understanding the rates of dry deposition of vehicle-related emis- sions and defining the transport and disper- sion of pollutants over long distances.

Summary of Research Recommendations

Given the state of knowledge outlined in this chapter, the following studies have the highest likelihood of yielding useful new data relevant to human exposure to vehicle-related emissions. Given limited re- sources, these topics should be considered in the following order.

HIGH PRIORITY

Recommendation 1
Open-Highway
Exposure

Determine, by physical modeling, the role of vehicular turbu- lence in the initial dispersion of pollutants. This research should focus on defining the factors influencing concentration fluctuations near and over the highway.

Recommendation 3
*Transport and
Dispersion in Street
Canyons*

Define the transport and dispersion of pollutants in street canyons through the use of chemical tracers. Ensure that many sites be investigated with widely varying building configurations and the observations be conducted over sufficient time to include many different meteorological conditions.

MODERATE PRIORITY

Recommendation 2
Buoyant Plume

Assess the importance of plume buoyancy on near-highway and street-canyon pollutant concentrations.

Recommendation 4
Physical Modeling

Develop physical models to simulate a variety of complex settings and atmospheric conditions.

Recommendation 6
*Regional-Scale
Transport and
Dispersion*

Assess the relative contribution of urban, ground-level emissions versus elevated point-source emissions to overall regional-scale pollutant levels. A dual tracer release conducted over a range of atmospheric conditions would have a high likelihood for success.

LOW PRIORITY

Recommendation 5
*Urban-Scale
Transport*

Develop an automated, remote-sensing wind-profiling system for use in defining wind flow within an urban atmosphere.

Recommendation 7
*Regional-Scale
Transport*

Define the most efficient method for augmenting the existing NWS rawinsonde measurement network to satisfy the data needs of regional-scale air pollution transport models.

References

Anderson, G. E. 1971. Mesoscale influences on wind fields, *J. Appl. Meteorol.* 10:377–386.

Anderson, G. E. 1973. A Mesoscale Windfield Analysis of the Los Angeles Basin, EPA-650/4-73-001, U.S. Environmental Protection Agency, Washington, D.C. 56 pp.

Angell, J. K., Dickson, C. R., and Hoecker, W. H. 1975. Relative diffusion within the Los Angeles basin as estimated by tetroon triads, *J. Appl. Meteorol.* 14:1490–1498.

Anthes, R. A., and Warner, T. T. 1978. Development of hydrodynamic models suitable for air pollution and other mesometeorological studies, *Mon. Weather Rev.* 106:1045–1078.

Aron, R. 1983. Mixing height—an inconsistent indicator of potential air pollution concentrations, *Atmos. Environ.* 17:2193–2197.

Benesh, F. 1978. Carbon Monoxide Hot Spot Guidelines, Vol. 5: User's Manual for Intersection-Midblock Model, EPA-450/3-78-037, U.S. Environmental Protection Agency, Washington, D.C.

Benson, P. E. 1979. CALINE3—A Versatile Dispersion Model for Predicting Air Pollutant Levels Near Highways and Arterial Streets, FHWA/CA/TL-79/23, NTIS:PB 220842, November 1979, 129 pp.

Benson, P. E. 1982. Modifications to the Gaussian vertical dispersion parameter, σ_z, near roadways, *Atmos. Environ.* 16:1399–1405.

Bonner, W. D. 1968. Climatology of the low-level jet, *Mon. Weather Rev.* 96:833–850.

Bornstein, R. D. 1975. The two-dimensional URBMET urban boundary layer model, *J. Appl. Meteorol.* 14:1459–1477.

Bowling, S. A. 1984. Meteorological Factors Responsible for High CO Levels in Alaskan Cities. U.S. Environmental Protection Agency, Environmental Research Laboratory, Corvallis, Ore.

Brice, R. M., and Roesler, J. F. 1966. The exposure to

Correspondence should be addressed to Perry J. Samson, Department of Atmospheric and Oceanic Science, University of Michigan, Ann Arbor, MI 48109-2143.

carbon monoxide of occupants of vehicles moving in heavy traffic, *J. Air Pollut. Contr.* 16:597–604.

Briggs, G. A. 1975. Plume rise predictions, In: *Lectures on Air Pollution and Environmental Impact Analysis* (D. A. Haugen, ed.), pp. 59–111, American Meteorological Society, Boston, Mass.

Bullin, J. A., and Maldonado, C. 1977. Modeling carbon monoxide dispersion from roadways, *Environ. Sci. Technol.* 11:1071–1076.

Cadle, S. H., Chock, D. P., Heuss, J. M., and Monson, P. R. 1976. Results of the General Motors Sulfate Dispersion Experiments, General Motors Res. Publ. GMT-2107, Warren, Mich. 140 pp.

Cadle, S. H., Chock, D. P., Monson, P. R., and Heuss, J. M. 1977. General Motors sulfate dispersion experiment: experimental procedures and results, *J. Air Pollut. Contr.* 27:33–38.

Calder, K. L. 1973. On estimating air pollution concentrations from a highway in an oblique wind, *Atmos. Environ.* 7:863–868.

Chang, T. Y., and Norbeck, J. M. 1983a. Wind shear effects on air column trajectories in urban air pollution models, *J. Air Pollut. Contr.* 33:488–491.

Chang, T. Y., and Norbeck, J. M. 1983b. Vehicular CO emission factors in cold weather, *J. Air Pollut. Contr.* 33:1188–1189.

Chock, D. P. 1977a. General Motors sulfate dispersal experiment—an overview of the wind, temperature, and concentration fields, *Atmos. Environ.* 11:553–559.

Chock, D. P. 1977b. Assessment of the EPA HIWAY model, *J. Air Pollut. Contr.* 27:39–45.

Chock, D. P. 1978a. An advection-diffusion model for pollutant dispersion near roadways, *J. Appl. Meteorol.* 17:976–989.

Chock, D. P. 1978b. A simple line source model for dispersion near roadways, *Atmos. Environ.* 12: 823–829.

Chock, D. P. 1983. Pollution dispersion near roadways—experiments and modeling, *Sci. Total Environ.* 25:111–132.

Chung, Y. S. 1977. Ground level ozone and regional transport of air pollutants, *J. Appl. Meteorol.* 16:1127–1133.

Clark, T. L., and Clarke, J. F. 1984. A Lagrangian study of the boundary layer transport of pollutants in the northeastern United States, *Atmos. Environ.* 18:287–298.

Coffey, P., and Stasiuk, W. 1975. Evidence of atmospheric transport of ozone into urban areas, *Environ. Sci. Technol.* 9:59–62.

Cortese, A. D., and Spengler, J. D. 1976. Ability of fixed monitoring station to represent personal carbon monoxide exposure, *J. Air Pollut. Contr.* 26:1144–1150.

Csanady, G. T. 1969. Diffusion in an Ekman layer, *J. Atmos. Sci.* 26:414–426.

Dabberdt, W. F. 1976. Multiple-tracer highway dispersion study. Proc IEEE Int Conf On Environ Sensing and Assessment, New York.

Dabberdt, W. F. 1977. Air Quality on and Near Roadways: Experimental Studies and Model Development, Project 2761, SRI International, Inc., Menlo Park, Calif.

Danard, M. B. 1972. Numerical modeling of carbon monoxide concentration near highways, *J. Appl. Meteorol.* 11:947–957.

Draxler, R. R., and Taylor, A. D. 1982. Horizontal dispersion parameters for long-range transport modeling, *J. Appl. Meteorol.* 21:367–372.

Drivas, P. J., and Shair, F. H. 1974. A tracer study of pollutant transport and dispersion in the Los Angeles area, *Atmos. Environ.* 8:1155–1164.

Eskridge, R. E., and Hunt, J. C. R. 1979. Highway modeling: Part 1: Prediction of velocity and turbulence fields in the wake of vehicles, *J. Appl. Meteorol.* 18:387–400.

Eskridge, R. E., and Rao, S. T. 1986. Turbulent diffusion behind vehicles: experimentally determined turbulence mixing parameters, *Atmos. Environ.*

Eskridge, R. E., and Thompson, R. S. 1982. Experimental and theoretical study of the wake of a block-shaped vehicle in a shear-free boundary flow, *Atmos. Meteorol.* 16:2821–2836.

Eskridge, R. E., Binkowski, F. S., Hunt, J. C. R., Clark, T. L., and Demerjian, K. L. 1979. Highway modeling, Part 2: Advection and diffusion of SF_6 tracer gas, *J. Appl. Meteorol.* 18:401–412.

Fay, J. A. 1975. Wake-Induced Dispersion of Automobile Exhaust Pollutants, Paper presented at the 68th Annual Meeting of the Air Pollution Control Assn, Abstracts p. 213.

Ferber, G. J. 1984. CAPTEX '83 completed: model evaluation workshop planned, *Bull. Amer. Meteorol. Soc.* 65:370–371.

Fernau, M. J., and Samson, P. J. 1985. Statistical Model of CO Levels in Anchorage, Alaska. University of Michigan.

Fowler, D. 1978. Dry deposition of SO_2 on agricultural crops, *Atmos. Environ.* 12:369–373.

Georgii, H. W., Busch, E., and Weber, E. 1967. Investigation of the Temporal and Spectral Distribution of the Emission Concentration of Carbon Monoxide in Frankfurt/Main, Rept. No. 11, Institute for Meteorology and Geophysics, Univ. of Frankfurt/Main.

Goodin, W. R., McRae, G. J., and Seinfeld, J. H. 1979. A comparison of interpolation methods for sparse data: application to wind and concentration fields, *J. Appl. Meteorol.* 18:761–771.

Green, N. J., Bullin, J. A., and Plasek, J. C. 1979. Dispersion of carbon monoxide from roadways at low wind speeds, *J. Air Pollut. Contr.* 29:1057–1061.

Haagenson, P. L. 1982. Review and evaluation of methods for objective analysis of meteorological variables, *Meteorol. Res.* 5:113–132.

Hanisch, J. L., Hart, B. R., Turetsky, W. S., et al. 1978. The Connecticut indirect source review: a methodology and model for evaluating CO concentrations, *Environ. Manage.* 2:127–133.

Heffter, J. L. 1965. The variation of horizontal diffusion parameters with time for travel periods of one hour or longer, *J. Appl. Meteorol.* 4:153–156.

Hesek, F. 1981. Numerical modeling of air pollution, *Meteorol. Zpra.* 34:144–147.

Hicks, B. B., and Liss, P. S. 1976. Transfer of SO_2

and other reactive gases across the air–sea interface, *Tellus* 28:348–354.

Holzworth, G. C. 1964. Estimates of mean maximum mixing depths in the contiguous United States, *Mon. Weather Rev.* 92:235–247.

Holzworth, G. C. 1967. Mixing depths, wind speeds and air pollution potential for selected locations in the United States, *J. Appl. Meteorol.* 6:1039–1044.

Hosker, R. P. 1984. Flow and diffusion near obstacles, In: *Atmospheric Science and Power Production* (D. Randerson, ed.), pp. 241–326, National Technical Information Service, Springfield, Va. NTIS DE84005177 (DOE/TIC-27601).

Hotchkiss, R. S., and Harlow, F. H. 1973. Air Pollution Transport in Street Canyons, EPA-R4-73-029, U.S. Environmental Protection Agency, Washington, D.C.

Hoydysh, W. G., and Chiu, H. H. 1971. An Experimental and Theoretical Investigation of the Dispersion of Carbon Monoxide in the Urban Complex, Paper No. 71-523, American Institute of Aeronautics and Astronautics.

Hoydysh, W. G., Griffiths, R. A., and Ogawa, Y. 1974. A Scale Model Study of the Dispersion of Pollution in Street Canyons, Paper presented at the 67th Annual Meeting of the Air Pollution Control Assn., Denver, Colo., June 9–13 1974.

Huebert, B. J., and Robert, C. H. 1985. The dry deposition of nitric acid to grass, *J. Geophys. Res.* 90:2085–2090.

Husar, R. B., Patterson, D. E., Paley, C. C., and Gillani, N. G. 1976. Ozone in hazy air masses, *Proceedings of the International Conference on Photochemical Oxidant Pollution and Its Control,* Raleigh, N.C., EPA-600/3-77-001, pp. 275–282.

Johnson, W. B., Ludwig, F. L., Dabberdt, W. F., and Allen, R. J. 1973. An urban diffusion simulation model for carbon monoxide, *J. Air Pollut. Contr.* 23:490–498.

Kahl, J. D., and Samson, P. J. 1986. Uncertainty in trajectory calculations due to low resolution meteorological data, *J. Clim. Appl. Meteorol.* 25:1816–1831.

Katen, P. C. 1977. Modelling Atmospheric Dispersion of Lead Particulates from a Highway, Colorado State Univ., Ft. Collins, Environmental Research Paper No. 11 (October 1977).

Keen, C. S., Lyons, W. A., and Schuh, J. A. 1979. Air pollution transport studies in a coastal zone using kinematic diagnostic analysis, *J. Appl. Meteorol.* 18:606–615.

Kennedy, I. M., and Kent, J. H. 1977. Wind tunnel modeling of carbon monoxide dispersal in city streets, *Atmos. Environ.* 11:541–547.

Kitabayashi, K., Sugawara, K., and Isomura, S. 1977. A Wind Tunnel Study of Automobile Exhaust Gas Diffusion in an Urban District, Paper presented at the Fourth International Clean Air Conference, May 16–20, 1977, Tokyo, Japan.

Kuo, Y. H., Skumanich, M., Haagenson, P. L., and Chang, J. S. 1985. The accuracy of trajectory models as revealed by the observing system simulation experiments, *Mon. Weather Rev.* 113:1852–1867.

Lane, D. D. 1976. Dispersion of pollutants in auto-

mobile wakes, *Am. Soc. of Civ. Engr., Environ. Eng. Div. J.* 102:571–580.

Leisen, P., and Sobottha, H. 1980. Simulation of the Dispersion of Vehicle Exhaust Gases in Street Canyons: Comparison of Wind Tunnel Investigations and Full-Scale Measurements, Paper presented at the IMA Conference on Modeling Dispersion in Transport Pollution, March 17–18, 1980, Southend, England.

Liu, M. K., and Seinfeld, J. H. 1975. On the validity of grid and trajectory models of urban air pollution, *Atmos. Environ.* 9:555–574.

Lurmann, F. 1979. User's Guide to the ELSTAR Photochemical Air Quality Simulation Model, Environmental Research and Technology Inc., Rept. No. P-5287-600, Westlake Village, Calif.

Lyons, W. A., and Cole, H. S. 1976. Photochemical oxidant transport: mesoscale lake breeze and synoptic-scale aspects, *J. Appl. Meteorol.* 15:733–743.

Malin, H. M., Jr. 1972. Computer model simulates air pollution over roads, *Environ. Sci. Technol.* 6:1071–1076.

Messina, A. D., Bullin, J. A., Nelli, J. P., and Noe, R. D. 1983. Estimates of Air Pollution Near Signalized Intersections, Texas Transportation Institute Rept. FHWA/RD-83/009, Texas A&M Univ., College Station, Tex.

Nicholson, S. 1975. A pollution model for street-level air, *Atmos. Environ.* 9:19–31.

Noll, K. E., Miller, T. L., and Claggett, M. 1978. A comparison of three highway line source dispersion models, *Atmos. Environ.* 12:1323–1329.

Pendergast, M. M. 1974. A Study of the Effect of the Urban Mesoclimate on Local and Regional Air Pollution Potential in SE Texas, Ph.D. thesis, Texas A&M Univ., College Station, Tex.

Petersen, W. B. 1980. User's Guide for HIWAY-2: A Highway Air Pollution Model, EPA-600/8-80-018, U.S. Environmental Protection Agency, Washington, D.C.

Petersen, W. B., and Allen, R. 1982. Carbon monoxide exposures to Los Angeles area commuters, *J. Air Pollut. Contr.* 32:826–833.

Petersen, W. B. 1984. Effects of Traffic Speed on the Ambient Pollutant Concentration Near Roadways, Paper No. 84-118.6 presented at the XXth Annual Meeting of the Air Pollution Control Assn., San Francisco, Calif.

Quickert, N. L., Dubois, L., and Wallworth, B. 1976. Characterization of an episode with elevated ozone concentrations, *Sci. Total Environ.* 5:79–93.

Ragland, K. W., and Peirce, J. J. 1975. Boundary layer model for air pollutant concentrations due to highway traffic, *J. Air Pollut. Contr.* 25:48–51.

Rao, S. T., 1984. Modeling carbon monoxide hot spots near roadway intersections, *Proceedings of the 10th North American Motor Vehicle Emissions Control Conference,* April 1–4, 1984, New York, N.Y.

Rao, S. T., and Keenan, M. T. 1980. Suggestions for improvement of the EPA-HIWAY model, *J. Air Pollut. Contr.* 30:247–256.

Rao, S. T., Chen, M., Keenan, M., Sistla, G., Peddada, R., Wotzak, G., and Kolak, N. 1978. Dispersion of Pollutants Near Highways: Experimental

Design and Data Acquisition Procedures, EPA-600/4-78-037, U.S. Environmental Protection Agency, Washington, D.C.

Rao, S. T., Keenan, M., Sistla, G., and Samson, P. 1979a. Dispersion of Pollutants Near Highways—Data Analysis and Model Evaluation, EPA-600/4-79-011, U.S. Environmental Protection Agency, Washington, D.C.

Rao, S. T., Sedefian, L., and Czapski, U. H. 1979b. Characteristics of turbulence and dispersion of pollutants near major highways, *J. Appl. Meteorol.* 18:283–293.

Rao, S. T., Sistla, G., Keenan, M. T., Wilson, J. S. 1980. An evaluation of some commonly used highway dispersion models, *J. Air Pollut. Contr.* 30:239–246.

Saffman, P. G. 1962. The effect of wind shear on horizontal spread from an instantaneous ground source, *Quart. J. Roy. Meteorol. Soc.* 88:382–393.

Samson, P. J., and Ragland, K. W. 1977. Ozone and visibility reduction in the Midwest: evidence for large scale transport, *J. Appl. Meteorol.* 16:1101–1106.

Schubert, J. F. 1976. A climatology of the mixed layer using acoustic methods, *Proceedings of the Third Symposium of Meteorological Instrumentation*, American Meteorological Society, Boston, Mass.

Schultz, P., and Warner, T. T. 1982. Characteristics of summertime circulations and pollutant ventilation in the Los Angeles basin, *J. Appl. Meteorol.* 21:672–682.

Sedefian, L., Rao, S. T., and Czapski, U. 1981. Effects of traffic-generated turbulence on near-field dispersion, *Atmos. Environ.* 15:527–536.

Sehmel, G. A. 1984. Deposition and resuspension, In: *Atmospheric Science and Power Production* (D. Randerson, ed.), pp. 533–583, National Technical Information Service, Springfield, Va. NTIS DE84005177 (DOE/TIC-27601).

Shair, F. H. 1974. Dispersion of an instantaneous cross-wind line source of tracer release, *Atmos. Environ.* 8:475–485.

Sheih, C. M. 1980. Numerical Simulation of Particle Dispersion from Emissions at a Highway, pp. 30–32, Argonne National Laboratory, Argonne, Ill., Rept. No. ANL-80-115, Pt. 4 (August 1980).

Sistla, G., Samson, P. J., Keenan, M., and Rao, S. T. 1979. A study of pollutant dispersion near highways, *Atmos. Environ.* 13:669–685.

Skinner, G. T., and Ludwig, G. R. 1976. Experimental Studies of CO Dispersion from a Highway Model, Rept. NA-5411-A-1 (August 1976), Buffalo, N.Y.

Smith, T. B., Blumenthal, D. L., Anderson, J. A., and Vanderpol, A. H. 1978. Transport of SO$_2$ in power plant plumes, *Atmos. Environ.* 12:605–611.

Sobottha, H., and Leisen, P. 1980. Pollutant dispersion of vehicle exhaust gases in street canyons, *Proceedings of the 5th International Clean Air Congress*, Buenos Aires.

Spicer, C. W., Joseph, D. W., Stickel, P. R., and Ward, G. F. 1979. Ozone sources and transport in the northeastern United States, *Environ. Sci. Technol.* 13:975–985.

Spicer, C. W., Joseph, D. W., Stickel, P. R. 1982. An investigation of the ozone from a small city, *J. Air Pollut. Contr.* 32:278–281.

Taylor, A. D. 1982. Puff growth in an Ekman layer, *J. Atmos. Sci.* 39:837–850.

Tyldesley, J. B., and Wallington, C. F. 1965. The effect of wind shear and vertical diffusion on horizontal dispersion, *Quart. J. Roy. Meteorol. Soc.* 91:158–174.

Wedding, J. B., Lombardi, D. J., and Cermak, J. E. 1977. A wind tunnel study of gaseous pollutants in street canyons, *J. Air Pollut. Contr.* 27:557–566.

Wesley, M. L., and Hicks, B. B. 1977. Some factors that affect the deposition rates of sulfur dioxide and similar gases on vegetation, *J. Air Pollut. Contr.* 27:1110–1116.

Wolff, G. T., Lioy, P. J., Wight, G. D., Meyers, R. E., and Cederwall, R. T. 1977. An investigation of long-range transport of ozone across the midwestern and eastern U.S., *Atmos. Environ.* 11:797–802.

Yamartino, R. J., and Wiegand, G. 1986. Development and evaluation of simple models for the flow, turbulence, and pollutant concentration fields within an urban street canyon, *Atmos. Environ.* 20:2137–2156.

Zimmerman, J. R., and McKenzie, K. W. 1974. Winter 73–74 Weather and Carbon Monoxide Air Pollution in Fairbanks, Alaska, National Oceanic and Atmospheric Administration Technical Memorandum NWS-11, Alaska Region, Anchorage, Alaska.

Zimmerman, J. R., and Thompson, R. S. 1975. "Users Guide for HIWAY. A Highway Air Pollution Model," EPA-650/4-74-008, U.S. Environmental Protection Agency, Washington, D.C.

Atmospheric Transformations of Automotive Emissions

ROGER ATKINSON
University of California, Riverside

Air Pollution, the Automobile, and Public Health. © 1988 by the Health Effects Institute. National Academy Press, Washington, D.C.

Components of Atmospheric Pollution

A wide spectrum of inorganic and organic chemical compounds are emitted from automotive use. These emissions arise from combustion as well as evaporative processes. They include the obvious water vapor and carbon dioxide (CO_2), as well as carbon monoxide (CO), oxides of nitrogen (NO_x), oxides and oxyacids of sulfur, reduced sulfur compounds, a wide variety of volatile organic compounds comprising fuel components and partially oxidized products of combustion, and particulate matter. The identities of these emissions and a quantitative understanding of their emission rates are the focus of "Automotive Emissions" (Johnson, this volume).

In highly urbanized regions, automotive emissions contribute a significant, and often major, fraction of the overall emission burden of NO_x, volatile organic compounds, and elemental carbon and/or particulate organic matter. For example, the 1979 mobile and stationary source contributions of NO_x, volatile reactive organic gases (ROG), oxides of sulfur (SO_x), total suspended particulate matter (TSP), CO, and lead (Pb) to the overall emissions in the Los Angeles South Coast Air Basin of California are given in table 1. In this particular urban air basin, mobile source emissions are predominantly automotive (since aircraft and ship emissions are relatively minor) and are major contributors to the overall emission inventory of NO_x, ROG, CO, and Pb.

Some of these emissions have a direct impact on the ecosystem, including human health. In addition, most of them can undergo chemical transformations in the atmosphere (see, for example, Atkinson and Lloyd 1984; Atkinson 1986), sometimes leading to the production of more toxic products. The possible chemical transformations and physical loss processes that occur in the atmosphere during transport of these primary automotive emissions from source to receptor are the main subjects of this chapter. The time scales of these atmospheric transformations and physical loss processes vary widely, with chemical life-

times ranging from ≤ 1 min for some highly reactive organic compounds to months or even years for other much more inert emissions (Atkinson 1986). For example, the Los Angeles urban plume has been identified by ambient air monitoring measurements at Niwot Ridge, Colorado, and the transit time estimated at approximately four days (Roberts et al. 1984).

To understand effects on health and to assess risk, it is necessary to know the identities, the ambient concentrations, and the distributions between gaseous and condensed phases of the chemical compounds impacting human receptors. Thus, it is necessary to determine the chemical and physical changes that primary automotive emissions undergo during their transport through the atmosphere, and the threats, if any, that the resulting products pose to human health.

It must be borne in mind that automotive emissions cannot be considered in isolation. Synergistic chemical and physical interactions occur between automotive emissions and emissions arising from, for example, stationary sources and vegetation, giving rise to a further multitude of product species. Clearly, changes in emission rates or chemical characteristics of these nonautomotive emissions can lead to changes in the photochemical reactivities of the overall atmospheric pollutant mixtures. The elucidation of the effect of automotive emissions on human health necessitates a complete knowledge of the emission inventories, the physical and chemical transformations, transport, and ambient atmospheric measurements of automotive, stationary source, and vegetative emissions, all combined within the framework of local, ur-

Table 1. Average Emission Rates of NO_x, ROG, SO_x, TSP, CO, and Pb in the South Coast Air Basin of California During 1979

Source	Emissions (tons/day)					
	NO_x	ROG	SO_x	TSP	CO	Pb
Mobile	837	853	73	91	7,060	9.1
Stationary	406	680	201	522	588	0.6
Total	1,243	1,533	274	613	7,650	9.7

SOURCE: Adapted with permission from South Coast Air Quality Management District 1982.

ban, or regional photochemical computer models.

This chapter assesses the atmospheric lifetimes of the various classes of automotive emissions, which, for compounds of low volatility, may vary markedly with their distribution between the gaseous and particulate phases. The state of knowledge about products formed by chemical reactions under atmospheric conditions, including indoor environments, is reviewed and discussed. In many cases, the products formed during the photodegradation of primary automotive emissions are not presently known, and studies are needed to determine the general nature, and toxicity to humans, of these products. A list of research recommendations to obtain the necessary data base about these atmospheric transformations is presented.

Physical and Chemical Transformations Under Atmospheric Conditions

Two decades of laboratory, environmental chamber, and ambient atmospheric measurements have revealed the physical and chemical processes that transform and/or remove chemical compounds emitted into the atmosphere. These atmospheric emissions are partitioned between the gas and particulate phases, and the atmospheric loss processes for both phases must be evaluated separately.

Physical Removal Processes

The physical removal processes can be defined as accretion (or coagulation) of particles, and dry and wet deposition of gases as well as particles. Removal of gases and particles at ground surfaces—including snow-covered ground and other moist surfaces—is known as dry deposition, whereas removal of these species by raindrops is referred to as wet deposition. These processes are dynamic, and we do not yet have a quantitative understanding of them (see, for example, Eisenreich et al. 1981; Graedel et al. 1982; Slinn 1982; Colbeck and Harri-

son 1985; Dolske and Gatz 1985; Jonas and Heinemann 1985; Ligocki et al. 1985a,b; Sehmel et al. 1985; Terry Dana et al. 1985; van Noort and Wondergem 1985).

Dry Deposition of Gases and Particles.

Gas-phase species and particles can be removed from the atmosphere by an overall process that involves downward transport from the atmospheric boundary layer to the ground surface. The complex atmospheric physical mechanisms that deliver gaseous and particulate species to the surface are generally combined with the chemical processes of mass transfer at the surface by use of a "deposition velocity" V_d. The dry deposition rate, F, is

$$F = V_d [C] \qquad (1)$$

where $[C]$ is the concentration of the species at some reference height (generally defined as 1 m). The deposition velocity depends on the specific gaseous chemical and/or particle species, the surface to which the species is being deposited, and the reference height. It also depends on the atmospheric stability, being highest during unstable conditions (see, for example, Cadle et al. 1985; Colbeck and Harrison 1985).

The deposition velocity is often defined by three "resistance" terms

$$V_d = (r_a + r_b + r_s)^{-1} \qquad (2)$$

where r_a is the resistance between the reference height and the laminar sublayer near the receiving surface; r_b is the laminar sublayer resistance; and r_s is the surface resistance, specific to each pollutant and surface type. For certain species (for example, gaseous nitric acid), the surface resistance r_s is effectively zero, and transport to the surface becomes rate determining (Huebert and Robert 1985).

The deposition velocities of particles depend on the particle size, exhibiting a minimum for particles of mean diameter of ~ 0.1 μm. It should also be noted that, for particles, a constant adsorption and desorption of chemicals occurs, characterized by their Henry's law properties. Thus there is a dynamic equilibrium between the gaseous and adsorbed (or particulate) phases which, in accordance with the Henry's law constants, depends on temperature, properties of the individual particles, vapor pressure, and liquid adsorption properties.

Wet Deposition of Gases and Particles (Rainout).

In addition to dry deposition, wet deposition can remove gaseous compounds and particles from the atmosphere. This process occurs during precipitation. Slinn and coworkers (1978) showed that a falling raindrop attains equilibrium with a gaseous chemical over a distance of ~ 10 m. As described by Eisenreich et al. (1981), the wet removal of gaseous chemicals arises from equilibrium partitioning, and a washout ratio, W, defining the scavenging efficiency of a gas-phase species, is given by

$$W = C_{rain}/C_{air} = RT/H \qquad (3)$$

where R is the gas constant, T is the temperature (°K), H is the Henry's law constant, and C_{rain} and C_{air} are the concentrations in rain and air, respectively. The deposition rate, F, is then given by

$$F = WJC_{air} \qquad (4)$$

where J is the precipitation rate.

Wet removal of gases is clearly most important for chemicals highly soluble in water, such as hydrogen peroxide, nitric acid, and phenols. Thus, following a precipitation event, the atmospheric concentrations of highly water soluble species may fall to near zero. For most gas-phase organic chemicals, however, it is likely that wet removal is of minor importance. Atlas and Giam (1981) calculated atmospheric residence times ranging from ~ 60 days for phthalates and hexachlorohexanes up to ~ 6 years for the polychlorinated biphenyl mixture Aroclor 1242.

Clearly, wet deposition is episodic. Thus, only average wet deposition velocities can be ascribed, and these are strong functions of the climatological conditions at the particular geographic location in question.

Chemical Removal Processes

Many chemical processes contribute to the removal of compounds emitted into the troposphere. For gas-phase chemicals, these removal processes involve

- photolysis during daylight hours;
- reaction with hydroxyl (OH) radicals during daylight hours;
- reaction with ozone (O_3) during daytime and nighttime;
- reaction with hydroperoxyl (HO_2) radicals during, typically, late daytime and early nighttime hours;
- reaction with the gaseous nitrate (NO_3) radical during nighttime hours;
- reaction with dinitrogen pentoxide (N_2O_5) during nighttime hours;
- reaction with NO_2 during daytime and nighttime hours; and
- reaction with gaseous nitric acid (HNO_3) and other species such as nitrous (HNO_2) and sulfuric (H_2SO_4) acids.

Additionally, the following processes are likely to contribute to the degradation of chemical compounds present in the adsorbed phase:

- photolysis;
- reaction with O_3;
- reaction with N_2O_5 during nighttime hours;
- reaction with NO_2, typically present throughout a full 24-hr period;
- reaction with H_2O_2;
- reaction with HNO_3, HNO_2, and H_2SO_4.

Synergism may be important in certain of these reactions involving adsorbed automotive emissions. For example, the presence of HNO_3 together with NO_2 may lead to enhanced nitration of adsorbed polycyclic aromatic hydrocarbons (PAHs) (see, for example, Pitts 1983).

In addition to photolysis and reactions of automotive emissions in the gaseous and adsorbed states with a variety of atmospherically important species, the chemical reactions of automotive emissions in rain, cloud, or fog water with other reactive components of these aqueous systems must be considered. This subject has recently received much attention because of an increasing emphasis on acid deposition (see, for example, Calvert 1984). Reactions of chemicals in the aqueous phase with reactive intermediates such as H_2O_2 and various radical and ionic species have been dealt with in some detail in connection with these acidic deposition studies (Graedel and Weschler 1981; Chameides and Davis 1982; Graedel and Goldberg 1983; Jacob and Hoffmann 1983; Chameides 1986; Graedel et al. 1986).

Photolysis. Automobile emissions can be removed from the atmosphere by photolysis. This process requires that a chemical compound absorb light in the actinic portion of the spectrum (that is, the wavelength region from ~ 290 to 1,000 nm) and, after absorption of a photon, undergo chemical change (Calvert and Pitts 1966). For most compounds, breakage of a chemical bond requires an energy in excess of ~ 40 kcal/mole (Benson 1976). Therefore, photolytic wavelengths of ≲ 700 nm are necessary. One fundamental tenet is that absorption of a single photon (referred to hereafter as $h\nu$) cannot photodissociate more than one molecule (Calvert and Pitts 1966).

Formation of Ozone. Ozone is formed in the troposphere from the photolysis of NO_2

$$NO_2 + h\nu \rightarrow NO + O(^3P) \qquad (5)$$

followed by reaction of the ground-state oxygen atom, $O(^3P)$, with O_2

$$O(^3P) + O_2 \xrightarrow{M} O_3 \qquad (6)$$

where M denotes a third body, air in this case. Tropospheric O_3 is also transported downward from the stratosphere (Logan 1985). In the clean troposphere, O_3 mixing ratios are typically 30 ± 10 parts per billion (ppb) at ground level (~ 7×10^{11} molecules/cm^3), and increase with altitude (Logan 1985). The relative contributions to tropospheric O_3 of photochemical formation and downward transport from the stratosphere are discussed by Logan (1985).

Formation of Hydroxyl Radicals. The OH radical is the major reactive species in the troposphere (Logan et al. 1981), and is formed by photolysis of O_3, photolysis of HNO_2, and reaction of the HO_2 radical

with nitric oxide (NO) (DeMore et al. 1985).

Photolysis of Ozone. Ozone photodissociates at wavelengths of < 319 nm to yield, in part, electronically excited oxygen atoms, $O(^1D)$,

$$O_3 + h\nu \rightarrow O(^1D) + O_2 \qquad (7)$$

which react with water vapor (eq. 8) or N_2 and O_2 (eq. 9)

$$O(^1D) + H_2O \rightarrow 2 OH \qquad (8)$$

$$O(^1D) + N_2, O_2 \rightarrow O(^3P) + N_2, O_2 \qquad (9)$$

For a relative humidity of ~ 50 percent at 298°K, ~ 0.2 OH radicals are formed for each $O(^1D)$ atom formed.

Photolysis of Nitrous Acid. Nitrous acid, which is present during nighttime hours in urban atmospheres (Platt et al. 1980a; Harris et al. 1982; Pitts et al. 1984a), is rapidly photolyzed at wavelengths of < 400 nm (eq. 10) during daylight hours to yield OH radicals (DeMore et al. 1985),

$$HNO_2 + h\nu \rightarrow OH + NO \qquad (10)$$

with a lifetime due to photolysis of ~ 10–15 min under noontime conditions.

Reaction of Hydroperoxyl Radicals with Nitric Oxide. Hydroperoxyl radicals, formed from the photodissociation of aldehydes and other photochemical processes (see below, and Atkinson and Lloyd 1984), react with NO to yield the OH radical

$$HO_2 + NO \rightarrow OH + NO_2 \qquad (11)$$

In the troposphere, under conditions where NO concentrations are less than 5×10^8 molecules/cm^3, HO_2 radicals are expected to react with HO_2 and alkyl peroxy (RO_2) radicals in competition with reaction with NO. Hence a knowledge of tropospheric NO concentrations is important for assessing the conversion of HO_2 to the more reactive OH radical (Logan et al. 1981; Crutzen 1982; Logan 1983).

Formation of Hydroperoxyl Radicals. Hydroperoxyl radicals are produced under tropospheric conditions from the photolysis of aldehydes (Atkinson and Lloyd 1984), for example from formaldehyde (HCHO),

$$HCHO + h\nu \begin{cases} \rightarrow H + HCO \\ \rightarrow H_2 + CO \end{cases} \quad (12)$$

followed by the rapid reactions of hydrogen atoms and HCO radicals with O_2

$$H + O_2 \xrightarrow{M} HO_2 \quad (13)$$

$$HCO + O_2 \rightarrow HO_2 + CO \quad (14)$$

The higher aldehydes also photodissociate to ultimately yield HO_2 radicals

$$RCHO + h\nu \rightarrow R\cdot + HCO$$
$$\downarrow O_2$$
$$HO_2 + CO \quad (15)$$

(Here and below, **R**, **R′**, **R″** . . ., represent unspecified groups and the dot (·) represents an incomplete chemical bond or unpaired electron.)

Additionally, HO_2 radicals are formed from reactions of alkoxy and α-hydroxy radicals, which are reactive intermediates produced during the photooxidation processes of most organic compounds (Atkinson and Lloyd 1984). For example,

$$\begin{array}{c} R \\ \diagdown \\ R' \end{array} CHO\cdot + O_2 \rightarrow RCOR' + HO_2 \quad (16)$$

$$R\dot{C}HOH + O_2 \rightarrow RCHO + HO_2 \quad (17)$$

Daytime tropospheric HO_2 radical concentrations are calculated to range from $\sim 10^8$ to 10^9 molecules/cm^3 (Hov and Isaksen 1979; Stockwell and Calvert 1983a).

Formation of Nitrate Radicals. The gaseous NO_3 radical has been shown to be an important constituent of nighttime atmospheres (see, for example, Winer et al. 1984; Atkinson et al. 1986b). Nitrate radicals are formed by the reactions

$$NO_2 + O_3 \rightarrow NO_3 + O_2 \quad (18)$$

$$NO_2 + NO_3 \xrightarrow{M} N_2O_5 \quad (19)$$

$$N_2O_5 \xrightarrow{M} NO_2 + NO_3 \quad (20)$$

with N_2O_5 being in relatively rapid (<1 min at 298°K and 760 torr total pressure) equilibrium with NO_2 and the NO_3 radical (DeMore et al. 1985).

Since NO_3 radicals photolyze at wavelengths between 400 and 650 nm (Graham and Johnston 1978; Magnotta and Johnston 1980),

$$NO_3 + h\nu \begin{cases} \rightarrow NO_2 + O(^3P) \\ \rightarrow NO + O_2 \end{cases} \quad (21)$$

with a lifetime of ~ 5 sec at noontime (Magnotta and Johnston 1980), and react rapidly with NO,

$$NO_3 + NO \rightarrow 2\,NO_2 \quad (22)$$

NO_3 radical concentrations are essentially negligible during daylight hours. After sunset, they can rise rapidly to levels of up to ~ 400 parts per trillion (ppt) ($\sim 1 \times 10^{10}$ molecules/cm^3) over continental areas (see, for example, Platt et al. 1984; Atkinson et al. 1986b). For example, at several semiarid and desert sites in southern California, Platt and coworkers (1984) consistently observed nighttime NO_3 radical concentrations of $\sim 2 \times 10^8$ to 2×10^9 molecules/cm^3.

Formation of Dinitrogen Pentoxide. As shown above (eq. 19 and 20), N_2O_5 is in equilibrium with NO_2 and the NO_3 radical. Maximum nighttime N_2O_5 concentrations of $\sim (2-3) \times 10^{11}$ molecules/cm^3 can be inferred from the equilibrium constant for these reactions, which is uncertain by a factor of $\pm \sim 1.2–1.5$ (Graham and Johnston 1978; Tuazon et al. 1984b; Burrows et al. 1985a; Perner et al. 1985); and concentrations greater than $\sim 2 \times 10^{10}$ molecules/cm^3 were calculated to be exceeded ~ 30 percent of the nights for which data are available (Atkinson et al. 1986b).

Formation of Gas-Phase Acids. Chemicals that are basic can react with gas-phase acids to form their salts. As presently understood, the major gas-phase acidic species are HNO_3 and HNO_2. Nitric acid is formed in the gas phase from the reaction of OH radicals with NO_2

$$OH + NO_2 \xrightarrow{M} HNO_3 \quad (23)$$

and can be formed, probably in the adsorbed phase, from the heterogeneous hydrolysis of N_2O_5

$$N_2O_5 + H_2O \xrightarrow{heterogeneous} 2\,HNO_3 \quad (24)$$

though this initially adsorbed HNO_3 may be desorbed back into the gas phase (Tuazon et al. 1983; Atkinson et al. 1986a).

Nitrous acid is formed from the reaction of OH radicals with NO

$$OH + NO \xrightarrow{M} HNO_2 \qquad (25)$$

although its rapid photolysis during daylight hours (eq. 10) leads to a low ambient daytime concentration. However, HNO_2 has been identified and measured in nighttime Los Angeles atmospheres at up to 8 ppb ($\sim 2 \times 10^{11}$ molecules/cm^3) (Harris et al. 1982). Indeed, nighttime HNO_2 levels of $\sim 2 \times 10^{10}$ to 2×10^{11} molecules/cm^3 are probably typical of many, if not most, urban environments. In environmental chambers and indoor environments, HNO_2 has been shown to be formed from the heterogeneous hydrolysis of NO_2 (Sakamaki et al. 1983; Pitts et al. 1984c, 1985d).

For automotive emissions associated with the particulate phase, reactions with NO_2, HNO_3, HNO_2, N_2O_5, and O_3 must be considered (see, for example, Pitts 1983). Many, if not most, of these reactions probably proceed by reaction of the adsorbed automotive emissions with adsorbed, rather than gas-phase, reactive atmospheric species. Additionally, photolysis of adsorbed automotive emissions also occurs and may be highly important (Behymer and Hites 1985).

■ **Recommendation 1.**   Study is required on the physical removal processes leading to wet and dry deposition of gases and particles. Investigations of the processes occurring on and in particulate and aerosol matter should focus on gas-to-particle conversion processes and the chemical processes that occur within aerosols (including fogs and clouds).

Atmospheric Lifetimes, Fates, and Products of the Atmospheric Transformations of Automotive Emissions

This section reviews and summarizes the present status of knowledge concerning the atmospheric loss processes and atmospheric lifetimes of automotive emissions and the products formed from them under atmospheric conditions. Reference is made whenever possible to existing reviews and/or evaluations, from which further details can be obtained.

Atmospheric Lifetimes

Data obtained during the past two decades have provided a comprehensive view of the chemical and physical removal processes that occur in the troposphere, and of the reaction rate constants for many of these processes. Table 2 lists the rate constants at room temperature (298°K, 77°F) for the known tropospheric chemical removal reactions for selected automotive emissions. The corresponding calculated lifetimes in the lower troposphere of these chemicals due to reaction with each of the atmospherically important reactive species listed in table 2 are given in table 3.

Although the individual rate constants are known to a reasonable degree of accuracy (sometimes to within ± 25 percent, and in most cases to within a factor of two), the calculated atmospheric lifetimes are much more uncertain because of the larger uncertainties about the ambient atmospheric concentrations of several of these key tropospheric species. For example, the ambient atmospheric OH radical concentrations at any given time and/or location are uncertain to a factor of at least five, and more likely ten (Hewitt and Harrison 1985). The tropospheric diurnally and annually averaged OH radical concentrations are better known, to within possibly a factor of two (Crutzen 1982), being $\sim 5 \times 10^5$ and $\sim 6 \times 10^5$ molecules/cm^3 in the northern and southern hemispheres, respectively. Similar arguments apply for the ambient nighttime tropospheric concentrations of the NO_3 radical and of N_2O_5 (Atkinson et al. 1986b).

In addition to these chemical loss processes of automotive emissions in the troposphere, physical loss processes must also be taken into account. Tables 4 and 5 give selected examples from the literature of dry deposition velocities and of washout ratios for several inorganic and organic species.

Table 2. Room Temperature Rate Constants at Atmospheric Pressure of Air for the Gas–Phase Reactions of Selected Automotive Emissions with Atmospherically Important Intermediate Species

Emission	Rate Constant [cm³/(molecule · sec)]				
	OH	O_3	NO_3	HO_2	N_2O_5
$NO_2{}^a$	1.1×10^{-11}	3.2×10^{-17}	1.2×10^{-12}	1.4×10^{-12}	
NO	6.6×10^{-12}	1.8×10^{-14}	3.0×10^{-11}	8.3×10^{-12}	
$HNO_2{}^a$	6.6×10^{-12}	$< 5 \times 10^{-19}$			
$HNO_3{}^b$	1.3×10^{-13}				
SO_2	9×10^{-13}	$< 2 \times 10^{-22}$	$< 7 \times 10^{-21}$	$< 1 \times 10^{-18}$	$< 4.2 \times 10^{-23}$
$NH_3{}^c$	1.6×10^{-13}				
$CH_3NH_2{}^c$	2.2×10^{-11}	2.1×10^{-20}			
HCN	3×10^{-14}				
H_2S	4.7×10^{-12}	$< 2 \times 10^{-20}$	$< 3 \times 10^{-14}$		
CH_3SH	3.3×10^{-11}		1×10^{-12}		
$H_2O_2{}^a$	1.7×10^{-12}		$< 2 \times 10^{-15}$		
Propane	1.2×10^{-12}	$< 6 \times 10^{-24}$			
n–Butane	2.5×10^{-12}	$< 1 \times 10^{-23}$	3.6×10^{-17}		
n–Octane	8.7×10^{-12}		9.9×10^{-17}		
1,2-Dichloroethane	2.2×10^{-13}				
1,2-Dibromoethane	2.5×10^{-13}				
Ethene	8.5×10^{-12}	1.8×10^{-18}	1.1×10^{-16}		
Propene	2.6×10^{-11}	1.1×10^{-17}	7.5×10^{-15}		
1-Butene	3.1×10^{-11}	1.1×10^{-17}	9.7×10^{-15}		
$trans$-2-Butene	6.4×10^{-11}	2.0×10^{-16}	3.8×10^{-13}	$< 4 \times 10^{-18}$	
Acetylene	7.8×10^{-13}	8×10^{-21}	$\leq 2.3 \times 10^{-17}$		
Propyne	6.1×10^{-12}	1.4×10^{-20}	9.4×10^{-17}		
Butadiyne	2×10^{-11}	$\sim 6 \times 10^{-20}$			
Formaldehyde^a	9.0×10^{-12}	$< 2 \times 10^{-24}$	5.8×10^{-16}	$\sim 1 \times 10^{-14}$	
Acetaldehyde^a	1.6×10^{-11}	$< 6 \times 10^{-21}$	2.4×10^{-15}		
Benzaldehyde^a	1.3×10^{-11}		2.0×10^{-15}		
Acrolein	2.0×10^{-11}	2.8×10^{-19}			
Crotonaldehyde	3.6×10^{-11}	9.0×10^{-19}			
Methyl vinyl ketone	1.9×10^{-11}	4.8×10^{-18}			
Acetone^a	2.3×10^{-13}				
2-Butanone^a	1.0×10^{-12}				
Dimethyl ether	3.0×10^{-12}		$< 3 \times 10^{-15}$		
Methanol	9×10^{-13}		$< 6 \times 10^{-16}$		
Ethanol	2.9×10^{-12}		$< 9 \times 10^{-16}$		
Formic acid	4.6×10^{-13}				
Methyl nitrite^a	$\sim 1.8 \times 10^{-13}$	1.3×10^{-20}			
Benzene	1.3×10^{-12}	7×10^{-23}	$< 2 \times 10^{-17}$		
Toluene	6.2×10^{-12}	1.5×10^{-22}	3.6×10^{-17}		
m-Xylene	2.5×10^{-11}	6×10^{-22}	1.3×10^{-16}		
1,2,4-Trimethyl-benzene	4.0×10^{-11}	1.3×10^{-21}	9.7×10^{-16}		
Phenol	2.8×10^{-11}		3.8×10^{-12}		
Naphthalene	2.2×10^{-11}	$< 2 \times 10^{-19}$			1.4×10^{-17}
2-Methylnaphthalene	5.2×10^{-11}	$< 4 \times 10^{-19}$			4.2×10^{-17}
2,3-Dimethylnaph-thalene	7.7×10^{-11}	$< 4 \times 10^{-19}$			5.7×10^{-17}
Phenanthrene	3.2×10^{-11}				
Anthracene	1.3×10^{-10}				

^{a} Photolysis also occurs at a significant rate (see table 3).
^{b} Also reacts with NH_3 to form NH_4NO_3.
^{c} Also reacts with gaseous HNO_3 to form nitrate salts.

Table 3. Calculated Atmospheric Lifetimes[a] for the Gas–Phase Reactions of the Selected Automotive Emissions with Atmospherically Important Intermediate Species[b]

Emission	Atmospheric Lifetime Due to Reaction with				
	OH	O_3	NO_3	HO_2	$h\nu$[c]
NO_2	2 days	12 hr	1 hr	2 hr	2 min
NO	4 days	1 min	3 min	20 min	
HNO_2	4 days	> 33 days			~ 10 min
HNO_3[d]	180 days				
SO_2[e]	26 days	> 200 yr	> 4.5 × 10⁴ yr	> 600 yr	
NH_3[f]	140 days				
CH_3NH_2[f]	12 hr	2 yr			
HCN	2 yr				
H_2S	5 days	> 2 yr	> 4 days		
CH_3SH	8 hr		1 hr		
H_2O_2	14 days		> 60 days		36 hr
Propane	19 days	> 7,000 yr			
n-Butane	9 days	> 4,500 yr	9 yr		
n-Octane	3 days		3 yr		
1,2-Dichloroethane	100 days				
1,2-Dibromoethane	90 days				
Ethene	3 days	9 days	3 yr		
Propene	11 hr	1.5 days	15 days		
1-Butene	9 hr	1.5 days	12 days		
trans-2-Butene	4 hr	2 hr	4 hr	> 150 yr	
Acetylene	30 days	6 yr	≥ 14 yr		
Propyne	4 days	3 yr	3.4 yr		
Butadiyne	1 day	~ 270 days			
Formaldehyde	3 days	> 2 × 10⁴ yr	210 days	23 days	4 hr
Acetaldehyde	1 day	> 7 yr	50 days		60 hr
Benzaldehyde	2 days		60 days		
Acrolein	1 day	60 days			
Crotonaldehyde	8 hr	18 days			
Methyl vinyl ketone	1 day	3 days			
Acetone	100 days				15 days
2-Butanone	23 days				
Dimethyl ether	7 days		> 40 days		
Methanol	26 days		> 190 days		
Ethanol	8 days		> 130 days		
Formic acid	50 days				
Methyl nitrite	~ 120 days	3 yr			8 min
Benzene	18 days	600 yr	> 16 yr		
Toluene	4 days	300 yr	9 yr		
m-Xylene	11 hr	75 yr	2 yr		
1,2,4-Trimethylbenzene	7 hr	35 yr	120 days		
Phenol	10 hr		20 min		
Naphthalene[e]	1 day	> 80 days			
2-Methylnaphthalene[e]	5 hr	> 40 days			
2,3-Dimethylnaphthalene[e]	4 hr	> 40 days			
Phenanthrene	9 hr				
Anthracene	2 hr				

[a] The time for the compound to decay to 37 percent of its original concentration.

[b] For concentrations of OH, 12-hr average of 1 × 10⁶ molecules/cm³ (Crutzen 1982); O_3, 24-hr average of 7 × 10¹¹ molecules/cm³ (Singh et al. 1978); NO_3, 12-hr average of 2 × 10⁸ molecules/cm³ (Platt et al. 1984); HO_2, 12-hr average of 10⁸ molecules/cm³ (Hov and Isaksen 1979).

[c] For solar zenith angle of 0°.

[d] Also reacts with NH_3 to form NH_4NO_3.

[e] Lifetimes due to gas-phase reaction with a 12-hr average concentration of N_2O_5 of 2 × 10¹⁰ molecules/cm³ (Atkinson et al. 1986b) are SO_2, > 7.5 × 10⁴ yr; naphthalene, ~80 days; 2-methylnaphthalene, ~ 30 days; 2,3-dimethylnaphthalene, ~ 20 days.

[f] Also reacts with gaseous HNO_3 to form nitrate salts.

Table 4. Dry Deposition Velocities for Several Inorganic and Organic Chemicals

Depositing Species	Mean Deposition Velocity (cm/sec)
O_3	0.3–0.5^a
	0.08–0.91^b
Particulate sulfur	0.17^a
Particles	
0.18-μm median diameter	0.16^a
0.25-μm median diameter	0.35^a
Calcium sulfate ($CaSO_4$) particles	
1-μm diameter	0.01^c
2-μm diameter	0.03^c
5-μm diameter	0.44^c
10-μm diameter	4.6^c
SO_2	2.1^a
HNO_3	2.5^a
Tetrachloroethene	$\sim 10^{-4d}$
Nitrobenzene	$\sim 10^{-4d}$
Polychlorinated biphenyls (PCBs)	~ 0.1–0.5^e

[a] From Dolske and Gatz 1985, with grass as the surface.
[b] From Colbeck and Harrison 1985, with grass as the surface.
[c] From Jonas and Heinemann 1985, with grass as the surface.
[d] From Sehmel et al. 1985.
[e] From Eisenreich et al. 1981.

For the particle-associated chemical species, the washout ratios W given in table 5 reflect the loss of the particles. Thus, as discussed by Eisenreich et al. (1981), the washout ratios W for aerosols are typically $\sim 10^5$ to 10^6, in comparison to values of $\sim 10^0$ to 10^4 for gaseous chemicals.

For particles, the atmospheric lifetimes due to dry deposition are of the order of several days for 0.1–1 μm diameter particles, and table 6 gives the average lifetimes of atmospheric particles as a function of their diameter. The dry deposition lifetimes of many organic compounds are also weeks or months (see, for example, Atlas and Giam 1981). However, for certain chemicals that have relatively slow gas-phase chemical loss rates, such as HNO_3 and SO_2, dry deposition can be the major loss process under typical atmospheric conditions. Because of the potential importance of the dry deposition atmospheric removal process, measurements of the deposition velocities of gaseous and particulate species need to be carried out for a variety of terrains. This is a difficult and time-consuming task, and emphasis must be given to extending the presently available experimental techniques and to developing and testing new experimental, and possibly theoretical, approaches (see, for example, Dolske and Gatz 1985).

Table 5. Washout Ratios for Selected Organic Chemicals

Phase	Chemical	Washout Ratio (W)
Gas	Ethene oxide	4–6^a
	Phenol	$(0.7$–$25) \times 10^{4b}$
	Nitrobenzene	$(2$–$4) \times 10^{3a}$
	Naphthalene	100–300^c
	Phenanthrene	$2,000$–$4,000^c$
	Pyrene	$3,000$–$9,000^c$
	Benz[a]anthracene	$7,000$–$22,000^c$
	Hexachlorobenzene	$1,500^d$
Particle	PCBs	$\sim (1$–$10) \times 10^{4d}$
	Particles	
	0.1–1.0-μm diameter	$\leq 10^{5d}$
	10-μm diameter	$\sim 10^{6d}$
	Tricosane through hexacosane	$\sim 2 \times 10^{4e}$

[a] From Terry Dana et al. 1985.
[b] From Leuenberger et al. 1985.
[c] From Ligocki et al. 1985a.
[d] From Eisenreich et al. 1981.
[e] From Ligocki et al. 1985b.

For nonpolar organic compounds, wet deposition appears to be of minor importance as an atmospheric loss process. However, for highly water-soluble gases such as HNO_3 and H_2O_2, wet deposition can be important (Jacob and Hoffmann 1983; Chang 1984), and in fog and cloud systems this process leads to removal of these and other compounds from the gas phase into the aqueous phase where reactions can occur that lead ultimately to the formation of acids and other oxygenated products.

Wet deposition rates are somewhat better understood, with the experimental results for gas-phase chemicals agreeing to within a factor of ~ 10 or better with theoretical expectations (see, for example, Ligocki et al. 1985a,b; Terry Dana et al. 1985). Further research is needed to provide a wider data base concerning the washout ratios of chemical compounds present in the gas phase and of aerosols and particles. These data will then allow the importance of this wet deposition process to be better evaluated, both as a loss process for primary automotive emissions as well as for the formation and deposition of acid species resulting from aqueous-phase reactions.

The major atmospheric loss process for most of the automotive emissions present in the gas phase is by daytime reaction with the OH radical. However, for certain classes of automotive emissions, photolysis, reaction with NO_3 radicals during nighttime hours, and reaction with O_3 can be important removal routes. Furthermore, reactions that are relatively minor removal processes may need to be considered if they generate products with potential health risks to humans. For example, the reactions of gas-phase N_2O_5 with PAHs appear to be of minor significance as a PAH loss process, but they form toxic nitropolycyclic aromatic hydrocarbons (nitro-PAHs) (Pitts et al. 1985b; Sweetman et al. 1986).

Clearly, a knowledge of the atmospheric loss processes and lifetimes for automotive emissions is important, since these lifetimes determine the geographic extent of the influence of the parent automotive emission. Thus, a short lifetime leads to local exposure, whereas a long lifetime leads to regional or global exposure at lower concentrations.

Table 6. Average Atmospheric Lifetimes for Particles Due to Dry Deposition

Diameter (μm)	Lifetime (days)
0.002	0.01
0.02	1
0.2	10
2	10
20	1
200	0.01

SOURCE: Adapted with permission from Graedel and Weschler 1981.

The atmospheric lifetimes of automotive emissions present in the particulate phase are less well known. Dry and wet deposition constitute the physical loss processes, and photolysis and/or reaction with gas-phase and coadsorbed reactive intermediates constitute the possible chemical loss processes. These chemical processes are substrate dependent, with photolysis, reaction with O_3, reaction with NO_2 and/or HNO_3, and reaction with N_2O_5 occurring (see, for example, Nielsen et al. 1983; Pitts 1983). However, due to the dependence of these loss processes on the nature of the substrate, it is presently impossible to cite any meaningful atmospheric lifetimes for adsorbed automotive emissions, except to remark that the reaction of PAHs with O_3 may lead to lifetimes on the order of hours, photolysis is probably slow, and reaction with N_2O_5—though slow as a loss process—leads to the formation of direct-acting mutagenic and possibly carcinogenic nitro-PAH products.

Atmospheric Transformations

Oxides of Nitrogen. Oxides of nitrogen emitted into the atmosphere as a result of automotive use comprise NO, NO_2, N_2O, HNO_2, and possibly HNO_3. N_2O has been shown to be chemically inert in the troposphere, being transported into the stratosphere where it photodissociates at wavelengths of < 220 nm (Liu et al. 1977),

$$N_2O + h\nu \rightarrow N_2 + O(^1D) \qquad (26)$$

and reacts with electronically excited oxygen atoms, $O(^1D)$, leading to formation of NO.

$$O(^1D) + N_2O \rightarrow 2\,NO \qquad (27)$$

Hence, tropospheric N_2O emissions are a major source of stratospheric NO, and the effect of increasing tropospheric N_2O emissions on stratospheric O_3 has been well documented (Liu et al. 1977).

In the troposphere, the other NO_x are interrelated by means of their reactions (Atkinson and Lloyd 1984; DeMore et al. 1985). At high mixing ratios [$\gtrsim$ 1 part per million (ppm)], NO is oxidized by O_2 to NO_2.

$$2\,NO + O_2 \rightarrow 2\,NO_2 \qquad (28)$$

Additionally, NO reacts rapidly with O_3 to yield NO_2.

$$NO + O_3 \rightarrow NO_2 + O_2 \qquad (29)$$

Nitrogen dioxide photolyzes rapidly at wavelengths of < 430 nm (with a lifetime of ~ 2 min at solar noon, 40°N latitude, summertime) to yield NO and an oxygen atom,

$$NO_2 + h\nu \rightarrow NO + O(^3P) \qquad (30)$$

with the $O(^3P)$ atom reacting with O_2 to yield O_3 (eq. 6). With this series of reactions, NO, NO_2, and O_3 are in a photostationary state,

$$(31)$$

with

$$[O_3] = \frac{k_{30}\,[NO_2]}{k_{29}\,[NO]}$$

where k_{29} and k_{30} are the rate constants for the reactions given in equations 29 and 30, respectively, and brackets [ ] signify concentrations. As shown below, this photostationary state is strongly affected by NO-to-NO_2 conversions caused by reactions involving organic compounds.

Further reactions of NO and NO_2 under atmospheric conditions involve the formation of NO_3 radicals and N_2O_5 (eq. 18–20). Removal of N_2O_5 by dry and wet deposition (eq. 24) may be an important nighttime NO_x sink in the lower troposphere (Heikes and Thompson 1983; Atkinson et al. 1986b).

However, measurements of the sticking coefficient of N_2O_5 to particles and of its dry deposition velocity are necessary before the importance of this nighttime NO_x removal process can be quantified. Thus, nighttime reactions involving NO_3 radicals and N_2O_5 need to be investigated, together with the health impacts of N_2O_5. Dinitrogen pentoxide is known from laboratory experiments to be in equilibrium with NO_3 radicals and NO_2, but it has not been observed in the ambient troposphere, although it is calculated to be present at mixing ratios of up to ~ 10–15 ppb (Atkinson et al. 1986b). Furthermore, it is well recognized that NO_3 and/or N_2O_5 are removed from the troposphere by presently unknown processes (Noxon et al. 1980; Platt et al. 1980b; Atkinson et al. 1986b). These NO_3 and/or N_2O_5 removal processes may be due to loss of NO_3 radicals or to loss of N_2O_5 (the latter by dry or wet deposition), and are important as nighttime loss processes for NO_x as well as for acid deposition. If N_2O_5 is hydrolyzed in the gas phase, presently viewed as unlikely (Atkinson et al. 1986a), significant gasphase HNO_3 concentrations could result in the presunrise hours. Conversely, heterogeneous processes may be, at least in part, an explanation of the highly acidic fogs and rains observed in and around Los Angeles (Waldman et al. 1982). Thus these heterogeneous loss processes of N_2O_5 and/or NO_3 require investigation, since they directly influence the gas-phase concentrations of NO_3 radicals and N_2O_5.

During daylight hours, NO_3 radicals photolyze rapidly (eq. 21) with a photolysis lifetime at solar noon of ~ 5 sec (Magnotta and Johnston 1980).

In addition, NO_3 radicals react with NO (eq. 22) sufficiently rapidly that NO and NO_3 cannot coexist at concentrations of a few ppt or higher.

Nitric oxide and NO_2 also react with HO_2 radicals (eq. 11, 32)

$$NO + HO_2 \rightarrow NO_2 + OH \qquad (11)$$

$$NO_2 + HO_2 \xrightarrow{M} HO_2NO_2 \qquad (32)$$

with the latter reaction being reversible due to the rapid (~ 10 sec lifetime at 298°K and

760 torr total pressure) back-decomposition of HO_2NO_2 (DeMore et al. 1985). Additionally, NO and NO_2 react with OH radicals (eq. 25, 23). The formation of HNO_2 is balanced during daytime hours (when OH radicals are present at appreciable concentrations) by its photolysis (eq. 10).

In addition to these homogeneous reactions of NO_x, heterogeneous reactions may also be atmospherically important. Thus, the formation of HNO_2 from the heterogeneous hydrolysis of NO_2,

$$2\,NO_2 + H_2O \rightarrow HNO_2 + (HNO_3) \qquad (33)$$

has been observed to occur in environmental chambers and in an indoor environment (Sakamaki et al. 1983; Pitts et al. 1984c, 1985d). This reaction may be a formation pathway for the HNO_2 observed during nighttime hours in a number of urban airsheds, at levels 3–5 percent of the NO_2 present (Perner 1980; Platt et al. 1980a; Harris et al. 1982; Pitts et al. 1984a), and in automobile exhaust (Pitts et al. 1984b). In environmental chambers, the formation rate of HNO_2 is first order with respect to the NO_2 concentration, and the potential coproduct HNO_3 has not been observed in the gas phase (Sakamaki et al. 1983; Pitts et al. 1984c). The HNO_2 formation rate in large-volume enclosures at $\sim$ 50 percent relative humidity is $\sim$ 0.1 ppb/min at an NO_2 concentration of 1 ppm (Sakamaki et al. 1983; Pitts et al. 1984c, 1985d). Surprisingly, if the observed nighttime HNO_2 concentrations and formation rates in the Los Angeles air basin are ascribed to this heterogeneous reaction, the calculated HNO_2 formation rates are similar to those observed in environmental chambers, despite the greatly different surface-to-volume ratios.

Thus, nighttime HNO_2 can arise from direct automotive emissions, heterogeneous hydrolysis of NO_2, and/or other processes. However, the relative contributions of direct emissions and heterogeneous formation of HNO_2 to observed nighttime HNO_2 levels need to be quantified. Furthermore, if HNO_2 formation from the heterogeneous hydrolysis of NO_2 is responsible for a major portion of the nighttime HNO_2 levels observed, the impor-

tance of this process during daylight hours needs to be investigated, especially since the photolysis of HNO_2 directly yields the key chain-carrying OH radical. In particular, research is needed to investigate whether this heterogeneous hydrolysis of NO_2 to HNO_2 is accelerated by light.

If so, this process could explain the "chamber-dependent" radical sources observed in environmental chambers (Carter et al. 1982), and would have a profound effect on the validity of chemical models developed for urban and regional airshed modeling. Present chemical mechanisms take into account the presence of chamber-dependent radical sources during their testing against environmental chamber experiments, but these radical sources are removed from the chemical mechanism for ambient atmospheric simulations. If HNO_2 (and hence, by photolysis, OH radicals) is produced at significant rates from NO_2, then the chemical computer models are being incorrectly applied under atmospheric conditions. The occurrence of this reaction would lead to significant underprediction of the reactivity of many present ambient atmospheric pollutant mixtures by computer modeling studies.

The observation of HNO_2 formation from NO_2 by heterogeneous hydrolysis is also of importance from an indoor air pollution viewpoint. For ppm mixing ratios of NO_2, HNO_2 formation rates of $\sim$ 0.1 ppb/min may be readily attained (Pitts et al. 1984c, 1985d), leading to indoor HNO_2 levels of up to $\sim$ 10 ppb, especially in the presence of mercury strip-lighting, which photolyzes HNO_2 slowly, if at all (Pitts et al. 1985d). Since HNO_2 can attain ppb mixing ratios in ambient atmospheres as well as indoor environments, it is imperative that its health effects be investigated.

In summary, these various nitrogen oxide species are readily interconverted in the lower atmosphere, but the major tropospheric loss process of NO_x occurs by the formation of gas-phase HNO_3 from the daytime reaction of OH radicals with NO_2 (the gas-phase HNO_3 being removed from the troposphere mainly by dry and wet deposition) and by wet and/or dry deposition of N_2O_5 during nighttime hours.

■ **Recommendation 2.** Further investigations of the transformations of NO_x under atmospheric conditions are needed. The kinetics and mechanism of gas-phase HNO_2 formation in indoor environments as well as ambient nighttime atmospheres should be investigated. Further research is also required on the nighttime reactions of the NO_3 radical, especially with respect to the products formed from its reactions with organic compounds, and on the heterogeneous and/or homogeneous reactions of N_2O_5.

Reduced Nitrogen Compounds. Reduced nitrogen compound emissions include ammonia (NH_3), hydrogen cyanide (HCN), and possibly their higher homologues such as the aliphatic and aromatic amines RNH_2, $RR'NH$ and $RR'R''N$, and the nitriles RCN, where R, R', R'' are alkyl or aryl groups.

The atmospheric reactions of these automotive emissions are not totally understood. These reactions are best considered separately as the amine and nitrile classes.

Amines. Reactions of amines with O_3 have been shown to be slow, and are of minor significance under atmospheric conditions (Atkinson and Carter 1984). The major atmospheric removal processes involve gas-phase reactions with the OH radical

$$OH + NH_3 \rightarrow H_2O + \dot{N}H_2 \quad (34)$$
$$OH + CH_3NH_2 \rightarrow H_2O + CH_3\dot{N}H \quad (35)$$

$$OH + (CH_3)_2NH \rightarrow H_2O + \begin{cases} (CH_3)_2\dot{N} \\ \dot{C}H_2NHCH_3 \end{cases} \quad (36)$$
$$OH + (CH_3)_3N \rightarrow H_2O + \dot{C}H_2N(CH_3)_2 \quad (37)$$

$$ (38) $$

and with gaseous HNO_3,

$$NRR'R'' + HNO_3 \rightarrow [NRR'R''H]^+NO_3^- \quad (39)$$

to form the corresponding nitrate salts. As shown in tables 2 and 3, the OH radical reactions are rapid, leading to lifetimes on the order of hours during daylight. In polluted atmospheres, however, reactions with gaseous HNO_3 could well be important, if not dominant, especially since this process continues during nighttime hours. Indeed, under certain conditions the reaction of ammonia with gaseous nitric acid is an important atmospheric loss process for NH_3 and HNO_3 (see, for example, Jacob et al. 1986). The resulting nitrate salts undergo dry and/or wet deposition.

The radical species ($\dot{N}RR'$) resulting from these OH radical reactions can react with O_2, NO, NO_2, or O_3.

$$\dot{N}RR' + O_2 \rightarrow products \quad (40)$$

$$\dot{N}RR' + NO \rightarrow \underset{\text{(nitrosamine)}}{RR'NNO} \quad (41)$$

$$\dot{N}RR' + NO_2 \begin{cases} \overset{a}{\rightarrow} \underset{\text{(nitramine)}}{RR'NNO_2} \\ \overset{b}{\rightarrow} RN=CHR'' + HNO_2 \quad (42) \end{cases}$$

$$\dot{N}RR' + O_3 \rightarrow products \quad (43)$$

To date, few data are available concerning the absolute or relative rate constants for these reactions. For the $\dot{N}H_2$ radical, the reaction rate constants are (Lesclaux 1984)

$NH_2 + O_2 \rightarrow products$
$$k_{40} < 3 \times 10^{-18} \text{ cm}^3/(\text{molecule·sec}),$$

$NH_2 + NO \rightarrow products$ (including $N_2 + H_2O$)
$$k_{41} \sim 2 \times 10^{-11} \text{ cm}^3/(\text{molecule·sec})$$

and

$NH_2 + NO_2 \rightarrow products$
$$k_{42} \sim 2 \times 10^{-11} \text{ cm}^3/(\text{molecule·sec}).$$

The only other data available for the subsequent reactions of $\dot{N}R_2$ radicals concern the dimethylamino (($CH_3)_2\dot{N}$) radical, for which Lindley and coworkers (1979) determined the rate constant ratios $k_{40}:k_{41}:k_{42a}:k_{42b} = 3.9 \times 10^{-7}:0.26:1.0:0.22$. For these aliphatic amines, the formation of carcinogenic nitrosamines, which photolyze rapidly (Tuazon et al. 1984a)

$$RR'NNO + h\nu \rightarrow RR'\dot{N} + NO \qquad (44)$$

and nitramines, which are longer-lived and react mainly with OH radicals (Tuazon et al. 1984a), can be important.

Nitriles. The available experimental data suggest that these reduced nitrogen compounds react mainly with OH radicals under tropospheric conditions. HCN reacts only slowly with OH radicals, with a room temperature rate constant at atmospheric pressure of 3×10^{-14} cm^3/(molecule·sec) (Fritz et al. 1984), by an addition process:

$$OH + HCN \rightarrow [HCN.OH] \rightarrow products \qquad (45)$$

The products of this reaction, as well as the subsequent reactions that occur under atmospheric conditions, are not known (Cicerone and Zellner 1983).

For the organic nitriles, the available data show that the OH radical reactions are more rapid than that for HCN, and that they probably occur by hydrogen-atom abstraction from the alkyl groups,

$$OH + CH_3CN \rightarrow H_2O + \dot{C}H_2CN \qquad (46)$$

which ultimately leads to the formation of CN radicals (Atkinson 1986).

Nitrites. Methyl nitrite is the most important compound of this class of organics, and photolysis is the only significant atmospheric loss process,

$$CH_3ONO + h\nu \rightarrow CH_3\dot{O} + NO$$
$$\downarrow O_2$$
$$HCHO + HO_2 \qquad (47)$$

resulting in a lifetime of $\sim$ 10–15 min at solar noon (Taylor et al. 1980).

Sulfur Oxides and Oxyacids. Sulfuric acid (H$_2$SO$_4$) exists almost entirely in the particulate phase under atmospheric conditions, and generally is neutralized by reaction with metal cations or ammonia to form salts such as Na$_2$SO$_4$, CaSO$_4$, (NH$_4$)$_2$SO$_4$, and NH$_4$HSO$_4$. Wet and dry deposition of particulate matter lead to its removal from the troposphere.

Sulfur dioxide is removed from the troposphere by gas- and aqueous-phase chemical reactions and by dry and wet deposition. With respect to chemical removal, gas-phase reaction with the OH radical is dominant,

$$OH + SO_2 \xrightarrow{M} HSO_3 \qquad (48)$$

followed by the formation of the chain-carrying HO$_2$ radical and H$_2$SO$_4$ (Stockwell and Calvert 1983b).

$$HSO_3 + O_2 \rightarrow HO_2 + SO_3$$
$$\downarrow H_2O$$
$$H_2SO_4 \qquad (49)$$

Reactions of SO$_2$ with the Criegee biradicals (RR'$\dot{C}O\dot{O}$) formed from alkene-O$_3$ reactions

$$RR'\dot{C}O\dot{O} + SO_2 \xrightarrow{H_2O} RR'CO + H_2SO_4 \qquad (50)$$

are of minor importance under most tropospheric conditions (Calvert and Stockwell 1983).

Since SO$_2$ is moderately soluble in the aqueous phase, it undergoes scavenging by fog, cloud water, and raindrops, leading to its inclusion in aqueous systems. Under these conditions, SO$_2$ is readily oxidized to sulfate by a variety of reactions, as discussed in detail by Graedel and Weschler (1981), Chameides and Davis (1982), Graedel and Goldberg (1983), Jacob and Hoffmann (1983), Chameides (1986), and Graedel et al. (1986).

Reduced Sulfur Compounds. Laboratory data show that the reduced sulfur-containing species (RSH) react with OH radicals during daytime hours and with NO$_3$ radicals during nighttime hours. For H$_2$S, the NO$_3$ radical reaction is slow (Wallington et al. 1986); therefore the dominant tropospheric removal process involves reaction with the OH radical (DeMore et al. 1985).

$$OH + H_2S \rightarrow H_2O + SH \qquad (51)$$

Although the SH radical apparently undergoes a series of reactions resulting in the formation of SO$_2$ (Thiemans and Schwartz 1978), the details of the reaction sequence are still not totally understood. The data of

Friedl et al. (1985) suggest that, under atmospheric conditions, SH radicals react with NO_2

$$SH + NO_2 \rightarrow HSO + NO \qquad (52)$$

and/or with O_3

$$\dot{S}H + O_3 \rightarrow HSO + O_2 \qquad (53)$$

with, typically, a lifetime of $\lesssim 1$ sec. The resulting HSO radicals react with O_2 to yield SO_2, by the intermediate formation of SO radicals. Reaction of HSO radicals with O_3 can apparently re-form SH radicals (Friedl et al. 1985).

For the aliphatic thiols, both NO_3 radical as well as OH radical reactions are important. For example, for CH_3SH

$$OH + CH_3SH \rightarrow \begin{bmatrix} OH \\ | \\ CH_3\dot{S}H \end{bmatrix} \rightarrow CH_3\dot{S} + H_2O \qquad (54)$$

$$\downarrow$$

$$HCHO, SO_2, \text{ and } CH_3SO_3H^{\cdot}$$

$$\uparrow$$

$$NO_3 + CH_3SH \rightarrow \begin{bmatrix} ONO_2 \\ | \\ CH_3\dot{S}H \end{bmatrix} \rightarrow CH_3\dot{S} + HNO_3 \qquad (55)$$

The reaction mechanisms and products formed under atmospheric conditions are, however, incompletely understood (Atkinson 1986; Mac Leod et al. 1986).

Oxides of Carbon. CO_2 is chemically stable under tropospheric conditions, and its removal from the troposphere involves transport to the stratosphere and absorption into the oceans.

CO is relatively long lived (~ 3 months), with its tropospheric lifetime determined by reaction with the OH radical (DeMore et al. 1985)

$$OH + CO \rightarrow HO\dot{C}O \rightarrow CO_2 + H$$
$$\downarrow_{O_2} HO_2 + CO_2 \qquad (56)$$

which generates the chain-carrying HO_2 radical.

Hydrogen Peroxide (H_2O_2) and Organic Peroxides. These compounds are readily absorbed into aqueous phases, including cloud and fog water (Jacob and Hoffmann 1983). In the gas phase, photolysis,

$$ROOH + h\nu \rightarrow OH + RO^{\cdot} \qquad (57)$$

where R represents either hydrogen or an alkyl radical, and reaction with OH radicals

$$OH + H_2O_2 \rightarrow H_2O + HO_2 \qquad (58)$$

$$OH + CH_3OOH \begin{array}{l} \nearrow H_2O + CH_3OO^{\cdot} \\ \searrow H_2O + \dot{C}H_2OOH \end{array}$$

$$\downarrow$$

$$HCHO + OH \qquad (59)$$

are the important chemical loss processes (DeMore et al. 1985; Atkinson 1986). The nighttime reaction of H_2O_2 with the NO_3 radical has been shown to be slow (Burrows et al. 1985b).

Alkanes. Alkanes, together with the aromatic hydrocarbons, comprise a major category of automotive emissions. Gasoline contains alkanes ranging in size from $\geq C_4$ through C_{12}–C_{15}, and diesel fuels from $\sim C_8$ through $\geq C_{20}$ (Carter et al. 1981). The atmospheric chemistry of the alkanes has been discussed by Carter and Atkinson (1985). Under tropospheric conditions, the alkanes react with OH radicals during daylight hours and with the NO_3 radical during nighttime hours, with the NO_3 reaction being of minor (< 10 percent) importance as an atmospheric loss process.

Both reactions proceed by hydrogen-atom abstraction from C—H bonds

$$\begin{Bmatrix} OH \\ NO_3 \end{Bmatrix} + RH \rightarrow \begin{Bmatrix} H_2O \\ HNO_3 \end{Bmatrix} + R^{\cdot} \qquad (60)$$

For the OH radical reaction, the contributions of hydrogen-atom abstraction at the various nonequivalent C—H bonds, and hence the distribution of the alkyl ($R^{\cdot}$) radicals formed, can be calculated using the estimation method of Atkinson (1986). Under atmospheric conditions, alkyl radicals react rapidly and exclusively with O_2 to yield alkyl peroxy ($RO_2^{\cdot}$) radicals.

$$R^{\cdot} + O_2 \xrightarrow{M} RO_2^{\cdot} \qquad (61)$$

Under tropospheric conditions, these $RO_2^{\cdot}$ radicals react with NO by two pathways,

$$RO_2^{\cdot} + NO \xrightarrow{M} \begin{cases} RO^{\cdot} + NO_2 \\ RONO_2 \end{cases} \quad (62)$$

with HO_2 radicals,

$$RO_2^{\cdot} + HO_2 \rightarrow RO_2H + O_2 \quad (63)$$

and with other $RO_2^{\cdot}$ radicals (including the same $RO_2^{\cdot}$ species).

$$RO_2^{\cdot} + R'O_2^{\cdot} \rightarrow \text{products (typically alkoxy}$$
$$\text{radicals, alcohols, and carbonyls)} \quad (64)$$

This competition between the reactions with NO and HO_2 and other $RO_2^{\cdot}$ radicals typically occurs for NO concentrations of the order of $(2–6) \times 10^8$ molecules/cm^3. Under polluted urban atmospheric conditions, and possibly for much of the lower troposphere in the eastern United States, reaction with NO is the dominant reaction pathway for $RO_2^{\cdot}$ radicals.

In addition, the reaction with NO (eq. 62) forms the corresponding alkoxy ($RO^{\cdot}$) radical and NO_2 or the corresponding alkyl nitrate. The yield of the alkyl nitrate increases with increasing pressure and with decreasing temperature. For secondary $RO_2^{\cdot}$ radicals at 298°K and 760 torr total pressure, the yields of the alkyl nitrates increase monotonically from ~ 0.04 for a C_3 alkane up to ~ 0.33 for a C_8 alkane (Carter and Atkinson 1985).

The resulting alkoxy radicals react under tropospheric conditions by unimolecular decomposition, unimolecular isomerization or with O_2, as shown, for example, for the 2-pentoxy radical:

$$\overset{\overset{\displaystyle O^{\cdot}}{\displaystyle |}}{CH_3CHCH_2CH_2CH_3}$$

decomposition / | \ isomerization

$$CH_3CHO + CH_3CH_2\dot{C}H_2 \quad | \quad CH_3CHOHCH_2CH_2\dot{C}H_2$$

$$\downarrow O_2$$

$$HO_2 + CH_3COCH_2CH_2CH_3 \quad (65)$$

The isomerization reactions are calculated to occur by 1,5-H-atom shifts, as shown. These isomerization reactions thus require a 4-carbon chain, and where this is available, isomerization is calculated to dominate over decomposition and/or reaction

with O_2. For a complete discussion of these alkoxy radical reactions, the review of Carter and Atkinson (1985) should be consulted.

The reactions that occur after alkoxy radical isomerization have not been demonstrated experimentally. However, it is expected that, for example, the reaction sequence for the 2-pentoxy radical will proceed by

$$\overset{\overset{\displaystyle O^{\cdot}}{\displaystyle |}}{CH_3CHCH_2CH_2CH_3} \rightarrow$$

$$CH_3CHOHCH_2CH_2\dot{C}H_2$$

$$\downarrow NO \dashrightarrow NO_2 \quad O_2$$

$$CH_3CHOHCH_2CH_2CH_2\dot{O}$$

$$\downarrow \text{second isomerization}$$

$$CH_3\dot{C}OHCH_2CH_2CH_2OH$$

$$\downarrow O_2$$

$$HO_2 + CH_3COCH_2CH_2CH_2OH \quad (66)$$

with this isomerization process giving rise to the bifunctional product pentan-4-one-1-ol.

Thus, for the alkanes the observed first-generation products are aldehydes, ketones, and alkyl nitrates, all of which react further under atmospheric conditions. The health effects of these first-generation products need to be assessed. For the alkanes containing $\geq C_4$ chains, the formation of δ-bifunctional compounds, of the general formula $RCOCH_2CH_2CHOHR'$, is also predicted. At the present time, however, no definitive evidence for the formation of these compounds has been reported. Whether this class of organic compounds is formed, and if so, what their health effects are, needs to be investigated.

Our understanding of alkane chemistry also suggests the possibility that δ-hydroxynitratoalkanes ($RCHOHCH_2CH_2$-$CH(ONO_2)R'$) could be formed. Investigations to determine whether or not these compounds are formed and, if so, their health effects, are also necessary. These studies apply mainly to the longer chain alkanes ($\geq C_5$) that are the major constituents of gasoline and, especially, diesel fuel.

For diesel fuel in which alkanes of $\geq C_{20}$ are present, many of the products are expected to be partitioned almost entirely into the particulate phase. Thus, the atmospheric concentrations and subsequent atmospheric transformations and health effects of these products should be studied.

■ **Recommendation 3.** The products arising from the OH radical–initiated reactions of alkanes—the major component of automobile emissions—require study. These products are likely to be distributed between the gas and particulate phases, and data are especially needed for the alkanes with eight or more carbon atoms.

Haloalkanes. The two major haloalkanes emitted from automotive use are 1,2-dichloroethane and 1,2-dibromoethane. These two haloalkanes react with the OH radical under atmospheric conditions with a room temperature rate constant of $\sim 2.5 \times 10^{-13}$ cm^3/(molecule·sec) (Atkinson 1986), corresponding to a lower tropospheric lifetime of ~ 100 days (tables 2 and 3).

The products of these reactions have not been experimentally determined, but a likely reaction sequence is,

$$OH + CH_2XCH_2X \rightarrow H_2O + \dot{C}HXCH_2X$$

$$HO_2 + \underline{CH_2XCXO} \xleftarrow{O_2} \dot{O}CHXCH_2X + NO_2$$

$$\dot{C}H_2X + \underline{HCXO}$$

$$HO_2 + \underline{HCXO} \tag{67}$$

where X represents bromine (Br) or chlorine (Cl) and the first-generation products are underlined. Further work is needed to elucidate the reaction routes of the 1,2-dihaloethanes under atmospheric conditions. Since their tropospheric lifetimes are long (~ 100 days), their degradation will take place under low NO mixing ratio conditions where reaction of $CH_2XCHXO\dot{O}$ radicals

with HO_2 and/or other $RO_2\dot{}$ radicals may occur. The products arising from these radical–radical reactions should be investigated.

Alkenes. The major alkenes present in gasoline and diesel fuels as well as in automotive exhaust are the smaller ones such as ethene ($CH_2{=}CH_2$), propene ($CH_3CH{=}CH_2$), 1-butene ($CH_3CH_2CH{=}CH_2$) and *cis*- and *trans*-2-butene ($CH_3CH{=}CHCH_3$). Laboratory research has shown that the alkenes are removed from the troposphere by reaction with OH radicals, NO_3 radicals, and O_3 (Atkinson and Lloyd 1984). These reaction pathways are discussed briefly below.

Hydroxyl Radical Reaction. These reactions are rapid, with the corresponding lifetimes for the smaller alkenes being given in table 3. All of these reactions proceed by OH radical addition to the $>C{=}C<$ double bond (Atkinson 1986),

followed by rapid addition of O_2 to yield the corresponding β-hydroxyperoxy radicals, for example,

Reaction with NO then yields the β-hydroxyalkoxy radical or the β-hydroxynitrate.

$$\text{(70)}$$

The β-hydroxynitrate formation pathway accounts for ~ 1–1.5 percent of the overall NO reaction at ~ 298°K and atmospheric pressure for propene (Shepson et al. 1985a). Thus, for the alkenes emitted from automotive use, formation of β-hydroxynitrates appears to be minor, accounting for only a few percent of the overall reaction pathways.

The β-hydroxyalkoxy radicals react with O_2, unimolecularly decompose, or unimolecularly isomerize (Atkinson and Lloyd 1984). This is shown, for example, for the $\dot{O}CH_2CHOHCH_2CH_3$ radical formed from internal OH radical addition to 1-butene.

$$\text{(71)}$$

The available data show that at room temperature, decomposition of the β-hydroxyalkoxy radicals dominates over their isomerization (Atkinson et al. 1985). Furthermore, only for the β-hydroxyalkoxy radical formed from ethene does the reaction with O_2 (eq. 72) compete with decomposition (eq. 73),

$$HOCH_2CH_2\dot{O} + O_2 \rightarrow HOCH_2CHO + HO_2 \quad (72)$$

$$HOCH_2CH_2\dot{O} \rightarrow HO\dot{C}H_2 + HCHO$$
$$\downarrow O_2$$
$$HCHO + HO_2 \quad (73)$$

with reaction 72 accounting for ~ 22 percent of the overall reaction for ethene at 298°K and 760 torr total air pressure (Niki et al. 1981a). Thus, for the alkenes with more than two carbons, these OH radical reactions lead, by ultimate cleavage of the $>C{=}C<$ double bonds, to the formation of aldehydes and/or ketones.

Nitrate Radical Reaction. Only for those alkenes more reactive than the 2-butenes does the NO_3 radical reaction become important under atmospheric conditions (Atkinson 1986). As presently understood, the initial reaction involves NO_3 radical addition to the $>C{=}C<$ double bond,

$$\text{(74)}$$

followed by reaction with O_2

$$\text{(75)}$$

The limited data available suggest that decomposition of the β-nitratoalkoxy radicals predominates, with minor amounts of dinitrates being formed by reaction of the β-nitratoalkoxy radicals with NO_2 (Bandow et al. 1980; Shepson et al. 1985a). However, because of the reported high toxicity of these dinitrates, further work is necessary to better define their yields from a variety of alkenes as a function of temperature and pressure. Furthermore, the reaction sequences that occur in the absence of NO, typically during nighttime hours when these NO_3 radical reactions are important, are not known (see Recommendation 2).

Reaction with Ozone. These reactions compete with the daytime OH radical reactions and the nighttime NO_3 radical reactions as a tropospheric loss process for the alkenes (table 3). These reactions proceed by initial O_3 addition to the $>C=C<$ double bond, followed by rapid decomposition of the resulting energy-rich (denoted as $[\ ]^{\ddagger}$) "molozonide,"

$$(76)$$

with the relative importance of pathways *a* and *b* assumed to be equal (see Atkinson and Carter 1984).

The major uncertainty concerns the fate under atmospheric conditions of the initially energy-rich biradicals $[RR'\dot{C}O\dot{O}]^{\ddagger}$, which can be collisionally stabilized or undergo unimolecular decomposition.

$$[RR'\dot{C}O\dot{O}]^{\ddagger} \xrightarrow{M} \begin{cases} RR'\dot{C}O\dot{O} \\ \text{decomposition} \end{cases} \quad (77)$$

Recent data of Hatakeyama and coworkers (1984) show that, at atmospheric pressure, the fraction of initially energy-rich Criegee biradicals which is stabilized decreases from ~ 0.39 for ethene (in agreement with the earlier data of Su et al. 1980, Kan et al. 1981, and Niki et al. 1981b) to ~ 0.18 for *trans*-2-butene, and to ~ 0.04 for cyclohex-

ene. The decomposition products, however, are not well known, and further work concerning these reactions is necessary.

The thermalized biradicals have been shown or postulated to react with a number of species, including H_2O (eq. 78), NO (eq. 79), NO_2 (eq. 80), SO_2 (eq. 81), CO (eq. 82), and aldehydes (eq. 83) (Calvert et al. 1978; Hatakeyama et al. 1981; Martinez and Herron 1981; Herron et al. 1982; Atkinson and Lloyd 1984).

$$R\dot{C}HO\dot{O} + H_2O \rightarrow RCOOH + H_2O \quad (78)$$

$$R\dot{C}HO\dot{O} + NO \rightarrow RCHO + NO_2 \quad (79)$$

$$R\dot{C}HO\dot{O} + NO_2 \rightarrow RCHO + NO_3 \quad (80)$$

$$R\dot{C}HO\dot{O} + SO_2 \xrightarrow{H_2O} RCHO + H_2SO_4 \quad (81)$$

$$R\dot{C}HO\dot{O} + CO \rightarrow \text{products} \quad (82)$$

$$(83)$$

In addition, biradicals such as $(CH_3)_2\dot{C}O\dot{O}$ may undergo unimolecular isomerization (Carter et al. 1986).

$$(84)$$

Again, further research is needed to elucidate the mechanisms and products of these reactions.

Oxygen-Containing Organics. The nonaromatic, oxygen-containing organic compounds that are either directly emitted as a result of automotive use or are formed as atmospheric transformation products include aldehydes, ketones, α,β-unsaturated carbonyls, carboxylic acids, esters, and ethers. Our current understanding of the atmospheric chemistry of these organic compounds is discussed below (see Atkinson 1986 for further details).

Aldehydes. These include aliphatic and aromatic aldehydes (unsaturated aldehydes such as acrolein and crotonaldehyde are dealt with below). These aldehydes are removed from the troposphere by photolysis and reaction with OH and NO_3

radicals (and, for formaldehyde, with HO_2 radicals). The reaction with NO_3 radicals is a relatively minor tropospheric loss process (table 3), but contributes to HO_2 radical and peroxyacetyl nitrate (PAN) formation during nighttime hours (Calvert and Stockwell 1983; Stockwell and Calvert 1983a; Cantrell et al. 1985, 1986). Thus the major loss processes involve photolysis and reaction with OH radicals. Recent studies (DeMore et al. 1985) show that the photodissociation quantum yields for the aliphatic aldehydes are lower than previously assumed, and for essentially all aldehydes, except HCHO, reaction with the OH radical represents the dominant daytime removal process (table 3).

For HCHO the OH radical reaction yields the HCO radical, which reacts rapidly with O_2 to form HO_2 and CO.

$$OH + HCHO \rightarrow H_2O + HCO$$
$$\downarrow O_2$$
$$HO_2 + CO \quad (85)$$

For the higher aldehydes, the $R\dot{C}O$ radical initially formed,

$$OH + RCHO \rightarrow H_2O + R\dot{C}O \quad (86)$$

rapidly adds O_2 to yield an acylperoxy radical:

$$R\dot{C}O + O_2 \rightarrow RC(O)O\dot{O} \quad (87)$$

Under polluted atmospheric conditions, these acylperoxy radicals then react with NO (eq. 88) or NO_2 (eq. 89),

$$RC(O)O\dot{O} + NO \rightarrow RC(O)\dot{O} + NO_2$$
$$\downarrow$$
$$R\cdot + CO_2 \quad (88)$$

$$RC(O)O\dot{O} + NO_2 \rightleftarrows RC(O)OONO_2$$
$$\text{Peroxyacyl nitrates} \quad (89)$$

with the peroxyacyl nitrates being in thermal equilibrium with NO_2 and the acylperoxy radicals. For PAN, the lifetime with respect to thermal decomposition (as observed in the presence of excess NO) is $\sim$ 45 min at 298°K (Atkinson and Lloyd 1984). PAN has been shown to be a direct-acting mutagen toward Ames *Salmonella typhimurium* strain TA100 (Kleindienst et al. 1985) and is phytotoxic; peroxypro-

pionyl nitrate (PPN) is even more phytotoxic. Thus, further studies concerning the health effects of these peroxyacyl nitrates need to be conducted.

Benzaldehyde is the simplest aromatic aldehyde, and is a primary automotive emission as well as being a product (see below) of the atmospheric transformations of toluene. Data concerning the atmospheric reactions of aromatic aldehydes are available only for benzaldehyde. For benzaldehyde, as for the aliphatic aldehydes, photolysis and OH radical reaction are the major loss processes (Atkinson and Lloyd 1984), with the nighttime NO_3 radical reaction being of minor importance (table 3). Reaction with the OH radical initiates a series of reactions leading to the formation of peroxybenzoyl nitrate (PBzN) and nitrophenols (Niki et al. 1979; Atkinson and Lloyd 1984).

$$OH + C_6H_5CHO \rightarrow H_2O + C_6H_5\dot{C}O$$
$$\downarrow O_2$$
$$\overset{O}{\underset{\parallel}{C_6H_5COO^{\cdot}}} \quad (90)$$

$$\overset{O}{\underset{\parallel}{C_6H_5COO^{\cdot}}} + NO_2 \rightleftarrows \overset{O}{\underset{\parallel}{C_6H_5COONO_2}} \text{ (PBzN)} \quad (91)$$

$$\overset{O}{\underset{\parallel}{C_6H_5COO^{\cdot}}} + NO \rightarrow \overset{O}{\underset{\parallel}{C_6H_5CO^{\cdot}}} + NO_2$$
$$\downarrow$$
$$C_6H_5^{\cdot} + CO_2 \quad (92)$$

$$C_6H_5\cdot + O_2 \rightarrow C_6H_5OO^{\cdot}$$
$$NO \rightarrow NO_2$$
$$C_6H_5O^{\cdot} \text{ (phenoxy radical)}$$
$$\overset{HO_2}{\swarrow} \qquad \overset{NO_2}{\searrow}$$
$$C_6H_5OH + O_2 \qquad o\text{- and } p\text{-nitrophenol} \quad (93)$$

Ketones. This class of organic emissions is exemplified by acetone and its higher homologues. As with the aldehydes, photolysis and reaction with the OH radical are the major atmospheric loss processes (Atkinson 1986). For acetone, the calculated lifetime of $\sim$ 15 days due to photolysis (Gardner et al. 1984) is significantly shorter than that of ~ 100 days due to OH radical reaction. For the other ketones, reaction with OH radicals appears to be the major tropospheric loss process, proceeding by, for example, for 2-butanone,

$$
\text{OH} + \text{CH}_3\text{CH}_2\text{COCH}_3 \begin{cases} \xrightarrow{\text{O}_2} \text{H}_2\text{O} \quad \overset{\text{OO}^{\cdot}}{\underset{|}{\text{+ CH}_3\text{CHCOCH}_3}} \\ \xrightarrow{\text{O}_2} \text{H}_2\text{O} \\ \quad + \text{CH}_3\text{CH}_2\text{COCH}_2\text{OO}^{\cdot} \\ \xrightarrow{\text{O}_2} \text{H}_2\text{O} \\ \quad + {}^{\cdot}\text{OOCH}_2\text{CH}_2\text{COCH}_3 \end{cases} \tag{94}
$$

with the top reaction being the major pathway (Atkinson 1986). Subsequent reaction of the radical formed in this reaction pathway with NO leads to the formation of acetaldehyde and the acetyl radical:

$$
\overset{\text{OO}^{\cdot}}{\underset{|}{\text{CH}_3\text{CHCOCH}_3}} + \text{NO} \rightarrow \text{NO}_2 + \overset{\text{O}^{\cdot}}{\underset{|}{\text{CH}_3\text{CHCOCH}_3}}
$$
$$
\downarrow
$$
$$
\text{CH}_3\text{CHO}
$$
$$
+ \text{CH}_3\dot{\text{C}}\text{O} \tag{95}
$$

Thus the major products from the atmospheric reactions of the ketones appear to be carbonyls and PAN precursors, although bifunctional oxygen-containing compounds may be formed in small amounts. The atmospheric degradation pathways need further investigation, and the health impacts of the products formed should be studied.

α,β-Unsaturated Carbonyls. These compounds, exemplified by acrolein, crotonaldehyde, and methyl vinyl ketone, are known to react with O_3, OH radicals, NO_3 radicals as well as undergo photolysis. Under atmospheric conditions, the OH radical reaction is the major loss process (Atkinson and Carter 1984). For the aldehydes, OH radical reaction can proceed by two reaction pathways: OH radical addition to the double bond and hydrogen-atom abstraction from the —CHO group (Atkinson 1986):

$$
\text{OH} + \text{CH}_3\text{CH}{=}\text{CHCHO} \rightarrow
$$
$$
\begin{cases} \rightarrow \overset{\text{OH}}{\underset{|}{\text{CH}_3\text{CH}\dot{\text{C}}\text{HCHO}}} \text{ and } \overset{\text{OH}}{\underset{|}{\text{CH}_3\dot{\text{C}}\text{HCHCHO}}} \\ \rightarrow \text{H}_2\text{O} + \text{CH}_3\text{CH}{=}\text{CH}\dot{\text{C}}\text{O} \end{cases} \tag{96}
$$

Analogous to the OH radical–initiated reactions with the alkenes and aliphatic aldehydes, respectively, these initial reactions are expected to be followed by the sequence of reactions shown below, where the first-generation products are underlined:

$$
\text{CH}_3\text{CHOH}\dot{\text{C}}\text{HCHO} + \text{O}_2 \rightarrow
$$
$$
\overset{\text{OO}^{\cdot}}{\underset{|}{\text{CH}_3\text{CHOHCHCHO}}}
$$
$$
\text{NO} \downarrow \text{NO}_2
$$
$$
\overset{\text{O}^{\cdot}}{\underset{|}{\text{CH}_3\text{CHOHCHCHO}}}
$$
$$
\downarrow
$$
$$
\text{CH}_3\dot{\text{C}}\text{HOH} + \underline{\text{(CHO)}_2} \text{ [glyoxal]}
$$
$$
\downarrow \text{O}_2
$$
$$
\underline{\text{CH}_3\text{CHO}} + \text{HO}_2 \tag{97}
$$

$$
\text{CH}_3\text{CH}{=}\text{CH}\dot{\text{C}}\text{O} + \text{O}_2 \rightarrow
$$
$$
\overset{\text{O}}{\underset{\|}{\text{CH}_3\text{CH}{=}\text{CHCOO}^{\cdot}}} \overset{\text{NO}}{\underset{\text{NO}_2}{\rightleftharpoons}} \text{CH}_3\text{CH}{=}\dot{\text{C}}\text{H} + \text{CO}_2
$$
$$
\text{NO}_2 \downarrow\uparrow \text{NO}_2
$$
$$
\overset{\text{O}}{\underset{\|}{\text{CH}_3\text{CH}{=}\text{CHCOONO}_2}} \tag{98}
$$

$$
\text{CH}_3\text{CH}{=}\dot{\text{C}}\text{H} + \text{O}_2 \rightarrow
$$
$$
[\text{CH}_3\text{CH}{=}\text{CHO}\dot{\text{O}}] \rightarrow \text{products} \tag{99}
$$

When the reaction of the vinyl radical ($\text{CH}_2{=}\dot{\text{C}}\text{H}$) with O_2 is used as an analogy (Slagle et al. 1984), the products from the reaction of O_2 with $\text{CH}_3\text{CH}{=}\dot{\text{C}}\text{H}$ may be CH_3CHO and HCO. Hence these α,β-unsaturated aldehydes are expected to ultimately give rise to α-dicarbonyls and carbonyls. For the α,β-unsaturated ketones, such as methyl vinyl ketone, the major atmospheric reaction with the OH radical occurs only by OH radical addition to the double bond:

$$OH + CH_2{=}CHCOCH_3 \xrightarrow{O_2}$$

$$\underset{HOCH_2\overset{|}{\overset{OO^{\cdot}}{C}HCOCH_3}}{} \text{ and } {}^{\cdot}\underset{OOCH_2\overset{|}{\overset{OH}{C}HCOCH_3}}{}$$

$$NO \overset{\rightarrow}{\rightarrow} NO_2 \qquad NO \overset{\rightarrow}{\rightarrow} NO_2$$

$$\underset{HOCH_2\overset{|}{\overset{O^{\cdot}}{C}HCOCH_3}}{} \qquad \underset{\overset{|}{\overset{OH}{O}}CH_2\overset{|}{C}HCOCH_3}{}$$

$$\underline{HOCH_2CHO} \qquad \underline{HCHO} \qquad (100)$$
$$+ CH_3\dot{C}O \qquad + CH_3CO\dot{C}HOH$$

$$\downarrow O_2$$

$$\underline{CH_3COCHO} + HO_2$$

where the first-generation intermediate products are again underlined. α-Dicarbonyls, together with aldehydes and hydroxyaldehydes, are expected to be formed.

Ethers. Under atmospheric conditions the aliphatic ethers, such as dimethyl and diethyl ether, react primarily with the OH radical. These reactions proceed by hydrogen-atom abstraction from the C—H bonds (Atkinson 1986), as shown below for diethyl ether (first-generation products are underlined).

$$OH + CH_3CH_2OCH_2CH_3 \rightarrow$$
$$H_2O + CH_3\dot{C}HOCH_2CH_3$$
$$\downarrow O_2$$
$$NO \overset{\rightarrow}{\rightarrow} NO_2$$

$$CH_3\overset{|}{\overset{O^{\cdot}}{C}}HOCH_2CH_3$$

$$O_2 \diagup \qquad \diagdown \qquad (101)$$

$$\underline{CH_3\overset{O}{\overset{||}{C}}OCH_2CH_3} + HO_2 \qquad \underline{CH_3CHO} + C_2H_5O^{\cdot}$$

$$\downarrow O_2$$

$$\underline{CH_3CHO} + HO_2$$

Few experimental data are available, but E.C. Tuazon (unpublished data) has shown that dimethyl ether yields mainly methyl-

formate ($HC(O)OCH_3$), presumably by the O_2 reaction route.

Alcohols. The reaction sequences for the simple aliphatic alcohols under atmospheric conditions are known (Atkinson 1986). These involve hydrogen-atom abstraction, mainly from the α-C—H bonds. For example,

$$OH + CH_3OH \begin{array}{c} \nearrow H_2O + \dot{C}H_2OH \\ \searrow H_2O + CH_3\dot{O} \end{array} \quad (102)$$

with the top reaction pathway accounting for $\sim$ 85 percent of the overall reaction at 298°K. Since the $\dot{C}H_2OH$ and $CH_3O^{\cdot}$ radicals react with O_2 to yield formaldehyde, the overall reaction can be written as

$$OH + CH_3OH \xrightarrow{O_2} H_2O + HCHO + HO_2 \quad (103)$$

The reaction pathways for ethanol are analogous, with acetaldehyde being the major product formed. For the higher alcohols, other, more complex, products are expected to be formed.

Carboxylic Acids. The carboxylic acids most likely to be emitted from automotive use are formic and acetic acids. The available data suggest that these carboxylic acids react with the OH radical under atmospheric conditions (Atkinson 1986). For formic acid, hydrogen-atoms are produced,

$$OH + H\overset{O}{\overset{||}{C}}OH \rightarrow\rightarrow H_2O + CO_2 + H \quad (104)$$

with the reaction probably proceeding through the initial formation of an addition complex. The reaction products formed from the reactions of OH radicals with the higher carboxylic acids are presently not known.

Esters. The esters, $RC(O)OR'$ also react primarily with the OH radical under atmospheric conditions (Atkinson 1986). The limited data available show that these reactions proceed by reaction with the alkoxy moiety.

$$OH + CH_3\overset{\overset{\displaystyle O}{\|}}{C}OCH_3 \rightarrow$$

$$CH_3\overset{\overset{\displaystyle O}{\|}}{C}O\dot{C}H_2 + H_2O$$

$$NO \overset{O_2}{\underset{\downarrow}{\rightarrow}} NO_2$$

$$CH_3\overset{\overset{\displaystyle O}{\|}}{C}OCH_2O\cdot \rightarrow CH_3\overset{\overset{\displaystyle O}{\|}}{C}O\cdot + HCHO$$

$$\overset{O_2}{\underset{\downarrow}{\quad}} \qquad \cdot CH_3 + CO_2$$

$$CH_3\overset{\overset{\displaystyle O}{\|}}{C}OCHO + HO_2 \qquad (105)$$

■ **Recommendation 4.** Investigations, under atmospheric conditions, of the reaction products for partially oxidized automotive emissions and their health impacts on humans are needed. This area of research includes the atmospheric transformations of methanol and ethanol, formaldehyde and acetaldehyde coemissions, and any other emissions associated with their use as alternative fuels. In addition, the atmospheric transformation products and associated health implications of aldehydes, ketones, α,β-unsaturated carbonyl compounds, carboxylic acids, and other products of incomplete combustion should be determined.

Monocyclic Aromatic Compounds. The aromatic hydrocarbons such as benzene, toluene, ethylbenzene, the xylenes, and the trimethylbenzenes are important constituents of gasoline and diesel fuel, as well as being major constituents of exhaust emissions. In addition, oxygen- and nitrogen-containing aromatic compounds, such as phenol, the cresols, and aromatic amines, may also be emitted.

The monocyclic aromatic hydrocarbons are removed from the atmosphere solely by reaction with the OH radical (table 3). These OH radical reactions proceed by two pathways: (1) a minor pathway involving hydrogen-atom abstraction from C—H bonds of, for benzene, the aromatic ring, or for alkyl-substituted aromatic hydrocarbons, the alkyl-substituted groups; and (2)

a major reaction pathway involving OH radical addition to the aromatic ring. For example, for toluene, these reaction pathways are

(106)

The hydrogen-atom abstraction pathway, top, leads mainly to the formation of aromatic aldehydes (Atkinson and Lloyd 1984)

Benzaldehyde   (107)

Subsequent reactions of these aromatic aldehydes with OH radicals lead to the formation of peroxybenzoyl nitrates and nitrophenols (see above, and Atkinson and Lloyd 1984). This hydrogen-atom abstraction pathway is minor, accounting for $\leqslant 10$ percent of the overall OH radical reaction for benzene and the alkyl-substituted aromatic hydrocarbons (Gery et al. 1985; Atkinson 1986).

The products arising from the OH radical addition reaction pathways are not well understood. Under atmospheric conditions, the initially formed OH-aromatic adduct is expected to react mainly with O_2, again by two reaction pathways. For example, for the toluene-OH adduct,

(108)

The hydrogen-atom abstraction reaction of the OH-aromatic adduct to yield phenolic compounds has been shown to be relatively minor, accounting for ~ 20 percent of the overall OH radical reaction mechanism for toluene (Atkinson and Lloyd 1984; Gery et al. 1985; Leone et al. 1985). The major reaction pathway involves other reactions of the OH-aromatic-O_2 adducts, and these have been shown to involve ring cleavage. Thus, the α-dicarbonyls glyoxal, methyl-glyoxal, and biacetyl have been identified and quantified from benzene and the methyl-substituted benzenes (Bandow et al. 1985; Bandow and Washida 1985a,b; Tuazon et al. 1986), and reaction pathways that lead to these products have been proposed (Atkinson et al. 1980; Killus and Whitten 1982). The reaction pathways that form α-dicarbonyls, phenolic products, and hydrogen-atom abstraction products, however, fail to account for ~ 30–50 percent of the overall reaction products. The recent semiquantitative or qualitative, but highly important, product studies of Dumdei and O'Brien (1984) and Shepson et al. (1984) have identified a variety of other bifunctional ring cleavage products from toluene and o-xylene, which include, from toluene, $CH_3COCOCH{=}CH_2$, $CHOCOCH{=}CH_2$, $CH_3COCH{=}CH_2$, $CH_3COCH{=}CHCH{=}CH_2$, $CHOC(OH){=}CHCHO$, and $CH_3COCH{=}CHCH{=}CHCHO$.

Much less information is available for the other aromatic compounds either directly emitted from automobiles, or formed as products from the primary aromatic emissions during their atmospheric transport. Indeed, most of the information has been derived from kinetic rather than direct product studies. For example, the phenolic compounds, which are known to be removed from the atmosphere primarily by chemical reaction with OH and NO_3 radicals (tables 2 and 3; Atkinson and Lloyd 1984) and by wet deposition from the gas phase (Leuenberger et al. 1985), react with OH radicals mainly by initial OH radical addition to the ring. However, the NO_3 radical reaction appears to proceed by hydrogen-atom abstraction from the substituent —OH group (Atkinson et al. 1984).

(109)

o- and p-nitrophenol

For other classes of monocyclic aromatic compounds, product data are not available. Product yields under atmospheric conditions are reliably known for only a few of the many aromatic hydrocarbons emitted from automotive use. Moreover, the health effects of most of these compounds are not known, although methylglyoxal has recently been reported to be mutagenic toward *Salmonella typhimurium* strain TA100 (Shepson et al. 1985b). Since the observed product yields typically account for ≲ 50 percent of the overall reaction products, an understanding of the remaining products and their health effects is necessary. This will include products formed in the particulate and gas phases.

■ **Recommendation 5.** The products arising from the OH radical–initiated reactions of the aromatic hydrocarbons, a major emission category from automotive use, need to be identified. These studies are important because of the relatively high reactivity of aromatic hydrocarbons, and will involve the identification and quantification of a plethora of bifunctional organic compounds, many of which will probably be present in low yield.

Polycyclic Aromatic Hydrocarbons (PAHs) and Their Derivatives and Analogues. A large number of these chemical com-

Table 7. Vapor Pressures at 298°K for a Series of PAHs

PAH	Vapor Pressure at 298°K (torr)
Naphthalene	8.0×10^{-2} [a]
Phenanthrene	1.2×10^{-4} [a]
Anthracene	6.0×10^{-6} [a]
Fluoranthrene	9.2×10^{-6} [a]
Pyrene	4.5×10^{-6} [a]
Benz[a]anthracene	2.1×10^{-7} [a]
Benzo[a]pyrene	5.6×10^{-9} [b]
Chrysene	6.4×10^{-9} [b]

[a] From Sonnefeld et al. 1983.
[b] From Yamasaki et al. 1984.

pounds, including PAHs (such as naphthalene, phenanthrene, anthracene, fluoranthene, pyrene, perylene, and benzo[a]pyrene), and their alkyl-substituted or oxygen-, sulfur-, and nitrogen-containing derivatives, as well as oxygen-, nitrogen- and sulfur-containing heterocyclic analogues, are, or may be, emitted from combustion sources. Although these compounds are relatively minor components of automotive emissions, they have assumed a "spotlight" position in automotive-related health risk assessments because of their potential toxicity.

The PAHs and their analogues and derivatives have relatively low vapor pressures (table 7), and are distributed between the gas and particulate phases, with this distribution being highly temperature dependent.

As presently understood, the atmospheric transformations of these PAHs and their derivatives are highly dependent upon the phase with which they are associated. The available data show that for the PAHs present in the gas phase, reaction with the OH radical predominates, leading to atmospheric lifetimes of a few hours or less (table 3). The nighttime reaction with N_2O_5 is of minor significance as a PAH loss process (table 3) (Pitts et al. 1985a), but may be important for the formation of nitro-PAHs (see below, and Arey et al. 1986).

Recent ambient atmospheric data from Norway, Denmark, and the United States show that 2-nitrofluoranthene and 2-ni-

tropyrene—nitro-PAHs not observed from combustion sources—are the major nitro-PAH components of particulate organic matter (POM) (Nielsen et al. 1984; Pitts et al. 1985c; Ramdahl et al. 1986). Since these two nitro-PAHs are not formed during the collection of POM, they must be formed in the atmosphere from the parent PAH during transport from source to receptor (Nielsen et al. 1984; Pitts et al. 1985c; Arey et al. 1986; Ramdahl et al. 1986; Sweetman et al. 1986; Zielinska et al. 1986). Indeed, it now appears that the majority of the nitro-PAHs present in ambient POM are formed via atmospheric transformations during transport from source to receptor.

Recent environmental chamber studies have shown that 2-nitrofluoranthene as well as 2-nitropyrene are formed from the gas-phase reactions of fluoranthene and pyrene with OH radicals in the presence of NO_x (Arey et al. 1986). 2-Nitrofluoranthene is also formed from the gas-phase reaction of N_2O_5 with fluoranthene (Sweetman et al. 1986; Zielinska et al. 1986).

Since many PAHs and their analogues and derivatives are partitioned primarily into the adsorbed phase under atmospheric conditions, a large number of experimental studies have been performed to delineate the reaction processes occurring for the adsorbed-phase compounds. However, most of these studies have been done using nonatmospherically realistic adsorbents such as glass fiber and Teflon-impregnated glass fiber and silica surfaces. The data obtained from these and from more realistic surfaces, such as carbon black and fly-ash, show that the reactions (including photolysis) are strongly dependent on the nature of the adsorbent species (see, for example, Pitts 1983; Behymer and Hites 1985).

Photolysis, reaction with O_3, NO_2, and/or HNO_3, and N_2O_5 have all been shown to lead to loss of PAHs on several substrates (see, for example, Pitts 1983; Pitts et al. 1985b, 1986). Certain of these reactions, in particular, those with NO_2 and HNO_3 and with N_2O_5, appear to be relatively slow under atmospheric conditions (Pitts et al. 1985b). However, because of the substrate dependence of these adsorbed-phase reactions (see, for example, Behymer and Hites 1985), no firm conclusions can be drawn about the importance of these reactions under atmospheric conditions. It does appear that photolysis and reaction with O_3 may be important for certain PAHs and their derivatives.

Clearly, a comprehensive and systematic investigation of the gas- and adsorbed-phase reactions of this class of automotive emissions is necessary before further risk assessment studies concerning these compounds can be carried out. This is a difficult research area because of the partitioning of many, if not most, of these emissions and their products between the gas and particulate phases, and because of the high potential for analytical artifacts during sampling with the currently available techniques.

■ **Recommendation 6.** The atmospheric transformation products of PAHs and their oxygen-, nitrogen-, and sulfur-containing analogues and homologues require study, in the gaseous and the adsorbed phases. In particular, the reaction pathways that lead to nitro-PAHs need to be quantitatively established. In addition, the atmospheric removal processes and resulting products of these nitro-PAHs should be studied further.

Particulate Matter. A variety of other chemical compounds, including metals such as Pb, are emitted from automotive use into the atmosphere in particulate form. As shown in table 6, particles are removed from the atmosphere at rates that depend markedly on the particle size. For particles of diameter 0.1–1 μm (the size that corresponds to most particles present in the atmosphere), dry and wet deposition processes lead to lifetimes of several days or more.

Since the metals emitted are expected to be present mainly in their oxidized form— for example, PbBrCl—chemical reactions are unlikely and their removal will occur principally by these physical processes. For organic chemicals emitted from automo-

tive use and present on ambient POM, reactions may occur during atmospheric transport. This topic has assumed importance because of the recent interest in acidic deposition and the role of aerosols in the formation of nitric, sulfuric, and organic acids. The reactions that can occur are complex and involve aqueous chemistry, gas-to-particle conversion, and heterogeneous reactions (see, for example, Graedel and Weschler 1981; Chameides and Davis 1982; Graedel and Goldberg 1983; Jacob and Hoffmann 1983; Chameides 1986; Graedel et al. 1986). However, this field is in a state of rapid change, and further research is necessary before a full understanding can be reached (see Recommendation 1).

Analytical Techniques

It has become apparent during the past several years that a major experimental initiative is necessary to develop new analytical techniques to allow the products of these complex atmospheric reactions to be identified and quantified. Fourier transform infrared absorption spectroscopy as well as gas chromatography/mass spectrometry (now often used in gas-phase studies) are subject to significant limitations when the organic products are complex and because of the possibility of the formation of artifacts. Similarly, for particle-associated chemicals, studies of their atmospheric degradation reactions and associated rates are often complicated by artifacts. Clearly, there is a need for new in situ analytical techniques allowing real-time analyses devoid of artifact product formation problems. This area is a major research topic in its own right and should be recognized as such. If this research area is not aggressively pursued, any advances in our knowledge about the atmospheric mechanisms and reaction products of automotive emissions may well become limited by the available analytical procedures. As research progresses in this area, it is also apparent that further techniques for studying low-volatility organics in the gas and the adsorbed phases must be developed.

■ **Recommendation 7.** A major research effort is needed to develop the necessary analytical techniques for identifying and quantifying the products of complex atmospheric reactions. Of prime importance is the development of nondestructive, nonintrusive, in situ analytical techniques that will allow the atmospheric transformations of gaseous and particulate-associated chemical species to be studied.

Summary

As a result of the last two decades of laboratory, computer modeling, and ambient atmospheric experiments, a large body of data now exists concerning the atmospheric loss processes and transformations of automotive emissions. However, significant gaps in our knowledge still remain, mainly about the products formed under atmospheric conditions.

As discussed in the sections above, the physical and chemical processes leading to the removal of automotive emissions from the atmosphere include

• wet and dry deposition of gases and particles;
• chemical reactions of gaseous automotive emissions with OH, NO_3, and HO_2 radicals, O_3, N_2O_5, and gaseous HNO_3;
• photolysis;
• reaction of particulate-associated organic compounds with a variety of gas- and adsorbed-phase species; and

• reactions in the aqueous phase with a variety of reactive species that are of importance in clouds, raindrops, and fog droplets, and lead to the formation of acidic precipitation.

Chemical reactions dominate the removal of most organic chemicals, with atmospheric lifetimes ranging from less than 1 min for highly reactive organic compounds reacting with the NO_3 radical during nighttime hours, to months or even years for the less reactive alkanes, haloalkanes, and substituted benzenes. Inorganic compounds emitted as a result of automotive use also exhibit a wide range of atmospheric lifetimes, with NO_2, HNO_2, and alkyl nitrites having photolysis lifetimes measured in minutes. In contrast, H_2O_2 and HNO_3 are readily removed at surfaces, and are predominantly removed from the gas phase by wet and/or dry deposition processes that can take several days or more.

The chemical transformations of automotive emissions lead to the formation of a wide variety of products. Many of these transformation products are unknown, and the health impacts on humans of those that are known have not been investigated. Future research programs must first require studies to determine the general chemical classes of products formed from the atmospheric transformations of automotive emissions. For those products suspected to threaten human health, additional work will then be necessary to better define their amounts and formation mechanisms.

Summary of Research Recommendations

Significant gaps still exist in our understanding of the physical and chemical transformations of automotive emissions that occur in the atmosphere during transport from source to receptor. The areas requiring further study are ranked in order of priority.

HIGH PRIORITY

Recommendation 2   Further investigations of the transformations of NO_x under atmospheric conditions are needed. This topic is important for indoor environments as well, since certain NO_x undergo important heterogeneous transformations. In particular, laboratory research has shown that, under conditions representative of certain indoor environments, NO_2 hydrolyzes on surfaces to yield gas–phase HNO_2 at significant rates. The kinetics and mechanism of this heterogeneous reaction should also be investigated in ambient nighttime atmospheres as should the reactions of the NO_3 radical (especially with respect to the products formed from its reactions with organic compounds) and the heterogeneous and/or homogeneous reactions of N_2O_5.

Recommendation 5	The products arising from the OH radical–initiated reactions of the aromatic hydrocarbons, a major emission category from automotive use, need to be identified. These studies are important because of the relatively high reactivity of aromatic hydrocarbons, and will involve the identification and quantification of a plethora of bifunctional organic compounds, many of which will probably be present in low yield.
Recommendation 6	The atmospheric transformation products of PAHs and their oxygen-, nitrogen-, and sulfur-containing analogues and homologues require study in the gaseous and the adsorbed phases. In particular, the reaction pathways that lead to nitro-PAHs need to be quantitatively established. In addition, the atmospheric removal processes and resulting products of these nitro-PAHs should be studied further. These studies will be difficult to perform because of the high potential for artifact formation.

MEDIUM PRIORITY

Recommendation 1	Study is required on the physical removal processes leading to wet and dry deposition of gases and particles. Investigations of the processes occurring on and in particulate and aerosol matter should focus on gas-to-particle conversion processes and the chemical processes that occur within aerosols (including fogs and clouds).
Recommendation 3	The products arising from the OH radical–initiated reactions of alkanes—the major component of automobile emissions—require study. These products are likely to be distributed between the gas and particulate phases, and data are especially needed for the alkanes with eight or more carbon atoms.
Recommendation 4	Investigations, under atmospheric conditions, of the reaction products for partially oxidized automotive emissions and their health impacts on humans are needed. This area of research includes the atmospheric transformations of methanol and ethanol, formaldehyde and acetaldehyde co-emissions, and any other emissions associated with their use as alternative fuels. In addition, the atmospheric transformation products and associated health implications of aldehydes, ketones, α,β-unsaturated carbonyl compounds, carboxylic acids, and other products of incomplete combustion should be determined.

Recommendation 7   A major research effort is needed to develop the necessary analytical techniques for identifying and quantifying the products of complex atmospheric reactions. Of prime importance is the development of nondestructive, nonintrusive, in situ analytical techniques that will allow the atmospheric transformations of gaseous and particulate-associated chemical species to be studied. This is clearly a long-term ideal, but utterly crucial in order to advance our current knowledge of the atmospheric transformations of automotive emissions.

References

Arey, J., Zielinska, B., Atkinson, R., Winer, A. M., Ramdahl, T., and Pitts, J. N., Jr. 1986. The formation of nitro-PAH from the gas-phase reactions of fluoranthene and pyrene with the OH radical in the presence of NO_x, *Atmos. Environ.* 20:2339–2345.

Atkinson, R. 1986. Kinetics and mechanisms of the gas phase reactions of the hydroxyl radical with organic compounds under atmospheric conditions, *Chem. Rev.* 86:69–201.

Atkinson, R., and Carter, W. P. L. 1984. Kinetics and mechanisms of the gas-phase reactions of ozone with organic compounds under atmospheric conditions, *Chem. Rev.* 84:437–470.

Atkinson, R., and Lloyd, A. C. 1984. Evaluation of kinetic and mechanistic data for modeling of photochemical smog, *J. Phys. Chem. Ref. Data* 13:315–444.

Atkinson, R., Carter, W. P. L., Darnall, K. R., Winer, A. M., and Pitts, J. N., Jr. 1980. A smog chamber and modeling study of the gas phase NO_x–air photooxidation of toluene and the cresols, *Int. J. Chem. Kinet.* 12:779–836.

Atkinson, R., Carter, W. P. L., Plum, C. N., Winer, A. M., and Pitts, J. N., Jr. 1984. Kinetics of the gas-phase reactions of NO_3 radicals with a series of aromatics of 296 ± 2 K, *Int. J. Chem. Kinet.* 16:887–898.

Atkinson, R., Tuazon, E. C., Carter, W. P. L. 1985. The extent of H-atom abstraction from the reaction of the OH radical with 1-butene under atmospheric conditions, *Int. J. Chem. Kinet.* 17:725–734.

Atkinson, R., Tuazon, E. C., Mac Leod, H., Aschmann, S. M., and Winer, A. M. 1986a. The gas-phase reaction of chlorine nitrate with water vapor, *Geophys. Res. Lett.* 13:117–120.

Atkinson, R., Winer, A. M., and Pitts, J. N., Jr. 1986b. Estimation of nighttime N_2O_5 concentrations from ambient NO_2 and NO_3 radical concentrations and the role of N_2O_5 in nighttime chemistry, *Atmos. Environ.* 20:331–339.

Atlas, E., and Giam, C. S. 1981. Global transport of organic pollutants: ambient concentrations in the remote marine atmosphere, *Science* 211:163–165.

Bandow, H., and Washida, N. 1985a. Ring-cleavage reactions of aromatic hydrocarbons studied by FT-IR spectroscopy. II. Photooxidation of *o*-, *m*-, and *p*-xylenes in the NO_x–air system, *Bull. Chem. Soc. Jpn.* 58:2541–2548.

Bandow, H., and Washida, N. 1985b. Ring-cleavage reactions of aromatic hydrocarbons studied by FT-IR spectroscopy. III. Photooxidation of 1,2,3-, 1,2,4- and 1,3,5-trimethylbenzenes in the NO_x–air system, *Bull. Chem. Soc. Jpn.* 58:2549–2555.

Bandow, H., Okuda, M., and Akimoto, H. 1980. Mechanism of the gas-phase reactions of C_3H_6 and NO_3 radicals, *J. Phys. Chem.* 84:3604–3608.

Bandow, H., Washida, N., and Akimoto, H. 1985. Ring-cleavage reactions of aromatic hydrocarbons studied by FT-IR spectroscopy. I. Photooxidation of toluene and benzene in the NO_x–air system, *Bull. Chem. Soc. Jpn.* 58:2531–2540.

Behymer, T. D., and Hites, R. A. 1985. Photolysis of polycyclic aromatic hydrocarbons adsorbed on simulated atmospheric particulates, *Environ. Sci. Technol.* 19:1004–1006.

Benson, S. W. 1976. *Thermochemical Kinetics*, 2nd ed., Wiley, New York.

Burrows, J. P., Tyndall, G. S., and Moortgat, G. K. 1985a. A study of the N_2O_5 equilibrium between 275 and 315 K and determination of the heat of formation of NO_3, *Chem. Phys. Lett.* 119:193–198.

Burrows, J. P., Tyndall, G. S., and Moortgat, G. K. 1985b. Absorption spectrum of NO_3 and kinetics of the reactions of NO_3 with NO_2, Cl and several stable atmospheric species at 298 K, *J. Phys. Chem.* 89:4848–4856.

Cadle, S. H., Dasch, J. M., and Mulawa, P. A. 1985. Atmospheric concentrations and the deposition velocity to snow of nitric acid, sulfur dioxide and various particulate species, *Atmos. Environ.* 19:1819–1827.

Calvert, J. G. (ed.) 1984. *SO₂, NO and NO₂. Oxidation Mechanisms: Atmospheric Considerations*, Butterworth, Boston.

Calvert, J. G., and Pitts, J. N., Jr. 1966. *Photochemistry*, Wiley, New York.

Calvert, J. G., and Stockwell, W. R. 1983. Acid generation in the troposphere by gas-phase chemistry, *Environ. Sci. Technol.* 17:428A–443A.

Calvert, J. G., Su, F., Bottenheim, J. W., and Strausz,

Correspondence should be addressed to Roger Atkinson, Statewide Air Pollution Research Center, University of California, Riverside, CA 92521.

O. P. 1978. Mechanism of the homogeneous oxidation of sulfur dioxide in the troposphere, *Atmos. Environ.* 12:197–226.

Cantrell, C. A., Stockwell, W. R., Anderson, L. G., Busarow, K. L., Perner, D., Schmeltekopf, A., Calvert, J. G., and Johnston, H. S. 1985. Kinetic study of the NO_3–CH_2O reaction and its possible role in nighttime tropospheric chemistry, *J. Phys. Chem.* 89:139–146.

Cantrell, C. A., Davidson, J. A., Busarow, K. L., and Calvert, J. G. 1986. The CH_3CHO–NO_3 reaction and possible nighttime PAN formation, *J. Geophys. Res.* 91:5347–5353.

Carter, W. P. L., and Atkinson, R. 1985. Atmospheric chemistry of alkanes, *J. Atmos. Chem.* 3: 337–405.

Carter, W. P. L., Ripley, P. S., Smith, C. G., and Pitts, J. N., Jr. 1981. Atmospheric chemistry of hydrocarbon fuels. Vol. 1: Experiments, results and discussion, Final Report ESL-TR-81-53, Engineering and Services Laboratory, Air Force Engineering and Services Center, Tyndall AFB, Fla. (November 1981).

Carter, W. P. L., Atkinson, R., Winer, A. M., and Pitts, J. N., Jr. 1982. An experimental investigation of chamber-dependent radical sources, *Int. J. Chem. Kinet.* 14:1071–1103.

Carter, W. P. L., Lurmann, F. W., Atkinson, R., and Lloyd, A. C. 1986. Development and testing of a surrogate species chemical reaction mechanism, EPA-600/3-86-031 (August 1986).

Chameides, W. L. 1986. Possible role of NO_3 in the nighttime chemistry of a cloud, *J. Geophys. Res.* 91:5331–5337.

Chameides, W. L., and Davis, D. D. 1982. The free radical chemistry of cloud droplets and its impact upon the composition of air, *J. Geophys. Res.* 87:4863–4877.

Chang, T. Y. 1984. Rain and snow scavenging of HNO_3 vapor in the atmosphere, *Atmos. Environ.* 18:191–197.

Cicerone, R. J., and Zellner, R. 1983. The atmospheric chemistry of hydrogen cyanide (HCN), *J. Geophys. Res.* 88:10689–10696.

Colbeck, I., and Harrison, R. M. 1985. Dry deposition of ozone: some measurements of deposition velocity and of vertical profiles to 100 metres, *Atmos. Environ.* 19:1807–1818.

Crutzen, P. J. 1982. The global distribution of hydroxyl, In: *Atmospheric Chemistry* (E. D. Goldberg, ed.), pp. 313–328, Springer-Verlag, New York.

DeMore, W. B., Margitan, J. J., Molina, M. J., Watson, R. T., Golden, D. M., Hampson, R. F., Kurylo, M. J., Howard, C. J., and Ravishankara, A. R. 1985. Chemical kinetics and photochemical data for use in stratospheric modeling, NASA Evaluation No. 7, Jet Propulsion Laboratory Publication 85-37 (July 1, 1985).

Dolske, D. A., and Gatz, D. F. 1985. A field intercomparison of methods for the measurement of particle and gas dry deposition, *J. Geophys. Res.* 90:2076–2084.

Dumdei, B. E., and O'Brien, R. J. 1984. Toluene

degradation products in simulated atmospheric conditions, *Nature* 311:248–250.

Eisenreich, S. J., Looney, B. B., and Thornton, J. D. 1981. Airborne organic contaminants in the Great Lakes ecosystem, *Environ. Sci. Technol.* 15:30–38.

Friedl, R. R., Brune, W. H., and Anderson, J. G. 1985. Kinetics of SH with NO_2, O_3, O_2 and H_2O_2, *J. Phys. Chem.* 89:5505–5510.

Fritz, B., Lorenz, K., Steinert, W., and Zellner, R. 1984. Rate of oxidation of HCN by OH radicals at lower temperatures, *Oxid. Commun.* 6:363–370.

Gardner, E. P., Wijayaratne, R. D., and Calvert, J. G. 1984. Primary quantum yields of photodecomposition of acetone in air under tropospheric conditions, *J. Phys. Chem.* 88:5069–5076.

Gery, M. W., Fox, D. L., Jeffries, H. E., Stockburger, L., and Weathers, W. S. 1985. A continuous stirred tank reactor investigation of the gas-phase reaction of hydroxyl radicals and toluene, *Int. J. Chem. Kinet.* 17:931–955.

Graedel, T. E., and Goldberg, K. I. 1983. Kinetic studies of raindrop chemistry. 1. Inorganic and organic processes, *J. Geophys. Res.* 88:10865–10882.

Graedel, T. E., and Weschler, C. J. 1981. Chemistry within aqueous atmospheric aerosols and raindrops, *J. Geophys. Res.* 19:505–539.

Graedel, T. E., Ayers, G. P., Duce, R. A., Georgii, H. W., Klockow, D. G. A., Morgan, J. J., Rodhe, H., Schneider, B., Slinn, W. G. N., and Zafiriou, O. C. 1982. Aqueous chemistry in the atmosphere, group report, In: *Atmospheric Chemistry* (E. D. Goldberg, ed.), pp. 93–118, Springer-Verlag, New York.

Graedel, T .E., Mandich, M. L., and Weschler, C. J. 1986. Kinetic model studies of atmospheric droplet chemistry. 2. Homogeneous transition metal chemistry in raindrops, *J. Geophys. Res.* 91:5205–5221.

Graham, R. A., and Johnston, H. S. 1978. The photochemistry of NO_3 and the kinetics of the N_2O_5–O_3 system, *J. Phys. Chem.* 82:254–268.

Harris, G. W., Carter, W. P. L., Winer, A. M., Pitts, J. N., Jr., Platt, U., and Perner, D. 1982. Observations of nitrous acid in the Los Angeles atmosphere and implications for predictions of ozone-precursor relationships, *Environ. Sci. Technol.* 16:414–419.

Hatakeyama, S., Bandow, H., Okuda, M., and Akimoto, H. 1981. Reactions of CH_2OO and $CH_2(^1A_1)$ with H_2O in the gas phase, *J. Phys. Chem.* 85: 2249–2254.

Hatakeyama, S., Kobayashi, H., and Akimoto, H. 1984. Gas-phase oxidation of SO_2 in the ozone-olefin reactions, *J. Phys. Chem.* 88:4736–4739.

Heikes, B. G., and Thompson, A. M. 1983. Effects of heterogeneous processes on NO_3, HONO and HNO_3 chemistry in the troposphere, *J. Geophys. Res.* 88:10883–10895.

Herron, J. T., Martinez, R. I., and Huie, R. E. 1982. Kinetics and energetics of the Criegee intermediate in the gas phase. I. The Criegee intermediate in ozone-alkene reactions, *Int. J. Chem. Kinet.* 14: 201–224.

Hewitt, C. N., and Harrison, R. M. 1985. Tropospheric concentrations of the hydroxyl radical—a review, *Atmos. Environ.* 19:545–554.

Hov, Ø., and Isaksen, I. S. A. 1979. Hydroxyl and peroxy radicals in polluted tropospheric air, *Geophys. Res. Lett.* 6:219–222.

Huebert, B. J., and Robert, C. H. 1985. The dry deposition of nitric acid to grass, *J. Geophys. Res.* 90:2085–2090.

Jacob, D. J., and Hoffmann, M. R. 1983. A dynamic model for the production of H^+, NO_3^- and SO_4^{2-} in urban fog, *J. Geophys. Res.* 88:6611–6621.

Jacob, D. J., Munger, J. W., Waldman, J. M., and Hoffmann, M. R. 1986. The H_2SO_4–HNO_3–NH_3 system at high humidities and in fogs. 1. Spatial and temporal patterns in the San Joaquin Valley of California, *J. Geophys. Res.* 91:1073–1088.

Jonas, R., and Heinemann, K. 1985. Studies on the dry deposition of aerosol particles on vegetation and plane surfaces, *J. Aerosol Sci.* 16:463–471.

Kan, C. S., Calvert, J. G., and Shaw, J. H. 1981. Mechanism of the ozone-ethene reaction in dilute N_2/O_2 mixtures near 1-atm pressure, *J. Phys. Chem.* 85:2359–2363.

Killus, J. P., and Whitten, G. Z. 1982. A mechanism describing the photochemical oxidation of toluene in smog, *Atmos. Environ.* 16:1973–1988.

Kleindienst, T. E., Shepson, P. B., Edney, E. O., and Claxton, L. D. 1985. Peroxyacetyl nitrate: measurement of its mutagenic activity using the *Salmonella/* mammalian microsome reversion assay, *Mutat. Res.* 157:123–128.

Leone, J. A., Flagan, R. C., Grosjean, D., and Seinfeld, J. H. 1985. An outdoor smog chamber and modeling study of toluene-NO_x photooxidation, *Int. J. Chem. Kinet.* 17:177–216.

Lesclaux, R. 1984. Reactivity and kinetic properties of the NH_2 radical in the gas phase, *Rev. Chem. Intermed.* 5:347–392.

Leuenberger, C., Ligocki, M. P., and Pankow, J. F. 1985. Trace organic compounds in rain. 4. Identities, concentrations and scavenging mechanisms for phenols in urban air and rain, *Environ. Sci. Technol.* 19:1053–1058.

Ligocki, M. P., Leuenberger, C., and Pankow, J. F. 1985a. Trace organic compounds in rain. II. Gas scavenging of neutral organic compounds, *Atmos. Environ.* 19:1609–1617.

Ligocki, M. P., Leuenberger, C., and Pankow, J. F. 1985b. Trace organic compounds in rain. III. Particle scavenging of neutral organic compounds, *Atmos. Environ.* 19:1619–1626.

Lindley, C. R. C., Calvert, J. G., and Shaw, J. H. 1979. Rate studies of the reactions of the $(CH_3)_2N$ radical with O_2, NO and NO_2, *Chem. Phys. Lett.* 67:57–62.

Liu, S. C., Cicerone, R. J., Donahue, T. M., and Chameides, W. L. 1977. Sources and sinks of atmospheric N_2O and the possible ozone reduction due to industrial fixed nitrogen fertilizers, *Tellus* 29:251–263.

Logan, J. A. 1983. Nitrogen oxides in the troposphere: global and regional budgets, *J. Geophys. Res.* 88:10785–10807.

Logan, J. A. 1985. Tropospheric ozone: seasonal

behavior, trends and anthropogenic influence, *J. Geophys. Res.* 90:10463–10482.

Logan, J. A., Prather, M. J., Wofsy, S. C., and McElroy, M. B. 1981. Tropospheric chemistry: a global perspective, *J. Geophys. Res.* 86:7210–7254.

Mac Leod, H., Aschmann, S. M., Atkinson, R., Tuazon, E. C., Sweetman, J. A., Winer, A. M., and Pitts, J. N., Jr. 1986. Kinetics and mechanisms of the gas phase reactions of the NO_3 radical with a series of reduced sulfur compounds, *J. Geophys. Res.* 91:5338–5346.

Magnotta, F., and Johnston, H. S. 1980. Photodissociation yields for the NO_3 free radical, *Geophys. Res. Lett.* 7:769–772.

Martinez, R. I., and Herron, J. T. 1981. Gas-phase reaction of SO_2 with a Criegee intermediate in the presence of water vapor, *J. Environ. Sci. Health* A16:623–636.

Nielsen, T., Ramdahl, T., and Bjørseth, A. 1983. The fate of airborne polycyclic organic matter, *Environ. Health Perspect.* 47:103–114.

Nielsen, T., Seitz, B., and Ramdahl, T. 1984. Occurrence of nitro-PAH in the atmosphere of a rural area, *Atmos. Environ.* 18:2159–2165.

Niki, H., Maker, P. D., Savage, C. M., and Breitenbach, L. P. 1979. Fourier transform infrared (FT-IR) studies of gaseous and particulate nitrogenous compounds, In: *Nitrogenous Air Pollutants* (D. Grosjean, ed.), pp. 1–16, Ann Arbor Press, Ann Arbor, Mich.

Niki, H., Maker, P. D., Savage, C. M., and Breitenbach, L. P. 1981a. An FT-IR study of mechanisms for the HO radical initiated oxidation of C_2H_4 in the presence of NO: detection of glycolaldehyde, *Chem. Phys. Lett.* 80:499–503.

Niki, H., Maker, P. D., Savage, C. M., and Breitenbach, L. P. 1981b. A FT-IR study of a transitory product in the gas-phase ozone-ethylene reaction, *J. Phys. Chem.* 85:1024–1027.

Noxon, J. F., Norton, R. B., and Marovich, E. 1980. NO_3 in the troposphere, *Geophys. Res. Lett.* 7:125–128.

Perner, D. 1980. HNO_2 in urban atmospheres and its photochemical significance, In: *Proceedings of the International Workshop on Test Methods and Assessment Procedures for the Determination of the Photochemical Degradation Behavior of Chemical Substances*, December 2–4, Berlin, pp. 159–173.

Perner, D., Schmeltekopf, A., Winkler, R. H., Johnston, H. S., Calvert, J. G., Cantrell, C. A., and Stockwell, W. R. 1985. A laboratory and field study of the equilibrium $N_2O_5 \rightleftarrows NO_3 + NO_2$, *J. Geophys. Res.* 90:3807–3812.

Pitts, J. N., Jr. 1983. Formation and fate of gaseous and particulate mutagens and carcinogens in real and simulated atmospheres, *Environ. Health Perspect.* 47:115–140.

Pitts, J. N., Jr., Biermann, H. W., Atkinson, R., and Winer, A. M. 1984a. Atmospheric implications of simultaneous nighttime measurements of NO_3 radicals and HONO, *Geophys. Res. Lett.* 11:557–560.

Pitts, J. N., Jr., Biermann, H. W., Winer, A. M., and Tuazon, E. C. 1984b. Spectroscopic identification and measurement of gaseous nitrous acid in dilute auto exhaust, *Atmos. Environ.* 18:847–854.

Pitts, J. N., Jr., Sanhueza, E., Atkinson, R., Carter, W. P. L., Winer, A. M., Harris, G. W., and Plum, C. N. 1984c. An investigation of the dark formation of nitrous acid in environmental chambers, *Int. J. Chem. Kinet.* 16:919–939.

Pitts, J. N., Jr., Atkinson, R., Sweetman, J. A., and Zielinska, B. 1985a. The gas-phase reaction of naphthalene with N_2O_5 to form nitronaphthalenes, *Atmos. Environ.* 19:701–705.

Pitts, J. N., Jr., Sweetman, J. A., Zielinska, B., Atkinson, R., Winer, A. M., and Harger, W. P. 1985b. Formation of nitroarenes from the reaction of polycyclic aromatic hydrocarbons with dinitrogen pentaoxide, *Environ. Sci. Technol.* 19:1115–1121.

Pitts, J. N., Jr., Sweetman, J. A., Zielinska, B., Winer, A. M., and Atkinson, R. 1985c. Determination of 2-nitrofluoranthene and 2-nitropyrene in ambient particulate organic matter: evidence for atmospheric reactions, *Atmos. Environ.* 19:1601–1608.

Pitts, J. N., Jr., Wallington, T. J., Biermann, H. W., and Winer, A. M. 1985d. Identification and measurement of nitrous acid in an indoor environment, *Atmos. Environ.* 19:763–767.

Pitts, J. N., Jr., Paur, H.-R., Zielinska, B., Arey, J., Winer, A. M., Ramdahl, T., and Mejia, V. 1986. Factors influencing the reactivity of polycyclic aromatic hydrocarbons adsorbed on filters and ambient POM with ozone, *Chemosphere* 15:675–685.

Platt, U., Perner, D., Harris, G. W., Winer, A. M., and Pitts, J. N., Jr. 1980a. Observations of nitrous acid in an urban atmosphere by differential optical absorption, *Nature* 285:312–314.

Platt, U., Perner, D., Winer, A. M., Harris, G. W., and Pitts, J. N., Jr. 1980b. Detection of NO_3 in the polluted troposphere by differential optical absorption, *Geophys. Res. Lett.* 7:89–92.

Platt, U. F., Winer, A. M., Biermann, H. W., Atkinson, R., and Pitts, J. N., Jr. 1984. Measurement of nitrate radical concentrations in continental air, *Environ. Sci. Technol.* 18:365–369.

Ramdahl, T., Zielinska, B., Arey, J., Atkinson, R., Winer, A. M., and Pitts, J. N., Jr. 1986. Ubiquitous occurrence of 2-nitrofluoranthene and 2-nitropyrene in air, *Nature* 321:425–427.

Roberts, J. M., Fehsenfeld, F. C., Liu, S. C., Bollinger, M. J., Hahn, C., Albritton, D. L., and Sievers, R. E. 1984. Measurements of aromatic hydrocarbon ratios and NO_x concentrations in the rural troposphere: observation of air mass photochemical aging and NO_x removal, *Atmos. Environ.* 18:2421–2432.

Sakamaki, F., Hatakeyama, S., and Akimoto, H. 1983. Formation of nitrous acid and nitric oxide in the heterogeneous dark reaction of nitrogen dioxide and water vapor in a smog chamber, *Int. J. Chem. Kinet.* 15:1013–1029.

Sehmel, G. A., Lee, R. N., and Horst, T. W. 1985. Hazardous air pollutants: dry-deposition phenomena, EPA-600/3-84-114 (January 1985).

Shepson, P. B., Edney, E. O., and Corse, E. W. 1984.

Ring fragmentation reactions in the photooxidations of toluene and o-xylene, *J. Phys. Chem.* 88:4122–4126.

Shepson, P. B., Edney, E. O., Kleindienst, T. E., Pittman, J. H., Namie, G. R., and Cupitt, L. T. 1985a. The production of organic nitrates from hydroxyl and nitrate radical reaction with propylene, *Environ. Sci. Technol.* 19:849–854.

Shepson, P. B., Kleindienst, T. E., Edney, E. O., Namie, G. R., Pittman, J. H., Cupitt, L. T., and Claxton, L. D. 1985b. The mutagenic activity of irradiated toluene/NO_x/H_2O/air mixtures, *Environ. Sci. Technol.* 19:249–255.

Singh, H. B., Ludwig, F. L., and Johnson, W. B. 1978. Tropospheric ozone: concentrations and variabilities in clean remote atmospheres, *Atmos. Environ.* 12:2185–2196.

Slagle, I. R., Park, J.-Y., Heaven, M. C., and Gutman, D. 1984. Kinetics of polyatomic free radicals produced by laser photolysis. 3. Reaction of vinyl radicals with molecular oxygen, *J. Am. Chem. Soc.* 106:4356–4361.

Slinn, W. G. N. 1982. Some influences of the atmospheric water cycle on the removal of atmospheric trace constituents. In: *Atmospheric Chemistry* (E. D. Goldberg, ed.), pp. 57–90, Springer-Verlag, New York.

Slinn, W. G. N., Hasse, L., Hicks, B. B., Hogan, A. W., Lal, D., Liss, P. S., Munnich, K. O., Sehmel, G. A., and Vittori, O. 1978. Some aspects of the transfer of atmospheric trace constituents past the air-sea interface, *Atmos. Environ.* 12:2055–2087.

Sonnefeld, W. J., Zoller, W. H., and May, W. E. 1983. Dynamic coupled-column liquid chromatographic determination of ambient temperature vapor pressures of polynuclear aromatic hydrocarbons, *Anal. Chem.* 55:275–280.

South Coast Air Quality Management District. 1982. Draft Air Quality Management Plan, Revision, Appendix No. IV-A, Final 1979 Emissions Inventory for the South Coast Air Basin.

Stockwell, W. R., and Calvert, J. G. 1983a. The mechanism of NO_3 and HONO formation in the nighttime chemistry of the urban troposphere, *J. Geophys. Res.* 88:6673–6682.

Stockwell, W. R., and Calvert, J. G. 1983b. The mechanism of the HO-SO_2 reaction, *Atmos. Environ.* 17:2231–2235.

Su, F., Calvert, J. G., and Shaw, J. H. 1980. An FT-IR spectroscopic study of the ozone-ethene reaction mechanism in O_2-rich mixtures, *J. Phys. Chem.* 84:239–246.

Sweetman, J. A., Zielinska, B., Atkinson, R., Ramdahl, T., Winer, A. M., and Pitts, J. N., Jr. 1986. A possible formation pathway for the 2-nitrofluoranthene observed in ambient particulate organic matter, *Atmos. Environ.* 20:235–238.

Taylor, W. D., Allston, T. D., Moscato, M. J., Fazekas, G. B., Kozlowski, R., and Takacs, G. A. 1980. Atmospheric photodissociation lifetimes for nitromethane, methyl nitrite and methyl nitrate, *Int. J. Chem. Kinet.* 12:231–240.

Terry Dana, M., Lee, R. N., and Hales, J. M. 1985.

Hazardous air pollutants: wet removal rates and mechanisms, EPA-600/3-84-113 (January 1985).

Thiemans, M. W., and Schwartz, S. E. 1978. The fate of HS radical under atmospheric conditions, paper presented at the 13th International Conference on Photochemistry, Clearwater Beach, Fla.

Tuazon, E. C., Atkinson, R., Plum, C. N., Winer, A. M., and Pitts, J. N., Jr. 1983. The reaction of gas phase N_2O_5 with water vapor, *Geophys. Res. Lett.* 10:953–956.

Tuazon, E. C., Carter, W. P. L., Atkinson, R., Winer, A. M., and Pitts, J. N., Jr. 1984a. Atmospheric reactions of *N*-nitrosodimethylamine and dimethylnitramine, *Environ. Sci. Technol.* 18:49–54.

Tuazon, E. C., Sanhueza, E., Atkinson, R., Carter, W. P. L., Winer, A. M., Pitts, J. N., Jr. 1984b. Direct determination of the equilibrium constant at 298 K for the $NO_2 + NO_3 \rightleftarrows N_2O_5$ reactions, *J. Phys. Chem.* 88:3095–3098.

Tuazon, E. C., Mac Leod, H., Atkinson, R., and Carter, W. P. L. 1986. α-Dicarbonyl yields from the NO_x-air photooxidation of a series of aromatic hydrocarbons in air, *Environ. Sci. Technol.* 20: 383–387.

van Noort, P. C. M., and Wondergem, E. 1985.

Scavenging of airborne polycyclic aromatic hydrocarbons by rain, *Environ. Sci. Technol.* 19: 1044–1048.

Waldman, J. M., Munger, J. W., Jacob, D. J., Flagan, F. C., Morgan, J. J., and Hoffmann, M. R. 1982. Chemical composition of acid fog, *Science* 218:677–680.

Wallington, T. J., Atkinson, R., Winer, A. M., and Pitts, J. N., Jr. 1986. Absolute rate constants for the gas-phase reactions of the NO_3 radical with CH_3SH, CH_3SCH_3, CH_3SSCH_3, H_2S, SO_2, and CH_3OCH_3 over the temperature range 280–350 K, *J. Phys. Chem.* 90:5393–5396.

Winer, A. M., Atkinson, R., and Pitts, J. N., Jr. 1984. Gaseous nitrate radical: possible nighttime atmospheric sink for biogenic organic compounds, *Science* 224:156–159.

Yamasaki, H., Kuwata, K., and Kuge, Y. 1984. Determination of vapor pressure of polycyclic aromatic hydrocarbons in the supercooled liquid phase and their adsorption on airborne particulate matter, *Nippon Kagaku Kaishi* 1324–1329.

Zielinska, B., Arey, J., Atkinson, R., Ramdahl, T., Winer, A. M., and Pitts, J. N., Jr. 1986. Reaction of dinitrogen pentoxide with fluoranthene, *J. Am. Chem. Soc.* 108:4126–4132.

Ambient Levels of Anthropogenic Emissions and Their Atmospheric Transformation Products

T. E. GRAEDEL
AT&T Bell Laboratories

Air Pollution, the Automobile, and Public Health. © 1988 by the Health Effects
Institute. National Academy Press, Washington, D.C.

Concentrations of Atmospheric Trace Constituents

The amount and sometimes the type of effect produced by an atmospheric constituent on a receptor—a human being, vegetation, a building material, for example—depends on its concentration. Relating the observed effect to its original cause, predicting effects in new circumstances, and controlling causes to limit the effects all require, in one way or another, an understanding of concentrations. In a comprehensive description, the concentration of a species reflects its emission flux (Johnson, this volume; Russell, this volume) modified by transport and turbulent diffusion (Sampson, this volume), and physical and chemical transformation and removal (Atkinson, this volume). An observed or statistically or mathematically predicted concentration is the starting point for estimating the flux delivered to a surface or to a tissue during respiration (Sexton and Ryan; Schlesinger; Sun, Bond, and Dahl; Overton and Miller; Ultman; all in this volume).

This paper reviews what is known about the measured concentrations of selected atmospheric trace constituents emitted or evolved from human activity, particularly the operation of motor vehicles. Two categories of environments, within which most of the exposure of human beings and much of the exposure of susceptible materials takes place, are emphasized: outdoor air in urban areas and indoor air within residential and nonindustrial buildings.

Actual concentrations are neither uniform over all space nor constant for all time, although some may be relatively uniform and stable for long times over large regions whereas others, especially concentrations of reactive species, those emitted intermittently, and those emitted at high concentration from a few widely spaced points, may be highly variable and/or nonuniform. Every actual measurement effectively averages the concentration over some finite volume and finite time interval, and every set of measurements is a distribution of such individual measurements at a sample of space/time regions distributed within the overall region of interest.

When the effect of an atmospheric species is approximately proportional to concentration as well as to duration of exposure, as it is in many cases, average concentration is a good measure of effective concentration. If long-term averages are measured, fewer measurements are required and they are less likely to be contaminated by occasional extreme conditions. For these reasons, long-term averages—often year-long averages—are considered where possible. In some cases, however, ambient air quality standards are established for extreme values (ozone is an example), so data appropriate to the standard are presented. In other cases, only a few "spot" measurements rather than an average of any sort are available. Since these give at least some idea of typical concentrations, they are presented with appropriate comments about their restricted validity. Finally, since the health effects of an atmospheric species may be a complex function of time and concentration, a table of extremes, ranges, and distributions is given.

Temporal and Spatial Patterns of Primary Gases

The primary gases are the principal gases other than water vapor and carbon dioxide (CO_2) that are directly emitted from combustion sources. Ambient air quality standards have been established in the United States and in several other countries for each primary gas. As a result, extensive measurement programs have been initiated to monitor concentrations for most of them, and substantial amounts of data are available. The quantity reported is usually the one named in the standard, most often a long-term average. For some species of gases and some purposes, however, that form may not be the best. For toxicologic purposes, for example, it is often important to know peak short-term concentrations whereas for other purposes it may be necessary to know the variability of concentrations from place to place and from time to time. These peaks and variations may go unreported in data summaries from governmental monitoring stations, yet they may be crucial to assessments of air quality effects on people and materials.

Even so, the data for the primary emitted gases are more extensive than those for any other atmospheric constituents. Despite occasional inadequacies, the data base is quite sufficient to represent atmospheric concentrations and predict their consequences. Atmospheric concentrations of several species, their ranges at different monitoring sites, and their long-term trends have been illustrated with box plots of data from the U.S. National Air Monitoring Stations (NAMS). The technique is shown in figure 1, where the 5th, 10th, and 25th percentiles of the data depict the concentration at the "cleaner" monitoring sites, the 75th, 90th, and 95th percentiles depict it at the "dirtier" sites, and the median and average describe the "typical" concentration. Although the average and the median both characterize typical behavior, the median has the advantage of not being affected by a few extremely high or low observations.

The major products of motor vehicle fuel combustion are, of course, water and CO_2, but their direct impact on human life is not significant. The totality of man-made CO_2 emissions, of which motor vehicle emissions make up only a small part, do have a measurable global effect, however, which is discussed in a later section.

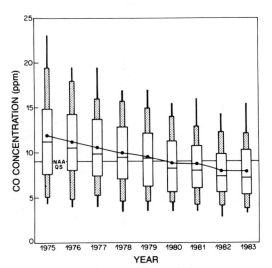

Figure 2. Trends in annual second highest nonoverlapping 8-hr average CO concentrations at 174 sites during 1975–1983. On this figure and others of its type, NAAQS indicates the U.S. National Ambient (outdoor) Air Quality Standard for the gas species. See figure 1 caption for explanation of plotting technique. (Adapted from U.S. Environmental Protection Agency 1985.)

Carbon Monoxide. About two-thirds of all carbon monoxide (CO) emissions come from transportation activities, with the combustion of solid waste and fuel providing most of the remainder (U.S. Environmental Protection Agency 1985). As a result, any reduction in CO emissions from automobiles is reflected directly in the measured CO concentrations.

Figure 2 shows the distribution of CO concentrations at 179 sites in the United States for nine years. The quantities reported are the second highest 8-hr averages measured in the over 1,000 non-overlapping 8-hr periods covering the year. Over the period 1975 to 1983, the median decreased from about 12 parts per million by volume (ppm) to about 7 ppm, well below the national ambient air quality standard (NAAQS) of 9 ppm. The concentrations at a number of sites exceed the standard, however, with a few at or above 15 ppm. For the foreseeable future, it appears likely that reductions in CO emission will be roughly offset by increases in the total number of vehicles and in the miles traveled per vehicle, so that little change in the CO concentration distribution is antici-

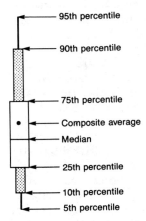

Figure 1. Technique used for box plots in subsequent figures in this paper. The 5th, 10th, and 25th percentiles depict the "cleaner" monitoring sites; the 75th, 90th, and 95th depict the "dirtier" sites; and the median and average represent typical concentration. (Adapted from U.S. Environmental Protection Agency 1985.)

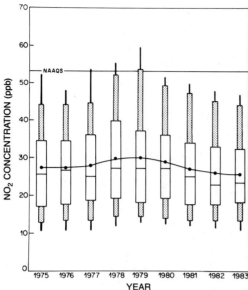

Figure 3. Trends in annual mean NO_2 concentrations at 177 sites during 1975–1983. See figure 1 caption for explanation of plotting technique. (Adapted from U.S. Environmental Protection Agency 1985.)

pated (U.S. Environmental Protection Agency 1985).

Oxides of Nitrogen. The emission flux of oxides of nitrogen (NO_x) is approximately equally divided between motor vehicles and stationary combustion activities (U.S. Environmental Protection Agency 1985). The total NO_x ($NO + NO_2$) emissions increased slightly in the late 1970s, decreased slightly in the early 1980s, and are now relatively stable. As a consequence, major changes in atmospheric concentrations of NO_x are not anticipated over the next few years.

Boxplots of annual mean nitrogen dioxide (NO_2) concentrations at 174 sites in the United States for nine years are shown in figure 3. The quantity reported is the annual mean concentration. The 95th, 90th, and 75th percentiles increase from 1975 to 1979, followed by a decrease from 1979 to 1983. The trend is less evident in the mean values of the annual concentrations and disappears entirely for the lower percentiles of the data. Most sites have annual mean concentrations of NO_2 in the range of 20–30 parts per billion by volume (ppb). Nearly all are below the NAAQS of 53

ppb. (Los Angeles is an exception; its NO_2 levels are in the upper 5 percent of values [not shown in the figure].) The detection technique generally used for NO_2 is sensitive to several other nitrogenous compounds as well, so the data are properly regarded as upper bounds to the true concentrations.

Hydrocarbons. The hydrocarbons (HC) found in the atmosphere comprise an extremely numerous and chemically diverse group of atmospheric compounds. They are of interest not only for their intrinsic properties but also because, with NO_x, they are precursors to ozone (O_3) and a variety of other atmospheric oxidants. Methane, the most abundant of the HCs, is of limited reactivity; as a consequence, data are often given for nonmethane hydrocarbons (NMHC) rather than for total HCs including methane.

Although there is an NAAQS for NMHC—240 ppb carbon by volume (ppbC), maximum 6–9 a.m. concentration, to be exceeded no more than once per year—the standard was established to serve only as a guide in assessing HC emission reductions needed to achieve O_3 standards. As such it has not been enforced, and only limited routine monitoring of NMHC has been performed. Thus the data on HC concentrations are less extensive than the data for some of the other species mentioned in this section.

Typical concentrations of total NMHC in urban areas are about 1–2 ppmC (Graedel and Schwartz 1977; Tilton and Bruce 1980). Rather than examining total concentrations, however, it is often more instructive to examine concentrations for individual compounds or groups of compounds that comprise the NMHC. Atmospheric HCs are often grouped into three classes for convenience of discussion: alkanes (aliphatic HCs characterized by a straight or branched carbon chain, generic formula C_nH_{2n+2}), alkenes (aliphatic HCs having one or more double bonds), and aromatic compounds (unsaturated cyclic HCs containing one or more rings). For the more reactive alkenes and aromatics, diagrams of typical concentrations in various

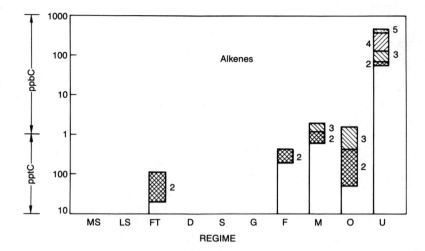

Figure 4. Approximate concentration ranges of alkenes in the following different atmospheric regimes: U = urban; O = oceanic; M = marshland; F = forest; G = grassland; S = steppes and mountains; D = desert (all of the previous measured within the boundary layer); FT = free troposphere (~5 km altitude); LS = lower stratosphere (15–20 km altitude); MS = middle stratosphere (~25 km altitude) (Graedel et al. 1986). Within the flags, the segments indicate the typical fraction of total concentration due to alkenes with the number of carbon atoms shown. For example, oceanic alkenes are typically comprised of about 65 percent ethylene (C = 2) and 35 percent propylene (C = 3).

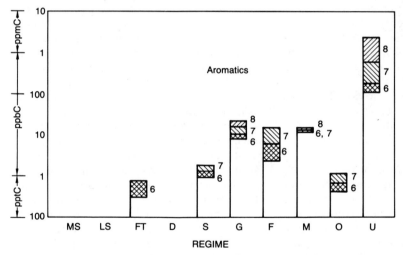

Figure 5. Approximate concentration ranges of the aromatic HCs benzene (6), toluene (7), and xylenes and ethylbenzene (8) in different atmospheric regimes. The regime code and the segmented division of the flags are explained in the caption of figure 4.

atmospheric regimes are available; they are reproduced in figures 4 and 5. In each case, concentrations in urban areas can be as high as about a thousand ppbC, and concentrations of several hundred ppbC are common. In more remote regions the measured

concentrations are sharply lower, reflecting lack of proximity to the principal sources as well as diminution of concentrations as a consequence of atmospheric reactions.

The U.S. Environmental Protection Agency (EPA) (1985) estimates that 37

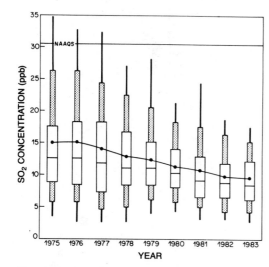

Figure 6. Trends in annual mean SO$_2$ concentrations at 286 sites during 1975–1983. (Adapted from U.S. Environmental Protection Agency 1985.)

percent of the volatile organic carbon compounds in the atmosphere come from motor vehicles, 37 percent from industrial activities, 15 percent from solid waste and miscellaneous, and 10 percent from volatilization of organic solvents. The total flux of these emissions decreased slightly over the period 1975–1983 and is now believed to be relatively stable.

Sulfur Dioxide. Sulfur dioxide (SO$_2$) is a trace gas of substantial concern because of its acid-forming potential in the atmosphere and because of its potential health effects. Its emission is dominated by fossil fuel combustion, with industrial activity being much less important and with motor vehicles emitting only a few percent of the total SO$_2$ flux. As increased controls and cleaner fuels have been used over the past decade, SO$_2$ concentrations have steadily decreased. This pattern is illustrated in figure 6 for annual mean concentrations measured at NAMS. As of 1983, typical annual mean concentrations were about 9 ppb.

Spatial and Temporal Variations in Concentration. Any effort to assess the effects of atmospheric trace gases on animate or inanimate objects must take into account the very great spatial and temporal variability of the trace gas concentrations. If the effects are cumulative and roughly propor-

tional to concentration and duration of exposure, the problem may not be severe. If, on the other hand, the effects are strong functions of peak intensity, some quantitative knowledge of variation from the mean value is necessary. The present review can offer no general rules to relate, for example, peak intensities to annual averages. What can be done is to illustrate some typical variations to provide some perspective on the magnitude of the problem.

Throughout the world, the concentrations of atmospheric trace gases demonstrate the degree to which sources of emissions are present in the vicinity of the monitoring sites, to what degree emissions from those sources are controlled, and to what degree the local meteorological situation influences the measured values. In figure 7, annual averages of SO$_2$ concentrations at various cities throughout the world are displayed. The ordinate on figure 7 is logarithmic, and the figure shows that the ratio of average SO$_2$ concentration at Milan (the site with the highest average annual concentration) and at Aukland (the site with the lowest) is about 16. The concentration also varies widely within cities because of local meteorological conditions. In a typical situation described by Johnson et al. (1973), circulation patterns in an urban street canyon concentrate emissions at certain locations within the canyon. The result is concentration patterns in which assessments made on one side of a city street can easily differ by 50 percent or more from similar assessments made on the other side.

The variation of concentration with time, as in space, can also be significant. In figure 8, the seasonal patterns in total HC (NMHC plus methane) concentration in Camden, New Jersey, are shown. The difference between the June value (the lowest) and the September value (the highest) is about 0.4 ppmC, or more than 20 percent of the mean value of about 1.9 ppmC. Diurnal differences can be important as well. In figure 9, 10 years of August total HC data are presented in hour-by-hour format. The difference between the 7 a.m. high value and the 4 p.m. low value is about 1.0 ppmC, that is, half the total concentration. Seasonal and diurnal fluctuations in concentration vary with location,

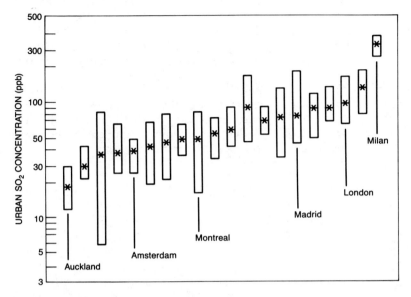

Figure 7. The range of annual averages of SO$_2$ concentrations measured at multiple sites within cities throughout the world. Several cities are identified here; further information is available in Bennett et al. (1985). The asterisk within each bar is the composite average for the city. (Adapted with permission from Bennett et al. 1985, and the American Chemical Society.)

species, and time of year, but those illustrated here are typical.

Because of the degree to which micrometeorology or meso- or synoptic-scale meteorology can influence species concentrations at any particular location, many species concentrations will tend to be in phase with each other. This situation has important consequences for studies attempting to link epidemiologic effects to atmospheric concentrations, for it requires that all potentially significant atmospheric variables be monitored simultaneously. Field measurements of such complexity have seldom been accomplished by epidemiologists.

■ **Recommendation 1.** No analytical instrument is readily available for routine monitoring of NMHC concentrations and concentration trends, although several techniques are available for potential incorporation into such an instrument. It is extremely important to achieve agreement on a satisfactory monitoring technique for NMHC (or some significant fraction thereof) and to begin to acquire data on a routine basis.

Selected Particle Constituents

Elemental Carbon. Carbon comprises 10 to 20 percent by weight of urban aerosols (Countess et al. 1980; Wolff et al. 1982b). Of this amount, nearly half is present as elemental carbon (soot), a consequence of incomplete combustion of fossil fuels. Diesel engines are the most significant of all the sources of elemental carbon (Wolff and Klimisch 1982). Extensive monitoring data are not available for elemental carbon, but the limited studies that have been performed suggest concentrations ranging from 1 to 35 μg/m^3 and averaging about 7 μg/m^3 (see, for example, Countess et al. 1980; Wolff 1985). The principal concerns with regard to soot are its efficient reduction of atmospheric visibility (Rosen et al. 1978), its potential as a catalytic oxidizer of SO$_2$ and other compounds in atmospheric droplets (Brodzinsky et al. 1980), and its ability to adsorb and concentrate toxic organic compounds and carry them into the lung (Sun, Bond, and Dahl, this volume).

Organic Carbon. Organic compounds are also a significant fraction of the urban atmospheric aerosol. The data of Grosjean and Friedlander (1975), Countess et al. (1980), and Wolff (1985) indicate that organics comprise between 2 and 40 μg/m^3 annually (mean value of 10–15 μg/m^3) of the aerosols in urban areas (see also National Academy of Sciences 1972). Motor vehicle emissions are responsible for perhaps somewhat less than half this amount.

The diurnal patterns of reactive aerosol constituents in urban areas have been demonstrated by the data of Grosjean and Friedlander (1975). In figure 10, these patterns are shown for a day in Pasadena, California. During this period, the change in concentration of total suspended particulates followed that of O_3, suggesting that smog chemistry is directly involved in the generation of the urban aerosol. Two features in the data are worth noting. The first is that the organic component of the aerosols is the largest of the four reactive com-

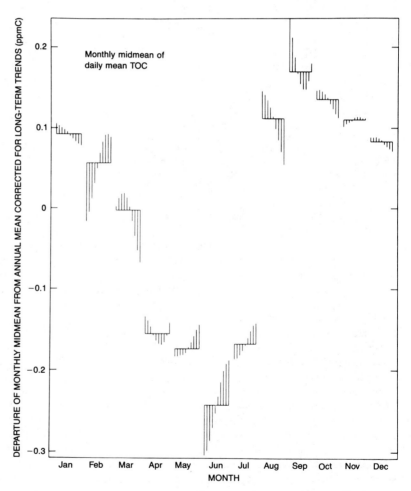

Figure 8. Derived seasonal dependence of the monthly midmeans of daily mean total organic carbon (TOC) concentrations (ppmC) in Camden, New Jersey, from 1968 to 1977. These data, with the long-term trend removed, are grouped into monthly sets, and the midmean of each set is determined. That value is plotted as a horizontal line. For each month, the excursions from the midmean are then determined for each of the years in the data sample and plotted as lines perpendicular to the midmean. The overall mean is approximately 1.9 ppmC. The seasonal trend for the entire time series can thus be seen, as well as the trend from year to year during each month. (Adapted with permission from Graedel and McRae 1982.)

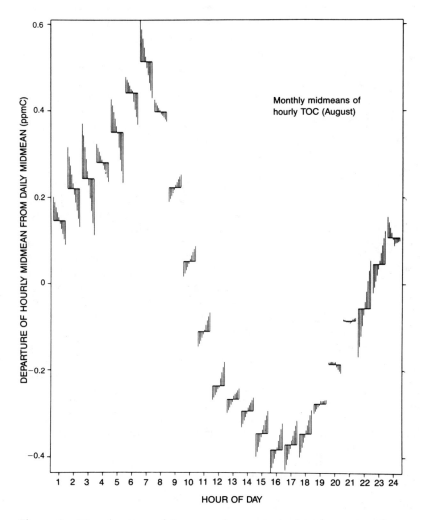

Figure 9. Diurnal pattern of the seasonal component of total organic carbon (TOC) data (ppmC) in Camden, New Jersey, for August days from 1968 to 1977. The overall August midmean is about 2.0 ppmC. The plotting technique is explained in the caption to figure 1. (Adapted with permission from Graedel and McRae 1982.)

ponents measured. (It is not always the largest component in all cities but is nearly always one of the major components.) The second is that the four aerosol constituents, all of which are produced by smog chemistry, comprise very large fractions of the total particle mass. The remainder of the mass, which largely consists of soot and soil dust, is important only during periods unfavorable for atmospheric chemistry.

Metals. The concentrations of metals in atmospheric aerosol particles are monitored routinely in the United States and in other

areas throughout the world. Most of the metals are primarily by-products of various industrial processes. Motor vehicles, however, have historically been major sources of lead (Pb) because of the use of Pb compounds as antiknock additives to gasoline. Manganese compounds are now becoming used for this purpose as well. It is therefore of interest to examine the atmospheric concentrations and trends of these two metals.

An NAAQS exists for Pb, and the result has been a careful study of its atmospheric abundance. Figure 11 shows the NAMS Pb

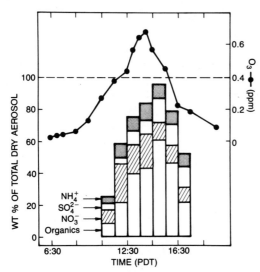

Figure 10. Diurnal patterns of nitrate, sulfate, and ammonium ions, total aerosol organics (as weight percent of total dry aerosol), and O_3 (-●-, ppm) in Pasadena, California, on July 25, 1973. The aerosol constituents were determined from seven 1-hr samples taken in late morning and throughout the afternoon. (Adapted with permission from Grosjean and Friedlander 1975, and the Air Pollution Control Association.)

data for the past decade. It is easy to see that the substantial reductions in leaded gasoline that have occurred during this period have been reflected in sharply decreasing atmospheric concentrations. As of 1983, the mean Pb level was about 0.3 $\mu g/m^3$.

Nitrate. The inorganic nitrate component of atmospheric aerosols is directly related to emissions of gaseous NO_x. Stationary combustion sources and motor vehicles may thus be supposed to be about equally responsible for aerosol nitrate; within urban areas, that attributable to motor vehicles probably dominates. Typical concentrations of nitrate fall within the range 1–10 $\mu g/m^3$, with a mean annual value of about 4 $\mu g/m^3$ (Graedel and Schwartz 1977; Harrison and Pio 1983). The diurnal behavior of urban aerosol nitrate is shown in figure 10. The largest concentrations tend to occur in the late morning, a circumstance that Grosjean and Friedlander (1975) attribute to a combination of rush-hour emissions of NO_x and rapid smog chemistry.

Sulfate. Urban aerosol sulfate concentrations arc typically in the range 1–20 $\mu g/m^3$,

with a mean annual value of about 7 $\mu g/m^3$ (Graedel and Schwartz 1977; Harrison and Pio 1983). The sulfate concentration in urban areas often increases during the day (see figure 10) as a result of the conversion of SO_2 to sulfate during periods of high photochemical activity. As with SO_2, stationary-source combustion of fossil fuel is the primary cause of aerosol sulfate in urban areas.

Ensemble Measurements of Suspended Particulate Matter. Total suspended particulate matter (TSP) in the atmosphere (that is, the atmospheric aerosol) is the most noticeable of the emittants from sources of atmospheric constituents. As such, it has been the object of substantial mitigation and reduction efforts over the years. As of 1983, emission fluxes from industrial processes and fuel combustion comprised about 30 percent each of the total emission flux. Motor vehicles and "solid waste and miscellaneous" contributed about 20 percent each (U.S. Environmental Protection Agency 1985).

Concentrations and trends in TSP over the past decade are illustrated in figure 12. A very significant reduction in the higher

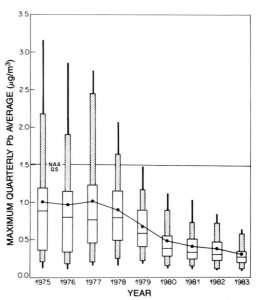

Figure 11. Trends in maximum quarterly Pb levels at 61 sites, 1975–1983. See figure 1 caption for explanation of plotting technique. (Adapted from U.S. Environmental Protection Agency 1985.)

percentiles of the data is seen, with the result that virtually all U.S. sites now meet the NAAQS for annual concentrations. The mean and percentiles near the median have fallen proportionately somewhat less than the higher percentiles; although, over the last two years represented in the data, noticeable reductions in these TSP concentrations occurred as well.

A feature that makes TSP data far from ideal indicators of health effects is that they are strongly influenced by high concentrations of very large particles (diameter> 15 μm), yet such particles are too large to be readily inhaled. Accordingly, many recent observations have been made with instruments designed to reject particles larger than 10 or 15 μm in aerodynamic diameter. The remaining particles are designated "inhalable particles" (IP). During normal breathing, these particles may travel to the bronchi and be retained there. Often the IP are further differentiated experimentally, the particles with aerodynamic diameter < 2.5 μm being known as the "respirable particles" (RPs). RPs travel as far as the lung parenchyma during normal breathing and may be retained there.

Extensive IP and RP data are not available, but it is clear that the chemical composition of larger particles is substantially different from that of the smaller ones. Elements residing on larger particles are emitted mainly by natural processes such as crustal erosion and sea spray. Those on smaller particles are commonly generated by high-temperature anthropogenic processes such as welding or soldering, smelting of metals, and combustion of fossil fuels (Milford and Davidson 1985). Thus sulfate (Wolff et al. 1985a), organic carbon (Wolff et al. 1982a), and elemental carbon (Wolff et al. 1982b) are among the species concentrated on RPs. The potential health effects of the RPs are much more significant than those of IPs, because of their chemical differences as well as the deeper respiratory system penetration that is characteristic of the smaller particles.

■ **Recommendation 2.** A difficult but essential job is to monitor the chemistry of atmospheric aerosol particles in much more detail than is now being done, concentrating especially on chemical differences as a function of particle size. Detailed organic analyses are particularly important.

Photochemical Products and Unregulated Emittants

The atmospheric species discussed thus far include those that are directly emitted from combustion sources and extensively monitored, either because they are the subject of a standard or as possible preparatory efforts in the establishment of a standard. In this section an attempt is made to discuss other compounds that may be of concern. One cannot begin to comment on all possible atmospheric compounds, however, because the total number thus far identified exceeds 2,800 (Graedel et al. 1986) and many are demonstrably unimportant. To render the discussion tractable, the following criteria have been used to select compounds for attention:

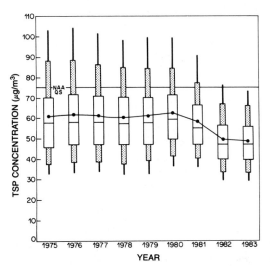

Figure 12. Trends in annual geometric mean total suspended particulate (TSP) concentrations at 1,510 sites, 1975–1983. See figure 1 caption for explanation of plotting technique. (Adapted from U.S. Environmental Protection Agency 1985.)

● High chemical reactivity
● Positive toxicologic test results
● Known effects, other than toxicologic, on humans
● Known effects other than those connected directly with human organisms.

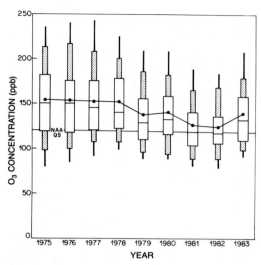

Figure 13. Trends in annual second highest daily maximum 1-hr O_3 concentrations at 176 sites, 1975–1983. See figure 1 caption for explanation of plotting technique. (Adapted from U.S. Environmental Protection Agency 1985.)

Inorganic Compounds. O_3 is the chemical species of most extensive interest in this group. It is not emitted directly from sources but is formed in the atmosphere by the gas-phase chemistry of NO_x, small organic molecules, and sunlight. O_3 is itself toxic and is the major oxidizing gas in photochemical smog mixtures.

The concentrations of O_3 at U.S. measurement sites are indicated in figure 13. The O_3 standard specifies the concentration at the second highest daily maximum during a year; therefore, the data in the figure correspond to near extreme, rather than average, values. The data suggest some decrease in concentrations since 1975 (perhaps as a result of a calibration change in O_3 monitors in 1979), but the increase in 1983 makes it clear that no monotonic trend exists. It is particularly significant that the majority of sites included in the data set have continued to exceed the NAAQS for O_3.

The influence of anthropogenic emittants on O_3 can be seen in the remote troposphere, far from populated regions. A careful data analysis by Logan (1985) suggests an increase of 20–100 percent in the annual average O_3 concentration in those regions since the 1940s. The increase is

attributed to photochemical production associated with increased levels of NO_x, HCs, and CO.

Other unregulated inorganic compounds worth comment are the oxy acids of nitrogen, which are among the final products of the chemistry of combustion-related emissions. Nitric acid (HNO_3) is the more abundant, with concentrations of a few ppb being common in urban areas. HNO_3 levels as high as a few tens of ppb are occasionally observed (Miller and Spicer 1975; Tuazon et al. 1981). Nitrous acid (HNO_2) concentrations are lower: they are often below 1 ppb, but occasionally peak near 10 ppb (Harris et al. 1982; Sjödin and Ferm 1985).

A fourth inorganic atmospheric species of great interest is hydrogen peroxide (H_2O_2), a product of atmospheric hydrocarbon chemistry. Its importance is due to its strong oxidizing capacity, which is thought to be of crucial importance to the chemistry of atmospheric water droplets. This species is exceptionally difficult to measure in the gas phase, so extensive data are not available. It appears, however, that urban concentrations of a few ppb are typical during the summer months (Kok et al. 1986).

Aldehydes. Formaldehyde and other aldehydes are common atmospheric constituents. They irritate the eyes and upper respiratory tract, but not generally at the outdoor concentrations experienced in urban areas. The aldehydes are directly emitted from a number of sources, including motor vehicles, but are also produced efficiently from small HCs by atmospheric reactions involved in smog chemistry.

Measurements of the atmospheric concentrations of formaldehyde are reasonably extensive, although no routine monitoring programs exist. Typical formaldehyde concentrations in urban areas in the summer are 5–20 ppb, although higher concentrations are occasionally observed (National Academy of Sciences 1981). Acetaldehyde concentrations are generally about a third those of formaldehyde (Hoshika 1977; Tanner and Meng 1984).

Outdoor urban concentrations of formaldehyde appear to reflect direct sources as well as atmospheric photochemistry. This is seen in figure 14, which shows diurnal formaldehyde concentrations in Newark, New Jersey, together with those for CO (an indicator of motor vehicle emissions) and O_3 (an indicator of photochemical activity). The signatures of both indicators appear to be reflected in the formaldehyde data, as is also the situation with other data sets illustrated by Cleveland et al. (1977). The published information also demonstrates clearly that formaldehyde has a strong seasonal pattern, though not as strong as that of O_3 (see figure 15). A similar seasonal pattern exists for acetaldehyde (Tanner and Meng 1984).

Polynuclear Aromatic Hydrocarbons. Poly-
nuclear aromatic hydrocarbons (PAH), a major fraction of atmospheric polycyclic

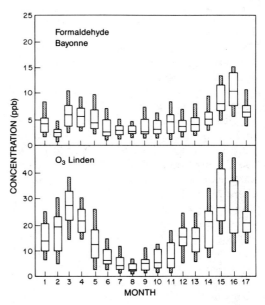

Figure 15. Monthly plots of daily averages of formaldehyde and O_3 in the New Jersey cities of Bayonne and Linden, respectively. Each index number corresponds to a month: 1 = May 1972; 2 = June 1972, . . . , 17 = September 1973. Unlike the boxplots shown earlier, the 5th and 95th percentiles and the composite average are omitted here. (Adapted from Cleveland et al. 1977.)

organic matter (POM), can be formed and/or directly released from any combustion process involving hydrocarbons. Power generation and refuse burning are the two largest sources; motor vehicle emissions are thought to contribute no more than a few percent of the total flux. In urban areas, of course, the motor vehicle contribution is proportionately higher.

Because of their intermediate vapor pressures, PAHs are generally present in the atmosphere in gaseous form and as particle components. Gas-phase PAH, of which fluorene, phenanthrene, and fluoranthene are among the more abundant, typically are present in urban air at concentrations of a few parts per trillion by volume (ppt) (Krstulovic et al. 1977; Cautreels and Van Cauwenberghe 1978). In the solid phase, where pyrene, benzo[*a*]pyrene, chrysene, and benzo[*ghi*]perylene are among the more abundant compounds, the PAH concentrations in urban areas are typically 10–20 ng/m^3 (the equivalent of about 1–2 ppt for a PAH of molecular weight 200).

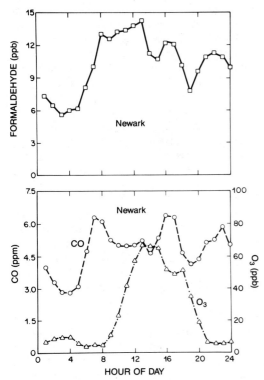

Figure 14. Formaldehyde, CO, and O_3 diurnal concentrations at Newark, New Jersey, for days within the periods June 1, 1973–August 31, 1973 and June 1, 1974–August 31, 1974 on which the daily maximum O_3 concentrations exceeded 100 ppb. (Adapted from Cleveland et al. 1977.)

Routine monitoring and speciation of PAHs are seldom performed, and any assessment of their abundance in specific urban areas generally requires special measurement programs.

Alcohols. Alcohols are of interest in this review because of the potential for rapidly increasing use in gasohol fuel blends for motor vehicles. If such use should occur, the unburnt alcohols that are emitted will participate in a number of atmospheric chemical processes including the production of organic acids.

To this point, few determinations of atmospheric alcohol concentrations have been performed. Those that have suggest a typical concentration of perhaps 5–10 ppb in urban areas (Bellar and Sigsby 1970; Hanst et al. 1975; Holzer et al. 1977; Snider and Dawson 1985). The alcohol concentrations do not follow regular diurnal patterns, and Hanst and his colleagues (1975) suggest that they are related to random emission from direct sources rather than to generation by atmospheric photochemistry. It is important to generate baseline data on the atmospheric compositions of the most abundant alcohols now, before the alcohol emission fluxes from motor vehicles increase markedly (if they do).

Nitro Compounds and Organic Nitrates.
Nitro compounds are those in which the NO_2 group is attached to an organic molecule. If the NO_2 group is attached to an oxygen atom, the compound is an organic nitrate. Many of these species are of interest because of their toxicity (National Academy of Sciences 1977) and because they have been shown to be harmful to certain types of vegetation (Taylor 1969).

The most widely measured of this group is peroxyacetyl nitrate (PAN). It is formed by chemical reactions involving small HCs and NO_x and its concentrations therefore tend to follow, in pattern if not in magnitude, those of indicators of smog activity such as O_3 (National Academy of Sciences 1977). Typical urban concentrations of PAN are 5–10 ppb (Spicer 1977; Tuazon et al. 1981). Since PAN is a mutagen and powerful eye irritant and can be measured

by relatively straightforward experimental techniques, its regular monitoring would seem to be advisable.

A number of other gas-phase nitrogen-containing organics occur in the atmosphere (Graedel et al. 1986), but their concentrations are very low. Most atmospheric nitro compounds occur in the solid phase at concentrations of a few tenths of a ng/m^3 in urban areas (Pitts et al. 1985b). Several direct mutagens are included in the species that comprise the solid-phase nitrated polynuclear aromatics.

Heterocyclic Organic Compounds.
Heterocyclic organic compounds are those in which an atom other than carbon is present in one or more ring structures. A number of such compounds have been shown to be potent mutagens (see Graedel et al. 1986), so their atmospheric concentrations are of interest. Few analyses have been made, however, either in the gas or aerosol phases. Heterocyclic nitrogen compounds, of which the most abundant are perhaps pyridine, indole, and quinoline, have urban concentrations of a few ppt (gas phase) or a few ng/m^3 (particle phase) (Dong et al. 1977; Ketseridis and Jaenicke 1978). Heterocyclic oxygen compounds, of which the most abundant appear to be furan, dioxane, coumarin, and their derivatives, have urban concentrations of about 10 ppt in the gas phase, and 0.10 ng/m^3 in the aerosol phase (Louw et al. 1977; Ketseridis and Jaenicke 1978; Harkov et al. 1983). In all cases, these concentration estimates are based on very limited data. If major concerns should arise concerning these substances, much additional information would be needed to ensure that a satisfactory picture of their urban concentrations was at hand.

■ **Recommendation 3.** An effort to develop a monitor for either formaldehyde or the aldehyde group with appropriate sensitivity (100 ppt or less), reliability, and an appropriately low cost, is a high priority for atmospheric chemists, particularly those addressing problems in urban air.

■ **Recommendation 4.** More attention should be given to monitoring unregulated, but potentially hazardous species, since extensive data will be required should regulatory action prove desirable. Among the species to which increased effort might be directed are PAN, nitric acid, hydrochloric acid, nitro–PAHs, and heterocyclic organics.

■ **Recommendation 5.** Given the possibility of sharply increased use of methanol and ethanol in motor vehicle fuels over the next decade or two, it is important to begin promptly to establish a satisfactory baseline against which future changes in atmospheric alcohol concentrations could be assessed.

Indoor Concentrations

Principal Trace Gases

Most people spend most of their lives indoors—in homes, offices, automobiles, and other enclosed places. As a consequence, the quality of indoor air may be of more concern from a public health standpoint than that of outdoor air quality. Many vulnerable materials—electronic components, fabrics, or works of art—are exposed to the atmosphere, mainly under cover. Nonetheless, the data base for assessment of indoor air quality is smaller than that for ambient air. Citywide average indoor air quality levels are clearly not appropriate or meaningful, because the available data indicate quite strongly that the concentrations of trace species in indoor locations vary greatly within cities, within neighborhoods, and within buildings.

For many trace gases, the concentrations in indoor air reflect indoor sources rather than infiltration of air from outdoors. For a few, injection of outdoor air appears to control indoor concentrations. In this section, measured indoor concentrations of species of interest are presented, as well as a summary of what is known about the controlling sources. Further information is available in the detailed and authoritative book *Indoor Pollutants* (National Research Council 1981), as well as in reports by Fanger and Valbjorn (1979), Aurand et al.

(1982), Spengler et al. (1982), and Berglund et al. (1986).

Carbon Dioxide. CO_2 is a natural constituent of the atmosphere, with typical concentrations of about 350 ppm in remote areas and higher concentrations near population centers. It is a principal product of fossil fuel combustion, and any indoor combustion source that is not well vented will give rise to heightened CO_2 levels.

Differences in building construction and indoor sources result in wide variations in CO_2 indoor concentrations. The minima generally correspond to that outdoors (Yocom 1982; Spengler and Sexton 1983), whereas maxima of 3,000 ppm are not too uncommon. Occasionally the peak values may reach the Occupational Safety and Health Administration exposure standard of 5,000 ppm (World Health Organization 1983).

Carbon Monoxide. Indoor concentrations of CO tend to be lower than those outdoors, but not markedly so (Ott and Flachsbart 1982). Urban indoor CO concentrations of a few ppm are thus common, and higher concentrations are sometimes seen. Two types of indoor environments are particularly likely to have high CO concentrations: indoor garages or buildings with attached indoor parking areas (Ott and Flachsbart 1982) and residences with improperly ventilated space heating equipment (National Research Council 1981). In these situations, concentrations as high as 100 ppm have been measured. High concentrations can also be produced as a consequence of cigarette smoking.

If obvious indoor sources are absent, indoor CO concentrations tend to follow outdoor CO concentrations but to be somewhat lower. The exposure of the population to CO thus has a lower limit set by the outdoor concentrations of CO. The concentration variations indoors lag behind those outdoors in time.

Oxides of Nitrogen. NO_x exists indoors as a consequence of both indoor and outdoor sources. Indoors, these sources consist largely of poorly ventilated products of burnt gas. Gas stoves are by far the most common problem. Somewhat elevated NO_x levels may also be seen in buildings with unvented space heaters, gas-fired hot water heaters, or gas-fired clothes dryers.

Many of the concentration dependencies have been studied by Moschandreas et al. (1980), whose data show that gas-heated residences generally have NO_x concentrations greater than those outdoors, whereas electrically heated residences tend to mimic outdoor NO_x conditions. Typical NO_x concentrations are 1–40 ppb (Spengler et al. 1979). Office buildings tend to be efficiently ventilated and to have heating systems well removed from office areas, so that the type of heating is often not a factor in their indoor NO_x concentrations.

Hydrocarbons. Indoor environments are rich sources of HCs and oxidized organic molecules, with several hundred different organic compounds known to be present in certain indoor environments (Jarke et al. 1981; Graedel et al. 1986). The most abundant HC compounds, at least in residences, are alkanes, alkylbenzenes, and terpenes (Mølhave and Møller 1979).

Indoor concentrations of gaseous NMHCs span a measured range of 0.006–40 ppm, with average concentrations in newer homes being of the order of 12 ppm and those in older homes being of the order of 1 ppm (Mølhave and Møller 1979; Wallace et al. 1982). Building materials and a host of other indoor sources appear to be responsible for many of the observed organic substances. In view of the observed infiltration of outdoor species such as CO and NO_x, it appears likely that outdoor concentrations will generally establish a lower bound to indoor NMHC concentrations of a few tenths of a ppm or higher in central urban areas.

Ozone. Except for occasional generation by electric arcing, photocopiers, and indoor ultraviolet light sources (Committee on Indoor Pollutants 1981), any O_3 indoors is present as a result of the injection or infiltration of outside air. O_3 is readily absorbed and destroyed on surfaces and, as a result, its indoor concentrations tend to be low. The National Research Council

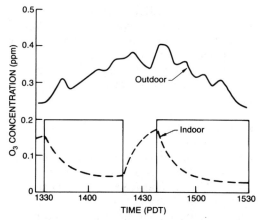

Figure 16. Relative indoor and outdoor concentrations of O_3 in the Spaulding Laboratory, California Institute of Technology, July 22, 1975. Boxed areas indicate times when the building's charcoal filtering system was in operation. (Adapted from Committee on Indoor Pollutants 1981.)

(1981) suggested an anticipated range for indoor O_3 of 20–200 ppb, with low values in this range being much more common than high ones.

The relationship between indoor and outdoor O_3 during smoggy conditions is shown in figure 16. This figure demonstrates that indoor O_3 reflects outdoor concentrations, but at a lower level, and that charcoal filtration is effective in removing O_3 from indoor environments. Similar findings in a number of geographical locations have been reviewed by Yocom (1982).

Sulfur Dioxide. In the absence of indoor sources, SO_2 concentrations are normally 0.3 to 0.7 times those outdoors, although air conditioning or filtration can reduce this factor substantially (Yocom 1982). Within a residence in which sulfur-containing fuel (such as coal, wood, or kerosene) is burned, however, inadequate venting of exhaust gases can result in high local SO_2 concentrations. Indoor SO_2 levels of a few ppb are typical (Spengler et al. 1979), but concentrations as high as several hundred ppb have been observed in special cases (World Health Organization 1983). As with outdoor SO_2, there is little or no evidence for a significant contribution by motor vehicle emissions.

Concentrations of Principal Trace Gases Within Motor Vehicles. A large number of people spend one or more hours a day within automobiles, buses, and other motor vehicles. Notwithstanding this fact, information on concentrations of trace gases within motor vehicles is sparse indeed. Some data on CO within automobiles indicate that these concentrations can exceed 100 ppm if vehicle exhaust products are allowed to intrude into the passenger compartment. Similar tendencies are suggested by the data of figure 17, which show high levels of interior CO in a vehicle moving slowly or halted in dense traffic. HCs show similar behavior (Mücke et al. 1984), and concentrations of NO_x and small particles would be expected to mimic those of CO as well. Additional data are needed to assess more fully the contribution of exposure in the passenger compartments of motor vehicles to total exposure burdens.

Personal Air Quality Monitors. A recent development in the assessment of exposure to atmospheric trace species is the introduction of personal exposure monitors. These instruments are designed to be carried or worn throughout one's normal activities, and thus are limited in power, size, and weight. In table 4 of their chapter in this volume, Sexton and Ryan summarize the direct personal monitoring studies that have been performed in the United States. The potential benefits from the use of such instruments is substantial, but much research is needed to develop suitable techniques for other species of interest and to miniaturize the equipment.

■ **Recommendation 6.** Personal exposure monitors have important future roles to play in health effects research. Instrument development is required, however, to improve portability, reliability, and the quantitative detection of many atmospheric species of interest not presently capable of being monitored in this way.

Minor Emittants and Products

Inorganic Compounds. As with outdoor air, one would like to have information on the indoor concentrations of nitric and nitrous acids, hydrogen peroxide, and perhaps other inorganic gases and aerosol constituents. To my knowledge, however, no such measurements have ever been made in unperturbed indoor environments. A related study of interest is that of Pitts et al.

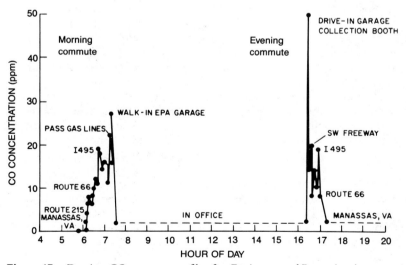

Figure 17. Daytime CO exposure profile of an Environmental Protection Agency employee. The employee's location in the Washington, D.C., area is given for points of interest on the diagram. The detector was not operated during the periods indicated by dotted lines. (Adapted with permission from Ott 1985, and the American Chemical Society.)

(1985a), who injected NO_2 into a humid indoor space and measured the formation of nitrous acid. The results suggested that indoor spaces with high concentrations of NO_2, such as can occur near gas stoves, might also be areas in which a few ppb of HNO_2 can be found. Much more study of the indoor nitrogen cycle in the presence of strong sources is needed.

Some data are available on the concentrations of major ions on indoor aerosols. Within fine particles (those with diameters <2.5 μm), Sinclair and coworkers (1985) found ammonium, sulfate, and nitrate to be predominant, whereas within coarse particles, they found calcium and nitrate concentrations to be highest. Typical average concentrations and peak values in office buildings were SO_4^{2-}, 0.1 and 0.3 $\mu g/m^3$; NO_3^-, 0.1 and 0.2 $\mu g/m^3$; NH_4^+, 0.2 and 0.3 $\mu g/m^3$; and Ca^{2+}, 0.05 and 0.2 $\mu g/m^3$.

Aldehydes. Formaldehyde (and to a much smaller extent, other aldehydes) is one of the major indoor air quality concerns. The current understanding of aldehyde sources is described in a recent report by the National Research Council (1981):

"The primary sources (of formaldehyde and other organic substances) are in the indoor environment itself—building materials, combustion appliances, tobacco smoke, and a large variety of consumer products. A buildup of formaldehyde may be exacerbated in buildings that have been subjected to energy-efficiency measures intended to reduce infiltration and, thus, energy consumption. Emission rates for formaldehyde and other organic pollutants emitted in the indoor environment are generally unknown."

More information on sources of aldehydes and their effects has been reported by the National Academy of Sciences (1981).

The indoor concentrations of formaldehyde—the only aldehyde for which any significant amount of data is available—vary greatly. They can be negligibly small in buildings that contain few or none of its common indoor sources. Conversely, in buildings such as new, well-insulated mo-

bile homes, concentrations may be as high as several ppm (National Research Council 1981; Hanrahan et al. 1985), although several tenths of a ppm is a more typical value (Bundesgesundheitsamt 1985; Sexton et al. 1986). Acetaldehyde concentrations are typically much lower (Wang 1975; DeBortoli et al. 1984). There is no indication that formaldehyde from infiltrated outdoor air plays any significant role in establishing indoor formaldehyde concentrations unless there are no indoor sources whatever.

Alcohols. Very limited data exist on the indoor concentrations of alcohols. Wang (1975) detected several alcohols in a college classroom and deduced that the sources were indoor rather than outdoor. Berglund and coworkers (1982) measured butanol concentrations in a school and attributed the presence of butanol to the vaporization of solvent from building materials. In both cases, concentrations of 5–50 ppb were observed.

Nitro Compounds and Organic Nitrates. Few studies of indoor nitro compounds have been conducted. Thompson and coworkers (1973) examined indoor and outdoor concentrations of gaseous peroxyacetyl nitrate (PAN) at several sites in the Los Angeles Basin. They found PAN levels of a few ppb, always lower indoors than outdoors, and attributed them to the infiltration of outdoor air. Seifert and coworkers (1984) detected amines and nitrosamines, the former at levels as high as a few hundred ppt, the latter at levels 10 to 100 times smaller. It has been suggested that the nitrosamines are formed in kitchens when NO_x and amines are simultaneously trapped in air-cleaning units.

Since condensed-phase nitro compounds are common outdoor constituents, one would expect to find them indoors as well, perhaps at much reduced concentrations. No studies of such species indoors have been performed.

Heterocyclic Organic Compounds. The only study identifying indoor heterocyclic compounds is that of Jarke et al. (1982). At homes in the Chicago area, they found many organic compounds, including furan, dioxane, and indole. No quantification of the concentrations was attempted, but the authors estimated their detection limit for these compounds was about 0.5 ppb. The sources of the compounds were not determined but might be supposed to be either the infiltration of outdoor air or the by-products of indoor fossil fuel combustion.

Polynuclear Aromatic Hydrocarbons. PAHs are readily detectable in the indoor environment, as a consequence of infiltration of outdoor air as well as from indoor combustion sources. Their combined concentrations indoors total perhaps 5–10 ng/m^3 (Butler and Crossley 1979; Sexton et al. 1985). Given PAH vapor pressures, these data imply as well that indoor equilibrium gaseous PAH concentrations will be around 1 ppt. Such levels are similar to, or slightly smaller than, outdoor PAH concentrations.

Suspended Particulate Matter. The concentration of suspended particulate matter in buildings without air filtration appears to be generally higher than it is outdoors. The National Research Council (1981) states a range of indoor TSP of 10–500 μg/m^3. Typical levels within most buildings are about 15–50 μg/m^3 (Sexton et al. 1984, 1985).

The chemical constituents of the indoor aerosols bear substantial resemblance to those outdoors. Organic carbon compounds make up perhaps 4–25 μg/m^3 of the total (Sexton et al. 1985); these compounds include phthalates, alkanes, fatty acids, and other oxygenated species (Weschler 1980, 1984; Weschler and Fong 1984). Elemental carbon accounts for some 10 percent of the total aerosol mass, or about 2–7 μg/m^3 (Sexton et al. 1985). Heavy metals, including aluminum, iron, copper, and zinc, are present at concentrations of a few tens or hundreds of ng/m^3 (Tosteson et al. 1982; Sexton et al. 1985). The ions common to outdoor aerosols are also found indoors (Sinclair et al. 1985). The principal sources of many of the organic and metallic compounds appear to be located within the buildings. In the case of Pb, however, most is present on fine particles and exists indoors as a result of infiltration of outdoor air.

■ **Recommendation 7.** Air quality researchers have only the most general ideas of indoor fluxes of trace species, the relative importance of indoor and outdoor sources to indoor species concentrations, and total exposures. A major effort should be made to acquire such data, without which no epidemiologic studies can hope to be authoritative. Special effort should be directed to the passenger compartments of automobiles, where total exposure is potentially quite high.

Emittants with Potential Global Influence

Carbon Dioxide

CO_2 is one of the principal products of the combustion of fossil fuels. Its emissions from motor vehicles are substantial but are small fractions of the global emission flux. CO_2 is not toxic at or near atmospheric concentrations, but its presence in the atmosphere has major effects on biogenic life cycles on the earth because it is a major absorber of the infrared radiation emitted toward space from the earth's surface. As a result, it is crucial to the establishment and maintainence of the planetary temperature structure.

The concentration of CO_2 at a remote atmospheric site is shown in figure 18. The upward trend is readily evident. It has been estimated that the atmospheric CO_2 concentration will double by the year 2030 or thereabouts, producing a global average temperature increase of about 1.5 to 4.5°C (National Research Council 1983). Despite considerable study, it appears unlikely that any global program for the reduction of CO_2 emissions will prove feasible. As a result, motor vehicle CO_2 emissions are unlikely to be controlled by law. It is possible that increasing amounts of CO_2, together with other radiation-absorbing gases, will change the total environment of

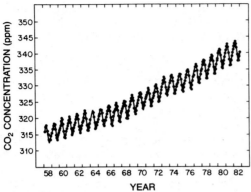

Figure 18. Concentration of atmospheric CO_2 at Mauna Loa Observatory, Hawaii, expressed as a mole fraction in parts per million of dry air. The dots depict monthly averages of visually selected data that have been adjusted to the center of each month. The curve represents the fit simultaneously to an exponential function, a spline function, and a linearly increasing seasonal cycle. (Adapted with permission from Bacastow et al. 1985, and the American Geophysical Union.)

the planet within two or three generations, and it may be prudent to keep that change as small as possible.

Carbon Monoxide

Most of the trace molecules emitted into the air are ultimately removed from it by reaction chains initiated by the hydroxyl (HO·) radical (Atkinson, this volume). It appears from the results of photochemical atmospheric models that the most impor-

tant reactant in controlling the global HO· concentration is CO, because of its abundance and its rapid reaction with HO·. As with CO_2, the atmospheric concentrations of CO are steadily increasing (figure 19). As was pointed out above, motor vehicles are responsible for about two-thirds of all CO emissions and are thus major factors in the global CO increase. Photochemical model studies (Levine et al. 1985) suggest that over the past 35 years the average HO· concentration has decreased by 25 percent. As a result, the ability of the atmosphere to cleanse itself has become increasingly inhibited.

Methane

Methane is an absorber of infrared radiation as well as a factor in controlling the atmospheric abundance of the HO· radical. As a consequence, its long-term trend is also of interest. A summary of atmospheric concentration measurements of methane over the past several years is given in figure 20. As with CO_2 and CO, methane concentrations are increasing, at about 1.2 percent per year. About a thousandth of the annual methane emissions are attributable to motor vehicles (Ehhalt and Schmidt 1978).

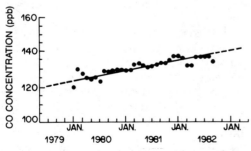

Figure 19. CO concentrations at Cape Meares, Oregon. Monthly concentrations are formed from the 440 to 2,200 measurements each month, and these averages are then combined to form 12-month moving averages. In this approach, seasonal cycles of a year or less disappear. The solid line is the trend calculated by linear least-squares techniques. (Adapted with permission from Khalil and Rasmussen 1984, © 1984 by AAAS.)

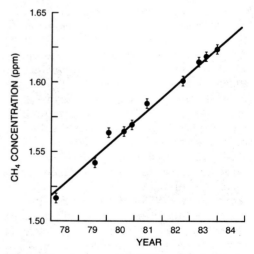

Figure 20. Average worldwide tropospheric concentrations of methane during the period 1978–1983. The solid line represents a least-squares fit with an increase of 0.018 ppm per year. (Adapted with permission from Blake and Rowland 1986, and D. Reidel Publishing Company.)

Summary

Data Adequacy

Data adequacy is taken here to mean that sufficient data are available to permit reasonable assessments of the effects of a given atmospheric constituent on animate and inanimate objects, and not that the concentrations are known on every street corner. By this definition, sufficient outdoor data are available for the "criteria species" CO, NO_2, SO_2, O_3, Pb, and TSP. For total or NMHCs the amount of available data is marginally adequate, in large part because no satisfactory routine monitoring instrument is available. Inside buildings a similar situation exists, with at least the approximate range of criteria species concentrations having been established. Within the passenger compartments of automobiles, very few concentration data have appeared in the literature.

In the case of atmospheric species for which ambient standards have not been established, the available outdoor data are generally inadequate to do more than infer order of magnitude concentrations and to produce some idea of the relative strengths of the potential sources of the compounds. For example, the data on formaldehyde and other small aldehydes are from very few sites and are not now being enhanced by any regular monitoring. As with NMHC, this is partly because no routine, reliable monitoring instrument is available. For methanol, ethanol, and manganese, species that may soon be emitted from motor vehicles at much higher rates, the data are extremely sparse. It is important that these compounds be included soon in routine monitoring programs in order to establish baseline concentrations for future reference. Other species for which more data are needed are those generally present in aerosol form which possess positive bioassay characteristics. Such compounds include the nitro derivatives of PAHs and several heterocyclic species.

Indoors, measurements have been restricted largely to the criteria species and to formaldehyde. Much greater characterization of trace species in the indoor environment is needed.

Trends

For CO, CO_2, and methane, the atmospheric concentrations show an increasing trend at global background locations, and stable or slightly decreasing peak levels in urban areas. For SO_2, Pb, and TSP, decreasing trends are seen. The concentrations of O_3 and NO_2 in most U.S. urban regions appear to be roughly stable on an annual basis (Los Angeles is the exception, having shown a 25 percent decrease over the last eight years); at global background sites the concentration of O_3, at least, appears to be increasing, thus implying also an increase in NO_2.

Concentrations

As is evident from the information above, concentrations of airborne species of interest show wide variations from site to site and time to time. Notwithstanding this complexity, it is useful to tabulate information from the literature on typical values of average and peak concentrations. Such data appear in table 1 for 21 species. In each case, an attempt is made to indicate the approximate state of measurement technology currently required. The ranges of values for urban areas are annual averages, if available. Peak values are given for urban outdoor environments, for indoor nonmanufacturing environments, and for the passenger compartments of automobiles. The availability of data generally decreases from left to right in the table. Many more data are extant on emitted gases and TSP than for other species shown.

For gaining a quick perspective on typical concentrations of trace species, graphical displays are often convenient. Such displays are presented here for species for which sufficient data are available to establish typical ranges of concentrations. Trace gases are displayed in figure 21. In most cases, the concentrations in remote areas are the lowest, those in outdoor urban air next highest, and those indoors highest of all. (The exception is O_3, which has roughly the same peak values in all three regimes.) The ordinate on figure 21 is logarithmic; remote and indoor concentrations differ by as much as four or five orders of magnitude in some cases.

Table 1. Typical Ranges and Peak Values for Gaseous and Particulate Species

Species	Measurement Capability	Concentration[a] Urban Range	Urban Peak	Indoor Peak	Auto Peak[b]	References
Emitted gases (ppb)						
CO	RM	$(3–15) \times 10^3$	4×10^4	1×10^5	3×10^4	National Research Council (1981); U.S. Environmental Protection Agency (1985); Mücke et al. (1984)
CO_2	RM	$(3–6) \times 10^5$	6×10^5	3×10^6	—	McRae and Graedel (1979); Spengler and Sexton (1983)
NO_x	RM	10–50	800	500	1×10^3	U.S. Environmental Protection Agency (1985); Mücke et al. (1984); Spengler and Sexton (1983)
NMHC	ET	$(1–5) \times 10^3$	1×10^4	3×10^4	2×10^3	DeBertoli et al. (1984); Mücke et al. (1984); Tilton and Bruce (1980)
SO_2	RM	3–20	300	20	—	National Research Council (1981); U.S. Environmental Protection Agency (1985)
Product gases (ppb)						
O_3	RM	90–210	350	200	—	National Research Council (1981); U.S. Environmental Protection Agency (1985); Tuazon et al. (1981)
HNO_2	ES	0.2–4(s)	8	5	—	Harris et al. (1982); Pitts et al. (1985b); Sjödin and Ferm (1985)
HNO_3	ES	1–5	30	—	—	Spicer (1977); Tuazon et al. (1981)
PAN	ES	5–10	25	—	—	Tuazon et al. (1981)
H_2O_2	ES	0.2–2(s)	—	—	—	Kok et al. (1986)
HCHO	ET	3–60	50	1×10^3	—	National Research Council (1981); National Academy of Sciences (1981); Tuazon et al. (1981)
Particle species ($\mu g/m^3$)						
Pb	RM	0.1–0.7	1.0	0.1	—	U.S. Environmental Protection Agency (1985); Tosteson et al. (1982)
EC	ET	1–15	35	—	—	Countess et al. (1980); Wolff (1985)
OC	ET	5–20	40	—	—	Countess et al. (1980); Wolff (1985)
NO_3^-	RM	1–10	15	0.7	—	Graedel and Schwartz (1977); Graedel et al. (1986)
SO_4^{2-}	RM	1–20	30	5	—	Graedel and Schwartz (1977); Graedel et al. (1986)
PAH	ES	$(5–10) \times 10^{-2}$(s)	0.1	10	—	Lahmann et al. (1984); Moschandreas et al. (1981); Seifert et al. (1983)
Nitro-PAH	ES	$(1–3) \times 10^{-4}$(s)	3×10^{-4}	—	—	Gibson (1982); Pitts et al. (1985a)

(*Table continued next page.*)

Table 1. *Continued*

| Species | Measure-ment Ca-pability | Concentration[a] | | | | References |
		Urban Range	Urban Peak	Indoor Peak	Auto Peak[b]	
TSP	RM	30–75	400	500	—	Bennett et al. (1985); U.S. Environmental Protection Agency (1985); Spengler and Sexton (1983)
IP	RM	5–80	120	—	—	Lioy et al. (1983); Wolff et al. (1985b)
RP	RM	10–75	210	500	150	Budiansky (1980); National Research Council (1981); DeBortoli et al. (1984)

[a] Averaging times are annual for urban range except shorter where noted by (s), daily for urban peak, an hour or two for indoor and automobile interior data.

[b] Concentrations as measured within the passenger compartment of an automobile.

NOTE: Abbreviations and symbols not elsewhere defined are EC = elemental carbon; OC = organic carbon; RM = routine monitoring; ET = event monitoring by technician; ES = event monitoring by scientist.

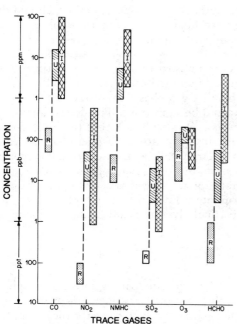

Figure 21. Typical concentration ranges of selected atmospheric trace gases in remote areas (R), urban areas (U), and indoors (I). Data for this display are from Noxon 1975; Spengler et al. 1979; Mølhave 1982; U.S. Environmental Protection Agency 1985; Wallace et al. 1985; and Graedel et al. 1986. Because of the form of the ambient air quality standards for CO and O_3, the urban data for those compounds represent upper extreme values (extracted from an annual data set) rather than mean values.

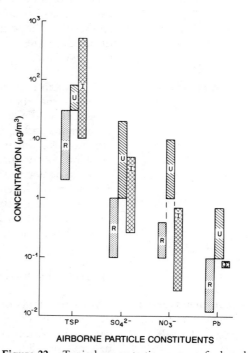

Figure 22. Typical concentration ranges of selected atmospheric airborne particle constituents in remote areas (R), urban areas (U), and indoors (I). The data for this display are from Rhodes et al. 1979; Graedel 1980; Sinclair et al. 1985; and the references given in table 1.

A similar display for particulate matter is given in figure 22. In general, the data for the three regimes are much more similar than was the case for the gases. Two caveats are worth noting, however: adequate trace metal data, except for Pb, are not extensive, and few indoor data exist for other than TSP. Limited studies of trace metal compositions throughout the world suggest that urban and remote concentration differences are substantial.

Summary of Research Recommendations

HIGH PRIORITY

Recommendation 7
Indoor Air Quality

The proportion of time spent indoors by most people is high, yet the available data for indoor air quality is quite sparse. This is particularly true of nonresidential environments such as automobile interiors, subway platforms, and the like. Air quality researchers have only the most general ideas of indoor fluxes of trace species, the relative importance of indoor and outdoor sources to indoor species concentrations, and total exposures. A major effort should be made to acquire such data, without which no epidemiologic studies can hope to be authoritative. Special effort should be directed to the passenger compartments of automobiles, where total exposure is potentially quite high.

Recommendation 3
Aldehyde Monitor

The alkanic aldehydes are unusual atmospheric constituents in the sense that they are directly emitted by sources as well as being produced in the atmosphere by gas-phase chemistry. Their effects on humans and animals could be significant. The determination of atmospheric aldehyde levels is a difficult task, but the limited data available show that much insight into atmospheric processes is likely to be derived from carefully collected, extended data records. An effort to develop a monitor for either formaldehyde or the aldehyde group with appropriate sensitivity (100 ppt or less), reliability, and an appropriately low cost, is a high priority for atmospheric chemists, particularly those addressing problems in urban air.

MODERATE PRIORITY

Recommendation 2
Unregulated Species—Particle Phase

Measurements of the total particulate loading of the atmosphere give only the crudest possible idea of the condensed-phase aerosol chemistry. A difficult but essential job is to monitor the chemistry of atmospheric aerosol particles in much more detail than is now being done, concentrating especially on chemical differences as a function of particle size. Detailed organic analyses are particularly important. It is likely that the most interesting and complex air quality problems in the next decade will relate to condensed-phase species; enhancing the level of analysis and depth of study of these species are thus matters of critical concern.

Recommendation 1
NMHC Monitor

Hundreds of NMHCs are present in the atmosphere. They are involved in the formation of O_3, PAN, formaldehyde, and other lachrymators, some of which are toxic, and some of which serve as precursors for such potentially hazardous compounds as the nitro-aromatics. Notwithstanding this central role, no analytical instrument is readily available for routine monitoring of their concentrations and concentration trends, although several techniques are available for potential incorporation into such an instrument. It is extremely important to achieve agreement on a satisfactory monitoring technique for NMHCs (or some significant fraction thereof) and to begin to acquire data on a routine basis.

Recommendation 4
Unregulated Species—Gas Phase

From the bioassay perspective, the complex chemical products of the compounds emitted to the atmosphere are often of the most concern. Perhaps 98 percent of monitoring efforts are directed at criteria species, however. It is important to recognize that many assessments having to do with atmospheric species cannot go forward unless supporting data for them are available. A few examples of crucial species listed earlier include PAN, nitric acid, hydrochloric acid, and others. It is important for atmospheric chemists to focus their thinking on unmet needs in species diversity and geographical diversity.

LOW PRIORITY

Recommendation 5
Alkanic Alcohol Monitoring

Atmospheric alcohols have not often been studied in the atmosphere, but the limited data suggest that in urban areas, at least, their concentrations rival those of many better known organic compounds. Once present in the atmosphere, the alcohols will react to produce aldehydes and organic acids, two groups of compounds potentially involved in a number of injurious interactions with animate and inanimate surfaces. Given the possibility of sharply increased use of methanol and ethanol in motor vehicle fuels over the next decade or two, it is important to begin promptly to establish a satisfactory baseline against which future changes in atmospheric alcohol concentrations could be assessed. The techniques now available, if perhaps not optimum, are satisfactory at least for a limited screening program.

Recommendation 6
Personal Exposure Monitors

The ultimate concern of the epidemiologist dealing with the effects of atmospheric species is not species concentrations at a monitoring site, but those encountered by human beings. Personal exposure monitors thus have important future roles to play in health effects research. Instrument development is required, however, to improve portability, reliability, and the quantitative detection of many atmospheric species of interest not presently capable of being monitored in this way.

Acknowledgments

I thank R. S. Freund, J. D. Sinclair, and C. J. Weschler for useful reviews of this paper, and K. Sexton and B. Seifert for providing considerable assistance in locating reference material dealing with indoor air quality measurements.

References

Aurand, K., Seifert, B., and Wegner, J. 1982. Luft-qualität in Innenräumen, Gustav Fischer Verlag, Stuttgart.

Bacastow, R. B., Keeling, C. D., and Whorf, T. P. 1985. Seasonal amplitude increase in atmospheric CO_2 concentration at Mauna Loa, Hawaii, 1951–1982. *J. Geophys. Res.* 90:10529–10540.

Bellar, T. A., and Sigsby, J. E., Jr. 1970. Direct gas chromatographic analysis of low molecular weight substituted organic compounds in emissions, *Environ. Sci. Technol.* 4:150–156.

Bennett, B. G., Kretzschmar, J. G., Akland, G. G., and de Koning, H. W. 1985. Urban air pollution worldwide, *Environ. Sci. Technol.* 19:298–304.

Berglund, B., Johansson, I., and Lindvall, T. 1982. A longitudinal study of air contaminants in a newly built preschool, *Environ. Int.* 8:111–115.

Berglund, B., Berglund, U., Lindvall, T., Spengler, J., and Sunden, J. (eds.) 1986. Indoor air, *Environ. Int.* 12: whole issue (nos. 1–4) pages 1–494.

Blake, D. R., and Rowland, F. S. 1986. World-wide increase in tropospheric methane, 1978–1983, *J. Atmos. Chem.* 4:43–62.

Brodzinsky, R., Chang, S. G., Markowitz, S. S., and Novakov, T. 1980. Kinetics and mechanism for the catalytic oxidation of sulfur dioxide on carbon in aqueous suspensions, *J. Phys. Chem.* 84:3354–3358.

Budiansky, S. 1980. Indoor air pollution, *Environ. Sci. Technol.* 14:1023–1027.

Bundesgesundheitsamt, Bundesanstalt für Arbeitsschutz, and Umweltbundesamt. 1985. *Formaldehyde*. Medizin Verlag, München.

Butler, J. D., and Crossley, P. 1979. An appraisal of relative airborne sub-urban concentrations of poly-cyclic aromatic hydrocarbons monitored indoors and outdoors, *Sci. Total Environ.* 11:53–58.

Cautreels, W., and Van Cauwenberghe, K. 1978. Experiments on the distribution of organic pollutants between airborne particulate matter and the corresponding gas phase, *Atmos. Environ.* 12:1133–1141.

Cleveland, W. S., Graedel, T. E., and Kleiner, B. 1977. Urban formaldehyde: observed correlation with source emissions and photochemistry, *Atmos. Environ.* 11:357–360.

Countess, R. J., Wolff, G. T., and Cadle, S. H. 1980.

The Denver winter aerosol: a comprehensive chemical characterization, *J. Air Pollut. Contr l Assoc.* 30:1194–1200.

DeBortoli, M., Knöppel, H., Pecchio, E., Peil, A., Rogora, L., Schauenburg, H., Schlitt, H., and Vissers, H. 1984. Integrating "real life" measurements of organic pollution in indoor and outdoor air of homes in northern Italy. *Proc. 3rd Int. Conf. Indoor Air Quality and Climate* 4:21–26.

Dong, M. W., Locke, D. C., and Hoffmann, D. 1977. Characterization of aza-arenes in basic organic portion of suspended particulate matter, *Environ. Sci. Technol.* 11:612–618.

Ehhalt, D. H., and Schmidt, U. 1978. Sources and sinks of atmospheric methane. *Pure Appl. Geophys.* 116:452–464.

Fanger, P. O., and Valbjorn, O. (eds.) 1979. *Indoor Climate*, Danish Bldg. Res. Inst., Copenhagen.

Gibson, T. L. 1982. Nitro derivatives of polynuclear aromatic hydrocarbon in airborne and source particulate matter, *Atmos. Environ.* 16:2037–2040.

Graedel, T. E. 1980. Atmospheric photochemistry, In: *Handbook of Environmental Chemistry* (O. Hutzinger, ed.), Vol. 2A, Springer, Heidelberg: pp. 107–143.

Graedel, T. E., and McRae, J. E. 1982. Total organic component data: a study of urban atmospheric patterns and trends, *Atmos. Environ.* 16:1119–1132.

Graedel, T. E., and Schwartz, N. 1977. Air quality reference data for corrosion assessment, *Mater. Perform.* 16(8):17–25.

Graedel, T. E., Hawkins, D. T., and Claxton, L. D. 1986. *Atmospheric Chemical Compounds: Sources, Occurrence, and Bioassay*, Academic Press, Orlando, Fla. 732 pp.

Grosjean, D., and Friedlander, S. K. 1975. Gas-particle distribution factors for organic and other pollutants in the Los Angeles atmosphere, *J. Air Pollut. Control Assoc.* 25:1038–1044.

Hanrahan, L. P., Anderson, H. A., Dally, K. A., Eckmann, A. D., and Kanarek, M. S. 1985. Formaldehyde concentrations in Wisconsin mobile homes. *J. Air Pollut. Control Assoc.* 35:1164–1167.

Hanst, P. L., Wilson, W. E., Patterson, R. K., Gay, B. W., Jr., Chaney, L. W., and Burton, C. S. 1975. A spectroscopic study of California smog, EPA Report 650/4-75-006, U.S. Environmental Protection Agency, Washington, D.C.

Harkov, R., Kebbekus, B., Bozzelli, J. W., and Lioy, P. J. 1983. Measurement of selected volatile organic compounds at three locations in New Jersey during the summer season, *J. Air Pollut. Control Assoc.* 33: 1177–1183.

Harris, G. W., Carter, W. P. L., Winer, A. M., Pitts, J. N., Jr., Platt, U., and Perner, D. 1982. Observations of nitrous acid in the Los Angeles atmosphere and implications for predictions of ozone-precursor relationships, *Environ. Sci. Technol.* 16: 414–419.

Harrison, R. M., and Pio, C. A. 1983. Major ion composition and chemical associations of inorganic atmospheric aerosols, *Environ. Sci. Technol.* 17: 169–174.

Holzer, G., Shanfield, H., Zlatkis, A., Bertsch, W., Jaurez, P., Mayfield, H., and Liebich, H. M. 1977.

Correspondence should be addressed to T. E. Graedel, AT&T Bell Laboratories, 600 Mountain Avenue, Murray Hill, NJ 07974.

Collection and analysis of trace organic emissions from natural sources, *J. Chromatogr.* 142:755–764.

Hoshika, Y. 1977. Simple and rapid gas-liquid-solid chromatographic analysis of trace concentrations of acetaldehyde in urban air, *J. Chromatogr.* 137:455–460.

Jarke, F. H., Dravnieks, A., and Gordon, S. M. 1981. Organic contaminants in indoor air and their relation to outdoor contaminants, *ASHRAE Trans.* 81 (Part 1):153–165.

Johnson, W. B., Ludwig, F. L., Dabberdt, W. F., and Allen, R. J. 1973. An urban diffusion simulation model for carbon monoxide, *J. Air Pollut. Control Assoc.* 23:490–498.

Ketseridis, G., and Jaenicke, R. 1978. Organische Beimengungen in atmosphärisher Reinluft: Ein Beitrag zur Budget-Abschatzung, In: *Organische Verunreinigungen in der Umwelt-Erkennen, Berwerben, Bermindern* (K. Aurand et al., eds.) E. Schmidt, Berlin: pp. 379–390.

Khalil, M. A. K., and Rasmussen, R. A. 1984. Carbon monoxide in the earth's atmosphere: increasing trend, *Science* 224:54–56.

Kok, G. L., Heikes, B. G., and Lazrus, A. L. 1986. Gas and aqueous phase measurements of hydrogen peroxide, paper presented at the National Meeting of the American Chemical Society, New York, N.Y., April 14, 1986.

Krstulovic, A. M., Rosie, D. M., and Brown, P. R. 1977. Distribution of some atmospheric polynuclear aromatic hydrocarbons, *Am. Lab.* 9(7):11–18.

Lahmann, E., Seifert, B., Zhao, L., and Bake, D. 1984. Immissionen von polycyclischen aromatischen kohlenwasserstoffen in Berlin (West), *Staub-Reinhalt. Luft* 44:149–157.

Levine, J. S., Rinsland, C. P., and Tennille, G. M. 1985. The photochemistry of methane and carbon monoxide in the troposphere in 1950 and 1985, *Nature* 318:254–257.

Lioy, P. J., Daisy, J. M., Reiss, N. M., and Harkov, R. 1983. Characterization of inhalable particulate matter, volatile organic compounds and other chemical species measured in urban areas in New Jersey. I. Summertime episodes, *Atmos. Environ.* 17:2321–2330.

Logan, J. A. 1985. Tropospheric ozone: seasonal behavior, trends, and anthropogenic influence, *J. Geophys. Res.* 90:10463–10482.

Louw, C. W., Richards, J. F., and Faure, P. K. 1977. The determination of volatile organic compounds in city air by gas chromatography combined with standard addition, selective subtraction, infrared spectrometry, and mass spectrometry, *Atmos. Environ.* 11:703–717.

McRae, J. E., and Graedel, T. E. 1979. Carbon dioxide in the urban atmosphere: dependencies and trends, *J. Geophys. Res.* 84:5011–5017.

Milford, J. B., and Davidson, C. I. 1985. The sizes of particulate trace elements in the atmosphere—a review, *J. Air Pollut. Control Assoc.* 35:1249–1260.

Miller, D. F., and Spicer, C. W. 1975. Measurement of nitric acid in smog, *J. Air Pollut. Control Assoc.* 25:940–942.

Mølhave, L. 1982. Indoor air pollution due to organic gases and vapours of solvents in building materials, *Environ. Int.* 8:117–127.

Mølhave, L., and Møller, J. 1979. The atmospheric environment in modern Danish dwellings—measurements in 39 flats, *Proc. 1st Int. Indoor Climate Symp.* 1:171–186.

Moschandreas, D. J., Zabransky, J., and Pelton, D. J. 1980. Comparison of Indoor-Outdoor Concentrations of Atmospheric Pollutants, GEOMET Report ES-823, Electric Power Research Institute, Palo Alto, Calif.

Moschandreas, D. J., Pelton, D. J., and Berg, D. R. 1981. The effects of woodburning on indoor pollutant concentrations. Paper No. 81-22.2 presented at the 74th Annual Meeting of the Air Pollution Control Assoc., Philadelphia, Pa.

Mücke, W., Jost, D., and Rudolf, W. 1984. Luftverunreinigungen in kraftfahrzeugen, *Staub-Reinhalt. Luft* 44:374–377.

National Academy of Sciences. 1972. *Particulate Polycyclic Organic Matter*, Washington, D.C., 361 pp.

National Academy of Sciences. 1977. *Ozone and Other Photochemical Oxidants*, Washington, D.C., 719 pp.

National Academy of Sciences. 1981. *Formaldehyde and Other Aldehydes.* National Academy Press, Washington, D.C., 340 pp.

National Research Council, Committee on Indoor Pollutants. 1981. *Indoor Pollutants*, National Academy Press, Washington, D.C.

National Research Council, Carbon Dioxide Assessment Committee. 1983. *Changing Climate*, National Academy Press, Washington, D.C., 496 pp.

Noxon, J. F. 1975. Nitrogen dioxide in the stratosphere and troposphere measured by ground-based absorption spectroscopy, *Science* 189:547–549.

Ott, W. R. 1985. Total human exposure, *Environ. Sci. Technol.* 19:880–886.

Ott, W., and Flachsbart, P. 1982. Measurement of carbon monoxide concentrations in indoor and outdoor locations using personal exposure monitors, *Environ. Int.* 8:295–304.

Pitts, J. N., Jr., Sweetman, J. A., Zielinska, B., Winer, A. M., and Atkinson, R. 1985a. Determination of 2-nitrofluoranthene and 2-nitropyrene in ambient particulate organic matter: evidence for atmospheric reactions, *Atmos. Environ.* 19:1601–1608.

Pitts, J. N., Jr., Wallington, T. J., Biermann, H. W., and Winer, A. M. 1985b. Identification and measurement of nitrous acid in an indoor environment, *Atmos. Environ.* 19:763–767.

Quackenboss, J. J., Kanarek, M. S., Spengler, J. D., and Letz, R. 1982. Personal monitoring for nitrogen dioxide exposure: methodological considerations for a community study, *Environ. Int.* 8:249–258.

Rhodes, R. C., Fair, D. H., Frazer, J. E., Long, S. J., Loseke, W. A., Wheeler, V. A., and Walling, J. F. 1979. Air Quality Data for Metals 1976 from the National Air Surveillance Networks. Report EPA-600/4-79-054. U.S. Environmental Protection Agency, Research Triangle Park, N.C.

Rosen, H., Hansen, A. D. A., Gundel, L., and Novakov, T. 1978. Identification of the optically absorbing component in urban aerosols, *Appl. Optics* 17:3859–3861.

Seifert, B., Ullrich, D., and Schmahl, H. J. 1983.

Occurrence of carcinogenic organic substances in kitchen air, In: *Proc. 6th World Cong. Air Quality* 2:177–179.

Seifert, B., Presher, K.-E., and Ullrich, D. 1984. Auftreten anorganischer und organischer substanzen in der luft von küchen und anderen wohnräumen. Inst. für Wasser-, Boden-, und Lufthygiene des Bundes gesundheitsantes, Berlin.

Sexton, K., Spengler, J. D., and Treitman, R. D. 1984. Effects of residential wood combustion on indoor air quality: a case study in Waterbury, Vermont, *Atmos. Environ.* 18:1371–1383.

Sexton, K., Liu, K.-S., Treitman, R. D., Spengler, J. D., and Turner, W. A. 1986. Characterization of indoor air quality in wood-burning residences, *Environ. Int.* 12:265–278.

Sexton, K., Liu, K.-S., and Petreas, M. X. 1986. Formaldehyde concentrations inside private residences: a mail-out approach to indoor air monitoring, *J. Air Pollut. Control Assoc.* 36:698–704.

Sinclair, J. D., Psota-Kelty, L. A., and Weschler, C. J. 1985. Indoor/outdoor concentrations and indoor surface accumulations of ionic substances, *Atmos. Environ.* 19:315–323.

Sjödin, A., and Ferm, M. 1985. Measurements of nitrous acid in an urban area, *Atmos. Environ.* 19:985–992.

Snider, J. R., and Dawson, G. A. 1985. Tropospheric light alcohols, carbonyls, and acetonitrile: concentrations in the southwest United States and Henry's law data, *J. Geophys. Res.* 90:3797–3805.

Spengler, J. D., and Sexton, K. 1983. Indoor air pollution: a public health perspective, *Science* 221:9–17.

Spengler, J. D., Ferris, B. G., Jr., Dockery, D. W., and Speizer, F. E. 1979. Sulfur dioxide and nitrogen dioxide levels inside and outside homes and the implications on health effects research, *Environ. Sci. Technol.* 13:1276–1280.

Spengler, J., Hollowell, C., Moschandreas, D., and Fanger, O. (eds.) 1982. Indoor air pollution, *Environ. Int.* 8:1–534.

Spicer, C. W. 1977. Photochemical atmospheric pollutants derived from nitrogen oxides, *Atmos. Environ.* 11:1089–1095.

Tanner, R. L., and Meng, Z. 1984. Seasonal variations in ambient atmospheric levels of formaldehyde and acetaldehyde, *Environ. Sci. Technol.* 18:723–726.

Taylor, O. C. 1969. Importance of peroxyacetyl nitrate (PAN) as a phytotoxic air pollutant, *J. Air Pollut. Control Assoc.* 19:347–351.

Thompson, C. R., Hensel, E. G., and Kats, G. 1973. Outdoor-indoor levels of six air pollutants, *J. Air Pollut. Control Assoc.* 23:881–886.

Tilton, B. E., and Bruce, R. M. 1980. Review of Criteria for Vapor-Phase Hydrocarbons. Report EPA-600/8-80-045. U.S. Environmental Protection Agency, Washington, D.C.

Tosteson, T. D., Spengler, J. D., and Weker, R. A. 1982. Aluminum, iron, and lead content of respirable particulate samples from a personal monitoring study, *Environ. Int.* 8:265–268.

Tuazon, E. C., Winer, A. M., and Pitts, J. N., Jr.

1981. Trace pollutant concentrations in a multiday smog episode in the California South Coast Air Basin by long path length Fourier transform infrared spectroscopy, *Environ. Sci. Technol.* 15:1232–1237.

U.S. Environmental Protection Agency. 1985. National Air Quality and Emissions Trends Report, 1983. Report EPA-450/4-84-029, Research Triangle Park, N.C.

Wallace, L., Zweidinger, R., Erickson, M., Cooper, S., Whitaker, D., and Pellizzari, E. 1982. Monitoring individual exposure. Measurements of volatile organic compounds in breathing-zone air, drinking water, and exhaled breath, *Environ. Int.* 8:269–282.

Wallace, L. A., Pellizzari, E. D., Hartwell, T. D., Sparacino, C. M., Sheldon, L. S., and Zelon, H. 1985. Personal exposures, indoor-outdoor relationships, and breath levels of toxic air pollutants measured for 355 persons in New Jersey, *Atmos. Environ.* 19:1651–1661.

Wang, T. C. 1975. A study of bioeffluents in a college classroom, *ASHRAE Trans.* 81(Part 1):32–44.

Weschler, C. J. 1980. Characterization of selected organics in size-fractionated indoor aerosols, *Environ. Sci. Technol.* 14:428–431.

Weschler, C. J. 1984. Indoor-outdoor relationships for nonpolar organic constituents of aerosol particles, *Environ. Sci. Technol.* 18:648–652.

Weschler, C. J., and Fong, K. J. 1984. Characterization of organic species associated with indoor aerosol particles, *Proc. 3rd Int. Conf. Indoor Air Quality and Climate*, 2:203–208.

Wolff, G. T. 1985. Characteristics and consequences of soot in the atmosphere, *Environ. Int.* 11:259–269.

Wolff, G. T., and Klimisch, R. L. (eds.) 1982. *Particulate Carbon: Atmospheric Life Cycle*, Plenum, New York, N.Y.

Wolff, G. T., Ferman, M. A., Kelly, N. A., Stroup, D. P., and Ruthkosky, M. S. 1982a. The relationships between the chemical composition of fine particles and visibility in the Detroit metropolitan area, *J. Air Pollut. Control Assoc.* 32:1216–1220.

Wolff, G. T., Groblicki, P. J., Cadle, S. H., and Countess, R. J. 1982b. Particulate carbon at various locations in the United States, In: *Particulate Carbon: Atmospheric Life Cycle* (G. T. Wolff and R. L. Klimisch, eds.), Plenum, New York, N.Y., pp. 297–315.

Wolff, G. T., Korsog, P. E., Kelly, N. A., and Ferman, M. A. 1985a. Relationships between fine particulate species, gaseous pollutants and meteorological parameters in Detroit, *Atmos. Environ.* 19:1341–1349.

Wolff, G. T., Korsog, P. E., Stroup, D. P., Ruthkosky, M. S., and Morrissey, M. L. 1985b. The influence of local and regional sources on the concentration of inhalable particulate matter in southeastern Michigan, *Atmos. Environ.* 19:305–313.

World Health Organization. 1983. Indoor Air Pollutants: Exposure and Health Effects, EURO Reports and Studies 78, Copenhagen, 42 pp.

Yocom, J. E. 1982. Indoor-outdoor air quality relationships: a critical review, *J. Air Pollut. Control Assoc.* 32:500–520.

Mathematical Modeling of the Effect of Emission Sources on Atmospheric Pollutant Concentrations

ARMISTEAD G. RUSSELL
Carnegie Mellon University

Air Pollution, the Automobile, and Public Health. © 1988 by the Health Effects
Institute. National Academy Press, Washington, D.C.

Development of Air Quality Models

When air pollution began to have a significant deleterious effect on human life, it became necessary to discover and understand the links between emission sources and the air quality deterioration and health effects they cause. Only after the impacts of sources have been assessed correctly will it be possible to devise and implement rational, convincing, and effective policies to improve air quality. Over $29 billion were spent in the United States in 1983 on air pollution abatement and control (Council on Environmental Quality 1984). If a fraction of that expense can be saved by better understanding the relation of air quality and health effects to emission sources, the monetary benefits will be tremendous. Knowledge of the relation between emissions by a source and pollutant concentrations in the air at later times and other places (that is, the source/receptor relationship) is essential to calculating the exposure of humans to these pollutants and hence to predicting the health impacts resulting from these source emissions. Mathematical models have evolved as the most practical means to relate source emissions to the subsequent air pollution concentrations.

Mathematical models integrate our knowledge of the chemical and physical processes of pollutant dynamics into a structured framework that can be used to explain the relationship between sources such as motor vehicle exhaust and the resulting impact on human health (figure 1). The multistep process begins with characterizing the emissions. The second step is to accurately determine the effects that atmospheric transport and chemical reactions have on pollutant concentrations. Mathematical models are ideally suited to this task. The next step is to correlate people's activities with pollutant concentrations and determine personal exposure. Exposure is related, through deposition in and absorption by the respiratory tract tissues, to dose. Finally, dose is related to health effects. Central to this process is the ability to accurately calculate the air quality contributions due to specific emission sources.

This chapter reviews the development and current status of air quality models. It differs from previous reviews in emphasizing the use of models in health-related studies. It also assesses the current state of air quality modeling technology. As a logical outcome, gaps in our current understanding are highlighted and research opportunities identified. Chemically reacting pollutant systems receive extra attention for two reasons: first, many of the signifi-

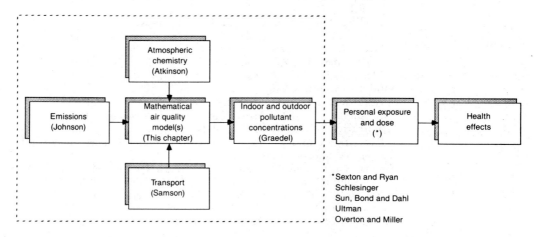

Figure 1. Steps required to link source emissions to health effects. Relevant chapters in this volume are given in parentheses. Central to the process is a mathematical model to predict pollutant concentrations as a function of emissions. Depending on the study, more than one model may be required, for example, to predict indoor pollutant concentrations. Up to this point, mathematical modeling studies have been limited almost exclusively to the steps within the boxed area.

cant components of automotive exhaust are very reactive and contribute to the formation of secondary products that are of as much, or more, concern as the original components; and second, air quality models that include descriptions of atmospheric chemistry are the most thorough and complete and will be the basis for future advanced models. By comparing our present knowledge with current needs, we can identify what these advances are likely to be. This chapter is intended for researchers interested in relating automotive emissions to the resulting health effects, not primarily for specialists in air quality modeling, and is organized to show how mathematical models are useful for providing critical information needed by the health effects community.

Components of Exposure

Human exposure to a pollutant, and its consequent impact on health, results from the simultaneous occurrence of two events—a pollutant concentration $c(x,t)$ at point x and time t, and the presence of people:

$$\text{Exposure} = f[P(x,t), c(x,t)]$$

where $P(x,t)$ represents the number of people at point x and time t inhaling a pollutant at concentration $c(x,t)$. Sexton and Ryan (this volume) explain in detail the three components of personal exposure: magnitude of the concentration, duration, and (if the exposure is a discrete event that recurs) frequency; or, more generally, the magnitude $c(x,t)$ of the concentration as a function of the path of the subject characterized by his or her position x at all times t for the duration of the time interval in which exposure takes place. This chapter discusses how air quality models can be used to determine how $c(x,t)$ depends on emission sources.

Source/Receptor Relationships

The most direct method for observing the effect of a single air pollution source is to eliminate it completely, but complete elimination is usually impractical or impossible.

A more feasible method is needed to predict the impacts of emission sources on air quality. Two distinctly different methods have been developed for making such predictions: mathematical models and physical models. A mathematical air quality model simulates pollutant evolution by interrelating symbolic descriptions of the important physical and chemical processes occurring in the atmosphere within a computational framework. A physical model simulates atmospheric processes with a scaled-down representation of the atmosphere in a laboratory setting. The most common example of a physical model is a smog chamber used to study atmospheric chemistry. Another example is wind tunnel testing using scale models of buildings to observe the transport of pollutants in city street canyons.

Mathematical models have a number of advantages over physical models when the question is to find out how much of each air pollutant at a given location is due to each particular emission source—a process called source apportionment. For example, smog chambers can only be used to study atmospheric chemical reactions in a fixed location and are not suited to simulate the effects of diffusion, changing spatial and temporal emission patterns, pollutant deposition at the ground, and varying meteorological conditions. On the other hand, by accurately describing the dynamics of pollutants as they travel from the many emission sites in a city to a sampling, or receptor, site, a mathematical model can keep track of the separate contributions of the sources of pollutants that influence air quality at a given location. The inputs to the calculation are the pollutant emission rates, and the output is the expected concentrations of the several atmospheric pollutants (figure 2).

Mathematical models used in air pollution analysis fall into two types: empirical/statistical and analytical/deterministic. In the former, the model statistically relates observed air quality data to the accompanying emission patterns, whereas chemistry and meteorology are included only implicitly (Seinfeld 1975). In the latter, analytical expressions describe the complex transport and chemical processes involving air pollutants. The pollutant concentrations

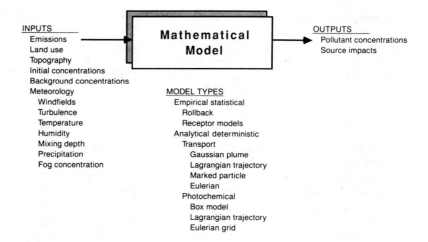

INPUTS
  Emissions
  Land use
  Topography
  Initial concentrations
  Background concentrations
  Meteorology
    Windfields
    Turbulence
    Temperature
    Humidity
    Mixing depth
    Precipitation
    Fog concentration

**Mathematical
Model**

OUTPUTS
  Pollutant concentrations
  Source impacts

MODEL TYPES
  Empirical statistical
    Rollback
    Receptor models
  Analytical deterministic
    Transport
      Gaussian plume
      Lagrangian trajectory
      Marked particle
      Eulerian
    Photochemical
      Box model
      Lagrangian trajectory
      Eulerian grid

Figure 2. Inputs, outputs, and types of models commonly used in air quality modeling studies.

are determined as explicit functions of the meteorology, topography, chemical transformation, and source characteristics, which are inputs to the calculation.

The subject matter of this chapter necessarily overlaps that of other chapters of this book. To minimize duplication, this chapter focuses on how mathematical models are used to predict pollutant concentrations as a function of emissions. Greatest attention is given to pollutants that are either known to be or suspected of being harmful to human health and to modeling on a scale appropriate to urban areas where pollutant concentrations and population densities are highest.

Our discussion begins with a section devoted to understanding the physical and chemical nature of the emissions, for these, in part, determine important characteristics that should be described by a mathematical model. Because of chemical reactions in the atmosphere, the dynamics of some automotive emissions and reaction products depend on the presence of other anthropogenic and natural sources, and it is often insufficient to consider one without the other. After the important emission source types have been identified, it is necessary to choose an appropriate model for each application. The different types of air quality models that are available are reviewed in the next section along with possible ad-

vances that could be made in their structure and application.

The section on modeling approaches presents our current understanding of the various individual physical and chemical processes (for example, transport, chemical reaction, dry deposition) that affect pollutant concentration in the atmosphere. A model's capabilities are determined by the level of detail at which each of the processes is described within the modeling framework. Many future advances in air quality modeling will come from better quantitative descriptions of individual processes, so a number of topics for fruitful research evolve from this section. The theoretical basis and accuracy of the complete model, each of its components, and the structure interrelating the components must be evaluated, as described in the succeeding section.

After a model has been evaluated, it is ready for use in conducting source apportionment, population exposure, and control strategy studies, as discussed in the next section. Studies of this type are of great interest, but few comprehensive control strategy studies have been conducted using state-of-the-art air quality models. Finally, a section addressing special topics and emerging issues in air quality modeling is followed by a summary of research recommendations.

Historical Perspective

The driving force behind the development of mathematical air quality models has been the Clean Air Act (American Meteorological Society 1981). Models have been used to demonstrate compliance with regulatory standards and to guide regulatory agencies toward possible emission control strategies for improving air quality. Air quality models motivated by the Clean Air Act are designed primarily to predict the concentrations of pollutants such as carbon monoxide (CO), nitrogen dioxide (NO_2), and ozone (O_3) that have been regulated by the federal government for many years, but not those of many trace toxic pollutants that are already of growing interest to health effects researchers and are likely to be subject to regulation in the future.

By the early 1970s, analytical models had been developed to the point that it was possible to predict the concentrations of pollutants such as CO that are largely determined by transport but not by atmospheric chemical reaction. The next step was to incorporate atmospheric chemistry into the model to describe the dynamics of pollutants, such as O_3 and NO_2 that are chemically active in the atmosphere (see, for example, Transportation Research Board 1976). By the early 1980s, photochemical airshed models had been developed that could accurately predict O_3 and NO_2 concentrations as a function of emissions. At present, a limiting factor in our ability to describe the dynamics of these two pollutants in an urban area is the availability of high-quality input data, not the model itself.

On the near horizon are models that describe aerosol processes in the atmosphere. So far, modeling studies have concentrated on specific aspects of the many different processes that control the size and composition of particulate matter in the atmosphere. Advances in this area are vital for providing better assessments of health impacts of emission sources.

The past decade has seen rapid development of empirical/statistical air quality models. Most models of the early 1970s assumed that basinwide air quality changed in direct proportion to total basinwide emissions. These "rollback" models were applied to basinwide emissions to predict concentrations of chemically inert as well as chemically reactive pollutants. Rollback models are limited in application because they ignore important effects due to the spatial distribution of emission source changes and atmospheric chemistry. Empirical receptor-oriented models that use the chemical composition of ambient pollution samples as a tracer for pollutant origin were introduced in the 1970s, but were initially applied in only a few cases. Because they accurately resolve source contributions to particulate matter concentrations, receptor models are now widely accepted as a replacement for rollback models.

Although there are still critical aspects of present models that could be improved, it is clearly time for more extensive use of models for explaining relationships between sources and health effects. A particularly pressing issue that can be studied using present models is the relationship between the nitrogen oxide emissions (NO and NO_2 and the sum is commonly symbolized schematically as NO_x) and organic gas emissions in the formation of O_3 (the O_3-precursor relationship—see Pitts et al. 1976; Chock et al. 1983; Pitts et al. 1983). If resources are provided, the next decade should see models that are able to describe the dynamics of aerosols and currently unregulated toxic gases and to resolve many current questions about sources and air quality.

An important but historically underused facet of mathematical models is that they collect and codify what is understood about the constituent processes in a large system such as the atmosphere. In cases where models fail to perform well, they then reveal what is not understood. In this way, evaluation of model performance directs our attention to fruitful problems and topics for further research.

Emission Source Characteristics

The composition of emissions from mobile sources is discussed in detail by Johnson, and atmospheric chemical transformations and transport are covered in chapters by Atkinson and Samson, respectively (all in this volume). It is important to realize that if the air quality model is to be an effective tool for predicting pollutant concentrations and health effects and devising strategies for controlling them, the essential characteristics of the sources must be retained within the model. For example, the dynamic behavior of power plant plumes is very different from that of automotive tail pipe emissions in that plumes are not immediately dispersed by the motion of and turbulence surrounding the source, but rise hundreds of meters because of thermal buoyancy. Likewise, the chemical composition of automotive emissions is quite different from that of power plant emissions. Consequently, it is useful to divide all sources into two categories: mobile and stationary. Most of the total mass of emissions from mobile sources comes from automobiles and trucks, but rail vehicles, ships, aircraft, motorcycles and off-the-road vehicles also make a contribution. Stationary sources are divided further into two classes: anthropogenic and natural emitters.

It is imprudent to neglect stationary sources when characterizing the impact of mobile source emissions. Chemical compounds emitted from stationary sources react extensively with automotive emissions to form various substances in the air. A classic example is the formation of O_3 in urban areas. NO_x emissions (primarily from automobiles, trucks, and stationary source combustion) react with hydrocarbons (HCs) from mobile and stationary sources to form O_3 and other photochemical oxidants (Atkinson, this volume).

Most mobile source emissions are generated by combustion, but other noncombustion releases occur. Significant quantities of HCs come from fuel evaporation, and particulate matter originates from tire wear, brake wear, and road dust. Auto exhaust contains NO, NO_2, CO, organics (commonly referred to as HCs), NH_3, and a variety of particulate species such as aerosol carbon, lead (especially in older vehi-

cles), and bromine. Near the source, the pollutants are rapidly mixed by turbulence generated mechanically from the movement of the automobiles. After initial mixing, the pollutants move away from the road by convection, and are further dispersed by atmospheric turbulence and transport.

Stationary sources, such as power plants and industrial complexes, and natural sources such as forest canopies, emit HCs, NO_x, sulfur oxides (SO_2 and SO_3, commonly called SO_x), NH_3, particulate matter, and CO. Large point sources often emit from tall stacks, and the momentum and buoyancy of the emitted gas can carry the pollutants above the mixed layer, reducing their local impact, but increasing their persistence in the atmosphere over long distances.

Organic compounds and NO_x emissions are both involved in reactions leading to the formation of O_3, NO_2, nitric acid (HNO_3), particulate nitrate (NO_3^-), peroxyacetyl nitrate (PAN), and other oxidized and nitrated organic compounds, and can increase the oxidation rate of sulfur dioxide (SO_2). Some of the compounds formed in the atmosphere by gas-phase reactions involving automotive exhaust compounds are mutagenic and potentially carcinogenic, for example nitroarenes (Pitts and Winer 1984), nitro-polycyclic aromatic hydrocarbons (nitro-PAHs) (Grosjean et al. 1983), and nitroxyperoxyalkyl nitrates and dinitrates (Bandow et al. 1980; Atkinson et al. 1984). Less effort has been devoted to developing mathematical models that will estimate concentrations and source contributions to the formation of these toxic trace species for a number of reasons: these species are not regulated, few data exist to quantify their ambient concentrations, and the chemistry leading to their formation is not completely understood. The necessary data are beginning to be assembled, and the use of mathematical models to study the formation and transport of trace, mutagenic, and carcinogenic organic compounds will become an important activity in the future.

Primary organic particulates, soot (also

called elemental carbon or graphitic carbon), lead, and bromine compounds do not participate extensively in the photochemical reactions but can be affected by gas-phase pollutants. Studies are beginning to elucidate the extent of formation of secondary atmospheric organic particulates and the conversion of compounds from one type to another while in the aerosol phase.

For modeling purposes, there are two distinct types of emissions: unreactive and reactive. Unreactive emissions include CO, lead, soot, and some fraction of the organic particulates. (CO participates in photochemical reactions, but its concentration is determined predominantly by direct CO emissions. Pollutants are referred to as unreactive if reactions do not appreciably affect their concentrations over the time scales being modeled.) Reactive pollutants include HCs, NO_x, and SO_2, which can react to form secondary pollutants such as O_3, PAN, and aerosol sulfates. As will be discussed in the next section, it is often more efficient and sometimes necessary to use different types of mathematical models to describe the dynamics of these two categories of pollutants.

Categories of Air Quality Models

Health effects can arise from exposure to a single pollutant species or from combined actions and interactions of a mixture of compounds the subject is exposed to. The health effects of short-term exposure to high concentrations may not be equivalent to those from longer contact with moderate levels of the pollutant of interest. These alternatives must be reflected in the choice of models used to establish connections between sources and ultimate health effects. First, the pollutants and the time and spatial scales of interest are defined, and then an appropriate model(s) is chosen. Models have been formulated in a number of ways. Each formulation involves certain approximations and has certain strengths and limitations. This chapter shows how models

can be used for relating health effects to sources. Consequently, limitations and strengths are stressed to assist in choosing the most effective models to best utilize the available resources.

If care is not exercised in choosing a model, one of two undesirable outcomes may ensue: a model may be chosen that by its formulation is incapable of doing the job (such as using a nonchemically reactive model to estimate the concentrations of O_3, PAN, and even NO_2), or a model is chosen that is more complex and time-consuming than is necessary (such as a photochemical airshed model to estimate elemental carbon or CO levels in an area heavily impacted by mobile source emissions).

Empirical/Statistical Models

Mathematical air quality models are of one of two types: empirical/statistical or deterministic (figure 2). Empirical/statistical models, such as receptor-oriented and rollback models, are based on establishing a relationship between historically observed air quality and the corresponding emissions. The linear rollback model is simple to use and requires few data, and for those reasons has been widely used (see, for example, Barth 1970; South Coast Air Quality Management District and Southern California Association of Governments 1982). Linear rollback models assume that the highest measured pollutant concentration is proportional to the basinwide emission rate, plus the background value; that is,

$$c_{max} = aE + c_b \qquad (1)$$

where c_{max} is the maximum measured pollutant concentration, E is the emission rate, c_b is the background concentration due to sources outside the modeling region, and a is the constant of proportionality. The constant a accounts for the dispersion, transport, deposition, and chemical reactions of the pollutant. Thus, the allowable emission rate, E_a, necessary to reach a desired ambient air quality goal, c_d, using the linear rollback model can be calculated from

$$\frac{E_a}{E_o} = \frac{c_d - c_b}{c_{max} - c_b} \qquad (2)$$

where E_o is the emission rate that prevailed at the time that c_{max} was observed. Presumably, pollutant concentrations at other times would also decrease toward background levels as emissions are reduced, and similar expressions can be written for relating annual mean concentrations to emission rates. Obviously this is a very simplified approach, and its application is limited. Nonlinear processes such as chemical reactions and spatial or temporal changes in the emission patterns are not accounted for explicitly in the rollback model formulation.

A second class of empirical/statistical models of continuing interest is the receptor-oriented model, used extensively for estimating source contributions to particulate matter concentrations in a number of geographic areas (Friedlander 1973; Heisler et al. 1973; Gartrell and Friedlander 1975; Gatz 1975, 1978; Gordon 1980; Watson et al. 1981; Cass and McRae 1983; Watson 1984; Hopke 1985). Nonreacting gases have also been tracked by receptor modeling methods (Yamartino 1983). Receptor models compare the measured chemical composition of particulate matter concentrations at a receptor site with the chemical composition of emissions from the major sources to identify the source contributions at ambient monitoring sites.

There are three major categories of receptor models: chemical mass balance, multivariate, and microscopic. Hybrid analytical and receptor (or combined source/receptor) models have been proposed and used, but further investigation into their capabilities is required.

Receptor models are powerful tools for source apportionment because of the vast amount of particulate species characterization data routinely collected at many sampling sites within the United States. Most of the information available is for elemental concentrations (for example, lead, nickel, aluminum) although recent measurements are leading to increased data on concentrations of compounds such as ionic species and carbon compounds. At a sampling (or receptor) site, the aerosol mass concentration of each species i is

$$c_i = \sum_{j=1}^{n} a_{ij}S_j \qquad i = 1, 2, \ldots m \qquad (3)$$

where c_i is the mass concentration of species i at the receptor site; S_j is the total mass concentration of all species emitted by source category j as found at the receptor site; a_{ij} is the fraction of the total mass from source j emitted as species i arriving at the sampling site; m is the total number of species measured; and n is the total number of sources. The mass concentration c_i measured at the receptor site of interest and the coefficients a_{ij} that describe the chemical composition for the major sources are the inputs from which S_j, the mass apportioned to source j, is determined. Because a_{ij} characterizes the source, it is referred to as the source fingerprint and should be unique to the source. When the chemical composition of the emissions from two source categories are similar, it is extremely difficult for receptor models to distinguish between the sources. The categories of receptor models are differentiated by the techniques used to determine S_j.

Chemical Mass Balance Methods. Given that the source fingerprints a_{ij} for each of n sources are known, and that the number of sources is less than or equal to the number of measured species ($n \le m$), an estimate for the solution to the system of equations in equation 3 can be obtained. If $m > n$, then the set of equations is overdetermined, and least-squares or linear programming techniques are used to solve for S_j. This is the basis of the chemical mass balance (CMB) method (Miller et al. 1972; Cooper and Watson 1980). If each source emits a particular species unique to it (commonly called a tracer species), then a very simple tracer technique can be used (Friedlander 1977). Examples of tracers commonly used are lead and bromine from mobile sources, nickel from fuel oil, and sodium from sea salt. Often the necessary condition to use the latter method—that each source have a tracer species unique to itself—is not met in practice.

Microscopic identification models are similar to the CMB methods except that more information is included that distinguishes the source of the aerosol. Such chemical or morphological data include particle size and individual particle composition and are often obtained by electron or optical microscopy.

Multivariate Models. Multivariate models, including factor analysis models (Henry and Hidy 1979, 1982; Hopke 1981, 1985), rely on finding the underlying structure of large sets of particulate air quality data in order to determine the sources of the aerosol. Models based on factor analysis are the most widely used. Multivariate models operate by identifying bundles of elements whose concentrations fluctuate together from day to day, implying that these bundles come from a single "source." When the composition of the hypothetical source is compared to the known composition of specific sources, it often becomes obvious what the group of cofluctuating chemical elements stand for. For example, lead and bromine concentrations are usually highly correlated because they are emitted primarily by the same sources (automobiles burning leaded gasoline). Thus, multivariate techniques identify groups of pollutants whose concentrations are correlated, and thus suggest the nature of the source. They do not rely on a detailed knowledge of the source fingerprint, a_{ij}, and can be used to refine estimates of the fingerprint.

Research intended to extend the power of receptor models for source apportionment is continuing, including development of methods to integrate measurement uncertainties into the analysis, incorporation of aerosol properties other than elemental composition, and inclusion of the effect of chemical reactions on secondary aerosol formation. Friedlander (1981) has proposed a method that includes a decay factor in the formulation of equation 3 to take into account the chemical transformation of aerosols such as PAHs. This method is limited to first-order decay and assumes a knowledge of the average pollutant residence time in the atmosphere. A more general technique that can be used to estimate the source contributions to secondary aerosol mass loadings using receptor modeling techniques would be of use.

Attempts to circumvent some of the limitations of receptor models include hybridization with source-oriented models that rely on mass emission rate data from the pollutant sources. Applications of this sort have met with varying success (Gartrell and Friedlander 1975; Pace 1979; Yamartino and Lamich 1979). Yamartino and Lamich used a hybrid model to identify areas with noninventoried emissions of CO. In theory, the source strengths of noninventoried or unknown emitters could be estimated using a hybrid technique, although uncertainty and sensitivity analyses need to be conducted on this type of model. Pace used a microinventory approach, assuming that most of the aerosol mass at a receptor is derived from nearby emitters, and was able to account for total suspended particulate concentrations (TSP) with a standard error of 17 percent. Note that hybrid models require additional data (that is, source strengths and meteorological data), but the prospects of added accuracy can justify the added effort. Hybrid models potentially could account for the secondary aerosols present in source apportionment studies. Further development and use of hybrid models is clearly warranted, since they potentially retain the strength of receptor-oriented as well as source-oriented (analytical) models.

■ **Recommendation 1.** Research should continue on the development of receptor models, especially on the hybridization of these models with other types of models. The inclusion of aerosol properties and formation should also be pursued.

Deterministic Models

Deterministic air quality models describe in a fundamental manner the individual processes that affect the evolution of pollutant concentrations. These models are based on solving the atmospheric diffusion/reaction equation, which is in essence the conservation-of-mass principle for each pollutant species (Lamb and Seinfeld 1973):

$$\frac{\delta c_i}{\delta t} + \overline{U} \cdot \nabla c_i = \nabla \cdot D_i \nabla c_i$$

$$+ R_i(c_1, c_2, c_3, \ldots c_n)$$

$$+ S_i(\overline{x}, t) \qquad i = 1, 2, 3, \ldots, n \qquad (4)$$

where c_i is the concentration of species i; $\overline{U}$ is the wind velocity vector; D_i is the molecular diffusivity of species i; R_i is the net production (depletion if negative) of species i by chemical reaction; S_i is the emission rate of i from sources; and n is the number of species. R can also be a function of the meteorological variables. In essence, this equation states that the time rate of change of a pollutant (term 1) depends on convective transport (2), diffusion (3), chemical reactions (4), and emissions (5). As discussed in the chapter on pollutant transport (Samson, this volume), the closure problem makes it necessary to approximate this equation, usually by K-theory (Lamb 1973):

$$\frac{\delta \langle c_i \rangle}{\delta t} + \langle \overline{U} \rangle \cdot \nabla \langle c_i \rangle = \nabla \cdot K \nabla \langle c_i \rangle$$

$$+ R_i(\langle c_1 \rangle, \langle c_2 \rangle, \ldots \langle c_n \rangle)$$

$$+ \langle S_i(x, t) \rangle \qquad i = 1, 2, 3, \ldots, n \qquad (5)$$

where the braces $\langle \rangle$ indicate an ensemble average, and K is the turbulent (eddy) diffusivity tensor. Pollutant dry deposition and ground level emissions enter the system as boundary conditions. Except for the simplest source distributions and chemical reaction mechanisms, $\langle S_i \rangle$ and R, there are no analytical solutions to equation 5. If equation 5 can be simplified for a particular application, it is usually advantageous to do so.

An examination of equation 5 shows that if there are no chemical reactions, $(R = 0)$, or if R is linear in $\langle c_i \rangle$ and uncoupled, then equation 5 forms a set of linear, uncoupled differential equations for determining the pollutant concentrations. This is the basis

of the transport only and transport with linear chemistry models (which, for brevity, will be called transport models). Transport models are suitable for studying the effects of CO sources and primary particulate emissions sources on air quality, but not for studying reactive pollutants such as O_3, NO_2, HNO_3, and secondary organic species. Transport of nonreactive pollutants is described in detail by Samson (this volume) and will be discussed here only briefly.

Lagrangian Models. There are two distinct reference frames from which to view pollutant dynamics. The most natural is the Eulerian coordinate system which is fixed at the earth's surface. In that case, a succession of different air parcels are viewed as being carried by the wind past an observer who is fixed to the earth's surface. The second is the Lagrangian reference frame in which the frame of reference moves with the flow of air, in effect maintaining the observer in contact with the same air parcel over extended periods of time. Because pollutants are carried by the wind, it is often convenient to follow pollutant evolution in a Lagrangian reference frame, and this perspective forms the basis of Lagrangian trajectory and Lagrangian marked-particle or particle-in-cell models. In a Lagrangian marked-particle model, the center of mass of parcels of emissions are followed, traveling at the local wind velocity, while diffusion about that center of mass is simulated by an additional random translation corresponding to the atmospheric diffusion rate (Lamb and Neiburger 1971; Cass 1981).

Lagrangian trajectory models can be viewed as following a column of air as it is advected in the air basin at the local wind velocity. Simultaneously, the model describes the vertical diffusion of pollutants, deposition, and emissions into the air parcel (figure 3). The underlying equation being solved is a simplification of equation 5:

$$\frac{\delta c_i}{\delta t} = \frac{\delta}{\delta z} K_{zz} \frac{\delta c_i}{\delta z} + S_i(t) + R(\overline{c}, t) \qquad (6)$$

Trajectory models require spatially and temporally resolved wind fields, mixing-

A

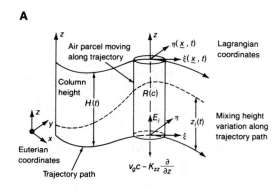

B

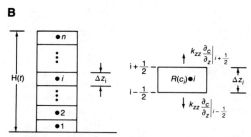

Figure 3. Schematic diagram of a Lagrangian trajectory model: (A) The column of air being modeled is advected at the local wind velocity along a trajectory path across the modeling region. Within the moving air parcel, the model describes the important processes affecting the pollutant (i) evolution and concentration (c) such as chemical reactions (R), deposition (V_g), emissions (E), and vertical diffusion (K_{zz}). (B) Vertical resolution is gained by dividing the column into a number of cells in the vertical direction.

height fields, deposition parameters, and data on the spatial distribution of emissions. Lagrangian trajectory models assume that vertical wind shear and horizontal diffusion are negligible. Other limitations of trajectory and Eulerian models are discussed by Liu and Seinfeld (1975).

Gaussian Plume Model. One of the basic and more widely used transport models based on equation 5 is the Gaussian plume model (figure 4). Gaussian plume models for continuous sources can be obtained from statistical arguments or can be derived by solving:

$$\overline{U}\frac{\delta c}{\delta x} = K_{yy}\frac{\delta^2 c}{\delta y^2} + K_{zz}\frac{\delta^2 c}{\delta z^2} \quad (7)$$

where $\overline{U}$ is the temporally and vertically averaged wind velocity; x, y, and z are the

distances in the downwind, crosswind, and vertical directions, respectively; and K_{yy} and K_{zz} are the horizontal and vertical turbulent diffusivities, respectively. For a source with an effective height H, with emission rate Q, and a reflecting (nonabsorbing) boundary at the ground, the solution is:

$$c(x, y, z) = \frac{Q}{2\pi\overline{U}\sigma_y(x)\sigma_z(x)}\exp\left[\frac{-y^2}{2\sigma_y{}^2(x)}\right]$$

$$\cdot\left[\exp\frac{-(z - H)^2}{2\sigma_z{}^2(x)} + \exp\frac{-(z + H)^2}{2\sigma_z{}^2(x)}\right] \quad (8)$$

This solution describes a plume with a Gaussian distribution of pollutant concentrations, where $\sigma_y(x)$ and $\sigma_z(x)$ are the standard deviations of the mean concentration in the y and z directions (figure 3). The standard deviations are the directional diffusion parameters, and are assumed to be related simply to the turbulent diffusivities, K_{yy} and K_{zz}. In practice, $\sigma_y(x)$ and $\sigma_z(x)$ are functions of x, $\overline{U}$, and the atmospheric stability as discussed by Samson (this volume), Gifford (1961), and Turner (1964, 1967).

Gaussian plume models are easy to use,

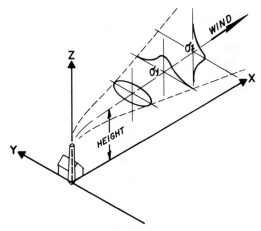

Figure 4. Diffusion of pollutants from a point source. Pollutant concentrations have separate Gaussian distributions in both the horizontal (y) and vertical (z) directions. The spread is parameterized by the standard deviations (σ) which are related to the diffusivity (K).

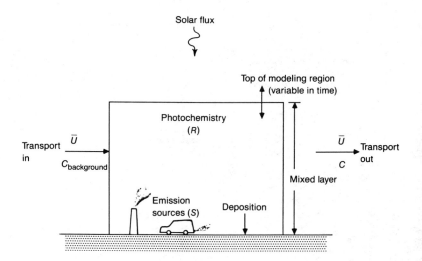

Figure 5. Schematic representation of a box model based on the conservation-of-mass equation. The stationary box allows pollutants to be advected into and out of the modeled region. The height of the modeling region can increase, accounting for an increase in the height of the mixed layer.

require relatively few input data, and are very quick computationally. Multiple sources are treated by superimposing the calculated contributions of individual sources to ambient concentrations at a given receptor site. It is possible to include the first-order chemical decay of pollutant species within the Gaussian plume framework. For chemically more complex situations, however, the Gaussian plume model simply fails to provide an acceptable solution. Because of its simplicity and because of its use by regulatory agents, the search for improvements to Gaussian plume models is still an active area of research.

Eulerian Models. Of the Eulerian models, the box model is the easiest to envision conceptually. Simply, the atmosphere over the modeling region is perceived as a well-mixed box, and the evolution of pollutants in the box is calculated following conservation-of-mass principles including emissions, deposition, chemical reactions, and a changing mixing (or inversion-base) height (figure 5).

Eulerian "grid" models are the most complex, but potentially the most powerful, air quality models, involving the least-restrictive assumptions, and are the most

computationally intensive. Grid models attempt to solve a finite approximation to equation 5, including temporal and spatial variation of the meteorological parameters, emission sources, and surface characteristics. Grid models divide the modeling region into a large number of cells, horizontally and vertically, that interact with each other by simulating diffusion, advection, and sedimentation (for particles) of pollutant species. Input data requirements for grid models are similar to those for Lagrangian trajectory models but, in addition, require data on background concentrations (boundary conditions) at the edges of the grid system used. Eulerian grid models produce pollutant concentration predictions throughout the entire airshed, which can be examined over successive time periods to observe the evolution of pollutant concentrations and how they are affected by transport and chemical reaction.

Modeling Chemically Reactive Compounds. A number of compounds, regulated as well as unregulated, are formed in the atmosphere by a series of complex, nonlinear chemical reactions. Often the compounds formed are more harmful than their precursors. In this case it is necessary to use models that not only describe pollutant transport, but also complex chemical transformation, $R(\bar{c},t)$ in equation 4. Examples of secondary pollutants are O_3, PAN, HNO_3, and many aerosols. Such models are also required to study the dynamics of chemically reactive primary pollutants such as benzene, and pollutants that are primary as well as secondary in origin, for example, NO_2 and formaldehyde. Addition of the capability to describe a series of interconnected chemical reactions greatly increases the computational requirements for computer storage as well as for time, and also the input data requirements. The increased computational demands arise because the evolution of some species must be followed simultaneously. One major difficulty encountered when numerically calculating the change of pollutant concentrations due to chemical reaction is that the characteristic lifetimes of the different pollutants are distributed over many orders of magnitude. Such systems are said to be computationally "stiff" and are generally time-consuming to solve. A suitable numerical solution scheme must be chosen when confronted by a stiff system. Some simplifications and procedures, described below, have been devised to help reduce the computational time, but the required computational time is still a deterrent to the widespread use of photochemical air quality models. Another major difficulty is that accurate, speciated emissions inventories for each of the many reactive air pollutants are needed. Such detailed emission inventories have been developed for only a few geographic areas, most notably Los Angeles, California (figure 6), and a chemically detailed regional inventory for the eastern United States.

Box, Lagrangian trajectory, and Eulerian grid models can be developed to include nonlinear chemical reactions. Box models, the first candidate, assume that the pollutants are mixed homogeneously within the modeling region, an assumption that is often inappropriate. Trajectory and grid models resolve pollutant dynamics on a much finer scale and have been used widely and with considerable success (Reynolds et al. 1973; Lloyd et al. 1979; Reynolds et al. 1979; Seinfeld and McRae 1979; Chock et

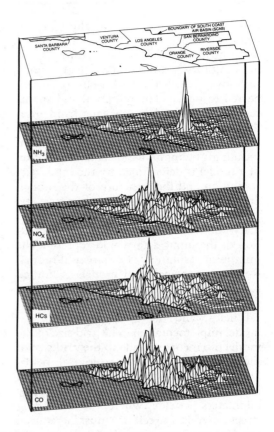

Figure 6. Emissions of NH_3, NO_x, HCs, and CO in the Los Angeles area during 1982. A spatially gridded, time-resolved, and speciated emissions inventory is necessary for conducting air quality modeling studies involving chemically reacting compounds. (Based on data from McRae 1981.)

al. 1981; Carmichael et al. 1986; Russell and Cass 1986). Chemically reacting models have received much attention because they are being used to plan air quality control programs in areas with photochemical smog problems and to study acid deposition. They will also provide the key to predicting (and hence controlling) the formation and dynamics of secondary aerosols and trace, but potentially harmful, gases in the atmosphere, such as PAN, HNO_3, and nitrous acid (HNO_2).

Temporal and Spatial Resolution of Empirical and Analytical Models

Short-term contact with high pollutant concentrations as well as chronic exposure to lower concentrations can affect health, and the effects can be different. The choice of air quality models to be used for assessing health risks should reflect the temporal scale over which the health effects are expected to occur. The temporal and spatial resolution of models can vary from minutes to a year and from several meters to hundreds of kilometers. The minimum meaningful temporal and spatial resolution of a model is determined by the input data resolution and the structure of the model. Statistical models generally rely on several years' worth of measurements of hourly or daily pollutant concentrations. The resolution of the input data would represent the minimum resolution of a statistical model. Resolution of analytical models is limited by the spatial and temporal resolution of the emissions inventory, the meteorological fields, and the grid size chosen for model implementation. The grid size of the model often corresponds to the grid size of the inventory and meteorological fields. For modeling urban air basins, the size of individual grid cells is on the order of a few kilometers per side, whereas for modeling street canyons, the cell size must be reduced to a few meters on each edge. The temporal resolution of urban models ranges from about 15 minutes to a few hours or days. Multiple time intervals can be combined to form pollutant concentration predictions for longer periods of time.

More than one model may be appropriate, if not necessary, for the analysis of a given problem. Choice of models will be influenced by available resources (time, computational facilities, and funds). A stepped approach is suggested, starting with simpler models (that is, Gaussian plume, rollback, or box models) for approximations, and building up to more sophisticated model formulations when greater precision is needed.

Modeling Approaches for Individual Processes

In general, models described in the previous section simply provide a framework for combining theoretical descriptions of individual physical and chemical processes. The model's ability to correctly predict pollutant dynamics and to apportion source contributions depends on the accuracy of the individual process descriptions, the accuracy of the input data, and the fidelity with which the framework reflects the true interactions of the processes.

Analytical models are composed of modules describing (depending on model type) pollutant transport, diffusion, chemical reactions, deposition and emissions, aerosol dynamics, and heterogeneous (for example, gas/aerosol) interactions. Problem areas in each of these process descriptions are discussed below. Transport-related processes, advection and diffusion, are described by Samson (this volume), and will be discussed here briefly from a computational viewpoint.

Turbulent Transport and Diffusion

Numerical schemes developed to calculate the rate of transport of pollutants suffer from numerical diffusion and dispersion (Roache 1976). Numerical diffusion and dispersion result from using a discrete approximation to the governing system of equations, and are manifested by the computed solution being artificially spread out and ripples being formed. Numerous numerical schemes have been developed to minimize the errors induced, including higher-order finite-difference, finite-element, particle-in-cell, filtered, and spectral methods. In reviewing the use of different advection routines for solving the atmospheric diffusion equation 5, McRae et al. (1982c), Chock and Dunker (1983), and Schere (1983) compare accuracy and computational requirements.

Closure of the atmospheric diffusion equation 5, can be accomplished by utilizing the K-theory, or gradient/diffusion, hypothesis (see Samson, this volume). K-theory is used to describe pollutant fluxes on scales smaller than the size resolvable by conventional wind velocity measurements, thus representing the many processes involved in turbulent diffusion. An obvious need when applying this theoretical treatment within an air quality model is some algorithm for establishing the value of the eddy diffusivity tensor, K. As a result of the large variety of processes involved, there are also a number of methods to parameterize the horizontal and vertical diffusion coefficients (Yu 1977). The usual limitation to the accuracy of diffusion calculations in a practical application is determined by the extent of measurements on the atmospheric structure taken during the period to be simulated. For most model applications, such as source apportionment studies, the number of observed factors relating to atmospheric turbulence are few and include only ground-level winds and temperatures, surface roughness, and cloud cover. At a few locations and times the inversion base (or mixing height), wind speeds aloft, and vertical temperature gradient may also be known. As the amount and accuracy of information characterizing atmospheric structure increases, confidence in model predictions of dispersion increases.

Complex Terrain: Street Canyons

Complex terrain represents an obstacle to modeling the transport of pollutants because large variations in the wind velocity occur over distances smaller than can be resolved by the wind sampling network. Classic examples are valleys and street canyons where the buildup of pollutants can be substantial. In urban areas, build-up of CO in street canyons is of interest and has been addressed by a number of authors (for example, Johnson et al. 1973), and transport in street canyons is discussed by Samson (this volume). These studies did not address the effect of chemical reactions on pollutant concentrations.

In regions subject to photochemical smog, modeling the transport and distribution of O_3 and the impact of automotive NO_x emissions in street canyons needs to be addressed for two reasons: to determine population exposure to these pollutants, and to explain the difference between predicted pollutant concentrations calculated when using a grid size much larger than the size of a street canyon and observed concentrations measured by air monitoring

stations that may be located within the influence of the street canyon (Nappo et al. 1982). As an example, air quality models now in use for studying the formation and transport of O_3 and NO_2 use grid sizes of about 5 km square, compared to a street canyon width of a few tens of meters.

■ **Recommendation 2.** Chemical interactions, especially of reactive pollutants, need to be included in street canyon models.

Removal Processes

Removal processes, particularly dry deposition and scavenging by rain and clouds, are a major factor in determining the dynamics and ultimate fate of pollutants in the atmosphere. (See also, Atkinson, this volume.) The potential for health and environmental impacts is thus closely tied to the physical processes removing pollutants from the atmosphere.

Dry Deposition. Dry deposition occurs in two steps: the transport of pollutants to the earth's surface, and the physical and chemical interaction between the surface and the pollutant. The first is a fluid mechanical process, the second is primarily a chemical process, and neither is completely characterized at the present time. The problem is confounded by the interaction between the pollutants and biogenic surfaces where pollutant uptake is enhanced or retarded by plant activity that varies with time (Hicks and Wesely 1981; Hicks et al. 1983). It is very difficult to measure the depositional flux of pollutants from the atmosphere, though significant advances have been made in the last 10 years. Accurate mathematical description of the depositional process has, as a result, been advancing rapidly over the same time span.

Many factors affect dry deposition, but for computational convenience air quality models resort to using a single quantity called the deposition velocity, v_d, to prescribe the deposition rate. The deposition velocity is defined such that the flux F_i of species i to the ground is

$$F_i = v_d c_i(z_r) \qquad (9)$$

where $c_i(z_r)$ is the concentration of species i at some reference height z_r, typically one to several meters. For a number of pollutants, v_d has been measured under various meteorological conditions and for a number of surface types. A basic problem with this parameterization is that it does not explicitly represent dry deposition as a complex linkage between turbulent diffusion in the surface boundary layer, molecular diffusion very near the surface, chemical reaction, and plant activity.

Early models used a value for v_d that remained constant throughout the day. However, measurements show that the deposition velocity increases during the day as the surface heating increases atmospheric turbulence and hence diffusion, and plant stomatal activity increases (Whelpdale and Shaw 1974; Wesely and Hicks 1977; Wesely et al. 1985). More recent models take this variation of v_d into account. In one approach, the first step is to estimate the upper limit for v_d in terms of the transport processes alone. This value is then modified to account for surface interaction, since the earth's surface is not a perfect sink for all pollutants. This has led to what is referred to as the resistance model (Wesely and Hicks 1977; Fowler 1978) that represents v_d as the analog of an electrical conductance

$$v_d = (r_a + r_b + r_s)^{-1} \qquad (10)$$

where r_a is the aerodynamic resistance controlled primarily by atmospheric turbulence, r_b is the resistance to transport in the fluid sublayer very near the plant surface, and r_s is the surface (or canopy) resistance (figure 7). Of the three resistances, r_a is essentially the same for all species, r_b is the same for gaseous species with the same diffusivities, though it can be considerably greater for aerosols, and r_s depends greatly on the surface affinity for the diffusing species. For example, HNO_3, which reacts rapidly with most surfaces, has a very low surface resistance, usually taken as zero (Huebert 1983; Huebert and Robert 1985; Walcek et al. 1986), whereas CO is not very reactive and has a high r_s. More recent models account for the variation of surface

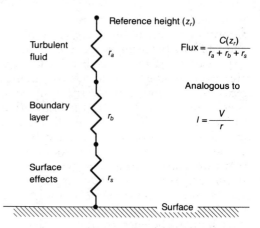

Figure 7. The resistance model of deposition showing the three regions over which deposition is depicted to take place. The total resistance to deposition is the sum of the three and is analogous to an electrical system of series resistors.

resistance and diurnal change in fluid mechanical transport. These parameterizations have been used to quantify the deposition flux of various compounds (McRae and Russell 1983; Walcek et al. 1986).

Less attention has been devoted to studying the deposition of aerosols and how to effectively model their rate of deposition. Major differences between the deposition of gases and aerosols are that aerosols have a much lower diffusivity, the rate of gravitational settling can be significant for larger particles, and the surface resistance for aerosols is not determined by species reactivity. Particulate deposition velocities have been measured for a number of species, leading to parameterization of deposition velocities (Liu and Ilori 1974; Sehmel and Hodgson 1974; Hicks 1977; Slinn and Slinn 1981; Wesely et al. 1985). More fundamental work has been conducted for deposition to smooth surfaces (Sehmel 1971, 1980; Reeks and Skyrme 1976), and should be expanded to nonideal surfaces.

It is important to better understand the processes leading to the deposition of atmospheric aerosols, so that the concentrations of these aerosols can be properly estimated and the related health effects assessed. Research in this area should follow two paths: experimental measurements of aerosol deposition in the environment, especially actual aerosol velocities near surfaces; and modeling and parameterization of the fundamental physical processes. Results from these studies can be used in refining models for the apportionment of aerosol contributions between different source types and may aid the improvement of models for aerosol deposition in human lungs.

■ **Recommendation 3.** Better characterization of the processes leading to dry deposition of chemically reactive pollutants and aerosols is needed.

Scavenging by Rain, Fog, and Clouds. Wet removal, or precipitation scavenging, can be effective in cleansing the atmosphere of pollutants, and depends on the intensity and size of the raindrops (Martin 1984). Fog and cloud droplets can also absorb gases, capture particles, and promote chemical reactions (Adewuyi and Carmichael 1982; Chameides and Davis 1982; Levine and Schwartz 1982; Munger et al. 1983; Graedel 1984; Kumar 1985). Current research into these processes is concentrating on more fundamental descriptions of the absorption of pollutants by droplets and chemical dynamics, taking into account the species solubility, reactivity, and the fluid mechanics of a falling drop (Schwartz and Frieberg 1981; Drewes and Hale 1982; Jacob and Hoffmann 1983; Jacob 1985). Precipitation scavenging is not as important on an urban scale as on a regional scale and is not included in most urban-scale models. Fog chemistry can be important to human health on an urban scale, as evidenced in London in 1952 when thousands of persons died during an episode of excess industrial air pollution and fog. (Seinfeld 1986); however, no attempt has been made to model the relationship between pollutant emissions and fog chemical dynamics in an urban area.

■ **Recommendation 4.** Research into the development of "emissions-to-fog chemistry" models is needed and would be valuable for determining source/health effects relationships in the instances where fog in urban areas may lead to compounds harmful to human health.

Representation of Atmospheric Chemistry Through Chemical Mechanisms

A complete description of atmospheric chemistry within an air quality model would require tracking the dynamics of many hundreds of compounds through thousands of chemical reactions. Many of these compounds affect human health. Atkinson (this volume) gives an account of the number and complexity of the interactions taking place and provides insight into how much is known (and unknown) about the rate and products of these reactions (see also Atkinson and Lloyd 1984). There are so many reactive species, particularly organic compounds, and reactions, that it is infeasible to incorporate an explicit statement of all reactions for each species within the chemical mechanism used by urban air pollution models, even if atmospheric chemistry were completely understood. Fortunately it is not necessary to follow every compound. Instead, a compact representation of the atmospheric chemistry, commonly called a chemical mechanism, is used. The chemical mechanism represents a compromise between using an exhaustive description of the chemistry and being computationally tractable. It is the principal method for modeling the dynamics of reactive compounds such as O_3, NO_2, hydroxy radicals, and PAN in air quality models. The level of chemical detail is balanced against the computational time that increases as the number of species and reactions increase. Instead of the hundreds of species present, chemical mechanisms use on the order of 50 species and about 100 reactions to accurately describe the principal features of atmospheric chemistry.

Three different types of chemical mechanisms have evolved in an attempt to simplify the HC (organic) chemistry: surrogate (Graedel et al. 1976; Dodge 1977), lumped (Falls and Seinfeld 1978; Atkinson et al. 1982), and carbon bond (Whitten et al. 1979; Killus and Whitten 1982). These mechanisms were developed primarily to study the formation of O_3 and NO_2 in photochemical smog but can be extended

Table 1.   Example of Lumping Alkane-OH Reactions

Explicit Reactions of Alkanes with OH:

$$C_2H_6 + OH \rightarrow C_2H_5^{\cdot} + H_2O$$
$$C_3H_8 + OH \rightarrow C_3H_7^{\cdot} + H_2O$$
$$\cdot$$
$$\cdot$$
$$\cdot$$
$$C_nH_{2n+2}^{\cdot} + OH \rightarrow C_nH_{2n+1}^{\cdot} + H_2O$$

$\left.\begin{array}{c} \\ \\ \\ \\ \\ \\ \end{array}\right\}$ Initial Reaction with OH

$$C_2H_5^{\cdot} + O_2 \rightarrow C_2H_5O_2^{\cdot}$$
$$C_3H_7^{\cdot} + O_2 \rightarrow C_3H_7O_2^{\cdot}$$
$$\cdot$$
$$\cdot$$
$$\cdot$$
$$C_nH_{2n+1}^{\cdot} + O_2 \rightarrow C_nH_{2n+1}O_2^{\cdot}$$

$\left.\begin{array}{c} \\ \\ \\ \\ \\ \\ \end{array}\right\}$ Alkyl Radical Oxidation Reaction

Lumped Representation:

$$\text{Alkane} + OH \rightarrow RO_2^{\cdot}$$

to compute the concentrations of other pollutants believed to be noxious.

Surrogate mechanisms use the chemistry of one or two compounds in each class of organics to represent the chemistry of all the species in that class; for example, the explicit chemistry of butane might be used to describe the chemistry for all the alkanes.

Lumped mechanisms are based on the grouping of chemical compounds into classes of similar structure and reactivity; for example, all alkanes are lumped into a single class whose reaction rates and products are based on a weighted average of the properties of all the alkanes present. Only the dominant chemical features and reactions of each lumped class are used to describe reaction steps. By taking advantage of the common features of the organics and free radicals, lumping allows one to greatly reduce the number of required species and steps needed to accurately describe the prevailing pollutant chemistry. For example, as illustrated in table 1, the various alkanes (C_nH_{2n+2}) react with OH in a similar manner to form alkyl radicals (C_nH_{2n+1}). The alkyl radical then reacts rapidly with O_2 to form an alkyl peroxy radical ($C_nH_{2n+1}O_2^{\cdot}$). (See Atkinson, this volume.) When expressed explicitly, this involves over 30 species and 20 reactions. This would lead to a mechanism too large

to be used in an air quality model. By lumping, the series of reactions can be reduced to one, and the number of required organic compounds is reduced to two. This is a tremendous savings in computational time while maintaining the necessary chemical detail. The carbon bond mechanism, a variation of a lumped mechanism, splits each organic molecule into functional groups using the assumption that the reactivity of the molecule is dominated by the chemistry of each functional group.

Leone and Seinfeld (1985) analyzed the performance of six chemical mechanisms by comparing, quantitatively, why they behave differently under identical conditions. They were able to identify critical areas that, when improved, would bring the predictions of each mechanism into much closer agreement. This analytical technique is suited to developing and testing new chemical mechanisms.

Given the importance of the chemical mechanism to the outcome of model evaluation, source apportionment, and control strategy studies, it is bothersome that the predictions of different chemical mechanisms do not always agree. Shafer and Seinfeld (1986) compared $NO_x/HC/O_3$ relationships, and the sensitivity of six chemical mechanisms to initial and boundary conditions. They found that the predicted

HC control needed to reduce O_3 concentrations from 0.4 to 0.12 parts per million (ppm) varied among the six mechanisms as did the sensitivities to perturbations in boundary and initial conditions. The effect of chemical mechanisms on model predictions, particularly the $NO_x/HC/O_3$ relationship, should be studied further. Different mechanisms should be embedded within a complete airshed model and the results compared.

New, or at least modified, reaction mechanisms will be required as the knowledge of atmospheric chemistry increases and as attention is turned toward less abundant, but potentially harmful, trace gases. For example, the chemistry of dinitrogen pentoxide (N_2O_5) and the NO_3 radical is just unfolding (Graham and Johnston 1978; Atkinson et al. 1984; Winer et al. 1984; Russell et al. 1985; Johnston et al. 1986), as are the reactions leading to nitroarenes, which are strong mutagens (Pitts and Winer 1984) and other organic compounds.

■ **Recommendation 5.** Research into the development of new chemical mechanisms is essential to advancing the accuracy and scope of air quality model predictions, especially as interest grows in the effects of noncriteria pollutants.

Aerosol Dynamics

Inclusion of a description of aerosol dynamics within air quality models is of primary importance because of the health effects associated with fine particles in the atmosphere (Schlesinger, as well as Sun, Bond, and Dahl, this volume), visibility deterioration, and the acid deposition problem. Although the effects of aerosols on health are not fully understood, it is known that aerosols can contain strongly mutagenic and toxic compounds such as PAHs, nitro-PAHs, and lead. Aerosol dynamics differ markedly from gaseous pollutant dynamics in that particles come in a continuous distribution of sizes and can coagulate, evaporate, grow in size by condensation, be formed by nucleation, or sediment out. Furthermore, the species mass concentra-

tion alone does not fully characterize the aerosol. The particle size distribution (which changes with time) and composition determine the fate of particulate air pollutants and their environmental and health effects. Particles of about 1 μm or smaller in diameter penetrate the lung most deeply and represent a substantial fraction of the total aerosol mass. The origin of these fine particles is difficult to identify because much of that fine particle mass is formed by gas–phase reaction and condensation in the atmosphere (figure 8).

Simulation of aerosol processes within an air quality model begins with the fundamental equation of aerosol dynamics which describes aerosol transport (term 2), growth (term 3), coagulation (terms 4 and 5), and sedimentation (Friedlander 1977):

$$\frac{\delta n}{\delta t} + \nabla \cdot \overline{U} n + \frac{\delta I}{\delta v} =$$

$$\frac{1}{2} \int_0^v \beta(\overline{v}, v - \overline{v}) n(\overline{v}) n(v - \overline{v}) d\overline{v}$$

$$- \int_0^\infty \beta(\overline{v}, v) n(\overline{v}) n(v) d\overline{v} - \nabla \cdot Cn \qquad (11)$$

where n is the particle size distribution function; $\overline{U}$ is the fluid velocity; I is the droplet current that describes particle growth and nucleation due to gas-to-particle conversion; v is the particle volume; β is the rate of particle coagulation; and C is the sedimentation velocity. The chemical composition of the aerosol also changes with size. Gelbard and Seinfeld (1980) present a framework for modeling the formation and growth of aerosols by sectioning the size distribution, n, into discrete ranges. Their sectional model can then follow the size and chemical composition of an aerosol as it evolves by condensation, coagulation, sedimentation, and nucleation. However, application of these methods to a simulation of the formation, growth, and transport of all the components of an urban aerosol from their emission sources using a fundamental description of the aerosol dynamics and chemistry has yet to be com-

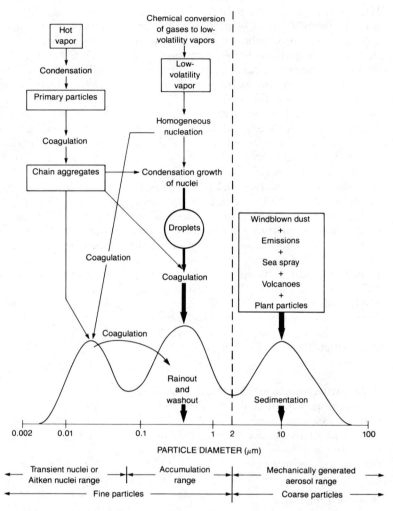

Figure 8. Example of size distribution of an urban aerosol showing the three modes containing much of the aerosol mass. The fine mode contains particles produced by condensation of low-volatility gases. The mid-range, or accumulation mode, results from coagulation of smaller aerosols and condensation of gases on preexisting particles. The largest aerosols are usually generated mechanically. The various processes leading to the different sizes ensure that aerosol composition will change with size. (Adapted with permission from Whitby and Cantrell 1976, and the Institute of Electrical and Electronic Engineers.)

pleted. Instead, as a first step, investigators have chosen to model the major individual aerosol components such as sulfates (Cass 1981; Carmichael et al. 1986), nitrates (Russell et al. 1983; Russell and Cass 1986), and carbonaceous aerosol (Gray 1986). These studies predicted aerosol mass and chemical composition, not the aerosol size distribution (that is, how the aerosol is distributed over specific size ranges). A primary reason

for not proceeding with the size-resolved calculation is the lack of adequate data for input and verification.

Middleton and Brock (1977) attempted to model the evolution of the aerosol size distribution and mass in Denver, Colorado, using as input a parameterized rate of condensable aerosol formation along with an inventory for primary aerosol emissions. They concluded that part of the disagreement between predictions and observations was due to errors in the aerosol emissions inventory. This problem is universal and will hinder any attempt to perform a full simulation of atmospheric aerosol dynamics. Construction of an accurate inventory of aerosol emissions will be an arduous task, although adoption of standards for particulate matter less than 10 μm in diameter (PM-10) should hasten inventory development, making it possible to conduct more accurate modeling studies.

The chemistry leading to the formation of some secondary organic aerosols has been clarified recently (Grosjean and Friedlander 1980; Grosjean 1984, 1985; Hatakeyama et al. 1985) to the point that it is now feasible to conduct preliminary modeling studies of secondary organic aerosol formation in the atmosphere. More research is required before it is possible to predict the formation of all the secondary organic particulates. Once procedures for modeling secondary organic aerosol formation have been developed and accurate field data become available, it should be possible to construct size-resolved and chemically resolved modeling programs for use in health effects, control strategy, and source apportionment work. The development of aerosol process models will be a very important area of research over the next decade.

Heterogeneous gas/aerosol interactions, such as the reaction between HNO_3 and sea salt (Duce 1969),

$$HNO_3(gas) + NaCl(solid \text{ or aqueous}) \rightarrow \\ HCl(gas) + NaNO_3(solid \text{ or aqueous})$$

have been included in very few modeling studies to date (Russell and Cass 1986). Pitts and Winer (1984) present evidence for heterogeneous reactions leading to the formation of very mutagenic, and possibly carcinogenic, nitro-PAHs (see also Atkinson, this volume). Study of gas/aerosol reaction rates under controlled laboratory conditions has been attempted in a few cases (Baldwin and Golden 1979; Jech et al. 1982). Model calculations by Chameides and Davis (1982) indicate that the presence of aerosols can affect concentrations of gaseous species. Dahneke (1983) presents an expression that can be used to estimate the reaction rate between aerosols and gases, given experimental measurements that characterize the fraction of collisions occurring between gases and aerosols that result in reaction. Additional research into methods for incorporating chemical reactions at aerosol surfaces into chemical mechanisms is warranted.

Development of aerosol process models incorporating gas-to-particle conversion of harmful compounds, heterogeneous reactions, and particle growth is perhaps the most critical research area for advancing air quality models to clarify relationships between sources and health effects. Given increased field data, our current understanding of processes governing the production and growth of aerosols is such that major advances in the use of aerosol process models should be realized in the next few years.

■ **Recommendation 6.** Size-resolved and chemically resolved measurements of atmospheric aerosols are needed to test and further develop aerosol process models.

Model Evaluation

An air quality model must be tested before it can be used confidently for a specific application, such as control strategy design or source apportionment. Confidence in model predictions is vital because of the large cost of implementing policy decisions based on them and because of the importance of the health and other effects that are influenced by the implementation of those policies. Model evaluation studies should determine the range of circumstances over which the model will perform adequately along with the accuracy of the inputs required to implement the model and, if possible, should identify and quantify the reasons for differences between predictions and observations, although this is often impractical, or impossible, because of uncertainties in input data.

There are three reasons why model predictions may not agree with observations:

modeling error, measurement error, and uncertainty inherent in model formulation (Fox 1984). Modeling error arises from incorrectly specifying input data or from model formulation problems due either to lack of a detailed understanding of the basic chemistry and physics or to the simplifications required to make the problem computationally tractable. Inherent uncertainties exist because concentration values measured at a single point in space are in part determined by a stochastic process (turbulent diffusion) and are being compared to a value predicted in a deterministic fashion for a large averaging volume. This will remain even if model predictions and measurements are error free.

Approaches for Testing Model Performance

At present, there are no formal standards or universally accepted tests used to validate model performance. One reason for this is that there are a wide variety of models developed for different purposes. For example, a model designed to predict annual average pollutant concentrations may not be easily compared to a model designed to predict hour-to-hour pollutant variations. Model evaluation procedures must account for the intended model application and formulation.

Some criteria have been suggested for measuring model performance (Brier 1975; Bowne 1980; American Meteorological Society 1981; Fox 1981). Fox (1981) identified three classes of performance measures:

1. Analysis of paired predicted versus observed concentrations for particular locations and times.
2. Ability of the model to predict observed peak concentrations.
3. Comparison of the cumulative frequency distributions of the unpaired predicted and observed concentrations.

Bencala and Seinfeld (1979) developed a computer program for performing statistical analyses. Table 2 lists some of the performance measures applied to the results of four photochemical models.

Each model's adherence to fundamental principles should be scientifically evaluated (Fox 1981). For models based on the atmospheric diffusion equation 5, this means that mass should be conserved and that physically unrealistic predictions such as negative concentrations do not occur. Graphic comparison of the predicted and observed concentrations together can be helpful in diagnosing the nature of the differences between observed and predicted pollutant levels. A final method for model evaluation involves comparing the results of one model against those of another, or to a particular case for which an analytical solution is available.

Data Requirements

The data requirements for conducting model evaluation studies differ greatly among model types. In many cases, lack of data is the major barrier to model evaluation and successful source impact studies. Acquiring the data can be an arduous task. For a Gaussian plume model, the required data could include as little as the mean wind velocity, source emission rate, atmospheric stability (and hence diffusivity), and effective source height (Weber 1982). At the other extreme, a large grid model that incorporates chemical kinetics requires considerably more information before it can be tested—millions of pieces of input data including (McRae and Seinfeld 1983):

1. Vertically resolved, three-dimensional wind fields for every hour of simulation;
2. An hourly emissions inventory for every species in each cell of the modeling region;
3. Hourly temperature, relative humidity, and mixing depth data for each cell;
4. Land use or surface roughness;
5. Vertically resolved initial concentration for every species in each cell;
6. Boundary conditions (concentrations) for each species;
7. Solar radiation data and cloud cover; and
8. Measured hourly ground level data for comparison against model predictions.

Table 2. Performance Statistics for the Caltech, SAI, LIRAQ, and ELSTAR Models

Performance Measure	Model			
	Caltech[a]	SAI[a]	LIRAQ[b]	ELSTAR[c]
Predicted peak ratio[d]				
O_3	0.80	0.71	0.94[e]	
NO_2	0.80	0.77		
Correlation coefficient between predicted and observed concentrations				
O_3	0.89	0.87	0.80	0.84
NO_2	0.67	0.64		0.49
Bias[f]				
O_3	0.002 (3%)	0.072[g]	−0.090[g]	0.017
NO_2	0.078 (11%)	0.024[g]		0.027
Timing of peak prediction (hr)[h]				
O_3	0	2		
NO_2	−2	0		

[a] Statistics are for a July 1974 evaluation period in the Los Angeles area. SAI = System Application, Inc.

[b] The Livermore Regional Air Quality (LIRAQ) model was not evaluated for NO_2 predictions.

[c] The Environmental Lagrangian Simulation of Transport and Atmospheric Reactions (ELSTAR) is a trajectory model that was tested using various trajectories for various days, each with different peak predicted concentrations. Hence, no single number for statistics involving peak predictions is included.

[d] (Maximum concentration predicted)/(Maximum observed).

[e] Average peak ratio.

[f] $\dfrac{1}{N} \sum\limits_{i=1}^{N} P_i - O_i$

where P_i is the ith predicted concentration and O_i is the corresponding observed concentration. Values in parentheses are percent of mean.

[g] Normalized bias is $\dfrac{1}{N} \sum\limits_{i=1}^{N} \dfrac{P_i - O_i}{O_i}$.

[h] Time of the predicted maximum minus the time of the observed maximum.

This list is not exhaustive, nor does every model application require all this information in the detail prescribed. Acquisition of the necessary data can be the major obstacle to a successful evaluation and application program. Nevertheless, the input data acquisition process is vital because the quality of the data ultimately limits the maximum possible quality of the model evaluation study results.

Meteorological data such as surface wind velocity, temperature, relative humidity, and cloud cover are more widely available than emissions inventories, though upper-level wind and temperature data are scarce. It is necessary to collect ambient air quality data, for specifying initial and boundary conditions as well as for comparison with model predictions. Often, the experimental data necessary for model evaluation are not available, and field studies must be executed specifically to collect the data required. Examples of field experiments conducted for such a purpose include the Los Angeles Reactive Pollutant Program (LARPP) (Zak 1982); a study to acquire regional HNO_3, aerosol NO_3^-, and PAN concentrations (Russell and Cass 1984); the Sulfate Regional Experiment (SURE) (Electric Power Research Institute 1981); Regional Air Pollutant Study (RAPS) (Schiermeier 1978); and a program designed to measure particulate carbon concentrations for use in an air quality model evaluation study (Gray 1986). Studies such as these are costly, time-consuming, and significantly increase the effort required to confirm model performance.

Data must be in a form compatible with the model. It may be necessary to interpolate pollutant concentration and meteorological data that are collected at a few discrete locations and times to develop continuous concentration and meteorological fields for model use. Some interpolation methods have been suggested for this purpose (see for example Goodin et al. 1979a). Particular care must be taken in developing wind fields from sparse data because the wind field should be mass consistent. Objective analysis procedures are used to reduce the divergence of interpolated wind fields (Endlich 1967; Dickerson 1978; Goodin et al. 1979b).

A field of input values generated by interpolation over a large geographic area from sparse data is intrinsically uncertain and leads to uncertainty in model predictions. Upper-level variables such as temperature structure (mixing depths), wind fields, and concentration data are particularly susceptible to this uncertainty (Russell and Cass 1986). Upper-level pollutant concentration data are also seldom available except from a few intensive measurement programs—for example, LARPP (Zak 1982) and RAPS (Schiermeier 1978) (see also Edinger 1973; Blumenthal et al. 1978). In the absence of measurements aloft, upper-level initial conditions must be estimated in a way that is consistent with the ground-level measurements and known chemical principles (Russell et al. 1986). Chemically reacting models also require that HC measurements, usually measured as total hydrocarbon concentration (THC), be split into the organic gas classes used by the model (Reynolds et al. 1979; McRae and Seinfeld 1983; Russell and Cass 1986). HC splitting factors either can be based on relative abundance of HCs in the emissions inventory or on detailed atmospheric chemical measurements (see, for example, Graedel 1978; Lamb et al. 1980; Grosjean and Fung 1984).

Analysis of Model Performance

Sensitivity/uncertainty analysis has been applied to estimating the effect that uncertainties in the inputs and reaction mechanisms have on model predictions (Falls et al. 1979; Dunker 1980, 1981; Seigneur et al.

1981; McRae et al. 1982b; Tilden and Seinfeld 1982). Tilden and Seinfeld (1982) present the sensitivity of O_3 and NO_2 predictions to variations in inputs of up to 50 percent, showing the complex relationship of the response. Dunker (1980, 1981) uses analysis of the partial derivatives to describe model response to scaling initial conditions, boundary conditions, and ground-level emissions. For small perturbations of input parameters, the model responded linearly, although nonlinearities were present for larger changes. Model sensitivity to initial conditions decreases with time, suggesting that multiday simulations should be conducted. Multiday simulations are necessary if control strategy or source apportionment calculations are planned. Otherwise, the initial conditions will dominate the results. However, for grid models, long simulations can become sensitive to uncertainties in boundary conditions. Modeling regions should be designed to minimize this effect over the area of most interest and also to capture the effect of inflow boundary conditions (McRae 1981). Sensitivity analysis should also be used to direct experimental research by identifying the model components, such as rate constants and physical parameterizations, that are major causes of uncertainties in predictions.

Extensive model evaluation studies have been conducted for a number of models beginning with the Gaussian plume models and continuing to the state-of-the-art urban and regional photochemical, air quality models. Turner (1964) used a multiple-source Gaussian model to predict 24-hr averaged concentrations of SO_2 in the Nashville, Tennessee, area. He included a first-order chemical decay of SO_2 to form sulfates. Fifty-eight percent of the predictions were within 30 $\mu g/m^3$ of the observations, and the root mean square (RMS) error was 95 $\mu g/m^3$. During the period, concentrations ranged from near zero to about 600 $\mu g/m^3$. The correlation coefficient between predictions and observations was 0.54. As evidence of the advancement in air quality modeling capabilities, compare this to the evaluation statistics of present-day photochemical models (table 2) that describe transport as well as reaction.

Gaussian plume models have also been used to estimate CO, NO_x, and particulate matter concentrations. More recent evaluations of Gaussian plume model performance have been made by Smith (1984) and Irwin and Smith (1984).

In an early application of a mass conservation model based on the numerical solution of equation 3 with simple chemical kinetics, Lamb (1968) calculated CO values in Los Angeles for September 23, 1968. The RMS error was 6.8 ppm, or 50 percent of the mean. Disagreement was ascribed to the lack of a vertically resolved wind field and the need for a more complete description of the chemistry, although present knowledge of emission levels and atmospheric chemistry would indicate that atmospheric production of CO is of lesser importance. Sklarew and coworkers (1972) used a particle-in-a-cell, Lagrangian model to examine the same set of data, reducing the RMS error to 2.7 ppm. They also compared model results to observations for NO_2. Agreement was disappointing for NO_2, presumably because of the need for a more accurate description of atmospheric chemistry.

Recently developed photochemical air quality models, in the Lagrangian trajectory as well as the Eulerian grid form, use more complete descriptions of atmospheric chemistry based on the condensed chemical mechanisms described in the section on Modeling Approaches for Individual Processes. Other improvements include more accurate descriptions of pollutant dry deposition, vertical transport, and more detailed input data. Examples of Lagrangian photochemical trajectory models include Environmental Lagrangian Simulation of Transport and Atmospheric Reactions (ELSTAR) (Lloyd et al. 1979), and the Caltech model (Seinfeld and McRae 1979; McRae et al. 1982a; Russell et al. 1983), which are vertically resolved, the kinetic model developed by Whitten and Hogo (1978), and Empirical Kinetic Modeling Approach (EKMA) developed for the EPA. Each of these models has been used to estimate the effect of emission controls on air quality.

Lloyd et al. (1979) tested the chemical mechanism of the ELSTAR model against smog chamber data. Then they used the data from the LARPP field study, which was specifically designed for testing Lagrangian models, to evaluate the model. Statistical comparison of predicted and observed O_3 and NO_2 concentrations is given in table 2. Seinfeld and McRae (1979) first tested the Caltech photochemical trajectory model in Los Angeles using data from a very smoggy (episode) day—June 27, 1974. Further evolution of the model included testing its capability to predict the formation of aerosol NO_3, PAN, and HNO_3 (Russell et al. 1983; Russell and Cass 1986). In order to reduce the effect of initial conditions, multiday simulations were used in the latter evaluation study.

A model for the long-range transport of nitrogen compounds (Bottenheim et al. 1984) also was developed to predict PAN and NO_3^- concentrations using the SURE data base. Predicted NO_3^- loadings agreed in magnitude with observations, but PAN predictions were generally high.

Lagrangian trajectory models can accurately predict pollutant concentrations and test emission control alternatives. They take relatively little time to execute on a computer (up to 500 or more times faster than grid models), but they produce pollutant concentration predictions only along a single air parcel trajectory. It is often desirable to study the areawide dynamics of pollutants, especially for population exposure calculations, and to present a more complete picture of the effects of source controls (for example, NO_x controls can have a very different effect on O_3 near the source than far away). Rather than run thousands of trajectory simulations, it is more efficient to use Eulerian grid models such as the System Applications, Inc. (SAI) Urban Airshed Model (Reynolds et al. 1973; Seigneur et al. 1983), the regional sulfate transport and reaction model (Carmichael and Peters 1984a,b), the Livermore Regional Air Quality (LIRAQ) model (MacCracken et al. 1978), and the Caltech Airshed Model (McRae et al. 1982a; McRae and Seinfeld 1983). Also, Eulerian grid models are subject to fewer fundamental constraints.

Both the Caltech and the SAI urban air

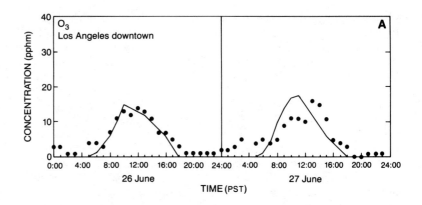

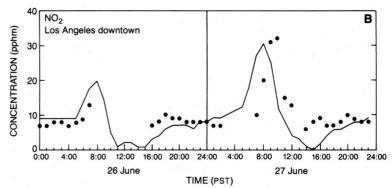

Figure 9. Plot of predicted (————) and observed (●) O_3 and NO_x concentrations (in parts per hundred million, pphm) at downtown Los Angeles during the June 26–27, 1974, modeling study showing the accuracy of model predictions. (Adapted with permission from McRae 1981.)

quality models are vertically resolved, as opposed to the vertically integrated LIRAQ model (Duewer et al. 1978; MacCracken et al. 1978) that has been used in San Francisco in two forms. LIRAQ I is used to model relatively nonreactive pollutant transport (for example, CO, SO_2). LIRAQ II, using a lumped chemical mechanism similar to that of Hecht et al. (1974), is used for computing photochemically reactive pollutant concentrations such as O_3.

Extensive statistical evaluations of the SAI and Caltech models were conducted using data for the June 26–27, 1974, smog episode in Los Angeles. McRae and Seinfeld (1983) calculated the uncertainty in the Los Angeles basin emissions data for the 1974 period to be ±20 percent for CO, ±15 percent for NO_x, and ±25 percent for reactive hydrocarbons (RHCs). Results of

the statistical analysis for these models is given in table 2. They applied the Fortran program developed by Bencala and Seinfeld (1979). Graphic results are shown in figure 9. For the June 26–27 period, both the Caltech and the SAI models tended to underpredict peak O_3 and NO_2 concentrations (table 2). Part of the disagreement between predicted and observed NO_2 concentrations can be ascribed to interference of HNO_3 and PAN with the measurement devices. Given the uncertainties in the meteorological and emissions data, the agreement is quite good. Input data quality is a definite limitation to model performance. Russell and coworkers (1986) updated the chemical mechanism and added the capability to predict ammonium nitrate aerosol concentrations within the Caltech model. They showed the model's ability to predict inorganic NO_3 and PAN, as well as O_3 and NO_2 concentrations. Extensive summaries of many recent model evaluation studies have been made by Dennis and Downton (1984) and Wagner and Ranzieri (1984).

Model evaluation is a vital part of any air quality modeling study. A major limitation is accurate input data, especially on unobserved, upper-level initial and boundary conditions, as well as meteorological parameters. Testing of the more advanced models has shown that they are capable of predicting O_3, NO_2, HNO_3, and PAN as well as nonreactive pollutant concentrations directly from data on meteorological conditions and pollutant emissions.

■ **Recommendation 7.** The most advanced air quality models should be compared against each other and against field experimental observations using a detailed and accurate set of input and verification data. Reasons for any discrepancies should be identified and conflicting findings reconciled.

Application of Air Quality Models

Analytical and receptor models are powerful tools for use in source apportionment, emission control strategy, and population exposure calculations. There is no doubt, however, that the full potential of the newer models has yet to be realized. The ultimate goal is their use in emission control strategy and health impact studies, of which exposure and source apportionment calculations are vital components.

Population Exposure Calculations

Advanced air quality models are powerful tools for use in exposure studies that seek to relate health effects to individual pollutant emission sources. These models can also provide a framework for predicting future exposures resulting from changing emissions. Advanced air quality models, however, have not yet been used widely for population exposure calculations. A preliminary demonstration of the potential for such use is contained in the 1982 Air Quality Management Plan (AQMP) for Los Angeles. Here the SAI urban airshed model was used to estimate the change in population exposure to O_3 that would result from a set of planned emissions reductions (South Coast Air Quality Management District 1982). That study presents a spatially resolved map of the change in population "dosage," defined as

$$D(x, y, K) = \sum_{t=1}^{T} P(x, y) c(x, y, t) \, F[c(x, y, t), K] \qquad (12)$$

where $P(x,y)$ is related to the local population density; F is a function that equals 1 if $c(x,y,t)$—the concentration at (x,y)—is greater than or equal to a threshold concentration K; and T is the number of hours in the simulation. Note that this is not the usual definition of dosage but is the definition used in that particular study. Thus dosage has units of ppm-person-hours and measures the cumulative amount of air pollutant to which a population is exposed above a threshold value, K. Likewise, they calculated "exposure," where exposure (E) is defined by

$$E(x, y, K) = \sum_{t=1}^{T} P(x, y) \, F[c(x, y, t), K] \qquad (13)$$

Again, this is not the usual definition of exposure, but is a measure of how long people are exposed to pollutant concentrations over a threshold value K. Units of exposure are typically person-hours. A finding of this study was that although emission reductions should decrease O_3 exposure in most portions of the basin, some locations would be adversely affected. The above calculation is a first step toward the development of integrated source/exposure studies.

Source Apportionment and Control Strategies

Receptor Models. Receptor models, by their formulation, are effective in determining the source contributions to particulate matter concentrations. The sources contributing to airborne particle loadings have been identified in Washington, D.C. (Gordon et al. 1981), St. Louis (Gatz 1978; Hopke 1981), Los Angeles (Gartrell and Friedlander 1975; Cass and McRae 1983), Portland, Oregon (Watson 1979), and Boston (Hopke et al. 1976; Alpert and Hopke 1980, 1981), as well as other areas, such as the desert (Gaarenstroom et al. 1977). In one effort, a number of researchers were convened to use various receptor models to examine the sources of the Houston aerosol (Stevens and Pace 1984).

Hopke (1981) used size-resolved data and factor analysis to analyze coarse and fine particle fractions in the St. Louis atmosphere. On the basis of data taken at a receptor station on the Washington University campus, he found that 15 percent of the fine particulate matter was from motor vehicles, although little of the coarse particle fraction was derived from mobile sources. An unusually high contribution from paint was attributable to a paint pigment factory in the city, thus showing how a receptor model can be used to identify unusual sources.

Gordon and coworkers (1981) used a chemical mass balance (CMB) technique and varied the number of elements used in the balance to test the sensitivity of the model's results to the choice of marker elements. They found that using nine carefully chosen chemical elements for their calculations gave results comparable to a similar analysis using data on 28 or more chemical components.

In one of the earlier applications of the CMB technique, Gartrell and Friedlander (1975) estimated the sources of particulate mass in Los Angeles atmosphere during the Aerosol Characterization Experiment (ACHEX) (Hidy 1975). In this case, mobile sources accounted for at least 6 percent of the aerosol mass at the Pomona receptor site. As noted previously, receptor models

are not directly suited for determining the source of secondary aerosols such as nitrates and secondary organics. According to the 1974 emissions inventory for the Los Angeles area (McRae and Seinfeld 1983), mobile sources are responsible for 62.3 percent of the NO_x emissions (precursor to NO_3^- aerosols). If one apportioned the measured NO_3^- to sources in proportion to their contribution to the basin-wide NO_x inventory, then the mobile source contribution in the study by Gartrell and Friedlander (1975) increases to 35 percent of the total aerosol mass. In addition, part of the unidentified organic compounds, ammonium, and water may be attributable to mobile sources. This is a rough calculation, indicating that source attribution of secondary aerosol species poses a problem for receptor models and a challenge for future research.

Core and coworkers (1981) combined the use of a receptor model developed by Watson (1979) and a source-oriented model as part of a particulate air quality control strategy analysis in Portland, Oregon. Using CMB techniques, they identified source contributions to the ambient aerosol and then used dispersion modeling to confirm those source contributions. Then, they compared the results obtained with the two models and revised the particulate emission inventory input into the source/dispersion model. Then, they used the revised emissions inventory in dispersion modeling of emission control strategy alternatives. This approach utilized the strengths of the two types of models. Receptor models are not suitable for predicting the outcome of arbitrary perturbations in some sources but not others. They are, however, good for determining the sources of particulate matter when an accurate emissions inventory is not available. Dispersion models, on the other hand, are well suited for modeling the impact of a wide variety of emissions changes that would result from changed emission control regulations but rely totally on an input emissions inventory, which may be uncertain or difficult to obtain. Figure 10 shows the mass apportionment of the aerosol in Portland, Oregon, and the aerosol emissions inventory

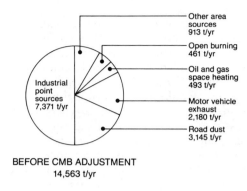

BEFORE CMB ADJUSTMENT
14,563 t/yr

Industrial point sources 7,371 t/yr
Other area sources 913 t/yr
Open burning 461 t/yr
Oil and gas space heating 493 t/yr
Motor vehicle exhaust 2,180 t/yr
Road dust 3,145 t/yr

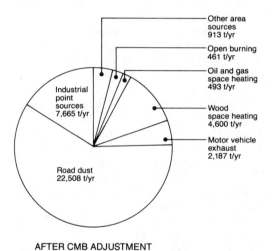

AFTER CMB ADJUSTMENT
38,827 t/yr

Industrial point sources 7,665 t/yr
Road dust 22,508 t/yr
Other area sources 913 t/yr
Open burning 461 t/yr
Oil and gas space heating 493 t/yr
Wood space heating 4,600 t/yr
Motor vehicle exhaust 2,187 t/yr

Figure 10. Emissions inventory of aerosol before and after using chemical mass balance modeling to improve the estimates of emission rates. (Adapted with permission from Core et al. 1981, and the American Chemical Society.)

before and after adjustment using the CMB receptor modeling study. Major deficiencies were identified and improved in the emission inventory for wood burning and road dust.

An extension of the simultaneous use of receptor and source models that merits investigation follows the above methodology except that a chemically reactive transport model is used to estimate the formation of secondary aerosols such as sulfate, NO_3^-, ammonium, and secondary organic carbon. Products of this research would

include estimates of source contributions to secondary aerosols, improved emissions-estimates for the aerosol precursors, and determination of the gross conversion rates of gases to aerosols (see Recommendations 1 and 4).

Source-Oriented Modeling Studies. Nonreactive, mass conservation models based on solving equation 3, including Gaussian plume models, have been used extensively for source apportionment, control strategy analysis, and source impact modeling of nonreactive pollutants such as CO, and of carbonaceous aerosol. Recently, Gray (1986) used a particle-in-cell model to estimate the sources that contribute to primary carbonaceous aerosol concentrations and further used the model to define optimal strategies for controlling aerosol carbon. Models of this type have been used to study the sources that contribute to secondary aerosol sulfate formation (Cass 1981).

Source apportionment, when applied to nonreactive pollutants, has a very clear meaning; that is, source apportionment means determining what proportion of pollutant measured at a receptor site was emitted from a given source. Source apportionment has a much more complex meaning when discussing secondary pollutants that are formed by series of complex atmospheric reactions, rather than being directly emitted from sources. These pollutants include O_3, PAN, and secondary aerosols. The reason is that an incremental change in the emissions of precursors to the formation of a secondary pollutant need not lead to a proportional change in the pollutant concentration, if any change at all results. For example, NO_x and HCs are precursors to the formation of O_3, but increasing the emissions of one precursor can have a very different result than increasing the other. In fact, decreasing NO_x emissions may increase local O_3 concentrations while at the same time decreasing O_3 concentrations downwind.

A common graphic representation of the relationship of maximum O_3 concentrations to initial precursor concentrations of NO_x and HCs is the O_3 isopleth diagram (figure 11).

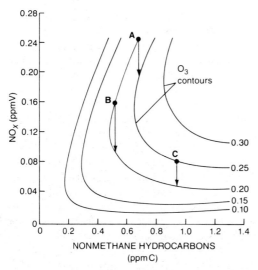

Figure 11. O_3 isopleth diagram showing the response of O_3 concentrations to changes in initial NO_x and nonmethane hydrocarbon concentrations (methane is not included because of its low reactivity). The varying response to NO_x reductions is dependent upon the particular initial concentrations. Initial NO_x concentration is given in parts per million by volume (ppmV) and HC as parts per million of carbon atoms (ppmC). Note that at one position on a curve, such as point C, a decrease in the NO_x concentration results in a much larger decrease in O_3 concentrations than a similar decrease at another position on the same isopleth, such as B. At point A, the graph indicates that decreasing NO_x would increase O_3 formation.

Analysis of the effect of emission control on the improvement in O_3 air quality is further complicated by the fact that the effect of controlling two emission sources together is not necessarily equal to the incremental improvement from controlling one source added to the incremental improvement from controlling the other source separately. This clouds the issue of definitively assessing the impact that a single source has on air quality in that its impact dynamically responds to changes in other sources and to varying meteorological conditions.

Mathematical modeling of photochemical air pollution can delineate the relationship that a source has on air quality. For example, a model that has undergone successful evaluation using the actual emissions inventory for the period studied can

have the inventory revised to exclude all emissions from a source. The second set of calculations using this perturbed inventory should simulate what the pollutant concentrations of the reactive as well as unreactive species would have been without that source. In this fashion, researchers can assess the impact individual sources have on air quality. A similar procedure is followed to estimate the improvement that can be expected from controlling source emissions to varying degrees, simulating implementation of control options or strategies.

Photochemical modeling studies that have examined the effect of specific sources on air quality are scarce. Notable examples are Chock et al. (1981), who used a trajectory model to study the impact of automotive emissions in Los Angeles, and Tesche et al. (1984), who used the SAI airshed model to evaluate emission controls proposed for the Los Angeles area.

The trajectory study of Chock and colleagues used the ELSTAR model described previously, in this case examining air quality changes in Los Angeles due to reductions in automotive emissions of HCs, NO_x, and CO. Two trajectories were modeled, each 8 or 10 hr long. Results indicate that drastic improvement (O_3 reduced from 0.20 to 0.04 ppm and NO_2 from 0.17 to 0.10 ppm for one of the trajectories) would result from reducing automotive NO_x emissions from 3.8 to 2.45 g/mile and HCs from 9.67 to 1.94 g/mile. Further NO_x reduction was found to be ineffective for controlling O_3. Pitts et al. (1983) commented on the conclusion that further NO_x reductions would not be beneficial, noting that short trajectories (of a few hours) are extremely sensitive to initial conditions, many of which are uncertain.

Tesche and colleagues present the results of a number of model calculations, depicting the result of 18 emission control possibilities in the Los Angeles basin. Among the conditions modeled were:

1. 1987 (AQMP) emissions inventory expected in the absence of further emission controls;
2. All elevated emissions removed from

case 1, above (that is, power plant emissions removed);

3. Refinery emissions removed from case 1, above;

4. Mobile source emissions removed from case 1, above;

5. No emissions at all.

In each of these calculations, the base case inventory used was the 1987 AQMP inventory, run 1 (South Coast Air Quality Management District and Southern California Association of Governments 1982). Of the three cases where a source type was removed, removing mobile sources (run 4) showed the greatest decrease in O_3 (lowered from 0.194 to 0.138 ppm). That study also provides a classic example of the nonlinearity of the photochemical system. The net improvement in O_3 for cases 2, 3, and 4 added together is 0.04 ppm. The reduction in NO_x and RHC emissions, when the three cases are combined, is 93 percent and 71 percent, respectively. However, model calculations indicate that a 100 percent emissions reduction should improve O_3 by 0.13 ppm, three times more than would be expected by addition of the effect of individual cases that add up to nearly a complete elimination of the emission sources.

In many of the simulations conducted by Tesche et al. (1984), NO_x control appears to be a relatively ineffective approach to controlling O_3. This is a point that is being debated in the scientific literature at present (Pitts et al. 1983). Trajectory simulations by Russell and Cass (1986) indicate that NO_x controls will reduce O_3, NO_2, PAN, and inorganic NO_3^- formation in the eastern portion of the Los Angeles basin. The debate on the effectiveness of NO_x controls is, perhaps, at present the most critical question to be answered by urban air quality models, and further research into the issue is critical for understanding how to control O_3 and related photochemically generated species. (See Recommendation 7.)

A motivation for developing advanced transport models and transport and reaction models is to create an ability to guide decisions regarding the most cost-effective

set of control techniques to obtain a desired air quality (that is, optimal control strategy development). In general, least-cost control strategy development attempts to solve the mathematical programming problem (Cass and McRae 1981):

Find $\bar{x}$ such that $C(\bar{x})$ is minimized, subject to $Q[E(x,t)M(x,t)] \leq S$

where $\bar{x}$ is a set of control measures that can be applied to the sources E, minimizing the cost C, such that the air quality Q, is less than or equal to a prescribed level S. M represents the changing meteorology, and t is time. Q and S may include a number of species.

Usually, blind application of the best available control technology to the largest sources, as is often proposed, is not the most cost-effective means for improving air quality. Cass and McRae (1981) showed that applying the best available technology to the largest sources first could cost about $70 million/yr to meet a 10 $\mu g/m^3$ sulfate level as contrasted with about $40 million for a least-cost strategy, or a savings of about half. Other studies have shown similar results. Kyan and Seinfeld (1974), following the work of Trijonis (1974), illustrate an economically optimized control strategy for photochemical pollutants.

The large data requirements and computational times make it expensive to test a large number of emission control strategies using the most advanced photochemical airshed models. Instead, realizing that the precursors of O_3 and NO_2 are HC and NO_x, least-cost control strategies can be estimated by identifying the least-cost approach to achieving various levels of HC and NO_x emissions and then using the advanced air quality models to identify the perturbed emissions level that will meet the desired air quality standard. Cass and McRae (1981) summarize the techniques for devising least-cost control strategies. One question not adequately addressed in the literature is whether or not an optimal strategy for reducing O_3 on high-episode days will be as effective at reducing O_3 on typical days.

Future Uses

As new technologies change pollutants and emission patterns, it is important to be able to answer, in advance, the question "What will be the probable effects of future emissions of novel substances?" Photochemical models are ideally suited for predicting the changes, a priori. Specific applications for models would be to test the effect of enlarging the fleet of diesel-powered vehicles on particulate and gas-phase pollutants, changing fuel compositions, or converting the vehicle fleet to methanol fuel.

One reason for advancing the technology base built into mathematical models is to be able to answer questions that will arise in the future. Given the lag time of several years between initiation of model development and proof of model performance, it is necessary to work continually on extending model capabilities. Development periods of four or more years can be expected. Typically, once a particular air quality problem has reached the point of public debate, the time scale allowed for technical analysis of the problem is shorter than the time needed to develop new modeling tools from scratch. Yet, without the appropriate tools for conducting a competent engineering analysis, inefficient—or worse, ineffective—costly decisions will be made. One current policy problem now awaiting completion of an advanced air quality model is that of acid deposition control.

In addition to source apportionment studies, air quality models can be used to identify potential areas of research by identifying gaps in our knowledge. Also, models can predict concentrations of trace gases that would be difficult to observe experimentally, thus alerting researchers to possible undetected problems.

It is clear that much time has been devoted to developing and evaluating advanced photochemical models (see earlier sections on Modeling Approaches for Individual Processes and Model Evaluation) and that much can be gained if these models are put into effective use by government regulatory agencies. One barrier to such use is manpower problem; there are very few organizations capable of conducting a source apportionment study that involves chemically reacting pollutant emissions (Cass and McRae 1981). Although further use of advanced models for the design of optimal control strategies, alone, would appear capable of identifying economic savings well in excess of the cost of conducting those studies, there is often no mechanism to pay for this necessary effort.

Special Topics and Emerging Issues in Air Quality Modeling

Previous sections of this chapter addressed the formulation and use of air quality models as they have generally been viewed in the past. As our knowledge of air pollutant transport and chemical reaction processes increases, new types of air quality models are being constructed. In many cases, problems that were computationally intractable can now be handled by faster computers. These areas that are undergoing rapid development present numerous opportunities for research. Aerosol process modeling that includes combined smog/fog cycles (as discussed in the section on Modeling Approaches for Individual Processes) is, perhaps, the most important emerging topic in modeling for health effects purposes. Other emerging issues and special topics are discussed below.

Modeling Large-Scale Processes

The advance in very powerful computers has made it possible to start thinking about modeling extremely large-scale transport and reaction systems in some detail, such as the problem of acid deposition in eastern North America. On yet a larger scale are the global circulation models which include a description of atmospheric chemistry to estimate how the chemical composition of the atmosphere will change with time because of increasing industrial and mobile source activity. Such models can help answer questions surrounding the increased emissions of novel substances such as fluorocarbons or long-term impacts of more mundane substances such as CO and CO_2. The questions addressed in this case often involve global-scale health effects such as the increase in skin cancer that would occur if greater amounts of ultraviolet solar radiation were to reach the earth's surface because of stratospheric O_3 depletion. The technical barriers facing the development of extremely large-scale models are essentially the same as for urban-scale models, magnified by problems of data collection on a global scale.

Modeling Small-Scale Processes

Often the information desired from modeling studies depends on processes that occur on spatial scales much smaller than the resolution of most urban air quality models. The modeling of NO_x air quality in street canyons involves small-scale processes of this sort. Introduction of point-source emissions into grid-based air quality models likewise involves a mismatch between the high concentrations that in fact do exist near the source versus the lower concentrations computed by a model that immediately mixes those emissions throughout a grid cell of several kilometers on each side. Because of computational time constraints, it is often impractical to fully describe the processes that take place on a scale smaller than the main model grid (subgrid scale), but one must be able to ensure that answers obtained from large-scale calculations are correct over the spatial averaging scale adopted by the model.

In urban-scale modeling, the usual grid dimensions are on the order of 1 to 10 km, with ground-level cell heights of 10 to 100 or more meters. Measured pollutant concentrations, against which model predictions are compared, however, are point values, taken a few meters above the ground, and these can be directly affected by nearby sources. This situation can frustrate comparisons to model predictions of directly emitted pollutants such as CO and NO or rapidly reacting secondary pollutants such as O_3 and NO_2. Concentrations of slowly reacting secondary pollutants such as HNO_3 would be less affected. One needs better methods to reconcile the differences between large-grid, volume-averaged predictions and point measurements (see Nappo et al. 1982).

How can the treatment of small processes be improved? Treatment of large point sources (for example, power plants) in urban-scale photochemical models can be handled in two general ways: (1) much like area sources where the emissions are mixed instantaneously throughout a cell, or (2) by separately treating an expanding, reacting plume that interacts with the atmosphere outside the plume, while maintaining its integrity. In an evaluation of the SAI urban airshed model using both approaches, Seigneur et al. (1983) found little difference for the regional dynamics of O_3 in Los Angeles. This may not be the case in all situations, especially when viewing the impact of major point sources in other geographic areas or when the concern is for air quality very near the source.

Point sources that dominate emissions in a specific area, such as offshore oil production platforms, may have to be treated in great detail. For reacting plumes—those containing NO_x, HCs, or possibly SO_x—the large concentration gradients that exist make the macro- and the micro-scale mixing processes important to the overall dynamics of pollutant evolution.

■ **Recommendation 8.** Additional research is needed on near-source dispersion and reactions of pollutants for inclusion in plume models.

Indoor/Outdoor Pollutant Relationships

Air quality models have traditionally dealt with calculating the effect of pollutant sources on *outdoor* air quality. However, much of the time people are indoors, either at home, at work, or in a car (National Research Council 1981). Thus, indoor pollutant concentrations make a major contribution to personal time-weighted pollutant exposure (Sexton and Ryan, this volume). Indoor air quality models currently are being developed to bridge the gap, relating the pollutant concentrations indoors to outdoor air quality, indoor emissions, ventilation rates, indoor transport, and indoor chemistry. Key questions are, "To what extent do pollutants derived from outdoor sources interact with indoor emissions, and what are the products of those interactions?"

As in outdoor situations, receptor-oriented and transport (source) models can be used to estimate source impacts on indoor air quality (Turk 1963; Shair and Heitner 1974; Borazzo et al. 1987; Sexton and Hayward 1987). Constraints and limitations on the two approaches indoors are similar to those discussed previously for outdoor applications. However, some unusual chemical constituents can be found at high concentrations indoors (for example, formaldehyde, radon, and tobacco smoke), as well as the traditional outdoor pollutants (CO, NO_2, O_3, and particulate matter).

The usual approach to indoor air quality modeling has been to apply a mass balance equation, similar to equation 5. For a single-compartment model this becomes (National Research Council 1981)

$$V \frac{dC}{dt} = q_0 C_0 (1 - F_0) + q_1 C (1 - F_1)$$
$$+ q_2 C_0 - (q_0 + q_1 + q_2) C$$
$$+ S - R \qquad (14)$$

where V is the volume of the structure (or room), q_0 and q_2 are the rates at which air is brought into the building from outdoors through a ventilation system or by infiltration, respectively (figure 12), and q_1 is the air recirculation rate. Both the makeup air (q_0) and the recirculated air may be filtered such that the pollutant concentration of the filtered air is $(1 - F)$ times that entering the unit. The characteristic filtration efficiencies are F_0 and F_1 for the makeup air filter and recirculated air filter, respectively. S represents indoor emission sources, and R is an indoor sink term. Multicompartment models involve a system of similar coupled differential equations. Mass balance models have been used to successfully relate indoor air quality to outdoor pollutant levels, especially for nonreactive gases such as CO. However,

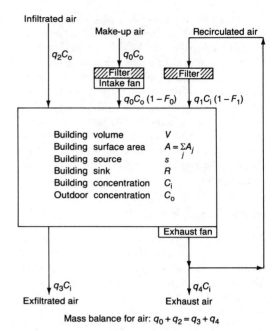

Figure 12. Indoor air quality model, including mass balance on pollutants and air. (Adapted with permission from Shair and Heitner 1974, and the American Chemical Society.)

agreement for reactive gases such as O_3 and SO_2 has not been as close (National Research Council 1981). Most indoor air quality models have yet to use as sophisticated a description of the chemical kinetics as have their outdoor counterparts.

A potential area for research involves studying the effect that chemistry has on indoor pollutants. The concentrations of some pollutants indoors will behave much differently than those outdoors because of the magnitude of the concentrations, artificial lighting, and the large surface areas for deposition. Another interesting question is "How will the pollutants emitted indoors, such as formaldehyde, react with vehicle-related pollutants drawn from outside, such as O_3 and NO_2?"

One factor that will complicate the use of indoor air quality models arises from the fact that different buildings (and rooms) vary tremendously in surface reactivity, humidity, ventilation, filtration, and diffusion rates. Input parameters for mass balance models should be measured for each individual building used in a study. Differences among buildings pose a problem for receptor models, too. Source signatures must be identified for each building.

Given the significant contribution that indoor pollutant concentrations add to personal exposure, it is evident that attention should be focused on determining the sources of pollutants found indoors. Realizing the critical role that outdoor pollutants play, indoor/outdoor air quality relationships should be further defined, and the sources of the indoor pollutants identified. By linking the results from outdoor air quality source apportionment studies with similar studies using models that relate indoor air quality to that outdoors, it should be possible to identify the effect of outdoor sources on indoor air quality, and the related human exposure, even for reactive gases. Results of combined indoor/outdoor studies can be used for setting outdoor air quality standards that consider the effect of outdoor air quality on indoor pollutant levels. Ultimately, indoor/outdoor air quality models can be used to devise optimal strategies for controlling emission sources in a way that is more directly related to human exposure.

■ **Recommendation 9.** Further research is needed into the use of models that relate indoor exposure to outdoor air quality. The procedures outlined by Sexton and Ryan (this volume) should be useful in guiding future studies of the relationship between emission sources and human exposure.

Conclusion

Mathematical models, statistical as well as deterministic, have evolved to become powerful tools for apportioning the impact of sources on certain aspects of air quality. Models can be used to study human exposure to air pollutants and to identify cost-effective control strategies. Their use for designing optimal emission control strategies, alone, could lead to large savings in emission control costs. Given the appropriate input data, air quality models can accu-

rately predict the concentrations of the regulated pollutants such as CO, O_3, and NO_2, as well as some of the noncriteria pollutants. A primary limitation on the accuracy of model results at present is not the model formulation, but the accuracy of the available input data.

Receptor-oriented (statistical) models use the large volume of data available on pollutant concentrations and use the underlying structure of a data set to separate the contribution of different emission sources to observed air quality. The most common types of receptor models use chemical mass balance and multivariate analysis techniques and have been used in a number of locations to identify and apportion source contributions at receptor sites. However, the assumptions involved in formulating receptor models limit their use for source impact research to studying nonchemically reacting systems. For control strategy development, other limitations exist. One area for promising research is the hybridization of receptor-oriented models with source-oriented (or analytical) models, thereby capturing the power of both methods.

Analytical models are composed of a number of modules each describing, mathematically, a physical or chemical process, such as transport, diffusion, deposition, and chemical reaction. This is particularly true of the advanced photochemical air quality models. Some research areas have been identified where model capabilities can be improved or expanded:

1. Advancing and testing the chemical mechanisms used to model air quality;
2. Inclusion of models of aerosol processes, including the chemical reactions leading to aerosol formation and heterogeneous reactions;
3. Models relating indoor and outdoor air quality;
4. Further use of air quality models in source apportionment and control studies, and in personal exposure research;
5. Improved description of pollutant deposition, both wet and dry.

In some of these areas, a better understanding of the underlying physical process is needed, requiring basic research into the actual physical phenomena involved. Deposition processes and some aspects of aerosol dynamics fall in this category. On the other hand, development of advanced chemical mechanisms is quite possible using our present knowledge of atmospheric chemistry.

Inclusion of aerosol processes within future air quality models was identified as a key area for future research, particularly because of the suspected health effects of small particles. The ability to relate particle size and composition to the original source(s) will be critical in future exposure and impact studies. By advancing air quality modeling methods now, we will be able to answer questions that now face us and be situated to address, in a timely manner, questions that arise in the future.

It is clear that models now can predict the dynamics of the regulated pollutants such as CO, NO_2, O_3, and some components of particulate matter directly from data on emissions and thus are well suited for defining source–air quality relationships for those pollutants. However, it is also clear that this capability has been extended to only a few of the many nonregulated pollutants that may be of interest to the health effects research community in the future. Inasmuch as regulation has been the principal driving force for model development, this is understandable. However, progress in expanding model capabilities could be encouraged if toxicologists and epidemiologists collaborated with physical scientists to specify the additional pollutants, concentrations, and averaging times of interest, so that air quality scientists could develop or modify models to suit the specific needs of the health effects research community and anticipate the demands likely to arise from future regulation. Clearly the research proposed here would involve a variety of disciplines. This cooperation would lead to a better understanding of the sources of the pollutants that impact human health.

Summary of Research Recommendations

Evaluating the present state of mathematical modeling as a means to relate emissions to air quality and consequently health effects points to a number of areas for promising research. However, advances in mathematical air quality models are ultimately limited by our understanding of the basic physics and chemistry being described within the model. In this regard, Samson and Atkinson (both this volume) have identified research that would enhance mathematical modeling of air quality by improving the understanding of the underlying physical and chemical processes on which such models are based.

We are currently able to describe mathematically the dynamics of unreactive pollutants in urban areas with a great deal of confidence. In addition, our ability to model NO_2 and O_3 is well advanced, though the issues that surround the effect of NO_x controls on O_3 air quality still should be resolved. Recommendations 5 and 7 (detailed below) would result in greatly increased confidence in model predictions and lead to answering major questions. Much of the limitation to developing a greater capability for defining source/air quality relationships is not due to the model itself, but rather to a lack of accurate data for use in the models.

Processes affecting the formation and growth of aerosols are not nearly as well understood as processes involving the gas–phase alone. The ability to model aerosol dynamics is, likewise, relatively undeveloped. This is understandable. It was necessary to develop gas-phase models before attempting a complete description of aerosol processes, because the formation and growth of aerosols is directly affected by gas-phase compounds, whereas the gas-phase is only slightly affected by aerosols. Presently, photochemical air quality models are able to provide the basis for an aerosol processes model. Because of the importance of inhalation of aerosols to human health, an aerosol process model is essential in determining source/health effects relationships. Recommendation 6, below, would lead to rapid development of a comprehensive aerosol process air quality model.

The final step in constructing a system for determining source/air quality relationships for use in exposure studies involves developing a comprehensive indoor air quality model, as described by Recommendation 9. The model envisioned would include gas-phase chemistry as well as aerosol dynamics, and hence relies on completing the first three projects.

Completion of the four high-priority research recommendations listed below is essential to an improved understanding of relationships between sources and health effects. A number of moderate- and lower-priority research recommendations arising from considerations in the text are listed next. Undoubtedly there are others whose urgency and importance will command attention as the field evolves. The following recommendations emphasize research efforts that will rapidly increase the capability to apply air quality models to describe the dynamics of air pollutants believed to be harmful to health, and to identify the sources of those pollutants.

HIGH PRIORITY

Recommendation 5
Construction of an
Advanced Chemical
Mechanism

Development of an accurate, condensed chemical mechanism would increase the confidence in using models to assess source impacts on air quality and could be used to examine the dynamics of compounds suspected of causing health problems. The mechanism should accurately reproduce smog chamber experiments when the expected wall radical source is included and agree with a large explicit "master" mechanism that includes a detailed description of atmospheric chemistry as it is now understood. As discussed by Leone and Seinfeld (1985), the concentration predictions from that condensed mechanism (including trace radical species) as well as the relative production routes of various species such as O_3 should be close to the predictions of an explicit mechanism over a variety of initial conditions and emission rates during the simulation. The condensed mechanism must be small enough to be used in an urban air quality model. The mechanism should then be incorporated into one of the advanced air quality models, and research Recommendation 7 then should be pursued.

Recommendation 7
Model Comparison
and Evaluation

The most advanced air quality models should be compared against each other and against field experimental observations, using a detailed and accurate set of input and verification data. Collection of the needed data is vital to air quality model development. Reasons for any discrepancies should be identified. Input data preparation would need to be well documented and open to review. A major issue to be addressed as part of this study concerns the effect of NO_x emissions on the formation of O_3 (Pitts et al. 1983). Previous modeling studies of the problem have been conducted with differing conclusions. It is very important to reconcile these conflicting findings, and this type of project is the most direct method to do so.

Recommendation 6
*Aerosol Process
Model Development*

The scientific knowledge currently exists that would permit development of models for basic atmospheric aerosol processes, but the atmospheric data needed to conduct preliminary tests of such a model are not available. What is required are size-resolved and chemically resolved aerosol measurements collected in a manner that can be fully utilized for model development. A three-step procedure is suggested:

a. Preliminary model calculations should be made using the limited data currently available to identify specific parameters that need to be well characterized during a large-scale aerosol measurement experiment.
b. A measurement program should be designed and conducted to obtain the data identified in step (a).
c. The results of steps (a) and (b) could then be used for more detailed model development and more thorough model testing. The model should include reactions leading to highly toxic compounds, such as PAH reactions with NO_x.

Recommendation 9
*Indoor Air Quality
Modeling*

Indoor air quality models complementary to outdoor air quality models are needed to relate indoor air quality and exposure to sources. Mathematical models are currently under development, along with characterization of important input parameters. Further work is needed, especially to advance model descriptions of gas-phase chemistry, deposition, and aerosol dynamics indoors. Receptor-oriented models have received less attention for indoor applications, although they could be a powerful tool for use in source apportionment studies. Results from indoor air quality studies that relate indoor pollutant concentrations to those outdoors can be combined with similar studies on outdoor air to help develop air quality standards and conduct source-related health impact studies.

MODERATE PRIORITY

Recommendation 3
*Pollutant Deposition
Modeling*

Dry deposition of chemically reactive air pollutants and aerosols is an area of current research interest. Given the importance to the fate and impact of pollutants, and as a vital part of any modeling studies, better characterization of the process leading to deposition would be valuable. This problem should be attacked using field experiments as well as laboratory analyses, complemented by derivation of new computer-based algorithms to be used for describing dry deposition processes based on fundamental physical principles. Laboratory analyses should focus on the mechanics of particle transport through boundary layers by making detailed particle velocity measurements near surfaces. Outdoor deposition measurements would benefit from improved instrumentation.

Recommendation 1
Receptor Modeling

Receptor models such as those using chemical mass balance techniques have proven to be very convenient tools for apportioning the contributions of sources to atmospheric particulate matter concentrations. Combining receptor and source models appears to have great potential. Further studies using hybrid or combined models will benefit from the strengths of both types of models. Also, it may be possible to add the ability to identify the sources of secondary aerosols when using receptor models.

Recommendation 2
Pollutant Dynamics in Street Canyons

Studies to date have concentrated on pollutant transport but not chemical interactions. Inclusion of chemical reactions within a street canyon model is important to determine near-source effects on the concentrations of pollutants such as NO_2 and O_3. A field study in which reactive pollutants such as O_3, NO, and NO_2 and a tracer are closely monitored in and above a street canyon would provide the data required for testing a chemically reactive street canyon air quality model.

LOWER PRIORITY

Recommendation 4
Fog Chemistry

Interactions between smog and fog droplets are known to increase fog acidity and acid deposition, although direct health effects are not well known. Smog/fog interactions will also affect the evolution of gas-phase pollutants. We should combine our knowledge of gas-phase and fog droplet chemistry into a single model to investigate how the interaction affects pollutant evolution in an urban atmosphere.

Recommendation 8
Reactive Plume and Subgrid Scale Modeling

Plumes may dominate pollutant concentrations in the near field, such as near a power plant or highway. Much of the work to date has considered chemically inert plumes, and the few reacting plume models have adopted extensive approximations. Given the reactivity of vehicular exhaust and the amount of time people spend on the road, it is important to gain a better understanding of the near-source dispersion and reaction of pollutants.

Acknowledgments

I thank Drs. Glen Cass and Ken Sexton for their comments during the preparation of this manuscript and am grateful for the many helpful comments of the reviewers.

Correspondence should be addressed to Armistead G. Russell, Department of Mechanical Engineering, Carnegie Mellon University, Pittsburgh, PA 15213.

References

Adewuyi, Y. G., and Carmichael, G. R. 1982. A theoretical investigation of gaseous absorption by water droplets from SO_2–HNO_3–NH_3–CO_2–HCl mixtures, *Atmos. Environ.* 16:719–729.

Alpert, D. J., and Hopke, P. K. 1980. A quantitative determination of sources in the Boston urban aerosol, *Atmos. Environ.* 14:1137–1146.

Alpert, D. J., and Hopke, P. K. 1981. A determination of the sources of airborne particles collected during the regional air pollution study, *Atmos. Environ.* 15:675–687.

American Meteorological Society. 1981. Air Quality Modeling and the Clean Air Act, Boston, Mass.

Atkinson, R., and Lloyd, A. C. 1984. Evaluation of kinetic and mechanistic data for modeling of photochemical smog, *J. Phys. Chem. Ref. Data* 13:315–444.

Atkinson, R., Lloyd, A. C., and Winges, L. 1982. An updated chemical mechanism for hydrocarbon/NO_x/SO_2 photooxidations suitable for inclusion in atmospheric simulation models, *Atmos. Environ.* 16:1341–1355.

Atkinson, R., Plum, C. N., Carter, W. P., Winer, A. M., and Pitts, J. R., Jr. 1984. Rate constants for the gas phase reactions of NO_3 radicals with a series of organics in air at 298° ± 1 K, *J. Phys. Chem.* 88:1210–1215.

Baldwin, A. C., and Golden, D. M. 1979. Heterogeneous atmospheric reactions: sulfuric acid aerosols as tropospheric sinks, *Science* 206:562–563.

Bandow, H., Okuda, M., and Akimoto, H. 1980. Mechanism of the gas-phase reactions of C_3H_6 and NO_3 radicals, *J. Phys. Chem.* 84:3604–3608.

Barth, D. S. 1970. Federal motor vehicle emission goal for carbon monoxide, hydrocarbon and NO_x based on desired air quality levels, *J. Air Pollut. Control Assoc.* 20:519–523.

Bencala, K. E., and Seinfeld, J. H. 1979. An air quality model performance assessment package, *Atmos. Environ.* 13:1181–1185.

Blumenthal, D. L., White, W. H., and Smith, T. B. 1978. Anatomy of a Los Angeles smog episode: pollutant transport in the daytime sea breeze regime, *Atmos. Environ.* 12:893–907.

Borazzo, J. E., Osborne, J. F., Fortmann, R. C., Keefer, R. L., and Davidson, C. I. 1987. Modeling and monitoring of CO, NO and NO_2 in a modern townhouse, *Atmos. Environ.* 21:299–312.

Bottenheim, J. W., Brice, K. A., and Anlauf, K. G. 1984. Discussion of a Lagrangian trajectory model describing long range transport of oxides of nitrogen, the incorporation of PAN in the chemical mechanism, and supporting measurements of PAN and nitrate species at rural sites in Ontario, Canada, *Atmos. Environ.* 18:2609–2620.

Bowne, N. E. 1980. Validation and performance criteria for air quality models, In: *Proceedings of the Second Joint Conference on Applications of Air Pollution Meteorology*, March 24–27, New Orleans, La., American Meteorological Society, pp. 614–626.

Brier, G. W., 1975. Statistical Questions Relating to the Validations of Air Quality Simulation Models, U.S. Environmental Protection Agency Report No. EPA-650/4-75-010, Research Triangle Park, N.C.

Carmichael, G. R., and Peters, L. K. 1984a. An Eulerian Transport/Transformation/Removal model for SO_2 and sulfate. I. Model development, *Atmos. Environ.* 18:937–952.

Carmichael, G. R., and Peters, L. K. 1984b. An Eulerian Transport/Transformation/Removal model for SO_2 and sulfate. II. Model calculation of SO_x transport in the eastern United States, *Atmos. Environ.* 18:953–967.

Carmichael, G. R., Kitada, T., and Peters, L. K.,

1986. A second generation model for regional-scale transport/chemistry/deposition, *Atmos. Environ.* 20:173–188.

Cass, G. R. 1981. Sulfate air quality control strategy design, *Atmos. Environ.* 15:1227–1249.

Cass, G. R., and McRae, G. J. 1981. Minimizing the cost of air pollution control, *Environ. Sci. Technol.* 15:748–757.

Cass, G. R., and McRae, G. J. 1983. Source receptor reconciliation of routine air monitor in data for trace metals—An emission inventory assisted approach, *Environ. Sci. Technol.* 17:129–139.

Chameides, W. L., and Davis, D. D. 1982. The free radical chemistry of cloud droplets and its impact upon the composition of rain, *J. Geophys. Res.* 87:4863–4877.

Chock, D. P., and Dunker, A. M. 1983. A comparison of numerical methods for solving the advection equation, *Atmos. Environ.* 17:11–24.

Chock, D. P., Dunker, A. M., Kumar, S., and Sloane, C. S. 1981. Effect of NO_x emission rates on smog formation in the California south coast air basin, *Environ. Sci. Technol.* 15:933–939.

Chock, D. P., Dunker, A. M., Kumar S., and Sloane, C. S. 1983. Reply to comment on "Effect of nitrogen oxide emissions in metropolitan regions," "Effect of NO_x emission rates on smog formation in the California south coast air basin," and "Effect of hydrocarbon and NO_x on photochemical smog formation under simulated transport conditions," *Environ. Sci. Technol.* 17:48–62.

Cooper, J. A., and Watson, J. G. 1980. Receptor-oriented methods of air particulate source apportionment, *J. Air Pollut. Control Assoc.* 30:1116–1125.

Core, J. E., Hanrahan, P. L., and Cooper, J. A. 1981. Air particulate control strategy development: a new approach using chemical mass balance methods, In: *Atmospheric Aerosols: Source/Air Quality Relationships* (E. S. Macias and P. K. Hopke, eds.), Symposium Series No. 167, American Chemical Society, Washington, D.C.

Council on Environmental Quality. 1984. Environmental Quality, the Fifteenth Annual Report, U.S. Government Printing Office, Washington, D.C.

Dahneke, B. 1983. Simple kinetic theory of Brownian diffusion in vapors and aerosols, In: *Theory of Dispersed Multiphase Flow* (R. Meyer, ed.), pp. 97–133, Academic Press, New York.

Dennis, R. L., and Downton, M. W. 1984. Evaluation of urban photochemical models for regulatory use, *Atmos. Environ.* 18:2055–2069.

Dickerson, M. H. 1978. MASCON—A mass-consistent atmospheric flux model for regions with complex terrain, *J. Appl. Meteorol.* 17:241–253.

Dodge, M. C. 1977. Combined Use of Modeling Techniques and Smog Chamber Data to Derive Ozone-Precursor Relationships, U.S. Environmental Protection Agency Report No. EPA-600/3-77-001a, Research Triangle Park, N.C.

Drewes, D. R., and Hale, J. M. 1982. SMICK—A scavenging model incorporating chemical kinetics, *Atmos. Environ.* 16:1717–1724.

Duce, R. A. 1969. On the source of gaseous chlorine

in the marine atmosphere, *J. Geophys. Res.* 74:4597–4599.

Duewer, W. H., MacCracken, M. C., and Walton, J. J. 1978. The Livermore regional air quality model: II. Verification and sample application in the San Francisco Bay area, *J. Appl. Meteorol.* 17:273–311.

Dunker, A. M. 1980. The response of an atmospheric reaction transport model to changes in input functions, *Atmos. Environ.* 14:671–679.

Dunker, A. M. 1981. Efficient calculation of sensitivity coefficients for complex atmospheric models, *Atmos. Environ.* 15:1155–1161.

Edinger, J. G. 1973. Vertical distribution of photochemical smog in the Los Angeles basin, *Environ. Sci. Technol.* 7:247–252.

Electric Power Research Institute. 1981. EPRI Sulfate Regional Experiment: Results and Implications, Report No. EPRI-EA-2165-SY-LD, Palo Alto, Calif.

Endlich, R. M. 1967. An iterative method for altering the kinematic properties of wind fields, *J. Appl. Meteorol.* 6:837–844.

Falls, A. H., and Seinfeld, J. H. 1978. Continued development of a kinetic mechanism for photochemical smog, *Environ. Sci. Technol.* 12:1398–1406.

Falls, A. H., McRae, G. J., and Seinfeld, J. H. 1979. Sensitivity and uncertainty of reaction mechanisms for photochemical air pollution, *Int. J. Chem. Kinet.* 11:1137–1162.

Fowler, D. 1978. Dry deposition of SO_2 on agricultural crops, *Atmos. Environ.* 12:369–373.

Fox, D. G. 1981. Judging air quality model performance, *Bull. Am. Meteorol. Soc.* 62:599–609.

Fox, D. G. 1984. Uncertainty in air quality modeling, *Bull. Am. Meteorol. Soc.* 65:27–36.

Friedlander, S. K. 1973. Chemical element balances and identification of air pollution sources, *Environ. Sci. Technol.* 7:235–240.

Friedlander, S. K. 1977. *Smoke, Dust and Haze*, Wiley-Interscience, New York, N.Y.

Friedlander, S. K. 1981. New developments in receptor modeling theory, In: *Atmospheric Aerosols: Source/Air Quality Relationships* (E. S. Macias and P. K. Hopke, eds.), pp. 1–20, Symposium Series No. 167, American Chemical Society, Washington, D.C.

Gaarenstroom, P. D., Perone, S. P., and Moyers, J. L. 1977. Application of pattern recognition and factor analysis for characterization of atmospheric particulate composition in southwest desert atmosphere, *Environ. Sci. Technol.* 11:795–800.

Gartrell, G., and Friedlander, S. K. 1975. Relating particulate pollution to sources: the 1972 California aerosol characterization study, *Atmos. Environ.* 9:279–299.

Gatz, D. F. 1975. Relative contributions of different sources of urban aerosols: application of a new estimation method to multiple sites in Chicago, *Atmos. Environ.* 9:1–18.

Gatz, D. F. 1978. Identification of aerosol sources in the St. Louis area using factor analysis, *J. Appl. Meteorol.* 17:600–608.

Gelbard, F., and Seinfeld, J. H. 1980. Simulation of multicomponent aerosol dynamics, *J. Colloid Interfac. Sci.* 78:485–501.

Gifford, F. A. 1961. Use of routine meteorological observations for estimating atmospheric dispersion, *Nucl. Saf.* 2:47–57.

Goodin, W. R., McRae, G. J., and Seinfeld, J. H. 1979a. A comparison of interpolation methods for sparse data: application to wind and concentration fields, *J. Appl. Meteorol.* 18:761–771.

Goodin, W. R., McRae, G. J., and Seinfeld, J. H. 1979b. An objective analysis technique for constructing three-dimensional, urban scale wind fields, *J. Appl. Meteorol.* 19:98–108.

Gordon, G. E. 1980. Receptor models, *Environ. Sci. Technol.* 14:792–800.

Gordon, G. E., Zoller, W. H., Kowalczyk, G. S., and Rheingrover, S. W. 1981. Composition of source components needed for aerosol receptor models, In: *Atmospheric Aerosol Source/Air Quality Relationships* (E. S. Macias, and P. K. Hopke, eds.), pp. 51–74, American Chemical Society, Symposium Series No. 167, Washington, D.C.

Graedel, T. E. 1978. *Chemical Compounds in the Atmosphere*, Academic Press, New York, N.Y.

Graedel, T. E. 1984. Effects of below cloud scavenging on raindrop chemistry over remote ocean regions, *Atmos. Environ.* 18:1835–1842.

Graedel, T. E., Farrow, L. A., Weber, T. A. 1976. Kinetic studies of the photochemistry of the urban troposphere, *Atmos. Environ.* 10:1095–1116.

Graham, R. A., Johnston, H. S. 1978. The photochemistry of NO_3 and the kinetics of the N_2O_5–O_3 system, *J. Phys. Chem.* 82:254–268.

Gray, H. A. 1986. Control of atmospheric fine primary carbon particle concentrations, Ph.D. thesis, California Institute of Technology, Pasadena, Calif.

Grosjean, D. 1984. Atmospheric reactions of orthocresol: gas phase and aerosol products, *Atmos. Environ.* 18:1641–1652.

Grosjean, D. 1985. Reaction of *o*-cresol and nitrocresol with NO_x in sunlight and with ozone-nitrogen dioxide mixtures in the dark, *Environ. Sci. Technol.* 19:968–974.

Grosjean, D., and Friedlander, S. 1980. Formation of organic aerosols from cyclic olefins and diolefins, In: *The Character and Origin of Smog Aerosols* (G. Hidy and P. K. Hueller, eds.), pp. 435–473, John Wiley and Sons, New York, N.Y.

Grosjean, D., and Fung, K. 1984. Hydrocarbons and carbonyls in Los Angeles air, *J. Air Pollut. Control Assoc.* 34:537–543.

Grosjean, D., Fung, K., and Harrison, J. 1983. Interactions of polycyclic aromatic hydrocarbons with atmospheric pollutants, *Environ. Sci. Technol.* 17:673–679.

Hatakeyama, S., Tanonaka, T., Weng, J., Bandow, H., Takagi, H., and Akimoto, H. 1985. Ozone-cyclohexene reaction in air: quantitative analysis of particulate products and the reaction mechanism, *Environ. Sci. Technol.* 19:935–942.

Hecht, T. A., Seinfeld, J. H., and Dodge, M. 1974. Further development of a generalized kinetic mechanism for photochemical smog, *Environ. Sci. Technol.* 8:327–339.

Heisler, S. L., Friedlander, S. K., and Husar, R. B. 1973. The relationship of smog aerosol size and chemical element distributions to source characteristics, *Atmos. Environ.* 7:633–649.

Henry, R. C., and Hidy, G. M. 1979. Multivariate analysis of particulate sulfate and other air quality variables by principal components. Part 1. Annual data from Los Angeles and New York, *Atmos. Environ.* 13:1581–1596.

Henry, R. C., and Hidy, G. M. 1982. Multivariate analysis of particulate sulfate and other air quality variables by principal components. II. Salt Lake City, Utah and St. Louis, Missouri, *Atmos. Environ.* 16:929–943.

Hicks, B. B. 1977. On the parameterization of aerosol fluxes to the Great Lakes, *J. Great Lakes Res.* 3:263–269.

Hicks, B. B., and Wesely, M. L. 1981. Heat and momentum characteristics of adjacent fields of soybeans and maize, *Boundary Layer Meteorol.* 20:175–185.

Hicks, B. B., Wesely, M. L., Coulter, R. L., Hart, R. L., Durham, J. L., Speer, R. E., and Stedman, D. H. 1983. An experimental study of sulfur deposition to grassland, In: *Precipitation Scavenging, Dry Deposition and Resuspension* (H. R. Pruppacher, R. G. Semonin, and W. G. N. Slinn, eds.), Elsevier, New York.

Hidy, G. M. 1975. Summary of the California aerosol characterization experiment, *J. Air Pollut. Control Assoc.* 25:1106–1114.

Hopke, P. K. 1981. The application of factor analysis to urban aerosol source resolution, In: *Atmospheric Aerosol: Source/Receptor Relationships* (E. S. Macias and P. K. Hopke, eds.), pp. 21–49, Symposium Series, No. 167, American Chemical Society, Washington, D.C.

Hopke, P. K. 1985. *Receptor Modeling in Environmental Chemistry.* John Wiley and Sons, New York, N.Y.

Hopke, P. K., Gladney, E. S., Gordon, G. E., Zoller, W. H., and Jones, A. G. 1976. The use of multivariate analysis to identify sources of selected elements in the Boston urban aerosol, *Atmos. Environ.* 10:1015–1025.

Huebert, B. J. 1983. Measurements of the dry deposition flux of nitric acid vapor to grasslands and forest, In: *Precipitation Scavenging, Dry Deposition and Resuspension* (H. R. Pruppacher, R. G. Semonin, and W. G. N. Slinn, eds.), Elsevier, New York.

Huebert, B. J., and Robert, C. H. 1985. The dry deposition of nitric acid to grass, *J. Geophys. Res.* 90:2085–2090.

Irwin, J., and Smith, M. 1984. Potentially useful additions to the rural model performance evaluation, *Bull. Am. Meteorol. Soc.* 65:559–568.

Jacob, D. J. 1985. Comment on "The photochemistry of a remote stratoform cloud," *J. Geophys. Res.* 90:5864.

Jacob, D. J., and Hoffmann, M. R. 1983. A dynamic model for the production of H^+, NO_3^- and SO_4^{2-} in urban fog, *J. Geophys. Res.* 88:6611–6621.

Jech, D. D., Easley, P. G., and Krieger, B. B. 1982. Kinetics of reactions between free radicals and surfaces (aerosols) applicable to atmospheric chemistry, In: *Heterogeneous Atmospheric Chemistry* (D. R. Schryer, ed.), pp. 107–121, American Geophysical Union, Washington, D.C.

Johnson, W. B., Ludwig, F. L., Dabberdt, W. F., and Allen, R. J. 1973. An urban diffusion simulation model for carbon monoxide, *J. Air Pollut. Control Assoc.* 23:490–498.

Johnston, H. S., Cantrell, C. A., and Calvert, J. G. 1986. Unimolecular decomposition of NO_3 to form NO and O_2 and a review of N_2O_5/NO_3 kinetics, *J. Geophys. Res.* 91:5159–5172.

Killus, J. P., and Whitten, G. Z. 1982. A New Carbon Bond Mechanism for Air Quality Simulation Modeling, U.S. Environmental Protection Agency Report No. EPA 60013-82-041, Research Triangle Park, N.C.

Kumar, S. 1985. An Eulerian model for scavenging of pollutants by raindrops, *Atmos. Environ.* 19:769–778.

Kyan, C. P., and Seinfeld, J. H. 1974. On meeting the provisions of the clean air act, *Am. Inst. Chem. Eng. J.* 20:118–127.

Lamb, R. G. 1968. An air pollution model of Los Angeles, M.S. thesis, University of California, Los Angeles.

Lamb, R. G. 1973. Note on application of K-theory to turbulent diffusion problems involving non-linear chemical reactions, *Atmos. Environ.* 7:257–263.

Lamb, R. G., and Neiburger, M. 1971. An interim version of a generalized urban air pollution model, *Atmos. Environ.* 5:239–264.

Lamb, R. G., and Seinfeld, J. H. 1973. Mathematical modeling of urban air pollution, *Environ. Sci. Technol.* 7:253–261.

Lamb, S. I., Petrowski, C., Kaplan, I. R., and Simonett, B. R. T. 1980. Organic compounds in urban atmospheres: a review of distribution, collection and analysis, *J. Air Pollut. Control Assoc.* 30:1098–1115.

Leone, J. A., and Seinfeld, J. H. 1985. A comparative analysis of chemical reaction mechanisms for photochemical smog, *Atmos. Environ.* 19:437–464.

Levine, S. Z., and Schwartz, S. E. 1982. In-cloud and below-cloud scavenging of nitric acid vapor, *Atmos. Environ.* 16:1725–1734.

Liu, B. Y. H., and Ilori, T. A. 1974. Aerosol deposition in turbulent pipe flows, *Environ. Sci. Technol.* 8:351–356.

Liu, M.-K., and Seinfeld, J. H. 1975. On the validity of grid and trajectory models of urban air pollution, *Atmos. Environ.* 9:555–574.

Lloyd, A. C., Lurmann, F. W., Godden, D. K., Hutchins, J. F., Eschenroeder, A. Q., and Nordsiek, R. A. 1979. Development of the ELSTAR Photochemical Air Quality Simulation Model and Its Evaluation Relative to the LARPP Data Base, Environmental Research and Technology Report No. P-5287-500, West Lake Village, Calif.

MacCracken, M. C., Wuebbles, D. J., Walton, J. J., Duewer, W. H., and Grant, K. E. 1978. The Livermore Regional Air Quality Model: I. Concept and Development, *J. Appl. Meteorol.* 17:254–272.

Martin, A. 1984. Estimated washout coefficients for

sulfur dioxide, nitric oxide, nitrogen dioxide and ozone, *Atmos. Environ.* 18:1955–1961.

McRae, G. J. 1981. Mathematical modeling of photochemical air pollution, Ph.D. thesis, California Institute of Technology, Pasadena, Calif.

McRae, G. J., and Russell, A. G. 1983. Dry deposition of nitrogen-containing species, In: *Deposition Both Wet and Dry* (B. B. Hicks, ed.), Acid Precipitation Series, Vol. 4, pp. 153–193, Butterworth, Boston.

McRae, G. J., and Seinfeld, J. H. 1983. Development of a second-generation mathematical model for urban air pollution. II. Evaluation of model performance, *Atmos. Environ.* 17:501–522.

McRae, G. J., Goodin, W. R., and Seinfeld, J. H. 1982a. Development of a second generation mathematical model for urban air pollution. I. Model formulation, *Atmos. Environ.* 16:679–696.

McRae, G. J., Goodin, W. R., and Seinfeld, J. H. 1982b. Numerical solution of the atmospheric diffusion equation for chemically reacting flows, *J. Comput. Phys.* 45:1–42.

McRae, G. J., Tilden, J. W., and Seinfeld, J. H. 1982c. Global sensitivity analysis in computational implementation of the fourier amplitude sensitivity test (FAST), *Comput. Chem. Eng.* 6:15–25.

Middleton, P. B., and Brock, J. R. 1977. Modeling the urban aerosol, *J. Air Pollut. Control Assoc.* 27:771–775.

Miller, M. S., Friedlander, S. K., and Hidy, G. M. 1972. A chemical element balance for the Pasadena aerosol, *J. Colloid Interfac. Sci.* 39:165–176.

Munger, J. W., Jacob, D. J., Waldman, J. M., and Hoffmann, M. R. 1983. Fogwater chemistry in an urban atmosphere, *J. Geophys. Res.* 88:5109–5121.

Nappo, C. J., Caneill, J. Y., Furman, R. W., Gifford, F. A., Kaimal, J. C., Kramer, M. L., Lockhart, T. J., Pendergast, M. M., Pielke, R. A., Randerson, D., Shreffler, J. H., and Wungard, J. C. 1982. The workshop on the representativeness of meteorological observations, June 1981, Boulder, Colorado, *Bull. Am. Meteorol. Soc.* 63:761–764.

National Research Council. 1981. *Indoor Pollutants,* National Academy Press, Washington, D.C.

Pace, T. G. 1979. An Empirical Approach for Relating Annual TSP Concentrations to Particulate Microinventory Emissions Data and Monitor Siting Characteristics, Report No. EPA-450/4-79-012, U.S. Environmental Protection Agency, Research Triangle Park, N.C.

Pitts, J. N., Jr., and Winer, A. M. 1984. Particulate and Gas Phase Mutagens in Ambient and Simulated Atmospheres, Final Report to the California Air Resources Board, Riverside, CA.

Pitts, J. N., Jr., Winer, A. M., Darnell, K. R., Doyle, G. J., and McAfee, J. M. 1976. Chemical Consequences of Air Quality Standards and of Control Implementation Programs: Roles of Hydrocarbons, Oxides of Nitrogen and Aged Smog in the Production of Photochemical Oxidant, Final Report to the California Air Resources Board, Contract No. 4-214, University of California, Riverside.

Pitts, J. N., Jr., Winer, A. M., Atkinson, R., and Carter, W. P. L. 1983. Comment on "Effect of nitrogen oxide emissions on ozone levels in metropolitan regions," "Effect of NO_x emission rates on smog formation in the California south coast air basin," and "Effect of hydrocarbon and NO_x on photochemical smog formation under simulated transport conditions." *Environ. Sci. Technol.* 17:54–57.

Reeks, M. W., and Skyrme, G. 1976. The dependence of particle deposition velocity on particle inertia in turbulent pipe flow, *J. Aerosol Sci.* 7:485–495.

Reynolds, S. D., Roth, P. W., and Seinfeld, J. H. 1973. Mathematical modeling of photochemical air pollution. I. Formulation of the model, *Atmos. Environ.* 7:1033–1061.

Reynolds, S. D., Tesche, T. W., and Reid, L. E. 1979. An Introduction to the SAI Airshed Model and Its Usage, Report No. SAI-EF79-31, Systems Applications, Inc. San Rafael, Calif.

Roache, P. J. 1976. *Computational Fluid Dynamics,* Hermosa Publishers, Albuquerque, N.M.

Russell, A. G., and Cass, G. R. 1984. Acquisition of regional air quality model validation data for nitrate, sulfate, ammonium ion and their precursors, *Atmos. Environ.* 18:1815–1827.

Russell, A. G., and Cass, G. R. 1986. Verification of a mathematical model for aerosol nitrate and nitric acid formation, and its use for control measure evaluation, *Atmos. Environ.* 20:2011–2025.

Russell, A. G., McRae, G. J., and Cass, G. R. 1983. Mathematical modeling of the formation and transport of ammonium nitrate aerosol, *Atmos. Environ.* 17:949–964.

Russell, A. G., McRae, G. J., and Cass, G. R. 1985. The dynamics of nitric acid production and the fate of nitrogen oxides, *Atmos. Environ.* 19:893–903.

Russell, A. G., McCue, K. F., and Cass, G. R. 1988. Mathematical modeling of the formation of nitrogen-containing air pollutants. I. Evaluation of an Eulerian photochemical model (in press).

Schere, K. L. 1983. An evaluation of several numerical advection schemes, *Atmos. Environ.* 17:1897–1907.

Schiermeier, F. A. 1978. Air monitoring milestones. RAPS field measurements are in, *Environ. Sci. Technol.* 12:644–651.

Schwartz, S. E., and Frieberg, J. E. 1981. Mass-transport limitation to the rate of reaction of gases in liquid droplets: Application to oxidation of SO_2 in aqueous solutions, *Atmos. Environ.* 15:1129–1144.

Sehmel, G. A. 1971. Particle diffusivities and deposition velocities over a horizontal smooth surface, *J. Colloid Interfac. Sci.* 37:891–906.

Sehmel, G. A. 1980. Particulate and gas dry deposition: A review, *Atmos. Environ.* 14:983–1011.

Sehmel, G. A., and Hodgson, W. H. 1974. Predicted dry deposition velocities, In: *Atmosphere-Surface Exchange of Particulate and Gaseous Pollutants,* pp. 399–419, Energy Research and Development Administration, NTIS CONF-740921.

Seigneur, C., Tesche, T. W., Roth, P. M., and Reid, L. E. 1981. Sensitivity of a complex urban air quality model to input data, *J. Appl. Meteorol.* 20:50–70.

Seigneur, C., Tesche, T. W., Roth, P. M., Liu, M-K. 1983. On the treatment of point source emissions in urban air quality modeling, *Atmos. Environ.* 17:1655–1676.

Seinfeld, J. H. 1975. *Air Pollution: Physical and Chemical Fundamentals*, McGraw-Hill, New York.

Seinfeld, J. H. 1986. *Atmospheric Chemistry and Physics of Air Pollution*, Wiley-Interscience, New York.

Seinfeld, J. H., and McRae, G. J. 1979. Use of models to establish source-receptor relationships and estimate relative source contributions of NO_x to air quality problems, for the technical Symposium on the Implications of a Low NO_x Vehicle Emission Standard, U.S. Environmental Protection Agency, Reston, Va., May 2–4.

Sexton, K., and Hayward, S. B. 1987. Source apportionment of indoor air pollution, *Atmos. Environ.* 21:407–418.

Shafer, T. B., and Seinfeld, J. H. 1986. Comparative analysis of chemical reaction mechanisms for photochemical smog. II. Sensitivity of EKMA to chemical mechanism and input parameters, *Atmos. Environ.* 20:487–499.

Shair, F. H., and Heitner, K. L. 1974. Theoretical model for relating indoor pollutant concentrations to those outside, *Environ. Sci. Technol.* 8:444–451.

Sklarew, R. C., Fabrick, A. J., and Prager, J. E. 1972. Mathematical modeling of photochemical smog using the PIC method, *J. Air Pollut. Control Assoc.* 22:865–869.

Slinn, S. A., and Slinn, W. G. N. 1981. Modeling of atmospheric particulate deposition to natural waters, In: *Atmospheric Pollutants in Natural Waters* (S. Eisenreich, ed.), Ann Arbor Science, Ann Arbor, Mich.

Smith, M. 1984. Review of the attributes and performance of 10 rural diffusion models, *Bull. Am. Meteorol. Soc.* 65:554–559.

South Coast Air Quality Management District. 1982. Results of the SAI airshed model evaluation of the 1987 impact of the draft 1982 AQMP revision short range control strategies, AQMP Technical Paper No. 9, South Coast Air Quality Management District, El Monte, Calif.

South Coast Air Quality Management District and Southern California Association of Governments. 1982. *Final Air Quality Management Plan*, 1982 Rev., El Monte, Calif.

Stevens, R. K., and Pace, T. G. 1984. Overview of the mathematical and empirical receptor models workshop (Quail Roost II), *Atmos. Environ.* 18:1499–1506.

Tesche, T. W., Seigneur, C., Oliver, W. R., and Haney, J. L. 1984. Modeling ozone control strategies in Los Angeles. *J. Environ. Eng.* 110:208–225.

Tilden, J. W., and Seinfeld, J. H. 1982. Sensitivity analysis of a mathematical model for photochemical air pollution, *Atmos. Environ.* 16:1357–1364.

Transportation Research Board. 1976. Assessing transportation-related air quality impacts, Special Report 167, National Academy of Sciences, Washington, D.C.

Trijonis, J. C. 1974. Economic air pollution control model for Los Angeles County in 1975, *Environ. Sci. Technol.* 8:811–826.

Turk, A. 1963. Measurements of odorous vapors in test chambers: theoretical, *ASHRAE J.* 5:55–58.

Turner, D. B. 1964. A diffusion model for an urban area, *J. Appl. Meteorol.* 3:83–91.

Turner, D. B. 1967. A Workbook of Atmospheric Dispersion Estimates, Public Health Service Publication No. 999-AP-26, U.S. Government Printing Office, Washington, D.C.

Wagner, K. K., and Ranzieri, A. J. 1984. Model performance evaluations for regional photochemical models in California, Paper No. 84-47.3, presented at the 77th Annual Meeting of the Air Pollution Control Association, San Francisco, Calif. (June).

Walcek, C. J., Brost, R. A., Chang, J. S., and Wesely, M. L. 1986. SO_2, sulfate, and HNO_3 deposition velocities computed using regional land use and meteorological data, *Atmos. Environ.* 20:949–964.

Watson, J. G. 1979. Chemical element balance to receptor methodology for assessing the sources of fine and total suspended particulate matter in Portland, Oregon, Ph.D. thesis, Oregon Graduate Center, Beaverton, Ore.

Watson, J. G. 1984. An overview of receptor modeling principles, *J. Air Pollut. Control Assoc.* 34:619–623.

Watson, J. G., Henry, R. C., Cooper, J. A., and Macias, E. S. 1981. The state of the art of receptor models relating ambient suspended particulate matter to sources, In: *Atmospheric Aerosol: Source/Air Quality Relationships* (E. S. Macias and P. K. Hopke, eds.), pp. 89–106, Symposium Series No. 167, American Chemical Society, Washington, D.C.

Weber, E. 1982. *Air Pollution, Assessment Methodology and Modeling*, Plenum, New York.

Wesely, M. L., and Hicks, B. B. 1977. Some factors that affect the deposition rates of sulfur dioxide and similar gases on vegetation, *J. Air Pollut. Control Assoc.* 27:1110–1116.

Wesely, M. L., Cook, D. R., Hart, R. L., and Speer, R. E. 1985. Measurements and parameterization of particulate sulfur dry deposition over grass, *J. Geophys. Res.* 90:2131–2143.

Whelpdale, D. M., and Shaw, R. W. 1974. Sulfur dioxide removal by turbulent transfer over grass, snow and water surfaces, *Tellus* 26:196–205.

Whitby, K., and Cantrell, B. 1976. Fine particles, Paper presented at the International Conference on Environmental Sensing and Assessment, Las Vegas, Nev., Institute of Electrical and Electronic Engineers.

Whitten, G. Z., and Hogo, H. 1978. A User's Manual for Kinetics Model and Ozone Isopleth Plotting Package, U.S. Environmental Protection Agency, No. EPA-600/3-78-014a, Research Triangle Park, N.C., 239 pp.

Whitten, G. Z., Hogo, H., Meldgin, M. J., Killus, J. P., and Bekowies, P. J. 1979. Modeling of simulated photochemical smog with kinetic mechanisms, Vol. I and II, U.S. Environmental Protection Agency Report No. EPA-600/3-79-001a, Research Triangle Park, N.C.

Winer, A. M., Atkinson, R., Pitts, J. N. 1984.

Gaseous nitrate radical—A possible nighttime atmospheric sink for biogenic compounds, *Science* 224:156–159.

Yamartino, R. J. 1983. Formulation and application of a hybrid source receptor model, In: *Receptor Models Applied to Contemporary Pollution Problems* (S. C. Dattner and P. K. Hopke, eds.), pp. 285–295, Air Pollution Control Assoc., Pittsburgh, Pa.

Yamartino, R. J., and Lamich, D. J. 1979. The formulation and application of a source finding algorithm, Fourth Symposium of Turbulence, Diffusion and Air Pollution, Jan. 15–18, Reno, Nev., pp. 84–88, American Meteorological Society, Boston.

Yu, T. W. 1977. A comparative study on parameterization of vertical turbulent exchange processes, *Monthly Weath. Rev.* 105:57–66.

Zak, B. D. 1982. Lagrangian studies of atmospheric pollutant transformations, *Adv. Environ. Sci. Technol.* 12:303–344.

Assessment of Human Exposure to Air Pollution: Methods, Measurements, and Models

KEN SEXTON
Health Effects Institute

P. BARRY RYAN
Harvard School of Public Health

Air Pollution, the Automobile, and Public Health. © 1988 by the Health Effects
Institute. National Academy Press, Washington, D.C.

Human Exposure: Introduction

Accurate estimates of human exposure to inhaled air pollutants are necessary for a realistic appraisal of the risks these pollutants pose and for the design and implementation of strategies to control and limit those risks. Except in occupational settings, such estimates are usually based on measurements of pollutant concentrations in outside (ambient) air, recorded with outdoor fixed-site monitors.

Indeed, compliance with existing National Ambient Air Quality Standards (NAAQS), intended to protect public health with an adequate margin of safety, depends exclusively on outdoor measurements of pollutants. But, such measurements are subject to biases because most people spend much more of their time indoors than out, and air pollutant concentrations are often much higher inside buildings than outside (National Research Council 1981; Spengler and Sexton 1983). In addition, available evidence indicates that personal exposure to many pollutants is not adequately characterized because the time people spend in different locations and their activities vary dramatically with age, gender, occupation, and socioeconomic status (National Research Council 1981; World Health Organization 1982, 1983; Yocum 1982; Spengler and Sexton 1983; Spengler and Soczek 1985).

In this chapter, the state of the art of air pollution exposure assessment is discussed with emphasis on gaps in our knowledge and the implications of those gaps for future research. First, important terms are defined, and then the methods available for monitoring exposure, the results of exposure assessment studies, and the models for exposure estimation are examined.

Definitions

Concentration, Exposure, and Dose

The concentration of a specific air pollutant is the amount of material per unit volume of air. Concentrations are most commonly expressed as mass per unit volume (for example, micrograms per cubic meter). Concentrations of pollutant gases may be reported as volume per unit volume (for example, parts per million by volume) and discrete particles as number per unit volume (for example, number of fibers per cubic centimeter).

Exposure refers to any contact between an airborne contaminant and a surface of the human body, either outer (for example, the skin) or inner (for example, respiratory tract epithelium). Thus, exposure requires the simultaneous occurrence of two events: a pollutant concentration at a particular place and time, and the presence of a person at that place and time (Duan 1982; Ott 1985). Exposure is expressed quantitatively by a description of the duration of the contact and the relevant pollutant concentration.

There is an important distinction between concentration and exposure. Concentration is a physical characteristic of the environment at a certain place and time, whereas strictly speaking, exposure describes an interaction between the environment and a living subject. Thus, a concentration in a room with people present is a surrogate measurement of exposure, but is valid only to the degree that it approximates the concentrations actually experienced by each individual in the room.

The distinction between exposure and dose is also important. As stated above, exposure is the pollutant concentration in the air at the point of contact between the body and the external environment. Dose is the amount of the pollutant that actually crosses one of the body's boundaries and reaches the target tissue.

The difference between exposure and dose is illustrated by considering two people, one sedentary and one vigorously active, in a room where the air pollutant concentration is constant. Both have the same nominal exposure. But because of faster and deeper breathing, the actual dose of air pollution delivered to lung tissues is greater in the active subject than in the sedentary subject.

Components of Exposure

Three aspects of exposure are important for determining related health consequences.

1. Magnitude: What is the pollutant concentration?

2. Duration: How long does the exposure last?

3. Frequency: How often do exposures occur?

Magnitude is an important exposure parameter because concentration typically is assumed to be directly proportional to dose and ultimately to the health outcome. But exposure implies a time component, and it is essential to specify the duration of an exposure. The health risks of exposure to a specific concentration for 5 minutes are likely to be different, all other factors being equal, than exposure to the same concentration for an hour. Similarly, the frequency of exposure or the time between subsequent exposures might have health implications. Whether a person is exposed once a week or several times a day can be an important determinant of air pollution injury.

A real-time air pollutant monitor carried by a person for 24 hr would provide a continuous exposure record for that period. Depending on the pollutant and the person's activities during that period, the record might show some intervals of zero exposure and some intervals of very high exposure. The full record would contain all exposure information for that day, but it is often too complex to work with, as well as too difficult and expensive to obtain.

It is common to rely on data summaries (averages) that depend on the capabilities of the available instruments. In most exposure studies, magnitude is defined as the average concentration over some specified time interval (for example, 1, 8, or 24 hr). Duration is the time (or average time) from the beginning to the end of a nonzero exposure, and frequency is the number of exposure episodes (of a specified duration) per unit of time.

Types of Exposure Information

Data on human exposure can be presented in several ways (Ott 1982, 1983-84, 1985). For an individual, i, a plot of exposure magnitude as a function of time, $C_i(t)$, typically covering a 24-hr period, is called an exposure profile. As shown in figure 1a,

additional data about the subject's activities can be combined with the exposure profile to show when, where, and how the highest-magnitude exposures occurred.

Integrating the function $C_i(t)$ with respect to time t for a specified time period yields the integrated exposure. The integration is represented graphically in figure 1b and shows a 24-hr integrated exposure of 960 parts per billion-hour (ppb-hr). The integrated exposure does not provide information about the pattern of exposure over subintervals of the averaging time, nor does it reflect the magnitude of short-term peaks in exposure. Figure 1c shows several examples of average exposure, t_a, arrived at by dividing the integrated exposure by the period of integration. The figure gives eight 3-hr averages, three 8-hr averages, and a single 24-hr average derived from the same 24-hr period of data.

In spite of the importance of these distinctions, it is common to refer to the average exposure—that is, the average concentration during a specific measurement period (for example, 24 hr)—as the exposure. In some instances, it is also common to refer to the average concentration measured by a fixed-site monitor as the exposure, even though no individual was actually in the vicinity of the instrument for the duration of the measurement period. The blurring of these distinctions, like those between weight and mass or between heat and temperature, causes little confusion for those well versed in the literature. For others, however, it is important to keep in mind that a measurement of air pollutant concentration is a surrogate for exposure only to the degree that it reflects actual concentrations experienced by people.

Individual Exposure Versus Population Exposure

The pollutant concentrations experienced by an individual during normal daily activities are referred to as personal or individual exposures. A personal exposure depends on the air pollutant concentrations that are present in the locations the person moves through as well as on the time spent in each location. Individual exposures for a speci-

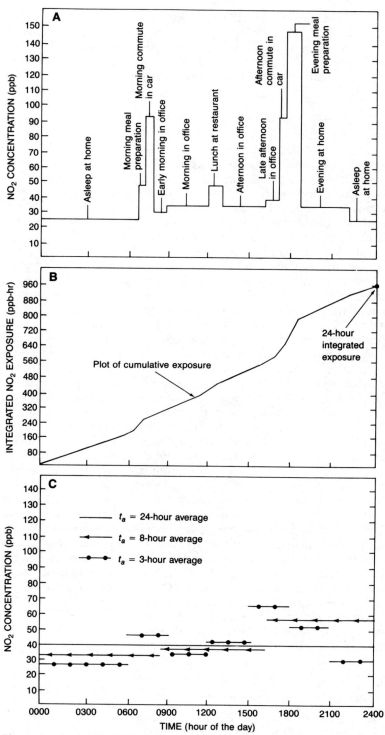

Figure 1. Examples of NO$_2$ exposure information: A: a 24-hr exposure profile and associated time–activity pattern data for a specific individual; B: a plot of cumulative exposure and the calculated 24-hr integrated exposure; C: calculated exposures averaged over 3 hr, 8 hr, and 24 hr.

fied group of people may vary widely because of their different time-activity patterns (Dockery and Spengler 1981; Quackenboss et al. 1982; Ott 1983–84; Sexton et al. 1983; Letz et al. 1984; Spengler et al. 1985; Stock et al. 1985; Wallace et al. 1985a).

Measuring any one person's exposure is a relatively straightforward procedure, but from a public health perspective it is important to determine the population exposure—the aggregate exposure for a specified group of people, such as a community or an occupational cohort. It is rarely necessary or desirable to measure the exposure of each member of the group. But some measure of the distribution of individual exposures is needed. This typically includes at least a measure of the central tendency (for example, mean exposure) and of its variability (for example, variance). An accurate and statistically valid characterization of even these simple descriptors of population exposure may require many personal exposure measurements.

The upper tail of the distribution is frequently of special interest, because it represents the segment of the population that has much higher-than-average personal exposures. Determination of the numbers and kinds of people who experience exceptionally high exposures can be critical for health risk assessment. This is especially true when the relationship between the pollutant dose and resultant health effects is highly nonlinear. Typically, more personal exposure measurements are needed to accurately estimate the tails of the distribution than are needed to estimate its mean and variance.

Methods

Basically, there are two general approaches to air pollution exposure assessment: (1) air monitoring, which depends on either direct measurements (personal monitors) or indirect measurements (fixed-site monitors combined with data on time-activity patterns), and (2) biological measurements that use biological markers to assess expo-

sure. In the past, questionnaires have also been used to estimate exposures, particularly in epidemiologic studies. Typically, questionnaires are used to categorize respondents into two or more groups (for example, exposed or unexposed, high exposure or low exposure). This is a qualitative, often retrospective, method for estimating air pollution exposure. It depends on a priori knowledge of exposures and their determinants to develop effective questionnaires (for example, high formaldehyde exposure for workers in certain industries, or high carbon monoxide [CO] and lead [Pb] exposure for traffic policemen, bus drivers, and toll collectors). Most often the information necessary to develop questionnaires is obtained from previous studies that used either air monitoring or biological monitoring to measure exposures. The questionnaire method is really a way to extend the results of prior air monitoring or biological measurements to a larger or different population and is not a separate approach.

Air Monitoring

Direct Approach to Exposure Assessment. A personal monitor is a small, lightweight device, such as a diffusion tube or a filter with a battery-operated pump, that can be carried or worn by a person during his or her normal daily activities. Personal monitors make it possible to measure exposures for an identified subset of the general population. Moreover, if study participants maintain records of their activities, then locations where highest concentrations occur as well as the nature of emission sources can often be inferred. The major impediment to this type of assessment has been the lack of suitable instruments.

Small, quiet, portable personal exposure monitors that are sensitive enough to measure ambient concentrations of some pollutants are now available (Lautenberger et al. 1981; Rose and Perkins 1982; Wallace and Ott 1982; Bartley et al. 1983; Underhill 1984). Pollutants that can be measured accurately with personal monitors include nitrogen dioxide (NO_2) (Palmes et al. 1976;

Palmes and Tomczyk 1979; Woebkenberg 1982; Yanagisawa and Nishimura 1982), respirable particles (Turner et al. 1979), formaldehyde (Geisling et al. 1982; Kring et al. 1984), sulfur dioxide (SO_2) (Coleman 1983; Kring et al. 1983), organic vapors (Feigley and Chastain 1982; Seifert and Abraham 1983; Vo-Dinh and Miller 1983; Compton et al. 1984; Reggin and Peterson 1985; Sheldon et al. 1985), and CO (Akland et al. 1985; Ott et al. 1986).

Personal monitors can be grouped into two general categories: integrated samplers that collect the pollutant over a specified time period and then are returned to the laboratory for analysis, and continuous samplers that use a self-contained analytical system to measure and record the pollutant concentration on the spot. Instruments in both categories can be either active or passive. Active monitors use a pump and a power source to move air past a collector or sensor. Passive monitors depend on diffusion to bring the pollutants into contact with the collector or sensor. Information about personal monitors is summarized in table 1 (Wallace and Ott 1982).

Most personal monitors available today are integrated samplers with sampling periods ranging from 8 hr to a week or more. Active integrated sampling devices are commonly used to obtain integrated exposure measurements over an 8- to 24-hr period. In general, they are bulky, noisy, and require frequent calibration to ensure the validity of the data they collect.

Passive samplers are simple, small, quiet, inexpensive, and easy to use; but at ambient concentrations normally require a longer sampling period (for example, 1 or 2 weeks) to collect enough material for analysis. A passive sampler, therefore, cannot be used to relate short-term exposures (minutes or hours) to specific events or sources. Moreover, passive samplers are affected by temperature, relative humidity, and air movement and tend to be less accurate than active monitors. They are most appropriate for large-scale surveys of population exposure, where pinpoint accuracy is not required and long-term exposures are of primary interest.

Although considerable progress has been

made in miniaturizing real-time analytical monitors, much work remains to be done. Continuous personal monitors have not been developed for most of the important pollutants (see table 1). Those that are available can provide data on measured concentrations as a function of time throughout the day. These data can be used to construct exposure profiles (see figure 1) and, when combined with time-activity information, can be used to relate short-term exposures to specific events and sources. Because they record a large number of real-time measurements, continuous personal monitors should log and store data to be most effective.

Participants are typically asked to maintain a detailed record of their time-activity patterns during the test period. The record is usually a log or diary documenting the subject's location and activity at particular times. Recently, a small microprocessor-based data logger was developed that automatically computes and stores times and average concentrations (Ott et al. 1986). The subject only records the type of activity engaged in and presses a button, and the instrument stores all other information electronically to be retrieved and analyzed later.

As Wallace and Ott (1982) pointed out, the direct measurement of exposures using personal monitors raises several methodological issues. Personal monitoring studies are complex, expensive, time-consuming, and labor-intensive. They present problems because they generally require the selection and recruitment of representative subjects; the distribution, maintenance, and retrieval of many monitors; either a laboratory analysis of many air samples returned from monitors in the field or calibration and validation of many real-time monitors; and the transcription and statistical analysis of data on pollutant concentrations and time-activity patterns. The problems raised by the three latter points are fairly obvious, but the difficulties associated with selecting and recruiting a test sample require amplification.

Personal exposure monitoring is, by its very nature, an intrusive event in the life of the study participants. The degree of incon-

Table 1. Personal Exposure Monitors Capable of Quantitative Pollutant Measurements at Ambient Concentrations

Monitor Type	Pollutants	Collection Method	Analytical Method
Integrated, Active	Respirable particles (sulfates, nitrates, metals)	Pump/stack filter (2 size fractions)	Microbalance chemical analysis PIXE
	Respirable particles (mass only)	Pump-impactor/precipitator	Piezoelectric
	Respirable particles (sulfates, nitrates, metals)	Pump/filter	Microbalance PIXE
	Sulfur dioxide, nitrogen dioxide, respirable particles	Pump/impingers/filter	Colorimetric gravimetric
	Nonpolar volatile organics	Pump/Tenax cartridge	Thermal desorption/GC–MS
	Organochlorine pesticides, polychlorinated biphenyls	Pump/polyurethane foam	GC
Integrated, Passive	Carbon monoxide	Diffusion	Electrochemical
	Nitrogen dioxide	Diffusion tube (TEA) adsorbent	Colorimetric
	Nitrogen dioxide	Badge/TEA	Colorimetric
	Nitrogen dioxide	Diffusion/dimethylsilicone filter/TEA	Colorimetric
	Nitrogen dioxide	Diffusion/TEA–impregnated filter	Colorimetric
	Formaldehyde	Permeable membrane	MBTH, pararosaniline
	Formaldehyde	Diffusion badge	Chromotropic acid
	Polynuclear aromatics	Diffusion badge	Room-temperature phosphorescence
	Vinyl chloride	Permeable membrane badge/activated charcoal	Solvent desorption/GC
	Radon	Plastic (records radiation damage)	Etching/microscopic examination
Continuous, Active	Carbon monoxide	Pump electrolyte	Sulfuric acid
	Carbon monoxide	Pump electrolyte	Solid polymer

NOTE: GC = gas chromatography; GC–MS = gas chromatography–mass spectrometry; MBTH = 3-methyl-2-benzothiozolinone; PIXE = proton-induced x-ray emission; TEA = triethanolamine.
SOURCE: Adapted with permission from Wallace and Ott 1982, and from the Air Pollution Control Association.

venience depends on the size, weight, appearance, and ease of operation of the monitor, as well as other aspects of the study, such as the need to fill out logs or diaries. The demands of the project protocol and the associated inconvenience may cause many people to refuse to cooperate. It is particularly difficult to get the cooperation of schoolchildren, non-English-speaking people, disadvantaged people, or those with low socioeconomic status. The response rate may be raised by offering incentives, but, even so, additional incentives for those that complete the study may be

necessary to forestall high dropout rates. In any case, simply wearing a monitor or filling out a log can cause the participant to change his or her behavior and consequently introduce bias (Sexton et al. 1986a; Ryan et al. 1987).

Direct personal monitoring is the most accurate means of exposure assessment, but it is also the most expensive. Large-scale personal monitoring studies are a recent development, so many survey design, logistic, and technical problems remain to be solved. More attention should be focused on these issues to make subsequent personal exposure studies more cost-effective.

Indirect Approach to Exposure Assessment.

The indirect approach estimates integrated exposure by combining measurements of pollutant concentrations at fixed sites (for example, outdoors at a busy intersection, indoors at home) with data logs and diaries about the times people spend in specific environments (Fugas et al. 1972; Fugas 1975; Dockery and Spengler 1981; Duan 1981, 1982; Ott 1982; Sexton et al. 1983, 1984b). The general form of the equation used to calculate time-weighted integrated exposure is

$$E_i = \sum_{j}^{J} C_j t_{ij} \qquad (1)$$

where E_i is the time-weighted integrated exposure for person i over the specified time period; C_j is the pollutant concentration in microenvironment j; t_{ij} is the aggregate time that person i spends in microenvironment j; and J is the total number of microenvironments that person i moves through during the specified time period.

A microenvironment is defined as a three-dimensional space where the pollutant level at some specified time is uniform or has constant statistical properties. Outdoors in a specific community, inside an individual motor vehicle, and inside a particular residence are examples of locations that can be defined, under appropriate conditions, as microenvironments. Examples of potentially important microenvironments for exposure assessment are given in table 2.

Several assumptions are implicit in the application of equation 1:

1. The concentration C_j in microenvironment j is assumed to be constant during the time t_{ij} that person i is there. This is not always the case. For example, it is likely that air pollution levels inside one's residence will vary substantially during the 14 to 16 hr/day that most people spend at home, because of variations in emission rates and air exchange rates.

2. The concentration C_j within microenvironment j and the time t_{ij} that person i spends there are assumed to be independent events. This assumption is not universally valid, however. Persons sensitive to pollutants like tobacco smoke and formaldehyde, or to noxious odors, such as those from paint and cleansing solutions, are likely to avoid microenvironments where concentrations of these pollutants are elevated.

3. The number of microenvironments necessary to characterize personal exposure adequately is assumed to be small, but in fact, it is not clear how many are necessary. Within the indoor residential environment, for example, the variability in short-term particle concentrations from activities such as cooking, smoking, and cleaning might necessitate the inclusion of several additional microenvironments in the model to comply with assumption 1 above.

4. The time-weighted integrated exposure, usually measured over 24 hr, is directly related to the health outcome. This may not be the case for adverse health effects due to short-term peak exposures (hours, minutes, or in some cases seconds) to pollutants such as formaldehyde, NO_2, or ozone (O_3).

The concept of a time-weighted integrated exposure is illustrated in figure 2. A unit width is indicated on the j axis for each of five microenvironments: indoors at home, indoors at work, indoors in other locations, in transit, and outdoors. The concentration of respirable particles (RSP) is displayed on the Y axis, and the fraction of time that person i spends in each microenvironment over the 24-hr period is plotted on the t axis. The volumes of the boxes shown in figure 2 represent contributions from each of the five microenvironments to the time-weighted integrated exposure. The contribution of each microenvironment is represented mathematically in the table at the bottom of figure 2.

Even though respirable particle concentration was low inside the home, it contributed significantly to the time-weighted exposure because this person spent 18 out of 24 hr there. Conversely, the respirable particle concentration outdoors made only a

Table 2. Potentially Important Microenvironments for Air Pollution Exposure Assessment

Microenvironments	Comments
Outdoors	
Urban	Metropolitan areas where air pollution levels are high as a result of a high density of mobile and stationary sources.
Suburban	Small- to medium-sized cities where air pollution levels tend to be lower than metropolitan areas, although transport of urban pollution can affect local air quality under certain conditions.
Rural	Agricultural communities and small towns with few major anthropogenic sources of air pollution. Air pollution levels tend to be low, although transport of urban and suburban pollution can affect local air quality under certain conditions.
Indoors—Occupational	
Industrial	Manufacturizing and production processes, such as those in petrochemical plants, pulp mills, power plants, and smelters.
Nonindustrial	Primarily service industries where workers are not involved in manufacturing and production processes, such as insurance companies, law offices, and retail sales outlets.
Indoors—Nonoccupational	
Residential	Single-family houses, apartments, mobile homes, condominiums
Commercial	Restaurants, retail stores, banks, supermarkets
Public	Post offices, courthouses, sports arenas, convention halls
Institutional	Schools, hospitals, convalescent homes
Indoors—Transportation	
Private	Automobiles, private airplanes
Public	Buses, subways, trains, commercial airplanes

minor contribution because this person was outdoors less than half an hour during the 24-hr period.

This illustrates the general problems associated with attempts to define the limits of microenvironments that are sufficiently homogeneous, to identify which among them are the significant contributors to integrated exposure, and to measure or estimate both the pollutant concentration C_j and the average time, t_{ij}, the subject spends in the microenvironment.

Better documentation of time-activity patterns, as well as more information about approximate indoor and outdoor pollutant concentrations would help investigators specify important microenvironments and choose fixed monitoring sites. In most cases, however, there is not enough information to determine which microenvironments are adequately defined, which can be bypassed or lumped with others, which should be subdivided, and which should have their limits altered to ensure accurate exposure estimates. Although the temporal

and spatial aspects of people's activity patterns are reflected separately in the time budgets and mobility patterns that sociologists, urban planners, economists, and transportation analysts use, these data are not in a form suitable for application to exposure assessment. Only in the past few years have both temporal and spatial aspects of people's everyday movements been investigated in conjunction with air pollution measurements (Spengler et al. 1980, 1985; Dockery and Spengler 1981; Dockery et al. 1981; Moschandreas 1981; Ott and Flachsbart 1982; Sega and Fugas 1982; Sexton et al. 1983, 1984b; Flachsbart and Brown 1985; Nagda and Koonz 1985; Wallace et al. 1985a,b).

Much of what is known about human time-activity patterns can be traced to two studies now more than a decade old (Szalai 1972; Chapin 1974). A summary of measured 24-hr time-activity patterns from these studies is provided in table 3. Both studies found that on most days people are inside their residences for an average of 65

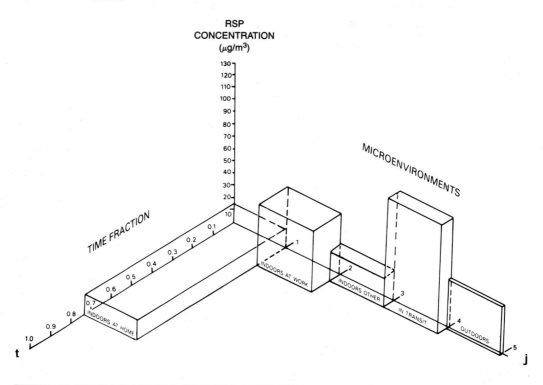

Microenvironment Type	RSP Concentration (C_j, $\mu g/m^3$)	Time Fraction[a] (t_{ij})	$C_j \times t_{ij}$ ($\mu g/m^3$)	Microenvironment Contribution to E_i (%)[b]
Indoors at Home	15	0.75	11.25	47
Indoors at Work	50	0.15	7.50	31
Indoors, Other	25	0.04	1.00	4
In Transit	90	0.04	3.60	15
Outdoors	40	0.02	0.80	3

$$E_i = \Sigma\, C_j \times t_{ij} = 24.15\ \mu g/m^3$$

[a] Fraction of 24 hr spent in each microenvironment.
[b] Percentage that each microenvironment contributes to the 24-hr, time-weighted, integrated exposure (E_i).

Figure 2. Examples of the relative contributions from specific microenvironments to an individual's time-weighted, integrated exposure to respirable particles (RSP).

to 70 percent of the time, and indoors at home, work, or elsewhere for more than 90 percent of the time. Although these values vary with age, gender, occupation, socio-economic status, and day of the week, it has become clear that indoor microenvironments must be taken into account for a realistic assessment of exposure to many air pollutants (National Research Council 1981; World Health Organization 1982, 1983; Yocum 1982; Spengler and Sexton 1983; Lebowitz et al. 1984; De Bortoli et al. 1985; Stock et al. 1985), including NO_2 (Quackenboss et al. 1982; Ryan et al. 1983; Sexton et al. 1983; Spengler et al. 1983), formaldehyde (Environmental Health Perspective 1984; Sexton et al. 1986a), CO (Jaeger 1981; Ott and Willits 1981; Ott and Flachsbart 1982; Ziskind et al. 1982), respirable particles (Spengler et al. 1981; Sexton et al. 1984a,b; Sexton et al. 1986b), radon (Nero and Lowder 1983), and organic va-

Table 3. Summary of Average Time-Activity Patterns for a 24-Hr Period

Location	Hours in Each Location	
	Chapin (1974)	Szalai (1972)
Indoors		
Home	16.03	16.75
Work	4.61	4.03
Other	1.31	1.63
Subtotal	21.95	22.41
Outdoors		
Home	0.27	0.23
Work	—	—
Other	0.27	0.12
Subtotal	0.54	0.35
In Transit		
All modes	1.16	1.25
Total	23.65[a]	24.01

[a] Shortfall from 24 hr not explained by the author.
SOURCE: Adapted with permission from World Health Organization 1982.

pors (Beall and Ulsamer 1981; Hollowell and Miksch 1981; Jarke et al. 1981; Miksch et al. 1982; Molhave 1982; Otson et al. 1983; Wallace et al. 1984; Andelman 1985; Wallace et al. 1985a; Sexton et al. 1986c).

In addition to the problems of identifying important microenvironments and of obtaining valid measurements of pollutant concentrations, the indirect approach suffers from the same problems as the direct approach: the selection and recruitment of a representative sample of people; the distribution, maintenance, and retrieval of many monitors; either a laboratory analysis of many samples returned from monitors in the field or calibration and validation of many real-time monitors; and the transcription and statistical analysis of data on pollutant concentrations and time-activity patterns.

Biological Monitoring

Air monitoring traditionally has been the principal means of exposure assessment. A major shortcoming of this approach is its failure to take account of factors such as respiration rate and depth of inspiration that may cause two individuals with the same measured exposure to receive vastly

different doses. Differences in dose at equivalent exposures, coupled with variations in individual susceptibility, introduce a large measure of uncertainty in the extrapolation from air pollutant measurements to the effects on human health. Thus there is an acute need for methods that provide better information about the interrelationships of exposure, dose, and health effects.

Biological monitoring is the measurement of environmental contaminants or their biological consequences after the contaminants have crossed one of the body's surfaces and entered tissues or fluids. There are two kinds: measurements of environmental contaminants or their metabolites and derivatives in body fluids or excreta (exposure markers); and measurements of biological responses in cells and tissues (exposure markers and effects markers). Examples of the first type include direct chemical analyses, immunoassays, and bioassays specific for mutagenicity; these methods can be used to measure chemicals in the blood, urine, breast milk, saliva, and semen. Examples of the second category include immunologic and chemical methods to detect and quantify covalently bound derivatives formed between activated chemicals and cellular macromolecules such as nucleic acids and proteins, as well as observations of mutation, sister chromatid exchange, and chromosome aberrations (Wogan and Gorelick 1985).

Biological measurements enable the development of exposure markers related qualitatively or quantitatively to measured air pollution concentrations (Goldstein 1981; Miller 1983; Berlin et al. 1984; National Institute of Environmental Health Science 1984; Wogan and Gorelick 1985; Ho and Dillion 1986). Exposure markers are not necessarily closely correlated with subsequent health effects for two reasons: first, the site and mechanism of toxic action associated with adverse health effects are not always fully understood; and second, some identified sites of toxic action are not accessible for analysis. For example, cotinine is a metabolite of nicotine that can be detected in the blood of infants whose mothers smoke as well as in the mothers themselves, but the role, if any, of this metabolite in the toxicity of tobacco smoke is unknown (National Institute of Environmental Health Science 1984).

Biological monitoring has three major advantages over environmental measurements for estimating health risks. First, only the pollutants that cross the boundary and enter the body are included in the analysis. Second, biological markers are more directly related to the biological processes from which the health consequences arise. And third, biological monitoring can serve as the basis for total risk estimates from multiple chemicals because it takes into account absorption by all routes and integrates exposures from all sources (Wogan and Gorelick 1985). However, the availability of biological exposure markers does not obviate the need for measurements of air pollution concentrations. They should complement air pollution measurements rather than replace them. For example, without data on the nature of the relationship between air pollution concentrations and their corresponding biological indicators, biological measurements by themselves are not sufficient to establish realistic air quality standards to protect public health.

Research Recommendations

Time-Activity Patterns. Available data on time budgets, activity patterns, and commuting behavior lack the specificity needed to make them useful for exposure estimation. For example, information is typically not available about the time spent in critical microenvironments, and about the presence of emissions sources such as smokers and unvented combustion appliances. Moreover, existing data are not coded so that important determinants of exposure, such as the amount of time spent indoors and outdoors, can be readily tabulated. In addition, available data analyses and summaries emphasize average values rather than distributions.

Studies are needed to investigate the spatial and temporal distribution of human populations as they relate to exposure. These studies could be carried out independently or in conjunction with exposure measurements. They should define distri-

butions of time spent in important microenvironments, air pollution sources present in these microenvironments, time of day that people are in specific locations, and differences in time-activity patterns according to demographic and socioeconomic factors. It is especially important to know, for example, whether potentially susceptible groups such as asthmatics, young children, and the elderly have time-activity patterns that differ substantially from those of the general population. It is also important to determine which population subgroups are likely to experience high exposures to certain air pollutants because of their specific time-activity patterns.

Two types of studies are needed: first, studies relating time-activity patterns (independent variable) to exposures (dependent variable); and second, studies relating factors such as age, gender, occupation, socioeconomic status, and pollutant susceptibility (independent variables) to time-activity patterns (dependent variable). Data of both types should be collected for representative samples that can support inferences for the general population. Studies should be designed to measure regional and seasonal differences because climate and weather affect the amount of time people spend outdoors as well as the characteristics of indoor microenvironments, through factors such as building weatherization, natural and forced ventilation, and the burning of wood, coal, and kerosene.

Prior to performing such studies, a consensus must be reached about what kinds of information are needed (important microenvironments and activities, presence of indoor pollution sources, etc.), how the information should be obtained (by trained interviewers, through self-administered questionnaires or diaries, or with automated trackers, etc.), and what statistical analyses and data summaries are most appropriate. Once the appropriate survey instruments have been designed and validated, they must be standardized to facilitate comparisons among studies and to enable results from many studies to be pooled effectively in a common data base.

■ **Recommendation 1. *Time-Activity Patterns.*** Studies should be undertaken to provide information on the spatial and temporal distributions of human populations as they relate to exposures. The goal of these studies should be to construct frequency distributions of time spent in important microenvironments, to identify the air pollution sources in those microenvironments, to specify the time of day that people are in particular microenvironments, and to determine the differences in time-activity patterns associated with demographic and socioeconomic factors.

Breathing Patterns. The relation among exposure, dose, and health outcome is complicated by the fact that respiration rate and mode (mouth or nose breathing, depth of inspiration) vary with a person's activity. The health effect of an inhaled pollutant is strongly influenced by how much actually crosses one of the body's boundaries and reaches a target tissue, which in turn is strongly affected by the manner and rate of a person's breathing. Respiration rate and mode are affected by whether a person is sedentary, standing, sitting, sleeping, exercising, or talking—conditions that most certainly are not statistically independent of time-activity patterns. Therefore, changes in respiration associated with these and other important human activities must be measured or otherwise taken into account before the health effects of air quality can be satisfactorily related to measurements of pollutant concentration.

■ **Recommendation 2.** *Breathing Patterns.* Respiration rate and mode are important determinants of air pollutant dose and therefore affect the health consequences of a measured exposure. The changes in respiration associated with repose, exercise, standing, sitting, sleeping, talking, and other important human activities should be measured or estimated.

Measurements

During the past 15 years, several studies have been undertaken to assess the extent and magnitude of human exposure to a variety of air pollutants. Most of these studies addressed the adequacy of outdoor measurements for exposure assessment. Although no comprehensive evaluation is available yet for any specific pollutant, the results from these preliminary investigations provide evidence of a consistent pattern. Outdoor measurements are weakly correlated, or not correlated at all, with individual exposure to most air pollutants, especially those with indoor sources. Furthermore, for airborne contaminants such as formaldehyde and many other volatile organic compounds, NO_2, CO, and respirable particles, individual exposures are significantly higher than measured outdoor concentrations alone would imply (World Health Organization 1982; Spengler and Sexton 1983; Ott 1985; Spengler and Soczek 1985).

The realization that outdoor monitors often are inappropriate to estimate air pollution exposure creates an anomalous situation. Compliance with National Ambient Air Quality Standards (NAAQS) is determined exclusively by measurements at outdoor, stationary monitoring sites. Thus, a substantial portion of the population residing in an area that is nominally in compliance with NAAQS might still be exposed to concentrations that are either above or below the standards, (Sexton et al. 1983; Ott 1983–84; Letz et al. 1984; Wallace et al. 1985a; Liu et al. 1986). Similarly, epidemiologic studies that rely on outdoor monitors to define exposures are subject to systematic and random bias, as well as to misclassification errors, which may lead to spurious conclusions about public health risks (Goldstein and Landovitz 1977a,b; Shy et al. 1978; Goldstein et al. 1979; Sexton et al. 1983; Spengler and Sexton 1983; Lippman and Lioy 1985; National Research Council 1985; Leaderer et al. 1986; Ozkaynak et al. 1986).

Most of the published exposure studies used either the direct or indirect approach, and focused on a small group of pollutants, including respirable particles, CO, NO_2, and, more recently, 20 to 40 individual volatile organic compounds (for example, toluene, benzene, and xylenes). A few studies also were carried out to examine exposures to Pb and O_3. Biological monitoring for environmental exposures is still

Table 4. Summary of Direct, Personal Monitoring Studies Carried Out in the United States

Pollutant	Individual Studies	Number of Subjects	Summary of Findings
RSP	Binder (1976)	20	Personal exposure did not correlate with outdoor measurements, and in most cases was substantially higher than outdoor concentrations would predict. Exposure to passive tobacco smoke was a major determinant of RSP exposure.
	Dockery and Spengler (1977)	22	
	Spengler et al. (1980)	46	
	Dockery and Spengler (1981)	37	
	Sexton et al. (1984a)	48	
	Spengler et al. (1985)	101	
CO	Cortese and Spengler (1976)	66	Time spent in transit was the primary determinant of personal exposure. Highest CO exposures were due primarily to motor vehicle exhaust.
	Jabara et al. (1980)	98	
	Ziskind et al. (1982)	9	
	Ackland et al. (1985)	1,083	
NO$_2$	Dockery et al. (1981)	9	Outdoor monitors overestimated exposures for people not exposed to indoor sources, but underestimated exposures for people who reside in homes with unvented combustion appliances.
	Quackenboss et al. (1982)	66	
	Silverman et al. (1982)	18	
	Quackenboss et al. (1986)	350	
VOCs	Wallace et al. (1982)	17	Outdoor measurements did not correlate well with personal exposures. Personal exposure and in-home concentrations tended to be higher than outdoor concentrations for many volatile organic compounds.
	Wallace et al. (1984)	12	
	Wallace et al. (1985a)	355	
Pb	Azar et al. (1975)	150	Highest Pb exposures were experienced by taxi drivers. All subjects except office workers experienced highest exposure at work.

NOTE: CO = carbon monoxide; NO$_2$ = nitrogen dioxide; Pb = lead; RSP = respirable particles; VOCs = volatile organic compounds.

in its infancy for most pollutants although wide-scale measurements of blood Pb levels are available (Annest et al. 1983).

Air Monitoring

Direct Approach to Exposure Estimation. Information about direct, personal monitoring studies carried out in the United States is summarized in table 4. The paucity of data on individual and population exposures is evident from the fact that fewer than 3,000 people have carried personal monitors for all pollutants combined. Most of these subjects were middle-class adult volunteers from urban areas.

Respirable particles. Personal monitoring studies indicate that individual exposure (24 hr) typically is higher than measured outdoor concentrations (see figure 3).

Furthermore, exposures correlated only weakly with outdoor concentrations. Exposure to environmental tobacco smoke accounted for a substantial portion of the difference between personal and outdoor values (Dockery and Spengler 1981; Sexton et al. 1984b). Since virtually all of the published studies on respirable particles focused solely on particle mass, there are insufficient data to evaluate the differences in particle composition (Colome et al. 1982; Sexton et al. 1986b,c).

Carbon monoxide. Elevated CO exposures have been associated with time spent in or near heavy traffic or in microenvironments affected by motor vehicle exhaust (for example, parking structures or lots, and gas stations). In-transit activity appears to account for most personal exposures (see table 5). Average CO exposures tend to be

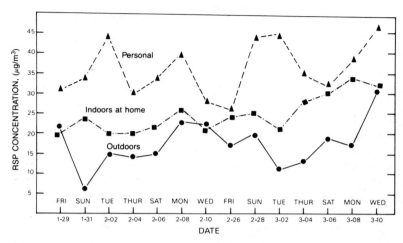

Figure 3. Daily variations in mean outdoor, indoor at home, and personal exposure to respirable particles (RSP) for 46 nonsmoking volunteers from 23 residences in Waterbury, Vt. (Sexton et al. 1984b).

below the NAAQS of 35 ppm/hr, but 1 to 3.5 percent of the subjects tested in Denver, Colorado, and Washington, D.C., received higher exposures (Akland et al. 1985). Therefore, outdoor monitors are likely to underestimate exposures for individuals who experience the highest concentrations.

Nitrogen dioxide. Outdoor NO_2 concentrations tend to be higher than measured personal exposure when there are no significant indoor sources (for example, unvented gas-fired appliances). When individuals are exposed indoors (for example, in residences with gas ranges or kerosene

heaters), outdoor measurements may significantly underestimate actual exposures (Quackenboss et al. 1982, table 6). Because people typically spend more time outdoors during the summer when natural ventilation in buildings also is highest, outdoor measurements are most likely to approximate individual exposure during the summer months (Spengler et al. 1983; Quackenboss et al. 1986).

Volatile organic compounds. The U.S. Environmental Protection Agency (EPA) has studied personal exposure to volatile organic compounds (VOCs) in several

Table 5. Summary of Mean CO Concentrations and Time Spent in Selected Microenvironments in Two U.S. Cities

Microenvironment	Denver, Colo.		Washington, D.C.	
	Mean Concn. (ppm)[a]	Time (min)[b]	Mean Concn. (ppm)[a]	Time (min)[b]
Indoors, parking garage	19	14	11	11
In transit, car	8	71	5	79
In transit, other vehicle	8	66	4	49
Outdoors, near roadway	4	33	3	20
In transit, walking	4	28	2	32
Indoors, restaurant	4	58	2	45
Indoors, office	3	478	2	428
Indoors, store	3	50	3	36
Indoors, residence	2	975	1	1,048

[a] Mean concentration in the specific microenvironment during the time that subjects were there.
[b] Median time spent by subjects in specific microenvironments.
SOURCE: Adapted with permission from Akland et al. 1985, and from the American Chemical Society.

Table 6. Comparison of Outdoor Concentrations, Indoor Concentrations, and Personal NO_2 Exposures in Portage, Wisconsin

		NO₂ Concn. ($\mu g/m^3$)		
Cooking Fuel	Sample Type	No. of Samples	Mean	Std. Dev.
Gas	Outdoors	9	5	1.8
	Indoors (kitchen)	9	67	29
	Personal[a]	33	34	16
Electric	Outdoors	10	12	4.5
	Indoors (kitchen)	10	8	4.9
	Personal[b]	33	14	6.4

[a] Occupants of the nine residences with gas stoves.
[b] Occupants of the 10 residences with electric stoves.
SOURCE: Adapted with permission from Quackenboss et al. 1982, and from Pergamon Press.

U.S. cities (Ott 1985; Wallace et al. 1985a,b). Findings from the Total Exposure Assessment Methodology (TEAM) project indicate that 11 compounds—chloroform, benzene, carbon tetrachloride, ethylbenzene, among others—are commonly present in indoor and outdoor air. Personal exposure to these chemicals was consistently higher than recorded outdoor concentrations, sometimes by an order of magnitude. In addition, nighttime indoor concentrations in residences were higher than matched outdoor samples (see table 7). Concentrations of these same compounds in exhaled breath correlated with personal exposure measurements (Wallace et al. 1985a).

Lead. Personal exposure for Pb is not well represented by outdoor measurements, since exposures occur primarily during commuting periods (Fugas et al. 1972). Highest Pb levels are encountered near streets with heavy traffic patterns. Therefore, individuals who spend a great deal of time in associated microenvironments, such as taxi drivers, experience the highest exposures (Azar et al. 1975).

Ozone. Few personal monitoring studies have been undertaken for O₃. Available

Table 7. Comparison of Indoor and Outdoor Concentrations of Selected Volatile Organic Compounds for 85 New Jersey Residences

	Concn. ($\mu g/m^3$)[a]			
	Outdoor Air		Indoor Air[b]	
Compound	Median	Maximum	Median	Maximum
Chloroform	0.7	22	2.9	220
1,1,1-Trichloroethane	4.2	40	16	880
Benzene	7.0	91	13	120
Carbon tetrachloride	0.8	14	1.4	14
Trichloroethylene	1.3	15	2.0	47
Tetrachloroethylene	2.6	27	5.6	250
Styrene	0.7	11	1.8	54
m,p-Dichlorobenzene (isomers)	0.8	13	2.8	920
Ethylbenzene	3.2	20	6.1	320
o-Xylene	3.0	27	5.0	46
m,p-Xylene (isomers)	9.9	70	16	120

[a] 12-hr averages.
[b] Bedroom.
SOURCE: Adapted with permission from Wallace et al. 1985a, and from Pergamon Press.

evidence suggests that outdoor O_3 values significantly overestimate personal exposure for most people because so much time is typically spent indoors where levels are substantially lower (Stock et al. 1985).

Indirect Approach to Exposure Estimation.

Given the diversity of microenvironments that people move through each day (see table 2), application of the indirect approach to exposure assessment is not straightforward. Its utility depends on identification of and sampling in the microenvironments with the greatest potential to influence human exposure. The costs and practical difficulties of monitoring in all, or even most, of the locations where people are likely to spend their time limits the scope of indirect measurements. Most studies are designed in conjunction with direct measurements and typically include sampling only of community outdoor air and of air inside private residences. A few studies have investigated in-vehicle exposure (Ott and Willits 1981; Flachsbart and Brown 1985).

Most air pollution measurements have been made in the outdoor community air microenvironment. Nevertheless, because most people are outside for such a small fraction of the time, and because the amount of air pollution that penetrates indoors is modified by building characteristics, outdoor measurements are of marginal value in estimating the actual exposures of humans to many airborne contaminants.

Information on important microenvironments and sources that affect exposure to air pollutants is given in table 8. In-home concentrations of NO_2 (Sexton et al. 1983; Quackenboss et al. 1986; Ryan et al. 1987), respirable particles (Dockery and Spengler 1981; Spengler et al. 1981, 1985; Sexton et al. 1984a,b), and many volatile organic compounds (Wallace et al. 1982, 1984; Wallace et al. 1985a) are the most important determinants of personal exposure for these pollutants. Outdoor or in-vehicle concentrations in areas of heavy traffic are the primary determinants of personal exposure for CO (Ott and Flachsbart 1982; Nagda and Koontz 1985) and Pb (Azar et al. 1975; Fugas 1975). Nonoccupational exposures

to formaldehyde are likely to occur primarily indoors, especially in newer manufactured residences, because of emissions from building materials and furnishings (Environmental Health Perspective 1984; Sexton et al. 1986a). Among the other pollutants for which indoor sources are the principal cause of personal exposure are environmental tobacco smoke, asbestos, radon, and various microorganisms (National Research Council 1981; Spengler and Sexton 1983).

Research Recommendation

Exposure Monitoring.

Past exposure assessment studies generally have investigated the magnitude of exposures compared to outdoor concentrations. Participants were often volunteers willing to undergo the inconvenience of personal or indoor monitoring. In only a few studies was a statistically random sample of participants selected (Akland et al. 1985; Wallace et al. 1985a; Sexton et al. 1986a; Ryan et al. 1987). These exploratory studies indicated that fixed-site, outdoor monitors do not adequately estimate human exposure to most air pollutants, and that accurate extrapolation to people beyond the sample population often is not possible. Furthermore, the high degree of variability in pollutant concentrations within individual buildings, within neighborhoods, and within cities will require substantial work to determine the distribution of exposures across a population (for example, frequency distribution for a city). In general, exposure assessment has not been not undertaken as part of a comprehensive and coordinated effort to define exposure distributions for a specific group of people.

■ **Recommendation 3. Exposure Monitoring.** Studies are required to provide data on human exposures and to investigate the link between measured exposures and adverse health effects. These efforts will require (a) the development of suitable instruments (for example, personal and indoor monitors) and measurement techniques (for example, noninvasive biological tests); (b) the application of appropriate

Table 8. Important Indoor and Outdoor Sources of Nonindustrial Air Pollution Exposure

Exposures Resulting Primarily from Outdoor Sources

Pollutant	Sources
Ozone (O_3)	Secondary product of photochemical reactions between hydrocarbons and nitrogen oxides (NO_x)
Sulfur dioxide (SO_2)	Stationary sources, such as coal-fired power plants
Particle-phase lead (Pb)	Mobile sources
Particle-phase sulfate	Secondary product of atmospheric reactions between SO_2 and other compounds
Pollens	Trees, grass, weeds, plants

Exposures Resulting Primarily from Indoor Sources

Pollutant	Sources
Formaldehyde	Particleboard, plywood, insulation, furnishings, adhesives, tobacco smoke
Asbestos	Insulation, fire retardants, texture paints, building construction materials
Radon	Underlying soil, building construction materials, well water
Pesticides	Consumer products (for example, insecticides, fungicides)
Microorganisms (bacteria, viruses, fungi)	Air-cleaning equipment, humidifiers, flush toilets, carpets, pets, plants, people
Environmental tobacco smoke	Cigarettes, cigars, pipes

Exposures Resulting from Both Indoor and Outdoor Sources

Pollutant	Indoor Sources	Outdoor Sources
Volatile organic compounds (VOCs)	Solvents, adhesives, synthetic building materials, aerosol sprays, pesticides, paint, tobacco smoke, cooking, metabolic processes	Mobile and stationary sources
Respirable particles (RSP)	Tobacco smoke, cooking, resuspended house dust, aerosol sprays, condensation of vapors, unvented combustion appliances	Mobile and stationary sources, secondary reactions in the atmosphere
Carbon monoxide (CO)	Unvented combustion appliances (gas-fired cooking stove, furnace, or hot-water heater, kerosene heater)	Primarily mobile sources
Nitrogen dioxide (NO_2)	Unvented combustion appliances (gas-fired cooking stove, furnace, or hot-water heater, kerosene heater)	Mobile and stationary sources, secondary reaction product in the atmosphere

statistical survey design methods; (c) the creation of extensive data bases on personal exposures, pollutant concentrations in important microenvironments, and human time–activity patterns; and (d) the development and application of appropriate models to estimate human exposure.

The respective roles of air monitoring (direct and indirect approaches) and biological monitoring in exposure assessment studies must be defined and detailed. Direct measurements of personal exposure are straightforward but costly, time-consuming, and labor-intensive. Indirect ap-

proaches, which combine data on pollutant concentrations in important microenvironments with time-activity information, are likely to be less costly, but also less accurate. Moreover, direct measurements are essential to validate the results of indirect (microenvironmental) exposure studies. Future studies should evaluate the utility of the indirect approach, which depends on the identification of the important microenvironments and the ability to measure pollutant concentrations in a representative sample, and determine to what extent the indirect approach complements direct exposure measurements.

Both the direct and the indirect measurement schemes should be coupled with a biological monitoring program to investigate the relationship between air concentrations, biological exposure markers, and related health consequences. This linkage is critical for regulatory purposes, since the airborne levels of a specific contaminant are typically the focus of health-based regulations.

Biological Monitoring

Most of the biological data on exposure markers are related to occupational environments. Lead workers, for example, are tested routinely for clinical signs and symptoms of Pb absorption, using markers such as porphyrins, δ-aminolevulinic acid, and aminolevulinate dehyratase. Biological markers for other metals such as cadmium, arsenic, mercury, selenium, tellurium, manganese, thallium, and zinc also are used widely (Ho and Dillion 1986). Biological tests for acute exposure to CO (carboxyhemoglobin in serum), trichloroethylene (trichloracetic acid in urine), and organophosphates (cholinesterase in serum) are well established (Miller 1983). The relevance of available analytical techniques for biological monitoring of occupationally exposed workers was reviewed by Linch (1974), Baselt (1980), and Lauwerys (1983).

Biological monitoring has not been used extensively outside the workplace. Among the exceptions are the analysis of adipose tissue and other body compartments to determine residues for chlorinated hydrocarbons (Hayes 1975), of human milk to assess exposure to organochlorine pesticides (Savage et al. 1981) and polychlorinated biphenyls (Rogan and Gladen 1983; Rogan et al. 1983), of urine for the presence of cotinine, a marker of exposure to environmental tobacco smoke (Jarvis et al. 1984), of plasma to assess exposure to benzo[a]pyrene (Hutcheon et al. 1983), and of exhaled breath to measure alveolar CO levels (Hartwell et al. 1984; Johnson 1984) and concentrations of selected volatile organic chemicals (Wallace et al. 1985a).

Techniques for the biological monitoring of human exposure to carcinogenic and mutagenic agents are a relatively recent

development (Goldstein 1981; Berlin et al. 1984; National Institute of Environmental Health Science 1984; Wogan and Gorelick 1985). Examples of techniques that have been used to detect human exposures to carcinogenic materials include determination of the degree of hemoglobin histidine alkylation as a measure of ethylene oxide exposure (Osterman-Golkar et al. 1976; Calleman et al. 1978); detection of mutagenic activity in the urine of anesthesiologists who use halogenated anesthetic gases (McCoy et al. 1977), of patients treated with anticancer drugs (Yamasaki and Ames 1977), and of cigarette smokers (Kreibel and Commoner 1980); the measurement of chromosome abnormalities and aberrations in workers exposed occupationally to Pb (Nordstrom et al. 1978a,b), vinyl chloride (Purchase et al. 1975; Szentesi et al. 1975; Hansteen et al. 1976), benzene (Tough et al. 1970), and organophosphates (Kiraly et al. 1976); and the use of radioimmunoassay to detect a metabolite of 4-aminobiphenyl (Johnson et al. 1980) or aflatoxin B (Sizaret et al. 1982) in the urine of exposed individuals.

Research Recommendation

Biological Markers of Exposure. Biological monitoring has evolved rapidly, especially for applications in the industrial workplace. Recent advances suggest that available biological measurement techniques can be applied to a wider range of environmental contaminants, including community and indoor air pollution. Studies are needed first to determine the associations among the air monitoring data, the biological markers of exposure, and the ultimate health outcome; and second to establish, prior to the general use of a particular biological measurement technique, its sources of error, the validity of sample collection methodology, the appropriateness of internal and/or external standards, and the adequacy of methods for quality control. In addition, specific studies might be undertaken to identify and monitor potential exposure markers in surviving animals throughout the course of large-scale chronic animal studies, to evaluate human materials including placentas, abortuses, and autopsy and biopsy reports as potential markers of exposure, and to reexamine the long-term or ongoing epidemiologic cohort studies to determine whether indicator or precursor lesions can be identified retrospectively (National Institute of Environmental Health Studies 1984).

■ **Recommendation 4.** *Biological Markers of Exposure.* Research is required to adapt available biological measurement techniques to community air pollution studies. Studies that better define the nature of the relationship between exposure and dose, and between dose and health outcome, are needed. It is especially important to establish in controlled human populations (a) the sources of error for a particular biological measurement technique, (b) the validity of sample collection methodology, (c) the appropriateness of internal and external standards, and (d) the adequacy of methods for quality control.

Modeling Human Exposure to Air Pollution

Models are useful tools to quantify the relationship between air pollutant exposure and important explanatory variables (for example, time-activity patterns), as well as to estimate exposures in situations where measurements are unavailable. In the United States, there are approximately 240 million people, about 82 million residences, and some 15 million commercial and public buildings. It is impractical to have everyone carry a personal monitor or to undertake pollutant measurements inside all buildings. Exposure models obviate the need for such extensive measurement programs by providing estimates of population exposures that are based on a small number of representative measurements. The challenge is to develop appropriate models that allow for extrapolation from relatively few exposure measurements to a much larger population.

Human exposure to air pollution can be thought of as a physical system in the same way that the forces acting on a falling object are a physical system. Newton's laws of motion might be selected as an appropriate model, for example, to describe the motion of an object falling in a gravitational field. Subsequent to selecting this model, it is necessary to verify it through experimentation. Experimental data might indicate that certain important determinants, such as air resistance, have been left out of the model. But even if all the critical parameters are included, some error is always associated with the predicted value because of measurement errors.

Statistical models are those for which predicted values are approximations subject to uncertainties introduced by factors deliberately omitted in the model, as well as by measurement errors. Techniques of statistical data analysis—a sophisticated form of curve-fitting—may be used instead of modeling the contributing physical, chemical, and biological processes, even if the processes are known and understood.

For simplicity and clarity, the following discussion distinguishes between statistical exposure models developed from a stochastic or probabilistic perspective, and physical exposure models, defined as those that are developed from an understanding of the underlying physical processes. This is an artificial distinction that focuses on the orientation, either statistical or physical, from which the model was constructed. In truth, all exposure models are statistical to some degree because the physical laws that constrain human activities and pollutant concentrations do not take into account absolutely every contributing process and factor. Physical-stochastic models, which combine the concrete aspects of the physical approach with the probabilistic aspects of the statistical approach, are also discussed. All three classifications of exposure models are compared in table 9.

Statistical Modeling

The statistical approach requires the collection of data on human exposures and the factors thought to be determinants of exposure. These data are combined in a sta-

Table 9. Comparison of Different Approaches to Air Pollution Exposure Modeling

Parameter	Model Type		
	Statistical	Physical	Physical-Stochastic
Method of formulation	Hypothesis testing	Physical laws	Physical laws and statistics
Required input	Collected data on human exposure	Knowledge of important parameters and their values in the system to be modeled	Knowledge of important parameters and their distributions in the systems to be modeled
Advantages	Makes use of real data in the model-building process	True model developed from a priori considerations	Model developed from a priori considerations; stochastic part allows uncertainty to contribute, reducing importance of research biases
Disadvantages	Requires data on hand for model building; extrapolation beyond data base is difficult	Includes researcher's biases; must be validated	Requires much knowledge of system; must be validated

tistical model, normally a regression equation or an analysis of variance (ANOVA), to investigate the relationship between air pollutant exposure (dependent variable) and the factors contributing to the measured exposure (independent variables) (see table 10). In principle, the results are applicable only to the data used to produce the model. If the study population constitutes a representative sample, however, then extrapolation of results to a broader group may be justified. Furthermore, selection of factors that influence exposure has a substantial effect on the outcome of the analysis. Spurious conclusions may be drawn, for example, from statistical models that include parameters that are correlated with, but not causally related to, air pollution exposure.

Initial statistical analysis of the data set usually focuses on defining the range of deviation associated with each variable and examining the bivariate and multivariate correlations between and among the independent variables. A lack of sufficient variability for one or more of the parameters in the model will limit the development of a useful predictive tool, whereas inclusion of independent variables that are highly collinear may obscure the importance of critical exposure determinants.

Once this type of exploratory analysis is completed, model building can begin. One of the simplest statistical models is the categorical comparison, in which exposure data are compared for two or more categories of a specific independent variable. An example of this approach is a comparison of NO_2 exposures inside residences with and without gas ranges (Speizer et al. 1980). Once it has been verified that NO_2 exposures are significantly higher in the homes with gas ranges, then it is feasible to divide the study population into two groups, people who live in homes with gas ranges (high exposure) and those who live in homes with electric ranges (low exposure) and compare symptom and illness rates on the basis of this simple dichotomous variable. Such categorical analyses often suggest additional avenues of investigation that further refine the nature of the relationship between exposure and important explanatory variables.

ANOVA provides a more quantitative method for categorical comparisons. A generalized linear model is developed that quantifies the relationship between the exposure and the selected explanatory variables in terms of the proportion of explained variance. The use of ANOVA allows for an investigation of the marginal effect of including new variables in the model, and thereby provides a means to determine the importance of identified exposure determinants.

Table 10. Different Approaches to Air Pollution Exposure Modeling

Model Type	Examples	References
Statistical	Various epidemiologic studies, such as Harvard Air Pollution/Lung Health Study	Ferris et al. (1979) Speizer et al. (1980)
Physical	National Exposure Model (NEM)	Biller et al. (1981, 1984) Johnson and Paul (1983a,b) Richmond and McCurdy (1985)
	Stepwise physical models which include physical parameters in stepwise regressions	Tosteson et al. (1982) Sexton et al. (1984b) Spengler et al. (1985) Quackenboss et al. (1986)
	Physical model of indoor air quality	Ryan et al. (1983) Sexton et al. (1983) Nazaroff and Cass (1986)
Physical-stochastic	Simulation of Human Air Pollution Exposure (SHAPE)	Ott (1981) Ott and Willits (1981) Ott (1983–1984)
	Simulation System (SIMSYS)	Letz et al. (1984) Ryan et al. (1986, 1987)

Statistical techniques such as factor analysis and cluster analysis can be used to elucidate the basic, underlying processes that determine air pollution exposure. These methods allow exposure to be partitioned into factors or clusters of correlated independent variables that tend to act together. Such analyses are useful for investigating correlations among independent variables and for understanding the relative contribution of specific factors or clusters of variables to the measured exposure.

Physical Modeling

The physical approach is based on the investigator's interpretation of the underlying processes that determine air pollution exposure. This interpretation is expressed as a quantitative description—mathematical formula, computer program, numerical tables, or graph—of the relationship between exposure and the determinants thought to be important. Since the model is chosen by the investigator, it may produce biased results because of the inadvertent inclusion of inappropriate parameters or the improper exclusion of critical determinants.

In the physical modeling approach, the modeler begins with certain a priori assumptions about the underlying physical processes that determine air pollution exposure. These assumptions are the basis for constructing a quantitative formulation that constitutes a physical exposure model. References and examples of the physical modeling approach are given in table 10.

A simple physical model can be constructed by assuming that personal exposure to air pollution is a strict function of the outdoor, or ambient, concentration. The mathematical form of this statement can be expressed as

$$E = f(C_{amb}) \qquad (2)$$

where E is exposure for a specific air contaminant, f denotes "a function of," and C_{amb} is the ambient (outdoor) concentration of the pollutant. This model would be most appropriate for air pollutants that result primarily from outdoor sources (see table 8).

An example of this basic model, which assumes that exposure can be approximated by a linear function, is

$$E = aC_{amb} + b \qquad (3)$$

where E is exposure, a is the slope of the line relating exposure to the ambient concentration, C_{amb} is the measured ambient concentration, and b is the exposure when the ambient concentration is zero. Several groups have combined this model with data about personal exposures and ambient concentrations to estimate values for a and b in equation 3 (Tosteson et al. 1982; Ryan et al. 1983; Sexton et al. 1983; Spengler et al. 1985). Further analysis has been carried out to delineate the relationship between the model parameters a and b and the physical processes such as the air-exchange rate, the first-order pollutant losses from physicochemical processes, and the indoor sources of air pollution (Ozkaynak et al. 1982; Ryan et al. 1983; Sexton et al. 1983; Letz et al. 1984).

The microenvironmental approach, discussed earlier in the Methods section under the Indirect Approach to Exposure Assessment, is a more complex model based on similar ideas. Pioneered by Fugas (1975), this approach assumes that a person's time-weighted, integrated exposure is the product of the air pollution concentration in identified microenvironments and the time spent in those microenvironments (see equation 1). Although this approach allows comparison of the contributions of selected microenvironments to the measured exposure, the identification and monitoring of pollutants in critical microenvironments is often difficult and expensive.

The NAAQS Exposure Model (NEM) is a physical model that uses the microenvironmental approach (Biller et al. 1981, 1984; Johnson and Paul 1983a; Richmond and McCurdy 1985). The NEM also incorporates the concept of a population cohort (a group of individuals having a statistical factor in common, such as, live in the same neighborhood or have the same commuting pattern)—an assumption that is analogous to the requirement for spatial and temporal uniformity of pollutant concentrations within a specific microenvironment. The model is designed to estimate the effect on population exposure that results from changes in air quality standards. The NEM has been applied to CO (Johnson and Paul 1983a,b), SO_2 (Biller et al. 1984), and O_3 (Richmond and McCurdy 1985).

A common shortcoming of the physical models described above is that while they do estimate expected exposure, they do not estimate the associated uncertainty. Evidence suggests that there is substantial variation in the time spent in various microenvironments (Sexton et al. 1984b; Clausing et al. 1986; Quackenboss et al. 1986), as well as in the pollutant concentrations within each microenvironment (Spengler et al. 1983; Sexton et al. 1984a; Akland et al. 1985; Sexton et al. 1986a,b). Letz and his colleagues (1984) attempted to estimate the uncertainty in predicted exposure by including estimates of the variance in each model parameter. The variance in predicted exposure is estimated by a Taylor-series expansion. Results of this approach correlate well with findings from personal monitoring studies.

Physical-Stochastic Modeling

The physical-stochastic approach can be thought of as a third type of exposure model, even though it is a computational method. It combines elements of both the physical and the statistical approaches to estimate exposure. A mathematical model that describes the physical basis for air pollution exposure is first constructed. Then a random or stochastic component that takes into account the imperfect knowledge of the physical parameters that determine exposure is introduced into the model. The inclusion of the random component limits the effect of investigator-induced bias and allows for estimates of population distributions of air pollutant exposure. Misleading results can still be produced if model parameters are selected ineptly. In addition, the required knowledge about distribution characteristics may be difficult and expensive to obtain.

By the introduction of a stochastic component into a physical model, the physical-stochastic approach attempts to account for the probabilistic nature of the physical processes that determine exposure. In this way, the inherent uncertainty associated with a mathematical abstraction of air pollution exposure is taken into account.

Two models, the Simulation of Human Air Pollutant Exposure (SHAPE) model and the Simulation System (SIMSYS) model, are representative of the physical-stochastic approach. Both models use the microenvironment concept discussed earlier and use similar statistical approaches. They differ primarily in their application and intended use.

The SHAPE model focuses on estimating personal exposures (Ott 1981, 1983–1984; Ott and Willits 1981). Statistical techniques are used to select the appropriate characteristics of the individuals in the study and the microenvironments through which they move. Time-activity data are generated by selecting the type of activity as well as the duration of activity from probability distributions. Air pollutant exposure is modeled as the sum of 1-min exposures that are experienced throughout the course of an individual's daily activities.

The SHAPE model has two distinct advantages: less-detailed information on timeactivity patterns is needed because one must know only the probability of going from one activity to another; and a small number of microenvironments (14 in the published version) is required to estimate exposure. Disadvantages of this approach include the potential bias introduced by the modeler's selection of relevant microenvironments; the need for accurate data on the probabilities of transitions between microenvironments and the time spent in specific microenvironments; and the difficulty of obtaining the distribution of pollutant concentrations in important microenvironments.

The SIMSYS model focuses on estimating the distribution of air pollutant exposures within a population, with emphasis on the contribution of specific microenvironments to the integrated exposure (Letz et al. 1984; Ryan et al. 1986, 1987). The SIMSYS model is based on a physical description of exposure similar to equation 3. Estimates of the probability distributions for the model parameters are obtained from the literature or from field studies. Basically, the SIMSYS approach is similar to the SHAPES model and therefore shares the same generic disadvantages. The advantage of this model is that it provides a means of evaluating the effects on human exposure of reducing air pollutant levels in specific microenvironments.

Source Apportionment

Before it is feasible to evaluate the adequacy and cost-effectiveness of air pollution control strategies, it is necessary to obtain more and better information about the relative contributions of indoor and outdoor emission sources to measured personal exposures. Models such as SHAPE and SIMSYS are useful tools that aid in understanding where and how exposures occur. They begin to address the issue of the extent to which public health is protected by the NAAQSs, which apply only to air outside buildings.

As pointed out by Sexton and Hayward (1986), informed decisions about appropriate resource allocation to control air pollution require more than just data on health effects. They depend also on adequate information about important emission sources (source identification), chemical and physical properties of emissions (emissions characterization), and the effects of important source categories on indoor and outdoor air quality, as well as on personal exposures (that is, source apportionment).

Although the major emission sources, indoors as well as outdoors, have been identified and work is progressing on the characterization of airborne discharges, the relative impact of indoor and outdoor emissions on personal and population exposures has not been addressed systematically and comprehensively.

Several types of source apportionment models have been applied to outdoor (ambient) air, but their application to air pollution inside buildings or to personal exposures is just beginning. Consequently, insufficient data are available to determine the relative contributions of indoor and outdoor sources to measured personal exposures. This lack of information seriously hinders attempts to evaluate the costs and benefits of alternative control options (Sexton and Hayward 1986).

Validation and Generalization

The models described in the preceding sections are mathematical abstractions of physical reality that may or may not provide adequate estimates of air pollution exposure. The only way to be sure that a model is capable of providing useful and accurate information is by validation— comparing model predictions with measurements independent of the measurements used to develop the model. Moreover, model validation is a necessary precondition for the generalization of model results to a different or larger population.

In the statistical modeling approach, data collection is an integral part of model construction. If the data are known to be from a statistically representative sample of the population, there is no need for further validation. If the results are to be extrapolated beyond the range for which the existing data base provides a statistical description, validation is necessary. The physical and physical-stochastic modeling approaches must be validated with actual data from separately conducted field studies. Care must be taken that the data used to validate a model are not biased with respect to crucial model parameters. The validation step is useful only to the degree that the sample population is representative of the group to which results will be extrapolated.

Research Recommendation

Exposure Modeling. Attempts to model human exposures to air pollutants are relatively recent. Models vary widely in complexity and have not been validated adequately. The lack of data on the variability and covariance of time-activity patterns among individuals is a critical hindrance to model development.

Perhaps the most pressing need associated with modeling human exposure is the necessity for the external review and validation of existing models. It is not clear, for example, whether current exposure models are adequate, or if a new generation of models needs to be developed. The validation of existing models, using data sets other than those from which they were generated, is essential to answer this question.

Source apportionment of ambient air pollution is a growing research field. Many investigators are now studying ambient air pollution to determine which pollution sources are affecting which receptor and to what degree. The work should go further and determine which sources most directly affect specific human populations. Future studies should focus on determining the relative contributions of indoor and outdoor emission sources to personal exposures.

■ **Recommendation 5.** *Exposure Modeling.* Research is needed to assess the adequacy of current exposure models through external review and validation. Validation of existing models is essential to determine whether these models are adequate or if a new generation of models should be developed. In addition, how to apportion contributions of specific emission sources to individual exposures requires further study. The relevance of existing models (outdoor air) for source apportionment of personal exposures and of indoor air pollution needs to be evaluated and new models need to be developed if existing models are shown to be deficient.

Summary and Conclusions

In its examination of the state of the art in air pollution exposure assessment, this chapter describes the general methods available to determine exposure, the published studies that report on measurements of actual exposures, and the models that are used to estimate individual and population exposures. The goals are to help the reader understand the rudiments of this emerging field and to highlight the critical areas where further research is needed. In addition, it attempts to impart an awareness of the importance of obtaining information about how, when, where, and why exposures occur.

Evidence accumulated over the past few years indicates that adequate estimates of individual and population exposures for most air pollutants, including regulated and unregulated substances, cannot be derived solely from measurements by traditional outdoor monitoring stations. Depending on the pollutant in question, exclusive reliance on outdoor measurements may over- or underestimate the magnitude, duration, and frequency of exposures for the general population, as well as for many potentially susceptible subgroups. Although the ramifications of these findings for the development and evaluation of air pollution control strategies have not been explored fully, it is clear that they raise policy issues that should be taken into account in future regulatory decisions (Sexton and Repetto 1982; Sexton 1986).

Perhaps the most important lesson to be drawn from this chapter is the realization that accurate estimation of human exposures is a prerequisite for realistic assessment of air pollution health risks. Quantitative risk assessment is rapidly becoming an integral part of the regulatory decisions that are aimed at protecting public health. Too often, however, the availability of suitable exposure data is taken for granted.

The generalized form of the equation used to estimate health risks from environmental contaminants is

$$\begin{array}{c} \text{Health Risk} \quad = \\ \text{(morbidity/mortality)} \end{array}$$

$$\begin{array}{cc} \text{Potency} & \times \quad \text{Exposure} \\ \text{(dose/response)} & \text{(concentration)} \end{array}$$

$$\begin{array}{c} \times \text{ Exposed Population} \\ \text{(number of people exposed)} \end{array}$$

Human exposure data are obviously crucial to the calculation of air pollution health risks since this information is needed to specify values for two terms in the equation: exposure (including magnitude, duration, and frequency) and exposed population. Moreover, exposure assessment is a critical element of epidemiologic studies, which are often used to develop values for the potency term in the equation. For example, epidemiologic studies that fail to account for indoor as well as outdoor exposures are prone to systematic and random bias and to the misclassification of exposures. Such errors can lead to spurious conclusions concerning dose/response relationships for airborne contaminants, and, ultimately, to inappropriate estimation of public health risks.

Summary of Research Recommendations

HIGH PRIORITY

Recommendation 1
Time-Activity
Patterns

Studies should be undertaken to provide information on the spatial and temporal distributions of human populations as they relate to exposure. The focus of these studies should be to construct the frequency distribution of time spent in important microenvironments, to identify the air pollution sources in those microenvironments, to specify the time of day that people are in particular microenvironments, and to determine the differences in time-activity patterns associated with demographic and socioeconomic factors.

Recommendation 3
Exposure
Monitoring

Studies are required to provide representative data on human exposures and to investigate the link between measured exposures and adverse health effects. These efforts will require (a) the development of suitable instruments (for example, personal and indoor monitors) and measurement techniques (for example, noninvasive biological monitoring); (b) an application of the appropriate statistical survey design methods; (c) the creation of extensive data bases on personal exposures, pollutant concentrations in important microenvironments, and time-activity patterns; and (d) the development and application of appropriate models to estimate human exposure.

Recommendation 4
Biological Markers of Exposure

Research is required to adapt available biological measurement techniques to community air pollution measurement and control. Studies that better define the nature of the relationship between exposure and dose and between dose and health outcome are needed. It is especially important to establish in controlled human populations (a) the sources of error for a particular biological measurement technique, (b) the validity of sample collection methodology, (c) the appropriateness of internal and external standards, and (d) the adequacy of methods for quality control.

Recommendation 5
Exposure Modeling

Research is needed to assess the adequacy of current exposure models through external review and validation. Validation of existing models is essential to determine whether these models are adequate or if a new generation of models should be developed. In addition, how to apportion the contributions of specific emissions sources to individual exposures requires further study. The relevance of existing models (outdoor air) for source apportionment of personal exposures and of indoor air pollution needs to be evaluated and new models need to be developed if existing models are shown to be deficient.

MEDIUM PRIORITY

Recommendation 2
Breathing Patterns

Respiration rate and mode (for example, mouth breathing versus nose breathing) are important determinants of air pollutant dose and therefore affect the health consequences of a measured exposure. The changes in respiration associated with repose, exercise, standing, sitting, sleeping, talking, or any other important human activity should be measured or estimated.

Acknowledgments

We thank the following people for their helpful comments on this manuscript: J. Bailar, J. Evans, and J. Spengler, Harvard University; I. Goldstein, Columbia University; W. Ott and L. Wallace, EPA; and A. Watson, HEI. J. Schwartz and G. Raisbeck provided editorial assistance. The manuscript was typed by M. E. Patten.

Correspondence should be addressed to Ken Sexton, U.S. Environmental Protection Agency, Office of Health Research, Washington, DC 20460, or P. Barry Ryan, Harvard School of Public Health, Department of Environmental Science and Physiology, 665 Huntington Avenue, Boston MA 02115.

References

Akland, G. G., Hartwell, T. D., Johnson, T. R., and Whitmore, R. W. 1985. Measuring human exposure to carbon monoxide in Washington, D.C., and Denver, Colorado, during the winter of 1982–1983, *Environ. Sci. Technol.* 19:911–918.

Andelman, J. B. 1985. Human exposures to volatile halogenated organic chemicals in indoor and outdoor air, *Environ. Health Perspect.* 62:313–318.

Annest, J. L., Pirkle, J. L., Makug, D., Neese, J. W., Bayse, D. D., and Kovar, M. G. 1983. Chronological trend in blood lead levels between 1976 and 1980, *New Engl. J. Med.* 308:1373–1377.

Azar, A., Snee, R. D., Habini, K. 1975. *Lead* (T. F. Griffen and J. H. Knelson, eds.), Academic Press, New York, N.Y.

Bartley, D. L., Doemeny, L. J., and Taylor, D. G. 1983. Diffusive monitoring of fluctuating concentrations, *Am. Ind. Hyg. Assoc. J.* 44:241–247.

Baselt, R. C. 1980. Biological Monitoring Methods for Industrial Chemicals, Biomedical Publications, Calif.

Beall, J. R., and Ulsamer, A. G. 1981. Toxicity of volatile organic compounds present indoors, *Bull. N.Y. Acad. Med.* 57:978–996.

Berlin, A., Draper, M., Hemminki, K., and Vainio, H. (eds.). 1984. *Monitoring Human Exposure to Carcinogenic and Mutagenic Agents,* IARC Scientific Publ. No. 59, Oxford University Press, New York, N.Y.

Biller, W. F., Feagans, T. B., Johnson, T. R., Duggen, G. M., Paul, R. A., McCurdy, T., and Thomas, H. C. 1981. A general model for estimating exposures associated with alternative NAAQS. Paper presented at the 74th Annual Meeting of the Air Pollution Control Association, Philadelphia, Pa., June 21–26, 1981.

Biller, W. F., Thomas, H. C., Jr., Stoechinus, T. E., and Paul, R. 1984. Estimator of short-term sulfur dioxide population exposures. Paper presented at the 77th Annual Meeting of the Air Pollution Control Association, San Francisco, Calif., June 24–28, 1984.

Binder, R. E. 1976. Personal exposure to respirable particles, *Arch. Environ. Health* 36:277–279.

Calleman, C. J., Ehrenberg, L., Jansson, B., Osterman-Golkar, S., Severback, D., Svensson, K., and Wachtmeister, G. A. 1978. Monitoring and risk assessment by means of alkyl groups in hemoglobin in persons occupationally exposed to ethylene oxide, *J. Environ. Pathol. Toxicol.* 2:427–442.

Chapin, F. S., Jr. 1974. *Human Activity Patterns in the City: Things People Do in Time and Space,* John Wiley & Sons, New York, N.Y.

Clausing, P., Mak, J. K., Spengler, J. D., and Letz, R. 1986. Personal NO$_2$ exposures of high school students, *Environ. Int.* 12:413–417.

Coleman, S. R. 1983. A tube diffusion dosimeter for sulfur dioxide, *Am. Ind. Hyg. Assoc. J.* 44:631–637.

Colome, S. D., Spengler, J. D., and McCarthy, S. 1982. Comparison of elements and inorganic compounds inside and outside of residences, *Environ. Int.* 8:197–212.

Compton, J. R., Dwiggins, G. A., Feigley, C. E., and Ludwig, D. A. 1984. The effect of square wave exposure profiles upon the performance of passive organic vapor monitoring badges, *Am. Ind. Hyg. Assoc. J.* 45:446–450.

Cortese, A. D., and Spengler, J. D. 1976. Ability of fixed monitoring stations to represent personal carbon monoxide exposure, *J. Air. Pollut. Contr. Assoc.* 26:1144–1150.

De Bortoli, M., Knoppel, H., Pecchio, E., Peil, A., Rogora, L., Schauenburg, H., Schlitt, H., and Vissers, H. 1985. Measurements of Indoor Air Quality and Comparison with Ambient Air, Commission of the European Communities, Luxembourg.

Dockery, D. W., and Spengler, J. D. 1981. Personal exposure to respirable particulates and sulfates, *J. Air Pollut. Contr. Assoc.* 31:153–159.

Dockery, D. W., and Spengler, J. D. 1977. Personal exposure to respirable particulates and sulfates versus ambient measurements, *Proceedings of the 70th Annual Meeting of the Air Pollution Control Association.* June 1977, Toronto, Canada. Paper no. 77-44.6.

Dockery, D. W., Spengler, J. D., Reed, M. P., and Ware, J. 1981. Relationship among personal, indoor, and outdoor NO$_2$ measurements, *Environ. Int.* 5:101–107.

Duan, N. 1981. Microenvironment Types: A Model for Human Exposures to Air Pollution, SIMS Technical Report No. 47, Department of Statistics, Stanford University, Palo Alto, Calif.

Duan, N. 1982. Models for human exposure to air pollution, *Environ. Int.* 8:305–309.

Environmental Health Perspectives. 1984. Report on the consensus workshop on formaldehyde, *Environ. Health Perspect.* 58:323–381.

Feigley, C. E., and Chastain, J. B. 1982. An experimental comparison of three diffusion samplers exposed to concentration profiles of organic vapors, *Am. Ind. Hyg. Assoc. J.* 43:227–234.

Ferris, B. G., Jr., Speizer, F. E., Spengler, J. D., Dockery, D. W., Bishop, Y. M. M., Wolfson, M., and Humble, C. 1979. Effects of sulfur oxides and respirable particles on human health: I. Methodology and demography of population in study, *Am. Rev. Respir. Dis.* 120:767–779.

Flachsbart, P. G., and Brown, D. E. 1985. Surveys of personal exposure to vehicle exhaust in Honolulu microenvironments, Report of the Department of Urban and Regional Planning, University of Hawaii at Manoa, December 1985.

Fugas, M. 1975. Assessment of total exposure to an air pollutant, *Proceedings of the International Symposium on Environmental Monitoring, Las Vegas, Nevada,* Vol. 2, pp. 38–45, Institute of Electrical and Electronic Engineers, Inc., New York, N.Y.

Fugas, M., Wilder, B., Paukovic, R., Hrsak, J., Steiner-Skreb, D. 1972. Concentration levels and particle size distribution of lead in the air of an urban and an industrial area as a basis for the calculation of population exposure. Pp. 961–968 in *Proceedings of the International Symposium of the Environmental Aspects of Lead,* Amsterdam, Holland.

Geisling, K. L., Tashima, M. K., Girman, J. R., Miksch, R. R., and Rappaport, S. M. 1982. A passive sampling device for determining formaldehyde in indoor air, *Environ. Int.* 8:153–158.

Goldstein, I. F. 1981. The use of biological markers in studies of health effects of pollutants, *Environ. Res.* 25:236–240.

Goldstein, I. F., and Landovitz, L. 1977a. Analysis of air pollution patterns in New York City—I. Can one station represent the large metropolitan area? *Atmos. Environ.* 11:47–52.

Goldstein, I. F., and Landovitz, L. 1977b. Analysis of air pollution patterns in New York City—II. Can one aerometric station represent the area surrounding it? *Atmos. Environ.* 11:53–57.

Goldstein, I. F., Fleiss, J. L., Goldstein, M., and Landovitz, L. 1979. Methodological problems arising from the choice of an independent variable in linear regression, with application to an air pollution epidemiological study, *Environ. Health Perspect.* 32:311–315.

Hansteen, I. L., Hillestad, L., and Thiis-Eveson, E. 1976. Chromosome studies in workers exposed to vinyl chloride, *Mutat. Res.* 38:112.

Hartwell, T. D., Clayton, C. A., Michie, R. M., Whitmore, R. W., Zelon, H. S., Jones, S. M., and Whitehurst, D. A. 1984. Study of Carbon Monoxide Exposure of Residents of Washington, D.C., and Denver, Colorado, Environmental Monitoring Systems Laboratory. U.S. Environmental Protection Agency, Research Triangle Park, N.C. EPA-600/S4-84-031, PB 84-183516.

Hayes, W. J. 1975. *Toxicology of Pesticides*, The Williams and Wilkins Co., Baltimore, Md.

Ho, M. H., and Dillion, H. K. 1986. Biological monitoring, *Environ. Sci. Technol.* 20:124–127.

Hollowell, C. D., and Miksch, R. R. 1981. Sources and concentrations of organic compounds in indoor environments, *Bull. N.Y. Acad. Med.* 57:962–977.

Hutcheon, D. E., Kantrowitz, J., Van Gelder, R. N., and Flynn, E. 1983. Factors affecting plasma benzo[a]pyrene levels in environmental studies, *Environ. Res.* 32:104–110.

Jabara, J. W., Keefe, T. J., Beaulieu, H. J., and Buchan, R. M. 1980. Carbon monoxide: dosimetry in occupational exposures in Denver, Colorado, *Arch. Environ. Health* 35:198–204.

Jaeger, R. J. 1981. Carbon monoxide in houses and vehicles, *Bull. N.Y. Acad. Med.* 57:860–872.

Jarke, F. H., Dravnieks, A., and Gordon, S. M. 1981. Organic contaminants in indoor air and their relation to outdoor contaminants, *ASHRAE Trans.* 87:153–165.

Jarvis, M., Tunstall-Pedoe, H., Feyerabend, C., Vesey, C., and Salloojee, Y. 1984. Biochemical markers of smoke absorption and self-reported exposure to passive smoking, *J. Epidemiol. Community Health* 38:335–339.

Johnson, T. 1984. A study of personal exposure to carbon dioxide in Denver, Colorado. Paper presented at the 79th Annual Meeting of the Air Pollution Control Association, San Francisco, Calif., June 24–29, 1984.

Johnson, H. J., Jr., Cernosek, S. F., Jr., Gutierrez-Cernosek, R. M., and Brown, L. L. 1980. Development of a radioimmunoassay procedure for 4-acetamidobiphenyl, a metabolite of the chemical carcinogen 4-aminobipheny, in urine, *J. Anal. Toxicol.* 4:86–90.

Johnson, T., and Paul, R. A. 1983a. The NAAQS Model (NEM) Applied to Carbon Monoxide. EPA 450/5-83-003. U. S. Environmental Protection Agency, Research Triangle Park, N.C.

Johnson, T., and Paul, R. A. 1983b. The NAAQS Model (NEM) Applied to Carbon Monoxide. EPA 450/5-83-004. U. S. Environmental Protection Agency, Research Triangle Park, N.C.

Kiraly, J., Czeizel, A., and Szentesi, I. 1976. Genetic studies on workers producing organophosphate insecticides, In: *Abstracts, 6th Annual Meeting of the European Environmental Mutagen Society, Gernrode, East Germany, Sept. 27–Oct. 1*, p. 109.

Kreibel, D., and Commoner, B. 1980. The mutage-

nicity of cigarette smokers. Paper presented at the American Public Health meeting, June 1980, New York, N.Y.

Kring, E. V., Henry, T. J., Damrell, D. J., and Bythewood, T. K. 1983. Laboratory and field comparison of three methods for monitoring sulfur dioxide in air, *Am. Ind. Hyg. Assoc. J.* 44:929–936.

Kring, E. V., Ansul, G. R., Basilio, A. N., McGibney, P. D., Stephens, J. S., and O'Dell, H. L. 1984. Sampling for formaldehyde in workplace and ambient air environments—additional laboratory and field verification of a passive air monitoring device compared with conventional sampling methods, *Am. Ind. Hyg. Assoc. J.* 45:318–324.

Lautenberger, W. J., Kring, E. V., and Morello, J. A. 1981. Theory of passive monitors, *Ann. Am. Conf. Gov. Ind. Hyg.* 1:91–99.

Lauwerys, R. R. 1983. Industrial Chemical Exposure: Guidelines for Biological Monitoring, Biomedical Publications, Calif.

Leaderer, B. P., Zagraniski, R. T., Berwick, M., Stolwijk, J. A. J. 1986. Assessment of exposure to indoor air contaminants from combustion sources: methodology and application, *Am. J. Epidemiol.* 124:275–289.

Lebowitz, M. D., Corman, G., O'Rourke, M. K., and Holberg, C. J. 1984. Indoor-outdoor air pollution, allergen, and meteorological monitoring in an arid southwest area, *J. Air Pollut. Contr. Assoc.* 34:1035–1038.

Letz, R., Ryan, P. B., and Spengler, J. D. 1984. Estimated distributions of personal exposure to respirable particles, *Environ. Monit. Assess.* 4:351–359.

Linch, A. L. 1974. *Biological Monitoring for Industrial Chemical Exposure Control*, CRC Press, Boca Raton, Fla.

Lippman, M., and Lioy, P. J. 1985. Critical issues in air pollution epidemiology, *Environ. Health Perspect.* 62:243–258.

Liu, K., Chang, B., Hayward, S. B., Kulasingam, G., and Sexton, K. 1986. Estimation of formaldehyde exposure for mobile home residents, *Proceedings of the 79th Annual Meeting of the Air Pollution Control Association*, Minneapolis, Minn.

McCoy, E. C., Hankel, R., Rosenkranz, H. S., Giuffrida, J. G., and Bizzari, D. V. 1977. Detection of mutagenic activity in the urine of anaesthesiologists: a preliminary report, *Environ. Health Perspect.* 21:221–223.

Miksch, R. R., Hollowell, C. D., and Schmidt, H. E. 1982. Trace organic chemical contaminants in office spaces, *Environ. Int.* 8:129–137.

Miller, S. 1983. A monitoring report, *Environ. Sci. Technol.* 17:343A–346A.

Molhave, L. 1982. Indoor air pollution due to organic gases and vapours of solvents in building materials, *Environ. Int.* 8:117–127.

Moschandreas, D. J. 1981. Exposure to pollutants and daily time budgets of people, *Bull. N.Y. Acad. Med.* 57:845–859.

Nagda, N. L., and Koontz, M. D. 1985. Microenvironment and total exposures to carbon monoxide for three population subgroups, *J. Air Pollut. Contr. Assoc.* 35:134–137.

National Institute of Environmental Health Science. 1984. Biochemical and cellular markers of chemical exposure and preclinical indicators of disease, In: Human Health and Environment—Some Research Needs, Report of the Third Task Force for Research Planning in Environmental Health Science.

National Research Council, Committee on Indoor Pollutants. 1981. *Indoor Pollutants*, National Academy Press, Washington, D.C.

National Research Council, Committee on the Epidemiology of Air Pollution. 1985. *Epidemiology and Air Pollution*, National Academy Press, Washington, D.C.

Nazaroff, W. W., and Cass, G. R. 1986. Mathematical modeling of chemically reactive pollutants in indoor air, *Environ. Sci. Technol.* 20:924–934.

Nero, A. V., and Lowder, M. M. (eds.). 1983. Indoor radon, *Health Phys.* 45:273–574.

Nordstrom, I., Beckman, G., Beckman, L., and Nordstrom, S. 1978a. Occupational and environmental risks in and around a smelter in northern Sweden. II. Chromosomal aberrations in workers exposed to arsenic, *Hereditas* 88:47–50.

Nordstrom, I., Beckman, G., Beckman, L., and Nordstrom, S. 1978b. Occupational and environmental risks in and around a smelter in northern Sweden. IV. Chromosomal aberrations in workers exposed to lead, *Hereditas* 88:263–267.

Osterman-Golkar, S., Ehrenberg, L., Segerback, D., and Hallstrom, I. 1976. Evaluation of genetic risks of alkylating agents. II. Hemoglobin as a dose monitor, *Mutat. Res.* 34:1–10.

Otson, R., Doyle, E. E., Williams, D. T., and Bothwell, P. D. 1983. Survey of selected organics in office air, *Bull. Environ. Contam. Toxicol.* 31:222–229.

Ott, W. R. 1981. Exposure estimates based on computer-generated activity patterns. Paper presented at the 74th Annual Meeting of the Air Pollution Control Association, Phildelphia, Pa., June 21–26, 1981.

Ott, W. R. 1982. Concepts of human exposure to air pollution, *Environ. Int.* 7:179–196.

Ott, W. R. 1983–1984. Exposure estimates based on computer-generated activity patterns, *J. Toxicol. Clin. Toxicol.* 21:97–128.

Ott, W. R. 1985. Total human exposure, *Environ. Sci. Technol.* 19:880–885.

Ott, W. R., and Flachsbart, P. 1982. Measurement of carbon monoxide concentrations in indoor and outdoor locations using personal exposure monitors, *Environ. Int.* 8:295–304.

Ott, W. R., and Willits, N. H. 1981. CO Exposures of Occupants of Motor Vehicles: Modeling the Dynamic Response of the Vehicle, Society of Industrial and Applied Mathematics Institute for Mathematics and Society Technical Report No. 48.

Ott, W. R., Rodes, C. E., Drago, R. J., Williams, C., Burmann, F. J. 1986. Automated data-logging personal exposure monitors for carbon monoxide, *J. Air Pollut. Contr. Assoc.* 36:883–888.

Ozkaynak, H., Ryan, P. B., Allen, G. A., and

Turner, W. A. 1982. Indoor air quality modeling: compartmental approach with reactive chemistry, *Environ. Int.* 8:461–471.

Ozkaynak, H., Ryan, P. B., Spengler, J. D., and Letz, R. 1986. Bias due to misclassification of exposures in epidemiologic studies, *Environ. Int.* 12:389–393.

Palmes, E. D., and Tomczyk, C. 1979. Personal sampler for NO_2, *Am. Ind. Hyg. Assoc. J.* 40: 588–591.

Palmes, E. D., Gunnison, A. F., DiMattio, J., and Tomczyk, C. 1976. Personal sampler for NO_2, *Am. Ind. Hyg. Assoc. J.* 37:570–577.

Purchase, I. F. H., Richardson, C. F., and Anderson, D. 1975. Chromosomal and dominant lethal effects of vinyl chloride, *Lancet* 2:410–411.

Quackenboss, J. J., Kanarek, M. S., Spengler, J. D., and Letz, R. 1982. Personal monitoring for nitrogen dioxide exposure: methodological considerations for a community study, *Environ. Int.* 8:249–258.

Quackenboss, J. J., Spengler, J. D., Kanarek, M. S., Letz, R., Duffy, C. P. 1986. Personal exposure to nitrogen dioxide: relationship to indoor/outdoor air quality and activity patterns, *Environ. Sci. Technol.* 20:775–783.

Reggin, R. M., and Peterson, B. A. 1985. Sampling analysis methodology for semivolatile and nonvolatile organic compounds in air, In: *Indoor Air and Human Health* (R. B. Gammage and S. V. Kaye, eds.), Lewis Publishers, Inc., Chelsea, Mich.

Richmond, H. M., and McCurdy, T. 1985. Estimating exposures and health risks for alternative ozone ambient air quality standards. Paper presented at the Society for Risk Assessment Annual Meeting, Alexandria, Va., October 6–9, 1985.

Rogan, W. J., and Gladen, B. C. 1983. Monitoring breast mild contamination to detect hazards from waste disposal, *Environ. Health Perspect.* 48:87–91.

Rogan, W. J., Gladen, B. C., McKinney, J. D., and Albro, P. W. 1983. Chromatographic evidence of polychlorinated biphenyl exposure from a spill, *J. Am. Med. Assoc.* 249:1057–1058.

Rose, V. E., and Perkins, J. L. 1982. Passive dosimetry state-of-the-art review, *Am. Ind. Hyg. Assoc. J.* 43:605–621.

Ryan, P. B., Spengler, J. D., and Letz, R. 1983. The effects of kerosene heaters on indoor pollutant concentrations: a monitoring and modeling study, *Atmos. Environ.* 17:1339–1345.

Ryan, P. B., Spengler, J. D., and Letz R. 1986. Estimating personal exposures to nitrogen dioxide, *Environ. Int.* 12:395–400.

Ryan, P. B., Soczek, M. L., Spengler, J. D., and Billick, I. H. In press. A survey methodology for characterization of residential NO_2 concentrations, *J. Air Pollut. Contr. Assoc.*

Savage, E. P., Keefe, T. J., Tessari, J. D., Wheeler, H. W., Applehans, F. M., Goes, E. A., and Ford, S. A. 1981. A national study of chlorinated hydrocarbon insecticide residues in human milk, USA. I. Geographic distribution of dieldrin, heptachlor, heptachlor epoxide, chlordane, oxychlordane, and mirex, *Am. J. Epidemiol.* 113:413–433.

Sega, K., and Fugas, M. 1982. Personal exposure

versus monitoring station data for respirable particles, *Environ. Int.* 8:259–263.

Seifert, B., and Abraham, H. J. 1983. Use of passive samplers for the determination of gaseous organic substances in indoor air at low concentration levels. *Int. J. Environ. Anal. Chem.* 13:237–253.

Sexton, K. 1986. Indoor air quality: an overview of policy and regulatory issues, *Sci. Technol. Human Values* 11:53–67.

Sexton, K., and Hayward, S. B. 1986. Source apportionment of indoor air pollution, *Atmos. Environ.* 21:407–418.

Sexton, K., and Repetto, R. 1982. Indoor air pollution and public policy, *Environ. Int.* 8:5–10.

Sexton, K., Letz, R., and Spengler, J. D. 1983. Estimating human exposure to nitrogen dioxide: an indoor/outdoor modeling approach, *Environ. Res.* 32:151–166.

Sexton, K., Spengler, J. D., and Treitman, R. D. 1984a. Effects of residential wood combustion on indoor air quality: a case study in Waterbury, Vermont, *Atmos. Environ.* 18:1371–1383.

Sexton, K., Spengler, J. D., and Treitman, R. D. 1984b. Personal exposure to respirable particles: a case study in Waterbury, Vermont, *Atmos. Environ.* 18:1385–1398.

Sexton, K., Liu, K., and Petreas, M. X. 1986a. Formaldehyde concentrations inside private residences: a mail-out approach to indoor air monitoring, *J. Air Pollut. Contr. Assoc.* 36:698–704.

Sexton, K., Liu, K., Treitman, R. D., Spengler, J. D., and Turner, W. A. 1986b. Characterization of indoor air quality in wood-burning residences, *Environ. Int.* 12:265–278.

Sexton, K., Webber, L. M., Hayward, S. B., and Sextro, R. G. 1986c. Characterization of particle composition, organic vapor constituents, and mutagenicity of indoor air pollutant emissions, *Environ. Int.* 12:351–362.

Sheldon, L. S., Sparacino, C. M., and Pellizzari, E. D. 1985. Review of analytical methods for volatile organic compounds in the indoor environment, In: *Indoor Air and Human Health* (R. B. Gammage and S. V. Kaye, eds.), Lewis Publishers, Inc., Chelsea, Mich.

Shy, C. M., Kleinbaum, D. G., and Morgenstern, H. 1978. The effect of misclassification of exposure status in epidemiological studies of air pollution health effects, *Bull. N.Y. Acad. Med.* 54:1155–1165.

Silverman, F., Corey, P., Mintz, S., Oliver, P., and Hosein, R. 1982. A study of effects of ambient urban air pollution using personal samplers: a preliminary report, *Environ. Int.* 8:311–316.

Sizaret, P., Malaveille, C., Montesano, R., and Frayssinet, C. 1982. Detection of aflatoxins and related metabolites by radioimmunoassay, *J. Natl. Cancer Inst.* 69:1375–1381.

Speizer, F. E., Ferris, B. G., Jr., Bishop, Y. M. M., and Spengler, J. D. 1980. Respiratory disease rates and pulmonary function in children associated with NO₂ exposure, *Am. Rev. Respir. Dis.* 121:3–10.

Spengler, J. D., and Sexton, K. 1983. Indoor air pollution: a public health perspective, *Science* 221:9–17.

Spengler, J. D., and Soczek, M. L. 1985. Evidence for improved ambient air quality and the need for personal exposure research, *Environ. Sci. Technol.* 18:268A–280A.

Spengler, J. D., Dockery, D. W., Reed, M. P., Tosteson, T., and Quinlan, P. 1980. Personal exposure to respirable particles. Paper presented at the 73rd Annual Meeting of the Air Pollution Control Association, Montreal, Quebec, Canada, June 23–27.

Spengler, J. D., Dockery, D. W., Turner, W. A., Wolfson, J. M., and Ferris, B. G., Jr. 1981. Longterm measurements of respirable sulfates and particles inside and outside homes, *Atmos. Environ.* 15:23–30.

Spengler, J. D., Duffy, C. P., Letz, R., Tibbits, T. W., and Ferris, B. G., Jr. 1983. Nitrogen dioxide inside and outside 137 homes and implications for ambient air quality standards and health effects research, *Environ. Sci. Technol.* 17:164–168.

Spengler, J. D., Treitman, R. D., Tosteson, T. D., Mage, D. T., and Soczek, M. L. 1985. Personal exposures to respirable particulates and implications for air pollution epidemiology, *Environ. Sci. Technol.* 19:700–707.

Stock, T. H., Kotchmar, D. J., Contart, C. F., Buffler, P. A., Holguin, A. H., Gehan, B. M., and Noel, L. M. 1985. The estimation of personal exposures to air pollutants for a community-based study of health effects in asthmatics—design and results of air monitoring, *J. Air Pollut. Contr. Assoc.* 35:1266–1273.

Szalai, A. 1972. *The Use of Time: Daily Activities of Urban and Suburban Populations in Twelve Countries,* Paris: Mouton, The Hague.

Szentesi, I., Hornyak, E., Ungvary, G., Czeizel, A., Bognar, Z., and Timar, M. 1975. High rate of chromosomal aberration in PVC workers, *Mutat. Res.* 37:313–316.

Tosteson, T., Spengler, J. D., and Weker, R. 1982. Aluminum, iron, and lead content of respirable particle samples from a personal monitoring study, *Environ. Int.* 8:265–268.

Tough, I. M., Smith, P. G., Court Brown, W. M., and Harnden, D. G. 1970. Chromosome studies on workers exposed to atmospheric benzene, *Eur. J. Cancer* 6:49–55.

Turner, W. A., Spengler, J. D., Dockery, D. W., and Colome, S. D. 1979. Design and performance of a reliable personal monitoring system for respirable particles, *J. Air Pollut. Contr. Assoc.* 29:747–749.

Underhill, D. W. 1984. Efficiency of passive sampling by adsorbents, *Am. Ind. Hyg. Assoc. J.* 45:306–310.

Vo-Dinh, T., and Miller, G. H. 1983. A new passive monitor for direct detection of PAH vapors, In: *Proceedings of the International Symposium on Polyaromatic Hydrocarbons,* October 26–28, 1983, Columbus, Ohio.

Wallace, L. A., and Ott, W. R. 1982. Personal monitors: a state-of-the-art survey, *J. Air Pollut. Contr. Assoc.* 32:601–610.

Wallace, L. A., Zweidinger, R., Erickson, M., Cooper, S., Whitaker, D., and Pellizzari, E. 1982. Monitoring individual exposure: measurements of volatile organic compounds in breathing-zone air, drinking water, and exhaled breath, *Environ. Int.* 8:269–282.

Wallace, L. A., Pellizzari, E. D., Hartwell, T., Rosenzweig, M., Erickson, M., Sparacino, C., and Zelon, H. 1984. Personal exposure to volatile organic compounds. I. Direct measurements in breathing zone air, drinking water, food, and exhaled breath, *Environ. Res.* 35:293–319.

Wallace, L. A., Pellizzari, E. D., Hartwell, T. D., Sparacino, C. M., Sheldon, L. S., and Zelon, H. 1985a. Personal exposures, indoor-outdoor relationships, and breath levels of toxic air pollutants measured for 355 persons in New Jersey, *Atmos. Environ.* 19:1651–1661.

Wallace, L. A., Pellizzari, E. D., and Gordon, S. M. 1985b. Organic chemicals in indoor air: a review of human exposure studies and indoor air quality studies, In: *Indoor Air and Human Health* (R. B. Gammage and S. V. Kaye, eds.), Lewis Publishers, Inc., Chelsea, Mich.

Woebkenberg, M. L. 1982. A comparison of three passive personal sampling methods for NO_2, *Am. Ind. Hyg. Assoc. J.* 43:553–561.

Wogan, G. N., and Gorelick, N. J. 1985. Chemical and biochemical dosimetry of exposure to genotoxic chemicals, *Environ. Health Perspect.* 62:5–18.

World Health Organization. 1982. Estimating Human Exposure to Air Pollutants, WHO Offset Publication No. 69, Geneva, Switzerland.

World Health Organization. 1983. Indoor Air Pollutants: Exposure and Health Effects, EURO Reports and Studies 78, Geneva, Switzerland.

Yamasaki, E., and Ames, B. N. 1977. Concentration of mutagens from urine by absorption with the non-polar resin XAD-2: cigarette smokers have mutagenic urine, *Proc. Nat. Acad. Sci. USA* 74:3555–3559.

Yanagisawa, Y., and Nishimura, H. 1982. A badge-type personal sampler for measurement of personal exposure to NO_2 and NO in ambient air, *Environ. Int.* 8:235–242.

Yocum, J. E. 1982. Indoor-outdoor air quality relationships: a critical review, *J. Air. Pollut. Contr. Assoc.* 32:500–520.

Ziskind, R. A., Fite, K., and Mage, D. T. 1982. Pilot field study: carbon monoxide exposure monitoring in the general population, *Environ. Int.* 8:283–293.

Biological Disposition of Airborne Particles: Basic Principles and Application to Vehicular Emissions

RICHARD B. SCHLESINGER
New York University Medical Center

Air Pollution, the Automobile, and Public Health. © 1988 by the Health Effects
Institute. National Academy Press, Washington, D.C.

The primary route of exposure to motor vehicle emissions is inhalation. The respiratory tract has a large internal surface area that is directly and continually exposed to 10,000 to 20,000 liters of ambient air inhaled daily, making it a potential target site for exhaust products. In addition, because the barrier between inhaled air and the pulmonary bloodstream is very thin, the respiratory tract is also an efficient portal of entry into the general circulation.

A large fraction of emissions is either directly released in particulate form or becomes adsorbed onto the surface of other ambient particles. The disposition of inhaled particles, and any adsorbed constituents, determines the dose delivered to target tissues. However, their ultimate fate and any potential hazard depend upon various interacting parameters: the physicochemical characteristics of the particles, the amount that actually deposits in the respiratory tract, and the rates and routes by which deposited material is cleared from the respiratory tract or translocated to other organs.

Particles derived from motor vehicles do not have unique properties that influence their deposition or clearance. Thus, their disposition can be assessed in general terms. This chapter is a review of the biological disposition of inhaled particulate matter in terms of the factors that influence and control its deposition, clearance, translocation, and ultimate retention. The fate of specific nonorganic particles found in automobile exhaust will be assessed as examples of the disposition of toxicologically relevant material.

Some of the information presented is based on studies with humans, but much is derived from experiments with laboratory animals. Since human experimentation is precluded in many instances and often yields only limited data, surrogate animal models are needed. However, extrapolation from animal studies requires information on similarities and differences between species that may influence the disposition of inhaled materials. Thus, an attempt has been made to interrelate and integrate human data with that obtained with experimental animals and, in some cases, even with in vitro systems.

The chapter is divided into five major

sections. The first describes the anatomy of the respiratory tract, since airway structure is a major determinant of particle disposition. The second section discusses the aspects of ventilation important in exposure assessment, including scaling for different species. The third section describes the physical mechanisms by which inhaled particles deposit in the respiratory tract, their controlling influences and modifying factors. It critically reviews the available data for total and regional deposition in humans and experimental animals and provides a comparative analysis of interspecies deposition patterns. The fourth section discusses the structure, physiology, kinetics, and modifiers of the mechanisms by which deposited particles are cleared from, or translocated and retained within, the respiratory tract. The fifth section discusses the fate of specific nonorganic particles of relevance to automobile exhaust toxicology, that is, diesel particles, metals, and sulfates. In all five sections, knowledge gaps are highlighted and recommendations for research to fill these gaps are presented.

Structure of the Respiratory Tract

The respiratory tract is divided into two sections according to function: one is concerned with transporting air from the external environment to the sites of gas exchange and consists of the upper respiratory tract and the tracheobronchial tree; the other, the pulmonary region, is involved in gas exchange.

Upper Respiratory Tract

This region originates at the nostrils and mouth and extends through the larynx; a diagram of the human upper respiratory tract is shown in figure 1. Air entering the nostrils passes first through the vestibule, the narrowest cross-section in the entire nasal region, before entering the main nasal passages. These consist of two airways separated almost symmetrically by the nasal septum. They are convoluted (due to

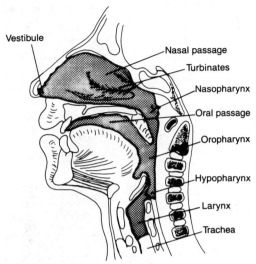

Figure 1. Diagram of the human upper respiratory tract.

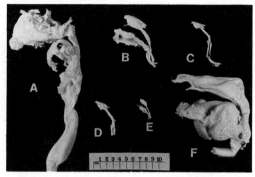

Figure 2. Silicone rubber replica casts of the naso-pharyngeal region of different species: (A) human; (B) rabbit; (C) rat; (D) guinea pig; (E) hamster; (F) baboon; scale in cm. (Adapted with permission from Patra 1986, and from Hemisphere Publishing Corporation.)

the folds of the nasal turbinates), downward-curving shelf-like structures, resulting in a large surface area and a relatively narrow distance between opposing airway walls. Here, exchange of heat and moisture modify the temperature and humidity of the inhaled air. The nasopharynx begins at the posterior end of the turbinates, where the septum ends and the nasal passage narrows and turns downward. Although the basic structure of the nasal airways is similar in humans and most other mammals, there are considerable interspecies differences in the relative position, shape, and size of individual components, as shown in figure 2. For example, the nasopharynx in the rat encompasses a greater percentage of the total length of the nasal passages than in the human, whereas that in rabbits and dogs is intermediate between rats and humans.

The oral passages begin at the mouth and are characterized by much greater inter- and intraindividual variation in shape and cross-section than the nasal passages. At the posterior of the mouth, inhaled air enters the oropharynx. The oro- and nasopharynx join to form the hypopharynx, an airway that extends to the entrance of the larynx. The latter extends to the trachea and has a variable cross-section depending upon the rate of airflow through it.

Tracheobronchial Tree

The tracheobronchial tree consists of airways from the trachea through the terminal bronchioles. The trachea divides into two main bronchi which then enter the lungs at the hilar region. These main bronchi further subdivide into smaller airways. Support for the trachea and bronchi are derived from cartilagenous rings or plates. As the bronchial tree proceeds distally, the cartilage eventually disappears, and these airways—the bronchioles—are supported by smooth muscle. In humans, the transition from bronchi to bronchioles occurs in airways of ~1-mm diameter.

Simplistically, the tracheobronchial tree can be considered to be a system of tubes connected at specific division points. In most cases, division is by dichotomy, whereby a single branch (the parent) gives

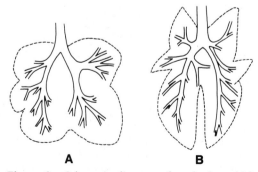

Figure 3. Schematic diagram of tracheobronchial tree branching patterns: (A) human lung; (B) monopodial system common in experimental animals.

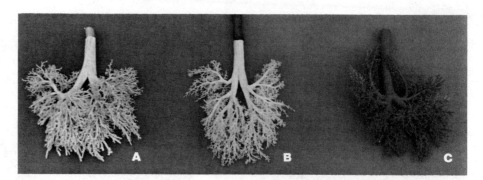

rise to two branches (the daughters). To describe this structure, the position of an individual airway is usually assigned a code number. There are two basic coding systems: the numbering of divisions up from distal end branches or, alternatively, down from the trachea. For example, in the often-used Weibel ordering system (Weibel 1963), each branching division is known as a generation; the trachea is generation 0, and each distal division increases by one number.

In a dichotomous branching system, the pattern can vary in terms of the degree of symmetry (figure 3). If both daughters have the same diameter and length, and branch from the parent at the same angle, the mode of division is known as regular or symmetrical. If the two daughters differ from each other in one or more dimensions, the mode of branching is termed irregular or asymmetrical, the extreme case of which is monopody. In a monopodial branching system, the larger-diameter daughter (major daughter) may not be easily distinguishable from the parent since the change in diameter and direction from the parent may be negligible.

A major difference in respiratory tract anatomy between humans and most other mammals commonly used in inhalation studies is in the pattern of bronchial airway branching. Figure 4 shows casts of the upper bronchial tree in humans and in a number of other species, and figure 5 presents a quantitative analysis that allows characterization of branching patterns. In a regular dichotomous branching system, the ratio of daughter diameters is 1, whereas in a perfect monopodial branching system, the ratio of major daughter diameter to

Figure 4. Silicone rubber replica casts of the tracheobronchial tree of different species. (A) human; (B) baboon; (C) dog; (D) rhesus monkey; (E) rabbit; (F) guinea pig; (G) rat; (H) hamster; (I) mouse. Both photos are reproduced here at the same scale, given in inches at the bottom. (Adapted with permission from Patra 1986, and from Hemisphere Publishing Corporation.)

parent diameter is 1. The human bronchial tree, at least for the first six generations, exhibits the most symmetrical branching of all of the species shown, whereas the dog's bronchial tree is almost ideally monopodal. The other species exhibit various degrees of irregularity. Recent qualitative observations on the tracheobronchial trees of two nonhuman primates—the rhesus monkey and the baboon—suggest a branching pattern that is more irregular than that of humans, but not to the extent of the experimental animal species shown in figure 5 (Patra 1986). But although there may be striking interspecies differences in the upper bronchial tree, the branching patterns in most mammals tend to approach more regular symmetry in distal conducting airways.

An important difference between regu-

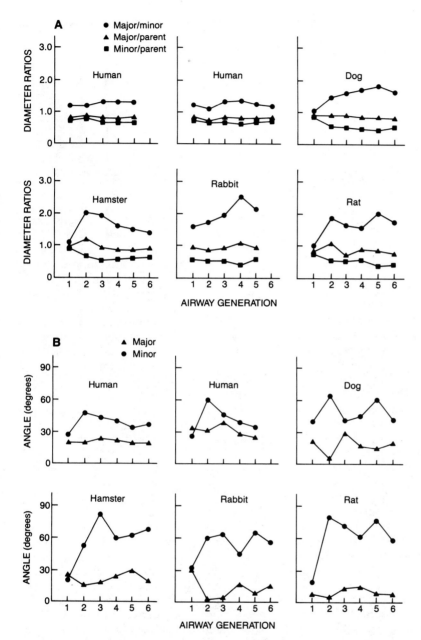

Figure 5. Morphometric relationships for the bronchial trees of different species. Each panel is derived from measurements of a single silicone rubber cast. (A) Ratios of airway diameters as a function of branching generation; (B) ratios of branching angles as a function of generation. (Adapted with permission from Schlesinger and McFadden 1981.)

lar and irregular dichotomous branching modes concerns the number of airways between the trachea and the terminal bronchioles. In a regular dichotomous branching system, the number of divisions and, therefore, the path length, between the trachea and the most distal conducting airways is the same along any pathway. In addition, all airways at any branch level have exactly the same dimensions. In an

Table 1. Airway Path Lengths

| Species | Number of Airway Divisions (Generations)[a] | | Reference |
	Mean (for entire lung)	Range of Means (for individual lobes)	
Human	17	—	Weibel (1963)
Human	15	14–17	Schum et al. (1976)
Rat	15	11–19	Raabe et al. (1977)
Hamster	14	10–18	Raabe et al. (1977)
Dog (beagle)	18	15–21	Schum et al. (1976)

[a] From trachea through terminal bronchioles.
SOURCE: Adapted with permission from Schlesinger and McFadden 1981.

irregular dichotomous system, the number of branch divisions from the trachea to each distal bronchiole is not the same along every pathway, and not all airways at a given branch level have the same dimensions. Table 1 presents the average number of branching generations from the trachea through the terminal bronchioles for various species. Humans have the narrowest range of branching generations, a reflection of the greater symmetry of their lungs.

Because of the complexity of airway branching structure, the geometry of the tracheobronchial tree has been represented by models; these are idealizations derived from experimental data, usually from measurements performed on castings prepared from actual lungs. One of the most widely used human structural models is the symmetrically dichotomous Weibel Model A (Weibel 1963). This is a 23-generation system, with generations 0–16 representing conducting airways. Although the assumption of regular dichotomy simplifies the treatment of morphometric data, the actual bronchial tree is asymmetrical, and a number of models of human airways that account for asymmetry have been described (Horsfield and Cumming 1968; Olson et al. 1970; Horsfield et al. 1971; Parker et al. 1971). In addition, Phalen and coworkers (1978) and Yeh and Schum (1980) developed structural models of the human lung which consist of "typical pathways," based on mean dimensions, for each lobe within the lung. Although the models were developed with symmetrical branching within each lobe, they do account for the asymmetry, and resultant variable path length, between different lobes. Most of the tra-

cheobronchial models have been based upon measurements made in only one lung. The very limited data base suggests that there is significant variability in airway dimensions between individuals (Nikiforov and Schlesinger 1985), but the only model that accounts for this is a statistical description of the tracheobronchial tree based upon the Weibel geometry (Soong et al. 1979).

Structural models of the bronchial tree have also been developed for experimental animals. These include symmetrical dichotomous models for the rabbit (Kliment 1974), the rat (Kliment 1973), and the guinea pig (Kliment et al. 1972) and typical pathway models for the dog, the rat, and the hamster (Yeh 1980).

Pulmonary Region

The pulmonary region extends from the respiratory bronchioles through the alveoli and contains airways involved in gas exchange between the air and blood (figure 6a). In the human lung, the final generation of airways that merely conduct air—the terminal bronchioles—branch into several generations of respiratory bronchioles, which are characterized by the presence of alveoli. The degree of alveolarization increases toward the lung periphery; when the airway becomes totally alveolarized, it is termed an alveolar duct. This may branch into other ducts, or into blind-ended alveolar sacs. The adult human lung contains ~ 375 million alveoli, the number varying with body size, and the average alveolar diameter is 250–300 μm. This results in a total alveolar surface area on the order of 150–180 m^2 (Weibel 1980).

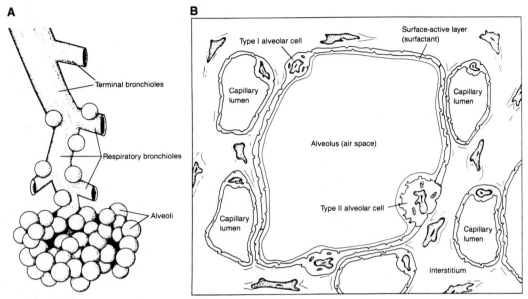

Figure 6. (A) Diagram of the human airways in the pulmonary region; (B) diagram of the cellular makeup and surrounding structures of the alveolus.

There are large interspecies differences in the gross structure of the pulmonary region (Gehr et al. 1981; Tyler 1983). The number of branching generations of respiratory bronchioles and alveolar ducts varies, and some species appear to have no respiratory bronchioles. The degree of alveolarization of the respiratory bronchioles also differs, as does alveolar size and total alveolar surface area, the latter increasing in direct proportion to body mass.

The alveolar surface is lined with a continuous layer of two distinct cell types (figure 6b). About 93–95 percent is covered by type I cells, which are characterized by a central nucleus surrounded by cytoplasm stretching out in thin winglike processes to form part of the alveolar wall. The remaining surface is covered by cuboidal-shaped type II cells, which are actually more numerous than the type I cells. The relative numbers of these cell types, as well as the percentage of the alveolar surface covered by each, are similar in humans and most other mammals (Crapo et al. 1983; Gehr 1984).

The alveoli are supported by a framework of connective tissue termed the interstitium. Capillary endothelial cells are joined through the interstitium to alveolar epithelial cells, to form the "alveolo-cap-illary membrane." This membrane is about 2 μm thick in humans, but appears to be thinner in most experimental animals (Meessen 1960; Crapo et al. 1983; Gehr 1984). The interstitium and associated structures form the part of the lung known as the parenchyma. This region also includes the pulmonary lymphatic vessels.

The lungs contain two lymphatic networks. One set (superficial or pleural network) is located within the connective tissue layer of the visceral pleura, whereas the other (deep or peribronchovascular network) consists of interconnecting vessels within the connective tissue surrounding both the airways (to the level of the respiratory bronchioles) and the pulmonary vascular system. A plexus of vessels connects the two sets. In both systems, the network begins as blind-ended capillaries and fluid flows toward the hilar region of the lung. Many larger lymphatic vessels are interspersed with nodes (encapsulated aggregates of lymphoid tissue); the most prominent of these are located along the trachea and main bronchi, and at branching sites between these airways. More diffuse lymphoid aggregates occur near the branching regions of smaller bronchi and bronchioles. Eventually, the entire pulmonary lymphatic system drains into the general venous circulation.

Research Recommendations

Quantitative anatomy—or morphometry—of the respiratory tract is essential for understanding the dosimetry of inhaled particles. The structure of the various components of the respiratory tract influences the airflow and, thus, the resultant pattern of particle deposition and the distribution of sites of potential damage. Morphometry must be assessed in humans as well as experimental animals, the latter so as to assist in the extrapolation of toxicologic data to humans. Data are available for normal adult humans and some other species, but critical gaps remain.

■ **Recommendation 1.** Variability in morphometry of the tracheobronchial and pulmonary regions in normal humans as well as experimental animals (including different strains) should be studied. Better statistical descriptions of interindividual variation at all levels of the respiratory tract are needed to validate conclusions drawn from current theoretical or empirical deposition models, which are generally based upon a single morphometric model.

■ **Recommendation 2.** Lung morphometry should be assessed in potentially "susceptible" subsegments of the human population: children, the elderly, and people with respiratory disease. Although data are becoming available on the morphometry of children's lungs at different ages, these are not yet sufficient to develop a comprehensive morphometric model describing growth of the tracheobronchial tree. No information exists at all for assessment of morphometric changes due to aging or disease.

■ **Recommendation 3.** Comparative morphometry of human and animal upper respiratory tracts should be assessed. Because of large interspecies differences in the nasopharyngeal region, more quantitative information is needed to allow better comparison with that in humans. For example,

rodents have essentially a straight pathway from the nostrils to the trachea, a situation radically different from that in humans and nonhuman primates. In humans, more detailed information on dimensions of the oral passages under different ventilatory conditions is also needed to assess particle removal by the upper respiratory tract.

■ **Recommendation 4.** Comparative structure and physiology of human and animal pulmonary lymphatic systems should be studied. This knowledge is needed for better comparisons of particle clearance by this route in humans and experimental animals.

Ventilation

Ventilatory Parameters

Ventilation is the movement of air in and out of the respiratory tract and is a factor in determining the amount of an exposure atmosphere that is actually inhaled. Ventilatory parameters also affect the deposition of particles once inhaled.

The amount of air inspired (or expired) during a normal breath is the tidal volume (V_T); it averages 450–600 ml in resting healthy males and slightly less for females. The fraction of the V_T that does not reach the alveolated airways—about 150–200 ml in resting males and 120–160 ml in females—is termed the anatomic dead space volume ($V_{D_{anat}}$).

Not all of the inspired air reaching the pulmonary region is equally effective in oxygenating the blood, since air may enter alveoli that are ventilated but poorly perfused. The portion of V_T that does not equilibrate with gas pressure in the pulmonary capillary blood is the alveolar dead space volume ($V_{D_{alv}}$). The total volume of inhaled air that does not participate in gas exchange, $V_{D_{anat}} + V_{D_{alv}}$, is termed the total or physiological dead space ($V_{D_{tot}}$).

During expiration, air within the tracheobronchial tree—largely from the previous inspiration—is expelled along with some alveolar air which is a mixture from a number of inspirations. Particles inhaled into the pulmonary region can therefore be exhaled over a number of breaths. Thus, the time available to deposit inhaled particles in the conducting airways is fairly short (a few seconds), whereas the residence time in pulmonary air may be longer (about a minute).

Total ventilation ($\dot{V}_E$), or minute volume (MV), is defined as the volume of air expired each minute and is equal to V_T times the breathing frequency (f). The average f during normal quiet breathing in adults is 11–17 breaths/min, and the resting $\dot{V}_E$ averages 5–10 liters/min. The $\dot{V}_E$ consists of anatomic dead space ventilation ($\dot{V}_{D_{anat}}$) and total alveolar ventilation ($\dot{V}_A$), the latter being the amount of air entering the pulmonary region each minute. The effective portion of $\dot{V}_A$ that participates in gas exchange is equal to $f(V_T - V_{D_{tot}})$.

Ventilation is affected by numerous exogenous factors such as altitude, ambient temperature, and smoking, as well as endogenous factors such as body size. Two of the major modifiers in any particular individual are physical activity and age.

Physical Activity. Healthy humans at rest normally breath through the nose, but when respiratory demand increases above a certain level there is a shift to oronasal (combined nose and mouth) breathing. Maximum inspiratory nasal airflow occurs at a $\dot{V}_E$ of 30–40 liters/min (Swift and Proctor 1977; Niinimaa et al. 1980), at which point ~40–60 percent of total airflow occurs through the nose. As respiratory demand increases further, the proportion of air entering the mouth increases, but even at high demand the oral pathway accounts for no more than 60 percent of the inhaled air (Swift and Proctor 1987).

With mounting respiratory demand, V_T and f increase, and the maximum volume of air that can be inhaled per minute, or the maximum voluntary ventilation, may rise to more than 10 times the resting ventilatory level. As breathing frequency increases, expiratory time diminishes, but inspiratory time remains relatively constant. Furthermore, respiratory pauses, the gaps between expiration and inspiration

which can occupy 25 percent of the breathing cycle in resting individuals, become shorter with increasing level of activity.

Growth and Aging. The volume of air in the lungs and the ventilatory capacity depend on body and lung size and, thus, increase with growth from childhood. In addition, the contribution of V_T and f to total ventilation also changes; V_T increases while f decreases until maturity is reached (Mauderly 1979).

Ventilatory function reaches a peak between the ages of 20 and 35 and then begins to decline. Although various models have been proposed to describe these changes, they differ in their assumptions about the rate of functional decline (Buist 1982). Furthermore, most of the reported data for age-related changes in lung function are derived from cross-sectional population studies and may not reflect the true aging process, especially since these studies may be measuring the heartiest survivors. The best way to avoid possible bias is to examine true aging patterns in longitudinal studies in which the same people are tested over a number of years. Such analyses are scarce, and those that do exist have measured only a few parameters (Fowler 1985).

Changes in lung function with aging are the result of deterioration of the lung tissue itself, a decrease in the strength of the respiratory muscles, and an increase in the stiffness of the thoracic cage. The time course varies from individual to individual and may be aggravated by chronic pollutant exposure. Some ventilatory indices are affected by age, whereas others are not. Figure 7 shows a diagram of the various divisions into which the volume of air in the lungs may be separated. With age, functional residual capacity (FRC) and residual volume (RV) increase, whereas vital capacity (VC), inspiratory capacity (IC), and expiratory reserve volume (ERV) decrease. Anatomic dead space ($V_{D_{anat}}$) increases with age because of a decrease in lung elasticity and a resultant increase in lung volume at the same pressure differentials.

Aging is associated with regional inequalities in the distribution of ventilation

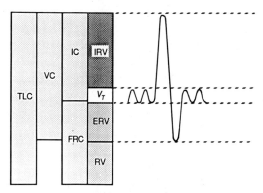

Figure 7. Diagram of subdivisions of lung volumes as measured with a spirometer. A typical spirometer tracing is shown on the right. TLC = total lung capacity, VC = vital capacity, RV = residual volume, FRC = functional residual capacity, IRV = inspiratory reserve volume, ERV = expiratory reserve volume, V_T = tidal volume, and IC = inspiratory capacity.

and a decrease in the uniformity of perfusion (Holland et al. 1968). Nonuniform mixing of inspired air may result when sections of the lungs communicate poorly with others and, because of this, some alveolar regions may not be continuously ventilated during normal tidal breathing. Nonuniform perfusion results in an increase in $V_{D_{alv}}$ which, together with the increase in $V_{D_{anat}}$, results in an aging-related rise in $V_{D_{tot}}$. Although this does not affect resting levels of $\dot{V}_E$, which show no major change with aging, the ability of the lungs to respond to increased activity is altered, and maximum voluntary ventilation declines by about 30 percent between ages 30 and 70.

Comparative Aspects of Ventilation

Since much of the toxicologic work with inhaled particles involves experimental animals, it is essential that their respiratory mechanics be quantitated. Various animal data exist (see, for example, Guyton 1947; Spell 1969), but the methods used to obtain them were not standardized, so there is much variability, even for similarly sized animals of the same species. "Representative" ventilatory values for a particular species are therefore difficult to specify, so generalized values based on scaling pro-

cedures are used. Scaling is based on the principle that respiratory mechanical properties may be related to body size or mass in some consistent fashion, even though there may be interspecies differences in the mechanisms that determine these properties. This allows quantitative comparisons of function between animals of different sizes, within or between species. Scaling makes use of dimensional or dimensionless parameters that either remain constant with body size or can be related to body size by some proportionality factor (Leith 1983). For example, $\dot{V}_E$ is proportional to body mass (M) raised to the ¾ power, whereas lung volumes, such as V_T, tend to vary with M to the first power. Similarly, breathing frequency is proportional to $M^{-1/4}$, whereas the ratio of $V_{D_{anat}}$ to V_T is independent of body size.

Stahl (1967), after an extensive literature search, developed predictive equations relating respiratory variables in mammals to body weight. These equations can be used to scale values between animals of different species as well as between individuals of different body weights within one species, as long as the animals are in comparable physiological states. Scaling is not a precise technique, however, and is only as good as the values upon which the exponents and proportionality factors are based. For example, many of these values have been obtained in anesthetized animals, in which actual lung volumes and ventilation may be less than normal (Sweeney et al. 1983).

Airflow Patterns

Patterns of airflow in the conducting airways are a major determinant of particle deposition sites. Basic principles of airflow are presented by Ultman (this volume). Aspects of airflow critical to particle deposition are addressed below.

Within straight tubes, two main types of flow may occur: laminar and turbulent. In laminar flow, gas molecules move in parallel as a smooth stream, with the highest velocity occurring at the center of this stream. The flow can be imagined as concentric layers of air sliding or telescoping lengthwise along each other, with no transverse mixing between layers. In turbulent flow, gas molecules are in an agitated state, and there is erratic mixing of concentric layers. Random secondary flows (eddies) are superimposed on the average longitudinal motion of flow velocity. Flow that is partially laminar and partially turbulent is termed transitional.

The type of flow that occurs depends upon the strength of the inertial forces in the moving air in relation to the frictional and viscous forces acting on it. For example, turbulence occurs when the former exceed the latter. Airflow may thus be described in terms of the ratio of inertial forces to viscous and frictional forces, which is expressed as the dimensionless Reynolds number (Re). The Reynolds number depends on the geometry of the conduit through which the air passes and the velocity of airflow, and flow characteristics change as Re passes certain critical values. Thus, for steady flow in a straight, smooth-walled, circular tube, flow will be laminar when Re is less than 2100, transitional when Re is between 2100 and 4000, and fully turbulent when it exceeds 4000 (Hinds 1982).

Within the respiratory tract, bends, bifurcations, constrictions, surface roughness and convolutions, and other features of airway shape that add inertial forces may lead to turbulent flow at a velocity lower than that at which turbulence would be initiated in a smooth, straight, obstacle-free tube having the same cross-section. Thus, flow instability and turbulence may occur in the upper respiratory tract and upper tracheobronchial tree at Reynolds numbers well below 2100 (West and Hugh-Jones 1959; Dekker 1961; Sekihara et al. 1968; Olson et al. 1973; Swift and Proctor 1977). Turbulence is also produced by the continuous acceleration and deceleration of air during the breathing cycle (Lakin and Fox 1974). But although turbulent flow generated in the upper airways upon inspiration may be propagated into a few generations of downstream bronchi, air velocity decreases with depth into the lung, and in the smaller conducting airways, flow is always laminar.

Because of structural differences between the tracheobronchial trees of humans and most other mammals, one would expect differences in resultant flow patterns. For example, the trachea of most mammals is much longer relative to its diameter than is the human trachea. Thus, any turbulence introduced by flow through the larynx is much less likely to persist into the downstream bronchi of nonhuman mammals. Unfortunately, there are few data on airflow patterns in the airways of most commonly used experimental animals (see, for example, Snyder and Jaeger 1983).

Research Recommendations

Ventilatory patterns and airflow dynamics are critical determinants of dose to the respiratory tract from inhaled particles. The following important gaps in our knowledge of ventilation in humans and in experimental animals should be filled.

■ **Recommendation 5.** Patterns and distribution of airflow in the tracheobronchial tree of healthy adult experimental animals and humans should be determined. This information is important for the development of deposition models and for the extrapolation of results of toxicologic studies to humans.

■ **Recommendation 6.** Effects of aging on ventilation in humans and experimental animals should be determined by use of longitudinal studies of humans and experimental animals involving numerous ventilatory parameters. In animals, a cross-correlation of age equivalencies between species should be performed, so that parameters of toxicologic studies may be better related to lung function in humans.

■ **Recommendation 7.** Ventilatory mechanics and airflow in children should be analyzed. Although data are available for some stages of growth, there is a gap between birth and ~9 years of age.

■ **Recommendation 8.** Flow patterns in the upper respiratory tracts of experimental animals and humans should be studied. Most experimental animals are obligate na-

sal breathers, so only their nasal passages
need be studied. But in humans, analyses of
the nature of flow in the oral passages
through the oropharynx, including the effects of speech and increased physical activity, are also needed.

Deposition of Inhaled Particles in the Respiratory Tract

The concentration of particles in ambient
air does not by itself define the dose delivered to the respiratory tract. To provide
such quantification, it is first necessary to
determine deposition sites—that is, regions
where inhaled particles initially contact airway surfaces. Deposition sites determine
the subsequent pathways for removal or
translocation and, as such, constitute a major contributor to the ultimate toxicologic
response.

Deposition Mechanisms and Controlling Factors

Specific Deposition Mechanisms. The
size of inhaled particles is a critical factor
affecting their deposition; thus, resultant
biological effects are, to some extent, par-
ticle-size dependent. Size may, however,
be expressed in various ways. For spherical
particles, actual measured diameter is un-
ambiguous, but for nonspherical or irregularly shaped particles some "effective" diameter is more appropriate. Such particles
are often described in terms of equivalent

spheres, on the basis of equal volume,
mass, or aerodynamic drag.

In order to compare deposition data obtained using particles of different materials,
a diameter that accounts for the factors
affecting deposition should be used; the
most common of these is aerodynamic
equivalent diameter (D_{ae}). This term accounts for shape and density and is defined
as the diameter of a spherical particle with
unit density that has the same terminal
settling velocity (see below) as the particle
in question. Particles that have higher than
unit density will have actual diameters
smaller than their D_{ae}.

The significant mechanisms by which
particles are deposited in the respiratory
tract are impaction, sedimentation, Brownian diffusion, interception, and electrostatic
precipitation (figure 8). The relative contribution of each depends on characteristics of
the particles as well as on ventilatory patterns and respiratory tract anatomy.

Impaction onto an airway surface occurs
when a particle's momentum prevents it
from changing course in an area where
there is a rapid change in the direction of
bulk airflow. It is the main deposition
mechanism for particles having $D_{ae} \gtrsim 0.5$
μm in the upper respiratory tract and at or
near tracheobronchial tree branching
points. The probability of impaction increases with increasing air velocity, rate of
breathing, particle density, and size.

Sedimentation is deposition due to gravity. When the gravitational force on a particle is balanced by the total forces due to
air buoyancy and air resistance, the inspired
particle will fall out of the airstream at a

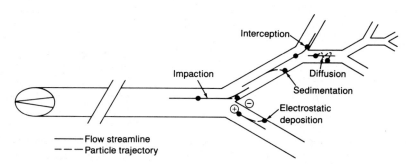

Figure 8. Mechanisms for particle deposition in the respiratory tract. (Adapted
with permission from Lippmann and Schlesinger 1984.)

constant rate, known as its terminal settling velocity. The probability of sedimentation increases with increasing residence time in the airway, particle size, and density, but decreases with increasing breathing rate. Sedimentation is important for particles with $D_{ae} \gtrsim 0.5$ μm in medium to small airways where air velocity is relatively low.

Submicrometer-size airborne particles, especially those with diameters ≤ 0.2 μm, have imparted to them a random motion due to bombardment by surrounding air molecules; this motion may then cause such particles to come into contact with the airway wall. The displacement sustained by the particle is a function of the diffusion coefficient, which is inversely related to particle cross-sectional area. Brownian diffusion is a major deposition mechanism in airways where bulk flow is low or no longer occurring, that is, in the bronchioli and alveoli. However, molecular-size particles may be deposited by diffusion in the upper respiratory tract, trachea, and larger bronchi.

As mentioned, particles with $D_{ae} \gtrsim 0.5$ μm are subject to impaction and sedimentation, whereas the deposition of particles ≤ 0.2 μm is diffusion dominated. Particles with diameters between these values are only minimally influenced by these mechanisms and tend to have prolonged suspension times in air. They may, thus, undergo little deposition, being carried out of the respiratory tract in the exhaled air.

Interception is a significant deposition mechanism for elongated particles, such as fibers, and occurs when the edge of the particle contacts the airway wall. The aerodynamic diameter of a fiber is related to its transverse diameter. Thus, fibers that are long (for example, 50–100 μm) but thin (for example, 0.5 μm) behave aerodynamically like small particles, penetrating into distal airways. Fiber shape is also important, since straight fibers penetrate more distally than do curly ones.

Some freshly generated particles can be electrically charged and may exhibit deposition greater than that expected on the basis of size alone. Electrostatic deposition results from image charges induced on the surface of the airways by charged particles and/or from space/charge effects, whereby repulsion of similarly charged particles causes increased migration toward the airway wall. The effect of charge on deposition increases with decreasing particle size and airflow rate. Since most ambient particles become neutralized naturally by air ions, electrostatic deposition is a minor contributor to particle collection by the respiratory tract.

Factors Controlling Deposition. An understanding of the extent and loci of particle deposition in the respiratory tract requires an appreciation of various controlling factors: characteristics of the inhaled particles, anatomy of the respiratory tract, and ventilation pattern.

Characteristics of Inhaled Particles. The major particle characteristic that influences deposition is size. However, particles are inhaled not singly, but as constituents of aerosols, which are suspensions of liquid or solid particles in a gas. The components of the particulate phase may differ, but even if this consists of a single material, a spectrum of particle sizes is often present. In general, the size distribution of particles in commonly encountered aerosols fits reasonably well with a lognormal distribution; that is, the logarithm of particle diameter is normally distributed. Such a distribution can be described by a geometric mean size (which is also the median diameter) and by an index of dispersion—the geometric standard deviation (σ_g). This latter is the ratio of the diameter at 84.1 percent (or 15.9 percent) cumulative probability, that is ± 1 standard deviation (SD) of the normal curve, to the diameter corresponding to 50 percent cumulative probability (figure 9). Depending upon the specific size parameter used to develop the distribution, the resultant median diameter may be count median (CMD, using the physical diameter of the particles), mass median (MMD, using the particle mass distribution relative to diameter), or aerodynamic mass median (MMAD, using aerodynamic equivalent diameter). If not directly measured, the MMD and MMAD may be calculated from the measured CMD for spherical particles.

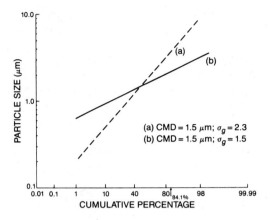

Figure 9. Cumulative frequency distribution plots of particle number for two polydisperse aerosols. Both aerosols have the same count median diameter (CMD) (50 percent probability), but they have different geometric standard deviations (σ_g). Because of the different σ_g, the percentage of particles with diameters greater or less than the CMD differs substantially for the two aerosols.

Radioactive or toxic aerosol size distributions are often expressed as activity median aerodynamic diameter (AMAD).

The deposition probability of particles with physical diameters $\gtrsim 0.5$ μm is governed largely by particle aerodynamic diameter, whereas deposition probability of smaller ones is governed by actual physical diameter. Thus, use of the MMAD parameter is appropriate only in describing aerosols in which most particles are physically $\gtrsim 0.5$ μm; the median size of aerosols containing particles with actual diameters less than this is usually expressed in terms of a diffusion diameter or actual physical size. Aerodynamic diameter is, therefore, the most appropriate unit for describing deposition by sedimentation and impaction, but not by diffusion.

The size distribution of an aerosol, which largely depends on its method of production, is characterized as monodisperse or polydisperse (heterodisperse). A monodisperse aerosol consists of particles of uniform size. Since, in reality, perfect monodispersity does not exist, an aerosol is considered monodisperse if the σ_g is <1.2 (Fuchs and Sutugin 1966). But use of this term in deposition analyses means that all of the particles are assumed to behave as if

they were exactly the same size, that is, the median size. In polydisperse aerosols, particles of widely differing sizes may be present, and the σ_g is ≥ 1.2.

If the σ_g of a polydisperse aerosol is <2, its total respiratory tract deposition will probably not differ substantially from that for a monodisperse aerosol having the same median size (Morrow 1981; Diu and Yu 1983). However, size distribution is important in determining the spatial pattern of initial dose. This is because the effect of size dispersion on regional deposition depends upon the sequential "filtering" action of each component of the respiratory tract, which in turn depends upon particle size. For example, as σ_g increases for aerosols with median sizes between 0.01 and about 0.07 μm, tracheobronchial deposition will likely increase, but pulmonary deposition will decrease because of less penetration into this region. On the other hand, as σ_g increases for aerosols with median sizes of 0.07 to about 1 μm, bronchial as well as pulmonary deposition will increase (Diu and Yu 1983).

A particle characteristic that may dynamically alter its size is hygroscopicity. Hygroscopic particles may grow substantially while they are still in transit in the respiratory tract and will be deposited according to their hydrated, rather than their initial dry size. The deposition pattern of specific hygroscopic aerosols can generally be related to their particle growth characteristics, if known.

Respiratory Tract Anatomy. Respiratory tract geometry affects particle deposition in various ways. For example, airway diameter sets the displacement required for a particle to contact a surface, whereas cross-section determines the air velocity for a given flow rate. Differences in pathway lengths in different lung lobes affect regional deposition. Lobes with the shortest average path length between the trachea and terminal bronchioles may have the highest concentration of deposited particles ≥ 1 μm in the alveoli. Regional differences become less obvious for submicrometer particles, which tend to deposit evenly in all lobes regardless of path length but in proportion to relative ventilation (Raabe et al. 1977).

Although humans differ from most other mammals in various aspects of respiratory tract anatomy, the implications of this for particle deposition have not been adequately appreciated. For example, alveolar size differs among species; since particles with $D_{ae} \gtrsim 0.5$ μm that reach the alveoli will be deposited primarily by sedimentation, and different-size alveoli have different characteristics as sedimentation chambers, the net result will be that the pulmonary region of various species will have different deposition efficiencies. Differences in deposition patterns affect the dosimetry of inhaled particles and the ability to use the results of toxicity tests in experimental animals for human risk assessments. In addition to interspecies differences, the data available indicate that size of tracheobronchial airways and alveoli vary considerably within species. Such variation is probably a major factor responsible for the observed differences in deposition efficiency among individuals of one species (Heyder et al. 1982).

Ventilatory Parameters and Mode of Inhalation. The pattern of respiration during particle exposure influences regional deposition sites and efficiencies. For example, high inhalation velocities enhance deposition by impaction but decrease that due to sedimentation and diffusion. Thus, a rise in flow rate, such as during increased physical activity, may shift regional deposition, increasing collection in the upper respiratory tract and central bronchi and reducing it in more distal conducting airways and the pulmonary region (Valberg et al. 1982; Morgan et al. 1984; Bennett et al. 1985). Increased flow velocities may also result in the development of turbulence, which tends to enhance particle deposition, the degree of potentiation depending on particle size (Schlesinger et al. 1982).

Tidal volume affects deposition by determining how deep into the lungs the inspired air penetrates. At a constant breathing frequency, increasing tidal volume deepens penetration of inhaled particles, thus increasing deposition in the smaller conducting airways and pulmonary region. Alterations in tidal volume can also dramatically affect total respiratory tract dep-

osition. For example, Schum and Yeh (1980) suggested that, in the rat, doubling tidal volume from 1.4 ml to 2.8 ml increases the deposition of a 1-μm (median D_{ae}) aerosol by seven times. Finally, the duration of respiratory pauses influences sedimentation or diffusion deposition by affecting particle residence time in relatively still air.

A major ventilatory change that occurs in humans when activity level increases is a switch from nasal to oronasal breathing. Since the nasal passages remove inhaled particles more efficiently than the oral passages, bypassing the nose increases the penetration of particles into the lungs. The actual magnitude of this increase is influenced by particle size, since larger particles are more effectively filtered in the nose than are smaller ones.

Measurement of Deposition

Measurement Techniques. Various techniques have been used to measure particle deposition in the respiratory tract of humans and experimental animals (Valberg 1985). Unfortunately, the use of different experimental methods and assumptions, especially in assessment of regional deposition, has resulted in large variations in reported values, even within the same species.

Total respiratory tract deposition has often been determined by a procedure that compares the concentration of test particles administered in inhaled air with that in collected exhaled air, the difference representing the total amount deposited. If assumptions are made about mixing and dead space, estimates of regional deposition can be derived from measurements of particle concentration in different volume fractions of the expired air, but such assumptions cannot be validated.

Specific particle characteristics may be used to measure deposition. Most commonly, radioactively tagged tracer particles are used with various types of detector systems. Total deposition is estimated by monitoring the thoracic and head regions immediately after exposure, whereas regional deposition is usually defined functionally on the basis of subsequent clearance. For example, it is often assumed that any particles remaining in the thorax 20 to 24 hr after exposure are in the pulmonary region, and particles that deposited in the tracheobronchial region were cleared from the lungs prior to this time. This is a reasonable assumption for healthy subjects but may not be for subjects with disease states where clearance is slower, and its use could result in an overestimation of pulmonary deposition and an underestimation of tracheobronchial deposition.

Deposition in the upper respiratory tract is inferred from measurements on the head immediately after exposure. Since this region clears rapidly to the stomach, even the first measurement may not accurately reflect actual deposition; accordingly, some investigators include an initial measurement of material in the gastrointestinal tract in their reported value for upper respiratory deposition. However, the upper respiratory tract, as defined in various studies, may include any or all of the following anatomic regions: nasopharynx, oropharynx, larynx, or upper trachea.

Another technique for deposition analysis in experimental animals is chemical and/or radiological assay of tissues or whole organs removed by dissection after exposure. Obtaining accurate deposition values requires immediate sacrifice, and the assumption of no particle translocation (except to the gastrointestinal tract) prior to or during dissection.

Experimental Deposition Assessment.
The species of choice for deposition analyses is the human. However, experimentation with human subjects is not always possible, and various animals are therefore used instead, with the ultimate goal of extrapolating the results to humans. If the results are to be valid, the extrapolation must take into account interspecies differences in total and regional deposition.

It is difficult to systematically compare deposition patterns obtained from reported studies in one species, and it is even harder to do this between species, because of variations in experimental protocols, measurement techniques, definitions of specific respiratory tract regions, and so on. For example, tests with humans are generally conducted under protocols that standardize tidal volume and breathing frequency (although the standardization parameters often differ in different laboratories), whereas those using experimental animals involve a wider variation in respiratory exposure conditions (for example, spontaneous breathing versus controlled breathing as well as various degrees of sedation). Much of the variability in the reported data for individual species is due to the lack of normalization for specific respiratory parameters during exposure.

In addition, experimental inhalation studies use different exposure techniques, such as nasal mask, oral mask, oral tube, or tracheal intubation. Regional deposition fractions are affected by the exposure route and delivery technique used (Wolfsdorf et al. 1969; Swift et al. 1977a). Even the specific size of the delivery device can affect inspired airflow rates, which influence the extent of deposition in the upper respiratory tract and the degree of particle penetration into the lungs (Heyder et al. 1980b).

Compilations of experimentally determined deposition values in humans and those experimental animals commonly used in inhalation toxicology studies are shown in figure 10. Not all deposition studies reported in the literature were included in this survey, since the objective was to make the intercomparisons as valid as possible. Thus, only studies where regional deposition values as a fraction of the amount of particles inhaled were provided, or could be derived, were included. Most studies describe regional fractions as a percentage of total deposition rather than in terms of amount of material inhaled and were, therefore, excluded. In addition, only studies using nonhygroscopic, nonviable, nonfibrous aerosols and reporting an aerodynamic or diffusion-related diameter were included. Most studies with humans used monodisperse aerosols, whereas many of those with experimental animals used polydisperse aerosols. Since some of these latter may have consisted of particles of widely different sizes, it is often difficult to evaluate deposition based upon the median size alone. However, it is necessary to include some of these studies, since a substantial amount of the existing data base is derived using such aerosols. Finally, although the tracer aerosols in some studies were not charge neutralized, data using these tracer aerosols were included. The presence of electrical charges could account for some of the variability between different studies using the same species and similar size particles.

Total Respiratory Tract. Figure 10a shows total respiratory tract deposition. In humans, nasal inhalation results in somewhat greater total deposition than oral exposures for particles with diameters >0.5 μm because the nasal passages collect larger particles more efficiently than the oral passages. There is little difference in total deposition between nasal or oral breathing for particles from 0.02 to 0.5 μm. With even smaller particles, total deposition should be greater with nose breathing than with mouth breathing, although the difference would be small, amounting to, for example, only about 5 percent for particles with diameters of 0.005 μm (Schiller et al. 1987).

Dogs and guinea pigs exhibit greater total deposition of 0.1–1-μm particles than do nasal-breathing humans. However, for particles >1 μm, deposition is less in dogs than in humans, but deposition in guinea pigs is similar to that in humans. On the other hand, both rats and hamsters generally show less total deposition than nasal-breathing humans.

In some cases, the data indicate that total deposition for the same size particle can be quite similar in experimental animals and humans. It therefore follows that deposition efficiency is independent of body (or lung) size (McMahon et al. 1977; Brain and Mensah 1983). However, different species exposed to the same size particles at the same exposure concentration will not receive the same initial mass deposition. If the total amount of deposition is divided by body (or lung) weight, smaller animals would receive greater initial particle burdens per unit weight per unit exposure time than would larger ones. For example, for 1-μm (D_{ae}) particles, it is predicted that the rat would receive an initial deposit 5 to 10 times that of humans, and the dog would

receive 3 times that of humans, if deposition was calculated on a per unit lung (or body) weight basis (Phalen et al. 1977).

Not all atmospheric particles to which an individual is exposed will be inhaled. The inspirable fraction is the portion of the ambient concentration that actually enters the upper respiratory tract. In humans, for example, the fraction for particles with D_{ae} <10 μm is greater than 80 percent, whereas that for particles ranging from 30 to 80 μm is about 50 percent (Vincent and Armbruster 1981). The probability of particles being inhaled into the respiratory tract depends upon particle size as well as the orientation of the individual to external air currents and the size of the entrance to the respiratory tract. Thus, inspirable fractions likely differ among species.

An additional point concerns hygroscopicity. If figure 10a is examined, it is evident that total deposition of hygroscopic particles < 0.5 μm inhaled by humans would tend to decrease if particles grow no larger than 0.5 μm, and deposition will only begin to increase if particle final diameter is >1 μm. Furthermore, since particles > 5 μm may grow minimally in one respiratory cycle, their deposition may not increase at all compared to nonhygroscopic particles (Ferron et al. 1987). On the other hand, deposition probability for 0.3 to 0.5-μm hygroscopic particles may change substantially.

Upper Respiratory Tract. Figure 10b shows upper respiratory tract deposition. There is substantial variability between species as well as large differences between individuals of the same species. Most experimental animals are obligate nasal breathers, and a large part of the intraspecies variability may be due to nasal geometry variation (Brain and Valberg 1979) as well as to different breathing patterns during exposure. Note the large intraspecies variability in deposition for particle sizes subject to impaction; this is probably responsible for a large portion of the intraspecies variation in total respiratory tract deposition (Stahlhofen et al. 1981a; Heyder et al. 1982).

In humans, nasal inhalation results in enhanced deposition compared to oral in-

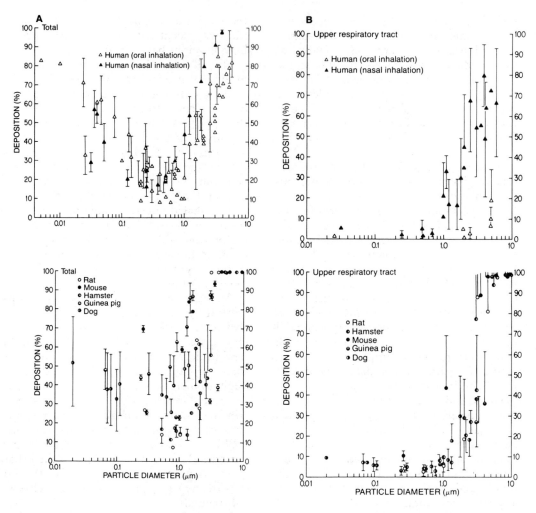

Figure 10. Deposition efficiency (that is, percentage deposition of amount inhaled) as a function of particle size for (A) total respiratory tract, (B) upper respiratory tract, (C) tracheobronchial tree, and (D) pulmonary region. All values are means (with standard deviations when available). Deposition efficiencies for experimental animals are based on nasal breathing. Particle diameters are aerodynamic (MMAD or AMAD) for those ≥ 0.5 μm and diffusion equivalent for those <0.5 μm. Data were compiled from Altshuler et al. 1957, 1966; Chan and Lippmann 1980; Craig and Buschbom 1975; Cuddihy et al. 1973; Foord et al. 1977; George and Breslin 1967; Giacomelli-Maltoni et al. 1972; Gibb and Morrow 1962; Heyder and Rudolf 1977; Heyder et al. 1973, 1975, 1980b, 1982;

halation. In all species shown, there is a rapid increase in deposition with increasing particle size about 1 μm, although the apparent "rate" of deposition increase with size is not the same. Thus, in humans, deposition appears to plateau somewhat for sizes >2 μm, whereas in rodents, deposition increases more rapidly.

Upper respiratory deposition of particles >1 μm is sometimes greater in nasal-

breathing humans than in the experimental animals. This is not necessarily expected, since the nasal passages of animals are more intricate than are those in humans and should therefore be more efficient particle collectors. However, the actual observations may be a reflection of exposure conditions. Many of the experimental animals were sedated or anesthetized and therefore breathed slower than fully awake animals.

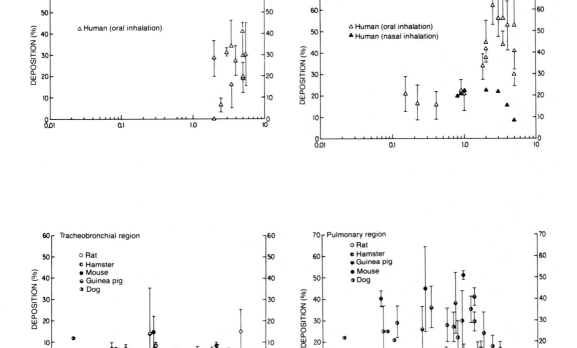

Figure 10 (*continued*).
Hounam et al. 1969; Johnson and Zeimer 1971; Kanapilly et al. 1982; Landahl et al. 1951, 1952; Lippmann 1970, 1977; Lippmann and Albert 1969; Lippmann and Altshuler 1976; Martens and Jacobi 1974; McMahon et al. 1977; Moores et al. 1980; Muir and Davies 1967; Palm et al. 1956; Pattle 1961; Raabe et al. 1977, 1987; Schiller et al. 1987; Stahlhofen et al. 1981a,b; Swift et al. 1977b; Tu and Knutson 1984; Wilson et al. 1985; Wolff et al. 1981, 1982; Yeh et al. 1980.

Since the dominant mechanism for deposition of particles >1 μm in the upper respiratory tract is impaction, low flow rates should reduce deposition efficiency. Inasmuch as smaller particles can penetrate the upper respiratory tract at all flow rates, deposition for these is similar in all species. If deposition were plotted in a manner that would normalize for flow, which is not possible for most of the experimental animal studies because of the lack of such data, the experimental animals would probably show greater deposition efficiency for larger particles than would humans at equivalent size/flow normalization parameters.

The extent of particle removal by the upper respiratory tract may vary depending upon whether an aerosol is mono- or polydisperse. For example, Thomas and Raabe (1978) compared the deposition in hamsters of a monodisperse and a polydisperse aerosol having similar median diameters (AMAD, 1.53 μm vs. 1.87 μm). The major difference was that the polydisperse aerosol deposited to a greater extent in the upper respiratory tract because of the presence of a certain percentage of larger particles that were effectively removed by impaction. Total respiratory tract deposition of the two aerosols, expressed as a percentage of inhaled amount, was the same. The less the deposition in the head, the greater is the amount available for removal in the lungs. Thus, the extent of removal in the upper respiratory tract may affect deposition patterns in distal regions.

Tracheobronchial Tree. Figure 10c shows tracheobronchial deposition; the amount of data available is less than for other regions. The figure indicates that the percentage of inhaled aerosol that is removed is greater in the oral-breathing human than in the nasal-breathing dog, hamster, or rat, at least in the limited region where particle sizes overlap. As mentioned above, a lower tracheobronchial deposition in experimental animals may be a reflection of greater upper respiratory tract deposition. On the other hand, the differences in regional deposition may be due to differences in flow in the upper bronchial tree and/or in airway branching patterns. In all cases, especially in the experimental animals, there is no well-defined trend relating deposition to particle size, unlike the situation in the other respiratory tract regions; on the contrary, fractional tracheobronchial deposition is relatively constant over a wide particle size range.

Pulmonary Region. Deposition in the pulmonary region is shown in figure 10d. In general, deposition in humans breathing orally increases as particle size decreases, after a minimum deposition is reached at about 0.5 μm. In nasal-breathing humans and experimental animals, deposition tends to decrease with increasing particle size.

The removal of particles in more proximal airways determines the shape of the pulmonary deposition curves. Increased upper respiratory and tracheobronchial deposition of particles ≥ 1 μm results in a reduction of pulmonary deposition that occurs more sharply in smaller animals than in humans. This is due not to reduced efficiency of pulmonary deposition of larger particles, but to the fact that only a small fraction of these large particles reach the lower respiratory tract. Similarly, nasal breathing in humans results in less pulmonary penetration of larger particles; thus, there is a lesser fraction of deposition for entering aerosol than for oral inhalation. In oral-breathing humans, the peak for pul-

monary deposition shifts upward to a larger-size particle compared to nasal breathing humans, and is more pronounced. On the other hand, with nasal breathing, there is a relatively constant pulmonary deposition over a wider size range, that is, 0.7–3 μm.

Pulmonary deposition is much less in hamsters and rats, which are similar to each other, than in dogs, guinea pigs, or humans. However, deposition in nasal-breathing humans is less than in these other species when available data for comparable size ranges are compared. Patterns are similar for oral inhalation, although the particle size for peak deposition is greater in humans than in guinea pigs or dogs. This is probably due to the more efficient removal of larger particles in the upper respiratory tract and tracheobronchial airways of these experimental animals.

Factors Modifying Deposition

Various factors cause deposition patterns to differ from those in normal healthy adults, the greatest contributors to the human data base. These include exposure to airborne irritants, lung disease, growth, and aging, all of which can affect particle deposition by changing ventilation patterns and/or airway geometry.

Bronchoconstriction induced by inhaling irritants increases impaction deposition in the upper airways. Thus, for example, cigarette smokers with no clinical disease exhibit somewhat greater tracheobronchial deposition of tracer particles ≥ 1 μm than do nonsmokers. As a consequence, smokers also exhibit a reduction in pulmonary region deposition compared to nonsmokers (Lippmann et al. 1972).

There are some deposition data in disease states. In humans with chronic bronchitis, an obstructive airway disease, tracheobronchial and upper respiratory tract deposition of particles $\geq 1\mu$m is quite variable, but greater than in healthy individuals (Thomson and Short 1969; Lippmann et al. 1972). Airway obstruction associated with lung disease in humans reduces the peripheral deposition of particles and may even entirely eliminate deposition in some parts of the lungs (Lourenco et al. 1972; Thomson and Pavia 1974). Total deposition has been found to be lower in rodents with enzyme-induced emphysema than in normal controls (Hahn and Hobbs 1979; Damon et al. 1983). This is probably due to an increase in alveolar size, resulting in greater distances to deposit on a surface and a concomitant reduction in pulmonary region deposition efficiency (Brain and Valberg 1979).

There are some data on the effects of fibrotic disease on deposition. Heppleston (1963) found that inhaled hematite particles deposited more distally in rats with coal- or silica-derived pneumoconiosis than in normal animals. On the other hand, Love and coworkers (1971) found no difference in the total respiratory tract deposition of 1-μm particles in coal workers with simple pneumoconiosis compared to normal people.

Most particle deposition studies in humans are performed with young to middle-aged adults, and few data are available on the growing lung. The available information is based on estimates of the influence of anatomic and ventilatory changes upon deposition during postnatal growth, obtained by using assumed respiratory parameters and mathematical particle deposition modeling techniques in conjuction with actual child lung morphometric values or scaled versions of available adult morphometric models (see, for example, Crawford 1982; Hofmann 1982; Martonen 1985; Phalen and Oldham 1985; Phalen et al. 1985; Phalen 1987). These studies suggest that the relative effectiveness of the major deposition mechanisms may differ at various times in the growth of the individual and that this, in turn, may alter the pattern of regional particle deposition. Thus, all age groups may not have the same distribution of deposition after exposure to the same particles. Although the results of the modeling studies are not consistent in terms of which regions or specific age groups differ, they all suggest that deposition efficiency, in at least some regions, is greater in children than adults. Taking into account the greater ventilation per unit body weight in children, the deposition fractions in certain regions could be well

above those measured in adults. Since there are also regional differences in clearance rates, it follows that the dose to specific lung compartments will vary with age from newborn to adult.

There are no systematic data that would allow an analysis of deposition in the aging lung, that is, between adulthood and senescence. There are also few data on deposition differences according to gender. Available evidence indicates that under equivalent inspiratory conditions, total respiratory tract and tracheobronchial deposition of particles with D_{ae} from 2.5 to 5 μm, inhaled orally at rest, is similar in men and women (Pritchard et al. 1987). However, women's smaller-diameter airways result in higher flow rates and, hence, more impaction in the upper respiratory tract, resulting in less deposition in the pulmonary region. This suggests that as particle size increases, women (and perhaps children) may be at less risk from material in pulmonary airways but at a greater risk from deposition in the upper respiratory tract and tracheobronchial tree.

Localized Patterns of Deposition

Specific patterns of enhanced local deposition within various regions of the respiratory tract are important to consider, since tissue dose depends on the surface density of deposited particles. The occurrence of nonuniform deposition suggests that the initial dose delivered to specific sites may be greater than that occurring if a uniform density of surface deposit is assumed. This is especially important for inhaled particles, such as irritants, that affect tissues on contact.

In the human upper respiratory tract, enhanced deposition may occur in the larynx, oropharyngeal bend, and soft palate (Swift 1981). Deposition is also nonuniform in the nasal passages; varying relative amounts occur in the anterior and posterior regions, depending largely on particle size (Itoh et al. 1985; Swift and Proctor 1987). The change in airflow direction at the vestibule in the nasal passages, together with the fact that it is an area of high velocity, produces locally enhanced deposition posterior to this region.

Studies in models and hollow casts of the human upper tracheobronchial tree have shown that deposition of aerosols >1 μm is not homogeneous. Entrance conditions produced by the larynx result in enhanced deposition in the upper trachea. At bronchial bifurcations, deposition is greatly enhanced relative to the rest of the airway length (Schlesinger et al. 1982; Cohen et al. 1987). This occurs by impaction during inspiration, although deposition is also enhanced downstream of bifurcations during exhalation (Schlesinger et al. 1983a). Enhanced deposition of submicrometric particles at bifurcations also occurs (Cohen et al. 1987); since this size aerosol is not subject to impaction, the effect is probably from turbulent diffusion.

The experimental conditions used in the numerous microdistribution studies varied widely, yet the relative distribution of enhancement among the airways was quite similar. This suggests that within the larger bronchi, local patterns of deposition may be fairly insensitive to particle size and airflow rate. Measurements in hollow human airway casts have also shown that the proportional distribution of deposition in specific airways is relatively constant over a wide range of particle sizes and overall deposition efficiencies (Schlesinger and Lippmann 1978). In addition, inhalation studies with rodents indicate that the distribution of deposition in the various lobes of the lungs is also relatively constant over a range of particle sizes and different total lung deposition efficiencies (Raabe et al. 1977).

There are few data on local deposition patterns for distal airways. Available information is based on examination by microscopy of tissues after in vivo exposures of experimental animals (Holma 1969; Brody and Roe 1983). These studies indicate that bronchiolar and alveolar duct bifurcations are preferential sites for deposition of a wide range of particles small enough to reach these regions.

Differences in the geometry of airways in humans and other species may result in differences in the microdistribution patterns of particle deposition, a factor that should be accounted for in extrapolation

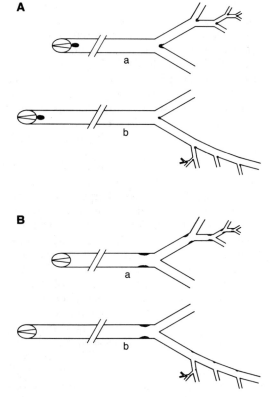

Figure 11. Location and relative intensity of enhanced tracheobronchial particle deposition for (A) inspiratory flow; and (B) expiratory flow, in humans (a) and nonprimate laboratory animals (b). (Adapted with permission from Lippmann and Schlesinger 1984.)

modeling. For example, nasal turbinates of rodents are more complex than those of primates and, as a result, the bulk of impaction-dominated deposition occurs more anteriorly in the nasal passages of rodents (Gooya and Patra 1986; Schreider 1986). Unfortunately, there are no other data to relate geometry to microdeposition. Speculated differences in the site and extent of localized deposition in the tracheobronchial tree are depicted in figure 11.

Mathematical Modeling

Mathematical models are needed to predict deposition sites and efficiencies since it is not possible to study all conditions of exposure experimentally. A mathematical model relates the main factors that control deposition to various geometric parameters and is used to predict the mean probability of particle deposition in the respiratory tract. Although most models have been designed for assessing deposition of spherical particles in humans, some have been developed for experimental animals (Kliment et al. 1972; Kliment 1973, 1974; Schreider and Hutchens 1979; Schum and Yeh 1980).

The first mathematical treatment of regional particle deposition in humans was performed by Findeisen (1935). This was later refined by others (Landahl et al. 1951, 1952; Beeckmans 1965; Task Group on Lung Dynamics 1966). Because of their very nature, these analytical models adopted assumptions and idealizations of almost all aspects of the respiratory tract and of particle dynamics. This simplification resulted in the loss of some important characteristics of the real system and often limited their ultimate usefulness and reliability. When compared to results from human experiments, these early models tended to overestimate pulmonary and underestimate tracheobronchial deposition (Mercer 1975). They were, however, very useful in quantitating the influence of various controlling parameters on deposition.

One of the major components of any deposition model, and one subject to the greatest oversimplification, is the representation of airway geometry. Most of the early predictive models made use of a very simple stylized lung structure. Recently, however, more realistic anatomic descriptions have become available, some developed specifically for use in deposition analyses and others easily adapted for such applications. As discussed earlier, some of these are symmetrical, others asymmetrical. Since the actual human bronchial tree is asymmetrical, and because the amount of deposition depends upon the path length over which the inhaled aerosol passes, realistic computations require consideration of asymmetry.

In order to assess the effect of anatomic structure, Yu and Diu (1982) calculated deposition in humans, using various symmetrical and asymmetrical geometries, and compared results to experimental data. They found that predicted total deposition did not differ greatly from geometry to geometry, and all compared reasonably well with experimental values. However, predicted regional deposition was quite sensitive to the particular geometry used. Since the anatomic models are generally derived from examination of single casts, they reflect, to some extent, actual interindividual structural variability. These results may provide the reason for variations from individual to individual in deposition experiments, even under identical breathing conditions. The sensitivity of predicted deposition to the specific anatomic model has also been noted by others (Martonen and Gibby 1982).

Another drawback of analytical models is the oversimplification of airflow pattern, a necessity since there are no exact expressions for flow dynamics in noncylindrical, tapering tubes that undergo repeated branchings and often have asymmetrical, nonlaminar flow profiles. Analytical models generally assume laminar flow, no disturbances produced by bifurcations, and uniform ventilation of airways. In addition, many assume a constant velocity of air during inspiration and expiration.

Any predictive model must also contain expressions for deposition probability. However, equations for deposition within a realistic geometry and flow pattern are not available. Thus, semiempirical expressions based upon analyses of simplified analogues of sections of the bronchial tree are often used. For example, impaction expressions are often obtained from the analytical solution to the equation of motion of a particle in ideal flow in a bent tube; diffusion expressions are obtained from the analytical solution for flow in an infinitely long, horizontal tube with ideal flow; and sedimentation expressions are obtained from the solution for deposition along a long, horizontal tube.

Recent mathematical deposition models have increased in sophistication and flexibility. Some allow for variations in air velocity, mode and pattern of breathing, polydispersity, hygroscopic growth, and even for changes in linear airway dimensions over the breathing cycle (see, for example, Taulbee and Yu 1975; Ferron

1977; Diu and Yu 1983; Egan and Nixon 1985). Some also include a more realistic treatment of the upper respiratory tract (Scott et al. 1978; Yu et al. 1981; Yu and Diu 1982). When predictions from these recent models are compared to experimental data, there is more often agreement for total than for regional deposition. However, given the complexity of the respiratory tract and inter-subject variability, this is not surprising.

A relatively recent approach to modeling in humans is dimensional analysis, in which deposition is related to some dimensionless parameter. Heyder and coworkers (1980a) formulated a parameter that is a function solely of particle size and flow rate and upon which total deposition was dependent; single parameters for regional deposition have been reported by Rudolf et al. (1987).

Because of the importance of drawing conclusions about humans from experiments with animals, special attempts have been made to develop methods for direct interspecies extrapolation. Stauffer (1975) used dimensional analysis to develop scaling factors based on particle physics and the assumption of a geometric similarity among all mammalian lungs. He predicted that interspecies particle deposition probabilities would be similar for sedimentation but a function of body weight for diffusion. McMahon and coworkers (1977) attempted to scale the collection efficiency of the respiratory tract in different species, based on inhalation studies in mice, hamsters, rats, rabbits, and dogs. They concluded that the overall collection efficiency of the lung would be independent of body size (this is essentially what is observed in figure 10a).

Research Recommendations

There is a considerable body of data on the deposition of inhaled aerosols in humans and experimental animals. We also have a fairly good understanding of some of the factors that control deposition. But our knowledge in certain critical areas is still not adequate.

■ **Recommendation 9.** Particle removal in the human upper respiratory tract should be assessed experimentally for oral and nasal breathing as should the influence of

breathing mode on deposition in other regions of the respiratory tract. Information on removal of particles in the upper tract will allow the prediction of appropriate starting concentrations for modeling particle transport into the lungs, since particle removal in the upper respiratory tract determines the concentration penetrating to distal regions. It is still not well defined how the specific route of entry affects regional deposition in humans, especially those with respiratory disease. No systematic studies have been done in which oral geometry, flow, and deposition of particles have been measured during natural oral breathing, that is, without an inhalation tube or mouthpiece placed in the anterior oral passages. These types of studies will enable better interpretation of the large available data base on deposition in humans, which was obtained with mouthpiece breathing. In addition, the effect of various degrees of oronasal breathing upon deposition should be assessed.

■ **Recommendation 10.** Microdistribution patterns of deposition should be studied under a wide range of exposure conditions. The nonuniformity of deposition in both the tracheobronchial tree and the pulmonary region may be important factors in ultimate dose. Further assessment of microdistribution is needed for incorporation of "enhancement functions" into deposition models.

■ **Recommendation 11.** Regional deposition efficiencies should be determined for ultrafine (<0.1 μm) particles in the human respiratory tract. Much of the lack of such data is due to the difficulty in generating monodisperse aerosols in this size range, as well as in accurately detecting the generated particles. More studies are needed that evaluate deposition in the upper respiratory tract where such small particles, if soluble, may be rapidly absorbed into the blood, as well as in the tracheobronchial tree, where increased deposition compared to larger submicrometer sizes may occur. The available theoretical models appear to be inconsistent in that predicted deposition for particles in this size range depends very much on the particular model being used.

■ **Recommendation 12.** The effects of anatomic variability on deposition should be analyzed systematically, and appropriate statistical descriptions developed for incorporation into deposition models. Theoretical predictions and in vivo studies show a dependence of regional deposition upon morphometry, and there is interindividual variability in structural characteristics of the human lungs. Yet the significance of this variability in affecting deposition is not currently known.

■ **Recommendation 13.** Effects of specific aspects of ventilation upon deposition should be determined. More data are needed with exercise breathing patterns, which may result in greater risk because of increased ventilation. More information is also needed on the relation of changes in tidal volume and breathing rate to the uniformity of deposition in the lungs. Studies should also be performed in experimental animals relating ventilation to deposition for use in extrapolation models.

■ **Recommendation 14.** Deposition in sensitive subsegments of the human population, such as children, the aged, and people with chronic lung disease (for example, emphysema and bronchitis), should be examined. Since children cannot be used in experimental studies, the development of deposition models based upon accurate ventilatory and morphometric information is critical. Although it is difficult to study deposition in individuals with lung disease, because of ventilatory and anatomic dysfunction which result in a large variability in deposition, more studies performed using well-controlled in vivo testing procedures and/or hollow airway cast systems would provide a better basis for assessing deposition in the compromised lung.

■ **Recommendation 15.** The deposition of hygroscopic particles in the human respiratory tract should be evaluated. Many important pollutant aerosols are hygroscopic, and there may be substantially greater deposition during inhalation as well as exhalation compared to dry particles of the same initial size, making mathematical predictions of deposition based on nonhygroscopic particles difficult. Most calculations of the growth of hygroscopic particles are based upon growth curves developed for sodium chloride or sulfate particles, and few data for other dynamic material exist. Studies of such deposition in the head, especially during nasal breathing, are needed.

■ **Recommendation 16.** Intercomparisons of regional deposition patterns among experimental animals (unsedated) and humans should be made using comparable monodisperse particles over a wide size range and comparable experimental techniques. Most of the regional deposition data that allow any cross-species comparison are for particles $> 1 \mu$m. In addition, there are no consistently applied methods for assessment of regional deposition in experimental animals and humans. "Calibration factors" need to be developed that may be used to relate results of toxicologic studies in experimental animals to human exposure assessment and health effects.

■ **Recommendation 17.** Models that allow calculation of deposition by airway generation should be expanded to other species. Coupled with data on ventilation and morphometry, this will allow better estimations of delivered dose in experimental animals.

Retention of Deposited Particles

Retention refers to the amount of particles remaining in the respiratory tract at specific times after exposure, and is the net result of deposition and clearance. Clearance is the physical removal from the respiratory tract of particles deposited on its surfaces.

Clearance Mechanisms: Basic Structure and Function

Particles are cleared from the respiratory tract by several different processes, some of which are regionally distinct as shown in table 2.

Table 2.   Respiratory Tract Clearance Mechanisms

Upper Respiratory Tract	Tracheobronchial Tree	Pulmonary Region
Mucociliary transport	Mucociliary transport	Macrophage transport
Sneezing	Coughing	Interstitial pathways
Nose wiping and blowing	Dissolution (for soluble particles)	Dissolution (for soluble and "insoluble" particles)
Dissolution (for soluble particles)		

Upper Respiratory Tract. The nasal passages beyond the vestibular region are lined with a ciliated epithelium overlaid by mucus (see Overton and Miller, this volume, figure 1). The mucus is produced by specialized epithelial cells and submucosal glands and consists of two layers: a low-viscosity hypophase that surrounds the cilia and within which they move, and a high-viscosity epiphase lying on top of the cilia (Lucas and Douglas 1934). The composition and characteristics of mucus are described in detail by Overton and Miller (this volume). Material depositing on the mucus is cleared by movement of the epiphase due to coordinated beating of the cilia.

The general flow of mucus in the ciliated nasal passages is toward the nasopharynx. In the region just distal to the nonciliated vestibule, however, mucous flow is forward, clearing deposited material to an area where sneezing, wiping, or blowing can occur. Soluble material deposited on the ciliated nasal epithelium will be accessible to underlying cells if diffusion through the mucus occurs at a rate faster than mucous flow. Since there is a rich vasculature in the nose, uptake into the blood can occur rapidly.

Insoluble particles deposited in the oral passages are cleared by swallowing into the gastrointestinal tract. Although there are no data on the clearance of soluble particles, oral tissue has the capacity for rapid systemic absorption (Swift and Proctor 1987). In the larynx, mucus moving toward the head from the trachea passes into the hypopharynx and is swallowed.

Tracheobronchial Tree. Most of the surface of the tracheobronchial tree through the terminal bronchioles is lined with ciliated epithelium overlaid by mucus, and insoluble particles are cleared primarily by the net movement of fluid toward the oropharynx. Some insoluble particles may traverse the tracheobronchial epithelium, entering the peribronchial region (Masse et al. 1974; Sorokin and Brain 1975), while soluble particles can be absorbed through the mucus into the circulation.

The bronchial surfaces are not homogeneous. For example, there are openings of daughter bronchi and islands of nonciliated squamous cells at bifurcations. Within these regions, the usual progress of mucous movement is interrupted (Hilding 1957), and clearance can be retarded. The efficiency with which nonciliated obstacles are passed depends on the traction of the mucous layer.

Pulmonary Region. A number of mechanisms and pathways contribute to clearance from the pulmonary region, but their relative importance is uncertain and depends to some extent on the physicochemical properties and amount of material deposited. The mechanisms involve either absorptive (dissolution) or nonabsorptive processes, which can occur simultaneously or at different times.

Nonabsorptive clearance processes, which are outlined in figure 12, are mediated primarily by alveolar macrophages. These large, mononuclear cells (figure 13) originate from precursors in bone marrow, reach the lung as circulating monocytes, and mature in the pulmonary interstitium, from which they traverse the epithelium to reach the alveolar surface. As macrophages move freely on alveolar surfaces, they phagocytize, transport, and detoxify deposited

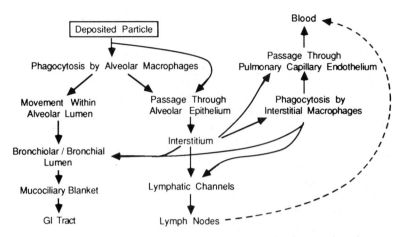

Figure 12. Flowchart of clearance pathways for particles depositing in the pulmonary region (dissolution is not included).

material, which they contact by chance or by directed motion due to chemotactic factors.

Macrophages normally comprise about 3 percent of total alveolar cells in healthy,

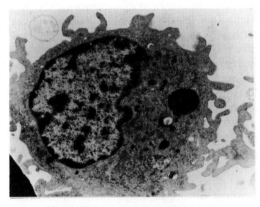

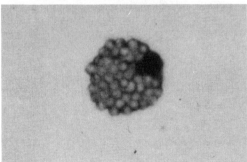

Figure 13. (Top) Electron micrograph (12,000×) of an alveolar macrophage. (Bottom) Light micrograph of an alveolar macrophage that has phagocytosed polystyrene latex particles. The dark area is the nucleus.

nonsmoking humans and in other mammalian species (Gehr 1984). However, the actual count can be influenced by particle deposition. Few particles may not result in an increase in cell number. Above a certain loading, however, macrophage numbers increase in proportion to particle number until a saturation point is reached (Brain 1971; Adamson and Bowden 1981). This increase is due to monocytic egress, proliferation of interstitial mononuclear cells, and/or to actual division of alveolar macrophages (Bowden and Adamson 1980; Blusse van Oud Alblas et al. 1983), and appears to be a generalized response that follows exposure to many types of particles, although the extent differs for particles of different composition (Adamson and Bowden 1981). Furthermore, the magnitude of the increase is more closely related to the number of deposited particles than to the total dose by weight, so equal masses of an identical material will stimulate more macrophages if the material is delivered as many small particles rather than as fewer large ones.

Particle-laden macrophages are cleared from the pulmonary region along a number of pathways. The primary route is by the mucociliary system, but the mechanism(s) by which cells reach it is not certain. One possibility is movement along the alveolar epithelium due to surface tension gradients between the alveoli and conducting airways; alternatively, locomotion could be directed along a density gradient, such as that produced by chemotactic factors released by macrophages actively ingesting deposited material (Kilburn 1968; Sorokin and Brain 1975; Ferin 1976). Another possible route to the mucociliary system involves passage through the alveolar epithelial wall into the interstitium (Brundelet 1965; Green 1973; Corry et al. 1984; Harmsen et al. 1985). Macrophages could then reach the surface of ciliated airways, perhaps through small collections of lymphatic tissue that exist at the alveolar/bronchiolar junction (Macklin 1955). Some of the cells that follow interstitial clearance pathways are probably resident interstitial macrophages that have ingested particles transported through the alveolar epithelium by endocytosis by type I pneumocytes (Brody et al. 1981; Bowden and Adamson 1984).

Particle-laden macrophages that do not clear by way of the bronchial tree may actively migrate within the interstitium to a nearby lymphatic channel or, along with uningested particles, be carried in the flow of interstitial fluid toward the lymphatic system. Passive entry into lymphatic vessels is fairly easy since endothelial cells are loosely connected with wide intercellular junctions (Lauweryns and Baert 1974); lymphatic endothelium has also been observed to actively engulf particles from the surrounding interstitium (Leak 1980). Deposited particles can then be translocated to the tracheobronchial lymph nodes, which often become reservoirs of retained material. Some particles subsequently appear in postnodal lymph, from which they enter the blood and may then translocate to extrapulmonary sites. Alternatively, uningested particles or macrophages in the interstitium may cross the alveolar capillary endothelium, entering the blood directly (Robertson 1980; Holt 1981; Raabe 1982). Whatever the route to the systemic circulation, particle-laden macrophages as well as free particles have been found in various extrapulmonary organs (Hourihane 1965; Pooley 1974; Lee et al. 1981; LeFevre et al. 1982).

Free particles in macrophages within the interstitium can end up in perivenous or subpleural sites, where they then become trapped. The migration and grouping of particles and macrophages can lead to the redistribution of deposits into focal aggregates in the lungs (Heppleston 1953).

The specific clearance route for particles depositing in the pulmonary region may depend upon loading. Earlier reports have suggested that at low-exposure concentrations, most particles are removed within macrophages via the bronchial mucociliary system, whereas at higher exposure levels, a larger proportion of free particles are translocated by the lymphatic system (Ferin 1977). More recently, however, researchers have found that the percentage of initial lung burden cleared by the lymphatic system after exposure to a high particle level is the same as that after exposure to a lesser burden (Snipes and Clem 1981; Snipes et al. 1983; Lehnert et al. 1986). Thus, some free particles are likely cleared by the lymphatics under most conditions of exposure.

The most important mechanism for absorptive clearance is dissolution. Free particles that dissolve in the alveolar fluid can diffuse through the epithelium and interstitium into the lymph or blood, whereas particles initially translocated to and trapped in interstitial sites may undergo dissolution there. Dissolution is a major clearance route even for particles usually considered to be relatively insoluble (Morrow et al. 1964; Mercer 1967). The factors affecting the solubility of deposited particles are poorly understood, although they are influenced by the particle's surface-to-volume ratio and other surface properties (Morrow 1973).

Some deposited material can dissolve after phagocytic uptake by macrophages. For example, certain metals dissolve within the acidic milieu of phagosomes. It is, however, not certain whether the dissolved material then leaves the cell. This is discussed further in a subsequent section.

Clearance Kinetics

Kinetic data are essential for determining the dosimetry of inhaled particles. Although the lungs may clear deposited material completely, the time frame over which clearance occurs determines the dose delivered to the lungs as well as to other organs. Tissue doses to the upper respiratory tract and tracheobronchial tree are often limited by the rapid clearance from these regions and are thus proportional to concentration and exposure duration. On the other hand, doses from material deposited in the pulmonary region depend much more on the characteristics of both the particle matrix and any associated materials.

Both the pulmonary region and the tracheobronchial clearance rates (that is, the fraction cleared per unit time) are well-defined functional characteristics of an individual human or experimental animal when repeated tests are performed under the same conditions (Gibb and Morrow 1962; Schlesinger et al. 1978; Lippmann et al. 1980; Bohning et al. 1982); but there is substantial interindividual variability. In addition, because of differences in experimental techniques and the fact that measured rates are strongly influenced by the specific methodology, comparisons between studies by different investigators are difficult to make.

Measurement Techniques. Methods for measuring clearance have been reviewed recently by Schlesinger (1985). Some of the techniques are identical to those used in assessing deposition, since the first measurement after aerosol inhalation is assumed to represent initial deposition.

The velocity of mucous transport in the nasal passages, trachea, and main bronchi can be measured directly by monitoring inert marker particles placed on the epithelium, or by measuring the movement of a bolus of radioactive particles selectively deposited within these airways or, in the case of the trachea and main bronchi, moving through them from more distal airways. The advantage of local velocity techniques is that they allow measurement in anatomically defined, albeit limited, airways. Because of this specificity, there is no doubt about whether clearance rates altered by toxicant exposure resulted from actual alterations in the mucociliary system or from a change in deposition pattern, a doubt not easily resolved when using whole-lung

clearance assays, as discussed below. However, local velocity techniques have a number of disadvantages. Some marker techniques are invasive, since particles may have to be selectively introduced into the airway of interest. Anesthesia, necessary in many cases, may affect the observed transport rates. Finally, the actual method of marker introduction can result in trauma to the airway.

In a number of studies aimed at assessing the effects of inhaled toxicants upon mucociliary clearance, alteration in tracheal transport rate has been used as the sole end point because it is easier to measure than is whole-lung clearance. However, an overall effect cannot necessarily be inferred from a change in this index, since toxicant-induced changes in bronchial clearance are not always associated with an alteration in tracheal transport rate. This could occur if the toxic particles are of a size too small to provide significant deposition within the trachea.

The most common technique to measure whole-lung clearance uses inhalation of radiolabeled (nonleaching) tracer particles. The total amount of radioactivity remaining in the lungs at selected intervals is then measured by external detector systems. The decline in emission rate, corrected for radioactive decay, represents clearance. Various types and configurations of mobile and/or fixed-scintillation detector systems have been used, and each has its own advantages and disadvantages in terms of spatial resolution and sensitivity.

One of the problems in using external monitors to assess tracheobronchial mucociliary clearance is that the observed clearance pattern depends on the initial deposition of the tracer particles. This is because the techniques are indirect, and clearance rate is proportional to transit pathways. For example, an apparent increase in clearance rate after toxicant exposure could be due to a proximal shift in deposition of the tracer rather than to an effect on the clearance system itself. This presents a special problem when different groups are being compared; for example, persons with chronic obstructive pulmonary disease tend to have greater central bronchial deposition than do healthy subjects for the same size tracer particles. There is, however, no basis for any kinetic distinction between mucous transport rates measured using different particles, as long as the deposition sites are the same. The rate of mucous transport in the trachea has been found to be independent of the shape, size, or composition of the insoluble marker used to measure it (van Antweiler 1958; Man et al. 1980). There is no a priori reason to assume that this does not hold true for transport in more distal airways as well, provided there is no toxic effect from the deposited material.

Pulmonary region clearance is relatively slow, and therefore measurements should be performed over, perhaps, several months. When radioactively tagged tracer particles are used, a nuclide having a long half-life is required. In addition, since the total dose to the subject should be minimized, especially if humans are used, long counting times may be required to obtain statistically reliable data. Health risks may therefore rule out such long-term radioactive tracer clearance studies.

A technique that avoids this problem is magnetopneumography (Cohen 1973). In this procedure, the subject is exposed to magnetizable particles and, at various times, a magnetic field is applied externally to the thorax in order to magnetize the bulk of the deposited particles. After the external field is removed, a remanent field remains, which is measured with an appropriate sensor (Freedman et al. 1982). Magnetopneumographic techniques have some advantages over radioaerosol techniques in terms of temporal and spatial resolution. Furthermore, certain information can be obtained using them that is not obtainable by other whole-animal in vivo techniques, for example, the assessment of in situ phagocytosis of tracer particles by macrophages (Brain et al. 1987). However, magnetopneumography has some significant drawbacks: all sources of external magnetic contamination on the subject or on the measurement apparatus must be removed; critical positioning is required since the measurements are highly sensitive to distance from the source; and there are difficulties in deducing actual particle distribution in the lungs from the data.

Fecal analysis is a technique for indirect monitoring of clearance using experimental animals that involves radioaerosols but not external monitoring. The radioactivity in feces collected at fixed intervals after exposure to tracer particles is measured. The fecal excretion activity curve presumably represents material cleared by the mucociliary system into the gastrointestinal tract and can thus be used to provide an index of tracheobronchial clearance. This technique assumes that all material cleared from the lungs is transported to the gastrointestinal tract and subsequently excreted in the feces. It is also very sensitive to feeding behavior of the animals; those that do not eat or do not excrete for a particular fraction of the sampling interval cannot be included in the analysis.

Clearance of deposited particles can also be assessed in experimental animals by serial sacrifice at various intervals after exposure, followed by microscopic, chemical, or radiological analysis. Various parts of the respiratory tract or the lungs as a whole can be sampled. The measured burden plotted as a function of time provides an index of clearance. Although microscopy can provide only a qualitative assessment of particle distribution and clearance from various regions, other techniques alone or combined with microscopy allow quantitative determination of the material retained regionally, and without interference from material in adjacent areas. Sacrifice techniques have the advantage of being very sensitive, but major disadvantages are that a large number of animals is needed for statistical reliability, the intraindividual variability in clearance cannot be examined, and the effects of toxicants on the course of clearance in the same individual on different occasions or under different conditions cannot be assessed.

Clearance Rates and Times. *Upper Respiratory Tract.* Nasal mucous flow rates are nonuniform. Regional velocities in the healthy adult human range from <2 to >20 mm/min (Proctor 1980), with the fastest flow in the midportion of the nasal passages; average velocities for the nasal passages as a whole range from 4 to 12 mm/min (Bang et al. 1967; Phipps 1981). The resultant mean transport time for insoluble particles over the nasal passages is about 10 to 13 min (Rutland and Cole 1981; Majima et al. 1983; Stanley et al. 1985). If soluble particles diffuse through the mucus within this time period, they become accessible to underlying epithelial cells.

Particles deposited in the anterior, non-ciliated portion of the nasal passages can be cleared slowly by mucous movement; a flow rate of 1 to 2 mm/hr has been suggested for fluid moved by traction from more distal cilia (Hilding 1963). Particles may take over 12 hr to be cleared by this mechanism and are usually more effectively removed by sneezing or wiping, in which case clearance may occur in under 30 min (Fry and Black 1973; Morrow 1977).

The velocity of mucous transport in the larynx has not been measured. However, it is probably about the same as that in the trachea (Swift and Proctor 1987).

Tracheobronchial Tree. Clearance of particles deposited on tracheobronchial airways occurs by the parallel processes of mucous transport and absorption. The fraction of deposited material cleared by either of these pathways is a function of its physicochemical properties, but because of the short time frame for mucociliary clearance, relatively insoluble material will be cleared solely by this route.

Mucous clearance occurs at different rates in different local regions; mucus moves fastest in the trachea, and progressively slower in more distal airways. Measured rates in the human trachea range from 4 to 20 mm/min, depending upon the experimental technique used (Yeates et al. 1981a). Anesthesia and invasive procedures affect transport, resulting in rates apparently slower than normal. Using noninvasive measurement procedures on unanesthetized, healthy nonsmokers, researchers have observed average tracheal mucous transport rates of 4.3 to 5.7 mm/min (Yeates et al. 1975, 1981b; Foster et al. 1980; Leikauf et al. 1981, 1984).

The mean mucous velocity in the human main bronchi has been found experimentally to be about 2.4 mm/min (Foster et al. 1980). Although rates of mucous movement in smaller airways cannot be measured directly, transport rates in human medium bronchi have been estimated at 0.2

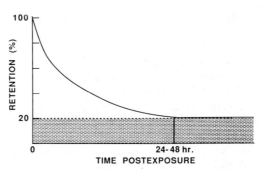

Figure 14. Schematic representation of tracheobronchial clearance after exposure to tracer particles. Particles remaining beyond 24 to 48 hr (shaded area) are assumed to have deposited in the pulmonary region.

to 1.3 mm/min, and those in the most distal ciliated airways as low as 0.001 mm/min (Morrow et al. 1967b; Yeates and Aspin 1978).

The total duration of bronchial clearance or some other time parameter is often used as an index of mucociliary function. In healthy, nonsmoking adult humans, 90 percent of insoluble particles depositing in the tracheobronchial tree will be cleared within 2.5 to 20 hr after deposition. The actual time depends on the individual subject and the size of the tracer aerosol used, which affects the depth of deposition and subsequent pathway length for removal (Albert et al. 1973). Clearance will be 99 percent completed by 48 hr after deposition (Bailey et al. 1985a).

In humans, normal tracheobronchial mucociliary clearance exhibits a two-phase pattern (figure 14): a short initial phase characterized by rapid clearance lasting a few hours, followed by a slower, second phase extending until 24 to 48 hr after exposure. These probably represent clearance of the tracer particles deposited in the "upper" and "lower" tracheobronchial tree, respectively. As the size of the tracer particles is reduced, resulting in more distal deposition, there is an increase in the fraction of total tracheobronchial clearance which is accounted for by the slower phase. Although some portion of the above clearance pattern may include rapid early clearance of material deposited in the pulmonary region, the contribution of this to the apparent bronchial clearance rate appears minimal.

Studies in rodents have shown that a small fraction of insoluble material is retained for prolonged periods within the upper respiratory tract or tracheobronchial tree (Patrick and Stirling 1977; Watson and Brain 1979; Gore and Patrick 1982). In humans, it has been estimated that the average residence time in bronchial tissue of insoluble particles derived from cigarette smoke is 3 to 5 months (Radford and Martell 1977). Soluble material may also be retained in ciliated airways for long periods because of binding to cells or specific macromolecules (Boecker et al. 1983).

The mechanism(s) underlying the long-term retention of insoluble particles is unknown. It may involve endocytosis by epithelial cells with subsequent translocation into deeper tissue or merely passive movement into the tissue (Sorokin and Brain 1975; Watson and Brain 1979; Gore and Patrick 1982). In addition, long-term tracheobronchial retention patterns for insoluble particles are not uniform. Enhanced retention occurs at bifurcation regions (Radford and Martell 1977; Henshaw and Fews 1984; Cohen et al. 1987), which may be the result of greater deposition as well as ineffective mucous clearance. Because of this nonuniformity, doses calculated using uniform surface retention density may be misleading.

Pulmonary Region. Particles are cleared from the pulmonary region by a number of pathways and mechanisms. Their effectiveness depends on the nature of the particles, but just what this dependence is has not been completely resolved. Consequently, the kinetics of clearance from the pulmonary region are not definitively understood, although particles deposited there generally remain longer than do those deposited in ciliated airways. Data on clearance rates in humans are limited, and those for experimental animals (and humans) vary widely because of different properties of the particles in the various studies. Many of these studies used high concentrations of particles, which may of itself have interfered with normal mechanisms, producing rates different from those that would occur at lower exposure levels.

Pulmonary region clearance data appear to fit an exponential model, and each component is believed to represent removal by a different mechanism or pathway (Casarett 1972). For example, an initial fast phase having a clearance half-time of about 2 to 6 weeks presumably represents rapid clearance by macrophages; an intermediate phase, with a half-time of months, may represent macrophage clearance by interstitial pathways; and a phase of prolonged clearance with a half-time of months to years represents removal by dissolution. Rates of removal by dissolution are extremely variable but likely dominate the long-term retention of relatively insoluble particles.

Rates that correspond to the various clearance phases can only be obtained if clearance is measured until all the deposited particles are removed from the lungs. This is usually not possible, and many studies are terminated when the radioactivity levels of retained particles fall below detectable limits. Clearance of inert insoluble particles in healthy, nonsmoking humans has been observed experimentally to consist of two phases: the first has a half-time measured in days, and the second in hundreds of days (Bailey et al. 1982; Bohning et al. 1982; Philipson et al. 1985). Table 3 summarizes data from numerous studies for the half-times of the longer, second phase of clearance. Wide variations in clearance times indicate a dependence upon the nature of the material being cleared. For example, when polydisperse aerosols are used, various size fractions clear by different routes and, thus, with varying rates (Snipes et al. 1984a,b). Different clearance rates have also been observed when using different-size particles (Morgan and Holmes 1980; Bailey et al. 1982); but if dissolution is accounted for, mechanical removal to the gastrointestinal tract and lymph nodes is independent of particle size (Snipes et al. 1983). There is also considerable intersubject variation in the clearance rates of similar particles, which increases with time postexposure (Bailey et al. 1985a; Philipson et al. 1985). This large difference in pulmonary region clearance kinetics among different individuals suggests that equivalent exposures to insoluble particles will result in differences in respiratory tract burdens.

Table 3. Long-Term Particle Clearance from the Pulmonary Region in Human Nonsmokers

Tracer Particle		Clearance Half-Time[a]	
Material	Size (μm)	(days)	Reference
Polystyrene latex	5	150–300	Booker et al. (1967)
Polystyrene latex	5	144–340	Newton et al. (1978)
Polystyrene latex	0.5	33–602	Jammet et al. (1978)
Polystyrene latex	3.6	296	Bohning et al. (1982)
Teflon	4	200–2,500	Philipson et al. (1985)
Aluminosilicate	1.2	330	Bailey et al. (1982)
Aluminosilicate	3.9	420	Bailey et al. (1982)
Iron oxide (Fe_2O_3)	0.8	62	Morrow et al. (1967a,b)
Iron oxide (Fe_2O_3)	0.1	270	Waite and Ramsden (1971)
Iron oxide (Fe_3O_4)	2.8	70	Cohen et al. (1979)

[a] Half-time of clearance for the slowest phase observed.

Even less is known about relative rates along specific pathways than about overall pulmonary region clearance kinetics. After deposition, the uptake of particles by alveolar macrophages is very rapid, unless the particles are cytotoxic (Lehnert and Morrow 1985; Naumann and Schlesinger 1986). The actual rate of subsequent clearance of these cells is not certain; perhaps 5 percent or less of their total number is translocated from the lungs each day (Masse et al. 1974; Lehnert and Morrow 1985).

Uningested particles may penetrate into the interstitium within a few hours after deposition (Sorokin and Brain 1975; Ferin and Feldstein 1978; Brody et al. 1981). The amount transported via transepithelial passage seems to increase as particle loading increases, especially when loading surpasses the level at which the number of macrophages saturate (Ferin 1977; Adamson and Bowden 1981). Similarly, a depression of phagocytosis by toxic particles may increase the number of free particles in the alveoli, enhancing removal by other routes. Free particles or those within alveolar macrophages reach the lymph nodes within a few days after deposition (Harmsen et al. 1985; Lehnert et al. 1987). However, most clearance by the lymphatic system is very slow (Sorokin and Brain 1975; Ferin 1976).

Soluble particles deposited in the pulmonary region are cleared rapidly by absorption through the epithelial surface into the blood, but there are few data on dissolution and transfer rates in humans. The rate does depend upon the size of the particle, with smaller ones clearing faster than larger ones. Some dissolved material may be retained in lung tissue because of binding with cellular components, preventing it from passing into the circulation (Cuddihy 1984).

Comparative Clearance Kinetics and Modeling. The retention of certain materials cannot be studied experimentally in humans, so experimental animals must be used. Since dosimetry depends upon clearance rates and routes, adequate toxicologic assessment necessitates relating clearance kinetics in animals to that in humans. Although the basic mechanisms of respiratory tract clearance are similar in humans and most other mammals, regional clearance rates vary substantially among species, even for similar particles deposited under comparable exposure conditions. It is likely that dissolution rates and rates of transfer of dissolved substances into blood are related solely to the properties of the material being cleared and are essentially independent of species (Cuddihy et al. 1979; Griffith et al. 1983; Bailey et al. 1985b). On the other hand, different rates of mechanical transport, such as macrophage clearance from the pulmonary region (Bailey et al. 1985b) or mucociliary transport in conducting airways (Felicetti et al. 1981), are found, resulting in species-dependent rate constants for these clearance pathways. Differences in regional (and perhaps total) clearance rates among species are probably due to these latter processes. Accordingly, respiratory tract clearance in humans can be predicted by using dissolution rates in experimental animals and mechanical clearance rates in humans, as long as lung damage or binding to lung molecules has not occurred (Bailey et al. 1985b).

Another approach used to predict clearance in humans is mathematical modeling. Various theoretical and empirical models have been developed to predict regional as well as total respiratory tract clearance of particles. Most of these models have been used for dosimetry of inhaled radionuclides (see, for example, International Commission on Radiological Protection 1959, 1972; Bailey and James 1979). In these models, fractional allocations are made between mechanical clearance processes and dissolution on the basis of properties of each specific material being assessed.

Mathematical models have also been developed that describe overall tracheobronchial clearance patterns by calculating mucous transport rates in each generation (Yeates and Aspin 1978; Lee at al. 1979; Yu et al. 1983). These models make various assumptions: for example, all mucus is produced in the terminal airways; no fluid reabsorption occurs; or the thickness of the mucous layer is constant in all airways. In addition, the overall clearance rates generated by some of these models are very sensitive to the rates assumed in the smallest airways; this is because this region has the slowest rate and a large surface area, and the models assume transport rates to be inversely proportional to surface area or circumference. Only limited testing of the accuracy of these models is possible because actual transport rates are not known for distal airways. Thus, predicted results are often compared to actual observations for total time of tracheobronchial clearance and to values for mucous transport rates in the trachea. Internal adjustments are made so that the predicted time is the same as that observed experimentally.

Factors Modifying Clearance

A number of host and environmental factors modify normal clearance patterns, affecting the dose delivered by exposure to

inhaled particles. These factors include aging, gender, work load, disease, and irritant inhalation exposure. In many cases, however, their exact role is not resolved.

The evidence for aging-related effects on mucociliary function in healthy individuals is contradictory, with studies showing either no change or a slowing in clearance with age after maturity (Goodman et al. 1978; Yeates et al. 1981a). One problem is that it is difficult to determine whether an apparent decrement in function is due to aging alone, or to long-term, low-level ambient pollutant exposure (Wanner 1977). There are no data for changes in overall pulmonary region clearance related to aging. Functional differences have been found between alveolar macrophages from mature and senescent mice (Esposito and Pennington 1983), although no age-related decline in human macrophage function has been seen (Gardner et al. 1981).

There are not sufficient data to assess changes in clearance in the growing lung. Nasal mucociliary clearance time in a group of children (average age 7 yr) has been found to be about 10 min (Passali and Ciampoli 1985), which is within the range for adults. There is one report of bronchial clearance in 12 yr olds, but this study was performed in hospitalized patients (Huhnerbein et al. 1984).

In terms of gender, no difference in nasal mucociliary clearance has been observed between male and female children (Passali and Ciampoli 1985), nor in tracheal transport rates in adults (Yeates et al. 1975). Slower bronchial clearance has been noted in male compared to female adults, but this was attributed to differences in lung size rather than inherent gender differences in transport velocities (Gerrard et al. 1986).

The effect of increased physical activity on mucociliary clearance is also unresolved, with the available data indicating either no change or an increase with exercise (Wolff et al. 1977; Pavia 1984). There are no data relating changes in pulmonary region clearance to increased activity levels, but Valberg and coworkers (1985) found that CO_2-stimulated hyperpnea had no effect on early pulmonary clearance and redistribution of particles.

Various diseases are associated with altered clearance. Nasal mucociliary clearance is prolonged in humans with chronic sinusitis, bronchiectasis, or rhinitis (Majima et al. 1983; Stanley et al. 1985), and with cystic fibrosis (Rutland and Cole 1981). Bronchial mucous transport may be impaired in people with bronchial carcinoma (Matthys et al. 1983), chronic bronchitis (Vastag et al. 1986), asthma (Pavia et al. 1985), and various acute respiratory infections (Lourenco et al. 1971b; Camner et al. 1979; Puchelle et al. 1980). In some of these conditions, coughing may enhance mucous clearance but is generally effective only if excess secretions are present.

Rates of pulmonary region particle clearance appear to be reduced in humans with chronic obstructive lung disease (Bohning et al. 1982), and the viability and functional activity of macrophages has been found to be impaired in human asthmatics (Godard et al. 1982). Reduced clearance from the pulmonary region of experimental animals with viral infections has also been observed (Cresia et al. 1973). On the other hand, Tryka and coworkers (1985) found increased pulmonary clearance in hamsters with interstitial fibrosis. Damon and coworkers (1983) observed no clearance difference in rats with emphysema. Hahn and Hobbs (1979), however, found that the copresence of inflammation resulted in prolonged retention. Inflammation may enhance the penetration of free particles and macrophages through the alveolar epithelium into the interstitium by increasing the permeability of the epithelium and the lymphatic endothelium (Corry et al. 1984).

Cigarette smoking in humans is associated with persistently slowed mucociliary clearance in both the nasal passages and the tracheobronchial tree (Lourenco et al. 1971a; Camner and Philipson 1972; Goodman et al. 1978; Stanley et al. 1984), and the extent of decline appears related to the amount of smoking (Vastag et al. 1986). Smokers can also exhibit specific clearance abnormalities, including intermittent retrograde mucous flow in the trachea and intermittent periods of stasis that alternate with abrupt drops in particle retention in the bronchi (Albert et al. 1971, 1973). The

rate of particle clearance from the pulmonary region also appears to be reduced in heavy cigarette smokers (Cohen et al. 1979; Bohning et al. 1982).

In addition to cigarette smoke, other inhaled irritants have an effect on mucociliary clearance function in humans as well as experimental animals (Wolff 1986). Single exposures to a particular material may increase or decrease the overall rate of tracheobronchial clearance, depending upon the exposure concentration. Although alterations in clearance rate following single exposures to moderate concentrations of irritants are transient—lasting <24 hr—repeated exposures may persistently retard clearance. The effects of irritant exposure may be enhanced by exercise, or by coexposure to other materials.

Acute and chronic exposures to inhaled irritants can also alter clearance from the pulmonary region. For example, nitrogen dioxide (NO_2), ozone (O_3), sulfuric acid (H_2SO_4), and some metals (for example, cadmium) have been shown to change the rate of tracer particle clearance (Ferin and Leach 1977; Oberdörster and Hochrainer 1980; Driscoll et al. 1986; Schlesinger and Gearhart 1986). Clearance may be accelerated or depressed, depending upon the specific material and/or length of exposure. Alterations in alveolar macrophage function may underly some of the observed changes, since numerous irritants have been shown to impair the functional properties of these cells (Gardner 1984).

Specific macrophage properties, which include phagocytosis and mobility, allow them to adequately perform their role in clearance. However, the relation between these characteristics and overall clearance is not certain. For example, in comparisons among a number of species, no positive correlation was found between macrophage mobility and clearance rate since slower movement was often associated with an acceleration in clearance (Metivier 1984; Naumann and Schlesinger 1986).

Research Recommendations

■ **Recommendation 18.** Interspecies comparisons of short-term (mucociliary) and long-term (pulmonary) clearance kinetics should be made, with an emphasis on mechanical processes, using equivalent insoluble particles and experimental techniques. These studies should assess the effects of exposure characteristics—for example, particle size and mass concentration—on retention patterns, and should examine the effects of differences in lung anatomy. The use of equivalent experimental conditions is essential to avoid differences in the results due to lack of standardization in measurement techniques and particle characteristics. Newer, nonradioactive experimental techniques should be used to expand the data base on long-term pulmonary region clearance kinetics in humans.

■ **Recommendation 19.** The long-term fate of particles in conducting airways should be examined. To determine whether specific cell types preferentially handle certain particles or to assess whether cells critical in clearance change during their normal activity, these studies should include morphological techniques to precisely locate sites of particle deposition and retention.

■ **Recommendation 20.** Studies should be undertaken on the pathways and circumstances by which macrophages and free particles are removed from alveoli to regions of potential long-term retention in the lung, such as peribronchial and perivascular sites, as well as lymph nodes. Effects of particle loading should be considered, since it is not clear how this affects transport by various routes.

■ **Recommendation 21.** Alveolar macrophages should be characterized. There is a need to study how alterations in cell functional properties are manifested in changes in overall clearance patterns, so as to determine whether or not specific functional changes induced by inhaled materials critically affect the cell's ability to participate in lung defense. The effects of various characteristics of exposure (for example, mass loading), on phagocytosis, mobility, release of chemotactic factors for neutrophils, or production of other mediators, requires further examination. In addition,

since macrophages are not homogeneous, and different subsets exhibit functional differences, more work is needed to characterize the heterogeneity of macrophages recovered after particle exposures.

■ **Recommendation 22.** In vitro effects on macrophages should be related to those produced in vivo. Dose to cells is better defined in in vitro studies, but calibration factors are needed to relate exposure concentration to actual target tissue dose in extrapolating in vitro to in vivo results.

■ **Recommendation 23.** The effects of exogenous factors on retention should be determined. For example, exercise may alter deposition pattern, but how clearance and ultimate retention are affected is not known.

■ **Recommendation 24.** Mucus should be characterized as to average and regional variations in thickness, physicochemical properties, and synthesis rates. Data are needed on humans as well as most experimental animals.

■ **Recommendation 25.** Fate of soluble particles in the tracheobronchial tree should be determined. There are uptake data for the nasal passages but not for other conducting airways.

■ **Recommendation 26.** Animal models that mimic human respiratory disease should be developed further. These models should be used to examine clearance and retention of potentially sensitive subsegments of the human population.

■ **Recommendation 27.** Clearance kinetics of the growing lung and the aging lung should be studied. Studies of mucociliary clearance and pulmonary region clearance are needed; when combined with deposition studies, a comprehensive picture of defense capabilities in the young and elderly segments of the population can be developed. Animal models can be used to some extent, since it is difficult in human populations to separate true aging effects from certain environmental influences, such as air pollution.

■ Recommendation 28. Gender-related differences in the disposition of inhaled particles in humans should be investigated. A comparison of the efficiency of defense mechanisms is needed to determine whether, as has been suggested, doses in males and females may indeed differ.

Disposition of Vehicular Particulate Emissions

This section addresses the fate of particles—specifically, carbon, metals, and sulfates—emitted in vehicular exhaust. Gasoline and diesel engines produce carbon, but much greater amounts are released by diesels. These particles contain adsorbed organic compounds, the fate of which is discussed by Sun, Bond, and Dahl (this volume). Metals present in fuel are released in the exhaust in amounts that differ with the type of fuel. Sulfates, primarily sulfuric acid, are produced by diesel and gasoline engines on approximately the same scale.

Diesel Exhaust Particles

Diesel exhaust particles have a diameter (MMD) of 0.2 to 0.3 μm and a dense carbonaceous core. They usually contain adsorbed organic matter, but this section discusses the general fate of inhaled diesel particles without regard to specific adsorbed components. Available data are based largely on inhalation studies using rodents exposed to diluted diesel exhaust containing particles, usually at predetermined concentrations, as well as various gases.

About 15 to 20 percent of diesel particles inhaled by rodents are initially deposited (Chan et al. 1981; Dziedzic 1981; Lee et al. 1983); this value is quite close to that for total respiratory tract deposition of similar-sized particles in humans (figure 10a). Although there are no actual experimental studies in humans, diesel particle deposition in different age groups under various breathing conditions has been estimated using mathematical modeling (Xu and Yu 1985). Results suggest that total and re-

gional deposition vary with age. For nasal breathing at rest, total and pulmonary deposition in infants and children are predicted to be greater than that in adults, with maximum deposition occurring at about two years of age. Because of its particle size, diesel particle deposition is predicted to be unaffected by the mode of inhalation or by the frequency of respiration but should increase with increasing tidal volume.

The clearance routes for diesel particles depend on their regional deposition. The fraction deposited in the tracheobronchial tree, about 40 to 50 percent of the initial deposit, is rapidly cleared by mucociliary transport in about one to two days (Chan et al. 1981; Lee et al. 1983), although there is some evidence that diesel particles depress mucociliary clearance rates (Battigelli et al. 1966). Most of the particles that reach the pulmonary region are phagocytized by macrophages (Barnhart et al. 1981; White and Garg 1981), and the numbers of these cells seem to increase in relation to the rate of particle entry into the lung rather than to the total cumulative exposure (Strom 1984; Mauderly et al. 1987). Although diesel particle exposures to levels up to about 2 mg/m^3 do not reduce viability of macrophages, phagocytic activity has been variously reported to be either depressed or unaltered (Chen et al. 1980a,b; Weller et al. 1980; Barnhart et al. 1981; Castranova et al. 1985).

Diesel particles can also be taken up by type I alveolar epithelial cells (Barnhart et al. 1981). Enhanced uptake probably occurs if the macrophages are overloaded, since the number of type I cells containing particles increases as particle concentration and exposure duration increase.

Following deposition, diesel particles are fairly evenly distributed throughout the pulmonary region. Gradually, within macrophages, the particles are moved from peripheral lung regions toward the terminus of the mucociliary transport system, from where they may be cleared via the tracheobronchial tree (White and Garg 1981). Clusters of particle-laden macrophages are often found at the distal ends of the terminal bronchioles after high- or low-level exposures (Barnhart et al. 1979; Puro 1980; Garg 1985). Aggregates of free particles have been observed in focal areas of the tracheobronchial tree, perhaps because of a general depression of mucociliary clearance, local accumulation in areas of inadequate transport, or longer-term retention after uptake by or through the epithelium.

Another clearance pathway from the pulmonary region is by the lymphatic system. Free diesel particles, as well as particle-laden macrophages, have been found in parenchymal lymphoid aggregates, lymphatic vessels, and mediastinal lymph nodes (Vostal et al. 1979). The amount cleared along this pathway increases with increasing duration and level of exposure (Chan et al. 1981; White and Garg 1981).

The kinetics of diesel particle clearance have been examined in rodents. However, exposure concentrations were generally high, and it is not known whether the kinetics are the same when exposure levels are lower. In rats, two phases of pulmonary clearance were observed, with half-times of 6 and 80 days, respectively (Lee et al. 1983). The faster clearance was ascribed to the mucociliary transport of particles deposited in proximal respiratory bronchioles, whereas the slower phase was ascribed to other alveolar removal processes. The kinetics were the same with exposure to either 7 mg diesel particulate/m^3 for 45 min, or 2 mg/m^3 for 140 min. Guinea pigs exposed to 7 mg/m^3 for 45 min showed little clearance from days 10 to 432 after exposure, even though initial deposition percentages and mucociliary clearance times were the same as in the rat. On the other hand, the clearance was found to occur at the same rate in rats and mice chronically exposed to 0.35 to 7.0 mg diesel particulate/m^3 (Henderson et al. 1982).

It has been suggested that the observed increase in macrophage numbers after exposure to diesel particles should increase the pulmonary clearance rate of this material relative to clearance in nonexposed controls (Lee 1981). Experiments do not always support this hypothesis; diesel particle exposure has been associated with depressed and accelerated clearance from the pulmonary region (table 4). Lung burden appears to be a critical factor affecting the overall efficiency of pulmonary clearance, perhaps by altering relative amounts cleared by different pathways. Diesel particles may be retained in the lungs for long periods of time after exposure, with the amount increasing with increasing deposition (Chan et al. 1981; Rudd and Strom 1981; Henderson et al. 1982). A long residence time provides an extended period for elution of adsorbed material.

The development of dust clusters and their residence times probably depend on exposure concentration and duration (Moore et al. 1978). Perhaps accumulation and persistence begin, or increase, when normal clearance processes are overloaded during chronic exposures; this could occur at exposure levels lower even than those used in most studies, but there are few data at realistic concentrations. Most of the data on diesel particle disposition were obtained from chronic studies in rodents using diluted exhaust with diesel particulate levels $\gtrsim 0.25$ mg/m^3. Furthermore, the observed rates and routes of clearance could have been affected by the various combustion products, many of which are irritants, found either in the gas phase of diesel exhaust or adsorbed on particle surfaces.

Metals

Many metals present in motor vehicle fuel are emitted in the exhaust (Lee and van Lehmden 1973). They are distributed in the atmosphere as individual particles or adsorbed onto the surfaces of other particles. Particle sizes range from submicrometer to about 2 to 3 μm, depending on the metal.

Once deposited in the respiratory tract, the disposition of a metal varies with its valence state and the compound containing it. The solubility of metals and their compounds in biological fluids strongly influences their biological availability, utilization, and toxicity. Insoluble forms of some metals can accumulate in the lungs over time because of continuous exposure and slow systemic absorption, whereas more soluble forms are rapidly absorbed into the blood and translocated to other organs. But

Table 4. Effects of Diesel Particles on Respiratory Tract Clearance

Species	Particle Mass Conc[a] (mg/m^3)	Duration	Effect	Reference
	Exposure Parameters			
Rat	0.2, 0.99, 4.1	7 hr/day, 5 days/week for 18 weeks	↓ pulmonary region clearance at 4.1 mg/m^3	Griffis et al. (1983)
Rat	0.35, 3.5, 7	7 hr/day, 5 days/week for 30 months	↓ pulmonary region clearance at 3.5 and 7 mg/m^3	Mauderly et al. (1987)
Mouse	0.35, 3.5, 7	7 hr/day, 5 days/week for 18 months	↓ pulmonary region clearance at 3.5 and 7 mg/m^3	Henderson et al. (1982)
Rat	0.25, 6	20 hr/day, 2 days/week 7–112 days	↓ pulmonary region clearance when burden ⩾ 6.5 mg/m^3	Chan et al. (1984)
Rat	8, 17	34–40 hr at 4–6-hr sessions	↓ tracheal clearance	Battigelli et al. (1966)
Rat	17	4 hr	↓ tracheal clearance	Battigelli et al. (1966)
Sheep	0.4–0.5 (resuspended particles)	0.5 hr	NC tracheal clearance	Abraham et al. (1980)
Rat	4.0	95 hr/week, for 19 months	↓ pulmonary region clearance	Muhle et al. (1987)
Hamster	4.0	95 hr/week for 19 months	↓ pulmonary region clearance	Muhle et al. (1987)
Rat	2	7 hr/day, 5 days/week 6 months	↑ pulmonary region clearance; NC bronchial clearance	Oberdörster et al. (1984)
Rat	4	daily, 16 months	↓ pulmonary region clearance	Heinrich et al. (1981)
Rat	0.35, 3.5, 7	7 hr/day, 5 days/week 30 months	NC tracheal clearance; ↓ pulmonary region clearance at 3.5 and 7 mg/m^3	Wolff et al. (1987)

[a] Concentration of particles in diluted exhaust.
NOTE: NC = no change; ↑ = significant acceleration of clearance; ↓ = significant retardation of clearance.

some soluble forms can actually undergo greater retention than insoluble ones because of binding to protein in the lungs. Unfortunately, most ambient measurements determine the total concentration of the metal and do not discriminate among different compounds or valence states. Thus, for dosimetric purposes, it is usually assumed that all compounds, whatever their source, will dissociate to some degree after deposition in the respiratory tract, releasing metal ions that will be absorbed and redistributed in the body in a similar manner.

The absorption efficiency for most metals is about 50 to 80 percent from the pulmonary region, and 5 to 15 percent from the upper respiratory tract and tracheobronchial tree (Natusch et al. 1974). This may reflect the more efficient extraction of metals from the smaller particles that would preferentially deposit in the pulmonary region, and/or the shorter residence time of particles depositing on the tracheobronchial tree. After absorption from the respiratory tract, these metals will be distributed rapidly to blood-rich organs and more slowly to other organs and to fat, and will very slowly equilibrate with poorly perfused tissues (Luckey and Venugopal 1977).

Metals that deposit in the pulmonary region have the greatest toxic potential because of the likelihood of extended residence times. The more insoluble the metal, the more likely it is to be cleared from this area by movement to the tracheobronchial tree, followed by swallowing. Systemic absorption from the respiratory tract is minimal in the course of this movement. In general, metals are absorbed less effectively from the gastrointestinal tract than from the lungs, in part because of differences in residence times (Natusch et al. 1975; Luckey and Venugopal 1977). Thus, material processed by the gastrointestinal tract often has less toxic impact, and may also be influenced by dietary factors.

Dissolution is a major clearance mechanism for metal particles deposited in the pulmonary region, and it may be enhanced if the particles are first phagocytized by macrophages. When deposited particles are ingested and subsequently exposed to the acidic environment of the phagosome, metal ions can be released. Studies of rabbit and human alveolar macrophages exposed in vitro to submicrometer manganese dioxide (MnO_2) particles have shown that the cells of both species were able to dissolve two to three times more of the material than was dissolved in the culture media within the same time (Lundborg et al. 1984, 1985). Dissolved metals can then leave the macrophage, and the lungs, at rates faster than their normal dissolution rate in lung fluid. The early clearance rate of a metal can therefore vary with the form in which it is inhaled. For example, MnO_2, which is insoluble in lung fluid, dissolves in the macrophage, but soluble manganese chloride ($MnCl_2$) probably dissolves extracellularly and is not ingested, so deposited Mn may clear at different initial rates depending upon its original state (Camner et

al. 1985). Furthermore, intracellular dissolution, by enhancing the release of metals into the cellular milieux, can be a mechanism for the local cytotoxic action of some phagocytosed metal particles, for example, lead (Pb) (DeVries et al. 1983). Intracellular dissolution must therefore be considered in models of the pulmonary clearance of particles and in assessment of mechanisms of differential toxicity of metal compounds.

Various characteristics of macrophages have been examined by performing bronchopulmonary lavage after in vivo exposure to metal particles or after direct in vitro exposures. Subchronic particle inhalations may or may not produce a nonspecific increase in macrophage number, depending upon the metal particle. For example, macrophages in rats increased after exposure to nickel oxide (NiO) but not nickel chloride ($NiCl_2$) or lead chloride ($PbCl_2$), whereas lead oxide (Pb_2O_3) exposure reduced the numbers of recovered cells (Bingham et al. 1972). Exposures of rabbit alveolar macrophages to soluble chlorides of cadmium (Cd^{2+}), Ni^{2+}, Mn^{2+}, and chromium (Cr^{3+}) resulted in significantly reduced viability, with Cd^{2+} being the most toxic (Waters et al. 1975). Except for Cd, these metals produced cell lysis in a roughly concentration-dependent manner and with a relative potency similar to that which characterized the change in viability. It is conceivable that the mode of cell death affects subsequent clearance pathways for a metal. If the cell dies after phagocytosis but does not lyse, the particle-laden cell may be cleared by the mucociliary system. Alternatively, phagocytosis-induced cell lysis would liberate particles for systemic absorption or reengulfment.

Examination of the viability of rabbit alveolar macrophages exposed in vitro to fly ash with adsorbed oxides of Pb, Ni, or Mn showed Pb to be the most cytotoxic, with Mn and Ni exhibiting somewhat lower toxicity (Aranyi et al. 1977). At any specific particle concentration, the effect on viability decreased with increasing particle size over a 2- to 8-μm range, even though the percentage of metal in the ash was the same for all sizes. This finding was ascribed both to the fact that bigger particles were ingested in fewer numbers than were smaller ones and that, once ingested, the larger particles presented less surface area to the phagosomal contents, resulting in less leaching of metal ions. Thus, the cytotoxicity of a metal at a specific concentration depends upon the size of its associated carrier particle and, conversely, the nature of the carrier particle influences the fate and effects of the compound it transports into the cell.

The rate of phagocytosis for equivalent-sized particles depends upon the nature of the adsorbed surface coating. Rabbit alveolar macrophages exposed to 5 μm Teflon particles coated with various metals phagocytosed those with carbon or Cr to a greater degree than those coated with Pb, Mn, or silver (Ag) (Camner et al. 1973, 1974). Direct in vitro exposure to Ni^{2+}, Cd^{2+}, vanadate (VO_3^-), Mn^{2+}, or Cr^{3+} impaired the phagocytic ability of macrophages (Graham et al. 1975; Waters et al. 1975; Castranova et al. 1980); in vivo exposure to MnO_2 also depressed phagocytosis (Bergstrom 1977). The effect may depend upon the concentration of the metal. In rats exposed to cadmium chloride ($CdCl_2$) at 1.5 or 5 mg/m^3, the lower level stimulated phagocytosis whereas the higher level depressed it (Greenspan and Morrow 1984). Reduced phagocytosis could result in increased lung burdens of the offending metal or of other deposited substances.

The effects of metals on various cytologic end points may not be equal. For example, Ni^{2+} impaired phagocytosis at a concentration much lower than that required to decrease viability, whereas Cd^{2+} and Cr^{3+} depressed phagocytosis as well as viability at comparable concentrations (Waters et al. 1975). Concentrations of VO_3^- that caused macrophage lysis did not reduce phagocytosis in the surviving cells (Graham et al. 1975). Thus, examination of viability as the only toxic end point may not be appropriate. Other functions may be more sensitive and may also play a role in determining the ultimate disposition of deposited particles.

Examples of the disposition of selected inhaled metals found in exhaust—namely Pb, Cd, Cr, Mn, Ni, and vanadium (V)—are discussed below.

Lead. Until recently, Pb was the metal of greatest concern, but as the amount of Pb in gasoline has been gradually reduced, so has the concern. About 20 to 60 percent of inhaled Pb particles deposit in the adult human respiratory tract (Nozaki 1966; Moore et al. 1980; Morrow et al. 1980). Most is rapidly cleared by absorption; lung retention half-times of 13 to 14 hr have been measured (Morrow et al. 1980). Of the Pb cleared to the gastrointestinal tract, only about 5 to 15 percent is subsequently absorbed in adults (Kehoe 1961; Goyer and Chisolm 1972; Baksi 1982), with the rest excreted in the feces. Gastrointestinal absorption is, however, greater in infants and children (Rabinowitz et al. 1976; Ziegler et al. 1978).

The total contribution of airborne Pb to total blood Pb levels is hard to determine, but the percentage of airborne Pb ultimately found in the blood is in the range of 7 to 40 percent (Patterson 1965; Rabinowitz et al. 1973, 1974; Manton 1977), and the biological half-time in blood is about 25 days (Baksi 1982). Absorbed Pb is excreted primarily in urine but also in bile and by exfoliation of epithelial tissue. About 25 to 40 percent of inhaled Pb is retained in the body, almost all in bone, where it accumulates slowly with age and continued exposure (Goyer and Chisolm 1972; Barry 1975; Gross et al. 1975).

Cadmium. The solubility of Cd salts varies widely, and the relative rates of clearance by specific pathways probably depend on specific form. For example, whereas soluble $CdCl_2$ and insoluble CdO_2 have similar long-term clearance rates, with a retention half-time of ~67 days (Oberdörster et al. 1979), a larger proportion of the oxide is cleared by an earlier fast phase, perhaps mediated by macrophages or mucociliary transport.

Although Cd is absorbed from the respiratory tract, the relation between exposure and uptake in humans is not known. On the basis of experimental animal studies, up to 30 percent appears to be absorbed, depending on the specific form of Cd (Friberg 1950). Absorption from the gastrointestinal tract is quite poor—less than 10 percent in experimental animals and adult humans—but there is increased absorption in young individuals (Rahola et al. 1972; Fleischer et al. 1974).

Once absorbed, Cd accumulates primarily in the liver and kidneys, which together account for about 50 percent of the total body burden (Fleischer et al. 1974). The primary route of excretion of absorbed Cd is urine, and its biological half-time in the human body is estimated at 19 to 38 years (Friberg et al. 1974).

Chromium. After deposition, the water-soluble salts of Cr are rapidly absorbed from the respiratory tract into the circulation, but the less-soluble, and more-toxic, forms remain primarily in the respiratory tract, where their concentration increases with age (Baselt 1982). Lesser amounts of Cr accumulate in skin, muscle, fat, and liver. There is low gastrointestinal absorption, only up to about 25 percent of the initial dose (Baselt 1982). Of the Cr that is absorbed, at least 80 percent is excreted in urine.

Manganese. The use of Mn additives as alternatives to Pb as antiknock ingredients in gasoline will probably result in an increase of Mn in exhaust emissions. A number of studies have examined the clearance of deposited particulate Mn from the lungs of humans and experimental animals (Morrow et al. 1964, 1967a; Maigetter et al. 1976; Bergstrom 1977; Drown et al. 1986). Unfortunately, the data are not directly comparable; residence times vary widely because of differences in particle size and resultant deposition, exposure duration, and concentration. However, like Cd, insoluble and soluble forms may clear at similar long-term rates, but dissimilar short-term rates.

Once it is absorbed, Mn is stored primarily in liver, kidneys, intestines, and pancreas, but does not accumulate with age; it is excreted primarily in bile. Injected radio-labeled Mn has been found to disappear from the human body at two rates: about 70 percent is removed with a half-time of 39 days, and 30 percent with a half-time of 4 days (Mahoney and Small 1968).

Nickel. Respiratory tract clearance mechanisms for Ni are not very effective, and lung levels remain high for years after exposures have ended (Torjussen and Anderson 1979; Williams et al. 1980). The passage of Ni across lung epithelium is slow, and studies in rodents show no significant removal by the lymphatic system (Williams et al. 1980). Because of this slow removal, the concentration of Ni within the respiratory tract increases with age, even during chronic exposure to low levels. In addition, the lung actually sequesters significant amounts of Ni because of binding to a variety of macromolecules; this may be an additional cause of Ni accumulation and toxicity. Of the Ni that reaches the gastrointestinal tract, more than 90 percent is excreted unabsorbed in the feces (Sunderman 1977). Aside from lungs, absorbed Ni tends to localize in connective tissue and kidney. Excretion of absorbed Ni is in the urine.

Vanadium. Vanadium concentrates primarily in fat, which can account for 90 percent of the total body burden, but also in bones and teeth (Schroeder et al. 1963; Myron et al. 1978). Of the other organs, the lungs contain the greatest concentration, but lung kinetics for V are not known. It is possible that insoluble forms may accumulate in the lungs with age, but this is also not known. Any V that reaches the gastrointestinal tract is excreted unabsorbed.

Sulfates

Sulfates in exhaust, primarily sulfuric acid (H_2SO_4) and its neutralization products with atmospheric ammonia, occur in ambient air as submicrometric aerosols. These particles are hygroscopic and their deposition depends upon their effective diameter within the respiratory tract which, in turn, depends upon the rate of particle growth. In guinea pigs and rats, total respiratory tract deposition of H_2SO_4 aerosols ranging in size from 0.4–1.2 μm (MMAD) increased with increasing initial droplet size (Dahl and Griffith 1983). Desposition models developed for hygroscopic sulfate particles (Martonen and Patel 1981) predict that total respiratory tract deposition efficiencies should be greater than those for nonhygroscopic particles only if the sulfate originated from dry particles with diameters greater than about 0.3 μm. Although the regional deposition of H_2SO_4 has not been studied experimentally in humans, predictive deposition models indicate that patterns of deposition for 0.5–1 μm (final size) H_2SO_4 particles are similar, with deposition concentrated in the distal conducting airways (Leikauf et al. 1984).

Sulfate is cleared from the lungs by diffusion (Charles et al. 1977), but the exact rate depends upon the inhaled concentration and associated cations. Using radioactive ^{35}S label, Dahl and coworkers (1983) studied the clearance of inhaled submicrometer H_2SO_4 from the lungs of experimental animals. For dogs, rats, and guinea pigs, they found that the half-time of sulfate clearance from all sites in the lungs ranged from 2 to 9 min, with smaller airways clearing faster than larger ones. The lungs cleared H_2SO_4 much faster than the nasal region, suggesting that clearance from the nose was not primarily by the blood. There were some interspecies differences; the dog cleared slower than the guinea pig, which cleared slower than the rat.

In humans as well as experimental animals, either acute or chronic H_2SO_4 exposure alters the bronchial mucociliary clearance rates of tracer particles. Acceleration or depression of clearance may occur, depending on the concentration and exposure regime. These effects, recently reviewed by Schlesinger (1986), are likely due to the deposition of hydrogen ion (H^+), rather than sulfate (SO_4^{2-}), on the airway surfaces. Sulfuric acid exposures of experimental animals have also been associated with alterations in the rate of clearance of tracer particles from the pulmonary region (Phalen et al. 1980; Naumann and Schlesinger 1986; Schlesinger and Gearhart 1986) and with changes in macrophage function (Naumann and Schlesinger 1986; Schlesinger 1987). Effects of H_2SO_4 on respiratory tract clearance are summarized in table 5.

Research Recommendations

■ **Recommendation 29.** Factors that control the bioavailability of material adsorbed onto particles should be examined. A major effort should be made to evaluate the effects of carrier particle characteristics, such as size, composition, surface area, and surface characteristics, on translocation and redistribution, intra- as well as extrapulmonary, of adsorbed nonorganic pollutants. Another aspect of this effort involves examination of modifiers of toxic action. For example, macrophages dissolve metals, but there are no comprehensive data to determine if this occurs for all metals of interest, or whether the extent varies significantly among animal species.

■ **Recommendation 30.** Studies should be undertaken at low concentrations of diesel exhaust which simulate actual human exposure. There is a need to determine whether all aspects of diesel particle disposition are the same for chronic exposures at low dose levels. In addition, studies of diesel particle retention beyond a 100-day postexposure observation time, to more completely assess long-term clearance, translocation, and body retention, should be performed.

■ **Recommendation 31.** Effects of particulate as well as gas-phase components of diesel exhaust should be studied. Because diesel exhaust is a complex mixture of particles and gases, it is essential that the effects of these components be separated to determine underlying toxicologic mechanisms.

■ **Recommendation 32.** In experimental animal systems, effects should be determined of concurrent exposures to more than one specific material on clearance pathways, retention patterns, and extrapulmonary disposition. This could involve coexposures to diesel exhaust with other components of ambient air, including cigarette smoke.

■ **Recommendation 33.** Effects of components of exhaust emissions on susceptible

Table 5. Effects of Sulfuric Acid on Respiratory Tract Clearance

Species	Mass Conc (mg/m³)	Particle Size (μm)	Duration	Effect	Reference
	Exposure Parameters				
Rat	1–100	1	6 hr	↑ tracheal clearance at 1 day after 100 mg/m³	Wolff (1986)
Guinea pig	1–27	0.8–0.9	6 hr	↓ tracheal clearance at 1 day after 1 mg/m³	Wolff (1986)
Sheep	14	0.1	0.3 hr	NC tracheal clearance	Sackner et al. (1978)
Sheep	4	0.1	4 hr	NC tracheal clearance	Sackner et al. (1978)
Human	1	0.5	2.5 hr	↑ bronchial clearance	Newhouse et al. (1978)
Human	0.1–1.0	0.5	1 hr	↑, ↓ bronchial clearance (concentration dependent), NC tracheal clearance	Leikauf et al. (1981, 1984)
Rabbit	0.1–2.2	0.3	1 hr	↑, ↓ bronchial clearance (concentration dependent)	Schlesinger et al. (1984)
Donkey	0.2–1.4	0.4	1 hr	↓ bronchial clearance (persistent in 2 of 4 animals), NC tracheal clearance	Schlesinger et al. (1978)
Mouse	1.5	0.6	4 hr	NC bronchial clearance	Fairchild et al. (1975)
Mouse	15	3.2	4 hr	↓ bronchial clearance	Fairchild et al. (1975)
Rat	3.6	1.0	4 hr	NC bronchial clearance ↓ pulmonary region clearance	Phalen et al. (1980)
Donkey	0.1	0.5	1 hr/day, 5 days/week up to 6 months	↓ bronchial clearance within 3 months	Schlesinger et al. (1979)
Rabbit	0.25–0.5	0.3	1 hr/day, 5 days/week up to 4 weeks	↑ bronchial clearance within 1 week	Schlesinger et al. (1983b)
Rabbit	1	0.3	1 hr	↑ pulmonary region clearance	Naumann and Schlesinger (1986)
Rabbit	0.25	0.3	1 hr/day, 5 days/week, up to 8 months	↑ pulmonary region clearance within 2 weeks	Schlesinger and Gearhart (1986)
Rabbit	0.5	0.3	2 hr/day, 14 days	↓ pulmonary region clearance	Schlesinger and Gearhart (1987)

NOTE: NC = no change; ↑ = significant acceleration of clearance; ↓ = significant retardation of clearance.

populations, such as those with respiratory disease, should be examined. This can be performed with animal models of specific human diseases. There is a need to use these models, and to develop others, so as to be able to study effects of exhaust products on specific sensitive individuals. Studies are also needed to assess the disposition of pollutants in exercising adults, the young, and the elderly. In addition, the role of concomitant stresses should be assessed in these individuals, as should the question of dose distribution in various extrapulmonary tissues as a function of age.

■ **Recommendation 34.** The effects of particulate emissions on clearance of other deposited particles should be examined. In addition to sulfates, other materials need investigation in this regard. Studies should assess underlying mechanisms of alteration; for example, changes in bronchial and alveolar epithelial permeability may affect ultimate clearance rates.

Summary

A basic goal of risk assessment is to relate dose to exposure. The deposition of inhaled particles on the internal surfaces of the airways defines the delivery rate to the initial contact site(s) and is controlled by various physical mechanisms that are influenced by particle characteristics, airflow patterns and rates, and respiratory tract anatomy. Biological effects are often more directly related to the quantitative pattern of deposition within specific sites than to total respiratory tract deposition. This is because regional deposition patterns determine the specific pathways and rates by which deposited particles are ultimately cleared and redistributed.

There are numerous data on regional deposition of inert particles in humans, but the risk of inhaling hazardous aerosols or chronic exposure protocols requires the use of experimental animals and interspecies extrapolation of the results. To adequately apply these results to human risk assessment it is essential to consider differences in regional deposition patterns. But different species exposed to the same aerosol may not receive identical doses in comparable respiratory tract regions and, thus, the use of a particular species influences the estimated initial lung dose, the subsequent translocation sites, and the relation of exposure to potential human health effects.

The toxic response from inhaled particles depends on both the amount of material deposited at target sites and the length of time this persists (that is, retention). Particles are cleared from their deposition sites by various routes and interacting processes. The specific pathway depends on the region of the respiratory tract where the material deposits, physicochemical properties of the material, and, perhaps, exposure concentration and duration.

The primary biological clearance mechanisms for insoluble particles are mucociliary transport in the nasal passages and tracheobronchial tree, and removal by resident macrophages from the pulmonary region. Residence time depends on route. Material deposited on the conducting airways is cleared within two days, although some long-term retention can occur. Particles deposited in the pulmonary region may remain for months to years or be retained indefinitely in various interstitial sites. Soluble particles, even those with relatively low solubility, can dissolve in the pulmonary region. Solubilized components can be retained in the lungs or be redistributed in the body, where they may be retained in extrapulmonary tissues or excreted. In the conducting airways, solubilization occurs only if the rate of dissolution is faster than the rate of removal by mucous transport.

Although clearance mechanisms are similar in humans and experimental animals, clearance rates may differ if mediated by biological processes, for example, mucous transport or macrophages. On the other hand, physical processes such as diffusion across epithelial barriers proceed at about the same rate in most species examined.

In addition to reviewing the principles and mechanisms that influence the deposition and clearance of particulate matter, the disposition of three classes of inorganic materials produced by vehicles were discussed: diesel particles, metals, and sulfates. Diesel exhaust contains carbonaceous particles onto which various, usually organic, materials are adsorbed. Only the fate of the matrix was discussed, although it is possible that it is affected by any associated materials. Almost all reported studies have been conducted using high concentrations of particles—at least 10 times that found in ambient air—and it is conceivable that disposition is not the same at lower levels of exposure. Inhaled diesel particles are cleared primarily by mucociliary transport and alveolar macrophages, but significant long-term retention in the lungs can occur.

The fate of inhaled metals depends largely on the particular metal, as well as its valence state or inhaled form. Some metals are cytotoxic to macrophages, whereas others alter the function of these cells without affecting their viability. Some metals accumulate in the lungs or extrapulmonary tissues after continuous exposure, whereas other metals reach a steady-state concentration unless exposure levels are very high.

Sulfates—primarily sulfuric acid—produce their main effect on the respiratory tract; response is probably related to the pH of the specific sulfate species. These materials alter the rates of clearance processes, both in conducting airways and in the pulmonary region, thus compromising the lung's defense capabilities.

Summary of Research Recommendations: Discussion

Many of the recommendations presented are for highly goal-oriented studies needed to expand or refine the data base on factors that control the disposition of inhaled particles. The precision of exposure assessment and risk analysis will improve from an enhanced understanding of these factors. In many cases, experimental animals should be used for studies that are not feasible in humans. In other cases, the use of physical models, such as airway casts, is appropriate.

A major area of concern is the disposition of inhaled particles in sensitive subsegments of the human population, such as children, the aged, and people with chronic respiratory diseases. Predictive models suggest that particle deposition will be greater in children than in adults and that children may receive a disproportionately large respiratory tract dose from inhaled toxicants in relation to their body mass. This is due largely to differences between children and adults in ventilation, general activity level, and lung morphometry, but such differ-

ences are ignored in exposure assessments. Since direct extrapolation from experimental inhalation studies using adults may not be valid, studies aimed directly at children are, therefore, needed. It is also important to determine whether the distribution of dose to extrapulmonary sites differs during growth.

Older individuals and people with chronic respiratory disease may also be more susceptible to effects of inhaled toxicants and/or may show differences in particle disposition compared to normal, healthy younger adults, the group upon which most of the current data are derived. Although some basic studies of particle deposition and clearance have been performed in people with chronic respiratory disease, results are difficult to interpret because of the large inherent variability in these individuals caused by ventilatory and anatomic irregularities. Expanding the data base depends on additional, well-controlled studies. Finally, to examine aging-related changes in particle deposition and clearance, longitudinal tests could be performed in healthy humans as they age.

The use of children in experimental studies is generally precluded. Thus, particle deposition in children is generally modeled, using predictive deposition equations coupled with empirically derived morphometric models. More data, therefore, are needed on growth-related morphometric and ventilatory changes in the lung. This information can be used to develop a better model of the growing lung which, in turn, will allow for more accurate predictions of particle disposition.

To assess clearance and retention characteristics of specific vehicular pollutants as a function of growth in vivo, animal models will be needed. For example, rodents ranging in age from newborn to adult could be exposed to particulate pollutants, and clearance or retention correlated with age. Along these same lines, animal models could be developed to assess how particle disposition in the aging or diseased lung compares to that in healthy young adult lungs.

Since certain types of information on particle disposition must be obtained from experimental animals, extrapolation to humans for use in risk estimates is necessary. However, each species has unique physiological and anatomic characteristics that serve to differentiate it from others, and these factors may play a role in determining species-specific responses to inhaled agents. Reliable interspecies comparisons can be performed only if appropriate data are available. Thus, some of the research recommendations are aimed at providing additional information on respiratory tract structure and function in experimental animals. It is essential that similarities and differences among these animals and humans be assessed. This information will allow selection of appropriate animal models that resemble humans for particular situations, or that have specific individual characteristics that are highly desirable for mechanistic studies. In addition, systematic studies of deposition and clearance in various experimental animal species and humans, wherein comparable techniques are used, are needed to provide cross-species calibration factors. Such studies will avoid the problems inherent when comparing data obtained from different laboratories and using various methodologies.

Another gap in the current data base on the disposition of vehicular-derived particles relates to exposure conditions. Despite the inherent problems, chronic exposures to toxicologically relevant materials at realistic concentrations are necessary. The disposition of particles may be concentration-dependent and, thus, risk assessments based upon studies at high concentrations may not be appropriate for assessing hazards at ambient levels. In addition, chronic studies are needed, since acute exposures are not always predictive of effects from longer-term exposures. Currently, there is no sound basis to extrapolate the effects of vehicular-derived particles from high to low concentrations, in terms of their disposition. Furthermore, the role of increased physical activity of exposed individuals in altering particle disposition needs to be determined, since enhanced ventilation may drastically alter dose.

In ambient situations, exposures to more than one material occur. It is necessary to determine whether retention kinetics of particles of a single material are adequate representations of that associated with joint exposures. Concomitant irritant exposures affect airway size, ventilation, clearance rates, and distribution of cells in the airways. Studies are needed to see if such effects alter the ultimate fate of vehicular-derived materials.

All of the research recommendations presented provide data that can be used as components of a dosimetric model; the more accurate the individual components of the model, the more reliable are the predictions. A complete model must include accurate input for regional deposition and clearance, species variability in ventilation and anatomy, as well as relevant properties of the inhaled particles.

Summary of Research Recommendations: Priorities

To meet the broad goals of a basic research initiative as discussed above, the following specific areas were proposed as requiring additional research. These are ranked in groups, depending upon their priority for improving risk estimations.

HIGH PRIORITY

These studies are essential in order to provide needed data for more accurate risk assessment.

Recommendation 1	Interindividual variability of dimensions of the upper respiratory tract, tracheobronchial tree, and pulmonary region for adult humans and experimental animals (including strain differences) should be assessed to provide statistical descriptions of morphometry at all levels of the respiratory tract.

Recommendation 12	The effects of interspecies anatomic variability on deposition should be analyzed systematically.
Recommendation 2	Morphometry of the human growing lung, the aging lung, and the diseased lung should be assessed.
Recommendations 5, 6, 7	Patterns and distribution of airflow in the nasal passages and tracheobronchial tree of experimental animals and humans, and in the oral passages of humans, should be determined. This should include assessments in the growing, aging, and diseased lung.
Recommendations 11, 16	Systematic study should be undertaken of regional deposition using a full range of particle sizes with comparable exposure conditions in humans and experimental animals; especially needed are studies with ultrafine (<0.1 μm) particles.
Recommendation 18	Interspecies comparison of clearance kinetics should be made, using comparable methods and particles, for assessment of mucociliary transport from conducting airways and mechanical transport from pulmonary airways.
Recommendations 13, 14, 23, 26, 27, 28	Effects of modifying factors on particle deposition, clearance, and retention should be studied. This should include examination of growing, aging, and diseased lungs, as well as of differences due to physical activity and gender. The development and use of appropriate animal models for these studies should be pursued.
Recommendation 29	The effects of carrier particle characteristics (for example, size, surface characteristics, mass concentration) on ultimate disposition of adsorbed material should be examined.
Recommendation 30	Chronic exposures to diesel exhaust products at realistic levels should be undertaken.
Recommendation 32	Coexposures to diesel exhaust products, or to diesel exhaust and other ambient pollutants, should be conducted.
Recommendation 33	Pollutant exposures in animal models of sensitive populations, such as the young, elderly, or diseased, should be performed.

MEDIUM PRIORITY

These studies will provide data to refine risk assessment.

Recommendation 3	Comparative morphometry of the upper respiratory tract in humans and experimental animals should be assessed.
Recommendation 9	The effects of breathing mode and of particle removal in the upper respiratory tract on regional deposition in animals and humans should be assessed systematically.

Recommendation 10	Nonuniform particle deposition (microdistribution) should be studied under a wide range of exposure conditions.
Recommendation 17	Models that allow calculation of deposition by airway generation should be expanded to other species.
Recommendation 19	Pathways and mechanisms of long-term retention in conducting airways (tracheobronchial tree and upper respiratory tract) should be examined and quantified.
Recommendation 20	Pathways of clearance from the pulmonary region to site(s) of long-term retention in the parenchyma should be studied.
Recommendations 15, 25	Regional deposition and ultimate fate of hygroscopic and soluble particles should be evaluated.
Recommendation 31	Effects of individual paticulate and gas-phase components of diesel exhaust should be studied.
Recommendation 34	Effect of vehicular particulate emissions on disposition of other inhaled particles should be examined.

LOW PRIORITY

These studies will provide information useful in fine tuning risk assessment, but are not critical to its development.

Recommendation 21	The relationship of changes in macrophage functional characteristics to particle clearance, including effects of exposure conditions, such as particle burden, should be characterized.
Recommendation 4	Comparative structure and physiology of human and laboratory animal pulmonary lymphatic systems should be studied.
Recommendation 8	Flow patterns in experimental animal nasal passages and human oral passages should be studied during different levels and types of activity.
Recommendation 22	The relationship between effects on macrophages in vivo and in vitro should be better defined.
Recommendation 24	The mucous layer should be characterized in various species.

References

Abraham, W., Kim, A., Januszkiewicz, M., Welker, M., Mingle, M., and Schreck, R. 1980. Effects of a brief low-level exposure to the particulate fraction of diesel exhaust on pulmonary function of conscious sheep, *Arch. Environ. Health* 35:77–80.

Adamson, I. Y. R., and Bowden, D. H. 1981. Dose response of the pulmonary macrophagic system to various particulates and its relationship to transepithelial passage of free particles, *Exp. Lung Res.* 2:165–175.

Albert, R. E., Lippmann, M., and Peterson, H. T., Jr. 1971. The effects of cigarette smoking on the kinet-

Correspondence should be addressed to Richard B. Schlesinger, Institute of Environmental Medicine, New York University Medical Center, 550 First Avenue, New York, NY 10016.

ics of bronchial clearance in humans and donkeys, In: *Inhaled Particles III* (W. H. Walton, ed.), vol. 1, pp. 165–180, Unwin Bros., Surrey, England.

Albert, R. E., Lippmann, M., Peterson, H. T., Jr., Sanborn, K., and Bohning, D. E. 1973. Bronchial deposition and clearance of aerosols, *Arch. Intern. Med.* 131:115–127.

Altshuler, B., Yarmus, L., Palmes, E. D., and Nelson, N. 1957. Aerosol deposition in the human respiratory tract, *Arch. Ind. Health* 15:293–303.

Altshuler, B., Palmes, E. D., and Nelson, N. 1966. Regional aerosol deposition in the human respiratory tract, In: *Inhaled Particles and Vapours II* (C. N. Davies, ed.), pp. 323–335, Pergamon Press, Oxford, England.

Aranyi, C., Andres, S., Ehrlich, R., Fenters, J. D., Gardner, D. E., and Waters, M. D. 1977. Cytotoxicity to alveolar macrophages of metal oxides adsorbed on fly ash, In: *Pulmonary Macrophage and Epithelial Cells* (C. L. Sanders, R. P. Schneider, G. E. Dagle, and H. A. Ragan, eds.), CONF-760927, pp. 58–65, National Technical Information Service, Springfield, Va.

Bailey, M. R., and James, A. C. 1979. In: *Biological Implications of Radionuclides Released from Nuclear Industries*, vol. 1, pp. 465–479, International Atomic Energy Agency, Vienna, Austria.

Bailey, M. R., Fry, F. A., and James, A. C. 1982. The long-term clearance kinetics of insoluble particles from the human lung, *Ann. Occup. Hyg.* 26:273–290.

Bailey, M. R., Fry, F. A., and James, A. C. 1985a. Long-term retention of particles in the human respiratory tract, *J. Aerosol Sci.* 16:295–305.

Bailey, M. R., Hodgson, A., and Smith, H. 1985b. Respiratory tract retention of relatively insoluble particles in rodents, *J. Aerosol Sci.* 16:279–293.

Baksi, S. N. 1982. Physiological effects of lead dusts, In: *Air Pollution—Physiological Effects* (J. J. McGrath and C. D. Barnes, eds.), pp. 281–310, Academic Press, New York.

Bang, B. G., Mukherjee, A. L., and Bang, F. B. 1967. Human nasal mucus flow rates, *Johns Hopkins Med. J.* 121:38–40.

Barnhart, M. I., Chen, S., and Puro, H. 1979. Input of diesel engine exhaust (DEE) particles on the structural physiology of the lung, Presented at International Symposium on the Health Effects of Diesel Engine Emissions, Cincinnati, Ohio (December 1979).

Barnhart, M. I., Chen, S., Salley, S. O., and Puro, H. 1981. Ultrastructure and morphometry of the alveolar lung of guinea pigs chronically exposed to diesel engine exhaust: six months' experience, *J. Appl. Toxicol.* 1:88–103.

Barry, P. S. I. 1975. A comparison of concentrations of lead in human tissues, *Br. J. Ind. Med.* 32:119–139.

Baselt, R. C. 1982. *Disposition of Toxic Drugs and Chemicals in Man*, Biomedical Publications, Davis, Calif.

Battigelli, M. C., Hengstenberg, F., Mannella, R. J., and Thomas, A. P. 1966. Mucociliary activity, *Arch. Environ. Health* 12:460–466.

Beeckmans, J. M. 1965. The deposition of aerosols in the respiratory tract. I. Mathematical analysis and comparison with experimental data, *Can. J. Physiol. Pharmacol.* 43:157–172.

Bennett, W. D., Messina, M. S., and Smaldone, G. C. 1985. Effect of exercise on deposition and subsequent retention of inhaled particles, *J. Appl. Physiol.* 59:1046–1054.

Bergstrom, R. 1977. Acute pulmonary toxicity of manganese dioxide, *Scand. J. Work Environ. Health* 3 (Suppl 1):1–41.

Bingham, E., Barkley, W., Zerwas, M., Stemmer, K., and Taylor, P. 1972. Responses of alveolar macrophages to metals. I. Inhalation of lead and zinc, *Arch. Environ. Health* 25:406–414.

Blusse van Oud Alblas, A., Mattie, H., and van Furth, R. 1983. A quantitative evaluation of pulmonary macrophage kinetics, *Cell Tissue Kinet.* 16:211–219.

Boecker, B. B., Hahn, F. F., Cuddihy, R. G., Snipes, M. B., and McClellan, R. O. 1983. Is the human nasal cavity at risk from inhaled radionuclides? In: *Proceedings of the 22nd Hanford Life Sciences Symposium: Life Span Radiation Effects Studies in Animals*, Richland, Wash. (September 1983).

Bohning, D. E., Atkins, H. L., and Cohn, S. H. 1982. Long-term particle clearance in man: normal and impaired, *Ann. Occup. Hyg.* 26:259–271.

Booker, D. V., Chamberlain, A. C., Rundo, J., Muir, D. C. F., and Thomson, M. L. 1967. Elimination of 5 μm particles from the human lung, *Nature* 215:30–33.

Bowden, D. H., and Adamson, I. Y. R. 1980. Role of monocytes and interstitial cells in the generation of alveolar macrophages. I. Kinetic studies of normal mice, *Lab. Invest.* 42:511–517.

Bowden, D. H., and Adamson, I. Y. R. 1984. Pathways of cellular efflux and particulate clearance after carbon instillation to the lung, *J. Pathol.* 143:117–125.

Brain, J. D. 1971. The effects of increased particles on the number of alveolar macrophages, In: *Inhaled Particles III* (W. H. Walton, ed.), vol. 1, pp. 209–223, Unwin Bros., Old Woking, England.

Brain, J. D., and Mensah, G. A. 1983. Comparative toxicology of the respiratory tract, *Am. Rev. Respir. Dis.* 128(Suppl.):S87–S90.

Brain, J. D., and Valberg, P. A. 1979. Deposition of aerosol in the respiratory tract, *Am. Rev. Respir. Dis.* 120:1325–1373.

Brain, J. D., Valberg, P. A., Gehr, P., and Bloom, S. B. 1987. Magnetic iron dust as a probe of particle clearance, phagocytosis, and particle cytotoxicity in the lungs, *Ann. Occup. Hyg.* (in press).

Brody, A. R., and Roe, M. W. 1983. Deposition pattern of inorganic particles at the alveolar level in the lungs of rats and mice, *Am. Rev. Respir. Dis.* 128:724–729.

Brody, A. R., Hill, L. H., Adkins, B., Jr., and O'Connor, R. W. 1981. Chrysotile asbestos inhalation in rats: deposition pattern and reaction of alveolar epithelium and pulmonary macrophages, *Am. Rev. Respir. Dis.* 123:670–679.

Brundelet, P. J. 1965. Experimental study of the dust

clearance mechanisms of the lung, *Acta Pathol. Microbiol. Scand.* 175:1–141.

Buist, A. S. 1982. Evaluation of lung function: concepts of normality, In: *Current Pulmonology* (D. H. Simmons, ed.), vol. 4, pp. 141–165, Wiley, New York.

Camner, P., and Philipson, K. 1972. Tracheobronchial clearance in smoking-discordant twins, *Arch. Environ. Health* 25:60–63.

Camner, P., Hellstrom, P. A., and Lundborg, M. 1973. Coating 5μ particles with carbon and metals for lung clearance studies, *Arch. Environ. Health* 27:331–333.

Camner, P., Lundborg, M., and Hellstrom, P. A. 1974. Alveolar macrophages and 5 μm particles coated with different metals. In vitro studies, *Arch. Environ. Health* 29:211–213.

Camner, P., Mossberg, B., Philipson, K., and Strandberg, K. 1979. Elimination of test particles from the human tracheobronchial tract by voluntary coughing, *Scand. J. Respir. Dis.* 60:56–62.

Camner, P., Curstedt, T., Jarstrand, C., Johannsson, A., Robertson, B., and Wiernik, A. 1985. Rabbit lung after inhalation of manganese chloride: a comparison with the effects of chlorides of nickel, cadmium, cobalt, and copper, *Environ. Res.* 38: 301–309.

Casarett, L. J. 1972. The vital sacs: alveolar clearance mechanisms in inhalation toxicology, In: *Essays in Toxicology* (W. Hayes, Jr., ed.), vol. 3, pp. 1–35, Academic Press, New York.

Castranova, V., Bowman, L., Reasor, M. J., and Miles, P. R. 1980. Effects of metallic ions on cellular and subcellular properties of rat alveolar macrophages, In: *Pulmonary Toxicology of Respirable Particles* (C. L. Sanders, F. T. Cross, G. E. Dagle, and J. A. Mahaffey, eds.), pp. 266–278, National Technical Information Service, Springfield, Va.

Castranova, V., Bowman, L., Reasor, M. J., Lewis, T., Tucker, J., and Miles, P. R. 1985. The response of rat alveolar macrophages to chronic inhalation of coal dust and/or diesel exhaust, *Environ. Res.* 36: 405–419.

Chan, T. L., and Lippmann, M. 1980. Experimental measurements and empirical modeling of the regional deposition of inhaled particles in humans, *Am. Ind. Hyg. Assoc. J.* 41:399–409.

Chan, T. L., Lee, P. S., Hering, W. E. 1981. Deposition and clearance of inhaled diesel exhaust particles in the respiratory tract of Fischer rats, *J. Appl. Toxicol.* 1:77–82.

Chan, T. L., Lee, P. S., Hering, W. E. 1984. Pulmonary retention of inhaled diesel particles after prolonged exposure to diesel exhaust, *Fundam. Appl. Toxicol.* 4:624–631.

Charles, J. M., Anderson, W. G., Menzel, D. B. 1977. Sulfate absorption from the airways of the isolated perfused lung, *Toxicol. Appl. Pharmacol.* 41:91–99.

Chen, S., Weller, M. A., and Barnhart, M. I. 1980a. Effects of Diesel Engine Exhaust on Pulmonary Alveolar Macrophages, GM Contract Publ. CR80-6/BI, General Motors Corp., Detroit, Mich.

Chen, S., Weller, M. A., and Barnhart, M. I. 1980b.

Effects of diesel engine exhaust on pulmonary alveolar macrophages, *Scan. Electron Microsc.* 3:327–338.

Cohen, B. S., Harley, N. H., Schlesinger, R. B., and Lippmann, M. 1987. Nonuniform particle deposition on tracheobronchial airways: implication for lung dosimetry, *Ann. Occup. Hyg.* (in press).

Cohen, D. 1973. Ferromagnetic contamination in the lungs and other organs of the human body, *Science* 180:745–748.

Cohen, D., Arai, S. F., and Brain, J. D. 1979. Smoking impairs long-term dust clearance from the lung, *Science* 204:514–517.

Corry, D., Kulkarni, P., and Lipscomb, M. F. 1984. The migration of bronchoalveolar macrophages into hilar lymph nodes, *Am. J. Pathol.* 115:321–328.

Craig, D. K., and Buschbom, R. L. 1975. The alveolar deposition of inhaled plutonium aerosols in rodents, *Am. Ind. Hyg. Assoc. J.* 36:172–180.

Crapo, J. D., Young, S. L., Fram, E. K., Pinkerton, K. E., Barry, B. E., and Crapo, R. O. 1983. Morphometric characteristics of cells in the alveolar region of mammalian lungs, *Am. Rev. Respir. Dis.* 128(Suppl.):S42–S46.

Crawford, D. J. 1982. Identifying critical human subpopulations by age groups: radioactivity and the lung, *Phys. Med. Biol.* 27:539–552.

Cresia, D. A., Nettesheim, P., and Hammons, A. S. 1973. Impairment of deep lung clearance by influenza virus infection, *Arch. Environ. Health* 26: 197–201.

Cuddihy, R. G. 1984. Mathematical models for predicting clearance of inhaled radioactive materials, In: *Lung Modelling for Inhalation of Radioactive Materials* (H. Smith and G. Gerber, eds.), pp. 167–175, Commission of the European Communities, Luxembourg.

Cuddihy, R. G., Brownstein, D. G., Raabe, O. F., and Kanapilly, G. M. 1973. Respiratory tract deposition of inhaled polydisperse aerosols in beagle dogs, *J. Aerosol Sci.* 4:35–45.

Cuddihy, R. G., Boecker, B. B., and Griffith, W. C. 1979. Modelling the deposition and clearance of inhaled radionuclides, In: *Biological Implications of Radionuclides Released from Nuclear Industries*, vol. II, pp. 77–89, International Atomic Energy Agency, Vienna, Austria.

Dahl, A. R., and Griffith, W. C. 1983. Deposition of sulfuric acid mist in the respiratory tracts of guinea pigs and rats, *J. Toxicol. Environ. Health* 12:371–383.

Dahl, A. R., Felicetti, S. A., and Muggenburg, B. A. 1983. Clearance of sulfuric acid–introduced ^{35}S from the respiratory tracts of rats, guinea pigs, and dogs following inhalation or instillation, *Fundam. Appl. Toxicol.* 3:293–297.

Damon, E. G., Mokler, B. V., and Jones, R. K. 1983. Influence of elastase-induced emphysema and the inhalation of an irritant aerosol on deposition and retention of an inhaled insoluble aerosol in Fischer-344 rats, *Toxicol. Appl. Pharmacol.* 67:322–330.

Dekker, E. 1961. Transition between laminar and turbulent flow in human trachea, *J. Appl. Physiol.* 16:1060–1064.

DeVries, C. R., Ingram, P., Walker, S. R., Linton,

R. W., Gutknecht, W. F., and Shelburne, J. D. 1983. Acute toxicity of lead particulates on pulmonary alveolar macrophages. Ultrastructural and micro analytical studies, *Lab. Invest.* 48:35–44.

Diu, C. K., and Yu, C. P. 1983. Respiratory tract deposition of polydisperse aerosols in humans, *Am. Ind. Hyg. Assoc. J.* 44:62–65.

Driscoll, K. E., Vollmuth, T. A., and Schlesinger, R. B. 1986. Early alveolar clearance of particles in rabbits undergoing acute and subchronic exposure to ozone, *Fundam. Appl. Toxicol.* 7:264–271.

Drown, D. B., Oberg, S. G., and Sharma, R. P. 1986. Pulmonary clearance of soluble and insoluble forms of manganese, *J. Toxicol. Environ. Health* 17:201–212.

Dziedzic, D. 1981. Differential counts of B and T lymphocytes in the lymph nodes, circulating blood, and spleen after inhalation of high concentrations of diesel exhaust, *J. Appl. Toxicol.* 1:111–114.

Egan, M. J., and Nixon, W. 1985. A model of aerosol deposition in the lung for use in inhaled dose assessments, *Radiat. Protect. Dosim.* 11:5–17.

Esposito, A. L., and Pennington, J. E. 1983. Effects of aging on antibacterial mechanisms in experimental pneumonia, *Am. Rev. Respir. Dis.* 128:662–667.

Fairchild, G. A., Kane, P., Adams, B., and Coffin, D. 1975. Sulfuric acid and streptococci clearance from respiratory tract of mice, *Arch. Environ. Health* 30:538–545.

Felicetti, S. A., Wolff, R. K., and Muggenburg, B. A. 1981. Comparison of tracheal mucous transport in rats, guinea pigs, rabbits, and dogs, *J. Appl. Physiol.* 51:1612–1617.

Ferin, J. 1976. Lung clearance of particles, In: *Air Pollution and the Lung* (C. F. Aharonson, A. Ben-Davis, and M. A. Klingberg, eds.), pp. 64–78, John Wiley and Sons, New York.

Ferin, J. 1977. Effects of particle content of lung on clearance pathways, In: *Pulmonary Macrophage and Epithelial Cells* (C. L. Sanders, R. P. Schneider, G. E. Dagle, and H. A. Ragan, eds.), pp. 414–423, National Technical Information Service, Springfield, Va.

Ferin, J., and Feldstein, M. L. 1978. Pulmonary clearance and hilar lymph node content in rats after particle exposure, *Environ. Res.* 16:342–352.

Ferin, J., and Leach, L. J. 1977. The effects of selected air pollutants on clearance of titanic oxide particles from the lungs of rats, In: *Inhaled Particles IV* (W. H. Walton, ed.), pt. 1, pp. 333–340, Pergamon Press, Oxford, England.

Ferron, G. A. 1977. The size of soluble aerosol particles as a function of the humidity of the air application to the human respiratory tract, *J. Aerosol Sci.* 8:251–267.

Ferron, G. A., Kreyling, W. G., and Haider, B. 1987. Influence of the growth of salt aerosol particles on the deposition in the lung, *Ann. Occup. Hyg.* (in press).

Findeisen, W. 1935. Uber das Absetzen Kleiner in der Luft suspendierten Teilchen in der menschlichen lung bei der Atmung, *Arch. Ges. Physiol.* 236:367–379.

Fleischer, M., Sarofim, A. F., Fassett, D. W. Ham-

mond, P., Shacklette, H. T., Nisbet, I. C. T., and Epstein, S. 1974. Environmental impact of cadmium: a review by the panel on hazardous trace substances, *Environ. Health Perspect.* 7:253–323.

Foord, H., Black, A., and Walsh, M. 1977. Pulmonary deposition of inhaled particles with diameters in the range of 2.5 to 7.5 μm, In: *Inhaled Particles IV* (W. H. Walton, ed.), pt. 1, pp. 137–149, Pergamon Press, Oxford, England.

Foster, W. M., Langenbach, E., and Bergofsky, E. H. 1980. Measurement of tracheal and bronchial mucus velocities in man: relation to lung clearance, *J. Appl. Physiol.* 48:965–971.

Fowler, R. W. 1985. Aging and lung function, *Age Ageing* 14:209–215.

Freedman, A. P., Robinson, S. E., and Green, F. Y. H. 1982. Magnetopneumography as a tool for the study of dust retention in the lungs, *Ann. Occup. Hyg.* 26:319–335.

Friberg, L. 1950. Health hazards in the manufacture of alkaline accumulators, with special reference to chronic cadmium poisoning, *Acta Med. Scand.* 138(Suppl.):240.

Friberg, L., Piscator, M., Nordberg, G. F., and Kjellstrom, T. 1974. *Cadmium in the Environment*, CRC Press, Cleveland, Ohio.

Fry, F. A., and Black, A. 1973. Regional particle deposition and clearance of particles in the human nose, *J. Aerosol Sci.* 4:113–124.

Fuchs, N. A., and Sutugin, A. G. 1966. Generation and use of monodisperse aerosols, In: *Aerosol Science* (C. N. Davies, ed.), Academic Press, New York.

Gardner, D. E. 1984. Alterations in macrophage functions by environmental chemicals, *Environ. Health Perspect.* 55:343–358.

Gardner, N. D., Lim, S. T. K., and Lawton, J. W. M. 1981. Monocyte function in ageing humans, *Mech. Ageing Dev.* 16:233–239.

Garg, B. D. 1985. Morphoquantitative analysis of pleural surface macrophage aggregates in the lungs of rats after a long term, low concentration exposure to diesel exhaust, Presented at Annual Meeting of American Association for Aerosol Research, Albuquerque, N. Mex. (November 1985).

Gehr, P. 1984. Lung morphometry, In: *Lung Modelling for Inhalation of Radioactive Materials* (H. Smith and G. Gerber, eds.), pp. 1–11, Commission of the European Communities, Luxembourg.

Gehr, P., Mwangi, D. K., Ammann, A., Maloiy, G. M. O., Taylor, C. R., and Weibel, E. R. 1981. Design of the mammalian respiratory system. V. Scaling morphometric pulmonary diffusing capacity to body mass: wild and domestic animals, *Respir. Physiol.* 44:61–86.

George, A. C., and Breslin, A. J. 1967. Deposition of natural radon daughters in human subjects, *Health Phys.* 13:375–378.

Gerrard, C. S., Gerrity, T. R., and Yeates, D. B. 1986. The relationships of aerosol deposition, lung size, and the rate of mucociliary clearance, *Arch. Environ. Health* 41:11–15.

Giacomelli-Maltoni, G., Melandri, C., Prodi, V., and Tarrone, G. 1972. Deposition efficiency of monodisperse particles in human respiratory tract, *Am. Ind. Hyg. Assoc. J.* 33:603–610.

Gibb, F. R., and Morrow, P. E. 1962. Alveolar clearance in dogs after inhalation of an iron-59 oxide aerosol, *J. Appl. Physiol.* 17:429–432.

Godard, P., Chaintreuil, J., Damon, M., Coupe, M., Flandre, O., de Paulet, A. C., and Michel, F. B. 1982. Functional assessment of alveolar macrophages: comparison of cells from asthmatic and normal subjects, *J. Allergy Clin. Immunol.* 70:88–93.

Goodman, R. M., Yergin, B. M., Landa, J. F., Golinvaux, M. H., and Sackner, M. A. 1978. Relationship of smoking history and pulmonary function tests to tracheal mucus velocity in nonsmokers, young smokers, ex-smokers and patients with chronic bronchitis, *Am. Rev. Respir. Dis.* 117:205–214.

Gooya, A., and Patra, A. 1986. Deposition of particles in a baboon nose cast, Presented at 39th Annual Conference on Engineering in Medicine and Biology Meeting, Baltimore, Md. (September 1986).

Gore, D. J., and Patrick, G. 1982. A quantitative study of the penetration of insoluble particles into the tissue of the conducting airways, *Ann. Occup. Hyg.* 26:149–161.

Goyer, R. A., and Chisolm, J. J. 1972. Lead, In: *Metallic Contaminants and Human Health*, pp. 57–95, Academic Press, New York.

Graham, J. A., Gardner, D. E., Waters, M. D., and Coffin, D. L. 1975. Effect of trace metals on phagocytosis by alveolar macrophages, *Infect. Immun.* 11: 1278–1283.

Green, G. M. 1973. Alveolobronchiolar transport mechanisms, *Arch. Intern. Med.* 131:109–114.

Greenspan, B. J., and Morrow, P. E. 1984. The effects of in vitro and aerosol exposures to cadmium on phagocytosis by rat pulmonary macrophages, *Fundam. Appl. Toxicol.* 4:48–57.

Griffis, L. C., Wolff, R. K., Henderson, R. F., Griffith, W. C., Mokler, B. V., and McClellan, R. O. 1983. Clearance of diesel soot particles from rat lung after a subchronic diesel exhaust exposure, *Fundam. Appl. Toxicol.* 3:99–103.

Griffith, W. C., Cuddihy, R. C., Boecker, B. B., Guilmette, R. A., Medinsky, M. A., and Mewhinney, J. A. 1983. Comparison of solubility of aerosols in lungs of laboratory animals, *Health Phys.* 45:233 (abstr.).

Gross, S. B., Pfitzer, E. A., Yeager, D. W., and Kehoe, R. A. 1975. Lead in human tissues, *Toxicol. Appl. Pharmacol.* 32:638–651.

Guyton, A. C. 1947. Measurement of the respiratory volumes of laboratory animals, *Am. J. Physiol.* 150:70–77.

Hahn, F. F., and Hobbs, C. H. 1979. The effect of enzyme-induced pulmonary emphysema in Syrian hamsters on the deposition and long-term retention of inhaled particles, *Arch. Environ. Health* 34:203–211.

Harmsen, A. G., Muggenburg, B. A., Snipes, M. B., and Bice, D. E. 1985. The role of macrophages in particle translocation from lungs to lymph nodes, *Science* 230:1277–1280.

Heinrich, V., Peters, L., Funcke, W., Pott, F., Mohr,

W., and Stober, W. 1981. Investigations of toxic and carcinogenic effects of diesel exhaust in long-term inhalation exposures of rodents. Presented at U.S. Environmental Protection Agency Diesel Emission Symposium, Raleigh, N.C. (October 1981).

Henderson, R. F., Wolff, R. K., Mauderly, J. L., and McClellan, R. O. 1982. Accumulation of diesel soot in lungs of rodents exposed in life span studies to diluted diesel exhaust, In: *Inhalation Toxicology Research Institute Annual Report 1981–1982*, Lovelace Biomedical and Environmental Research Institute, Albuquerque, N.Mex., LMF-102, National Technical Information Service, Springfield, Va.

Henshaw, D. L., and Fews, A. P. 1984. The microdistribution of alpha emitting particles in the human lung, In: *Lung Modelling for Inhalation of Radioactive Materials* (H. Smith and G. Gerber, eds.), pp. 199–208, Commission of the European Communities, Luxembourg.

Heppleston, A. G. 1953. Pathological anatomy of simple pneumoconiosis in coal workers, *J. Pathol. Bacteriol.* 66:235–246.

Heppleston, A. G. 1963. Deposition and disposal of inhaled dust, *Arch. Environ. Health* 7:548–555.

Heyder, J., and Rudolf, G. 1977. Deposition of aerosol particles in the human nose, In: *Inhaled Particles IV* (W. H. Walton, ed.), pt. 1, pp. 107–125, Pergamon Press, Oxford, England.

Heyder, J., Gebhart, J., Heigwer, G., Roth, C., and Stahlhofen, W. 1973. Experimental studies of the total deposition of aerosol particles in the human respiratory tract, *J. Aerosol Sci.* 4:191–208.

Heyder, J., Armbruster, L., Gebhart, J., Grein, E., and Stahlhofen, W. 1975. Total deposition of aerosol particles in the human respiratory tract for nose and mouth breathing, *J. Aerosol Sci.* 6:311–328.

Heyder, J., Gebhart, J., Rudolf, G., and Stahlhofen, W. 1980a. Physical factors determining particle deposition in the human respiratory tract, *J. Aerosol Sci.* 11:505–515.

Heyder, J., Gebhart, J., and Stahlhofen, W. 1980b. Inhalation of aerosols: particle deposition and retention, In: *Generation of Aerosols* (K. Willeke, ed.), pp. 65–103, Ann Arbor Science Publishers, Ann Arbor, Mich.

Heyder, J., Gebhart, J., Stahlhofen, W., and Stuck, B. 1982. Biological variability of particle deposition in the human respiratory tract during controlled and spontaneous mouth breathing, *Ann. Occup. Hyg.* 26:137–147.

Hilding, A. C. 1957. Ciliary streaming in the bronchial tree and the time element in carcinogenesis, *N. Engl. J. Med.* 256:634–640.

Hilding, A. C. 1963. Phagocytosis, mucus flow and ciliary action, *Arch. Environ. Health* 6:61–71.

Hinds, W. C. 1982. *Aerosol Technology*, Wiley Interscience, New York.

Hofmann, W. 1982. Mathematical model for the postnatal growth of the human lung, *Respir. Physiol.* 49:115–129.

Holland, J., Milic-Emili, J., Macklem, P. T., and Bates, D. V. 1968. Regional distribution of pulmonary ventilation and perfusion in elderly subjects, *J. Clin. Invest.* 47:81–92.

Holma, B. 1969. Scanning electron microscopic observation of particles depositing in the lung, *Arch. Environ. Health* 18:330–339.

Holt, P. F. 1981. Transport of inhaled dust to extrapulmonary sites, *J. Pathol.* 133:123–129.

Horsfield, K., and Cumming, G. 1968. Morphology of the bronchial tree in man, *J. Appl. Physiol.* 24: 373–383.

Horsfield, K., Dart, G., Olson, D. E., Filley, G., and Cumming, G. 1971. Models of the human bronchial tree, *J. Appl. Physiol.* 31:207–217.

Hounam, R. F., Black, A., and Walsh, M. 1969. Deposition of aerosol particles in the nasopharyngeal region of the human respiratory tract, *Nature* 221:1254–1255.

Hourihane, D. O. B. 1965. A biopsy series of mesotheliomata and attempts to identify asbestos with some of the tumors, *Ann. N.Y. Acad. Sci.* 132:647–673.

Huhnerbein, J., Otto, J., and Thal, W. 1984. Untersuchungsergebnisse der mukoziliaren Clearance bei lungengesunden Kindern, *Padiat. Grenzgeb.* 23: 437–443.

International Commission on Radiation Protection. 1959. *Report of Committee II on Permissible Dose for Internal Radiation*, Publ. 2, Pergamon Press, Oxford, England.

International Commission on Radiation Protection. 1972. *The Metabolism of Compounds of Plutonium and Other Actinides*, Publ. 19, Pergamon Press, Oxford, England.

Itoh, H., Smaldone, G. C., Swift, D. L., and Wagner, H. N. 1985. Mechanisms of aerosol deposition in a nasal model, *J. Aerosol Sci.* 16:529–534.

Jammet, H., Drutel, P., Parrot, R., and Roy, M. 1978. Etude de l'epuration pulmonaire chez l'homme apres administration d'aerosols de particules radioactives, *Radioprotection* 13:143–166.

Johnson, R. F., Jr., and Zeimer, P. L. 1971. The deposition and retention of inhaled [152–154]europium oxide in the rat, *Health Phys.* 20:187–193.

Kanapilly, G. M., Wolff, R. K., DeNee, P. B., and McClellan, R. O. 1982. Generation, characterization and inhalation deposition of ultrafine aggregate aerosols, *Ann. Occup. Hyg.* 26:77–91.

Kehoe, R. A. 1961. The metabolism of lead in man in health and disease: the metabolism of lead under abnormal conditions, *J. Roy. Inst. Public Health Hyg.* 24:101–120.

Kilburn, K. H. 1968. A hypothesis for pulmonary clearance and its implications, *Am. Rev. Respir. Dis.* 98:449–463.

Kliment, V. 1973. Similarity and dimensional analysis evaluation of aerosol deposition in the lungs of laboratory animals and man, *Folia Morphol.* 21: 59–66.

Kliment, V. 1974. Dichotomical model of respiratory airways of the rabbit and its significance for the construction of deposition models, *Folia Morphol.* 22:286–290.

Kliment, V., Libich, J., and Kaudersova, V. 1972. Geometry of guinea pig respiratory tract and application of Landahls' model of deposition of aerosol particles, *J. Hyg. Epidemiol. Microbiol. Immunol.* 16:107–114.

Lakin, M. B., and Fox, V. G. 1974. Transient flow characteristics in an idealized bronchial bifurcation, *Respir. Physiol.* 21:101–117.

Landahl, H. D., Tracewell, T. N., and Lassen, W. H. 1951. On the retention of airborne particulates in the human lung, *Arch. Ind. Hyg. Occup. Med.* 3: 359–366.

Landahl, H. D., Tracewell, T. N., and Lassen, W. H. 1952. Retention of airborne particulates in the human lung, III, *Arch. Ind. Hyg. Occup. Med.* 6:508–511.

Lauweryns, J. M., and Baert, J. H. 1974. The role of the pulmonary lymphatics in the defenses of the distal lung: morphological and experimental studies of the transport mechanisms of intratracheally instilled particles, *Ann. N.Y. Acad. Sci.* 221:244–275.

Leak, L. V. 1980. Lymphatic removal of fluids and particles in the mammalian lung, *Environ. Health Perspect.* 35:55–76.

Lee, K. P., Barras, C. E., Griffith, F. D., and Waritz, R. S. 1981. Pulmonary response and transmigration of organic fibers by inhalation exposure, *Am. J. Pathol.* 102:314–322.

Lee, P. S. 1981. Clearance of inhaled titanium dioxide dust in the control of diesel exposed rats, Presented at Twentieth Annual Meeting of Society of Toxicology, San Diego, Calif. (March 1981).

Lee, P. S., Gerrity, T. R., Hass, F. T., and Lourenco, R. V. 1979. A model for tracheobronchial clearance of inhaled particles in man and a comparison with the data, *IEEE Trans. Biomed. Eng.* 26:624–630.

Lee, P. S., Chan, T. L., and Hering, W. E. 1983. Long-term clearance of inhaled diesel exhaust particles in rodents, *J. Toxicol. Environ. Health* 12: 801–813.

Lee, R. E., Jr., and van Lehmden, D. J. 1973. Trace metal pollution in the environment, *J. Air Pollut. Control Assoc.* 23:853–857.

LeFevre, M. E., Green, F. H. Y., Joel, D. D., and Laqueur, W. 1982. Frequency of black pigment in livers and spleens of coal workers. Correlation with pulmonary pathology and occupational information, *Human Pathol.* 13:1121–1132.

Lehnert, B. E., and Morrow, P. E. 1985. Association of [59]iron oxide with alveolar macrophages during alveolar clearance, *Exp. Lung Res.* 9:1–16.

Lehnert, B. E., Valdez, Y. E., and Stewart, C. C. 1986. Translocation of particles to the tracheobronchial lymph nodes after lung deposition: kinetics and particle-cell relationships, *Exp. Lung Res.* 10:245–266.

Lehnert, B. E., Valdez, Y. E., and Hyler, S. 1987. Translocation of particles to the pleural space and hilar lymph nodes, and the distribution of particles in the lung free cells following their deposition in the lung, *Ann. Occup. Hyg.* (in press).

Leikauf, G., Yeates, D. B., Wales, K. A., Albert, R. E., and Lippmann, M. 1981. Effects of sulfuric acid aerosol on respiratory mechanics and mucociliary particle clearance in healthy nonsmoking adults, *Am. Ind. Hyg. Assoc. J.* 42:273–282.

Leikauf, G. D., Spektor, D. M., Albert, R. E., and Lippmann, M. 1984. Dose-dependent effects of submicrometer sulfuric acid aerosol on particle clearance from ciliated human lung airways, *Am. Ind. Hyg. Assoc. J.* 45:285–292.

Leith, D. E. 1983. Comparative mammalian respiratory mechanics, *Am. Rev. Respir. Dis.* 128(Suppl.): S77–S82.

Lippmann, M. 1970. Deposition and clearance of inhaled particles in the human nose, *Ann. Otol. Rhinol. Laryngol.* 79:1–10.

Lippmann, M. 1977. Regional deposition of particles in the human respiratory tract, In: *Handbook of Physiology, Reactions to Environmental Agents* (D. H. K. Lee, H. L. Falk, and S. D. Murphy, eds.), vol. 9, pp. 213–232, American Physiological Society, Bethesda, Md.

Lippmann, M., and Albert, R. E. 1969. The effect of particle size on the regional deposition of inhaled aerosols in the human respiratory tract, *Am. Ind. Hyg. Assoc. J.* 30:257–275.

Lippmann, M., and Altshuler, B. 1976. Regional deposition of aerosols, In: *Air Pollution and the Lung* (E. F. Aharonson, A. Ben-David, and M. A. Klingberg, eds.), pp. 25–48, Wiley, New York.

Lippmann, M., and Schlesinger, R. B. 1984. Interspecies comparison of particle deposition and mucociliary clearance in tracheobronchial airways, *J. Toxicol. Environ. Health* 13:441–469.

Lippmann, M., Albert, R. E., and Peterson, H. T., Jr. 1972. The regional deposition of inhaled aerosols in man, In: *Inhaled Particles III* (W. H. Walton, ed.), pp. 105–120, Unwin Bros., Surrey, England.

Lippmann, M., Yeates, D. B., and Albert, R. E. 1980. Deposition, retention and clearance of inhaled particles, *Br. J. Ind. Med.* 37:337–362.

Lourenco, R. V., Klimek, M. F., and Borowski, C. J. 1971a. Deposition and clearance of 2 μ particles in the tracheobronchial tree of normal subjects—smokers and non-smokers, *J. Clin. Invest.* 50:1411–1420.

Lourenco, R. V., Stanley, E. D., Gatmaitan, B., and Jackson, G. G. 1971b. Abnormal deposition and clearance of inhaled particles during upper respiratory and viral infections, *J. Clin. Invest.* 50:62a.

Lourenco, R. V., Loddenkemper, R., and Cargan, R. W. 1972. Patterns of distribution and clearance of aerosols in patients with bronchiectasis, *Am. Rev. Respir. Dis.* 106:857–866.

Love, R. G., Muir, D. C. F., and Sweetland, K. F. 1971. Aerosol deposition in the lungs of coalworkers, In: *Inhaled Particles III* (W. H. Walton, ed.), vol. I, pp. 131–139, Unwin Bros., Surrey, England.

Lucas, A. M., and Douglas, L. C. 1934. Principles underlying ciliary activity in the respiratory tract. II. A comparison of nasal clearance in man, monkey and other mammals, *Arch. Otolaryngol.* 20:518–541.

Luckey, T. D., and Venugopal, B. 1977. *Metal Toxicity in Mammals. 1. Physiological and Chemical Basis for Metal Toxicity*, Plenum Press, New York.

Lundborg, M., Lind, B., and Camner, P. 1984. Ability of rabbit alveolar macrophages to dissolve metals, *Exp. Lung Res.* 7:11–22.

Lundborg, M., Eklund, A., Lind, B., and Camner, P. 1985. Dissolution of metals by human and rabbit alveolar macrophages, *Br. J. Ind. Med.* 42:642–645.

Macklin, C. C. 1955. Pulmonary sumps, dust accumulation, alveolar fluid and lymph vessels, *Acta Anat.* 23:1–21.

Mahoney, J. P., and Small, W. J. 1968. Studies on manganese. III. The biological half-life of radiomanganese in man and factors which affect this half-life, *J. Clin. Invest.* 47:643–653.

Maigetter, R. Z., Ehrlich, R., Fenters, J. D., and Gardner, D. E. 1976. Potentiating effects of manganese dioxide on experimental respiratory infections, *Environ. Res.* 11:386–391.

Majima, Y., Sakakura, Y., Matsubara, T., Murai, S., and Miyoshi, Y. 1983. Mucociliary clearance in chronic sinusitis: related human nasal clearance and in vitro bullfrog palate clearance, *Biorheology* 20:251–262.

Man, S. F. P., Lee, T. K., Gibney, R. T. N., Logus, J. W., and Noujaim, A. A. 1980. Canine tracheal mucus transport of particulate pollutants: comparison of radiolabeled corn pollen, ragweed pollen, asbestos, silica, and talc to Dowex® anion exchange particles, *Arch. Environ. Health* 35:283–286.

Manton, W. I. 1977. Sources of lead in blood: identification by stable isotopes, *Arch. Environ. Health* 32:149–159.

Martens, A., and Jacobi, W. 1974. Die in vivo Bestimmung der Aerosolteilchendeposition in Atemtrakt bei Mund-bzur. Nasenatmung, In: *Aerosols in Physik, Medizin und Technik*, pp. 117–121, Gesellschaft fur Aerosolforschung, Bad Soden.

Martonen, T. B. 1985. The effect of age on regional aerosol deposition in man, Presented at Annual Meeting of American Association for Aerosol Research, Albuquerque, N.Mex. (November 1985).

Martonen, T., and Gibby, D. 1982. Computer models of aerosol deposition in two human tracheobronchial geometries, *Comp. Biomed. Res.* 15:425–433.

Martonen, T. B., and Patel, M. 1981. Computation of ammonium bisulfate aerosol deposition in conducting airways, *J. Toxicol. Environ. Health* 8:1001–1014.

Masse, R. 1971. Etude cytologique comparee de l'influence du plutonium et de la silica inhales sur le comportement du macrophage alveolaire, In: *Inhaled Particles III* (W. H. Walton, ed.), vol. 1, pp. 247–257, Unwin Bros., Surrey, England.

Masse, R., Ducousso, R., Nolibe, D., Lafuma, J., and Chretien, J. 1974. Passage transbronchique des particules metalliques, *Rev. Fr. Mal. Respir.* 1:123–129.

Matthys, H., Vastag, E., Kohler, D., Daikeler, G., and Fischer, J. 1983. Mucociliary clearance in patients with chronic bronchitis and bronchial carcinoma, *Respiration* 44:329–337.

Mauderly, J. L. 1979. Effect of age on pulmonary structure and function of immature and adult animals and man, *Fed. Proc.* 38:173–177.

Mauderly, J. L., Jones, R. K., Henderson, R. F., Wolff, R. K., Pickrell, J. A., McClellan, R. O., and Gillett, N. A. 1987. Relationship of lung structural and functional changes to accumulation of diesel exhaust particles, *Ann. Occup. Hyg.* (in press).

McMahon, T. A., Brain, J. D., and LeMott, S. R. 1977. Species difference in aerosol deposition, In: *Inhaled Particles IV* (W. H. Walton, ed.), pt. 1, pp. 23–33, Pergamon Press, Oxford, England.

Meessen, H. 1960. Die pathomorphologic der Diffusion und Perfusion, *Verh. Dtsch. Ges. Pathol.* 44:98–106.

Mercer, T. T. 1967. On the role of particle size in the dissolution of lung burdens, *Health Phys.* 13:1211–1221.

Mercer, T. T. 1975. The deposition model of the Task Group on Lung Dynamics: a comparison with recent experimental data, *Health Phys.* 29:673–680.

Metivier, H. 1984. Animal data on clearance, In: *Lung Modelling for Inhalation of Radioactive Materials* (H. Smith and G. Gerber, eds.), pp. 77–89, Commission of the European Communities, Luxembourg.

Moore, M. R., Meredith, P. A., and Goldberg, A. 1980. Lead and heme biosynthesis, In: *Lead Toxicity* (R. L. Singhal and J. A. Thomas, eds.), pp. 79–117, Urban and Schwarzenberg, Munich.

Moore, W., Orthoefer, J. G., Burkart, J. K., and Malanchuk, M. 1978. Preliminary findings on the deposition and retention of automotive diesel particulate in rat lungs, In: *Proceedings of the 71st Annual Meeting of the Air Pollution Control Association*, Houston, Tex. (June 1978).

Moores, S. R., Black, A., Lambert, B. E., Lindop, P. J., Morgan, A., Pritchard, J., and Walsh, M. 1980. Deposition of thorium and plutonium oxides in the respiratory tract of the mouse, In: *Pulmonary Toxicology of Respirable Particles* (C. L. Sanders, F. T. Cross, G. E. Dagle, and J. A. Mahaffey, eds.), pp. 103–118. National Technical Information Service, Springfield, Va.

Morgan, A., and Holmes, A. 1980. Concentrations and dimensions of coated and uncoated asbestos fibers in the human lung, *Br. J. Ind. Med.* 37:25–32.

Morgan, W. K. C., Ahmad, D., Chamberlain, M. J., Clague, H. W., Pearson, M. G., and Vinitski, S. 1984. The effect of exercise on the deposition of an inhaled aerosol, *Respir. Physiol.* 56:327–338.

Morrow, P. E. 1973. Alveolar clearance of aerosols, *Arch. Intern. Med.* 131:101–108.

Morrow, P. E. 1977. Clearance kinetics of inhaled particles, In: *Respiratory Defense Mechanisms* (J. D. Brain, D. F. Proctor, and L. M. Reid, eds.), pt. II, pp. 491–543, Marcel Dekker, New York.

Morrow, P. E. 1981. Aerosol factors affecting respiratory deposition, In: *Proceedings of International Symposium on Deposition and Clearance of Aerosols in the Human Respiratory Tract*, Bad Gleichenberg, Austria (May 1981).

Morrow, P. E., Gibb, F. R., and Johnson, L. 1964. Clearance of insoluble dust from the lower respiratory tract, *Health Phys.* 10:543–555.

Morrow, P. E., Gibb, F. R., and Gazioglu, K. M. 1967a. The clearance of dust from the lower respiratory tract of man: an experimental study, In: *Inhaled Particles and Vapours II* (C. N. Davies, ed.), pp. 351–358, Pergamon Press, London.

Morrow, P. E., Gibb, F. R., and Gazioglu, K. M. 1967b. A study of particulate clearance from the human lung, *Am. Rev. Respir. Dis.* 96:1209–1221.

Morrow, P. E., Beiter, H., Amato, F., and Gibb, F. R. 1980. Pulmonary retention of lead: an experimental study in man, *Environ. Res.* 21:373–384.

Muhle, H., Bellman, B., and Heinrich, U. 1987. Overloading of lung clearance after chronic exposure of experimental animals to particles, *Ann. Occup. Hyg.* (in press).

Muir, D. C. F., and Davies, C. N. 1967. The deposition of 0.5 μm diameter aerosols in the lungs of man, *Ann. Occup. Hyg.* 10:161–174.

Myron, D. R., Zimmerman, T. J., and Schuler, T. R. 1978. Intake of nickel and vanadium by humans. A survey of selected diets, *Am. J. Clin. Nutr.* 31:527–531.

Natusch, D. F. S., Wallace, J. R., and Evans, C. A., Jr. 1974. Toxic trace elements: preferential concentration in respirable particles, *Science* 183:202–204.

Natusch, D. F. S., Wallace, J. R., and Evans, C. A. 1975. Concentration of toxic species in submicrometer size airborne particles—the lung as a preferential absorption site, *Am. Inst. Chem. Eng. Symp. Series No.* 147, vol. 71.

Naumann, B. D., and Schlesinger, R. B. 1986. Assessment of early alveolar particle clearance and macrophage function following an acute inhalation of sulfuric acid mist, *Exp. Lung Res.* 11:13–33.

Newhouse, M. T., Dolovich, M., and Obminski, G. 1978. Effect of TLV levels of SO_2 and H_2SO_4 on bronchial clearance in exercising man, *Arch. Environ. Health* 33:24–32.

Newton, D., Fry, F. A., Taylor, B. T., Eagle, M. C., and Shorma, R. C. 1978. Interlaboratory comparison of techniques for measuring lung burdens of low energy protein emitters, *Health Phys.* 35:751–771.

Niinimaa, V., Cole, P., Mintz, S., and Shephard, R. J. 1980. The switching point from nasal to oronasal breathing, *Respir. Physiol.* 42:61–71.

Nikiforov, A. I., and Schlesinger, R. B. 1985. Morphometric variability of the human upper bronchial tree, *Respir. Physiol.* 59:289–299.

Nozaki, K. 1966. Method for studies on inhaled particles in human respiratory system and retention of lead fume, *Ind. Health* 4:118–122.

Oberdörster, G., and Hochrainer, D. 1980. Lung clearance of Fe_2O_3 and $CdCl_2$ during chronic CdO inhalation, In: *Aerosols in Science, Medicine and Technology, Proceedings of the 8th Conference on Aerosol Research*, Schmallenberg, Germany.

Oberdörster, G., Baumert, H.-P., Hochrainer, D., and Stoeber, W. 1979. The clearance of cadmium aerosols after inhalation exposure, *J. Am. Ind. Hyg. Assoc.* 40:443–450.

Oberdörster, G., Green, F. Y. H., and Freedman, A. P. 1984. Clearance of $^{59}Fe_2O_3$ particles from the lungs of rats during exposure to coal mine dust and diesel exhaust, *J. Aerosol Sci.* 15:235–237.

Olson, D. E., Dart, G. A., and Filley, G. F. 1970. Pressure drop and fluid flow regime of air inspired into the human lung, *J. Appl. Physiol.* 28:482–494.

Olson, D. E., Sudlow, M. F., and Horsfield, K. 1973. Convective patterns of flow during inspiration, *Arch. Intern. Med.* 131:51–57.

Palm, P. E., McNerney, J. M., and Hatch, T. 1956.

Respiratory dust retention in small animals. A comparison with man, *Arch. Ind. Health* 13:355–365.

Parker, H., Horsfield, K., and Cumming, G. 1971. Morphology of distal airways in the human lung, *J. Appl. Physiol.* 31:386–391.

Passali, D., and Ciampoli, M. B. 1985. Normal values of mucociliary transport time in young subjects, *Int. J. Pediatr. Otorhinolaryngol.* 9:151–156.

Patra, A. L. 1986. Comparative anatomy of mammalian respiratory tract: the nasopharyngeal region and the tracheobronchial tree, *J. Toxicol. Environ. Health* 17:163–174.

Patrick, G., and Stirling, C. 1977. The retention of particles in large airways of the respiratory tract, *Proc. Roy. Soc. London, Ser. V* 198:455–462.

Patterson, C. C. 1965. Contaminated and natural lead environments of man, *Arch. Environ. Health* 11:344–360.

Pattle, R. E. 1961. The retention of gases and particles in the human nose, In: *Inhaled Particles and Vapours* (C. N. Davies, ed.), pp. 302–309, Pergamon Press, Oxford, England.

Pavia, D. 1984. Lung mucociliary clearance, In: *Aerosols and the Lung* (S. W. Clarke and D. Pavia, eds.), pp. 127–155, Butterworths, London.

Pavia, D., Bateman, J. R. M., Sheahan, N. F., Agnew, J. E., and Clarke, S. W. 1985. Tracheobronchial mucociliary clearance in asthma: impairment during remission, *Thorax* 40:171–175.

Phalen, R. F. 1987. Particle deposition predictions for infants, children and adolescents, *Ann. Occup. Hyg.* (in press).

Phalen, R. F., and Oldham, M. J. 1985. Predicted particle deposition efficiency of the newborn's nose, Presented at Annual Meeting of American Association for Aerosol Research, Albuquerque, N.Mex. (November 1985).

Phalen, R., Kenoyer, J., and Davis, J. 1977. Deposition and clearance of inhaled particles: comparison of mammalian species, In: *Proceedings of the Annual Conference on Environmental Toxicology*, vol. 7, pp. 159–170, AMRL-TR-76-125, National Technical Information Service, Springfield, Va.

Phalen, R. F., Yeh, H. C., Schum, G. M., and Raabe, O. G. 1978. Application of an idealized model of morphometry of the tracheobronchial tree, *Anat. Rec.* 190:167–176.

Phalen, R. F., Kenoyer, J. L., Crocker, T. T., and McClure, T. R. 1980. Effects of sulfate aerosols in combination with ozone on elimination of tracer particles inhaled by rats, *J. Toxicol. Environ. Health* 6:797–810.

Phalen, R. F., Oldham, M. J., Beaucage, C. B., Crocker, T. T., and Mortensen, J. D. 1985. Postnatal enlargement of human tracheobronchial airways and implications for particle deposition, *Anat. Rec.* 212:368–380.

Philipson, K., Falk, R., and Camner, P. 1985. Long-term lung clearance in humans studied with teflon particles labeled with chromium-51, *Exp. Lung Res.* 9:31–42.

Phipps, R. J. 1981. The airway mucociliary system, In: *International Review of Physiology: Respiratory Physiology* (J. G. Widdicombe, ed.), vol. 23, pp. 213–259, University Park Press, Baltimore, Md.

Pooley, F. 1974. Locating fibers in the bowel wall, *Environ. Health Perspect.* 9:235.

Pritchard, J. N., Jane-Jefferies, S., and Black, A. 1987. Regional deposition of 2.5 to 5.0 μm polystyrene microspheres inhaled by women, *Ann. Occup. Hyg.* (in press).

Proctor, D. F. 1980. The upper respiratory tract, In: *Pulmonary Diseases and Disorders* (A. P. Fishman, ed.), pp. 209–223, McGraw-Hill, New York.

Puchelle, E., Zahm, J. M., Girard, F., Bertrand, A., Polu, J. M., Aug, F., and Sadoul, P. 1980. Mucociliary transport in vivo and in vitro—relations to sputum properties in chronic bronchitis, *Eur. J. Respir. Dis.* 61:254–264.

Puro, H. 1980. Light microscopic findings in lungs of rats and guinea pigs exposed to diesel exhaust, GM Research Laboratories Publ. CR80-7/BI, General Motors Corp., Detroit, Mich.

Raabe, O. G. 1982. Deposition and clearance of inhaled aerosols, In: *Mechanisms in Respiratory Toxicology* (H. Witschi and P. Nettesheim, eds.), pp. 27–76, CRC Press, Boca Raton, Fla.

Raabe, O. G., Yeh, H. C., Newton, G. J., Phalen, R. J., and Velazquez, D. J. 1977. Deposition of inhaled monodisperse aerosols in small rodents, In: *Inhaled Particles IV* (W. H. Walton, ed.), pt. 1, pp. 3–20, Pergamon Press, Oxford, England.

Raabe, O. G., Ali-Bayati, M. A., Rasolt, A., and Teague, S. V. 1987. Regional deposition of inhaled monodisperse coarse and fine aerosol particles in small laboratory animals, *Ann. Occup. Hyg.* (in press).

Rabinowitz, M. B., Wetherill, G. W., and Kopple, J. D. 1973. Lead metabolism in the normal human: stable isotope studies, *Science* 182:725–727.

Rabinowitz, M. B., Wetherill, G. W., and Kopple, J. D. 1974. Studies of human lead metabolism by use of stable isotope tracers, *Environ. Health Perspect.* 7:145–153.

Rabinowitz, M. B., Wetherill, G. W., and Kopple, J. D. 1976. Kinetic analysis of lead metabolism in healthy humans, *J. Clin. Invest.* 58:260–270.

Radford, E. P., and Martell, E. A. 1977. Polonium-210: lead 210 ratios as an index of residence times of insoluble particles from cigarette smoke in bronchial epithelium, In: *Inhaled Particles IV* (W. H. Walton, ed.), pt. 2, pp. 567–580, Pergamon Press, Oxford, England.

Rahola, T., Aaran, R. K., and Miettinen, J. K. 1972. Half-time studies of mercury and cadmium by whole body counting, In: *Assessment of Radioactive Contamination in Man*, International Atomic Energy Agency, Vienna, Austria.

Robertson, B. 1980. Basic morphology of the pulmonary defense system, *Eur. J. Respir. Dis.* 61(Suppl. 107):21–40.

Rudd, C. J., and Strom, K. A. 1981. A spectrophotometric method for the quantitation of diesel exhaust particles in guinea pig lung, *J. Appl. Toxicol.* 1:83–87.

Rudolf, G., Gebhart, J., Heyder, J., Scheuch, G., and Stahlhofen, W. 1987. Mass deposition from inspired polydisperse aerosols, *Ann. Occup. Hyg.* (in press).

Rutland, J., and Cole, P. J. 1981. Nasal mucociliary clearance and ciliary beat frequency in cystic fibrosis compared with sinusitis and bronchiectasis, *Thorax* 36:654–658.

Sackner, M. A., Ford, D., Fernandez, R., Cipley, J., Perez, D., Kwoka, M., Reinhart, M., Michaelson, E. D., Schreck, R., and Wanner, A. 1978. Effects of sulfuric acid aerosol on cardiopulmonary function of dogs, sheep and humans, *Am. Rev. Respir. Dis.* 118:497–510.

Schiller, C. F., Gebhart, J., Heyder, J., Rudolf, G., and Stahlhofen, W. 1987. Deposition of monodisperse insoluble aerosol particles in the 0.005 to 0.2 μm size range within the human respiratory tract, *Ann. Occup. Hyg.* (in press).

Schlesinger, R. B. 1985. Clearance from the respiratory tract, *Fundam. Appl. Toxicol.* 5:435–450.

Schlesinger, R. B. 1986. The effects of inhaled acids on respiratory tract defense mechanisms, *Environ. Health Perspect.* 63:25–38.

Schlesinger, R. B. 1987. Functional assessment of rabbit alveolar macrophages following intermittent inhalation exposures to sulfuric acid mist, *Fundam. Appl. Toxicol.* 8:328–334.

Schlesinger, R. B., and Gearhart, J. M. 1986. Early alveolar clearance in rabbits intermittently exposed to sulfuric acid mist, *J. Toxicol. Environ. Health* 17:213–220.

Schlesinger, R. B., and Gearhart, J. M. 1987. Intermittent exposures to mixed atmospheres of nitrogen dioxide and sulfuric acid: effect on particle clearance from the respiratory region of rabbit lungs, *Toxicology* 44:309–319.

Schlesinger, R. B., and Lippmann, M. 1978. Selective particle deposition and bronchogenic carcinoma, *Environ. Res.* 15:424–431.

Schlesinger, R. B., and McFadden, L. 1981. Comparative morphometry of the upper bronchial tree in six mammalian species, *Anat. Rec.* 199:99–108.

Schlesinger, R. B., Lippmann, M., and Albert, R. 1978. Effects of short-term exposures to sulfuric acid and ammonium sulfate aerosols upon bronchial airway function in the donkey, *Am. Ind. Hyg. Assoc. J.* 39:275–286.

Schlesinger, R. B., Halpern, M., Albert, R. E., and Lippmann, M. 1979. Effects of chronic inhalation of sulfuric acid mist upon mucociliary clearance from the lungs of donkeys, *J. Environ. Pathol. Toxicol.* 2:1351–1367.

Schlesinger, R. B., Gurman, J. L., and Lippmann, M. 1982. Particle deposition within bronchial airways: comparisons using constant and cyclic inspiratory flow, *Ann. Occup. Hyg.* 26:47–64.

Schlesinger, R. B., Concato, J., and Lippmann, M. 1983a. Particle deposition during exhalation: a study in replicate casts of the human upper tracheobronchial tree, In: *Aerosols in the Mining and Industrial Work Environment* (B. Liu and V. Marple, eds.), pp. 165–176, Ann Arbor Science Publishers, Ann Arbor, Mich.

Schlesinger, R. B., Naumann, B. D., and Chen, L. C. 1983b. Physiological and histological alterations in the bronchial mucociliary clearance system of rabbits following intermittent oral or nasal inhalation of sulfuric acid mist, *J. Toxicol. Environ. Health* 12:441–465.

Schlesinger, R. B., Chen, L. C., and Driscoll, K. E. 1984. Exposure-response relationship of bronchial mucociliary clearance in rabbits following acute inhalations of sulfuric acid mist, *Toxicol. Lett.* 22:249–254.

Schlesinger, R. B., Vollmuth, T. A., Naumann, B. D., and Driscoll, K. E. 1986. Measurement of particle clearance from the alveolar region of the rabbit respiratory tract, *Fundam. Appl. Toxicol.* 7:256–263.

Schreider, J. P. 1986. Comparative anatomy and functions of the nasal passages, In: *Toxicology of the Nasal Passages* (C. S. Barrow, ed.), pp. 1–25, McGraw-Hill, New York.

Schreider, J. P., and Hutchens, J. O. 1979. Particle deposition in the guinea pig respiratory tract, *J. Aerosol Sci.* 10:599–607.

Schroeder, H. A., Balassa, J. J., and Tipton, I. H. 1963. Abnormal trace metals in man—vanadium, *J. Chronic Dis.* 16:1047–1071.

Schum, G. M., and Yeh, H. C. 1980. Theoretical evaluation of aerosol deposition in anatomical models of mammalian lung airways, *Bull. Math. Biol.* 42:1–15.

Schum, G. M., Duggan, M. T., and Yeh, H. C. 1976. Tracheobronchial anatomy: species differences in branching patterns, In: *Inhalation Toxicology Research Institute Annual Report 1975–1976*, Lovelace Biomedical and Environmental Institute, Albuquerque, N.Mex.

Scott, W. R., Taulbee, D. B., and Yu, C. P. 1978. Theoretical study of nasal deposition, *Bull. Math. Biol.* 40:581–603.

Sekihara, T., Olson, D. E., and Filley, G. F. 1968. Airflow regimes and geometrical factors in the human airways, In: *Current Research in Chronic Respiratory Disease: Proceedings of the Eleventh Aspen Emphysema Conference,* pp. 103–114, U.S. Department of Health, Education and Welfare, Washington, D.C.

Snipes, M. B., and Clem, M. F. 1981. Retention of microspheres in the rat lung after intratracheal instillation, *Environ. Res.* 24:33–41.

Snipes, M. B., Boecker, B. B., and McClellan, R. O. 1983. Retention of monodisperse or polydisperse aluminosilicate particles inhaled by dogs, rats, and mice, *Toxicol. Appl. Pharmacol.* 69:345–362.

Snipes, M. B., Boecker, B. B., and McClellan, R. O. 1984a. Respiratory tract clearance of inhaled particles in laboratory animals, In: *Lung Modelling for Inhalation of Radioactive Materials* (H. Smith and G. Gerber, eds.), pp. 63–71, Commission of the European Communities, Luxembourg.

Snipes, M. B., Chavez, G. T., and Muggenburg, B. A. 1984b. Disposition of 3-, 7-, and 13-μm microspheres instilled into lungs of dogs, *Environ. Res.* 33:333–342.

Snyder, B., and Jaeger, M. J. 1983. Lobar flow patterns in a hollow cast of canine central airways, *J. Appl. Physiol.* 54:749–756.

Soong, T. T., Nicolaides, P., Yu, C. P., and Soong, S. C. 1979. A statistical description of the human tracheobronchial tree geometry, *Respir. Physiol.* 37:161–172.

Sorokin, S. P., and Brain, J. D. 1975. Pathways of clearance in mouse lungs exposed to iron oxide aerosols, *Anat. Rec.* 181:581–626.

Spell, K. E. 1969. Comparative studies in lung mechanics based on a survey of literature data, *Respir. Physiol.* 8:37–57.

Stahl, W. R. 1967. Scaling of respiratory variables in mammals, *J. Appl. Physiol.* 22:453–460.

Stahlhofen, W., Gebhart, J., and Heyder, J. 1981a. Biological variability of regional deposition of aerosol particles in the human respiratory tract, *Am. Ind. Hyg. Assoc. J.* 42:348–352.

Stahlhofen, W., Gebhart, J., Heyder, J., Philipson, K., and Camner, P. 1981b. Intercomparison of regional deposition of aerosol particles in the human respiratory tract and their long-term elimination, *Exp. Lung Res.* 2:131–139.

Stanley, P. J., MacWilliam, L., Greenstone, M. A., Daly, C., and Cole, P. J. 1984. Prolonged nasal mucociliary clearance in healthy smokers, *Thorax* 39:239 (abstr).

Stanley, P. J., Wilson, R., Greenstone, M. A., Mackay, I. S., and Cole, P. J. 1985. Abnormal nasal mucociliary clearance in patients with rhinitis and its relationship to concomitant chest disease, *Br. J. Dis. Chest* 79:77–82.

Stauffer, D. 1975. Scaling theory for aerosol deposition in the lungs of different mammals, *J. Aerosol Sci.* 6:223–225.

Strom, K. A. 1984. Response of pulmonary cellular defenses to the inhalation of high concentrations of diesel exhaust, *J. Toxicol. Environ. Health* 13:919–944.

Sunderman, F. W., Jr., 1977. A review of the metabolism and toxicology of nickel, *Ann. Clin. Lab. Sci.* 7:377–398.

Sweeney, T. D., Brain, J. D., Tryka, A. F., and Godleski, J. J. 1983. Retention of inhaled particles in hamsters with pulmonary fibrosis, *Am. Rev. Respir. Dis.* 128:138–143.

Swift, D. L. 1981. Aerosol deposition and clearance in the human upper airways, *Ann. Biomed. Engl.* 9:593–604.

Swift, D. L., and Proctor, D. F. 1977. Access of air to the respiratory tract, In: *Respiratory Defense Mechanisms* (J. D. Brain, D. F. Proctor, and L. M. Reid, eds.), pt. 1, pp. 63–93, Marcel Dekker, New York.

Swift, D. L., and Proctor, D. F. 1987. A dosimetric model for particles in the respiratory tract above the trachea, *Ann. Occup. Hyg.* (in press).

Swift, D. L., Cobb, J. A. C., and Smith, J. C. 1977a. Aerosol deposition in the dog respiratory tract, In: *Inhaled Particles IV* (W. H. Walton, ed.), pt. 1, pp. 237–245, Pergamon Press, Oxford, England.

Swift, D. J., Shanty, F., and O'Neill, J. T. 1977b. Human respiratory tract deposition of nuclei particles and health implications, Presented at American Nuclear Society Winter Meeting, San Francisco (November–December 1977).

Task Group on Lung Dynamics. 1966. Deposition and retention models for internal dosimetry of the human respiratory tract, *Health Phys.* 12:173–207.

Taulbee, D. B., and Yu, C. P. 1975. A theory of aerosol deposition in the human respiratory tract, *J. Appl. Physiol.* 38:77–85.

Thomas, R. L., and Raabe, O. G. 1978. Regional deposition of ^{137}Cs-labelled monodisperse and polydisperse aluminosilicate aerosols in Syrian hamsters, *Am. Ind. Hyg. Assoc. J.* 39:1009–1018.

Thomson, M. L., and Pavia, D. 1974. Particle penetration and clearance in the human lung, *Arch. Environ. Health* 29:214–219.

Thomson, M. L., and Short, M. D. 1969. Mucociliary function in health, chronic obstructive airway disease and asbestosis, *J. Appl. Physiol.* 26:535–539.

Torjussen, W., and Anderson, I. 1979. Nickel concentrations in nasal mucosa, plasma and urine in active and retired nickel workers, *Ann. Clin. Lab. Sci.* 9:289–298.

Tryka, A. F., Sweeney, T. D., Brain, J. D., and Godleski, J. J. 1985. Short-term regional clearance of an inhaled submicrometric aerosol in pulmonary fibrosis, *Am. Rev. Respir. Dis.* 132:606–611.

Tu, K. W., and Knutson, E. O. 1984. Total deposition of ultrafine hydrophobic and hygroscopic aerosols in the human respiratory system, *Aerosol Sci. Technol.* 3:453–465.

Tyler, W. S. 1983. Comparative subgross anatomy of lungs, *Am. Rev. Respir. Dis.* 128(Suppl.):S32–S36.

Valberg, P. A. 1985. Determination of retained lung dose, In: *Toxicology of Inhaled Materials* (H. P. Witschi and J. D. Brain, eds.), pp. 57–91, Springer-Verlag, Heidelberg.

Valberg, P. A., Brain, J. D., Sneddon, S. L., and LeMott, S. R. 1982. Breathing patterns influence aerosol deposition sites in excised dog lung, *J. Appl. Physiol.* 53:824–837.

Valberg, P. A., Wolff, R. K., and Mauderly, J. L. 1985. Redistribution of retained particles. Effect of hyperpnea, *Am. Rev. Respir. Dis.* 131:273–280.

van Antweiler, H. 1958. Uber die Function des Flimmerepithels der Luftwege, insbesondere under Staubbelastung, *Beitr. Silikose-Forsch.* 3:509.

Vastag, E., Matthys, H., Zsamboki, G., Kohler, D., and Daikeler, G. 1986. Mucociliary clearance in smokers, *Eur. J. Respir. Dis.* 68:107–113.

Vincent, J. H., and Armbruster, L. 1981. On the quantitative definition of the inhalability of airborne dust, *Ann. Occup. Hyg.* 24:245–248.

Vostal, J. J., Chan, T. L., Garg, B. D., Lee, P. L., and Strom, K. S. 1979. Lymphatic transport of inhaled diesel particles in the lungs of rats and guinea pigs exposed to diluted diesel exhaust, In: *Proceedings of the International Symposium on Health Effects of Diesel Engine Emission*, U.S. Environmental Protection Agency, Cincinnati, Ohio (December 1979).

Waite, D. A., and Ramsden, D. 1971. The inhalation of insoluble iron oxide particles in the sub-micron range, pt. I, chromium 51 labelled aerosols, AEEW-R740, Atomic Energy Authority, United Kingdom.

Wanner, A. 1977. Clinical aspects of mucociliary transport, *Am. Rev. Respir. Dis.* 116:73–125.

Waters, M. D., Gardner, D. E., Aranyi, C., and

Coffin, D. L. 1975. Metal toxicity for rabbit alveolar macrophages in vitro, *Environ. Res.* 9:32–47.

Watson, A. Y., and Brain, J. D. 1979. Uptake of iron oxide aerosols by mouse airway epithelium, *Lab. Invest.* 40:450–459.

Weibel, E. R. 1963. *Morphometry of the Human Lung*, Academic Press, New York.

Weibel, E. R. 1980. Design and structure of the human lung, In: *Pulmonary Diseases and Disorders* (A. P. Fishman, ed.), pp. 224–271, McGraw-Hill, New York.

Weller, M. A., Chen, S., and Barnhart, M. I. 1980. Acid Phosphatase in Alveolar Macrophages Exposed in vivo to Diesel Engine Exhaust, GM Contract Publ. CR80-5/BI, General Motors Corp., Detroit, Mich.

West, J. B., and Hugh-Jones, P. 1959. Patterns of gas flow in the upper bronchial tree, *J. Appl. Physiol.* 14:753–759.

White, H. J., and Garg, B. D. 1981. Early pulmonary response of the rat lung to inhalation of high concentration of diesel particles, *J. Appl. Toxicol.* 1:104–110.

Williams, S. J., Holden, K. M., Sabransky, M., and Menzel, D. B. 1980. The distributional kinetics of Ni^{2+} in the rat lung, *Toxicol. Appl. Pharmacol.* 55:85–93.

Wilson, F. T., Jr., Hiller, F. C., Wilson, J. G., and Bone, R. C. 1985. Quantitative deposition of ultrafine stable particles in the human respiratory tract, *J. Appl. Physiol.* 58:223–229.

Wolff, R. K. 1986. Effects of airborne pollutants on mucociliary clearance, *Environ. Health Perspect.* 66:222–237.

Wolff, R. K., Dolovich, M. B., Obminski, G., and Newhouse, M. T. 1977. Effects of exercise and eucapnic hyperventilation on bronchial clearance in man, *J. Appl. Physiol.* 43:46–50.

Wolff, R. K., Kanapilly, G. M., DeNee, P. B., and McClellan, R. O. 1981. Deposition of 0.1 μm chain aggregate aerosols in beagle dogs, *J. Aerosol Sci.* 12:119–129.

Wolff, R. K., Kanapilly, G. M., Chang, Y. S., and McClellan, R. O. 1982. Deposition of 0.1 μm aggregate and spherical $^{67}Ga_2O_3$ particles inhaled by beagle dogs, In: *Annual Report of the Inhalation Toxicology Research Institute 1981–1982* (M. B. Snipes, T. C. Marshall, and B. S. Martinez, eds.), Lovelace Biomedical and Environmental Research Institute, Albuquerque, N.Mex., National Technical Information Service, Springfield, Va.

Wolff, R. K., Henderson, R. F., Snipes, M. B., Griffith, W. C., Mauderly, J. L., Cuddihy, R. G., and McClellan, R. O. 1987. Alterations in particle accumulation and clearance in lungs of rats chronically exposed to diesel exhaust, *Fundam. Appl. Toxicol.* 9:154–166.

Wolfsdorf, J., Swift, D. L., and Avery, M. E. 1969. Mist therapy reconsidered: an evaluation of the respiratory deposition of labelled water aerosols produced by jet and ultrasonic nebulizers, *Pediatrics* 43:799–808.

Xu, G. B., and Yu, C. P. 1985. Deposition of inhaled diesel emission particulates in different human age groups at various respiratory conditions, Presented at Annual Meeting of the American Association for Aerosol Research, Albuquerque, N.Mex. (November 1985).

Yeates, D. B., and Aspin, M. 1978. A mathematical description of the airways of the human lungs, *Respir. Physiol.* 32:91–104.

Yeates, D. B., Aspin, M., Levison, H., Jones, M. T., and Bryan, A. C. 1975. Mucociliary tracheal transport rates in man, *J. Appl. Physiol.* 39:487–495.

Yeates, D. B., Gerrity, T. R., and Garrard, C. S. 1981a. Particle deposition and clearance in the bronchial tree, *Ann. Biomed. Eng.* 9:577–592.

Yeates, D. B., Pitt, B. R., Spektor, D. M., Karron, G. A., and Albert, R. E. 1981b. Coordination of mucociliary transport in human trachea and intrapulmonary airways, *J. Appl. Physiol.* 51:1057–1064.

Yeh, H. C. (ed.) 1980. Respiratory Tract Deposition Models—Final Report, DOE Research and Development Report, LF-72, National Technical Information Service, U.S. Department of Commerce, Springfield, Va.

Yeh, H., and Schum, G. M. 1980. Models of human lung airways and their application to inhaled particle deposition, *Bull. Math. Biol.* 42:461–480.

Yeh, H. C., Barr, E. B., and Esparza, D. C. 1980. Deposition of inhaled dual aerodynamically similar aerosols in Syrian hamsters, In: *Respiratory Tract Deposition Models: Final Report to NIEHS* (H. C. Yeh, ed.), Inhalation Toxicology Research Institute, Lovelace Biomedical and Environmental Research Institute, Albuquerque, N.Mex., National Technical Information Service, Springfield, Va.

Yu, C. P., and Diu, C. K. 1982. A comparative study of aerosol deposition in different lung models, *Am. Ind. Hyg. Assoc. J.* 43:54–65.

Yu, C. P., Diu, C. K., and Soong, T. T. 1981. Statistical analysis of aerosol deposition in nose and mouth, *Am. Ind. Hyg. Assoc. J.* 42:726–733.

Yu, C. P., Hu, J. P., Leikauf, G., Spektor, D., and Lippmann, M. 1983. Mucociliary transport and particle clearance in the human tracheobronchial tree, In: *Aerosols in the Mining and Industrial Work Environments* (V. A. Marple and B. Y. H. Liu, eds.), pp. 177–184, Ann Arbor Science Publishers, Ann Arbor, Mich.

Ziegler, E. E., Edwards, B. B., Jensen, R. L., Mahaffey, K. R., and Fomon, S. J. 1978. Absorption and retention of lead by infants, *Pediatr. Res.* 12:29–34.

Biological Disposition of Vehicular Airborne Emissions: Particle-Associated Organic Constituents

JAMES D. SUN
JAMES A. BOND
ALAN R. DAHL
Lovelace Biomedical and Environmental Research Institute

Air Pollution, the Automobile, and Public Health. © 1988 by the Health Effects
Institute. National Academy Press, Washington, D.C.

Significance of Carrier Particles

Public concern has been aroused in recent decades over damage to human health from inhaling man-made particles, and there is overwhelming evidence that this concern is not misplaced (Committee on Biological Effects of Atmospheric Pollutants 1972). The concentration of man-made airborne particles is highest near industrialized areas and areas where the density of motor vehicle traffic is high. Among the major sources of man-made particles polluting the atmosphere are electrical power plants (Chrisp et al. 1978; Fisher et al. 1979); industries that burn fossil fuels (Lofroth 1978); and vehicles powered by internal combustion engines that burn either gasoline (Wang et al. 1978) or diesel fuel (Clark and Vigil 1980).

It has been suggested that the high incidence of human cancers in urban areas near industry and high-density traffic may be associated with inhaling organic pollutants, and other deleterious health effects have been attributed to organic pollutants as well (Committee on Biological Effects of Atmospheric Pollutants 1972). Most studies of the toxicologic consequences of inhaled organic pollutants reported in the literature have been performed using pure compounds. However, many organic pollutants are adsorbed on relatively inert and insoluble particles (Williams and Swarin 1979; Hanson et al. 1985). Consequently, for a complete evaluation of the toxicity and human health risks of inhaled organic pollutants, studies with pure organic compounds must be complemented with studies of organic compounds adsorbed onto particles.

What organic pollutant is carried on a particle, how much, and how it later becomes separated from the particle are affected by physical and chemical properties of both the organic pollutant and the "carrier" particle. Conditions for adsorption of organic pollutants onto particles are particularly favorable with the components emitted in automotive exhaust. The distribution of organic pollutants in the body can be quite different from the distribution of inhaled organic pollutants not attached to particles and is determined by the chemical characteristics of the particle as well as the organic pollutant. These differences are particularly evident with respect to the rate and path of clearance of inhaled organic pollutants from the respiratory tract.

A few inhalation studies have been performed with organic compounds adsorbed onto particles. These investigations have been limited to the potential carcinogenic effects of inhaled particle-associated polycyclic aromatic hydrocarbons (PAHs). However, other disease processes and chemical classes are also important.

Many of the airborne particle-associated compounds of toxicologic concern are not directly toxic but produce adverse effects when activated metabolically. Thus, the simple presence of such a potentially toxic compound within a tissue or organ does not necessarily cause a health problem; the tissue or organ in question must be capable of transforming the compound into toxic metabolites.

In addition to tissues and organs, pulmonary alveolar macrophages may play an important role in the disposition and metabolic fate of inhaled organic pollutants. Most organic compounds are relatively soluble in the lungs and can be cleared by direct absorption through the pulmonary epithelium into blood. In contrast, deep lung clearance of relatively insoluble particles, such as those that carry adsorbed organic pollutants, depends primarily on the phagocytic activity of pulmonary macrophages and their eventual translocation to the lymphatic system. Adsorbed soluble organic pollutants may remain with these insoluble particles instead of dissolving immediately and may be cleared from the lungs by the same mechanisms that clear insoluble particles. Translocation of these potentially toxic organic compounds and their metabolites to the lymphatic system may have an effect on the immune system.

This chapter first reviews certain relevant characteristics of polluting airborne particles, emphasizing the importance of particle-associated organic compounds. The biological fate of inhaled particle-associated organic compounds is largely determined by two factors: their specific site of deposition among the different tissues that exist in the respiratory tract and their distribution and metabolism by these and other tissues of the body. The disposition and clearance of particles per se is treated extensively by Schlesinger, and the disposition, absorption, and metabolism of vapors and gases are discussed by Overton and Miller and by Ultman (all in this volume). The concepts treated by those authors are essential to an understanding of biological fate of organic compounds adsorbed on inhaled particles.

This chapter proceeds with a discussion of the disposition of inhaled particle-associated organic compounds, including the effects of particle association on lung clearance and bioavailability of these organic compounds. This is followed by a discussion of the paths and mechanisms by which particle-associated inhaled organic compounds produce a biological effect. The bulk of relevant research in these areas up to now has dealt with the metabolism of chemical carcinogens and, to a lesser extent, immunologic effects of inhaled particles and particle-associated organic compounds. These sections are followed by a summary and an overview of the research needed to provide essential information for evaluating potential health risks from inhaled particle-associated organic compounds.

Characteristics of Particle-Associated Air Pollutants

Hundreds of chemical compounds have been identified in spark-ignited and diesel automotive exhausts. Behymer and Hites (1984) provide a more extensive discussion on this subject. At least four major classes of organic compounds are found associated with particles in automotive exhausts:

1. aliphatic hydrocarbons and their oxidation products (alcohols, aldehydes, carboxylic acids);
2. aromatic compounds, including heterocycles, and their oxidation products (phenols to quinones);
3. alkyl-substituted aromatic compounds

and their oxidation products (alkylphenols, alkylquinones, aromatic carboxaldehydes, and carboxylic acids); and
4. nitroaromatic compounds (nitro-polycyclic aromatic hydrocarbons or nitro-PAHs).

Other chemical classes also exist but are of lesser importance in terms of their prevalence and/or current knowledge of their toxicologic significance.

Exhaust particles from diesel and spark-ignition engines carry the same types of organic compounds (Behymer and Hites 1984). Some of the organic compounds carried by particles from other combustion processes such as conventional and fluidized-bed coal burning (Natusch 1978; Hanson et al. 1981) and cigarette smoke (Guerin 1980) are similar, but have important differences. Like automotive exhaust, cigarette smoke and effluents from burning coal contain PAHs, but they have far less of the nitro-PAHs that are the major toxic components of concern in automotive exhaust. On the other hand, cigarette smoke contains highly toxic alkaloids, nitrosamines, and aromatic amines that are present in negligible amounts in automotive exhaust. These dissimilarities arise from differences in the composition of the burning material as well as from differences in the mode of burning. For example, cigarette smoke has components arising from pyrolysis and vaporization of tobacco preceding the burning cone as well as components resulting from the oxidation of tobacco and cigarette paper (Guerin 1980).

Quantitative data about individual compounds present in air samples collected from near highway tunnels, oil- and gas-fired electrical power plants, industrial boilers, coke ovens, oil refineries, and coal tar heaters, have been published in a recent report by Daisey et al. (1986). Table 1 shows the reported concentrations of benzo[a]pyrene (BaP) in particulate matter.

Toxic chemical compounds ranging in volatility from nitrogen dioxide (Hanson et al. 1985) to PAHs such as BaP (Williams and Swarin 1979) have been identified on airborne particles from many sources, including automotive emissions. Classed

Table 1. Concentrations of Benzo[*a*]pyrene in Particulate Matter from Various Sources

Source	ppm B*a*P in Particulate Matter	References
Tunnel samples	66–500	Daisey et al. (1986)
Coal burning		
Residential, anthracite	10–20	Sanborn et al. (1983)
Residential, bituminous	240–600	Sanborn et al. (1983)
Power plants	0.0007	Bennett et al. (1979)
Oil-burning power plants	0.005	Bennett et al. (1979)
Residential wood burning		
Fireplaces	3–141	Dasch (1982)
Wood stoves	213–870,900	Truesdale and Cleland (1982) Knight et al. (1983)
Coke plant	1,400–5,800	Bjorseth et al. (1978)
Soil	0.1–2.3	Wang and Meresz (1982) Butler et al. (1984)

SOURCE: Adapted with permission from Daisey et al. 1986, and from the Air Pollution Control Association.

according to toxicity, particle-associated compounds include direct-acting mutagens such as pyrene-3,4-dicarboxylic acid anhydride (Rappaport et al. 1980), indirect carcinogens and mutagens such as B*a*P (Williams and Swarin 1979), and irritants such as acrolein and sulfur dioxide (Hanson et al. 1985), as well as others.

Among the four major classes of organic compounds found in automotive exhausts, the ones that have been studied most are the nitro-PAHs, such as nitropyrene (NP) and dinitropyrene, and the unsubstituted PAHs such as B*a*P. These compounds and others of these classes command attention because so many of them are known to be mutagens and/or carcinogens. But organic compounds carried on inhaled particles can produce other toxic effects that can also be as important as cancer.

Inhalation of Airborne Particles

Disposition of Inhaled Particle-Associated Organic Compounds

Particles carrying adsorbed organic compounds are deposited in the respiratory tract by the same mechanisms and according to the same principles as particles without adsorbed compounds. Inhaled aerosols of B*a*P or NP alone, B*a*P or NP adsorbed onto gallium oxide (Ga_2O_3) or diesel ex-

haust particles, uncoated Ga_2O_3 particles alone, or diesel exhaust particles alone are all deposited along the respiratory tract in about the same pattern (Chan et al. 1981; Sun et al. 1982, 1983, 1984; Wolff et al. 1982).

Inhaled particles are cleared rapidly from the upper respiratory tract and tracheobronchial regions by the ciliated epithelium, which sweeps deposited particles upward for eventual removal by expectoration and then ingestion. In most measurements of this clearance mechanism, inorganic particles such as Ga_2O_3 have been used (Wolff et al. 1982), but there is no reason to believe that clearance occurs in a different manner for pure organic aerosols or particles with adsorbed organic compounds. Available evidence supports this inference (Schlesinger, this volume).

Lipid and water soluble compounds and their metabolites can be absorbed also through the mucous membranes and into blood or lymph. Because mucociliary and absorption mechanisms clear material rapidly, it is difficult to distinguish between the two experimentally. Studies have been performed in which rats were treated with radiolabeled organic particles or with the same organic materials adsorbed onto inorganic particles, either by intratracheal instillation (Sun and McClellan 1984) or by inhalation (Sun et al. 1982, 1983, 1984). In these experiments, no radiolabeled particles were detected in the stomach at any time after exposure of the animals to pure organic particles, but substantial amounts of organic constituents adsorbed onto inorganic particles were found in the stomach. One can infer that pure organic particles deposited in these regions clear primarily by absorption through the respiratory tract epithelium into the blood. These examples show that organic compounds associated with particles and the same compounds in pure form can clear by different routes and thus can expose different tissues of the body.

Slowly dissolving inorganic particles do not clear from the pulmonary region as fast as from the upper respiratory tract. The main reason is that the lower respiratory tract lacks ciliated airways, and insoluble particles are cleared by the action of phagocytic macrophages moving slowly to the lymphatic system. The long-term half-times of lung clearance of inhaled Ga_2O_3 or diesel engine soot particles in rats is about 65 days (Chan et al. 1981; Griffis et al. 1982; Wolff et al. 1982).

In contrast, lipid or water soluble compounds clear quickly from the pulmonary region, principally by dissolution and absorption into the blood. Inhaled BaP (Mitchell 1982; Sun et al. 1982), NP (Sun et al. 1983; Bond et al. 1986b), aminoanthracene (Mitchell et al. 1984), phenanthridone (Dutcher and Mitchell 1983), and dibenzo[c,g]carbazole (Bond et al. 1986a) are almost completely cleared during the short-term phase of respiratory tract clearance (figure 1). Attempts to measure lung clearance times of these chemicals yielded half-times that measured as short as hours rather than weeks or months.

Particle-associated organic compounds clear from the upper respiratory tract (short-term phase) about as fast as pure organic particles (Sun et al. 1982, 1983). However, the pulmonary clearance rates (long-term phase) of particle-associated organic compounds are longer for many particle-associated compounds than for inhaled pure organic compounds. The cause is presumed to be the tenacity with which the organic material is bound to the slowly cleared "carrier" particles. These binding

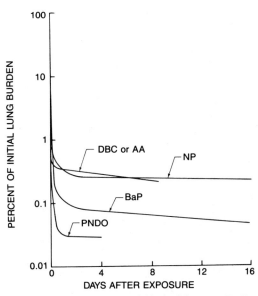

Figure 1. Lung clearance of benzo[a]pyrene (BaP), nitropyrene (NP), aminoanthracene (AA), phenanthridone (PNDO), or dibenzo[c,g]carbazole (DBC) in rats after acute exposures.

strengths very likely vary with particle type and organic material, and probably govern long-term clearance rates of particle-associated organic compounds. This increased long-term retention of particle-associated organic compounds in the pulmonary regions of the respiratory tract is believed to be of toxicologic importance.

■ **Recommendation 1.** Additional research is needed on the actual toxic effects of increased lung retention of inhaled particle-associated organic compounds.

Sun and coworkers (1982, 1983) compared the clearance rates of inhaled BaP and NP aerosols from the lungs with those of the same PAHs adsorbed onto Ga_2O_3 (figure 2). They reported that BaP adsorbed onto Ga_2O_3 cleared slower than BaP inhaled as a pure aerosol. However, NP on Ga_2O_3 and pure NP aerosols cleared at about the same rate. The data suggest that Ga_2O_3-associated organic material clears from lungs primarily by dissolution and direct absorption into the blood. In the pulmonary regions, the long-term retention of BaP or NP adsorbed onto diesel engine exhaust particles is as much as 230

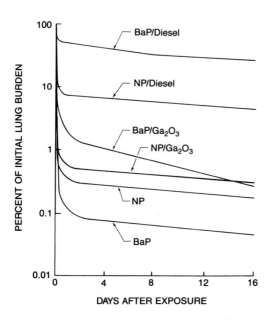

Figure 2. Lung clearance of benzo[a]pyrene (BaP) or nitropyrene (NP) adsorbed onto gallium oxide particles (Ga_2O_3) or diesel engine exhaust particles and of BaP and NP in pure form in rats after acute exposure.

times greater than that of BaP or NP alone (Sun et al. 1984; Bond et al. 1986b) (figure 2). Sun and McClellan (1984) further support the finding that particle-associated organics are cleared more slowly from the lungs than are pure organic aerosols. They operated a diesel engine on ^{14}C-radiolabeled fuel to create radiolabeled exhaust in which the majority of the ^{14}C was with the organic compounds associated on the carbonaceous core particles. Then, they intratracheally instilled this radiolabeled soot into rats and measured the clearance rate of ^{14}C from lungs. For comparison, they extracted the radiolabeled organic compounds from the soot and instilled it into the lungs of rats. These particle-free organic compounds cleared much faster than the particle-associated organic compounds.

Clearance of particle-associated organics from lungs appears to be governed by factors related to the binding properties of the particles on which the organic material is adsorbed. Chemical composition as well as physical properties of the particle surface appear to play a part, although the exact

mechanisms are unknown. For example, when Henry and Kaufman (1973) measured clearance of intratracheally instilled BaP coated on carbon, aluminum oxide, and ferric oxide particles in hamsters, they found that the BaP cleared substantially slower when carried by carbon than when coated on metal oxide particles of similar size and shape. There was little difference between the clearance rates of BaP coated on aluminum oxide and BaP coated on ferric oxide. On the other hand, irregularly shaped particles or particles having a high degree of porosity have a greater surface area per unit mass than smooth spherical particles. An amount of material that would make a loosely bound layer several molecules thick on a smooth particle is carried in a more tightly bound monomolecular layer on a rough or porous particle (Gregg and Sing 1982).

After these organic compounds and their metabolites are cleared from the respiratory tract, they are widely distributed to many tissues in the body (Sun et al. 1982, 1983, 1984; Bond et al. 1986a,b). Bond and coworkers (1986b) used ^{14}C-NP associated with diesel exhaust particles to study what form this organic material has after reaching nonrespiratory tract tissues. Within 1 hr after exposure, a large proportion of the ^{14}C had cleared from the lungs to other tissues and more than 90 percent of the ^{14}C in these tissues was associated with NP metabolites.

Bioavailability of Particle-Associated Organic Compounds

Vostal (1983) has postulated that particle-associated compounds must be eluted off the particle and made available for various cellular metabolic processes before a toxic response can result. Other researchers (Brooks et al. 1980; King et al. 1981) have reported that the mutagenic components associated with diesel engine exhaust particles are removed by various physiological fluids, tissue homogenates, and serum. They found that, in the presence of such biological fluids, the mutagenic activity of these organic compounds was reduced, suggesting that organic constituents in the media are bound to proteins or metabolized. Similarly, King and coworkers (1983) found that when pulmonary alveolar macrophages were incubated with diesel engine exhaust particles, amounts of organic compounds and mutagenic activity decreased measurably from the amount originally associated with these particles, suggesting that organics were removed from phagocytized particles. Collectively, these studies suggest that particle-associated organics become "bioavailable" to respiratory tract cells, allowing metabolic processes to occur.

Increased toxicity to lung tissue caused by the long-term retention of particle-associated organics has been illustrated by Saffiotti et al. (1964), Creasia and Nettesheim (1974), and Henry et al. (1975). Their studies showed that BaP is retained in the lungs longer and results in a higher incidence of lung carcinomas when intratracheally instilled on iron oxide particles than when instilled alone. But how and why prolonged retention of particle-associated organics in lungs increases the toxic potential of those compounds has not been thoroughly investigated. One way that particle association might make organic compounds more toxic is by facilitating their uptake into lipid bilayers and microsomes, primary sites for cellular metabolism of chemicals. Lakowitz and coworkers (1980) reported that BaP adsorbed on particles was transported into model membranes composed of phosphatidylcholine dipalmitoyl faster than suspensions of BaP microcrystals. The degree of enhanced uptake varied with particle type. Those researchers also found that four types of asbestos particles facilitated transport into membranes to a greater degree than particulate hematite, silica, titanium dioxide, porous glass, or talc. However, transport was not enhanced when BaP and particles were added to their assay system as a simple mixture. Bevan and coworkers (1981) reported that transport was enhanced for particle-associated dibenzoanthracene, benzoperylene, and 3-methylcholanthrene, but not for dibenzocarbazole adsorbed onto particles.

Automobile exhaust particles have not been used to characterize the transport of particle-associated organics. However, Bevan and Worell (1985) used BaP associated with carbon black particles, which have physical and chemical characteristics similar to those of exhaust particles, and observed enhanced transport of this PAH into phospholipid vesicles.

Bevan and Manger (1985) measured the rates of uptake into rat liver microsomes of BaP adsorbed onto asbestos and iron oxide particles and, using the *Salmonella typhimurium* microsome assay, they measured the microsomal metabolism of BaP and the mutagenicity of the metabolites. They correlated enhanced uptake of particle-associated BaP into microsomes with higher rates of production of mutagenic metabolites.

Formation of adducts to DNA is an important marker of effective dose to target tissues of organic compounds such as those found in vehicular exhaust. The interaction of reactive organic chemicals or their metabolites with DNA is important in the overall carcinogenic response of tissues to inhaled chemicals. Weinstein and coworkers (1984, 1985) have discussed some of the potential mechanisms involved in the carcinogenic response of the respiratory tract. For example, rats chronically exposed to high levels of diesel engine exhaust have higher levels of DNA adducts in their lungs than rats exposed only to air (Wong et al. 1986). Taken as a whole, these data suggest that inhaled diesel engine exhaust may initiate a carcinogenic response. Indeed, Mauderly and coworkers (1987) observed an increased incidence of lung tumors in rats exposed to diesel exhaust. These results also suggest that measurement of DNA adduct formation may be an indicator for estimating exposure dose.

We are in the early stages of understanding how particle association affects the toxicity of inhaled organic compounds. A few inhalation studies suggest that the particle-associated organic compounds are more toxic than the same organic compounds in pure aerosol form, probably because they are retained in the lungs longer. However, in all of these studies the exposure was brief and at relatively high concentrations. Prolonged exposures at lower concentrations would be more relevant. Under such conditions, equilibrium concentrations of the inhaled organic compounds and their metabolites in various organs and tissues might be more important than their clearance rates from the lungs.

To better evaluate the toxic potential of particle association of inhaled organic compounds, it will be necessary to understand how these insoluble particles influence the biological fate of these organics. The specific site at which these inhaled organics are initially deposited along the respiratory tract is probably a crucial factor that determines their biological fate. Particle association may influence the deposition characteristics of these inhaled organic compounds to a high degree in terms of where the organics are carried along the respiratory tract. Related to this is the rate at which particle-associated organic compounds are dissolved off these particles after deposition and become available for clearance, metabolism, and toxic action. It is likely that, if these organic compounds never desorb from their "carrier" particles, they will be biologically inert.

■ **Recommendation 2.**   The effect of carrier particles on the delivery of adsorbed compounds to specific regions of the respiratory tract needs to be determined.

■ **Recommendation 3.**   Desorption rates of adsorbed compounds from inhaled particles should be quantified.

Toxicity of Inhaled Organic Compounds

Varieties of Toxic Responses

A wide variety of toxic effects may result from the inhalation of chemicals. Carbon monoxide (CO), for example, avidly binds to hemoglobin, and its inhalation can lead to asphyxiation. The irritant gas sulfur dioxide (SO_2) causes bronchoconstriction; and the oxidant gases nitrogen dioxide (NO_2) and ozone (O_3), when inhaled at sufficiently high concentrations or for a prolonged period, can cause chronic lung disease. Inhalation of other substances, such as cocaine or the active ingredient of marijuana, tetrahydrocannabinol, on the other hand, can cause toxic effects that are more systemic in nature. This section focuses on those toxic responses of great public concern that may result from inhalation of organic compounds associated with automotive exhaust.

People have coexisted with automotive exhaust for more than seven decades. During this time, most concerns over its acute toxicity have focused on the accidental or intentional deaths caused by inhaling high concentrations of CO that occur when a vehicle is operated in closed quarters. But as the number of automobiles on the road increases, especially in densely populated areas, the concern over the long-term effects of chronic exposures to automotive exhaust has grown.

A significant discovery about the organic compounds emitted in automotive exhaust is that many are mutagenic and/or carcinogenic; thus, carcinogenicity is one focus of this section; the other focus is immunotoxicologic effects. As discussed earlier in this chapter, many of the organic compounds in automotive exhaust are not released in their pure form, but are adsorbed onto the insoluble, carbonaceous soot of automotive exhaust. Clearance of insoluble particles from lungs depends on the phagocytic activity of pulmonary alveolar macrophages and their eventual translocation to the lymphatic system (Schlesinger, this volume), where they may have toxic effects on the immune system. The remainder of this chapter will therefore address the inhalation of particle-associated organic compounds as it pertains to chemical carcinogenesis and possible effects on the immune system.

Metabolism of Chemical Carcinogens

Most carcinogens associated with automotive exhaust and other air pollutants are procarcinogens that must be transformed to reactive metabolites by metabolic activation before they can produce a carcinogenic effect. Chemical metabolism may proceed through several stages to produce proximate carcinogens and finally ultimate carcinogens. In many instances, proximate and ultimate carcinogens are metabolized by enzymes to inactive metabolites. But the reactive metabolites interact with cells in many ways, most notably by binding covalently with macromolecules such as RNA, DNA, and proteins. The balance between the rate of formation of reactive metabolites and the rate of formation of inactive metabolites plays a crucial role in determining the levels of reactive metabolites. Usually, ultimate carcinogens comprise only a small fraction of the total metabolic products of a chemical. A number of reviews have been published relating to the metabolism or metabolic activation of chemical carcinogens (Miller and Miller 1966, 1981; Dipple et al. 1985) and other toxic, but noncarcinogenic, chemicals that may be associated with particles (Magee 1974; Nelson et al. 1977; Boyd et al. 1980).

Enzymes responsible for most of the metabolic conversions of procarcinogens are part of the mixed-function oxygenase system (Gelboin et al. 1970, 1972). This enzyme complex responsible for biological activation of PAHs is found in most mammalian tissues. It is NADPH-dependent and catalyzes the incorporation of molecular oxygen into substrate molecules. The enzyme complex has an absolute requirement for three components: a hemoprotein referred to as cytochrome P-450; a flavoprotein referred to as NADPH cytochrome *c* (or P-450) reductase; and a phospholipid, typically phosphatidylcholine. Lu (1976) has reviewed the components and properties of this enzyme system; Guengerich and MacDonald (1984) have reviewed the mechanisms by which chemicals are metabolized by cytochrome P-450; and Hodgson and coworkers (1980) and Testa and Jenner (1976) have summarized the different enzyme systems responsible for biotransformation of procarcinogens.

Some classes of procarcinogens adsorbed on airborne particles include PAHs, nitro-PAHs, aromatic amines, nitrosamines, and *N*- or *S*-containing heterocycles. In cells, some PAHs are enzymatically converted to epoxide intermediates which can spontaneously rearrange to phenols, be converted enzymatically to *trans*-dihydrodiols via epoxide hydrolase, be reduced back to the parent compound via epoxide reductase, be conjugated enzymatically or nonenzymatically with glutathione via glutathione *S*-epoxide transferase, or react directly with cellular macromolecules. The *trans*-dihydrodiols may be further oxidized to the diol epoxides which also react with cellular macromolecules (see figure 3 for some of the known pathways for metabolism of BaP, a frequently studied PAH). Nitro-PAHs are metabolized by "nitro-reductases" to *N*-hydroxy intermediates that can be metabolized to *N*-sulfated acetamides (as in the case for *N*-hydroxy-acetyl amino fluorene) by the acetyl transferases and sulfo transferases (Miller and Miller 1981) (see figure 4 for some known pathways for metabolism of NP, a typical nitro-PAH). (For a detailed discussion of the metabolism of PAHs and nitro-PAHs see Hecht, this volume.) Aromatic amines are also metabolized to *N*-hydroxy intermediates, possibly by the flavin-containing monooxygenases and the mixed-function monooxygenases. Nitrosamines are metabolized by cytochrome P-450 monooxygenases to unstable hydroxy (α) carbon compounds that decompose to the highly electrophilic *N*-monoalkylnitrosamines (Farrelly and Stewart 1982) (see figure 4 for some known pathways for aminopyrene metabolism, a typical aromatic amine).

Biotransformation of carcinogens occurs in the nose (Dahl et al. 1985), lung, skin, intestine, placenta, kidney, testes, adrenals, and liver (Jenner and Testa 1980). The liver is believed to be responsible for the major quantitative contribution to the metabolism of xenobiotics (organic chemicals foreign to the body), with a few exceptions noted below. However, relative levels of activating and detoxifying enzymes vary in different tissues, so that the amounts of ultimate carcinogenic metabolites formed in different tissues is not necessarily proportional to the amount of initial carcinogen metabolized there. This may explain, in part, the localization of certain chemically induced tumors outside the liver.

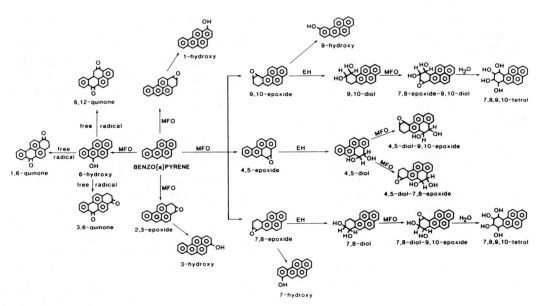

Figure 3. Pathways of benzo[*a*]pyrene metabolism, where MFO = the mixed-function oxygenase enzyme system, and EH = epoxide hydrolase. Hydroxylated metabolites can undergo conjugation reactions that result in the formation of glucuronide, sulfate, or mercapturic acid derivatives of the metabolite.

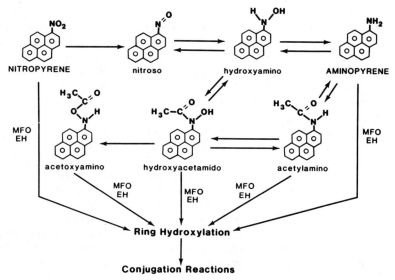

Figure 4. Pathways of nitropyrene and aminopyrene metabolism, where MFO = the mixed-function oxygenase enzyme system, and EH = epoxide hydrolase. Conjugation reactions result in the formation of glucuronide, sulfate, or mercapturic derivatives of the metabolite.

Respiratory Tract Metabolism of Particle-Associated Carcinogens.

With few exceptions, the quantitative formation of metabolites by respiratory tract tissues is less than that by the liver. However, since the respiratory tract is directly exposed to particle-associated carcinogens, their activation by respiratory tract tissues may still have an important role in the pathogenesis of carcinogen-induced lesions in these tissues.

Carcinogen metabolism has been studied in different preparations of respiratory tract tissues (including nasal tissue), cultured trachea and bronchus, the perfused lung, isolated lung cells, and pulmonary alveolar macrophages. These studies focused on pure compounds, in particular PAHs, rather than particle-associated carcinogens. References to the liver, a major organ for metabolism of endogenous as well as exogenous compounds, are introduced for comparison with various respiratory tract tissues throughout the following review.

Nasal Tissue. The metabolism of xenobiotics in nasal tissue has been generally neglected, although some reports have been published (Dahl et al. 1982, 1987; Bond 1983a,b; Dahl and Hadley 1983; Hadley and Dahl 1983; McNulty et al. 1983; Brittebo and Ahlman 1984; Casanova-Schmitz et al. 1984; Dahl 1986). Nasal tissue metabolism is important because many environmental pollutants contain known carcinogens adsorbed onto particles of sizes that deposit in the nasopharyngeal region of the respiratory tract (Task Group on Lung Dynamics 1966; Natusch and Wallace 1974).

It is believed that nasal tissue contains the full complement of enzymes necessary for the metabolism of xenobiotics, but further work is needed to substantiate this belief. Xenobiotic-metabolizing enzymes known to be in the nasal cavity include cytochrome P-450 and flavin-containing monooxygenases, aldehyde dehydrogenases, epoxide hydrolases, glutathione transferases, UDP-glucuronyl transferases, and carboxyl esterases (for a review, see Dahl 1986, 1987). In general, enzymes in the nasal tissue have turnover rates comparable to those in liver.

The nasal tissue metabolism of only two PAHs—BaP and NP—has been thoroughly investigated (Bond 1983a,b). Those in vitro studies using nasal tissue homogenates indicated that nasal tissue can metabolize these compounds to phenols, quinones, dihydrodiols, and tetrols. In vivo studies in hamsters have also shown that BaP is metabolized by nasal tissue (Dahl et al. 1985). The profile of BaP metabolites produced in hamster noses is nearly identical to that measured using nasal tissue homogenates, suggesting that in vitro models for nasal tissue metabolism of PAHs may predict the metabolic profile in the intact animal.

Further research on nasal tissue metabolism of chemicals is necessary to adequately characterize nasal tissue enzymes. It is important to determine the capacity of the nose to activate and inactivate the various chemicals associated with pollutants. Kinetic studies are also required to provide information on concentrations of pollutants that may "saturate" nasal tissue enzymes. These studies may provide clues about the mechanisms involved in nasal tissue carcinogenesis.

There are no reports in the literature of nasal tissue metabolism of particle-associated organics. This deficiency must be addressed in future studies if we are to understand the role of nasal tissue in the overall biological fate of particle-associated carcinogens. Other research is needed to determine where the reactive metabolites and procarcinogens produced in nasal tissue are translocated and at what rates. Dahl and coworkers (1985) have shown that nasal tissue metabolites can be swallowed, exposing the alimentary tract to potentially reactive metabolites. Translocation of these reactive metabolites to other tissues requires further study.

Trachea and Bronchi. A considerable amount of research has been done on bronchial metabolism of PAHs (for a review, see Autrup 1982) and BaP in particular (Harris et al. 1974, 1976, 1977; Jeffrey et al. 1977; Yang et al. 1977; Daniel et al. 1983). The data indicate that bronchial tissue is capable of metabolizing BaP to several compounds including phenols, dihydrodiols, and quinones. The profiles of organic soluble metabolites of BaP in the respiratory tissues from humans and experimental animals are similar. Evidence also suggests that bronchial tissue contains phase II enzymes such as UDP-glucuronyl transferase, aryl sulfatase, and glutathione transferase. These enzymes are responsible for the detoxification and elimination of reactive, toxic metabolites. Additional research indicates that bronchial tissue metabolizes PAHs to compounds that are capable of binding to DNA. The BaP-DNA adducts detected in human bronchial tissue are similar to those found in cultured tissue from experimental animals. Human bronchial epithelial cells also activate 7,12-dimethylbenz[a]anthracene, 3-methylcholanthrene, and dibenz[a,h]anthracene to metabolites that covalently bind to DNA. In general, there is a positive correlation between the level of covalent binding of these PAHs to bronchial DNA and their carcinogenic potency in experimental animals (Huberman and Sachs 1977).

Metabolism of PAHs in the trachea has not been studied as extensively as that in bronchi. Cohen and Moore (1976) observed that ethyl acetate–soluble metabolites from the culture media of rodent tracheal and bronchial cultures were quantitatively similar. Other investigators (Moore and Cohen 1978; Cohen et al. 1979) have shown that BaP is metabolized to oxidative and conjugated metabolites in cultured rodent trachea. Kaufman and coworkers (1973) demonstrated that these metabolites were covalently bound to tracheal DNA. Data obtained from these studies of rodent trachea showed BaP metabolite profiles similar to those found in short-term organ cultures of human bronchus.

Research on metabolism of chemicals in the trachea and bronchi is needed in the subject areas described above for nasal tissue. There is no information available on the capacity of these tissues to metabolize particle-associated carcinogens. Whether these particles can even penetrate the different types of cells in these tissues remains to be determined and should be of high priority in terms of research needs.

Pulmonary Tissues. BaP and other PAHs are readily metabolized by lung homogenates (Hundley and Freudenthal 1977;

Prough et al. 1977, 1979), by perfused lungs (Ball et al. 1979; Warshawsky et al. 1980; Smith and Bend 1981; Bond and Mauderly 1984; Bond et al. 1984, 1985), lung slices (Stoner et al. 1978), cultured type II alveolar cells (Devereux and Fouts 1981; Jones et al. 1983; Sivarajah et al. 1983), and Clara cells (Boyd 1980; Jones et al. 1983) (see also reviews by Hook and Bend 1976; Philpot et al. 1977; Boyd 1980; Jenner and Testa 1980; Philpot and Wolf 1981). However, metabolism of PAH in terms of amount per unit of incubation time is slower in these lung preparations than in the liver. The spectrum of these lung systems produced different spectra of metabolites, but in all cases lung cells or intact lungs metabolized PAHs to intermediates capable of covalently binding to DNA. In addition to the in vitro studies, several studies have demonstrated metabolism of PAHs—BaP as well as NP—in lungs following inhalation (Mitchell 1983; Sun et al. 1984; Bond et al. 1986b).

Warshawsky and coworkers (1978, 1983, 1984) and Schoeny and Warshawsky (1983) have shown that the presence of particulate matter in the perfused lung can enhance the metabolic activation of BaP. In particular, they found that the levels of dihydrodiols were higher in lungs from animals preexposed in vivo to particles than in animals that had not received particle preloading. They further showed that when perfused lungs were exposed to BaP and particles, extracts of lung macrophages were consistently mutagenic. Törnquist and coworkers (1985) demonstrated that the presence of urban air particles increases the residence time of BaP in lungs and alters the BaP metabolic profile in a way that enhances binding to DNA. However, in contrast, Bevan and Manger (1985) demonstrated that BaP hydroxylase activity in rat liver microsomes was slightly inhibited when BaP was adsorbed onto particulate matter.

Studies in intact animals have shown that following exposure to particle-associated PAHs (either BaP or NP), metabolites of these chemicals can be found in the lungs (Sun et al. 1984; Bond et al. 1986b). In animals exposed to BaP on particles, metabolites (phenols as well as quinones) were measured in lungs as long as 20 days after the end of a 1-hr exposure to BaP. These studies, however, did not determine if these metabolites were formed by lung cells per se or by alveolar macrophages.

■ **Recommendation 4.** There are large differences in rates of metabolism at different sites in the respiratory tract. Although these sites contain many of the enzymes necessary for overall metabolism of "pure" PAHs, additional research is necessary to more fully characterize respiratory tract metabolism of inhaled particle-associated xenobiotics.

A key issue is whether organic compounds associated with particulate matter must first be "desorbed" prior to being acted upon by tissue enzymes. Studies using laboratory animals indicate that PAHs associated with particulate matter are retained longer in lungs than "pure" PAHs. The increased lung retention of the particle-associated PAHs might result in a larger or more protracted "dose" to the tissue than inhalation of pure PAHs. But if organic compounds must first be "desorbed" from the particles, then doses depend initially on the rates of desorption in the different portions of the respiratory tract. If desorption rates are relatively slow, then clearance of the particles up the mucociliary escalator and to the lymph nodes may remove the particle-associated organics prior to desorption. Such clearance of the particles would decrease the effective dose of the compounds to respiratory tract tissues, but could increase the dose to other tissues.

■ **Recommendation 5.** Future research endeavors should determine whether desorption of organic compounds from particles is required before metabolic activation can occur, and how desorption characteristics affect the metabolic fate, rate of metabolite formation, and the distribution of particle-associated organic compounds.

Pulmonary Alveolar Macrophages. The role of pulmonary alveolar macrophages in metabolizing inhaled organic compounds has not been studied as extensively as other cells, but its importance should not be overlooked. Some particles deposited in lungs are phagocytized by macrophages, some macrophages with engulfed particles remain in the lung for extended periods of time (see Schlesinger, this volume), and slow release of organic compounds and their metabolites from these macrophages subjects surrounding tissues to extended exposure to potentially toxic or carcinogenic reactive metabolites.

PAHs have been studied in human and laboratory animal pulmonary alveolar macrophages (McLemore et al. 1977; Autrup et al. 1978; Palmer et al. 1978; Marshall et al. 1979; Bond et al. 1984) as they have in other tissues. BaP has been used as a model compound in these studies. Although the amount of PAH metabolized per unit of incubation time (metabolic rate) is lower in pulmonary alveolar macrophages than in other portions of the respiratory tract, macrophages nevertheless do activate BaP to reactive intermediates that bind to DNA (Harris et al. 1978; Romert and Jenssen 1983). These metabolites are released into the surrounding medium, and it has been demonstrated that the metabolites formed by macrophages are capable of being taken up by surrounding respiratory tract tissue (Palmer et al. 1978). Studies have also shown that macrophages obtained from smokers have greater capacity to metabolize xenobiotics than macrophages from nonsmokers (Cantrell et al. 1973).

Limited data from a few studies of the capacity of pulmonary alveolar macrophages to metabolize particle-associated BaP or dimethylbenz[a]anthracene (DMBA) (Tomingas et al. 1971; Autrup et al. 1979; Bond et al. 1984; Palmer and Creasia 1984; Greife et al. 1986) suggest that when BaP or DMBA is coated on a particle (for example, diesel exhaust or urban air particles), pulmonary alveolar macrophages can engulf the particle and metabolize it to several compounds including the proximate carcinogen. However, these studies do not indicate whether macrophages can metabolize particle-bound BaP or DMBA, since after the particle has been engulfed, BaP or DMBA may have been removed from the particle before metabolism occurred.

Although macrophages can metabolize PAHs associated with particles, it remains to be determined whether site-specific metabolism of PAHs is the major contributing factor in the overall carcinogenic response or whether the metabolites released from macrophages play any role in contributing to the total "dose" to tissue. Evidence from in vitro and in vivo studies indicates that different portions of the respiratory tract can metabolize PAHs to compounds that bind to "critical" macromolecules. However, whether metabolites produced in macrophages and released in different portions of the respiratory tract bind to critical macromolecules of other tissues is unknown.

■ **Recommendation 6.** The importance of macrophage metabolism in the activation of particle-associated organics and the contribution of metabolic products to tissue dose need to be determined.

Effects on the Immune System

Organic compounds associated with inhaled particles affect the immune system in at least two important ways: they affect the translocation of antigen from the lung to the lung-associated lymph nodes, and they directly affect lymphoid cells in these tissues.

Table 2. Decrease in Phagocytizing Pulmonary Alveolar Macrophages as a Result of In Vitro Exposures to Particle Extracts

Particle Source	IC$_{50}$[a] (μg/mL)
Rural road	471
City street	270
City roof	124
Auto tunnel	143
Gasoline engine	5

[a] IC$_{50}$ is the concentration of extract needed to decrease the number of phagocytizing pulmonary alveolar macrophages to 50 percent.
SOURCE: Adapted with permission from Romert et al. 1985, and from Pergamon Journals, Ltd.

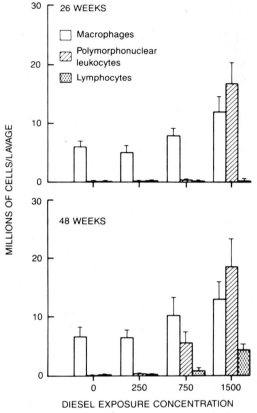

Figure 5. Cell populations in lavage fluid from rats exposed to different concentrations of diesel exhaust particles for 26 weeks or 48 weeks. The data are means ± standard deviations (n = 6). (Adapted with permission from Strom 1984.)

Translocation Mechanisms. Particles and antigen deposited in the broncho-alveolar region of the lungs can be transported to the lung-associated lymph nodes where immune responses are produced. Macrophages probably participate in this transport (Harmsen et al. 1985); therefore, any toxicant associated with the particle that kills macrophages or inhibits phagocytosis affects the capacity of the lung to clear antigen from the lung to lung-associated lymph nodes.

Toxic effects on pulmonary alveolar macrophages have been observed from organic chemicals associated with particles from a variety of sources including automotive emissions (Romert et al. 1983, 1985) (see table 2). Undiluted automotive exhaust is highly toxic to pulmonary alveolar macrophages in vitro. However, Strom (1984) found that, despite this toxicity, the effect on overall phagocytosis in vivo may be small because the number of pulmonary alveolar macrophages in rats increased in response to exposure to diesel exhaust. After exposure of rats to high concentrations of diesel exhaust, phagocytic neutrophils were recruited into the lung, thereby potentially further increasing the overall phagocytic capacity (figure 5). This capacity to recruit polymorphonuclear (PMN) leukocytes (that is, neutrophils, eosinophils, and basophils) appears to be a common phenomenon after exposure to particles. It has been observed, for instance, after exposure to cigarette smoke particles, but not to the vapors (Kilburn and McKenzie 1975). The contribution of particle-associated organic compounds to the recruitment of these phagocytic cells, however, is not known. Also, if a portion of the organics originally adsorbed on the particles are transported with them to the lymph nodes, then toxic metabolites may be formed from them in the lymphatic system.

■ **Recommendation 7.** Research should be pursued on the effects of particle-associated materials on lung phagocyte recruitment and activity, clearance rates to the lymph nodes, and translocation of particle-associated organic compounds to the lymphatic system.

Lymphoid Cells. Cells responsible for immunity to antigens deposited in the lung are affected by some but not all of the compounds that often are associated with particles. For example, BaP instilled in hamster lungs suppressed the splenic systemic humoral immune response, whereas treatment with benzo[e]pyrene had no effect on immune response (Zwilling 1977). The effect of particle association of BaP, however, was not examined in that study.

BaP instilled in the lungs can also alter the induction of immunity in the lung-associated lymph nodes in response to a particulate antigen (Schnizlein et al. 1982). For example, the numbers of lung-associated lymph node IgM as well as IgG anti-

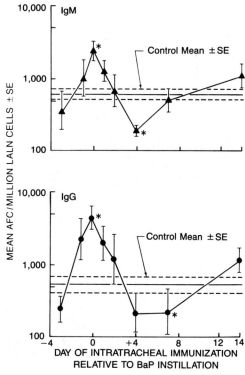

Figure 6. The number of IgM and IgG antisheep red blood cell (SRBC) antibody-forming cells (AFC) per million lung-associated lymph node (LALN) cells seven days after intratracheal immunization of rats with 10^6 SRBC at various times relative to the instillation of 1 mg BaP into their lungs. (*) Denotes significance ($p < 0.05$) as determined by Student's *t* test. (Adapted with permission from Schnizlein et al. 1982.)

sheep red blood cell antibody-forming cells increased in rat lungs instilled with BaP if the rats were immunized at the time of BaP instillation, but decreased if the rats were immunized four days after BaP exposure (figure 6). This relationship raises interesting questions about BaP (or compounds with similar effects) adsorbed onto particles. BaP adsorbed onto Ga_2O_3 (Sun et al. 1982) or diesel engine exhaust particles (Sun et al. 1984) cleared from the lung more slowly than pure BaP. A more protracted dose of BaP to the lymph nodes might thus result from particle-associated BaP than from pure BaP. At present, no information is available that would predict whether the immune response to sheep red blood cells (SRBC) instilled four days after instillation of particle-associated BaP

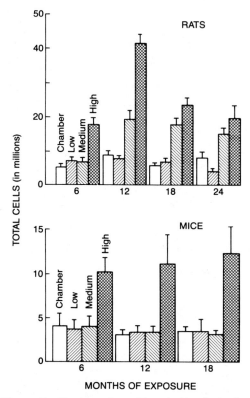

Figure 7. Total number of lymphoid cells in the lung-associated lymph nodes from rats (top panel) or mice (bottom panel) exposed to different levels of diesel engine exhaust for 6, 12, 18, or 24 months. Data are presented as geometric means ± 1 SE. (Adapted with permission from Bice et al. 1985.)

would be like the response seen either when coadministered with pure compound or administered after a four-day delay.

Automotive exhaust also affects the lung-associated lymph nodes. The particulate fraction of the exhaust, along with associated organic compounds, may contribute to the effects. In rats and mice exposed to diesel exhaust at 0.35, 3.5, or 7.0 mg/m³ of soot particles for 7 hr/day, 5 days/week, the number of cells present in lung-associated lymph nodes increased significantly (figure 7) (Bice et al. 1985). After immunization with SRBC, rats and mice exposed at the 7.0 mg/m³ level had elevated numbers of IgM antibody-forming cells.

Combustion aerosols from nonautomotive sources that have been studied with regard to immune responses include coal fly ash and cigarette smoke. Fly ash had no effect on the response to antigenic challenge in mice immunized with live bacillus Calmette-Guerin organisms, or on the ability of pulmonary alveolar macrophages to function in T-lymphocyte mutagenesis assays (Zarkower et al. 1982). These negative findings might have resulted from the low concentrations of toxic organic compounds associated with coal fly ash (Hanson et al. 1981). On the other hand, cigarette smoke, which has relatively high concentrations of particle-associated organic toxicants (Guerin 1980) has significant effects on immune responses. For example, smokers had increased numbers of germinal centers in their carinal lymph nodes, suggesting either that increased translocation of antigen from the lungs occurred in smokers or that the lymphoid cells were directly stimulated by tobacco smoke (Soutar 1977).

The response of the immune system to inhaled particles varies with the toxicity of the particles, their source, and other factors such as the timing between exposure and the measurement of effects. At present, there are not enough data to explain the observed effects rationally. Knowledge of the chemical composition of inhaled aerosols, their rates of release from particles, and the metabolic fate of specific compounds are of key importance. Studies with lymph node tissue show that rat lymph nodes have low levels of the constitutive enzymes responsible for aryl hydrocarbon hydroxylase (AHH) activities (Ciaccio and De Vera 1975). Intraperitoneal injection of BaP, however, increased the lymph node AHH activity by a factor of 11.

The presence of AHH activities, and possibly of other enzyme activities, in lymph nodes shows that organic compounds on particles that clear to the lymph nodes may be metabolized in that tissue. As a result, electrophilic metabolites formed in the lymph nodes may produce toxic effects at that site.

■ **Recommendation 8.** The translocation of particle-associated organic compounds and their eventual metabolism to reactive metabolites in lymph nodes requires investigation.

Summary

All portions of the respiratory tract contain enzymes capable of metabolizing xenobiotics (foreign organic chemicals) including PAHs. With few exceptions, metabolism of xenobiotics by these enzymes has been studied in vitro in tissues of various laboratory animals and humans. Significant progress has been made in our understanding of the comparative metabolism of the different anatomic portions of the respiratory tract. Although people are exposed to many different chemicals associated with particles, most studies have involved only pure chemicals, often BaP.

Inhaled particle-associated organic compounds are deposited in all areas of the respiratory tract. Association with particles can affect the deposition sites and retention times of such organic materials and creates opportunities for metabolic activation that would be different for the pure organic material. The few studies on the effects of particle association on clearance of organic compounds from lungs indicate that the rates of removal are much slower than those for the pure compounds and that the clearance mechanisms may be different. Thus, particle-associated organic compounds may pass through different paths, or at different rates or concentrations, or to different tissues, or be exposed to different metabolic environments, than pure compounds.

Particle-associated organics are transported into microsomal membranes more readily than are pure compounds, and are susceptible to metabolic activation for a longer time. This increased activation may make the lungs a target organ for organics associated with particles that in pure form would be cleared rapidly and with less toxic effect.

Inhaled particle-associated organic compounds have two kinds of effects on the immune system. First, they can impair phagocytosis by specific types of phagocytes, altering the clearance of inhaled antigens from the lung to lymphoid tissue. For example, extracts from particles generated by a gasoline engine decreased the number of phagocytizing macrophages in an in vivo assay (Romert et al. 1985). Second, they can affect lymphoid cells in the lung-associated lymph nodes. Effects on the number of antibody-forming cells and the quantity of antibody produced in the lung-associated lymph nodes has been demonstrated, for example, in rats after inhalation of diesel exhaust at high (7 mg particles/m^3) concentrations (Bice et al. 1985). Compounds frequently associated with particles, such as BaP, have similar effects on lymph node cells. Enzymes that metabolize BaP and similar compounds to reactive (electrophilic) compounds are known to occur in lymph nodes and to be inducible by BaP. This is evidence that particle-associated compounds can have effects on the immune system, but more research is required to establish the toxicologic importance of such effects.

Summary of Research Recommendations

HIGH PRIORITY

Because of the limited amount of information available concerning the toxicology of inhaled particle-associated organic compounds, further research is needed. Outlined below are the most critical areas of study that would provide the information needed for better estimations of the potential human health risks of inhaled atmospheric pollutants.

Recommendation 2   Little information is available on the specific regions and/or cell types in the respiratory tract where inhaled organic compounds adsorbed onto particles become bioavailable. These particles may act as "carriers" of these compounds to sites or cell types that would be different from those to which inhaled pure forms of the same organic materials would be deposited. Such studies would determine if these carrier particles affect the delivery of adsorbed organic compounds to "critical" cell types.

Recommendations 3, 5

Information is needed to determine the rate of desorption of compounds from particles after deposition in the respiratory tract. Such information would provide insight into the rate of delivery of these organic constituents to respiratory tract tissues as well as to other tissues in the body. Studies in this area will determine if particle association of potentially toxic organic compounds causes these inhaled chemicals to be delivered to "critical" cell types at an exposure rate that would ultimately result in a deleterious effect.

MODERATE PRIORITY

Studies using new methodologies that better determine the degree of toxicity and the mechanisms of action of inhaled particle-associated organic compounds need to be developed and conducted.

Recommendation 1

Particle association of organic compounds appears to increase the long-term retention of those compounds in lungs. At present, it is usually accepted that the increased retention of inhaled toxic/carcinogenic compounds in lungs will increase the deleterious effects of those compounds. This may occur by maintaining the concentration of the organic compound(s) in lungs above some "critical" dose for extended periods. Alternatively, longer lung retention of particle-associated organic compounds may actually make those compounds less toxic/carcinogenic. In this latter case, the slow release of compounds from particles may result in keeping the concentration below the "critical" dose, which may be a level where metabolic detoxification or repair can effectively occur. The toxicologic effects of particle-associated organic compounds and the mechanism of those effects need to be tested in long-term inhalation studies using organic compound(s) that are retained substantially longer in the lungs when adsorbed onto particles than when inhaled in pure form.

Recommendation 4

Little is known about the capacity of respiratory tract tissues to metabolize organic compounds that are adsorbed on particles. This is particularly true for nasal tissue and the tracheobronchial regions of the respiratory tract. The "effective" dose of a carcinogen is thought to be determined by the amount of reactive metabolite bound to critical cellular macromolecules such as DNA. The key issue that remains to be addressed is whether particle-associated carcinogens can penetrate cells and be accessible to various metabolizing enzymes. Related to this is whether metabolism of particle-associated carcinogens occurs in these regions of the respiratory tract, and, if so, whether reactive metabolites can then be translocated to other tissues and at what rates. These are important areas of research that need to be addressed for the different anatomic regions and cell types of the respiratory tract.

LOW PRIORITY

Studies that investigate secondary or partial effects of particle-associated organic compounds that deposit in the respiratory tract are needed.

Recommendation 6	In vitro data indicate that macrophages can metabolize pure as well as particle-associated organic compounds. However, it is not clear whether macrophages contribute to the overall "dose" to tissue in terms of formation of reactive chemical species that can interact with portions of the respiratory tract or other tissues. Since large numbers of macrophages are recruited following a "particle" insult, the importance of macrophage metabolism in the activation of particle-associated toxins/carcinogens needs to be determined.
Recommendation 7	As part of the lung's normal defense mechanisms, phagocytizing cells are recruited to anatomic areas where inhaled particles have been deposited. An area needing further study is the effect that particle-associated organic compounds may have on this recruitment process. Related to this is the need to further investigate the potential toxicity that these particle-associated organic compounds may have on phagocytizing cells and the role these cells may play in the overall clearance of these organic compounds from the different regions of the respiratory tract.
Recommendation 8	Following phagocytosis, particles and associated organic compounds are translocated to lung-associated lymphoid tissue. This translocation may allow for the metabolism of these organic compounds in the lymph nodes. The degree to which this may occur, the metabolic profile and characteristics of lymphoid tissue metabolism of particle-associated organic compounds, and the effect of these metabolites on the lung's normal immunologic response require investigation.

Acknowledgments

The authors give special thanks to Mary Jo Waltman for providing technical assistance for this chapter. The authors also acknowledge Drs. D. E Bice, R. G. Cuddihy, R. L. Hanson, R. F. Henderson, T. R. Henderson, C. H. Hobbs, J. L. Mauderly, R. O. McClellan, M. A. Medinsky, C. E. Mitchell, P. J. Sabourin, M. B. Snipes, and R. K. Wolff for their critical reviews.

Portions of the research were supported by the Office of Health and Environmental Research under U.S. Department of Energy Contract No. DE-AC04-76EV01013.

Correspondence should be addressed to James D. Sun, James A. Bond, or Alan R. Dahl, Inhalation Toxicology Research Institute, Lovelace Biomedical and Environmental Research Institute, P.O. Box 5890, Albuquerque, NM 87185.

References

Autrup, H. 1982. Carcinogen metabolism in human tissues and cells, *Drug Metab. Rev.* 13(4):603–646.
Autrup, H., Harris, C. C., Stoner, G. D., Selkirk, J. K., Schafer, P. W., and Trump, B. F. 1978.

Metabolism of (^{3}H) benzo[a]pyrene by cultured human bronchus and cultured human pulmonary alveolar macrophages, *Lab. Invest.* 38(3):217–224.

Autrup, H., Harris, C. C., Schafer, P. W., Trump, B. F., Stoner, G. D., and Hsu, I. C. 1979. Uptake of benzo[a]pyrene–ferric oxide particulates by human pulmonary macrophages and release of benzo[a]pyrene and its metabolites, *Proc. Soc. Exp. Biol. Med.* 161:2800–2804.

Ball, L. M., Plummer, J. L., Smith, B. R., and Bend, J. R. 1979. Benzo[a]pyrene oxidation, conjugation and disposition in the isolated perfused rabbit lung: role of the glutathione S-transferases, *Med. Biol.* 57:298–305.

Behymer, T. D., and Hites, R. A. 1984. Similarity of some organic compounds in spark-ignition and diesel engine particulate extracts, *Environ. Sci. Technol.* 18:203–206.

Bennett, R. L., Knapp, K. T., Jones, P. W., Wilkerson, J. E., and Strup, P. E. 1979. Measurement of polynuclear aromatic hydrocarbons and other hazardous organic compounds in stack gases, In: *Polynuclear Aromatic Hydrocarbons* (P. W. Jones and P. Leber, eds.), pp. 419–428, Ann Arbor Science Publishers, Inc., Ann Arbor, Mich.

Bevan, D. R., and Manger, W. E. 1985. Effect of particulates on metabolism and mutagenicity of benzo[a]pyrene, *Chem.-Biol. Interact.* 56:13–28.

Bevan, D. R., and Worrell, W. J. 1985. Elution of benzo[a]pyrene from carbon blacks into biomembranes in vitro, *Toxicol. Environ. Health* 15:697–710.

Bevan, D. R., Riemer, S. C., and Lakowicz, J. R. 1981. Effects of particulate matter on rates of membrane uptake of polynuclear aromatic hydrocarbons, *Toxicol. Environ. Health* 8:241–250.

Bice, D. E., Mauderly, J. L., Jones, R. K., and McClellan, R. O. 1985. Effects of inhaled diesel exhaust on immune responses after lung immunization, *Fundam. Appl. Toxicol.* 5:1075–1086.

Bjorseth, A., Bjorseth, O., and Fjeldstad, P. E. 1978. PAHs in the work atmosphere. II. Determination in a coke plant, *Scand. J. Work Environ. Health* 4:224.

Bond, J. A. 1983a. Some biotransformation enzymes responsible for PAH metabolism in rat nasal turbinates; effects on enzyme activities of in vitro modifiers and intraperitoneal and inhalation exposure of rats to inducing agents, *Cancer Res.* 43:4804–4811.

Bond, J. A. 1983b. Bioactivation and biotransformation of 1-nitropyrene in liver, lung and nasal tissue of rats, *Mutat. Res.* 124:315–324.

Bond, J. A., and Mauderly, J. L. 1984. Metabolism and macromolecular covalent binding of (^{14}C-1-nitropyrene in isolated perfused and ventilated rat lungs, *Cancer Res.* 44:3924–3929.

Bond, J. A., Butler, M. M., Medinsky, M. A., Muggenburg, B. A., and McClellan, R. O. 1984. Dog pulmonary macrophage metabolism of free and particle-associated (^{14}C)benzo[a]pyrene, *Toxicol. Environ. Health* 14:181–189.

Bond, J. A., Mauderly, J. L., Henderson, R. F., and McClellan, R. O. 1985. Metabolism of 1-(^{14}C)nitropyrene in respiratory tract tissue of rats exposed to diesel exhaust, *Toxicol. Appl. Pharmacol.* 79:461–470.

Bond, J. A., Ayres, P. H., Medinsky, M. A., Cheng, Y.-S., Hirshfield, D., and McClellan, R. O. 1986a. Disposition and metabolism of ^{14}C-dibenzo[c,g]carbazole aerosols in rats after inhalation, *Fundam. Appl. Toxicol.* 7:76–85.

Bond, J. A., Sun, J. D., Medinsky, M. A., Jones, R. K., Yeh, H. C. 1986b. Deposition, metabolism and excretion of ^{14}C-1-nitropyrene and ^{14}C-1-nitropyrene coated on diesel exhaust particles as influenced by exposure concentration, *Toxicol. Appl. Pharmacol.* 85:102–117.

Boyd, M. R. 1980. Biochemical mechanisms in chemical-induced lung injury: roles of metabolic activation, In: *Critical Reviews in Toxicology* (L. Goldberg, ed.), Vol. 7, pp. 103–176, CRC Press, Inc., Boca Raton, Fla.

Boyd, M. R., Buckpitt, A. R., Jones, R. B., Statham, C. N., and Longo, N. S. 1980. Metabolic activation of toxins in extrahepatic target organs and target cells, In: *The Scientific Basis of Toxicity Assessment* (H. R. Witschi, ed.), p. 141, Elsevier/North-Holland, New York.

Brittebo, E. B., and Ahlman, M. 1984. Metabolism of a nasal carcinogen, phenacetin, in the mucosa of the upper respiratory tract, *Chem.-Biol. Interact.* 50:233–245.

Brooks, A. L., Wolff, R. K., Royer, R. E., Clark, C. R., Sanchez, A., and McClellan, R. O. 1981. Deposition and biological availability of diesel particles and their associated mutagenic chemicals, *Environ. Int.* 5:263–267.

Butler, J. D., Butterworth, V., Kellow, S. C., and Robinson, H. G. 1984. Some observations of the PAH content of surface soils in urban areas, *Sci. Total Environ.* 33:75.

Cantrell, E. T., Warr, G. A., Busbee, D. L., and Martin, R. R. 1973. Induction of aryl hydrocarbon hydroxylase in human pulmonary alveolar macrophages by cigarette smoking, *J. Clin. Invest.* 52(8):1881–1884.

Casanova-Schmitz, M., David, R. M., and Heck, H. D. 1984. Oxidation of formaldehyde and acetaldehyde by NADH-dependent dehydrogenases in rat nasal mucosal homogenates, *Biochem. Pharmacol.* 33(7):1137–1142.

Chan, T. L., Lee, P. S., and Hering, W. E. 1981. Deposition and clearance of inhaled diesel exhaust particles in the respiratory tract of Fischer rats, *J. Appl. Toxicol.* 1:77–82.

Chrisp, C. E., Fisher, G. L., and Lammert, J. E. 1978. Mutagenicity of filtrates from respirable coal fly ash, *Science* 199:73–75.

Ciaccio, E. I., and De Vera, H. 1975. Effect of benzo[a]pyrene and chlorpromazine on aryl hydrocarbon hydroxylase activity from rat tissues, *Biochem. Pharmacol.* 25:985–987.

Clark, C. R., and Vigil, C. 1980. Influence of lung and liver homogenates on the mutagenicity of diesel exhaust particulate extracts, *Toxicol. Appl. Pharmacol.* 56:110–115.

Cohen, G. M., and Moore, B. P. 1976. Metabolism of (^{3}H)benzo[a]pyrene by different portions of the respiratory tract, *Biochem. Pharmacol.* 25:1623–1629.

Cohen, G. M., Marchok, A. C., Nettesheim, P.,

Steele, V. E., Nelson, F., Huang, S., and Selkirk, J. K. 1979. Comparative metabolism of benzo[a]pyrene in organ and cell cultures derived from rat tracheas, *Cancer Res.* 39:1980–1984.

Committee on Biological Effects of Atmospheric Pollutants. 1972. *Particle Polycyclic Organic Matter*, pp. 1–361, National Academy of Sciences, Washington, D. C.

Creasia, D. A., and Nettesheim, P. 1974. Respiratory cocarcinogenesis studies with ferric oxide: a test case of current experimental models, In: *Experimental Lung Cancer, Carcinogenesis and Bioassays* (E. Karbe and J. F. Park, eds.), p. 234, Spenger-Verlag, Frankfurt, West Germany.

Dahl, A. R. 1986. The role of nasal xenobiotic metabolism in toxicology, In: *Current Topics in Pulmonary Pharmacology and Toxicology* (Mannfred A. Hollinger, ed.), Vol. 1, pp. 143–164, Elsevier, New York.

Dahl, A. R. 1987. The effect of cytochrome P-450–dependent metabolism and other enzyme activities on olfaction, In: *Molecular Neurobiology of the Olfactory System* (in press).

Dahl, A. R., and Hadley, W. M. 1983. Formaldehyde production promoted by rat nasal cytochrome P-450–dependent monooxygenases with nasal decongestants, essences, solvents, air pollutants, nicotine and cocaine as substrates, *Toxicol. Pharmacol.* 67:200–205.

Dahl, A. R., Hadley, W. M., Hahn, F. F., Benson, J. M., and McClellan, R. O. 1982. Cytochrome P-450–dependent monooxygenases in olfactory epithelium of dogs; possible role in tumorigenicity, *Science* 216:57–59.

Dahl, A. R., Coslett, D. S., Bond, J. A., and Hesseltine, G. R. 1985. Exposure of the hamster alimentary tract to benzo[a]pyrene metabolites produced in the nose, *J. Nat. Cancer Inst.* 75:135–139.

Dahl, A. R., Bond, J. A., P-Fischer, J., Sabourin, P. J., and Whaley, S. J. 1987. Effects of the respiratory tract on inhaled materials (and vice versa), *Fundam. Appl. Toxicol.* (in press).

Daisey, J. M., Cheney, J. L., and Lioy, P. J. 1986. Profiles of organic particulate emissions from air pollution sources: status and needs for receptor source apportionment modeling, *J. Air Pollut. Control Assoc.* 36:17–33.

Daniel, F. B., Schut, H. A. J., Sandwisch, D. W., Schenck, K. M., Hoffmann, C. O., Patrick, J. R., and Stoner, G. D. 1983. Interspecies comparisons of benzo[a]pyrene metabolism and DNA-adduct formation in cultured human and animal bladder and tracheobronchial tissues, *Cancer Res.* 43:4723–4729.

Dasch, J. M. 1982. Particulate and gaseous emissions from wood burning fireplaces, *Environ. Sci. Technol.* 16:639.

Devereux, T. R., and Fouts, J. R. 1981. Xenobiotic metabolism by alveolar type II cells isolated from rabbit lung, *Biochem. Pharmacol.* 30:1231–1237.

Dipple, A., Michejda, C. J., and Weisburger, E. K. 1985. Metabolism of chemical carcinogens, *Pharmacol. Ther.* 27(3):265–296.

Dutcher, J. S., and Mitchell, C. E. 1983. Distribution

and elimination of inhaled phenanthridone in Fischer-344 rats, *J. Toxicol. Environ. Health* 12:709–719.

Farrelly, J. G., and Stewart, M. L. 1982. The metabolism of a series of methylalkylnitrosamines, *Carcinogenesis* 3(11):1299–1302.

Fisher, G. L., Chrisp, C. E., and Raabe, O. G. 1979. Physical factors affecting the mutagenicity of fly ash from a coal-fired power plant, *Science* 204:879–881.

Gelboin, H. V., Wiebel, F., and Diamond, L. 1970. Dimethylbenzanthracene tumorigenesis and aryl hydrocarbon hydroxylase in mouse skin: inhibition by 7,8-benzoflavone, *Science* 170(954):169–171.

Gelboin, H. V., Wiebel, F. J., and Kinoshita, N. 1972. Microsomal aryl hydrocarbon hydroxylases: on their role in polycyclic hydrocarbon carcinogenesis and toxicity and the mechanism of enzyme induction, *Biochem. Soc. Symp.* 34:103.

Gregg, S. J., and Sing, K. S. W. 1982. *Adsorption, Surface Area and Porosity*, 2nd Ed., Academic Press, New York.

Greife, R., Schoeny, R., and Warshawsky, D. 1986. Alveolar macrophage metabolism of benzo[a]pyrene with presence of the co-carcinogen ferric oxide, In: *Proceedings of the American Association of Cancer Research*, Vol. 27, p. 118, abstr. 464, March 1986.

Griffis, L. C., Wolff, R. K., Henderson, R. F., Griffith, W. C., Mokler, B. V., and McClellan, R. O. 1982. Clearance of diesel soot particles from rat lung after a subchronic diesel exhaust exposure, *Fundam. Appl. Toxicol.* 3:99–103.

Guengerich, F. P., and MacDonald, T. L. 1984. Chemical mechanisms of catalysis by cytochromes P-450: a unified view, *Acc. Chem. Res.* 17:9–16.

Guerin, M. R. 1980. Chemical composition of cigarette smoke, In: *Banbury Report: A Safe Cigarette?* (G. B. Gori and F. G. Bock, eds.), pp. 191–204, Cold Spring Harbor Laboratory, Cold Spring Harbor, N.Y.

Hadley, W. M., and Dahl, A. R. 1983. Cytochrome P-450–dependent monooxygenase activity in nasal membranes of six species, *Drug Metab. Dispos.* 11:275–276.

Hanson, R. L., Weissman, S. H., Carpenter, R. L., Newton, G. J., and Rothenberg, S. J. 1981. Sampling and chemical characterization of coal combustion effluents from an experimental fluidized-bed combustor, In: *Coal Conversion and the Environment* (D. D. Mahlum, R. H. Gray, and W. D. Felix, eds.), pp. 52–66, Proceedings of the 20th Annual Hanford Life Sciences Symposium at Richland, Wash., Technical Information Center.

Hanson, R. L., Dahl, A. R., Rothenberg, S. J., Benson, J. M., Brooks, A. L., and Dutcher, J. S. 1985. Chemical and biological characterization of volatile components of environmental samples after fractionation by vacuum line cryogenic distillation, *Arch. Environ. Contam. Toxicol.* 14:289–297.

Harmsen, A. G., Muggenburg, B. A., Snipes, M. B., and Bice, D. E. 1985. The role of macrophages in particle translocation from lungs to lymph nodes, *Science* 230:1277–1280.

Harris, C. C., Genta, V. M., Frank, A. L., Kaufman, D. G., Barrett, L. A., McDowell, E. M., and

Trump, B. F. 1974. Carcinogenic polynuclear hydrocarbons bind to macromolecules in cultured human bronchi, *Nature* 252:68–69.

Harris, C. C., Frank, A. L., van Haaften, C., Kaufman, D. G., Connor, R., Jackson, F., Barrett, L. A., McDowell, E. M., and Trump, B. F. 1976. Binding of (^{3}H)benzo[a]pyrene to DNA in cultured human bronchus, *Cancer Res.* 36:1011–1018.

Harris, C. C., Autrup, H., Stoner, G., Yang, S. K., Leutz, J. C., Gelboin, H. V., Selkirk, J. K., Connor, R. J., Barrett, L. A., Jones, R. T., McDowell, E., and Trump, B. F. 1977. Metabolism of benzo[a]pyrene and 7,12-dimethylbenz[a]anthracene in cultured human bronchus and pancreatic duct, *Cancer Res.* 37:3349–3355.

Harris, C. C., Hsu, I. C., Stoner, G. D., Trump, B. F., and Selkirk, J. K. 1978. Human pulmonary alveolar macrophages metabolize benzo[a]pyrene to proximate and ultimate mutagens, *Nature* 272:633–634.

Henry, M. C., and Kaufman, D. G. 1973. Clearance of benzo[a]pyrene from hamster lungs after administration on coated particles, *J. Nat. Cancer Inst.* 51(6):1961–1964.

Henry, M. C., Port, C. D., and Kaufman, D. G. 1975. Importance of physical properties of benzo[a]pyrene–ferric oxide mixtures in lung tumor induction, *Cancer Res.* 35:207–217.

Hodgson, E., Kulkarni, A. P., Fabacher, D. L., and Robacker, K. M. 1980. Induction of hepatic drug metabolizing enzymes in mammals by pesticides: a review, *J. Environ. Sci. Health* 15(6):723–754.

Hook, G. E. R., and Bend, J. R. 1976. Pulmonary metabolism of xenobiotics, *Life Sci.* 18:279.

Huberman, E., and Sachs, L. 1977. DNA binding and its relationship to carcinogenesis by different polycyclic hydrocarbons, *Int. J. Cancer* 19(1):122–127.

Hundley, S. G., and Freudenthal, R. I. 1977. A comparison of benzo[a]pyrene metabolism by liver and lung microsomal enzymes from 3-methylcholanthrene–treated rhesus monkeys and rats, *Cancer Res.* 37(9):3120–3125.

Jeffrey, A. M., Weinstein, I. B., Jennette, K. W., Grzeskowiak, K., and Nakanishi, K. 1977. Structures of benzo[a]pyrene–nucleic acid adducts formed in human and bovine bronchial explants, *Nature* 269:348–350.

Jenner, P., and Testa, B. (eds.). 1980. *Concepts in Drug Metabolism*, Pt. A, Marcel Dekker, New York.

Jones, K. G., Holland, J. F., Foureman, G. L., Bend, J. R., and Fouts, J. R. 1983. Xenobiotic metabolism in Clara cells and alveolar type II cells isolated from lungs of rats treated with β-naphthoflavone, *J. Pharmacol. Exp. Therap.* 225:316–319.

Kaufman, D. G., Genta, V. M., Harris, C. C., Smith, J. M., Sporn, M. B., and Saffiotti, U. 1973. Binding of ^{3}H-labeled benzo[a]pyrene to DNA in hamster tracheal epithelial cells, *Cancer Res.* 33:2837–2841.

Kilburn, K. H., and McKenzie, W. 1975. Leukocyte recruitment to airways by cigarette smoke and particle phase in contrast to cytotoxicity of vapor, *Science* 189:634–637.

King, L. C., Kohan, M. J., Austin, A. C., Claxton, L. D., and Huisingh, J. L. 1981. Evaluation of the release of mutagens from diesel particles in the presence of physiological fluids, *Environ. Mutagen.* 3:109–121.

King, L. C., Loud, K., Tejada, S. B., Kohan, M. J., and Lewtas, J. 1983. Evaluation of the release of mutagens and 1-nitropyrene from diesel particles in the presence of lung macrophages in culture, *Environ. Mutagen.* 5:577–588.

Knight, C. V., Graham, M. S., and Neal, B. S. 1983. Polynuclear aromatic hydrocarbons and associated organic emissions for catalytic and noncatalytic wood heaters. In: *Polynuclear Aromatic Hydrocarbons: Formation, Metabolism and Measurement* (M. Cooke and A. J. Dennis, eds.), pp. 689–710, Battelle Press, Columbus, Ohio.

Lakowicz, J. R., Bevan, D. R., and Riemer, S. C. 1980. Transport of a carcinogen, benzo[a]pyrene, from particulates to lipid bilayers: a model for the fate of particle-adsorbed polynuclear aromatic hydrocarbons which are retained in the lungs, *Biochim. Biophys. Acta* 629:243–258.

Lofroth, G. 1978. Mutagenicity assay of combustion emissions, *Chemisphere* 7:791–798.

Lu, A. Y. 1976. Liver microsomal drug-metabolizing enzyme system: functional components and their properties, *Fed. Proc.* 35(13):2460–2463.

Magee, P. N. 1974. Activation and inactivation of chemical carcinogens and mutagens in the mammal, *Essays Biochem.* 10:105–136.

Marshall, M. V., McLemore, T. L., Martin, R. R., Jenkins, W. T., Snodgrass, D. R., Corson, M. A., Arnott, M. S., Wray, N. P., and Griffin, A. C. 1979. Patterns of benzo[a]pyrene metabolism in normal human pulmonary alveolar macrophages, *Cancer Lett.* 8:103–109.

Mauderly, J. L., Jones, R. K., Griffith, W. C., Henderson, R. F., and McClellan, R. O. 1987. Diesel exhaust is a pulmonary carcinogen in rats exposed chronically by inhalation, *Fundam. Appl. Toxicol.* 9:208–221.

McLemore, T. L., Martin, R. R., Busbee, D. L., Richie, R. C., Springer, R. R., Toppell, K. L., and Cantrell, E. T. 1977. Aryl hydrocarbon hydroxylase activity in pulmonary macrophages and lymphocytes from lung cancer and noncancer patients, *Cancer Res.* 37:1175–1181.

McNulty, M. J., Casanova-Schmitz, M., and Heck, H. D. 1983. Metabolism of dimethylamine in the nasal mucosa of the Fischer-344 rat, *Drug Metab. Dispos.* 11:421–425.

Miller, E. C., and Miller, J. A. 1966. Mechanisms of chemical carcinogenesis: nature of proximate carcinogens and interactions with macromolecules, *Pharmacol. Rev.* 18:805.

Miller, E. C., and Miller, J. A. 1981. Mechanisms of chemical carcinogenesis, *Cancer* 47:1055–1064.

Mitchell, C. E. 1982. Distribution and retention of benzo[a]pyrene in rats after inhalation, *Toxicol. Lett.* 11:35–42.

Mitchell, C. E. 1983. The metabolic fate of benzo[a]pyrene in rats after inhalation, *Toxicology* 28:65–73.

Mitchell, C. E., Henderson, R. F., and McClellan,

R. O. 1984. Distribution, retention, and fate of 2-aminoanthracene in rats after inhalation, *Toxicol. Appl. Pharmacol.* 75:52–59.

Moore, B. P., and Cohen, G. M. 1978. Metabolism of benzo[a]pyrene and its major metabolites to ethyl acetate–soluble and water–soluble metabolites by cultured rodent trachea, *Cancer Res.* 38:3066–3075.

Natusch, D. F. S. 1978. Potentially carcinogenic species emitted to the atmosphere by fossil-fueled power plants, *Environ. Health Perspect.* 22:79.

Natusch, D. S., and Wallace, J. R. 1974. Urban aerosol toxicity: the influence of particle size, *Science* 186:695–699.

Nelson, S. D., Boyd, M. R., and Mitchell, J. R. 1977. Role of metabolic activation in chemical-induced tissue injury, In: *Drug Metabolism Concepts* (D. M. Jerina, ed.), ch. 8, American Chemical Society, Washington, D.C.

Palmer, W., and Creasia, D. 1984. Metabolism of 7,12-dimethylbenz[a]anthracene by alveolar macrophages containing ingested ferric oxide, aluminum oxide or carbon particles, *J. Environ. Pathol. Toxicol. Oncol.* 5(4):261–270.

Palmer, W. G., Allen, T. J., and Tomaszewski, J. E. 1978. Metabolism of 7,12-dimethylbenz[a]anthracene by macrophages and uptake of macrophage-derived metabolites by respiratory tissues in vitro, *Cancer Res.* 38:1079–1084.

Philpot, R. M., and Wolf, C. R. 1981. The properties and distribution of the enzymes of pulmonary cytochrome P-450–dependent monooxygenase systems, reviews, In: *Biochemical Toxicology* (E. Hodgson, J. R. Bend, and R. M. Philpot, eds.), Vol. 3, pp. 51–76, Elsevier/North-Holland, New York.

Philpot, R. M., Anderson, M. W., and Eling, T. M. 1977. Uptake, accumulation, and metabolism of chemicals by the lung, In: *Metabolic Functions of the Lung* (Y. S. Bakhle and J. R. Vane, eds.), ch. 5, pp. 123–172, Marcel Dekker, New York.

Prough, R. A., Sipal, Z., and Jakobsson, S. W. 1977. Metabolism of benzo[a]pyrene by human lung microsomal fractions, *Life Sci.* 21:1629–1636.

Prough, R. A., Patrizi, V. W., Okita, R. T., Masters, B. S. S., and Jakobsson, S. W. 1979. Characteristics of benzo[a]pyrene metabolism by kidney, liver, and lung microsomal fractions from rodents and humans, *Cancer Res.* 39:1199–1206.

Rappaport, S. M., Wang, Y. Y., Wel, E. T., Sawyer, R., Watkins, B. E., and Rapoport, H. 1980. Isolation and identification of a direct-acting mutagen in diesel-exhaust particulates, *Environ. Sci. Technol.* 14:1505–1509.

Romert, L., and Jenssen, D. 1983. Rabbit alveolar macrophage-mediated mutagenesis of polycyclic aromatic hydrocarbons in V79 Chinese hamster cells, *Mutat. Res.* 111(2):245–252.

Romert, L., Bernson, V., and Pettersson, B. 1983. Effects of air pollutants on the oxidative metabolism and phagocytic capacity of pulmonary alveolar macrophages, *J. Toxicol. Environ. Health* 12:417–427.

Romert, L., Bernson, V., and Pettersson, B. 1985. The evaluation of air sample extracts on the phagocytosis of alveolar macrophages and studies of mac-

rophage-mediated mutagenesis in co-cultivated V79 Chinese hamster cells, *Environ. Int.* 11:341–346.

Saffiotti, U., Borg, S. A., Grote, M. I., and Karp, D. B. 1964. Retention rates of particulate carcinogens in lungs, *Chicago Med. School Quart.* 2.24: 10–17.

Sanborn, C. R., Cooke, M., Bresler, W. E., and Osborne, M. C. 1983. Characterization of emissions of PAHs from residential coal-fired space heaters, Paper No. 83-54.4 presented at the 76th Annual Meeting of the Air Pollution Control Association, Atlanta, Ga.

Schnizlein, C. T., Bice, D. E., Mitchell, C. E., and Hahn, F. F. 1982. Effects on rat lung immunity by acute lung exposure to benzo[a]pyrene, *Arch. Environ. Health* 37:201–206.

Schoeny, R., and Warshawsky, D. 1983. Mutagenicity of benzo[a]pyrene metabolites generated on the isolated perfused lung following particulate exposure, *Teratogen. Carcinogen. Mutagen.* 3:151–162.

Sivarajah, K., Jones, K. G., Fouts, J. R., Devereux, T., Shirley, J. E., and Eling, T. E. 1983. Prostaglandin synthetase and cytochrome P-450–dependent metabolism of benzo[a]pyrene 7,8-dihydrodiol by enriched populations of rat Clara cells and alveolar type II cells, *Cancer Res.* 43:2632–2636.

Smith, B. R., and Bend, J. R. 1981. Metabolic interactions of hydrocarbons with mammalian lung, reviews, In: *Biochemical Toxicology* (E. Hodgson, J. R. Bend, and R. M. Philpot, eds.), pp. 77–122, Elsevier/North-Holland, New York.

Soutar, C. A. 1977. Abnormalities of the tracheobronchial lymph nodes in smokers and subjects with chronic bronchitis: a necropsy study of the distribution of immunoglobulins, *Thorax* 32:397–405.

Stoner, G. D., Harris, C. C., Autrup, H., Trump, B. F., Kingsbury, E. W., and Myers, G. A. 1978. Explant culture of human peripheral lung. I. Metabolism of benzo[a]pyrene, *Lab. Invest.* 38:685.

Strom, K. A. 1984. Response of pulmonary cellular defenses to the inhalation of high concentrations of diesel exhaust, *J. Toxicol. Environ. Health* 13:919–944.

Sun, J. D., and McClellan, R. O. 1984. Respiratory tract clearance of ^{14}C-labeled diesel exhaust compounds associated with diesel particles or as a particle-free extract, *Fundam. Appl. Toxicol.* 4:388–393.

Sun, J. D., Wolff, R. K., and Kanapilly, G. M. 1982. Deposition, retention, and biological fate of inhaled benzo[a]pyrene adsorbed onto ultrafine particles and as a pure aerosol, *Toxicol. Appl. Pharmacol.* 65:231–244.

Sun, J. D., Wolff, R. K., Aberman, H. M., and McClellan, R. O. 1983. Inhalation of 1-nitropyrene associated with ultrafine insoluble particles or as a pure aerosol: a comparison of deposition and biological fate, *Toxicol. Appl. Pharmacol.* 69:185–198.

Sun, J. D., Wolff, R. K., Kanapilly, G. M., and McClellan, R. O. 1984. Lung retention and metabolic fate of inhaled benzo[a]pyrene associated with diesel exhaust particles, *Toxicol. Appl. Pharmacol.* 73:48–59.

Task Group on Lung Dynamics. 1966. Disposition

and retention models for internal dosimetry of the human respiratory tract, *Health Phys.* 12:173–207.

Testa, B., and Jenner, P. 1976. *Drug Metabolism: Chemical and Biochemical Aspects*, Marcel Dekker, New York.

Tomingas, R., Dehnen, W., Lange, H. U., Beck, E. G., and Manojlovic, N. 1971. The metabolism of free and soot-bound benzo[a]pyrene by macrophages from guinea pigs in vitro, *Zentralblat. Bakteriol. Parasitenkd. Infektionskr. Hyg. Abt. 1: Orig. Reihe B* 155:159–167.

Törnquist, S., Wiklund, L., and Toftgard, R. 1985. Investigation of absorption, metabolism kinetics and DNA-binding of intratracheally administered benzo[a]pyrene in the isolated, perfused rat lung: a comparative study between microcrystalline and particulate adsorbed benzo[a]pyrene, *Chem.-Biol. Interact.* 54:185–198.

Truesdale, R. S., and Cleland, J. G. 1982. Residential stove emissions from coal and other alternate fuels, In: *Residential Wood and Coal Combustion*, pp. 115–128, Air Pollution Control Association, Pittsburgh, Pa.

Vostal, J. J. 1983. Bioavailability and biotransformation of the mutagenic component of particulate emissions present in motor exhaust samples, *Environ. Health Perspect.* 47:269–281.

Wang, D. T., and Meresz, O. 1982. Occurrence of potential uptake of polynuclear aromatic hydrocarbons of highway traffic origin by proximally grown food crops, In: *Polynuclear Aromatic Hydrocarbons: Physical and Biological Chemistry* (M. Cooke, A. J. Dennis, and G. L. Fisher, eds.), pp. 885–896, Battelle Press, Columbus, Ohio.

Wang, Y. Y., Rappaport, S. M., Sawyer, R. F., Talcott, R. E., and Wei, E. T. 1978. Direct-acting mutagens in automobile exhaust, *Cancer Lett.* 5:39–47.

Warshawsky, D., Niemeier, R. W., and Bingham, E. 1978. Influence of particulates on metabolism of benzo[a]pyrene in the isolated perfused lung, In: *Carcinogenesis, Vol. 3: Polynuclear Aromatic Hydrocarbons* (P. W. Jones and R. I. Freudenthal, eds.), pp. 347–360, Raven Press, New York.

Warshawsky, D., Bingham, E., and Niemeier, R. W. 1980. The effects of n-dodecane pretreatment on the metabolism and distribution of benzo[a]pyrene in the isolated perfused rabbit lung, *Life Sci.* 27(20): 1827–1837.

Warshawsky, D., Bingham, E., and Niemeier, R. W. 1983. Influence of airborne particulate on the metabolism of benzo[a]pyrene in the isolated perfused lung, *Toxicol. Environ. Health* 11:503–517.

Warshawsky, D., Bingham, E., and Niemeier, R. W. 1984. The effects of a cocarcinogen, ferric oxide, on the metabolism of benzo[a]pyrene in the isolated perfused lung, *Toxicol. Environ. Health* 14:191–209.

Weinstein, I. B., Gattoni-Celli, S., Kirschmeier, P., Lambert, M., Hsiao, W., Backer, J., and Jeffrey, A. 1984. Initial cellular targets and eventual genomic changes in multistage carcinogenesis, *IARC Sci. Publ.* 56:277–297.

Weinstein, I. B., Arcoleo, J., Lambert, M., Hsiao, W., Gattoni-Celli, S., Jeffrey, A. M., and Kirschmeier, P. 1985. Mechanisms of multistage chemical carcinogenesis and their relevance to respiratory tract cancer, *Carcinogenesis* 8:395–409.

Williams, R. L., and Swarin, S. J. 1979. Benzo[a]pyrene Emissions from Gasoline and Diesel Automobiles, SAE Technical Paper Series No. 790419, Society of Automotive Engineers, Warrendale, Penn.

Wolff, R. K., Griffis, L. C., Hobbs, C. H., and McClellan, R. O. 1982. Deposition and retention of ultrafine $^{67}Ga_2O_3$ aggregate aerosols in rats following whole body exposures, *Fundam. Appl. Toxicol.* 2:195–200.

Wong, D., Mitchell, C., Wolff, R. K., Mauderly, J. L., Jeffrey, A. M. 1986. Identification of DNA damage as a result of exposure of rats to diesel engine exhaust, *Carcinogenesis* 7:1595–1597.

Yang, S. K., Gelboin, H. V., Trump, B. F., Autrup, H., and Harris, C. C. 1977. Metabolic activation of benzo[a]pyrene and binding to DNA in cultured human bronchus, *Cancer Res.* 37:1210–1215.

Zarkower, A., Eskew, M. L., Scheuchenzuber, W. J., and Graham, J. A. 1982. Effects of fly ash inhalation on murine immune function: changes in macrophage-mediated activities, *Environ. Res.* 29:83–89.

Zwilling, B. S. 1977. The effect of respiratory carcinogenesis on systemic humoral and cell-mediated immunity of Syrian Golden hamsters, *Cancer Res.* 37:250–252.

Transport and Uptake of Inhaled Gases

JAMES S. ULTMAN
Pennsylvania State University

Air Pollution, the Automobile, and Public Health. © 1988 by the Health Effects
Institute. National Academy Press, Washington, D.C.

When a pollutant gas is breathed, its transport through the highly branched tracheobronchial tree results in a unique internal distribution of concentration and uptake rates. A major factor complicating the study of these processes is the fact that gases with different molecular properties can exhibit different internal dose distributions. Highly water-soluble gases, such as sulfur dioxide (SO_2), are essentially removed by the upper airways, the primary site of respiratory defense. A highly reactive gas of only moderate solubility, such as ozone (O_3), can reach the tracheobronchial tree, where it reacts with the protective mucous layer and eventually damages underlying tissue in the small bronchioles. A gas that has limited aqueous activity but is highly reactive with hemoglobin, such as carbon monoxide (CO), is able to penetrate further to the respiratory zone and diffuse to the pulmonary circulation in quantity.

The guiding philosophy presented in this chapter is that direct measurements on human volunteers can be minimized by using mathematical models to anticipate uptake rates and identify target tissues of potentially hazardous species. An ideal model enables one to calculate dose distribution from basic anatomic and physicochemical parameters and to estimate the effects of pathological airway derangements and changes in ventilatory demand. But in many situations it is more practical to use models that simply extrapolate measurements from laboratory animals to humans. In either case, it is necessary to validate proposed models with experiments, and for that reason both theoretical and laboratory methodologies are discussed in this paper.

The role of physicochemical factors in the absorption of gas species from a flowing airstream is of central concern to chemical engineers who design industrial gas separation equipment. Many of the same concepts used to develop gas absorption or chromatographic separation processes can be applied with some modification to the lung. Two widely read sources on this subject are Bird et al. (1960), a fundamental treatise on fluid flow and diffusion phenomena, and Treybal (1980), a textbook that describes conventional design methods. Complementing this engineering knowledge is a wealth of physiology literature concerning the influence of lung anatomy and mechanics on the distribution of foreign gases (for example, Engel and Paiva 1985). In this chapter, we will integrate these separate fields as Hills (1974) has done for normal respiratory gases. Although the material here is intended to encompass all pollutant gases—those already recognized as well as those still unrecognized—an encyclopedic review has not been attempted. Rather, basic concepts and methods are stressed to provide a general overview, and concrete examples with respect to three species—SO_2, O_3, and CO—have been used to illustrate the spectrum of solubility and reactivity effects.

This article begins with two sections devoted to issues at the heart of mathematical model development: first, lung anatomy and its influence on aerodynamics; and second, fundamentals of diffusion and chemical reaction and their characterization in terms of mass transfer coefficients. The next two sections address the structure of alternative mathematical models and their validation by experimental measurements.

Implications of Lung Anatomy

This section, in presenting an overview of lung anatomy and its influence on aerodynamics and hemodynamics, focuses on the human respiratory system. Although most of the concepts are also valid for animals (see Schlesinger, this volume), the numerical values used to illustrate specific anatomic and functional features are typical of normal adult humans. As a supplement to these data, table 1 has been included to indicate the difference in scale between the lungs of humans and animals.

The pulmonary airways consist of three distinct functional units. The upper airways, extending from the nares and lips to the larynx, are the primary sites at which the temperature and humidity of inspired air are equilibrated to body conditions. The conducting airways, a branched–tube net-

Table 1. Comparisons Among Anatomical Measurements Related to Breathing in Humans and Laboratory Animals

Anatomical Measurement	Human	Large Dog	Small Dog	Monkey	Rabbit	Small Rat	Mouse
Body weight (kg)	74	22.8	11.2	3.71	3.6	0.14	0.023
Total lung volume (ml)	4,341	1,501	736	184	79	6.3	0.74
Total alveolar surface (m²)	143	90	41	13	5.9	0.39	0.068
Total capillary surface (m²)	126	72	33	12	4.7	0.41	0.059
Total capillary volume (ml)	213	119	50	15	7.2	0.48	0.084
Mean air–blood barrier thickness (μm)	2.22	1.42	1.64	1.52	1.51	1.42	1.25

SOURCE: Adapted with permission from Gil (1982, tables 1 and 2). Copyright CRC Press, Boca Raton, Fla.

work originating at the trachea and extending to include terminal bronchioles, provide an orderly yet compact expansion of flow cross-section. The respiratory zone, composed of all the alveolated airways and airspaces, is responsible for oxygenation of and carbon dioxide removal from the pulmonary capillaries. Taken together, the conducting airways and respiratory zone are referred to as the lower respiratory tract, and since they contain no gas exchange surface, the upper and conducting airways combined are often called the anatomic dead space.

The respiratory zone, with a volume of about 3 liters compared to the 160 ml of anatomic dead space, contains most of the functional residual gas volume, and yet the total path length of 6 mm along alveolated airways is only a fraction of the 40 cm between the nose and the terminal conducting airways. In other words, the total cross-sectional area in the respiratory zone is much greater than in the dead space. However, this total is the sum of the cross-sections of a large number of small passages in which the axial diffusion distance is very short. Therefore, longitudinal mixing of fresh inspired air with residual gas is favored in the respiratory zone, whereas bulk flow with little mixing of fresh and residual air occurs in the dead space (Gomez 1965).

The walls of the upper and conducting airways are composed of an inner mucosal layer, a submucosal layer, and an outer adventitia. The mucosal layer, sometimes called the mucous membrane, contains ciliated epithelium and mucus-producing goblet cells, with additional mucus-producing glands located in the submucosal layer. In the lumen of these airways, cilia are surrounded by a low-viscosity periciliary fluid, above which high-viscosity mucus floats. The cilia beat in an organized unidirectional fashion, propelling this mucous blanket from the lower respiratory tract to the oropharynx with a daily clearance of about 100 ml.

Blood is supplied to the lower respiratory tract through the bronchial arteries, which arise from the aorta and deliver oxygen to conducting airways, and through the pulmonary circulation that perfuses the respiratory zone via the pulmonary artery. During rest, the bronchial circulation constitutes only 1 percent of the total 5 liter/min of systemic arterial blood flow, and it feeds sparsely distributed capillaries in the submucosal layer of bronchial tissue. In contrast, the entire output of the right heart enters the pulmonary circulation to supply the dense meshwork of capillaries in the walls of the alveoli.

Upper Airways

During quiet breathing, a typical tidal volume is 500 ml at a frequency of 12 breaths/min, corresponding to a minute volume of 6 liters/min. Under these conditions, air normally enters the nose and flows by way of the nasopharynx, pharynx, and larynx to the trachea. In the nose, the convoluted well-perfused surface of the nasal turbinates with its large specific surface area of 8 cm^2/ml provides an efficient site for absorption of soluble gases. Alternatively, air may follow a path from the mouth directly to the pharynx, thereby depriving the inspirate of purification by the nose. Most individuals breathe exclusively through their nose when minute volume is below 34 liters, but during vigorous exercise when ventilatory demand is larger, 45 to 60 percent of respired air is inspired orally (Niinimaa et al. 1980, 1981). The presence of a nasal obstruction or even a disease of the lower respiratory tract such as asthma can also exaggerate oral breathing, making the lung more likely to be exposed to pollutant gases.

The geometry of upper airways is irregular and variable, depending on the point of air access, on the breathing pattern, and on the state of mucous membrane engorgement by nasal blood flow. Perhaps this explains why no mathematical model of its anatomy has been established. Roughly speaking, upper airway volume is 50 ml, with a path length of 17 cm from lips to glottis and 22 cm from nares to glottis. As respired gas flows through the upper airways, it encounters a series of sudden contractions and expansions in cross-section leading to corresponding accelerations and decelerations in gas velocity (figure 1). This is quite different from the steady flow of gases in straight tubes of constant cross-section.

The nature of flow in a tube is characterized by the Reynolds number (*Re*), the dimensionless ratio of inertial force to frictional resistance:

$$Re = \dot{V}d/A\nu \qquad (1)$$

where *d* is tube diameter, $\dot{V}$ is volumetric flow rate, *A* is cross-sectional area available for flow, and *ν* is kinematic viscosity. In sufficiently long straight tubes, flow is fully developed and the Reynolds number uniquely determines whether the velocity field is laminar (*Re* < 2100), turbulent (*Re* > 4000), or in transition between these two regimes (4000 > *Re* > 2100). In laminar flow, material moves along fixed axial

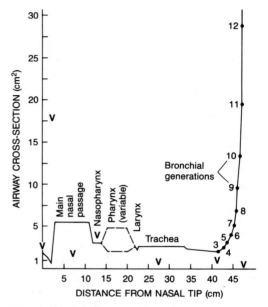

Figure 1. Cross-sectional area available for gas flow. Axial velocity (m/sec for flow of 200 ml/sec) during quiet respiration is also shown at selected "V" (velocity) points. (Adapted with permission from Swift and Proctor 1977, p. 68, by courtesy of Marcel Dekker, Inc.)

streamlines so that radial transport can only occur by molecular diffusion. In turbulent flow, the mixing action of turbulent eddies facilitates radial transport of gas species to the tube wall, a favorable condition for gas absorption.

Because of disturbances created by geometric irregularities, flow in the upper airways is never fully developed. For example, the inspired airstream that is propelled through the glottic constriction into the trachea forms a confined jet immediately downstream of the larynx. Large velocity gradients within the jet generate turbulent eddies at Reynolds numbers well below the fully developed transition value of 2100 (Simone and Ultman 1982), and these eddies can increase radial mixing as gas flows downstream, even during quiet inspiration. Also, the rapid deceleration of air passing from the ostium internum to the turbinated structures can induce turbulence in the nose (Swift and Proctor 1977).

In addition to the promotion of turbulence, three other phenomena unique to the upper airways may enhance gas absorption.

First, fresh mucus swept into the tubinates from the paranasal sinuses supplements the capacity of the nasal passages for airborne pollutants. Second, the nasal mucosa and submucosa are well perfused with blood flowing countercurrent to the mucus flow, and this is a favorable configuration for gas absorption (Brain 1970). Third, because increased blood flow can elicit an erectile response in surrounding tissue, vasodilation by cholinergic pollutants may increase gas absorption both by increasing capillary surface and by reducing the lateral diffusion distance in the nasal passages.

Conducting Airways

The conducting airways, also referred to as the tracheobronchial tree, are a series of more or less dichotomously branched tubes originating at the larynx and extending just proximal to the point where alveolated airways arise. Those branches containing cartilage are defined as bronchi and generally occupy the first 10 or so generations. The more distal cartilage-free branches are defined as bronchioli. Although volume changes in the trachea and major bronchi are minimized by stiff cartilage rings embedded in the airway walls, the smaller bronchi and the bronchioli undergo isotropic volume changes in proportion to changes in lung volume (Hughes et al. 1972).

Raabe (1982) reviewed several numerical models of the tracheobronchial tree, which may be divided into two categories: the so-called symmetric models, in which a regular dichotomy of branching exists at every generation and all branches in a given generation have the same diameters and the same lengths, and the asymmetric models, which account for the unequal distribution of lung lobes—three on the right and two on the left—or for irregularities within the lobes themselves. Whereas a symmetric model is an idealization wherein the geometry of all transport paths from larynx to terminal bronchioles is identical, in the more realistic asymmetric model, airway diameters and lengths and even the number of branch divisions may differ along alternative paths. For example, in Horsfield and Cumming's asymmetric model (1968),

there are between 8 and 25 branchings with a corresponding distribution in path lengths from 7 to 23 cm between the trachea and the respiratory lobules. On the other hand, in Weibel's symmetric model A (1963, p. 136) there are 16 branchings with an equal length of 13 cm along all conducting airway paths.

A symmetric model is a convenient starting point for the analysis of diffusion and flow because all transport paths are equivalent. Weibel's widely used model A portrays the conducting airways as an expanding network of dichotomously branching tubes, wherein generation number, z, increases from zero at the trachea to 16 at the terminal bronchioli and 23 at the terminal alveolar sacs. Each generation contains 2^z branches. Figure 2 summarizes the important geometric features of this model (top) and their impact on gas flow (bottom). Although the diameters of individual branches generally decrease with increasing z, beyond the eighth generation the total surface available for gas absorption, S, is a strongly increasing function of longitudinal distance, y. Increases in the summed cross-section available for flow, A, parallel those for S so that axial velocity as well as branch Reynolds number fall dramatically as gas is transported distally. During quiet respiration, the velocity is sufficiently low that Re does not exceed the upper limit of 2100 for fully developed laminar flow in any branch, but during moderate exercise, Re values in excess of 2100 indicate the possibility of turbulent flow in the trachea and first four generations.

The entry length fraction, L/L_e, is the ratio of actual airway length, L, to length, L_e, required for flow to become fully developed (Olson et al. 1970). Entry length fractions are less than unity proximal to the seventh generation, implying that velocity fields within the trachea and major bronchi do not reach fully developed behavior, even during quiet breathing. This explains how it is possible for random velocity fluctuations to be observed in bronchial airway casts at inspiratory flows well within the laminar Reynolds number range (Dekker 1961; Olson et al. 1973). Such turbulence was undoubtedly due to the propagation of incompletely dissipated eddies from the upper airway segment of the

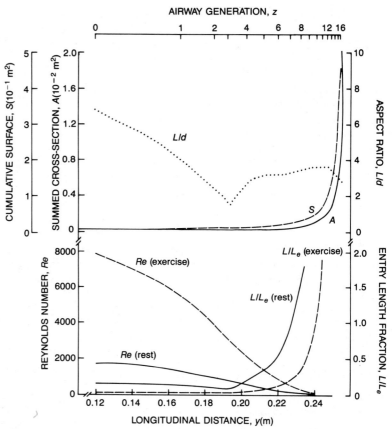

Figure 2. Geometric and aerodynamic characteristics of a symmetric tracheobronchial model. All values were derived for Weibel's model A (Weibel 1963) scaled to
a functional residual capacity of 3 liters. Entry length fraction was computed with
the equation of Olson et al. (1973). Flows of 0.4 and 1.6 liters/sec have been used for
rest and exercise conditions, respectively.

cast. Ordered flow disturbances consisting
of a paired vortex pattern during inspiration and a quadruple vortex pattern during
expiration have also been reported in isolated bronchi at Reynolds numbers within
the laminar range (Schroter and Sudlow
1969). The radial mixing brought about by
these lateral flow circulations aids in transport of gas species toward the airway walls.

Accompanying the decrease in branch
diameters between trachea and terminal
bronchioles, there is a progressive decrease
in airway wall thickness and tissue perfusion rate. There are also fewer mucus-
secreting elements and less ciliated epithelium, which reduces mucus velocity and
thickness. Estimates of some relevant mucus, tissue, and blood flow parameters are
presented in table 2.

Respiratory Zone

The respiratory zone is defined by the
presence of regions containing membranous outpouchings—the alveoli—normally
responsible for gas exchange. An individual
alveolus has a characteristic diameter of
about 0.2 mm, and its shape may be modeled in various ways: a truncated sphere, a
truncated cone, or a cylindrical wedge
(Weibel 1963, p. 60). The fine structure of
the alveolar wall consists of little more than
a shell-like mesh of pulmonary capillaries
enclosed in a layer of airway epithelium
covered by a thin surfactant film, and
therefore the blood-gas barrier imposed by
the alveolar membrane is only about 2 μm
thick. Because the pulmonary capillaries
are short, typically 10 μm long, there is

Table 2. Conducting Airway Parameters[a]

Airway	Diameter/Length[b] (10^{-2} m)	Mucus Thickness[c] (10^{-6} m)	Tissue Thickness[d] (10^{-4} m)	Blood Flow[e] (10^{-7} m³/sec)	Mucus Velocity[c] (10^{-4} m/sec)
Trachea	1.6/12	7	8	1.7	2.5
Main bronchi	1.0/6.0	7	5	0.6	1.7
Lobar bronchi	0.4/3.0	7	2	0.2	0.50
Segmental bronchi	0.2/1.5	7	1	0.09	0.067
Subsegmental bronchi	0.15/0.5	7	0.75	0.2	0.010
Terminal bronchioles	0.06/0.3	4.2	0.30	1.7	0.00033

[a] Parameter values are based on a single airway branch.
[b] Landahl (1950).
[c] National Research Council (1977, table 7-3).
[d] Computed as 5% of airway diameter (DuBois and Rogers 1968).
[e] DuBois and Rogers (1968, table 2).

extensive branching in the capillary mesh, and the associated pulmonary blood flow resembles a moving sheath of fluid interrupted by regularly spaced posts of tissue (Rosenquist et al. 1973).

When packed together to form the lung parenchyma, the alveoli give a honeycomb appearance, with a specific surface area greater than 200 cm²/ml airspace volume. The edges of contact between adjacent alveoli, called alveolar septa, protrude into the airspace, thereby partitioning gas and impeding axial diffusion near the airway wall. Because their walls are so thin and have such a large specific surface, alveoli are well suited to normal gas exchange and probably to the absorption of air pollutants as well. To illustrate this, consider that an erythrocyte, which normally resides in the pulmonary capillaries for only 1 sec, has sufficient time to come to equilibrium with alveolar oxygen (O_2) and carbon dioxide (CO_2) partial pressures.

The fraction of airway wall occupied by alveoli increases with distal distance from the first alveolated airway to the blind-ended alveolar sacs that terminate every path of the tracheobronchial tree. As was the case for the conducting airways, both symmetric and asymmetric models have been proposed for the respiratory zone. In Weibel's model A, there are seven symmetric generations of respiratory airways, and along the 6-mm path leading from the first

respiratory bronchioli to the alveolar sacs, the air-tissue surface available for gas exchange increases from 0.16 to 39 m² per generation. Simultaneously, the flow cross-section increases from 0.03 to 1.2 m² with a concomitant drop in branch Reynolds number from 0.5 to 0.01 during quiet breathing. More recent work by Hansen and Ampaya (1975) indicates that Weibel's model significantly underestimates both tissue surface and flow cross-section, but in either case it is reasonable to conclude that the influence of gas flow is negligible and transport in the respiratory zone occurs principally by diffusion.

The distal airway model of Parker et al. (1971) is an example of an asymmetric model of the respiratory zone. It is a dichotomously branching network in which the total number of branchings varies from three to eight along different airway paths. In addition, this model incorporates distributions of alveoli, from 7 to 25 alveoli per duct (mean 15.9) and from 8 to 15 alveoli per sac (mean 9.6), rather than the fixed values of 20 alveoli in each duct and 17 in each sac of Weibel's symmetric model.

Nonuniformities in Ventilation and Perfusion

If the lungs were perfectly symmetric and mechanical tissue properties and forces were uniform among generations, then air

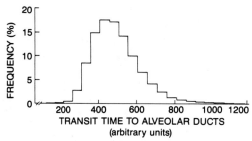

Figure 3. Frequency distribution of transit time from the carina to the alveolar ducts. (Adapted with permission from Horsfield and Cumming 1968, p. 379, and from the American Physiological Society.)

flow and blood flow to all sibling branches would also be uniform. In this idealized situation, the local concentration and uptake rate of inhaled pollutants could vary only with respect to longitudinal position along the equivalent airway transport paths. In the real lung, however, both small-scale (intraregional) and large-scale (interregional) inhomogeneities exist, and they may impose a difference in dose between equivalent structural units. For example, two terminal bronchioles located in different lobes of the lung might experience different pollutant exposure levels.

Two specific types of nonuniformities normally present in the lung have been well documented. First, because of anatomic asymmetries, there is a natural distribution of path lengths between the larynx and the respiratory zone, with shorter path lengths containing less gas volume than longer path lengths. If the rate of alveolar expansion distal to these paths is uniform, the transit times for fresh inspired gas to reach the alveoli by different paths will differ (figure 3), and there will be differences in time of exposure among equivalent structural units. Second, under the influence of gravity, when the body is upright, there is more ventilation and more blood flow near the bottom of the lungs than near the top (figure 4). In addition to the effects caused individually by the distribution of ventilation, $\dot{V}_A$, and perfusion, $\dot{Q}$, the $\dot{V}_A/\dot{Q}$ ratio has special importance regarding gas uptake in the respiratory zone. That is, if a pollutant is to continuously absorb into the respiratory zone, it must first reach an alveolus and ultimately be removed by the pulmonary circulation. Therefore, alveoli that receive little local ventilation (that is, $\dot{V}_A/\dot{Q}$ is small) are as incapable of continuously absorbing soluble gases as are adequately ventilated alveoli that receive an

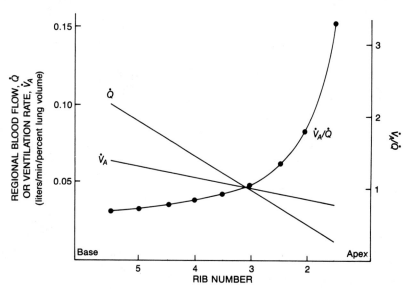

Figure 4. Ventilation/perfusion ratio, $\dot{V}_A/\dot{Q}$, from base to apex of the normal upright human lung. (Adapted with permission from West 1977, p. 202, and from Academic Press, Orlando, Fla.)

abnormally small blood flow (that is, $\dot{V}_A/\dot{Q}$ is large). This implies that gas uptake is maximized at some intermediate value of the ventilation/perfusion ratio.

Fundamentals of Mass Transport

A complete description of mass transport consists of three basic elements: the mass conservation equation, rate equations for diffusion and chemical reaction, and thermodynamic equilibrium expressions. The mass conservation equation balances the rate at which a species may appear with the rate of its accumulation in a spatial region of interest often called the control volume. Given that the permissible modes of appearance are convection, diffusion, and chemical reaction, the mass conservation equation has the form

$$\begin{array}{l} \text{Rate of} \\ \text{Accumulation} \end{array} =$$

$$\begin{array}{ccc} \text{Net Input} & \text{Net Input} & \text{Net Rate of} \\ \text{Rate by} + & \text{Rate by} + & \text{Production} \quad (2) \\ \text{Bulk Flow} & \text{Diffusion} & \text{by Reaction} \end{array}$$

The mathematical expression of mass conservation is a differential field equation applicable to any three-dimensional time-varying problem (Bird et al. 1960, p. 559), but in practice the general equation must be simplified by selecting a set of physically realistic assumptions. This will be considered in detail in the next section, which is devoted to mathematical modeling.

In this section, attention is focused on the fundamentals of thermodynamic equilibria and rate expressions for diffusion and chemical reaction. The formulation of mass transfer coefficients to describe transverse transport and the use of retention parameters and mixing coefficients to characterize longitudinal transport are also described.

Thermodynamic Equilibria

Phase equilibrium refers to the distribution of a species between adjacent immiscible phases. Suppose, for example, that a liquid phase containing a physically dissolved species X is placed in contact with gas containing an arbitrary partial pressure p_x of the same solute species. Thereafter, molecules of X will continually redistribute between the liquid and gas phases, altering p_x until it reaches a stationary value p_x^* referred to as the equilibrium partial pressure or the gas tension. In this equilibrium state, the ratio of the molar concentration C_x of species X in the liquid phase to the corresponding value of p_x^* in the gas phase is a parameter, α_x, called the Bunsen solubility coefficient:

$$\alpha_x = C_x/p_x^* \qquad (3)$$

Generally, α_x is a function of temperature only, its value being equivalent to the molar volume of the liquid divided by Henry's Law constant. Therefore, when temperature is fixed, the dissolved species concentration is a linear function of gas tension. The solubility coefficients of most foreign gases in mucus, tissue, and blood are not precisely known and must be approximated by values measured in readily available solvents such as water (table 3) and hydrocarbon oils.

Chemical reaction equilibrium can impose additional constraints on solute concentrations. During chemical reactions, molecular combinations and decompositions can both occur such that a dissolved pollutant species is reversibly bound to another endogenous species. Typical of such reactions are the aqueous ionization of SO_2 to form bisulfite; the ionization of CO_2 to form bicarbonate; and the binding of O_2, CO_2, or CO to hemoglobin. To exemplify the underlying reaction equilibrium, consider the ionization of SO_2 in water given by the stoichiometric equation

$$SO_2 + H_2O \rightleftharpoons H^+ + HSO_3^- \qquad (4)$$

Although HSO_3^- also undergoes a weak dissociation to SO_3^{2-}, equation 4 is a sufficiently accurate model of the overall process (Schroeter 1966, p. 17).

As the concentration of SO_2 is depleted by the formation of HSO_3^-, the law of mass action dictates that the forward reaction rate must slow down. Simultaneously, the concentration of bisulfite builds up, and this leads to an increase in reverse reaction rate. This progressively decreasing forward

Table 3. Estimates of Physical Properties for Various Gases

Gas	Diffusivity in Air[a] D_x (10^{-5} m²/sec)	Aqueous Diffusivity[b] D_x (10^{-9} m²/sec)	Aqueous Solubility[c] α_x (kg-mol/m³/Pa)	First-Order Reactivity[d] k_r (10^3/sec)
SO_2	1.15	2.14	9.13×10^{-6}	0
O_3	2.11	3.06	5.62×10^{-8}	$1.2(m):50(t):5(b)$
CO	2.17	3.06	8.09×10^{-9}	0

[a] Chapman-Enskog equation (Bird et al. 1960, p. 510).
[b] Wilke-Chang equation (Bird et al. 1960, p. 515).
[c] Computed from Henry's Law coefficients (National Research Council 1977, table 7-1).
[d] Miller et al. (1985, table 3) for mucus (m), tissue (t), and blood (b).

rate and increasing reverse rate continues until the two rates are equal and the concentrations of all species become stationary. In this equilibrium state, the concentrations of reactants and products possess a specific algebraic relationship. For the SO_2 ionization reaction,

$$(K_e)_{SO_2} = C_{HSO_3} C_H / C_{SO_2} \quad (5)$$

where $(K_e)_{SO_2}$, the reaction equilibrium constant specific to the ionization of SO_2, is a function of temperature (Pearson et al. 1951).

In pure water, where hydroxyl ion concentration is low, electroneutrality dictates that bisulfite and hydrogen ion concentrations are approximately equal, and the summed concentration of SO_2 in the unreacted form C_{SO_2} and in the reacted form C_{HSO_3} can then be expressed as

$$C_{SO_2} + C_{HSO_3} = $$
$$\alpha_{SO_2} p^*_{SO_2} + [(K_e)_{SO_2} \alpha_{SO_2} p^*_{SO_2}]^{1/2} \quad (6)$$

This equation illustrates the ability of a reversible chemical reaction to increase the capacity of a solution for gaseous solute. To emphasize the importance of this, figure 5a compares the physical solubility of SO_2 to the corresponding level of bisulfite ions. Clearly, the concentration of reacted species is far greater than the concentration of physically dissolved species, and the relationship of bisulfite concentration to gas tension is nonlinear. In mucus and tissue, increases in hydrogen ion concentration occurring with SO_2 ionization are probably suppressed by the buffering action of other solutes, and as equation 5 indicates, the bisulfite concentration would be even larger than in pure water.

To provide a unified treatment of the dissolved and reversibly bound forms of a soluble gas, a reactive capacitance coefficient, β_x, analogous to the solubility coefficient, can be defined as the derivative of the reacted solute concentration, C_{xr}, with respect to gas tension:

$$\beta_x = dC_{xr}/dp^*_x \quad (7)$$

Unlike α_x, which is independent of p^*_x, the reactive capacitance coefficient for a reversibly bound solute species is generally a decreasing function of gas tension (figure 5b).

Diffusion and Reaction Rates

Transport of any species occurs by a combination of diffusion and convection. Whereas convection is the translation of molecules at the mean flow velocity, diffusion is the transport that occurs in response to a concentration gradient whether or not any net flow occurs. Fick's First Law is the general rate equation that accounts for convection and diffusion (Bird et al. 1960, p. 502), and for the special case of one-dimensional diffusion of species X in the absence of flow, it reduces to

$$\dot{M}_x = -D_x S d(C_x/dz) \quad (8)$$

where $\dot{M}_x$ is the mass transport rate of species X; dC_x/dz is the concentration gradient in the diffusion direction, z; S is the surface area perpendicular to z; D_x is the molecular diffusion coefficient; and the negative sign indicates that diffusion is along the path of decreasing concentration. Strictly speaking, equation 8 is valid only for solutions composed of two species, but Fick's Law may also be applied to multi-

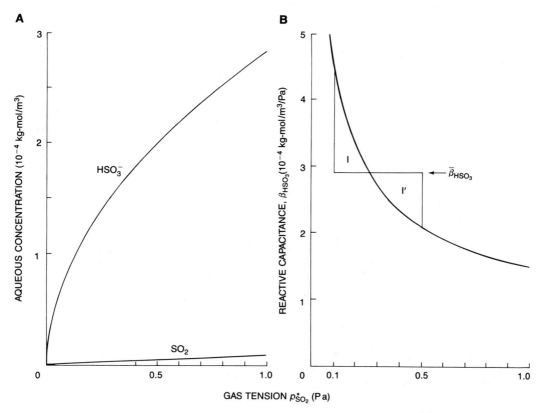

Figure 5. Solubility of sulfur dioxide at 37°C in pure water. A: The Physical solubility of SO_2 and its equilibrium concentration as reacted SO_2 (that is, HSO_3^-) are compared. B: In the graphical construction of average reactive capacitance at SO_2 tensions between 0.1 and 0.5 Pa, a horizontal line is drawn so that the areas I and I′ are equal. Note that 1 Pa is equivalent to 21.5 ppm of SO_2 when total pressure is 1 atm.

component solutions when the solute species of interest is present at sufficiently low concentration or partial pressure relative to the solvent species (Bird et al. 1960, p. 571). Chang et al. (1975) analyzed situations in the airway lumen where multicomponent diffusion effects may be important, but the existence of similar phenomena in mucus and tissue have not been investigated.

The lung consists of gas, tissue, and blood regions, and the customary assertion that phase equilibrium applies at the interfaces between regions leads to corresponding discontinuities in species concentration. This mathematical inconvenience can be circumvented by substituting, in Fick's Law, the gas tension gradient for the concentration gradient.

$$\dot{M}_x = -\alpha_x D_x S(dp_x^*/dz) \qquad (9)$$

The product, $\alpha_x D x$, sometimes called Krogh's constant of diffusion (Dejours 1981), indicates that the diffusion rate of gases with a small diffusion coefficient can be significant when compensated by a sufficiently large solubility.

A useful form of equation 9 results for the case of steady-state diffusion through a planar barrier of thickness l. Then the transport rate $\dot{M}_x$ is constant and equation 9 can be integrated with the result that

$$\dot{M}_x = (\alpha_x D_x/l) S(p_{x_1}^* - p_{x_2}^*) \qquad (10)$$

where $p_{x_1}^*$ and $p_{x_2}^*$ are the gas tensions at the two sides of the barrier. Thus, the tension of an inert gas is linearly distributed in the diffusion direction, and the proper driving force for diffusion is the gas tension difference (figure 6a).

If a gas species undergoes reversible

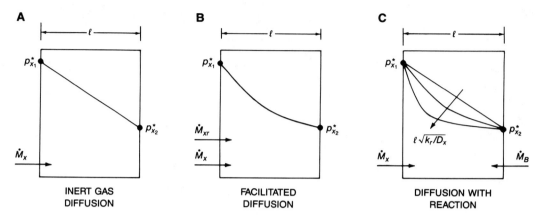

Figure 6. Schematic representation of steady-state diffusion through a stationary barrier of thickness l. A: The gas tension distribution is linear for an inert gas. B: When a species simultaneously diffuses in physically dissolved ($\dot{M}_x$) and reversibly reacted forms ($\dot{M}_{xr}$), the distribution is steepest at the highest gas tensions where the reactive capacitance is smallest. C: When irreversible reaction occurs with a biological substrate B, the distribution becomes increasingly nonlinear as the reaction rate constant increases and the diffusion coefficient of pollutant X decreases.

chemical reaction, its transport rate is altered because it simultaneously diffuses in physically dissolved and reacted forms. When the reaction rate is so fast that chemical equilibrium is closely approached throughout the diffusion barrier, the reactive capacitance coefficient can be utilized with Fick's Law for steady-state conditions to obtain

$$\dot{M}_x = [(\alpha_x D_x + \bar{\beta}_x D_{xr})/l]S(p^*_{x_1} - p^*_{x_2}) \quad (11)$$

where $\bar{\beta}_x$ is the integral average reactive capacitance as illustrated by the graphical construction in figure 5b, and D_{xr} is the diffusion coefficient of the reacted species. The term $(\alpha_x D_x + \bar{\beta}x D_{xr})$ is an extended Krogh constant that accounts for the increase in diffusion brought about by reversible reaction. The existence of such facilitated diffusion has been used to explain enhanced diffusion of oxygen in hemoglobin solutions (Keller and Friedlander 1966) and might also be an important factor in the transport of foreign gases through mucus and tissue (figure 6b).

When chemical reaction occurs at a finite rate or is not reversible, reactive capacitance can no longer relate the concentration of reacted species to the gas tension. Instead, a reaction rate expression must be used in conjunction with Fick's Law. Sup-

pose, for example, that each molecule of pollutant X absorbed from the airway lumen reacts in mucus with b molecules of a biological substrate B that is supplied from underlying tissue (figure 6c). That is,

$$X + bB \rightleftharpoons cC \quad (12)$$

where species C represents reaction products and c represents the number of molecules of C. Then the net rate of depletion of species X per unit volume of mucus can be formulated as

$$R_x = k_{r_f} (C_B)^b (C_x - C_{x_e}) \quad (13)$$

where k_{r_f} is the rate constant of the forward reaction, C_x and C_B are the reactant concentrations, and C_{x_e} represents the concentration of pollutant that would be present if the system was in reaction equilibrium at the prevailing concentrations of biological materials B and C.

Unfortunately, few detailed kinetic data have been collected for inhaled pollutants and, therefore, the rate constants necessary to use reaction expressions such as equation 13 cannot be evaluated. In modeling bronchial uptake, some investigators (McJilton et al. 1972; Miller et al. 1978) have circumvented this problem by assuming that reaction of pollutants with biological substrates is instantaneous and irreversible. In

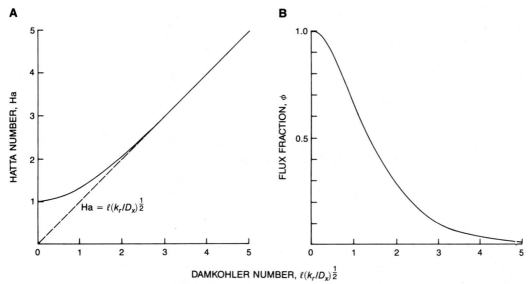

Figure 7. Simultaneous diffusion and first-order reaction in a planar stationary barrier. A: The Hatta number is the ratio of uptake rate in the presence and the absence of chemical reaction. B: The flux fraction ϕ is the ratio of efflux from the barrier to influx into the barrier.

that case, chemical reaction occurs at a thin plane whose position within the mucous layer is determined by the relative diffusional influx of pollutant and substrate.

A more general approach is to assume that biological substrate is so abundant that C_B is essentially constant, and equation 13 can be approximated by first-order kinetics (Miller et al. 1985):

$$R_x = k_r(C_x - C_{x_e}) \quad (14)$$

where k_r is a modified reaction rate constant. The diffusion equation can then be formulated and solved for two alternative cases—fast irreversible reaction (that is, $C_{x_e} = 0$) or slow reversible reaction. For both of these situations the uptake rate is given by (Hobler 1966)

$$\dot{M}_x = Ha(\alpha_x D_x/l)S(p_{x_1}^* - p_{x_2}^*) \quad (15)$$

where the Hatta number, Ha, is a dimensionless parameter that depends upon the dimensionless ratio of reaction to diffusion time, $l[k_r/D_x]^{1/2}$, known as a Damköhler number (figure 7a).

This equation is similar to the inert gas uptake rate with the addition of the Hatta number to provide a proportional correction for the effect of chemical reaction. As

Damköhler number approaches zero, corresponding to inert gas behavior, Ha reduces to unity and equation 15 becomes identical to equation 10. When Damköhler number is greater than zero, Ha is greater than unity, indicating an enhancement of diffusion by chemical reaction. Above a value of 3.5, Hatta and Damköhler numbers are essentially equal and are proportional to the barrier thickness so that the diffusion rate is independent of l. In other words, when Damköhler numbers are larger than 3.5, the relative reaction time is so short that most of species X is converted to product before it can diffuse through the barrier.

In analyzing the transport of an air pollutant through a chemically reactive barrier, it is important to assess the fraction of absorbed gas that completely penetrates. For the special case of first-order kinetics, figure 7b indicates how the ratio of reactant species efflux to influx, ϕ, depends upon Damköhler number. The fraction of influx that is converted to reaction products is $(1 - \phi)$. Using estimates of the reaction rate constant, aqueous diffusivity (table 3), and barrier thickness (table 2), a value of $\phi = 0.025$ was computed for O_3 diffusion through the mucous blanket of the trachea and first four bronchial generations. This indicates that more than 97 percent of the O_3 absorbed at the airway wall is scavenged by mucus. On the other hand, because the mucous layer on the terminal bronchioli is much thinner, the associated value of ϕ was 0.14, indicating that 14 percent of the absorbed O_3 is expected to penetrate the mucus and reach underlying tissue. Throughout the tracheobronchial tree, the estimated value of ϕ for tissue was nearly zero, implying that airway walls are sufficiently thick to deplete all the O_3 that reaches them.

■ **Recommendation 1.** There is a critical need to quantify the chemical interaction of specific pollutants with mucus, tissue, and blood. Besides determining solubility and diffusion coefficients, it is essential to determine the coefficients in reaction rate equations.

Individual Mass Transfer Coefficients

The uptake formulations reveal that transport rate normal to a diffusion barrier is generally proportional to the product of the solubility, surface area, and gas tension difference across the diffusion barrier:

$$\dot{M}_x = \alpha_x k_m S(p_{x_1}^* - p_{x_2}^*) \qquad (16)$$

where the proportionality constant, k_m, is the individual mass transfer coefficient of species X. Table 4 summarizes some useful formulas for k_m with entry 1 listing the

stationary barrier models represented by equations 10, 11, and 15.

It is also important to consider the uptake that occurs in the presence of flow, as in the airway lumen and the bloodstream. In these situations convection is transverse to the principal direction of absorption, and a diffusion barrier, known as the concentration boundary layer, appears in the moving fluid. In the airway lumen, a boundary layer is formed where air is in contact with the mucous blanket, and in blood vessels a boundary layer is formed where blood is in contact with the endothelial cell lining. In general, as velocity increases, the boundary-layer thickness decreases and uptake rate increases.

Entry 2 in table 4 is a commonly used boundary-layer equation that correlates individual mass transfer coefficients with fluid velocity appearing in the Reynolds number and with the dimensionless ratio of kinematic viscosity to diffusion coefficient, the so-called Schmidt number (Sc). To use this equation, it is necessary to substitute specific values of the correlation constants (c, m, n, q), depending on conduit geometry and flow conditions. A collection of correlation constants pertaining to diffusion within the airway lumen is given in table 5. Using these constants, gas phase k_{m_g} values have been estimated in each conducting airway generation of Weibel's model A. The resulting distributions (figure 8) generally exhibit a maximum value between the eighth and twelfth generation, indicating that the mucus lining these airways is

Table 4. Useful Formulas for Individual Mass Transfer Coefficients, k_m

Description	Formula
(1) Stationary barrier of thickness l^a	
(a) Inert gas	$(k_m l / D_x) = 1$
(b) Reactive gas at equilibrium	$(k_m l / D_x) = 1 + (\bar{\beta}_x D_{xr} / \alpha_x D_x)$
(c) Reactive gas, first-order kinetics	$(k_m l / D_x) = Ha$
(2) Boundary-layer diffusion in tube of	
diameter d and length L^b	$(k_m d / D_x) = c Re^m Sc^n (d/L)^q$
(3) Surface renewal^c	
(a) Inert gas, turnover time τ	$k_m = [D_x / \tau]^{1/2}$
(b) Reactive gas, first-order kinetics	$k_m = [k_r D_x]^{1/2}$

^{a} Equations 10, 11, and 15 of text.
^{b} Hills (1974, p. 58).
^{c} Astarita (1967, pp. 7, 37).

Table 5. Correlation Constants for Mass Transfer Coefficients in the Airway Lumen, k_{m_g}

Conduit Geometry, Flow Conditions	$(k_{m_g}d/D_x) = cRe^m Sc^n(d/L)^q$			
	c	m	n	q
(1) Upper airways, nasal breathing[a]				
(a) Inspiration	0.028	0.854	0.854	0
(b) Expiration, $Re < 7800$	0.0045	1.08	1.08	0
(c) Expiration, $Re > 7800$	0.310	0.585	0.585	0
(2) Upper airways, oral breathing[a]				
(a) Inspiration	0.035	0.804	0.804	0
(b) Expiration, $Re < 12{,}000$	0.0006	1.269	1.269	0
(c) Expiration, $Re > 12{,}000$	0.094	0.704	0.704	0
(3) Lower respiratory tract[a]				
(a) Inspiration	0.0777	0.726	0.726	0
(b) Expiration	0.0589	0.752	0.752	0
(4) Straight tubes, developed flow[b]				
(a) $L/d < 0.1\, ReSc$, $Re < 4000$	1.86	0.333	0.333	0.333
(b) $L/d > 0.1\, ReSc$, $Re < 4000$	0.500	1.0	1.0	1.0
(c) $L/d > 0.1\, ReSc$, $Re > 4000$	0.026	0.800	0.333	0

[a] Nuckols (1981). Re is evaluated at tracheal conditions.
[b] Bird et al. (1960).

particularly susceptible to attack by airborne pollutants. The computations also indicate that the increased respiratory flow present during exercise triples k_{m_g} values, but the difference in mass transfer coefficients among the three pollutants is relatively minor.

Entry 3 in table 4 is an alternate formulation for k_m that might be useful in mucus. Many investigators believe that the mucous blanket is composed of a mosaic of 1- to 100-μm patches (Jeffery and Reid 1977). As these patches flow, it is possible for underlying mucus to be swept to the surface, thereby exposing a fresh sink for pollutant gas. With gaps between patches spaced as close as 1 μm, and at the highest mucus velocity of 1 cm/min, the characteristic time for mucus turnover, τ, could be as short as 6 msec. Higbie's penetration theory and Danckwert's surface renewal model (Astarita 1967) have been developed for predicting mass transfer coefficients in such situations, but unfortunately, existing data on the structure and fluid mechanics of mucus are insufficient to perform reliable computations. On the other hand, because mucus velocity is at least four orders of magnitude less than the adjacent gas velocity, it is reasonable to apply stationary barrier models (table 4, entry 1) as a first approximation.

The estimation of k_m values in the blood spaces is an exceptionally difficult problem. First, diffusion processes depend on geometry that varies widely between the erectile vessels of the nasal passages, the sparsely distributed capillaries of the bronchial circulation, and the densely perfused plate-and-post structure of the pulmonary capillary beds. Second, quantitative anatomic models of vascular structure in the bronchial tissue and the upper airways do not exist. Last, blood is a two-phase material composed primarily of plasma and red blood cells, a fact that must be considered when analyzing diffusion within the small blood vessels.

Overall Mass Transfer Coefficients

In transport across a bronchial (or alveolar) wall, four major barriers to diffusion are encountered: the flowing gas, the mucus (or surfactant-hypophase), the tissue, and the blood layers (figure 9). However, gas tensions at air/mucus, mucus/tissue, and tissue/blood interfaces are usually unknown, and it is necessary to compute uptake rate from the gas-phase partial pressure, p_{x_g}, and blood gas tension, $p_{x_b}^*$.

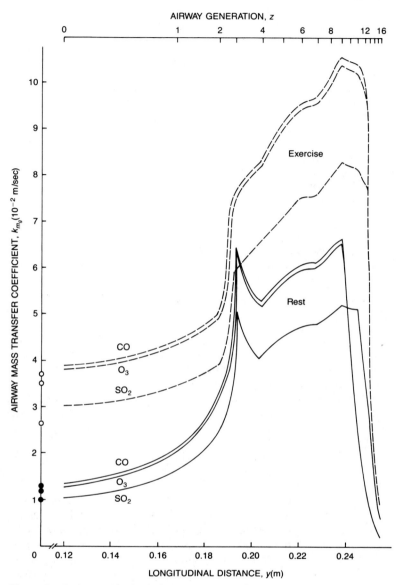

Figure 8.   Estimates of individual mass transfer coefficients in the conducting airways for three common pollutants, based on a straight tube correlation (table 5, entry 4). Flows of 0.4 and 1.6 liters/sec have been used to simulate quiet respiration (solid curves, closed circles) and exercise conditions (broken curves, open circles), respectively. Points on the ordinate are average values computed for the upper airways (table 5, entry 1).

By separately formulating diffusion rate in each of the barriers and applying continuity of uptake to combine these four equations, an absorption rate based on the overall driving force $(p_{x_g} - p_{x_b}^*)$ is obtained:

$$\dot{M}_x = K_m S(p_{x_g} - p_{x_b}^*) \qquad (17)$$

where K_m is the overall mass transfer coef-

ficient. In the sense that finite values of blood gas tension reduce the driving force for uptake, $p_{x_b}^*$ is often referred to as a diffusion backpressure.

When the flux of species X is equal in all diffusion layers, the overall mass transfer coefficient has the well-known form (Treybal 1980, p. 109)

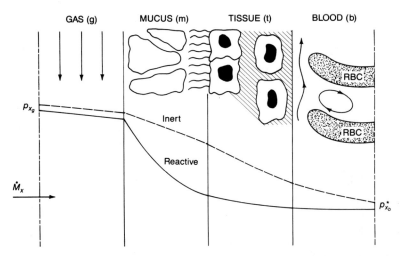

Figure 9. Schematic representation of the four-layer model of diffusion through the bronchial wall. For convenience the blood layer has been drawn as a continuous planar diffusion barrier when, in fact, the bronchial capillaries are sparsely distributed throughout the tissue layer.

$$K_m = [RT/k_{m_g} + 1/(\alpha_x k_m)_m + 1/(\alpha_x k_m)_t + 1/(\alpha_x k_m)_b]^{-1} \quad (18)$$

so that K_m can be constructed from individual mass transfer coefficients and solubility coefficients in gas (subscript g), mucus (subscript m), tissue (subscript t), and blood (subscript b). Since K_m is a harmonic mean, its value will be controlled mainly by the layer having the smallest value of diffusion conductance $\alpha_x k_m$ or, in other words, the largest diffusion resistance, $1/\alpha_x k_m$.

For species that undergo irreversible reaction, the flux is progressively reduced from layer to layer, and the flux fraction, ϕ, must be introduced as a correction factor in the formulation of K_m. If the species reacts in both mucous and tissue layers, then

$$K_m = [RT/k_{m_g} + 1/(\alpha_x k_m)_m + \phi_m/(\alpha_x k_m)_t + \phi_m \phi_t/(\alpha_x k_m)_b]^{-1} \quad (19)$$

where the presence of fractional ϕ values increases K_m relative to the prediction of equation 18.

To examine the implications of these equations, mass transfer coefficients were estimated for the common pollutants, CO, SO_2, and O_3, using the anatomic constants in table 2. For all three species, individual

gas-phase coefficients were computed at quiet breathing conditions, using entry 2 in table 4. Individual mucus and tissue coefficients were calculated from the formulas in table 4 by assuming that CO transport occurs by pure diffusion (entry 1a); SO_2 undergoes instantaneous and reversible reaction to bisulfite ion (entry 1b); and O_3 undergoes irreversible first-order reaction (entry 1c). The overall mass transfer coefficients (with omission of the blood-layer contribution) were then determined by using equation 18 for CO and SO_2 and equation 19 for O_3.

The resulting values of the overall mass transfer coefficients (table 6) decrease from the species SO_2 to O_3 to CO. This ordering is consistent with the National Research Council report (1977, p. 282) that SO_2 is almost completely absorbed within the upper airways; O_3 is transported beyond the upper airways, its uptake being most significant in the distal conducting airways and proximal respiratory branches; and CO absorption occurs deeper within the respiratory zone. A comparison of the values of $\alpha_x k_m$ shows that the controlling diffusion resistance is in gas and tissue layers for SO_2, in the mucous layer for O_3, and in the tissue layer for CO. Thus, it is clear that mucus provides a protective function for an irreversibly reacting gas such as O_3, but its

Table 6. Estimates of the Mass Transfer Coefficient, $\alpha_x k_m$ $(10^{-10}$ kg-mol/m^2/Pa/sec), in a Conducting Airway Model

Airway	SO$_2$[a]				CO				O$_3$[b]			
	Gas	Mucus	Tissue	Overall	Gas	Mucus	Tissue	Overall	Gas	Mucus	Tissue	Overall
Trachea	32	3,970	35	17	57	0.035	0.00031	0.00031	52	1.1	28	1.04
Main bronchi	54	3,970	56	27	91	0.035	0.00050	0.00049	89	1.1	28	1.05
Lobar bronchi	136	3,970	139	68	227	0.035	0.0012	0.0012	223	1.1	28	1.05
Segmental bronchi	272	3,970	278	133	454	0.035	0.0025	0.0023	445	1.1	28	1.06
Subsegmental bronchi	415	3,970	371	187	692	0.035	0.0033	0.0030	679	1.1	28	1.06
Terminal bronchioli	20	6,620	927	20	14	0.059	0.0083	0.0073	13	1.1	4.9	0.84

[a] A $\bar{\beta}_{SO_2}$ value of 0.0013 kg-mol/m^3/Pa was used. This corresponds to the capacitance of pure water in contact with a gas containing 1 ppm SO$_2$ at 1 atm total pressure. It was also assumed that $D_{SO_2} = D_{HSO_3}$.
[b] The tissue coefficient was divided by ϕ_m to account for flux reduction by chemical reaction.

barrier effects have little influence on the absorption of inert or soluble species. The relative diffusion resistances also indicate that increases in respiratory flow tending to diminish the gas-phase mass transfer coefficient should have little effect on O$_3$ and CO uptake, but would probably increase SO$_2$ absorption measurably.

■ **Recommendation 2.** A serious effort is needed to analyze mass transfer through individual diffusion barriers, particularly the mucous layer, the bronchial wall, and the alveolar capillary network.

Longitudinal Gas Transport

Friction imposes drag on the layer of gas moving next to the wall of an airway. Viscosity communicates some of this drag to adjacent layers. As a result, gas near the center of an airway moves downstream faster than gas near the airway wall. This situation can give rise to both radial and axial variations in species concentrations. To avoid considering variations in both spatial directions, it is often convenient to utilize a concentration, C_x, that has been averaged over the airway cross-section. In that case, Fick's Law for the total rate of longitudinal transport, $\dot{N}_x$, is expressed by the sum of a bulk transport and an axial mixing term as follows:

$$\dot{N}_x \quad = \quad \gamma_x \dot{V} C_x \quad - \quad \delta_x S(dC_x/dy) \quad (20)$$

$$\underset{\text{of Transport}}{\text{Total Rate}} \qquad \underset{\text{Bulk Flow}}{\text{Rate of}} \qquad \underset{\text{Mixing}}{\text{Rate of}}$$

where $\dot{V}$ is volumetric flow rate through the airway, y is longitudinal position, γ_x is a retention coefficient, and δ_x is an overall longitudinal mixing coefficient.

Bulk transport refers to the longitudinal motion of a gas species that occurs because the entire gas mixture has a finite volumetric flow. To understand the need to use a retention coefficient in expressing bulk transport, consider a concentration front that divides inspired air containing a foreign gas species from residual air initially devoid of foreign gas (figure 10). When the foreign gas is chemically inert and insoluble in tissue, it is confined to the airway lumen, and the concentration front is displaced at a volumetric rate identical to the bulk flow rate of the entire gas mixture. In that case, it is not necessary to use a retention coefficient (that is, $\gamma_x = 1$). On the other hand, when a soluble gas is inspired, then reversible absorption into the airway wall causes a reduction in its mean axial displacement rate, and the retention coefficient is less than unity (Dayan and Levenspiel 1969). Finally, when a gas species is chemically reactive, then slower-moving molecules near the airway walls are preferentially and

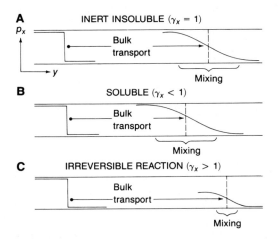

A   INERT INSOLUBLE ($\gamma_x = 1$)

B   SOLUBLE ($\gamma_x < 1$)

C   IRREVERSIBLE REACTION ($\gamma_x > 1$)

Figure 10. Wash-in of foreign gas. Bulk transport refers to the longitudinal displacement of the concentration front in a flowing gas stream, and mixing is characterized by the increase in the width of the front.

permanently absorbed from the airstream, leaving a greater fraction of faster-moving molecules in the vicinity of the airway centerline. In that case, the mean rate of translation of the concentration front increases above $\dot{V}$, and the retention coefficient has a value greater than unity (Sankarasubramanian and Gill 1973).

In contrast to bulk transport, which has been visualized in terms of the axial displacement of a concentration front, longitudinal mixing is characterized by the axial spreading or dispersion of transported species about the front. Three basic mechanisms may be responsible for longitudinal mixing: axial diffusion is due to the Brownian motion of individual molecules and is always present, whether or not there is flow; mixing by asymmetry is due to the nonuniform distribution of transport path lengths and to the unequal flow along these paths; and convective diffusion is due to the presence of a velocity profile which causes a progressive separation between the faster-moving molecules near the airway centerline and the slower-moving molecules near the airway wall. Details regarding the mathematical formulation of the mixing coefficients, δ_x, corresponding to these mechanisms, can be found in a recent review paper (Ultman 1985).

Mathematical Models

Modeling is the process by which the behavior of a system of interest is represented by a simplified mathematical formulation to serve a specific purpose such as correlation of data collected under a variety of conditions; scale-up or extrapolation of performance from small to large systems; or the prediction of behavior in completely new situations. An integral part of the modeling process is the selection of physical and mathematical assumptions, approximations, and simplifications to make the governing equations perspicuous and tractable. This can lead to a variety of alternative models of varying complexity and utility (Himmelblau and Bishoff 1968). Selection of a suitable model must be made and justified within the context of the purpose it is to serve and the data required for its validation and implementation.

A common simplification is to ignore spatial variation in species concentration in one or more dimensions, in the most extreme case assuming that, in designated regions, concentration is independent of position and is a function of time only. This corresponds to the physical assumption that materials are well mixed within each region. The governing material balances for such compartment models are ordinary differential equations. Although compartment models require few parameter values—an advantage in scale-up calculations—they cannot realistically incorporate detailed effects of geometry or fluid dynamics. To do so is a more ambitious task that requires a model in which concentration varies continuously along one or more principal directions instead of being represented by a single value in each of a few well-mixed regions. Such a distributed-parameter model is governed by partial rather than ordinary differential equations. Because its parameters can be closely related to basic material properties and anatomic features, distributed-parameter models are capable of predicting system behavior.

In modeling pulmonary gas uptake, it is expedient to make assumptions that uncouple the mass transport processes in the lungs from those in other organ systems. For example, in treating the uptake of O_3, it has been commonly assumed that rapid irreversible reaction occurs in the mucus, bronchial tissue, or both, so that the O_3 has no opportunity to reach other organs. And in classic models applied to the uptake of CO, it has been assumed that the extraordinary affinity of hemoglobin for CO suppresses its blood gas tension. Thus, the molar concentration of bound CO reaching the systemic circulation may be significant, but its backpressure on the pulmonary diffusion process is negligible. Both of these approximations make sense in the analysis of acute exposure, where lung tissue and blood do not have the opportunity to build up a significant store of pollutant and where chemical reaction is not limited by the availability of biological substrates. However, the modeling of chronic exposure should consider such effects, and in so doing, must also consider the influence of metabolism and excretion in other organs (see, for example, Haggard 1924; Kety 1951).

Compartment Models

The oldest mathematical model of the lungs is attributed to Bohr (1891), who suggested that the airways be idealized by two serial compartments: a dead space representing the upper and conducting airways in which gas is transported by bulk flow with complete absence of mixing, and an alveolar region in which mixing is so complete that composition is uniform. Whereas the Bohr model predicts that expired gas is composed of a dead space of constant composition followed immediately by alveolar gas also of constant composition, modern gas monitoring has shown that in addition to pure dead-space gas (phase I), there is a finite transition in composition (phase II) to the alveolar gas sample (phase III) which exhibits a gradually sloping plateau as expiration continues (see, for example, Fowler 1948).

To analyze gas uptake, the Bohr airway model must be coupled to a model of pulmonary circulation that simulates the progressive equilibration of capillary blood with alveolar gas. A realistic approach is to neglect backmixing in the direction of

blood flow and to combine the uptake rate given by equation 17 with a steady-state material balance along a differential length, dy, of capillary (figure 11a):

$$K_m S(p_{x_g} - p_{x_b}^*)(dy/L) = \dot{Q}(dC_x + dC_{xr}) \quad (21)$$

$$\underset{\text{into Blood}}{\text{Uptake Rate}} \qquad \underset{\text{by Blood Flow}}{\text{Net Output Rate}}$$

where L is capillary length, $\dot{Q}$ is volumetric rate of blood flow, p_{x_g} is the (constant) alveolar partial pressure, $p_{x_b}^*$ is the blood gas tension, and C_x and C_{xr} are the molar concentrations in blood of species X in physically dissolved and chemically combined forms, respectively. To solve this equation, it is necessary to provide relationships between C_x, C_{xr}, and $p_{x_b}^*$. For inert gases, equation 3 constitutes this relationship, but the situation is more complicated for reactive gases.

Consider, for example, the case of a reversibly reacting gas that undergoes a reaction sufficiently rapid that equilibrium occurs at all points in the capillary. Then equation 7 and a differential form of equation 3 must both be applied to equation 21 with the result:

$$(K_m S/\dot{Q})(y/L) =$$
$$\int_{p_{x_v}^*}^{p_{x_b}^*(L)} (\alpha_{x_b} + \beta_{x_b})/(p_{x_g} - p_{x_b}^*) dp_{x_b}^* \quad (22)$$

where $p_{x_v}^*$ and $p_{x_b}^*(L)$ are the equilibrium partial pressures of species X in venous and arterial blood, respectively. To perform this so-called Bohr integration (Hills 1974, p. 64), the exact dependence of reactive capacitance on gas tension must be known. For purposes of discussion, however, β_x can be fixed at an appropriate mean value, $\bar{\beta}_x$, thereby allowing analytical integration of equation 22 (Piiper et al. 1971).

The resulting gas tension distributions (figure 11b) depend upon a single dimensionless parameter,

$$Bo = K_m S/[\dot{Q}(\alpha_{x_b} + \bar{\beta}_{x_b})] \quad (23)$$

the ratio of characteristic convection time to diffusion time. For small values of this

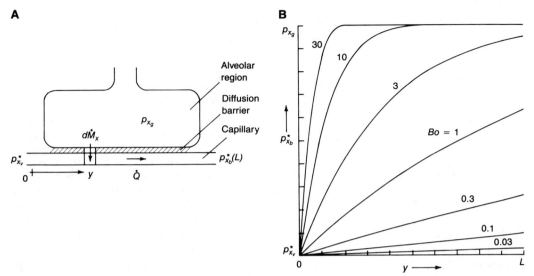

Figure 11. Alveolar uptake model. A: Schematic diagram of model. B: Longitudinal capillary gas tension distributions resulting from Bohr integration. (Adapted with permission from Piiper and Scheid 1980, p. 138, and from Academic Press, Orlando, Fla.)

ratio, diffusion across the parenchymal tissue is slow, and there is insufficient time for end-capillary blood to become equilibrated with alveolar gas. For large values of the Bo ratio, however, the relative perfusion rate is small enough to ensure complete equilibration. This opposition of convection and diffusion effects is also apparent in the overall uptake equation derived from the Bohr integration (Piiper and Scheid 1980).

$$\dot{M}_x = K_m S[(1 - e^{-Bo})/Bo](p_{x_g} - p_{x_v}^*) \quad (24)$$

Equation 24 suggests that the overall driving force for alveolar uptake is the difference between alveolar partial pressure p_{x_g} and venous tension $p_{x_v}^*$ and the quantity in square brackets corrects the overall mass transfer coefficient for any added resistance due to a perfusion limitation. When Bo is small, this correction factor approaches unity so that the uptake is determined by $K_m S$ alone (that is, it is diffusion limited), but when Bo is large, the correction term approaches $1/Bo$ and uptake is proportional to $\dot{Q}(\alpha_{x_b} + \bar{\beta}_{x_b})$ (that is, perfusion limited).

Although envisioned by physiologists as a means of evaluating pulmonary function, CO uptake should be regarded as the most extensively studied prototype of pollutant absorption. Carbon monoxide has a low

aqueous solubility, but its reactive capacitance with hemoglobin is so large that the value of Bo is only 0.02, implying diffusion-limited uptake. Moreover, because of its strong affinity for hemoglobin, CO gas tension in pulmonary capillary blood is much below its alveolar partial pressure. In the extreme case where CO venous tension can be disregarded entirely, equation 24 indicates that $K_m S$ may be computed as the ratio of uptake rate to alveolar partial pressure, $\dot{M}_x/p_{x_g}$, a quantity called the pulmonary diffusing capacity. And even though carboxyhemoglobin reaction may not be sufficiently rapid to justify this assumption of negligible CO backpressure, it is still possible to equate $K_m S$ with $\dot{M}_x/p_{x_g}$ when the reaction kinetics are first order with respect to CO concentration (Roughton and Forster 1957). In that case, $K_m S$ is composed of a hemoglobin reaction capacity, $(\alpha_x k_r V)_b$, in addition to a true alveolar membrane-diffusing capacity, $(K_m S)'$, as given by

$$K_m S = [1/(K_m S)' + 1/(\alpha_x k_r V)_b]^{-1} \quad (25)$$

where V_b is blood volume in the capillary bed, and k_{r_b} is the reaction rate constant of species X in blood. For CO, the membrane diffusion resistance $1/(K_m S)'$ is typically three times the reaction limitation $1/(\alpha_x k_r V)_b$.

A popular determination of diffusing capacity is the single-breath method in which the decline in expired alveolar composition during a known breath-holding interval is measured. In this situation, it is appropriate to apply an unsteady-state differential material balance to the alveolar compartment.

$$\frac{1}{RT}[d(V_g p_{x_g})/dt] + \dot{M}_x = 0 \qquad (26)$$

Accumulation   Uptake
Rate   Rate

An important feature of the Bohr model is that inspiration and expiration are, in effect, occurring simultaneously and the alveolar volume, V_g, is constant. After introducing the additional assumption that venous gas tension is negligible, equations 24 and 26 can be integrated to obtain

$$p_{x_g}(t_b)/p_{x_g}(0) =$$
$$\exp\{-[K_m S(1 - e^{-Bo})RT/V_g Bo]t_b\} \qquad (27)$$

where $p_{x_g}(t_b)$ is the alveolar partial pressure of species X after breath-holding time t_b. This formula, which accounts for both perfusion and diffusion effects, is a generalization of the more traditional Krogh equation in which a diffusion limitation for CO is assumed.

Three important limitations in the Bohr model have been identified in previous applications. First, the dynamics occurring during periods of inspiration and expiration are not included in the conventional Bohr model and, therefore, the effect of breathing pattern cannot be evaluated. This restriction may be overcome by using a dynamic Bohr model (see, for example, Murphy 1969; Graham et al. 1980) in which input and output by gas flow explicitly appear as terms in the differential material balance. Second, the Bohr model fails to anticipate imperfect mixing phenomena, such as those responsible for phase II and sloping phase III behavior of expired single-breath concentration curves. These mixing dynamics can be simulated by using multicompartment models in which parallel ventilation

paths are provided (Robertson et al. 1950). Third, the Krogh equation predicts a linear semilogarithmic relationship between expired alveolar concentration and breath-holding time, whereas the relationship for highly soluble gases has been shown to be curved (Cander and Forster 1959). Models incorporating permeable bronchial compartments in addition to the alveolar region would be useful in simulating such data.

The model of Saidel et al. (1973), which overcomes the first two of these three limitations, exemplifies the potential flexibility of multicompartment analysis. Their five-compartment model (figure 12a) can simulate stratified and regional concentration inhomogeneities, nonuniformities in gas flow distribution, and distension of airway volumes, and it can incorporate any desired breathing pattern. The general mass balance equation for a compartment j in Saidel's model is similar to that of the Bohr model, equation 26, with addition of terms to account for input and output by gas flow:

$$\frac{1}{RT}[d(V_j p_{x_j})/dt] =$$

Accumulation
Rate

$$(\dot{V}/RT)\sum_k (f_{kj}p_{x_k} - f_{jk}p_{x_j}) - \dot{M}_{x_j} \qquad (28)$$

Net Input Rate by   Uptake
Gas Flow   Rate

where $\dot{V}$ is the time-dependent tracheal flow; V_j is compartment volume; p_{x_j} is species partial pressure; f_{jk} is the fraction of tracheal flow from compartment j to an adjacent compartment k; and $\dot{M}_{x_j}$ is uptake rate through the walls of compartment j. In the two alveolar compartments, $\dot{M}_{x_j}$ was related to p_{x_j} by equation 17, and in the three bronchial compartments, $\dot{M}_{x_j}$ was neglected. The five simultaneous differential equations that result cannot be solved analytically, but must instead be integrated using well-established computer algorithms.

In simulating the single-breath CO uptake test, Saidel et al. (1973) assumed that the CO backpressure was zero. They also

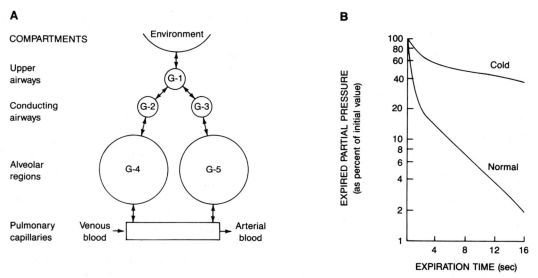

Figure 12. Multicompartment model of CO uptake. A: Schematic diagram of the model. B: Simulation of CO partial pressure expired from compartment G-1. (Adapted with permission from Saidel et al. 1973, p. 484, and from the National Center for Scientific Research, Paris.)

assumed that CO uptake from the two alveolar compartments was governed by identical values of $K_m S$. And instead of the conventional technique of changing breath-holding time to visualize the effect of uptake on expired alveolar composition, these investigators simulated expiration phases of varying times. Figure 12b illustrates the resulting CO uptake in the presence of chronic obstructive pulmonary disease (COPD) as compared to the more rapid uptake in a healthy lung. The dramatic difference between the two simulations is merely the result of adjusting the f_{jk} parameters to account for the inhomogeneity of alveolar ventilation in the presence of COPD.

It would be useful to extend Saidel's model for general application to pollutant gases. The principal element required to complete the formulation is a rate expression for bronchial absorption, analogous to equation 24 for alveolar absorption. DuBois and Rogers (1968) modeled bronchial uptake of nonreactive foreign gases as diffusion from a well-mixed airway compartment to a well-mixed capillary compartment across a homogeneous tissue barrier (figure 13, inset). The use of a well-mixed compartment implies that pulmonary capillaries are randomly oriented, thus provid-

ing extensive backmixing of blood; and the assignment of a diffusion resistance to the tissue alone implies that diffusion through the gas boundary layer and mucous blanket is relatively rapid. In addition, the analysis was restricted to acute exposure where recirculation of pollutant to afferent capillary blood is negligible.

Under these conditions, a steady-state material balance around capillary compartment j results in the following relationship between uptake rate and airway partial pressure:

$$\dot{M}_{x_j} = K_{m_j} S_j p_{x_j} \qquad (29)$$

where S_j is the contact area between tissue and capillary compartments, and $K_{m_j} S_j$ may be viewed as a bronchial diffusing capacity. According to the DuBois-Rogers model,

$$K_{m_j} = (1/\alpha_{x_t} k_{m_j} + S_j/\alpha_{x_b} \dot{Q}_j)^{-1} \qquad (30)$$

so that K_{m_j} is analogous to an overall mass transfer coefficient that results from the sum of a tissue diffusion resistance characterized with an individual mass transfer coefficient, k_{m_j}, and a capillary perfusion resistance arising from the finite rate of blood flow, $\dot{Q}_j$. Using the stationary barrier model (table 4, entry 1a) to estimate values of k_{m_j}, DuBois and Rogers predicted the

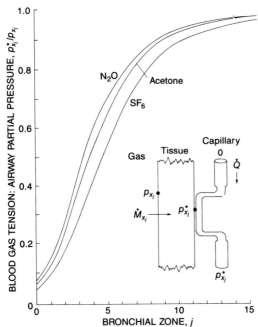

Figure 13. Bronchial uptake model. Partitioning of pollutant between capillary blood, $p_{x_j}^*$, and airway lumen, p_{x_j}, in the conducting airway generations. (Adapted with permission from DuBois and Rogers 1968, pp. 40, 42, and from Elsevier Science Publishers.)

partitioning of gas tension between capillary blood and the airway lumen, $p_{x_j}^*/p_{x_j}$, in each of 15 conducting airway generations of Weibel's model A. When capillaries were assumed to envelop the outside of the bronchial wall, $p_{x_j}^*/p_{x_j}$ values distal to the tenth generation are at least 0.95 (figure 13), indicating that blood gas tension is within 5 percent of airway gas partial pressure. In other words, soluble gas penetration from the airway lumen through the bronchial wall is nearly complete. However, in more proximal generations where the tissue layer is thicker, there is a more significant departure from complete penetration.

Recognizing that the spatial distribution of blood vessels can influence the penetration of pollutants through tissue, DuBois and Rogers also investigated models in which perfusion occurred within the bronchial wall. Generally speaking, they found that capillaries located near the airway lumen are effective in removing absorbing

species before they can penetrate far into tissue, and deep tissue damage is thereby averted. For example, when perfusion was assumed to be uniform throughout the bronchial wall, $p_{x_j}^*/p_{x_j}$ never exceeded 0.25.

Hori and Suzuki (1984) presented a model of bronchial uptake of reactive gases wherein diffusion from an airway compartment to tissue and to capillary blood occurred in parallel. A physical basis for this model would be a bronchial surface composed of coexisting capillary and tissue regions. Although this may be a reasonable picture of the upper airways where a rich network of superficial capillaries exists, it would seem to be less appropriate in bronchial airways. The overall mass transfer coefficient for the Hori-Suzuki model was formulated as

$$K_m = [RT/k_{m_g} + 1/(\alpha_{x_b} k_{m_b}) (1 + \theta)]^{-1} \quad (31)$$

where θ, the ratio of $\alpha_x k_m$ values between tissue and blood, reflects the relative affinity of the pollutant species for these two competing absorbent phases.

In estimating uptake in the conducting airways during inspiration, Hori and Suzuki analyzed the steady flow of polluted air through a series of 16 well-mixed airway compartments corresponding to the bronchial generations of Weibel's model A. Utilizing equation 31 in an appropriate steady-state material balance, uptake rates for each compartment were computed, and then summed to determine accumulated uptake. Since constant values of 0.0181 and 1.41 cm/sec were specified for k_{m_b} and k_{m_g}, respectively, and θ was fixed at 6.21 regardless of the diffusing species, uptake values were a function of blood solubility only. The results of these computations (figure 14a) indicate that accumulated uptake is negligible when $\alpha_{x_b} < 10^{-9}$ kg-mol/m^3/Pa, but when $\alpha_{x_b} > 10^{-5}$, absorption into the conducting airways is essentially 100 percent.

Distributed-Parameter Upper Airway Models

The upper and lower respiratory tracts have often been modeled separately as distributed-parameter models, but rarely to-

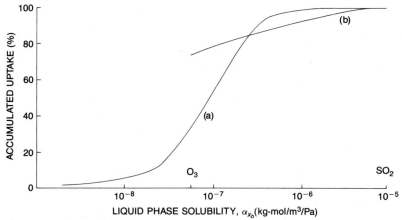

Figure 14. Comparison of model predictions of total lung uptake. Based on data from a: Hori and Suzuki (1984); b: McJilton et al. (1972).

gether. There are at least two reasons why this separation is natural. First, the upper airways are adapted to perform some functions not shared with the lower respiratory tract: filtering out particulate material and warming and moistening inspired air before it reaches the lower airways. Second, the geometry of the upper airways is complex, irregular, and asymmetric, making it difficult to represent concisely, but the tracheobronchial tree is regular and symmetric in a way that makes it possible to represent a vast number of interconnecting branches with a simple and tractable idealization.

When air is inhaled simultaneously through the nose and mouth, analyzing uptake rate through the upper airways requires separate models for nasal and for oral absorption coupled according to the distribution of airflow through the two channels. As applied to absorption during a steady inspiratory flow, this has been expressed formally (Oulrey et al. 1983; Kleinman 1984) as:

$$p_{x_o}/p_{x_i} = f_n(p_{x_o}/p_{x_i})_n + (1 - f_n)(p_{x_o}/p_{x_i})_m \quad (32)$$

where p_{x_o}/p_{x_i}, sometimes known as tracheal penetration, is the ratio of pollutant partial pressures in the trachea and inhaled air; $(p_{x_o}/p_{x_i})_n$ and $(p_{x_o}/p_{x_i})_m$ are the separate tracheal penetrations for air access through the nose and through the mouth, respectively; and f_n is the fraction of tracheal flow that enters through the nose. Whereas values of

f_n are available from direct flow measurements (Niinimaa et al. 1981), the nasotracheal and orotracheal penetrations have been formulated with distributed-parameter models.

To explain the penetration of a variety of soluble vapors during steady inspiratory flow, Aharonson et al. (1974) modeled the nose as a tortuous conduit surrounded by a combined mucous/tissue layer that is enveloped by a sheath of capillaries (figure 15a). Neglecting longitudinal mixing in the airstream, a steady-state material balance was written around a differential length, dy, of the conduit.

$$-\frac{\dot{V}}{RT}\frac{dp_{x_g}}{dy} = K_m(y)a(y)(p_{x_g} - p_{x_b}^*)$$

$$(33)$$

Net Input Rate   Uptake
by Gas Flow   Rate

where the overall mass transfer coefficient $K_m(y)$ and airway perimeter $a(y)$ are both functions of longitudinal position, y. By assuming that gas tension of pollutant throughout the blood is negligible, this ordinary differential equation was integrated from the airflow inlet to the outlet with the result

$$p_{x_o}/p_{x_i} = \exp(-K_m SRT/\dot{V}) \quad (34)$$

where $K_m S$ is the value of $K_m(y)a(y)$ integrated over the length of the conduit. A

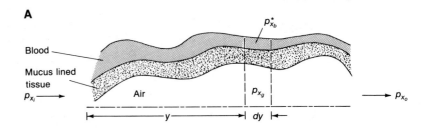

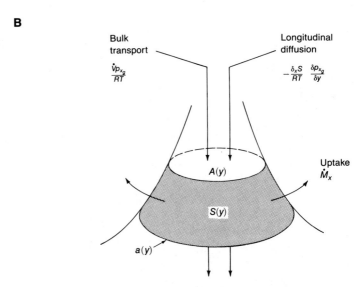

Figure 15. Geometry of distributed-parameter models. A: Upper airways (adapted with permission from Aharonson et al. 1974, p. 655, and from the American Physiological Society). B: Symmetric tracheobronchial tree.

similar model was utilized both by Oulrey et al. (1983) and by Kleinman (1984) to correlate SO_2 absorption data obtained in the upper airways. They alternatively applied equation 34 to the nose and to the mouth by using separate values of $K_m S$.

As Brain (1970) pointed out, the nose functions much as an industrial gas-liquid scrubber, and general design methods for such devices are common in chemical engineering literature (see, for example, Treybal 1980, pp. 300–313). These methods facilitate integration of the differential gas-phase material balance (eq. 33) without the need to neglect the possible buildup of pollutant backpressure in the liquid absorbent. For example, when a nonreactive soluble pollutant is absorbed with a parallel flow of initially pure liquid, then

$$p_{x_o}/p_{x_i} = \{F + \exp\left[(-K_m SRT/\dot{V})(1 + F)\right]\}$$
$$/\{1 + F\} \quad (35)$$

where F is a dimensionless ratio defined in terms of the gas-liquid flow ratio $\dot{V}/\dot{Q}$ as

$$F = \dot{V}/RT\alpha_{x_b}\dot{Q} \quad (36)$$

On the basis of an estimated respiratory air-to-blood flow ratio of $\dot{V}/\dot{Q} = 10$ (La Belle et al. 1955), gases of low solubility ($10^{-6} > \alpha_{x_b} > 10^{-8}$ kg-mol/m^3/Pa) should have nasal F values of one or more, whereas pollutants of very high solubility ($\alpha_{x_b} > 10^{-5}$ kg-mol/m^3/Pa) may have F values considerably less than one. In the latter case, equation 35 reduces to the Aharonson model (eq. 34), implying that backpressure created by absorbed pollutant is negligible. On the other hand, when F is greater than one, there is significant back-pressure, and equation 35 predicts penetration ratios, p_{x_o}/p_{x_i}, which are larger than those expected from the Aharonson model.

Distributed-Parameter Lower Airway Models

Whereas the development of upper airway models has been limited, distributed-parameter modeling of gas transport in the tracheobronchial tree has received more attention. A primary motivation has been to understand airway mixing phenomena such as the slope of the phase III alveolar plateau or the dependence of anatomic dead space on respiratory flow and breath-holding time. Rauwerda (1946) was the first to combine a spatially varying anatomic geometry with the diffusion equation in order to analyze the longitudinal distribution of inert gas within the airways. He assumed that summed airway cross-section increased linearly with longitudinal distance, and he ignored the contribution of individual airway dimensions and branching characteristics. Major improvements in Rauwerda's model have included more realistic representation of summed airway cross-section (LaForce and Lewis 1970); the incorporation of convective bulk transport due to breathing pattern (Cumming et al. 1971); the use of a concentric two-zone model of respiratory airways to account for the stagnant gas pockets between alveolar septa (Scherer et al. 1972; Paiva 1973); and the analysis of diffusion and convection between parallel units (Paiva and Engel 1979) to simulate ventilation inhomogeneities.

These analyses all share the philosophy that, first, the key variable along a given transport path is the longitudinal distribution of partial pressure resulting from simultaneous diffusion and convection, and, second, alternative transport paths are alike enough that the tracheobronchial tree can be represented as a single airway with summed cross-section that expands with distal distance from the trachea (figure 15b). McJilton et al. (1972) followed the same philosophy in developing the first distributed-parameter model of reactive gas uptake, beginning with

a material balance on a differential section of the airway lumen given by

$$\frac{A}{RT}\frac{\partial p_{x_g}}{\partial t} = -\frac{\dot{V}}{RT}\frac{\partial p_{x_g}}{\partial y}$$

$$\text{Accumulation} \qquad \text{Net Input Rate}$$
$$\text{Rate} \qquad \text{by Gas Flow}$$

$$+\frac{1}{RT}\frac{\partial}{\partial y}\left(D_x A\frac{\partial p_{x_g}}{\partial y}\right) - a\left(\frac{\dot{M}_x}{S}\right) \qquad (37)$$

$$\text{Longitudinal} \qquad \text{Uptake}$$
$$\text{Diffusion Rate} \qquad \text{Rate}$$

where p_{x_g} represents partial pressure averaged over the cross-section A available for flow, a is the perimeter available for gas absorption, and both A and a are increasing functions of distal longitudinal distance, y.

In solving this equation, McJilton et al. used a finite-difference algorithm in which respiratory flow, $\dot{V}$, varied with time in sinusoidal fashion, and in which the airway domain was divided into 25 longitudinal zones patterned after the 23 generations of Weibel's model A. Each airway zone was bounded by a tissue layer coated with an inner liquid film of either mucus in conducting zones or surfactant in respiratory zones. The principal assumptions in specifying absorption between the gas, liquid, and tissue layers were that the diffusion resistance of the gas boundary layer within the airway lumen is negligible; transport through the liquid film occurs by steady-state diffusion; and chemical reaction occurs exclusively at the liquid/tissue interface, where pollutant is instantaneously consumed. Since there is no accumulation or depletion of pollutant within the liquid film, the influx of pollutant from the airway lumen is identical to tissue dose. And because pollutant concentration is negligible at the liquid/tissue interface, the uptake flux, $\dot{M}_x/S$, could be expressed as the product of diffusion conductance in the liquid film, $\alpha_x k_m$, with the pollutant partial pressure in the airway lumen, p_{x_g}. In performing computations, $\alpha_x k_m$ values were estimated using the stationary barrier equation (table 4, entry 1a) in conjunction with appropriate values of aqueous-phase solubility and diffusion coefficient and liquid film thickness of 10 μm in upper generations, 3–5 μm in alveolar ducts, and 0.3 μm in the alveoli.

Predicted values of the local dose, expressed as mass absorbed per unit bronchial surface area per breath, are shown in figure 16 for O_3 and SO_2. The O_3 dose is uniformly distributed within the conducting airway segments and then declines sharply in the terminal airspaces distal to the twentieth model segment. For SO_2, a much more soluble gas, the uniform portion of the dose distribution is at a higher level but does not span as many airway generations as in the case of O_3. Thus, the integrated doses of these two pollutants are similar. This explains why predicted values of accumulated uptake are relatively insensitive to solubility, experiencing an increase of only 30 percent for a 100-fold increase in α_{x_b} (figure 14b).

Compared to Hori and Suzuki's (1984) steady-state inhalation model (figure 14a), the McJilton model predicts a greater total uptake for a low-solubility gas such as O_3, but less uptake for a higher-solubility gas such as SO_2. The former result is due to absorption in the respiratory zone, which was ignored in the Hori-Suzuki model and which is an important consideration for gases of low solubility. The latter result is probably due to mass transfer coefficient values: those used by McJilton et al. were far smaller than those of Hori and Suzuki.

A major shortcoming of the McJilton model is the localization of chemical reaction at the liquid/tissue interface. Therefore, Miller et al. (1978) developed a model of O_3 transport in which the site of pollutant reaction could vary within the mucus. In particular, the mucous layer was modeled as a stagnant barrier into which O_3 diffuses from the airway lumen and reactive biological substrates simultaneously diffuse from underlying tissue. At the reaction plane where the O_3 and substrates meet, they are instantaneously consumed by chemical reaction. And the depth of penetration of this plane into the mucous layer is limited by the rate of O_3 absorption relative to the rate of substrate supply. By definition, tissue exposure is nonzero only when

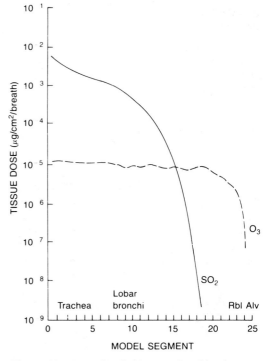

Figure 16. Dose distributions predicted by the McJilton model. Environmental concentration is 1,000 μg/m^3, tidal volume is 500 ml, and breathing frequency is 15 breaths/min. Rbl refers to respiratory bronchioles; Alv to alveolus. (Adapted with permission from McJilton et al. 1972.)

O_3 diffusion is sufficiently rapid that the reaction plane reaches the mucus/tissue interface, at which point all pollutant is consumed by the tissue. Other additions to this model include the use of a mass transfer coefficient, k_{m_g}, and a mixing coefficient, δ_x, to better describe lateral and longitudinal transport in the airway gas.

The tissue doses predicted by Miller et al. in Weibel's geometric model generally increase from a value of zero in the trachea, where the reaction plane cannot completely penetrate the thick mucous layer, to a maximum value in the 17th generation where the mucous layer is thinnest (figure 17). In more distal generations, the dose declines because absorption is spread out over the rapidly increasing alveolar surface area and also because the surfactant layer is assumed to be unreactive with O_3. In spite of the marked decline in tissue dose per unit area

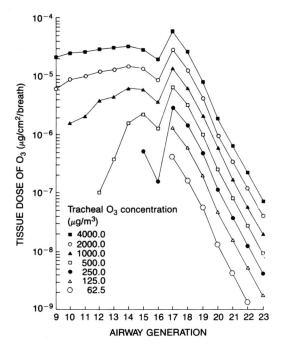

Figure 17. Ozone dose distribution predicted by the model of Miller et al. Tidal volume is 500 ml and breathing frequency is 15 breaths/min. (Adapted from Miller et al. 1978, p. 91.)

from conducting airways to respiratory airspaces, the total mass of O_3 absorbed into the respiratory zone was found to be five times that absorbed into conducting airways.

The model of McJilton et al. predicts a tissue dose strictly proportional to tracheal concentration, but this relation in the Miller model is nonlinear because of the moving reaction plane. For example, consider the influence of tracheal concentration on dose distribution in conducting airways as predicted by the Miller model. Proximal to the sixteenth generation, there is no tissue exposure at tracheal concentrations below 60 $\mu g/m^3$. But at progressively larger concentrations, more and more conducting airway tissue is exposed to pollutant, beginning with the more distal generations and progressing toward the trachea. Thus, the total uptake in conducting airway tissue increases with tracheal concentration in a disproportionate fashion.

The most significant finding in the analyses of both McJilton et al. and Miller et al. is that certain tissues receive a much higher

dose than others, indicating that O_3 damage might be localized. The predictions of the two models are somewhat different, however. McJilton's model exhibits a uniform dose throughout the conducting airways that is considerably greater than O_3 exposure in the respiratory zone. On the other hand, the distribution curves computed by Miller et al. have dosage peaks in the respiratory bronchioli that are many times larger than doses in adjacent conducting airways and respiratory airspaces alike. Such sharply defined peaks indicate that the respiratory bronchioli are particularly susceptible to O_3 damage, and evidence from histological studies on primates (Dungworth et al. 1975) supports this conclusion. Miller et al. (1983) have applied the same model to NO_2 uptake, and the shape of resulting dose distribution curves is similar to those of O_3.

Miller et al. (1985) sought to improve their previous model for O_3 transport by accounting for the finite rate of O_3 reaction in surfactant, tissue, and blood, as well as in mucus. The bronchial wall was modeled by adjacent layers of stagnant mucus and solid tissue, and the respiratory airway wall was modeled by a series of stagnant surfactant, solid tissue, and stationary blood layers. Quasi-steady transport by simultaneous diffusion and irreversible chemical reaction was assumed to occur through each layer. The use of first-order reaction kinetics implying a constant concentration of substrate precluded a limitation on pollutant flux by the supply of biological species.

Rather, the O_3 flux was restricted by boundary conditions at the edge of the (infinitely thick) bronchial tissue layer and at the centerline of the pulmonary blood layer. The equations that ultimately described this model were linear, and therefore, when normalized by the tracheal concentration, the predicted dose distributions were independent of tracheal concentration.

A primary objective in carrying out computations with this model was to determine the sensitivity of dose predictions to several parameters including the gas-phase mass transfer coefficient, the longitudinal mixing coefficient, the mucus reac-

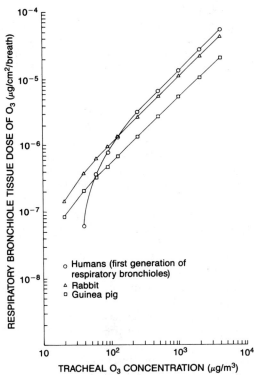

Figure 18. Comparison of respiratory bronchial tissue dose in humans and laboratory animals. (Adapted from Miller et al. 1978, p. 95.)

tory airway tissue to O_3 is aggravated by increasing levels of exercise.

An important application of these models is the comparison of dose distributions in laboratory animals to those in humans. Miller et al. (1978, 1983) utilized their model with morphometric and physicochemical data available for guinea pigs and rabbits as well as for humans. From the viewpoint of O_3 dose absorbed into respiratory bronchioles (figure 18), the model predicts that humans are more like rabbits than guinea pigs, at least at tracheal concentrations above 100 $\mu g/m^3$. It must be borne in mind, however, that the validity of such predictions depends strongly on the accuracy of the many anatomic, equilibrium, diffusion, and reaction parameters used in the model.

■ **Recommendation 3.** To analyze total uptake of pollutants and to predict dose distribution, mathematical models that account for ventilation and perfusion limitations, including their regional distribution, should be developed and validated.

Experiments

In this section, experimental methods that can be used to validate proposed transport models and estimate relevant thermodynamic, diffusion, and reaction parameters are discussed. Where published data exist, they are compared with the predictions of models already presented in the Mathematical Models section. Three categories of experimental methods are described: in vitro methods that use physical models or isolated tissue preparations; in vivo methods that, because of a high degree of risk or invasiveness, use animals as subjects of study; and in vivo techniques that are suitable for use on human subjects.

In Vitro Methods

Physical models fabricated from stock materials or from casts of actual lung airways offer the opportunity to evaluate mathematical models in the absence of undesired

tion coefficient, and the minute volume of breathing. As has been established in table 6, the gas-phase diffusion resistance of O_3 is relatively small because of its low solubility. Therefore, it is not surprising that dose predictions were insensitive to the gas-phase mass transfer coefficient. Similarly, the magnitude of the longitudinal mixing coefficient had little effect on O_3 uptake throughout the respiratory zone, which is consistent with previous mathematical analyses of O_2 and CO_2 exchange (Pack et al. 1977). On the other hand, dose predictions were quite sensitive to the mucus reaction rate constant, k_r. As the value of k_r was increased, the dose to bronchial mucus became larger and the bronchial tissue dose decreased dramatically. This demonstrates the effectiveness of mucus in removing O_3 before it can damage underlying tissue. Simulations performed at progressively increasing tidal volumes and breathing frequencies implied that exposure of respira-

biological variables. Physical models were used to study gas transport as early as 1915, when Henderson et al. (1915) introduced smoke into a tube in order to visualize airflow patterns and concluded that individual gas species can penetrate well beyond their tidal volume because of nonuniform longitudinal convection. A continuing series of such investigations, using inert tracer gases in both upper airway casts (Simone and Ultman 1982) and tracheobronchial network models (see, for example, Scherer et al. 1975; Ultman and Blatman 1977), have added considerably to the understanding of longitudinal convective-diffusion processes and to the evaluation of longitudinal mixing coefficients.

Of greater importance than longitudinal mixing is the lateral diffusion of absorbable species in the vicinity of airway walls. Nuckols (1981) applied the analogy between heat and mass transfer (Bird et al. 1960, p. 642) to infer gas-phase mass transfer coefficients from thermal measurements in an adult upper airway cast as well as in a symmetric tracheobronchial model. Measurements were made in both steady inspiratory and steady expiratory flows, and the resulting correlations of the Sherwood number, $k_{m_g}d/D_x$, show a strong dependence on both Reynolds and Schmidt numbers (table 5).

Using the napthalene wall sublimation technique, Hanna and Scherer (1986) determined local mass transfer coefficients in an adult upper airway cast supplied with a fixed inspiratory airflow of 12 liters/min. The distribution in k_{m_g} values from the external nares to the larynx was highly nonuniform, with the largest value occurring at the entrance of the nasopharynx where there is an abrupt convergence of airflow from the turbinates.

Neither Nuckols nor Hannah and Scherer attempted to simulate absorption processes within the airway wall. Ichioka (1972), however, measured SO_2 absorption from a flowing gas stream into moistened filter paper intended to represent the upper-airway mucosa. In each experiment, a straight tube, 60 cm long × 0.9 cm diameter, was exposed to a steady flow of a mixture of 5 ppm SO_2 in nitrogen for 15 min, and the cylindrical sections of filter paper lining the tube were moistened with either distilled water or with 1 percent bovine serum albumin (BSA) solution. Ichioka's data are presented in terms of the inlet-to-outlet partial pressure ratio, p_{x_i}/p_{x_o}, allowing comparison to the Aharonson model of uptake into the upper airways (figure 19). According to equation 34, a chord drawn from the origin to any data point on a plot of $\ln(p_{x_i}/p_{x_o})$ versus S/V has a slope equal to RTK_m, and therefore the plot is linear when the overall mass transfer coefficient is constant. Clearly this is not the case for Ichioka's measurements. The slopes of the curves progressively decrease, implying that K_m decreases as the transit time, S/V, increases.

To interpret these results, consider the overall mass transfer coefficient as being composed of diffusion resistances due to the gas boundary layer (subscript g) and to the filter paper (subscript l):

$$RTK_m = [1/k_{m_g} + 1/RT(\alpha_x k_m)_l]^{-1} \quad (38)$$

According to this equation, the largest K_m values occur when the resistance of the filter paper is negligible, and RTK_m is well approximated by k_{m_g}. The p_{x_i}/p_{x_o} curve for this gas-phase limited diffusion has been constructed in figure 19 using standard correlations for k_{m_g} (table 5, entry 4). Because the experimental p_{x_i}/p_{x_o} values are consistently smaller than the gas-limited predictions, the wet filter paper must constitute a significant impediment to diffusion. This is especially true for long transit times, when absorption is so great that a significant backpressure of SO_2 is built up at the gas liquid interface. Judging from typical values of k_{m_g} and RTK_m calculated from the slopes of the dotted chords in figure 19, equation 38 indicates that the filter paper contributes 33 percent of the overall diffusion resistance.

Unlike the gas-limited curve, the data correlations in figure 19 depend upon flow rate, but curiously this effect is different for distilled water and for BSA solution. At any given S/V value, the BSA data suggest that K_m is larger at a gas flow of 3 liters/min than it is at 1 liter/min, whereas the opposite is true for distilled water. This differ-

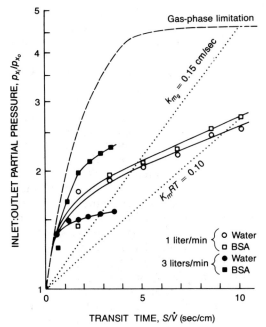

Figure 19. Measurements of Ichioka (1972) for SO_2 absorption from a nitrogen stream flowing through a straight-tube model of the upper respiratory tract. Dotted chords illustrate the graphical determination of RTK_m, and the solid curves have been drawn through the data by eye. The broken curve representing gas-phase limited diffusion has been computed using entry 4 in table 5.

ence is most likely due to a variation in SO_2 chemistry between the two liquids. For example, BSA solution may have a buffering capacity for hydrogen ions produced in the bisulfite reaction (eq. 4), thereby increasing the reaction affinity for SO_2. Lacking this buffering capacity, distilled water creates higher SO_2 backpressures, which reduce the driving force for absorption.

Beyond certain limits of minimum scale and maximum geometric complexity, it is impractical to construct physical models. Rather, an isolated organ preparation should be used, which more accurately represents the biochemistry and mechanics. Postlethwait and Mustafa (1981) studied NO_2 uptake by an isolated perfused rat lung that was supplied with a reciprocating flow of the pollutant-in-air mixture to the trachea. These investigators were able to measure both total uptake and the partitioning of reaction products between pa-

renchymal tissue and perfusate. They demonstrated that NO_2 is absorbed into tissue primarily as NO_2^-, which is then converted to NO_3^- by erythrocytes in the vascular space. Although not pursued in this particular study, the effects of perfusion and ventilation rate on absorption could be measured in an isolated lung preparation. By analyzing such data in conjunction with a suitable mathematical model such as equation 24, global values of mass transfer coefficients could be estimated and the relative importance of perfusion and ventilation limitations could be determined.

It might also be desirable to quantify reaction dynamics between pollutants and biological substrates by using isolated tissue samples. In such small-scale systems, the confounding effects of diffusion resistance can be eliminated by using high convection rates and thin samples. Rasmussen (1984) reviewed various systems for culturing lung cells for this purpose. As he points out, the unique problem in achieving a realistic model is that the tissue sample must simultaneously be exposed to aqueous nutrient medium on the endothelial side and pollutant gas flow on the epithelial side. One possible solution to this problem is to culture cells in growth medium on the outside of permeable hollow fiber tube bundles (Knazek et al. 1972) and then pass the pollutant gas through the lumen of the bundles. Such a bioreactor has the added advantage of a very high surface-to-volume ratio for pollutant uptake.

In Vivo Animal Experiments

The most popular and fruitful approach in studying absorption dynamics has been to ventilate an anesthetized animal through a specially designed face mask or cannula from which a pollutant air mixture can be directed to either the nose or the mouth. Then, by measuring pollutant concentration at the gas inlet as well as in the trachea, it is possible to infer pollutant penetration along a path from nose to trachea or from mouth to trachea. Figure 20 shows tracheal concentration monitored in response to the onset of various O_3 levels in a steady airflow entering the nose (solid lines) or the

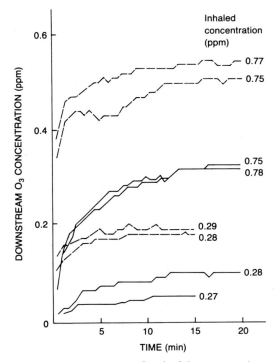

Figure 20. Time course of tracheal O_3 concentration in response to the introduction of a fixed concentration into the nose (solid curves) or the mouth (broken curves) of a dog. O_3 entered the upper airways at a steady airflow of 3.5 liters/min and at the concentration labeled on the curves. (Adapted from Yokoyama and Frank 1972, p. 134, with permission of the Helen Dwight Reed Educational Foundation. Published by Heldref Publications. © 1972.)

mouth (broken lines). In this experiment, the trachea was surgically isolated from the lower respiratory tract, which was separately ventilated with clean air (Yokoyama and Frank 1972). For each curve, there is a finite time during which O_3 concentration rises from its initial value of zero toward a steady-state level, and this rise time appears to be longer for the nose than for the mouth. In principle, a dynamic model of these data would yield useful information about the underlying transport processes. However, the slow-responding devices that are typically used to measure pollutant concentration are incapable of monitoring the true transient behavior.

Therefore, all previous studies have been restricted to the analysis of steady-state p_{x_i}/p_{x_o} values. In figure 21, p_{x_i}/p_{x_o} data for airflow between the nose and the trachea of

dogs has been plotted for several noxious gases. The use of transit time for expressing the observed variation with flow is consistent with the treatment of in vitro data in figure 19. Because SO_2 absorption by the nose is so great, it has to be graphed separately using a compressed ordinate. At the other extreme, CO is so poorly absorbed by the upper airways (Vaughan et al. 1969) that it does not appear in figure 21.

It is obvious that none of the data curves in figure 21 are linear. Instead there is a consistent trend of increasing slope with increasing airflow rate. It has been suggested that a higher airflow stimulates greater blood flow, causing distension of blood vessels and thereby increasing the effective area available for absorption. Simultaneously, increased blood flow would more effectively wash out the capillary bed and diminish vascular backpressure (Aharonson et al. 1974). Then again, similar curvature was exhibited by the in vitro data (figure 19), where no physiological response was possible. An explanation that conforms to both in vivo and in vitro data is that the gas-phase mass transfer coefficient, which is directly related to airflow, makes a substantial contribution to the overall absorption process.

To explore this hypothesis further, tracheal penetration data obtained in dogs at two steady inspiratory flows and along nasotracheal as well as orotracheal paths have been assembled in table 7. By applying equation 34 to these data, RTK_mS values have been computed for O_3 and for SO_2 at every experimental condition. For O_3, these values are of the order of 100 cm^3/sec and, assuming an upper airway surface of 100 cm^2, the order of magnitude of RTK_m is 1 cm/sec. But the range of individual k_{m_g} values predicted by the upper airway correlations in table 5 is from 1 to 3.5 cm/sec. Therefore, gas-phase resistance is indeed a significant factor in the overall absorption of O_3 and in the uptake of the even more soluble SO_2.

In characterizing the flow behavior of the data in table 7, RTK_mS is assumed to vary as $\dot{V}^m$, and the flow sensitivity parameter m was computed from each data pair. For SO_2, the orotracheal mass transfer coeffi-

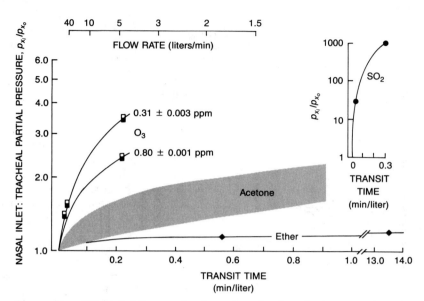

Figure 21. Effect of airflow on the absorption of various foreign gases along the nasotracheal path of the upper airways. (Adapted with permission from Aharonson et al. 1974, p. 656, and from the American Physiological Society.)

cient is much less sensitive to flow than is the nasotracheal coefficient. For O_3, however, there is an astonishing similarity of flow sensitivity for the nasotracheal and the orotracheal paths, and at both inlet concentrations reported. With the exception of the orotracheal value for SO_2, the values of m are all the same order of magnitude but somewhat less than the 0.854 value reported for k_{m_g} in physical models (table 5, entry 1a). The smaller flow dependence of the in vivo data is probably due to diffusion resistances in tissue and mucous layers that are not directly affected by gas flow.

Animal data also suggest that both O_3 and

SO_2 uptake are sensitive to atmospheric concentration. This effect is portrayed by the p_{x_o}/p_{x_i} tracheal penetration values in figure 22. The SO_2 data (Strandberg 1964) were obtained from free-breathing rabbits who inspired a pollutant mixture of known composition from a head chamber. Tracheal samples were obtained at peak inspiratory flow (I) and peak expiratory flow (E), the former sample being representative of upper airway absorption and the latter indicative of lower airway uptake. The O_3 data were obtained on dogs by applying subatmospheric pressure to a tracheostomy, thereby withdrawing pollutant mix-

Table 7. Comparison of Absorption Characteristics of the Nose and Mouth

		Nasotracheal			Orotracheal		
Inspired Concentration	Steady Flow (liters/min)	p_{x_o}/p_{x_i}	RTK_mS $(10^{-4}\text{m}^3/\text{sec})$	m	p_{x_o}/p_{x_i}	RTK_mS $(10^{-4}\text{m}^3/\text{sec})$	m
1 ppm SO_2[a]	3.5	0.001	4.03	0.70	0.0044	3.17	−0.12
	35	0.032	20.1		0.660	2.42	
26–34 ppm O_3[b]	3.5–6.5	0.283	1.05	0.52	0.665	0.340	0.42
	35–45	0.631	3.07		0.884	0.822	
78–80 ppm O_3[b]	3.5–6.5	0.408	0.747	0.49	0.732	0.260	0.47
	35–45	0.733	2.07		0.902	0.687	

[a] Frank et al. (1969).
[b] Yokoyama and Frank (1972).

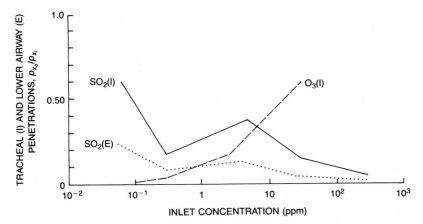

Figure 22. The concentration dependence of SO_2 penetration (Strandberg 1964) in rabbits and O_3 penetration (Vaughan et al. 1969) in dogs. Both inspiratory (I) and expiratory (E) samples were obtained in the trachea.

tures from a reservoir into the nose and through the upper airways (Vaughan et al. 1969). Tracheal concentration was determined by sampling from a tracheostomy tube while the dog breathed spontaneously through the caudal portion of the tracheostomy.

These data reveal that penetration of SO_2 through the upper airways is inversely related to its inlet concentration, but penetration of O_3 is directly related to concentration. It is possible to attribute the behavior of O_3 to a saturation limitation in chemical reaction rate as concentration increases. Such reaction kinetics are common for a variety of biochemical reactions and are often described by the Michaelis-Menten equation (Mahler and Cordes 1968, p. 153). It is difficult to conceive of a purely physical explanation for the concentration dependence of SO_2 penetration. Rather, it has been postulated that short exposures to high levels of SO_2 stimulate mucus secretion, thereby reducing penetration as compared to exposures at low concentrations (Brain 1970).

Far less information is available for absorption in the lower airways than for uptake in the upper airways. In experiments where the lower airways of dogs were surgically isolated from the upper airways, penetration of O_3 to the respiratory airspaces estimated as the ratio of expired-to-inspired partial pressures was

from 0.15 to 0.20, depending on the mechanical ventilation rate and the inlet concentration (Yokoyama and Frank 1972). And in free-breathing rabbits (figure 22), the analogous ratio for SO_2 was from 0.2 to 0.4. Because of the naturally reversing respiratory flow in the latter experiments, some desorption of pollutant may have occurred during expiration, as pollutant-depleted air from distal airways passed over the pollutant-rich tissue and mucus in more proximal airways. Therefore, the expiration-to-inspiration partial pressure ratios may be somewhat larger than the actual lower airway penetrations.

An important factor to consider in the design of uptake experiments is exposure time of the animal to the pollutant. Whereas an acute exposure results in data relevant to transport in a healthy animal lung, chronic exposure can result in anatomic and functional derangements that further affect the absorption process. All the investigations cited above utilized short exposures, usually less than an hour. However, Moorman et al. (1973) measured O_3 absorption into the upper airways of dogs that were chronically exposed for 8 to 24 hr, and compared the results with data from acutely exposed dogs. Their results show that in dogs, penetration through the nasotracheal path is generally greater during chronic exposure than during acute exposure. These investigators hypothesized that decreased mucus flow due to chronic exposure was responsible for the increased penetration.

■ **Recommendation 4.** Using a consistent experimental protocol, total uptake of selected pollutants should be measured in different animal species and then used to develop basic rules of extrapolation.

In Vivo Human Subject Studies

Extensive research into the transport of foreign gases in the human lung has been directed toward the development of noninvasive tests of pulmonary function. To a large extent, progress in this area has been stimulated by the development of reliable fast-responding gas analyzers. Literature on the use of inert insoluble gases for the characterization of gas mixing and distribution is extensive (Engel and Paiva 1985).

One of the most widely used lung function tests employing inert insoluble gas is the multibreath wash-out (Fowler et al. 1952). In this measurement, the end-expired nitrogen fraction is recorded for a series of regular breaths following a change in inspired gas mixture from room air to pure oxygen. If the lungs behaved in accordance with the static Bohr model, then a semilogarithmic plot of expired nitrogen fraction versus breath number would be a straight line having a slope and intercept from which dead space and alveolar volume could be determined. However, for patients with diseased lungs, this plot is far from linear, and even in normal subjects, expired nitrogen fraction exhibits multiexponential rather than single-exponential decay. Such nonideal behavior has been successfully simulated using multicompartment models with ventilation inhomogeneities between parallel regions (Robertson et al. 1950).

Most research on reactive foreign gases has been devoted to the uptake of CO. Because the absorption of this species is limited by diffusion through the alveolar membranes, decreased CO uptake can serve as an indicator of parenchymal tissue abnormalities. In the single-breath method, CO diffusing capacity is computed from the ratio of final-to-initial alveolar concentration measured during a series of breath-holding periods of known duration (Apthorp and Marshall 1961). The diffusion-limited form of equation 27 (that is, $Bo = 0$) is appropriate for this calculation, and it predicts that a semilogarithmic plot of alveolar concentration ratio versus breath-holding time has a slope proportional to K_mS. Roughton and Forster (1957) recognized that pulmonary diffusing capacity is composed of a true membrane-diffusing capacity and a hemoglobin reaction capacity that is proportional to capillary blood volume (eq. 25). Moreover, they developed a method of estimating the capillary blood volume by measuring the change in reaction capacity at different levels of inspired O_2. Although the method is fraught with difficulties, primarily because a unique value of alveolar concentration must be inferred from expired gas analysis, the single-breath CO uptake procedure as standardized by Ogilvie et al. (1957) is still in use.

There has also been considerable interest in soluble nonreactive gases such as acetylene. Since the absorption of acetylene is perfusion-limited, it can be used as an indicator of pulmonary blood flow, $\dot{Q}$. Acetylene uptake has been measured by the same breath-holding procedure developed for CO, and the data were then analyzed with equation 27. For a moderately soluble gas, the Bo parameter is large enough that a semilogarithmic plot of expired alveolar concentration versus breath-holding time should have a slope proportional to $\dot{Q}$. To study in detail the influence of solubility on uptake, Cander and Forster (1959) performed single-breath experiments using five different nonreactive gases including acetylene. Their results depart from the theory in two important ways, particularly for the most soluble gases, ethyl ether and acetone (figure 23).

First, the "percent of initial alveolar concentration" does not extrapolate to the expected value of 100 percent at zero breath-holding time. This was attributed to an initially rapid absorption of the foreign gas into parenchymal tissue, thereby causing an instantaneous drop in alveolar partial pressure. By extending the mathematical model

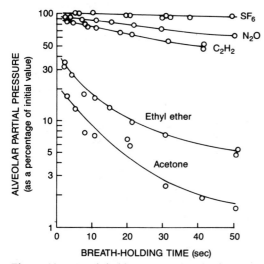

Figure 23. Breath-holding uptake data for five inert gases of increasing solubility obtained in a series of single breaths. (Adapted with permission from Cander and Forster 1959, p. 544, and from the American Physiological Society.)

to account for this, Cander and Forster were able to estimate reasonable values of the tissue volume. Second, the semilogarithmic plots are not linear. Instead, as breath-holding time increases, the alveolar concentration data become progressively larger than expected from a linear extrapolation of the initial data. Cander and Forster attributed this result to the contamination of expired alveolar gas with foreign gas that was initially absorbed into conducting airway tissues. Although no formal model was proposed, this behavior could undoubtedly be predicted by a multicompartment simulation that accommodates desorption processes during expiration.

By combining individual measurements into one multiple-gas test, it is possible to simultaneously estimate several transport parameters for the same subject. Moreover, since the parameters are measured under a single set of conditions, their values and the relations among their values may be more reliable than if each was measured under conditions that must necessarily differ, even if only slightly. An example of such an approach is the work of Sackner et al. (1975) who performed a series of breath-holding experiments on subjects inspiring a gas mixture containing helium (to assess alveolar volume), CO (to evaluate mem-

brane-diffusing capacity and capillary blood volume), and acetylene (to determine blood flow). Whereas Sackner's study used a single-compartment static model to analyze the uptake data, Saidel et al. (1973) used a more sophisticated multicompartment model to elucidate both ventilation distribution and uptake dynamics.

These investigators carried out parameter estimation experiments in two stages. They first performed multibreath wash-out measurements in which uptake of the nitrogen test gas is negligible. By matching simulations (eq. 28) to these data, flow fraction parameters, f_{kj}, governing distribution of volume and ventilation among the four distensible lower-airway compartments (figure 12a) could be evaluated for each subject. Then, the subjects were administered a steady-state uptake test in which a dilute CO–air mixture was inspired and the end-tidal concentration and uptake of CO was measured during consecutive breaths. By using the f_{kj} ventilation parameters already established from the nitrogen wash-out data, simulations of CO uptake data could be performed to estimate the parenchymal diffusion parameters, $K_m S$.

In humans as in animals, it is also possible to study uptake in the upper airways independent of the lower airways. For example, Speizer and Frank (1966) described an experiment in which cooperating subjects inhaled a 15-ppm mixture of SO_2–air into the nose, while inspiratory and expiratory samples were automatically withdrawn through a nasal sampling tube just inside the nares and a pharyngeal sampling tube was inserted through the mouth. By comparing concentrations between the nose and pharynx, it was clear that SO_2 penetration beyond the upper airways was only 1 percent during inspiration. And because the pharyngeal concentration was on the order of 0.4 ppm during both inspiration and expiration, it appears that the lower airways neither absorbed nor desorbed a detectable quantity of SO_2. However, the expired nasal concentration was 2 ppm, five times larger than the pharyngeal value, indicating that desorption from the nasal passages was promoted during expiration by the flow of air that had been stripped of SO_2 during inspiration. This temporal countercurrent exchange process enhances the protective capability of the upper airways. That is, in addition to preventing pollutants from reaching lower airways, countercurrent exchange reduces pollutant loading in the upper airway mucosa.

Up to this point, the discussion has focused on physicochemical problems, namely, absorption rates and internal concentration distributions of foreign gases. From a medical point of view, however, these physicochemical descriptions are useful only if they correspond to functional abnormalities. The connection between characterization of local dose on the one hand, and graded response of lung function on the other, has clearly been established for SO_2. In particular, it is well known that acute exposure to SO_2 leads to a reversible increase in airway flow resistance because of bronchoconstriction mediated by neural chemical sensors (Frank 1970). Because these sensors are particularly abundant near the larynx and carina, it should be possible to correlate local dose at these sites with airway resistance. Amdur (1966) used published SO_2 penetration values obtained by Strandberg (1964) on free-breathing rabbits in conjunction with her own airway flow-resistance values measured in guinea pigs and graphed the logarithm of airway resistance increase versus the logarithm of tracheal concentration. This local dose/response plot was linear with a slope of 0.6, corresponding to an approximate square root dependence of airway resistance increase on tracheal SO_2 concentration.

Going one step further, Kleinman (1984) hypothesized that if changes in airway resistance are directly related to the dose of SO_2 reaching the postpharyngeal airways, then apparent differences in response that have been observed during rest, exercise, free breathing, and breathing through a mouthpiece (figure 24a) can be explained by the dependence of upper airway penetration on flow rate and on the point of air access. Kleinman presented a quantitative analysis in which equation 32 was used to convert the inhaled dose rate, $\dot{V}p_{x_i}$, of the nine data points in figure 24a into their

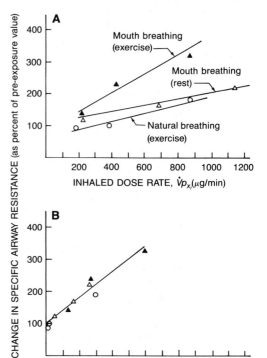

Figure 24. Dose/response data for increased airway resistance following exposure to SO_2. A: Correlation with respect to inhaled dose. B: Improved correlation with respect to postpharyngeal dose. (Adapted with permission from Kleinman 1984, pp. 33, 35, and from the Air Pollution Control Association.)

corresponding postpharyngeal dose rate, $\dot{V}p_{x_o}$. The flow fraction entering the nose, f_n, was evaluated as a function of airflow by assembling available human subject data; and the nasotracheal penetration, $(p_{x_o}/p_{x_i})_n$, and orotracheal penetration, $(p_{x_o}/p_{x_i})_m$, were both modeled by a formula similar to equation 34, with separate values of $K_m S$ for the nose and for the mouth estimated on the basis of mixed data from dogs and man. When the change in airway resistance is replotted against the predicted values of postpharyngeal dose (figure 24b), the dose/response correlation is considerably improved relative to the use of inhaled dose (figure 24a).

Oulrey et al. (1983) performed a similar analysis of SO_2 dose/response, but used a larger data set composed of 23 grouped measurements of airway resistance. They concluded that the increase in specific airway resistance is best correlated with the square of postpharyngeal penetration.

■ **Recommendation 5.** Noninvasive pulmonary function tests such as the CO uptake method should be extended to the evaluation of pollutant transport in humans.

Summary

Mathematical models can serve several purposes in the analysis of pollutant gas uptake. In order of increasing importance, a model can be used to correlate experimental measurements made under differing environmental or respiratory conditions; to extrapolate data obtained on laboratory animals to those values expected in humans; and to estimate sites and rates of uptake under conditions for which no data are available. There is no single model that best serves all these tasks. Rather, there may be a different model appropriate to each. For example, in data correlation and in extrapolation, a compartment model can incorporate important physiologic phenomena (for example, geometric asymmetry, airflow nonuniformities, ventilation/perfusion inequalities) within a mathematical framework that requires little detailed information for its development and only modest computational power for its implementation. On the other hand, a distributed-parameter model, which requires more elaborate input and more complex numerical algorithms to solve, is capable of predicting outcomes from first principles.

Whatever the nature of the model selected, it is assembled from four basic building blocks. First, an idealized geometry accounting for the structure of the airways, tissue, and blood spaces must be decided upon. Then material balance equations describing the time-dependent and possibly spatially distributed transport of pollutant are formulated. At the foundation of these material balances are the basic thermodynamic equilibrium, diffusional flux, and chemical reaction rate equations. Finally, using a specified set of pulmonary function parameters as forcing functions (for example, respiratory and pulmonary blood flows), the material balance equations are solved to provide a numerical simulation of pollutant uptake.

Experimental measurements are necessary to provide the geometric and physicochemical data required as inputs to a model and also to validate predictions by the model. Ideally, this is accomplished by a combination of separate experiments. For example, basic thermodynamic and reaction rate data can be obtained from in vitro systems such as isolated perfused lungs or excised tissue samples, whereas predicted uptake rates might be verified with noninvasive measurements in human subjects or invasive measurements in intact animals.

Clearly, an adequate quantification of pollutant gas transport and uptake is an interdisciplinary problem. Its solution requires the modeling skills of engineers and physicists, as well as the biological expertise of biochemists, toxicologists, and physiologists. And if some scientists should choose to straddle two or more disciplines, then so much the better!

Summary of Research Recommendations

Recommendation 1 *Basic Property Data*	*Objective.* There is a critical need to quantify the chemical interaction of specific pollutants with mucus, tissue, and blood. Besides determining solubility and diffusion coefficients, it is essential to determine the coefficients in reaction rate equations. The difficulty of this task is complicated by the fact that the associated thermodynamics are undoubtedly nonideal, and nonlinear concentration effects are likely.

Motivation. The descriptions of pollutant chemistry that are contained in this chapter were intended to be illustrative of quantitative methods, but were not completely accurate. In particular, physical properties in biological media were represented by aqueous values, and in the absence of chemical rate data, it was necessary to assume either instantaneous reaction or first-order kinetics.

To develop reliable mathematical models and provide sound interpretation of absorption data, basic property data remain to be established, even for the most common pollutants. For example, the explanation of concentration effects in O_3 and SO_2 uptake data is still somewhat speculative, largely because of our ignorance of the underlying chemistry.

Approaches. The use of isolated tissue preparations, such as tissue cultures or excised organ segments, could provide more direct measurements of biochemical properties than are possible in the entire organ. For example, by incorporating such a preparation into a flow-through or batch reactor, it is possible to determine a reaction rate expression using the well-established engineering principles of reactor design.

Recommendation 2
Individual Mass Transfer Coefficients

Objective. A serious effort is needed to analyze mass transport through individual diffusion barriers, particularly the mucous layer, the bronchial wall, and the alveolar capillary network. Undoubtedly, the sparsely perfused bronchial wall will require different mass transfer theory than the richly perfused plate-and-post structure of the alveolar walls. And the analysis of diffusion through the mucous blanket, because it may have a discontinuous dynamically changing conformation, poses unique challenges.

Motivation. At the core of any mathematical model of pollutant uptake are the individual mass transfer coefficients for the diffusion barriers. The values of mass transfer coefficients presented in this chapter were merely estimates. Considerable refinement is necessary.

Approaches. Although some physical modeling may be appropriate, it is also possible to perform computations based on existing geometric and hydrodynamic data. DuBois and Rogers (1968) have illustrated the application of diffusion theory to the bronchial wall; mechanical engineers have reported methods for analyzing transport in interrupted flows similar to those in the pulmonary circulation (Wieting 1975); and chemical engineers have developed a surface renewal theory to deal with dynamically changing interfaces such as the gas/mucus boundary (Astarita 1967).

Recommendation 3
Ventilation and Perfusion Effects

Objective. To analyze total uptake of pollutants and to predict dose distribution, mathematical models that account for ventilation and perfusion limitations, including their regional distribution, should be developed and validated.

Motivation. It is clear that there is a nonuniform distribution of ventilation and perfusion, even in a normal lung. And the possibility of diffusion and perfusion limitations exists for all pollutant gases. These interrelated phenomena have not been systematically investigated for pollutant gases, and yet it seems likely that they will have an impact on uptake distribution.

Approaches. A combination of measurement and mathematical modeling is necessary. The use of isolated perfused lung preparations could allow the measurement of total uptake under conditions where the overall ventilation/perfusion ratio is controlled. Moreover, intravascular tracer methods previously developed for determining ventilation/perfusion ratios in humans (Wagner 1981) might also be applied to an isolated lung. In analyzing data, distributed-param eter models may be unnecessarily complicated. Multicompartment lumped-parameter models are probably more appropriate.

Recommendation 4
Extrapolation Modeling

Objective. Using a consistent experimental protocol, total uptake of select pollutants should be measured in different animal species and then used to develop basic rules of extrapolation. More specifically, these uptake data could be correlated using known interspecies differences in lung volume, surface area, breathing rate, and rate-limiting mass transfer coefficients within the framework of an appropriate mathematical model.

Motivation. There is virtually no information in the literature that allows prediction of uptake by the human lung from data obtained in smaller laboratory animals. This represents a critical problem in setting air quality standards in cases where measurements on humans do not exist or cannot be taken.

Approaches. Yokoyama (1984) described an enclosure, similar to a closed-circuit metabolic chamber, in which the total O_3 uptake by a free-breathing rat could be monitored, without the need to anesthetize the animal. Using such a device, or possibly several chambers of different sizes, it would be possible to amass total uptake data on a series of different animal species. This data base could be analyzed with a simple compartment model that treats the animal and the chamber as two separate subsystems.

Recommendation 5
Noninvasive Methods

Objective. Noninvasive pulmonary function tests such as the CO uptake method should be extended to the evaluation of pollutant gas transport in humans. The data from such experiments, particularly when several indicator gases are used simultaneously, can be analyzed with an appropriate mathematical model to extract considerable information about regional inhomogeneities in uptake rate and dose.

Motivation. To date, most pollutant uptake data have been obtained in animals using protocols that required heavy sedation, and in some cases, extreme surgical procedures. Moreover, these measurements were made in the absence of complementary tests that characterize other important functional features such as the distribution of ventilation.

Approaches. It would be useful to extend the methodology of Saidel et al. (1973) to soluble and reactive pollutants. Naturally, the multicompartment model used by these investigators to simulate CO uptake must be generalized to include absorption into the upper airway and conducting airway compartments. Also, because of limitations in gas analyzer response and the potential health hazard during continual exposure, it may not be practical to use the steady-state uptake technique; the single-breath, breath-holding method may be a better choice.

Acknowledgment

This work was supported in part by National Institutes of Health Grant HL-20347.

References

Aharonson, E. F., Menkes, H., Gurtner, G., Swift, D. L., and Proctor, D. F. 1974. Effect of respiratory airflow rate on removal of soluble vapors by the nose, *J. Appl. Physiol.* 37:654–657.

Amdur, M. O. 1966. Respiratory absorption data and SO_2 dose-response curves, *Arch. Environ. Health* 12:729–732.

Apthorp, G. H., and Marshall, R. 1961. Pulmonary diffusing capacity: a comparison of breath-holding and steady-state methods using carbon monoxide, *J. Clin. Invest.* 40:1775–1784.

Astarita, G. 1967. *Mass Transfer with Chemical Reaction,* Elsevier, New York.

Bird, R. B., Stewart, W. E., and Lightfoot, E. N. 1960. *Transport Phenomena,* Wiley, New York.

Bohr, C. 1891. Uber die lungenatmung, *Skand. Arch. Physiol.* 2:236–268.

Brain, J. D. 1970. The uptake of inhaled gases by the nose, *Ann. Otol. Rhinol. Laryngol.* 79:529–539.

Cander, L., and Forster, R. E. 1959. Determination of pulmonary parenchymal tissue volume and pulmonary capillary blood flow in man, *J. Appl. Physiol.* 14:541–551.

Chang, H. K., Tai, R. C., and Farhi, L. E. 1975. Some implications of ternary diffusion in the lung, *Respir. Physiol.* 23:109–120.

Cumming, G., Horsfield, K., and Preston, S. B. 1971. Diffusion equilibrium in the lungs examined by nodal analysis, *Respir. Physiol.* 12:329–345.

Dayan, J., and Levenspiel, O. 1969. Dispersion in smooth pipes with adsorbing walls, *Ind. Eng. Chem. Fundam.* 8:840–842.

Dejours, P. 1981. *Principles of Comparative Respiratory Physiology,* Elsevier, New York, ch. 5.

Dekker, E. 1961. Transition between laminar and turbulent flow in human trachea, *J. Appl. Physiol.* 16:1060–1064.

DuBois, A. B., and Rogers, R. M. 1968. Respiratory factors determining the tissue concentrations of inhaled toxic substances, *Respir. Physiol.* 5:34–52.

Dungworth, D. L., Castleman, W. L., Chow, C. K., Mellick, P. W., Mustafa, M. G., Tarkington, B., and Tyler, W. S. 1975. Effect of ambient levels of ozone on monkeys, *Fed. Proc.* 34:1670–1674.

Engel, L. A., and Paiva, M. (eds.) 1985. *Gas Mixing and Distribution in the Lung,* Dekker, New York.

Fowler, W. S. 1948. Lung function studies: II. The respiratory dead space, *Am. J. Physiol.* 154:405–416.

Correspondence should be addressed to James S. Ultman, Pennsylvania State University, Department of Chemical Engineering, 106 Fenske Laboratory, University Park, PA 16802.

Fowler, W. S., Cornish, E. R., Jr., and Kety, S. S. 1952. Lung function studies: VIII. Analysis of alveolar ventilation by pulmonary N_2 clearance curves, *J. Clin. Invest.* 31:40–50.

Frank, R. 1970. The effects of inhaled pollutants on nasal and pulmonary flow-resistance, *Ann. Otol. Rhinol. Laryngol.* 79:540–546.

Frank, N. R., Yoder, R. E., Brain, J. D., and Yokoyama, E. 1969. SO_2(^{35}S labeled) absorption by the nose and mouth under conditions of varying concentration and flow, *Arch. Environ. Health* 18:315–322.

Gil, J. 1982. Comparative morphology and ultrastructure of the airways, In: *Mechanisms in Respiratory Toxicology* (H. W. Witschi and P. Nettesheim, eds.), ch. 1, CRC Press, Boca Raton, Fla.

Gomez, D. M. 1965. A physico-mathematical study of lung function in normal subjects and in patients with obstructive diseases, *Med. Thorac.* 22:275–294.

Graham, B. L., Dosman, J. A., and Cotton, D. J. 1980. A theoretical analysis of the single breath diffusing capacity for carbon dioxide, *IEEE Trans. Biomed. Eng.* BME-27 4:221–227.

Haggard, H. W. 1924. The absorption, distribution, and elimination of ethyl ether; I. Amount of ether absorbed in relation to concentration inhaled and its fate in the body, *J. Biol. Chem.* 59:737–751.

Hanna, L. M., and Scherer, P. W. 1986. Measurement of local mass transfer coefficients in a cast model of the upper respiratory tract, *Trans. Am. Soc. Mech. Eng.* 108:12–18.

Hansen, J. E., and Ampaya, E. P. 1975. Human air spaces, sizes, areas and volumes, *J. Appl. Physiol.* 38:990–995.

Henderson, Y., Chillingworth, F. P., and Whitney, J. L. 1915. The respiratory dead space, *Am. J. Physiol.* 38:1–19.

Hills, B. A. 1974. *Gas Transfer in the Lung,* Cambridge University Press, Cambridge, Mass.

Himmelblau, D. M., and Bishoff, K. B. 1968. *Process Analysis and Simulation: Deterministic Systems,* Wiley, New York.

Hobler, T. 1966. *Mass Transfer in Absorbers,* ch. 5, Pergamon, New York.

Hori, Y., and Suzuki, S. 1984. Estimation of the rate of absorption of atmospheric pollutants in respiratory conducting airways by mass transfer theory, *J. Jpn. Soc. Air Pollut.* 4:263–270.

Horsfield, K., and Cumming, G. 1968. Morphology of the bronchial tree in man, *J. Appl. Physiol.* 24:373–383.

Hughes, J. M., Hoppin, F. G., and Mead, J. 1972. Effect of lung inflation on bronchial length and diameter in excised lungs, *J. Appl. Physiol.* 32:25–35.

Ichioka, M. 1972. Model experiments on absorbability of the airway mucous membrane of SO_2 and NO_2 gases, *Bull. Tokyo Med. Dent. Univ.* 19:361–372.

Jeffery, P. K., and Reid, L. M. 1977. The respiratory mucous membrane, In: *Respiratory Defense Mechanisms, Part I* (J. D. Brain, D. F. Proctor, and L. M. Reid, eds.), ch. 7, Dekker, New York.

Keller, K. H., and Friedlander, S. K. 1966. Investigation of steady-state oxygen transport in hemoglobin solution, *Chem. Eng. Progr. Symp. Ser.* 62(66): 88–95.

Kety, S. S. 1951. The theory and applications of the exchange of inert gas at the lungs and tissues, *Pharmacol. Rev.* 3:1–41.

Kleinman, M. T. 1984. Sulfur dioxide and exercise: relationship between response and absorption in upper airways, *Air Poll. Control Assoc. J.* 34:32–36.

Knazek, R. A., Gullino, P. M., Kohler, P. O., and Dedrick, R. L. 1972. Cell culture on artificial capillaries: An approach to tissue growth in vitro, *Science* 178:65–67.

La Belle, C. W., Long, J. E., and Christofano, E. E. 1955. Synergistic effects of aerosols-particulates as carriers of toxic vapors, *Arch. Ind. Health* 11:297–304.

LaForce, R. C., and Lewis, B. M. 1970. Diffusional transport in the human lung, *J. Appl. Physiol.* 28:291–298.

Landahl, H. D. 1950. On the removal of airborne droplets by the human respiratory tract. I. The lung, *Bull. Math. Biophys.* 12:43–56.

Mahler, H. R., and Cordes, E. H. 1968. *Basic Biological Chemistry,* Harper and Row, New York.

McJilton, C. E., Thielke, J., and Frank, R. 1972. A model to predict the uptake of ozone and other pollutant gases by the respiratory system (unpublished manuscript–personal communication from R. Frank).

Miller, F. J., Menzel, D. B., and Coffin, D. L. 1978. Similarity between man and laboratory animals in regional pulmonary deposition of ozone, *Environ. Res.* 17:84–101.

Miller, F. J., Graham, J. A., Overton, J. H., and Myers, E. T. 1983. Pulmonary dosimetry of nitrogen dioxide in animals and man, National Technical Information Service PB83-243394:1–10.

Miller, F. J., Overton, J. H., Jaskot, R. H., and Menzel, D. B. 1985. A model of the regional uptake of gaseous pollutants in the lung. I. The sensitivity of the uptake of ozone in the human lung to lower respiratory tract secretions and to exercise, *J. Toxicol. Environ. Health* 79:11–27.

Moorman, W. J., Chmiel, J. J., Stara, J. F., and Lewis, T. R. 1973. Comparative decomposition of ozone in the nasopharynx of Beagles—acute vs chronic exposure, *Arch. Environ. Health* 26:153–155.

Murphy, T. W. 1969. Modeling of lung gas exchange—mathematical models of the lung: The Bohr model, static and dynamic approaches, *Math. Biosci.* 5:427–447.

National Research Council. 1977. Respiratory transport and absorption, In: *Ozone and Other Photochemical Oxidants,* ch. 7, National Academy of Sciences, Washington, D.C.

Niinimaa, V., Cole, P., Mintz, S., and Sheppard, R. J. 1980. The switching point from nasal to oronasal breathing, *Respir. Physiol.* 42:61–71.

Niinimaa, V., Cole, P., Mintz, S., and Sheppard, R. J. 1981. Oronasal distribution of respiratory airflow, *Respir. Physiol.* 43:69–75.

Nuckols, M. L. 1981. Heat and mass transfer in the human respiratory tract, *Proc. 34th Annu. Conf. Eng. Med. Biol.* 23:34.

Ogilvie, C. M., Forster, R. E., Blakemore, W. S., and Morton, J. W. 1957. A standardized breath-holding technique for the clinical measurement of the diffusing capacity of the lung for carbon monoxide, *J. Clin. Invest.* 36:1–17.

Olson, D. E., Dart, G. A., and Filley, G. F. 1970. Pressure drop and fluid flow regime of air inspired into the human lung, *J. Appl. Physiol.* 28:482–494.

Olson, D. E., Sudlow, M. F., Horsfield, H., and Filley, G. F. 1973. Convective patterns of flow during inspiration, *Arch. Intern. Med.* 131:51–57.

Oulrey, B., Moore, J., and Husman, P. 1983. Bronchoconstrictive effects—a combined statistical analysis of seven published studies, In: State of California Air Resources Board, "Public Hearing to Consider Amendments to Section 70100(i) and 70200, Title 17, California Administrative Code, Regarding the Short-term (One-hour) State Ambient Air Quality Standard for Sulfur Dioxide and Measurement Method." Appendix B.

Pack, A., Hooper, M. B., Nixon, W., and Taylor, J. C. 1977. A computational model of pulmonary gas transport incorporating effective diffusion, *Respir. Physiol.* 29:101–124.

Paiva, M. 1973. Gas transport in the human lung, *J. Appl. Physiol.* 35:401–410.

Paiva, M., and Engel, L. A. 1979. Pulmonary interdependence of gas transport, *J. Appl. Physiol. Respir. Environ. Exer. Physiol.* 47:296–305.

Parker, H., Horsfield, K., and Cumming, G. 1971. Morphology of distal airways in the human lung, *J. Appl. Physiol.* 31:386–391.

Pearson, D. A., Lundberg, L. A., West, F. B., and McCarthy, J. L. 1951. Absorption of sulfur dioxide in water in a packed tower, *Chem. Eng. Progr.* 47:257–264.

Piiper, J., and Scheid, P. 1980. Blood-gas equilibration in lungs, In: *Pulmonary Gas Exchange* (J. B. West, ed.), ch. 5, Academic Press, New York.

Piiper, J., Dejours, P., Haab, P., and Rahn, H. 1971. Concepts and basic quantities in gas exchange physiology, *Respir. Physiol.* 13:292–304.

Postlethwait, E. M., and Mustafa, M. G. 1981. Fate of inhaled nitrogen dioxide in isolated perfused rat lung, *J. Toxicol. Environ. Health* 7:861–872.

Raabe, O. G. 1982. Deposition and clearance of inhaled aerosols, In: *Mechanisms in Respiratory Toxicology* (H. W. Witschi and P. Nettesheim, eds.), ch. 2, CRC Press, Boca Raton, Fla.

Rasmussen, R. E. 1984. In vitro systems for exposure of lung cells to NO_2 and O_3, *J. Toxicol. Environ. Health* 13:397–411.

Rauwerda, P. E. 1946. Unequal ventilation of different parts of the lungs and the determination of cardiac output. Ph.D. dissertation, University of Groningen, Netherlands.

Robertson, J. S., Siri, W. E., and Jones, H. B. 1950. Lung ventilation patterns determined by analysis of nitrogen elimination rates: use of the mass spectrometer as a continuous gas analyzer, *J. Clin. Invest.* 29:577–590.

Rosenquist, T. H., Bernick, S., Sorbin, S. S., and

Fung, Y. C. 1973. The structure of the pulmonary interalveolar microvascular sheet, *Microvasc. Res.* 5:199–212.

Roughton, F. J. W., and Forster, R. E. 1957. Relative importance of diffusion and chemical reaction rates in determining rate of exchange of gases in the human lung, with special reference to true diffusing capacity of pulmonary membrane and volume of blood in the lung capillaries, *J. Appl. Physiol.* 11:290–302.

Sackner, M. A., Greeneltch, D., Heiman, M. S., Epstein, S., and Atkins, N. 1975. Diffusing capacity, membrane diffusing capacity, capillary blood volume, pulmonary tissue volume, and cardiac output measured by a rebreathing technique, *Am. Rev. Respir. Dis.* 111:157–165.

Saidel, G. M., Militano, T. C., and Chester, E. H. 1973. A theoretical basis for assessing pulmonary membrane transport, *Bull. Physio-Path. Respir.* 9:481–496.

Sankarasubramanian, R., and Gill, W. N. 1973. Unsteady convective diffusion with interphase mass transfer, *Proc. Roy. Soc. (London)* 333A:115–132.

Scherer, P. W., Shendalman, L. H., and Greene, N. M. 1972. Simultaneous diffusion and convection in single-breath lung washout, *Bull. Math. Biophys.* 34:393–412.

Scherer, P. W., Shendalman, L. H., Greene, N. M., and Bouhuys, A. 1975. Measurement of axial diffusivities in a model of the bronchial airways, *J. Appl. Physiol.* 38:719–723.

Schroeter, L. C. 1966. *Sulfur Dioxide. Application in Foods, Beverages and Pharmaceuticals,* Pergamon Press, New York.

Schroter, R. C., and Sudlow, M. F. 1969. Flow patterns in models of the human bronchial airways, *Respir. Physiol.* 7:341–355.

Simone, A. F., and Ultman, J. S. 1982. Longitudinal mixing by the human larynx, *Respir. Physiol.* 49:187–203.

Speizer, F. E., and Frank, N. R. 1966. The uptake and release of SO_2 by the human nose, *Arch. Environ. Health* 12:725–728.

Strandberg, L. G. 1964. SO_2 absorption in the respiratory tract: studies on the absorption in rabbit, its dependence on concentration and breathing phase, *Arch. Environ. Health* 9:160–166.

Swift, D. L., and Proctor, D. F. 1977. Access of air to the respiratory tract, In: *Respiratory Defense Mechanisms, Part I* (J. D. Brain, D. F. Proctor, and L. M. Reid, eds.), ch. 3, Dekker, New York.

Treybal, R. E. 1980. *Mass-Transfer Operations,* 3rd ed., McGraw-Hill, New York.

Ultman, J. S. 1985. Gas transport in the conducting airways, In: *Gas Mixing and Distribution in the Lung* (L. A. Engel and M. Paiva, eds.), ch. 3, Dekker, New York.

Ultman, J. S., and Blatman, H. S. 1977. Longitudinal mixing in pulmonary airways. Analysis of inert gas dispersion in symmetric tube network models, *Respir. Physiol.* 30:349–367.

Vaughan, T. R., Jennelle, L. F., and Lewis, T. R. 1969. Long-term exposure to low levels of air pollutants: Effects on pulmonary function in the Beagle, *Arch. Environ. Health* 19:45–50.

Wagner, P. D. 1981. Estimation of distributions of ventilation/perfusion ratios, *Ann. Biomed. Eng.* 9:543–556.

Weibel, E. 1963. *Morphology of the Lung,* Academic Press, New York.

West, J. B. 1977. Gas exchange, In: *Regional Differences in the Lung* (J. B. West, ed.), ch. 5, Academic Press, New York.

Wieting, A. R. 1975. Empirical correlations for heat transfer and flow friction characteristics of rectangular offset-fin plate-fin heat exchangers, *J. Heat Transf. Trans. ASME* 101:488–490.

Yokoyama, E. 1984. A simple method for measuring the respiratory uptake of carbon monoxide in unanesthetized rats: an application to rats acutely exposed to ozone, *Arch. Environ. Health* 39:375–378.

Yokoyama, E., and Frank, R. 1972. Respiratory uptake of ozone in dogs, *Arch. Environ. Health* 25:132–138.

Dosimetry Modeling of Inhaled Toxic Reactive Gases

JOHN H. OVERTON
FREDERICK J. MILLER
U.S. Environmental Protection Agency

Applications of Dosimetry Modeling

This chapter reviews modeling of the absorption of inhaled toxic reactive gases in the respiratory tracts of humans and animals. It focuses on our knowledge and understanding of the processes and factors influencing absorption and on mathematical dosimetry models and their use. The processes and factors considered are mainly those associated with the fluids and tissues of the respiratory tract. (A discussion of transport in airway lumens and air spaces is presented by Ultman in this volume.) Consideration of reactive gases such as carbon monoxide, which are transferred out of the respiratory tract and require a consideration of processes and factors outside this region, are beyond the scope of the review.

Mathematical dosimetry models—models that predict the uptake and distribution of absorbed gases in the respiratory tract—facilitate the integration of our knowledge and understanding of the physical, chemical, and biological processes involved in absorption. For example, the gathering of information to develop models identifies areas where information is missing; performing sensitivity studies with models can be used to determine the more important parameters and processes as well as to indicate those needing further research; and comparing predicted results to experimental data can be used to focus attention on needed theoretical and experimental research.

Furthermore, since knowledge of dose is an essential component of quantitative risk assessment, dosimetry modeling has an important role in the extrapolation of animal toxicologic results to humans. Dosimetry models can be used to estimate exposure levels that result in the same dose in different animals of the same or different species for use in comparing toxicologic effects, as well as to assist in experimental design. Predicted regional or local doses can be correlated with observed health effects, thereby allowing the prediction of effects in situations for which experiments are not feasible. Ultimately, in order to assess human health effects, models can be used to establish general principles and

guidelines for the evaluation and integration of the results of clinical, epidemiologic, and animal studies.

The review begins with a discussion of the physiological and chemical factors that must be quantified for dosimetry modeling. These factors are considered relative to anatomical models and to the characteristics of the liquid lining of the upper and lower respiratory tracts and their associated blood and tissue. Physicochemical processes such as solubility, chemical reactions, molecular diffusion, and convective transport are explained, and needs relative to dosimetry modeling are outlined. Several dosimetry models are surveyed and a discussion of their major features and assumptions is provided. Examples of uses of the models are illustrated, and the importance of and need for experimental data to be used in the dosimetry models are discussed.

Anatomical, Physiological, and Chemical Considerations

This section discusses some of the biological aspects of the mammalian respiratory tract that must be understood and quantified in order to develop dosimetry models. Anatomical models are discussed as well as the physical, chemical, and structural characteristics of respiratory tract fluids and tissues. For more information about respiratory tract structure, see the chapters by Schlesinger and by Ultman, this volume.

Anatomical Models

Knowledge of the dimensions of the lumen of the airways and of the air spaces of an animal's respiratory tract are important for a number of reasons. For example, unequal path lengths to equivalent morphological areas in lungs may result in significantly different doses; incorrect surface areas can result in erroneous predictions of uptake; and an incorrect tracheobronchial volume would result in erroneous estimates of the quantity of gas delivered by convection to the pulmonary region.

Respiratory tract dimensions useful to dosimetry modeling are an important facet of anatomical models. One type of model organizes the many branching airways of the lower respiratory tract (airways distal to and including the trachea) into sequential segments (also called generations, and other groupings have been used as well). Associated with each segment are idealized airways, in that all model airways of a given sequential segment are assumed to be the same size. Each model airway has the average length and diameter of all or some of the actual or real airways associated with the segment. The pulmonary region can be characterized by specifying the volume and surface area of the average alveoli as well as the number of alveoli per airway.

Probably the best known lower-respiratory-tract anatomical model (LRT model) is the one developed by Weibel (1963) for humans. The branching structure of the airways is assumed to be dichotomous and there is only one unique model path from the trachea to a given terminal unit (alveolar sac). However, all of the model paths (~ 8.4 million) are equivalent, requiring that only one path be considered for dosimetry modeling.

More recently, anatomical models have been developed that take into account some of the actual variability in different paths. The models of Yeh and Schum (1980) for human lungs and of Yeh et al. (1979) for rat lungs are examples. In addition to a model for the whole lung(s) of each species, these investigators also reported lobar models. With more detailed anatomical models such as these, the effects of intralung differences on predicted uptake and on dose at equivalent but differently located morphological sites can be investigated.

Most LRT models are not based on morphometric measurements of lung volumes during normal breathing (see, for example, Weibel 1963; Yeh et al. 1979; Yeh and Schum 1980). To use such data for dosimetry modeling, the data should be modified to better represent the lung size for the breathing conditions being considered. Procedures for modifying reported lung dimensions to those experienced during a breathing cycle are needed.

Upper respiratory tract (airways proximal to the trachea) dimensions also have been reported in terms of sequential segments along the path of air flow. Dimensions of the upper respiratory tracts of several animals are given by Schreider and Raabe (1981b). Measurements of cross-sectional areas and perimeters along the air path allow for estimating local volumes, surface areas, and other parameters, such as the gas-phase mass transfer coefficient, necessary for dosimetry modeling. By contrast, only the length, volume, and surface area of a species upper respiratory tract can be reported (see, for example, Swenberg et al. 1983). Other models (see, for example, Kliment 1973; Schreider and Hutchens 1980) have more than one segment but report only the lengths and volumes. With surface areas for each segment, even these simplistic upper-respiratory tract anatomical models (URT models) could prove useful in dosimetry modeling.

Liquid Lining of the Respiratory Tract

Upper Respiratory Tract and Tracheobronchial Region.
The epithelium of the upper respiratory tract of mammals is covered by a continuous two-layer liquid (Morgan et al. 1984) of viscous mucus (the epiphase) overlying a serous, or periciliary, fluid (the hypophase) in which the cilia move in a coordinated fashion (Lucas and Douglas 1934). For more information on the major functions of the mucous/serous/cilia system, see Kaliner et al. (1984).

In the tracheobronchial region, the liquid lining is similar in structure to that of the upper respiratory tract (see figure 1). Proceeding distally, from a thickness of 10 to 15 μm in the trachea, the mucous layer decreases in thickness to where, in the smallest bronchioles of healthy animals, there is no mucus (Gil and Weibel 1971). However, where there is mucus, it may not form a continuous layer. The periciliary layer is about as thick as the cilia are long, 4 to 6 mm, depending on location. Whether or not this layer is thinner in the regions not occupied by groups of ciliated cells apparently is not discussed in the literature.

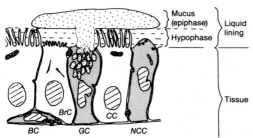

Figure 1. Diagram of the airway epithelium. Luminal or superficial cells include ciliated (CC), goblet or mucous (GC), nonciliated serous (NCC), and brush (BrC); a basal cell (BC) is also represented. The epiphase or mucous layer is depicted as discontinuous, being made up of flakes, as it is viewed by some researchers. The hypophase is composed of a low-viscous periciliary fluid in which cilia beat or move in such a way as to propel mucus toward the glottis. (Adapted with permission from Jeffery and Reid 1977, p. 198, by courtesy of Marcel Dekker, Inc.)

There are two basic concepts as to the extent to which mucus covers the periciliary layer. Some researchers see the mucus forming a continuous blanket (Luchtel 1978; Kaliner et al. 1984); whereas others view the mucous layer as being made up of discrete flakes and droplets (figure 1; see Jeffery and Reid 1977; van As 1977). In either case, reported values for mucous thickness suggest large thickness variations in airways of the same generation as well as in a given airway. Thus, there could be selective locations in airways of the same generation that are at different risk from exposure to inhaled gases. From the viewpoint of dosimetry modeling and predicting the quantities of absorbed gas by a given compartment (for example, mucous layer, periciliary layer, epithelial layer), the thickness distribution of mucus, as a function of location, is of considerable importance.

Going from the tracheobronchial region to the pulmonary region, the liquid lining probably is continuous with the thinner lining of the pulmonary region; however, no quantitative description of the transition was found in the literature. Since this transition region is a place of maximum tissue damage from gases such as nitrogen dioxide (NO_2) and ozone (O_3), an understanding of lining thickness here could be important in quantifying centriacinar uptake.

Most of what is known about the chemical constituents of the tracheobronchial region liquid lining comes from lavage data from patients with pulmonary diseases such as asthma, cystic fibrosis, and chronic bronchitis. Only recently have data been obtained from patients without lung problems (see, for example, Woodward et al. 1982). However, techniques are still needed to collect the periciliary fluid since most of the present approaches recover the mucus of the epiphase (Boat and Cheng 1980). Lavage fluid is separated into two components, the insoluble constituents (gel portion) and the soluble materials (sol portion). Whether or not this latter watery substance is chemically similar to the periciliary fluid is not known. The gel portion, often considered equivalent to mucus, has four major constituents: glycoprotein (2–3 percent), lipids (0.3–0.5 percent), proteins (0.1–0.5 percent), and water (95 percent) (Lopez-Vidriero and Reid 1980). However, disease may modify proportions of the constituents.

Pulmonary Region. The pulmonary epithelium is considered to be covered, at least in part, by an acellular lining composed of a serous fluid, possibly serum transudate, from 0.01 to as much as several microns thick, but on the average about 4 percent of the air-to-blood distance (Weibel 1973). Covering this fluid and separating it from air is a 0.002- to 0.01-mm-thick surface active monomolecular film called surfactant (Clements and Tierney 1965).

Hills (1982) has suggested the possibility of a discontinuous pulmonary liquid lining, the pulmonary surfaces being largely dry. Obviously, the uncertainty in the physical nature of the pulmonary acellular lining presents problems to dosimetry modeling and to the interpretation of toxic effects that are similar to problems posed by the uncertainty in the nature of the tracheobronchial region liquid lining.

Information on the chemical composition of the pulmonary region liquid lining comes mainly from analysis of the insoluble fraction of pulmonary lavage material. This fraction, corresponding to the monolayer surfactant material, is 80–90 percent lipid, 10–20 percent protein, and 1–2 percent carbohydrates (Sahu and Lynn 1977). The unsaturated fatty acid composition of the insoluble part of pulmonary lavage fluid and their estimated effective concentrations are reported by Miller et al. (1985). The serous fluid, because of its greater thickness, probably has more influence on gas absorption than the thinner surfactant layer; however, the fluid is seldom analyzed. Further characterization (chemical and physical) of the thicker serous layer is needed.

Lung Tissue and Blood

Most of the epithelium of the upper respiratory tract is pseudostratified and columnar. Luminal cell types are goblet, ciliated, and nonciliated, with their relative abundance dependent on location (Jeffery and Reid 1977; Mygrind et al. 1982). Columnar cells (ciliated and unciliated) are covered by 300 to 400 microvilli up to 2 μm long, helping to increase exchange processes across the epithelium as well as preventing dryness (Mygrind et al. 1982).

In the tracheobronchial region, cells are either ciliated or nonciliated. Nonciliated cells can be further classified as secretory (serous, Clara, goblet) or nonsecretory (brush, intermediate). The relative numbers and types of cells depend on location in the respiratory tract as well as on species (see, for example, Castleman et al. 1975). Figure 1 illustrates some of these cells and their relationship to the liquid lining.

Respiratory bronchioles, if present, are transitional airways, and their cellular makeup reflects this. The cells change in nature proceeding distally from the terminal bronchioles. For example, in monkeys, the cells are initially cuboidal and are replaced toward the distal end by squamous-type cells similar to alveolar type I cells (see, for example, Castleman et al. 1975). In addition, the respiratory bronchioles have outpockets of alveoli whose number increases distally from the bronchioles.

Tissue of the alveolar septum (figure 2) is, for the most part, a three-layered structure composed of the alveolar epithelium, an intermediate interstitium, and the capillary endothelium. The thickness of this structure is from 0.4 μm to less than 0.8 μm, depending on location and species. The capillary endothelium is made up of simple squamous cells that are thin and cover large areas. Alveolar epithelium is composed mainly of type I and type II cells. Type I cells are similar to endothelial cells, with broad thin cytoplasmic sheets extending from a bulkier nuclear region. This thin (0.1- to 0.2-μm thick) cell facilitates gas exchange (Burri 1985) since it covers from 90 to 97 percent of the alveolar surface area. Type II cells are cuboidal and are believed to be the source of surfactant (Burri 1985).

Capillaries are an integral part of the pulmonary alveolar structure (figure 2). Blood flowing through the capillaries is composed of plasma and blood cells in about equal proportion. Gases not depleted by reactions in the air/blood barrier may

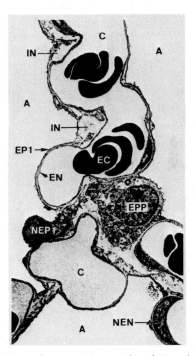

Figure 2. Electron micrograph of interalveolar septa, including alveolar capillaries (C) containing erythrocytes (EC), endothelial cells and nuclei (EN and NEN, respectively), type I and II epithelial cells (EP1 and EPP, respectively), the interstitial space (IN), and the alveolar air space (A). (Adapted with permission from Gehr et al. 1978, and Elsevier Science Publishers.)

penetrate to capillary blood and react further with blood components.

The chemical makeup of tissue and blood is not much different than that of the liquid lining—mainly water with traces of glycoproteins, lipids, proteins, as well as smaller molecules. However, the relative amounts or concentrations of the molecular components of the major constituents can be very different (Miller et al. 1985).

Physical and Chemical Factors Affecting Absorption of Reactive Gases

In general, the absorption of gases is affected by diffusion, convection, and, if relevant, chemical reactions in the gas phase (lumen and air spaces) as well as by solubility, diffusion, convection, and chemical reactions in the liquids and tissues of the respiratory tract. In this section, the concepts and nature of solubility, diffusion, convection, and chemical reactions are discussed as they apply to dosimetry modeling. (For a discussion of gaseous transport in the lumen of the airways, see Ultman, this volume.)

Solubility

Solubility refers to the ability of a medium to absorb a gas. There are many different definitions for solubility, including Bunsen's, Kuenen's, Ostwald's, and Henry's laws (Clever and Battino 1975). In equilibrium conditions, these definitions relate quantities of a gas in the gas phase to quantities of the gas in a liquid. For dosimetry modeling, one of the most convenient definitions of solubility is Henry's law, expressed as $C_g = HC_l$, where C_g and C_l are, respectively, the gas- and liquid-phase molar concentrations of the absorbed gas, and H is Henry's law constant for this particular formulation. This law only applies to the free or uncombined form of a trace gas in solution and can be used to quantify the concentration of the molecular form of the trace gas in a liquid, even if the absorbed gas is involved in chemical reac-

tions. The constant is a function of temperature and the molecular properties of the liquid and the gas; however, for constant temperature and the ranges of ambient concentrations of trace gases, the coefficient H can be considered constant.

Henry's law constants have been determined for many gases in water (see, for example, Altman and Dittmer 1971; National Research Council 1977). Often these values are used as approximations for in vivo values. Altman and Dittmer (1971) give data on Henry's law constants for a few gases, such as oxygen (O_2), carbon dioxide (CO_2), and NO_2, in water and in several biological fluids and tissues. In general, there is not much variation in the constant for a given gas among the various substances represented. For most cases, the use of the water value would seem justifiable. Although incomplete, the data suggest that the value of Henry's law constant is not influenced much by different biological tissues and liquids, indicating that a known value for one tissue or liquid may be a good approximation for missing values.

Henry's law can be extended to trace gases in equilibrium in two different media with a common interface. In this situation, the ratio of the concentrations in the two media is the ratio of the Henry's law constant of the two gases. This ratio is called the distribution coefficient or the partition coefficient.

Altman and Dittmer (1971) give a table of partition coefficients for several biological tissues and fluids. The values are close to one (to within 15 percent), suggesting that animal fluids and tissues are very similar with respect to Henry's law constant.

Molecular Diffusion

Molecular diffusion is a result of the random motions of molecules, an action that redistributes molecules so that there is a net flow from regions of high concentration to regions of lower concentration (Danckwerts 1970). The process often is described in terms of the diffusional flux, which is the net rate of transfer (due to random motion) of mass or molecules across a plane perpendicular to a given direction. Mathematically, the flux is expressed as $F = -D (dc/dx)$, where dc/dx is the concentration gradient along the given direction; and F is the diffusional flux (Danckwerts 1970). The formula is often referred to as Fick's law and can be used to describe diffusion in gases, liquids, and tissues. D is the molecular diffusion coefficient and is defined by the above equation; its units are (length)^2/time and its value depends on the properties of the molecule and the medium.

According to Sherwood et al. (1975), the diffusion of small molecules such as O_2 and CO_2 in solutions of biological proteins is approximately the same as in polymer solutions; however, diffusion in polymers does not follow a simple pattern. Nevertheless, the diffusion coefficients of small molecules in dilute polymer solutions having constituents similar to those of biological fluids and tissues are probably similar to, but smaller than, the value for diffusion in water. These conclusions most likely apply for small molecules other than O_2 and CO_2, such as O_3, NO_2, and formaldehyde (HCHO). Unfortunately, for the larger biological molecules, diffusion coefficients may be difficult to estimate from water values, and measurements in biological substances may be necessary.

The molecular diffusion coefficients of a few gases such as O_2 and CO_2 in biological fluids and tissues have been measured. Values are generally less than in water. For example, the values for O_2 in water, ox serum, frog muscle, dog connective tissue, and rat lung tissue are, respectively, 3×10^{-5}, 1.7×10^{-5}, 1.2×10^{-5}, 0.97×10^{-5}, and 2.3×10^{-5} cm^2/sec at 37°C (Altman and Dittmer 1971).

Convection in Respiratory Tract Fluids

There are two lung fluids in which convection may play a role in the absorption of reactive gases—the liquid lining of the upper respiratory tract and the tracheobronchial region, and capillary blood. Both fluids are in motion, and absorption may be enhanced by the removal of absorbed gases from a location or the replenishing of biochemical reactants.

In modeling, to account for convection, the flow rates of the fluids are needed. In some cases, mucous flow rates in the upper respiratory tract (see, for example, Morgan et al. 1984) and airways of the tracheobronchial region have been measured (see, for example, Iravani and van As 1972). In addition, rates in all airways have been estimated on the basis of clearance data from selected airways, anatomical data, and other assumptions (see, for example, van As 1977; Miller et al. 1978). For example, Velasquez and Morrow (1984) applied kinetic equations to data on particle retention in five airway zones (based on airway diameter) of guinea pigs to estimate transport rates in each airway generation. The calculated mucociliary (particle) rates ranged from 0.001 mm/min in the distal bronchioles to approximately 8 mm/min in the trachea. However, mucociliary rates are not necessarily the same as the liquid lining flow rates, and further data or assumptions must be used to estimate convection velocities.

Pulmonary capillary blood flows have been measured by Horimoto et al. (1981), among others, as well as theoretically calculated by Zhuang et al. (1983). However, capillary blood flow measurements for the upper respiratory tract and the tracheobronchial region were not found in the literature; no doubt, reasonable estimates could be obtained if needed. On the other hand, because the air/blood barrier in these regions is much thicker than it is in the pulmonary region, reactive gases (within the scope of this chapter) will not reach the capillaries proximal to the pulmonary region.

Heck and coworkers (1983) demonstrated that either HCHO or, most probably, its reaction products were transferred to tissues and fluids outside the upper respiratory tract. Also, NO_2 products are known to be transferred out of the lung (Postlethwait and Mustafa 1981). Presumably, capillary blood flow is a major factor in the transfer, suggesting that in developing dosimetry models the effect on gas absorption by the removal of reactants and products from the lung by blood must be considered relative to its effect on absorption.

Chemical Reactions

Chemical reactions occur as a result of the collision of molecules whereby the interaction of the colliding molecules (reactants) results in one or more different molecules (products). For modeling purposes, the rates of reaction or the rates of change of concentrations (for example, moles per liter per second) are important because the rates are the means whereby the loss and gain of chemical species are quantified. In general, rates are a complicated function of the local concentration of each of the reactants and products involved.

Theoretically, all of the chemical species involved and the reaction rate constants are needed to characterize a reacting system for modeling purposes. In practice, however, complete information is not always available; even if reaction rates are known, the products or the product formation rates, or both, may not have been measured. This poses no problem in modeling systems of reacting compounds if the unknown products do not react significantly with the known species. Even if this is not the case, approximations are often possible that will allow the modeling of absorption of the reacting gases.

Much of the results of mathematical modeling of the absorption and chemical reaction of gases in thin films, found in chemical engineering and mass transfer books (see, for example, Astarita 1967; Danckwerts 1970), are applicable to the thin layers of the liquid lining, tissues, and capillaries of the lung. The film theory models are most relevant since turbulence is not expected to be an important transport mechanism. Disturbances due to cilia and blood motion may prove the exception; however, these processes can be taken into account.

The main constituent of lung tissue and fluids is water, 85–95 percent; thus, the chemistry of absorbed gases in water is of interest. Many toxic gases react with water, affecting the absorption rate as well as creating products that may, in turn, react with the absorbed gases and the biological constituents. For example, O_3, NO_2, sulfur dioxide (SO_2), ammonia (NH_3), and

CO_2 react with water as well as interact in water (see Durham et al. 1981, 1984). These reactions lead to the formation of sulfate, bisulfite, nitrite, and nitrate molecules, and have the potential for changing pH. Thus, water reactions could have an indirect or direct adverse effect on tissue and fluids.

With the introduction of biochemical constituents, the reacting system becomes more complex. However, depending on conditions, some of the reactions will be more important than others. If necessary, the important reactions can be determined by chemical kinetic modeling to gain information that will allow simplification of the system for use in dosimetry modeling by keeping only the important reactions.

HCHO, NO_2, and O_3 are toxic reactive gases derived from mobile sources. Their reactions with lung constituents are briefly discussed as examples of the types of reactions that should be considered in formulating dosimetry models.

Formaldehyde. HCHO is a by-product of normal body metabolism, and in small quantities is not toxic. It is extremely reactive, even with itself, and the unhydrated form reacts rapidly with water (Gerberich et al. 1980; National Research Council 1981; Madestau 1982). It is highly reactive with amines and reacts with proteins, amino acids, nucleic acid, and histones as well (National Research Council 1981). HCHO has also been found to damage DNA (Ballenger 1984), implying that DNA is directly attacked by HCHO or reacts with HCHO reaction products. The reactive compounds are thought to bind to specific sites on single-stranded DNA. HCHO does not react with double-stranded DNA (Swenberg et al. 1983). A two-step mechanism has been suggested for the reaction of HCHO with amino groups, such as those that compose proteins and nucleic acid. The first step is fast and reversible; the second is irreversible, forming a stable product (Swenberg et al. 1983).

Nitrogen Dioxide. NO_2, as well as other oxides of nitrogen such as NO, is toxic.

The reactions of NO_2 with water lead to nitrous acid (HNO_2), which may form nitrosamines that are carcinogenic; also, NO_2 reacts with unsaturated fatty acids to form radicals leading to the autooxidation of unsaturated fatty acids and to HNO_2 (Pryor 1981). According to Postlethwait and Mustafa (1981), over 70 percent of the NO_2 absorbed by a ventilated perfused rat lung was converted to nitrite (NO_2^-). They also concluded that nonwater substances (that is, biological constituents) played the major role in the conversion.

Ozone. According to Menzel (1976), O_3 is the most toxic of the oxidizing air pollutants, its toxic effects being a result of its oxidative properties. Although it reacts with almost all classes of biological substances, biochemical, physiological, and morphological evidence indicates that cellular membranes are the site of toxicity (Menzel 1984), suggesting lipids as a major target.

Olefins are particularly sensitive to O_3, the Criegee mechanism being the accepted mechanism of reaction (Menzel 1976; Pryor et al. 1983). In this process, O_3 attacks the carbon double bond in the unsaturated fatty acid. The generation of free radicals also plays a role in toxicity; however, the nonradical reaction is considered dominant for most unsaturated fatty acids (Pryor et al. 1983). Although vitamin E is known to protect unsaturated fatty acids as well as entire animals against some of the effects of O_3 by scavenging radicals, the vitamin does not interfere with the nonradical Criegee process (Pryor et al. 1983).

Rate constants for the reaction of O_3 with some biological unsaturated fatty acids are known, for example, in carbon tetrachloride (Razumovskii and Zaikov 1972); however, in vivo rate constants in biological substances are for the most part lacking. For example, the extent to which O_3 is able to penetrate membranes and react with unsaturated fatty acids in lipids is unknown (Pryor et al. 1983).

According to Pryor and coworkers (1983), O_3 also reacts with amino acids and proteins. In water the only amino acids found to react with O_3 are, in order of decreasing reactivity: cysteine, tryptophan

or methionine, tyrosine, hystidine, cystine, and phenylalanine. Although the mechanisms for damage to proteins are not well known, there is evidence for a radical path and, possibly, for a nonradical Criegee process. The reactions of O_3 with sugars and nucleic acids have not been studied; however, sugar reactions are expected to be slow.

Dosimetry Modeling

By dosimetry models we mean mathematical or experimental models that predict, simulate, or are used to explain the quantitative uptake or absorption of gases in specific regions or locations. The formulation of such a model for inhaled gases requires information on the physical, biological, and chemical properties of the respiratory tract, as discussed previously, as well as an understanding of the nature of gas transport in the lumen and air spaces (as discussed by Ultman, this volume). The processes and features modeled are very complex, and essential information is often missing. By its very nature, a model is a simplified representation of a real process or object; complex processes and geometries are reduced to their essences with some aspects omitted and others retained.

Using a model to explore the effects of assumptions, combined with comparisons of simulation results or predictions with experimental data leads to more useful models and, more important, to a better understanding of the chemical, physical, and biological processes modeled. A survey of mathematical dosimetry models and their basic features is presented. This is followed by a discussion of results obtained by using models to predict the uptake and distribution of toxic gases in the respiratory tract. Emphasis is placed on examples that illustrate the sensitivity of predicted results to uncertainties in physical, chemical, and biological factors. Finally, the relationship between dosimetry modeling and experimental data is considered. Dosimetry models constructed from experimental equipment are considered by Ultman (this volume).

Upper Respiratory Tract and Total Respiratory Tract Models

We are not aware of any URT or total respiratory tract dosimetry model that has been developed to predict the absorption of toxic reactive gases. The two URT models discussed below were developed to analyze experimental data and to estimate parameters, although they could be used with modifications for predicting dose or uptake.

Chang and coworkers (1983) devised a very simple dosimetry model that they used to better understand species differences in nasal toxicity due to HCHO. This model, for all its simplicity, embodies most of the principles of the most complex dosimetry models, including the use of species-defining characteristics such as ventilatory and anatomical parameters. The authors defined the "dose" available for deposition on the nasal surface as the HCHO concentration times the minute volume divided by the nasal cavity surface area. Their "dose" is probably a good estimate since the upper respiratory tract absorbs most of the HCHO. Chang and coworkers (1983) concluded that the use of their "dose" helped to understand species differences in nasal toxicity.

Aharonson and coworkers (1974) developed a model based on the assumptions of mass balance, approximate steady state, and that the flux of gas to the air/liquid lining interface is proportional to the trace gas-phase partial pressure. They applied the model data on the uptake of acetone, ether, O_3, and SO_2 in the upper respiratory tract of dogs to estimate the dependence of the effective mass transfer coefficient on flow rates. They concluded that the transfer coefficients of the four gases increased with increasing airflow rate.

■ **Recommendation 1.** In order to better understand toxic effects in the upper respiratory tract, dosimetry models for this region are needed. These models should be designed so that they augment present LRT models in order to relate lower respiratory tract predictions to ambient concentrations. Simple empirical models, similar to that of Aharonson et al. (1974), may be sufficient if

toxic effects only in the lower respiratory tract are of interest.

Lower Respiratory Tract Models

In the model of McJilton and coworkers (1972; see also Morgan and Frank 1977; National Research Council 1977), absorption and transport in the lumen and air spaces are based on a one-dimensional differential equation that accounts for convection, molecular diffusion, and the loss of gas by wall absorption. On the assumption that transfer is controlled by the liquid lining, the flux of gas to the air/liquid lining interface is defined in terms of a liquid-lining mass transfer coefficient. Chemical reactions are not considered in the liquid lining, and the transfer coefficient is based on the physical properties of O_3 and the lining (Henry's law constant, molecular diffusion coefficient, and lining thickness) and the requirement that the O_3 concentration at the liquid lining/tissue interface be zero. This latter requirement is based on the assumption that O_3 reacts instantaneously with the tissue constituents and cannot penetrate to any significant depth in the tissue compartment (National Research Council 1977).

The model is used in conjunction with the airway model of Weibel (1963) to simulate the uptake of O_3 and SO_2 in humans. Weibel's anatomical model defines the radii, length, surface areas, and volumes of the lumen and air spaces of each of its 24 generations. A sinusoidal breathing pattern is used, but model lung size is assumed to be constant during "breathing". The liquid lining thickness depends on the generation: "10 μm in the upper generations, 3–5 μm in the alveolar ducts, and 0.3 μm in the alveoli" (National Research Council 1977).

The differential equation developed to take into account the above factors is solved using finite difference methods. The local dose (mass per unit area gained by the airway surface) is computed for each generation.

One of the major deficiencies in the model developed by McJilton and coworkers (1972) is the lack of chemical reactions in the liquid lining. Miller (1977) and, later, Miller and coworkers (1978) developed an O_3 dosimetry model to address this limitation. The formulation of this model is similar to McJilton's model with respect to how lumen and air space transport is modeled and in the use of anatomical models, mass transfer coefficients, and distally decreasing liquid lining thickness. However, differences do exist. For example, the model of Miller and co-workers uses an axial dispersion coefficient to account for air velocity inhomogeneities in place of axial molecular diffusion used by McJilton and coworkers. Furthermore, no assumption is made as to whether the radial flux is limited by the liquid lining. Instead, a gas-phase mass transfer coefficient is calculated on the basis of an approximate radial O_3 concentration profile and combined with the liquid-phase transfer coefficient to obtain an overall mass transfer coefficient. Although such enhancements make the model developed by Miller et al. (1978) more physiologically sophisticated than McJilton's, the additions have a minor effect on simulation results. Nevertheless, such enhancements are necessary to determine what processes and factors are important.

Chemical reactions in the liquid lining are accounted for by assuming that the reaction rates of O_3 with biochemical constituents are so fast that the reactions can be characterized by an "instantaneous reaction regime" similar to that discussed by Astarita (1967). In order to model this description, the production rates of biochemical reactants in each generation are required. These are estimated by assuming that the production rates decrease distally, by using tracheal mucous flow rate data, by using data on the surface area of each tracheo-bronchial generation, and by specifying the concentration of the reacting biochemical constituents and their stoichiometry of reaction with O_3 (Miller 1977; Miller et al. 1978). Mucous transport rates, such as those estimated by Velasquez and Morrow (1984), would have been helpful in modeling the production of biochemical reactants throughout the tracheobronchial region. The concentration of O_3 was assumed to be zero at the liquid/tissue interface for reasons similar to that given by McJilton et al. (1972). Thus, O_3 did not penetrate into a tissue compartment beyond the liquid/tissue interface—all tissue absorption took place at the interface.

A second model of Miller and co-workers (1985) differs from the first, mainly in how chemical reactions are modeled and in the inclusion of tissue and pulmonary blood compartments where reactions take place. The reactions of O_3 with the biological constituents of the liquid lining, tissue, and blood compartments are assumed to be second order; however, the concentrations of the biological constituents remain constant during the time of simulation (Miller et al. 1985). Thus, the model uses pseudo first-order reactions to account for chemical reactions. The major reactions considered in estimating an effective first-order rate constant are those of O_3 with the unsaturated fatty acids. The reactions of O_3 with the amino acids and constituents other than unsaturated fatty acids are assumed relatively ineffective (as far as dosimetry is concerned) and are not included in the estimations of the net concentration or of the effective rate constant.

Conceptually, in this latest model, transport in the gas phase is essentially the same as in the original (Miller et al. 1978) model. However, a recent modification of the model by Overton and Graham (1985) takes into account varying lung dimensions during the breathing cycle. This modification causes negligible changes in simulation results compared to results without the modification.

Mockros et al. (1985) developed a mathematical simulation model to investigate transport, absorption, and chemical reactions of toxic gases in the lower respiratory tract. This model and its predictions of lower respiratory tract uptake of O_3 in humans and rabbits is similar to the instantaneous reaction regime model and predictions of Miller et al. (1978).

Influence of Anatomical and Physiological Factors

Figure 3 is an example of a simulation using the model of McJilton and associates (see Morgan and Frank 1977); it illustrates the effect on predicted results of a modification to an anatomical model. The purpose of the simulations was to explore the effects on O_3 uptake in humans with modified lung

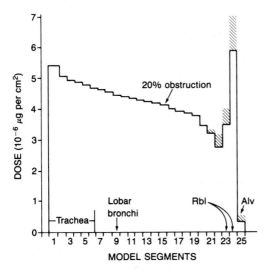

Figure 3. Results of a simulation using the model of McJilton and associates. Plotted for two simulations are the simulated doses of O_3 for humans versus model segment. The position of the trachea, lobar bronchi, respiratory bronchioles (rbl), and alveoli (alv) are indicated. The distance along the airpath from the trachea distally to the end of the airway model was divided into 25 segments, not necessarily corresponding to generations. Simulated "normal" O_3 uptake is illustrated by the heavy black line. Shaded areas correspond to dose increases using the same anatomical model, except that the number of airways distal to the seventh Weibel generation (model segment 15) is reduced by 20 percent. (Adapted with permission from Morgan and Frank 1977, p. 183, by courtesy of Marcel Dekker, Inc.)

geometry, such as might occur with disease. To simulate a pulmonary mechanical defect, the number of airways distal to the seventh Weibel generation (segment 15) was reduced by 20 percent. The shaded areas correspond to the predicted doses that resulted from modifying the anatomical geometry. The effect of the "pulmonary mechanical defect" is to increase the tissue dose in segments distal to the obstruction or defect. The major increases in dose occur in the pulmonary segments, where for one respiratory bronchiole segment the increase is as much as 17 percent. Both simulation curves have the same general shape. Dose is relatively high in the trachea, decreasing distally to the bronchioles where, at the respiratory bronchioles, there is a sharp increase in dose followed by an even sharper decline in dose.

Overton and Miller (1985) applied the first-order chemical reaction model of Miller and coworkers (1985) to different anatomical models of laboratory animals. The results shown in figure 4a were obtained using the anatomical model of Kliment (1973) for a 160-g rat in which several sequential generations are grouped into nine zones. Figure 4b is a plot of dose(s) versus generation using the airway model of Yeh et al. (1979) for a 330-g rat. The corresponding curves in both figures have the same basic shapes, the sharp peak in the first alveolated generation in figure 4b being the major exception. However, according to Overton and Miller (1985), the major differences between results are due to the effect of ventilatory parameters on percent uptake and on the difference in predicted percent uptake. For a constant minute volume the total uptake of the 160-g rat decreased from 93 percent at 80 breaths/min to only 87 percent at 140 breaths/min. On the other hand, at 80 breaths/min the uptake for the 330-g rat model was 74 percent (19 percentage points

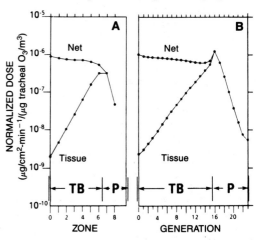

Figure 4. Use of the first-order chemical reaction regime dosimetry model of Miller and coworkers to explore the effects of anatomical models on predicted O_3 dose in rat lungs is illustrated for generations of the tracheobronchial (TB) and pulmonary (P) regions; dose is normalized to tracheal concentration. (A) Using the anatomical model of Kliment (1973), for a 160-g rat, dose is plotted according to zone (that is, sequential generations of airways), for a tidal (intake) volume of 0.7 ml at 144 breaths/min. (B) Using the anatomical model of Yeh et al. (1979), for a 330-g rat, dose is plotted according to airway generation for a tidal volume of 1.84 ml at 105 breaths/min. Based on Overton and Miller (1985).

less than for the 160-g rat), decreasing to 50 percent (37 percentage points less) at 140 breaths/min. Using allometric equations to scale the 330-g rat to 160 g did not reduce the Yeh et al. (1979) rat to a Kliment rat as far as percentage uptake and sensitivity to ventilatory parameters were concerned, a further indication of the importance of anatomic models in predicting uptake.

■ **Recommendation 2.** Anatomical models should be developed for different subpopulations—diseased, healthy, young, old, and so on—of humans and laboratory animals. These models should accurately reflect dimensions associated with physiological conditions.

The effects of various ventilatory parameters on tissue dose are illustrated in figure 5. The simulations were performed using the first-order chemical reaction model of Miller et al. (1985) in conjunction with Weibel's (1963) anatomical model for the purpose of estimating the effects of exercise on tissue dose in humans. The four curves presented have the same general shape: Tissue dose increases distally to some generation in the pulmonary region and then rapidly decreases. The location of the peak tissue dose depends on the ventilatory parameters; the higher the tidal volume the more distal the peak. In addition, as the ventilatory parameters increase, so does the quantity of O_3 absorbed in the pulmonary region. For the largest minute volume, the pulmonary absorption is 13.6 times as much as for the lowest minute volume (resting state) for the same length of time. On the other hand, for the same increase in ventilatory parameters, the tracheobronchial absorption increases only by a factor of 1.4. In another sensitivity study using a rat anatomical model, Overton and Miller (1985) showed that percent uptake was sensitive to tidal volume for a given minute volume.

■ **Recommendation 3.** Much of the ventilatory data on laboratory animals comes from anesthetized or restrained, quiet animals. Data based on restrained or anesthetized animals should be shown to be sufficient for modeling purposes, or data that

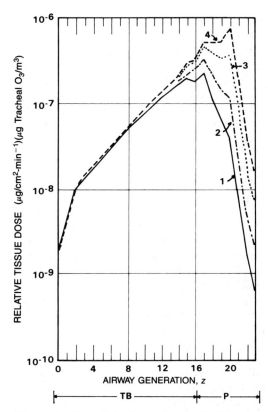

Figure 5. Use of the first-order chemical reaction regime dosimetry model of Miller and coworkers to explore the effects of changes in tidal volume (V_T) and breathing frequency (f) on predicted tissue dose of O_3 in tracheobronchial (TB) and pulmonary (P) segments of human lungs is illustrated. Curve 1: V_T = 500 ml; f = 15 breaths/min. Curve 2: V_T = 1,000 ml; f = 15 breaths/min. Curve 3: V_T = 1,750 ml; f = 20.3 breaths/min. Curve 4: V_T = 2,250 ml; f = 30 breaths/ min. Tissue dose has been normalized to the tracheal concentration; to obtain dose (mg/cm²-min^{-1}), multiply figure values by the tracheal concentration (mg/m³). (Adapted with permission from Miller et al. 1985, and Academic Press, Inc.)

correspond to more realistic conditions should be obtained.

There are several important gaps in our understanding of the liquid lining. There is not yet agreement as to whether the epiphase (mucous layer) of the liquid lining in the tracheobronchial region is continuous. More important, the local values of epiphase and hypophase thickness throughout the airways are needed. Likewise, the extent and thickness of the pulmonary liquid lining is in dispute. A region of importance for NO_2 and O_3 is the centriacinar region where the liquid lining makes a transition from the tracheobronchial region to the pulmonary region; little is known about how the thickness of the liquid lining changes in going from one region to another.

Miller and coworkers (1985) illustrated the importance of liquid lining thickness on predicted tissue dose. Their study indicated that a wide variation of tissue doses can be predicted using the range of liquid lining thickness reported in the literature. For example, halving or doubling the liquid lining thickness in the trachea increased or decreased, respectively, the tracheal tissue dose by a factor of more than 10 relative to the control simulation results. Another example by Miller et al. (1985) illustrates the effect of liquid lining thickness in the first generation of respiratory bronchioles of humans; a decrease in thickness by a factor of 10 resulted in a threefold increase in respiratory bronchiole tissue dose.

■ **Recommendation 4.** The thickness of the liquid lining of human and experimental animal respiratory tracts should be accurately characterized. This includes determining the thickness distribution of the mucous layer (epiphase) and of the underlying hypophase as a function of airway and morphological location, and determining how the liquid lining varies in thickness in going from the terminal bronchioles into and through the first alveolated duct.

Influence of Physicochemical Factors

The dosimetry models of Miller et al. (1978, 1985) were developed to simulate the uptake of O_3. However, the models will simulate the uptake of any gas whose properties conform to the theoretical assumptions of the models. On the assumption that the properties of NO_2 meet the criterion, Miller and coworkers (1982) used the original model (with the instantaneous reaction regime) in conjunction with Weibel's (1963) anatomical model to investigate the effects of Henry's law constant and of the mucous production rate on NO_2 uptake in humans. At the time of the simulations, the value of Henry's law constant was uncer-

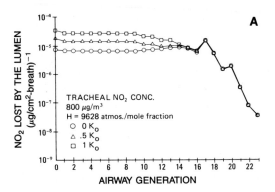

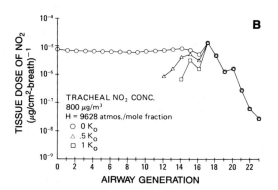

Figure 6. The instantaneous reaction regime dosimetry model of Miller and coworkers is used in conjunction with Weibel's (1963) anatomical model to simulate the effect of three mucous production rates (K_o) on the predicted absorption of 800 $\mu g/m^3$ NO_2 in human lungs. H (Henry's law constant) = 9,628 atm/mole fraction; 0 K_o (○), 0.5 K_o (△), and 1 K_o (□) indicate simulations with no mucous production, one-half of standard, and standard mucous production rates, respectively. The tidal volume and respiration frequency are 500 ml and 15 breaths/min, respectively. (A) Predicted NO_2 "lost by the lumen" or the net quantity NO_2 absorbed per unit area per breath for each generation according to generation number. (B) Predicted tissue dose of NO_2 (quantity of NO_2 absorbed per unit area of tissue per breath) by generation number. (Adapted with permission from Miller et al. 1982, and Elsevier Science Publishers.)

tain, and the necessary quantitative data on the reactions of NO_2 with biological constituents were missing (and still are). The results are shown in figures 6 and 7.

Figures 6a and 6b are dose profiles for the net quantity of NO_2 absorbed in each generation and the tissue dose, respectively, for three different mucous production rates (K_o). Mucous production rates in each generation are indicated by 0 (none), 0.5 (half of standard), and 1 (standard). As the mucous production rate increases, the amount

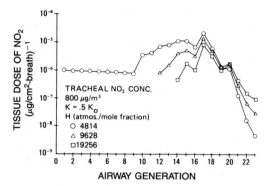

Figure 7. The instantaneous reaction regime dosimetry model of Miller and coworkers is used to simulate the effect of three values of Henry's law constant on the predicted absorption (tissue dose) of 800 $\mu g/m^3$ NO_2 in human lungs. Tissue dose is plotted according to airway generation, using a one-half standard (0.5 K_o) mucous production rate and Henry's law constants (atm/mole fraction) of 4,814 (O); 9,628 (△); and 19,256 (□). The tidal volume and respiration frequency were 500 ml and 15 breaths/min, respectively. (Adapted with permission from Miller et al. 1982, and Elsevier Science Publishers.)

of NO_2 absorbed in each generation increases in the tracheobronchial region, but remains the same in the pulmonary region (figure 6a). However, the NO_2 tissue dose (figure 6b) decreases in the tracheobronchial region as the mucous production rate increases, but no change occurs in the pulmonary region where the net and tissue doses are essentially the same.

The discontinuity of the tissue dose profiles for the two largest mucous production rates is a result of the instantaneous reaction regime—either NO_2 penetrates to the tissue or it does not. In the more recent model, in which first-order chemical reactions are assumed, a similiar sensitivity analysis was performed for O_3 (Miller et al. 1985). No discontinuities occurred; however, depending on the mucous chemical rate constant, tissue dose in the trachea could be significantly less than the net dose. Otherwise, increasing the rate constant had the same qualitative effect on the net and tissue dose curves as increasing the mucous production rate did in the older model.

Figures 6 and 7 illustrate the predicted results of a dosimetry model in which the instantaneous reaction regime was used to approximate chemical reactions. The first-order chemical reaction model of Miller et al. (1985) resulted in the predicted tissue

doses illustrated in figures 4 and 5. Although there are similarities in the tissue dose profiles, there are also differences. For example, for O_3, when using Weibel geometry and similar ventilatory parameters, the instantaneous reaction regime results in 60 percent uptake (Miller et al. 1978; Mockros et al. 1985); whereas, the first-order reaction model predicts 89 percent uptake (Miller et al. 1985). These differences in predicted uptake illustrate the importance of knowing the kinetic mechanisms, rate constants, reactant concentrations, and possibly biochemical reactant production rates.

Unfortunately, quantitative chemical information is, for the most part, lacking; most information available is from studies of chemicals such as amino acids and olefins, which can be major biochemical constituents. If the reaction chemistry of a molecule is known, then it is necessary to know if it reacts in the same way when it is part of a larger molecule. Such information will make data obtained in nonbiological situations more useful to dosimetry modeling.

■ Recommendation 5. The in vivo kinetic mechanisms of the important reactions of toxic gases with biological substances of the respiratory tract should be determined as well as kinetic reactions under nonbiological conditions that are applicable in vivo. This would make the present data bases more useful.

■ Recommendation 6. The local concentrations of bioreactants in the respiratory tract should be identified for different human and animal subpopulations (that is, according to age, gender, health conditions, exposure to toxic gases, and so on).

The discussion on the values of Henry's law constant and diffusion coefficients in biological tissues and fluids indicates that known values for water or for biological substances may be good estimates for missing data. Unfortunately, values for water or other fluids are not always available or they are highly uncertain. The following sensitivity study, using the instantaneous reaction regime model of Miller and co-workers (1982), shows that a factor of two in the uncertainty of Henry's law constant may result in more than a factor of two difference in predicted tissue dose.

In figure 7, tissue doses of NO_2 are plotted according to airway generations for three values of Henry's law constant. On the figure, 9,628 atm/mole fraction corresponds to a value used in the literature. The other two values are one-half and twice this value. Increasing the constant has a similar effect on tracheobronchial tissue dose as does increasing the mucous production rate (figure 6); the higher Henry's law constant, the lower the tissue dose. A different effect occurs in the pulmonary region (generations 17–23). The three curves are not only separated, but cross over at the 19th or 20th generation where the lowest value of Henry's law constant results in the lowest tissue dose. However, the general shape of the tissue dose profiles from the 15th generation distally is independent of the values of the Henry's law constants used; that is, the curves increase from the 15th generation to peak at the 17th, and then decrease.

■ **Recommendation 7.** Dosimetry models should be used to determine the importance of the values of the Henry's law constant and liquid- and tissue-phase molecular diffusion coefficients to predictions. Then, if necessary, the value of the parameters should be measured in vivo.

Importance of Experimental Data

The importance of experimental uptake data to dosimetry modeling cannot be overemphasized. With the appropriate type of data, dosimetry modelers can obtain an idea of the reliability of their models and infer which processes have been modeled correctly and which ones need improvements. As previously discussed, the in vivo values of physical, biological, and chemical parameters often are not well known. By obtaining data from appropriately designed experiments, the values of some of these parameters could be estimated or refined, resulting in more reliable predictions. Furthermore, if a parameter's value is considered the same for more than one animal species, then values determined for one species can be applied to other species, extending the usefulness of the data.

Methods of model validation and parameter estimation are needed to provide a link between experimental data and dosimetry models, as well as to provide guidelines for experimental designs. The techniques used to compare experimental and predicted data will determine, to a large extent, the type of data needed. Unfortunately, most dosimetry experiments have not been designed for the purpose of validation and estimation. Thus, the extent to which the present data base can be used profitably with dosimetry models is limited.

■ **Recommendation 8.** Methods of model validation and parameter estimation applicable to dosimetry models of toxic reactive gases should be identified or developed. Experimental dosimetry data should be obtained for model validation and parameter estimation. Evaluation methods and experimental designs should be developed hand in hand, leading to optimal experimental and evaluation methods. Integral to and very much a part of this process is the necessary continuation of the development and refinement of dosimetry models.

Summary

This chapter focuses on the physical, chemical, and biological processes and factors involved in the absorption of reactive gases. Emphasis is placed on the importance of these factors in developing dosimetry models, with special consideration being given to the role of lung fluids and tissues. Several dosimetry models are discussed and illustrations of predicted results presented to demonstrate the application of the models to the uptake of NO_2 and O_3, and to demonstrate the use of models in determining the effects of physical, chemical, and biological parameters on dosimetry predictions. Gaps in our knowledge and understanding of the processes of dosimetry are pointed out and research recommendations made to increase our understanding of the processes and to enhance the development of dosimetry models.

Summary of Research Recommendations

HIGH PRIORITY

Recommendation 8	Identify or develop methods of model validation and parameter estimation and obtain experimental dosimetry data for model validation and estimation.

MEDIUM PRIORITY

Recommendation 1	Dosimetry models should be developed that encompass the entire respiratory tract.
Recommendation 2	Anatomical models should be developed for different subpopulations (for example, diseased, healthy, young, old) humans and laboratory animals.
Recommendation 4	The thickness of the liquid lining of human and laboratory animal respiratory tracts should be accurately characterized.
Recommendation 5	The in vivo kinetic mechanisms of the important reactions of toxic gases with biological substances of the respiratory tract should be determined.
Recommendation 6	The local concentrations of bioreactants in the respiratory tract should be identified for different human and animal subpopulations, that is, according to age, gender, health conditions, exposure to toxic gases, and so on.

LOW PRIORITY

Recommendation 3	Ventilatory data from laboratory animals that correspond to realistic conditions should be obtained.
Recommendation 7	Values of Henry's law constants and of liquid- and tissue-phase molecular diffusion coefficients that are applicable to respiratory tract tissues and fluids should be obtained.

References

Aharonson, E. F., Menkes, H., Gurtner, G., Swift, D. L., and Proctor, D. F. 1974. Effect of respiratory airflow rate on removal of soluble vapors by the nose, *J. Appl. Physiol.* 27:654–657.

Altman, P. L., and Dittmer, D. S. 1971. Respiration and Circulation, Federation of American Societies for Experimental Biology, Bethesda, Md.

Astarita, G. 1967. *Mass Transfer with Chemical Reaction*, p. 187, Elsevier, New York.

Ballenger, J. J. 1984. Some effects of formaldehyde on the upper respiratory tract, *Laryngoscope* 94:1411–1413.

Boat, T. F., and Cheng, P. W. 1980. Biochemistry of airway mucus secretions, *Fed. Proc.* 39:3067–3074.

Burri, P. H. 1985. Morphology on respiratory function of the alveolar unit, *Int. Arch. Allergy Appl. Immunol.* (Suppl. 1) 76:2–12.

Castleman, W. L., Dungworth, D. L., and Tyler, W. S. 1975. Intrapulmonary airway morphology in

Correspondence should be addressed to John H. Overton or Frederick J. Miller, Toxicology Branch, Inhalation Toxicology Division, Health Effects Research Laboratory, U.S. Environmental Protection Agency, Research Triangle Park, NC 27711.

three species of monkeys: a correlated scanning and transmission electron microscopic study, *Am. J. Anat.* 142:107–121.

Chang, J. C. F., Gross, E. A., Swenberg, J. A., and Barrow, C. S. 1983. Nasal cavity deposition, histopathology, and cell proliferation after single or repeated formaldehyde exposure in B6C3F1 mice and F-344 rats. *Toxicol. Appl. Pharmacol.* 68:161–176.

Clements, J. A., and Tierney, D. F. 1965. Alveolar instability associated with altered surface tension, In: *Handbook of Physiology* (W. D. Fenn and H. Rahn, eds.), Vol. II, ch. 69, pp. 15–84, American Physiological Society, Washington, D.C.

Clever, H. L., and Battino, R. 1975. The solubility of gases in liquids, In: *Solutions and Solubilities* (M. R. J. Dack, ed.), pp. 379–385, John Wiley and Sons, New York.

Danckwerts, F. R. S. 1970. *Gas-Liquid Reactions*, McGraw-Hill, New York.

Durham, J. L., Overton, J. H., Jr., and Aneju, V. P. 1981. Influence of gaseous nitric acid on sulfate production and acidity in rain, *Atmos. Environ.* 15:1059–1068.

Durham, J. L., Barnes, H. M., and Overton, J. H., Jr. 1984. Acidification of rain by oxidation of dissolved sulfur dioxide and the absorption of nitric acid, In: *Chemistry of Particles, Fogs and Rain* (J. L. Durham, ed.), Vol. 2, pp. 197–236, Butterworths, Boston.

Gehr, P., Bachofen, M., and Weibel, E. R. 1978. The normal human lung: ultrastructure and morphometric estimation of diffusion capacity, *Respir. Physiol.* 32:121–140.

Gerberich, H. R., Stautzenberger, A. L., and Hopkins, W. C. 1980. Formaldehyde, In: *Kirk-Othmer Encyclopedia of Chemical Technology,* 3rd Ed., Vol. II, pp. 231–250, John Wiley and Sons, New York.

Gil, J., and Weibel, E. R. 1971. Extracellular lining of bronchioles after perfusion—fixation of rat lungs for electron microscopy, *Anat. Rec.* 169:185–200.

Hales, J. M., and Sutter, S. L. 1973. Solubility of sulfur dioxide in water at low concentrations, *Atmos. Environ.* 1:997–1001.

Heck, H. d'A., Chin, T. Y., and Schmitz, M. C. 1983. Distribution of [^{14}C] formaldehyde in rats after inhalation exposure, In: *Formaldehyde Toxicity* (J. E. Gibson, ed.), pp. 26–37, Hemisphere Publishing Corp., Washington, D.C.

Hills, B. A. 1982. Liquid layer lining the lung, *J. Appl. Physiol.: Respir. Environ. Exercise Physiol.* 52(5):1383–1385.

Horimoto, M., Koyama, T., Kikuchi, Y., Kakiuchi, Y., and Murao, M. 1981. Effect of transpulmonary pressure on blood-flow velocity in pulmonary microvessels, *Respir. Physiol.* 43:31–41.

Iravani, J., and van As, A. 1972. Mucus transport in the tracheobronchial tree of normal and bronchitic rats, *J. Pathol.* 106:81–93.

Jeffery, P. K., and Reid, L. E. 1977. The respiratory mucous membrane, In: *Respiratory Defense Mechanism, Part I* (J. D. Brain, D. F. Proctor, and L. M. Reid, eds.), pp. 193–245, Marcel Dekker, Inc., New York.

Kaliner, M., Marom, Z., Patow, C., and Shelhamer, J. 1984. Human respiratory mucus, *J. Allergy Clin. Immunol.* 73(3):318–323.

Kliment, V. 1973. Similarity and dimensional analysis, evaluation of aerosol deposition in the lungs of laboratory animals and man, *Folia Morphol.* 21:59–64.

Lopez-Vidriero, M. T., and Reid, L. 1980. Respiratory tract fluid—chemical and physical properties of airway mucus, *Eur. J. Respir. Dis.* (Suppl.) 110:21–26.

Lucas, A. M., and Douglas, L. C. 1934. Principles underlying ciliary activity in the respiratory tract, *Arch. Otolaryngol.* 20:528–541.

Luchtel, D. L. 1978. The mucus layer of the trachea and major bronchi in the rat. *Scan. Electron. Microsr.* 11:1089–1098.

Madestau, L. 1982. Formaldehyde, In: *McGraw-Hill Encyclopedia of Science and Technology,* 5th Ed., pp. 669–670, McGraw-Hill, New York.

McJilton, C., Thielke, J., and Frank, R. 1972. Ozone uptake model for the respiratory system, In: *Abstracts of Technical Papers*, American Industrial Hygiene Conference, May 14–19, 1972, paper no. 45.

Menzel, D. B. 1976. The role of free radicals in the toxicity of air pollutants (nitrogen dioxide and ozone), In: *Free Radicals in Biology* (W. A. Pryor, ed.), Vol. II, pp. 181–202, Academic Press, New York.

Menzel, D. B. 1984. Ozone: an overview of its toxicity in man and animals, In: *Fundamentals of Extrapolation Modeling of Inhaled Toxicants: Ozone and Nitrogen Dioxide* (F. J. Miller and D. B. Menzel, eds.), pp. 3–24, Hemisphere Publishing Corp., Washington, D. C.

Miller, F. J. 1977. A mathematical model of transport and removal of ozone in mammalian lungs, Ph.D. thesis, North Carolina State University, Raleigh, N.C.

Miller, F. J., Menzel, D. B., and Coffin, D. L. 1978. Similarity between man and laboratory animals in regional pulmonary deposition of ozone, *Environ. Res.* 17:84–101.

Miller, F. J., Overton, J. H., Meyers, E. T., and Graham, J. A. 1982. Pulmonary dosimetry of nitrogen dioxide in animals and man, In: *Air Pollution by Nitrogen Dioxide* (T. Schneider and L. Grant, eds.), Proceedings of U.S.-Dutch International Symposium on NO$_x$, Maastrict, Netherlands, pp. 377–386, Elsevier Biomedical Press, Amsterdam.

Miller, F. J., Overton, J. H., Jr., Jaskot, R. H., and Menzel, D. B. 1985. A model of the regional uptake of gaseous pollutants in the lung. I. The sensitivity of the uptake of ozone to lower respiratory tract secretions and exercise, *Toxicol. Appl. Pharmacol.* 79:11–27.

Mockros, L. F., Grotberg, J. B., and Sheth, B. V. 1985. Absorption of noxious gases in pulmonary airways. Abstract, 38th ACEMB Meeting, Chicago, Ill., Sept. 30–Oct. 2, 1985.

Morgan, K. T., Jrang, X-Z., Patterson, D. L., and Gross, E. A. 1984. The nasal mucociliary apparatus, *Am. Rev. Respir. Dis.* 130:275–281.

Morgan, M. S., and Frank, R. 1977. Uptake of pollutant gases by the respiratory system, In: *Res-*

piratory Defense Mechanisms. Part I (J. D. Brain and D. F. Proctor, eds.), pp. 157–189, Marcel Dekker, Inc., New York.

Mygrind, N., Pederson, M., and Nielson, M. H. 1982. Morphology of the upper airway epithelium, In: *The Nose. Upper Airway Physiology and the Atmospheric Environment* (D. F. Proctor and I. Andersen, eds.), pp. 245–278, Elsevier Biomedical Press, Amsterdam.

National Research Council. 1977. *Ozone and Other Photochemical Oxidants*, National Academy Press, Washington, D.C.

National Research Council. 1981. *Formaldehyde and Other Aldehydes*, National Academy Press, Washington, D.C.

Overton, J. H., Jr., and Graham, R. C. 1985. The influence of upper respiratory tract models on simulated lower respiratory tract uptake of O_3, In: *Program and Abstracts of the 1985 Annual Meeting of the American Associates for Aerosol Research,* Nov. 18–22, Abstract no. 9P7, p. 223.

Overton, J. H., Jr., and Miller, F. J. 1985. Sensitivity of mathematical model ozone (O_3) dosimetry to anatomical and ventilatory parameters of laboratory animals, In: *The Toxicologist*, 5(1), Abstract no. 489, p. 122, Society of Toxicology, Akron, Ohio.

Postlethwait, E. M., and Mustafa, M. G. 1981. Fate of inhaled nitrogen dioxide in isolated perfused rat lung, *J. Toxicol. Environ. Health* 7:861–872.

Pryor, W. A. 1981. Mechanism and detection of pathology caused by free radicals. Tobacco smoke, nitrogen dioxide, and ozone, In: *Environmental Health Chemistry. The Chemistry of Environmental Agents as Potential Human Hazards* (J. D. McKinney, ed.), pp. 445–466, Ann Arbor Science, Ann Arbor, Mich.

Pryor, W. A., Dooley, M. M., and Church, D. F. 1983. Mechanism for the reaction of ozone with biological molecules: the source of the toxic effects of ozone, *Adv. Mod. Environ. Toxicol.* 5:7–19.

Razumovskii, S. D., and Zaikov, G. E. 1972. Effect of structure of an unsaturated compound on rate of its reaction with ozone, *J. Gen. Chem. USSR* 8(3): 468–472. (English translation from *Zh. Org. Cheskoi Khim.* 8(3):464–468.)

Sahu, S., and Lynn, W. S. 1977. Lipid composition of secretions from patients with asthma and patients with cystic fibrosis, *Am. Rev. Respir. Dis.* 115:233–239.

Schreider, J. P., and Hutchens, J. O. 1980. Morphology of the guinea pig respiratory tract, *Anatom. Rec.* 196:313–321.

Schreider, J. P., and Raabe, O. G. 1981a. Structure of the human respiratory acinus, *Am. J. Anat.* 162: 221–232.

Schreider, J. P., and Raabe, O. G. 1981b. Anatomy of the nasal-pharyngeal airway of experimental animals, *Anatom. Rec.* 200:195–205.

Sherwood, T. K., Pigford, R. L., and Wilke. 1975. *Mass Transfer*, McGraw-Hill, New York.

Swenberg, J. A., Gross, E. A., Martin, J., and Popp, J. A. 1983. Mechanisms of formaldehyde toxicity, In: *Formaldehyde Toxicity* (J. E. Gibson, ed.), pp. 26–37, Hemisphere Publishing Corp., Washington, D.C.

van As, A. 1977. Pulmonary airway clearance mechanisms: a reappraisal. *Am. Rev. Respir. Dis.* 115: 721–726.

Velasquez, D. J., and Morrow, P. E. 1984. Estimation of guinea pig tracheobronchial transport rates using a compartmental model, *Exp. Lung Res.* 7:163–176.

Weibel, E. R. 1963. *Morphometry of the Human Lung*, Academic Press, New York.

Weibel, E. R. 1973. Morphological basis of alveolar-capillary gas exchange, *Physiol. Rev.* 53(2):419–495.

Woodward, H., Horsey, B., Bhavanandan, V. P., and Davidson, E. A. 1982. Isolation, purification, and properties of respiratory mucus glycoproteins, *Biochemistry* 21:694–701.

Yeh, H. C., and Schum, G. M. 1980. Models of human lung airways and their application to inhaled particle deposition, *Bull. Math. Biol.* 42:461–480.

Yeh, H. C., Schum, G. M., and Duggan, M. T. 1979. Anatomic models of the tracheobronchial and pulmonary region of the rat, *Anatom. Rec.* 195:483–492.

Zhuang, F. Y., Fung, Y. C., and Yen, R. T. 1983. Analysis of blood flow in cat's lung with detailed anatomical and elasticity data, *J. Appl. Physiol.: Respir. Environ. Physiol.* 55(4):1341–1348.

Part III: Biological Effects

Epidemiologic Studies of Effects of Oxidant Exposure on Human Populations

EDDY A. BRESNITZ
KATHLEEN M. REST
The Medical College of Pennsylvania

Air Pollution, the Automobile, and Public Health. © 1988 by the Health Effects
Institute. National Academy Press, Washington, D.C.

Role of Epidemiology in Air Pollution Research

Epidemiology is the primary research discipline that allows investigators to examine the long-term effects of air pollution on public health. Controlled human studies in exposure chambers generate useful hypotheses and allow maximum quantification of dose/response relationships, but they can only address acute effects and short-term changes in functional parameters. The assessment of long-term exposures and chronic health effects in humans must necessarily fall to the epidemiologist. In epidemiologic investigations, causal relationships between exposure and effect are usually plausibly inferred by the strength of the association, the consistency of data, the specificity of results, the temporality of observations, the demonstration of a biological gradient, and the plausibility and coherence of results (Hill 1965).

In air pollution research, animal, chamber, and epidemiologic studies have served to investigate a variety of health outcomes. Mortality studies have often followed on the heels of major air pollution episodes. Morbidity studies have looked at respiratory as well as nonrespiratory end points. The former include acute effects such as asthma and infection, chronic effects such as chronic obstructive pulmonary disease (COPD), and long-term effects such as lung cancer and accelerated decline in lung function. Research on nonrespiratory effects has usually focused on nonoxidant exposures (for example, on lead, organic solvents, and carbon monoxide), and has investigated neurotoxic effects, heart disease, and leukemia.

When epidemiology is used to study these outcomes in air pollution research, several technical and methodological issues arise. These include selection of appropriate study design and study population, assessment of exposure, definition and assessment of adverse health effects, control of bias and confounding variables, and analysis of data. Many studies done to date have been flawed by their method of addressing these difficult problems, leaving their results and conclusions open to question. Yet the weight of the aggregate evidence suggests that air pollutants do cause adverse health effects at certain levels, and that further epidemiologic studies are needed to quantify dose/response and to assess the consequences of long-term exposure to air pollutants at low levels.

The use of epidemiology in air pollution research is examined in this chapter. The focus of the chapter is on photochemical oxidants, mainly ozone (O_3) and nitrogen dioxide (NO_2), and their effects on the respiratory system, exclusive of lung cancer. A review of selected studies illustrates how investigators have addressed important technical and methodological issues, and suggests how such issues might be better addressed in the future. Important scientific and methodological knowledge gaps are then identified, and several approaches for closing these gaps are recommended.

Prior and Ongoing Studies—Designs, Findings, and Problems

Researchers have made considerably more progress in studying the acute effects of exposure to short-term, high levels of O_3 and NO_2 than in studying the effects of long-term, low-level oxidant exposure, or repeated episodic oxidant exposure at peak levels. The contribution of these exposure patterns to such respiratory diseases as COPD, asthma, and pulmonary infection remains a research gap. The effect of oxidants on the normal rate of decline in pulmonary function with age is also an unresolved question.

The studies required to fill these gaps are difficult to accomplish because of several factors: the generally low level of ambient (outdoor) exposures in the United States; the relative infrequency of some chronic respiratory diseases; the multifactorial nature of these chronic respiratory diseases; the lack of sensitivity and specificity of some of the tools used to detect physiological dysfunction or disease; the poor characterization of the actual composition of pollutants over time; the difficulty in estimating biologically effective dose; and the ethical problems inherent in conducting controlled studies of long-term exposures. Epidemiologic observational studies to date have generally suffered from one or more of these problems, precluding definitive, *quantitative* conclusions about causal relationships between exposure to oxidants and adverse health effects.

In table 1, the most commonly used types of epidemiologic study design are described. The cross-sectional study has been the most frequently used observational design in epidemiologic studies of air pollution. Studies with this design determine the prevalence of particular health outcomes in a population at one or more points in time and correlate them with some concurrent measure of exposure. The risk for disease development (incidence) and the temporal sequence of exposure and disease cannot always be determined in this type of study. Although cross-sectional studies have been useful for generating hypotheses, comparing health outcomes in communities exposed to high and low levels of air pollution, and studying the acute effects of short-term exposures, they have not been very helpful in assessing dose/response effects of long-term exposure.

A cross-sectional study may be repeated in populations over time to assess trends in the prevalence of specific outcomes of interest. A study done this way is called a secular trend or time-series analysis. In air pollution research, the outcomes under study may be correlated with serial measurements of ambient air pollutants. Parallel changes in the prevalence of symptoms or mean pulmonary function and average air pollutant levels would suggest an association between the exposure and the outcomes.

A closer examination of several cross-sectional studies that attempt to assess the relationship between exposure to oxidants and respiratory morbidity illustrates the problems and limitations of this type of study.

Cross-Sectional Studies of Pulmonary Function and Symptoms

Healthy Populations. Shy and coworkers (1970a) compared the ventilatory function of 987 second-grade children who lived in Chattanooga, Tennessee, in four geographic areas that varied in their average 24-hr levels of ambient NO_2 and particulates (one high-NO_2 area, one high-particulate area, and two control areas). Measurements were taken from stationary air pollutant monitoring sites during the 65-day study period.

Assessment of mean weekly height-adjusted forced expiratory volume in the first 0.75 sec ($FEV_{0.75}$) of the spirogram, averaged over each month of the study for children of the same sex, showed that the $FEV_{0.75}$ was statistically significantly lower in the high-NO_2 area compared to the two control areas in both months of the study. However, the differences were very small and clinically not very significant.

This study has several methodological problems. First, the investigators did not adjust the results for parental smoking and indoor sources of NO_2 which may have been associated both with decreased lung function in the children (Speizer et al. 1980) and levels of outdoor NO_2. That is, parental smoking may have confounded this association. Second, the levels of total suspended particulates (TSP) and sulfates (SO_4) were highest in the high-NO_2 area. These other air pollutants may interact with NO_2 or confound the relationship between lung function and the oxidant levels. Finally, the technique used to measure NO_2 was subsequently shown to be invalid. Thus, technical as well as methodological problems affected the results of the study.

In the second phase of the study, Shy and coworkers (1970b) investigated the trend in the incidence of acute respiratory illness among all families in the study having a child in the second grade. A total of 4,043 study subjects, comprising 871 families of the children who had participated in the ventilatory testing portion of the study, agreed to participate in this phase.

Temporal variations of self-reported respiratory illness rates were similar in each study area, and rates among smokers did not differ from rates in nonsmokers. Absolute illness rates over the entire study period were consistently and significantly

Table 1. Epidemiologic Research Designs

Study Type	Description	Advantages	Disadvantages
Cross-sectional (also called prevalence study)	• Examines presence or absence of exposure and effects at same point in time. • Sometimes repeated observations will be made over time and combined to compare trends in the population. • Is essentially the first phase of any cohort study.	• Relatively easy and quick to perform. • Especially suited for studying subclinical health effects for which records do not exist, and for studying effects that can be quantitated (e.g., pulmonary function) and that can vary over time. • Also useful for studying relatively frequent diseases that have long duration. • Generates useful etiologic hypotheses for analytical research.	• Does not permit cause/effect inferences. • Individuals who demonstrate an effect or have a disease (prevalent cases) may not be representative of all individuals who have same effect/disease. • Temporal relationship between exposure and effect may be difficult to ascertain.
Cohort	• Individuals (exposed and unexposed) are selected for observation and followed over time. • Study proceeds from suspected cause to effect (i.e., subjects selected on basis of exposure). • Data may be collected prospectively (i.e., outcome and exposure data unknown at beginning of study but collected as study proceeds). OR • Data may be collected retrospectively (i.e., exposures and outcome have already occurred at beginning of study and data are historical in nature).	• Excellent method for studying effects of rare exposures. • Permits observation of multiple effects. • Quality control in data collection is more manageable. • Yields incidence rates as well as relative risk. • Permits observation of entire exposure to disease state continuum.	• Requires long follow-up period. • Very costly. • Requires large number of subjects. • Problems with attrition and loss to follow-up. • Changes in exposure, lifestyle factors during course of study may make findings irrelevant or difficult to interpret.
Case-control	• Individuals with particular diseases or conditions (cases) are selected and compared to individuals without the disease or condition (controls). They are compared with respect to the exposure of interest. • Study proceeds from effect to cause (i.e., subjects selected on basis of disease). • Exposure data are collected retrospectively.	• Excellent method for studying rare diseases with long latency periods. • Relatively quick and inexpensive. • Requires small number of subjects. • Existing records may be available for use. • Permits investigation of multiple causal factors.	• Exposure data may not exist, may be inadequate, or may be available only through recall. • Validation of exposure data may be difficult or impossible. • Selection of proper control group may be difficult. • Control of extraneous variables may be incomplete. • Incidence rates cannot be ascertained.

higher in all family segments in the high-NO_2 areas compared to the two control areas, especially during the interinfluenza periods. The differences in illness rates be- tween the high-particulate area versus the two control areas were smaller and less consistent. There was no consistent differ- ence in the severity of illnesses among the different areas, nor was there a dose/response effect on the rates of illnesses observed in the subjects whose children attended the three schools in the high-NO_2 area.

In addition to the study design weaknesses discussed earlier, the major problem in this part of the study was the failure to relate symptoms with incidence of true infection. Self-reporting of disease was accepted as evidence of respiratory illness. Although the investigators assessed the severity of the illnesses through follow-up telephone interviews, there was no seroepidemiologic evidence of infection. Moreover, the investigators were unable to discriminate between the effects of short-term peak levels and long-term averages in pollutant exposures, nor did they adjust for differences in indoor exposure to NO_2. Also, the absolute differences in rates were small despite being statistically significant. The authors concluded that exposure to higher levels of ambient NO_2 and to particulates (to a lesser degree) increased the incidence of acute respiratory illnesses. It is not possible to conclude that infection(s) was the cause of these illnesses in this secular trend analysis.

Cohen et al. (1972) studied 441 nonsmoking, middle-aged Seventh Day Adventists living in either a high- or a low-oxidant pollution area in California. Air pollutant concentrations were measured at four sampling stations in each area; only total oxidants were measured. The mean percent daily maximum hourly concentration of oxidants in the two areas varied, but the annual mean value of oxidants was identical in both areas. Of the study subjects, 76 percent completed spirometry and maximum expiratory flow-volume curves and 97 percent completed a Medical Research Council questionnaire assessing symptoms. There were no differences in age, socioeconomic status, occupation, ethnic background, and length of residence in the two different areas. Maximum hourly averages of exposure to pollutants other than oxidants between the two groups were also the same. There was no significant difference from expected values in either respiratory symptoms or pulmonary function adjusted for age, gender, and height in either group.

This cross-sectional study did well in adjusting for some variables that may confound or bias any apparent association between the level of ambient oxidants and the health outcome of interest. Unfortunately, the differences in the pulmonary function measurements were small and the likelihood of detecting statistically significant differences in outcomes was correspondingly diminished. Moreover, the overall prevalence of chronic bronchitis was less than two percent. Limiting the study to nonsmokers, however, strengthened the conclusions about lack of an effect on previously healthy individuals. It did not address the question of a possible increased risk of morbidity secondary to low-level, long-term exposures among individuals with or without preexisting respiratory disease.

Other, more recent cross-sectional studies have investigated the effects of short-term, low levels of O_3 exposure on differences in pulmonary function in children (Lippmann et al. 1983) and exercising adults (Selwyn et al. 1985). These studies found a significant correlation between one or more measures of pulmonary function derived from spirometry and some measure of maximum O_3 concentration.

There was a significant negative correlation between a single day's peak 1-hr O_3 concentration and the peak expiratory flow rate in 39 healthy children between the ages of 7 and 13 (Lippmann et al. 1983). The O_3 level was less than 0.1 parts per million (ppm) on most days (28 of 32) of the study. The magnitude of the relationship, however, may have been underestimated because of a selection bias; that is, children who had positive histories of respiratory illness and did not participate in spirometric tests on four or more days were not included in the analysis.

A study of spirometry performed in 24 healthy adult runners before and after a 3-mile run correlated changes in airflow during 28 separate runs with trackside measurements of a time-weighted average of maximum O_3 concentrations over 15-min periods (Selwyn et al. 1985). Increasing O_3

concentrations were statistically significantly correlated with greater pre/post run differences in the volume of air expired during the first second (FEV_1) and the forced midexpiratory flow ($FEF_{25-75\%}$). Although these findings suggest a direct relationship between increasing biological dose and pulmonary function, this association did not remain statistically significant after adjustment for relative humidity.

The University of California at Los Angeles (UCLA) population study (Detels et al. 1979, 1981, 1982; Rokaw et al. 1980) is a longitudinal investigation of COPD that examines the relationship between long-term exposure to photochemical oxidants and the prevalence of pulmonary function decrements and respiratory symptoms. The initial survey in this (or any) longitudinal study can be analyzed as a cross-sectional study.

One phase of the study (Detels et al. 1981) compared 3,192 white adult residents of Lancaster, California, a low-pollution area, to 2,369 residents of Glendora, California, a high-oxidant area. The age, gender, race, and income distributions were similar in the two cities. Daily maximum hourly concentrations of specific air pollutants were measured at stationary air monitoring stations. The air monitoring station used to estimate exposure in Glendora was three times further from the center of the community than the one used in Lancaster.

Pulmonary function tests included spirometry and the single-breath nitrogen test. The National Heart, Lung, and Blood Institute respiratory questionnaire was used to assess the prevalence of symptoms. Testing in the two areas was separated by a three-year interval.

The results were reported by smoking status and were age-adjusted to the 1970 U.S. population census. The prevalence of cough, sputum production, and wheezing was higher in the high-pollution area compared to the low-pollution area among never smokers and smokers alike, although the differences were smaller among smokers. Although many of the differences were statistically significant, the absolute differences were small.

The mean percents predicted for the forced vital capacity (FVC), the FEV_1, and the $FEF_{25-75\%}$ were minimally lower in Glendora men, but there were no differences between women in the two areas. Differences in mean values were greater between smoking strata in both areas. The prevalence of study subjects whose FEV_1 and FVC were less than 50 percent predicted was higher in Glendora residents, in male and female smokers and nonsmokers. Tests primarily associated with small airway function showed little or no differences between the study subjects in two areas.

This study has several limitations. First, the measurement of exposure most likely underestimated the true exposure in the high-oxidant area because of the distance of the monitoring site from the community. Second, pulmonary function testing and questionnaire administration were not performed concurrently in the two areas because of financial constraints. Changes in smoking habits could have affected performance of pulmonary function tests and thus the study's findings. However, the investigators stratified and compared study populations on smoking habit. Third, questionnaire responses were not tested for reliability or validity. Fourth, there was no adjustment for potential confounders such as occupation and indoor air pollutants, or for effect modifiers such as daily activity. Finally, all levels of pollutants were elevated in the Glendora area.

Therefore, it is not possible to specify the pollutant(s) that best explain(s) the differences between the two groups, even if those differences could be attributed solely to air pollutants and not to unanalyzed confounders. The authors' conclusion that long-term exposure to high concentrations of photochemical oxidants, NO_2, and SO_4 results in respiratory impairment is tempered by these limitations.

Confounders and Effect Modifiers. The previous discussion refers to several potential study confounders such as parental smoking and levels of other pollutants. A confounder is an extraneous variable that is a risk factor for the disease or symptoms being studied and is associated with the exposure of interest, but is not a consequence of that exposure. The effect of a confounding variable may be to mask the underlying association or to explain partially (or wholly) the apparent effects of the exposure of interest. For example, the prevalence of smokers in a given geographic area may vary with the degree of air pollution in that area (for example, more blue-collar workers in an industrial area). In this case, smoking is a confounder; it is a risk factor for COPD, and it is associated with the levels of air pollution.

There are two general classes of potential confounders that must be considered in air pollution research: environmental confounders such as other air pollutants, occupational exposures, smoking, and meteorological variables; and personal confounders such as allergies and respiratory infections. The validity of any study is affected by the appropriate control of confounders, either in the design phase or the analysis phase of the study.

The investigators in the Chattanooga studies did not consider the effect of parental smoking and indoor sources of NO_2 on pulmonary function. This was a serious omission in view of recent evidence suggesting an association between exposure to sidestream smoke as well as to NO_2 from gas cooking and respiratory symptoms or impaired pulmonary function (Keller et al. 1979a,b; Tager et al. 1979; Speizer et al. 1980; Comstock et al. 1981; Ware et al. 1984).

An effect modifier is a variable that does not, by itself, cause the effect under study but modifies the effect of the risk factor(s) under study. Use of medications, socioeconomic status, and activity levels may affect either the extent of exposure or the expression of dysfunction. Failure to account for these variables may lead to underestimation of exposure or disease or both.

Although confounders, effect modifiers, and other potential sources of bias must be considered in all epidemiologic studies, specific variables are not uniformly important across all studies. For example, smoking and socioeconomic status may be important confounders or effect modifiers in prospective or retrospective cohort studies of the relationship between oxidant exposure and COPD, but they may be essentially immaterial in cross-sectional studies that correlate oxidant air pollution data with hospital admissions (Bates and Sizto 1983).

Sensitive Populations. The impact of ambient oxidant exposure on respiratory morbidity in sensitive populations is potentially greater than the impact on healthy populations. High-risk groups include people with one or more of the following: (1) specific diseases such as asthma, COPD, bronchitis, allergies, and sensitive airways; (2) specific exposures such as cigarette smoke and toxic fumes; and (3) biological factors such as age and α-1-proteinase inhibitor deficiency.

Several cross-sectional studies of asthmatics have evaluated the association of oxidant exposure and the frequency and severity of asthma attacks. Table 2 summarizes those studies. Many of these studies suffer from the same methodological problems encountered in some of the previously cited studies on healthy populations. The studies are especially difficult to compare in view of the variability or lack of definition for an attack of bronchospasm. (See also Bromberg, this volume.) However, the study by Holguin et al. (1985) deserves special comment in view of its overall excellence.

Holguin and coworkers studied 42 medically stable, nonsmoking, well-defined asthmatics without other pulmonary diseases in Houston during periods of high photochemical oxidant levels in 1981. Subjects maintained 12-hr logs, twice daily recording symptoms, location, activity, and use of medications. The maximum value of self-administered peak flow determinations was recorded for each 12-hr shift. Quality control was maintained by prestudy instruction of the participants and weekly review of the daily diaries and calibration of the peak-flow meters by trained technicians.

Exposures for O_3, NO_2, pollen, temperature, and relative humidity were determined by air monitoring stations located less than 2.5 miles from all study subjects. An hourly exposure estimate was calcu-

Table 2. Studies of Aggravation of Existing Asthma by Photochemical Oxidant Pollution

Av. Max. Conc. Range (ppm)	Pollutant	Study Description	Results and Comments	Reference
0.13[a]	Oxidants	Daily records of times of onset and severity of asthma attacks of 137 asthmatics residing and working in Pasadena, California, between September 3 and December 9, 1956; daily maximum hourly average oxidant levels (KI) from LA-APCD.	Of the 3,435 attacks reported, <5% were associated with smog, and most of these occurred in the same individuals; time-lagged correlations were lower than concurrent correlations; mean number of patients having attacks on days when oxidant levels were >0.25 ppm was significantly higher than days when levels were <0.25 ppm.	Schoettlin and Landau (1961)
0.01–0.37	O_3	Daily diaries for symptoms and medication of 45 asthmatics (aged 7–72 yr) residing in Los Angeles from July 1974 to June 1975; daily average concentration of O_3, NO, NO_2, SO_2, and CO by LA-APCD within the subjects' residential zone.	No significant relationship between pollutants and asthma symptoms; increased number of attacks at >0.28 ppm in a very small number of subjects; other factors such as animal dander and other pollutants may be important.	Kurata et al. (1976)
Not reported	O_3	Daily log for symptoms, medication, and hospital visitation of 80 children with asthma (aged 8–15 yr) in the Chicago area during 1974–75; air quality data on SO_2, CO, PM; partial data for O_3, pollen, and climate.	Bad weather and high levels of SO_2, CO, and PM exerted a minor influence on asthma, accounting for only 5–15% of the total variance; high levels of O_3 increased both the frequency and the severity of asthmatic attacks; pollen density during fall and winter temperature variations had no influence; no exposure data given for quantitative treatment.	Khan (1977)
0.004–0.235	O_3	Daily symptom rates in 82 asthmatic and allergic patients compared to 192 healthy telephone company employees in New Haven, Connecticut, from July to September 1976; average maximum hourly levels of O_3 and average daily values for SO_2, TSP, SO_4^{2-}, pollen, and weather were monitored within 0.8 km of where the subjects were recruited.	Maximum oxidants were associated with increased daily prevalence rates for cough, and eye and nose irritation in heavy smokers and patients with predisposing illnesses; pH of particulate was also associated with eye, nose, and throat irritation, whereas suspended sulfates were not associated with any symptoms. Questionable exposure	Zagraniski et al. (1979)

(Table continued next page.)

Table 2. *Continued*

Av. Max. Conc. Range (ppm)	Pollutant	Study Description	Results and Comments	Reference
			assessment, use of prevalence rather than incidence data, and lack of correction for dropout rates limit the usefulness of this study for developing quantitative exposure/ response relationships.	
0.03–0.15	Oxidants	Statistical analysis (repeated-measures design) of CHESS data on daily attack rates for juvenile and adult asthmatics residing in six Los Angeles area communities for 34-week periods (May–December) during 1972–75; daily maximum hourly averages for oxidants (KI) were monitored by LA-APCDs, 24-hr averages for TSP, RSP, SO_x, NO_x, SO_2, and NO_2 were monitored by the EPA, and meteorological conditions were monitored within 1 to 8 miles of homes in each community.	Daily asthma attack rates increased on days with high oxidant and particulate levels and on cool days; presence of attack on the preceding day, day of week, and day of study were highly significant predictors of an attack. Questionable exposure assessment including lack of control for medication use, pollen counts, respiratory infections, and other pollutants and possible reporting biases limit the usefulness of this study for developing quantitative exposure/response relationships.	Whittemore and Korn (1980)
0.07–0.39[a]	O_3	Emergency room visits and hospital admissions for children with asthma symptoms during periods of high and low air pollution in Los Angeles from August 1979 to January 1980; daily maximum hourly concentrations of O_3, SO_2, NO, NO_2, HCs, and COH; weekly maximum hourly concentrations of SO_4^{2-} and TSP; biweekly allergens and daily meterological variables from regional monitoring stations.	Asthma positively correlated with COH, HCs, NO_2, and allergens on same day and negatively correlated with O_3 and SO_2; asthma positively correlated with NO_2 on days 2 and 3 after exposure; correlations were stronger on day 2 for most variables; nonsignificant correlations for SO_4^{2-} and TSP. No indication of increased symptoms or medication use during high-pollution period; however, peak flow decreased (no differentiation of pollutants). Factor analysis suggested possible synergism between NO, NO_2, RH, and wind speed; O_3, SO_2, and temperature; and allergens and wind speed. Presence of	Richards et al. (1981)

(*Table continued next page.*)

Table 2. *Continued*

Av. Max. Conc. Range (ppm)	Pollutant	Study Description	Results and Comments	Reference
			confounding variables, lack of definitive diagnoses for asthma, and questionable exposure assessment limit the quantitative interpretation of this study.	
0.03–0.12	O_3	Admissions to 79 acute-care hospitals in southern Ontario for the months of January, February, July, and August in 1974, 1976–78. Hourly average concentrations of particulate COH, O_3, SO_2, NO_2, and daily temperature from 15 air sampling stations within the region.	Excess respiratory admissions associated with SO_2, O_3, and temperature during July and August with 24- and 48-hr lag; only temperature was associated with excess respiratory admissions and total hospital admissions for January and February. Lack of sufficient exposure analysis limits the quantitative use of this study.	Bates and Sizto (1983)
0.02–0.16[a]	O_3	Fifty-one asthmatics (aged 7–55 yr) exposed to ambient air from May to October 1981 in Houston, Texas; attack status determined each 12-hr period using log forms; individual environmental exposure estimates for the 12-hr period (1-hr max used for health data analysis) from regression model describing relationship of fixed-site monitors (within 2.5 miles) to microenvironment; multiple logistic regression model of Whittemore and Korn (1980) used for 42 subjects along with data on O_3, NO_2, temperature, humidity, pollen, and attack status.	Increased probability of asthma attack associated with increased O_3 and decreased temperature (and previous asthma attack); definition of asthma was specific for each individual; good control of confounders except for pollen levels. Quality control of questionnaires was excellent; peak flow changes were measured but not reported; personal monitors consistently indicated underestimates of exposures determined by fixed-site monitors.	Holguin et al. (1985)

[a] Represents average maximum hourly concentrations during low (minimum) and high (maximum) pollution periods.

NOTE: CHESS = Community Health Environmental Surveillance System; COH = coefficients of haze; HC = hydrocarbon; KI = potassium iodine method; LA-APCD = Los Angeles Air Pollution Control District; PM = particulate matter; RH = relative humidity; RSP = respirable suspended particulates; TSP = total suspended particulates.

SOURCE: Adapted from U.S. Environmental Protection Agency 1986.

lated for each subject for each hour of every 12-hr period, and the 12-hr exposure estimate was defined as the maximum hourly estimate during the period. Time spent indoors was not included in the estimate. O_3 and NO_2 were the only exposure variables that covaried significantly during the daytime.

The study developed individual attack definitions after all health and activity data had been collected but before the analysis of air monitoring data began. Attack definitions included self-reported symptoms, a decrease in expiratory peak flow, and an increase in the use of asthma-specific medication (Holguin et al. 1983). A logistic regression model (Whittemore and Korn 1980) was used to estimate the effect of the measured variables on the risk of an attack. The probability of an attack was significantly associated with an attack on the previous day, increasing O_3 levels, and decreasing temperature. NO_2, pollen count, and relative humidity had little effect on the risk of an attack. The average ambient O_3 level in the 12-hr periods during which the study subjects experienced attacks was less than 0.05 ppm in all instances.

The strength of this study lies in its excellent assessment of individual exposure and its effects, its careful definition of an asthmatic attack *unique* for each individual, and the quality of its statistical analysis. Nevertheless, that study and the other cross-sectional studies discussed here only *suggest* associations between increased levels of ambient oxidants and decrements in pulmonary function, higher prevalence of respiratory symptoms, and more frequent asthma attacks. They cannot and do not indicate whether these effects are temporary, have any long-term *clinical* significance, or lead to increased morbidity with time. Longitudinal cohort studies are best suited to assess these issues.

Prospective Cohort Studies

The prospective cohort method identifies study subjects on the basis of individual exposure and then follows them forward in time to assess disease development in ex-

posed and unexposed groups. The ability to identify exposure prior to disease development makes this design theoretically more powerful than cross-sectional studies in assessing causality in disease/exposure relationships. The magnitude of the risk factor is measured by obtaining the ratio of the disease incidence rates in the exposed and unexposed groups (the relative risk). The cohort design is particularly good for studying many outcome measures simultaneously. Prospective cohort studies allow for better quality and control of data collection.

The disadvantages of the cohort design include potentially long follow-up periods with the attendant problems of loss to follow-up and migration out of the geographic area; expense of maintaining follow-up; changing levels of exposure; and the low incidence of certain diseases, such as COPD, necessitating large groups of study subjects.

The Six-Cities study (Ferris et al. 1979) is an ongoing cohort study of the respiratory health effects of air pollution on children and adults. The cities in the study were selected on the basis of their historical levels of air pollution. They represent a range of ambient SO_2 and particulate exposure levels above and below the current ambient air quality standard. The main goal of the study is to assess the health effects of long-term exposure to these pollutants. The investigators are collecting data on SO_2, NO_2, O_3, TSP, and mass respirable particles by use of personal monitors (where possible) as well as central site monitors.

The published reports to date have focused on the health effects of ambient SO_2 and TSP. The Six-Cities study has supported earlier findings by Spengler and coworkers (1979) that indoor sources of NO_2 contribute more to total NO_2 exposure than outdoor, ambient levels. Ferris and coworkers have not yet published their analyses of the respiratory health effects of ambient oxidants. However, the study may provide important information on the effects of oxidants on the rate of decline in pulmonary function and on the prevalence of symptoms in smokers and nonsmokers if the cities vary in their level of oxidants.

That study and the UCLA population studies are essentially the only ongoing prospective cohort studies in air pollution research. The UCLA study discussed previously should provide additional longitudinal information about accelerated changes in pulmonary function among long-term residents of the high-pollution areas compared to the low-pollution areas. Repeat pulmonary function tests in the participants from Glendora and Lancaster will permit comparison of changes in decline of pulmonary function. An accelerated rate of decline in the high-pollutant community will suggest an effect of long-term, increased levels of air pollutants. Separating the effects of oxidants from the effects of other pollutants may be difficult.

■ **Recommendation 1.** A cohort study of the effects of long-term exposure to oxidants on respiratory morbidity in children, normal adults, and sensitive populations should be done if the Six-Cities study and the UCLA population studies cannot assess these issues.

Biological Markers. Use of early markers of lung injury would shorten considerably the follow-up period necessary to assess the development of disease in any cohort study. A shorter follow-up period would decrease the problems of loss to follow-up, migration, cost, investigator turnover, and low incidence of disease. Several investigators have proposed measuring hydroxyproline (Yanagisawa et al. 1986), antiproteinase fragments (Johnson and Travis 1979), elastin-derived peptides (Kucich et al. 1984), and antibodies to collagen (Michaeli and Fudenberg 1974) as early markers of emphysema. Cohort studies that assessed cumulative exposure or delivered dose, followed subjects for an increase in one or more of these markers as evidence of early disease, and controlled for potential confounders such as smoking and occupation could yield valuable information on threshold levels and dose/response relationships between exposure to O_3 and NO_2 and the development of emphysema. Unfortunately, to date there is no evidence that establishes these substances as valid

markers of early injury or disease. In fact, these substances may only be markers of exposure and reflect the activity of adaptive physiological processes.

Retrospective Studies

Two other study designs may be useful in studying the relationship of respiratory morbidity to oxidant exposure: the historical (retrospective) cohort study and the case-control study. In both designs, information on confounders and past exposures of interest is ascertained by a variety of methods including direct interviews, abstraction of medical records, and the use of existing data bases. These designs are particularly good for studying rare diseases and chronic diseases with long latency periods. They can be performed relatively quickly and often with minimal cost. Moreover, these methods can identify more than one risk factor in the same set of data. Their major limitation is determining the validity and reliability of the information on past exposures. Both designs have been used to study and imply causal relationships between disease and exposure to drugs, infectious agents, chemicals, and smoking. Neither design has been used in air pollution research, most likely because of the difficulty of accurately ascertaining past exposures to pollutants.

■ **Recommendation 2.** Future epidemiologic studies of air pollution should include the use of retrospective methods to investigate the relationship between oxidant exposure and respiratory morbidity if more accurate methods of assessing past cumulative exposure can be developed.

Generic Issues

Exposure Assessment

Estimating exposure, prospectively as well as retrospectively, is a difficult task. In most epidemiologic studies of air pollution, exposure has been measured by stationary air pollution monitors put into place for public health and regulatory purposes, not for research purposes. The distance of these monitoring stations from the population or area under study has varied a great deal.

Area Monitors. A few studies have included better-placed area monitors to supplement and make more precise the data obtained from regional monitoring stations (Kagawa and Toyama 1975; Holguin et al. 1985; Selwyn et al. 1985). Well-placed area monitors can yield a more precise assessment of exposure, especially when supplemented with time-activity data.

Personal Monitors. Elsewhere in this volume, Sexton and Ryan review the status of exposure assessment in air pollution research, and suggest several approaches for improving the quality of this assessment process. For the most part, their recommendations focus on the relationship between personal monitoring and ambient air quality measurements. Because there are no sensitive biological markers of exposure to NO_2 and O_3 as there are for carbon monoxide and lead, personal monitors currently provide the most accurate assessment of exposure.

Investigators should consider using personal monitors in all future cohort studies of air pollution as is being done in the Six-Cities study (Ferris et al. 1983). Results from personal monitoring should be correlated with exposure assessment determined by questionnaires, stationary samplers in the home (for NO_2), and measurements from air monitoring stations. Only a sample of the subjects need personal monitors when the size of the study groups is large. By adjusting the results from personal monitors for time and activity patterns as determined by personal activity logs, investigators can more accurately assess total personal exposure and better correlate this measure with early markers of effect. This information should provide more accurate assessment of dose/response effects at low levels of exposures. Separate assessments for average total exposure and peak exposure to oxidants should also be made.

■ **Recommendation 3.** Future epidemiologic studies should use personal monitors

and a greater number of better-placed area monitors to assess true exposure in at least a sample of the individuals under study.

Exposure Questionnaires. Standardized questionnaires for exposure assessment would be useful in ensuring comparability of data across studies. They should collect qualitative and quantitative information on indoor and outdoor activities, long-term residence, all potential confounders (personal as well as environmental), and effect modifiers. Each question should be tested for reliability and validity in different populations at risk. Measures of reliability and validity would allow for adjustment of responses among different subgroups where needed. A standardized exposure questionnaire should be validated against estimates of exposure determined by personal and well-placed environmental monitors. A questionnaire that correlated well with other measures would obviate the need to assess actual exposure in every individual.

■ **Recommendation 4.** Valid and reliable questionnaires that help assess personal exposure to air pollutants should be developed and used in future epidemiologic studies.

Effects Assessment

Questionnaires and Diaries. Questionnaires have been the mainstay of assessing symptoms in epidemiologic air pollution studies. Variations of questionnaires developed by the Medical Research Council (MRC) (1960) and the Division of Lung Disease in the National Heart, Lung, and Blood Institute (Ferris 1978) have been the most frequently used. Although many of the questions in these instruments have been tested for reliability and validity, most epidemiologic studies of air pollution have not actually assessed the accuracy of the responses that these questions have elicited in their own studies. Validity could be assessed by comparing symptoms recorded by the patient with responses to questions posed to the patient's spouse or a close family member.

The use of diaries to collect health effects data may be preferable to periodic use of questionnaires or personal interviews. Methodological studies of the use of diaries to collect health data indicate their potential for collecting accurate information (National Center for Health Services Research 1980). A 1973–74 Survey Research Laboratory study (University of Illinois) found that households of low as well as high educational levels were willing and able to use diaries, that they reported 14 percent more medical events than did households during recall interviews, that a ledger format was superior to a journal format, and that compensated households reported higher levels of events than uncompensated households (National Center for Health Services Research 1980).

Future epidemiologic studies could use diaries to collect data on specific symptoms, days of restricted activity, physician and emergency room visits, purchase/use of medicine, or days the respondent felt ill but performed usual activities. Statistical analysis of data obtained through diaries may be problematic if the instrument or instructions use a rating system that depends on the previous day's response (for example, better than yesterday). If such a problem is not eliminated by the design of the diary, special care will be required during data analysis. As with other data collection methods, validation studies should precede widespread use.

■ **Recommendation 5.** Valid and reliable questionnaires or diaries that assess health effects should be further developed to facilitate epidemiologic studies in air pollution. Ongoing cohort studies (Six-Cities, UCLA) could be used to develop and test such data collection instruments.

Spirometry and Single-Breath Tests. Spirometry is the most practical test to assess pulmonary function and the one most often used in epidemiologic studies. Pulmonary function changes from predicted values are reflected in several parameters, including average absolute differences in flow rates and lung volumes, variation in rates of decline of flow rates, and increases in airway resistance and airway response to bronchoconstrictors.

Spirometry is a well-standardized test for determining changes in rates of expiratory flow. Its ease of administration, reliability, and validity favor its continued use as a tool for measuring functional impairment of the lung. The forced expiratory volume in one second (FEV_1) has a low coefficient of variation. The forced expiratory flow rate between 25 and 75 percent of the vital capacity ($FEF_{25-75\%}$) is derived from the forced vital capacity and is a good measure of obstruction in small airways. The $FEF_{25-75\%}$ varies widely among individuals and therefore may be overly sensitive in detecting minor degrees of airflow obstruction if the lower limit of normal is defined as the conventional 80 percent predicted. In a cohort study, however, the rate of decline would be less affected by this variability, and the measure may have better predictive value for early COPD.

The single-breath nitrogen or oxygen test has also been used in epidemiologic studies to assess small airway function. The measures that are derived from this test have high coefficients of variation. The test is somewhat more difficult to administer than spirometry, but it may be more sensitive than the $FEF_{25-75\%}$ in detecting small airways obstruction in its reversible stage, prior to the development of COPD. However, evidence of small airways disease does not yet have predictive value for the development of COPD. Becklake and Permutt (1979) have indicated that measures of small airway function, derived either from spirometry or the single-breath oxygen test, may eventually be useful for predicting the development of COPD in smokers. It will be important to assess these tests in nonsmokers who are long-term residents of high- and low-oxidant environments. The UCLA population study and the Six-City study may yield information on changes in small airway function that are associated with varying ambient oxidant levels.

■ **Recommendation 6.** Longitudinal cohort studies should assess small airway function in nonsmokers. If residents of high-oxidant environments have decreases in small airway function compared to residents of low-oxidant environments, the cohorts should be followed for a sufficient amount of time to assess the development of COPD.

Bronchial Provocation Testing. In addition to using changes in spirometric values as outcome measures, some investigators have focused on bronchial reactivity as a possible predictor of progression to COPD among sensitive individuals. Barter and Campbell (1976) have theorized that, among smokers, individuals with sensitive airways are at special risk for progression to COPD, specifically chronic bronchitis and emphysema. Methacholine challenge is often used to assess bronchial sensitivity; histamine and cold air challenge have also been used. In a methacholine challenge test, airflow and airway conductance are measured serially after the inhalation of progressively increasing concentrations of methacholine, a bronchoconstrictor. An individual is classified as being sensitive if airflow or conductance falls a specified percent after the inhalation of a relatively low concentration of methacholine. This test serves as the gold standard for diagnosing bronchial hyperreactivity. (For additional discussion, see Bromberg, this volume.)

Recently, Chan-Yeung and coworkers (1986) surveyed an occupational cohort of 1,392 males for respiratory symptoms and the degree of hyperresponsiveness to methacholine inhalation. The prevalence of bronchial hyperresponsiveness was 12 percent among 694 asymptomatic individuals with normal lung function. Among the symptomatic participants, hyperresponders were more than three times as likely to have chronic bronchitis than the nonhyperresponders.

Asthmatics may also be at greater risk for accelerated decline in lung function. A study comparing asthmatic smokers and asthmatic nonsmokers to normal subjects without a history of smoking or respiratory illness, showed a significant correlation between rate of decline of FEV_1 and the level of bronchial responsiveness to histamine (Woolcock et al. 1986). Smoking and asthma were independent risk factors for an accelerated decline in FEV_1. It is possible

that sensitive individuals without clinical asthma, exposed repetitively to high peak levels or to chronically elevated low levels of oxidants, are at higher risk for developing an accelerated decline in lung function and COPD.

■ **Recommendation 7.** Small panels of individuals with sensitive airways should be studied carefully to assess whether oxidants are risk factors for various outcome measures of respiratory morbidity.

Data Bases. The utility of extant data bases for collecting data on the health effects of air pollution remains relatively unexplored. There are several population-based data bases that, if modified, may provide useful information on the effects of oxidants on health. Data collected by the National Health and Nutrition Examination Survey, the National Health Interview Survey, health maintenance organizations, preferred provider organizations, and insurance companies are some examples of potentially useful data sources.

A survey of existing data bases is needed to determine which are most likely to produce useful information on the health effects of air pollutants. Potentially useful data bases would probably require modification to ensure collection of information on potential confounders. Some data bases may serve as sampling frames for longitudinal, prospective cohort studies needed to test specific hypotheses associating exposure to oxidants with a specific health outcome (for example, development of early COPD).

■ **Recommendation 8.** Extant data bases should be explored for their use in studying the effects of air pollution on health.

Hospital and Physician Records. Data on health service utilization have been used as indices of morbidity in populations. In air pollution research, hospital and physician records may be helpful in evaluating threshold effects and the effects of exposure on sensitive groups.

The use of hospital statistics as an indirect measure of community morbidity has several limitations. Statistics such as number of visits to the emergency room, number of admissions, and length of stay can be affected by illness behavior, the demographic characteristics of the population in the catchment area, the availability of beds, and changes in the definitions of disease or coding practices (Bennett 1981). Most of these problems also apply to data based on outpatient visits to physicians' offices. As a result, comparison of illness rates based on encounters with the health care system can be problematic. The lack of a gold standard for diagnosing many respiratory illnesses makes the interpretation of studies that utilize these measures all the more difficult.

Nevertheless, several investigators have utilized hospital data to assess the effects of air pollutants on respiratory morbidity (Sterling et al. 1966, 1967; Bates and Sizto 1983). Bates (1985) has argued that despite limitations on hospital data, systematic diagnostic bias would be unlikely in studies that involved large populations over long periods of time and many different institutions and physicians. With such caveats, hospital data may be used efficiently and appropriately, especially in cross-sectional studies.

Specific Issues and Study Approaches

A number of questions remain unanswered in oxidant air pollution research. Still lacking is quantitative information on the threshold levels of effects, dose/response relationships, effects on sensitive groups, and the quantification of disease burden on public health. In particular, the effect of long-term oxidant exposure on human health remains a significant and unanswered question. Does such exposure increase the risk of or exacerbate existing COPD, bronchial hyperresponsiveness (for example, asthma), or pulmonary infection? Several approaches may be useful in examining these questions. Each approach will have its strengths and weaknesses; no one study by itself will produce conclusive ev-

idence of causality. Nevertheless, the consistency of findings from studies done in different populations, during different periods of time, and using different study designs would allow valid causal inference.

Chronic Obstructive Pulmonary Disease

Cohort Methods. The two ongoing cohort studies (Six-Cities and UCLA) are collecting morbidity data based on pulmonary symptoms and pulmonary function test results. These and any future cohort studies should look carefully at nonsmoking young adults who are long-term residents (more than 10–20 years) of the specified geographic areas to assess the relationship between accelerated decline in lung function and exposure to oxidants. Exposure should be assessed periodically by using personal monitors and time-activity logs in at least a sample of the study population. Information on confounders and effect modifiers should be elicited periodically; any differences in the prevalence of these variables between the exposure groups must be adjusted in the analysis. Information from questionnaires will require reliability and validity testing. To assess the future clinical significance in any statistically increased rates of decline in high-exposure groups, prolonged follow-up of these cohorts will be necessary.

Retrospective Methods. A case-control study comparing individuals with a specific respiratory illness (for example, chronic bronchitis) to healthy controls for estimated measures of exposure collected retrospectively might yield dose/response and threshold response information on the effects of oxidants on the disease under study.

One of the data bases mentioned earlier could be useful in assessing the role of oxidants (or other air pollutants) in the etiology of COPD, using a case-control study of nonsmokers with COPD. For example, most states now have large data bases consisting of all patients enrolled in Medicaid. Individual information such as age, gender, race, county, diagnoses, drugs dispensed, procedures, and place of visit are recorded in these data bases. Use of these data would allow comparisons of patients who have a recorded diagnosis of COPD, chronic bronchitis, or emphysema to controls without these diagnoses for county of residence.

Air pollution monitoring data in each county should provide a qualitative estimate of exposure. Personal contact with the medical providers for selected cases and controls would be necessary to estimate the prevalence of smoking in each group and the length of residence in each area. If the COPD group lived in higher air pollution areas than the control group, the study would suggest air pollution as a risk factor for the development of COPD. Arkansas and Ohio are two states worthy of study, given the high mortality rates for chronic bronchitis in certain portions of each state (National Research Council 1985).

■ **Recommendation 9.** Investigators should do case-control studies using existing health service data bases to assess the relationship between specific diseases and long-term residence in high-pollution areas.

Cross-Sectional Methods. Cross-sectional studies and secular trend analyses using the Medicaid data bases could also assess the prevalence of chronic respiratory disease and, perhaps, the frequency of various acute respiratory diseases in counties with different average levels of air pollutants. A rough estimate of a dose/response relationship may be evident in this type of analysis, which would also give information on other effects of air pollution exposure. The diagnoses recorded by the physician in the office or hospital would be used as the outcome variable of interest.

Cross-sectional studies can also be used to study the effects of point-source emissions on respiratory health. New roads and traffic patterns as well as industrialization of previously rural areas may produce new areas of high air pollution. If possible, pulmonary function measurements and morbidity surveys should be performed before and after the development of a new point source of oxidant pollutants. Changes in the mean levels and predicted rates of

decline in pulmonary function or in the prevalence of morbidity outcomes would implicate the point source as a causal factor.

A new point source may develop in Spring Hills, Tennessee, where the construction of a model automobile manufacturing plant is planned. General Motors expects to employ approximately 6,000 workers at the new plant. Construction and operation of the plant is expected to increase and change traffic patterns in the community. The study outlined in the preceding paragraph could identify changes in pulmonary function or morbidity outcomes in relation to specific areas of high traffic volume or congestion.

To strengthen the study, a nearby community that is unaffected by changing traffic patterns can serve as an unexposed control group. The two areas can then be compared on various measures of respiratory morbidity in the community (hospitalizations, visits to emergency rooms, and so on) and personal morbidity (symptoms, pulmonary function).

Such a study would be quasi experimental in that both communities would have a similar initial exposure to air pollutants, but a different level of exposure after the construction of the plant.

■ **Recommendation 10.** New point-source emissions should be identified and their effects on air pollution and respiratory morbidity should be studied longitudinally.

Asthma/Bronchial Hyperresponders

Future epidemiologic studies of air pollution should include those groups at high risk for developing adverse effects. Studying high-risk groups will enhance our ability to detect increases in the risk of exacerbating preexisting diseases in populations exposed to low ambient levels of pollutants.

Cohort Methods. The cohort design would be appropriate for assessing the effects of oxidant exposure on the morbidity of sensitive and normal individuals as determined by methacholine challenge. The National Heart, Lung, and Blood Institute recently funded a large multicenter randomized clinical trial in smokers to assess whether smoking cessation will effectively modify the course of COPD at a stage when mild dysfunction has already occurred (Gunby 1984). A subsample of smokers with hyperreactive airways will also be followed over a period of five years. This study is being organized in eight cities across the United States and may be a good source of hyperreactive patients who are exposed to different ambient levels of air pollutants.

Retrospective Methods. Alternatively, a case-control design can be used to compare the past oxidant exposure of bronchial hyperresponders and nonresponders. Assuming development of a valid questionnaire for past exposures, higher exposure levels in sensitive individuals would suggest that oxidants either increase sensitivity or exacerbate the sensitive state. If oxidants have one or both of these effects, exposure to oxidants may contribute to the increased prevalence of chronic bronchitis in hyperresponders and to the accelerated decline in lung function in asthmatics noted above.

Cross-Sectional Methods. The study design used by Holguin and coworkers (1985) to study asthmatics should be applied to other sensitive groups, such as people with COPD, who may be at increased risk for the exacerbation of symptoms. The design could be strengthened by including healthy people in the study group and by using personal monitors to better assess actual exposures to NO_2. Studies should focus on smokers as well as nonsmokers with sensitive airways as determined by methacholine challenge testing.

■ **Recommendation 11.** A variety of design methods should be used to study nonsmoking and smoking individuals with bronchial hyperresponsiveness to assess whether long-term residence in high-oxidant environments causes accelerated rates of decline in pulmonary function, more frequent encounters with the health care

system, and increased severity of pulmonary symptoms.

Healthy Cohort Effect. Investigators must be cautious in comparing morbidity rates in groups of high-risk individuals living in different air pollution environments. Standardized morbidity rates may be equal in populations with extremes in their exposure to oxidants, yet the high-oxidant environment may still be a risk factor for exacerbation of disease that is not evident because of a selection bias. This bias is analogous to the "healthy worker effect" seen in occupational epidemiologic studies where disease rates in occupational cohorts appear artificially low because of the "healthiness" of working populations (Monson 1980; Sterling and Weinkam 1986). This may be termed the "healthy cohort" effect in epidemiologic air pollution studies.

For example, asthmatics who are highly susceptible to oxidants and other pollutants may migrate out of a high-air-pollution environment, leaving behind those asthmatics—the "healthy cohort"—who may tolerate high exposures. Comparison of the remaining asthmatics (or a representative sample) with a panel of asthmatics in a low-exposure environment may show similar asthma attack rates that suggest no increased risk in high-pollution areas. Clearly, the selection of relatively healthy asthmatics is a form of selection bias that may obscure the risk of exposure to environmental pollutants.

The magnitude of the healthy cohort effect is substantial and affects all age groups. For example, an analysis of data from the Household Interview Survey of the National Center for Health Statistics showed that employed individuals had lower standardized morbidity rates of chronic respiratory ailments compared to unemployed people (Sterling and Weinkam 1985). Unemployed males were four times as likely to have chronic respiratory ailments as employed males. Although the magnitude of the healthy person effect is unlikely to be so large in morbidity studies on air pollutants, the size of the effect should be estimated in all air pollution studies. This may be done by asking the study subjects whether they have ever moved for health reasons.

■ **Recommendation 12.** All epidemiologic studies should analyze the outcome measure of interest by stratifying groups on their length of residence in the study areas.

Respiratory Infection

Pennington (this volume) indicated that there is little experimental evidence of a relationship between automotive emissions and the risk of respiratory infections. To date, there have been no epidemiologic studies assessing the incidence or severity of infection in populations exposed to high oxidant levels. Several studies have looked at the prevalence of symptoms that may indicate either infection or a direct irritant effect of oxidants (Shy et al. 1970a,b; Pearlman et al. 1971; Speizer et al. 1980; Comstock et al. 1981). Investigators have not used either sputum cultures or serology to assess the presence of true infection.

A recent review examined the relationship between respiratory illness in childhood and chronic airflow obstruction in adulthood (Samet et al. 1983). The authors decided that results from different studies on this question conflicted and that the data to date were inconclusive. However, the studies cited in the review support the conclusion that a history compatible with lower respiratory infection is associated with impaired pulmonary function in children. This association may be causal or noncausal and may be mediated by a host factor such as atopy (allergy that is probably hereditary) or an environmental factor such as parental smoking or oxidant exposure.

Epidemiologic investigation of the relationship between the risk of respiratory infection and exposure to oxidants would be difficult to conduct. Selection of the organisms to study would be the first of several difficult decisions. Even limiting the study to viral organisms would not necessarily narrow the spectrum of tests needed to assess true infection. Antibody levels for

Table 3. Sample Size Requirements for a Cohort Study in a Low-Oxidant Area

Relative Risk	Frequency of Disease	Total Sample Size
1.5	0.5	104
	0.4	176
	0.2	534
	0.1	1,254
	0.05	2,690
2	0.5	18
	0.4	40
	0.2	148
	0.1	362
	0.05	794
3	0.2	40
	0.1	112
	0.05	254
	0.01	1,404

NOTE: $\alpha = 0.05$ (1-tail), $\beta = 0.20$.

several viral antigens would be needed to ensure complete assessment for the most likely pathogens.

Cohort Methods. After deciding on the organism(s) to evaluate, investigators could study cohorts of high-risk populations (for example, chronic bronchitics) residing in high- versus low-level oxidant areas. Serial measurements of antibody titers could be compared to assess whether the high-oxidant group had higher conversion rates (fourfold rise) than the low-oxidant group. Such a study would presuppose that the right organism(s) has been identified a priori, that the populations have had the same likelihood of exposure to these organisms, that they are susceptible to infection with these organisms, and that the prevalence of exposure is high enough to ensure a sufficient incidence of clinical disease with a fourfold rise in antibody titers in both groups. Culture of organisms would not necessarily increase the ability to detect differences in infection rate between the two groups (Tager and Speizer 1975).

It is easy to calculate the sample size of previously uninfected subjects required to detect a significant clinical difference between the high- and low-oxidant groups. Table 3 shows how the required sample size varies depending on the frequency of infection in the low-oxidant group and the relative risk that the investigator would like to detect with a reasonable degree of certainty ($\beta = 0.20$) (see also Schlesselman 1974).

For example, 267 subjects would be required in each area (534 total) to detect a 50 percent increase in infection rate in the high-oxidant group, assuming that 20 percent of the low-oxidant group developed disease on exposure to the organism of interest. Note that the sample size requirements are minimum estimates and do not account for loss to follow-up, failure to develop clinical disease on exposure to organism, and the likelihood of being exposed to the organism. All these factors would increase the sample size needed to do the study. However, if infection with *any one of several* identified organisms was the outcome measure of interest, the sample size required to detect a difference in the incidence of infection between the high- and low-oxidant group would decrease.

Retrospective Methods. A case-control study of the relationship between respiratory infection and oxidant exposure would also be difficult to do. However, if one could specify the organism(s) of interest in advance and, on a prospective basis, identify cases as they are presented to the hospital (or to a physician), one could then compare them to controls selected on the basis of diagnoses not conceivably related to oxidant exposure. Clearly, the investigator would require a reliable and valid instrument for assessing past exposure to oxidants and the presence of potential confounders. The Medicaid data base described previously may be an excellent resource for doing this type of study.

■ **Recommendation 13.** Seroepidemiologic surveys should be done in cohorts that have different exposures to oxidants. More than one organism should be selected as a suspected respiratory pathogen to maximize the likelihood of detecting differences in infection rate among the cohorts.

Attributable Risk

Air pollution research efforts have focused primarily on acute effects in attempts to address issues of threshold effects, dose/response relationships, and disease incidence and prevalence in sensitive groups. Investigators have paid little attention to assessing the potential impact of improved air quality standards on reductions in health risks. Since most of the diseases associated with exposure to oxidants (and other pollutants) have multifactorial origins, it would be useful to know how reducing average pollutant levels might proportionally reduce morbidity. The attributable risk of exposure to oxidants is the proportion of disease in the population that can be theoretically prevented by eliminating exposure to oxidants.

Morganstern and Bursic (1982) outlined a mathematical method for estimating the potential impact of a public health policy on the health status of a target population. Investigators applying this method to ambient oxidants would have to assume that (1) the distribution of oxidant levels in the population is known; (2) oxidants are true determinants of the disease(s) in question (as measured by the relative risks); (3) a reduction in ambient oxidant levels would reduce the risk of disease(s) in previously exposed persons compared to the risk of disease(s) in previously unexposed persons; (4) there are no significant secular trends in the risk of disease(s) because of changes unrelated to the reduction in ambient oxidant levels; and (5) there are no significant changes in other risk factors that would interact with oxidants to synergistically increase the risk of disease(s). Because of the latter assumption, such models must be applied cautiously to air pollution research. When there are several, noninteracting causes of adverse health effects, the simple addition of the attributable risks associated with each causal factor will yield the total benefit in risk reduction that is theoretically achievable. However, in the case of interacting or synergistic pollutants, attributable risks must be viewed more carefully. When synergism exists, no single etiologic factor can be said to "cause" a percentage of the disease that is numerically equivalent to the attributable risk. For example, it is true that smoking cessation could prevent x percent of disease in a target population, but it may also be true that the elimination of other occupational or environmental toxins could prevent y percent of the disease in the same population. The addition of x percent and y percent can yield a number much greater than 100 percent in the case of synergistic risk factors (Ashford 1985; Sterling 1985).

With these caveats, studies that assess attributable risk of oxidant exposure would be highly useful to policy makers and regulators, as well as of great interest to the general public, the automotive industry, and public health professionals. Mathematical modeling using data collected from previous studies could provide some estimate of the amount of morbidity that could be detected in populations with different diseases and potential exposures. In addition, mathematical modeling could help estimate the expected degree of interaction between air pollutants and other risk factors.

■ **Recommendation 14.** The potential attributable risk for disease caused by exposure to oxidants and other air pollutants should be determined and expressed in a meaningful way.

Summary

Epidemiologic studies in air pollution research are necessary to assess the effects of long-term oxidant exposure on human health. Most studies done to date have utilized the cross-sectional design, which is best suited for correlating current levels of exposure with the prevalence of various measures of acute respiratory morbidity. Review of selected air pollution studies shows that higher oxidant exposures are associated with a higher prevalence of respiratory symptoms, more frequent asthma attacks, greater decreases from expected pulmonary function, and more frequent utilization of health care services. The long-term and chronic effects of low-level exposure to oxidants is essentially unknown. The major gaps in oxidant air pollution research include information on threshold levels of effects and dose/response relationships, long-term effects on sensitive groups, and the quantification of disease burden on public health.

Investigators must address several methodological issues when conducting epidemiologic research in this area. They involve developing more specific tools for assessing personal exposure and adverse health effects; improving assessment and control of confounders and effect modifiers; identifying sensitive populations for more detailed study; and assessing potential biases in population selection.

Several approaches for addressing these methodological issues and for answering important research questions have been addressed in this paper. Cross-sectional, case-control, and cohort studies can be used to examine the relationship between oxidant exposure and increased risk for or exacerbation of existing COPD, asthma, or pulmonary infections. Each approach has its strengths and weaknesses and will be appropriate in different situations. The consistency and reproducibility of findings from different epidemiologic studies are necessary to allow causal inference between oxidant exposure and adverse health effects.

Summary of Research Recommendations

Specific Research Issues
HIGH PRIORITY

Studies that will provide data on the long-term, chronic health effects of exposure to low levels of photochemical oxidants are of highest priority.

Recommendation 1	A cohort study of the effects of long-term exposure to oxidants on respiratory morbidity in children, normal adults, and sensitive populations should be done if the Six-Cities study and the UCLA population studies cannot assess these issues.
Recommendation 6	Longitudinal cohort studies should assess small airway function in nonsmokers. If residents of high-oxidant environments have decreases in small airway function compared to residents of low-oxidant environments, the cohorts should be followed for a sufficient amount of time to assess the development of COPD.
Recommendation 7	Small panels of individuals with sensitive airways should be studied carefully to assess whether oxidants are risk factors for various outcome measures of respiratory morbidity.
Recommendation 11	A variety of design methods should be used to study nonsmoking and smoking individuals with bronchial hyperresponsiveness to assess whether long-term residence in high-oxidant environments causes accelerated rates of decline in pulmonary function, more frequent encounters with the health care system, and increased severity of pulmonary symptoms.

MODERATE PRIORITY

Existing data bases may be an untapped resource of information on health effects. New environments may also provide an opportunity to explore exposure/effect relationships.

Recommendation 8	Extant data bases should be explored for their use in studying the effects of air pollution on health.
Recommendation 10	New point-source emissions should be identified and their effects on air pollution levels and respiratory morbidity should be studied longitudinally.

LOW PRIORITY

Causal hypotheses are strengthened if a variety of study designs, methods, and data sources are used and yield similar results.

Recommendation 2	Future epidemiologic studies of air pollution should include the use of retrospective methods to investigate the relationship between oxidant exposure and respiratory morbidity if accurate methods of assessing past cumulative exposure can be developed.
Recommendation 9	Investigators should do case-control studies using existing health service data bases to assess the relationship between specific diseases and long-term residence in high-pollution areas.
Recommendation 13	Seroepidemiologic surveys should be done in cohorts that have different exposures to oxidants. More than one organism should be

selected as a suspected respiratory pathogen to maximize the likelihood of detecting differences in infection rate among the cohorts.

Recommendation 14   The potential attributable risk for disease caused by exposure to oxidants and other air pollutants should be determined and expressed in a meaningful way.

Generic Issues

Methodological innovation will greatly enhance the ability of epidemiologic studies to investigate dose/response relationships, long-term effects of low-level exposure, and attributable risk. Such methodological research deserves the highest priority.

Recommendation 3   Future epidemiologic studies should use personal monitors and a greater number of better-placed area monitors to improve the assessment of true exposure in at least a sample of the individuals under study.

Recommendation 4   Valid and reliable questionnaires that help assess personal exposure to air pollutants should be developed and used in future epidemiologic studies.

Recommendation 5   Valid and reliable questionnaires or diaries that assess health effects should be developed further to facilitate epidemiologic studies in air pollution. Ongoing cohort studies (Six-Cities, UCLA) could be used to develop and test such data collection instruments.

Recommendation 12   All epidemiologic studies should analyze the outcome measure of interest by stratifying groups on their length of residence in the study areas.

Acknowledgment

The authors wish to thank the following individuals for their careful reviews and thoughtful comments: Nicholas Ashford, Ph.D., J. D.; Peter Gann, M.D.; Murray Gilman, M.D.; Tee Guidotti, M.D.; Sandy Norman, Ph.D.; and Brian Strom, M.D. The authors also appreciate the unflagging clerical support of Karen Menna and Frances Seeds.

Correspondence should be addressed to Eddy A. Bresnitz, Division of Occupational and Environmental Health, Department of Community and Preventive Medicine, The Medical College of Pennsylvania, 3300 Henry Avenue, Philadelphia, PA 19129.

References

Ashford, N. A. 1985. Unpublished comments on the OTA Report on Smoking-Related Death and Financial Costs.

Barter, C. E., and Campbell, A. H. 1976. Relationship of constitutional factors and cigarette smoking to decrease in 1-second forced expiratory volume, *Am. Rev. Respir. Dis.* 113:305–314.

Bates, D. V. 1985. The strength of the evidence relating air pollutants to adverse health effects, Carolina Environmental Essay Series VI, University of North Carolina Institute for Environmental Studies, Chapel Hill, N.C.

Bates, D. V., and Sizto, R. 1983. Relationship between air pollution levels and hospital admissions in Southern Ontario, *Can. J. Public Health* 74:117–133.

Becklake, M. R., and Permutt, S. 1979. Evaluation of tests of lung function for "screening" for early detection of chronic obstructive lung disease, In: *The Lung in the Transition Between Health and Disease*

(P. T. Macklem and S. Permutt, eds.), pp. 345–387, Marcel Dekker, New York.

Bennett, A. E. 1981. Limitations of the use of hospital statistics as an index of morbidity in environmental studies, *J. Air Pollut. Control Assoc.* 31:1276–1278.

Chan-Yeung, S., Vedal, S., and Enarson, D. 1986. Asthma, asthma-like symptoms, chronic bronchitis and the degree of bronchial hyperresponsiveness in epidemiologic surveys (abs.), *Am. Rev. Respir. Dis.* 133:A155.

Cohen, C. A., Hudson, A. R., Clausen, J. L., and Knelson, J. H. 1972. Respiratory symptoms, spirometry, and oxidant air pollution in nonsmoking adults, *Am. Rev. Respir. Dis.* 105:251–261.

Comstock, G. W., Meyer, M. B., Helsing, K. J., and Tockman, M. S. 1981. Respiratory effects of household exposures to tobacco smoking and gas cooking, *Am. Rev. Respir. Dis.* 12:143–148.

Detels, R., Rokaw, S. N., Coulson, A. H., Tashkin, D. P., Sayre, J. W., and Massey, R. J. 1979. The UCLA population studies of chronic obstructive respiratory disease. I. Methodology and comparison of lung function in areas of high and low pollution, *Am. J. Epidemiol.* 109:33–58.

Detels, R., Sayre, J. W., Coulson, A. H., Rokaw, S. N., Massey, F. J., Tashkin, D. P., and Wu, M. M. 1981. The UCLA population studies of chronic obstructive respiratory disease. IV. Respiratory effect of long-term exposure to photochemical oxidants, nitrogen dioxide, and sulfates on current and never smokers, *Am. Rev. Respir. Dis.* 124:673–680.

Detels, R., Tashkin, D. P., Simmons, M. S., Carmichael, H. E., Sayre, J. W., Rokaw, S. N., and Coulson, A. H. 1982. The UCLA population studies of chronic obstructive respiratory disease, *Chest* 82:630–638.

Ferris, B. G. 1978. Epidemiology standardization project, *Am. Rev. Respir. Dis.* 118(Suppl.):7–53.

Ferris, B. G., Speizer, F. E., Spengler, J. D., Dockery, D., Bishop, Y. M. M., Wolfson, M., and Humble, C. 1979. Effects of sulfur oxides and respirable particles on human health. Methodology and demography of populations in study, *Am. Rev. Respir. Dis.* 120:767–779.

Ferris, B. G., Dockery, D. W., Ware, J. H., Speizer, F. E., and Spiro, R. 1983. The six-city study: examples of problems in analysis of the data, *Environ. Health Perspect.* 52:115–123.

Gunby, R. 1984. Intervention trial begins with young smokers, *J. Am. Med. Soc.* 252:2802–2803.

Hill, A. B. 1965. The environment and disease: association or causation, *Proc. Roy. Soc. Med.* 58: 295–300.

Holguin, A. H., Contant, C. F., Noel, L. M., Mei, M., Buffler, P. A., and Hsi, B. P. 1983. Defining an asthmatic attack in epidemiologic studies, In: *International Symposium on the Biomedical Effects of Ozone and Related Photochemical Oxidants* (S. D. Lee, M. G. Mustafa, and M. A. Mehlman, eds.), pp. 527–538, Princeton Scientific Publishers, Inc., Princeton, N.J.

Holguin, A. H., Buffler, P. A., Contant, C. F., Stock, T. H., Kotchmar, D. J., Hsi, B. R., Jenkins, D. E.,

Gehan, B. M., Noel, L. M., and Mei, M. 1985. The effects of ozone on asthmatics in the Houston area, In: *Transactions of the APCA International Specialty Conference: Evaluation of the Scientific Basis for Ozone/Oxidant Standards* (S. D. Lee, ed.), pp. 262–280, Air Pollution Control Association, Pittsburgh, Pa.

Johnson, D., and Travis, J. 1979. The oxidative inactivation of human alpha-1-proteinase inhibitor, *J. Biol. Chem.* 254:4022.

Kagawa, J., and Toyama, T. 1975. Photochemical air pollution: its effects on respiratory function of elementary school children, *Arch. Environ. Health* 30:117–122.

Keller, M. D., Lanese, R. R., Mitchell, R. I., and Cote, R. W. 1979a. Respiratory illness in households using gas and electricity for cooking. I. Survey of incidence, *Environ. Res.* 19:495–503.

Keller, M. D., Lanese, R. R., Mitchell, R. I., and Cote, R. W. 1979b. Respiratory illness in households using gas and electricity for cooking. II. Symptoms and objective findings, *Environ. Res.* 19:504–515.

Khan, A. U. 1977. The role of air pollution and weather changes in childhood asthma, *Ann. Allergy* 39:397–400.

Kucich, U., Abrams, W. R., Christner, P., Rosenbloom, J., Kimbel, P., and Weinbaum, G. 1984. Molecular weight distribution of elastin peptides in plasmas from human non-smokers, smokers and emphysema patients, *Am. Rev. Respir. Dis.* 129:307.

Kurata, J. H., Glovsky, M. M., Newcomb, R. L., and Easton, J. G. 1976. Multifactorial study of patients with asthma. 2: Air pollution, animal dander, and asthma symptoms, *Ann. Allergy* 37:398–409.

Lippmann, M., Lioy, P. J., Leikauf, G., Green, K. B., Baxter, D., Morandi, M., and Pasternack, B. S. 1983. Effects of ozone on the pulmonary function of children. In: *International Symposium on the Biomedical Effects of Ozone and Related Photochemical Oxidants* (S. D. Lee, M. G. Mustafa, and M. A. Mehlman, eds.), pp. 423–446, Princeton Scientific Publishers, Inc., Princeton, N.J.

Medical Research Council, Committee on the Aetiology of Chronic Bronchitis. 1960. Standardized questionnaires on respiratory symptoms, *Brit. Med. J.* 2:1665.

Michaeli, D., and Fudenberg, H. H. 1974. Antibodies to collagen in patients with emphysema, *Clin. Immunol. Immunopathol.* 3:187–192.

Monson, R. R. 1980. *Occupational Epidemiology*, CRC Press, Inc., Boca Raton, Fla.

Morganstern, H., and Bursic, E. S. 1982. A method for using epidemiologic data to estimate the potential impact of an intervention on the health status of a target population, *J. Commun. Health* 7:292–307.

National Center for Health Services Research. 1980. Health Care Surveys Using Diaries, p. 286, Publication No. PHS 80-3279, U.S. Department of Health and Human Services, National Center for Health Services Research, Hyattsville, Md.

National Research Council. 1985. *Epidemiology and Air Pollution*, National Academy Press, Washington, D.C.

Pearlman, M. E., Finklea, J. F., Creason, J. P., Shy,

C. M., Young, M. M., and Horton, H. S. M. 1971. Nitrogen dioxide and lower respiratory illness, *Pediatrics* 47:391–398.

Richards, W., Azen, S. P., Weiss, J., Stocking, S., and Church, J. 1981. Los Angeles air pollution and asthma in children, *Ann. Allergy* 47:348–354.

Rokaw, S. N., Detels, R., Coulson, A. H., Sayre, J. W., Tashkin, D. P., Allwright, S. S., and Massy, F. J. 1980. The UCLA population studies of chronic obstructive lung disease. III. Comparison of pulmonary function in three communities exposed to photochemical oxidants, multiple primary pollutants, or minimal pollutants, *Chest* 78:252–262.

Samet, J. M., Tager, I. B., and Speizer, F. E. 1983. The relationship between respiratory illness in childhood and chronic air flow obstruction in adulthood, *Am. Rev. Respir. Dis.* 127:508–523.

Schlesselman, J. J. 1974. Sample-size requirements in cohort and case-control studies of disease, *Am. J. Epidemiol.* 99:381–384.

Schoettlin, C. E., and Landau, E. 1961. Air pollution and asthmatic attacks in the Los Angeles area, *Public Health Rep.* 76:545–548.

Selwyn, B. J., Stock, T. H., Hardy, R. J., Chan, F. A., Jenkins, D. E., Kotchmar, D. J., and Chapman, R. S. 1985. Health effects of ambient ozone exposure in vigorously exercising adults, In: *Transactions of the APCA International Specialty Conference: Evaluation of the Scientific Basis for Ozone/Oxidant Standards.* (S. D. Lee, ed.), pp. 281–296, Air Pollution Control Association, Pittsburgh, Pa.

Shy, C. M., Creason, J. P., Pearlman, M. E., McClain, K. E., Benson, F. B., and Young, M. M. 1970a. The Chattanooga school children study: effects of community exposure of nitrogen dioxide. I. Methods, description of pollutant exposure and results of ventilatory function testing, *J. Air Pollut. Control Assoc.* 20(8):539–545.

Shy, C. M., Creason, J. P., Pearlman, M. E., McClain, K. E., Benson, F. B., and Young, M. M. 1970b. The Chattanooga school study: effects of community exposure to nitrogen dioxide. II. Incidence of acute respiratory illness, *J. Air Pollut. Control Assoc.* 20(9):582–588.

Speizer, F. E., Ferris, B., Jr., Bishop, Y. M. M., and Spengler, J. 1980. Respiratory disease rates and pulmonary function in children associated with NO₂ exposure, *Am. Rev. Respir. Dis.* 121:3–10.

Spengler, J. D., Ferris, B. G., Jr., and Dockery, D. W. 1979. Sulfur dioxide and nitrogen dioxide

levels inside and outside homes and the implication on health effects research, *Environ. Sci. Technol.* 13:1276–80.

Sterling, T. D., 1985. Letter to the Honorable Don Sundquist, U.S. House of Representatives. Comments on OTA Report on Smoking Related Death and Financial Costs (September 20).

Sterling, T. D., and Weinkam, J. J. 1985. The "healthy worker effect" on morbidity rates, *J. Occup. Med.* 27:477–482.

Sterling, T. D., and Weinkam, J. J. 1986. Extent, persistence, and constancy of the healthy worker or healthy person effect by all and selected causes of death, *J. Occup. Med.* 28:348–353.

Sterling, T. D., Phair, J. J., Pollack, S. V., Schurnsky, D. A., and Degroot, I. 1966. Urban hospital morbidity and air pollution. A first report, *Arch. Environ. Health* 15:362–374.

Sterling, T. D., Pollack, S. V., and Phair, J. H. 1967. Urban hospital morbidity and air pollution. A second report. *Arch. Environ. Health* 15:362–374.

Tager, I. B., and Speizer, F. E. 1975. Role of infection in chronic bronchitis, *N. Engl. J. Med.* 292:563–571.

Tager, I. B., Weiss, S. T., Rosner, B., and Speizer, F. E. 1979. Effect of parental cigarette smoking on the pulmonary function of children, *Am. J. Epidemiol.* 110:15–26.

U.S. Environmental Protection Agency. 1986. Air Quality Criteria for Ozone and Other Photochemical Oxidants, Report EPA/600/8-84/020EF, volume V, Research Triangle Park, N.C.

Ware, J. H., Dockery, D. W., Spiro, A., Speizer, F. E., and Ferris, B. G. 1984. Passive smoking, gas cooking, and respiratory health of children living in 6 cities, *Am. Rev. Respir. Dis.* 129:366–374.

Whittemore, A. S., and Korn, E. L. 1980. Asthma and air pollution in the Los Angeles area, *Am. J. Public Health* 70:687–696.

Woolcock, A. J., Peat, J. K., and Cullen, K. J. 1986. The effect of asthma on rate of decline of lung function (abs.), *Am. Rev. Respir. Dis.* 133:A156.

Yanagisawa, Y., Nishmura, H., Osaka, F., and Kasuga, H. 1986. Personal exposure and health effect relation for NO₂ with urinary hydroxyproline to creatinine ratio as indicator, *Arch. Environ. Health* 41:41–48.

Zagraniski, R. T., Leaderer, B. P., and Stolwijk, J. A. J. 1979. Ambient sulfates, photochemical oxidants, and acute health effects: an epidemiological study, *Environ. Res.* 19:306–320.

Biochemical and Cellular Interrelationships in the Development of Ozone-Induced Pulmonary Fibrosis

JEROLD A. LAST
University of California, Davis

Air Pollution, the Automobile, and Public Health. © 1988 by the Health Effects
Institute. National Academy Press, Washington, D.C.

Difficulties in Relating Ozone Exposure to Lung Disease

The focus of this chapter is on the evidence that exposure to ozone (O_3) can cause pulmonary fibrosis, generally considered to be a chronic lung disease. A description of the underlying mechanisms of lung injury that might be predictive of other adverse health effects provides a background to the discussion of O_3-induced pulmonary fibrosis.

True chronic experiments have rarely been performed with O_3. Examinations of the development of lung fibrosis, a disease widely perceived as chronic, have been made primarily on the basis of data from experiments performed in an acute time frame, usually days or weeks. Long-term data are generally products of earlier experiments not performed under present-day standards of care and precision. Animal hygiene in those experiments was not always adequate to prevent outbreaks of pneumonia and other confounding disease.

It may be that the human disease analogous to that caused by O_3 is not chronic lung fibrosis but acute forms of pulmonary fibrosis such as the adult respiratory distress syndrome. The many similarities between the acute disease in humans and the responses of animals to O_3 are also discussed in this chapter.

The misconception that O_3 can cause emphysema deserves special note. Early work of Stokinger and coworkers (1957) suggested this to be the case. However, in those studies O_3 was routinely generated from air rather than oxygen, and there was concurrent exposure to nitrogen dioxide (arising from oxidation of atmospheric nitrogen), a known emphysema-provoking agent. Intercurrent animal illness might also have influenced those results.

Perhaps the frequency of occurrence of chronic obstructive pulmonary disease (COPD) vis-à-vis fibrosis tempted investigators to find a solution to this much more prevalent disease in humans. Although data are lacking that suggest that exposure to O_3, by itself, can cause emphysema, without large-scale, careful, lifetime exposures of animals to O_3, the possibility that exposure to O_3 may contribute to the develop-

ment of COPD in humans cannot be totally ruled out.

Another difficulty lies in trying to distinguish between an "effect" and an "adverse health effect." Acute exposure to O_3, especially at higher-than-ambient concentrations, causes changes in a wide variety of measurable biological parameters. Some of these changes are probably relevant to the mechanism(s) of action of O_3. However, some are not—they are more likely to be epiphenomena (for example, changes in barbiturate-induced sleeping time) or artifacts (for example, sleeplessness and lack of appetite in rats for 1–2 days after the start of exposures to levels of O_3 that presumably are irritating enough to cause "sore throats"). Interpretation of whether measured changes, even statistically significant changes, are "adverse health effects" may be a particularly difficult problem when human pulmonary function tests are performed on volunteers and one or two interdependent parameters out of 20 are shown to change slightly.

Such "experiments of nature" as epidemiologic studies of the risk of breathing polluted atmospheres in, for example, the Los Angeles air basin, have essentially nothing to tell us thus far. This is true for several reasons. Most important, from the perspective of this chapter, the clinical recognition of pulmonary fibrosis is difficult, except for the most severe manifestations of this disease. Routine spirometry (simple pulmonary function testing), such as may be performed on large populations, is a very insensitive technique for scoring lung fibrosis. There have been very few studies (and no large-scale ones) on autopsies of people routinely exposed to O_3 and dying of random causes. Costs of such a large-scale study might be prohibitive.

Confounding effects of cigarette smoking further complicate large-scale epidemiologic studies, as do definitions of appropriate control groups. The relationship between outcome in an individual and actual dose of O_3 to that individual is never known; personal monitoring of individuals in a large-scale study probably has not been performed for O_3. Finally, epidemiologic studies are most powerful when the disease at risk is a rare one, and fibrosis is not rare. All of these limitations also pertain to determination of whether O_3 exposure increases the risk of cancer.

Finally, the characteristics of particularly susceptible populations at risk for pulmonary fibrosis are unknown. Intuitively—and limited animal data support this intuition—it is hypothesized that neonates and youngsters whose lungs are still developing might be one such sensitive population. Populations receiving higher-than-ordinary doses of O_3, such as joggers and other outdoor exercisers and athletes, and obligate mouth breathers, might constitute another. Definition of such potentially susceptible populations has profound impact on the proper design and interpretation of animal exposure studies, but this is an area where little has been done experimentally.

Some may argue that mathematical modeling of calculated dose delivered to tissue sites can replace such actual experimentation, but this is not yet possible. Not enough is known about lung structure, target cells and molecules, thickness and reactivity of protective layers at different sites in the lung, or the role of secondary reactions in amplification of injury, to interpret such modeling exercises, either biologically or for regulatory purposes. The requisite data to refine such models to the point where they may replace animal experiments will most probably not be available in the near future. In the meantime, measurement of equivalence of biological responses may be the best indicator for interpretation of exposure/dose relationships in humans vis-à-vis animal experiments.

Response to Fibrogenic Insult

When the lung is exposed to O_3 or other fibrogenic agents administered systemically or by inhalation, the critical response seems to be an inflammatory influx of macrophages to sites of injury, with accompanying edema. This influx may be responsible for the release of chemotactic factors that, in turn, are responsible for the recruitment of fibroblasts to these sites (Reiser and Last 1979). These fibroblasts would then synthesize the collagen that constitutes the fibrotic "scar." Macrophages may release substances that stimulate fibroblasts to produce collagen, or to produce more collagen than their baseline synthesis levels (Reiser and Last 1979).

Other cells, including lymphocytes, granulocytes, and eosinophils, also seem able to release substances chemotactic for one another (Sobel and Gallin 1979). Some of these same inflammatory cell types also seem to be able to release factors that stimulate fibroblast mitogenesis (Nathan et al. 1980), which in turn increases the capacity for collagen synthesis at such sites. The interplay of chemotactic, mitogenic, and collagen synthesis-stimulating factors in the etiology of organ fibrosis and the regulation of cell infiltration and turnover in the damaged lung is poorly understood but presently under very active investigation. It will be discussed more thoroughly below.

Acute Lung Injury and Edema

The earliest quantifiable response of the lung to many types of injury, including exposure to O_3, is pulmonary edema. At high levels of injury, this response is easy to quantify; the animal is sacrificed and the lung weighed. An increase in lung wet weight over weight of lungs from matched control animals exposed to filtered air can be equated with the increase of fluid that constitutes edema. To control for differences in animal size, data are expressed as the ratio of lung weight to body weight. However, the best way to quantify edema in most animal models—that is, the ratio of wet weight to dry weight of the lung—is not appropriate with O_3 (Cross et al. 1981).

The edema fluid accumulating in lungs of rats exposed to less than 1 part per million (ppm) of O_3 has a wet weight to dry weight ratio similar to lung and to blood serum; therefore, pulmonary edema induced by O_3 has little or no effect on the lung wet weight: dry weight ratio, even in lungs whose wet weight has doubled. Measurement of wet weight (either directly or normalized to body weight) alone, however, can be used to quantify O_3-induced edema. Such measurements show concentration/response behavior between about 0.5 and 1.0 ppm O_3, but these measurements are insensitive compared to quantification by other methods.

Two sensitive methods deserve special mention: accumulation of serum proteins in lung lavage fluid (Hu et al. 1982; Guth et al. 1986) and tracer transport from blood to lavage fluid (and vice versa) (Alpert et al. 1971; Guth et al. 1986). Both techniques have been used to examine effects of exposure of laboratory animals to O_3 at low concentrations. Under the assay conditions used in those studies, the increased protein content of lavage fluid from lungs of rats exposed to O_3 was due almost completely to movement of serum albumin from blood to broncho-alveolar and pulmonary interstitial fluid, as determined by electrophoretic analysis of proteins in the lavage fluid. More recent studies (Warren et al. 1986; Warren and Last 1987) document a positive response by this assay at 0.12 ppm O_3, the current peak hourly National Ambient Air Quality Standard.

Increased permeability of the lung to serum proteins can be equated with pulmonary edema, a recognized adverse health effect. Increased permeability has been observed in animal experiments at O_3 concentrations routinely encountered, or even routinely exceeded, as peak hourly values in southern California smog episodes. At a given concentration of O_3, rats as obligate nose breathers probably receive a lower total dose to the peripheral lung than do humans. Further work is necessary to ascertain the reversibility of these changes suggestive of pulmonary edema and the long-term significance of a succession of short-term episodes of pulmonary edema. Current experiments in humans, using labeled aerosols of diethylenetriaminepentaacetic acid (DTPA) to probe for permeability changes of the lung after exposure to O_3 in humans (see, for example, Gellert et al. 1985) and rats (for example, Minty and Royston 1985) should be correlated with similar animal work.

■ **Recommendation 1.** Using experimental animals, anatomic regions in the lung associated with tracer transport from the blood to lavage fluid and vice versa after O_3 exposure should be localized and correlated with the bulk movement of albumin into the airspaces.

■ **Recommendation 2.** To facilitate interpretation of human studies, sensitive lavage-based assays of protein content need to be correlated with DTPA aerosol transport assays in animals.

■ **Recommendation 3.** Human studies using DTPA aerosols or other tracers in conjunction with concentration/response assays for edema should be undertaken.

There are two important barriers to the accumulation of fluid in the lung in pulmonary edema. The first, the endothelial cell permeability barrier, is thought to be intrinsically leaky. Fluid is constantly crossing this barrier and constantly being pumped from the interstitial region back to the vasculature to maintain homeostasis. O_3 does not appear to damage endothelial cells directly since it is too chemically reactive to traverse the distance from the airways to the vasculature and thereby to reach these cells. However, products of O_3 reaction with epithelial cells or mucous constituents might well have long enough lifetimes to affect cells deeper in the lung or elsewhere in the body.

The other barrier—the epithelial cell layer with its tight junctions between cells—is thought to prevent fluid movement between the vasculature (and interstitial space) and the airways. It is thus easy to visualize the relationship between damage to epithelial cells in the small airways and centriacinar regions of the lung and pulmonary edema.

Transition from Pulmonary Edema to Cellular Inflammation

The first lung cells to encounter inhaled or intratracheally instilled fibrogenic agents are the epithelial cells lining the respiratory tract. Oxidant gases such as nitrogen dioxide (NO_2) and O_3 have long been known to damage epithelial cells directly; much has been published on the mechanisms by which such oxidant gases may exert their cytotoxic effects (see, for example, Mustafa and Tierney 1978). Systemically administered fibrogenic agents such as paraquat, radiation, and possibly bleomycin may also exert a direct cytotoxic effect on lung parenchymal cells through generation of free radicals in vitro, and thus these agents may share common pathways of injury and fibrogenesis with the oxidant gases.

There appears to be a wide range of response of the type I and II pneumonocytes to such agents. In general, type I pneumonocytes appear to be more sensitive to damage than type II pneumonocytes, although the much greater surface area of type I cells (about 95 percent of normal alveolar epithelium) may merely present a correspondingly larger target for injury.

Exposure of an animal to a pneumotoxic agent often results in a characteristic pattern of injury and repair in which type II cells rapidly begin proliferating to repair the epithelial lesions resulting from injury and death of type I cells. Despite their greater cytoplasmic differentiation, type II cells apparently serve as progenitor cells of type I pneumonocytes (Haschek and Witschi 1979).

Cellular damage may also result in the local release of soluble mediators thought to be chemotactic for phagocytotic cells. Such mediators could include products from fibrin, fibronectin, albumin, prostaglandins, leukotrienes, and a vast array of poorly characterized chemoattractants and other factors.

Present concepts of the critical sequence of events in lung injury suggest that the response of the lung to cellular damage and/or to pulmonary edema is the movement of alveolar macrophages and (perhaps) the recruitment of leukocytes to sites of damage within or near the lung epithelial surface. This process may also contribute to increased airway reactivity. For most models of lung damage, the neutrophil is a ubiquitous participant in this inflammatory response.

O_3 is unusual as a pneumotoxicant; the inflammatory response of the lung to O_3 seems to be predominantly macrophagic, and there is as yet no evidence for a role of the neutrophil in this process, especially in the critical early stages. It is tempting to speculate that the lack of a neutrophilic response to O_3 may have mechanistic significance, especially with regard to lung fibrosis and, perhaps, to the failure to observe an emphysematous response to O_3. This would be consistent with the presumed role of neutrophil elastase in the pathogenesis of emphysema, as discussed by Wright (this volume).

Accumulation of alveolar macrophages at sites of lung damage can be beneficial or detrimental. If the response is limited, dogma suggests that the lung repairs itself. However, the long-term consequences to the lung of multiple episodes of this type and whether the lung maintains its putative ability for self-repair in the face of repeated limited insults are unknown. If the macrophagic response is more exuberant, then the alveolar macrophages elaborate factors that stimulate the proliferation of lung fibroblasts. Such fibroblast proliferation gives rise to increased numbers of interstitial lung cells producing collagen, the hallmark of pulmonary fibrosis. This process is not well understood. Specifically, the extent of macrophage accumulation at sites of lung damage cannot be used to determine whether lung repair or progression to fibrosis occurs. However, macrophage accumulation elicited by "inert" particles (for example, fly ash) does not result in lung fibrosis.

The concept of cellular damage is often invoked in discussions of toxicity but seldom defined in biochemical terms. There seems to be a general consensus as to the mechanism of action of O_3, which is assumed to initiate lipid peroxidation in lungs by reacting with unsaturated fatty acids of cell membranes. This putative mechanism of action, although widely accepted, has not been proven. Neither is the chemical basis for generation of free radicals by these reactions completely straightforward. It is likely that if free radicals and active oxygen species do play a role in cellular damage elicited by O_3, macrophages and leukocytes participating in the inflammatory response might well be the source of such active oxygen species.

The initiation of lipid peroxidation in the lung is a possible mechanism for oxidant damage by gases such as O_3 and NO_2 (Fridovich 1976; Waling 1963). In recent years a considerable body of evidence has been amassed in support of the hypothesis that lung damage may be mediated in large part by reactions involving free radicals. Such radicals may be directly generated by reaction of toxicants with target molecules (for example, O_3 or NO_2 reaction with polyunsaturated fatty acids), or indirectly as a result of phagocytotic activity by alveolar macrophages or neutrophils. These cells release free radicals such as superoxide and hydroxyl during increased metabolism which is often termed "oxidative burst."

Evidence for the role of free radicals in lung damage includes a wide variety of observations. Uncontrolled clinical trials have shown increased survival of patients suffering from adult respiratory distress syndrome after they are treated with vitamin E (Wolf and Seeger 1982). Treatment with various hydroxyl radical scavengers protected rats from pulmonary edema induced by high doses of thiourea or from otherwise lethal levels of gamma irradiation (Fox et al. 1983). Recent studies have shown protection against hyperoxia by superoxide dismutase or catalase stabilized by encapsulation in liposomes (Turrens et al. 1984).

Production of oxygen radicals in rat lungs during hyperoxia has been directly demonstrated by Freeman and Crapo (1981), and in vitro studies by Martin (1984) support this observation. Numerous studies, among them Crapo et al. (1978) and Mustafa and Tierney (1978), have reported increases in the activity of free radical–scavenging enzymes in lungs of animals surviving insult with O_3, NO_2, or other toxicants, thereby indirectly supporting this hypothesis. In vitro experiments also support the concept that O_3 can directly injure cells by a mechanism involving free radicals (see, for example, Morgan and Wenzel 1985).

This is a field of intense investigation at present, with new observations reported in practically every issue of the relevant journals. The linkages by which early damage are connected to "late" pulmonary fibrosis have not been defined and remain an area for future studies.

■ **Recommendation 4.** Animal studies should be done to determine the role of free radicals and active oxygen species in lung injury.

The Middle Phase

The middle phase, or the events that occur between injury to cells and deposition of excess collagen, is the component of the lung's response to injury that we understand the least. The recruitment of alveolar macrophages to sites of injury and the subsequent proliferation of fibroblasts are key cellular events in the pathogenesis of pulmonary fibrosis. In addition, there are published studies (reviewed by Haschek and Witschi 1979) suggesting an important role for the alveolar type II cells in the determination of the balance between cell repair and fibrogenesis in the damaged lung.

Epithelial Cells. Alveolar type II cell proliferation may represent a critical period in terms of repair (Haschek and Witschi 1979). Disruption of the normal sequence of events during this period may favor the development of pulmonary fibrosis. For example, systemic administration of butylated hydroxytoluene to mice results in widespread necrosis of type I cells within 24 hr. On days 2 and 3, most dividing cells are type II pneumonocytes. If mice are exposed to high (70 percent) concentrations of oxygen (O_2) during this period, type II cells are severely affected, whereas dividing interstitial cells are not. Inhibition of the epithelial cell proliferation allows the interstitial cells to proliferate relatively unchecked, resulting in increased collagen production and pulmonary fibrosis. Animals exposed to x rays instead of O_2 showed similar results (Witschi et al. 1980).

These experiments also showed that the timing of the second insult was critical. If it did not coincide with the period of reepi-thelialization, then the enhanced fibrosis was not observed. Other investigators have found a similar enhancement of fibrosis in rats exposed to bleomycin followed by high levels of O_2 (Rinaldo et al. 1982; Tryka et al. 1982). Interestingly, paraquat, which by itself can induce irreversible fibrosis, has been observed to destroy type I as well as type II cells (Vejeyaratnam and Corrin 1971; Smith and Heath 1973; Skill-rud and Martin 1984). These data suggest that epithelial cell control of fibroblast proliferation may be an important early mechanism in pulmonary fibrosis of various etiologies (Bowden 1984).

■ **Recommendation 5.** Studies of the linkages between epithelial cell damage and repair and changes in populations of pulmonary macrophages and interstitial cells should be undertaken in whole animals.

Pulmonary Alveolar Macrophages. Many investigators have allocated a central role to the pulmonary alveolar macrophage in the initiation of fibrosis. The role of the macrophage in inflammatory and immunologic processes is enormously complex and the topic is reviewed periodically (Hocking and Golde 1979; Nathan et al. 1980).

A related topic is the effects of fibrogenic agents on pulmonary alveolar macrophages in terms of their recruitment and functional response. The time course of macrophage accumulation in relation to a fibrogenic stimulus appears to vary with the insult and, to some extent, with the experimental design. In many models of fibrosis, macrophages begin accumulating extremely rapidly after exposure to the fibrogenic agent.

Brody and coworkers (1985) hypothesized that complement may play a key role in particle-induced macrophage migration. They found that rats depleted of complement by cobra venom and mice genetically deficient in complement had far less macrophage accumulation following asbestos exposure. In addition, the molecular weight of the chemotactic factor from lavage fluid was consistent with its being C_{5a}. These researchers did not speculate on the source of the complement component, ex-cept to point out that edema fluid (serum transudate) is an important potential source of lower-molecular-weight components of complement in the damaged lung.

Other investigators have shown that macrophages themselves may participate in macrophage recruitment. Dauber and Daniele (1980) found that guinea pig lung macrophages secreted a chemoattractant for macrophages as well as for neutrophils and lymphocytes. Kagan and coworkers (1983b) found that alveolar macrophages from rats exposed to asbestos appear to secrete a chemoattractant for macrophages. That study did not, however, examine the effects of this chemoattractant on other effector cells. Partial characterization suggested it to be a protein and heterogeneous with regard to molecular weight. Circulating immune complexes, whose etiology is unknown, are also believed to stimulate macrophage accumulation (Hunninghake et al. 1981).

The direct effects of fibrogenic agents on macrophages also seem to vary considerably, depending on the specific agent and the experimental design. The role of the macrophage in fibrogenesis has often been envisioned as passive; that is, the macrophage may release various soluble mediators that play a role in fibrogenesis. More recent studies suggest that macrophages may respond to a fibrogenic agent in various ways.

Silica-exposed rat lungs showed numerous silica-containing macrophages in lavage fluid as well as in situ immediately after exposure, with high percentages of silica-containing macrophages remaining six weeks later (Brody et al. 1982; Warheit et al. 1984). Ultrastructural examination revealed that the silica-containing macrophages appeared normal, and functional studies (measurement of oxygen consumption and phagocytosis) of lavaged macrophages revealed no abnormalities.

A decrease in phagocytic capacity is not necessarily synonymous with functional impairment. For example, Tryka and coworkers (1984a) found that alveolar macrophages lavaged from lungs of hamsters exposed to bleomycin and O_2 at time points ranging up to 120 days increased in number but had a decreased amount of cell surface antigen, indicating that they were relatively immature. Those researchers attribute the markedly decreased phagocytic capability at least partly to the decrease in surface antigen. Other data from that study suggest that fibrosis may be associated with increased macrophage turnover.

Once macrophages have been recruited, they are capable of releasing many mediators that modulate the behavior of other effector cells in the lung. Kazmierowski et al. (1977) observed that macrophages obtained from normal primate lungs secreted at least two chemotactic factors. One factor had properties consistent with its identity as the complement component C_{5a}, and was chemotactic for neutrophils, mononuclear cells, and eosinophils. The other factor had an apparent molecular weight of about 5000, did not appear to be a complement component, and was a specific chemoattractant for neutrophils.

Similarly, Merrill and coworkers (1980) found that human alveolar macrophages secreted two chemoattractants for neutrophils, and Dauber and Daniele (1980) found that macrophages from guinea pig lung secreted chemoattractants for macrophages, neutrophils, and lymphocytes. A phagocytic stimulus increased the release of the neutrophil chemoattractant.

■ **Recommendation 6.** Basic research in cell culture systems should be performed to examine the biochemical basis of cell–cell communication and the molecular nature of various mediators, released by leukocytes and macrophages from damaged lungs, that enhance or prolong the cellular inflammatory response.

Macrophages may also modulate effector cells in another way. Stimulated macrophages are capable of releasing arachidonic acid from cell phospholipid pools. Further metabolism through the cyclooxygenase pathway results in the production of the prostaglandins PGE_2 and PGI_2, among other products. In vitro data show that PGE_2 and PGI_2 suppress a variety of neutrophil, macrophage, and lymphocyte functional responses. They may also partic-

ipate in the induction of suppressor T cells. In marked contrast, metabolic products of arachidonic acid resulting from the lipooxygenase pathway appear to enhance the inflammatory response (Hunninghake et al. 1984).

■ **Recommendation 7.** Mediators possibly released by damaged lung epithelial cells or derived from damaged matrix components, which maintain and amplify lung injury after acute cellular or organ damage, should be characterized.

Macrophage secretion of effector cell chemoattractants appears to be increased in fibrosis. Schoenberger and coworkers (1982) observed that asbestos stimulates alveolar macrophages to increase neutrophil migration to the lung. Wesselius and coworkers (1984) found that macrophages lavaged from exposed rat lungs between 5 and 20 days after bleomycin instillation secreted increased amounts of neutrophil chemoattractant as compared with macrophages from controls. Since bleomycin did not directly stimulate macrophages, the mechanism for the increased macrophage secretion is unclear. The authors postulate that the stimulus may result from cell injury caused by bleomycin. In addition, data from a study of macrophage-derived chemoattractants from hamsters instilled with bleomycin suggest that macrophages may be regulating the sequence of effector cell migration following injury (Kaelin et al. 1983).

The interaction between macrophages and fibroblasts is particularly complex since macrophages appear to have the capacity to stimulate fibroblasts as well as to suppress them. Stimulation of fibroblast proliferation alone is also complex. Stiles and coworkers (1979) proposed dividing growth factors into competence factors (which provide a signal early in the G_1 phase of the cell cycle) and progression factors (which provide a signal later in G_1), stimulating the cell to replicate.

Bitterman and coworkers (1982) showed that alveolar macrophages secrete a growth factor (alveolar macrophage–derived growth factor, mol. wt. = 18,000) for fibroblasts that is distinct from other described growth factors. The alveolar macrophage–derived growth factor is believed to function as a progression factor and to stimulate fibroblasts to produce their own progression factor.

Macrophages secrete fibronectin, a large glycoprotein known to mediate cell/matrix interactions through a variety of functions including its chemotactic properties (Hunninghake et al. 1984). Fibronectin is believed to act as a competence factor for fibroblast growth (Bitterman et al. 1983). In addition, fibronectin is chemotactic for fibroblasts. Indeed, macrophage-derived fibronectin is 1,000-fold more potent as a chemoattractant than is plasma fibronectin (Rennard et al. 1982).

Macrophages also appear to be capable of suppressing fibroblast growth. Elias and coworkers (1985) found that supernatants from normal human alveolar macrophages inhibit growth of log-phase fibroblasts. Their study showed that the inhibitory capacity of the supernatant was directly related to its capacity to stimulate fibroblast prostaglandin production. They separated macrophage subpopulations by density gradients and found that the factor(s) appeared to be preferentially elaborated by smaller and denser macrophages. In some respects this factor resembles those described by Korn et al. (1980) and Clark et al. (1982).

Investigation of the effects of specific fibrogenic agents on macrophage regulation of fibroblast growth illustrates the complexities involved. For example, Lugano and coworkers (1984) found that 2 and 14 days after silica exposure, lavaged macrophages depressed fibroblast proliferation, whereas at 42 days macrophages stimulated fibroblast proliferation. Clark and others (1982) discovered that macrophages from hamsters instilled with bleomycin had a greater suppressive effect on fibroblast proliferation and collagen synthesis compared with control macrophages. They found that suppression was associated with increased PGE_2 and intracellular cAMP, and that fibroblast suppressive activity decreased in the first days after bleomycin instillation, and then increased after 8 days. This suppressive activity may

be a mechanism for modulating fibroblast proliferation and fibrosis following fibrogenic exposure.

Recently published data of Schmidt et al. (1984) suggest that interleukin-1 may play a role in fibroblast proliferation. Although other researchers have not found that interleukin-1 has fibroblast-stimulating properties, Schmidt and coworkers point out differences in experimental design that might account for this discrepancy. Peripheral blood monocytes were used in that particular study, and it should be noted that alveolar macrophages are also capable of secreting interleukin-1. No detailed studies have been reported with macrophages or effector cells from animals exposed to O_3. Such work should be done to test whether (and which of) these pathways might be operative in the lungs of animals exposed to O_3.

Broncho-alveolar lavage was used to obtain macrophages in many of the studies discussed above. Given the heterogeneity of macrophages, it is unclear if the populations present in lavage fluid accurately reflect the populations actually present in lung tissue, particularly in disease states. The problem is compounded by the fact that there is no consensus as to the appropriate functional and/or structural definitions of such macrophage subpopulations. Brain and coworkers (1977), Mason (1977), and Lum and coworkers (1983) have reviewed some of the potential problems in studying lavaged macrophages.

In detailed morphometric analyses of centriacinar macrophages in situ and pulmonary macrophages lavaged from control and O_3-exposed rats, Lum and coworkers (1983) observed significant differences in most parameters studied between the lavaged and in situ macrophages in the control as well as the O_3-exposed rats. These data suggest that, at least in this model of fibrosis, lavaged macrophages are not representative of macrophages present at the sites of greatest lung damage. Alternatively, it is questionable whether interstitial macrophages, and those resident in airways and therefore accessible by lavage, are a homogeneous population in equilibrium or are somehow "different." Clearly, similar problems may exist in studying any of the effector cells obtained by lavage.

Fibroblasts. Presumably the cells responsible for synthesizing the "abnormal" collagen of pulmonary fibrosis—whether it be abnormal in amount, location, or type—are the fibroblasts. Although they often are perceived as passive target cells for the effector cells and their mediators, fibroblasts may on occasion play a more active role in directing the course of fibrosis.

In some cases fibroblasts may directly interact with the fibrogenic agent. Several workers have examined the effects of bleomycin on fibroblasts. Sterling and co-workers (1982) found that collagen synthesis increased in human fetal lung fibroblasts exposed to bleomycin for 48 hr, but degradation also increased. They also found that polysomes from bleomycin-treated fibroblasts synthesized twice as much collagen as control polysomes, but noncollagen protein synthesis was not affected. Similar results have been reported by Clark et al. (1980) and Phan et al. (1985).

In an examination of the effects of bleomycin, hyperoxia, and the presence of lung macrophages on collagen synthesis by human WI-38 fibroblasts, Robin and Juhos (1983) found that bleomycin alone directly stimulated collagen synthesis, as measured by hydroxyproline in the culture dishes. The addition of hyperoxia and/or the presence of lung macrophages did not further increase collagen synthesis, and hyperoxia alone significantly decreased collagen synthesis. However, hyperoxia in the presence of lung macrophages increased collagen synthesis about the same extent as did bleomycin.

There is also evidence that bleomycin affects fibroblast proliferation. In a study of the effects of in vitro and in vivo exposure to bleomycin on growth characteristics of fibroblasts, Absher and coworkers (1984) observed that both types of bleomycin exposure appeared to decrease growth of fibroblasts in comparison with controls. In a similar system of in vitro exposure, Phan and coworkers (1985) found that bleomycin exposure had no effect on growth.

However, since that study examined isolated fibroblasts 14 days after bleomycin instillation, its authors suggest that the difference in timing may account for the apparently discrepant findings. That is, during the first week after instillation, bleomycin toxicity may impair fibroblast growth, whereas during the second week, recovery may be occurring. Furthermore, this recovery may involve recruitment or selection of a population of fibroblasts with different growth characteristics. Whether these provocative findings in the bleomycin system accurately model events in lungs of animals exposed to O_3 remains to be tested.

Fibroblasts are believed to play a role in epithelial cell growth and function. Fibroblast pneumonocyte factor has been isolated from glucocorticoid-treated fibroblasts and has been shown to stimulate disaturated phosphatidylcholine synthesis in whole lung in vitro (Smith 1979) and in isolated type II pneumonocytes (Smith 1978). A small somatomedin-like growth factor specific for pneumonocytes is believed to be secreted by fibroblasts after pneumonectomy (Smith et al. 1980). Pulmonary fibroblasts exposed to hyperoxia in vitro secrete an epithelial cell growth factor as well as a lipid-synthesis-stimulating factor (Tanswell 1983). These particular factors appear to differ from any of the others previously described.

Fibroblasts may also affect effector cells. Although their predominant function is apparently production of collagen, other matrix components, and mucopolysaccharides, they also secrete biologically active products such as C_{1q} and interferon (Al-Adnani and McGee 1976). Cultured fibroblasts produce a factor chemotactic for monocytes as well as for neutrophils (Sobel and Gallin 1979). In addition, fibroblast culture fluid is capable of generating chemotactic activity from human serum, probably by cleaving C_{5a} from C_5. Fibroblasts are capable of producing macrophage migration inhibition factor (Tubergen et al. 1972). Whether such "fibrokines" play a role in pulmonary fibrosis has yet to be demonstrated.

Collagen itself is another fibroblast product with chemotactic properties. Type I collagen and its isolated chains are chemotactic for monocytes but not neutrophils (Stecher 1975; Postlethwaite and Kang 1976). In contrast, rat collagen is chemotactic for rat neutrophils in vivo (Chang and Houck 1970).

■ **Recommendation 8.** Examination of factors released by inflammatory cells that modulate collagen synthesis or fibroblast proliferation, especially in response to O_3, might help to define the mechanism(s) underlying the transition from the damaged, inflamed lung to the fibrotic lung.

Lung Fibrosis

Lung fibrosis, as defined clinically, refers to interstitial fibrosis as is seen in the later stages of idiopathic pulmonary fibrosis (also called cryptogenic fibrosing alveolitis in the literature of the United Kingdom). In this disease the hallmark of pulmonary fibrosis as seen by the pathologist is increased focal staining of collagen fibers in the alveolar interstitium. Despite earlier misconceptions, it seems clear that fibrotic lungs from humans with either acute or chronic pulmonary fibrosis contain increased amounts of collagen as evaluated biochemically, in agreement with the histological findings.

In the normal lung, interstitial collagen fibers are thought to provide structural matrix or scaffolding upon which the lung cells are assembled. These fibers are also thought to be responsible for the limits to which the alveoli can be stretched during inhalation or to which they can relax during expiration. The deposition of additional collagen in the fibrotic lung is presumably responsible for the increased stiffness of these lungs, whereby the volume to which they can expand at a given distending pressure is decreased as compared with normal values.

Unfortunately, pure interstitial fibrosis does not generally occur in lungs damaged by toxicants. In many ways such toxicants cause disease that more closely resembles adult or infant respiratory distress syndrome than chronic fibrosis. Excess lung collagen is usually observed not only in the alveolar interstitium, but also throughout the centriacinar region, including the alveolar ducts and respiratory bronchioles. The relationship between increased collagen deposition around small airways and lung mechanics is not understood, either theoretically or empirically.

■ **Recommendation 9.** The reversibility of excess collagen deposition in the fibrotic lung, as reflected by increased hydroxyproline content of the lung, should be determined.

There are at least 13 genetically distinct collagen types known to occur in all mammals, of which at least seven have been found in normal lungs or synthesized by isolated lung cells. Two types predominate in the lung, representing about 90 percent or more of the total lung collagen. Type I and III collagens are the major interstitial components and are found in the normal lungs of all mammals in an approximate ratio of 2:1.

This ratio is altered in fibrotic lungs. It is not known whether shifts in collagen types, as compared with absolute increases in collagen content, account for the increased stiffness of fibrotic lungs. Type III collagen is much more compliant than is type I; thus, an increasing proportion of type I relative to type III collagen might result in a stiffer lung as is observed in pulmonary fibrosis. Changes in collagen cross-linking in fibrotic lungs may also contribute to the increased stiffness observed. In addition, because type I collagen is the material that stains histologically as "collagen," whereas type III collagen does not, an increased proportion of type I relative to type III collagen would be appreciated histologically as an "increased amount of lung collagen." Therefore, it is unclear whether the observed increase in stainable collagen is due solely to the increase in collagen content of the lungs observed biochemically, or whether altered collagen types or cross-linking might also contribute to the histological changes seen.

Some types of pulmonary fibrosis, including that induced by O_3, involve abnormalities in the type of collagen being made.

Although the elaboration of recruitment and proliferation factors by effector cells in the damaged lung might account for the accumulation of fibroblasts and the increased deposition of collagen in the fibrotic lung, they do not in themselves account for some of the qualitative abnormalities found in the collagen of fibrotic lungs. For example, there is an increase in type I collagen relative to type III collagen in idiopathic pulmonary fibrosis (Seyer et al. 1976). Similar shifts have been demonstrated in lungs of adults and infants dying of acute respiratory distress syndrome (Last et al. 1983b; Shoemaker et al. 1984). In the acute diseases, pulmonary fibrosis develops very rapidly, within weeks, as is the case with animals after short-term exposure to high levels of O_3.

Increased collagen type I:type III ratios have also been observed in newly synthesized collagen in several animal models of acute pulmonary fibrosis (Reiser and Last 1981; Haschek et al. 1982), including the lungs of rats exposed to high concentrations of O_3. A similar shift in collagen type ratios was also observed in experimental fibrosis induced by several other agents, including paraquat and bleomycin (Reiser and Last 1981), and butylated hydroxytoluene with and without supplemental O_2 (Haschek et al. 1982). The latter observations were made in mouse lung using in vivo assays. Three weeks after a fibrogenic insult with bleomycin, the shift in collagen type ratios could be detected in total lung collagen (Reiser and Last 1983).

A shift in collagen type I:type III ratios has also been demonstrated in several in vitro systems. Clark and coworkers (1980) found that fibroblasts exposed to bleomycin in vitro not only had increased collagen synthesis rates, but also synthesized more type I collagen when compared with controls. Similarly, Phan and coworkers (1985) observed that collagen synthesized by fibroblasts exposed either in vivo or in vitro to bleomycin had an increased ratio of type I to type III collagen.

Although the mechanism for this shift in collagen types is unknown, there are many possible explanations. Clones of fibroblasts responsive to recruitment/proliferation factors may preferentially synthesize type I collagen as compared with the action of fibroblasts normally present. In fact, Kelley and coworkers (1981) found that exposure to lung macrophages caused cultured fibroblasts to alter the ratios of type I and III collagens being synthesized. Those researchers suggest that macrophages may exert an influence on collagen type ratios by selectively stimulating a subpopulation of fibroblasts with a predetermined collagen phenotype.

Alterations in the extracellular matrix, resulting from inflammatory mediators secreted by various effector cells, might also cause the fibroblasts to switch the collagen phenotype being synthesized. There is ample evidence from in vitro experiments that alteration of culture conditions can alter collagen phenotypes (Daniel 1976; Deshmukh and Kline 1976; Mayne et al. 1976; Deshmukh and Sawyer 1977; Hata and Peterkofsky 1977; Smith and Niles 1980). The possible role of alterations in composition of the basement membrane secondary to cellular damage or killing in these processes remains to be defined.

■ **Recommendation 10.** In addition to collagen, the content and structure of extracellular matrix components in the fibrotic lung should be examined.

Collagen associated with fibrosis may also be abnormal with respect to cross-linking. Alterations in cross-links in experimental silicosis (Last 1985; Reiser and Gerriets 1985) and in bleomycin-induced fibrosis (Reiser et al. 1986) have recently been described. Recent data (Reiser et al. 1987) suggest that similar changes in cross-linking are detectable in lungs of monkeys exposed intermittently to 0.5 ppm O_3, 8 hr/day for 1 year. As in the case of alterations in collagen type ratios, it is unclear if the mechanisms can be ascribed to changes in the clones of fibroblasts actively synthesizing collagen or to changes in the milieu that secondarily affect the nature of the collagen being made by a given population of lung fibroblasts.

Lung elastin degradation is thought to be an important component of the pathophysiology of emphysema, although there is no known relationship between lung elastin metabolism and pulmonary fibrosis. Assays analogous to the quantification of lung collagen synthesis rate can be used to measure elastin synthesis in lungs of rats or other laboratory animals exposed to O_3 (Dubick et al. 1981).

Since changes in lung compliance (elasticity) have been reported in rats (Bartlett et al. 1974) and rabbits (Frank et al. 1979) exposed to O_3, such short-term techniques for evaluating putative changes in lung elastin content or cross-linking are of obvious interest. Lung mince techniques for studying rates of elastin biosynthesis, utilizing antibodies to soluble precursors of elastin, are currently under development. Such techniques might allow investigators to look at lung elastin synthesis independently of cross-linking reactions and therefore in a much shorter time frame with less complex experiments.

Models of Exposure

Experimental Approaches

Choice of Animal Models. Animal toxicology studies must remain the major source of information, on a phenomenological as well as a mechanistic level, on whether exposure to ambient concentrations of O_3 causes pulmonary fibrosis. In this section, the techniques used for assessing the fibrogenic potential of O_3 are addressed.

The choice of animal model is complicated by the wide range in species response to O_3. For example, with common laboratory animals, LD_{50} doses for 1-hr exposures range from 21 ppm in mice and rats to 52 ppm in guinea pigs. Birds show greater resistance to the edemagenic effects of O_3 than mammals, consistent with their lack of alveoli and other important differences in lung structure; for example, turkeys survive comparable exposures to 417 ppm (Melton 1982).

Since animals are exposed by inhalation, choosing an animal with a respiratory system similar to the human one is particularly desirable. The respiratory system of monkeys most closely resembles that of humans. However, availability and cost of animals and the necessity for special facilities for housing monkeys and performing long-term exposures clearly limit the use of this model. Ethical considerations with regard to monkeys as experimental animals include the confinement of the primate in small exposure chambers for prolonged periods of time. Rats are widely used, although fundamental differences in respiratory anatomy (for example, lack of respiratory bronchioles) and function (rats are obligate nose breathers) can complicate extrapolation of effects to humans.

It is of great interest to know whether, and to what extent, the fibrogenic effects of O_3 observed in rats also occur in monkeys. In lung biopsy specimens obtained from monkeys exposed to O_3, Last and co-workers (1981) observed increased rates of collagen synthesis over lung biopsy specimens obtained from the same animals exposed only to air. This experimental design, in which each animal serves as its own control, is a practical way to perform such exposure experiments using small numbers of valuable animals without the necessity of sacrificing them.

In an earlier study, Last and others (1979) found that although some variation occurred in the response of individual animals to O_3 exposure, by the criterion of an "average response" calculated after one week of exposure to 1.2 ppm O_3, monkeys appeared to be quite a bit more sensitive to high levels of O_3 than did rats. This suggests that quantitative data from rodent exposures might underestimate actual risks of fibrogenesis to humans breathing O_3.

This finding may well reflect the important difference between concentration and dose, since monkeys and humans breathing by the oronasal route probably receive a greater dose of O_3 to the deep lung for a given concentration of O_3 than do nose-breathing rats. One may assume that the equivalence of quantifiable biological response may be equated with equivalence of dose in these experiments, suggesting a means of extrapolation for human risk assessment.

Finally, histological studies (Freeman et

al. 1974; Schwartz et al. 1976; Last et al. 1979; Castleman et al. 1980) have also demonstrated marked similarities (within the context of their anatomical differences) between monkeys and rats in terms of the cellular response of their lungs to O_3 exposure. In both species, increased collagen deposition has been observed in the terminal bronchiolar/alveolar duct region and in the alveolar interstitium, that is, in the centriacinar regions of the lung. Thus, it seems likely that a common pathogenetic mechanism underlies the response of these two different species to O_3 exposure.

Exposure. Generation of a gas available in high purity as a compressed "tank gas"— for example, sulfur dioxide (SO_2), O_2, or NO_2—is relatively straightforward, and metering and dilution produce appropriate concentrations for exposure. Exposures requiring generation of toxicant in situ may be more difficult. For example, O_3 is usually generated by passing an electrical discharge through pure oxygen. If air is used as a source of oxygen, then the resultant O_3 will be contaminated with oxides of nitrogen arising from oxidation of nitrogen in the air.

Monitoring and quantifying gaseous pollutants requires either expensive detectors needing frequent calibration (and usually a computer to process the tremendous amount of data generated) or very labor-intensive wet chemical analysis procedures after exhaust gases from the chambers are bubbled through traps. The current method of choice for analysis of O_3 is ultraviolet (UV) photometry. Earlier methods usually involved quantification of O_3 by trapping chamber exhaust gases in solutions of iodine/potassium iodide. Such values are about 20 percent higher than those observed by ultraviolet photometry.

Exposure chambers must allow for rapid attainment of desired concentrations of toxicants, maintenance of desired levels homogeneously throughout the chamber, adequate capacity for experimental animals, and minimal accumulation of undesired products of animal metabolism (usually heat and carbon dioxide). A major concern, especially with regard to exposure to acid aerosols (see below), has been the putative buildup of ammonia in chambers because of microbial action upon animal excreta. Thus, maximal loading factors and sanitation must also be considered in chamber usage. Finally, concern for the environment and for facility personnel safety suggests prudence in how chambers are exhausted.

Lung Morphology. Histological and morphological studies have been the predominant techniques used for assessing fibrogenicity of O_3. For example, exposure of experimental animals to high concentrations of O_3 (1 ppm and above) has been reported to produce pulmonary fibrosis as defined by morphological criteria (Stokinger et al. 1957). Others have suggested that exposure to O_3 at lower levels may cause pulmonary fibrosis in dogs and rats (Freeman et al. 1973, 1974; Last et al. 1979).

The terminal bronchiolar and proximal alveolar duct region is an area in which a fibroblastic response to O_3 would be expected since several studies have demonstrated O_3-induced epithelial cell injury and proliferation, as well as a moderate inflammatory response, in this location (Stephens et al. 1974; Schwartz et al. 1976). Mathematical modeling based on the known aqueous solubility of O_3 has also suggested that this area of the lung receives the maximum dose when O_3 is inhaled (Miller et al. 1978; see also Ultman, and Overton and Miller, this volume).

Analysis of Lung Lavage. It seems reasonable to assume that pulmonary edema and/or inflammation are obligatory precursors of fibrosis. Thus, markers of early events generally are chosen to reflect lung edema or cellular changes in the lung. The most popular of these types of assays have quantified various parameters in lung lavage from animals exposed to pneumotoxic substances. Generally, lungs of exposed and control animals are washed with multiple small volumes of isotonic saline. Many parameters may be evaluated from such lung washings. This technique has the further appeal of allowing direct comparisons with data accessible from normal human volunteers or from patients undergoing bronchopulmonary lavage for therapeutic purposes.

Differential cell counts after centrifugation of the lavage fluids and/or determination of various enzyme activities in the supernatant fluids after removal of cells have been the most widely used of these approaches. Henderson and coworkers (1978) advocated measurement of lactate dehydrogenase as a sensitive indicator of lung damage, and several groups have taken this approach with various toxicants, including silica (Moores et al. 1981; Sjostrand and Rylander 1984), dichloroethylene (Forkert et al. 1982), and a veritable potpourri of toxicants (Roth 1981; Beck et al. 1983). In later work, Henderson and coworkers (1979a,b), as well as others (Morgan et al. 1980; Kagan et al. 1983a; Sykes et al. 1983; Guth and Mavis 1986), advocated measurement of cytological and enzymatic profiles rather than of lactate dehydrogenase alone.

Current emphasis seems to be on measurement of polymorphonuclear leukocytes, macrophages, and monocytes (and their phagocytotic capabilities) in the cellular fraction, and of lactate dehydrogenase (and its substituent isoenzymes), N-acetylglucosaminidase, acid or alkaline phosphatase, other lysosomal hydrolases, lavagable total protein and/or albumin, and sialic acid. Although such measurements have often been the basis of mechanistic interpretations, we really do not have a rigorous theoretical understanding of any of these parameters.

Analysis of Lung Collagen. Biochemical approaches for assessing O_3 fibrogenicity have also been used. Last and coworkers (1979) measured collagen synthesis rates in lungs from rats subjected to short-term O_3 exposure at levels ranging from 0.4 to 1.6 ppm (UV photometric analysis). Collagen synthesis rates were significantly elevated at all levels of O_3 exposure and corresponded with increases in histological lesions in alveolar ducts. These results are consistent with the increased lung collagen content (see below) observed by biochemical analysis in rats exposed to 0.4 ppm O_3

(UV photometric analysis) for up to 180 days (Last and Greenberg 1980).

Fibrotic collagen may be distinguished from normal lung collagen by quantitative analysis of certain characteristic cross-links (Reiser and Last 1987). Such new methods may allow us to distinguish collagen synthesis associated with normal lung repair from abnormal deposition of collagen associated with fibrosis, and to distinguish between reversible and irreversible events in lung collagen metabolism. These methods may also be made very sensitive; they remain to be validated for this purpose. This is an area of future research that deserves a high priority.

■ **Recommendation 11.** The relationship between the presence of specific abnormal collagen cross-links in the lung and the reversibility of fibrosis should be examined.

Chronic Exposure

In rats exposed to O_3 for 180 days (0.4 ppm O_3 delivered for 23.5 hr/day), Last and Greenberg (1980) observed increased amounts of lung protein and of lung collagen (hydroxyproline) throughout the study. Collagen synthesis rates measured in lung minces were also elevated, and the increased lung hydroxyproline content at 180 days persisted 2 months postexposure breathing filtered air. The observed biochemical changes were consistent with concurrent morphological observations of the occurrence of mild pulmonary fibrosis.

A study of long-term effects of exposure to lower concentrations of O_3 in juvenile cynomolgus monkeys showed that lung collagen content increased significantly after exposure to 0.64 ppm O_3 for 8 hr/day for 1 year (Last et al. 1984b). These findings could be correlated with an increase in the length of their respiratory bronchioles, as evaluated morphometrically (Tyler et al. 1985).

These long-term experiments suggest an additional mechanism that may be operative in O_3 exposure. Exposure to O_3 may accelerate normal aging or growth-related processes (Last and Greenberg 1980). Exposed and control rats reached the same endpoint collagen content after six months, with the exposed animals reaching this value more rapidly (at three months). It seems likely that the changes in hydroxyproline content of the lungs from exposed rats observed over the first three months of the study were related to excessive deposition of collagen (fibrosis) in their lungs (as observed morphologically), but a role for accelerated aging (or growth) of the lungs in O_3-exposed rats cannot be ruled out as an alternative explanation of these observations. One disadvantage of rats as an experimental animal for chronic studies is that rats continue to grow throughout their entire lifetime, and lung growth is not "complete" at maturity as in humans.

■ **Recommendation 12.** The response of the developing lung versus the mature lung in animals exposed to O_3 should be studied to ascertain whether the developing lung is more susceptible to damage, fibrosis, or change.

Progression of Lung Injury After Cessation of Exposure

What are the consequences of allowing animals to "recover" during a postexposure period? In another set of experiments, intermittent exposure to high levels of O_3 (0.64 or 0.96 ppm) for 8 hr/day elicited the same increases in lung collagen content of adult rats as did a continuous exposure protocol of 23.5 hr/day (Last et al. 1984b). This finding emphasizes the importance of not assuming that effects of O_3 exposure may be estimated by a simple concentration × time relationship. A six-week postexposure "recovery period" breathing filtered air exacerbated the increase in lung collagen content appreciated immediately after cessation of exposure. This result suggests that not only are these effects irreversible, at least in this time frame, but that they are also progressive.

The mechanism for exacerbating damage, or repairing such damage, to the lung by a postexposure period breathing filtered air is not obvious. Before any meaningful

hypothesis can be presented, further experiments examining cellular changes within the lung during such a postexposure period are necessary. Examination of the time course and cellular components of reepithelialization of the alveolar ducts and walls during the postexposure period would be especially important in this regard.

Haschek and Witschi (1979) stressed the importance of this component of lung repair in potentiation by O_2 of pulmonary fibrosis after lung damage with butylated hydroxytoluene. Further, to avoid misinterpretation of data because of altered growth rates in exposed and control animals, use of appropriate (pair-fed) controls in experiments with growing animals exposed to high levels of O_3 is obligatory. This specific area of research on postexposure effects of O_3 is an important one for further study.

■ **Recommendation 13.** Lung structure and biochemistry over long postexposure periods should be studied in detail.

Synergistic Interactions

The design of experiments to assess fibrogenicity of a given agent becomes much more complex when the possibility of synergistic (or antagonistic) actions between agents is considered. For example, Haschek and coworkers found that exposure of mice to 70 percent O_2 enhances pulmonary fibrosis previously induced by intraperitoneal injection of butylated hydroxytoluene (Haschek and Witschi 1979; Haschek et al. 1983). They concluded that the severe fibrosis seen after the combination of butylated hydroxytoluene and O_2 is the result of synergism, and proposed the following mechanism: when butylated hydroxytoluene causes lung damage, there is an initial phase of epithelial proliferation. If mice are exposed to 70 percent O_2 during this period, the epithelial type II cells are either inhibited from dividing or killed. Damage to the epithelial cells may then lead to uninhibited interstitial cell growth. This specific interaction between butylated hydroxytoluene and O_2 seems to occur only in mice.

A similar synergism apparently occurs with bleomycin and 70 percent O_2 in rats, hamsters, and other laboratory rodents (Tryka et al. 1984b) and is not specific to the mouse. Other such two-agent models also seem to work in multiple species (Haschek et al. 1983); such combinations include drugs and O_2, radiation and O_2, and cytotoxic agents and bleomycin.

In another study of synergism between particulates and gases, McJilton and Charles (1976) examined the effects on guinea pigs of exposure to sodium chloride aerosols and SO_2. When the mixture was at a high humidity, those researchers found decreases in airway flow resistance. Normally, SO_2 does not penetrate deep into the lung, and McJilton and Charles proposed that the highly soluble SO_2 dissolved in droplets and thus was able to "piggyback" into the lower respiratory tract. Ellison and Waller (1978) have reviewed this topic.

In a study of oxidant gases plus SO_2, Gardner and coworkers (1977) found that a protocol of O_3 and sulfuric acid (H_2SO_4) aerosol presented sequentially was more toxic than either agent alone, based on mortality rates of mice exposed to *Streptococcus pyogenes*. Juhos and coworkers (1978) found evidence for synergism between O_3 and H_2SO_4, based on very limited histologic evaluations of rat lungs. Last and Cross (1978), using several biochemical criteria and studying the effects upon tracheal mucous-producing cells, reported a synergism between O_3 and H_2SO_4 aerosols at relatively low concentrations of each. Hazuka and Bates (1975) observed synergism between O_3 and sulfate aerosols and suggested that it was responsible for decreased maximal flow rates observed during human exposure to O_3 and SO_2 or H_2SO_4 at near-ambient levels.

In a recent study of the effects of ammonium sulfate aerosols in combination with O_3 or NO_2 on collagen metabolism in rat lungs, Last and coworkers (1983a) found that ammonium sulfate aerosols alone had no effect on collagen synthesis rates; however, they significantly potentiated the effects of the oxidant gases. Guth and coworkers (1986) found that lavagable protein content increased significantly in lungs of rats continuously exposed for 1–2 days to 0.12 ppm O_3, and, more important, Warren and Last (1987) saw elevated collagen synthesis when rats were exposed to 0.12 ppm O_3 plus H_2SO_4 aerosol. The latter researchers obtained similar results with rats exposed to 0.2 ppm $O_3 \pm 40$ $\mu g/m^3$ of acid aerosol. We interpret these results as being indicative of a prefibrotic response of the lung to injury.

Furthermore, responses in this and other assays in rats exposed to 0.2 ppm O_3 for 8 hr/night and in rats exposed continuously have been similar. That is, the increase in lung collagen synthesis rate is not contingent upon continuous exposure of rats to O_3, but may occur under an intermittent exposure regimen that models human diurnal exposures as well. These results must be confirmed, especially in another species. Last and coworkers (1984a) suggested a hypothetical mechanism for this synergy involving increases in the stability of free radicals generated in situ. Since the mechanisms of injury elicited by individual pneumotoxins are so poorly understood, it is hardly surprising that mechanisms underlying synergistic interactions remain highly speculative.

■ **Recommendation 14.** A rational basis should be developed for prediction of synergistic or antagonistic interactions of pollutant mixtures by systematically examining binary and ternary combinations of the pollutants. The possibility that agents known to affect epithelial cell turnover might interact synergistically with O_3 deserves special attention.

Determination of Human Risk

What are the implications of these experiments for those charged with setting ambient air quality standards? Clearly, the traditional data base used for setting such standards is heavily skewed in the direction of detecting acute, short-term effects such as reflex bronchoconstriction in human subjects undergoing controlled experimental exposures. However, increased airway resistance may occur in response to release of stored histamine from airway mast cells in response to signals from irritant receptors (Dixon and Mountain 1965) and may have little or no long-term consequences to health. The inflammatory response of the lung to O_3 may also contribute to increased airway reactivity after exposure.

The potential adverse health effects of air pollution that constitute the major concern, however, are chronic effects from intermittent, long-term, low-level exposures: cancer, emphysema, pulmonary fibrosis, and chronic obstructive lung disease. Controlled human exposures are of absolutely no value for assessing these types of risks. To date, epidemiologic studies have also been of little value in assessing these risks (Committee on Medical and Biological Effects of Environmental Pollutants 1979), partly because of the overwhelming impacts of smoking and occupational exposures on the incidence of these diseases in our population, and the uncertainty about individual doses as compared with exposures of entire populations.

It is not practical to look at each pollutant and every possible combination of pollutants in long-term dose/response experiments that may require inhalation exposures for six months, or a year, or an entire lifetime (three years or more in rats or mice). Thus the need is for short-term assays that probe for potential structural changes in the lung, such as those for collagen and elastin metabolism described above. Such short-term assays may detect potential adverse health effects of air pollutants that cannot be ascertained by controlled human exposures or other currently used methodology for risk assessment. However, the regulatory climate until now has tended to ignore animal inhalation toxicology experiments in favor of data from controlled human exposures (Whittenberger 1985).

■ **Recommendation 15.** Greater use should be made of animal models of susceptible populations.

We also seem to be allowing ourselves to be lulled into a sense of security with the current ambient air quality standards, on the basis of the belief that we adapt to pollutants upon continued exposure. The concept of adaptation comes from the attenuation of reflex bronchoconstriction in controlled human exposures to, for example, O_3 upon continued exposure. There are no data suggesting that the lung can adapt to continued exposure to O_3 when the assay for effects is based upon structural changes in the lung rather than upon transient responses such as reflex bronchoconstriction or localized inflammation.

It is only by designing experiments that correlate data between short-term assays for structural change (to evaluate dose/response characteristics of the lung) with selected long-term chronic exposures that include detailed examination of lung structure that we will be able to evaluate properly the true risks of exposure to ambient air pollutants and their mixtures.

■ **Recommendation 16.** Long-term (lifetime) exposures of rodents should be performed under realistic protocols with sophisticated assays at termination. Such studies should also examine tumor incidence and, to be meaningful, would require very large groups of animals.

Identification of Susceptible Populations

The wording of the 1977 version of the Clean Air Act raises the question of potential susceptible populations in relation to any attempt to evaluate actual or hypothetical human risks of exposure to O_3. Are there identifiable susceptible populations from the standpoint of potential lung fibrosis? First, and most important, are very young children and infants, whose lungs are still developing and who would seem from animal experiments to be at risk of their lungs developing differently in atmospheres containing O_3. "Differently" in this context relates to the higher collagen content found in lungs of young rats than in lungs of matched, pair-fed controls (Last et al. 1984b) and the elongated alveolar ducts or respiratory bronchioles observed in rats or monkeys after exposure to O_3 (Tyler et al. 1985). It is not known if these observations constitute adverse health effects, hence, use of the word "differently."

Second, and obvious, is the population that breathes polluted outdoor air rather than conditioned indoor air by the nature of their professions, by choice, or by necessity during episodes of pollution, especially in conjunction with strenuous exercise (which increases the inhaled dose to the lungs—the prototypical "freeway jogger").

Since fibrosis is an insidious disease and the lung has a substantial reserve capacity to protect itself from the physiological consequences of fibrosis, a third population at risk includes people whose lungs are already severely compromised by chronic obstructive pulmonary disease, emphysema, congenital disease, or fibrosis.

A fourth group at risk would be those with diseases of the small airways, the focal site of O_3-induced fibrosis. The largest group with small airways pathology would be cigarette smokers. If the cellular mechanisms of lung inflammation and injury discussed earlier pertain in human populations, subjects with chronic lung inflammation (alveolitis) should be at increased risk. This group would include workers and others with allergic alveolitis and patients (such as those with idiopathic pulmonary fibrosis) who are subject to recurrent episodes of inflammatory infiltration of the distal lung with leukocytes and macrophages.

Summary

Inhalation of O_3 by experimental animals causes pulmonary edema after exposures of short duration and results in peribronchiolar and centriacinar pulmonary fibrosis after longer exposures. This discussion has stressed possible linkages between these outcomes, including the probable central role of the alveolar macrophage as an effector cell. The phenomenon and the underlying mechanisms of pathogenesis of pulmonary edema are probably well enough known for these observations to be extrapolated to assessment of human risks upon exposure to a given concentration of O_3, and sensitive clinical methods applicable to humans are available to compare directly with such estimates. However, the reversibility of such changes and whether increased fluid in the bronchial and alveolar airspaces is "damage" in the regulatory sense (that is, "an adverse health effect") are unknown.

At the other end of the process, much is being learned about the biochemistry of collagen in the fibrotic lung. Little or nothing is known of the cellular mechanisms responsible for excessive synthesis and deposition of fibrotic collagen, or whether such changes are reversible. As far as O_3 is concerned, nothing is known of the corresponding cellular and biochemical mechanisms in humans.

The events linking acute and chronic responses of the lung to O_3 are poorly understood and should be a primary focus of future research, in cell biology as well as in biochemistry. Most of the methodology necessary for such studies in animals is probably available.

Other topics of interest are progressive versus self-limited injury to the lung, reversibility of damage, and the potential for synergism between mixtures of air pollutants, especially O_3 and respirable aerosols or particulates. Recent studies in various animal models have suggested that certain combinations of agents, which alone cause limited or undetectable lung injury, can cause progressive pulmonary fibrosis.

These models have certain features in common with the continuing changes observed in lungs of rats and monkeys exposed to O_3 followed by periods where they breathe only filtered pure air. Rats exposed to O_3 in combination with various acidic aerosols, which by themselves apparently cause no lung damage at the concentrations tested, have suffered enhanced acute lung damage. The significance of these observations to potential chronic, irreversible changes in rat lungs, or to potential human health effects, remains to be determined.

Summary of Research Recommendations

Acute Phase of Injury

In the acute injury phase, the greatest need is for a better understanding of the pathophysiological significance of apparent alterations in broncho-alveolar epithelial permeability in animals and humans acutely exposed to O_3.

HIGH PRIORITY

Recommendation 2   To facilitate interpretation of human studies, sensitive lavage-based assays of protein content need to be correlated with diethylenetriaminepentaacetic (DTPA) acid aerosol transport assays in animals.

MEDIUM PRIORITY

Recommendation 1   Using experimental animals, anatomic regions in the lung associated with tracer transport from blood to lavage fluid, and vice versa, after O_3 exposure, should be localized and correlated with bulk movement of albumin into airspaces.

Recommendation 3   Human studies using DTPA aerosols or other tracers in conjunction with concentration/response assays for edema should be undertaken.

LOW PRIORITY

Recommendation 4   Animal studies should be done to determine the role of free radicals and active oxygen species in lung injury.

Middle Phase of Injury

The greatest needs for research in the middle phase of injury/ fibrosis are for a better understanding of the basis of cellular changes that perpetuate and amplify inflammation and cellular damage in lungs of animals exposed to O_3.

HIGH PRIORITY

Recommendation 5   Studies of the linkages between epithelial cell damage and repair and changes in populations of pulmonary macrophages and interstitial cells should be undertaken in whole animals.

Recommendation 6   Basic research in cell culture systems should be performed to examine the biochemical basis of cell-cell communication and the molecular nature of various mediators, released by leukocytes and macrophages from damaged lungs, that enhance or prolong the cellular inflammatory response.

MEDIUM PRIORITY

Recommendation 7   Mediators possibly released by damaged lung epithelial cells or derived from damaged matrix components (arachidonic acid metabolites, chemotactic factors, cytokines), which maintain and amplify lung injury after acute cellular or organ damage should be characterized.

Recommendation 8   Examination of factors released by inflammatory cells that modulate collagen synthesis or fibroblast proliferation, especially in response to O_3, might help to define the mechanism(s) underlying the transition from the damaged, inflamed lung to the fibrotic lung.

Late Phase of Injury

In the fibrotic phase, or late stages of injury, the greatest need is to understand whether the increased collagen synthesis observed in lungs of rats and monkeys acutely exposed to O_3 results in pulmonary fibrosis.

HIGH PRIORITY

Recommendation 9   The reversibility of excess collagen deposition in the fibrotic lung, as reflected by increased hydroxyproline content of the lung, should be determined.

Recommendation 11   The relationship between the presence of specific abnormal collagen cross-links in the lung and the reversibility of fibrosis should be examined.

Recommendation 12   The response of the developing lung versus the mature lung in animals exposed to O_3 should be studied to ascertain whether the developing lung is more susceptible to damage, fibrosis, or change.

MEDIUM PRIORITY

Recommendation 10   In addition to collagen, the content and structure of other extracellular matrix components in the fibrotic lung should be examined.

Other Considerations
HIGH PRIORITY

Recommendation 13   Detailed study of lung structure and biochemistry over long postexposure periods should be made to allow better understanding of progression of injury after exposure to O_3 is terminated.

Recommendation 14   A rational basis should be developed for prediction of synergistic or antagonistic interactions of pollutant mixtures by systematically examining binary and ternary combinations of the pollutants. The possibility that agents known to affect epithelial cell turnover might interact synergistically with O_3 deserves special attention.

Recommendation 15   Greater use should be made of animal models of susceptible populations.

Recommendation 16   Descriptive, subjective pathology is of little or no value. Long-term (lifetime) exposures of rodents under realistic protocols with sophisticated assays at termination are required, with morphometric and detailed biochemical examination of structural components at the least. Such studies should also examine tumor incidence and, to be meaningful, would require very large groups of animals.

References

Absher, M., Hildebran, J., Trombley, L., Woodcock-Mitchell, J., and Marsh, J. 1984. Characteristics of cultured lung fibroblasts from bleomycin-treated rats: comparisons with in vitro exposed normal fibroblasts, *Am. Rev. Respir. Dis.* 129:125–129.

Al-Adnani, M. S., and McGee, J. O'D. 1976. C1q production and secretion by fibroblasts, *Nature* 263:145–146.

Alpert, S. M., Schwartz, B. B., Lee, S. D., and Lewis, T. R. 1971. Alveolar protein accumulation. A sensitive indicator of low level oxidant toxicity, *Arch. Intern. Med.* 128:69.

Bartlett, J. D., Faulkner, C. S., and Cook, K. 1974. Effects of chronic ozone exposure on lung elasticity in young rats, *J. Appl. Physiol.* 37:92–96.

Beck, B. D., Gerson, B., Feldman, H. A., and Brian, J. D. 1983. Lactate dehydrogenase isoenzymes in hamster lung lavage fluid after lung injury, *Toxicol. Appl. Pharmacol.* 71:59–71.

Bitterman, P. B., Rennard, S. I., Hunninghake, G. W., and Crystal, R. G. 1982. Human alveolar macrophage growth factor for fibroblasts. Regulation and partial characterization, *J. Clin. Invest.* 70:806–822.

Bitterman, P. B., Rennard, S. I., Adelberg, S., and Crystal, R. G. 1983. Role of fibronectin as a growth factor for fibroblasts, *J. Cell Biol.* 97:1925–1932.

Bowden, D. H. 1984. Unraveling pulmonary fibrosis: the bleomycin model, *Lab. Invest.* 50(5):487–488.

Brain, J. D., Godleski, J. J., and Sorokin, S. P. 1977. Quantification, origin and fate of pulmonary macrophages, In: *Respiratory Defense Mechanisms, Part II* (J. D. Brain, D. F. Proctor, and L. M. Reid, eds.), pp. 849–892, Marcel Dekker, New York.

Brody, A. R., Roe, M. W., Evans, J. N., and Davis, G. S. 1982. Deposition and translocation of inhaled silica in rats. Quantification of particle distribution, macrophage participation, and function, *Lab. Invest.* 47(6):533–542.

Brody, A. R., Hill, L. A., and Warheit, D. B. 1985. Induction of early alveolar injury by inhaled asbestos and silica, *Fed. Proc.* 44:2596–2601.

Castleman, W. L., Dungworth, D. L., Schwartz, L. W., and Tyler, W. S. 1980. Acute respiratory bronchiolitis, *Am. J. Pathol.* 98:811–840.

Chang, C., and Houck, J. C. 1970. Demonstration of the chemotactic properties of collagen, *Proc. Soc. Exp. Biol. Med.* 134:22–26.

Clark, J. G., Starcher, B. C., and Uitto, J. 1980. Bleomycin-induced synthesis of type I procollagen by human lung and skin fibroblasts in culture, *Biochim. Biophys. Acta* 631:359–370.

Clark, J. G., Kostal, K. M., and Marino, B. A. 1982. Modulation of collagen production following bleomycin-induced pulmonary fibrosis in hamsters, *J. Biol. Chem.* 257(14):8098–8105.

Committee on Medical and Biologic Effects of Environmental Pollutants. 1979. *Ozone and Other Photochemical Oxidants*, pp. 416–438, National Academy of Sciences, Washington, D.C.

Crapo, J. D., Sjostrom, K., and Drew, R. T. 1978. Tolerance and cross-tolerance using NO_2 and O_2. I. Toxicology and biochemistry, *J. Appl. Physiol.* 44:364–369.

Cross, C. E., Parsons, G., Gorin, A. B., and Last, J. A. 1981. Pulmonary edema, In: *Mechanisms in Respiratory Toxicology* (H-P. Witschi and P. Nettesheim, eds.), p. 219, CRC Press, West Palm Beach, Fla.

Daniel, J. C. 1976. Changes in type of collagen synthesized by chick fibroblasts in vitro in the presence of 5-bromodeoxyuridine, *Cell. Differ.* 5:247–253.

Dauber, J. H., and Daniele, R. P. 1980. Secretion of chemotaxins by guinea pig lung macrophages. I. The spectrum of inflammatory cell responses, *Exp. Lung Res.* 1:23–32.

Deshmukh, K., and Kline, W. G. 1976. Characterization of collagen and its precursors synthesized by rabbit-articular-cartilage cells in various culture systems, *Eur. J. Biochem.* 69:117–123.

Deshmukh, K., and Sawyer, B. D. 1977. Synthesis of collagen by chondrocytes in suspension culture: modulation by calcium, 3':5'-cyclic AMP, and prostaglandins, *Proc. Natl. Acad. Sci. U.S.A.* 74(9):3864–3868.

Dixon, J. R., and Mountain, J. T. 1965. Role of histamine and related substances in development of tolerance to edemagenic gases, *Toxicol. Appl. Pharmacol.* 7:756–766.

Dubick, M. A., Rucker, R. B., Last, J. A., Lollini, L. O., and Cross, C. E. 1981. Elastin turnover in murine lung after repeated ozone exposure, *Toxicol. Appl. Pharmacol.* 58:203–210.

Elias, J. A., Rossman, M. D., Zurier, R. B., and Daniele, R. P. 1985. Human alveolar macrophage inhibition of lung fibroblast growth. A prostaglandin-dependent process, *Am. Rev. Respir. Dis.* 131:94–99.

Correspondence should be addressed to Jerold A. Last, Department of Internal Medicine and California Primate Research Center, University of California, Davis, CA 95616.

Ellison, J. M., and Waller, R. E. 1978. A review of sulphur oxides and particulate matter as air pollutants with particular references to effects on health in the United Kingdom, *Environ. Res.* 17:302–325.

Forkert, P. G., Custer, E. M., Alper, A. J., Ansari, G. A. S., and Reynolds, E. S. 1982. Lactate dehydrogenase activity in mouse lung following 1,1-dichloroethylene: index of airway injury, *Exp. Lung Res.* 4:67–77.

Fox, R. B., Harada, R. N., Tate, R. M., and Repine, J. E. 1983. Prevention of thiourea-induced pulmonary edema by hydroxyl radical scavengers, *J. Appl. Physiol.* 55:1456–1459.

Frank, R., Brain, J. D., Knudson, D. E., and Kronmal, R. A. 1979. Ozone exposure, adaptation, and changes in lung elasticity, *Environ. Res.* 19:449–459.

Freeman, B. A., and Crapo, J. D. 1981. Hyperoxia increases oxygen radical production in rat lungs and lung mitochondria, *J. Biol. Chem.* 256:10986–10992.

Freeman, G., Stephens, R. J., Coffin, D. L., and Stara, J. F. 1973. Changes in dogs' lungs after long-term exposure to ozone, *Arch. Environ. Health* 26:209–216.

Freeman, G., Juhos, L. T., Furiosi, N. J., Mussenden, R., Stephens, R. J., and Evans, M. J. 1974. Pathology of pulmonary disease from exposure to interdependent ambient gases (nitrogen dioxide and ozone), *Arch. Environ. Health* 29:203–210.

Fridovich, I. 1976. In: *Free Radicals in Biology* (W. A. Pryor, ed.), Vol. I, pp. 239–277, Academic Press, New York.

Gardner, D. E., Miller, J. F., Illing, J. W., and Kirtz, J. M. 1977. Increased infectivity with exposure to ozone and sulfuric acid, *Toxicol. Lett.* 1:59–74.

Gellert, A. R., Perry, D., Langford, J. A., Riches, P. G., and Rudd, R. M. 1985. Bronchoalveolar lavage fluid proteins and their relationship to pulmonary epithelial permeability, *Chest* 88(5):730–735.

Guth, D. J., and Mavis, R. D. 1986. Biochemical assessment of acute nitrogen dioxide toxicity in rat lung, *Toxicol. Appl. Pharmacol.* 84:304–314.

Guth, D. J., Warren, D. L., and Last, J. A. 1986. Comparative sensitivity of measurements of lung damage made by bronchoalveolar lavage after short-term exposure of rats to ozone, *Toxicology* 40:131–143.

Haschek, W. M., and Witschi, H. 1979. Pulmonary fibrosis: a possible mechanism, *Toxicol. Appl. Pharmacol.* 51:475–487.

Haschek, W. M., Klein-Szanto, A. J. P., Last, J. A., Reiser, K. M., and Witschi, H. P. 1982. Long-term morphologic and biochemical features of experimentally induced lung fibrosis in the mouse, *Lab. Invest.* 46:438–449.

Haschek, W. M., Reiser, K. M., Klein-Szanto, A. J. P., Kehrer, J. P., Smith, L. H., Last, J. A., and Witschi, H. P. 1983. Potentiation of butylated hydroxytoluene-induced acute lung damage by oxygen. Cell kinetics and collagen metabolism, *Am. Rev. Respir. Dis.* 127:28–34.

Hata, R-I., and Peterkofsky, B. 1977. Specific changes in the collagen phenotype of BALB 3T3 cells as a result of transformation by sarcoma viruses or a chemical carcinogen, *Proc. Natl. Acad. Sci. U.S.A.* 74(7):2933–2937.

Hazuka, M., and Bates, D. V. 1975. Combined effects of ozone and sulphur dioxide on human pulmonary function, *Nature* 257:50–51.

Henderson, R. F., Damon, E. G., and Henderson, T. R. 1978. Early damage indicators in the lung. I. Lactate dehydrogenase activity in the airways, *Toxicol. Appl. Pharmacol.* 44:291–297.

Henderson, R. F., Rebar, A. H., and Denicola, R. B. 1979a. Early damage indicators in the lung. IV. Biochemical and cytologic response of the lung to lavage with metal salts, *Toxicol. Appl. Pharmacol.* 51:129–135.

Henderson, R. F., Rebar, A. H., Pickrell, J. A., and Newton, J. F. 1979b. Early damage indicators in the lung. III. Biochemical and cytological response of the lung to inhaled metal salts, *Toxicol. Appl. Pharmacol.* 50:123–136.

Hocking, W. G., and Golde, D. W. 1979. The pulmonary-alveolar macrophage, *N. Engl. J. Med.* 301:580–587.

Hu, P. C., Miller, F. J., Daniels, M. J., Hatch, G. E., Graham, J. A., Gardner, D. E., and Selgrade, M. J. 1982. Protein accumulation in lung lavage fluid following ozone exposure, *Environ. Res.* 29:377–388.

Hunninghake, G. W., Gadek, J. E., and Lawley, T. J. 1981. Mechanisms of neutrophil accumulation in the lungs of patients with idiopathic pulmonary fibrosis, *J. Clin. Invest.* 68:259–269.

Hunninghake, G. W., Garrett, K. C., Richerson, H. B., Fantone, J. C., Ward, P. A., Rennard, S. I., Bitterman, P. B., and Crystal, R. G. 1984. Pathogenesis of the granulomatous lung diseases, *Am. Rev. Respir. Dis.* 130:476–496.

Juhos, L. T., Evans, M. J., Mussenden-Harvey, R., Furiosi, N. J., Lapple, C. E., and Freeman, G. 1978. Limited exposure of rats to H_2SO_4 with and without ozone, *J. Environ. Sci. Health* 13:33–47.

Kaelin, R. M., Center, D. M., Bernarod, J., Grant, J., and Snider, G. L. 1983. The role of macrophage-derived chemoattractant activities in the early inflammatory events of bleomycin-induced pulmonary injury, *Am. Rev. Respir. Dis.* 128:132–137.

Kagan, E., Oghiso, Y., and Hartmann, D. 1983a. The effects of chrysotile and crocidolite asbestos on the lower respiratory tract: analysis of bronchoalveolar lavage constituents, *Environ. Res.* 32:382–397.

Kagan, E., Oghiso, Y., and Hartmann, D. P. 1983b. Enhanced release of a chemoattractant for alveolar macrophages after asbestos inhalation, *Am. Rev. Respir. Dis.* 128:680–687.

Kazmierowski, J. A., Gallin, J. J., and Reynolds, H. Y. 1977. Mechanism for the inflammatory response to primate lungs. Demonstration and partial characterization of an alveolar macrophage-derived chemotactic factor with preferential activity for polymorphonuclear leukocytes, *J. Clin. Invest.* 59:273–281.

Kelley, J., Trombley, L., Kovacs, E. J., Davis, G. S., and Absher, M. 1981. Pulmonary macrophages alter the collagen phenotype of lung fibroblasts, *J. Cell Physiol.* 109:353–361.

Korn, J. H., Halushka, P. V., and LeRoy, E. J. 1980. Mononuclear cell modulation of connective tissue function. Suppression of fibroblast growth by endogenous prostaglandin production, *J. Clin. Invest.* 65:543–554.

Last, J. A. 1985. Changes in the collagen pathway in fibrosis, *Fundam. Appl. Toxicol.* 5:210–218.

Last, J. A., and Cross, C. E. 1978. A new model for health effects of air pollutants: evidence for synergistic effects of mixtures of ozone and sulfuric acid aerosols on rat lungs, *J. Lab. Clin. Med.* 91:328–339.

Last, J. A., and Greenberg, D. B. 1980. Ozone-induced alterations in collagen metabolism of rat lungs. II. Long-term exposures, *Toxicol. Appl. Pharmacol.* 55:108–114.

Last, J. A., Greenberg, D. B., and Castleman, W. L. 1979. Ozone-induced alterations in collagen metabolism of rat lungs, *Toxicol. Appl. Pharmacol.* 51: 247–258.

Last, J. A., Hesterberg, T. W., Reiser, K. M., Cross, C. E., Amis, T. C., Gunn, C., Steffey, E. P., Grandy, J., and Henrickson, R. 1981. Ozone-induced alterations in collagen metabolism of monkey lungs: use of biopsy-obtained lung tissue, *Toxicol. Appl. Pharmacol.* 60:579–585.

Last, J. A., Gerriets, J. E., and Hyde, D. M. 1983a. Synergistic effects on rat lungs of mixtures of oxidant air pollutants (ozone or nitrogen dioxide) and respirable aerosols, *Am. Rev. Respir. Dis.* 128:539–544.

Last, J. A., Siefkin, A., and Reiser, K. M. 1983b. Type I collagen content is increased in lungs of patients with adult respiratory distress syndrome, *Thorax* 38:364–368.

Last, J. A., Hyde, D. M., and Chang, D. P. Y. 1984a. A mechanism of synergistic lung damage by ozone and a respirable aerosol, *Exp. Lung Res.* 7:223–235.

Last, J. A., Reiser, K. M., Tyler, W. S., and Rucker, R. B. 1984b. Long-term consequences of exposure to ozone. I. Lung collagen content, *Toxicol. Appl. Pharmacol.* 72:111–118.

Lugano, E. M., Dauber, D. H., Elias, J. A., Bashey, R. I., Jimenez, S. A., and Daniele, R. P. 1984. The regulation of lung fibroblast proliferation by alveolar macrophages in experimental silicosis, *Am. Rev. Respir. Dis.* 129:767–771.

Lum, H., Tyler, W. S., Hyde, D. M., and Plopper, C. G. 1983. Morphometry of in situ and lavaged pulmonary alveolar macrophages from control and ozone-exposed rats, *Exp. Lung Res.* 5:61–75.

Martin, W. J. 1984. Neutrophils kill pulmonary endothelial cells by a hydrogen-peroxide-dependent pathway. An in vitro model of neutrophil-mediated lung injury, *Am. Rev. Respir. Dis.* 130:209–213.

Mason, R. J. 1977. Metabolism of alveolar macrophages, In: *Respiratory Defense Mechanisms, Part II* (J. D. Brain, D. F. Proctor, and L. M. Reid, eds.), pp. 893–926, Marcel Dekker, New York.

Mayne, R., Vail, M. S., Mayne, P. M., and Miller, E. J. 1976. Changes in type of collagen synthesized as clones of chick chondrocytes grow and eventu-

ally lose division capacity, *Proc. Natl. Acad. Sci. U.S.A.* 73(5):1674–1678.

McJilton, C. E., and Charles, R. J. 1976. Influence of relative humidity on functional effects of an inhaled sulphur dioxide aerosol mixture, *Am. Rev. Respir. Dis.* 113:163–169.

Melton, C. E. 1982. Effects of long-term exposure to low levels of ozone: a review, *Aviat. Space Environ. Med.* 53:105–111.

Merrill, W. M., Naegel, G. P., Matthay, R. A., and Reynolds, H. Y. 1980. Alveolar macrophage-derived chemotactic factor. Kinetics of in vitro production and partial characterization, *J. Clin. Invest.* 65:268–276.

Miller, F. J., Menzel, D. B., and Coffin, D. L. 1978. Similarity between man and laboratory animals in regional pulmonary deposition of ozone, *Environ. Res.* 17:84–101.

Minty, B. D., and Royston, D. 1985. Cigarette smoke induced changes in rat pulmonary clearance of ^{99m}TcDTPA. A comparison of particulate and gas phases, *Am. Rev. Respir. Dis.* 132:1170–1173.

Moores, S. R., Black, A., Evans, J. C., Evans, N. H., Holmes, A., and Morgan, A. 1981. The effect of quartz, administered by intratracheal instillation, on the rat lung. II. The short-term biochemical response, *Environ. Res.* 24:275–285.

Morgan, A., Moores, S. R., Holmes, A., Evans, J. C., Evans, N. H., and Black, A. 1980. The effect of quartz, administered by intratracheal instillation, on the rat lung. I. The cellular response, *Environ. Res.* 22:1–12.

Morgan, D. L., and Wenzel, D. G. 1985. Free radical species mediating the toxicity of ozone for cultured rat lung fibroblasts, *Toxicology* 36:243–257.

Mustafa, M. G., and Tierney, D. F. 1978. Biochemical and metabolic changes in the lung with oxygen, ozone, and nitrogen dioxide toxicity, *Am. Rev. Respir. Dis.* 118:1061–1090.

Nasr, A. N. M. 1967. Biochemical aspects of ozone intoxication: a review, *J. Occup. Med.* 9:589–596.

Nathan, C. F., Henry, W. M., and Cohn, Z. A. 1980. The macrophage as an effector cell, *N. Engl. J. Med.* 303:622–626.

Phan, S. H., Varani, J., and Smith, D. 1985. Rat lung fibroblast collagen metabolism in bleomycin-induced pulmonary fibrosis, *J. Clin. Invest.* 76:241–247.

Postlethwaite, A. E., and Kang, A. H. 1976. Collagen and collagen peptide-induced chemotaxis of human blood monocytes, *J. Exp. Med.* 143:1299–1307.

Reiser, K. M., and Gerriets, J. E. 1985. Experimental silicosis: mechanisms of acute and chronic lung changes, In: *Silica, Silicosis and Cancer: Controversy in Internal Medicine* (D. F. Goldsmith, D. M. Winn, and C. M. Shy, eds.), Praeger, Philadelphia.

Reiser, K. M., and Last, J. A. 1979. Silicosis and fibrogenesis: fact and artifact, *Toxicology* 13:51–72.

Reiser, K. M., and Last, J. A. 1981. Pulmonary fibrosis in experimental acute respiratory disease, *Am. Rev. Respir. Dis.* 123:58–63.

Reiser, K. M., and Last, J. A. 1983. Type V collagen: quantitation in normal lungs and in lungs of rats with bleomycin-induced pulmonary fibrosis, *J. Biol. Chem.* 258:269–275.

Reiser, K. M., and Last, J. A. 1987. A molecular marker for fibrotic collagen in lungs of infants with respiratory distress syndrome, *Biochem. Med. Metabol. Biol.* 37:16–21.

Reiser, K. M., Tryka, A. F., Lindenschmidt, R. C., Last, J. A., and Witschi, H. P. 1986. Changes in collagen crosslinking in bleomycin-induced pulmonary fibrosis, *J. Biochem. Toxicol.* 1:83–91.

Reiser, K. M., Tyler, W. S., Hennessey, S. M., Dominguez, J. J., and Last, J. A. 1987. Long-term consequences of exposure to ozone. II. Structural alterations in lung collagen of monkeys, *Toxicol. Appl. Pharmacol.* 89:314–322.

Rennard, S., Hunninghake, G., Davis, W., Mortiz, E., and Crystal, R. 1982. Macrophage fibronectin is 1000-fold more potent as a fibroblast chemoattractant than plasma fibronectin (abstract), *Clin. Res.* 30:356.

Rinaldo, J., Goldstein, R. H., and Snider, G. L. 1982. Modification of oxygen toxicity after lung injury by bleomycin in hamsters, *Am. Rev. Respir. Dis.* 126:1030–1033.

Robin, E. D., and Juhos, L. T. 1983. Some implications of augmented collagen levels with bleomycin exposure or hyperoxic exposure of lung fibroblasts: fibrosis as altered phenotypic expression, *Trans. Assoc. Am. Physicians* 96:412–416.

Roth, R. A. 1981. Effect of pneumotoxicants on lactate dehydrogenase activity in airways of rats, *Toxicol. Appl. Pharmacol.* 57:69–78.

Schmidt, J. A., Oliver, C. N., Lepe-Zuniga, J. L., Green, I., and Gery, I. 1984. Silica-stimulated monocytes release fibroblast proliferation factors identical to interleukin 1. A potential role for interleukin 1 in the pathogenesis of silicosis, *J. Clin. Invest.* 73:1462–1472.

Schoenberger, C. I., Hunninghake, G. W., Kawanami, O., Ferrans, V. J., and Crystal, R. G. 1982. Role of alveolar macrophages in asbestosis: modulation of neutrophil migration to the lung after acute asbestos exposure, *Thorax* 37:803–809.

Schwartz, L. W., Dungworth, D. L., Mustafa, M. G., Tarkington, B. K., and Tyler, W. S. 1976. Pulmonary responses of rats to ambient levels of ozone, *Lab. Invest.* 34:565–578.

Seyer, J. M., Hutcheson, E. T., and Kang, A. H. 1976. Collagen polymorphism in idiopathic chronic pulmonary fibrosis, *J. Clin. Invest.* 57:1498–1507.

Shoemaker, C. T., Reiser, K. M., Goetzman, B. W., and Last, J. A. 1984. Elevated ratios of type I/III collagen in the lungs of chronically ventilated neonates with respiratory distress, *Pediatr. Res.* 8: 1176–1180.

Sjostrand, M., and Rylander, R. 1984. Enzymes in lung lavage fluid after inhalation of silica dust, *Environ. Res.* 33:307–311.

Skillrud, D. M., and Martin, W. J., II. 1984. Paraquat-induced injury of type II alveolar cells. An in vitro model of oxidant injury, *Am. Rev. Respir. Dis.* 129:995–999.

Smith, B. D., and Niles, R. 1980. Characterization of collagen synthesized by normal and chemically

transformed rat liver epithelial cell lines, *Biochemistry* 19:1820–1825.

Smith, B. T. 1978. Fibroblast-pneumonocyte factor: intercellular mediator of glucocorticoid effect on fetal lung, In: *Intensive Care in the Newborn* (L. Stern, ed.), Vol. II, pp. 65–86, Masson, New York.

Smith, B. T. 1979. Lung maturation in the fetal rat: acceleration by injection of fibroblast-pneumonocyte factor, *Science* 204:1094–1095.

Smith, B. T., Galaugher, W., and Thurlbeck, W. M. 1980. Serum from pneumonectomized rabbits stimulates alveolar type II cell proliferation in vitro, *Am. Rev. Respir. Dis.* 121:701–707.

Smith, P., and Heath, D. 1973. The ultrastructure and time sequence of the early stages of paraquat lung in rats, *J. Pathol.* 114(4):177–184.

Sobel, J. D., and Gallin, J. I. 1979. Polymorphonuclear leukocyte and monocyte chemoattractants produced by human fibroblasts, *J. Clin. Invest.* 63:609–618.

Stecher, V. J. 1975. The chemotaxis of selected cell types to connective tissue degradation products, *Ann. N.Y. Acad. Sci.* 289:177–189.

Stephens, R. J., Sloan, M. F., Evans, M. J., and Freeman, G. 1974. Early responses of lung to low levels of ozone, *Am. J. Pathol.* 74:31–58.

Sterling, K. M., Jr., DiPetrillo, T. A., Kotch, J. P., and Cutroneo, K. R. 1982. Bleomycin-induced increase of collagen turnover in IMR-90 fibroblasts: an in vitro study of connective tissue restructuring during lung fibrosis, *Cancer Res.* 42:3502–3506.

Stiles, C. D., Capone, G. T., Sher, C. D., Antoniades, H. N., Van Wyk, J. J., and Pledger, W. J. 1979. Dual control of cell growth by somatomedins and platelet derived growth factor, *Proc. Natl. Acad. Sci. U.S.A.* 76:1279–1283.

Stokinger, H. E., Wagner, W. D., and Dobrogorski, O. J. 1957. Ozone toxicity studies. III. Chronic injury to lungs of animals following exposure at a low level, *Arch. Environ. Health* 16:514–522.

Sykes, S. E., Morgan, A., Moores, S. R., Holmes, A., and Davison, W. 1983. Dose-dependent effects in the subacute response of the rat lung to quartz. I. The cellular response and the activity of lactate dehydrogenase in the airways, *Exp. Lung Res.* 5:229–243.

Tanswell, A. K. 1983. Cellular interactions in pulmonary oxygen toxicity in vitro: I. Hyperoxic induction of fibroblast factors which alter growth and lipid metabolism of pulmonary epithelial cells, *Exp. Lung Res.* 5:23–36.

Tryka, A. F., Skornik, W. A., Godleski, J. J., and Brain, J. D. 1982. Potentiation of bleomycin-induced lung injury by exposure to 70% oxygen, *Am. Rev. Respir. Dis.* 126:1074–1079.

Tryka, A. F., Godleski, J. J., and Brain, J. D. 1984a. Alterations in alveolar macrophages in hamsters developing pulmonary fibrosis, *Exp. Lung Res.* 7:41–52.

Tryka, A. F., Godleski, J. J., and Brain, J. D. 1984b. Differences in effects of immediate and delayed hyperoxia exposure on bleomycin-induced pulmonary injury, *Cancer Chemother. Rep.* 68:759–764.

Tubergen, D. G., Feldman, J. D., Pollock, E. M., and

Lerner, R. A. 1972. Production of macrophage migration inhibition factor by continuous cell lines, *J. Exp. Med.* 135:255–266.

Turrens, J. F., Crapo, J. D., and Freeman, B. A. 1984. Protection against oxygen toxicity by intravenous injection of lysosome-entrapped catalase and superoxide dismutase, *J. Clin. Invest.* 73:87–95.

Tyler, W. S., Tyler, N. K., Barstow, T., Magliano, D., Hinds, D., and Young, M. 1985. Persistence of ozone lesions in monkeys with growing lungs during a 6 month postexposure period, *Am. Rev. Respir. Dis.* 131:A169.

Vejeyaratnam, G. S., and Corrin, B. 1971. Experimental paraquat poisoning: a histological and electron-optical study of the changes in the lung, *J. Pathol.* 103:123–129.

Waling, C. 1963. Chemistry of the organic peroxides, *Radiat. Res., Suppl.* 3:3–16.

Warheit, D. B., Chang, L. Y., Hill, L. H., Hook, G. E. R., Crapo, J. D., and Brody, A. R. 1984. Pulmonary macrophage accumulation and asbestos-induced lesions at sites of fiber deposition, *Am. Rev. Respir. Dis.* 129:301–310.

Warren, D. L., and Last, J. A. 1987. Synergistic interaction of ozone and respirable aerosols on rat lungs. III. Ozone and sulfuric acid, *Toxicol. Appl. Pharmacol.* 88:203–216.

Warren, D. L., Guth, D. J., and Last, J. A. 1986. Synergistic interaction of ozone and respirable aerosols on rat lungs. II. Synergy between ammonium sulfate aerosol and various concentrations of ozone, *Toxicol. Appl. Pharmacol.* 84:470–479.

Wesselius, L. J., Catanzaro, A., and Wasserman, S. I. 1984. Neutrophil chemotactic activity generation by alveolar macrophages after bleomycin injury, *Am. Rev. Respir. Dis.* 129:485–490.

Whittenberger, J. L. 1985. Report of the workshop on environmentally related nononcogenic lung disease, *Environ. Res.* 38:417–469.

Witschi, H. P., Haschek, W. M., Meyer, K. R., Ullrich, R. L., and Dalbey, W. E. 1980. A pathogenetic mechanism in lung fibrosis, *Chest* 78(2): 395–399.

Wolf, H. R. D., and Seeger, H. W. 1982. Experimental and clinical results in shock lung treatment with vitamin E, *Ann. N.Y. Acad. Sci.* 393:392–408.

Relation of Pulmonary Emphysema and Small Airways Disease to Vehicular Emissions

JOANNE L. WRIGHT
University of British Columbia

Air Pollution, the Automobile, and Public Health. © 1988 by the Health Effects
Institute. National Academy Press, Washington, D.C.

The ability of gasoline and diesel engine emissions to cause pulmonary disease in humans depends not only on the character and components of emissions but also on the structure of the lung and the adequacy of its defense mechanisms. This defense system can be modulated by other factors such as age, general health, or the presence of specific lung diseases.

To understand the effects of emissions, it is necessary to understand the mechanisms of damage and their relationship, if any, with such lesions as emphysema or small airways disease. In this chapter, lung structure is described as it pertains to disease processes, and then methods of diagnosis and mechanisms thought to be involved in the production of pulmonary diseases are outlined.

The lesions of emphysema and small airways disease produced by tobacco smoke inhalation in humans are similar to those that appear in animals experimentally exposed to emissions. To investigate the effects of emissions on human lungs it is necessary to use animal models. A review of similarities and disparities between animal models and humans is followed by a discussion of similarities between human lung disease and diseases produced by manipulation of animal models.

Finally, a summary description of the lesions that have been identified in animals exposed to emissions or emission components is followed by a discussion of areas in which knowledge is lacking and recommendations for further experiments.

Lung Anatomy and Defense Mechanisms

Lung anatomy has been elegantly described by Nagaishi et al. (1972). Grossly, the lung is divided into lobes and segments. The smallest structures are the secondary lobules, consisting of lung tissue confined by lobular septa.

Embedded within the lung tissue is the bronchial tree, a complex structure formed of approximately 15 million branches. The more proximal airways are supported by cartilage rings or plates, whereas the more distal airways are muscular. There are approximately 13 divisions from the trachea to the membranous bronchioles, characterized by a complete wall formed of fibromuscular tissue. At approximately the eighteenth generation, respiratory bronchioles are identified. These airways are alveolated; that is, they have alveoli budding from the wall. The membranous and respiratory bronchioles less than 2 mm in internal diameter have been termed "the small airways." The respiratory bronchioles branch and form alveolar ducts which in turn divide to form alveolar sacs and finally alveoli. The airway branching patterns are well described by Horsfield (1976), and are reviewed by Schlesinger in this volume.

Airway Cells

The mucus-secreting cells and ciliated epithelial cells are the most important cellular structures in lung defense mechanisms (Gail and Lenfant 1983). The epithelial mucous cells and serous cells secrete some of the components of airway mucus, but the major source of mucus is the submucosal mucoserous glands. The ciliated epithelial cells, which occur from the trachea to the respiratory bronchioles, contribute to the mucociliary transport system. These airway components are known to react to, and be damaged by, inhaled particles and gases.

In addition to epithelial cells, macrophages and polymorphonuclear neutrophils contribute to lung defenses. Neutrophils originate in the blood and migrate onto airway surfaces. Macrophages originate as blood monocytes and enter the pulmonary interstitium where they mature. A continually renewing population of matured cells migrate onto pulmonary surfaces. Under conditions of inflammation, the numbers and functional capacities of cell types increase. (For a more detailed discussion of cellular components and their products see Last, this volume.)

■ **Recommendation 1.** In vitro cell experiments should be directed to basic cell biology wherein pollutant constituents are introduced to cell cultures of epithelial and/or inflammatory cells.

Alveoli

There are two types of alveolar epithelial cells. Type I are attenuated cells covering 93 percent of the alveolar surface. They are highly susceptible to injury and their death is followed by proliferation of type II cells which then differentiate into type I cells. Type II cells are responsible for the synthesis of the phospholipid and protein components of surfactant. These cells are thought to play a role in initiation of fibrotic repair reactions in the lung.

The lung matrix is composed of collagen, elastin, glycosaminoglycans, and fibronectin (Turino 1985). Collagen and elastin provide the tissue structural elements and together form what have been loosely termed the lung "scleroproteins." Types I and III are the most common of the multiple subtypes of collagen. The balance of these two types appears to be important in lung repair reactions. The other substances appear to induce and modulate the responses of the matrix to injury.

Elastin is formed as a soluble protein which is secreted into the extracellular space. The soluble elastin is converted into mature elastic fiber by a cross-linking process that is mediated by the enzyme lysine oxidase. The signal that stimulates elastin synthesis is unknown. Elastin degradation releases peptides which can be measured in the plasma, urine, or broncho-alveolar lavage fluid.

■ **Recommendation 2.** Studies should be undertaken to ascertain whether the peptides present in body fluids can be measured accurately, and whether the measured parameters do, in fact, relate to the degradation and repair process.

Deposition of Particulates

Deposition in the lung of the particulate components of inhaled material depends on the mean diameter of the particles and the distribution of particle diameters. There are a number of experimental techniques for measuring deposition (Brain and Valberg 1979). Particles less than 1 μm in diameter may reach the alveoli, but particles less than 5 μm in diameter are deposited in the airways by processes of sedimentation and inertial impaction. A large fraction of the particles emitted by gasoline and diesel engines is less than 5 μm in diameter.

The main defense mechanism in the airways is the mucociliary escalator. In the alveoli, it is the macrophage. The balance between particle deposition and action of the defenses can be altered by pathological changes in the airways or in the lung parenchyma. These changes affect the mechanisms of deposition and clearance. (For a review of deposition and clearance, see Schlesinger, this volume.)

Pathologic Conditions: Concepts and Quantification

The definitions and pathophysiological concepts of emphysema and small airways disease have been formulated primarily in the context of their relationship to tobacco smoke. It has been postulated that tobacco smoke produces three separate but highly interrelated pathologic conditions: emphysema, small airways disease, and mucus hypersecretion. Furthermore, because tobacco smoke is the only source of air pollution where there is a large data base, it is often used as a model for studying the effects of inhaled pollutants such as automotive emissions.

Emphysema

Emphysema has been defined as a condition of the lung characterized by abnormal, permanent enlargement of airspaces distal to the terminal bronchioles, accompanied by destruction of their walls and without obvious fibrosis. Destruction is defined as nonuniformity in the pattern of respiratory airspace enlargement, so that the orderly appearance of the acinus and its components is disturbed and may be lost (Snider et al. 1985).

There are three anatomic subtypes of emphysema, of which two—panacinar and centriacinar emphysema—are pertinent to this discussion. In panacinar emphysema, all components of the acinus are equally involved, whereas in centriacinar emphysema, destruction is centered around the respiratory bronchiole.

Panacinar Emphysema. Panacinar emphysema has been observed in humans with α-1-proteinase inhibitor (α-1-PI) deficiency. Use of papain, pancreatic, neutrophil, or bacterial elastase in animal models has produced anatomic lung destruction identical to panacinar emphysema in humans (Snider et al. 1985). The realization of these facts led to a theory of proteolysis/antiproteolysis imbalance as a pathogenic factor in pulmonary emphysema.

This theory has immediate bearing on the question of effects of auto emissions on the lungs. For instance, what are the effects of emissions on a population with altered levels of α-1-PI? And what are the effects on α-1-PI of direct emissions that alter the balance of proteinase and antiproteinase?

In relation to proteinase, researchers have shown that cigarette smokers have an increased number of neutrophils, both in the peripheral blood (Sparrow et al. 1984) and within the lung, as seen by broncho-alveolar lavage (Hunninghake and Crystal 1983). Smoke is chemotactic (that is, an attractant) for macrophages and neutrophils, and the neutrophilic response can be amplified by macrophage-secreted chemotactic factors and stimulated by a complement that has been activated by cigarette smoke. The idea that these neutrophils show increased enzymatic activity is more controversial.

Although elastase-like activity has been found in the lavage of smokers, it was metallo-enzyme in type, implicating the pulmonary macrophage as its source. This is not surprising, since macrophages in the lumens of respiratory bronchioles are a prominent feature in young healthy cigarette smokers (Niewoehner et al. 1974). This response may reflect early disease, and may be important in the pathogenesis of emphysema.

■ **Recommendation 3.** Cell damage and its relation to inflammation should be studied through further research on exposure-related increases in neutrophils and macrophages, including collagen and elastin biomechanics, and the dose/time relationship between exposure and neutrophil and macrophage increase.

There are several proteinase inhibitors in the respiratory tract, including low molecular weight (MW) bronchial mucus inhibitor, α-2-macroglobulin, and α-1-PI. It was thought that α-1-PI was the most important inhibitor, but this may not be entirely true (Janoff 1985). This is certainly an area that needs further investigation in regard to smoking as well as to the effects of emissions. Cigarette smoke has been shown to oxidize a methionyl residue on α-1-PI, thus rendering it essentially inactive. This method of inactivation may be important with regard to emissions, since nitrogen dioxide (NO_2) is able to inactivate α-1-PI in an in vitro situation, presumably through an oxidative mechanism.

■ **Recommendation 4.** The relation of oxidants to the proteinase/antiproteinase balance should be studied.

Centriacinar Emphysema. There are two morphologically similar subtypes of centriacinar emphysema, both applicable to this discussion. Exposure to coal dusts results in the dilatation of respiratory bronchioles with abundant collections of dust particles. Whatever the pathophysiological mechanism involved in this lesion, it is conceivable that it may be related to, or amplified by, exposure to the emissions from the machinery used in mining or processing. In recent years, the bulk of investigation has been directed toward the proteinase/antiproteinase hypothesis for the causation of the second subtype of centriacinar emphysema found in people who smoke cigarettes.

Quantification of Pulmonary Destruction. Indirect methods of quantifying pulmonary destruction include pulmonary function tests and radiographic examinations. Both methods can detect abnormalities in established disease but neither is particularly useful in the identification of early disease.

Direct methods of estimation and/or quantification of lung parenchymal destruction can be performed on several levels. Most accurate are the methods based on analysis of tissue parameters. However, these methods can only be applied to experimental animal models and human autopsy or surgical resection material, and the resulting data represent a single point in a cross-sectional study. These methods are not applicable to case-control longitudinal studies.

■ **Recommendation 5.** Pulmonary mechanics should be assessed on excised lung specimens obtained from human autopsies or surgical specimens and results related to data collected from live subjects. Morphological examinations of human lung tissues could duplicate methods used to investigate effects of tobacco smoke.

Gross and subgross methods involve either whole lung slices or Gough sections in which the degree of destruction is ranked by comparison with a standard grading panel (Thurlbeck et al. 1970). These methods are only truly applicable to inflated human lungs and are by their very nature imprecise. Since gross estimation of emphysema requires a fully established lesion, methods in which the earliest phases of disease can be defined and quantified must be considered.

Measurements of tissue using microscopy include calculation of mean linear intercept (Dunnill 1962), destructive index (Saetta et al. 1985a), and analysis of alveolar attachments (Saetta et al. 1985b). These measurements are more precise than those obtained by gross and subgross methods, and they reflect airspace enlargement (mean linear intercept), or alveolar destruction (destructive index, alveolar attachments). They can be performed on human and animal tissues alike. A minor disadvantage is that the methodology requires lungs inflated to a standard pressure.

Measurements of destructive index and alveolar attachments show relatively good separations of smoking and nonsmoking populations. Since mean linear intercept

reflects dilatation of the airspace rather than destruction per se, it is affected by aging. However, it may be more sensitive than other methods to scleroprotein alteration, and it should not be abandoned.

An image analysis system can perform detailed measurements of alveolar surface area and surface density to provide information relating both to alveolar space dilatation and destruction. This is a labor-intensive technique, however, and it does not yield data of greater value or accuracy than the techniques mentioned previously.

Transmission or scanning electron microscopy techniques are applicable to assess substructural alteration such as changes of cell types, nuclear and/or cytologic alterations, and collagen and elastin structural changes. Such techniques may identify extremely early lesions in the microstructure. Disadvantages include the potential for sampling error because of the small sample size, and the necessity for special fixation and preparation. Potential advantages include applications for biopsy technique and use in longitudinal studies.

■ **Recommendation 6.** Examination should be made of epithelial cells, collagen, and elastin by electron microscopy to document possible progression of emission-related disease during a recovery period after direct exposure ceases.

Biochemical techniques can be direct, using portions of lung, or indirect, using lavage fluid or urine. These techniques can be used to assay hydroxyproline and fibronectin and lung tissue collagen types. Measurements of by-products of elastin turnover are more exciting, because of their potential use in longitudinal studies. Urinary desmosine correlates well with lung destruction in an animal model, but its applications in human investigations have been less promising (Janoff 1985). However, elastin peptides in human blood have been shown to separate nonsmokers from smokers, and measurements of lysyl oxidase as an indicator of elastin synthesis have shown that cigarette smoke inhibits elastin repair after initial damage with instilled elastase (Janoff 1985).

Small Airways Disease

The concept of small airways disease as a major cause of airflow obstruction in cigarette smokers arose from early work showing that peripheral airways resistance was markedly increased in patients with chronic obstructive pulmonary disease (Hogg et al. 1968). This was initially thought to be due to abnormalities in airways of less than 2-mm internal diameter and was associated with inflammation and fibrosis of airway walls as well as with epithelial changes (Cosio et al. 1977).

These initial reports were followed by pathological description and semiquantitative analysis of airways in the lungs of patients with varying smoking histories, degrees of emphysema, and abnormalities of pulmonary function (Cosio et al. 1980; Wright et al. 1984). The most important components of small airways disease now appear to be an inflammatory response and a fibrotic repair reaction.

The broncho-alveolar lavage fluid of smokers shows greater numbers of polymorphonuclear neutrophils and macrophages. Some of these cells must represent an airway inflammatory response. The pathological abnormalities of inflammation and fibrosis relate to perturbations of the pulmonary function tests, including flow rates and elastic recoil.

Pathological parameters in the small airways can be assessed by semiquantitative grading schemes or by direct morphometric analysis. The grading methodology is subjective but easy to perform and has the advantage of being usable on noninflated lungs (Wright et al. 1985). In this method, various degrees of abnormality are defined and displayed as a poster format (figure 1). The test-case airways are compared to these standards, and grade score is assessed. This method isolates individual parameters such as inflammation and fibrosis while also giving an overall estimation of abnormality.

Morphometric analysis is more precise but requires standard inflation pressures of fixation and some technical expertise. This method allows the investigator to measure directly the airway diameter and wall thicknesses, and in addition, to quantify the types and numbers of inflammatory cells in the airway walls and lumens.

Small airways disease can be assessed indirectly with specialized pulmonary function tests including nitrogen wash-out curve and forced expiratory flow of 25–75 liters/sec (FEF_{25-75}). It would be difficult to detect minor degrees of injury in this fashion, but the tests would be suitable for longitudinal studies (Buist et al. 1984). Furthermore, the tests could be used to document established disease.

■ **Recommendation 7.** Some of the more detailed pulmonary function tests should be used to identify progressive dysfunction.

Mucus Hypersecretion

Cigarette smoking increases the size of the bronchial mucous glands (Reid 1960) and the proportion of glands in the bronchial wall and also causes goblet cell metaplasia (Mitchell et al. 1976). The Reid index measures the thickness of the bronchial mucous glands compared to the thickness of the bronchial wall measured from perichondrium to basement membrane. This thickness correlates with the presence of chronic bronchitis (Reid 1960). The proportion of mucous glands in the bronchial wall can be estimated either by a point-counting technique or by direct measurement of the areas.

Although these changes are associated with chronic cough and sputum production (a process formerly referred to as chronic bronchitis) with their attendant nonesthetic qualities and psychological disabilities, they have not been associated with airflow obstruction (Fletcher et al. 1976) and physical disabilities. Since mucus hypersecretion appears to be a nonspecific response to an irritant, and since this does not produce pulmonary function abnormalities, it will not be considered further.

Animal Models of Human Disease

It is important to ascertain whether lung disease produced in animals by manipulations such as the instillation or inhalation of

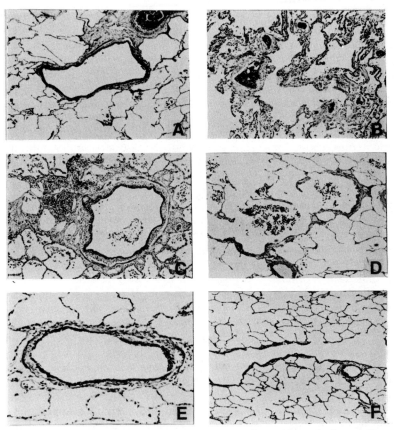

Figure 1. Type of poster format applicable in the semiquantitative grading technique for assessing pathological parameters in the small airways: top, normal respiratory (A) and membranous bronchioles (MB) (B); middle, grade III intralumenal macrophages in the respiratory bronchioles (RB) (C) and grade III intramural inflammation in the membranous bronchioles (D) (In the MB, the grading is based on overall cellularity rather than the inflammatory cell types.); bottom, grade III fibrosis of both the RB (E) and MB (F). (Adapted with permission from Wright et al. 1985. *Arch. Pathol. Lab. Med.* 109:163–165. Copyright 1985 American Medical Association.)

various materials is similar to human disease. Animal models have been used extensively to test specific hypotheses related to the pathogenesis of emphysema and small airways disease.

Comparisons of animal and human lung structure have concentrated on two areas: airway structure and airway surface epithelial cells and glands. The casts of the tracheobronchial trees of humans have a distinctive, almost spherical shape with a relatively symmetrical branching pattern. In contrast, most other mammals have elongated casts with tapered monopodial airways and small lateral branches.

Small airways and terminal units in species other than rodents are similar to human lung structure. In rodents, the respiratory bronchiole is either absent or very short. This may be significant since the respiratory bronchiole appears to be a target area for injury after inhalation of dusts and fumes.

Various mammalian species, including humans, have different cell populations in the normal surface epithelium, and even within an individual species there are changes dependent upon maturation. There are also species variations in the total volume of submucosal glands as well as in their distribution within the airways. In addition, differences in the histochemical staining of the epithelial cells and glands indicate different mucus secretions among the species.

Emphysema

Elastin/Collagen Destruction.

Elastin/collagen destruction in animal models has been studied primarily by using inflammatory cell increases, elastase administration, or reduction of α-1-PI to test the relationships between proteinases and antiproteinases. Administration of elastases, either pancreatic or leukocytic (endogenous or exogenous), has produced airspace destruction morphologically similar to panacinar emphysema (Janoff 1985).

Electron microscopy of enzyme-induced emphysema shows elastic fiber disruption, with beading and irregularity of the fibers (Kuhn et al. 1976). Biochemical analysis of elastase-instilled lungs has shown a rapid and marked decrease in total elastin content with a lesser decrease in collagen (Karlinsky et al. 1983). After a postexposure recovery period, however, lung elastin content returned to normal limits.

Alterations in physiological parameters show a dose/response effect (Raub et al. 1982), and age at exposure appears important since the effects of elastase have been shown to be greater in younger animals (Goldstein 1982). Administration of cigarette smoke to animals previously given elastase has augmented destruction (Hoidal and Niewoehner 1983). Chronic administration of the oxidant chloramine-T to interfere with the function of α-1-PI produces similar results. Cigarette smoke itself, in high doses, has been shown to produce enlargement of airspaces and abnormalities in pulmonary function (Huber et al. 1981; Heckman and Dalbey 1982). These studies have also described, but not quantified, changes in the small airway walls.

■ **Recommendation 8.** Research should be undertaken on the elastin and collagen degradation/repair balance to determine whether repair is affected by pollutants, and if so, by what mechanism. Studies should map the time course of collagen and elastin degradation and repair, relating data to type and concentration of exposure as well as ascertaining the possibility of alteration or repair by pollutants or other oxidants.

Elastin/Collagen Formation.

Lysyl oxidase, a copper-requiring enzyme that mediates the conversion of lysine to the elastin-specific cross-links desmosine and isodesmosine, is involved in the process of elastin/collagen formation. Elastin synthesis occurs rapidly after elastase-induced emphysema, and there is an associated increase in lysyl oxidase activity (Janoff 1985). Decreased lysyl oxidase activity results in poorly formed elastin and in lung lesions morphologically and physiologically identical to panlobular emphysema in the blotchy mouse (Snider et al. 1985).

■ **Recommendation 9.** Pollutant interference with lysyl oxidase and impairment of elastin resynthesis should be investigated as a possible destructive mechanism.

Dietary insufficiencies have also produced pulmonary abnormalities. Protein/calorie starvation produces enlargement of the airspaces, but physiological alterations are different from those seen in human emphysema (Janoff 1985).

When exposed to cigarette smoke, experimental animals with elastase-induced emphysema show a greater degree of lung destruction. These data, in conjunction with in vitro studies showing that cigarette smoke can inhibit the activity of lysyl oxidase (Janoff 1985), suggest that interference with elastin synthesis contributes to the alteration of the lung structure as seen in emphysema.

Administration of cadmium chloride to experimental animals produces inflammation and a granulation tissue response followed by airspace enlargement. Administration of the lathyrogen beta amino propionitrile (BAPN) appears to limit the fibrosis (Niewoehner and Hoidal 1982). The mechanism for disease is unclear in this model, but it appears to involve some balance between destruction and repair reaction with ultimate fibrosis.

The physiological requirements for the diagnosis of emphysema in humans are very controversial. Since these physiological requirements have not yet been established, only the morphology of human and animal emphysema can be compared. This comparison itself may be difficult, since the lungs of the various species are not identical. Although physiological abnormalities are often found in animal models, the presence or absence of emphysema is determined by whether airspace enlargement can be documented.

Within these limits, instillation or inhalation of elastase is an excellent model for panacinar emphysema; however, there is no workable model for centriacinar emphysema at the present time. Some animal manipulations produce airspace enlargement as well as fibrosis. Although these manipulations cannot be used as models for emphysema, they are appropriate to assess conditions where the balance of lung destruction and fibrogenic reaction are important.

■ **Recommendation 10.** Research should be undertaken on the balance between fibrosis and destruction and its relation to particulates/oxidants.

Small Airways Disease

Two studies utilizing acid inhalation in either acute or a short-term exposure correlate airways inflammation with airways dysfunction (Baile et al. 1982; Peters and Hyatt 1986). This model suggests that any inhaled substance that results in an inflammatory response is capable of causing airflow obstruction. These data are similar to those obtained from human studies of the association between inflammation and airflow obstruction in cigarette smokers.

Use of an acid nebulization technique in a dog model allowed researchers to ascertain whether acute airway inflammation could be detected by pulmonary function tests. Inhalation of a weak hydrochloric solution was associated with discernable inflammation in the airways and with abnormalities in the tests for small airway function (Baile et al. 1982).

The acid inhalation technique was also used with a semiquantitative grading technique to evaluate the airways. Researchers found inflammation and fibrosis of the cartilaginous as well as the noncartilaginous airways. These changes correlated with decreased dynamic compliance and increased slope of phase III of the nitrogen wash-out

curve, as well as with a decrease in flow rates (Peters and Hyatt 1986).

Emphysema and Small Airways Disease: Relations to Vehicular Emissions

Data on pathophysiology of disease due to cigarette smoking can help document the presence of disease and the progression from early to well-developed disease and provide information on the mechanisms of disease. Recent reports suggest key physiological alterations relevant to production of disease in animals.

The exhaust from diesel engines contains most of the pollutants present in the emissions of gasoline engines, but differs in that it has a large particulate component. When possible, studies involving effects of diesel emissions are treated separately.

Lung Disease Produced in Animals by Vehicular Emissions

It is important first to ascertain whether lung diseases produced in animals by inhalation of vehicular emissions are similar to human diseases suspected to be caused by emissions. A comparison of the physiological responses of NO_2-induced disease in several animal models to those of human emphysema (figure 2) showed that the animals had similar shifts in flow/volume and pressure/volume curves as well as in lung volumes (Mauderly 1984). The author concluded that animals provide data applicable to humans. When he compared acute responses to NO_2, the magnitude of the effects was different, but all subjects showed irritant effects at various doses.

An earlier study examined the morphology of normal and experimentally induced emphysema in animals and compared these data to those on emphysema in horses and in humans (figure 3). Dilatation of airspaces occurred in all species. Papain or continuous NO_2 exposure with peak increases or followed by intermittent exposure, caused similar lung destruction in all of the rodent models (Port et al. 1977). The morphology

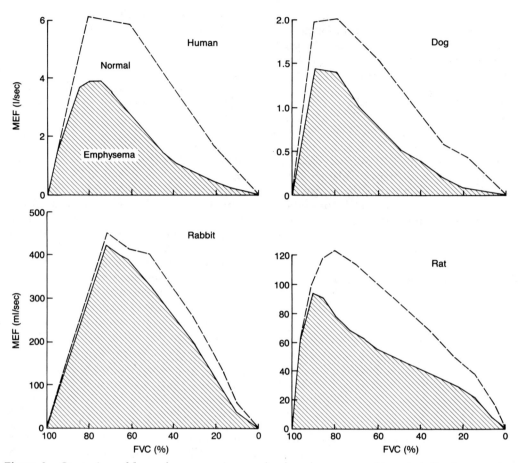

Figure 2. Comparison of flow-volume curves in normal and emphysematous subjects, including humans, dogs, rabbits, and rats. Flow rates of all species were reduced to a similar extent by emphysema. Maximum Expiratory Flow (MEF); Forced Vital Capacity (FVC). (Adapted with permission from Mauderly 1984, and Hemisphere Publishing Corp.)

of emphysema in rodents differs from that in horses and humans, but the models are certainly applicable for research on pathophysiological relationships.

Evidence of Emphysema and/or Small Airways Disease.

Several studies have provided evidence of emphysema and/or small airways disease by demonstrating changes in physiological as well as morphological parameters after the administration of various concentrations of NO_2 (Freeman et al. 1968; Freeman et al. 1972; Kleinerman and Niewoehner 1973; Coffin and Stokinger 1977; Evans and Freeman 1980).

NO_2 appears to exert its major damage on the bronchioles and alveolar ducts, and this seems related to a concentration gradient (Menzel 1980). Loss of cilia is a subtle airway change which occurs early in acute experiments (Evans 1984), and is seen after long-term exposures as low as 0.3 ppm (Nakajima et al. 1980). Not surprisingly, these structural abnormalities are associated with delayed mucociliary clearance (Giordano and Morrow 1972).

Bronchiolar epithelial cell damage from NO_2 appears to be repaired by proliferation of nonciliated epithelial cells—cells that act as the progenitors for both ciliated and nonciliated cells. Type I damage is repaired by type II cell proliferation (Evans 1984).

Physiological Changes. Pulmonary function tests in animals exposed to NO_2 con-

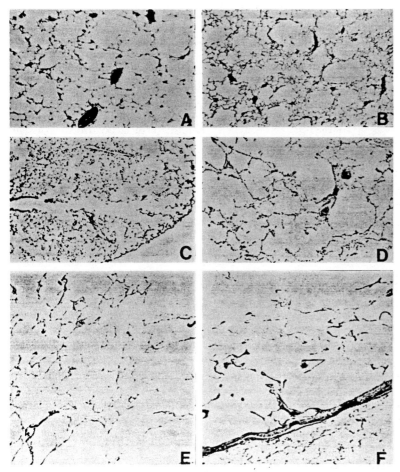

Figure 3. Comparison of the light microscopic findings from (A) NO$_2$-exposed rat, (B) papain-exposed hamster, (C) NO$_2$-exposed mouse, (D) emphysematous human, and (E, F) emphysematous horse. All lungs show dilatation of the airspaces. (Reproduced with permission from Port et al. 1977, and Hemisphere Publishing Corp.)

sistently indicate airflow obstruction. A large portion of this obstruction appears to be due to airway disease and is potentially reversible (Kleinerman and Niewoehner 1973; Kleinerman et al. 1976). An early study concluded that the pulmonary function changes in the rabbit model were due to bronchiolitis (Davidson et al. 1967).

In other studies of bronchiolar abnormalities related to exposure to NO$_2$, researchers have found massively enlarged collagen fibrils underlying bronchioles and associated with thickened basement membranes in rats (Stephens et al. 1971), and increased thickness of the basement membrane with increased collagen in the interstitium in squirrel monkeys (Bils 1976). In the latter study, basement membrane thickness and collagen in the interstitium increased even during recovery. In contrast, a study by Buell (1970) indicated that the morphological parameters of collagen and elastin denaturation following exposure to NO$_2$ were no longer present after a 7-day recovery period.

Biochemical Changes. Biochemical findings in animal exposure studies suggest that the degradation or repair of collagen and elastin do not necessarily occur at the same time and that they can therefore be individually influenced by exogenous or endogenous factors (Hacker et al. 1976; Kleinerman 1979).

Hydroxylysine in either urine or lavage

fluid has been suggested as a marker of lung injury from NO_2 exposure (Kelley et al. 1986). This may represent a potential mechanism for identifying early disease or following the time course of tissue remodeling.

■ Recommendation 11. Degradation products such as elastin, peptides, urinary desmosine, and hydroxylysine should be quantified for use in animal and human studies.

Investigators have examined the effects of NO_2 on lung lipids and suggested that NO_2 damage is mediated by the oxidative properties and the free-radical potential of the gas. Specifically, lipid peroxidation of the cell membrane could act on the endothelial as well as the epithelial cells (Mustafa et al. 1980). Other investigators have speculated that the changes in lung lipids of rat models could be related to increased air trapping (Arner and Rhoades 1973).

In early work on proteolysis/antiproteolysis, Rynbrandt and Kleinerman (1977) examined proteolytic activity in lung extracts or in lavage fluid of animals exposed to NO_2. This is an important concept, but the data are suspect due to defective methodology, and the work should be repeated using modern analytical techniques. To my knowledge, no estimates of the amount of α-1-PI in lavage fluids have been reported in NO_2-exposed animals.

■ Recommendation 12. Short-term animal experiments to determine pathobiology of disease related to emissions should be designed to measure biochemical parameters in lung tissue and to count and type inflammatory cells in lavage or tissue.

The increased numbers of macrophages seen in NO_2-exposed animals would potentially add to the proteolytic burden of the lung. These cells increase by migration into the alveolar space as well as by active division (Evans et al. 1973). The dividing cells appear to be localized near the opening of bronchioles, the site of destruction in centriacinar emphysema. Neutrophilic leukocytes also appear to increase in numbers during the initial phases of exposure (Gardner et al. 1977), but there are no

long-term data regarding the biomechanics of the proteolytic cell.

Among other factors affecting injury is dose dependency, with high concentrations causing greater injury than lesser concentrations. In addition, short-term exposure to a higher concentration has a greater effect than long-term exposure to a lower concentration (Gardner et al. 1979). This indicates that a concentration/time coefficient is not an adequate method of expressing exposure, although a small spike (5 ppm) superimposed on a low-level baseline may have minimal effect (Gregory et al. 1983).

Effects of other gases and compounds are particularly important in conjunction with automotive emissions. In one study, NaCl did not alter epithelial or hematological response to NO_2 (Furiosi et al. 1973). In another study suggestions of a synergism between NO_2 and ozone (O_3) were found during biochemical analyses (Mustafa et al. 1984). At higher doses of O_3, effects of NO_2 appeared to be overshadowed.

■ Recommendation 13. Dose dependency should be quantified and extrapolated to lower doses.

■ Recommendation 14. The degree of synergism between components and their relative importance in the causation of disease should be investigated.

At equivalent doses, older animals developed more severe disease than did young animals (Evans et al. 1977). In contrast, a mild injury during the newborn period was found to result in physiological and morphometric changes suggestive of mild emphysema (Lam et al. 1983). There is evidence that repair is delayed in older animals (Cabral-Anderson et al. 1977), and another review cites studies suggesting that exposure during infancy results in amelioration of effects upon reexposure (Evans 1984).

■ Recommendation 15. Long-term, multidisciplinary animal studies should be designed to document abnormalities produced by exposure to emissions at various points in the lifespans of the subjects.

The area of nutrition is interesting as it relates to the proteolysis/antiproteolysis hypothesis. Animals with vitamin C deficiency are more sensitive to NO_2, even at levels as low as 1 ppm (Selgrade et al. 1981). In contrast, animals with the highest vitamin E content in their diet had the best survival after exposure (Fletcher and Tappel 1973), suggesting that vitamin E induced a protective effect against acute toxicity as well as systemic effects such as weight changes. General body nutrition also appears to be important (Coffin and Stokinger 1977; Evans and Freeman 1980).

■ **Recommendation 16.** Experiments should be conducted to analyze effects of nutrition, stress, infection, exercise, and co-contaminants such as diesel particulates or cigarette smoke on disease.

Multidisciplinary Studies. The group of experiments initiated by the Division of Air Pollution of the University of Cincinnati and finalized at University of California at Davis exemplifies the sort of multidisciplinary approach necessary for investigating the effects of auto emissions (Vaughan et al. 1969; Lewis et al. 1974; Orthoefer et al. 1976; Hyde et al. 1978; Gillespie et al. 1980).

The dog model (beagles) was chosen for the similarity of response between this animal and humans. For example, histological similarities between the lungs of dogs exposed to cigarette smoke and the lungs of a group of cigarette smokers appeared to exhibit a rough dose/response relationship (Hammond et al. 1970).

The study objective was to examine the effects of long-term exposure to environmentally realistic levels of whole auto exhaust and its individual components. A further objective was to assess irreversible chronic effects, which might provide valuable data on the potential toxic effects of auto exhaust on humans (Stara et al. 1980). The levels of pollutants to which the animals were exposed were comparable to those measured in Chicago, Cincinnati, Washington, D.C., and Los Angeles. In addition, irradiation chambers were constructed to simulate the effects of sunlight on emissions. The animals were randomized, and all efforts were made to reduce chamber variation.

A limited battery of pulmonary function tests was performed at 18 months, followed by more extensive testing at 36 and 61 months. No difference between control animals and exposed animals was found at 18 months with the less sensitive tests, but progressive abnormalities were found in exposed animals at 36 and 61 months. The changes were most marked in the groups exposed to high levels of nitrogen oxides (NO_x) and raw and irradiated exhaust, and suggested abnormalities both of airways and lung parenchyma. Two years postexposure, the results of the pulmonary function tests showed continued deterioration.

Tissues were analyzed after the animals had been allowed to recover for approximately 3 years in an indoor environment. There was no difference between the control and exposed groups in the amount of hydroxyproline. However, there were increases of the tissue enzyme prolyl hydroxylase in the exposed group, indicative of an increased potential to make collagen and perhaps suggestive of a dynamic repair and regeneration sequence.

Morphometric and qualitative scanning electron microscopic analysis of the lung tissue (Hyde et al. 1978) showed decreased alveolar surface density with airspace enlargement, as well as an increase in the size and number of the alveolar pores, centered on the respiratory bronchioles and alveolar ducts in the groups treated with NO_x and sulfur oxides (SO_x). Since these measurements were made 3 years after exposure, disease present is at least nonreversible and may well be progressive, as the pulmonary function tests suggest. This is an important point in view of the progressive nature of destruction in animals given intratracheal elastase.

An additional question is whether actual lung destruction occurred in these experiments, since alteration of the lung scleroproteins could also produce the morphometric changes noted above. Participants formulated an alternate hypothesis that emissions resulted in excessive and premature aging of the lung (Stara et al. 1980), concluding that exposure to specific air pollutants produced pulmonary injury and loss of pulmonary function which continues following termination of exposure. They further suggest that "due to the ubiquitous nature of nitrogen and sulfur oxides and auto exhaust (per se and photochemically reacted) in the ambient air of urban communities, the chronic cardiopulmonary changes . . . denote serious potential health hazards to the populace of certain communities." This statement underscores the importance of these multidisciplinary studies, and stresses the value of chronic studies.

Animal Exposure to Diesel Exhaust

In studies involving the administration of diesel exhaust to animals, researchers have found abnormalities that appear to be focused on the centriacinar space (Wiester et al. 1980; Barnhart et al. 1981; Vostal et al. 1981; Plopper et al. 1983). Several of these investigators have noted small foci of fibrosis in relation to large clusters of macrophages (Wiester et al. 1980; Vostal et al. 1981). Some (Barnhart et al. 1981) have localized these changes to the bronchioles. Others have found the bronchiolar fibrosis to be progressive, even after a 6-month recovery period; they also found a trend toward increases in total lung collagen, with a twofold increase in newly synthesized collagen (Hyde et al. 1985).

Pulmonary function testing in diesel-exposed animals has produced conflicting results. Studies (Pepelko 1981) have shown dose-dependent decreases in vital capacity (VC) and diffusing capacity for carbon monoxide ($D_{L_{CO}}$) associated with loss of elastic recoil pressure in hamsters. Cat models have shown decreases in VC and $D_{L_{CO}}$ accompanied by decreased maximum expiratory flow, but also accompanied by a decrease in total lung capacity (TLC). This suggests a combination restrictive/obstructive pattern.

A rat model showed evidence of decreased compliance, reduced $D_{L_{CO}}$, and inhomogeneity of airways emptying, without evidence of airflow obstruction (Mauderly et al. 1983). A progressive increase in hydroxyproline peptides as well as

in total lung collagen appeared to be a function of increasing time as well as increasing exhaust concentrations.

Tracheal muscle histamine dose/response curves documented in rats exposed to diesel exhaust and/or coal dust for 2 years demonstrate an additive effect from coal dust and diesel exhaust (Fedan et al. 1985). These data may reflect a role of the immune system. A shortcoming of this study is the lack of any morphological, physiological, or biochemical data relating to the lungs of these animals.

There is also evidence of nonspecific dysfunction in exposed animals, with decreased spontaneous forced locomotor activity (Pepelko 1981). Data from the same laboratory suggest that exposure during the first week of life affected adult learning abilities.

■ **Recommendation 17.** Studies should be undertaken to determine whether the effects of diesel exhaust are different from the effects of gasoline exhaust in regard to disease progression, the effects of age at exposure, and additional applied stress.

Summary

It is apparent that exposure to gasoline or diesel emissions or their components has the potential to produce disease. Animal models of emphysema and small airways disease, with some exceptions, are very similar to the same diseases in humans.

Most histopathological, morphometric, and electron microscopic data on effects of emissions and their components have been derived from animal models which provide evidence of alveolar wall destruction with loss of surface area, and of abnormalities of the airways. Investigations of lung disease and tobacco smoke have utilized fairly sophisticated measurements, and these methods can be, and in some areas have been, easily utilized in emission protocols. Simple histological descriptions are no longer justified.

The pulmonary disease produced in experimental animal models by vehicular emissions appears to involve, as a first step, inflammation and destruction at the level of the respiratory bronchioles. The lesions produced by exposure to emissions may destroy the alveolar parenchyma and result in emphysema. Lesions may also act on the airways to produce fibrosis and subsequent airway narrowing and airflow obstruction.

These morphological reactions to emissions are very similar to those produced by cigarette smoke, leading to speculation that pulmonary disease caused by emissions involves the same mechanisms as disease induced by cigarette smoke.

The hamster model of Hoidal and Niewoehner (1982) suggests that the particulate component of smoke is important, not only in inducing inflammation in the bronchioles, but also in activating macrophages. NO_2 itself results in macrophage accumulation, but in that case, the chemotactic stimulus may be the injured epithelial cells. If both of these mechanisms are functional, one would suspect that the addition of particulates would amplify the inflammatory response, perhaps making diesel exhaust dangerous. Indeed, this could be expected to happen with the addition of any other injurious agent—perhaps some of the other components of gasoline exhaust, for instance.

Vehicular emissions and their components, like cigarette smoke, have oxidant properties. Therefore emissions can injure the lung through direct damage to the cell membrane as well as through inactivation of the antiproteolytic system. With inactivation of antiproteinase as well as an increase in the source of proteinases, the proteolysis/antiproteolysis balance might be tipped, resulting in lung destruction.

Other data suggest that the antioxidants selenium and vitamins C and E may be important in lung defense. Furthermore, emissions produce an inflammatory response involving neutrophils as well as macrophages, with evidence for collagen and elastin degradation and repair. Finally, there is good correlation of disease (emphysema and airway fibrosis and inflammation) with abnormalities of pulmonary function.

Summary of Research Recommendations: Discussion

There are two main questions to ask about the relation of pulmonary disease to vehicular emissions. Although both questions are studied in relation to cigarette smoke, data are still sparse.

First, does *significant* disease occur, and if so, what is it? This is difficult to determine in a human population, as data on cigarette smoking have shown. Second, what is the cellular basis of this disease?

Other questions appropriate to the study of pulmonary disease in relation to vehicular emissions are: What is the role of the proteolysis/antiproteolysis balance in pollutant exposure? What agent induces a response by the inflammatory cells, and does it have one or multiple components? What roles do the particulates and the immune system play? What is the role of the antiproteinases, and are they affected by the oxidants in pollutants—and which oxidants are more important?

Pathogenesis of Pulmonary Disease

The similarity of damage caused by vehicular emission exposure to damage caused by cigarette smoke has implicated the proteolysis/antiproteolysis balance in the pathogenesis of pulmonary disease. Measurement of elastase and antiproteinases in lavage fluids has been useful in investigating damage related to tobacco smoke and to the adult respiratory distress syndrome, and the feasibility of using this procedure to analyze proteolysis/antiproteolysis imbalance is an important subject for further research (Recommendation 2).

The inflammatory cells stimulated by cigarette smoke are a source of proteinases and represent a possible mechanism for lung destruction. In addition, since cigarette smoke, acting as an oxidant, is able to inhibit α-1-PI by binding to a methionine residue, it is possible that the oxidants in emissions can act in a similar fashion. This in itself would not be harmful unless proteolytic factors (that is, neutrophil and macrophage enzymes) were also increased, either by increasing the amount of enzyme per cell, or by increasing the number of cells. Both of these facts are true for cigarette smokers, but it is not clear which component of smoke is responsible (Recommendation 4). Some information suggests that gasoline and diesel exhaust exposure increase neutrophils as well as macrophages. This suggestion needs further clarification relative to dose and time as well as to collagen and elastin biomechanics (Recommendation 3).

The elastin and collagen degradation/repair balance is another mechanism whose operation elicits such questions as: Is repair affected by pollutants, and if so, by which component and by what mechanism? The same primary agent can produce fibrosis as well as destruction (Niewoehner and Hoidal 1982). That study was directed at cadmium, but it is very tempting to suppose that model would be applicable to disease produced by emissions. Further work on this hypothesis could use the lathyrogen BAPN. This manipulation would allow assessment of any destructive potential.

Collagen can be subtyped, and it is important to determine whether any subtype in particular is affected. Hydroxyproline or hydroxylysine from lavage can also be quantified and may be indicators of lung injury and repair. Fibronectin, which may prove important in both degradation and repair, can be quantified as well. Measurement of collagen subtypes or hydroxyproline would allow mapping of the dose response and the time course of the fibrotic response (Recommendation 8).

Initial data have suggested that in the early stages of emission injury, collagen and elastin degrade, but in different time frames. Studies should map the time course of degradation and repair and determine whether repair could be altered by pollutants or other oxidants such as cigarette smoke. These data should be further quantified and related to type and concentration of exposure.

Other studies should attempt to quantify degradation products such as elastin, peptides, urinary desmosine, and hydroxylysine for use in long-term animal and human investigations (Recommendation 11).

Cigarette smoke is thought to impair elastin resynthesis by interfering with lysyl oxidase. It is not known if this results from air pollutant exposure, but it could be an important mechanism underlying any destruction that occurs (Recommendation 9).

How could these data relate to the centriacinar versus panacinar locations of destruction in the lung? In α-1-PI deficiency or elastase administration the imbalance is a generalized event. In cigarette smoke-induced emphysema, the imbalance is localized in the area of the proteinase excess— the respiratory bronchiole. This localization might be expected to occur also during pollutant-induced lung destruction, where the respiratory bronchioles do appear to be abnormal, and there is an inflammatory infiltrate of macrophages.

What is the balance between fibrosis and destruction, and is this related to particulates rather than oxidants (Recommendation 10)? This has been suggested by work on diesel exhaust (Mauderly et al. 1986) which shows that there is focal fibrosis in the alveoli as well as in the bronchiolar wall. There may be some differences in the effects of gasoline and diesel emissions. Such possible differences should be investigated in regard to disease progression, effects of age at exposure, and additional applied stress (Recommendation 17).

Destruction of alveolar parenchyma has not been a universal finding; indeed, foci of alveolar fibrosis have been described (Mauderly et al. 1986). Perhaps if there is no imbalance of the proteolytic system, and there is particle stimulation of the interstitial cells, there will be fibrosis. But if there is imbalance of proteolysis, the collagen/elastin repair mechanism is hampered, and destruction results.

Although there is evidence for dose dependency of injury, this should be further quantified and related to the lower doses (Recommendation 13). Determining which emission components are most injurious and the degree of synergism between components active in the causation of disease is also important (Recommendation 14).

Some data suggest progression of disease even after the insult has been discontinued.

Such progression should be fully documented. Experiments could count and type inflammatory cells in lavage or in tissue. With electron microscopy, examinations for early abnormalities in collagen, elastin, and epithelial cells (Recommendation 6) could be extended to analyze the repair reaction and to identify agents that amplify or ameliorate the response. The data obtained would help to clarify whether damage was related to the oxidant components of the emissions.

Abundant evidence suggests abnormalities of airflow and lung volumes in animals exposed to emissions. Although measurements of lung volume, compliance, and resistance are important, their normal range varies widely, so it is important to analyze indicators of flow such as the pressure/volume curve, as well as flow/volume curves and tests such as the nitrogen washout curve indicating inhomogeneity of airflow. Mauderly's group (1984) has used these techniques to great advantage in their experimental models and have been able to identify subtle changes in pulmonary function.

Pathobiology

In vitro cell experiments should be directed to basic cell biology. Such studies might focus on introducing pollutant constituents, gaseous as well as particulate, to cell cultures of epithelial and/or inflammatory cells (Recommendation 1). Epithelial cells could then be examined for damage and production of inflammatory cell chemotactic factors. The inflammatory cells could also be examined for evidence of activation with production of oxygen radicals. To obtain obvious abnormalities, a dose range should initially be broad, with some concentrations fairly high. These experiments should measure biochemical parameters including α-1-PI and elastase in lavage fluid, elastin, desmosine, collagen type, and lysyl oxidase in lung tissue. These data would suggest which components of pollutants are important and whether combinations have greater effects than single exposures. They could also form the basis for short- and long-term animal experiments.

Animal Studies

Short-term animal experiments should be designed to determine the pathobiology of disease related to emissions (Recommendation 12). Since age at exposure may be a factor in disease, research should document abnormalities produced by exposure beginning at various points in the lifespans of animals studied. Researchers should analyze the effects of additional stress such as diesel particulates, cigarette smoke, infection, and nutritional deficiencies on the production of pulmonary disease (Recommendation 16).

The influence of particle size and composition as well as the gaseous components of the emissions need analysis. Brain and Valberg (1979), as well as others, have shown that there are multiple factors, including age and health of the experimental subject, that influence the deposition and subsequent management of a particulate aerosol. These workers have also described various methods to assess deposition, and these factors should be investigated.

Other necessary data include those on morphological abnormalities at multiple points over a long-term exposure, particularly at the lower doses. The animal model chosen should therefore have a long lifespan. Long-term animal studies should be multidisciplinary and should include physiological, morphological, and biochemical analyses (Recommendation 15). Study hypotheses should be related to the pathophysiology of disease and directed toward a low concentration range of emissions including single or multiple components. Long-term experiments could answer the question of severity of disease or tolerance to emissions in relation to exposure at an early age.

Since there are some data suggesting disease progression very similar to the postexposure progression of elastase-induced emphysema, it is important to analyze a recovery period as well as several points during exposure.

Sacrifice at various time points would provide information relating collagen and elastin breakdown and repair to parenchymal destruction and production of airway abnormalities. These data would be correlated with lung function. If proven helpful in the short term, biochemical estimations of elastin products would be useful to monitor disease without the necessity for sacrifice. Similarly, certain types of pulmonary function could be performed on a routine basis without sacrifice, following the general procedure of Mauderly (1984). These methods would allow a longitudinal study and avoid potential artifact produced by tracheostomy.

Data from such long-term animal experiments would provide a basis for long-term epidemiologic studies on humans. They would help to develop a scheme to suggest important variables and allow for appropriate testing and analysis of human data. The data obtained from such experiments could also be important in assessment of early disease and of the transition between health and disease in a pollutant-exposed population.

Although this list of recommendations treats the areas of histopathology, biochemistry, and physiology separately, researchers performing animal experiments should collect these data and correlate them. In particular, the data provided by the groups of dog model (beagle) studies at the University of California at Davis have evinced the need for study of the mechanisms of pulmonary disease. This can only be accomplished by carefully planned multidisciplinary studies.

Human Studies

Physiological and morphological investigations form a major tool for human studies. Although there are some difficulties in performing the more specialized physiological tests in a community or workplace environment, these difficulties are not insurmountable, and the additional data provided are extremely valuable. These specialized tests can certainly be performed in the short-term-exposure laboratory setting (Recommendation 7).

Researchers performing structural/functional correlations on human autopsies can perform lung mechanics on excised specimens and relate results to data collected on the live subject. Morphological examinations for human exposures could also be performed on autopsy and surgical specimens, duplicating the methods used for tobacco smoke investigations (Recommendation 5).

Summary of Research Recommendations: Priorities

The research recommendations are resummarized here by experimental technique rather than by subject area, for the purpose of focusing on a research program. In the following listing, the needed experiments are ranked from high to low priority within each category.

In Vitro Experiments
HIGH PRIORITY

Recommendation 1   Pollutant constituents should be introduced to cell cultures of epithelial and/or inflammatory cells.

MEDIUM PRIORITY

Recommendation 3   Cell damage and its relation to inflammation should be studied through further research on exposure-related increases in neutrophils and macrophages, including collagen and elastin biomechanics, and the dose/time relationship between exposure and neutrophil and macrophage increase.

| Recommendation 8 | Research should be undertaken on the elastin and collagen degradation/repair balance to determine whether repair is affected by pollutants, and if so, by what mechanism. Studies should map the time course of collagen and elastin degradation and repair, relating data to type and concentration of exposure as well as ascertaining the possibility of alteration or repair by pollutants or other oxidants. |

LOW PRIORITY

| Recommendation 4 | The relation of oxidants to the proteinase/antiproteinase balance should be studied. |

In Vivo Experiments
HIGH PRIORITY*

| Recommendation 2 | Studies should be undertaken to ascertain whether the peptides present in body fluids can be measured accurately, and whether the measured parameters do, in fact, relate to the degradation and repair process. |

| Recommendation 12 | Experiments should be performed to determine the pathobiology of emissions-related diseases by measuring biochemical parameters in lung tissue and by counting and typing inflammatory cells in lavage or tissue. |

| Recommendation 15 | Long-term, multidisciplinary animal studies should be designed to document abnormalities produced by exposure to emissions at various points in the lifespans of subjects. |

MEDIUM PRIORITY

| Recommendation 9 | Pollutant interference with lysyl oxidase and impairment of elastin resynthesis should be investigated. |

| Recommendation 10 | Research should be undertaken on the balance between fibrosis and destruction and their relation to particulates/oxidants. |

| Recommendation 11 | Degradation products such as elastin, peptides, urinary desmosine, and hydroxylysine should be quantified. |

| Recommendation 13 | Dose dependency should be quantified and extrapolated to lower doses. |

LOW PRIORITY

| Recommendation 6 | Epithelial cells, collagen, and elastin should be examined by electron microscopy to document possible progression of emis- |

---

* Experiments are meant to be short term unless explicitly stated otherwise.

sion-related disease after the cessation of direct exposure, in both short-term and long-term studies.

Recommendation 14	The degree of synergism between components and their relative importance in causing disease should be investigated.
Recommendation 16	Experiments should be conducted to analyze the effects of nutrition, stress, infection, exercise, and co-contaminants such as diesel particulates and cigarette smoke on pulmonary disease.
Recommendation 17	Short-term and long-term studies should be undertaken to determine whether the effects of diesel exhaust are different from the effects of gasoline exhaust in regard to disease progression, the effects of age at exposure, and additional applied stress.

Human Studies
HIGH PRIORITY

Recommendation 5	Pulmonary mechanics should be assessed on excised lung specimens obtained from human autopsies and results related to data collected from live subjects. Morphological examinations of human autopsies could duplicate methods used to investigate the effects of tobacco smoke.
Recommendation 7	Some of the more detailed pulmonary function tests should be used to identify progressive dysfunction.

References

Arner, E. C., and Rhoades, R. A. 1973. Long-term nitrogen dioxide exposure, *Arch. Environ. Health* 26:156–160.

Baile, E. M., Wright, J. L., Pare, P. D., and Hogg, J. C. 1982. The effect of acute small airway inflammation on pulmonary function in dogs, *Am. Rev. Respir. Dis.* 126:298–301.

Barnhart, M. I., Salley, S. O., Chen, S.-T., and Puro, H. 1981. Morphometric ultrastructural analysis of alveolar lungs of guinea pigs chronically exposed by inhalation to diesel exhaust (DE), *Develop. Toxicol. Environ. Sci.* 10:183–200.

Bils, R. F. 1976. The connective tissues and alveolar walls in the lungs of normal and oxidant-exposed squirrel monkeys, *J. Cell Biol.* 70:318.

Brain, J. D., and Valberg, P. A. 1979. Deposition of aerosol in the respiratory tract, *Am. Rev. Respir. Dis.* 120:1325–1373.

Buell, G. C. 1970. Biochemical parameters in inhalation carcinogenesis, In: *Inhalation Carcinogenesis*

Correspondence should be addressed to J. L. Wright, Pulmonary Research Laboratory, St Paul's Hospital, 1081 Burrard St., Vancouver, British Columbia, Canada V6Z 1Y6.

(M. G. Hanna, P. Nettesheim, and J. R. Gilbert, eds.), pp. 209–225, USAEC Division of Technical Information, Oak Ridge, Tenn.

Buist, S., Vollmer, W., and Johnson, L. 1984. Does the single breath N_2 test identify the susceptible individual? *Chest* 85:10S.

Cabral-Anderson, L. J., Evans, M. J., and Freeman, G. 1977. Effects of NO_2 on the lungs of aging rats. I: Morphology, *Exp. Mol. Pathol.* 27:353–365.

Coffin, D. L., and Stokinger, H. E. 1977. Biological effects, In: *Air Pollution. Vol. II: The Effects of Air Pollution* (A. C. Stern, ed.), Academic Press, New York, N.Y.

Cosio, M., Ghezzo, H., Hogg, J. C., Corbin, R., Loveland, M., Dosman, J., and Macklem, P. T. 1977. The relations between structural changes in small airways and pulmonary function tests, *N. Engl. J. Med.* 298:1277–1281.

Cosio, M. G., Hale, K. A., and Niewoehner, D. E. 1980. Morphologic and morphometric effects of prolonged cigarette smoking on the small airways, *Am. Rev. Respir. Dis.* 122:265–271.

Davidson, J. T., Lillington, G. A., Haydon, G. B., and Wasserman, K. 1967. Physiologic changes in the lungs of rabbits continuously exposed to nitrogen dioxide, *Am. Rev. Respir. Dis.* 95:790–796.

Dunnill, M. S. 1962. Quantitative methods in the study of pulmonary pathology, *Thorax* 17:320–328.

Evans, M. J. 1984. Oxidant gases, *Environ. Health Perspect.* 55:85–95.

Evans, M. J., and Freeman, G. 1980. Morphological and pathological effects of NO_2 on the rat lung, In: *Nitrogen Oxides and Their Effects on Health* (S. D. Lee, ed.), pp. 243–264, Ann Arbor Science Publishers, Ann Arbor, Mich.

Evans, M. J., Cabral, L. J., Stephens, R. J., and Freeman, G. 1973. Cell division of alveolar macrophages in rat lung following exposure to NO_2, *Am. J. Pathol.* 70:199–208.

Evans, M. J., Cabral-Anderson, L. J., and Freeman, G. 1977. Effects of NO_2 on the lungs of aging rats. II: Cell proliferation, *Exp. Mol. Pathol.* 27:366–376.

Fedan, J. S., Frazer, D. G., Moorman, W. J., Attfield, M. D., Franczak, M. S., Kosten, C. J., Cahill, J. F., Lewis, T. R., and Green, F. H. Y. 1985. Effects of a two-year inhalation exposure of rats to coal dust and/or diesel exhaust on tension responses of isolated airway smooth muscle, *Am. Rev. Respir. Dis.* 131:651–655.

Fletcher, B. L., and Tappel, A. L. 1973. Protective effects of dietary alpha-tocopherol in rats exposed to toxic levels of ozone and nitrogen dioxide, *Environ. Res.* 6:165–175.

Fletcher, C., Peto, R., Tinken, C., and Speizer, F. E. 1976. *The Natural History of Chronic Bronchitis and Emphysema*, pp. 70–105, Oxford University Press, New York, N.Y.

Freeman, G., Crane, S. C., Stephens, R. J., and Furiosi, N. J. 1968. Pathogenesis of the nitrogen dioxide–induced lesion in the rat lung: a review and presentation of new observations, *Am. Rev. Respir. Dis.* 98:429–443.

Freeman, G., Crane, S. C., Furiosi, N. J., Stephens, R. J., Evans, M. J., and Moore, W. D. 1972. Covert reduction in ventilatory surface in rats during prolonged exposure to subacute nitrogen dioxide, *Am. Rev. Respir. Dis.* 106:563–579.

Furiosi, N. J., Crane, S. C., and Freeman, G. 1973. Mixed sodium chloride aerosol and nitrogen dioxide in air, *Arch. Environ. Health* 27:405–408.

Gail, D. B. and Lenfant, C. J. M. 1983. Cells of the lung: biology and clinical implications, *Am. Rev. Respir. Dis.* 127:366–387.

Gardner, D. E., Coffin, D. L., Pinigin, M. A., and Sidorenko, G. I. 1977. Role of time as a factor in the toxicity of chemical compounds in intermittent and continuous exposures. Part I: Effects of continuous exposure, *J. Toxicol. Environ. Health* 3:811–820.

Gardner, D. E., Miller, F. J., Blommer, E. J., and Coffin, D. L. 1979. Influence of exposure mode on the toxicity of NO_2, *Environ. Health Perspect.* 30:23–29.

Gillespie, J. R., Berry, J. B., White, L. L., and Lindsay, P. 1980. Effects of pulmonary function of low-level nitrogen dioxide exposure, In: *Nitrogen Oxides and Their Effects on Health* (S. D. Lee, ed.), pp. 231–241, Ann Arbor Science Publishers, Ann Arbor, Mich.

Giordano, A. M., and Morrow, P. E. 1972. Chronic low-level nitrogen dioxide exposure and mucociliary clearance, *Arch. Environ. Health* 25:443–449.

Goldstein, R. H. 1982. Response of the aging hamster lung to elastase injury, *Am. Rev. Respir. Dis.* 125:295–298.

Gregory, R. E., Pickrell, J. A., Hahn, F. F., and Hobbs, C. H. 1983. Pulmonary effects of intermittent subacute exposure to low-level nitrogen dioxide, *J. Toxicol. Environ. Health* 11:405–414.

Hacker, A. D., El Sayed, N., Mustafa, M. G., Ospital, J. J., and Lee, S. D. 1976. Effects of short-term nitrogen dioxide exposure on lung collagen synthesis, *Am. Rev. Respir. Dis.* 113:107.

Hammond, E. C., Kirman, D., and Garfinkel, I. 1970. Effects of cigarette smoking on dogs, *Arch. Environ. Health* 21:740–753.

Heckman, C. A., and Dalbey, W. E. 1982. Pathogenesis of lesions induced in rat lung by chronic tobacco smoke inhalation, *J. Nat. Cancer Inst.* 69:117–129.

Hogg, J. C., Macklem, P. T., and Thurlbeck, W. M. 1968. Site and nature of airway obstruction in chronic obstructive lung disease, *N. Engl. J. Med.* 278:1355–1360.

Hoidal, J. R., and Niewoehner, D. E. 1982. Lung phagocyte recruitment and metabolic alterations induced by cigarette smoke in humans and in hamsters, *Am. Rev. Respir. Dis.* 126:548–552.

Hoidal, J. R., and Niewoehner, D. E. 1983. Cigarette smoke inhalation potentiates elastase-induced emphysema in hamsters, *Am. Rev. Respir. Dis.* 127:478–481.

Horsfield, K. 1976. Lung morphology, *Recent Adv. Respir. Med.* 1:123–154.

Huber, G. L., Davies, P., Zwilling, G. R., Pochay, V. E., Hinds, W. C., Hicholas, V. K., Mahajan, V. K., Hayashi, M., and First, M. W. 1981. A morphological and physiologic bioassay for quantifying alterations in the lung following experimental chronic inhalation of tobacco smoke, *Bull. Eur. Physiopathol. Respir.* 17:269–327.

Hunninghake, G. W., and Crystal, R. G. 1983. Cigarette smoking and lung destruction: accumulation of neutrophils in the centri-lobular form of hypertrophic emphysema and its relation to chronic bronchitis, *Thorax* 12:219–235.

Hyde, D., Orthoefer, J., Dungworth, D., Tyler, W., Carter, R., and Lum, H. 1978. Morphometric and morphologic evaluation of pulmonary lesions in beagle dogs chronically exposed to high ambient levels of air pollutants, *Lab. Invest.* 38:455–469.

Hyde, D. M., Plopper, C. G., Weir, A. J., Murname, R. D., Warren, D. L., Last, J. A., and Pepelko, W. E. 1985. Peribronchiolar fibrosis in lungs of cats chronically exposed to diesel exhaust, *Lab. Invest.* 52:192–206.

Janoff, A. 1985. Elastases and emphysema: current assessment of the proteinase-antiproteinase hypothesis, *Am. Rev. Respir. Dis.* 132:417–433.

Karlinsky, J., Fredette, J., Davidovits, G., Catanese, A., Snider, R., Faris, B., Snider, G. L., and Franzblau, C. 1983. The balance of lung connective tissue elements in elastase-induced emphysema, *J. Lab. Clin. Med.* 102:151–162.

Kelley, J., Hemenway, D., and Evans, J. N. 1986. Hydroxylysine as a marker of lung injury in nitrogen dioxide exposure, *Am. Rev. Respir. Dis.* 133:A86.

Kleinerman, J. 1979. Effects of nitrogen dioxide on elastin and collagen contents of lung, *Arch. Environ. Health* 34:228–232.

Kleinerman, J., and Niewoehner, D. 1973. Physiological, pathological and morphometric studies of long term nitrogen dioxide exposures and recovery in hamsters, *Am. Rev. Respir. Dis.* 107:1081.

Kleinerman, J., Rynbrandt, R., and Sorensen, J. 1976. Chronic obstructive airways disease in cats produced by NO$_2$, *Am. Rev. Respir. Dis.* 113:107.

Kuhn, C., Shiu-Yeh, Y., Chraplyvy, M., Linder, H. E., and Senior, R. 1976. The induction of emphysema with elastase, *Lab. Invest.* 34:372–380.

Lam, C., Kattan, M., Collins, A., and Kleinerman, J. 1983. Long-term sequelae of bronchiolitis induced by nitrogen dioxide in hamsters, *Am. Rev. Respir. Dis.* 129:1020–1023.

Lewis, T. R., Moorman, W. J., Yang, Y.–Y., and Stara, J. F. 1974. Long-term exposure to auto exhaust and other pollutant mixtures: effects on pulmonary function in the beagle, *Arch. Environ. Health* 29:102–106.

Mauderly, J. L. 1984. Respiratory function responses of animals and man to oxidant gases and to pulmonary emphysema, *J. Toxicol. Environ. Health* 13:345–361.

Mauderly, J. L., Benson, J. M., Bice, D. E., Carpenter, R. L., Evans, M. J., Henderson, R. F., Jones, R. K., McClellan, R. O., Pickrell, J. A., Redman, H. C., Shami, S. G., and Wolff, R. F. 1983. Inhalation Toxicology Research Institute Annual Report, pp. 305–316, LMF-107 Inhalation Toxicology Research Institute, Lovelace Biomedical and Environmental Research Institute, Albuquerque, N.M.

Mauderly, J. L., Bice, D. E., Gillett, N. A., Henderson, R. F., and Wolff, R. K. 1986. Differences in the responses of developing and adult rats to inhaled diesel exhaust, *Am. Rev. Respir. Dis.* 133:A84.

Menzel, D. B. 1980. Pharmacological mechanisms in the toxicity of nitrogen dioxide and its relation to obstructive respiratory disease, In: *Nitrogen Oxides and Their Effects on Health* (S. D. Lee, ed.), pp. 199–216, Ann Arbor Science Publishers, Ann Arbor, Mich.

Mitchell, R. S., Stanford, R. E., Johnson, J. M., Silvers, G. W., Dart, G., George, M. S. 1976. The morphologic features of the bronchi, bronchioles and alveoli in chronic airways obstruction: a clinic pathological study, *Am. Rev. Respir. Dis.* 114:137–145.

Mustafa, M. G., Faeder, E. J., and Lee, S. D. 1980. Biochemical effects of nitrogen dioxide on animal lungs, In: *Nitrogen Oxides and Their Effects on Health* (S. D. Lee, ed.), pp. 161–171, Ann Arbor Science Publishers, Ann Arbor, Mich.

Mustafa, M. G., El Sayed, N. M., von Dohlen, F. M., Hassett, C. M., Postlethwait, E. M., Quinn, C. L., Graham, J. A., and Gardner, D. E. 1984. A comparison of biochemical effects of nitrogen dioxide, ozone, and their combination in mouse lung, Toxicol. Appl. Pharmacol. 72:82–90.

Nagaishi, C., Nagasawa, N., Yamashita, M., Okada, Y., and Inoba, N. 1972. Functional Anatomy and Histology of the Lung, University Park Press, Baltimore, Md.

Nakajima, T., Oda, H., Kusumoto, S., and Nogami, H. 1980. Biological effects of nitrogen dioxide and nitric oxide, In: *Nitrogen Oxides and Their Effects on Health* (S. D. Lee, ed.), pp. 121–139, Ann Arbor Science Publishers, Ann Arbor, Mich.

Niewoehner, D. E., and Hoidal, J. R. 1982. Lung fibrosis and emphysema: divergent responses to a common injury, *Science* 217:359–360.

Niewoehner, D. E., Kleinerman, J., and Rice, D. P. 1974. Pathologic changes in the peripheral airways of young cigarette smokers, *N. Engl. J. Med.* 291:755–758.

Orthoefer, J. G., Bhatnagar, R. S., Rahman, A., Yang, Y. Y., Lee, S. D., and Stara, J. F. 1976. Collagen and prolyl hydroxylase levels in lungs of beagles exposed to air pollutants, *Environ. Res.* 12:299–305.

Pepelko, W. E. 1981. EPA studies on the toxicological effects of inhaled diesel engine emissions, *Develop. Toxicol. Environ. Sci.* 10:121-142.

Peters, S. G., and Hyatt, R. E. 1986. A canine model of bronchial injury induced by nitric acid, *Am. Rev. Respir. Dis.* 133:1049–1054.

Plopper, C. G., Hyde, D. M., Weir, A. J. 1983. Centriacinar alterations in lungs of cats chronically exposed to diesel exhaust, *Lab. Invest.* 49:391–399.

Port, C. D., Ketels, K. V., Coffin, D. L., and Kane, P. 1977. A comparative study of experimental and spontaneous emphysema, *J. Toxicol. Environ. Health* 2:589–604.

Raub, J. A., Mercer, R. R., Miller, F. J., Graham, J. A., O'Neil, J. J. 1982. Dose response of elastase-induced emphysema in hamsters, *Am. Rev. Respir. Dis.* 125:432–435.

Reid, L. 1960. Measurement of the bronchial mucous gland layer: a diagnostic yardstick in chronic bronchitis, *Thorax* 15:132–141.

Rynbrandt, D., and Kleinerman, J. 1977. Nitrogen dioxide and pulmonary proteolytic enzymes: effect on lung tissue and macrophages, *Arch. Environ. Health* 32:165–172.

Saetta, M., Shiner, R. J., Angus, G. E., Kim, W. D., Wang, N.-S., King, M., Ghezzo, H., and Cosio, M. G. 1985a. Destructive index: a measurement of lung parenchymal destruction in smokers, *Am. Rev. Respir. Dis.* 131:764–769.

Saetta, M., Ghezzo, H., Kim, W. D., King, M., Angus, G. E., Wang, N.-S., and Cosio, M. G. 1985b. Loss of alveolar attachments in smokers: a morphometric correlate of lung function impairment, *Am. Rev. Respir. Dis.* 132:894–900.

Selgrade, M. K., Mole, M. L., Miller, F. J., Hatch, G. E., Gardner, D. E., and Hu, P. C. 1981. Effect of NO$_2$ inhalation and vitamin C deficiency on protein and lipid accumulation in the lung, *Environ. Res.* 6:422–437.

Snider, G. L., Kleinerman, J., Thurlbeck, W. M., and Bengali, Z. H. 1985. The definition of emphysema: report of a National Heart, Lung, and Blood Institute, Division of Lung Diseases workshop, *Am. Rev. Respir. Dis.* 132:182–185.

Sparrow, D., Glynn, R. J., Cohen, M., and Weiss,

S. T. 1984. The relationship of the peripheral leukocyte count and cigarette smoking to pulmonary
function among adult men, *Chest* 86:383–386.

Stara, J. F., Dungworth, D. L., Orthoefer, J. G., and
Tyler, W. S. 1980. Long-Term Effects of Air Pollutants: In Canine Species, series no. 8, Environmental Protection Agency, Office of Research
and Development, EPA-600/8–80-014, Cincinnati,
Ohio.

Stephens, R. J., Freeman, G., and Evans, M. J. 1971.
Ultrastructural changes in connective tissue in lungs
of rats exposed to NO_2, *Arch. Intern. Med.*
127:873–883.

Thurlbeck, W., Dunnil, M. S., and Hurtung, W.
1970. A comparison of other methods of measuring
emphysema, *Human. Pathol.* 1:215–226.

Turino, G. M. 1985. The lung parenchyma—a dynamic matrix, *Am. Rev. Respir. Dis.* 132:1324–1334.

Vaughan, T. R., Jr., Jennelle, L. F., and Lewis, T. R.
1969. Long-term exposure to low levels of air
pollutants: effects on pulmonary function in the
beagle, *Arch. Environ. Health* 19:45–50.

Vostal, J. J., White, H. J., Strom, K. A., Siak, J.-S.,
Chen, K.-C., and Dziedzic, D. 1981. Response of
the pulmonary defense system to diesel particulate
exposure: toxicological effects of emissions from
diesel engines, *Develop. Toxicol. Environ. Sci.*
10:201–221.

Wiester, M. J., Iltis, R., and Moore, W. 1980. Altered
function and histology in guinea pigs after inhalation of diesel exhaust, *Environ. Res.* 22:285–297.

Wright, J. L., Lawson, L. M., Pare, P. D., Kennedy,
S., Wiggs, B., and Hogg, J. C. 1984. The detection
of small airways disease, *Am. Rev. Respir. Dis.*
129:989–994.

Wright, J. L., Cosio, M., Wiggs, B., and Hogg, J. C.
1985. A morphologic grading scheme for membranous and respiratory bronchioles, *Arch. Pathol. Lab.
Med.* 109:163–165.

Asthma and Automotive Emissions

PHILIP A. BROMBERG
University of North Carolina

Air Pollution, the Automobile, and Public Health. © 1988 by the Health Effects
Institute. National Academy Press, Washington, D.C.

Urban air pollution is characterized by increased levels of ozone (O_3) and nitrogen dioxide (NO_2) resulting from photochemical reactions of automotive emissions. Lung damage caused by exposure to oxidant gases is well recognized. For instance, inhalation of 50–100 parts per million (ppm) NO_2 causes acute pulmonary edema in humans, and survivors may develop a chronic progressive obstructive bronchiolitis several weeks after apparent recovery from the acute edema. O_3 is also capable of producing acute pulmonary edema at a concentration of less than 10 ppm.

Levels of NO_2 and O_3 in ambient air generally do not exceed 1 ppm. Therefore, the toxic effects of inhalation of high levels of these gases may not be relevant for low-level ambient exposures. Nevertheless, several population surveys have found a relation between episodes of oxidant air pollution and adverse respiratory health effects including exacerbations of asthma. More chronic effects of oxidant air pollution on respiratory function have also been suggested (see Bresnitz and Rest, this volume; U.S. Environmental Protection Agency 1986).

The notion that individuals with preexisting respiratory disease might be especially susceptible to effects of ambient air pollution has been strengthened by the relatively recent observation that, unlike normal subjects, asthmatic individuals with mild disease are highly susceptible to the experimental inhalation of low concentrations of sulfur dioxide (SO_2) during exercise. During brief exposures, asthmatic subjects have developed acute, sometimes symptomatic, increases in specific airflow resistance (SR_{aw}) attributed to bronchoconstriction (Sheppard et al. 1981; Bethel et al. 1983).

In this chapter, the clinical forms of asthma and their pathogenesis are described, and the results from controlled exposure studies of the effects of inhalation of pollutant gases relevant to automotive emissions on normal and asthmatic subjects are summarized. These findings are discussed with respect to the mechanisms of asthma, and key questions about possible interactions between asthma and automotive air pollution whose answers have so far been elusive are formulated.

Asthma: Definition, Demography, and Clinical Spectrum

Definition

Asthma is a common paroxysmal disorder of the airways but is difficult to define because of its complex pathogenesis, multiple etiologic factors, and clinical overlap with other airway diseases. It is characterized by recurrent episodes of diffuse airways obstruction associated with wheezing, coughing, and breathlessness (dyspnea). These episodes are largely reversible, especially with pharmacological therapy, but anatomic and physiological evidence of persistent diffuse airways disease can be found on careful investigation even during apparently quiescent periods. One important indication of persistent abnormality, even in asymptomatic asthmatics, is nonspecific bronchial hyperreactivity, that is, enhanced airways response to bronchoconstrictive stimuli that are neither specific antigens operating through immunologic mechanisms nor specific chemicals in the occupational environment (for example, toluene diisocyanate).

Demography

A 1970 U.S. Public Health Service survey of national health suggested that the prevalence of asthma was about 3 percent (Wilder 1973). Regional population surveys in Tecumseh, Michigan (Broder et al. 1962, 1974a,b), and Tucson, Arizona (Lebowitz et al. 1975), suggest a prevalence of 4–8 percent. In the Public Health Service survey, three-fifths of the asthmatics had consulted a physician for asthma during the previous year, one-fifth had visited their physician at least five times during the year, and one-sixth felt the disease limited their activity (Bonner 1984).

Prevalence is highest among children (males $\gg$ females) and a second peak occurs in older adults (males $>$ females). This later peak may include patients with relatively irreversible airways obstruction. In young to middle-aged adults, women may be somewhat more at risk than men. In young children, asthma (wheezing) associated with viral respiratory tract infections is a common problem. This association (Busse 1985) and the demonstration of IgE-type antibody, particularly to respiratory syncytial virus, in such patients (Welliver et al. 1986) suggest that respiratory tract infections can provoke an asthmatic state in some children. Of the various "types" of asthma, allergen-related (extrinsic) asthma occurs more commonly in somewhat older children and in young adults than does intrinsic (no identifiable allergen or specific antibody) asthma (Bonner 1984). Intrinsic asthma increases in importance in older adults and is a form of the disease that can manifest increasingly irreversible airways obstruction even in the face of chronic treatment with large doses of corticosteroids.

Clinical Spectrum

The clinical severity of asthma (either the background level of airways obstruction or the intensity and the frequency of the episodes) varies greatly, not only among affected individuals, but at different times in the same individual. Some asthmatic children seem to grow out of their disease. On the other hand, relatively asymptomatic asthma can flare so severely that it becomes life-threatening or even fatal. This variability is reflected in the intensity of therapy and the number of physician contacts. Some asthmatics use only an occasional inhalation of a bronchodilator aerosol; others use aerosols more frequently, either as required or on a regular basis. Other asthmatics use long-acting oral bronchodilators supplemented with one or more inhaled agents. Some affected individuals require intermittent or chronic corticosteroid therapy. Life-threatening, severe episodes require hospitalization and often involve a period of assisted ventilation.

Nonspecific Bronchial Hyperreactivity.

Although patients with recently developed asthma may have normal airway reactivity between attacks, the presence of nonspecific bronchial hyperreactivity is considered an important, but not specific, feature of asthma (Hargreave et al. 1981). Various stimuli have been used to elicit a bronchoconstrictive response. Quantitative inhalation challenge with aerosolized histamine and methacholine has gained widespread acceptance and use as an investigative and diagnostic tool. Individual reactivities to histamine and methacholine challenge correlate well with the presence of clinically severe asthma as judged by degree of symptoms, medication requirements, occurrence of spontaneous early morning increase in airflow obstruction, presence of more-or-less continuous airflow obstruction, and general instability of the airways despite chronic use of multiple bronchodilator medications. Such unstable patients with exquisite nonspecific bronchial hyperreactivity are at risk for life-threatening asthmatic crises. A sharp decrease in nonspecific bronchial hyperreactivity correlates with clinical improvement (Hargreave et al. 1985b; Woolcock et al. 1985).

Inhalation of cold dry air is also commonly used as a bronchoconstrictive stimulus, but its quantitation is more difficult, its mode of action is less well understood than that of specific drugs, and its effects are subject to the development of decreased response on repeated application (Hargreave et al. 1985a).

Nonspecific bronchial hyperreactivity to various challenges is an index of the intrinsic responsiveness of elements in the airway wall (smooth muscle, submucosal blood vessels and glands, afferent nerve endings and efferent nerve ganglia, mast cells, and possibly neuroendocrine cells), and presumably reflects or mimics certain events in "naturally occurring" asthma. Of course, multiple mediators are involved in asthma. Even with respect to histamine or methacholine, the naturally occurring release of these autacoids probably causes changes in local concentrations at receptor sites that are not precisely mimicked by inhalation challenge.

Other Obstructive Airways Disease. Diffuse airways disease associated with the impairment of forced expiratory flow rates

and increased airflow resistance is not limited to asthma. In addition to at least 6 million asthmatics, it has been estimated that 7.5 million people in the United States have chronic bronchitis and 2 million have emphysema (U.S. Department of Health and Human Services 1979). Etiologies vary among the obstructive airways diseases. Cigarette smoking is a major risk factor for chronic nonspecific bronchitis as well as for emphysema. Chronic airways infection with suppurative destruction of airway walls characterizes diffuse bronchiectasis. In some cases, a specific underlying reason, such as dysmotile cilia syndromes, immunoglobin deficiencies, autoimmune diseases, cystic fibrosis, or hereditary deficiency of the α-1-proteinase inhibitor can be identified. In others, the etiology is obscure and may be related to remote diffuse lung injury or infection occurring in childhood.

Chronic airways obstruction generally is much less reversible in these diseases than in asthma, but to varying degrees. Airways obstruction associated with cystic fibrosis frequently manifests considerable reversibility in response to bronchodilators. Obstruction is partially reversible with β-adrenergic or muscarinic agonists in some patients with chronic bronchitis. Airways hyperreactivity to bronchoconstrictor drugs also occurs in some chronic bronchitics. Unusual reactivity to bronchodilators or bronchoconstrictors may be associated with relatively rapid deterioration of pulmonary function in chronic bronchitis (Barter and Campbell 1976).

An important unresolved question is whether there are specific risk factors for the development of irreversible obstructive airways disease in some asthmatics. In view of the extensive inflammation and epithelial damage that appears to be characteristic of asthma (see below), it is surprising that most patients with asthma retain the feature of substantial reversibility of airways obstruction for long periods.

Patients with chronic obstructive pulmonary disease (COPD) and increased airways reactivity may not be considered "asthmatics," but they should not be neglected in studies of the health effects of air pollution. Such disease groups would include patients with chronic nonspecific bronchitis with reactive airways, cystic fibrosis, and children with a history of bronchopulmonary dysplasia or of respiratory infections associated with wheezing.

Asthma: Pathogenesis

Extrinsic and Intrinsic Asthma

The variability of the clinical features among different asthmatics and in a given individual at different times or periods of life has prompted efforts to classify the disease in a meaningful fashion. One approach differentiates extrinsic asthma, in which antigens (allergens) affect tissue cells—primarily mast cells sensitized by specific IgE antibody (see Chemical Mediators)—from intrinsic asthma, which occurs without an identifiable antigen or a specific IgE antibody.

Thus, extrinsic asthma occurs commonly among allergic (atopic) individuals who have (usually) multiple specific IgE-class antibodies to airborne antigens associated with pollens, mold spores, animal fur, or various insects (for example, mites found in house dust, cockroaches). The proclivity toward the development of IgE antibodies against common inhaled antigens (allergens) characterizes atopic individuals and is importantly influenced by genetic factors. Asthma exacerbations are provoked by inhaled antigen (for example, laboratory workers handling certain animals or insects) and tend, in the case of pollen and spore allergy, to be seasonal and regional.

In many atopic individuals, clinical manifestations are limited to the skin or the nasal mucosa (allergic rhinitis) and the conjunctivae (allergic conjunctivitis). These patients often have some symptoms compatible with asthma that are sufficiently mild to go unrecognized. They also may have a modest increase in nonspecific bronchial reactivity to histamine or methacholine. Indeed, Fish and Norman (1985) have found substantial hyperreactivity to inhaled prostaglandin (PG) $F_{2\alpha}$ in atopic, nonasthmatic subjects. Atopic persons may constitute a useful population for study of the effects of exposure to air pollutants since the nasal epithelium is a good model for the large-airways epithelium. Basic immunologic mechanisms in the allergic bronchial mucosa are also present in the nasal mucosa, which is readily accessible to experimental manipulation as well as to environmental exposures.

Intrinsic asthma implies the absence of atopy, that is, no evidence of specific IgE-class antibodies against common airborne antigens in the patient's environment. These asthmatics have perennial rather than seasonal symptoms. Exacerbations seem to be triggered by poorly characterized respiratory tract infections, which on the basis of serologic evidence and in the absence of a bacterial infecting agent, are assumed to be viral. A subclass of these individuals has nasal polyps and demonstrates an unusual proclivity to precipitation of asthma shortly after ingesting acetylsalicylic acid (aspirin) or other nonsteroidal antiinflammatory agents—all of which inhibit cyclooxygenase activity and therefore affect cellular arachidonate metabolism.

Asthma also can develop in workers (probably nonatopic) exposed to chemicals such as toluene diisocyanate, plicatic acid (western red cedar wood dust), and others. Exquisite airways sensitivity to inhalation of the specific offending agent is often present, although specific antibodies may not be demonstrable. Some of these individuals develop chronic perennial asthma despite prolonged avoidance of the sensitizing chemical. Such cases of asthma do not fit clearly into either the extrinsic or the intrinsic category.

Early and Late Reactions to Allergens

Recent descriptions of asthma pathophysiology emphasize the discovery that tissues bearing IgE-sensitized mast cells, such as skin, nasal mucosa, and airways, display a complex series of responses to specific antigen (Gleich 1982; Cockcroft 1983). A rapid, "early" response is dominated by local microvascular dilatation, edema, and in the airways, smooth-muscle contraction, all of which are mediated by substances released from sensitized mast cells as well as other sensitized cells. This reaction regresses spontaneously, or regression can be induced with β-adrenergic agonists.

Several hours later, a second local reaction occurs, again associated with bronchospasm in the airway, but now with substantial cellular infiltration by eosinophils, neutrophils, and basophils. These cells are thought to be attracted by chemotactic factors released by the activated mast cells. For reasons that are not clear, this "late" reaction is not reversed by β-adrenergic agonists but can be diminished or prevented by pretreatment with corticosteroids.

Bronchial Hyperreactivity

Late reactions following antigen challenge in asthmatic subjects are associated with enhancement of nonspecific bronchial hyperreactivity. Marked enhancement of bronchial hyperreactivity may last for weeks and is associated with the emergence of more severe clinical manifestations of asthma. The patient develops bronchoconstriction on exposure to a wide variety of stimuli. If no further exposure to the provoking agent (antigen or chemical) occurs, the bronchial hyperreactivity gradually recedes to baseline levels along with the clinical symptoms. The intensity of the late response to antigen correlates with the subsequent degree and duration of increased nonspecific bronchial hyperreactivity (Cartier et al. 1982; Cockcroft 1983).

Airways inflammation constitutes a central feature of both the late reaction to specific antigens and the genesis of nonspecific bronchial hyperreactivity. The specific features of the inflammatory process required for the development of hyperreactivity are far from understood. For example, among patients with chronic bronchitis, only a fraction appear to exhibit nonspecific bronchial hyperreactivity although all presumably have some degree of airways inflammation.

The prolonged state of bronchial hyperreactivity provoked by a single exposure to a specific antigen or to certain occupationally related chemicals suggests the possibility that the airways of so-called intrinsic asthmatics may be sensitized to nonspecific stimuli in a similar fashion by intermittent exposure to some undefined specific agent(s).*

The intensity of the bronchoobstructive response of a sensitized individual exposed to a particular antigen will therefore depend not only on the specific IgE-mediated discharge of mediators from local mast cells (and possibly other cells), but also on the degree of coexistent nonspecific bronchial reactivity (Cockcroft et al. 1979). A corollary to this notion is that any stimulus (for example, O_3) that increases nonspecific bronchial reactivity should also enhance reaction to inhaled antigen, even in the absence of any change in immunologic status. Furthermore, to the extent that air pollutants can provoke bronchoconstriction in a "nonspecific" manner, highly bronchoreactive asthmatics should be more sensitive to such exposures.

Controlled Exposure Studies

While field studies and practical experience indicate an association between air pollution and exacerbation of asthma, confirmation of the association and its physiological manifestations have been sought in controlled studies. The following section assesses the status of this research to date.

Normal Subjects

Ozone and Nitrogen Dioxide. Although NO_2 and O_3 are both classified as oxidants, and are relatively insoluble in aqueous media and thus able to penetrate into the lower

*Note, however, asthmatic children who are perennially exposed to the relevant antigen, and who receive repetitive antigen injections in hyposensitization therapy, tend to outgrow their asthma and show a decrease in IgE and IgG antibodies. In contrast, children who are intermittently (seasonally) exposed to another allergen, such as rye-grass pollen, continue to demonstrate seasonal asthma associated with rises in IgG and IgE antibodies (Hill et al. 1981).

airways, their mechanisms of action on the airways may well be quite different.

At concentrations <1 ppm, effects of acute NO_2 inhalation on the respiratory system of normal subjects have not been demonstrated. In contrast, acute (2-hr) exposure of exercising adults to levels of O_3 as low as 0.12 ppm causes a reduction of mean vital capacity accompanied by cough (McDonnell et al. 1983). Similar findings, which can be accounted for specifically by the O_3 content of the air, have resulted from exposure of exercising adults to Los Angeles air pollution under controlled conditions (Avol et al. 1984), in active children exposed to about 0.10 ppm O_3 in a summer camp (Lippman et al. 1983), and in children exposed to 0.12 ppm O_3 in an environmental chamber (McDonnell et al. 1985a).

Decreases in vital capacity do not appear to be caused by changes in mechanical properties of the lung (Beckett et al. 1985; Hazucha et al. 1986), but are probably due to neurally mediated involuntary inhibition of inspiration. Bronchoconstriction is not a prominent feature of the response of normal individuals to O_3; nor does the degree of bronchoconstriction correlate with the vital capacity decrease (McDonnell et al. 1983). In addition, inhaled atropine failed to prevent O_3-induced changes in vital capacity (Beckett et al. 1985). Unmyelinated airway sensory nerves (C-fibers), which contain and can release substance P and other physiologically active neuropeptides, could mediate the inhibition of inspiration, the subjective airway sensations, and a more shallow, rapid pattern of breathing. Since the stimulation of airway C-fibers in dogs also causes reflex bronchoconstriction (Roberts et al. 1981; Coleridge and Coleridge 1986) as well as reflex tracheal gland secretion (Davis et al. 1982), the relative weakness of the human bronchoconstriction response to inhaled O_3 is surprising. Substance P or other neuropeptides may also provoke other features of the acute airways response to O_3, such as mucous secretion (Coles et al. 1984) and increased epithelial permeability and ion transport (Al-Bazzaz et al. 1985).

Among a group of normal subjects, there is a substantial range of vital capacity responses to a given O_3 exposure (McDonnell et al. 1983). In a single individual, however, the response to O_3 exposure is relatively reproducible (McDonnell et al. 1985b). Whether this between-subject variability is attributable to different tissue doses of O_3 in different individuals under the same exposure conditions or to biological factors affecting individual responsiveness remains to be determined.

A feature of repeated daily experimental exposures to O_3 is the initial enhancement (Hackney et al. 1977; Farrell et al. 1979; Bedi et al. 1985; Folinsbee and Horvath 1986) but eventual disappearance of the vital capacity response and the associated subjective sensations (Farrell et al. 1979; Folinsbee et al. 1980; Horvath et al. 1981). Among the possible explanations for this phenomenon (often termed "tolerance" or "adaptation") is the depletion of neuropeptides from repeatedly stimulated C-fibers.

A more subtle feature of the response to O_3, first described 20 years ago in guinea pigs (Easton and Murphy 1967), and then shown in dogs (Lee et al. 1977), sheep (Abraham et al. 1980), and humans (Golden et al. 1978), is a transient increase, lasting hours to perhaps a day in humans, in the bronchoconstrictive response to parenteral (Easton and Murphy 1967; Gordon and Amdur 1980; Gordon et al. 1984; Murlas and Roum 1985a) as well as inhaled histamine and cholinergic agonists (Lee et al. 1977; Golden et al. 1978; Holtzman et al. 1979; DiMeo et al. 1981; Holtzman et al. 1983; Roum and Murlas 1984). This response also appears to undergo an adaptive suppression on repeated O_3 exposure (DiMeo et al. 1981).

Mixtures of Ozone and Other Pollutants.

Synergy between O_3 and other pollutants in causing respiratory effects during environmental chamber exposures has not been proven. Hazucha and Bates (1975) described synergism between SO_2 and O_3 in normal subjects, but other researchers were not able to duplicate their results (Bell et al. 1977; Bedi et al. 1979; Kleinman et al. 1981; Folinsbee et al. 1985). Stacy and coworkers (1983) noted increased mean changes of respiratory parameters when

low concentrations of acid aerosols were mixed with 0.4 ppm O_3, but these differences were not statistically significant. Using a sequential rather than simultaneous exposure protocol, Kulle and colleagues (1982) found no effect of sulfuric acid (H_2SO_4) aerosol (100 $\mu g/m^3$) following an exposure to 0.3 ppm O_3.

■ **Recommendation 1.** The uptake profile of pollutant gases in different regions of the airways should be measured by sampling and analyzing inspired air at different airway levels. Variables to be explored with such systems include concentration of pollutants, ventilatory parameters, duration of exposure, and presence of disease (for example, asthma, chronic bronchitis).

■ **Recommendation 2.** These data should be compared with the predictions of currently available mathematical models.

■ **Recommendation 3.** The uptake profiles should be examined for their ability to account for some of the variability in vital capacity response to O_3 exposure observed among individuals.

Asthmatic Subjects

Ozone. Controlled studies have yet to demonstrate that O_3 dramatically affects lung function in asthmatic subjects, atopic nonasthmatic subjects, or patients with COPD. The most pertinent publications on asthmatics (Linn et al. 1978, 1980; Silverman 1979; Koenig et al. 1985), persons with smoking-related COPD (Linn et al. 1982, 1983; Solic et al. 1982; Hackney et al. 1983; Kehrl et al. 1983, 1985; Kulle et al. 1984), and atopic subjects (Holtzman et al. 1979) have been reviewed in the U.S. Environmental Protection Agency's (1986) most recent revision of the air quality criteria document for O_3 and other photochemical oxidants.

Unfortunately, deficiencies in the experimental design in some of these studies preclude making a definitive statement about the effects of O_3 on pulmonary function in asthmatic subjects. Among these deficiencies are incomplete characterization of the nature of the obstructive airways process, little information on nonspecific bronchial reactivity status and no testing of the effect of exposure on bronchial reactivity, absence of airflow resistance (R_{aw}) measurements to assess airways response, no control of the use of medications by asthmatic subjects, possible investigator avoidance of subjects with severe asthma, and inadequate levels of ventilation (exercise) during the exposure (this was a critical variable in the demonstration of exquisite responsiveness of asthmatics to SO_2 exposure). Furthermore, no studies of asthmatic subjects appear to have been performed using prolonged single exposures or repeated daily exposures. Nor has possible synergism between O_3 and other relevant pollutants been studied in concurrent or sequential exposure protocols.

Nitrogen Dioxide. Orehek and coworkers (1976) were the first to claim that some asthmatics develop increased airways reactivity to inhaled bronchoconstrictor drug challenge after exposure to only 0.1 ppm NO_2. An editorial by Dawson and Schenker (1979) provides a succinct evaluation of our understanding of the effects of NO_2 inhalation as of the late 1970s. Since then, other groups have explored the effect of low levels of NO_2 (<0.5 ppm) on lung function and bronchial reactivity in asthmatics.

Koenig and colleagues (1985) examined atopic asthmatic adolescents exposed to 0.12 ppm NO_2 by mouthpiece for 1 hr at rest and found no changes in lung function during or after the exposure. Other researchers (Roger et al. 1985; Bauer et al. 1986) independently found evidence that bronchoconstriction in asthmatics following exercise was enhanced in the presence of 0.3 ppm NO_2. In contrast, Linn and Hackney (1984) and Linn and others (1985) observed no effect on SR_{aw} of exposure to 4 ppm NO_2 (!) for 75 min with light or heavy exercise. The route of inhalation (oronasal versus oral), the duration of exposure, and the number as well as intensity of exercise stints may be significant variables (Kulle 1982; Bauer et al. 1985). Bylin and colleagues (1985) found enhanced nonspecific bronchial reactivity to aerosolized histamine in asthmatics exposed to 0.5 ppm

NO_2 for 20 min. Kleinman and coworkers (1983) found no acute effect on respiratory mechanics after a 2-hr exposure, which included intermittent light exercise, to 0.2 ppm NO_2. The enhancement of nonspecific bronchial reactivity to aerosolized methacholine was, however, borderline in terms of group means. Ahmed and colleagues (1982) reported increased airways reactivity to aerosolized carbachol in resting normal subjects as well as in asthmatics following a 1-hr exposure to 0.1 ppm NO_2, despite the absence of any effects on baseline lung function. However, in a careful study, Hazucha and coworkers (1983) were unable to confirm the Orehek finding of increased airways reactivity following a 0.1-ppm NO_2 exposure. In addition, Roger and colleagues (1986) failed to demonstrate enhanced methacholine reactivity following exposure of exercising asthmatics to NO_2 levels as high as 0.6 ppm. Finally, Ahmed and colleagues (1983) challenged ragweed-sensitive asthmatic subjects with specific antigen immediately and 24 hr after a 1-hr exposure to 0.1 ppm. No effect of the exposure on specific bronchial reactivity was found, but the exposure conditions were extremely mild. None of the other studies cited above appear to have examined *specific* bronchial reactivity.

Evidence suggests that the airways of exercising and perhaps even resting asthmatic subjects are affected by exposure to <0.5 ppm NO_2. However, when all the reports are considered, a coherent picture fails to emerge. Even within a particular laboratory, there may be difficulty in reproducing an observation (D. Horstman, personal communication). The reasons for this inconsistency are presently obscure. Some of the problems mentioned for O_3 exposure studies may also apply to NO_2.

Need for Further Research

The preceding discussion suggests that, in contrast to the experience with SO_2, initial controlled exposure studies have failed to demonstrate that asthmatics exhibit unusual sensitivity to acute exposure to O_3. Although the situation is less clear with respect to NO_2, the effects claimed have been modest in extent. However, a conclusion that asthmatics do *not* constitute a sensitive subgroup in terms of possible adverse health effects of air pollution related to automotive emissions would be premature and possibly incorrect. Reasons for this posture have already been alluded to and include:

1. Two types of field studies suggest that asthmatics do exhibit special sensitivity to ambient oxidant air pollution. In one, "panels" of asthmatic and control subjects have been selected and their health status monitored in relation to ambient air composition over time; the best example is the analysis of Whittemore and Korn (1980). In the other, the respiratory health of an entire community is assessed in terms of emergency room visits, hospital admissions, and physician visits, and correlated with environmental air quality. The best study of that type is the survey of almost 6 million individuals in southern Ontario who receive medical care from a national health service with a computerized data base (Bates and Sizto 1983; Bates 1985).

2. The known bronchoconstrictive power of SO_2 and H_2SO_4 aerosols in asthmatics supports the possibility that experimental exposure to certain combinations of pollutants, including oxidants, could demonstrate that asthmatics constitute an oxidant-susceptible population. The work of Bates and Sizto (1983), and of others, in which environmental air composition is analyzed should be invaluable in designing the types of multiple pollutant exposures that could be tested in controlled exposure chambers.

3. Severe asthmatics and chronic bronchitics with reactive airways have undergone little or no systematic study.

4. As previously noted, the clinical severity of asthma is related to the degree of nonspecific bronchial hyperreactivity. Ozone inhalation reproducibly causes an enhancement of bronchial reactivity and therefore may increase clinical manifestations of an underlying asthmatic condition under the appropriate circumstances.

5. Despite inconsistencies, there is significant evidence suggesting increased sensitivity of asthmatics to the effects of NO_2 inhalation.

Additional support for the plausibility of significant interactions between oxidant air pollution and asthma may emerge from a more detailed comparison of the pathophysiology of asthma and of the effects of oxidants, primarily O_3 which has been studied more extensively than NO_2, on the airway tissues.

Before proceeding, however, it may be useful to anticipate some questions and issues relevant to the goals in this chapter.

In extrinsic asthma, inhaled antigen must gain access to antibody-sensitized cells in the airways, particularly the submucosal mast cells. In order to reach underlying mast cells or immunecompetent cells, antigens must penetrate the epithelial barrier. Acute O_3 exposure increases respiratory epithelial permeability and should enhance the ability of inhaled antigens to reach critical cells in the submucosa. Thus, acute or repeated oxidant pollutant exposures of sensitized individuals might modify the subsequent early and/or late airways responses to specific challenge. In addition, increased epithelial permeability would allow egress of submucosal albumin onto the airways surface where it could alter the viscoelastic properties of the surface liquid and impair mucociliary clearance.

Some important effector mechanisms in asthma involve stimulation of sensory nerves and neurally mediated reflexes; release of chemical mediators, including arachidonate metabolites, from mast cells and possibly other cells; recruitment of inflammatory cells to the airways; and damage to airways epithelium. Ozone is known to have all of these effects. In addition, some extracellular defense mechanisms against proteolytic enzymes, for example, α-1-proteinase inhibitor, are oxidant sensitive. By causing such overlapping effects in the same tissue, short-term oxidant exposure could lead to some acute enhancement of asthma mechanisms.

Asthma: Pathophysiology

The hallmark of acute asthma is widespread decrease in the diameter of airway passages, due, in part, to contraction of circularly arrayed airways smooth muscle. There is also edema of the submucosal tissues, and airway lumens are often plugged by tenacious, albumin-rich secretions containing mucins, intact and degenerating inflammatory cells, and sheets of airways epithelium. Eosinophils are prominent among the inflammatory cells (Hogg 1985).

Assessment of Airways Size

Descriptions of the caliber or geometry of the airways are, at best, incomplete. The conducting airways are a complex branching structure that includes over 20 branch points before a fully alveolated region of lung parenchyma is reached. In addition, the caliber of the airways is affected by lung elastic recoil which is a function of lung volume and is continually changing during respiration.

Several techniques are available that can be used to estimate airway size. The simultaneous measurement of the pressure gradient between the alveoli and the airway orifice and of airflow either during spontaneous breathing or panting allows an empirical measurement of R_{aw}, which is defined as the ratio between pressure gradient and flow. This relation varies with lung volume and depends on airflow rate. Nevertheless, this ratio is commonly used as a descriptor of overall airways geometry.

Another indirect approach to assessing airways caliber is the forced expiratory spirogram (volume expired versus time) which is equivalent in informational content to the flow-expired volume relation. Since the inspiration that precedes the forced expiratory maneuver may temporarily dilate constricted airways or, in asthmatics, provoke increased bronchoconstriction, the size of the inspiration can be reduced and partial forced expiratory maneuvers performed in order to avoid these confounding factors. The quantitative changes observed with R_{aw} and spirometric measurements in patients with airways narrowing may be poorly correlated. Thus, careful attention to selection of measurement techniques is necessary when probing for relatively small effects on airways function.

The direct relation between lung volume (and lung elastic recoil) and intrathoracic airways caliber is well known. Recent reports suggest that the dose/response relation of the airways to inhaled bronchoconstrictors is quite sensitive to the lung volume at which R_{aw} measurements are made (Martin et al. 1986). This factor will need to be considered in assessing bronchial reactivity.

Distribution of Airflow Resistance Along the Airways

The longitudinal distribution of the pressure changes between the airway opening and alveoli is complex. The larynx represents a significant problem in clinical studies whereas an endotracheal tube poses problems in intubated experimental animals. During quiet breathing or panting, a major pressure drop occurs in the larynx and the large airways of normal individuals or mild asthmatics. In normal adults, the small airways contribute little to total airflow resistance but become a major component of the elevated airflow resistance in persons with diffuse airways disease. In addition, changes in the small airways are particularly important in the pathogenesis of COPD. Significant pathological changes in the very small airways have been described in primates chronically exposed to moderate levels of O_3 (Tyler et al. 1985).

The caliber of the upper airways and the trachea can be measured with radiological or acoustic reflection techniques. Using tantalum dust as an experimental contrast material, researchers have obtained good visual resolution of the intrapulmonary bronchi (Hahn et al. 1976; Smith et al. 1979; Shioya et al. 1987). A variety of approaches have been devised to fractionate resistance between the "large" and "small" airways. Some of these (for example, analysis of the frequency dependence of respiratory impedance using random-noise forced oscillation, analysis of the gas-density dependence of forced expiratory flow) are noninvasive. A more direct, though invasive, approach uses a wedged bronchoscope to isolate smaller airways. This technique has been applied to the study of O_3

effects on peripheral lung airflow resistance and histamine reactivity in dogs by Menkes and his colleagues (Gertner et al. 1983a,b,c). Little or no attempt has been made in studies of pollutant effects in asthmatic subjects to fractionate airways resistance changes or, more specifically, to examine small airways function.

Mechanisms of Bronchoconstriction

Fundamental mechanisms intrinsic to smooth muscle that regulate its contraction and relaxation have been reviewed by Kamm and Stull (1985), Rasmussen (1986), and Russell (1986). However, it is not known whether asthma is associated with changes in these mechanisms. It is, for example, conceivable that the hypertrophied smooth muscle in the airways of severe asthmatics might show differences in composition of one or more of the molecules in the contractile or regulatory apparatus, or in properties of membrane calcium channels, or in the quantity or type of membrane receptors. Such abnormalities, and others, could produce an abnormal contractile response to normal stimuli.

More attention has been focused on abnormalities of mechanisms exogenous to the muscle cells that might stimulate excessive contraction and/or impair relaxation of normally functioning smooth muscle. These mechanisms fall into two broad classes: neural and chemical.

Neural Mechanisms. The airway wall elements, including smooth muscle, are innervated by several different kinds of nerves (for review, see Nadel and Barnes 1984). Best known are the postganglionic parasympathetic fibers. These fibers are cholinergic and depolarize the muscle membrane by reaction of acetylcholine released from the nerve endings with specific muscarinic muscle membrane receptors. Normal individuals exhibit parasympathetically mediated airways smooth-muscle tone which can be blocked by atropine or ipratropium bromide, resulting in a significant decrease in airways resistance. The preganglionic fibers in the airway wall are also cholinergic, but their synapse with the postganglionic nerve cell bodies is nicotinic (rather than muscarinic) and can be blocked by agents such as hexamethonium. The ganglion cells are subject to other neural and chemical influences that also influence their excitability. Detailed study of the anatomy and physiology of these important tissue ganglia is just beginning.

There appears to be little sympathetic innervation of the airways despite the abundance of adrenergic receptors on various cellular elements in the airway walls. On smooth muscle, these receptors are normally of the beta type, and their activation causes muscle relaxation. Under some circumstances, however, α-adrenergic constrictive responses have been demonstrated which can be blocked by appropriate inhibitors. An abnormality of adrenergic receptor function, in particular β-adrenergic blockade, was proposed 20 years ago by Szentivanyi (1968) as an important mechanism underlying asthma. Asthmatic patients often exhibit extraordinary sensitivity to orally or even topically (ocularly) administered β-adrenergic blockers which can precipitate a serious exacerbation of the asthmatic state. On the other hand, nonasthmatic individuals fail to develop bronchial hyperreactivity or asthma when treated chronically with β-adrenergic blocking agents.

Less well understood is the nonadrenergic, noncholinergic inhibitory system (see Barnes 1984) whose neurotransmitter(s) remain(s) to be firmly identified. Vasoactive intestinal peptide (VIP) is an attractive candidate for this role. Airway smooth muscle preparations from appropriate species including humans, when electrically stimulated, develop a transient relaxation followed by constriction. Constriction can be blocked by atropine pretreatment; this unveils a relatively prolonged relaxation that cannot be prevented by β-adrenergic blockade. The neural nature of this response is shown by its blockade by the neuronal sodium ion channel blocker, tetrodotoxin. It is possible that the postganglionic cholinergic fibers are also VIP-ergic, with the polypeptide transmitter modulating the effect of the "classical"—that is, acetylcholine—transmitter. Recently, it has been suggested, along lines similar to that of the β-adrenergic blockade hypothesis, that asthmatics have a blockade or a deficiency in the nonadrenergic inhibitory system.

Sensory nerves are abundant in the airways at all levels and provide normal reflex mechanisms (Coleridge and Coleridge 1986) which can alter airways function, especially by causing bronchoconstriction. So-called irritant receptors (rapidly adapting stretch receptors) are stimulated by mechanical, chemical (for example, histamine), and possibly osmotic events, and are susceptible to superficial stimuli. Irritants such as ammonia, cigarette smoke, diethyl ether, and phenyldiguanide are not potent stimulants, however (Sampson and Vidruk 1975). From the subglottal airways, these impulses ascend in myelinated fibers to the brain where several synaptic reflexes may initiate cough or diffuse bronchoconstriction. Such reflexes allow localized stimulation of the airways or of the larynx to give rise to diffuse motor responses.

Reflex bronchoconstriction can be blocked by inhalation of muscarinic antagonists. Bronchoconstriction evoked by histamine or antigen challenge is partially blocked by atropine pretreatment, indicating a significant role for reflex mechanisms as well as for more direct, chemically mediated effects on smooth muscle. Atropine pretreatment will block bronchoconstriction evoked by O_3 exposure (Beckett et al. 1985) and in asthmatics, by SO_2 exposure, indicating the important role of reflex cholinergic muscarinic neural mechanisms in the genesis of some of the acute effects of inhaled pollutants on airways function. There have been some differences of opinion as to the importance of cholinergically mediated reflexes in the genesis of bronchospasm in various situations. This may be due to experimental difficulties in achieving adequate concentrations of muscarinic antagonist at the critical site with certain stimuli (Holtzman et al. 1983; Sheppard et al. 1983; Boushey 1985).

Classical irritant receptors respond to histamine but not to methacholine (Vidruk et al. 1977). Therefore, O_3-induced hyperreactivity to inhaled methacholine cannot be explained simply by a sensitization of irritant receptors. Furthermore, Sampson and coworkers (1978) failed to find any enhancement of irritant receptor response to histamine following O_3 exposure in dogs.

The airways are also abundantly innervated by nonmyelinated or poorly myelinated sensory nerves, termed C-fibers. These nerves arborize extensively and supply not only the epithelium but also smooth muscle, submucosal glands, blood vessels, nerve ganglion cells in the airway wall, and other elements (Lundberg et al. 1984). These fibers can be stimulated by SO_2, histamine, prostaglandins, bradykinin, and capsaicin (Coleridge et al. 1965), a chemical found in hot peppers (Coleridge and Coleridge 1986), which in turn causes reflex bronchoconstriction and tracheal gland mucous secretion (Roberts et al. 1981; Davis et al. 1982). Coleridge and colleagues (1978) have pointed out that airway C-fibers are highly sensitive to chemical irritants, whereas the so-called "irritant" receptors are more sensitive to mechanical stimuli.

Sensory C-fibers synthesize several peptides including substance P (Iversen 1982; Hua et al. 1985; Besson and Chaouch 1987), which when released cause local smooth-muscle constriction, increased microvascular permeability, and other effects. The terminal arborization of these sensory nerves allows them to subserve antidromic (nonsynaptic) transmission by releasing biologically active peptides at sites other than the point of stimulation. Thus, capsaicin (Lundberg et al. 1983a) or toxicants such as cigarette smoke (Lundberg et al. 1983b) cause an antidromic reflex leading to motor effects such as bronchoconstriction, submucosal protein-rich edema, and gland secretion. Since O_3 inhalation by humans characteristically causes a sensation of large airways irritation and pain on deep inspiration, it is possible that the gas stimulates this sensory neural system and provokes antidromic reflexes as well as causing involuntary pain inhibition of deep inspiration and consequent impairment of spirometric performance (Hazucha et al. 1986; Bromberg 1987). Neuropeptides released from

C-fiber endings may also influence mast cells and immunecompetent cells and might therefore modulate immunologic processes in the airways (Payan et al. 1984).

The mechanism of the impairment of vital capacity by O_3 inhalation is probably mediated by neural pathways. Airway C-fibers are likely candidates for this role and deserve further investigation, particularly since stimulation of such fibers also leads to motor effects on airway glands and vessels mediated by axon reflexes and release of neuropeptides, as well as to bronchoconstriction and gland secretion by synaptic reflexes.

■ **Recommendation 4.** Specific neuropeptides should be assayed in airways surface liquid after O_3 exposure, and the phenomenon of "tolerance" to the O_3 effect on vital capacity should be explored along these lines. Highly sensitive neuropeptide assays will be required, and rapid inhibition of peptidase activity may also be necessary to prevent hydrolysis of peptides in the sample.

■ **Recommendation 5.** Ozone responsiveness in human subjects should be compared to the responses to known stimulants of the airway C-fibers, such as capsaicin (Collier and Fuller 1984; Fuller et al. 1985).

■ **Recommendation 6.** Animals prepared so as to render their airway C-fiber systems nonfunctional should be used to examine the role of this sensory system in O_3 effects on epithelial permeability, mucous secretion, epithelial ion transport, bronchial reactivity, and airways inflammation.

Chemical Mediators. Chemical mediators exert profound effects on smooth muscle and other effector elements in the airway wall, by direct combination with specific membrane receptors and by neural reflex mechanisms.

Major attention has been devoted to mediators associated with the strategically located airway mast cell (Robinson and Holgate 1985). This key cell synthesizes and stores certain mediators in granules. When stimulated, the cell discharges the granules and also releases other mediators that are synthesized de novo. Kaliner (1985) has categorized many known mast cell–derived mediators. Preformed mediators that are rapidly released from the granules include histamine, kininogenase, and chemotactic factors for eosinophils (ECF-A) and for neutrophils (NCF). Substances that remain associated with the discharged granule matrix include heparin and proteolytic enzymes. Mediators that are synthesized de novo from cell phospholipids (Dvorak et al. 1983) include platelet-activating factor (acetylglyceryletherphosphorylcholine) and a number of metabolites of arachidonate: PGD_2; slow-reacting substance of anaphylaxis (SRS-A, consisting largely of leukotrienes C_4, D_4, E_4); hydroxyeicosatetraenoic (HETES) and hydroperoxyeicosatetraenoic (HPETES) acids; thromboxane A_2; and leukotriene B_4.

All of these mediators cause bronchoconstriction, and some are extremely potent and capable of inducing prolonged bronchoconstriction. Many of the mast cell products are capable of increasing vascular permeability. Mucous secretion can be stimulated by leukotrienes D_4 and C_4. The chemotactic factors ECF-A and NCF are presumably responsible for the cellular inflammatory response involved in the late reaction. In addition, leukotriene B_4 and HETES may play a role in neutrophil recruitment.

Products liberated and secreted by recruited neutrophils and eosinophils undoubtedly play major roles in causing chronic changes after a single antigen exposure. The effects of eosinophil-derived proteins are particularly interesting and have been summarized by Gleich et al. (1985). Platelet activation has also been suggested to occur during antigen-induced airway reactions in asthmatics (Knauer et al. 1981).

Because the mast cell is coated with IgE, the cell can be stimulated to secrete by presentation of specific antigen. Mast cells can also be stimulated by nonantigenic means, such as hyperventilation of cold dry air. In addition to binding IgE, the mast cell membrane appears to have receptors for a variety of autacoids that stabilize (for example, β-adrenergic agents) or destabilize (for example, adenosine) the cell. The direct effect of air pollutants on mast cell secretion has not been studied. Indirect effects on mast cells by mediators released from pollutant-exposed airway epithelial cells or nerves are also possible.

Although an important role of the airways mast cell in allergic asthma is generally conceded, it has been difficult to demonstrate the release of all the mediators in vivo in allergen- or hyperventilation-provoked reactions (Deal et al. 1980; Nagy et al. 1982; Lee et al. 1983a,b). Peripheral blood analysis is limited by severe dilution problems. Nasal and bronchial lavage provide an alternative approach (Metzger et al. 1985a,b; Peters et al. 1985; Wasserman 1985). Furthermore, a number of important mediators (for example, arachidonate derivatives) can be derived from other airway cells, so the presence of a particular substance in the airway or the blood does not necessarily prove its mast cell origin.

In addition to mast cell and other airway cell products that may alter smooth-muscle tone, circulating endogenous adrenergic agonists can activate β-adrenergic receptors on smooth muscle and cause relaxation. α-Adrenergic receptors may be present, particularly on the muscle of chronically inflamed airways, and their stimulation, on the other hand, could cause constriction (Walden et al. 1985).

Ozone Exposure

Nonspecific Bronchial Hyperreactivity. Acutely increased airway response to histamine following exposure of rodents to O_3 was observed about 20 years ago (Easton and Murphy 1967). A similar effect was subsequently shown in dogs (Lee et al. 1977), humans (Golden et al. 1978), and sheep (Abraham et al. 1980). In human subjects, bronchial hyperreactivity has also been induced by inhaled methacholine; the degree is relatively modest and lasts hours rather than days (Holtzman et al. 1979).

It was attractive to postulate that the hyperreactivity was due to increased rate of uptake of the bronchoconstrictors across a damaged airway epithelium. However parenteral administration of the bronchoconstrictor agent in O_3-exposed guinea pigs elicits hyperreactivity even more reliably than does inhalation (Gordon et al. 1984; Roum and Murlas 1984), showing that other mechanisms must also exist. Such mechanisms may include increased bronchial submucosal blood flow and vascular permeability resulting in increased local presentation of parenterally administered drug; liberation of chemical or neurochemical mediators that sensitize airways smooth muscle to the effects of bronchoconstrictors; or decreased secretion of a tonic bronchodilating substance from airways epithelium. However, the most carefully studied hypothesis links bronchial hyperreactivity to airways inflammation.

The potential role of inflammation as a prerequisite for the development of bronchial hyperreactivity in dogs exposed to O_3 has been documented in a series of papers from the Cardiovascular Research Institute summarized in the next paragraph. These studies should be viewed in the larger context of the general mechanisms underlying all bronchial hyperreactivity, including that observed in asthma and after antigen challenge of sensitive subjects (Boushey et al. 1980; Fabbri 1985).

Holtzman and colleagues (1983) showed that post-O_3 (2 ppm, 2 hr) hyperreactivity to aerosolized acetylcholine was found only in dogs with neutrophil invasion of the airways mucosa and epithelium; the regression of inflammation coincided with the disappearance of hyperreactivity. O'Byrne and coworkers (1984b) showed that although severe hydroxyurea-induced neutropenia did not affect baseline bronchial reactivity to inhaled acetylcholine, after O_3 exposure (3 ppm, 2 hr), bronchial reactivity failed to increase and neutrophil infiltration of the epithelium did not occur. In nonleukopenic dogs, bronchoalveolar lavage fluid after O_3 exposure contained increased numbers of neutrophils and epithelial cells, but only epithelial cell numbers were increased in the hydroxyurea-pretreated leukopenic animals. O'Byrne and coworkers (1984a) further showed that indomethacin treatment, which had no effect on pre-O_3 baseline bronchial reactivity, prevented O_3-induced bronchial hyperreactivity despite the presence of neutrophil infiltration of the epithelium. This finding suggests that cyclooxygenase products of arachidonate are essential to bronchial hyperreactivity but not essential to the airway neutrophilic inflammatory response after exposure to high levels of O_3. Whether these putative cyclooxygenase products are derived from the neutrophils or from airways epithelial cells is unclear. In addition, leukotrienes produced by neutrophils might provoke epithelial cells, or other cells, to release cyclooxygenase products that are directly responsible for the induction of hyperreactivity.

Although these studies used high levels of O_3, their relevance to human exposures is supported by the observation that inhalation of 0.5 ppm O_3 by resting subjects provoked an increase in neutrophils in nasal lavage liquid (Graham et al. 1985). Furthermore, when Seltzer and colleagues (1986) exposed exercising subjects to 0.4 and 0.6 ppm O_3, increases in methacholine reactivity occurred 1 hr postexposure. Analysis of bronchoalveolar lavage fluid showed that, especially at the higher dose, the percent of neutrophils, but not epithelial cells, increased substantially. The lavage fluid was also examined for the presence of arachidonate oxidation products, and an increase in cyclooxygenase products, but not 5-lipoxygenase products, was found.

Roum and Murlas (1984) used parenteral acetylcholine to assess bronchial reactivity in guinea pigs and found it, at least in this species, to be a more sensitive and reproducible method than challenge with inhaled aerosolized acetylcholine. Unlike the dog model, increased airway responsiveness, as well as epithelial mucin discharge and ciliary damage following O_3 exposure, was observed prior to neutrophilic infiltration of the mucosa (Murlas and Roum 1985a). In neutrophil-depleted guinea pigs, these investigators (1985b) failed to inhibit any of these post-O_3 effects. Thus, neutrophils are not essential to the genesis of O_3-induced hyperreactivity in guinea pigs although there is no doubt that O_3 exposure is a potent stimulus to neutrophilic inflammation. Pretreatment with an inhibitor of leukotriene synthesis (U-60257) abolished the post-O_3 bronchial hyperresponsiveness (Murlas and Lee 1985). Pretreatment with indomethacin, a cyclooxygenase inhibitor, not only failed to prevent post-O_3 hyperreactivity, but actually potentiated the effect of "subthreshold" O_3 exposure (Murlas et al. 1986). Interestingly, these indomethacin effects are the opposite of what O'Byrne and coworkers (1984a) reported in O_3-exposed dogs but were similar to the effects observed in ovalbumin-sensitive guinea pigs challenged with histamine (Brink et al. 1981). In addition, indomethacin pretreatment somewhat enhanced the direct bronchoconstrictive effect of a 15-min exposure to 3.0 ppm O_3 (Lee and Murlas 1985).

The role of inflammatory cells in the genesis of post-O_3 bronchial hyperreactivity therefore remains uncertain. Whether differences between the dog and guinea pig models are attributable to species differences, anesthesia, inhalation as opposed to parenteral administration of the bronchoconstrictor, or other factors is not clear. Using guinea pigs, Thompson and coworkers (1986) studied the relation of neutrophilic infiltration to toluene diisocyanate–induced airway hyperresponsiveness to inhaled acetylcholine. Cyclophosphamide or hydroxyurea treatment abolished the inflammatory cell infiltration, but only hydroxyurea inhibited the toluene diisocyanate–induced airway hyperresponsiveness. Ozone exposure can cause a neutrophilic inflammatory response in the airways of humans. Whether this response is essential to the development of bronchial hyperreactivity in O_3-exposed normal human subjects is not known. Whether asthmatic subjects would exhibit a similar or altered inflammatory response to O_3 inhalation, and what effect this response may have on bronchial reactivity to specific as well as nonspecific challenge, is not known.

Specific Airways Reactivity. Studies in sensitized rodents (Matsumura 1970a,b; Matsumura et al. 1972; Osebold et al. 1980; Gershwin et al. 1981) suggest that O_3 exposure results in increased reactivity to challenge with inhaled or intravenous specific antigen. Abraham and coworkers (1983a), however, were unable to confirm this finding in *Ascaris suum*-sensitive sheep following a 2-hr exposure to 0.5 ppm O_3. The authors suggested that O_3 exposure might have degranulated airway mast cells (Dixon and Mountain 1965), thus making them less responsive to subsequent antigen challenge. Alternatively, O_3-induced mucus accumulation on airway surfaces could have impaired antigen penetration.

There is presently a significant discrepancy between an apparent increase in symptoms of asthma observed clinically in populations exposed to increased levels of ambient O_3, and the apparent absence of any particularly striking effects of short-term O_3 exposure on asthmatic subjects in experimental exposure chambers. In addition, the effects of NO_2 exposure on asthmatics have been inconsistent. Apart from the possibility that low levels of NO_2 are indeed "inert," and that despite its irritant effects, O_3 simply has no special effect on asthmatic subjects, several possibilities are suggested.

First, experimental conditions may not adequately mimic natural exposures. Variables to consider include duration of exposure; number of days of exposure; presence of other air pollutants, especially acid sulfate aerosols, which may interact with O_3 in causing health effects; presence of other nonspecific or specific (antigenic) airborne substances. Thus:

■ **Recommendation 7.** Exercising asthmatics should be experimentally exposed to relevant O_3 and NO_2 levels for periods up to 8 hr, with monitoring of airways caliber.

■ **Recommendation 8.** Asthmatics should be exposed experimentally for 8 hr to O_3 or NO_2 for two or three consecutive days, with monitoring of airways caliber.

■ **Recommendation 9.** Exposure chamber atmospheres that mimic the acid sulfate content and particle size distribution in ambient pollution should be created. First, it will be necessary to analyze in greater detail the composition of ambient atmospheres associated with increased symptomatology in patients with asthma and other respiratory diseases.

■ **Recommendation 10.** Asthmatic subjects should be exposed to such chamber atmospheres, with monitoring of airways caliber. The study of atmospheres containing more than one pollutant will greatly complicate the design of such experiments and will require some choices to be made of concentrations of the pollutants, time course of the concentration of each pollutant, and particle size range of a particulate pollutant (for example, acid sulfate aerosol).

■ **Recommendation 11.** Response of airways of extrinsic asthmatics to pollutants, especially NO_2, should be assessed in the presence versus the absence of chronic low-level exposure to specific allergens.

Second, striking pollutant effects may be limited to asthmatics and subjects who have particularly high levels of nonspecific bronchial hyperreactivity. Such asthmatics are more likely to be clinically "unstable" and will therefore present ethical as well as experimental design problems. Thus:

■ **Recommendation 12.** The effect of experimental pollutant exposures on nonspecific airways reactivity in asthmatic subjects selected to display a range of baseline reactivities should be measured.

■ **Recommendation 13.** The effect of experimental pollutant exposures on airways function should be measured in patients with COPD (for example, nonspecific chronic bronchitis, cystic fibrosis), in whom increased bronchial reactivity is present.

■ **Recommendation 14.** Pollutant effects on airways should be measured in extrinsic asthmatics in whom a transient state of marked bronchial hyperreactivity has been induced by a single antigen inhalation challenge. Individual subject responses could then be assessed over time at several levels of baseline bronchial reactivity.

Third, the provocative agents used to assess the nonspecific bronchial reactivity response to O_3 and NO_2 exposure have been limited mostly to acetylcholine congeners and histamine. Other provocative agents might reveal more dramatic effects. Effects on *specific* reactivity also should be assessed. Thus:

■ **Recommendation 15.** The effect of oxidant pollutant exposures of asthmatics on bronchial reactivity to stimuli such as SO_2, cold dry air, nonisotonic aerosolized solutions, and certain mast cell–derived mediators should be explored.

■ **Recommendation 16.** The effect in allergic rhinitis patients or in extrinsic asthmatics of experimental pollutant exposures on postexposure nasal or bronchial reactivity ("early" and "late" phases) to specific antigen challenge should be ascertained.

Respiratory Epithelium: Asthma and Ozone

It is apparent why efforts to understand the pathophysiology of asthma have focused on the components of airway walls, including mast cells, smooth muscle, inflammatory cells, and neural elements. More recently, however, the role of the airway epithelium has attracted attention (Hogg and Eggleston 1984). Increased understanding of the normal functions of the epithelium has led to a more precise definition of its structure; its role in mucociliary transport and in regulating the volume and composition of the airways surface liquid; its role as a barrier between the submucosal and surface liquid compartments; and its ability to produce arachidonate metabolites. In addition, an epithelial-derived smooth muscle relaxing activity has been found (Flavahan and Vanhoutte 1985; Flavahan et al. 1985; Frossard and Muller 1986). Furthermore, the abundant sensory nerve system in the epithelium can alter

airways function antidromically and by means of synaptic reflexes.

In asthmatics, but not normal individuals, cough and/or bronchoconstriction are provoked by inhalation of aerosols of hypo or hypertonic solutions (Sheppard et al. 1983), and the anionic composition of aerosolized isotonic solutions is a significant variable in causing airway responses (Eschenbacher et al. 1984). These findings strongly suggest that changes in airway surface liquid composition are detected by adjacent elements, that is, epithelial cells, surface basophiloid cells and macrophages, or intraepithelial nerve endings. Indeed, although respiratory heat loss has been suggested as the mechanism of exercise-induced bronchospasm, the loss of surface water and the development of hypertonicity is an attractive alternative. An important area of research in airways physiology and disease will be to define the driving forces and fluxes of water across the respiratory epithelium, the pathways of water movement, and the role of submucosal blood flow in relation to water flux across the airway surface. The response of individual airway cells to osmotic gradients (for example, secretion of mediators, ion transport, volume regulation) can be investigated with disaggregated cells in suspension and with intact epithelial sheets. Along these lines of investigation, histamine secretion by circulating basophils was shown by Findlay et al. (1981) to be provoked by hyperosmolarity, especially in conjunction with immunologic stimulation. Eggleston and coworkers (1984, 1987) showed that human lung mast cells as well as basophils released histamine in response to hyperosmolar conditions in vitro. Lee and colleagues (1982a,b) found elevated serum levels of a mast cell–derived neutrophil chemotactic factor in association with exercise-induced bronchospasm. Possible differences between the response of airways epithelium from asthmatics and normals to hypo- and hypertonic conditions remain to be studied.

Equally important to consider is that highly reactive gases such as O_3, NO_2, or SO_2 should react with the surface components of the air/airway interface, including the epithelium. Although direct chemical analysis of such reactions is not yet available, there is every reason to believe that appropriate studies will confirm this conjecture. Indeed, it will be important to develop techniques for analysis of tissue dosimetry for gases in relation to ambient exposure levels and ventilatory patterns (see Overton and Miller, and Ultman, this volume). Whether exposure of asthmatics or normal individuals to pollutants alters subsequent airway responses to osmotic or ionic stimuli has not been investigated.

Permeability

The epithelium is important in segregating the airway surface liquid from the submucosal interstitial fluid. By measuring the rate of uptake of polar probe molecules of various sizes instilled or aerosolized onto the respiratory surface, it is possible to quantify this barrier function in vivo (Hogg et al. 1979). Indeed, following inhalation antigen challenge of A. suum-sensitive rhesus monkeys, the resulting hyperreactivity to inhaled histamine was observed to be accompanied by increased uptake of instilled histamine (Boucher et al. 1979), horseradish peroxidase, and other probe molecules (Boucher et al. 1977). Airways hyperpermeability, assessed in this manner, has also been demonstrated in guinea pigs after acute exposure to histamine, methacholine, diethyl ether (Boucher et al. 1978), cigarette smoke (Boucher et al. 1980), or platelet-activating factor (PAF) (M. Ivanick, R. Schreiber, S. Wyrick, P. Bromberg, and V. Ranga, unpublished observations).

Hyperpermeability has been observed following O_3 exposure. Increased uptake of probe molecules occurred in guinea pigs exposed to low levels (1 or 0.3 ppm for 3 hr) of the oxidant (Davis et al. 1980; Bromberg et al. 1984). If, however, the trachea was excised after exposure and mounted as a cylinder in a bath so as to isolate and perfuse both the lumenal and submucosal surfaces of the tissue, changes in permeability were not found (V. Ranga, M. Ivanick, and P. Bromberg, unpublished observations). Whether this was due to the washout of mediators into the lumenal bathing

solution, loss of innervation, absence of circulating blood cells that contribute to inflammation, or other factors, is not known. Increased concentrations of serum albumin, another marker of permeability, have been reported in the bronchoalveolar lavage liquid from guinea pigs 15 hr after exposure to only 0.26 ppm O_3 (Hu et al. 1982). Kehrl and coworkers (1987) have recently shown that normal humans exposed for 2 hr to a regimen of 0.4 ppm O_3 with intermittent vigorous exercise develop an increased uptake rate of aerosolized ^{99m}Tc-diethylenetriaminepentaacetic acid (DTPA) deposited on the respiratory surfaces, suggesting increased epithelial permeability. Bromberg and colleagues (1984) found evidence of "adaptation" of the epithelial permeability response after four consecutive daily exposures of guinea pigs to 1.0 ppm O_3 for 3 hr. This observation is particularly interesting because of its possible parallel with the "adaptation" of the human response to O_3 after consecutive daily exposures.

The transepithelial pathways for molecular movements following oxidant exposure probably involve changes in intercellular tight junctions, but such changes have been difficult to demonstrate by electron microscopy which is used to examine small regions of the tight junction. Abundant horseradish peroxidase has been observed between epithelial cells in suitably prepared trachea from O_3-exposed guinea pigs. However, this marker has also been observed intracellularly, suggesting that intracellular penetration may also occur (V. Ranga and P. Bromberg, unpublished observations).

The mechanisms whereby a variety of agents, including O_3, increase airways epithelial permeability, are unknown. Many of these agents also cause bronchoconstriction. Modulation of epithelial permeability by substance P and other neuropeptides released from unmyelinated sensory nerve fibers is an attractive possibility but has not been examined. Alternatively, the presence of immunologically competent mast cells on the airway surface (Tomioka et al. 1984) provides a way in which mediators that affect epithelial permeability could be released immediately after inhalation of various agents including antigen. The possible role of epithelial cell products in regulating tight junctional integrity remains to be explored.

Do stable asthmatics have hyperpermeable airways? The presence of increased albumin levels in asthmatic bronchial secretions suggests that they do. Buckle and Cohen (1975) observed more rapid absorption of ^{125}I–human serum albumin from the nasal mucosa of patients with allergic rhinitis. Furthermore, bronchial biopsies taken from intralobar as well as more proximal large airways in eight asthmatic patients, clinically ranging from "mild" to "severe," and who were hyperreactive to histamine, showed extensive epithelial damage, especially to ciliated cells (Laitinen et al. 1985). Areas of gross denudation of epithelium were also observed. These observations show that evidence of extensive epithelial damage in bronchial asthma is not limited to necropsy (Dunnhill 1971, 1975; Cutz et al. 1978). The presence of desquamated sheets of epithelium and clusters of columnar epithelium (Creola bodies) in sputum from asthma patients is also well established (Naylor 1962). Exposed sensory nerve endings and mast cells have also been noted by Laitinen and coworkers (1985) in their biopsy material.

It is therefore surprising that a study by Elwood and colleagues (1983) failed to demonstrate increased uptake of aerosolized inhaled ^{99m}Tc-DTPA in clinically stable asthmatics who were hyperreactive to inhaled methacholine. Conversely, cigarette smokers with some evidence of small airways disease, but normal FEV_1, failed to exhibit bronchial hyperreactivity to inhaled histamine, but experienced increased ^{99m}Tc-DTPA uptake (Kennedy et al. 1984). Thus, although an acute allergic reaction in airways causes a sharp increase in airways epithelial permeability, the state of the airways epithelial barrier function and its relation to bronchial hyperreactivity in chronic asthma remains unclear.

Mucous Production and Secretion

Excessive mucous production is a feature of asthma as well as of other airway diseases.

In the bronchi, the relative contributions of the surface secretory cells, which are often increased in number in chronic inflammatory airways disease, as compared to the submucosal glands, which also are increased in volume in airways disease, are difficult to determine. In smaller airways, the glands are much less numerous, and the surface epithelial secretory cells must play the dominant role.

Gland secretion is under autonomic control. An isotonic fluid is produced that contains not only mucin macromolecules but also proteins such as secretory IgA, lysozyme, lactoferrin, and an antiprotease. Cholinergic stimulation greatly augments gland secretion volumes but does not alter viscosity and protein concentration. β-Adrenergic stimulation, on the other hand, increases the viscosity and protein content out of proportion to flow rate. The proteins are secreted by serous gland cells which degranulate and release lysozyme in response to α-adrenergic and cholinergic agonists, and to substance P. β-Adrenergic and cholinergic agents cause marked degranulation of mucous-type gland cells. Secretion is also increased by several prostaglandins (for example, PGD_2) and products of 5-lipoxygenase oxidation of arachidonate, such as leukotriene C_4 and D_4 (Basbaum 1986).

The surface goblet cells do not appear to be innervated and are not stimulated by neurotransmitters. They are, however, stimulated by certain proteases (Klinger et al. 1984) and possibly by other secretagogues (Kaliner et al. 1984). The secretory activity of surface epithelial cells in small airways, such as Clara cells, is just coming under scrutiny.

Acute O_3 exposure produces marked loss of stainable mucin in the surface epithelium and an increased layer of surface mucus of the airways of guinea pigs. Other irritants, such as cigarette smoke, also cause goblet cell secretion. The mechanism underlying this effect is unknown, and it has not been studied using quantitative measures of mucin macromolecule secretion. The effect of O_3 on the surface goblet cells of normal subjects or individuals with increased numbers of surface epithelial goblet cells (for example, asthmatics and chronic bronchitics) is as yet unknown. The only known effect of air pollutants on small airways epithelial function is the replacement of the normal small airways epithelium by goblet-type secretory cells not normally present in this region (Reid et al. 1983). Pollutant-induced damage to airways epithelium appears preferentially to involve ciliated cells. The repair response includes a burst of mitotic activity followed by differentiation, possibly into a different cell type. In chronic injury, goblet cells commonly replace ciliated cells (for review, see U.S. Environmental Protection Agency 1986).

Ion Transport

The salt and water composition as well as the volume of the airway surface liquid is regulated by transepithelial ion transport. Volume absorption can be driven by active, electrogenic sodium ion absorption whereas volume secretion can be driven by active sodium-dependent electrogenic chloride ion secretion. These processes depend on the presence of certain ion-selective pumps, channels, and transporters in the basolateral and apical portions of the epithelial cell membrane. These membrane regions are functionally isolated from one another by the tight junctions near the apical margins of the epithelial cells. The active ion transports generate transepithelial electrical potentials and currents that can be measured in vivo (Knowles et al. 1986) as well as in vitro (Al-Bazzaz 1986; Widdicombe 1986). Epithelial chloride secretion is stimulated by mediators that increase c-AMP levels (Smith et al. 1982), increased levels of intracellular ionized calcium, and in the canine trachea, substance P (Al-Bazzaz et al. 1985). Inhibitors of cellular energy metabolism generally decrease chloride secretion (Stutts et al. 1984).

Whether these ion-transporting mechanisms and associated water movements are disturbed in asthma, and how this might contribute to the pathophysiology of asthma, is not yet clear. In cystic fibrosis, the airway epithelial cells have markedly diminished apical membrane chloride permeability as well as increased sodium ion

absorption. Excessive absorption of salt and water by the epithelium in this disease and how it relates to the viscosity of mucous-containing secretion have been discussed by Knowles et al. (1986).

Guinea pig trachea, excised following acute exposure to O_3, has shown increased active ion transport, which produced an increased transepithelial potential difference in spite of a modest increase of electrical conductance (Stutts and Bromberg 1987). However, in vivo measurements failed to reveal similar changes of electrical potential in canine and guinea pig tracheas (P. Bromberg and M. Knowles, unpublished observations). These findings suggest that, although O_3 may alter cellular ion transport mechanisms in airways epithelium, concomitant increases in epithelial permeability and electrical conductance prevent an increase in transepithelial potential difference from being observed in vivo. The absence of such permeability changes in trachea mounted in vitro might be due to dilution or wash-out of mediator substances.

Mucociliary Clearance

How these complex processes of macromolecular secretion and salt and water transport interact with the ciliary apparatus to produce the integrated process of mucociliary transport remains mysterious. Various pollutant exposures have been reported to depress, enhance, or have no effect on mucociliary clearance by the airways. Asthmatics may have a slow tracheal mucous transport velocity, which is further slowed by antigen challenge. Impaired mucociliary transport in *A. suum*–sensitive sheep has been observed to last for several days after a single antigen challenge. This response appeared to be inhibitable by cromolyn pretreatment, suggesting a mechanism involving release of mast cell mediators. Other studies suggest that lipoxygenase oxidation products of arachidonate may suppress tracheal mucous transport, whereas other mediators may enhance it. The inhibitory effects of antigen challenge do not appear to be attributable to direct effects of mediators on ciliary movement but may result from increased secretion of mucous glycoproteins stimulated by lipoxygenase oxidation products of arachidonate (Wanner 1986).

■ **Recommendation 17.** The effects of exposure to O_3, or O_3 plus other air pollution components, on respiratory epithelial cell function should be probed in greater detail. These functions include mucociliary clearance, permeability to molecules including albumin and antigens, secretion of mucins and of other specific macromolecules, airway surface liquid composition and volume control, and release of mediator substances.

Experimental Models

Numerous experimental models are available that can be used to pursue the research recommendations presented. Descriptive as well as some mechanistic studies can be performed in human subjects, but some investigations may be better suited by animal models or by cultured cells, human as well as animal.

Human Studies

The limitations on the study of asthmatic humans in relation to air pollutant effects are significant but not prohibitive. The effects of acute and of recurrent exposure to relevant pollutants can be studied in subpopulations defined by immunologic, clinical, functional, and bronchial hyperreactivity criteria. In addition to asthma of different types and degrees of severity, other types of reactive airways diseases can be studied in humans. The possible effect of exposure to a particular pollutant on subsequent reaction to antigenic and nonantigenic (including another pollutant) substances can be explored. The design of such complex studies should take into account the results of field surveys that correlate the time course of air pollution levels with increased respiratory complaints.

The subglottic airways are accessible to the fiberoptic bronchoscope, and it is even possible to isolate an airway segment for antigen provocation and local lavage (Metzger et al. 1985a,b). Some aspects of airways epithelial function can be assessed. Mucociliary clearance is measured using radioactively labeled particles or radiologically opaque Teflon discs. Biopsies can be obtained for histologic and other studies. Airway surface liquid can be sampled without dilution using filter paper strips (Boucher et al. 1981). Alternatively, the fluid can be washed out and urea used as a suitable volume marker (Rennard et al. 1986). By use of sampling catheters, the profile of pollutant gas concentration along the airways during inspiration and expiration, and gas uptake in different airway regions, can be measured to provide information about dosimetry in human subjects.

The airways caliber can be assessed noninvasively by various techniques, and a variety of inhaled drugs have been used to assess airways smooth muscle and neural function. New imaging techniques based on nuclear magnetic resonance may be developed to study metabolism as well as anatomy of the airway tissues.

The airways epithelium is well modeled by the nasal epithelium, and patients with specific nasal allergies are available for study. The nasal epithelium can be exposed to inhaled gases and to antigens. Nasal surface liquid can be obtained for analysis. The epithelium can be studied in situ by electrophysiological techniques and can be biopsied for histologic evaluation. The behavior of airways smooth muscle is not, however, modeled by the nose.

Thus, substantial mechanistic as well as descriptive studies can be performed in appropriate human subjects. These studies will, however, need to be precisely designed and targeted to maximize the information obtained, and this will require careful attention to the results obtained in nonhuman studies.

Laboratory Animal Models

One of the problems in developing a suitable animal model for asthma is the fact that asthma is a chronic disease with demonstrable histologic and functional abnormalities even during periods of apparent health (see reviews by Wanner and Abraham 1982 and Hirshman 1985). In addition to persistent nonspecific bronchial hyperreactivity, asthmatic airways also exhibit chronic changes including smooth muscle hypertrophy, submucosal gland hypertrophy, increased numbers of epithelial goblet cells, and a thickened epithelial basement membrane. These changes presumably result from chronic activity of the underlying disease. Mediators that stimulate acute smooth muscle contraction or gland secretion may, over time, also cause hypertrophy of these structures. Chronic shedding of epithelial cells results in increased mitosis of the stem cells and may result in a relative increase in numbers of goblet cells and in a change in basement membrane properties. These changes may contribute to the persistent nonspecific bronchial hyperreactivity observed in asthmatic patients. Increased airway glands and smooth muscle, and, speculatively, reorganization of the smooth muscle into a single-unit type of structure with features of a syncytium are important in this regard. The tendency of asthmatics to develop transient bronchoconstriction, rather than the normal bronchodilation, when airways are passively dilated by a deep inspiration, is consistent with the behavior of single-unit muscle. However, isolated strips of bronchial smooth muscle from airways of asthmatics have failed to demonstrate exceptional reactivity to histamine.

To the extent that an animal model of asthma fails to exhibit baseline bronchial hyperreactivity or to manifest smooth muscle hypertrophy and chronic inflammatory changes in the airways, the model is not faithful to human asthma.

The fact that many cases of asthma do not appear to involve extrinsic antigen(s) may reduce the applicability of animal models in which bronchoconstriction is acutely evoked by antigen challenge. Sheep and dogs, for example, exhibit spontaneous sensitivity to *A. suum,* probably because of prior infestation with related worms. Other antigens can be used after a period of active immunization (for example, guinea pigs with ovalbumin, rabbits with antigen prepared from a mold, *Alternaria tenuis*). The product of crossing the basenji dog with the greyhound (the B-G dog) (see review by Hirshman 1985) exhibits an exceptional ease of sensitization to inhaled antigens and very marked changes in R_{aw} and lung compliance following antigen challenge. Interestingly, the B-G dog exhibits spontaneous nonspecific bronchial hyperreactivity to a variety of inhaled drugs, including methacholine, histamine, and citric acid, and responds to β-adrenergic blockade by an increase in airways resistance. The B-G dog and the allergic mongrel dog have been reported to release histamine and SRS-A during antigen challenge. In the B-G dog, in vitro release of histamine by leukocytes correlated with airways response to antigen challenge. The B-G dog also responds with SRS-A to a nonantigenic, citric acid aerosol.

An interesting rabbit model with early and late airway responses to *A. tenuis* antigen inhalation has been described by Larsen (1985). Both reactions are passively transferrable with IgE-containing serum and require neither cellular immune mechanisms nor IgG. Indeed, the presence of specific IgG antibody seems to blunt the response to antigen challenge in this model. The IgE-mediated early and late reactions respond to β-adrenergic agonists, cromolyn sodium, and corticosteroids similarly to those in human allergic asthmatics. Airways cellular infiltrate is seen in the early response. Finally, passively sensitized rabbits showed airways hyperreactivity to histamine 3 days after exposure to antigen which provoked a late asthmatic response.

Wanner and colleagues have extensively studied airways function in conscious sheep, including sheep allergic to *A. suum*. They investigated the in vivo effects on airways of a variety of inhaled substances, including O_3 and other pollutants, as well as specific antigens. Following antigen challenge, Abraham and coworkers (1983b) demonstrated a late-phase pulmonary response in allergic sheep. In addition, like asthmatic humans, these sheep exhibited increased airway smooth muscle tone after β-adrenergic blockade with propranolol (Wanner and Abraham 1982). Abraham and colleagues (1984) reported that although no immediate effect on specific lung resistance occurred after a 2-hr exposure to 1.0 ppm O_3, resistance doubled by 6.5–8 hr later. Some increase in leukotriene B_4 levels in bronchoalveolar lavage liquid was also found. Whether *A. suum*-sensitive sheep would exhibit similar or exaggerated delayed changes in lung resistance following O_3 exposure would be of interest.

Thus, a number of promising intact animal models of asthma are available. Tissue from the airways of such animals can be obtained for in vitro study and in vivo protocols. In addition, drugs that would be unsuitable in human subjects can be used.

Cells in Culture

The development of in vitro cell and organ culture techniques has greatly expanded the potential use of lung cells from humans as well as animals for research (for review, see Schiff 1986). For example, Friedman and colleagues (1985, 1986) studied the effects of O_3 exposure and of x-ray radiation on cultured bovine pulmonary artery endothelial cells. Several groups are studying the effect of O_3 exposure in vivo and in vitro on animal and human pulmonary alveolar macrophages in culture.

Crandall and colleagues (1987) examined the effects of NO_2 exposure on cultured rodent alveolar epithelial cells. Alink and coworkers (1983) described toxic effects of O_3 on cultured cells of the human A-549 alveolar type II cell line. Good techniques have been developed for isolating and culturing airway epithelial cells for functional as well as morphological studies (Yankaskas et al. 1985; Wu 1986). Bovine tracheal cells in culture have been observed to alter their arachidonate metabolism when exposed to O_3 (Leikauf et al. 1986). Van Scott and coworkers (1987) and Devereux and Fouts (1980) have isolated and cultured Clara cells which are particularly numerous in the small airways. Human bronchial tissue has been successfully explanted in organ culture (Lechner et al. 1986). Methods have been developed to expose cultured cells to pollutant gases in vitro in such a manner as to minimize the interference of the culture medium on the contact between gas and cells.

In view of the importance of airways epithelium in air pollution toxicology as well as asthma, the potential of cultured respiratory epithelial cells to evaluate precise dose/response relations for reactive gases and airways in various species, including humans, should be stressed. Pulmonary alveolar macrophages could serve as a marker for the alveolar region of the lung in measuring dose/response relations in vivo as well as in vitro.

■ **Recommendation 18.** Direct measurements of pollutant uptake (that is, covalent chemical reaction) by cells or tissues can be attempted using "labeled" gases such as $^{18}O_3$. Cultured pulmonary endothelial cells, alveolar macrophages, and epithelial cells could be used. Such studies would be particularly useful if they could be correlated to specific pollutant effects on the system under study.

■ **Recommendation 19.** Although not a good representative of the conducting airways, the pulmonary alveolar macrophage, as obtained by bronchoalveolar lavage following in vivo pollutant exposure, may exhibit functional changes that could be correlated with in vitro dose/response studies of cultured alveolar macrophages. Such data would provide some information on parenchymal tissue dosimetry.

Summary of Research Recommendations: Discussion

Controlled exposures of asthmatics to oxidant air pollutants have thus far largely failed to provoke reproducible airway responses. It therefore seems necessary to continue phenomenological or "descriptive" studies in the expectation that clearcut findings that can be correlated with data from ongoing field studies will emerge. Asthmatic subjects and persons with other diseases associated with bronchial hyperreactivity should be tested using NO_2 as well as O_3.

With regard to field studies, it may be useful to digress and present several suggestions despite the fact that this general area is addressed by Bresnitz and Rest (this volume). In studies of large populations (for example, Bates and Sizto 1983) careful attention to the air quality measurements is required. In addition to quantifying ambient gaseous and inorganic particulate pollutants, as well as temperature and relative humidity across a substantial area, it is necessary to ensure that all pollutant species are examined and that the measurements are frequent enough to assess the pattern of fluctuation of each component of interest. Indeed, these air quality measurements may point the way to developing controlled exposure protocols. In addition, if asthma is the health effect of concern, it is desirable to quantify common airborne particulate antigens. The possible presence of confounders, such as epidemics of respiratory infections, must also be considered.

If not prohibited by considerations of privacy and ethics, one should identify individuals within the study population whose respiratory health status "drives" any overall population correlations observed between asthma and air quality. Such individuals could be invited to join a panel of asthmatics for a prospective study. They could also be clinically characterized in depth to define the features of a putatively sensitive subpopulation. Finally, they might serve as subjects for controlled exposures.

In studies of panels of asthmatics, air quality monitoring might include personal monitors and home monitors in an effort to obtain the most precise exposure data. Again, common airborne allergens should be measured and evidence of respiratory tract infection sought. The respiratory status of panel members should be characterized in detail.

In addition to descriptive studies, studies aimed at enhancing our understanding of the mechanisms underlying the multiple airways effects of oxidant pollutants, especially O_3, should be considered. Such investigations should clarify the relation of automotive air pollution to relevant pathophysiological mechanisms in asthma and related diseases. Even if no relation to asthma were to emerge, a better understanding of this issue is essential to a rational health policy.

Some of these mechanistic studies can be performed in human subjects (for example, mucociliary clearance, airways surface liquid composition and volume, analysis of bronchial washes or of bronchoalveolar lavage samples for mediators and macromolecules). The nasal mucosa can serve as a model for the airways. However, to the extent that invasive procedures are required (bronchoscopy), there will be obvious limitations, especially in asthmatic subjects. Sensitive and specific analytical techniques for certain cellular secretory products will be needed, and their development may have to await further progress in lung biology.

Some investigations may be better suited by animal models, or by cultured cells, human as well as animal. For example, the bioelectric properties of respiratory epithelium are being explored in considerable detail using cultured cells as well as in vivo techniques. Macromolecular secretion can also be studied in cultured epithelial cell layers. Alveolar macrophages are particularly abundant in bronchoalveolar lavage liquid and can be cultured and exposed to air pollutants.

Excised perfused lungs allow for control of the composition of the perfusate as well as of the inspired gas or airways fluid. The neural elements are disrupted, however, and the bronchial circulation is absent. In situ isolated lobe preparations are therefore preferable to study permeability effects of O_3 exposure for various probe molecules including specific antigens. Intact animal models of antigen-induced asthma can be used for a variety of studies of pollutant effects.

Summary of Research Recommendations: Priorities

Descriptive Studies

HIGH PRIORITY

Recommendation 9   Exposure chamber atmospheres that mimic the acid sulfate content and particle size distribution in ambient pollution should be created. First, it will be necessary to analyze in greater detail the composition of ambient atmospheres associated with increased symptomatology in patients with asthma and other respiratory diseases.

Recommendation 10   Asthmatic subjects should be exposed to such chamber atmospheres, with monitoring of airways caliber. The study of atmospheres containing more than one pollutant will greatly complicate the design of such experiments and will require some choices to be made of concentrations of the pollutants, time course of the concentration of each pollutant, and particle size range of a particulate pollutant (for example, acid sulfate aerosol).

Recommendation 11   Response of airways of extrinsic asthmatics to pollutants, especially NO_2, should be assessed in the presence versus the absence of chronic low-level exposure to specific allergens.

Recommendation 16   The effect in allergic rhinitis patients or in extrinsic asthmatics of experimental pollutant exposures on postexposure nasal or bronchial reactivity ("early" and "late" phases) to specific antigen challenge should be ascertained.

MEDIUM PRIORITY

Recommendation 7   Exercising asthmatics should be exposed to relevant O_3 and NO_2 levels for periods up to 8 hr, with monitoring of airways caliber.

Recommendation 13   The effect of experimental pollutant exposures on airways function should be measured in patients with COPD (for example, nonspecific chronic bronchitis, cystic fibrosis), in whom increased bronchial reactivity is present.

Recommendation 15   The effect of oxidant pollutant exposures of asthmatics on bronchial reactivity to stimuli such as SO_2, cold dry air, nonisotonic aerosolized solutions, and certain mast cell–derived mediators should be explored.

LOW PRIORITY

Recommendation 8   Asthmatics should be exposed experimentally for 8 hr to O_3 or NO_2 for two or three consecutive days, with monitoring of airways caliber.

| Recommendation 12 | The effect of experimental pollutant exposures on nonspecific airways reactivity in asthmatic subjects selected to display a range of baseline reactivities should be measured. |

| Recommendation 14 | Pollutant effects on airways should be measured in extrinsic asthmatics in whom a transient state of marked bronchial hyperreactivity has been induced by a single antigen inhalation challenge. Individual subject responses could then be assessed over time at several levels of baseline bronchial reactivity. |

Mechanistic Studies
HIGH PRIORITY

| Recommendation 1 | The uptake profile of pollutant gases in different regions of the airways should be measured by sampling and analyzing inspired air at different airway levels. Variables to be explored with such systems include concentration of pollutants, ventilatory parameters, duration of exposure, and presence of disease (for example, asthma, chronic bronchitis). |

| Recommendation 2 | These data should be compared with the predictions of currently available mathematical models. |

| Recommendation 3 | The uptake profiles should be examined for their ability to account for some of the variability in vital capacity response to O_3 exposure observed among individuals. |

| Recommendation 4 | Specific neuropeptides should be assayed in airways surface liquid after O_3 exposure, and the phenomenon of "tolerance" to the O_3 effect on vital capacity should be explored along these lines. Highly sensitive neuropeptide assays will be required, and rapid inhibition of peptidase activity may also be necessary to prevent hydrolysis of peptides in the sample. |

| Recommendation 6 | Animals prepared so as to render their airway C-fiber systems nonfunctional should be used to examine the role of this sensory system in O_3 effects on epithelial permeability, mucous secretion, epithelial ion transport, bronchial reactivity, and airways inflammation. |

MEDIUM PRIORITY

| Recommendation 5 | O_3 responsiveness in human subjects should be compared to the responses to known stimulants of the airway C-fibers, such as capsaicin. |

| Recommendation 17 | The effects of exposure to O_3, or O_3 plus other air pollution components, on respiratory epithelial cell function should be probed in greater detail. These functions include mucociliary clearance, permeability to molecules including albumin and anti- |

gens, secretion of mucins and of other specific macromolecules, airway surface liquid composition and volume control, and release of mediator substances.

Recommendation 18   Direct measurements of pollutant uptake (that is, covalent chemical reaction) by cells or tissues can be attempted using "labeled" gases such as $^{18}O_3$. Cultured pulmonary endothelial cells, alveolar macrophages, and epithelial cells could be used. Such studies would be particularly useful if they could be correlated to specific pollutant effects on the system under study.

Recommendation 19   Although not a good representative of the conducting airways, the pulmonary alveolar macrophage, as obtained by bronchoalveolar lavage following in vivo pollutant exposure, may exhibit functional changes that could be correlated with in vitro dose/response studies of cultured alveolar macrophages. Such data would provide some information on parenchymal tissue dosimetry.

References

Abraham, W. M., Januszkiewicz, A. J., Mingle, M., Welker, M., Wanner, A., and Sackner, M. A. 1980. Sensitivity of bronchoprovocation and tracheal mucous velocity in detecting airway responses to O_3, *J. Appl. Physiol.* 48(5):789–793.

Abraham, W. M., Yerger, L., Marchette, B., and Wanner, A. 1983a. The effect of ozone on antigen-induced bronchospasm in allergic sheep, In: *Advances in Modern Environmental Toxicology, Vol. 5* (S. D. Lee, M. G. Mustafa, and M. A. Mehlman, eds.), Vol. 5, pp. 193–203, Princeton Scientific Publ. Inc., Princeton, N.J.

Abraham, W. M., Delehunt, J. C., Yerger, L., and Marchette, B. 1983b. Characterization of a late phase pulmonary response after antigen challenge in allergic sheep, *Am. Rev. Respir. Dis.* 128(5):839–844.

Abraham, W. M., Stevenson, J. S., Chapman, G. A., Yerger, L. D., Codias, E., Hernandez, A., and Sielczak, M. W. 1984. Ozone induces late bronchial responses after antigen challenge in allergic sheep, *Physiologist* 27:239 (abstr.).

Ahmed, T., Marchette, B., Danta, I., Birch, S., Dougherty, R. L., Schreck, R., and Sackner, M. A. 1982. Effect of 0.1 ppm NO_2 on bronchial reactivity in normals and subjects with bronchial asthma, *Am. Rev. Respir. Dis.* 125:A152.

Ahmed, T., Danta, I., Dougherty, R. L., Schreck, R., and Sackner, M. A. 1983. Effect of NO_2 on specific bronchial reactivity to ragweed antigen in subjects with allergic asthma, *Am. Rev. Respir. Dis.* 127:A160.

Correspondence should be addressed to Philip A. Bromberg, Center for Environmental Medicine, School of Medicine, The University of North Carolina, Chapel Hill, NC 27514.

Al-Bazzaz, F. J. 1986. Regulation of salt and water transport across airway mucosa, *Clin. Chest Med.* 7(2):259–272 (Review).

Al-Bazzaz, F. J., Kelsey, J. G., and Kaage, W. D. 1985. Substance P stimulation of chloride secretion by canine tracheal mucosa, *Am. Rev. Respir. Dis.* 131(1):86–89.

Alink, G. M., Rietjens, I. M. C. M., van der Linden, A. M. A., and Temmink, J. H. M. 1983. Biochemical and morphological effect of ozone on lung cells in vitro, In: *Advances in Modern Environmental Toxicology* (S. D. Lee, M. G. Mustafa, and M. A. Mehlman, eds.), Vol. 5, pp. 449–458, Princeton Scientific Publ. Inc., Princeton, N.J.

Avol, E. L., Linn, W. S., Venet, T. G., Shamoo, D. A., and Hackney, J. D. 1984. Comparative respiratory effects of ozone and ambient oxidant pollution exposure during heavy exercise, *J. Air Pollut. Control Assoc.* 34(8):804–809.

Barnes, P. J. 1984. The third nervous system in the lung: physiology and clinical perspectives, *Thorax* 39:561–567.

Barter, C. E., and Campbell, A. H. 1976. Relationship of constitutional factors and cigarette smoking to decrease in 1-second forced expiratory volume, *Am. Rev. Respir. Dis.* 113:305–314.

Basbaum, C. B. 1986. Regulation of airway secretory cells, *Clin. Chest Med.* 7(2):231–235.

Bates, D. V. 1985. The strength of the evidence relating air pollutants to adverse health effects, Carolina Environmental Essay Series VI, Institute for Environmental Studies, University of North Carolina, Chapel Hill.

Bates, D. V., and Sizto, R. 1983. Relationship between air pollution levels and hospital admissions in Southern Ontario, *Can. J. Public Health* 74:117–133.

Bauer, M. A., Utell, M. J., Morrow, P. E., Speers, D. M., and Gibb, F. R. 1985. Route of inhalation influences airway responses to 0.30 ppm NO_2 in asthmatic subjects, *Am. Rev. Respir. Dis.* 131:A171.

Bauer, M. A., Utell, M. J., Morrow, P. E., Speers, D. M., and Gibb, F. R. 1986. Inhalation of 0.3 ppm nitrogen dioxide potentiates exercise-induced bronchospasm in asthmatics, *Am. Rev. Respir. Dis.* 134(6):1203–1208.

Beckett, W. S., McDonnell, W. F., Horstman, D. H., and House, D. E. 1985. Role of the parasympathetic nervous system in acute lung response to ozone, *J. Appl. Physiol.* 59(6):1879–1885.

Bedi, J. F., Folinsbee, L. J., Horvath, S. M., and Ebenstein, R. S. 1979. Human exposure to sulfur dioxide and ozone: absence of a synergistic effect, *Arch. Environ. Health* 34:233–239.

Bedi, J. F., Dreshler-Parks, D. M., and Horvath, S. M. 1985. Duration of increased pulmonary function sensitivity to an initial ozone exposure, *Am. Ind. Hyg. Assoc. J.* 46(12):731–734.

Bell, K. A., Linn, W. S., Hazucha, M., Hackney, J. D., and Bates, D. V. 1977. Respiratory effects of exposure to ozone plus sulfur dioxide in Southern Californians and Eastern Canadians, *Am. Ind. Hyg. Assoc. J.* 38(12):696–706.

Besson, J. M., and Chaouch, A. 1987. Peripheral and spinal mechanisms of nociception, *Physiol. Rev.* 67:67–118.

Bethel, R. A., Erle, D. J., Epstein, J., Sheppard, D., Nadel, J. A., and Boushey, H. A. 1983. Effect of exercise rate and route of inhalation on sulfur-dioxide-induced bronchoconstriction in asthmatic subjects, *Am. Rev. Respir. Dis.* 128(4):592–596.

Bonner, J. R. 1984. The epidemiology and natural history of asthma, *Clin. Chest Med.* 5(4):557–565.

Boucher, R. C., Pare, P. D., Gilmore, N. J., Moroz, L. A., and Hogg, J. C. 1977. Airway mucosal permeability in the *Ascaris suum*-sensitive rhesus monkey, *J. Allergy Clin. Immunol.* 60(2):134–140.

Boucher, R. C., Ranga, V., Pare, P. D., Inoue, S., Moroz, L. A., and Hogg, J. C. 1978. Effect of histamine and methacholine on guinea pig tracheal permeability to HRP, *J. Appl. Physiol.* 45(6):939–948.

Boucher, R. C., Pare, P. D., and Hogg, J. C. 1979. Relationship between airway hyperreactivity and hyperpermeability in *Ascaris*-sensitive monkeys, *J. Allergy Clin. Immunol.* 64(3):197–201.

Boucher, R. C., Johnson, J., Inoue, S., Hulbert, W., and Hogg, J. C. 1980. The effect of cigarette smoke on the permeability of guinea pig airways, *Lab. Invest.* 43(1):94–100.

Boucher, R. C., Stutts, M. J., Bromberg, P. A., and Gatzy, J. T. 1981. Regional differences in airway surface liquid composition, *J. Appl. Physiol.* 50(3):613–620.

Boushey, H. A. 1985. Role of the vagus nerves in bronchoconstriction in humans, *Chest* 87:197S–201S.

Boushey, H. A., Holtzman, M. J., Sheller, J. R., and Nadel, J. A. 1980. Bronchial hyperreactivity, *Am. Rev. Respir. Dis.* 121(2):389–413 (Review).

Brink, C., Duncan, P. G., and Douglas, J. S. 1981. The response and sensitivity to histamine of respiratory tissues from normal and ovalbumin-sensitized guinea pigs: effects of cyclooxygenase and lipoxygenase inhibition, *J. Pharmacol. Exp. Ther.* 217(3):592–601.

Broder, I., Barlow, P. P., and Horton, R. J. M. 1962. The epidemiology of asthma and hay fever in a total community, Tecumseh, Michigan, *J. Allerg.* 33:513–523, 524–531.

Broder, I., Higgins, M. W., Mathews, K. P., and Keller, J. B. 1974a. Epidemiology of asthma and allergic rhinitis in a total community, Tecumseh, Michigan. III. Second survey of the community, *J. Allergy Clin. Immunol.* 53(3):127–138.

Broder, I., Higgins, M. W., Mathews, K. P., and Keller, J. B. 1974b. Epidemiology of asthma and allergic rhinitis in a total community, Tecumseh, Michigan. IV. Natural history, *J. Allergy Clin. Immunol.* 54(2):100–110.

Bromberg, P. A. 1987. Mechanism of effects of ozone inhalation on lung function: a hypothesis, In: *Fourth Health Effects Institute Annual Conference,* p. 37 (abstr.), Health Effects Institute, Cambridge, Mass.

Bromberg, P. A., Ivanick, M., and Ranga, V. 1984. Effect of ozone (O_3) on airways epithelial permeability to polar solutes: demonstration of adaptation following repeated exposure, *Physiologist* 27:276 (abstr.).

Buckle, F. G., and Cohen, A. B. 1975. Nasal mucosal hyperpermeability to macromolecules in atopic rhinitis and extrinsic asthma, *J. Allergy Clin. Immunol.* 55(4):213–221.

Busse, W. W. 1985. The precipitation of asthma by upper respiratory infections, *Chest* 87:44S–48S.

Bylin, G., Lindvall, T., Rehn, T., and Sundin, B. 1985. Effects of short-term exposure to ambient nitrogen dioxide concentrations on human bronchial reactivity and lung function, *Eur. J. Respir. Dis.* 66(3):205–217.

Cartier, A., Thomson, N. C., Frith, P. A., Roberts, R., and Hargreave, F. E. 1982. Allergen-induced increase in bronchial responsiveness to histamine: relationship to the late asthmatic response and change in airway caliber, *J. Allergy Clin. Immunol.* 70(3):170–177.

Cockcroft, D. W. 1983. Mechanism of perennial allergic asthma, *Lancet* 2(8344):253–256.

Cockcroft, D. W., Ruffin, R. E., Frith, P. A., Cartier, A., Juniper, E. F., Dolovich, J., and Hargreave, F. E. 1979. Determinants of allergen-induced asthma: dose of allergen, circulating IgE antibody concentration, and bronchial responsiveness to inhaled histamine, *Am. Rev. Respir. Dis.* 120(5):1053–1058.

Coleridge, H. M., and Coleridge, J. C. G. 1986. Reflexes evoked from tracheobronchial tree and lungs, In: *Handbook of Physiology,* Section 3, The Respiratory System. Vol. II. Control of Breathing, Part 1, Chap. 12. (N. S. Cherniack and J. G. Widdicombe, eds.), pp. 395–429, American Physiological Society, Bethesda, Md.

Coleridge, H. M., Coleridge, J. C. G., and Luck, J. C. 1965. Pulmonary afferent fibers of small diameter stimulated by capsaicin and by hyperinflation of the lungs, *J. Physiol. London* 179:248–262.

Coleridge, H. M., Coleridge, J. C., Baker, D. G.,

Ginzel, K. H., and Morrison, M. A. 1978. Comparison of the effects of histamine and prostaglandin on afferent C-fiber endings and irritant receptors in the intrapulmonary airways, *Adv. Exp. Med. Biol.* 99:291–305.

Coles, S. J., Neill, K. H., and Reid, L. M. 1984. Potent stimulation of glycoprotein secretion in canine trachea by substance P, *J. Appl. Physiol.* 57(5):1323–1327.

Collier, J. C., and Fuller, R. W. 1984. Capsaicin inhalation in man and the effects of sodium cromoglycate, *Br. J. Pharm.* 81:113–117.

Crandall, E. D., Postlethwait, E. M., and Cheek, J. M. 1987. Effects of nitrogen dioxide on alveolar epithelial transport properties, In: *Fourth Health Effects Institute Annual Conference,* p. 38 (abstr.), Health Effects Institute, Cambridge, Mass.

Cutz, E., Levison, H., and Cooper, D. M. 1978. Ultrastructure of airways in children with asthma, *Histopathology* 2(6):407–421.

Davis, B., Roberts, A. M., Coleridge, H. M., and Coleridge, J. C. 1982. Reflex tracheal gland secretion evoked by stimulation of bronchial C-fibers in dogs, *J. Appl. Physiol.* 53(4):985–991.

Davis, J. D., Gallo, J., Hu, E. P. C., Boucher, R. C., and Bromberg, P. A. 1980. The effects of ozone on respiratory epithelial permeability, *Am. Rev. Respir. Dis.* 121:231 (abstr.).

Dawson, S. V., and Schenker, M. B. 1979. Editorial: health effects of inhalation of ambient concentrations of nitrogen dioxide, *Am. Rev. Respir. Dis.* 120:281–292.

Deal, E. C., Jr., Wasserman, S. I., Soter, N. A., Ingram, R. H., Jr., and McFadden, E. R., Jr. 1980. Evaluation of role played by mediators of immediate hypersensitivity in exercise-induced asthma, *J. Clin. Invest.* 65(3):659–665.

Devereux, T. R., and Fouts, J. R. 1980. Isolation and identification of Clara cells from rabbit lung, *In Vitro* 16:958–968.

DiMeo, M. J., Glenn, M. G., Holtzman, M. J., Sheller, J. R., Nadel, J. A., and Boushey, H. A. 1981. Threshold concentrations of ozone causing an increase in bronchial reactivity in humans and adaptation with repeated exposures, *Am. Rev. Respir. Dis.* 124:245–248.

Dixon, J. R., and Mountain, J. T. 1965. Role of histamine and related substances in development of tolerance to edemagemic gases, *Toxicol. Appl. Pharmacol.* 7:756–766.

Dunnhill, M. S. 1971. The pathology of asthma, In: *Identification of Asthma* (R. Porter and J. Birch, eds.), Ciba Foundation Study Group No. 38. Churchill Livingstone, Edinburgh.

Dunnhill, M. S. 1975. The morphology of the airways in bronchial asthma, In: *New Directions in Asthma* (M. Stein, ed.), pp. 213–221, American College of Chest Physicians, Park Ridge, Ill.

Dvorak, A. M., Dvorak, H. F., Peters, S. P., Shulman, E. S., MacGlashan, D. W., Jr., Pyne, K., Harvey, V. S., Galli, S. J., and Lichtenstein, L. M. 1983. Lipid bodies: cytoplasmic organelles important to arachidonate metabolism in macrophages and mast cells, *J. Immunol.* 131(6):2965–2976.

Easton, R. E., and Murphy, S. D. 1967. Experimental ozone preexposure and histamine, *Arch. Environ. Health* 15:160–166.

Eggleston, P. A., Kagey-Sobotka, A., Schleimer, R. P., and Lichtenstein, L. M. 1984. Interaction between hyperosmolar and IgE-mediated histamine release from basophils and mast cells, *Am. Rev. Respir. Dis.* 130(1):86–91.

Eggleston, P. A., Kagey-Sobotka, A., and Lichtenstein, L. M. 1987. A comparison of the osmotic activation of basophils and human lung mast cells, *Am. Rev. Respir. Dis.* 135(5):1043–1048.

Elwood, R. K., Kennedy, S., Belzberg, A., Hogg, J. C., and Pare, P. D. 1983. Respiratory mucosal permeability in asthma, *Am. Rev. Respir. Dis.* 128(3):523–527.

Eschenbacher, W. L., Boushey, H. A., and Sheppard, D. 1984. Alteration in osmolarity of inhaled aerosols causes bronchoconstriction and cough, but absence of a permeant anion causes cough alone, *Am. Rev. Respir. Dis.* 129(2):211–215.

Fabbri, L. M. 1985. Airway inflammation and asthma. Importance of arachidonate metabolites for airway hyperresponsiveness, *Prog. Biochem. Pharmacol.* 20:18–25 (Review).

Farrell, B. P., Kerr, H. D., Kulle, T. J., Sauder, L. R., and Young, J. L. 1979. Adaptation in human subjects to the effects of inhaled ozone after repeated exposures, *Am. Rev. Respir. Dis.* 119(5):725–730.

Findlay, S. R., Dvorak, A. M., Kagey-Sobotka, A., and Lichtenstein, L. M. 1981. Hyperosmolar triggering of histamine release from human basophils, *J. Clin. Invest.* 67(6):1604–1613.

Fish, J. E., and Norman, P. S. 1985. Responsiveness to prostaglandin $F_{2\alpha}$ in atopic and non-atopic subjects, *Chest* 87:206S–207S.

Flavahan, N. A., and Vanhoutte, P. M. 1985. The respiratory epithelium releases a smooth muscle relaxing factor, *Chest* 87:189S–190S.

Flavahan, N. A., Aarhus, L. L., Rimele, T. J., and Vanhoutte, P. M. 1985. Respiratory epithelium inhibits bronchial smooth muscle tone, *J. Appl. Physiol.* 58(3):834–838.

Folinsbee, L. J., and Horvath, S. M. 1986. Persistence of the acute effects of ozone exposure, *Aviat. Space Environ. Med.* 57(12):1136–1143.

Folinsbee, L. J., Bedi, J. F., and Horvath, S. M. 1980. Respiratory responses in humans repeatedly exposed to low concentrations of ozone, *Am. Rev. Respir. Dis.* 121(3):431–439.

Folinsbee, L. J., Bedi, J. F., and Horvath, S. M. 1985. Pulmonary response to threshold levels of sulfur dioxide (1.0 ppm) and ozone (0.3 ppm), *J. Appl. Physiol.* 58(6):1783–1787.

Friedman, M., Madden, M. C., Saunders, D. S., Gammon, K., White, G. C., II, and Kwock, L. 1985. Ozone inhibits prostacyclin synthesis in pulmonary endothelium, *Prostaglandins* 30(6):1069–1083.

Friedman, M., Ryan, U. S., Davenport, W. C., Chaney, E. L., Strickland, D. L., and Kwock, L. 1986. Reversible alterations in cultured pulmonary artery endothelial cell monolayer morphology and albumin permeability induced by ionizing radiation, *J. Cell. Physiol.* 129(2):237–249.

Frossard, N., and Muller, F. 1986. Epithelial modulation of tracheal smooth muscle responses to antigenic stimulation, *J. Appl. Physiol.* 61(4):1449–1456.

Fuller, R. W., Dixon, C. M., and Barnes, P. J. 1985. Bronchoconstrictor response to inhaled capsaicin in humans, *J. Appl. Physiol.* 58(4):1080–1084.

Gershwin, L. J., Osebold, J. W., and Zee, Y. C. 1981. Immunoglobulin E–containing cells in mouse lung following allergen inhalation and ozone exposure, *Int. Arch. Allergy Appl. Immunol.* 65(3):266–277.

Gertner, A., Bromberger-Barnea, B., Dannenberg, A. M., Jr., Traystman, R., and Menkes, H. 1983a. Responses of the lung to 1.0 ppm ozone, *J. Appl. Physiol.* 55(3):770–776.

Gertner, A., Bromberger-Barnea, B., Traystman, R., Berzon, D., and Menkes, H. 1983b. Responses of the lung periphery to ozone and histamine, *J. Appl. Physiol.* 54(3):640–646.

Gertner, A., Bromberger-Barnea, B., Traystman, R., and Menkes, H. 1983c. Effects of ozone on peripheral lung reactivity, *J. Appl. Physiol.* 55(3):777–784.

Gleich, G. J. 1982. The late phase of the immunoglobulin E-mediated reaction: a link between anaphylaxis and common allergic disease? *J. Allergy Clin. Immunol.* 70(3):160–169 (Review).

Gleich, G. J., Loegering, D. A., and Adolphson, C. R. 1985. Eosinophils and bronchial inflammation, *Chest* 87(1):10S–13S.

Golden, J. A., Nadel, J. A., and Boushey, H. A. 1978. Bronchial hyperirritability in healthy subjects after exposure to ozone, *Am. Rev. Respir. Dis.* 118(2):287–294.

Gordon, T., and Amdur, M. O. 1980. Effect of ozone on respiratory response of guinea pigs to histamine, *J. Toxicol. Environ. Health* 6:185–195.

Gordon, T., Venugopalan, C. S., Amdur, M. O., and Drazen, J. M. 1984. Ozone-induced airway hyperreactivity in the guinea pig, *J. Appl. Physiol.* 57(4):1034–1038.

Graham, D., Henderson, F., and House, D. 1985. An increase in nasal lavage polymorphonuclear leukocytes in humans exposed to ozone, *Am. Rev. Respir. Dis.* 131:A172.

Hackney, J. D., Linn, W. S., Mohler, J. G., and Collier, C. R. 1977. Adaptation to short-term respiratory effects of ozone in men exposed repeatedly, *J. Appl. Physiol.* 43(1):82–85.

Hackney, J. D., Linn, W. S., Fischer, A., Shamoo, D. A., Anzar, U. T., Spier, C. E., Valencia, L. M., and Venet, T. G. 1983. Effects of ozone in people with chronic obstructive lung disease (COLD), In: *Advances in Modern Environmental Toxicology* (S. D. Lee, M. G. Mustafa, and M. A. Mehlman, eds.), Vol. 5, pp. 205–211, Princeton Scientific Publ. Inc., Princeton, N.J.

Hahn, H. L., Graf, P. D., and Nadel, J. A. 1976. Effect of vagal tone on airway diameters and on lung volume in anesthetized dogs, *J. Appl. Physiol.* 41(4):581–589.

Hargreave, F. E., Ryan, G., Thomson, N. C., O'Byrne, P. M., Latimer, K., Juniper, E. F., and

Dolovich, J. 1981. Bronchial responsiveness to histamine or methacholine in asthma: measurement and clinical significance, *J. Allergy Clin. Immunol.* 68:347–355.

Hargreave, F. E., O'Byrne, P. M., and Ramsdale, E. H. 1985a. Mediators, airway responsiveness, and asthma, *J. Allergy Clin. Immunol.* 76(2):272–276.

Hargreave, F. E., Sterk, P. J., Ramsdale, E. H., Dolovich, J., and Zamel, N. 1985b. Inhalation challenge tests and airway responsiveness in man, *Chest* 87:202S–206S.

Hazucha, M. J., and Bates, D. V. 1975. Combined effect of ozone and sulfur dioxide on human pulmonary function, *Nature (London)* 257:50–55.

Hazucha, M. J., Ginsberg, J. F., McDonnell, W. F., Haak, E. D., Jr., Pimmel, R. L., Salaam, S. A., House, D. E., and Bromberg, P. A. 1983. Effects of 0.1 ppm nitrogen dioxide on airways of normal and asthmatic subjects, *J. Appl. Physiol.* 54(3):730–739.

Hazucha, M. J., Bates, D. V., and Bromberg, P. A. 1986. Mechanisms of action of ozone on the human lung, *Am. Rev. Respir. Dis.* 133:A214.

Hill, D. J., Hosking, C. S., Shelton, M. J., and Turner, M. W. 1981. Growing out of asthma: clinical and immunological changes over 5 years, *Lancet* 2:1359–1362.

Hirshman, C. A. 1985. The basenji-greyhound dog model of asthma, *Chest* 87:172S–178S.

Hogg, J. C. 1985. The pathology of asthma, *Chest* 87:152S–153S.

Hogg, J. C., and Eggleston, P. A. 1984. Is asthma an epithelial disease? *Am. Rev. Respir. Dis.* 129(2):207–208.

Hogg, J. C., Pare, P. D., and Boucher, R. C. 1979. Bronchial mucosal permeability, *Fed. Proc.* 38(2):197–201.

Holtzman, M. J., Cunningham, J. H., Sheller, J. R., Irsigler, G. B., Nadel, J. A., and Boushey, H. A. 1979. Effect of ozone on bronchial reactivity in atopic and nonatopic subjects, *Am. Rev. Respir. Dis.* 120(5):1059–1067.

Holtzman, M. J., Fabbri, L. M., Skoogh, B. E., O'Byrne, P. M., Walters, E. H., Aizawa, H., and Nadel, J. A. 1983. Time course of airway hyperresponsiveness induced by ozone in dogs, *J. Appl. Physiol.* 55(4):1232–1236.

Horvath, S. M., Gliner, J. A., and Folinsbee, L. J. 1981. Adaptation to ozone: duration of effect, *Am. Rev. Respir. Dis.* 123:496–499.

Hu, P. C., Miller, F. J., Daniels, M. J., Hatch, G. E., Graham, J. A., Gardner, D. E., and Selgrade, M. K. 1982. Protein accumulation in lung lavage fluid following ozone exposure, *Environ. Res.* 29(2):377–388.

Hua, X. Y., Theodorsson, N. E., Brodin, E., Lundberg, J. M., and Hokfelt, T. 1985. Multiple tachykinins (neurokinin A, neuropeptide K and substance P) in capsaicin-sensitive sensory neurons in the guinea-pig, *Regul. Pept.* 13(1):1–19.

Iversen, L. L. 1982. Substance P, *Br. Med. Bull.* 38(3):277–282 (Review).

Kaliner, M. 1985. Mast cell mediators and asthma, *Chest* 87:2S–5S.

Kaliner, M., Marom, Z., Patow, C., and Shelhamer,

J. 1984. Human respiratory mucus, *J. Allergy Clin. Immunol.* 73(3):318–323 (Review).

Kamm, K. E., and Stull, J. T. 1985. The function of myosin and myosin light chain kinase phosphorylation in smooth muscle, *Annu. Rev. Pharmacol. Toxicol.* 25:593–620 (Review).

Kehrl, H. R., Hazucha, M. J., Solic, J. J., and Bromberg, P. A. 1983. The acute effects of 0.2 and 0.3 ppm ozone in persons with chronic obstructive lung disease (COLD), In: *Advances in Modern Environmental Toxicology* (S. D. Lee, M. G. Mustafa, and M. A. Mehlman, eds.), Vol. 5, pp. 213–225, Princeton Scientific Publ., Inc., Princeton, N.J.

Kehrl, H. R., Hazucha, M. J., Solic, J. J., and Bromberg, P. A. 1985. Responses of subjects with chronic obstructive pulmonary disease after exposures to 0.3 ppm ozone, *Am. Rev. Respir. Dis.* 131(5):719–724.

Kehrl, H. R., Vincent, L. M., Kowalsky, R. J., Horstman, D. H., O'Neil, J. J., McCartney, W. H., and Bromberg, P. A. 1987. Ozone exposure increases respiratory epithelial permeability in humans, *Am. Rev. Respir. Dis.* 135(5):1124–1128.

Kennedy, S. M., Elwood, R. K., Wiggs, B. J., Pare, P. D., and Hogg, J. C. 1984. Increased airway mucosal permeability of smokers. Relationship to airway reactivity, *Am. Rev. Respir. Dis.* 129(1): 143–148.

Kleinman, M. T., Bailey, R. M., Chang, Y. T., Clark, K. W., Jones, M. P., Linn, W. S., and Hackney, J. D. 1981. Exposures of human volunteers to a controlled atmospheric mixture of ozone, sulfur dioxide and sulfuric acid, *Am. Ind. Hyg. Assoc. J.* 42(1):61–69.

Kleinman, M. T., Bailey, R. M., Linn, W. S., Anderson, K. R., Whynot, J. D., Shamoo, D. A., and Hackney, J. D. 1983. Effects of 0.2 ppm nitrogen dioxide on pulmonary function and response to bronchoprovocation in asthmatics, *J. Toxicol. Environ. Health* 12:815–826.

Klinger, J. D., Tandler, B., Liedtke, C. M., and Boat, T. F. 1984. Proteinases of Pseudomonas aeruginosa evoke mucin release by tracheal epithelium, *J. Clin. Invest.* 74(5):1669–1678.

Knauer, K. A., Lichtenstein, L. M., Adkinson, N. F., Jr., and Fish, J. E. 1981. Platelet activation during antigen-induced airway reactions in asthmatic subjects, *N. Engl. J. Med.* 304(23):1404–1407.

Knowles, M. R., Stutts, M. J., Yankaskas, J. R., Gatzy, J. T., and Boucher, R. C., Jr. 1986. Abnormal respiratory epithelial ion transport in cystic fibrosis, *Clin. Chest Med.* 7(2):285–297 (Review).

Koenig, J. Q., Covert, D. S., Morgan, M. S., Horike, M., Horike, N., Marshall, S. G., and Pierson, W. E. 1985. Acute effects of 0.12 ppm ozone or 0.12 ppm nitrogen dioxide on pulmonary function in healthy and asthmatic adolescents, *Am. Rev. Respir. Dis.* 132(3):648–651.

Kulle, T. J. 1982. Effects of NO_2 on pulmonary function in normal healthy humans and subjects with asthma and chronic bronchitis, In: *Air Pollution by Nitrogen Oxides* (T. Schneider and L. Grant, eds.), pp. 477–486, Elsevier, Amsterdam.

Kulle, T. J., Kerr, H. D., Farrell, B. P., Sauder, L. R., and Bermel, M. S. 1982. Pulmonary function and bronchial reactivity in human subjects with exposure to ozone and respirable sulfuric acid aerosol, *Am. Rev. Respir. Dis.* 126(6):996–1000.

Kulle, T. J., Milman, J. H., Sauder, L. R., Kerr, H. D., Farrell, B. P., and Miller, W. R. 1984. Pulmonary function adaptation to ozone in subjects with chronic bronchitis, *Environ. Res.* 34(1):55–63.

Laitinen, L. A., Heino, M., Laitinen, A., Kava, T., and Haahtela, T. 1985. Damage to the airway epithelium and bronchial reactivity in patients with asthma, *Am. Rev. Respir. Dis.* 131(4):599–606.

Larsen, G. L. 1985. The rabbit model of the late asthmatic response, *Chest* 87:184S–188S.

Lebowitz, M. D., Knudson, R. J., and Burrows, B. 1975. Tucson epidemiologic study of obstructive lung diseases. I: Methodology and prevalence of disease, *Am. J. Epidemiol.* 102(2):137–152.

Lechner, J. F., Stoner, G. D., Yoakum, G. H., Willey, J. C., Grafstrom, R. C., Masui, T., LaVeck, M. A., and Harris, C. C. 1986. In vitro carcinogenesis studies with human tracheobronchial tissues and cells, In: *In Vitro Models of Respiratory Epithelium* (L. V. Schiff, ed.), Chap. 6, pp. 143–159, CRC Press, Boca Raton, Fla.

Lee, H. K., and Murlas, C. 1985. Ozone-induced bronchial hyperreactivity in guinea pigs is abolished by BW 755C or FPL 55712 but not by indomethacin, *Am. Rev. Respir. Dis.* 132:1005–1009.

Lee, L. Y., Bleecker, E. R., and Nadel, J. A. 1977. Effect of ozone on bronchomotor response to inhaled histamine aerosol in dogs, *J. Appl. Physiol.* 43(4):626–631.

Lee, T. H., Brown, M. J., Nagy, L., Causon, R., Walport, M. J., and Kay, A. B. 1982a. Exercise-induced release of histamine and neutrophil chemotactic factor in atopic asthmatics, *J. Allergy Clin. Immunol.* 70(2):73–81.

Lee, T. H., Nagy, L., Nagakura, T., Walport, M. J., and Kay, A. B. 1982b. Identification and partial characterization of an exercise-induced neutrophil chemotactic factor in bronchial asthma, *J. Clin. Invest.* 69(4):889–899.

Lee, T. H., Assoufi, B. K., and Kay, A. B. 1983a. The link between exercise, respiratory heat exchange, and the mast cell in bronchial asthma, *Lancet* 1(8323):520–522.

Lee, T. H., Nagakura, T., Papageorgiou, N., Iikura, Y., and Kay, A. B. 1983b. Exercise-induced late asthmatic reactions with neutrophil chemotactic activity, *N. Engl. J. Med.* 308(25):1502–1505.

Leikauf, G. D., Ueki, I. F., Nadel, J. A., and Widdicombe, J. H. 1986. Release of cyclooxygenase products from cultured epithelium derived from human and dog trachea, *Fed. Proc.* 44:1920 (abstr.).

Linn, W. S., and Hackney, J. 1984. Short-Term Human Respiratory Effects of NO_2: Determination of Quantitative Dose-Response Profiles. Phase II. Exposure of Asthmatic Volunteers to 4 ppm NO_2, CAPM 48-83, Final Report, Coordinating Research Council, Atlanta, Ga.

Linn, W. S., Buckley, R. D., Spier, C. E., Blessey, R. L., Jones, M. P., Fischer, D. A., and Hackney,

J. D. 1978. Health effects of ozone exposure in asthmatics, *Am. Rev. Respir. Dis.* 117(5):835–843.

Linn, W. S., Jones, M. P., Bachmayer, E. A., Spier, C. E., Mazur, S. F., Avol, E. L., and Hackney, J. D. 1980. Short-term respiratory effects of polluted ambient air: a laboratory study of volunteers in a high-oxidant community, *Am. Rev. Respir. Dis.* 121(2):243–252.

Linn, W. S., Fischer, D. A., Medway, D. A., Anzar, U. T., Spier, C. E., Valencia, L. M., Venet, T. G., and Hackney, J. D. 1982. Short-term respiratory effects of 0.12 ppm ozone exposure in volunteers with chronic obstructive pulmonary disease, *Am. Rev. Respir. Dis.* 125(6):658–663.

Linn, W. S., Shamoo, D. A., Venet, T. G., Spier, C. E., Valencia, L. M., Anzar, U. T., and Hackney, J. D. 1983. Response to ozone in volunteers with chronic obstructive pulmonary disease, *Arch. Environ. Health* 38(5):278–283.

Linn, W. S., Solomon, J. C., Trim, S. C., Spier, C. E., Shamoo, D. A., Venet, T. G., Avol, E. L., and Hackney, J. D. 1985. Effects of exposure to 4 ppm nitrogen dioxide in healthy and asthmatic volunteers, *Arch. Environ. Health* 40(4):234–239.

Lippman, M., Lioy, P. J., Leikauf, G., Green, K. B., Baxter, D., Morandi, M., and Pasternack, B. S. 1983. Effects of ozone on the pulmonary function of children, In: *Advances in Modern Environmental Toxicology* (S. D. Lee, M. G. Mustafa, and M. A. Mehlman, eds.), Vol. 5, pp. 423–446, International Symposium on the Biomedical Effects of Ozone and Related Photochemical Oxidants, Princeton Scientific Publ., Inc., Princeton, N.J.

Lundberg, J. M., Brodin, E., and Saria, A. 1983a. Effects and distribution of vagal capsaicin-sensitive substance P neurons with special reference to the trachea and lungs, *Acta Physiol. Scand.* 119(3):243–252.

Lundberg, J. M., Martling, C. R., Saria, A., Folkers, K., and Rosell, S. 1983b. Cigarette smoke-induced oedema due to activation of capsaicin-sensitive vagal afferents and substance P release, *Neuroscience* 10(4):1361–1368.

Lundberg, J. M., Hokfelt, T., Martling, C. R., Saria, A., and Cuello, C. 1984. Substance P–immunoreactive sensory nerves in the lower respiratory tract of various mammals including man, *Cell Tissue Res.* 235(2):251–261.

Martin, J. G., Dong-Jie, D., and Macklem, P. T. 1986. Effects of lung volume on methacholine-induced bronchoconstriction in normal subjects, *Am. Rev. Respir. Dis.* 133:A15.

Matsumura, Y. 1970a. The effects of ozone, nitrogen dioxide, and sulfur dioxide on the experimentally induced allergic respiratory disorder in guinea pigs. I. The effect of sensitization with albumin through the airway, *Am. Rev. Respir. Dis.* 102:430–437.

Matsumura, Y. 1970b. The effects of ozone, nitrogen dioxide, and sulfur dioxide on the experimentally induced allergic respiratory disorder in guinea pigs. II. The effects of ozone on the absorption and the retention of antigen in the lung, *Am. Rev. Respir. Dis.* 102:438–447.

Matsumura, Y., Mizuno, K., Miyamoto, T., Suzuki,

T., and Oshima, Y. 1972. The effects of ozone, nitrogen dioxide, and sulfur dioxide on experimentally induced allergic respiratory disorder in guinea pigs. IV. Effects on respiratory sensitivity to inhaled acetylcholine, *Am. Rev. Respir. Dis.* 105(2):262–267.

McDonnell, W. F., Horstman, D. H., Hazucha, M. J., Seal, E., Jr., Haak, E. D., Salaam, S. A., and House, D. E. 1983. Pulmonary effects of ozone exposure during exercise: dose-response characteristics, *J. Appl. Physiol.* 54(5):1345–1352.

McDonnell, W. F., Chapman, R. S., Leigh, M. W., Strope, G. L., and Collier, A. M. 1985a. Respiratory responses of vigorously exercising children to 0.12 ppm ozone exposure, *Am. Rev. Respir. Dis.* 132:875–879.

McDonnell, W. F., Horstman, D. H., Abdul-Salaam, S., and House, D. E. 1985b. Reproducibility of individual responses to ozone exposure, *Am. Rev. Respir. Dis.* 131:36–40.

Metzger, W. J., Moseley, P., Nugent, K., Richerson, H. B., and Hunninghake, G. W. 1985a. Local antigen challenge and bronchoalveolar lavage of allergic asthmatic lungs, *Chest* 87:S155–S156.

Metzger, W. J., Nugent, K., Richerson, H. B., Moseley, P., Lakin, R., Zavala, D., and Hunninghake, G. W. 1985b. Methods for bronchoalveolar lavage in asthmatic patients following bronchoprovocation and local antigen challenge, *Chest* 87:S16–S19.

Murlas, C., and Lee, H. K. 1985. U-60257 inhibits O_3 induced bronchial hyperreactivity in the guinea pig, *Prostaglandins* 30:563–572.

Murlas, C., and Roum, J. H. 1985a. Sequence of pathological changes in the airway mucosa of guinea pigs during ozone-induced bronchial hyperreactivity, *Am. Rev. Respir. Dis.* 131:314–320.

Murlas, C., and Roum, J. H. 1985b. Bronchial hyperreactivity occurs in steroid-treated guinea pigs depleted of leukocytes by cyclophosphamide, *J. Appl. Physiol.* 58:1630–1637.

Murlas, C., Lee, H. K., and Roum, J. H. 1986. Indomethacin increases bronchial reactivity after exposure to subthreshold ozone levels, *Prostaglandins Leukotrienes Med.* 21(3):259–268.

Nadel, J., and Barnes, P. 1984. Autonomic regulation of the airways, *Ann. Rev. Med.* 35:451–467.

Nagy, L., Lee, T. H., and Kay, A. B. 1982. Neutrophil chemotactic activity in antigen-induced late asthmatic reactions, *N. Engl. J. Med.* 306:497–501.

Naylor, B. 1962. The shedding of the mucosa of the bronchial tree in asthma, *Thorax* 17:69.

O'Byrne, P. M., Walters, E. H., Aizawa, H., Fabbri, L. M., Holtzman, M. J., and Nadel, J. A. 1984a. Indomethacin inhibits the airway hyperresponsiveness but not the neutrophil influx induced by ozone in dogs, *Am. Rev. Respir. Dis.* 130(2):220–224.

O'Byrne, P. M., Walters, E. H., Gold, B. D., Aizawa, H. A., Fabbri, L. M., Alpert, S. E., Nadel, J. A., and Holtzman, M. J. 1984b. Neutrophil depletion inhibits airway hyperresponsiveness induced by ozone exposure, *Am. Rev. Respir. Dis.* 130(2):214–219.

Orehek, J., Massari, J. P., Gayrard, P., Grimaud, C., and Charpin, J. 1976. Effect of short-term, low-level nitrogen dioxide exposure on bronchial sensitivity of asthmatic patients, *J. Clin. Invest.* 57(2): 301–307.

Osebold, J. W., Gershwin, L. J., and Zee, Y. C. 1980. Studies on the enhancement of allergic lung sensitization by inhalation of ozone and sulfuric acid aerosol, *J. Environ. Pathol. Toxicol.* 3(5/6):221–234.

Payan, D. G., Levine, J. D., and Goetzl, E. J. 1984. Modulation of immunity and hypersensitivity by sensory neuropeptides, *J. Immunol.* 132(4):1601–1604 (Review).

Peters, S. P., Naclerio, R. M., Togias, A., Schleimer, R. P., MacGlashan, D. W., Kagey-Sobotka, A., Adkinson, N. F., Norman, P. S., and Lichtenstein, L. M. 1985. In vitro and in vivo model systems for the study of allergic and inflammatory disorders in man. Implications for the pathogenesis of asthma, *Chest* 87:162S–164S.

Rasmussen, H. 1986. The calcium messenger system, *New Engl. J. Med.* 314:1094–1101, 1164–1170.

Reid, L., Bhaskar, K., and Coles, S. 1983. Control and modulation of airway epithelial cells and their secretions, *Exp. Lung Res.* 4(2):157–170.

Rennard, S. I., Basset, G., Lecossier, D., O'Donnell, K. M., Pinkston, P., Martin, P. G., and Crystal, R. G. 1986. Estimation of volume of epithelial lining fluid recovered by lavage using urea as marker of dilution, *J. Appl. Physiol.* 60(2):532–538.

Roberts, A. M., Kaufman, M. P., Baker, D. G., Brown, J. K., Coleridge, H. M., and Coleridge, J. C. 1981. Reflex tracheal contraction induced by stimulation of bronchial C-fibers in dogs, *J. Appl. Physiol.* 51(2):485–493.

Robinson, C., and Holgate, S. T. 1985. Mast-cell dependent inflammatory mediators and their putative role in bronchial asthma, *Clin. Sci.* 68:103–112.

Roger, L. J., Horstman, D. H., McDonnell, W. F., Kehrl, H. R., Seal, E., Chapman, R. S., and Massaro, E. J. 1985. Pulmonary effects in asthmatics exposed to 0.3 ppm NO_2 during repeated exercise, *Toxicologist* 5:70.

Roger, L. J., Horstman, D. H., and Kehrl, H. R. 1986. Absence of change in sensitivity to inhaled methacholine in asthmatics exposed to NO_2, *Toxicologist* 6:8.

Roum, J. H., and Murlas, C. 1984. Ozone-induced changes in muscarinic bronchial reactivity by different testing methods, *J. Appl. Physiol.* 57:1783–1789.

Russell, J. A. 1986. Tracheal smooth muscle, *Clin. Chest Med.* 7(2):189–200.

Sampson, S. R., and Vidruk, E. H. 1975. Properties of "irritant" receptors in canine lung, *Respir. Physiol.* 25:9–22.

Sampson, S. R., Vidruk, E. H., Bergren, D. R., Dumont, C., and Lee, L. Y. 1978. Effects of ozone exposure on responsiveness of intrapulmonary rapidly adapting receptors to bronchoactive agents in dogs, *Fed. Proc.* 37:712 (abstr.).

Schiff, L. V. (ed.) 1986. *In Vitro Models of Respiratory Epithelium,* CRC Press, Boca Raton, Fla.

Seltzer, J., Bigby, B. G., Stulbarg, M., Holtzman, M. J., Nadel, J. A., Ueki, I. F., Leikauf, G. D., Goetzl, E. J., and Boushey, H. A. 1986. O_3-induced change in bronchial reactivity to methacholine and airway inflammation in humans, *J. Appl. Physiol.* 60(4):1321–1326.

Sheppard, D., Nadel, J. A., and Boushey, H. A. 1981. Inhibition of sulfur dioxide–induced bronchoconstriction by disodium cromoglycate in asthmatic subjects, *Am. Rev. Respir. Dis.* 124:257–259.

Sheppard, D., Rizk, N. W., Boushey, H. A., and Bethel, R. A. 1983. Mechanism of cough and bronchoconstriction induced by distilled water aerosol, *Am. Rev. Respir. Dis.* 127(6):691–694.

Shioya, T., Munoz, N. M., and Leff, A. R. 1987. Effect of resting smooth muscle length on contractile response in resistance airways, *J. Appl. Physiol.* 62(2):711–717.

Silverman, F. 1979. Asthma and respiratory irritants (ozone), *Environ. Health Perspect.* 29:131–136.

Smith, P., Stitik, F., Smith, J., Rosenthal, R., and Menkes, H. 1979. Tantalum inhalation and airway responses, *Thorax* 34(4):486–492.

Smith, P. L., Welsh, M. J., Stoff, J. S., and Frizzell, R. A. 1982. Chloride secretion by canine tracheal epithelium: I. Role of intracellular cAMP levels, *J. Membr. Biol.* 70(3):217–226.

Solic, J. J., Hazucha, M. J., and Bromberg, P. A. 1982. The acute effects of 0.2 ppm ozone in patients with chronic obstructive pulmonary disease, *Am. Rev. Respir. Dis.* 125(6):664–669.

Stacy, R. W., Seal, E., Jr., House, D. E., Green, J., Roger, L. J., and Raggio, L. 1983. A survey of effects of gaseous and aerosol pollutants on pulmonary function of normal males, *Arch. Environ. Health* 38(2):104–115.

Stutts, M. J., and Bromberg, P. A. 1987. Effects of ozone on airway epithelial permeability and ion transport, *Toxicol. Lett.* 35:315–319.

Stutts, M. J., Gatzy, J. T., and Boucher, R. C. 1984. Metabolic inhibition of bronchial ion transport, *Fed. Proc.* 43:829.

Szentivanyi, A. 1968. The beta adrenergic theory of the atopic abnormality in bronchial asthma, *J. Allergy* 42:203.

Thompson, J. E., Scypinski, L. A., Gordon, T., and Sheppard, D. 1986. Hydroxyurea inhibits airway hyperresponsiveness in guinea pigs by a granulocyte-independent mechanism, *Am. Rev. Respir. Dis.* 134(6):1213–1218.

Tomioka, M., Ida, S., Shindoh, Y., Ishihara, T., and Takishima, T. 1984. Mast cells in bronchoalveolar lumen of patients with bronchial asthma, *Am. Rev. Respir. Dis.* 129(6):1000–1005.

Tyler, W. S., Tyler, N. K., Barstow, T., Magliano, D., Hinds, D., and Young, M. 1985. Persistence of ozone lesions in monkeys with growing lungs during a 6 month post-exposure period, *Am. Rev. Respir. Dis.* 131:A169.

U.S. Department of Health and Human Services. 1981. *Current estimates from the National Health Interview Survey: 1979,* In: Vital and Health Statistics, Ser. 10, No. 136, DHHS Publ. No. PHS 81-1564, Public Health Service, Office of Health Research,

Statistics and Technology, National Center for Health Statistics; Hyattsville, Md.

U.S. Environmental Protection Agency. 1986. Air Quality Criteria for Ozone and Other Photochemical Oxidants, EPA Report No. EPA-600/8-84-020dF, Environmental Criteria and Assessment Office, Research Triangle Park, N.C.

Van Scott, M., Hester, S., and Boucher, R. C. 1987. Ion transport by rabbit non-ciliated bronchiolar epithelial (Clara) cells in culture, *Proc. Natl. Acad. Sci. USA* 84:5498–5500.

Vidruk, E. H., Hahn, H. L., Nadel, J. A., and Sampson, S. R. 1977. Mechanisms by which histamine stimulates rapidly adapting receptors in dog lungs, *J. Appl. Physiol.* 43(3):397–402.

Walden, S. M., Britt, E. J., Permutt, S., and Bleecker, E. R. 1985. The effects of β-adrenergic and antihistiminic blockade on conditioned cold air and exercise-induced asthma, *Chest* 87:195S–197S.

Wanner, A. 1986. Mucociliary clearance in the trachea, *Clin. Chest Med.* 7(2):247–258.

Wanner, A., and Abraham, W. M. 1982. Experimental models of asthma, *Lung* 160:231–243.

Wasserman, S. I. 1985. Mast cell mediators in the blood of patients with asthma, *Chest* 87:13S–15S.

Welliver, R. C., Sun, M., Rinaldo, D., and Ogra,

P. L. 1986. Predictive value of respiratory syncytial virus-specific IgE responses for recurrent wheezing following bronchiolitis, *J. Pediatr.* 109(5):776–780.

Whittemore, A. S., and Korn, E. L. 1980. Asthma and air pollution in the Los Angeles area, *Am. J. Public Health* 70:687–696.

Widdicombe, J. H. 1986. Ion transport by tracheal epithelial cells in culture, *Clin. Chest Med.* 7(2):299–305.

Wilder, C. S. 1973. Prevalence of selected respiratory conditions, *Vital Health Stat.* 10:1–49.

Woolcock, A. J., Yan, K., Salome, C. M., Sedgewick, C. J., and Peat, J. 1985. What determines the severity of asthma? *Chest* 87:209S–213S.

Wu, R. 1986. In vitro differentiation of airway epithelial cells, In: *In Vitro Models of Respiratory Epithelium* (L. V. Schiff, ed.), Ch. 1, pp. 1–26, CRC Press, Boca Raton, Fla.

Yankaskas, J. R., Cotton, C. U., Knowles, M. R., Gatzy, J. T., and Boucher, R. C. 1985. Culture of human nasal epithelial cells on collagen matrix supports. A comparison of bioelectric properties of normal and cystic fibrosis epithelia, *Am. Rev. Respir. Dis.* 132(6):1281–1287.

Effects of Automotive Emissions on Susceptibility to Respiratory Infections

JAMES E. PENNINGTON

Brigham and Women's Hospital and Harvard Medical School

Air Pollution, the Automobile, and Public Health. © 1988 by the Health Effects
Institute. National Academy Press, Washington, D.C.

The Hypothesis

Respiratory infection is the most common type of infection occurring in the United States. It is estimated that between 200 million and 600 million respiratory infections occur in this country each year, resulting in a loss of more than 150 million work days (Monto and Ullman 1974; Garibaldi 1985). Costs associated with respiratory infections are enormous, with $10 billion expended each year for the common cold alone (Dixon 1985). Mortality from most types of respiratory infections is low, with morbidity and economic factors the major concerns. However, for pneumonia, which accounts for 1 to 2 percent of respiratory infections in adults (Garibaldi 1985), mortality also becomes an issue: pneumonia is currently the sixth most frequent cause of death in the United States (Garibaldi 1985). Thus, any factor that favorably or unfavorably influences the incidence or severity of respiratory infections is of enormous importance to our population.

The hypothesis considered in this chapter is that exposure to automotive emissions can result in greater susceptibility to respiratory infections. This hypothesis is worthy of careful study for a number of reasons. First, if increased susceptibility to respiratory infection can be documented among populations heavily exposed to mobile source emissions (for example, traffic police, tollbooth attendants, mechanics), then specific vaccination programs, respiratory function monitoring, and occupational counseling for these high-risk individuals can be initiated. Second, if emissions are clearly linked to risk of infection, then more extensive studies can be undertaken to identify the most hazardous component(s) of automotive emissions; and, if specific components are identified, they could be taken into account in future engine design. Finally, short-term and long-term noninfectious sequelae have been linked to respiratory infections. For example, protracted bronchial hyperreactivity sometimes follows an acute episode of influenza (Empey et al. 1976), and childhood respiratory infections have been linked to an increased risk of adult lung disease (Kattan 1979). Thus, prevention of respiratory infections associated with automotive emissions might reduce subsequent noninfectious pulmonary diseases.

In organizing a method for assessment of the basic hypothesis, a natural starting point is agreement about the constituents of automotive emissions and about the concentration ranges relevant to human exposures; some components (for example, nitrogen dioxide, acrolein) exist in low amounts in automotive emissions but in exceedingly high amounts elsewhere (for example, blasting areas, siloes, cigarette smoke). The emission components relevant to this discussion and their upper limits of exposure recommended by the National Ambient Air Quality Standards (NAAQS) of the U.S. Environmental Protection Agency, are listed in table 1. Recently, aldehydes have been added to this list in anticipation of future use of methanol fuel. In that case, acrolein, formaldehyde, and acetaldehyde would increase in importance as components of automotive emissions. However, on the basis of studies of Los Angeles smog, Tuazon and coworkers (1981) estimated that formaldehyde levels deriving from automotive emissions would be considerably less than 1 ppm, and levels of other aldehydes would be even lower.

What methods, then, are available for analysis of the hypothesis that automotive emissions increase susceptibility to respiratory infection? The approaches taken include epidemiologic studies, and experimental studies using animal models. Although chamber studies using nitrogen dioxide (NO_2), ozone (O_3), and other rel-

Table 1. Major Pollutants Related to Automotive Emissions

Component	Air Quality Standard (ppm)
Nitrogen dioxide (NO_2)	≤ 0.05/year
Ozone (O_3)	≤ 0.12 (1-hr average)
Carbon monoxide (CO)	9 ppm/8 hr
	35 ppm/1 hr
Total suspended particulates (TSP)	260 $\mu g/m^3$/24 hr

Table 2. Ambient Concentrations of NO_2

Setting	Concn. (ppm)
Rural air	0.01
Urban air	0.03–0.12
Los Angeles freeway (usual)	0.15–0.45
Los Angeles freeway, highest recorded (1962)	1.3
Cigarette smoke exiting cigarette	1.0–5.0
Mine shafts immediately after dynamite blast	250
Siloes with advanced decomposition of ensilage	500

evant gases on humans have been reported, those studies have been directed toward range finding for acute toxicity and effects on bronchial hyperreactivity (Ferris 1978). Virtually no data on the relationship between pollutants and the risk of infection have been published.

The discussion that follows reviews epidemiologic and experimental approaches to this problem. Since data on NO_2 are extensive and the approaches taken in NO_2 studies appear relevant to the question raised here, the background information presented below emphasizes findings with NO_2 exposures. Table 2 lists ambient concentrations of NO_2 in various settings which may be useful in the experimental design of some of the studies discussed below. After the background discussion, gaps in our understanding are identified, and suggestions for future studies that might improve our ability to prove or disprove the basic hypothesis are presented.

Background

Determinants of Susceptibility

Frequency of infections as well as severity, as measured by physician visits, hospitalization, absenteeism from work or school, and so on, should be considered relevant indicators of susceptibility. Health-impaired populations (for example, the immunosuppressed or chronic lung disease patients) are important to consider, as are infants and the elderly, in whom developing or senescent lungs may be factors in susceptibility. Also to be considered is the specific type of infection, that is, viral versus bacterial, especially as it relates to available vaccines. And certain anatomic locations in the lung and conducting airways may be at greater risk of infection than others. For example, is ciliary dysfunction with associated bronchitis/sinusitis a major determinant, or is the alveolar parenchymal region more prone to infection if stressed by automotive emissions? Naturally, if certain locations in the respiratory tract appear to be more susceptible, then research could be directed at analysis of the most relevant local defense mechanisms.

Occurrence of Respiratory Infections

In considering the contribution of emissions to respiratory infections, it is worthwhile to review the types of infectious agents that commonly cause respiratory tract infection in the United States. In a large prospective study over a 6-yr period, approximately 14,600 cases of respiratory infections were documented in 4,905 residents of Tecumseh, Michigan (Monto and Ullman 1974). Microbiological monitoring was carried out using throat cultures for Group A streptococcus, respiratory viral agents, and mycoplasma. Cultures were only obtained if available within 2 days after onset of symptoms. Of the isolates, 82 percent were viral, 13.3 percent were Group A streptococcal, and 4.7 percent were "other," which included mycoplasma. Respiratory infections were far more frequent during childhood (five to six per year in newborns to 2-year-olds) and decreased steadily with age. By adulthood, one to two infections occurred per year. This study and others (for example, Garibaldi 1985) emphasize the predominance of upper respiratory sites for infection, with bronchitis and pneumonia accounting for only 10 percent of the episodes.

Although upper respiratory viral infections are more common, the morbidity and mortality associated with bronchitis and pneumonia are far greater (Pennington 1983). The major agents causing serious lower respiratory infection are *Mycoplasma pneumoniae, Legionella* species, influenza A and B, parainfluenza type 1, respiratory syncytial virus, and bacteria, especially *Streptococcus pneumoniae* ("pneumococcus") and *Hemophilus influenzae*. In certain high-risk individuals such as alcoholics, diabetics, and the elderly, enteric gram-negative bacilli also may cause pneumonia.

Thus, in selecting appropriate infectious agents for surveillance studies of respiratory infection, a broad-based cultural and/or serologic approach is necessary. In selecting relevant infectious agents for use in animal models of respiratory infection, a wider range of infectious agents could be considered.

Lung Defense Mechanisms

Requisite to an analysis of effects of inhaled pollutants on lung defenses against infection is a thorough understanding of the respiratory defense system that is operative under normal conditions. Numerous reviews describe the lung's complex and remarkably effective defense against infection (Green 1970; Kaltreider 1976). Table 3 summarizes the basic components of this system. Specific defects in this defense system appear to predispose to specific types of infections (Reynolds 1983; Pennington

Table 3. Host Defense Mechanisms in the Human Respiratory Tract

Upper airways
  Nasopharyngeal filtration
  Mucosal adherence
  Bacterial "interference"
  Saliva (proteases, lysosome)
  Secretory IgA
Epiglottis
Lower airways and alveoli
  Cough reflex
  Mucociliary clearance
  Humoral factors
    Immunoglobulins
    Complement
  Cells and cell products
    Bronchus-associated lymphoid tissue
    Lymphocytes
    Alveolar macrophages
    Polymorphonuclear leukocytes
    Cytokines (interleukin-1, interleukin-2, interferons)

1984). For example, impaired mucociliary clearance (which can occur with ciliary dyskinesia, cystic fibrosis, bronchiectasis) results in bacterial sinusitis and bronchitis, generally caused by encapsulated bacteria such as *H. influenzae*, and the pneumococcus. Local deficiency in immunoglobulins such as IgG_2 and IgG_4 are associated with recurrent bacterial bronchitis and bronchopneumonia. Absent or impaired cough reflex from neurological disease results in aspiration pneumonias.

Cellular defenses are particularly important in the lower respiratory tract. For example, immunosuppressive drugs may reduce alveolar macrophage and local lymphocyte function, resulting in opportunistic pneumonias, such as *Pneumocystis carinii*, and cryptococcal pneumonia (Pennington 1985). Impaired polymorphonuclear leukocyte recruitment to the lungs during myelosuppression correlates with an increased risk for aerobic gram-negative bacillary pneumonia (Pennington 1985). All of these relationships have been well described, and numerous other components of the lung defense system are now being investigated. For example, investigations undertaken by Bukowski et al. (1984) on the local importance of natural killer (NK) cells in antiviral activity and by Ennis et al. (1978) on the importance of other cytotoxic lymphocyte populations in lung defenses are being actively pursued. Likewise, the role of various cell-derived immune modulators—interleukin-1, interleukin-2, and interferons—in local pulmonary defenses is a subject of great interest (Pinkston et al. 1983; Wewers et al. 1984; Robinson et al. 1985).

The functional integration of this complex system is not completely understood and in some cases is controversial. For example, some debate exists about the importance of alveolar macrophages versus neutrophils in bacterial defenses; alveolar macrophages appear to be critical resident phagocytes for the surveillance against low numbers of bacteria arriving in the lung, but recruitment of neutrophils into the lung is necessary for larger bacterial challenges. Furthermore, although secretory immunoglobin A (IgA) is generally considered the primary humoral immunoglobulin involved in local respiratory defenses, recent information suggests that IgG is a more potent humoral factor in the lower respiratory tract (Reynolds 1983). Finally, the lymphocyte-directed cellular immune response has been traditionally viewed as the mainstay against facultative intracellular pathogens such as mycobacteria but of little importance against acute infection with pyogenic organisms such as *Pseudomonas aeruginosa*. Recent investigations, however, suggest that this distinction may not be clear-cut. It thus appears that for most infections, a combination of mechanical, secretory, and cellular defenses is needed for optimal lung defense.

Long-Term Effects of Infection

Although most respiratory infections are not fatal, long-term adverse sequelae may result from childhood respiratory infections (Burrows et al. 1977; Kattan 1979; Pullan and Hey 1982). Long-term effects of infections may be particularly severe if infection-related lung injury occurs during infancy or early childhood—a critical period in lung development. Sequelae may take the form of asthma in children or adolescents or chronic airway obstruction in adults. However, a review of the relevant studies by Samet et al. (1983) suggests that evidence for these associations was incomplete and that many such studies suffer from recall bias. Nevertheless, the potential importance of such associations with childhood respiratory infections offers considerable incentive to pursue this analysis.

Epidemiologic Approach

The ideal method to test the hypothesis that links automotive emission exposure and increased susceptibility to respiratory infections would be to document the frequency and/or severity of respiratory infections in individuals with greater exposure to emissions. Numerous epidemiologic studies have suggested that exposure to NO_2 or other air pollutants increases the incidence of respiratory illness, and, in some cases, actual infection (discussed below). To date, no firm link between mobile source emission exposures and respiratory infection has been established because of methodological difficulties including insufficient use of diagnostic tests; unreliable techniques for collection of data (for example, recall, questionnaires); and inaccurate measurements of ambient gas concentrations. Since exposure to air pollutants may cause respiratory irritation and increased bronchial hyperreactivity, the use of symptoms such as cough, sputum production, sore throat, and wheezing as indicators of infection is much less accurate than viral or mycoplasmal serologies, or sputum and throat cultures. The use of questionnaires to record respiratory symptoms as a marker of infection is particularly problematic. Methodological problems must be taken into account in the design of future epidemiologic studies, even though some of the problems may prove insurmountable. The following discussion highlights important considerations in the design and implementation of epidemiologic investigations that explore the link between emissions and infection (see also Bresnitz and Rest, this volume).

Documentation and Measurement of Respiratory Infection

Past Studies. Numerous epidemiologic studies have attempted to demonstrate an association between NO_2 exposure and respiratory illness. In some cases direct evidence of respiratory infection was sought, but in most cases infection was simply implied by association with cough, coryza (head cold, inflammation of nasal mucous membranes), or sore throat. In no studies were serologic assays performed, nor, with rare exception, were throat or sputum cultures obtained.

Several of these studies illustrate this diagnostic difficulty. In one questionnaire-based study Speizer and Ferris (1973a,b) compared respiratory symptoms of a group of urban traffic police exposed to automotive emissions with symptoms of a group of suburban police. No differences in symptoms were noted, and the incidence of respiratory infections could not be determined in either group. Similarly, a large-scale study was undertaken in groups of Chattanooga, Tennessee, schoolchildren residing in geographic areas with high versus low levels of NO_2 (Shy et al. 1970a,b). Respiratory symptoms were monitored by biweekly postcard survey, and positive responses were followed up by direct questioning. In addition, teams of parents using spirometers measured gas volumes of second-grade children in each locality. Initial analysis of these data suggested that children in the high-NO_2 area experienced more frequent respiratory symptoms, were more prone (according to history) to "influenza" when exposed during an epidemic, and had lower $FEV_{0.75}$ values. A subsequent study in this area by Pearlman et al. (1971) also suggested that an increased incidence of childhood bronchitis was associated with increased NO_2 exposure. Later analysis of study design and of the method used for NO_2 measurements, however, rendered the findings from these studies less conclusive than originally thought (Ferris 1978).

Another relevant subject in investigating the health effects of NO_2 exposure deals with the levels of NO_2 inside homes with gas cooking appliances (Melia et al. 1977, 1979; Keller et al. 1979; Speizer et al. 1980). In these studies, indoor exposure to NO_2 was used as a method for isolated analysis of an ambient pollutant gas. Results of these studies conflict, again impaired by the use of questionnaires as well as the use of contemporary fuel exposures to assess past events.

Practical suggestions for improving diagnostic methods in future studies are more difficult to devise than might be imagined, because such methods of documenting and measuring respiratory infections generally suffer from insensitivity, impracticality, and expense.

Personal History and Physical Examination. A personal history taken during an acute respiratory illness is useful but extremely time-consuming. Even with careful questioning, differentiation between allergic and viral rhinitis or between asthmatic and infectious cough may be impossible. The presence of fever, purulent sputum, pleuritic chest pain, and sore throat are informative but are absent in many cases of respiratory infection (Glezen and Denny 1973; Monto and Ullman 1974). Of course, questionnaires, letters, infrequent telephone surveys, and patient recall are all even less accurate, although much less expensive, than personal histories.

As with history taking, individual physical examinations of patients with respiratory complaints are labor and cost intensive. Also, symptoms of most bacterial infections are evident on examination (exudative pharyngitis, purulent sputum production, or even septic appearance), but viral or mycoplasmal infections may be difficult to differentiate from hypersensitivity because of the similarity of symptoms (for example, coryza, erythematous throat, wheezing, rales).

Thus, taken together a physical examination and a personal history can provide useful concurrent diagnostic information. These data would clearly be superior to the data collected using questionnaire surveys, but the costs of these combined procedures may preclude their use for large-scale studies.

Nonmicrobiological Laboratory Methods. Hematologic or radiographic tests do not provide conclusive evidence of respiratory infection. An elevated white blood cell count may suggest bacterial infection, and infiltrative patterns on chest x rays may suggest lung infection (among other possibilities). However, these tests are expensive and offer little diagnostic advantage over more specific microbiological tests discussed previously. If it is critical, however, that pneumonia be differentiated from bronchitis, then chest x rays would be necessary.

Microbiological Methods. Although widely considered the methods of choice for diagnosis of respiratory infections (McIntosh 1985), microbiological methods are surprisingly insensitive. For example, in large prospective studies of respiratory infections in pediatric practices, routine use of throat swab cultures for viral, mycoplasmal, and bacterial isolation yielded etiologic agents in less than 30 percent of the cases (Henderson et al. 1979; Murphy et al. 1981). Similar results were reported in the Tecumseh study, in which throat cultures were collected from patients with symptoms for 2 days or less (Monto and Ullman 1974). In several studies (reviewed by Pennington 1983) of adults who required hospitalization for community-acquired pneumonia, specific etiologic agents were found in only 60 to 70 percent of the cases.

In attempts to define etiologies of respiratory infection, several studies used throat swab cultures to determine whether the pathogen was a virus, mycoplasma, or group A streptococcus (Glezen and Denny 1973; Monto and Ullman 1974; Henderson et al. 1979; Murphy et al. 1981), because respiratory infections caused by these agents are extremely common and because most infected patients cannot produce sputum for culture. But the use of throat cultures to isolate respiratory pathogens has certain drawbacks: first, most viral pathogens are excreted for brief periods, early in infection, and often in low titers; second, viral, mycoplasmal, and chlamydial cultures require specialized and expensive methods, which are not available in most diagnostic laboratories; and third, in large population studies the logistics for obtaining proper specimens early in the illness may be difficult. Bacterial culture techniques are widely available, but most respiratory infections are nonbacterial.

Although serologic tests to detect most nonbacterial respiratory pathogens are available (table 4), these tests are expensive, require acute and convalescent specimens (except for direct immunofluorescent preparations), and may remain negative in infants with infection. Furthermore, antigenic heterogeneity for rhinovirus, the most common cause of respiratory infections, precludes serologic evaluation. In short, serologic diagnosis is expensive, time-consuming, and incomplete.

Severity of Respiratory Infections. In addition to documenting the incidence of respiratory infections, a measurement of the severity of infection may be useful. Clinicians tend to measure severity in subjective as well as objective terms. For data collection, objective parameters are preferable. Measures such as duration of symptoms and fever, time lost from work or school, time bedridden, whether sputum was produced, are all useful to gauge severity. Questions such as how sore is your throat, how bad is your cough, are particularly unreliable. In addition, some measure of the overall effect of the respiratory infection may be useful in determining severity. Were physician visits needed, how much did the entire illness cost, were symptoms residual (for example, wheezing and coughing for weeks to months), can all be useful questions.

Other Epidemiologic Variables

Age. The highest risk of respiratory infection occurs during the first year of life and decreases steadily until adulthood (Monto and Ullman 1974). On the other hand, risks of pneumonia and mortality from respiratory infection increase with advancing age (Pennington 1983). Such variables may be useful in targeting groups for epidemiologic study.

Season. Most studies indicate that the peak incidence of respiratory infections occurs during winter months (Glezen and Denny 1973).

Table 4. Serologic Diagnosis of Common Respiratory Infections

Agent	Tests
Viral	
Respiratory syncytial virus	CF, IF
Influenza A and B	CF, IF
Parainfluenza	HI, IF
Adenovirus (group)	CF, IF
Mycoplasma pneumoniae	CF
Legionella sp.	IFA, IF
Chlamydia	IFA

NOTE: CF = complement fixation; HI = hemagglutination inhibition; IF = immunofluorescence (direct smear); IFA = indirect fluorescent antibody.

Socioeconomic Status and Family Setting.
The Tecumseh study (Monto and Ullman 1974) showed that the incidence of respiratory infections increased with education but decreased with family income. Superior reporting by well-educated subjects may partially account for this finding. In addition, larger families and rural life may increase the risk of infection (Glezen and Denny 1973). Additional factors, such as crowded family quarters, parental smoking, and use of day-care centers, must be considered in evaluation of epidemiologic data.

Underlying Diseases. Increases in incidence and/or severity of respiratory infections have been well described for certain high-risk individuals (Pennington 1983). These groups include individuals with chronic bronchitis, cystic fibrosis, alcoholism, malnutrition, or immunosuppressed states arising, for example, from organ transplants and cancer chemotherapy. Recent data suggest that bacterial pneumonia is more frequent in patients with acquired immune deficiency syndrome (AIDS) (Polsky et al. 1986). If limited resources dictate studies of only selected, high-risk populations, these may be the best groups to study.

Experimental Approach

Experimental Infections

A link between respiratory mucosal damage from inhalation of O_3 and increased susceptibility to localized infections was postulated by Miller and Ehrlich (1958). They reasoned that mucous membranes, known to be important in antiinfective defenses, might be damaged by O_3 during high-altitude flight. To prove their hypothesis, they exposed mice in inhalation chambers to 4 ppm O_3 for 3 hr and then monitored survival rates after exposures to aerosolized *Klebsiella pneumoniae*. Mortality was significantly higher among the O_3-exposed mice than among nonexposed but infected control groups. This work was the first in a series of studies by Miller and

Ehrlich, as well as by others, examining the many variables that pertain to the effects of pollutant exposure and susceptibility to experimental infection as measured by survival rates.

Later work using this model dealt with NO_2 rather than O_3 (Purvis and Ehrlich 1963; Ehrlich 1966; Ehrlich and Henry 1968). Ehrlich and coworkers concluded that acute exposures (for example, 1 to 2 hr) to NO_2 at levels below 3.5 ppm do not affect survival from *Klebsiella* challenge. In contrast, they found that chronic NO_2 exposure ($\geq$ 3 months) of 0.5 ppm increased mortality from *Klebsiella* challenge (Ehrlich 1966; Ehrlich and Henry 1968). Recent attempts to duplicate these results have been unsuccessful. McGrath and Oyervides (1983, 1985) found that mice exposed for 3 or 8 months to 0.5 ppm NO_2 were not significantly different from controls in their resistance to aerosol challenge with *Klebsiella*. In a subsequent study, they exposed mice to 0.5, 1.0, and 1.5 ppm NO_2 for 3 months, and again no decrease in resistance to aerosolized *Klebsiella* challenges was found. Acute exposures (3 days) to 5 ppm NO_2, however, did decrease survival rates (McGrath and Oyervides 1985). These authors speculated that the older NO_2 monitoring devices previously used by Ehrlich and coworkers may have provided imprecise data during their chronic exposure studies.

Inhalation studies using squirrel monkeys exposed to NO_2 indicate that NO_2 exposures in the 5- to 10-ppm range adversely affect lung resistance to *Klebsiella* (Henry et al. 1969) and influenza virus inocula (Henry et al. 1970). However, in subsequent studies Fenters et al. (1973) found that the antiviral defenses of monkeys exposed to 1 ppm NO_2 for more than a year were not significantly impaired.

Yet another system of analysis uses quantitative bacteriological monitoring of lung tissues in animals exposed to inhaled gases. This methodology provides an in vivo evaluation of microbicidal function in respiratory tissues. In one typical study by Goldstein et al. (1973), mice were challenged with aerosolized *Staphylococcus aureus* and were then exposed to various concentrations of NO_2 (0 to 14.8 ppm) for 4-hr periods. Their lungs were removed and quantities of remaining viable bacteria were determined. Animals exposed to NO_2 levels of 1.9 ppm killed pulmonary bacteria as well as control groups, but bactericidal capacity was reduced in groups exposed to $\geq$ 3.8 ppm NO_2.

The method of analysis—survival rates or assays of intrapulmonary microbicidal capacity—most indicative of the clinical setting is speculative. In either model system, exposure to rather high levels of NO_2 or O_3 is required before significant impairment of lung resistance to infection can be demonstrated. Thus, it might be concluded that NO_2 only impairs lung defenses at levels at or above 1.0 ppm, and that automotive emissions probably do not impair resistance to infection. An equally valid possibility, however, is that these methods of evaluation are not sensitive enough to detect impairment of lung defenses that might occur during exposures to NO_2 in lower concentrations.

Several methodological concerns can be identified when reviewing past studies of animal models of experimental infection. First, in most cases, healthy animals were used for the exposure studies. If impaired hosts are involved, such as immunosuppressed animals, lung-injured animals, or even neonatal or senescent animal hosts, adverse effects of pollutant gas exposure on lung defenses may occur at much lower concentrations. Second, in most animal studies, rather unusual respiratory pathogens were used (for example, *Klebsiella*, *S. aureus*, *Streptococcus pyogenes*). More clinically relevant choices of bacterial pathogens might include *S. pneumoniae* or *H. influenzae*. And, although some work using respiratory viruses has been reported (Henry et al. 1970; Fenters et al. 1973), far more emphasis on the use of viruses, Mycoplasma, and *Legionella* sp. in animal models might produce more definitive conclusions regarding the effects of inhaled gases on lung defenses. Furthermore, the combined or synergistic effects of viral plus bacterial infections, as described by Astry and Jakab (1983), may be a more relevant method for analysis of emission effects on

lung infection. Finally, it is quite possible that acutely overwhelming lungs with bacterial challenges simply does not simulate the pathogenesis of most human lung infections, and that a more detailed analysis of various components of the lung defense apparatus would be necessary to detect adverse effects of realistically low gas concentrations.

Studies of Lung Defense Mechanisms

Careful analysis of past studies describing the importance of specific lung defense components in resistance against specific types of infection is critically important to test the hypothesis that automotive emissions impair lung defenses. For example, careful epidemiologic studies such as those discussed above may demonstrate that individuals working or living in urban areas with high levels of ambient NO_2 have an increased risk for viral respiratory infection but not bacterial infections. In that case, investigative priorities could be placed on analyses of the effects of emission components on antiviral defense mechanisms such as NK cells, cytotoxic lymphocytes, and local and systemic interferon responses. In other words, the first clue regarding subtle but important effects of emissions on specific components of lung defenses may be suggested indirectly rather than by directly analyzing the specific component.

In the meantime, the effects of NO_2 and O_3 on alveolar macrophages and certain other components of lung defenses have been analyzed by Kavet and Brain (1975) and Green et al. (1977). In general, these researchers first exposed animals to gas inhalants, and then collected bronchoalveolar cells and fluids by lavage. In a few instances, they exposed lung cells or tracheal tissues in vitro to gases. In consideration of future research recommendations, a review of these past findings is likely to be helpful.

Using histologic methods, Freeman et al. (1968) found that 2-ppm NO_2 exposures resulted in decreased ciliary integrity of the tracheal epithelium. Schiff (1977), on the other hand, found no morphologic damage to cilia in hamster tracheal explants after daily in vitro exposures to 2 ppm NO_2 for up to 3 weeks, although decreased ciliary beat frequency as well as increased susceptibility of tissues to influenza A viral inoculations were observed.

Effects of gas exposures on alveolar macrophages have also been studied. For example, Vassallo and coworkers (1973) found that alveolar macrophages obtained from rabbits exposed to 10 ppm NO_2 demonstrated reduced phagocytosis of bacteria, and Gardner et al. (1969) found that higher NO_2 levels resulted in reduced numbers of pulmonary macrophages. Others have shown that O_3 exposure reduces alveolar macrophage hydrolases and other antibacterial metabolic functions (Hurst et al. 1970; Gardner et al. 1971; Hurst and Coffin 1971). In another series of studies, Valand et al. (1970) and Acton and Myrvik (1972) acutely exposed rabbits to 25 ppm NO_2, and then evaluated alveolar macrophages for antiviral defenses. In healthy rabbits, the alveolar macrophages obtained after parainfluenzal challenges produced interferon; however, NO_2 exposures suppressed this normal response. Williams et al. (1972) found that NO_2 exposures suppressed uptake of virus by macrophages. Voisin and coworkers (1977) observed that after 30-min in vitro exposures to 0.1 ppm NO_2 alveolar macrophages showed reduced bactericidal capacity as well as reduced cellular adenosine-triphosphate. The relationship between in vitro NO_2 concentrations and the in vivo situation, however, remains speculative. Finally, in more recent work, Greene and Schneider (1978) exposed baboons to 2 ppm NO_2 for 8 hr/day, 5 days/week for 6 months and then obtained alveolar macrophages by lavage. They found that the capacity for alveolar macrophage response to a lymphokine (migration inhibition factor) generated from autologous lymphocytes was decreased among NO_2-exposed animals. In summary, the recurrent finding resembles that in the infectivity studies: exposure to levels of NO_2 or O_3 above 2 ppm is necessary for adverse effects to be observed.

Effects of inhaled gas exposure on systemic humoral antibody response have also been studied. Holt et al. (1979) exposed mice to 10 ppm NO_2 daily for 30 weeks. At various intervals during the 30-week exposure period, they monitored serum antibody response to a T-cell independent antigen (polyvinyl pyrrolidone) and a T-cell dependent antigen (red blood cells), finding that antibody response to red blood cells, but not to polyvinyl pyrrolidone, was blunted after prolonged NO_2 exposure. In contrast, Fenters et al. (1973) found slight increases in serum antibody response to influenza virus in squirrel monkeys exposed to 1 ppm NO_2/day for over 1 year. Fujimaki and Shimizu (1981) found that exposure to 5 ppm NO_2 for 12 hr did not decrease antibody responses to red blood cells in mice, but exposures to 20 ppm and 40 ppm did reduce antibody-forming capability. Thus, it appears that animals exposed to NO_2 levels far above those expected to result from automotive emissions have a reduced capacity for primary antibody response to certain antigens. Studies of the effects of gas exposure on local antibody production by pulmonary tissues have not been reported.

Gaps in Knowledge and Research Recommendations

Epidemiologic Study Design

Diagnostic Specificity. The epidemiologic information currently available is not sufficient either to prove or disprove the hypothesis that exposure to automotive emissions increases susceptibility to respiratory infection. The most direct method to prove this hypothesis would be to demonstrate clearly that individuals exposed to higher levels of automotive emissions experience more frequent and/or more severe respiratory infections than individuals exposed to lower levels. Reliance on symptoms (cough, sore throat, "colds going to chest," sputum production) rather than on specific diagnostic methods as indicators of respiratory infection should be avoided.

Difficulties encountered in such studies

may include lack of availability of certain serologic tests in local laboratories, poor patient compliance in collection of paired (acute and convalescent) serologic specimens, high costs of such a meticulous diagnostic approach, and a high incidence of false-negative cultures reported for community-acquired respiratory infections. These problems may require that larger rather than smaller populations be studied. They may also dictate that more specific target groups be identified for prospective study.

High-Risk Populations. An attempt has been made to analyze special respiratory risks to children during NO_2 exposures (reviewed above), but a much broader view of high-risk populations should be taken. For example, the difficulty in finding consistent data in the childhood studies reported to date may be related to the fact that school-aged children are not among the groups more sensitive to low-level NO_2 exposures. The elderly or infants may, in fact, be far more susceptible to inhaled air pollutants. Likewise, patients whose immune systems are compromised because of drug therapy or AIDS may be especially vulnerable to low levels of NO_2, O_3, carbon monoxide (CO), or aldehyde. It is noteworthy that the lung is the most frequent target organ for infectious complications among immunocompromised patients, regardless of underlying disease (Pennington 1985). Further susceptibility to respiratory infections for such patients residing in congested urban areas with high traffic density would be meaningful to document. Other high-risk groups deserving study would include patients with chronic lung diseases (for example, cystic fibrosis, chronic bronchitis, emphysema) or even other chronic medical ailments. Even if an epidemiologic survey in a general population is not economically feasible using the diagnostic methods described above, an analysis in these high-risk populations would be worthwhile.

Difficulties that might be encountered in this type of study include rapid patient attrition due to underlying illness, multiple concurrent illnesses, and alterations in medications coinciding with infectious episodes.

■ **Recommendation 1.** Epidemiologic survey in high-risk populations exposed to varying levels of automotive emissions should be conducted.

These surveys should adhere to the following minimum set of requirements:

a. Use of population with defined exposure. Prospective and accurate ambient air analysis should document elevated levels of relevant emission component(s) in the environment for study.

b. Use of defined group. Cost considerations argue for selection of specific study groups of persons at high risk for respiratory infection, in particular the elderly, infants, patients with chronic lung diseases, and immunocompromised patients. One problem to be expected with groups suffering from chronic lung disease is the high rate of chronic symptoms not associated with, but potentially confused with, acute infection.

c. Use of adequate history. Retrospective questionnaires should not be used. Rather, prospective and frequent telephone surveys, followed up by visits to the clinic for symptomatic patients, are recommended.

d. Use of diagnostic tests for infection. At a minimum, throat swabs for respiratory viral cultures and bacterial cultures should be collected within 2 days of onset of illness. In addition, acute and convalescent sera should be evaluated for *Mycoplasma pneumoniae* and *Legionella* sp. antibody titers. If sputum is produced, it should be cultured. These methods can be expected to yield specific diagnoses in 20 to 30 percent of cases. Although considerably more expensive, viral serologies could also be done (see table 4), increasing diagnoses by at least another 20 to 30 percent.

e. Measures of severity. Factors discussed above dealing with severity of infection and overall impact of illness (that is, duration of symptoms and fever, absenteeism from work or school, time bedridden, sputum production, need for physician visits, cost of illness, residual symptoms) should be routinely recorded.

Experimental Studies

Animal Versus Human Studies. Before discussing areas where specific experimental studies could provide more useful information regarding effects of inhaled gases on lung defenses, it is important to point out the lack of relevant data derived from human studies. In recent years a methodology has been developed to collect cells and fluids from the human lower respiratory tract safely and rapidly with the flexible fiberoptic bronchoscope (Hunninghake et al. 1979). Much valuable information is now available regarding normal bronchoalveolar cell populations and the fluid-phase constituents of lower respiratory secretions (Hunninghake et al. 1979). Further analyses using specimens obtained bronchoscopically from patients with lung diseases such as sarcoid, idiopathic interstitial pneumonitis or asthma have identified cell population alterations that correlate with specific diseases and with the status of disease activity (Weinberger et al. 1978; Hunninghake et al. 1979; Pinkston et al. 1983; Rankin et al. 1984).

To date, bronchoscopic analysis has not been used in humans to evaluate the effects of inhaled gases on various components of the lung defense system. Ethical considerations have, perhaps, slowed acceptance of this approach because the risks attendant with inhalation chamber exposures to NO_2 or O_3 in humans are unknown. But, numerous chamber studies have been performed to develop guidelines for air quality recommendations (reviewed by Ferris 1978). Those using low-level, short-term exposures appear to be safe and may be considered favorably by human studies committees. On the other hand, it could be argued that until the basic hypothesis has been disproved, it may not be wise to expose humans even to low levels of inhaled pollutants. After all, a basic premise of such investigations is that low-level exposure induces subtle but important defects in local defenses, and that such defects will be detected only by careful analytical methodology.

Yet another method of human study might use experimental viral infections in association with inhalant exposures. The use of healthy subjects for experimental viral infections such as influenza and respiratory syncytial virus is an accepted practice (see, for example, Bell et al. 1957; Johnson et al. 1961; Feery et al. 1979), particularly to evaluate vaccine efficacy. Combining chamber exposures with experimental viral infections may be a useful method to evaluate the effects of emissions on respiratory pathogenicity of these agents. For example, the effect of inhalants on severity and duration of infection, and on minimum size of infective inoculum, could be determined. Again, ethical considerations must be addressed in designing such studies. Furthermore, the dose range maintained for inhaled gas exposures should be low enough to address the issue of relevance to automotive emissions.

In summary, studies should be designed for human subjects as well as animal models. Although dose-response studies can be carried out over much broader ranges of pollutant concentration in animals, the more direct clinical relevance of human data argues for strong consideration of at least some analyses being carried out in human subjects. Ideally, subjects exposed to high levels of automotive emissions could be compared with subjects from rural or other areas of low-level exposure, but, if this is not possible, chamber exposure methods could be used.

Infectivity Models. Despite the large body of information regarding susceptibility to infection in animals exposed acutely or chronically to inhaled NO_2 or O_3, questions regarding lung defenses during experimental infection remain unanswered. For example, little or no information exists about defenses against several extremely common respiratory pathogens, including mycoplasma, respiratory syncytial virus, *Legionella* sp., *S. pneumoniae*, and *H. influenzae*. Experimental models of infection with each of these pathogens have been reviewed by Pennington (1986) and could be used for analyses similar to those for *Klebsiella* and influenza virus.

■**Recommendation 2.** Perform experimental studies in animals using common respiratory pathogens.

Altered-Host Studies. The influence of altered-host status on lung susceptibility to infection after exposure to components of automotive emissions is poorly understood. Ethical considerations preclude bronchoscopic analyses in high-risk patients exposed to pollutant gases, but numerous animal models of immunosuppression (Pennington 1985), chronic lung damage (Snider et al. 1986), or extremes of age (Sherman et al. 1977; Esposito and Pennington 1983) exist from which valuable information may be obtained.

■**Recommendation 3.** Animal models should be used to determine whether altered-host status or extremes of age influence the lung's susceptibility to infection after exposure to automotive emission components. Survival or lung clearance during experimental infections could be evaluated as well as the additive influences of gas exposure plus underlying host defect on lung defense components (for example, alveolar macrophages, inflammatory reaction in airways after infection, bacterial adherence to respiratory mucosa).

Antiviral Defense Mechanisms. Judging from current epidemiologic data, the most likely source for increased infection in individuals exposed to high levels of automotive emissions is viral. As discussed above, this conclusion is by no means proven, but sore throat, coryza, and cough without sputum are the symptoms most frequently noted in such studies. As such, high priority must be placed on analysis of the effects of inhaled emission components on antiviral defenses in the respiratory tract. Some early data suggested that NO_2 exposure suppresses interferon production by alveolar macrophages (Valand et al. 1970; Acton and Myrvik 1972) and decreases lung resistance to influenza (Henry et al. 1970) or parainfluenza (Williams et al. 1972). The NO_2 levels used in those studies, however, were high (5 to 25 ppm). Using the more sophisticated methodologies now available

for analysis of antiviral activities (Sissons and Oldstone 1985), it may be possible to detect adverse effects on lung defenses at levels of gas exposure relevant to ambient concentrations. Such studies may be ideal for human subjects because low levels of gas (for example, NO_2) could be used in chamber exposures, followed by collection of large numbers of cells by bronchoscopic lavage. This approach may be especially useful because obtaining enough of the appropriate cell population (for example, NK cells, K cells, cytotoxic T cells) may be impossible in small animals but feasible in humans (Pinkston et al. 1983; Robinson et al. 1984).

Naturally, animal studies of antiviral defense mechanisms also could be carried out in experimentally exposed groups, allowing more convenient dose-response studies. If mice are used, pooling of lung specimens would probably be necessary to obtain enough cells for well-controlled, replicate assays. If larger animals such as guinea pigs and rabbits are used, difficulties in obtaining appropriate reagents for analysis (for example, monoclonal antibodies for NK cells) may be encountered.

Especially important in these studies would be the analysis of numbers and function of pulmonary NK cells (Stein-Streilein et al. 1983; Bukowski et al. 1984), cytotoxic T lymphocytes (Ennis et al. 1978), antibody-dependent cellular cytotoxicity (using alveolar macrophages and K cells) (Hunninghake and Fauci 1977; Kohl et al. 1977), and interferon α, β, and γ production by lung cell populations (Robinson et al. 1985). Since available data show no viral antibody responses in animals exposed to relevant levels of NO_2, it is less likely that studies of antibody response will be helpful. Analysis of pulmonary secretory antibody levels may be useful, however.

■**Recommendation 4.** Components of respiratory antiviral defense mechanisms should be analyzed with respect to impairment from exposure to automotive emission products.

Immunologic Modulators (Cytokines). A rapid expansion has occurred recently in our understanding of how various immunologic cell populations communicate and modulate cellular responses to infectious (and other antigenic) agents. Researchers have shown that interleukin-1 (Wewers et al. 1984), interleukin-2 (Pinkston et al. 1983), and various interferons (Robinson et al. 1985), are produced by pulmonary cells and may be affected by various disease states, including infection (Lamontagne et al. 1985) and sarcoidosis (Pinkston et al. 1983; Wewers et al. 1984; Robinson et al. 1985), or by immunosuppressive drugs (Salomon et al. 1985). Little is known regarding the effects of automotive emission exposure on these important cytokines.

■ **Recommendation 5.** Animal and/or human lung cell populations should be evaluated for the effect of inhaled gases on the production of interleukin-1 by alveolar macrophages, interleukin-2 by T lymphocytes, and the production of various interferons by alveolar macrophages and lymphocytes. Availability of enough bronchoalveolar lymphocytes may be a limiting factor, and, for that reason, human specimens may be especially valuable.

Alveolar Macrophage Functions. Information on alveolar macrophage functions with regard to the viral defense mechanisms and cytokines discussed above has increased considerably in recent years (Hunninghake et al. 1979, 1985). For example, expression of the type 2 histocompatibility antigen determinant on the macrophage surface is now known to be critical for alveolar macrophage processing of antigenic and infectious challenges (Mason et al. 1982). In fact, the role of alveolar macrophages in accessory cell functions is becoming increasingly clear (Toews et al. 1984). Also, the capacity of interferon γ treatment to activate alveolar macrophages for microbicidal activity has been closely associated with their defense function (Schaffner 1985). Furthermore, the capacity for respiratory burst (for example, superoxide anion production) is known to be directly related to alveolar macrophage microbicidal capacity (Hoidal et al. 1979; Pen-

nington 1985). Other metabolic activities, such as the production of chemotactic leukotrienes (leukotriene B_4) (Martin et al. 1984), as well as complement components (Pennington et al. 1979), and a low-molecular-weight neutrophil-activating factor (Pennington et al. 1985), have been identified for alveolar macrophages. Finally, the intrinsic motility of alveolar macrophages (Pennington and Harris 1981), plus their capacity to produce neutrophil chemotactic factors (Merrill et al. 1980), clearly relate to lung defense activities. It is safe to say that virtually no information is available regarding the effects of automotive emission components on these important alveolar macrophage products and functions, so the possibility exists that one or more of them may be significantly impaired by low and relevant levels of air pollutants.

■ **Recommendation 6.** The effects of automotive emission components on alveolar macrophage function should be determined.

Mucosal Binding. Several studies already mentioned in the Background section have described morphologic and functional defects in mucociliary apparatus of airway mucosa in intact animals or explanted tissues exposed to NO_2. These defects might adversely affect mucociliary clearance of potential infectious agents, but another mechanism—altered mucosal binding properties—might also predispose to infection. Increased mucosal binding affinity for pathogenic bacteria occurs under a number of stress conditions including surgery (Woods et al. 1981a), necessity for intensive care unit management (Woods et al. 1981b), and malnutrition (Niederman et al. 1984). Increased adherence of gram-negative bacilli to airway mucosa predisposes to subsequent respiratory infections in certain individuals (Johanson et al. 1972). The impact of emission exposures on binding affinity of airway mucosa for potential pathogenic infectious agents is unknown. The adherence properties of airway cells obtained from animals or humans experimentally exposed to relevant levels of automotive emission components should be studied. Such assays could be performed by using radiolabeled, gram-negative bacilli and quantitating their binding affinity for buccal or tracheobronchial cells removed by scraping and placed into tissue cultures (Niederman et al. 1984; Woods et al. 1981a, b), or tracheal explant cultures (Ramphal and Pyle 1983). These assays are relatively insensitive due to high background radioactivity, and it may be impossible to detect very subtle alterations in mucosal binding properties.

■ **Recommendation 7.** Mucosal cell binding affinity for respiratory pathogens should be studied after exposure to relevant concentrations of emission components.

Summary

The basic hypothesis under consideration is that exposure to automotive emissions results in increased susceptibility to respiratory infections. The rationale for exploring this hypothesis is that if it is true, then special programs for vaccination, clinical monitoring, and occupational counseling of high-risk groups should be undertaken. Current data are insufficient to test this hypothesis, but numerous studies suggest that it may be true. For operational purposes, the components of automotive emissions of interest include nitrogen oxides (especially NO_2), CO, O_3, and particulates. In addition, interest in aldehydes has increased in expectation that methanol fuel sources may be used in the near future.

Two basic research approaches have been used in past studies to explore this hypothesis. One approach has involved epidemiologic surveys of populations exposed to varying levels of known components of automotive emissions. Frequency and severity of respiratory symptoms, as well as performance on spirometry testing, were monitored and compared but results are conflicting. Outdated methods for monitoring ambient gas levels, as well as lack of serologic or cultural tests for diagnosing infection, are justified criticisms of these

studies. Future epidemiologic studies should take these problems into account. Also, future studies may wish to identify and focus attention on high-risk populations such as chronic lung disease patients, the immunosuppressed, and the elderly.

The second approach has been to expose animals to inhaled gases, commonly NO_2 or O_3, and then to study the effects of exposure on lung defenses against infection. Numerous studies have demonstrated decreased survival from experimental bacterial or viral infection and decreased capacity to kill bacteria in the lungs of animals exposed to pollutant gases. Levels of $NO_2 \geq 0.5$ ppm (often much higher), however, were required to demonstrate these adverse effects. Infectivity studies with animal models may not be sufficiently sensitive to identify more subtle defects in lung defense which might result from lower levels of gas exposures.

To address this possibility, individual components of the lung defense system have been evaluated in specimens obtained from exposed animals. To date, most studies have focused on alveolar macrophages, although some studies have examined morphologic effects of gases on mucociliary tissues and effects on systemic antibody responses. As in the infection models, impaired defenses have been identified only after high (that is, ≥ 2.0 ppm) and often prolonged NO_2 exposure. It is fair to point out, however, that newer assays have been developed (for example, NK cell function, interleukin-1 and -2 production, cell migration) that may be better suited to detect subtle, yet potentially important, defects in lungs exposed to low-level emission components. Furthermore, it is now safe and ethical to obtain human lung specimens using the flexible fiberoptic bronchoscope. Thus, future chamber studies with human subjects could be followed by an evaluation of bronchoscopic specimens. In planning such studies, it should be kept in mind that adverse effects on one component of lung defense might be compensated for by augmented activity of other defense systems. The complexity of this type of analysis cannot be overstated.

Summary of Research Recommendations

HIGH PRIORITY

On the basis of current information, the following studies are most likely to yield useful data and, given limited funding resources, most highly recommended.

Recommendation 1
Epidemiologic survey in high-risk populations exposed to varying levels of automotive emissions or components. The design of such surveys should include populations with defined exposure, specific study groups, adequate histories, diagnostic tests for infection, and measures of severity.

Recommendation 4
Evaluation of components of respiratory antiviral defense mechanisms. This analysis could be done in humans or animals exposed to emission components under controlled conditions. Studies most likely to yield useful new information would be numbers and function of pulmonary NK cells and cytotoxic lymphocytes; interferon production by alveolar macrophages and lymphocytes; and antibody-dependent cellular cytotoxic activity of Fc receptor-bearing lung cells (that is, alveolar macrophages and K lymphocytes).

MODERATE PRIORITY

Studies that include recent developments in methodology may allow detection of more subtle defects, particularly at relevant emission exposure levels.

Recommendation 5
Assays of immunologic modulator production by alveolar macrophages (interleukin-1, interferon) and pulmonary lymphocytes (interleukin-2, interferon). As before, chamber studies with human subjects or animal models could be used.

Recommendation 3
Altered-host infection models. The influence of altered-host status on lung susceptibility to emission components is important. Animals at extremes of age, immunosuppressed, or with experimentally induced chronic lung damage, should be studied to evaluate the effects of low-level exposures on lung defenses against infection.

Recommendation 6
Specialized alveolar macrophage functions, such as chemotaxis, production of complement and chemotactic factors, accessory cell function, and capacity for respiratory burst. Chamber studies with human subjects or animal models could be used.

LOW PRIORITY

These studies are similar to studies already performed which have shown negative results at low emission levels (infection model), or utilize relatively insensitive methodologies. However, if funding is available, the indicated modifications of study may provide new information.

Recommendation 7	Mucosal binding affinity. Human or animal respiratory cell binding affinity for respiratory pathogens should be studied after exposure to relevant concentrations of emission components.
Recommendation 2	Infection models using common human pathogens. Studies using viral agents, mycoplasma, *Legionella* sp., *S. pneumoniae*, and *H. influenzae* for experimental infections in exposed animals might offer more relevant information than the numerous past studies using *Klebsiella* and *S. aureus*.

References

Acton, J. D., and Myrvik, Q. N. 1972. Nitrogen dioxide effects on alveolar macrophage, *Arch. Environ. Health* 24:48–52.

Astry, C. L., and Jakab, G. J. 1983. The effects of acrolein exposure on pulmonary antibacterial defenses, *Toxicol. Appl. Pharmacol.* 67:49–54.

Bell, J. A., Ward, T. G., Kapikian, A. Z., Shelokov, A., Reichelderfer, T. E., and Huebner, R. J. 1957. Artificially induced Asian influenza in vaccinated and unvaccinated volunteers, *J. Am. Med. Assoc.* 165:1366–1373.

Bukowski, J. F., Woda, B. A., and Welsh, R. M. 1984. Pathogenesis of murine cytomegalovirus infection in natural killer cell-depleted mice, *J. Virol.* 52:119–128.

Burrows, B., Knudson, R. J., and Lebowitz, M. D. 1977. The relationship of childhood respiratory illness to adult obstructive airway disease, *Am. Rev. Respir. Dis.* 115:751–760.

Dixon, R. E. 1985. Economic costs of repiratory tract infections in the United States, *Am. J. Med.* 78(Suppl. 6B):45–51.

Ehrlich, R. 1966. Effect of nitrogen dioxide on resistance to respiratory infection, *Bacteriol. Rev.* 30:604–614.

Ehrlich, R., and Henry, M. C. 1968. Chronic toxicity of nitrogen dioxide. I. Effect on resistance to bacterial pneumonia, *Arch. Environ. Health* 17:860–865.

Empey, D. W., Laitinen, L. A., Jacobs, L., Gold, W. M., and Nadel, J. A. 1976. Mechanisms of bronchial hyperreactivity in normal subjects after upper respiratory tract infection, *Am. Rev. Respir. Dis.* 113:131–139.

Ennis, F. A., Wells, M. A., Butchko, G. M., and Albrecht, P. 1978. Evidence that cytotoxic T cells are part of the host's response to influenza pneumonia, *J. Exp. Med.* 148:1241–1250.

Esposito, A. L., and Pennington, J. E. 1983. Effects of aging on antibacterial mechanisms in pneumonia, *Am. Rev. Respir. Dis.* 128:662–667.

Feery, B. J., Evered, M. G., and Morrison, E. I. 1979. Different protection rates in various groups of volunteers given subunit influenza virus vaccine in 1976, *J. Infect. Dis.* 139:237–241.

Fenters, J. D., Findlay, J. C., Port, C. D., Ehrlich, R., and Coffin, D. L. 1973. Chronic exposure to nitrogen dioxide. Immunologic, physiologic, and pathologic effects in virus-challenged squirrel monkeys, *Arch. Environ. Health* 27:85–89.

Ferris, B. G. 1978. Health effects of exposure to low levels of regulated air pollutants. A critical review, *J. Air Pollut. Control Assoc.* 28:482–497.

Freeman, G., Crane, S. C., Stephens, R. J., and Furiosi, N. J. 1968. Environmental factors in emphysema and a model system with NO_2, *Yale J. Biol. Med.* 40:566–575.

Fujimaki, H., and Shimizu, F. 1981. Effects of acute exposure to nitrogen dioxide on primary antibody response, *Arch. Environ. Health* 3:114–119.

Gardner, D. E., Holzman, R. S., and Coffin, D. L. 1969. Effect of nitrogen dioxide on pulmonary cell population, *J. Bacteriol.* 98:1041–1043.

Gardner, D. E., Pfitzer, E. A., Christian, R. T., and Coffin, D. L. 1971. Loss of protective factor for alveolar macrophages when exposed to ozone, *Arch. Intern. Med.* 127:1078–1084.

Garibaldi, R. A. 1985. Epidemiology of community-acquired respiratory tract infections in adults, *Am. J. Med.* 78(Suppl. 6B):32–37.

Glezen, W. P., and Denny, F. W. 1973. Epidemiology of acute lower respiratory disease in children, *N. Engl. J. Med.* 288: 498–505.

Goldstein, E., Eagle, C., and Hoeprich, P. D. 1973. Effect of nitrogen dioxide on pulmonary bacterial defense mechanisms, *Arch. Environ. Health* 26:202–204.

Green, G. M. 1970. The J. Burns Amberson lecture—In defense of the lung, *Am. Rev. Respir. Dis.* 102:691–703.

Green, G. M., Jakab, G. J., Low, R. B., and Davis, G. S. 1977. Defense mechanisms of the respiratory membrane, *Am. Rev. Respir. Dis.* 115:479–514.

Greene, N. S., and Schneider, S. L. 1978. Effects of NO_2 on the response of baboon alveolar macrophages to migration inhibitory factor, *J. Toxicol. Environ. Health* 4:869–880.

Correspondence should be addressed to James E. Pennington, Clinical Research Department, Cutler Laboratories, P.O. Box 1986, Berkeley, CA 74701.

Henderson, F. W., Collier, A. M., Denny, F. W., Semor, R. J., Sheaffer, C. I., Conley, W. G., and Christian, R. M. 1979. The etiologic and epidemiologic spectrum of bronchiolitis in pediatric practice, *J. Pediatr.* 95:183–190.

Henry, M. C., Ehrlich, R., and Blair, W. H. 1969. Effect of nitrogen dioxide on resistance of squirrel monkeys to *Klebsiella pneumoniae* infection, *Arch. Environ. Health* 18:580–587.

Henry, M. C., Findlay, J., Spangler, J., and Ehrlich, R. 1970. Chronic toxicity of NO_2 in squirrel monkeys. III. Effect on resistance to bacterial and viral infection, *Arch. Environ. Health* 20:566–570.

Hoidal, R. J., Fox, R. B., and Repine, J. E. 1979. Defective oxidative metabolic responses in vitro of alveolar macrophages in chronic granulomatous disease, *Am. Rev. Respir. Dis.* 120:613–618.

Holt, P. G., Finlay-Jones, L. M., Keast, D., and Papadimitrou, J. M. 1979. Immunological function in mice chronically exposed to nitrogen oxides (NO_x), *Environ. Res.* 19:154–162.

Hunninghake, G. W., and Fauci, A. S. 1977. Immunologic reactivity of the lung. III. Effects of corticosteroids on alveolar macrophage cytotoxic effector cell function, *J. Immunol.* 118:146–150.

Hunninghake, G. W., Gadek, J. E., Kawanami, O., Ferrans, V. J., and Crystal, R. G. 1979. Inflammatory and immune processes in the human lung in health and disease: Evaluation by bronchoalveolar lavage, *Am. J. Pathol.* 97:149–206.

Hunninghake, G. W., Fick, R. B., and Nugent, K. W. 1985. Pulmonary host defenses: Cellular factors, In: *Advances in Host Defense Mechanisms* (J. I. Gallin and A. S. Fauci, eds.), pp. 89–113, Raven Press, New York, N.Y.

Hurst, D. J., and Coffin, D. L. 1971. Effect of ozone on lysosomal hydrolases in vitro, *Arch. Intern. Med.* 127:1059–1063.

Hurst, D. J., Gardner, D. E., and Coffin, D. L. 1970. Effect of ozone on acid hydrolases of pulmonary macrophages, *J. Reticuloendothel. Soc.* 8:288–300.

Johanson, W. G., Jr., Pierce, A. K., Sanford, J. P., and Thomas, G. D. 1972. Nosocomial respiratory infections with gram-negative bacilli. The significance of colonization of the respiratory tract, *Ann. Intern. Med.* 77:701–706.

Johnson, K. M., Chanock, R. M., Rifkind, D., Kravetz, H. M., and Knight, V. 1961. Respiratory syncytial virus. IV. Correlation on virus shedding, serologic response, and illness in adult volunteers, *J. Am. Med. Assoc.* 176:663–667.

Kaltreider, H. B. 1976. Expression of immune mechanisms in the lung, *Am. Rev. Respir. Dis.* 113:347–379.

Kattan, M. 1979. Long-term sequelae of respiratory illness in infancy and childhood, *Pediatr. Clin. N. Am.* 26:525–535.

Kavet, R. I., and Brain, J. D. 1975. Reaction of the lung to air pollutant exposure, *Life Sci.* 15:849–861.

Keller, M. D., Lanese, R. R., Mitchell, R. I., and Cote, R. W. 1979. Respiratory illness in households using gas and electric cooking, *Environ. Res.* 19:495–515.

Kohl, S., Starr, S. E., Oleske, J. M., Shore, S. L.,

Ashman, R. B., and Nahmias, A. J. 1977. Human monocyte–macrophage-mediated antibody-dependent cytotoxicity to herpes simplex virus–infected cells, *J. Immunol.* 118:729–735.

Lamontagne, L., Gauldie, J., Stadnyk, A., Richards, C., and Jenkins, E. 1985. In vivo initiation of unstimulated in vitro interleukin-1 release by alveolar macrophages, *Am. Rev. Respir. Dis.* 131:326–330.

Martin, T. R., Altman, L. C., Albert, R. K., and Henderson, W. R. 1984. Leukotriene B_4 production by the human alveolar macrophage: A potential mechanism for amplifying inflammation in the lung, *Am. Rev. Respir. Dis.* 129:106–111.

Mason, R., Austyn, J., Brodsky, F., and Gordon, S. 1982. Monoclonal anti-macrophage antibodies: human pulmonary macrophages express HLA-DR (Ia-like) antigens in culture, *Am. Rev. Respir. Dis.* 125:586–593.

McGrath, J. J., and Oyervides, J. 1983. Response of NO_2-exposed mice to klebsiella challenge, *Adv. Mod. Environ. Toxicol.* 5:475–485.

McGrath, J. J., and Oyervides, J. 1985. Effects of nitrogen dioxide on resistance to *Klebsiella pneumoniae* in mice, *J. Am. Coll. Toxicol.* 4:227–231.

McIntosh, K. 1985. Diagnostic virology, In: *Virology* (B. N. Fields, D. M. Knipe, R. M. Chanock, J. L. Melnick, B. Roizman, and R. E. Shope, eds.), pp. 309–322, Raven Press, New York, N.Y.

Melia, R. J. W., Florey, C. du V., Altman, D. S., and Swan, A. V. 1977. Association between gas cooking and respiratory disease in children, *Brit. Med. J.* 2:149–152.

Melia, R. J. W., Florey, C. du V., and Chinn, S. 1979. The relation between respiratory illness in primary school children and the use of gas for cooking. I. Results from a national survey, *Int. J. Epidemiol.* 8:333–338.

Merrill, W. W., Naegel, G. P., Matthay, R. A., and Reynolds, H. Y. 1980. Alveolar macrophage-derived chemotactic factor. Kinetics of in vitro production and partial characterization, *J. Clin. Invest.* 65:268–276.

Miller, S., and Ehrlich, R. 1958. Susceptibility to respiratory infections of animals exposed to ozone. I. Susceptibility to *Klebsiella pneumoniae*, *J. Infect. Dis.* 103:145–149.

Monto, A. S., and Ullman, B. M. 1974. Acute respiratory illness in an American community. The Tecumseh study, *J. Am. Med. Assoc.* 227:164–169.

Niederman, M. S., Merrill, W. W., Ferranti, R. D., Pagano, K. M., Palmer, L. B., and Reynolds, H. Y. 1984. Nutritional status and bacterial binding in the lower respiratory tract in patients with chronic tracheostomy, *Ann. Intern. Med.* 100:795–800.

Pearlman, M. E., Finklea, J. F., Creason, J. P., Shy, C. M., Young, M. M., and Horton, R. J. M. 1971. Nitrogen dioxide and lower respiratory illness, *Pediatrics* 47:391–398.

Pennington, J. E. 1983. Community-acquired pneumonia and acute bronchitis, In: *Respiratory Infections: Diagnosis and Management* (J. E. Pennington, ed.), pp. 125–134, Raven Press, New York, N.Y.

Pennington, J. E. 1984. Respiratory tract infections: intrinsic risk factors, *Am. J. Med.* 76:34–41.

Pennington, J. E. 1985. Immunosuppression of pulmonary host defenses, In: *Advances in Host Defense Mechanisms* (J. I. Gallin and A. S. Fauci, eds.), pp. 141–164, Raven Press, New York, N.Y.

Pennington, J. E. 1986. Use of animal models to evaluate antimicrobial therapy for bacterial pneumonias, In: *Experimental Models in Antimicrobial Chemotherapy* (M. A. Sande and O. Zak, eds.), pp. 237–256, Academic Press, London.

Pennington, J. E., and Harris, E. A. 1981. Influence of immunosuppression on alveolar macrophage chemotactic activities in guinea pigs, *Am. Rev. Respir. Dis.* 123:299–304.

Pennington, J. E., Matthews, W. J., Marino, J. T., and Colten, H. R. 1979. Cyclophosphamide and cortisone acetate inhibit complement biosynthesis by guinea pig bronchoalveolar macrophages, *J. Immunol.* 123:1318–1321.

Pennington, J. E., Rossing, T. H., Boerth, L. W., and Lee, T. H. 1985. Isolation and partial characterization of a human alveolar macrophage-derived neutrophil activating factor, *J. Clin. Invest.* 75:1230–1237.

Pinkston, P., Bitterman, P. B., and Crystal, R. G. 1983. Spontaneous release of interleukin-2 by lung T lymphocytes in active pulmonary sarcoidosis, *N. Engl. J. Med.* 308:793–800.

Polsky, B., Gold, J. W. M., Whimbey, E., Dryjanski, J., Brown, A. E., Schiffman, G., and Armstrong, D. 1986. Bacterial pneumonia in patients with the acquired immunodeficiency syndrome, *Ann. Intern. Med.* 104:38–41.

Pullan, C. R., and Hey, E. N. 1982. Wheezing, asthma, and pulmonary dysfunction 10 years after infection with respiratory syncytial virus in infancy, *Brit. Med. J.* 284:1665–1669.

Purvis, M. R., and Ehrlich, R. 1963. Effect of atmospheric pollutants on susceptibility to respiratory infection. II. Effect of nitrogen dioxide, *J. Infect. Dis.* 113:72–76.

Ramphal, R., and Pyle, M. 1983. Adherence of mucoid and nonmucoid *Pseudomonas aeruginosa* to acid-injured tracheal epithelium, *Infect. Immun.* 41:345–351.

Rankin, J., Snyder, P. E., Schachter, N., and Matthay, R. A. 1984. Bronchoalveolar lavage: its safety in subjects with mild asthma, *Chest* 85:723–728.

Reynolds, H. Y. 1983. Normal and defective respiratory host defenses, In: *Respiratory Infections: Diagnosis and Management* (J. E. Pennington, ed.), pp. 1–23, Raven Press, New York, N.Y.

Robinson, B. W. S., Pinkston, P., and Crystal, R. G. 1984. Natural killer cells are present in the normal human lung but are functionally impotent, *J. Clin. Invest.* 74:942–950.

Robinson, B. W. S., McLemore, T. L., and Crystal, R. G. 1985. Gamma interferon is spontaneously released by alveolar macrophages and lung T lymphocytes in patients with pulmonary sarcoidosis, *J. Clin. Invest.* 75:1488–1495.

Salomon, D. R., Boerth, L. W., Rocher, L. L., and Pennington, J. E. 1985. Dose-dependent reduction in alveolar macrophage populations during cyclosporine treatment, *Clin. Res.* 33:581A.

Samet, J. M., Tager, I. B., and Speizer, F. E. 1983. The relationship between respiratory illness in childhood and chronic air-flow obstruction in adulthood, *Am. Rev. Respir. Dis.* 127:508–523.

Schaffner, A. 1985. Therapeutic concentrations of glucocorticoids suppress the antimicrobial activity of human macrophages without impairing their responsiveness to gamma interferon, *J. Clin. Invest.* 76:1755–1764.

Schiff, L. J. 1977. Effect of nitrogen dioxide on influenza virus infection in hamster tracheal organ culture, *Proc. Soc. Exp. Biol. Med.* 156:546–549.

Sherman, M., Goldstein, E., Lippert, W., and Wennberg, R. 1977. Neonatal lung defense mechanisms: a study of the alveolar macrophage system in neonatal rabbits, *Am. Rev. Respir. Dis.* 116:433–440.

Shy, C. M., Creason, J. P., Pearlman, M. F., McClain, K. E., Benson, F. B., and Young, M. M. 1970a. The Chattanooga school children study. Effects of community exposure to nitrogen dioxide. I. Methods, description of pollutant exposure and results of ventilatory testing, *J. Air Pollut. Control Assoc.* 20:539–545.

Shy, C. M., Creason, J. P., Pearlman, M. F., McClain, K. E., Benson, F. B., and Young, M. M. 1970b. The Chattanooga school children study. Effects of community exposure to nitrogen dioxide. II. Incidence of acute respiratory illness, *J. Air Pollut. Control Assoc.* 20:582–586.

Sissons, J. G. P., and Oldstone, M. B. A. 1985. Host response to viral infections, In: *Virology* (B. N. Fields, D. M. Knipe, R. M. Chanock, J. Melnick, B. Roizman, and R. E. Shope, eds.), pp. 265–279, Raven Press, New York, N.Y.

Snider, G. L., Lucey, E. C., and Stone, P. J. 1986. Animal models of emphysema, *Am. Rev. Respir. Dis.* 133:149–169.

Speizer, F. E., and Ferris, B. G., Jr. 1973a. Exposure to automobile exhaust. I. Prevalence of respiratory symptoms and disease, *Arch. Environ. Health* 26: 313–318.

Speizer, F. E., and Ferris, B. G., Jr. 1973b. Exposure to automobile exhaust. II. Pulmonary function measurement, *Arch. Environ. Health* 26:319–324.

Speizer, F. E., Ferris, B. G., Jr. Bishop, Y. M. M., and Spengler, J. 1980. Respiratory disease rates and pulmonary function in children associated with NO_2 exposure, *Am. Rev. Respir. Dis.* 121:3–10.

Stein-Streilein, J., Bennett, M., Mann, D., and Kumar, V. 1983. Natural killer cells in mouse lung: surface phenotype, target preference, and response to local influenza virus infection, *J. Immunol.* 131:2699–2704.

Toews, G. B., Vial, W. C., Dunn, M. M., Guzzetta, P., Nunez, G., Stastny, P., and Lipscomb, M. F. 1984. The accessory cell function of human alveolar macrophages in specific T cell proliferation, *J. Immunol.* 132:181–186.

Tuazon, E. C., Winer, A. M., and Pitts, J. N. 1981. Trace pollutant concentrations in a multiday smog episode in the California South Coast Air Basin by long path length Fourier transform infrared spectroscopy, *Environ. Sci. Technol.* 15:1232–1237.

Valand, S. B., Acton, J. D., and Myrvik, Q. N. 1970. Nitrogen dioxide inhibition of viral-induced resistance in alveolar monocytes, *Arch. Environ. Health* 20:303–309.

Vassallo, L. L., Domm, B. M., Poe, R. H., Duncombe, M. L., and Gee, J. B. L. 1973. NO_2 gas and NO_2 effects on alveolar macrophage phagocytosis and metabolism, *Arch. Environ. Health* 26:270–274.

Voisin, C., Aerts, C., Jakubczak, E., Houdret, J. L., and Tonnel, A. B. 1977. Effects of nitrogen dioxide on alveolar macrophages surviving in the gas phase, *Bull. Euro. Physiopathol. Respir.* 13:137–144.

Weinberger, S. E., Kelman, J. A., Elson, N. A., Young, R. C., Reynolds, H. Y., Fulmer, J. D., and Crystal, R. G. 1978. Bronchoalveolar lavage in interstitial lung disease, *Ann. Intern. Med.* 89:459–466.

Wewers, M. D., Rennard, S. I., Hance, A. J., Bitterman, P. B., and Crystal, R. G. 1984. Normal human alveolar macrophages obtained by bronchoalveolar lavage have a limited capacity to release interleukin-1, *J. Clin. Invest.* 74:2208–2218.

Williams, R. D., Acton, J. D., and Myrvik, Q. N. 1972. Influence of nitrogen dioxide on the uptake of parainfluenza-3 virus by alveolar macrophages, *J. Reticuloendothel. Soc.* 11:627–636.

Woods, D. E., Straus, D. C., Johanson, W. G., Jr., and Bass, J. A. 1981a. Role of salivary protease activity in adherence of gram-negative bacilli to mammalian buccal epithelial cells in vivo, *J. Clin. Invest.* 68:1435–1440.

Woods, D. E., Straus, D. C., Johanson, J. G., Jr., and Bass, J. A. 1981b. Role of fibronectin in the prevention of adherence of *Pseudomonas aeruginosa* to buccal cells, *J. Infect. Dis.* 143:784–790.

Assessment of Carcinogenicity: Generic Issues and Their Application to Diesel Exhaust

DAVID G. KAUFMAN
University of North Carolina

Air Pollution, the Automobile, and Public Health. © 1988 by the Health Effects Institute. National Academy Press, Washington, D.C.

Despite our limited understanding of carcinogenesis, practical concerns in the "real world" confront us with the need to assess the potential significance of diesel exhaust as a human carcinogen. Such an assessment requires progressing from fragmentary theoretical insights into the process of carcinogenesis to estimates of the human risk posed by diesel exhaust. Confounding this effort is the fact that diesel exhaust is an imprecisely characterized and inconsistently constituted product composed of chemicals that may trigger carcinogenesis individually, cooperatively, or even sequentially.

Researchers are now confronting the difficulties of understanding the etiology and pathogenesis of multifactorial, multistep disease processes, and they are just beginning to recognize general principles that may operate in most typical cases of cancer. There is awareness of the relationship between the dose of carcinogens and the resulting tumor response, and recognition of the importance of the metabolism of a carcinogen into reactive intermediates that may cause damage. Cellular mechanisms such as DNA replication may provide opportunities for carcinogens to transform genetic information, and targets in DNA may include specific genes or sites at which chromosomes are prone to breakage. Enhancing factors, such as promoters, may increase the likelihood of cancer development. Variations in human susceptibility to cancer make evaluation of the activity of specific carcinogens difficult, although it is clear that certain human tissues or certain individuals are more susceptible to cancer than others. Certain familial tendencies or acquired illnesses are also thought to predispose people to cancer.

In this chapter, the evidence on the carcinogenicity of diesel engine exhausts and the methods used to make quantitative risk estimates from these data are evaluated. Specific evidence concerning carcinogenesis of diesel exhaust in experimental systems is reviewed, and relationships between this information and reviews in other chapters are identified. Current knowledge as well as areas of ignorance influence efforts to estimate human risks by extrapolation from the experimental data on animals. A discussion of these issues serves as an outline for making such estimates in the future.

Mechanisms of Carcinogenesis Relevant to Assessment of Mobile Source Emissions

Chemical carcinogenesis is a very complex topic. Thus, this review is selective in its consideration of carcinogenesis, focusing on several general concepts rather than on specific details. The constructive role of studying cancer development in animal models is considered, and certain aspects of the general principles operating in most typical cases of carcinogenesis are examined. The review also touches on unusual cases that appear not to fit the typical pattern of cancer development. It considers the evidence for and the problems associated with evaluating a disease that develops as the result of a multistep process. Finally, the factors that define individual variations in susceptibility are discussed, and features of carcinogen metabolism and translocation are reviewed.

Experimental Models in Chemical Carcinogenesis

Experimental animal models have been employed to reproduce tumors of the histologic types and organs of origin that commonly occur in humans. Such models permit direct experimental study of factors that influence the development of the most common cancers in humans and the mechanisms of action of particular carcinogens. Examples of valuable animal models and their applications are listed in table 1. Some unique insights have been derived from comparisons of the properties of animal tissues in which the tumor response is a good model for the human disease, to tissues of other species in which the response is very different from that of humans.

Studies of particular tissues have been facilitated by using organ cultures (Saffiotti

Table 1. Examples of Valuable Animal Models of Human Carcinogenesis

Rodent Species	Type of Human Cancer Modeled	References	Application
Hamster	Lung	Saffiotti et al. (1968)	Dose/ response
Hamster	Pancreatic	Pour (1984) Scarpelli et al. (1984)	Metabolic mechanisms
Rat	Breast	Huggins et al. (1961) Gullino et al. (1975)	Hormonal influences
Rat	Colon	Ward et al. (1973) Reddy et al. (1974)	Dietary influences
Mouse	Skin	Berenblum and Shubik (1947)	Promotion

and Harris 1979). In this technique, pieces of intact tissue representative of the sampled organ are grown in culture. Many features of the tissue that exist in vivo, including the interrelationship between the epithelial components and the supportive cells, are preserved. Such cultures can be used to assess morphological features, macromolecular synthesis, and responses to hormones as well as capacity to metabolize carcinogens and to repair DNA damage. Use of organ cultures has been a principal approach used for analysis of properties related to carcinogenesis in human tissues.

Some of the attractive features of organ cultures—for example, their maintenance of natural relationships between epithelial and supportive cells—are related to some of their major shortcomings. In contrast to cell cultures, organ cultures cannot be propagated, and material from an individual human subject is rapidly exhausted. Cell culture overcomes this limitation because cells may be propagated in culture. However, the very process of propagation exerts a selective pressure, and the cell type that emerges may be unlike that predominating in the intact tissue. Nevertheless, isolated cells have proven very useful in the study of common and unique features of carcinogenesis under far more controlled environmental conditions than is possible in an intact animal.

Although direct experimentation with the objective of inducing carcinogenesis is clearly unethical in humans, a broader, deeper information base is needed on the properties of human cells and tissues that

relate to carcinogenesis. This goal has been approached by undertaking culture studies of human cells and tissues obtained at immediate autopsies or from surgical specimens (Harris and Trump 1983).

Studying the properties of human tissues in vitro allows examination of the human diversity in cancer development. For example, in vitro techniques can be used to explore the individual variability in metabolizing carcinogens, repairing DNA damage, responding to various hormones, and perhaps even to determine the degree to which various nutrients serve as cofactors in carcinogenesis.

Although in vitro carcinogenesis with human cells in culture is rather new, transformation of normal cells to neoplastic ones has been accomplished with a number of cell types. Results from such studies permit the direct comparison of the stages in the presumed multistep process of carcinogenesis in humans and in animals. For example, the apparently greater difficulty in transforming human cells than animal cells may parallel the comparative susceptibility to cancer of these various species. If the determinants of the various stages in carcinogenesis are successfully characterized in human cells, it may be possible to develop improved methods for early detection of preneoplastic or early neoplastic lesions.

Some human tissues have been maintained as viable xenotransplants in nude mice (Valerio et al. 1981). Such models are an ethically acceptable method for in vivo study of the process of carcinogenesis in human tissue (Shimosato et al. 1980). This model may provide for direct comparisons of features of carcinogenesis between humans and experimental animals that are commonly used in bioassays. Such information would clearly be valuable in determining the risks to humans of agents demonstrated as carcinogenic in animal bioassays.

Role of DNA Replication and Repair

It is a well-recognized clinical observation that cancer typically occurs in tissues that have a high rate of cell proliferation or in tissues in which cell proliferation occurs in response to injury. Conversely, cancer is extremely rare in adult tissues or cell types in which cell proliferation does not occur. It was the opinion of classical pathologists that chronic irritation or injury was the etiologic factor for the development of cancer. Subsequently, a variety of specific carcinogenic etiologic agents have been recognized. Nonetheless, cell proliferation plays a significant role in the evolution of cancers (Grisham et al. 1983). This is well illustrated in the case of liver cancers induced in rats by chemical carcinogens. Typical liver carcinogens at effective doses are also hepatotoxic, and they induce restorative hyperplasia to replace cells lost as the result of the toxicity.

The influence of cell proliferation as a contributing factor in the development of cancer presumably results from effects on the mitotic process and on DNA synthesis. Replicating DNA is vulnerable for a variety of reasons. First, replicating DNA is affected to a greater extent by chemical carcinogens than is nonreplicating DNA (Cordeiro-Stone et al. 1982). Second, replication of DNA that contains carcinogen adducts may cause incorporation of incorrect nucleotides at sites of altered or excised bases. Third, some carcinogens may modify nucleotide precursors, and altered precursors may be incorporated into DNA. Fourth, DNA replication itself occurs with a low, but nonzero, error rate. Situations that increase cell replication are likely to cause mutations strictly as the result of these errors.

Mammalian cells have a number of mechanisms to repair DNA damage and to reduce the likelihood of errors during DNA replication. Treatments of cells or animals with chemical carcinogens or radiation cause the onset of DNA repair processes. In studies in which cell proliferation has been inhibited and DNA repair has been allowed to remove some or most carcinogen-induced DNA adducts, the transforming effects of the carcinogen damage have been reduced (Ikenaga and Kakunaga 1977). In contrast, in patients with defective DNA repair processes, such as the genetically determined syndrome known as xeroderma pigmentosum, increased incidences of tumors have been observed (Setlow 1978; Hanawalt and Sarasin 1986). Thus, DNA repair processes appear to be protective against tumor development, whereas defects of DNA repair appear to be associated with increased risks of cancer.

There appears to be a critical interrelationship between the repair and replication of DNA as factors in the etiology of cancer (Kakunaga 1975). If DNA replication proceeds within a damaged region prior to repair, there is a substantial risk of error-making during replication, which may cause a mutation to occur as the result of alteration of the base sequence of the complementary DNA strand. Of course, this does not occur if the repair of the damage precedes replication. Consequently, the relationship in time of the repair and replication of DNA may be a major determinant of the potential for the occurrence of mutations and also, presumably, of carcinogenesis.

Genetic Effects of Carcinogen Damage to DNA

Chemical carcinogens have been shown to produce a variety of types of DNA damage that can lead to genetic effects on cells (table 2) (Sarma et al. 1975; Drake and Baltz 1976; Singer and Grunberger 1983). Point mutations and frameshift mutations can alter the regulatory or coding regions of genes. On a larger scale, carcinogens can directly affect chromatids and chromosomes (Evans 1983). By still unknown mechanisms, carcinogen damage can cause the exchange of

Table 2. Genetic Effects of Carcinogen Damage to DNA

Point mutations—transitions and transversions
Frameshift mutations—small deletions or additions
Mutations at "hot spots"
Chromosomal breakage at "fragile sites"
Recombinations and rearrangements
Sister chromatid exchanges
Translocations of portions of chromosomes
Gene amplification
Aneuploidy

DNA segments between sister chromatids, and chromosomal breakage that leads to large deletions or transposition of chromosomal segments to other chromosomes. Presumably, such damage may lead to failures of mitotic division with unequal distribution of chromosomes between daughter cells, resulting in abnormal DNA content. DNA damage is also thought to be one mechanism for the amplification of segments of DNA.

The significance of many or all of these forms of damage to DNA does not concern the chemical composition of this molecule but rather its content of genetic information. Valuable insights about these genetic effects, particularly with regard to oncogenes, have arisen from recent studies in viral carcinogenesis and molecular biology. Investigations of the mechanism of cell transformation by oncogenic retroviruses have shown that their transforming genes, designated as oncogenes, are derived from the coding regions of cellular precursor genes known as proto-oncogenes (Bishop 1983). Proto-oncogenes are believed to play an important, though as yet unknown, role in normal cellular function or differentiation because they are highly conserved in widely divergent species—from yeast to humans. Recent studies have shown that proto-oncogenes can acquire transforming activity as the result of genetic alterations that affect their DNA sequence or place them under abnormal genetic regulation by chromosomal rearrangements, insertion of promoters, or gene amplification (Weinberg 1985; Barbacid 1986). The number of known retroviral oncogenes is quite limited—about two dozen. Even when the proto-oncogenes from which they are derived and the closely related cellular genes

(for example, N-myc and N-ras) are added to the sum, the total of retrovirus-related oncogenes is still small. Although further studies of human and animal tumors have identified additional genes with transforming activity, it is not yet possible to estimate the number of cellular genes that have transforming activity induced by genetic alteration.

It is well known that mutations occur at exceptionally high rates at specific sites in DNA of viruses and other prokaryotic organisms. This nonhomogeneous effect is recognized for spontaneous mutations as well as mutations induced by radiation or chemicals. The location of these so-called "hot spots" relates to the specific form of radiation or chemical carcinogen that induces the mutations. DNA sequence as well as the structural features of DNA, including bending and association with proteins, appears to influence the spectrum of hot spots. Clearly the DNA sequence in higher organisms such as mammals is far less completely defined, and the means for cataloging the spectra of hot spots in DNA of these organisms are very limited. Nonetheless, some evidence suggests that there are sites selectively affected by carcinogens where mutations occur at high frequency.

Fragile sites are locations in chromosomes that are particularly prone to breakage. When cell growth conditions are altered, such as through deprivation of thymidine and folic acid, chromosomes have been found to break consistently at the same sites. These sites are closely related to sites where chromosomal rearrangements occur in human cancers (Yunis and Soreng 1984), suggesting that structural peculiarities that make these sites prone to breakage may be important factors in the development of cancers. Another notable point is the chromosomal location of these fragile sites relative to several of the known proto-oncogenes. Although the power of the scientific methods used to compare the locations of these sites is not great, the apparent statistical relationship within the experimental error of the methods suggests that some very important feature of cancer development is related to the structure of DNA at these sites.

Techniques for identifying subregions (bands) within chromosomes now allow abnormal chromosomes in cancer cells to be examined with far greater resolution and specificity than previously possible. Surveys of the chromosomal banding patterns of a wide spectrum of cancer cells have shown some consistent patterns of chromosomal abnormalities for many different types of cancer (Sandberg 1983; Mitelman 1986). For some cancers—for example, Burkitt's lymphoma—there is a very high degree of consistency in the type of alteration observed. Most Burkitt's lymphomas show balanced translocations of portions of specific chromosomes. In other cases, such as the development of the Philadelphia chromosome (loss of a portion of the long arms of chromosome 22) in chronic myelogenous leukemia, the appearance of the chromosomal abnormality accompanies the chronic phase of the disease.

Another very common feature of cancer cells is the development of aneuploidy, with cells having more or less than the normal diploid number of chromosomes or an abnormal DNA content. In fact, the large size and the hyperchromaticity characteristic of cancer cell nuclei are largely due to the increased DNA content of typical aneuploid cells. Aneuploidy is presumed to develop as a consequence of unequal mitotic divisions during the evolution of cancer cells. The presence of abnormal mitotic figures is one feature of cancers used to arrive at a pathological diagnosis. One of the consequences of the abnormal chromosomal content of cancer cells is that particular genes are present in low or high copy number. One can speculate how the loss of a normal inhibiting function can occur with the loss of a chromosome in hypodiploid cancer cells. Hyperdiploid cells can greatly overexpress particular gene products, or they may generate insufficient inhibitory activity to balance the high copy number of some cancer-related gene.

Atypical Carcinogens

A number of substances very different from the typical chemical carcinogens have been shown to be carcinogenic in humans and in experimental animals. With atypical carcinogens, carcinogenesis can be induced by physical agents and chemicals that do not directly alter DNA. The differences between these atypical carcinogens and the common carcinogens challenge the classical concepts of carcinogenesis and demand the development of theories of carcinogenesis that can include their mode of action.

For some time, asbestos has been recognized as carcinogenic, first in humans (Doll 1955) and later confirmed in experimental animals (Wagner et al. 1973). When directly instilled into the pleural cavity of experimental animals it has been shown to produce tumors like those that follow asbestos exposure in humans (Wagner et al. 1973). The critical property of asbestos best associated with carcinogenicity is the physical dimensions of fibers (Stanton and Wrench 1972) rather than the chemical composition of the asbestos or the substances adsorbed on it. This was confirmed by showing that glass fibers, prepared in length and width comparable to asbestos fibers, were also carcinogenic.

The cellular response to asbestos fibers and other foreign bodies involves the foreign-body inflammatory reaction wherein the fibers are surrounded by macrophages and fibroblasts (Brand et al. 1975). Current hypotheses suggest that the inflammatory cells or epithelial cells produce reactive forms of oxygen molecules which may affect the DNA of the epithelial cells, and this damage is fundamental to the carcinogenic process. Others suggest that asbestos acts by affecting chromosomal segregation during mitosis. On a practical level, asbestos is relevant to the topic of mobile source emissions. It is known that asbestos exposure is associated with mesothelial cancer in humans. However, in individuals in whom asbestos exposure is combined with cigarette smoking, the risk of cancer is greatly increased, and the leading type of cancer is bronchogenic carcinoma (Selikoff et al. 1968). It is conceivable that individuals who have been exposed to asbestos will represent a group at increased risk from the combined effects of asbestos and mobile source emissions.

A number of studies have shown that

unusual substances, functioning as atypical carcinogens, can produce cancers in experimental animals. Plastic films have been shown to produce tumors when implanted into animals. However, when the films were sufficiently fenestrated, or when they were ground to a powder, the material was not carcinogenic. Several metals, in the forms of ores, refinery process by-products, and ions and salts, have been shown to be carcinogenic in humans or experimental animals (International Agency for Research on Cancer 1980). Examples include various forms of arsenic, chrome, and nickel.

A number of reports in recent years have noted that chemicals, including therapeutic agents that cause proliferation of peroxisomes, are carcinogenic (Reddy et al. 1980). Unlike chemicals such as phenobarbital, these agents appear to function as complete carcinogens rather than just as promoters. Investigations of examples of this class of chemicals have shown that they do not form adducts with DNA.

Several other chemicals and drugs are carcinogenic in animals or humans, but are not known to interact with or form adducts with DNA. Among these are agents that affect enzymes involved in the metabolism of DNA precursors or that more directly affect DNA precursor pools. These properties make some of these chemicals effective therapeutic agents for treating cancer. The action of some of these agents is believed to be a consequence of imbalances in DNA precursor pools, which also cause mutations.

Promotion, Cocarcinogenesis, and Enhancement

Exposure to carcinogens is not the only determinant of cancer development. Other substances or other processes can influence the risk for cancer development, particularly when they complement exposures to carcinogens or act on animals that have a high spontaneous tumor incidence. These factors must be recognized when the observational data derived from carcinogenicity tests in animals are being interpreted mechanistically. The terms "enhancers" and "enhancement" describe effects that include those typically classified as promoters or cocarcinogens but without attribution of a mechanism of action.

Promotion is defined operationally, based on classical experiments in which tumors were induced in mouse skin by a two-step treatment protocol (Berenblum 1975). The first treatment involved the application of a subcarcinogenic dose of a strong carcinogen to the mouse skin, followed by a prolonged series of applications of a noncarcinogenic agent. The combined treatments produced a tumor response, whereas the same dose of the carcinogen or the second agent, which has come to be known as a promoter when used alone, was ineffective or vastly weaker. The two steps of the treatment process were designated as initiation and promotion, and have come to be interpreted as separate events or processes in the evolution of cancers.

In contrast to the separate application of initiator and promoter, cocarcinogens are agents that enhance the development of cancers when administered concurrently with a carcinogen. Cocarcinogens act through a variety of mechanisms. They may modify the metabolism of carcinogens to yield a greater quantity or proportion of ultimate carcinogenic metabolites. They may act by causing cell or tissue toxicity with accompanying accelerated cell proliferation; this in turn may increase the risk of malignant transformation. They may also act by interfering with normal defense mechanisms that function to counteract the detrimental effects of carcinogens.

Enhancing or inhibiting effects from exposure to a wide variety of substances (for example, certain constituents of mobile source emissions), not just exposure to carcinogens (such as the possibly carcinogenic constituents of such mobile source emissions), determine the tumor response. Our understanding of these effects and the interactions between substances is very limited.

More-specific enhancing effects, in some cases affecting individual tissues, may come from exposures to noncarcinogens that exert a promoting or cocarcinogenic effect. Individual genetics, prior or concurrent medical conditions, and diet all may contribute to an individual's specific risk from a given level of exposure. Such enhancers presumably increase the effects of other carcinogens. It is particularly important to ascertain whether mobile source emissions contain constituents with enhancing activity.

Mobile source emissions may represent a serious public health problem if they enhance carcinogenesis initiated by other exposures. This is an important general problem that requires further attention. Different methods of bioassay from those used to detect carcinogens will have to be developed to determine whether these emissions have enhancing activity. Such methods are needed to explore the possibilities that enhancing activities are specific in augmenting the activity of particular types of carcinogens or that their activity differs according to tissue sites at which they act.

It is clear that standard carcinogenicity bioassays are not pure tests for either cancer-initiating activity or activity as a complete carcinogen. They are phenomenologic studies that associate excess cancers with particular treatments, but they do not indicate the mechanism by which the cancers are produced. Particularly in the case of tumors in tissues with a high spontaneous tumor incidence in untreated animals, increases in the incidence of tumors may reflect a toxic or promoting activity of the tested chemical.

If this effect is not the result of toxicity associated with high exposure levels, then the result may be a demonstration of promotion activity. Such a conclusion might distinguish these compounds from standard carcinogens, but it does not indicate that they are without risk. In view of the hazard posed by chemicals of this type, it is important to develop methods to demonstrate how they cause tumors. Since some of the constituents of diesel exhaust may also have this kind of activity, it would be useful to have the means of identifying and quantitating these chemicals.

Promoting or enhancing activity may involve a number of organs and tissues other than the skin. For this reason it will be necessary to evaluate the differences in enhancing effects in different tissues. For example, 12-O-tetradecanoylphorbol-13-

acetate (TPA) is a good promoter for mouse skin, but is not a good promoter for rat liver; conversely, phenobarbital is very effective in rat liver, but not in skin. The possibility also exists that enhancing activity relates to the type of initiator. The type of initiator may determine which tissues or organs will be sensitized to promoter or enhancer effects. It is entirely conceivable that the broad spectrum of chemicals in diesel exhaust contains enhancing agents with distinctive patterns of organ selectivity.

To test these hypotheses it will be necessary to examine the enhancing activity of materials such as diesel exhaust following an initiating treatment with any of a variety of carcinogens with a range of organ specificities. In this manner it may be possible to develop a standard panel of animal test models that would have the ability to detect and quantify promoters and enhancers that are active in any of a number of tissues.

■ **Recommendation 1.**   The role of promoters and enhancers in human carcinogenesis should be determined.

Multistep Processes

On the basis of a variety of clinical and experimental evidence, the development of cancer is believed to be a multistep process (Armitage 1985). Clinical experience has shown that the incidence of most cancers rises with age and most are seen to pass through premalignant stages prior to the development of clinically overt cancer (Doll 1971). The most thoroughly studied case is that of the multistep evolution of squamous cell carcinoma of the uterine cervix. The validity of a multistep interpretation is attested to by the fact that clinical intervention at an early stage vastly reduces the incidence of the overt, invasive tumors of this type. A similar sequence of premalignant lesions of the bronchial epithelium precedes invasive lung cancer (Auerbach et al. 1961). One study followed uranium miners with repetitive sputum cytologies for many years (Schreiber et al. 1974). Progressive changes in cytologic findings proceeded from squamous metaplasia through various stages of dysplasia, in situ carcinoma, and invasive carcinoma over the course of several years. Subsequently, comparable sequences of epithelial lesions have been found to precede overt cancers in a number of sites (Farber 1984).

Initiation and promotion in the mouse skin bioassay is an example of carcinogenesis as a two-step process (Berenblum 1975). Comparatively recent studies have shown that the process of promotion itself can be divided into stages (Slaga et al. 1980). In the case of the evolution of tumors of rat liver, cancer is believed to be the end result of a process in which foci or areas of enzyme-altered hepatocytes and neoplastic nodules precede malignant tumors (Farber 1980). In the respiratory tract of hamsters, carcinogen treatment has been found to cause a progressive sequence of histologic alterations that culminate in invasive, malignant tumors (Saffiotti and Kaufman 1975). These lesions demonstrate a spectrum of morphological changes very close in appearance to the lesions of the respiratory tract seen in humans. In fact, the evolving lesions shed cells analogous to those observed in the cytology preparations from the uranium miners cited above (Schreiber et al. 1974).

Methods to study the transformation of cells by chemical carcinogens in tissue culture have been available for about two decades. These studies first were successful in rodent embryo and fibroblast cells. More recently, similar results have been achieved using rodent epithelial cells, and in the past few years human cells have also been transformed with chemical carcinogens. A number of morphological, biological and phenotypic changes have often been observed in these in vitro transformation systems as the cells progress from the original cell population to demonstrably malignant cells. In some cases, for example in studies using Syrian hamster embryo cells, a specific sequence of changes in the culture has been linked to malignancy (Barrett and Ts'o 1978; Smith and Sager 1982). With cultured rat tracheal epithelial cells, a sequence of progressive changes in the biological behavior of carcinogen-treated cells has been observed (Nettesheim and Barrett 1984). In this system, the cultured cells can be evaluated for their relationship to the morphological alterations observed in vivo by allowing them to repopulate a transplanted rat trachea which had been deepithelialized (Klein-Szanto et al. 1982). Other evidence of the multistep nature of transformation found in this system is a two-step transformation involving an initiating carcinogen treatment followed by in vitro promotion with TPA (Steele et al. 1980). Studies of the transformation of human cells in vitro have also shown that several distinct alterations occur consistently and in a generally similar order (Kakunaga et al. 1983).

Variations in Susceptibility

Rates of development of spontaneous benign and malignant tumors vary in animals of different species and strains (Grasso and Hardy 1974). Some animals are highly resistant to tumor development, and even after a long life few will die with tumors. In contrast, some species and strains of animals have a very high incidence of cancers, in some cases 100 percent. In these species and strains, the type and quantity of these background tumors are characteristic of the animal and are presumed not to be the result of unusual exposures to environmental factors. Among the animals typically chosen for carcinogenesis bioassays, mice have an exceptionally high incidence of liver tumors, particularly in males. In female rats, mammary tumors are very common. Treatment of these animals with chemical carcinogens results in tumors at various locations and of types that depend on the activity and dose of the carcinogen as well as the route of administration and other factors. Commonly, these treatments also affect the incidence and multiplicity of the tumors characteristic of the untreated animals, indicative of the sensitivity of these tissues to transformation.

The human population, in comparison, appears to have a relatively low background level of cancer, as determined from cancer incidence data for certain low-risk groups in underdeveloped nations or in specific populations such as Mormons or Seventh Day Adventists in the United

Table 3. Factors Affecting Human Susceptibility to Carcinogenesis

Exposures to carcinogens

Diet composition and nutritional status

Personal habits including cigarette smoking and alcohol consumption

Determinants of geographic variations in cancer development

Genetic diseases or heterozygous carrier states

Acquired illnesses and infections

Unknown factors determining familial predispositions

Variations in metabolic activation or inactivation of carcinogens

States in which there are religious restrictions on smoking or certain dietary practices. Despite the low overall cancer rates in these groups, certain cancers are seen and these may represent the background tumors of humans. These include leukemias and lymphomas, soft-tissue sarcomas, skin tumors, and a low rate of tumors of several epithelial tissues. Above this background, several factors appear to affect the susceptibility of humans to the development of cancers (table 3).

The incidence in humans of tumors of various organs differs by country and even by population group within countries. In the United States, cancer of the lung is the most common significant cancer in males and females, whereas in Egypt and Japan, cancers of the urinary bladder and stomach, respectively, are the most common. Within the United States, there appear to be geographic differences in incidence of tumors of various organs (Pickle et al. 1987). Clearly, a large proportion of these tumors are induced rather than spontaneous and are of environmental origin.

In contrast to the general population, there are individuals who are genetically predisposed to the development of cancers. Examples of genetic diseases associated with a high incidence of cancer include xeroderma pigmentosum, ataxia telangiectasia, familial retinoblastoma, Fanconi's anemia, Gardner's syndrome, familial polyposis coli, and many others. Studies have shown that for the recessively inherited genetic disease ataxia telangiectasia, close relatives who do not have the disease,

but are heterozygous carriers, also have an elevated cancer risk (Swift et al. 1976). In fact, the heterozygous carriers of the ataxia telangiectasia trait may represent up to a few percent of the human population. At present, the biological basis for these genetic diseases and their link to cancer are unknown. However, it is known that there are defects of DNA repair functions, presumably different defects, for xeroderma pigmentosum, ataxia telangiectasia, and Fanconi's anemia. Familial retinoblastoma has been shown to be consistent with a deletion or mutation of chromosome 13. Familial polyposis coli and Gardner's syndrome are associated with abnormalities of cellular growth control.

These observations provide clues to possible steps in the presumed multistep process of malignant transformation. To the extent that carrier states for these diseases are common in the population, these genetic traits may be factors that influence individual risk for developing cancer (Swift et al. 1976). It is likely that other genetically determined factors may influence cancer risks even if they do not yield recognized genetic diseases. For example, there may be genetically determined influences on the rate or route of metabolism of chemicals. The racial differences in alcohol metabolism illustrate that such differences occur in the human population. Individual variations in other factors, such as those affecting responses to injury, may also influence cancer risk. Knowledge about such factors is limited at present but may be an important and useful area for continued research.

There are a variety of illnesses and infections that predispose people to the development of cancer. For examples, hepatitis B and schistosomiasis of the bladder are important factors in the causation of liver and bladder cancers, respectively. Certain lymphomas are associated with infectious diseases (for example, human T-cell lymphotrophic virus types I and III, or Epstein-Barr virus), and colon cancers are associated with ulcerative colitis. These diseases and conditions cause a high level of cell proliferation in specific target cell populations which may predispose to cancer development in the affected tissue.

Some tumors of the lung are associated with scars of the parenchyma. There has also been speculation that other lung conditions predispose people to lung cancer development (Kuschner 1985). Although it is likely that these conditions predispose people to lung cancer because of increased cell proliferation rates, it is also possible that these conditions affect the capacity of the lung to clear exogenous materials, including potential environmental carcinogens.

Omitting the known genetic diseases that predispose to cancer development, and in the absence of acquired diseases that are associated with cancer, there are still a number of families with an unusually high incidence of cancer. In most of these families there is variable penetrance of tumor risk with less than 100 percent incidence of cancer in these populations. In some cases there are distinct patterns of tumor development with particular organs affected to unusual extents and with different tumors predominating in males and females. It is unclear whether these occurrences are primarily the result of unrecognized genetic diseases or a heterozygous carrier state for a recessive genetic disease. Alternatively, these families may develop these cancers because of elusive environmental factors passed socially from generation to generation, such as diet or personal habits.

Clinical and experimental evidence suggest that there are important differences in susceptibility to cancer among individuals in the human population. This could be a very significant factor in efforts to control specific types of cancer, including any related to exposure to diesel exhausts. The population of susceptible individuals may account for a disproportionate share of particular types of cancer. It may be possible to significantly change the overall incidence of specific types of cancer by identifying susceptible individuals and concentrating cancer prevention activities on this population. It might be possible to identify individuals for whom specific types of carcinogens or diesel exhaust represent a particular hazard and protect them from such hazards.

We have limited knowledge of biological and enzymatic factors that determine these states of unusual susceptibility. More data are needed about the range of variation of metabolic processes, DNA repair processes, constitutive and induced cell proliferation rates, and responsiveness to hormones in tissues from human subjects. These data are needed for each tissue in which cancer is common, and this information should be obtained where possible to determine the variability according to age, gender, genetic background, and personal factors such as diet, therapeutic drug use, and personal habits (for example, cigarette smoking). Accomplishing these goals will require development of methods to obtain human tissues in an acceptable manner. Furthermore, human subjects will have to be chosen scientifically so that they are representative of the population as a whole or the subpopulations that appear to be at unusual risk. If these studies are successful, the next step will be development of methods to test these characteristics in samples of tissue that can be obtained from normal individuals with little or no risk.

On the basis of epidemiologic observations that there is a range of responses within populations apparently exposed to the same levels of a carcinogenic substance, it is conceivable that there are individual factors that are major determinants of risk. Identifying the portion of the population at exceptional risk and concentrating protective efforts on that population might have a major impact on the overall cancer incidence at particular tissue-specific sites. Such an approach has proven notably effective in reducing myocardial infarction rates in individuals with genetic abnormalities of low-density lipoprotein metabolism and in individuals with acquired coronary artery disease.

Developing methods to determine the elements of individual risk will require great attention. The development of appropriate and acceptable methods for obtaining cells from various body sites with little or no risk should be included in this method. Methods to test various cellular characteristics that have been associated with the development of cancer will also have to be devised. Such factors as the capacity for carcinogen metabolism, DNA repair, oncogene activation, or chromosomal breakage might all be included among the individual characteristics evaluated. It will be particularly important to determine how to scale methods down to fit the number of cells that can be made available for testing. The availability of monoclonal antibodies, enzyme-linked immunosorbent assay (ELISA) techniques, in situ nucleic acid hybridization, chromosomal banding methods, and flow cytometry offer many new and unexplored approaches to determining these individual characteristics within the numbers of cells that may be obtained from persons without demonstrable disease.

■ **Recommendation 2.** Methods should be developed to identify individuals at high risk.

Metabolic Conversion and Carcinogen Activation

Many chemical carcinogens are inert and unreactive in their native form. Cells have a wide variety of metabolic enzymes which may inactivate and modify endogenous as well as exogenous chemicals, including those that are potentially harmful. In many cases these metabolic processes increase the water solubility of these chemicals as a way of improving the body's ability to excrete them. The process of metabolic alteration proceeds via a number of reactions which often produce a variety of products.

Some of these products are reactive and interact with cellular constituents to produce damage (Miller 1970). An example is benzo[a]pyrene (BaP) metabolism: spontaneously reactive diol-epoxide intermediates are formed as minor products whereas the majority are converted into unreactive excretion products. The diol-epoxides are the major, and perhaps the exclusive, proximate carcinogenic forms of the parent carcinogen. Additional examples are provided in chapters by Hecht and by Marnett (both in this volume).

Metabolic processes are highly regulated and responsive to the environmental conditions of the organism and their constituent cells. In most cases the metabolic pathways are induced or activated by substrates. At the same time, many of the substrates metabolized by the same enzyme systems compete for metabolism, particularly before inductive mechanisms increase metabolic capacity. Since these metabolic processes use molecular oxygen, NADH, or other cofactors in the reactions, factors that influence the supply of these cofactors may also modify metabolic rates. These factors include oxidative poisons, alcohol, dietary constituents, and a variety of drugs.

Some carcinogens, even if administered systemically, produce tumors in specific target tissues (Merletti et al. 1984). For most of these chemicals the organ specificity is a manifestation of the metabolic properties of the particular tissue. For example, certain N-nitrosamine compounds with propyl, butyl, or phenyl substituents produce tumors in the esophagus or urinary bladder of rats with great selectivity. In contrast, diethylnitrosamine principally produces tumors in the liver and lungs of rats, and dimethylnitrosamine under appropriate conditions primarily produces tumors of the liver and kidneys.

Differences in specificity are presumed to relate to the absence of certain enzymes in particular species or organs where some of these compounds are ineffective. In other cases, particular metabolic pathways may show exceptional affinity for specific forms of a given class of compounds in a particular tissue and lesser activity for other forms. This results in large rate differences in metabolism that correlate with the spectrum of distribution of tumors.

Qualitative Assessments of Carcinogenicity

Epidemiologic Evaluation

The most powerful and convincing evidence for the carcinogenicity of a substance is the demonstration of its association with an excess of cancer in epidemiologic studies. Epidemiologic studies vary in type and design, with different levels of scientific power, but all seek to compare the rates of cancer in a defined test group versus that in a control group. These studies may be broadly based, as in the case of geographically based investigations of differences in cancer incidence, or they may focus on large population groups such as those of a certain religion with particular constraints on foods and habits, or groups with particular ethnic or genetic backgrounds. Studies of this type are likely to provide only the most general information. More-specific information is derived from experiments that focus more narrowly on particular chemicals or occupational conditions of exposure.

The general approach in epidemiologic studies involves a comparison of the risk for the development of cancer in a defined experimental group to that in an appropriate and scientifically defensible control group (or two or more control groups) (see Bresnitz and Rest, this volume). Trends in cancer incidence may also be determined by comparisons between cohorts selected as a function of graded intervals of time of exposure or rates of exposure. Trends of cancer incidence with time or extent of exposure support an association between the trend-related factors and cancer development. Statistically significant differences in cancer incidence between experimental and control groups indicate probable, causal relationships to cancer development.

The extent of the increase in incidence and the strength of the statistical significance are used to weigh the importance of the relationship. In cases where there is good epidemiologic evidence linking an environmental exposure to cancer development and there are good quantitative data on exposure of the human subjects, it is possible to make reasonable predictions of the risks associated with future exposures at similar dose levels. Such information is available for very few chemicals or environmental exposures. In most cases, evidence for carcinogenicity is derived from experimental data obtained from carcinogenicity bioassays performed with rodents.

Bioassays in Experimental Animals

For most chemicals there is no epidemiologic evidence to evaluate carcinogenicity.

Epidemiologic evidence may lack the power to adequately assess the carcinogenicity of the substance because of insufficient numbers of subjects, confounding factors, or inadequate documentation of exposure. In these cases the only evidence for carcinogenicity may be derived from the results of long-term tests in experimental animals.

The carcinogenicity of substances is determined from experiments in which test animals, typically rodents, are exposed to the subject chemical, most commonly for long periods of time at high doses that approach the maximally tolerated dose. Rodents are the preferred animals because of their comparatively small size, their rather short life span, and their sensitivity to the development of cancer. Since the life span of these animals generally is two to three years, it is necessary to expose the animals to high enough doses of the test substance to compensate for the comparatively short period of treatment in the rodents relative to the lifetime of humans.

The philosophical basis for testing at comparatively high dose levels is the belief that carcinogenesis is an unusual form of chronic toxicity. Very high doses of chemicals given either acutely or chronically are most commonly toxic and kill the test animals without causing tumors. Chemicals tested for carcinogenicity are usually selected because they are thought to have a high probability of being carcinogenic; even among these selected chemicals, many or most do not show a carcinogenic response, or they show an equivocal response. The exception to this is the group of known human carcinogens which have all proven to be carcinogenic in animal tests. This activity of human carcinogens in animal experiments is one of the observations that supports this method of identification of carcinogens.

Long-term carcinogenicity test methods have evolved over time. Earlier studies tended to be done in a single species, often with very small numbers of animals, with incompletely defined experimental methods and materials, and with incomplete methods for pathological evaluation of the test animals. Nonetheless, in cases in which clear strong evidence of carcinogenicity resulted, these studies still serve as data for assessment of the substance.

More recently, particularly through the efforts of the National Toxicology Program and with the adoption of the Good Laboratory Practices Act, better documented and more thoroughly designed carcinogenicity tests have become the standard of performance. Potential carcinogens are tested in at least two species, usually rats and mice, in sufficient numbers to allow adequate statistical evaluation of the results, and usually with several dose rates. Pathology is thoroughly evaluated in the test animals and is subject to independent verification. The overall results are also reviewed by an independent group of experts who are familiar with tests of this type.

Analysis of the results of carcinogenesis bioassays involves statistical as well as biological evaluations. Statistical analyses compare the incidence and multiplicity of tumors between experimental and control groups. Test results are analyzed for statistical significance. In cases where multiple experimental groups are tested at different dose levels, analyses are included to seek statistically significant dose-related trends. Where possible, control group results are compared to the historical record of previous control groups of the same species, strain, and gender, maintained under similar conditions.

Biological considerations include examination of the study for possible confounding factors such as the route or method of administration, group size, and survival. Statistically significant results that stand without biological reservation are included in the qualitative assessment of carcinogenicity. Results that show the development of unusual tumors are given greater weight than those that show an increase of tumors that occur at a high spontaneous background level in controls. Additional weight is given to cases where tumors arise with exceptional rapidity or with unusually high incidence. All of these factors gain strength if the pattern of results is consistent between species. The International Agency for Research on Cancer (IARC) has been the leading international source for assess-

Table 4. IARC Classification of Evidence of Carcinogenicity

Classification	Nature of Evidence
No evidence	No studies of the material.
Inadequate	Experimental data are scientifically compromised or are inconclusive.
Limited	Positive evidence for carcinogenicity is limited to one species and is not exceptional.
Sufficient	Positive results in two species, particularly with evidence of dose/response relationships, and if there are multiple confirmatory results.

NOTE: IARC = International Agency for Research on Cancer.

ing carcinogenicity of chemicals, environmental mixtures, and occupational exposures. The evaluation approach used by IARC incorporates the considerations noted above and concludes with a judgment about the strength of the evidence for the carcinogenicity to animals of the substance of exposure (table 4).

Results showing carcinogenicity can be judged "sufficient" if they show an unusual type of tumor rarely seen in the species, particularly if these tumors occur with unusually high incidence, or if their latency period (time to first tumor) is exceptionally short. This assessment considers the strength of the evidence that the compound is carcinogenic (qualitatively), but does not assess the strength or potency of the substance as a carcinogen.

Cell proliferation, either constitutive or induced in response to injury, is an essential characteristic of tissues with a high risk of developing cancer. The information presented earlier indicates that cell proliferation, and in particular DNA synthesis, may have a crucial role in the various mechanisms of carcinogenesis. Thus, factors that increase cell proliferation rates may potentiate the development of cancer.

In long-term bioassays for carcinogenicity, the doses of chemicals are chosen in relation to the highest dose found to have a minimal effect in terms of subacute toxicity, that is, no reduction of survival in the subacute phase up to several weeks following treatment. The doses chosen, therefore, do not sufficiently compromise the function of any vital tissue or organ so as to appreciably reduce the survival of test animals during the subacute phase. This does not imply that the dose of the test agent has no effect on the tissues or organs of the test animals. Certainly, the goal is to demonstrate a carcinogenic effect of the substance if this is a property of the compound. Since the doses used to test for this activity are generally much higher than doses usually experienced by the human population, it is conceivable that toxicity may be a significant factor in tumor development.

A chemical that tests positively in such a bioassay may require cell toxicity and the resulting increased cell proliferation of regenerative processes in order to evoke a tumor response. Thus, increased cell proliferation may be an essential feature of the apparent carcinogenicity of the chemical in the bioassay. This may not occur under realistic conditions of exposure at levels to which humans are subjected.

In species and strains of animals used for bioassays, there are some tissues in which there is a notable spontaneous incidence of tumors. In these tissues it may suffice to produce an increased rate of cell proliferation, strictly as a matter of cell toxicity and regeneration, to demonstrate an increased tumor incidence. Thus the toxicity of the high dose of chemical may be modulating (accelerating) the development of tumors. These tumors may have developed spontaneously, but more slowly. Alternatively, these tumors may have arisen from cells that were already initiated as an intrinsic property of the tissue of the specific animal, although they would not have been expressed as tumors without the accelerating (promoting) effect of the chemical. In such cases, it is conceivable that the bioassay that yielded an increased incidence of tumors did so only because the high dose of the test chemical was toxic in these susceptible tissues and increased the cell proliferation rate.

It is possible that in such a circumstance cellular protective mechanisms may prevent toxicity at the doses of these compounds to which humans are exposed, and may also prevent stimulation of cell proliferation. Thus a chemical that produced an apparent positive carcinogenic result in an animal bioassay might not represent any appreciable carcinogenic risk at realistic levels of exposure in humans.

For this reason some recommendations have been made to test chemicals for carcinogenicity at levels at which humans are exposed. Testing at doses that produce no detectable toxicity might be more justifiable. Still, the underlying hypothesis for this line of argument is largely untested. Whether such mechanisms represent a tangible factor in the positive test for carcinogenicity of any compounds is unknown. In order to identify chemicals accurately as carcinogens in bioassays, it is essential to distinguish true positive results from apparent positives based on this confounding effect for compounds which in fact are only false positives.

To assess this possibility it may be desirable to undertake studies utilizing doses selected for the extent of their toxicity, ideally including some specific dose that causes no toxicity. If it were possible to determine that a correlation existed between tissue toxicity and tumorigenesis, then it might be possible to provide some correction for cancer risks based on the degree of toxicity in the tissue as determined even from short-term in vivo studies. Such an effort would identify more clearly the kinds of compounds that might require testing in larger numbers or at lower doses to relate carcinogenicity risks in humans to results in animal studies.

■ **Recommendation 3.**   The role of toxicity in carcinogenesis should be evaluated.

Short-Term Tests in Vivo and in Vitro

Short-term tests have come to be used in the qualitative assessment of carcinogenicity and in the prioritization of chemicals for in vivo testing. Typical assays evaluate the mutagenicity of the test chemical either directly or following metabolic activation, usually using the S9 microsome-enriched fraction from livers of rats whose metabolism has been induced with agents such as Aroclors. Other assays evaluate the ability of the compounds to produce large-scale disruption of DNA, specifically chromosomal damage, including the induction of sister chromatid exchanges, aneuploidy, and micronuclei.

Induction of DNA repair is another type of end point used to assess a chemical's ability to cause DNA damage. Perhaps the short-term test that most closely approaches a test of carcinogenicity in vivo is evaluation of the ability of a chemical to transform cells in vitro. A variety of cells and organisms—from prokaryotic organisms, to plants and insects, and to mammalian cells (including human cells)—are used in these studies. Assessment of chemical activity often is based on results obtained in multiple test systems. The more systems in which the chemical has shown activity, the greater the evidence of its potential for carcinogenicity.

These assessments can be used to prioritize chemicals for testing in long-term carcinogenicity studies in vivo. Results of short-term tests may also be used to confirm the results of carcinogenicity testing in vivo. The short-term tests generally cannot be used alone to determine carcinogenicity because validation studies for these tests are not sufficiently accurate and precise. Although most carcinogens have been shown to be genotoxic, not all are genotoxic. Consequently, these tests are prone to false negatives.

Other tests are being evaluated for their ability to detect carcinogens that act via nongenotoxic mechanisms or through mechanisms that only generate genotoxic mediators indirectly in specific cells. At present, this area must be regarded as one in development rather than a defined area that can be used routinely for regulatory purposes.

Methods for Quantitative Extrapolations to Human Cancer Risk

In this section, the methods used to estimate quantitatively the risk for human cancer development resulting from "real-

Table 5. Factors Involved in Quantitative Estimates of Human Risk

Estimates of quantitative risk in experimental animals
Extrapolation among species
Extrapolations among routes of administration and exposure
Extrapolation to dose levels of human exposure

world" exposures to substances judged qualitatively to be carcinogenic are considered. In the event that sound epidemiologic evidence for carcinogenicity exists and adequate data on exposure rates in study populations are available, it may be possible to directly estimate cancer risks for humans. Because such data are not available in most cases, extrapolations must be made from results of animal studies (table 5). Both the procedures and philosophy of the methods in use and alternative methods that have been proposed to make these extrapolations are considered in this section.

Estimation of Quantitative Risk in Laboratory Animals

The first step in quantitative evaluation of carcinogenicity is to determine the quantitative tumor response in long-term studies in animals. Such studies must determine and verify tumor responses in various organs and tissues of test animals and controls. Results are analyzed to determine whether there is an excess incidence of tumors, cataloged on a site-by-site basis, in experimental animals compared to controls, and whether these excesses meet statistical tests of significance. Any errors in the diagnosis of tumors will profoundly affect calculated incidences and subsequent extrapolations. Judgments regarding inclusion of animals dying prematurely or otherwise lost from the study will also affect the population size (denominator) in the incidence calculation.

The type of tumor used in the quantitation presents another problem. Some tumors are widely diagnosed by similar criteria and are not controversial. Pathological diagnosis can pose a far more difficult problem, particularly when attempts are made to distinguish an early neoplastic lesion from a nonneoplastic but undifferentiated area of tissue repair. Also, only a single observation is made of the lesion. Lack of the extensive natural history detail and clinical follow-up known for most human tumors complicates diagnosis of many types of tumors in animal bioassays and influences their quantitative reliability. In most cases only a single positive result is evaluated, and questions of quantitative reproducibility are not considered. Consequently, there is no way to estimate accurately the error in derived estimates of tumor incidence, nor is this error propagated through the quantitative extrapolation to yield the probable error in the risk estimate.

Extrapolations Among Species

Carcinogenicity studies have traditionally been performed in strains and species of animals sensitive to tumor development. Rodents are commonly used because they can be inbred with high uniformity, have a comparatively short life-span, and are comparatively inexpensive to maintain. The choice of sensitive species and strains facilitates the qualitative detection of carcinogenic activity, but raises problems with regard to quantitative aspects of the tumor response.

The principal issue is whether the assay is determining an inherent carcinogenic activity of the material tested or whether it is influencing the rate or extent of occurrence of tumors that occur spontaneously in the animal subjects, perhaps facilitated by the toxicity of the compound and the extent of the resulting regenerative process. This is not a problem in judging the qualitative carcinogenicity of a substance, because increase of tumor incidence by any mechanism is consistent with the definition of carcinogenicity. The problem does arise when this type of data is used for quantitative assessments of cancer risks in humans.

Human sensitivity to carcinogens is not well understood. We do not know how the human population compares with typically used experimental animals with regard to sensitivity to carcinogens. It is not known whether small or large proportions of the human population are at high risk from carcinogens. For that matter, there are only fragmentary data comparing sensitive rodent strains with more-resistant strains. Sensitivity and resistance are probably variables dependent on the types of carcinogens considered. Since human exposures are to complex mixtures of chemicals rather than to a single pure compound, our general lack of knowledge regarding interactions among carcinogens and toxins raises a further, major complication. This problem is of particular concern in considering the risk from complex mixtures such as diesel exhaust.

Extrapolations Among Routes of Administration or Exposure

Carcinogenicity tests are used to evaluate the activity of substances; they are not necessarily designed to reproduce precisely the conditions of human exposure. Most commonly, potential carcinogens are tested by being incorporated into the diet of experimental animals and administered throughout life. In some cases, but less frequently, the substance is added to the drinking water or directly instilled into the stomach by gavage. Skin painting (with or without subsequent applications of promoters) and intraperitoneal injections are still the routes of administrations in test systems with skin papillomas or lung adenomas, respectively, as end points in mice. Rarely, other less common routes of administration are used, such as inhalation, subcutaneous injections, or intratracheal administrations. In most cases, the route of administration is chosen for practical reasons: to simplify the methods of the study and to limit costs, not because of the similarity to the typical exposure route of the human population.

Because long-term tests often involve routes of administration dissimilar to those of typical human exposure, methods have been developed to relate experimental results in animals to estimates of human cancer risk. If humans are usually exposed to a compound by inhalation, and the carcinogenesis bioassay utilizes ingestion, there must be a way to relate these studies to compensate for peculiarities of tumor localization or number resulting from the route of administration. The practical solution to this problem has been to make an evaluation by analogy with another compound.

The test chemical is compared to a reference compound that has been tested by the same route (for example, ingestion), but also has been tested by the route of administration most typical of human exposure (for example, inhalation). Studies using the same route of administration (that is, ingestion) compare the potency of the two chemicals. The tumor response for the test chemical is related to an equivalent dose of the reference compound yielding the same tumor response. On the basis of this information, the tumor response to the test chemical via the normal route of human exposure is extrapolated to the tumor response of the equivalent (calculated) dose of the reference chemical by that route of administration.

Essential to making this a workable means of extrapolation is to develop a body of information concerning the carcinogenicity of sufficiently diverse reference chemicals tested by various routes of administration. The selection of the reference chemical to be used for extrapolation would be based on similarities between this chemical and the test substance with regard to chemical structure or metabolic mechanisms that convert it to an active carcinogen. Although this approach offers a reasonable and pragmatic solution when practiced with adequate biological insight, its reasonableness does not ensure its correctness or accuracy, since it has not been adequately studied and validated by rigorous scientific tests. Even allowing for substantial safety factors in applying this method, it is unclear whether the conclusions generated are sufficiently protective or vastly too restrictive when used in extrapolations to human cancer risks.

The extrapolation process presently used to make quantitative risk assessments requires that several estimates be made to adjust for differences between human exposures and the conditions of animal experimentation. Examples include corrections for differences in the route of administration, dose rate, and species. These correction factors usually exaggerate the estimates of human risk or the relationship to humans so that the extrapolations do not underestimate the actual human risk. The corrections are applied to the best estimate of the actual excess of tumors in the test group as compared to appropriate controls. Extrapolated estimates of excess cancer deaths in the human population are calculated from the application of these correction factors to the quantitative estimate of excess tumors.

The problem with this process concerns the validity and verification of estimation and extrapolation. The approach is logical and offers a well-intentioned attempt to maximize the protection of the population at risk. However, the elements of the extrapolation process are all estimates based on inadequate amounts of data and incorporating a wide safety margin. Furthermore, these estimates and extrapolations have been accepted and used without adequate testing and validation.

To extrapolate from available data concerning the carcinogenicity of substances tested in animals to quantitative estimates of cancer risk in the human population is difficult and problematic. Much essential data are not available to support the methods that are used for correcting differences in doses; in species, including the use of sensitive species; in routes of administration; and between the test compound and the chemical on which extrapolation methods are based.

Distinguishing experimental tumors induced by complete carcinogens from tumors promoted by the biological environment of the particular animal tissue would improve significantly our current methods of extrapolating and quantitating risks. These two different development mechanisms of specific types of tumors may determine the choice between different methods for extrapolation to low doses.

Additional information is also required for understanding the implications of the differences in carcinogenicity between species. To what extent is the sensitivity of the commonly used test animals a function of differences in carcinogen metabolism or defense mechanisms such as DNA repair capacity? Only with such information can it be concluded with confidence that the means of correcting for differences in routes of administration for chemicals of different types have been established. Additional data and a sound scientific validation study would allow greater confidence in making these corrections for route of administration.

Extrapolation methods, although a logical and responsible approach, are still largely untested. For example, extrapolations concerning diesel exhaust have used topside coke oven emissions as a comparison standard. Objective tests of the validity of the process of extrapolation by analogy must be undertaken. If the currently used extrapolation methods are precise or even reasonably effective, then they should be used to make testable estimates of human cancer risks, or, conversely, predict effects in experimental animals from human data. A rigorous scientific test is needed to validate the currently used method for risk extrapolation.

■ **Recommendation 4.** Critical data should be gathered for quantitative assessments.

Extrapolation to Dose Levels of Human Exposure

For practical reasons—basically the need to produce tumors within the short life span of comparatively small numbers of rodents—carcinogenicity tests are usually performed with high doses of the test substances. These dose levels are usually quite different from the levels to which humans are exposed. It is therefore necessary to extrapolate from the high dose levels of the animal test to the much lower levels of human exposure. Unfortunately, little is known about the effects of carcinogens at the low dose rates typical of human exposures. A particular concern about high doses is that their effect may be more than that of a pure carcinogen. Doses may be sufficiently high to cause toxicity in the

target tissue; cell necrosis and restorative regeneration with elevated cell proliferation may potentiate the intrinsic carcinogenic effect of the agent. At the low doses at which humans are exposed, this toxicity may not occur, and tumor induction, if it occurs at all, may be at vastly lower rates.

Carcinogens generally show a cancer-causing potential in relation to the dose of the chemical administered. At high doses, the tumor response is usually directly related to dose, with progressively more tumors as the dose is increased. The nature of dose/response relationships at low doses is the subject of considerable debate. For most studies, the experimental design—principally the numbers of animals used—is suited to detect strongly or moderately carcinogenic substances.

There are usually too few animals in an experiment to show a statistically significant, small increase in tumors caused by a less carcinogenic substance or a low dose of a more active substance. As a consequence, typical tests to detect carcinogens have produced little information about tumor responses at low dose levels. The dearth of information on this topic has made it impossible to determine the shape of the dose/response curve at low doses, to ascertain whether there is a certain minimal dose (threshold dose) below which there is *no effect*, or whether there is an effect at any level down to zero exposure.

In order to obtain further information about the shape of the dose/response curve at lower exposures, the National Center for Toxicological Research undertook a study using large numbers of animals and extending the dose range lower by an order of magnitude (Cairns 1980). This study (ED_{01}) used two strong carcinogens—diethylnitrosamine and 2-acetylaminofluorene—and evaluated their effects in C3H × BL6 hybrid mice (Gaylor 1980; Littlefield et al. 1980). Study results suggested that for spontaneously occurring liver tumors in this strain of mice, the induced tumor incidence was linear to the dose of carcinogens administered. In the case of bladder tumors, where the carcinogen is assumed to function both as the initiator and the promoter, tumor incidences decreased rapidly as the dose was reduced, with what could be interpreted as a threshold effect. It is possible that in the case of spontaneously occurring tumors, the animal strain or the conditions of maintenance provide sufficient promoting stimulus to produce cancers in proportion to the extent of initiation by carcinogen treatment. If this interpretation is accepted, then this study provides a measure of support for both the linear extrapolation model and, at the other extreme, the threshold model.

In spite of this large-scale effort, our knowledge has advanced only modestly. The results do not provide enough information to predict the shape of the dose/response curve for a particular chemical at the very low doses of human exposures. In fact, the results show that each model may have some measure of validity, and the appropriate extrapolation may relate to the carcinogenic agent and the species (and strain) at risk. Such a conclusion would greatly increase the complexity of achieving reliable extrapolations of animal carcinogenicity data to human cancer risk estimates. It would point out the deficiencies of adoption of any general and universal extrapolation method for human cancer risk of all chemicals.

The effects of small doses of carcinogens have been the subject of considerable conjecture, because of the absence of critical data concerning carcinogenic effects at very low doses and a still quite incomplete knowledge of the mechanistic details of carcinogenesis. The development of strongly entrenched opposing positions reflects the lack of real knowledge in this area and the significant practical consequences resulting from the use of either of the extremes of extrapolation.

It is not germane here to consider the relative merits of the various extrapolation analyses that have been made. Various methods of extrapolation yield estimates of cancer risk in animals differing by as much as several orders of magnitude. When these uncertainties are combined with uncertainties about the comparative sensitivity of humans, the accuracy of the estimate of tumor incidence in bioassays, and the effects of mixed exposures, the estimation of

cancer risks in humans is clearly far from accurate or reliable.

Carcinogen treatments or exposures can occur under a wide variety of circumstances. Human exposures to a particular carcinogen may occur as the result of a single incident, as in the case of atomic bomb survivors, or may result from chronic or episodic exposures at lower levels. Experimental studies also have varied the mode of carcinogen exposure. These range from single treatments with either short-lived or persistent substances, to multiple discrete exposures with any of a variety of intervals during a treatment period of weeks to months; or continuous treatments, usually by incorporation into the animals' food or drinking water.

Results of tumor development following discrete treatment protocols with the same substance but using different treatment schedules have been used to compare the effects of dose fractionation. Single treatments have generally been found to be less effective than the same total dose given over multiple treatments. More fractionated doses tend to produce lower toxicity, allowing more of the treated animals to survive long enough to develop tumors. Protocols that administer the carcinogen in few doses over a short interval early in the animals' life-span can result in reductions in the latency period, but these protocols may result in a lower overall incidence of tumors in the treated population.

Exposure levels, particularly as they apply to the individuals at risk, are poorly defined in the case of most human exposures to carcinogenic or potentially carcinogenic substances. Estimates of average levels of exposure are usually based on current measurements of ambient concentrations in air, water, food, and so on, and these are extrapolated back in time for the duration of the study period. Estimates of concentrations may be stratified according to the specific type of work classification, the amounts of various kinds of food consumed, and so on. Although such approaches are very limited in defining individual exposures, they may be the best way of evaluating these parameters in an entirely retrospective manner. More desirable would be careful individual measurements of the presence and action of the substance in the target tissues of the subjects throughout the course of the epidemiologic study.

It would be valuable to have data on exposure at the target tissues in order to estimate the activity of environmental compounds as carcinogens in the organs of human subjects. Such data would also vastly improve our understanding of experimental studies in animals. If these data were available for experimental and epidemiologic studies, it might be possible to predict cancer risks better, and these estimates of risks might be provided on an individual basis. With such information, it might be possible to monitor work conditions or even modify the design of internal combustion engines or their fuel, in order to minimize levels of harmful agents, or to reduce their levels to acceptable limits.

One of the major limitations of most epidemiologic investigations of environmental factors as determinants of cancer is the inadequacy of quantification of exposure to specific agents. Consequently, these studies have had to rest on estimates of exposure levels that are based on such parameters as geographic location, classification of job or other function, and retrospective estimates as averages based on current measurements. These limitations all reduce the power of epidemiologic studies to detect positive carcinogenic effects, or, circumscribe the statistical range for which a no-effect result can be substantiated for negative studies.

Performing dosimetry studies on experimental subjects would improve the power and specificity of epidemiologic studies of the carcinogenic effects of environmental exposures. Recent scientific advances have made this goal far more feasible. It is now possible to detect the adducts formed by the binding of carcinogens to DNA or other macromolecules within cells, using specific antibodies and sensitive methods for amplifying the signals they produce. ELISA methods can provide this amplification and have been shown to detect environmental levels of some substances.

Since these methods are largely untested as a means of environmental monitoring,

tests are needed to verify their validity and discriminatory power. Such tests could be undertaken using well-studied animal model systems in which quantified tumor responses have been observed with carcinogens for which these modern techniques exist. If these tests in experimental animals prove the power and sensitivity of this approach under the controlled conditions of the laboratory, these methods should be applied to the study of epidemiologically defined groups to determine whether they offer distinct improvements to the estimation of exposures.

Such a test could be conducted on groups for which conventional environmental monitoring of the workplace, food contamination, cigarette smoking, and so on, can be used as a quantifiable basis for comparison with experimental results. In the event that these methods are shown to be reliable as discriminators of exposures for a few well-defined environmental hazards, then these methods could be used to gain further insights about suspected human carcinogens in a prospective study. If this type of study demonstrates that exposure levels at the target tissue are highly correlated with the tumor incidence rates, then this approach should be used routinely to augment or replace conventional methods of environmental monitoring.

■ **Recommendation 5.** Acceptable methods should be developed for dosimetry in humans.

Experimental Evidence on Carcinogenicity of Diesel Exhaust

In this section the information available about the carcinogenicity of diesel exhaust emissions is reviewed, and the data from animal studies and human epidemiologic studies and the risk extrapolation based on these data are described. Other studies that describe the results of short-term tests that provide additional insight about the potential carcinogenic activity of diesel exhaust emissions are also reviewed.

Table 6. Short-Term Tests Used to Assess
Diesel Engine Exhaust

Reverse mutations in *Salmonella typhimurium* bioassay
(Ames assay)
D3 recombinogenic assay in *Saccharomyces cerevisiae*
Forward mutagenesis in mouse L5178Y lymphoma
cells
Mutagenesis in Chinese hamster ovary (CHO) cells
DNA fragmentation by alkaline elution in Syrian
hamster embryo (SHE) cells
DNA adduct detection using the ^{32}P-postlabeling
technique
Sister chromatid exchange in CHO cells
Sister chromatid exchange in human lymphocytes
treated in vitro
Sister chromatid exchange in fetal hamster liver
exposed transplacentally
Cell focus transformation assay in SHE cells
Cell focus transformation assay in BALB/c mouse
3T3 cells

Short-Term Tests of Activity of Diesel Emissions

Short-term in vitro assays that detect mutagenesis or other activities that suggest
carcinogenic potential (table 6) have been
used to evaluate diesel engine exhaust
(Lewtas et al. 1981). In these studies, diesel
exhaust particulates from four different engines were compared with positive control
substances including BaP, and extracts
from cigarette smoke, coke oven emissions, roofing tar, and gasoline engine
emissions.

Each of these substances was tested for
mutagenic activity in the *Salmonella typhimurium* bioassay (Ames assay) (Claxton
1981). The assay was performed with and
without the addition of induced rat liver
microsomes (S9 fraction) as a source of
mammalian metabolizing activity. Each of
these test substances was mutagenic in the
presence of S9, but extracts of roofing
tar and cigarette smoke were inactive in
its absence. In comparisons of uniform
amounts of extracted organic material, the
diesel exhausts produced a 20-fold range of
activity, with the exhaust from a Nissan
engine being most active, those from Oldsmobile and Volkswagen engines being intermediate, and that from a Caterpillar
engine being least active. The addition of
S9 to diesel exhausts did not increase mu-

tagenic activity, but reduced it in most
cases.

Mitchell and coworkers (1981) tested
these substances using the D3 recombinogenic assay in *Saccharomyces cerevisiae*, forward mutagenesis in mouse L5178Y lymphoma cells, and sister chromatid exchange
in Chinese hamster ovary (CHO) cells. The
recombinogenic assay was insufficiently
sensitive to detect activity, but positive
results were found with the other two
assays. Most of the extracts proved positive
in both assays in the presence of a metabolic
activation system. Interestingly, most
compounds induced sister chromatid exchange and were mutagenic in the absence
of an exogenous activation system; this
suggests that the extracts contain direct-
acting mutagens and DNA-damaging
agents.

These compounds were also assessed for
genotoxicity by examining mutagenesis in
CHO cells and by studying DNA fragmentation by alkaline elution and cell transformation by focus assay in Syrian hamster
embryo (SHE) cells (Casto et al. 1981).
These assays, which depend on the test cells
themselves for metabolic activation, gave
far fewer positive results. Extracts of
Volkswagen and Datsun diesel engine exhausts were positive in the CHO mutagenesis assay, whereas extracts from coke oven
emissions or from gasoline engine exhaust
were active in this assay and in the SHE cell
DNA fragmentation assay. Studies of
transformation of SHE cells by these compounds were negative.

In other studies, these compounds were
assayed for their ability to transform mouse
BALB/c 3T3 cells when tested with and
without an S9 activating system from livers
of rats treated with Aroclor-1254 (Curren
et al. 1981). Although there were no clear
dose/response relationships, qualitative assessments were possible and an approximate ranking of activity could be achieved.
By this evaluation, emissions from a coke
oven and gasoline engine were most active
and approximately equal in activity, Nissan
diesel was less active, and roofing tar was
least active. Extracts from Oldsmobile and
Caterpillar diesel exhausts were inactive.

Several concurrent studies have examined the mutagenic activity of extracts of diesel engine exhaust particulates and the ability of these extracts to induce sister chromatid exchange. Nachtman and colleagues (1981) tested extracts from diesel exhaust particulates of unspecified origin using the Ames *Salmonella* bioassay. They found the extracts to be mutagenic without an induced rat liver S9 metabolic activating system. Addition of the S9 fraction reduced activity, presumably by competing with bacteria for binding of the reactive compound. Treatment of the extract with reducing agents greatly diminished mutagenic activity; addition of the induced S9 fraction restored the activity. This was interpreted as suggestive of the presence of nitroarene mutagens in the diesel exhaust particulates.

Pitts and coworkers (1982) studied extracts of exhaust particulates from a Nissan diesel engine. They demonstrated the presence of several nitroarene compounds in the extracts and showed the mutagenicity of some of these chemicals in the Ames assay. By comparing a standard *Salmonella* strain to one deficient in nitroreductase activity, it was possible to recognize the sizable contribution of active nitroarene mutagens to the total mutagenic activity of the extracts.

Li and Royer (1982) examined the effect of diesel exhaust extracts on mutagenesis induced by BaP and N-methyl-N'-nitro-N-nitrosoguanidine (MNNG) in CHO cells, testing exhaust extracts from five different brands of diesel engine. CHO cells were treated with the strong mutagen (BaP or MNNG) plus diesel exhaust, and the mutation frequencies were compared to the sum of the mutation frequencies in cells treated with each component separately. For MNNG alone and for BaP together with an exogenous metabolic activating system, the diesel exhaust extracts potentiated mutagenesis frequencies in the CHO cells. The results were interpreted as showing the presence of comutagens or cocarcinogens in the extracts of the particle fraction of diesel exhaust.

The ability of diesel emissions to induce sister chromatid exchanges has been studied in human lymphocytes in vitro (Lockard et al. 1982) and in fetal hamster liver following transplacental exposure (Pereira et al. 1982). Both systems used a Nissan engine to generate exhausts. In the former study, cells were exposed to an organic extract of diesel exhaust particulates; in the latter study, hamsters were exposed transplacentally to exhaust emissions, exhaust particulate material, or to an organic extract of the particulates. In both studies the diesel exhaust extracts produced a dose-related increase in sister chromatid exchanges.

Wong and colleagues (1986) exposed F344 rats to diesel exhausts by inhalation for 30 months using methods that had been shown previously to produce lung tumors in these animals. They evaluated DNA extracted from lung tissue of diesel-exposed and control rats for damage by the use of the [^{32}P]-postlabeling technique to detect the presence of adducted nucleosides. This technique had the sensitivity to distinguish exposed from unexposed rats, based on the presence and quantity of adducts detected.

Data on Carcinogenic Activity of Diesel Exhaust Emissions

Kotin and coworkers (1955) tested the carcinogenicity of extracts of diesel engine exhaust particulates. The concentrations of polynuclear aromatic hydrocarbons (PAHs) in extracts of exhaust particulates were found to vary according to the conditions of operation of the engine. Extracts were tested for carcinogenicity by skin painting on the backs of C57 black and strain A mice. The extracts produced notable systemic toxicity in both mouse strains. In both strains the extract induced the formation of skin tumors, whereas none developed in controls. There were few tumors in the C57 black mice, but 50 to 85 percent of the strain A mice developed skin tumors. In one of the groups a large proportion of the tumors were carcinomas rather than papillomas.

Nesnow and colleagues (1981, 1982, 1983a) and Slaga and coworkers (1981) showed that extracts from Nissan, Oldsmobile, Volkswagen, and Caterpillar diesel engines and from a Mustang gasoline engine acted as initiators in two-stage carcinogenesis when promoted with TPA in the skin of SENCAR mice. Exhaust extracts were prepared according to established protocols using exhausts produced under carefully defined engine operating conditions and with a uniform fuel; these extracts were tested parallel to a number of other substances which were positive controls and scaling factors (extracts from cigarette smoke, coke oven emissions, roofing tar, and so on) (Lewtas et al. 1981). The exhaust extracts from Nissan and Volkswagen engines produced skin papillomas in proportion to dose. The Caterpillar engine extract was regarded as active although no clear dose response was evident. The Oldsmobile and Volkswagen exhausts as well as the gasoline exhaust produced a weak response. In each case, the total tumor-initiating activity exceeded the activity of the BaP content of the exhaust samples. This indicates that there are components of the mixture in addition to BaP that are involved in the total tumor-inducing activity. A later study by this same group (Nesnow et al. 1983b) showed that Nissan diesel exhaust extract was a complete carcinogen as well as an initiator, but not a tumor promoter, in SENCAR mouse skin.

The carcinogenic activity of diesel engine exhaust was evaluated in a separate study using the lung adenoma in strain A mice as an end point (Orthoefer et al. 1981). Exposures were to diesel exhaust by inhalation or to an exhaust particulate fraction by intraperitoneal injection for a period of 8 weeks followed by sacrifice at 26–30 weeks. Other animals were exposed by inhalation for up to 7 months. None of these treatments produced a notable increase in lung adenoma incidence or multiplicity as compared to controls. In separate experiments, low doses of urethane were added to the treatment protocol for diesel-exposed and for unexposed mice. In these studies, there was a significant increase in the number of adenomas as a result of the combined treatments with urethane and diesel exhaust.

Diesel exhaust has been studied for its carcinogenicity by inhalation exposure in F344 rats (McClellan et al. 1986a). Inhalation exposures at soot concentrations of 0.35, 3.5, and 7.0 mg/m^3 were carried out on a 7-hr/day, 5-day/week schedule for up to 30 months; exhaust was generated by an Oldsmobile engine operating under defined conditions. Rats chronically exposed to diesel exhaust developed areas of pulmonary fibrosis and accumulations of soot, particularly at the highest exposure level. The rates of clearance of soot particles were shown to be significantly impaired at the two higher dose levels (Wolff et al. 1986). A significant increase in the incidence of lung tumors, particularly adenocarcinomas and squamous cell carcinomas, was found in the group of rats exposed to the highest concentrations of diesel exhaust (McClellan et al. 1986a,b; Mauderly et al. 1987).

In a separate series of studies, rats, mice, and hamsters were exposed to diesel exhausts by inhalation (Heinrich et al. 1982, 1985, 1986). Hamsters were studied to detect whether exposure to diesel exhaust affected tumor responses in animals treated with proven respiratory tract carcinogens: diethylnitrosamine (DEN) was injected subcutaneously, or dibenzo[a,h]anthracene was instilled intratracheally (Heinrich et al. 1982, 1985). Inhalation exposures were to exhaust from a Daimler-Benz diesel engine, either whole exhaust or exhaust with the particulate fraction removed; control hamsters breathed air. Lung tumors developed in too few hamsters for meaningful evaluation. Papilloma of the larynx and trachea induced by DEN were potentiated by exposure to diesel exhaust, either with or without particles. Focal proliferations in the lung periphery were seen in all groups exposed to total exhaust; these lesions were also seen in groups treated with the known carcinogen and the particulate-free exhaust.

In a separate study, Heinrich and coworkers (1986) exposed rats, mice, and hamsters to total diesel exhaust (from a different engine), particle-free exhaust, or clean air by inhalation. No lung tumors were observed among hamsters, but there was an increased incidence of bronchiolo-alveolar hyperplasia and emphysematous lesions in animals exposed to total exhaust. In mice, exposure to diesel exhaust with or without particles increased the incidence of lung adenocarcinomas as compared to controls. Bronchiolo-alveolar hyperplasias were far more common in mice exposed to total exhaust as compared to particle-free exhaust or controls. Differences between control rats and those exposed to particle-free exhaust were not significant. Rats exposed to total exhaust had several lesions that were significantly increased, including bronchiolo-alveolar adenomas, squamous cell tumors (one low-grade squamous cell carcinoma and eight benign keratinizing cysts), bronchiolo-alveolar metaplasias, and bronchiolo-alveolar hyperplasias. Other animals were exposed to the exhausts and also were treated with a known carcinogen to evaluate the potentiating effect of the inhalation exposures. For hamsters and mice, there were no effects and inconsistent effects, respectively. Total diesel exhaust potentiated the induction, by dipentyl nitrosamine, of malignant lung tumors, adenocarcinomas as well as squamous cell carcinomas.

Diesel exhaust has also been evaluated for its ability to initiate the induction of foci of altered hepatocytes in rat liver. Pereira and colleagues (1981) exposed rats to diesel exhausts generated by a Nissan engine operating under defined conditions. They also performed two-thirds partial hepatic resections on the rats and subjected them to the promoting effects of maintenance on a choline-deficient diet. At the end of the experiment, the rats were sacrificed and their livers were evaluated for the presence of the presumptive preneoplastic lesions, γ-glutamyltranspeptidase–positive foci. The researchers found no increase in the number of positive foci in the diesel-exhaust treated rats. It is possible that this assay is insufficiently sensitive to detect carcinogenic or initiating activity, for doses of the magnitude that could be received by inhalation in the brief period of restorative hyperplasia in the liver.

Epidemiologic studies of carcinogenicity of diesel exhaust in humans are limited in number. Studies by Hueper (1955), Raffle (1957), Kaplan (1959), Waxweiller et al. (1973), Menck and Henderson (1976), Heino et al. (1978), Luepker and Smith (1978), and Wegman and Peters (1978) have been reviewed by Schenker (1980). These studies were all regarded as flawed because of limited durations of exposure or follow-up, because of small numbers of subjects, or because of uncertainties about the extent of diesel exhaust exposures or the types of tumors that developed. In four of these studies (Raffle 1957; Kaplan 1959; Waxweiller et al. 1973; and Menck and Henderson 1976) researchers found no increases in lung cancer incidence that could be associated with diesel exhaust exposure. Each of the other four studies reported an excess of lung cancers in the subject groups as compared to control groups. However, flaws make it impossible to reach a definitive conclusion regarding the carcinogenicity of diesel exhaust from these studies.

Harris (1983) reviewed the study of London transport workers by Raffle (1957) and updated information concerning that study to include the years 1950–1974. From these data he observed that there was no increased incidence of lung cancer, but by introducing corrections for uncertainties and confounding factors, he calculated an upper-limit estimate of potential risk for lung cancer attributable to diesel exhaust. Hall and Wynder (1984) investigated the potential carcinogenicity of diesel engine exhaust exposure using a case-control analysis. They compared lung cancer patients and controls without tobacco-related disease for occupational exposures to diesel exhaust as judged from job classification. Although the study showed a strong relationship between smoking and lung cancer, it did not show an association between exposure to diesel exhaust and lung cancer.

Quantitative Assessment of the Cancer Risk of Diesel Exhaust in Humans

Some of the abovementioned studies have been used to assess the potential carcinogenicity of diesel engine emissions for the human population. Rather than the step-by-step extrapolation approach described earlier, the evaluation rested on earlier calculations for other compounds and the approach used to make this assessment was an extrapolation by comparisons and analogy (Albert et al. 1983; Lewtas et al. 1983). Results of skin carcinogenesis and skin initiation in SENCAR mice were compared for diesel exhaust extracts and extracts from gasoline engine exhaust, cigarette smoke condensate, roofing tar vapors, and coke oven emissions. Similarly, results of short-term assays for genotoxicity were compared for these same compounds. From these data the comparative potency of diesel exhaust was estimated on the basis of the most active diesel extract. To relate these values to estimates of risk for the human population, estimates of human lung cancer risk were made for coke oven emissions, roofing tar, and cigarette smoke condensate. For each of these compounds there were epidemiologic data relating exposure to human cancer and experimental data in test systems identical to that for diesel exhausts. The risk per unit quantity for diesel exhausts was extrapolated by determining the human risk on the basis of a unit quantity of organic extractable material. The estimated unit risk obtained for human lung cancer was $0.02–0.60 \times 10^{-4}$ (lung cancers)/μg exhaust particulates/m^3 of air.

To understand this estimate, it is important to recognize the inherent assumptions of the method (Lewtas et al. 1983). The method assumes that the relative potency of carcinogens in one carcinogenesis assay is directly proportional to that in another bioassay. Further, this comparability is assumed to apply across biological systems and species. This assumes that the bioavailability of the active compounds at the target tissue is proportional even when extrapolations are made between species and between routes of exposure. As Cuddihy and McClellan (1983) note, the estimates derived by this method suggest that extracts of diesel exhaust particles are not "orders of magnitude more potent than other emissions."

This risk assessment has a number of limitations. These studies and extrapolations are not based on whole, fresh exhausts; the exhausts are not acting on a population exposed to a myriad of other carcinogens and active compounds unrelated to diesel exhaust; and the exhausts are acting on a homogeneous population where genetic factors, prior illness, and personal habits do not influence the susceptibility to these insults. Also, this assessment offers an estimate of risk strictly for lung cancer, although cancers in other sites might also be affected. The risk estimate is also provided as specific risk per unit of exposure. This specific risk is not very dissimilar to those for the other materials to which it was compared experimentally. Thus, the comparison with the specific risk for gasoline engines is somewhat misleading when one considers the fact that diesel engines may generate one to two orders of magnitude more particles than a gasoline engine with a catalytic converter.

More recently, the comparative potency approach has been used to assess the human cancer risk associated with diesel exhaust in a more comprehensive analysis that includes estimates of population exposures (Cuddihy et al. 1984; McClellan 1986). Those analyses used previously reported estimates of specific risk of lung cancer development (lung cancers/μg diesel particulates/m^3 of air/year) (Albert et al. 1983; Cuddihy et al. 1983; Harris 1983). Exposure estimates were based on environmental concentrations in various locations and distribution of the population according to concentration levels and assuming a 20 percent proportion of diesel-powered light-duty vehicles. Analyses were also based on the estimates of risk from epidemiologic studies. From these data, calculated values for excess lung cancer deaths per year ranged from 100 to 3,500, a range attributable to an increase to a 20 percent light-duty diesel-powered fleet. As with the earlier estimate noted above, there are numerous potential sources of error in these calculated risk values. Despite these limitations the risk estimate offers a starting point for determining the overall potential for changes in the rate of cancer deaths as a result of increasing the use of diesel engines in the U.S. transportation fleet.

Too little generic information exists on the carcinogenicity of the gaseous and particulate emissions of mobile sources. Studies have been performed on representative emissions generated by particular sources operating under specific conditions. Scientifically, it is not clear to what extent such results apply to different engines operating with different fuel or other different conditions. Further, it is not clear how these results relate to the same engine operating under other conditions or even to the same engine operating under presumably identical conditions at a different time. Differences between individual engines or operating conditions can result in the generation of emissions with quantitative and even qualitative differences in the products formed. These differences in turn can be the major determinant of the activity of the emission in carcinogenicity tests. This situation is quite unlike the testing of a pure chemical, in which case there is a reasonable assurance of repeatability upon retesting.

Given that there is no standard mixture for mobile source emissions, the question arises whether these mixtures can be evaluated on the basis of quantity and activity of certain "sentinel" constituents. If these most active components could be monitored and minimized, then optimum engines and operating conditions might be selected. Although this idea has merit as a comparative measure, the assessment of the actual quantitative risk at any level of these sentinel compounds may be difficult to determine. A further complication is the fact that little is known about the possibility of interactions between carcinogenic compounds. It is unlikely that the risk associated with the mixture of sentinel compounds is simply the sum of the effects of the individual compounds. This uncertainty is further amplified when the numerous other constituents of emissions are considered as influencing the activities of the sentinel compounds.

Further bioassays are needed to provide a sufficient body of information about the carcinogenicity of diesel exhausts in experimental animals. In view of the variability of diesel engine exhausts due to engine design, conditions of operation, and the fuel used, it is necessary to perform more studies to evaluate the influence of these variables on tumor responses. Exposure to diesel alone should be complemented by studies in which diesel exhaust exposures follow initiating carcinogen treatment in each of several organs or tissues. Exposures should not be limited to a particular fraction of diesel exhaust condensate or to the particulate material, since the complete exhausts may have additive or inhibitory effects that would otherwise not be detected.

■ **Recommendation 6.** Additional studies should be performed on the carcinogenicity of diesel exhaust.

Assessment of the carcinogenicity of mixtures poses two conflicting problems. The first concerns the nature of mixtures and the fact that each mixture represents a unique case. The second concerns attempts to evaluate mixtures by dividing them into their constituents. In such cases it is difficult to determine how to reconstitute the effects of the individual constituents into the effects of the total mixture.

Mixtures such as the diverse combustion products in mobile source emissions can be administered to experimental animals and tested for carcinogenicity. However, these mixtures are not intentionally formulated with precise analytical procedures. They are the products of a process or source that may not have exceptional reproducibility. Thus, the emissions from two different diesel engines may have quantitative differences in the products of combustion. Even the same engine, operating under slightly different, or even unmodified conditions, may yield mixtures of products with some quantitative differences. Despite the slight quantitative differences in the various combustion products, it is likely that the mixtures will have similar qualitative effects: they will prove to be carcinogenic or they will not. The magnitude of the carcinogenic response may be affected by the actual composition of the mixtures.

Under any condition of operation or engine design, diesel exhaust is a mixture of many chemicals. The composition of this mixture is highly variable and depends on such factors as the engine, fuel, and operating conditions. It is conceivable that the interaction of the components of this mixture is the determinant of overall carcinogenicity of the complete mixture. Therefore, the only currently valid method to determine the carcinogenic activity of each form of diesel exhaust is by a separate animal bioassay.

However, animal bioassay is not practical for evaluating modifications of diesel engine design or other aspects of their operation as they affect carcinogenicity. It would be valuable to have some method of estimating changes in carcinogenic activity based on knowledge about the changes in composition of diesel exhaust. This would require a better understanding of the interactions of components of complex mixtures in causing cancer. To learn how constituents of mixtures interact in carcinogenesis, it will be necessary to determine how carcinogenicity changes with the variation of the concentration of individual components. Choices of chemicals to study would presumably be based on the activity of the isolated compound or its relative abundance in the diesel exhaust mixture. In addition, constituents that have demonstrated or are suspected of enhancing (or inhibiting, for that matter) the action of carcinogens (for example, promoters or enhancers) will also need to be studied.

This problem will require additional study to evaluate the effects of variation in the composition of diesel exhaust on the tumorigenicity of other carcinogens; this serves as a model of the multiple complex exposures associated with human environments and lifestyles. Optimally it would be possible to achieve a reasonable estimate for the complex exhaust mixture that is based on measurements of the concentrations of a certain small number of index compounds. This hypothesis and experimental approach should be tested and validated. If it is found to predict certain levels of carcinogenic activity, the predictions should be tested by performing animal bioassays.

■ **Recommendation 7.** Methods should be developed for assessing the carcinogenicity of mixtures.

Summary

This chapter reviews information on mechanisms of carcinogenesis and considers factors that influence the rate of tumor formation. It also considers the criteria for identifying a chemical qualitatively as a carcinogen, and methods that have been used to extrapolate from these data to quantitative estimates of cancer risk in humans. Finally, data have been reviewed regarding the qualitative assessment of diesel exhaust as a carcinogen and the extrapolations made using these data to estimate human cancer risk from diesel exhaust exposure.

The review of current knowledge about carcinogenesis points out great advances that have been made in our understanding of cancer and also reflects the vast remaining ignorance. Cancer development has come to be recognized as a slowly progressing, multifactorial, multistep process. Cell proliferation is known to have one or several roles in the process, and abnormalities of the control of this process are fundamental features of cancer cells. Factors such as chronic injury or toxicity (for example, toxicity that is produced by high, but nonlethal doses of administered drugs) can result in elevated rates of cell proliferation with the attendant increase in cancer risk. Some chemicals are known to act as carcinogens by direct effects on DNA; in some cases, specific mutations induced by chemicals have been identified. Other genes whose effects are recognized in their absence or altered state in certain genetic diseases predisposing to cancer have been localized cytogenetically, and efforts are in progress to isolate the genes. Studies of atypical carcinogens that do not have a direct mutagenic effect have suggested alternate pathways or separate steps in the pathway to the development of cancer. Similarly, studies of the role of promoters in carcinogenesis have pointed out the multistep nature of the process, and the potential influence of factors that may act by selecting cells with abnormal properties. The human population is diverse in its genetic background and its exposures to harmful materials. Many chemicals require enzymatic activation to become reactive ultimate carcinogenic metabolites; these metabolic processes may be included among the factors influenced by genetics and environmental exposures, that determine the individual variations in susceptibility to cancer. The list of issues about the process of carcinogenesis considered in this section was necessarily incomplete, limited by space constraints rather than the exhaustion of important facts.

The section on qualitative evaluations of carcinogenicity and quantitative estimates of cancer risks in humans considered the criteria for designating a chemical as a carcinogen and how these data are quantified and extrapolated to estimates of human cancer risk. The section shows that there are reasonably well-defined criteria for judging whether a chemical is a carcinogen. That is not to say that this evaluation is not without problems. Weak carcinogenic responses may be difficult to distinguish from background levels of tumors. Increased rates of tumors may be observed as the result of exposure to promoters of carcinogenesis; on the basis of a positive tumor response, the promoter is classified formally as a carcinogen. However, the tumor response produced by promoters may be critically dependent on dose and may even have a threshold level for activity. Consequently, promoters may become classified as carcinogens although their mode of action may be quite different from that of strong mutagenic carcinogens. The review of methods for quantitative risk assessment includes discussions of the several factors that must be considered in extrapolating estimates of human cancer risk. The discussion of extrapolation to humans from the species used in the carcinogenicity tests includes consideration of differences between species in the sensitivity to tumor formation in particular organs. Also noted are the considerations that must be given to account for differences in dose and in the route of exposure to a compound when extrapolating from animal tests to estimates of human risk.

The review of data specifically concerned with diesel engine exhaust emissions demonstrates that these exhausts have biological activity. Short-term tests have shown that diesel engine exhausts are mutagenic and can cause chromosomal damage. The activity in these studies was influenced by the source of the emissions tested, for example, the type of engine used. A variety of studies have evaluated the activity of diesel engine exhaust as a complete carcinogen and as an initiator or promoter of carcinogenesis. Some of the studies failed to produce positive results or were equivocal. Positive results, however, have been found in inhalation studies and in mouse skin painting and lung adenoma formation assays. The results of these studies have been used with current, though admittedly imperfect, risk extrapolation methods, and values for projected human cancers have been calculated. The risk for diesel engine exhaust was calculated to be comparable to the approximate range found for other carcinogenic human exposures such as coke oven emissions and roofing tar. Within the limitations of these estimates, diesel engine exhausts do not appear to be notably more active than these other materials. Review of the epidemiologic studies of the risk of diesel engine exhausts shows that exposure to these exhausts does not cause a strong effect like cigarette smoking. However, because of the limitations of the studies, it is difficult to conclude conversely that the carcinogenic activity is negligible or absent.

Summary of Research Recommendations: Priorities, Purposes, and Responsibilities

Many factors must be considered in developing a research plan that sets priorities for the pursuit of the various recommended research goals. For example, these priorities might be selected on the basis of the unique mission of the Health Effects Institute, or they might be viewed on the basis of the more general need for furthering our knowledge about how to make quantitative risk assessments. From a practical point of view, it might be preferable to place the highest priority on goals that will require the longest time to accomplish or that are not getting adequate attention and support from other sources. It is also reasonable to place highest priority on goals that might significantly affect the cancer risks that might be attributable to diesel engine exhaust, even without accomplishing all of the proposed research goals.

If accomplishment of the unique mission of the Health Effects Institute is the perspective from which priorities are determined, then highest priority must be given to performing additional studies on the carcinogenicity of diesel exhaust (Recommendation 6) and developing methods for assessing the carcinogenicity of mixtures (Recommendation 7). It is unlikely that other sources or organizations will place comparable emphasis on the direct study of diesel exhaust as a carcinogen. The issue of the carcinogenicity of mixtures is a more general problem, but it is essential for the evaluation of diesel exhaust, although only a secondary concern in the evaluation of many other materials.

If the view is taken that adequate assessment of the hazards of diesel exhaust will not be possible without obtaining more knowledge about how to make quantitative risk assessments in general, then priorities might be set somewhat differently. In this case, the highest priority might be placed on evaluating the role of toxicity in carcinogenesis (Recommendation 3) and gathering critical data for quantitative assessments (Recommendation 4). By learning how to make quantitative risk assessments that account for effects of toxicity, and which involve extrapolations to low doses, among routes of administration, and among species, in a manner more firmly founded on scientific knowledge, better estimates of human cancer risks in general will become possible, and this will benefit the assessments of diesel exhaust.

Another basis for setting priorities might be consideration of practical issues. For example, priorities could be set so that the research goals all might be accomplished in the shortest time. From this perspective,

highest priority might be placed on goals that require the longest time to accomplish. Thus, priority might be given to evaluating the role of toxicity in carcinogenesis (Recommendation 3), gathering critical data for quantitative assessments (Recommendation 4), and developing methods for assessing the carcinogenicity of mixtures (Recommendation 7). Each of these is a complex problem that will require the performance of long-term studies to accomplish, and may require several such studies in sequence. By beginning these studies at the earliest time and phasing in other goals later, it might be possible to have the more complete body of knowledge with which to make scientific risk assessments at the earliest time.

Another perspective is to place the highest priority on goals that are not receiving adequate attention and support from other sources. It could be argued that many or most of the Research Recommendations are not being pursued with the vigor that might be desired. The conclusion from this, however, is that all of the Recommendations should be given a high priority. This point of view may be accurate but it does not contribute to a practical plan.

Another view might be predicated on the idea that early availability of certain critical knowledge might make it possible to affect the cancer risks from diesel exhaust significantly even before all of the needed information for scientific risk assessments has been obtained. A possible scenerio that might fit this perspective would place the highest priority on developing methods to identify individuals at high risk (Recommendation 2) and developing acceptable methods for dosimetry in humans (Recommendation 5). For example, if one could identify the individuals who were at high risk for the development of cancer if they are excessively exposed to diesel engine exhaust, then it would be possible to focus preventive health measures on this group. If it were possible to carry out dosimetry on exposed individuals, then preventive measures might be developed that would reduce exposure and risk.

Summary of Research Recommendations: A Research Plan

From the preceding discussion, it is clear that there are many ways to assign priorities for the pursuit of the various Research Recommendations. The following plan considers these different perspectives in defining a preferred set of priorities.

HIGH PRIORITY

No other organization will commit comparable attention or resources to the study of diesel engine exhaust, and therefore this must be done by the Health Effects Institute. One would hope that research on the scientific problems in making critical extrapolations in quantitative risk assessment and in validating the process would be widely supported and actively pursued. Unfortunately, this need has been clear for some time, yet there has been less progress in solving this problem than might have been expected. Accomplishment of the following two goals will provide the most urgently needed information to perform better assessments of the human risks resulting from diesel exhausts.

Recommendation 4	Critical data should be gathered for quantitative assessments.
Recommendation 6	Additional studies should be performed on the carcinogenicity of diesel exhaust.

MEDIUM PRIORITY

Development of methods for human dosimetry may benefit from investigator-initiated basic research and even from the research of commercial enterprises. Therefore, the pursuit of these goals may be given somewhat lower priority. A similar lower priority may be given to the evaluation of the carcinogenicity of mixtures. This problem is not unique to the assessment of diesel exhausts and knowledge may be gained from studies supported by other regulatory programs.

Recommendation 5	Acceptable methods should be developed for dosimetry in humans.
Recommendation 7	Methods should be developed for assessing the carcinogenicity of mixtures.

LOW PRIORITY

The remaining recommendations are important but are generic goals that would improve our general ability to make risk assessments. These issues touch on basic research that is being pursued in investigator-initiated studies. Investigations of this type may also be pursued by other agencies that are required to make risk assessments. Thus, although these are important goals, they may deserve lower priority in this program.

| Recommendation 1 | The role of promoters and enhancers in human carcinogenesis should be determined. |

| Recommendation 2 | Methods should be developed to identify individuals at high risk. |

| Recommendation 3 | The role of toxicity in carcinogenesis should be evaluated. |

Acknowledgments

The author thanks Dianne Shaw for excellent technical editing and Brigitte Cooke for skillful secretarial assistance.

References

Albert, R. E., Lewtas, J., Nesnow, S., Thorslund, T. W., and Anderson, E. 1983. Comparative potency method for cancer risk assessment: application to diesel particulate emissions, *Risk Anal.* 3:101–117.

Armitage, P. 1985. Multistage models of carcinogenesis, *Environ. Health Perspect.* 63:195–201.

Auerbach, O., Stout, A. P., Hammond, E. C., and Garfinkel, L. 1961. Changes in bronchial epithelium in relation to cigarette smoking and in relation to lung cancer, *N. Engl. J. Med.* 265:253–267.

Barbacid, M. 1986. Oncogenes and human cancer: cause or consequence? *Carcinogenesis* 7:1037–1042.

Barrett, J. C., and Ts'o, P. O. P. 1978. Evidence for the progressive nature of neoplastic transformation in vitro, *Proc. Natl. Acad. Sci. USA* 75:3297–3301.

Berenblum, I. 1975. Sequential aspects of chemical carcinogenesis: skin, In: *Cancer: A Comprehensive Treatise* (F. F. Becker, ed.), pp. 323–344, Plenum Press, New York.

Berenblum, I., and Shubik, P. 1947. A new, quantitative, approach to the study of the stages of chemical carcinogenesis in the mouse's skin, *Br. J. Cancer* 1:383–391.

Bishop, J. M. 1983. Cellular oncogenes and retroviruses, *Ann. Rev. Biochem.* 52:301–354.

Brand, K. G., Buoen, L. C., Johnson, K. H., and Brand, I. 1975. Etiological factors, stages, and the role of the foreign body in foreign body tumorigenesis: a review, *Cancer Res.* 35:279–286.

Cairns, T. 1980. The ED_{01} study: introduction, objectives, and experimental design, *J. Environ. Pathol. Toxicol.* 3:1–7.

Casto, B. C., Hatch, G. G., and Huang, S. L. 1981. Mutagenic and carcinogenic potency of extracts of diesel and related environmental emissions: in vitro mutagenesis and oncogenic transformation, *Environ. Int.* 5:403–409.

Claxton, L. D. 1981. Mutagenic and carcinogenic potency of diesel and related environmental emissions: *Salmonella* bioassay, *Environ. Int.* 5:389–391.

Cordeiro-Stone, M., Topal, M. D., and Kaufman, D. G. 1982. DNA in proximity to the site of replication is more alkylated than other nuclear DNA in S Phase 10T1/2 cells treated with *N*-methyl-*N*-nitrosourea, *Carcinogenesis* 3:1119–1127.

Cuddihy, R. G., and McClellan, R. O. 1983. Evaluating lung cancer risks from exposures to diesel engine exhaust, *Risk Anal.* 3:119–123.

Cuddihy, R. G., Griffith, W. C., and McClellan, R. O. 1984. Health risks from light-duty diesel vehicles, *Environ. Sci. Technol.* 18:14A–21A.

Curren, R. D., Kouri, R. E., Kim, C. M., and Schechtman, L. M. 1981. Mutagenic and carcinogenic potency of extracts from diesel related environmental emissions: simultaneous morphological transformation and mutagenesis in BALB/c 3T3 cells, *Environ. Int.* 5:411–415.

Doll, R. 1955. Mortality from lung cancer in asbestos workers, *Br. J. Ind.* 12:81–86.

Doll, R. 1971. The age distribution of cancer: implications for models of carcinogenesis, *J. Roy. Soc. Med.* 134:133–166.

Drake, J. W., and Baltz, R. H. 1976. The biochemistry of mutagenesis, *Ann. Rev. Biochem.* 45:11–37.

Evans, H. J. 1983. Effects on chromosomes of carcinogenic rays and chemicals, In: *Chromosome Mutations and Neoplasia* (J. German, ed.), pp. 253–279, A. R. Liss, New York.

Farber, E. 1980. The sequential analysis of liver cancer induction, *Biochim. Biophys. Acta* 605:149–166.

Farber, E. 1984. Chemical carcinogenesis: a current biological perspective, *Carcinogenesis* 5:1–5.

Gaylor, D. W. 1980. The ED_{01} study: summary and conclusions, *J. Environ. Pathol. Toxicol.* 3:179–183.

Grasso, P., and Hardy, J. 1974. Strain differences in natural incidence and response to carcinogens, In: *Mouse Hepatic Neoplasia* (W. H. Butler and P. M. Newberne, eds.), pp. 111–132, Elsevier Press, Amsterdam.

Grisham, J. W., Kaufmann, W. K., and Kaufman, D. G. 1983. The cell cycle and chemical carcinogenesis, *Surv. Synth. Pathol. Res.* 1:49–66.

Gullino, P. M., Pettigrew, H. M., and Grantham, F. H. 1975. *N*-Nitrosomethylurea as mammary gland carcinogen in rats, *J. Nat. Cancer Inst.* 54:401–414.

Hall, N. E., and Wynder, E. L. 1984. Diesel exhaust

Correspondence should be addressed to David G. Kaufman, Department of Pathology, School of Medicine, University of North Carolina, Brinkhous-Bullitt Building, 228H, Chapel Hill, NC 27514.

exposure and lung cancer: a case-control study, *Environ. Res.* 34:77–86.

Hanawalt, P. C., and Sarasin, A. 1986. Cancer-prone hereditary diseases with DNA processing abnormalities, *Trends Genet.* 2:124–129.

Harris, C. C., and Trump, B. F. 1983. Human tissues and cells in biomedical research, *Surv. Synth. Pathol. Res.* 1:165–171.

Harris, J. E. 1983. Diesel emissions and lung cancer, *Risk Anal.* 3:83–100.

Heino, M., Ketola, R., and Makela, P. 1978. Work conditions and health of locomotive engineers. I. Noise, vibration, thermal climate, diesel exhaust constituents, ergonomics, *Scand. J. Work Environ. Health* 4:3–14.

Heinrich, U., Peters, L., Funcke, W., Pott, F., Mohr, U., and Stober, W. 1982. Investigation of toxic and carcinogenic effects of diesel exhaust in long-term inhalation exposure of rodents, In: *Toxicologic Effects of Emissions from Diesel Engines* (J. Lewtas, ed.), pp. 225–242, Elsevier Science Publishing, Inc., Amsterdam.

Heinrich, U., Pott, F., Mohr, U., and Stober, W. 1985. Experimental methods for the detection of the carcinogenicity and/or cocarcinogenicity of inhaled polycyclic-aromatic-hydrocarbon-containing emissions, In: *Carcinogenesis, A Comprehensive Survey. Vol. 8, Cancer of the Respiratory Tract: Predisposing Factors* (M. J. Mass, D. G. Kaufman, J. M. Siegfried, V. E. Steele, and S. Nesnow, eds.), Vol. 8, pp. 131–146, Raven Press, New York.

Heinrich, U., Muhle, H., Takenaka, S., Ernst, H., Fuhst, R., Pott, F., Mohr, U., and Stober, W. 1986. Chronic effects on the respiratory tract of hamsters, mice and rats after long-term inhalation of high concentrations of filtered and unfiltered diesel engine emissions, *J. Appl. Toxicol.* 6:383–397.

Hueper, W. C. 1955. A Quest into the Environmental Causes of Carcinoma of the Lung. Public Health Monograph No. 36, U.S. Department of Health, Education and Welfare, Public Health Service.

Huggins, C., Grand, L. C., and Brillantes, F. P. 1961. Mammary cancer induced by a single feeding of polynuclear hydrocarbons and its suppression, *Nature* 189:204–207.

Ikenaga, M., and Kakunaga, T. 1977. Excision of 4-nitroquinoline 1-oxide damage and transformation in mouse cells, *Cancer Res.* 37:3672–3678.

International Agency for Research on Cancer. 1980. *IARC Monographs on the Evaluation of the Carcinogenic Risks of Chemicals to Humans, Vol. 23, Some Metals and Metallic Compounds*, IARC, Lyon, France.

Kakunaga, T. 1975. The role of cell division in the malignant transformation of mouse cells treated with 3-methylcholanthrene, *Cancer Res.* 35:1637–1642.

Kakunaga, T., Crow, J. D., Hamada, H., Hirakawa, T., and Leavitt, J. 1983. Mechanisms of neoplastic transformation of human cells, In: *Human Carcinogenesis* (C. C. Harris and H. N. Autrup, eds.), pp. 371–399, Academic Press, New York.

Kaplan, I. 1959. Relationship of noxious gases to carcinoma of the lung in railroad workers, *J. Am. Med. Assoc.* 171:2039–2043.

Klein-Szanto, A. J. P., Terzaghi, M., Mirkin, L. D., Nartin, D., Shiba, M. 1982. Propagation of normal human epithelial cell populations using an in vivo culture system, *Am. J. Pathol.* 108:231–239.

Kotin, P., Falk, H. L., and Thomas, M. 1955. Aromatic hydrocarbons. III. Presence in particulate phase of diesel-engine exhausts and the carcinogenicity of exhaust extracts, *AMA Arch. Ind. Hyg. Occup. Med.* 11:113–120.

Kuschner, M. 1985. Perspective on pathologic predisposition to lung cancer in humans, In: *Carcinogenesis, A Comprehensive Survey. Vol. 8, Cancer of the Respiratory Tract: Predisposing Factors* (M. J. Mass, D. G. Kaufman, J. M. Siegfried, V. E. Steele, and S. Nesnow, eds.), pp. 17–21, Raven Press, New York.

Lewtas, J., Bradow, R. L., Jungers, R. H., Harris, B. D., Zweidinger, R. B., Cushing, K. M., Gill, B. E., and Albert, R. E. 1981. Mutagenic and carcinogenic potency of extracts of diesel and related environmental emissions: study design, sample generation, collection and preparation, *Environ. Int.* 5:383–387.

Lewtas, J., Nesnow, S., and Albert, R. E. 1983. A comparative potency method for cancer risk assessment: clarification of the rationale, theoretical basis and application to diesel particulate emissions, *Risk Anal.* 3:133–137.

Li, A. P., and Royer, R. E. 1982. Diesel-exhaust-particle extract enhancement of chemical-induced mutagenesis in cultured Chinese hamster ovary cells: possible interaction of diesel exhaust with environmental carcinogens, *Mutat. Res.* 103:349–355.

Littlefield, N. A., Farmer, J. H., and Gaylor, D. W. 1980. Effects of dose and time in a long-term, low-dose carcinogenic study, *J. Environ. Pathol. Toxicol.* 3:17–34.

Lockard, J. M., Kaur, P., Lee-Stephens, C., Sabharwal, P. S., Pereira, M. A., McMillan, L., and Mattox, J. 1982. Induction of sister-chromatid exchanges in human lymphocytes by extracts of particulate emissions from a diesel engine, *Mutat. Res.* 104:355–359.

Luepker, R. V., and Smith, M. C. 1978. Mortality in unionized truck drivers, *J. Occup. Med.* 20:677–682.

Mauderly, J. L., Jones, R. K., Griffith, W. C., Henderson, R. F., and McClellan, R. O. 1987. Diesel exhaust is a pulmonary carcinogen in rats exposed chronically by inhalation, *Fundam. Appl. Toxicol.* 9:208–221.

McClellan, R. O. 1986. Health effects of diesel exhaust: a case study in risk assessment, *Am. Ind. Hyg. Assoc. J.* 47:1–13.

McClellan, R. O., Bice, D. E., Cuddihy, R. G., Gillett, N. A., Henderson, R. F., Jones, R. K., Mauderly, J. L., Pickrell, J. A., Shami, S. G., and Wolff, R. K. 1986a. Health effects of diesel exhaust, In: *Aerosols: Research, Risk Assessment and Control Strategies* (S. D. Lee, T. Schneider, L. D. Grant, and P. J. Verkerk, eds.), pp. 597–615, Lewis Publishers, Inc., Chelsea, Mich.

McClellan, R. O., Mauderly, J. L., Jones, R. K., Henderson, R. F., and Wolff, R. K. 1986b. Lung tumor induction in rats by chronic exposure to diesel exhaust, Abstract of lecture presented at the Second International Aerosol Conference, Berlin, West Germany, September 22–26, 1986.

Menck, H. R., and Henderson, B. E. 1976. Occupational differences in rates of lung cancer, *J. Occup. Med.* 18:797–801.

Merletti, F., Heseltine, E., Saracci, R., Simonato, L., Vainio, H., and Wilbourn, J. 1984. Target organs for carcinogenicity of chemicals and industrial exposures in humans: a review of results in the IARC Monographs on the Evaluation of the Carcinogenic Risk of Chemicals to Humans, *Cancer Res.* 44:2244–2250.

Miller, J. A. 1970. Carcinogenesis by chemicals: an overview—G. H. A. Clowe's Memorial Lecture, *Cancer Res.* 30:559–576.

Mitchell, A. D., Evans, E. L., and Jotz, M. M. 1981. Mutagenic and carcinogenic potency of extracts of diesel and related environmental emissions: in vitro mutagenesis and DNA damage, *Environ. Int.* 5: 393–401.

Mitelman, F. 1986. Clustering of chromosomal breakpoints in neoplasia, *Cancer Genet. Cytogenet.* 19:67–71.

Nachtman, J. P., Xiao-bai, X., Rappaport, S. M., Talcott, R. E., and Wei, E. T. 1981. Mutagenic activity in diesel exhaust particulates, *Bull. Environ. Contam. Toxicol.* 27:463–466.

Nesnow, S., Triplett, L. L., and Slaga, T. J. 1981. Tumorigenesis of diesel exhaust, gasoline exhaust and related emission extracts on SENCAR mouse skin, In: *Short-Term Bioassays in the Analysis of Complex Environmental Mixtures II* (M. D. Waters, S. S. Sandhu, J. L. Huisingh, L. Claxton, and S. Nesnow, eds.), pp. 277–297, Plenum Publishing Corp., New York.

Nesnow, S., Triplett, L. L., and Slaga, T. J. 1982. Comparative tumor-initiating activity of complex mixtures from environmental particulate emissions on Sencar mouse skin, *J. Nat. Cancer Inst.* 68:829–834.

Nesnow, S., Triplett, L. L., and Slaga, T. J. 1983a. Mouse skin tumor initiation-promotion and complete carcinogenesis bioassays: mechanisms and biological activities of emission samples, *Environ. Health Perspect.* 47:255–268.

Nesnow, S., Triplett, L. L., and Slaga, T. J. 1983b. Mouse skin carcinogenesis: application to the analysis of complex mixtures, In: *Short-Term Bioassays in the Analysis of Complex Environmental Mixtures III* (M. D. Waters, S. S. Sandhu, J. N. Lewtas, L. Claxton, N. Chernoff, and S. Nesnow, eds.), pp. 367–390, Plenum Publishing Corp., New York.

Nettesheim, P., and Barrett, J. C. 1984. Tracheal epithelial cell transformation: a model system for studies on neoplastic progression, *Crit. Rev. Toxicol.* 12:215–239.

Orthoefer, J. G., Moore, W., Kraemer, D., Truman, F., Crocker, W., and Yang, Y. Y. 1981. Carcinogenicity of diesel exhaust as tested in strain A mice, *Environ. Int.* 5:461–471.

Pereira, M. A., Shinozuka, H., and Lombardi, B. 1981. Test of diesel exhaust emissions in the rat liver foci assay, *Environ. Int.* 5:455–458.

Pereira, M. A., McMillan, L., Kaur, P., Gulati, D. K., Sabharwal, P. S. 1982. Effect of diesel exhaust emissions, particulates and extract on sister chromatid exchange in transplacentally exposed fetal hamster liver, *Environ. Mutagen.* 4:215–220.

Pickle, L. P., Mason, T. J., Howard, N., Hoover, R., and Fraumeni, J. F. 1987. *Atlas of U.S. Cancer Mortality Among Whites: 1950–1980*, pp. 1–149, National Institutes of Health, Bethesda, Md.

Pitts, J. N., Lokensgard, D. M., Harger, W., Fisher, T. S., Mejia, V., Schuler, J. J., Scorziell, G. M., and Katzenstein, Y. A. 1982. Mutagens in diesel exhaust particulate: identification and direct activities of 6-nitrobenzo[a]pyrene, 9-nitroanthracene, 1-nitropyrene and 5H-phenanthro [(4,5-BCD)] pyran-5-one, *Mutat. Res.* 103(3-6):241–249.

Pour, P. M. 1984. Histogenesis of exocrine pancreatic cancer in the hamster model, *Environ. Health Perspect.* 56:229–243.

Raffle, P. A. B. 1957. The health of the worker, *Br. J. Ind. Med.* 14:73–80.

Reddy, B. S., Weisburger, J. H., Narisawa, T., Wynder, E. L. 1974. Colon carcinogenesis in germfree rats with 1,2-dimethylhydrazine and N-methyl-N'-nitro-N-nitrosoguanidine, *Cancer Res.* 34:2368–2372.

Reddy, J. K., Azarnoff, D. L., and Hignite, C. E. 1980. Hypolipidemic hepatic peroxi-some proliferators form a novel class of chemical carcinogens, *Nature* 283:397–398.

Saffiotti, U., and Harris, C. C. 1979. Carcinogenesis studies on organ cultures of animal and human respiratory tissue, In: *Carcinogens: Identification and Mechanisms of Action* (A. C. Griffin and C. R. Shaw, eds.), pp. 65–82, Raven Press, New York.

Saffiotti, U., and Kaufman, D. G. 1975. Carcinogenesis of laryngeal carcinoma, *Laryngoscope* 85: 454–457.

Saffiotti, U., Cefis, F., and Kolb, L. H. 1968. A method for the experimental induction of bronchogenic carcinoma, *Cancer Res.* 28:104–124.

Sandberg, A. A. 1983. A chromosomal hypothesis of oncogenesis, *Cancer Genet. Cytogenet.* 8:277–285.

Sarma, D. S. R., Rajalakshmi, S., and Farber, E. 1975. Chemical carcinogenesis: interactions of carcinogens with nucleic acids, In: *Cancer, A Comprehensive Treatise* (F. F. Becker, ed.), Vol. I, pp. 235–287, Plenum Press, New York.

Scarpelli, D. G., Rao, M. S., and Reddy, J. K. 1984. Studies of pancreatic carcinogenesis in different animal models, *Environ. Health Perspect.* 56:219–227.

Schenker, M. B. 1980. Diesel exhaust—an occupational carcinogen? *J. Occup. Med.* 22:41–46.

Schreiber, H., Saccomanno, G., Martin, D. H., and Brennan, L. 1974. Sequential cytological changes during development of respiratory tract tumors induced in hamsters by benzo[a]pyrene–ferric oxide, *Cancer Res.* 34:689–698.

Selikoff, I. J., Hammond, E. C., and Churg, J. 1968. Asbestos exposure, smoking and neoplasia, *J. Am. Med. Assoc.* 204:106–112.

Setlow, R. B. 1978. Repair deficient human disorders and cancer, *Nature* 271:713–715.

Shimosato, Y., Kodama, T., Tamai, S., and Kameya, T. 1980. Induction of squamous cell carcinoma in human bronchi transplanted into nude mice, *Gann* 71:402–407.

Singer, B., and Grunberger, D. 1983. *Molecular Biology of Mutagens and Carcinogens*, pp. 45–219, Plenum Press, New York.

Slaga, T. J., Fischer, S. M., Nelson, K., and Gleason, G. L. 1980. Studies on the mechanism of skin tumor promotion: evidence for several stages in promotion, *Proc. Nat. Acad. Sci. USA* 77:3659–3663.

Slaga, T. J., Triplett, L. L., and Nesnow, S. 1981. Mutagenic and carcinogenic potency of extracts of diesel and related environmental emissions: two-stage carcinogenesis in skin tumor sensitive mice (Sencar), *Environ. Int.* 5:417–423.

Smith, B. L., and Sager, R. 1982. Multistep origin of tumor-forming ability in Chinese hamster embryo fibroblast cells, *Cancer Res.* 42:389–396.

Stanton, M. F., and Wrench, C. 1972. Mechanisms of mesothelioma induction with asbestos and fibrous glass, *J. Nat. Cancer Inst.* 49:797–821.

Steele, V. E., Marchok, A. C., and Nettesheim, P. 1980. Enhancement of carcinogenesis in cultured respiratory tract epithelium by 12-O-tetradecanoyl-phorbol-13-acetate, *Int. J. Cancer* 26:343–348.

Swift, M., Sholman, L., Perry, M., and Chase, C. 1976. Malignant neoplasms in the families of patients with ataxia-telangiectasia, *Cancer Res.* 36:209–215.

Valerio, M. G., Fineman, E. L., Bowman, R. L., Harris, C. C., Stoner, G. D., Autrup, H., Trump, B. F., McDowell, E. M., and Jones, R. T. 1981. Long-term survival of normal adult human tissues as xenografts in congenitally athymic nude mice, *J. Nat. Cancer Inst.* 66:849–858.

Wagner, J. C., Berry, G., and Timbrell, V. 1973. Mesotheliomata in rats after inoculation with asbestos and other materials, *Br. J. Cancer* 28:173–185.

Ward, J. M., Yamamato, R. S., and Brown, C. A. 1973. Pathology of intestinal neoplasms and other lesions in rats exposed to azoxymethane, *J. Nat. Cancer Inst.* 51:1029–1039.

Waxweiller, R. J., Wagner, J. K., and Archer, W. C. 1973. Mortality of potash workers, *J. Occup. Med.* 15:406–409.

Wegman, D. H., and Peters, J. M. 1978. Oat cell cancer in selected occupations, *J. Occup. Med.* 20:793–796.

Weinberg, R. A. 1985. The action of oncogenes in the cytoplasm and nucleus, *Science* 230:770–776.

Wolff, R. K., Henderson, R. F., Snipes, M. B., Sun, J. D., Bond, J. A., Mitchell, C. E., Mauderly, J. L., and McClellan, R. O. 1986. Lung retention of diesel soot and associated organic compounds, Abstract of lecture presented at the International Symposium on Toxicological Effects of Emissions from Diesel Engines, Tsukuba Science City, Japan, July 26–28, 1986.

Wong, D., Mitchell, C., Wolff, R. K., Mauderly, J. L., and Jeffrey, A. M. 1986. Identification of DNA damage as a result of exposure of rats to diesel engine exhaust, *Proc. Am. Assoc. Cancer Res.* 27:84.

Yunis, J. J., and Soreng, A. L. 1984. Constitutive fragile sites and cancer, *Science* 226:1199–1204.

Potential Carcinogenic Effects of Polynuclear Aromatic Hydrocarbons and Nitroaromatics in Mobile Source Emissions

STEPHEN S. HECHT
American Health Foundation

Air Pollution, the Automobile, and Public Health. © 1988 by the Health Effects
Institute. National Academy Press, Washington, D.C.

Historical Perspective and Directions for the Future

Pott first observed an association between soot and cancer in 1775 (Pott 1775), and by the early twentieth century it was clear that soot, coal tar, and pitch could cause cancer in humans (International Agency for Research on Cancer 1985a). In the 1930s, pioneering studies by Kennaway and Hieger (1930) and Cook et al. (1933) established that polynuclear aromatic hydrocarbons (PAHs) were carcinogenic components of pitch. Although the following 50 years have brought major advances in our understanding of the mechanisms by which PAHs can cause cancer, our ability to assess the health effects—and in particular the potential carcinogenic effects—of PAHs in humans, remains incomplete. The aspects of PAH carcinogenesis summarized in this chapter include epidemiologic studies that may link PAHs to human cancer, carcinogenicity assays in laboratory animals, key features of PAH metabolic activation and detoxification, and the effects of modifiers on these processes.

In contrast to the PAHs, known for half a century, nitro-substituted PAHs (nitro-PAHs) have only recently been recognized as environmental carcinogens, whose presence in diesel exhaust is of particular concern (Schuetzle 1983). Although less is known about their health effects than about those of PAHs, it is clear that some nitro-PAHs are potent mutagens and carcinogens.

Using these studies as a base, we identify significant gaps that detract from our ability to assess PAH carcinogenicity in humans, summarize data on the carcinogenic effects and metabolism of nitro-PAHs in laboratory animals, and suggest directions for future research.

Recent rapid progress in research on PAHs and nitro-PAHs is due to improved analytical and spectroscopic techniques as well as to major advances in molecular biology. Those new techniques, which may soon permit the measurement of an effective biological dose of a carcinogen for humans, combined with animal experiments, may make it possible to develop indicators of individual susceptibility to PAH or nitro-PAH carcinogenesis. These exciting developments represent an important frontier in chemical carcinogenesis research, and their application to assessing the health risks of PAHs and nitro-PAHs are a focus of this chapter.

Evaluation of Polyaromatic Hydrocarbon and Nitro-Polyaromatic Hydrocarbon Carcinogenicity

Role of Polyaromatic Hydrocarbons as Human Carcinogens

Since the PAHs to which humans are exposed always occur in a mixture of many compounds, some potentially carcinogenic, assessing PAHs as human carcinogens by epidemiologic studies is difficult. The International Agency for Research on Cancer (IARC) has evaluated carcinogenicity of such mixtures and published results in the *IARC Monographs on the Evaluation of the Carcinogenic Risk of Chemicals to Humans*. Evidence for carcinogenicity from studies in humans is categorized as follows: (1) "sufficient evidence" indicates a causal relation between the agent and human cancer; (2) "limited evidence" indicates a credible causal relation without excluding other explanations; (3) "inadequate evidence" indicates either that pertinent data are few or that the available data do not exclude a presumed chance association; and (4) "no evidence" indicates that adequate studies are available and these show no evidence of carcinogenicity. Other working groups have also considered the possible roles of PAH-containing mixtures in human cancer etiology. Studies of sources, other than automotive, implicate PAHs as human carcinogens and are discussed below. Evidence for the role of mobile sources is discussed in Carcinogenicity of Nitro-PAHs.

Tobacco Smoke. Sufficient evidence now indicates that tobacco smoke is carcinogenic to humans, that malignant tumors of the respiratory tract and upper digestive system are causally related to smoking tobacco, and that malignant tumors of the bladder, pancreas, and renal pelvis specifically are causally related to cigarette smoking (International Agency for Research on Cancer 1986). The 1982 report by the Surgeon General of the United States concluded that "cigarette smoking is the major single cause of cancer mortality in the United States. Tobacco's contribution to all cancer deaths is estimated to be 30 percent" (U.S. Department of Health and Human Services 1982a). Evidence increasingly suggests that passive exposure to tobacco smoke, as in polluted indoor environments, may increase the risk of lung cancer (International Agency for Research on Cancer 1986).

PAHs are important carcinogenic constituents of the particulate phase of cigarette smoke in concentrations ranging from 1 to 60 ng/cigarette (U.S. Department of Health and Human Services 1982b), and their tumorigenic activities are enhanced by other agents in tobacco smoke. It is very likely that they are involved in the cancer-causing properties of tobacco smoke, but their role is difficult to assess because of the many other toxic and carcinogenic constituents of tobacco smoke, such as nitrosamines whose concentrations exceed those of PAHs (Hoffmann and Hecht 1985), volatile aldehydes, and aromatic amines.

Coal Tars, Shale Oils, and Soots.
An IARC Working Group (1985a) concluded that occupational exposure to coal tars is causally associated with incidence of skin cancer in humans, and that coal tar pitches are carcinogenic in humans. They noted that a cohort study of U.S. roofers indicated a greater risk of lung cancer and other cancers. Among the 10,000 compounds that may be present in coal tars, PAHs occur in concentrations ranging from 0.1 to 10 percent in high-temperature coal tars, and certainly contribute largely to the observed carcinogenic properties of coal tars. Other IARC groups (1985b,c) have concluded that there is sufficient evidence that shale oils and soot are carcinogenic to humans and contain relatively high levels of PAHs.

Coal Gasification and Coke Production.
The IARC (1984c) has concluded that certain exposures in the retort houses of older coal gasification processes and in the coke production industry are carcinogenic in humans. The relative risk of lung cancer is as high as 16-fold in topside coke oven workers with 15 years or more of exposure. Consideration of duration and location of employment in the plant has shown a dose/response relationship for lung cancer. In an association between mortality from lung cancer and exposure to coal tar pitch volatiles, PAHs are again likely prominent causative agents (Redmond 1983).

Aluminum Reduction and Iron and Steel Founding.
Although an IARC Working Group (1984b) found only limited evidence of increased incidence of cancer in aluminum production and iron and steel foundry workers, some evidence links aluminum production to bladder cancer. Levels of total PAHs in aluminum production range from <1 to 2,800 $\mu g/m^3$, determined by personal sampling for 2–6 hr at various sites. PAHs or their metabolites have also been detected in the urine of exposed workers (Becher and Bjorseth 1983; International Agency for Research on Cancer 1984b).

Mineral Oils.
Mineral oils used in such occupations as mulespinning, metal machining, and jute processing have been found to be carcinogenic to humans, and exposures have consistently been linked to cancer of the skin, and particularly of the scrotum. The levels of PAHs present in such oils vary, depending on source and processing (International Agency for Research on Cancer 1984a).

Urban Pollution.
In contrast to certain occupational exposures, epidemiological evidence indicates that after correction for smoking and occupation, exposure to general air pollution (defined as a body of contaminated air extending over a population area of appreciable size) has little, if any, effect on rate of death from lung cancer (Hammond and Garfinkel 1980).

Conclusion.
Although none of the studies mentioned here has specifically incriminated PAHs as causative agents of cancer, the overall data, together with the results

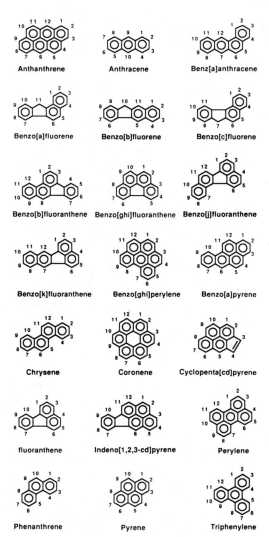

Figure 1. Structures of some PAHs commonly detected in exhaust emissions.

of bioassays described below, strongly indicate that PAHs can cause cancer in humans.

Tumorigenicity of Polyaromatic Hydrocarbons in Laboratory Animals

Figure 1 illustrates the structures of the PAHs commonly detected in exhaust emissions (International Agency for Research on Cancer 1983). Methylchrysenes as well as methyl- and dimethylphenanthrenes are also detectable. To evaluate carcinogenic potential, these compounds have been tested most extensively on mouse skin, since such assays are convenient and inexpensive. Some compounds have been tested by other protocols, including subcutaneous injection in mice and rats, intrapulmonary injection in rats, and intratracheal instillation in hamsters. Tables 1 and 2 summarize an IARC Working Group evaluation of carcinogenicity of these PAHs in laboratory animals.

Mouse Skin Bioassays. Two protocols are used. The first, the initiation/promotion protocol, consists of the application of a single large dose or series of smaller doses of the PAH to the skin, followed by repeated application of the tumor promoter 12-O-tetradecanoylphorbol-13-acetate (TPA). (A tumor promoter is a substance that does not itself induce tumors, but when applied after a tumor initiator, for example, a PAH, enhances its activity.) The second, or complete carcinogenicity protocol, consists of repeated applications of the PAH to the skin. Frequently, the results of both protocols agree (LaVoie et al. 1979).

Among the unsubstituted PAHs, significant activity is observed only in compounds with four or five aromatic rings. Benzo[*a*]pyrene (B*a*P) and benzo[*b*]fluoranthene are the most tumorigenic, followed by benzo[*j*]fluoranthene. Weak tumorigenicity has been observed for benz-[*a*]anthracene, benzo[*k*]fluoranthene, chrysene, cyclopenta[*cd*]pyrene, and indeno-[1,2,3-*cd*]pyrene. Methyl substitution can significantly alter tumorigenicity. As shown in table 2, 5-methylchrysene is a potent tumorigen, with activity similar to that of B*a*P, and 1,4- and 4,10-dimethylphenanthrene also are relatively strong tumor initiators. For a detailed review of structure/activity relationships among methylated PAHs, see Hecht et al. (1988).

Respiratory Tract Bioassays. These protocols are relevant to the problem of human respiratory exposure to PAHs, and include intratracheal instillation, lung implantation, and inhalation. Intratracheal instillation of PAHs has been used most extensively with the Syrian golden hamster

Table 1. Representative Tumor-Initiating Activity of PAHs on Mouse Skin and IARC Evaluations of Carcinogenicity of Parent PAHs Commonly Detected in Exhaust Emissions

Compound	Dose (nmole)	TBA (%)	T/A	Reference	IARC Evaluation of Carcinogenicity in Laboratory Animals[a]
	Representative Assays of Tumor-Initiating Activity on CD-1 Mouse Skin				
Anthanthrene	910	7	0.1	Hoffmann and Wynder (1966)	Limited
Anthracene	10,000	14	0.1	Scribner (1973)	None
Benz[a]anthracene	2,000	23	0.3	Wood et al. (1977)	Sufficient
Benzo[a]fluorene	4,630	10	0.2	LaVoie et al. (1981a)	Inadequate
Benzo[b]fluorene	4,630	20	0.4	LaVoie et al. (1981a)	Inadequate
Benzo[c]fluorene	4,630	25	0.3	LaVoie et al. (1981a)	Inadequate
Benzo[b]fluoranthene	119	60	2.3	LaVoie et al. (1982b)	Sufficient
Benzo[ghi]fluoranthene	NT[b]				Inadequate
Benzo[j]fluoranthene	119	30	0.6	LaVoie et al. (1982b)	Sufficient
Benzo[k]fluoranthene	119	5	0.1	LaVoie et al. (1982b)	Sufficient
Benzo[ghi]perylene	910	7	0.1	Hoffmann and Wynder (1966)	Inadequate
Benzo[a]pyrene	119	85	4.9	LaVoie et al. (1982b)	Sufficient
Benzo[e]pyrene	6,000	14	0.1	Buening et al. (1980)	Inadequate
Chrysene	4,390	55	0.6	Hecht et al. (1974)	Limited
Coronene	1,670	30		Van Duuren et al. (1968)	Inadequate
Cyclopenta[cd]pyrene	2,500	37	0.5	Wood et al. (1980)	Limited
Fluoranthene	4,950	3	0	Hoffmann et al. (1972)	None[c]
Indeno[1,2,3-cd]pyrene	900	17	0.3	Hoffmann and Wynder (1966)	Sufficient
Perylene	3,970	5	0.1	El-Bayoumy et al. (1982)	Inadequate
Phenanthrene	5,620	0	0	LaVoie et al. (1981b)	Inadequate
Pyrene	4,950	5	0.1	El-Bayoumy et al. (1982)	None
Triphenylene	NT				Inadequate

[a] From International Agency for Research on Cancer (1983).
[b] It was not active as a complete carcinogen on mouse skin (Wynder and Hoffmann 1959).
[c] A study has shown that fluoranthene is tumorigenic in newborn mice (Busby et al. 1984).
NOTE: NT = not tested as a tumor initiator; T/A = tumors per animal; TBA = tumor-bearing animals.

because it has no spontaneous lung tumors and is resistant to pulmonary infection and inflammation. Tumors are induced by instillation of the PAH and a carrier, for example, ferric oxide, (Fe_2O_3) or in a suspension in saline (Stinson and Saffiotti 1983). Squamous cell carcinoma of the trachea and bronchi induced in a high percentage of Syrian golden hamsters given intratracheal instillations of BaP closely resemble human tumors.

Although the carcinogenic effects of BaP in the hamster respiratory tract have been studied, investigations of other PAHs have been few. Sellakumar and Shubik (1974) found that when either benzo[b]fluoranthene, dibenz[a,h]anthracene, benz[a]anthracene, or pyrene was instilled with Fe_2O_3 in the hamster trachea, the incidence of respiratory tumors was insignificant.

However, in this model system dibenzo[a,i]pyrene and dibenzo[c,g]carbazole were highly tumorigenic. In experiments with other species, Hirao and coworkers (1980) found that BaP in saline caused lung cancer in rabbits, and Yoshimoto and coworkers (1977) observed that BaP with a carrier induced tumors in mice and rats.

Stanton and coworkers (1972) implanted PAHs dissolved in a mixture of beeswax and trioctanoin in rat lungs and observed the development of epidermoid lung carcinoma. Deutsch-Wenzel and coworkers (1983) tested a variety of PAHs using this protocol, and found that BaP was the most tumorigenic PAH of those commonly detected in exhaust emissions, while anthanthrene, benzo[b]fluoranthene, and indeno[1,2,3-cd]pyrene were moderately tumorigenic. However, the tumorigenicity

Table 2. Representative Tumor-Initiating Activity on Mouse Skin and IARC Evaluations of Carcinogenicity of Methylated Chrysenes and Phenanthrenes

Parent System	Representative Assays of Tumor-Initiating Activity on Mouse Skin				IARC Evaluation of Carcinogenicity in Laboratory Animals[a]
	Methyl Isomer	Dose (nmole)	TBA (%)	T/A	
Chrysene[b]	1	4,130	30	0.3	Inadequate
	2	4,130	42	0.7	Limited
	3	4,130	70	1.3	Limited
		1,240	20	0.4	
		410	15	0.2	
	4	4,130	35	0.5	Limited
	5	4,130	85	4.8	Sufficient
		1,240	100	8.0	
		410	100	5.5	
		100	90	5.2	
		33	80	3.9	
	6	4,130	35	0.6	Limited
Phenanthrene[c]	1	5,210	0	0	Inadequate
	2	5,210	0	0	NC
	3	5,210	0	0	NC
	4	5,210	0	0	NC
	9	5,210	0	0	NC
	1,4	4,850	100	5.3	NC
		1,460	80	3.2	NC
	1,9	4,850	0	0	NC
	2,7	4,850	5	0.1	NC
	3,6	4,850	0	0	NC
	4,5	4,850	5	0.1	NC
	4,9	4,850	10	0.1	NC
	4,10	4,850	55	1.5	NC
		1,460	35	1.6	NC

[a] From International Agency for Research on Cancer (1983).
[b] Hecht et al. (1974).
[c] LaVoie et al. (1981b, 1982a).
NOTE: NC = not considered; T/A = tumors per animal; TBA = tumor-bearing animals.

of anthanthrene was higher than expected on the basis of mouse skin studies.

Kendrick and coworkers (1974) found that instillation of PAHs into subcutaneous tracheal transplants in rats results in hyperplasia, dysplasia, squamous metaplasia, and squamous cell carcinoma. Benzo[a]pyrene, but not benzo[e]pyrene (BeP), was carcinogenic in that system. However, Topping and coworkers (1981) found that BeP was cocarcinogenic for the connective tissue, but not the tracheal epithelium.

Because of their expense, inhalation experiments with PAHs have been limited. In one experiment, 2 of 21 rats housed in fresh air and 5 of 21 rats housed in an atmosphere containing sulfur dioxide (SO_2) developed squamous cell carcinoma of the lung (Laskin et al. 1970). In another experiment, Thyssen and coworkers (1981) observed exposure-related neoplasms in the nasal cavity, larynx, pharynx, esophagus, and forestomach of Syrian golden hamsters exposed to BaP.

Dose/Response Relationships. Clear dose/response relationships for BaP-induced tumor formation have been demonstrated by using the Syrian golden hamster intratracheal instillation model and the mouse skin initiation/promotion protocol. The yield of respiratory tract tumors, tumor latency, and tumor multiplicity related to dose in hamsters treated with BaP and Fe_2O_3 (Saffiotti et al. 1972a,b), or BaP in buffer and physiological saline (Ketkar et al. 1979), depended on the number of administrations, the dosage per administration, and the fractionation of doses.

Similar effects have been observed on mouse skin (Saffiotti and Shubik 1956). In the mouse skin initiation/promotion protocol, a linear dose/response relationship for papilloma formation in Sencar mice ranged between 100 and 600 nmole/mouse; at higher doses the tumor response leveled off (Ashurst et al. 1983). The effects of differing initiating doses of 7,12-dimethylbenz[a]-anthracene showed that linear extrapolation from high doses may lead to underestimation of low-dose tumor risks (Stenbäck et al. 1981).

Effects of Mouse Strain. Aryl hydrocarbon (Ah)-responsive mice (for example, C57BL/6N) are more susceptible to tumor induction from subcutaneous injection of BaP, 3-methylcholanthrene, or dibenz-[a,h]anthracene than are Ah-nonresponsive mice (for example, DBA/2N). These differences have been attributed to the differing abilities of these mice to metabolize the PAHs to their ultimate carcinogenic forms. Inducibility of the cytochrome P-450 enzymes that metabolize PAHs appears to be related to PAH tumorigenicity (Nebert 1981). The *Ah* gene codes for a cytosolic receptor which can bind the inducer, such as 3-methylcholanthrene. The inducer/receptor complex is translocated into the nucleus and in some way initiates increased P-450 synthesis.

Modifiers of Polyaromatic Hydrocarbon Carcinogenesis

The influence of modifiers is perhaps one of the most important but least well understood areas of cancer induction by PAHs. Modifiers can be broadly classified as either promoters, cocarcinogens, or inhibitors of carcinogenesis. Promoters are generally noncarcinogenic substances which, when applied subsequently to PAHs, will cause tumors. In the case of the most widely studied promoter, TPA, tumors can be induced on mouse skin even if the TPA is

applied one year after administration of 7,12-dimethylbenz[a]anthracene (Van Duuren et al. 1975).

This observation supports the concept that initiation, even by a single dose of a PAH, is essentially irreversible, consistent with a change in DNA. Exposure to a single dose of a PAH can initiate cells but may not cause tumors, but exposure of initiated cells to multiple doses of a promoter can lead to tumor development. Thus, promotion may be important in determining whether or not environmental exposure to PAHs results in tumor development. TPA is an exceptionally effective tumor promoter for experimental studies, but it does not occur in significant quantities in the environment.

The most important known environmental tumor promoters are tobacco smoke and diet. The tumor-promoting activity of tobacco smoke and its condensate has been clearly demonstrated in studies using PAHs as initiators in either the mouse skin or the intratracheal instillation models (Hoffmann et al. 1978). Epidemiologic studies indicate that cessation of cigarette smoking leads to a lower risk for lung cancer, consistent with reversibility of promotion (Wynder and Hoffmann 1979). Extensive studies with the 7,12-dimethylbenz-[a]anthracene–induced Sprague-Dawley rat breast tumor model have demonstrated that a high-fat diet can promote breast tumor development. This is also in agreement with some epidemiologic studies (National Research Council, Committee on Diet, Nutrition, and Cancer 1982).

Cocarcinogens are defined as substances that enhance tumorigenicity when administered simultaneously with a carcinogen. Important environmental cocarcinogens include tobacco smoke, polyphenols, and PAHs. The cocarcinogenicity of tobacco smoke is due to its neutral polar and weakly acidic fractions (Hoffmann et al. 1978). Investigations of the compounds responsible for this activity have led to the identification of a number of cocarcinogens that are also environmental or dietary constituents. Important among these is catechol, which is strongly cocarcinogenic with BaP on mouse skin (Van Duuren and Goldschmidt 1976; Hecht et al. 1981). Compounds related to catechol occur widely in the diet, and conjugates of catechol are excreted in normal human urine (Carmella et al. 1982).

The cocarcinogenic activities of PAHs are particularly important because PAHs always occur as mixtures in the environment. Pyrene, benzo[e]pyrene, and fluoranthene are all essentially nontumorigenic, but they all enhance the tumorigenicity of BaP (Van Duuren and Goldschmidt 1976; Hoffmann et al. 1978). Other cocarcinogens include benzo[ghi]perylene, decane, undecane, 4,4'-dichlorostilbene, 1-methylindole, and 9-methylcarbazole (Van Duuren and Goldschmidt 1976; Hoffmann et al. 1978). Such interactions have to be considered when assessing human risk for cancer development upon exposure to PAHs.

A broad spectrum of compounds, many of them dietary, are capable of inhibiting PAH tumorigenesis. These include a variety of phenols, phenolic antioxidants such as butylated hydroxyanisole (BHA) and butylated hydroxytoluene (BHT), naturally occurring dietary indoles and isothiocyanates, various flavones, selenium salts, protease inhibitors, retinoids, and carotenes (Slaga and DiGiovanni 1984; Wattenberg 1985).

Inhibition has been observed in various model systems including mouse skin, mouse forestomach, mouse lung, and rat breast. A number of different protocols have been used, and the timing of administration of inhibitor versus administration of PAH can be critical in determining whether inhibition of tumorigenesis is observed. The identification of naturally occurring and synthetic inhibitors of carcinogenesis offers promise for prevention of PAH carcinogenesis. However, the influences of chemopreventive agents on PAH carcinogenesis are complex. For example, BHA is carcinogenic under certain conditions, and in some experimental protocols, BHA as well as BHT can act as tumor promoters (Ito et al. 1985).

Carcinogenicity of Nitro-Polyaromatic Hydrocarbons

Whereas the carcinogenic activities of PAHs have been evaluated extensively, re-search on nitro-PAHs is relatively limited. The observation that nitro-PAHs appear to account for a major portion of the direct-acting mutagenicity of diesel exhaust particulates has caused interest in their potential carcinogenic effects. No epidemiologic studies are available on the potential carcinogenicity of the nitro-PAHs that have been identified in diesel exhaust; these include 1-nitropyrene, dinitropyrenes, hydroxynitropyrenes, methyl nitropyrenes, 3-nitrofluoranthene, 2-nitrofluorene, 9-nitroanthracene, and 6-nitrobenzo[a]pyrene (Schuetzle 1983).

With respect to occupational exposure to diesel exhaust, "excess risk of cancer of the lung, or of any other site, has not been convincingly demonstrated" (National Research Council, Health Effects Panel of the Diesel Impacts Study Committee 1981). However, the Committee found fault with the studies that have been performed and called for additional carefully controlled studies of populations occupationally exposed to diesel engine exhaust.

Indeed, two recent studies of motor exhaust–related occupations and bladder cancer, based on the National Bladder Cancer Study, indicated that males usually employed as truck drivers or deliverymen had a statistically significant 50 percent increase in risk of bladder cancer (Silverman et al. 1983, 1986). The authors speculated that nitro-PAHs might be involved as causative agents. PAHs as well as nitro-PAHs are present in motor exhaust of various types.

Experimental studies of nitro-PAH tumorigenicity are summarized in table 3. 1-Nitropyrene, one of the predominant nitro-PAHs in diesel exhaust, induces tumors at the site of subcutaneous application as well as in the breast in newborn CD rats. It also is moderately tumorigenic in the A/J mouse lung adenoma assay. However, it appears to be inactive in adult CD rats and is weakly active or inactive in a number of other experimental models.

In contrast to 1-nitropyrene, the dinitropyrenes are highly tumorigenic. They have induced high incidences of subcutaneous tumors in rats (Ohgaki et al. 1984, 1985) and in mice (Tokiwa et al. 1985), and 1,6-dinitropyrene has caused lung carcinomas in 90–100 percent of Syrian golden

Table 3. Nitro-PAH Tumorigenicity Assays

Compound	Test System	Result	Reference
1-Nitropyrene	Newborn CD rat	Subcutaneous and mammary tumors	Hirose et al. (1984)
	A/J mouse	Lung adenomas	El-Bayoumy et al. (1984a)
	Mouse skin	Insignificant activity	El-Bayoumy et al. (1982) Nesnow et al. (1984)
	F344/Du Crj rat	Insignificant activity	Ohgaki et al. (1985)
	Newborn mouse	Liver tumors	Wislocki et al. (1985)
	BALB/c mouse	Insignificant activity	Tokiwa et al. (1984)
	CD rat	Insignificant activity	Imaida et al. (1985)
2-Nitropyrene	CD rat	Insignificant activity	Imaida et al. (1985)
4-Nitropyrene	CD rat	Mammary tumors	Imaida et al. (1985)
	Newborn mouse	Liver and lung tumors	Wislocki et al. (1985)
3-Nitrofluoranthene	A/J mouse	Lung adenomas	El-Bayoumy et al. unpublished observations
6-Nitrochrysene	Mouse skin	Skin tumors	El-Bayoumy et al. (1982)
	Newborn mouse	Lung and liver tumors	Busby et al. (1984) Wislocki et al. (1985)
6-Nitrobenzo[a]pyrene	Mouse skin	Insignificant activity	El-Bayoumy et al. (1982)
	Newborn mouse	Liver tumors (weak)	Wislocki et al. (1985)
3-Nitroperylene	Mouse skin	Skin tumors	El-Bayoumy et al. (1982)
7-Nitrobenz[a]anthracene	Newborn mouse	Liver tumors (weak)	Wislocki et al. (1985)
1,3-Dinitropyrene	F344/Du Crj rat	Subcutaneous tumors	Ohgaki et al. (1985)
	Newborn mouse	Liver tumors	Wislocki et al. (1985)
1,6-Dinitropyrene	F344/Du Crj rat	Subcutaneous tumors	Ohgaki et al. (1985)
	Syrian golden hamster	Lung tumors and leukemia	Takayama et al. (1985)
	BALB/c mouse	Subcutaneous tumors	Tokiwa et al. (1984)
	Newborn mouse	Liver tumors	Wislocki et al. (1985)
1,8-Dinitropyrene	F344/Du Crj rat	Subcutaneous tumors	Ohgaki et al. (1984)
2-Nitrofluorene	Holtzman rat	Forestomach carcinoma	Miller et al. (1955)
5-Nitroacenaphthene	Wistar rat	Ear duct, mammary, small intestine tumors	Takemura et al. (1974)
	Syrian golden hamster	Cholangiomas	Takemura et al. (1974)
	F344 rat	Ear duct, mammary, lung tumors	National Cancer Institute (1978)
	B6C3F₁ mouse	Liver tumors	

hamsters treated by intratracheal instillation (Takayama et al. 1985). In the newborn mouse, 6-nitrochrysene was exceptionally tumorigenic (Busby et al. 1984; Wislocki et al. 1985) whereas a number of other nitro-PAHs were weakly tumorigenic or inactive.

Taken together, these studies show that nitro-PAHs are carcinogenic in laboratory animals. Their activities depend greatly on structure and on the model system used. Further studies are necessary to define more clearly the structural requirements for tumorigenicity of nitro-PAHs as well as the most appropriate model systems for their bioassay.

Metabolic Activation and Detoxification of Polyaromatic Hydrocarbons and Nitro-Polyaromatic Hydrocarbons

Absorption and Distribution upon Inhalation

Clearance of BaP from the respiratory tract is slower when it is associated with particles than when it is in a pure form (Stinson and Saffiotti 1983). The major route of BaP excretion is in the feces (Heidelberger and Weiss 1951). The results of a comparative inhalation study by Sun and coworkers

(1982) suggests that most of a BaP aerosol was cleared by absorption into blood followed by biliary and fecal excretion, whereas in animals exposed to particle-associated BaP, a substantial amount of the lung clearance occurred by mucociliary clearance and ingestion. These studies demonstrate the importance of particles to the in vivo fate of inhaled BaP, and they are clearly relevant to evaluating health risks of environmental exposure to PAHs.

Using a similar approach, Sun and co-workers (1983) studied the fate of 1-nitropyrene. In contrast to the results obtained with BaP, no apparent differences in lung retention were observed between pure and particle-associated material. However, the rats exposed to 1-nitropyrene coated on particles excreted the majority of the dose in the feces whereas those exposed to the 1-nitropyrene aerosol excreted most of the dose in the urine. The excretion in the feces is consistent with mucociliary clearance and subsequent ingestion, since 1-nitropyrene and its metabolites administered by gavage are excreted primarily in the feces (El-Bayoumy and Hecht 1984a).

Benzo[a]pyrene

It is now generally accepted that modification of DNA is a key step in the initiation of the carcinogenic process. PAHs, like many other carcinogens, do not themselves react covalently with DNA, but require metabolic activation to reactive species. Thus, they are considered procarcinogens. Metabolites that are on the pathway to reaction with DNA are called proximate carcinogens and those that react with DNA are termed ultimate carcinogens. The latter are electrophiles and are formed as intermediates in the normal response of the organism to foreign compounds, which is generally to convert them to more polar forms that are readily excreted (see LaVoie and Hecht 1981).

Major metabolic transformations of BaP are summarized in figure 2. The metabolism of BaP has been extensively investigated in various systems, and reviews of its metabolic activation and detoxification are available (Gelboin 1980; LaVoie and Hecht

1981; Conney 1982; International Agency for Research on Cancer 1983). The pathway BaP → BaP-7,8-epoxide → BaP-7,8-diol → BaP-7,8-diol-9,10-epoxide is generally considered to be the major activation pathway in BaP-induced tumorigenesis. All other metabolic pathways illustrated in figure 2 are generally thought to be detoxification routes.

These generalizations are useful for considering the effects of BaP in various tissues and species, but it is becoming increasingly clear that they are oversimplifications. There are probably aspects of BaP metabolism other than diol epoxide formation that contribute to its tumorigenic activity. DNA adducts are also formed from metabolites other than the diol epoxides, such as the 4,5-epoxide of 9-hydroxy-BaP, and a number of unidentified BaP/DNA adducts are produced in cultured rat mammary cells in vitro and following direct application of BaP to rat mammary glands in vivo (Phillips et al. 1985). In addition, major unidentified material that elutes rapidly from chromatographic columns and is indicative of unknown DNA adducts has been observed in virtually all studies of BaP/DNA adduct formation.

Factors that Influence Metabolism.

The metabolism of BaP is extraordinarily complex, and alterations in one of the many pathways could affect BaP-induced tumorigenesis. The initial oxidation to arene oxides is controlled by the cytochrome P-450 system. The existence of cytochrome P-450 in multiple forms that differ in their capacity to catalyze oxidation of BaP at different positions is well established (Coon 1981; Conney 1982; Gelboin 1983). The distribution of these forms is dependent on the tissue of interest, the strain and species, and to a great extent on the effects of numerous inducers. Inducibility can be controlled by genetic or environmental factors. Thus it is clear that the production of particular arene oxides and phenolic metabolites of BaP in a given system will depend on a great many contributing factors.

The fate of the arene oxide intermediates is controlled to a large extent by the en-

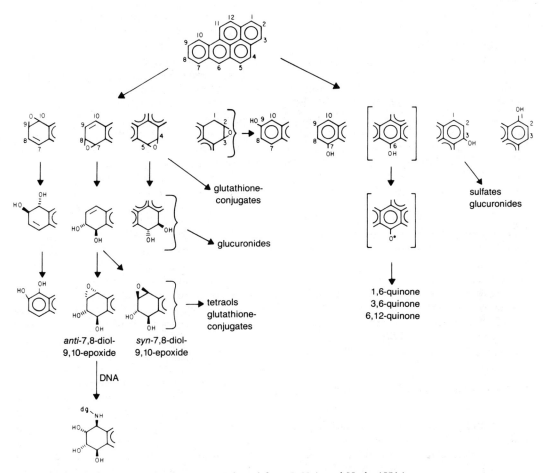

Figure 2. Metabolism of benzo[a]pyrene. (Adapted from LaVoie and Hecht 1981.)

zyme epoxide hydrolase. Its tissue concentrations depend on species, inducer pretreatment, and on the presence of inhibitors such as 1,1,1-trichloropropene oxide (Oesch 1980). Conjugation of phenolic and dihydrodiol metabolites catalyzed by glucuronyl transferases, and detoxification of epoxides and dihydrodiol epoxides by multiple forms of glutathione-S-transferases are, like the cytochrome P-450 and epoxide hydrolase activities, dependent on multiple factors.

The effects on metabolic activation and detoxification pathways of modifiers of BaP tumorigenesis are complex, but they do provide insights on potential mechanisms of cocarcinogenesis or inhibition. Studies of the cocarcinogens catechol, fluoranthene, pyrene, benzo[e]pyrene, and

BHA, taken together with related studies, are for the most part consistent with the concept that formation of 7,8-diol-9,10-epoxides is an important activation pathway in BaP-induced tumorigenesis.

Metabolism in Human Tissues. The metabolism of BaP has been extensively studied in various subcellular, cellular, and organ culture systems from human tissues (International Agency for Research on Cancer 1983). These studies have shown that the basic pattern of BaP metabolism is qualitatively similar in human tissues and in laboratory animal tissues, but major quantitative differences can occur. The formation of DNA adducts via the 7,8-diol-9,10-epoxide pathway is regularly observed in human tissues.

Comparisons among tissues have shown that the highest mean level of DNA adduct formation occurs in the human bronchus, with intermediate levels in the trachea and the esophagus, and the lowest levels in the colon. Major interindividual variations in DNA binding—75-fold in the bronchus, 100-fold in the esophagus, and 135-fold in the colon—have been observed (Autrup et al. 1980). Similar variations have also been observed in studies on levels of aryl hydrocarbon hydroxylase activity (formation of 3-hydroxy-BaP) in human cells and tissues (Gelboin 1983). The large interindividual variations are not surprising in view of the complexity of BaP metabolism and the multiple effects of genetic factors and environmental modifiers as discussed above. A correlation has been observed, however, among DNA binding levels of BaP from tissues of the same individual, suggesting that samples from one tissue might be indicative of metabolism in other tissues (Autrup et al. 1980).

Other Polyaromatic Hydrocarbons

The discovery that 7,8-diol-9,10-epoxide formation is a major activation pathway for BaP (Sims et al. 1974) led to the "bay region hypothesis" of PAH activation as proposed by Jerina and Daly (1977). A "bay region" is the three-sided portion of a PAH delineated by the 4-5 positions of phenanthrene (figure 1). Calculations of electronic distribution predicted that a diol epoxide metabolite with one carbon terminus of the epoxide ring in the bay region would have exceptional reactivity with nucleophiles because of stabilization of the resulting carbonium ion. Thus, it was predicted that chrysene-1,2-diol-3,4-epoxide and benz[a]anthracene-3,4-diol-1,2-epoxide should be major ultimate carcinogens of chrysene and benz[a]anthracene.

Extensive studies have shown that such bay region dihydrodiol epoxides are in fact major ultimate carcinogens of numerous PAHs including chrysene, benz[a]anthracene, dibenz[a,h]anthracene, benzo[c]phenanthrene, 3-methylcholanthrene, dibenzo[a,i]pyrene, and dibenzo[a,h]pyrene (Conney 1982). It has become clear, however, that stereochemical factors are also important in determining whether or not a particular diol epoxide isomer has exceptional tumorigenic activity.

Among the compounds in figure 1 other than BaP, benzo[b]fluoranthene (BbF) is the most potent carcinogenic PAH commonly detected in exhaust emissions. It has a bay region at the 1-12 positions and its 9,10-dihydrodiol has been shown to have tumorigenic activity similar to that of BbF (LaVoie et al. 1982b). However, the 9,10-dihydrodiol does not appear to be formed extensively in rat liver or mouse skin. The 1,2- and 11,12-dihydrodiol metabolites of BbF do not show significant tumorigenicity (Geddie et al. 1987).

Bay region dihydrodiol epoxide metabolites cannot be formed from some of the other commonly encountered tumorigenic PAHs in exhaust emissions, such as fluoranthene, benzo[j]fluoranthene, benzo[k]fluoranthene, cyclopenta[cd]pyrene, and indeno[1,2,3-cd]pyrene. The 9,10- and 8,9-dihydrodiol metabolites of benzo[j]fluoranthene and benzo[k]fluoranthene, respectively, do not show tumorigenic activity greater than that of the parent hydrocarbons, suggesting that other pathways are involved in their activation (LaVoie et al. 1982b). The 3,4-epoxide of cyclopenta[cd]pyrene and 4,5-epoxide of indeno[1,2,3-cd]pyrene seem to be important in their metabolic activation (Wood et al. 1980; Rice et al. 1985). The 2,3-dihydrodiol and 2,3-diol-1,10b-epoxide of fluoranthene are important in its activation to a mutagen (LaVoie et al. 1982c) but the basis for its tumorigenicity in newborn mice is not known.

The bay region hypothesis alone does not explain the exceptional tumorigenicity of 5-methylchrysene (table 2) since all methylchrysenes can form bay region dihydrodiol epoxides. The high tumorigenicity of 5-methylchrysene can, however, be explained by the exceptional tumorigenicity and DNA binding properties of its anti-1,2-diol-3,4-epoxide which has a methyl group and an epoxide ring in the same bay region (Hecht et al. 1985; Melikian et al. 1985). Such bay region dihydrodiol epoxides are the major ultimate carcin-

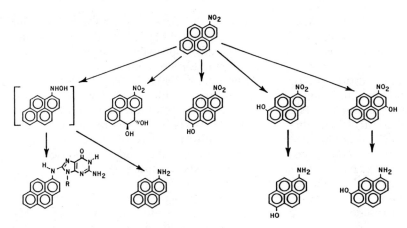

Figure 3. Metabolism of 1-nitropyrene.

ogens of numerous methylated PAHs (Hecht et al. 1988).

Nitro-Polyaromatic Hydrocarbons

1-Nitropyrene. Figure 3 illustrates some principal known metabolic pathways of 1-nitropyrene. The potential complexity of 1-nitropyrene metabolism, compared to PAH metabolism, is greatly increased by the presence of the nitro group. Extensive studies have shown that nitro-reduction is a key feature of 1-nitropyrene metabolism in *Salmonella typhimurium*. A major DNA adduct is formed by reaction of the hydroxylamine with the C-8 position of deoxyguanosine as illustrated in figure 3 (Beland et al. 1985). In the absence of an exogenous activating system, oxidative metabolism of 1-nitropyrene does not occur in *S. typhimurium*.

In mammalian systems the metabolism of 1-nitropyrene is complex. Experiments with 9,000 × *g* supernatant fractions of rat liver, and mouse liver and lung have demonstrated that 1-nitropyrene-3-ol, 1-nitropyrene-6-ol, and 1-nitropyrene-8-ol, as well as 1-nitropyrene-4,5-dihydrodiol, are major metabolites. In contrast to the PAH metabolites, these metabolites are still mutagenic in *S. typhimurium*, probably because of the nitro group, and thus may not represent detoxification pathways (El-Bayoumy and Hecht 1983; El-Bayoumy et al. 1984a). Rats can reduce 1-nitropyrene and these metabolites in vivo, resulting in ex-

cretion in the feces and bile of 1-aminopyrene, 1-amino-6-hydroxypyrene, and 1-amino-8-hydroxypyrene. Glucuronide and sulfate conjugates of these metabolites, as well as conjugates of the 1-nitropyrenols, have been detected in bile and urine (El-Bayoumy and Hecht 1984a). Intestinal microflora are important in the reductive metabolism of 1-nitropyrene in vivo (El-Bayoumy et al. 1984b). These results indicate that the pattern of DNA adduct formation from 1-nitropyrene in vivo should be complex. There has been one report of the in vivo formation in rat mammary tissue of the C-8 adduct shown in figure 3, along with other unknown adducts (Stanton et al. 1985). Further studies are necessary to elucidate the major activation and detoxification pathways of 1-nitropyrene metabolism.

Other Nitro-Polyaromatic Hydrocarbons.
The position of the nitro group and the nature of the ring system has a major influence on the metabolism of nitro-PAHs. Comparative studies of 5-nitroacenaphthene and 1-nitronaphthalene have shown that 5-nitroacenaphthene, which is carcinogenic, undergoes ring oxidation followed by nitro-reduction, but that under identical conditions, 1-nitronaphthalene, which is not carcinogenic, does not undergo significant amounts of nitro-reduction (El-Bayoumy and Hecht 1982a,b). The major metabolite of the potent tumorigen 6-nitrochrysene formed in rat liver in

vitro is 1,2-dihydro-1,2-dihydroxy-6-nitrochrysene. Formation of the corresponding 7,8-dihydrodiol is not observed, indicating that the 6-nitro group inhibits oxidation at the adjacent position (El-Bayoumy and Hecht 1984b). Inhibition of dihydrodiol formation has similarly been observed in studies of 6-nitrobenzo[*a*]pyrene, although an 8,9-dihydrodiol has been detected as a metabolite of 7-nitrobenz-[*a*]anthracene (Fu et al. 1982; Fu and Yang 1983). Nitro-reduction alone does not generally appear to be a major metabolic pathway for either 6-nitrochrysene or 6-nitrobenzo[*a*]pyrene. These results contrast to those observed with 1-nitropyrene (El-Bayoumy and Hecht 1983, 1984b), and with the dinitropyrenes that appear to be activated at least partially by nitro-reduction followed by *O*-acetylation (Beland et al. 1985).

Our present knowledge of the mechanisms of nitro-PAH metabolic activation is inadequate. The general aspects of activation and detoxification are not clearly understood, even for 1-nitropyrene which is the most extensively studied nitro-PAH. Further research is needed in this area. Given the advanced state of technology and experience with other carcinogens, these studies should proceed rapidly.

Research Problems Relating to the Potential Carcinogenic Effects of Polyaromatic Hydrocarbons and Nitro-Polyaromatic Hydrocarbons in Humans

Individual Dosimetry

The wide interindividual variation in PAH metabolism observed in studies on human tissue samples or cells is a result of the extraordinary complexity of the metabolic pathways and the many factors, genetic as well as environmental, that control the levels of the various enzyme activities involved. DNA adduct formation and protein adduct formation are significant end points of these processes because they reflect the generation of electrophilic inter-

mediates in PAH metabolism, and the persistence of PAH/DNA adducts in replicating cells is probably one important determinant for initiation of carcinogenesis. Although extensive studies of DNA adduct and protein adduct formation and persistence after single administrations of various doses of PAHs have been performed, little if any information is available concerning these end points under conditions of chronic PAH treatment (Stowers and Anderson 1985).

It will be essential to perform such studies in animals treated with doses of PAHs known to result in differing tumor incidences and to determine the relationship, if any, between DNA adduct and protein adduct levels and tumor development. These chronic administration experiments are important because they are more closely related to conditions of human exposure than are the acute administration protocols. The availability of sensitive assays for DNA adducts and protein adducts, without using labeled PAHs, will make these chronic studies feasible. The results of these investigations will be important in forming a baseline for interpretation of analogous data obtained from measurements of DNA adducts and protein adducts in humans.

■ **Recommendation 1.** The structures of the major DNA adducts and protein adducts formed from representative PAHs and nitro-PAHs should be determined in laboratory animals.

For dosimetry studies to be undertaken, the structures of appropriate DNA adducts and protein adducts must be known. These studies should initially focus on representative PAHs and nitro-PAHs: BaP, benzo[*b*]fluoranthene, fluoranthene, 1-nitropyrene, 1,6-dinitropyrene, and 6-nitrochrysene are recommended based on current knowledge of their environmental occurrence and carcinogenicity in laboratory animals. This list could change as further data become available. Among these six compounds, the structures of major DNA adducts formed in vivo are known only for BaP. Blood protein adducts are formed via BaP-7,8-diol-9,10-

epoxide (Santella et al. 1986; Shugart 1986). Further studies are required to characterize the DNA adducts and blood protein adducts of the other five compounds.

■ **Recommendation 2.** Methods should be developed for determining individual uptake and metabolic activation of representative PAHs and nitro-PAHs.

Although our understanding of the processes involved in tumor development is incomplete, there is no question that the metabolic generation of specific reactive PAH or nitro-PAH metabolites is one important feature of the process. These metabolites bind to DNA and protein. The measurement of DNA adducts provides a biologically significant end point which bypasses the many variables involved in individual exposure to, uptake of, and metabolism of PAHs. A drawback of DNA adducts as dosimeters is that they are removed from various cells at different rates depending on repair mechanisms and on normal cellular turnover.

In contrast, adducts with proteins such as hemoglobin have a more predictable and longer lifetime. The lifetime of hemoglobin in humans is four months and thus PAH/hemoglobin adducts could provide a measure of chronic exposure (Calleman et al. 1978; Garner 1985). Although such adducts may not be biologically significant per se, they do provide a cumulative measure of individual exposure to PAHs and activation of PAHs to electrophiles. Thus, methods should be developed to measure DNA adducts and protein adducts of the six compounds listed above.

Several methods are available for sensitive detection of PAH/DNA adducts. Immunoassay techniques have been developed and applied to the analyses of BaP/DNA adducts in human tissue samples and in peripheral blood lymphocytes (Harris et al. 1985; Santella et al. 1985). These studies have indicated the presence of these adducts in lung cancer patients, coke oven workers, roofers, and foundry workers, but not in noncancer patients.

Other methods for measuring DNA adducts include synchronous fluorescence spectrophotometry (Harris et al. 1985) and fluorescence-line-narrowed spectra (Heisig et al. 1984). An alternative method that shows great promise for a variety of PAH/DNA adducts is the ^{32}P-postlabeling technique (Randerath et al. 1985). Although some of these techniques are still in the developmental stage and require refinement before being applied routinely to a variety of representative PAHs and nitro-PAHs, they are generally promising.

Methods for assessing formation of PAH/protein adducts are being developed (Santella et al. 1986; Shugart 1986). On the basis of results obtained with 4-aminobiphenyl (Green et al. 1984), the measurement of nitro-PAH/hemoglobin adducts is likely to be feasible. The further development of these methods should certainly be a focus of future research.

In addition to the measurement of PAH/DNA adducts and PAH/protein adducts, there are a number of other methods for assessing carcinogen activation which may be appropriate as an adjunct to the approaches described above. These include the measurement of urinary or fecal metabolites (Becher and Bjorseth 1983), the use of monoclonal antibodies to type human tissues for individual cytochrome P-450s (Gelboin 1983), the detection of antibodies to PAH/DNA adducts (Harris et al. 1985), and the use of blood cells to metabolically activate BaP in vitro (Gelboin 1983).

■ **Recommendation 3.** Under conditions of chronic administration of PAHs or nitro-PAHs to laboratory animals, the relationship between DNA adduct or protein adduct formation and tumor development should be determined.

In order to assess the relationship of PAH or nitro-PAH adducts with DNA or protein to the risk for cancer development, it will be necessary to perform chronic studies in laboratory animals. The most desirable route of administration for such studies would be inhalation, but practical considerations prohibit extensive use of this model. As a compromise, concurrent use of the Syrian golden hamster intratracheal instillation model is recommended. Groups of hamsters should be treated chronically with a range of doses of the six representative compounds listed above with Recommendation 1, as well as mixtures of these compounds.

DNA adducts should be measured in respiratory tract tissue, in cheek pouch or other nonrespiratory tissues, and in peripheral blood lymphocytes. Blood protein adducts should also be quantified. Parallel inhalation studies with selected compounds should be carried out to validate the results obtained by intratracheal instillation. The results should be evaluated in light of the tumor incidence in the various groups. These data will provide the basis for interpreting data that eventually will be obtained from human populations potentially exposed to PAHs and nitro-PAHs.

These measurements in humans, taken together with the data from laboratory animals, should allow determination of individual PAH dose and may provide markers for assessing individual risk for cancer development. Such markers would be invaluable in epidemiologic studies. The lack of data relating levels of DNA adducts and protein adducts to cancer risk in laboratory animals, under conditions of chronic PAH or nitro-PAH treatment, is probably the most significant gap in our present ability to assess risk in humans, given the fact that methods for making these measurements are becoming available.

■ **Recommendation 4.** Pilot studies should be undertaken on individuals potentially exposed to PAHs or nitro-PAHs in order to determine the feasibility of monitoring DNA adducts and protein adducts in humans.

Application to humans is the goal of developing methods for measuring individual dosimetry of PAHs and nitro-PAHs. Therefore, it is essential that the feasibility of these methods be assessed. As mentioned above, studies of this type are already ongoing for some PAH/DNA adducts and aromatic amine/protein adducts. When assays have been shown in animal studies to have the requisite sensitivity, they should be applied to groups of 25–50 individuals who are supposed to have been exposed to PAHs or nitro-PAHs, and to corresponding numbers of controls. The results should be carefully cross-checked with other methods, to ensure their validity. For example, PAH/DNA adducts could be measured by immunoassay and by ^{32}P-postlabeling. If both methods are valid, the results should agree. Multilaboratory collaborative studies should be undertaken for cross checking of results. Parallel assays should also be made to aid in identification of the exposure source, for example, urinary or salivary nicotine or cotinine as monitors for tobacco exposure.

Bioassays in Laboratory Animals

Although extensive evaluations of PAH tumorigenicity have been performed using the mouse skin bioassay system, and structure/activity relationships are fairly well understood, limited data are available on induction of respiratory tumors by PAHs. Assays by intratracheal instillation have been performed with only a few PAHs that are found in mobile source emissions (Sellakumar and Shubik 1974). Several other PAHs have been tested by lung implantation with results not entirely in agreement with expectations based on mouse skin studies (Stanton et al. 1972; Deutsch-Wenzel et al. 1983). Inhalation experiments are practically nonexistent. It is possible that the extensive reliance on mouse skin assays could give a distorted perspective of the potential importance of particular PAHs in respiratory carcinogenesis.

■ **Recommendation 5.** Inhalation bioassays and intratracheal instillation bioassays of selected PAHs should be performed.

Although inhalation experiments on PAHs are necessary, practical considerations demand that they be somewhat limited. Of the PAHs present in mobile source emissions, BaP and benzo[b]fluoranthene are the most tumorigenic on mouse skin. They are also tumorigenic in the rat lung implantation system. It would be important to determine the comparative carcinogenicity of these two hydrocarbons by inhalation studies in rats or Syrian golden hamsters. Fluoranthene and pyrene are the two most prevalent cocarcinogenic hydrocarbons in mobile exhaust emissions, according to mouse skin assays. Their cocarcinogenicity with BaP in inhalation assays should be tested.

Intratracheal instillation bioassays have been performed on only a limited number of PAHs present in mobile source emissions. These studies should be extended at least to some of the more prevalent or carcinogenic components such as anthanthrene, benzo[ghi]fluoranthene, benzo[j]fluoranthene, benzo[ghi]perylene, 5-methylchrysene, cyclopenta[cd]pyrene, fluoranthene, indeno[1,2,3-cd]pyrene, and pyrene. The intratracheal instillation model would probably also be the most suitable for testing mixtures of PAHs as they occur in mobile source emissions.

Bioassays of nitro-PAHs have been limited, and structure/activity relationships are not yet predictable. Extensive systematic studies on nitro-PAH carcinogenicity in several model systems are required. In addition, it is possible that carcinogenic activities of PAH mixtures could differ significantly from expectations based solely on activities of the components of the mixtures. Many examples of cocarcinogenic or inhibitory activities of one PAH upon another are known. It would be important to assess the respiratory carcinogenicity of PAH/nitro-PAH mixtures, using relative concentrations similar to those observed in mobile source emissions.

■ **Recommendation 6.** The tumorigenicity of environmental nitro-PAHs should be evaluated.

The most appropriate system for bioassays of nitro-PAHs is probably intratracheal instillation, at least for screening the activities of the numerous nitro-PAHs that are present in diesel exhaust. Studies with 1,6-dinitropyrene have shown that it does induce lung tumors when applied by intratracheal instillation in hamsters (Takayama et al. 1985). The nitro-PAH showing the greatest activity in other model systems, such as 1,6-dinitropyrene and 6-nitrochrysene, should be chosen for inhalation ex-

periments to complement those already in progress with 1-nitropyrene.

Limited data are available on the effects of modifiers of PAH carcinogenesis in respiratory tract carcinogenesis models. It would be important to determine the effects of such cocarcinogens as catechol, or of dietary inhibitors, on PAH carcinogenesis in the respiratory tract. In addition, only limited data are available on the potential cocarcinogenicity and tumor-promoting activities of compounds to which humans are extensively exposed.

■ **Recommendation 7.** Bioassays should be performed to discover environmental modifiers of PAH and nitro-PAH carcinogenicity.

To study environmental modifiers of PAH and nitro-PAH carcinogenicity, relatively inexpensive assay systems such as mouse skin, mouse forestomach, A/J mouse lung, or the newborn mouse should be used, with BaP, benzo[b]fluoranthene, fluoranthene, 1-nitropyrene, 1,6-dinitropyrene, and 6-nitrochrysene as representative carcinogens. This work should focus on exploring structural analogues of known cocarcinogens, such as catechol, or analogues of known chemopreventive agents such as p-methoxyphenol, indole-3-carbinol, benzyl isothiocyanate, or sodium selenite and related organoselenium compounds. Emphasis should be on those compounds to which humans are exposed in relatively high concentrations.

Mechanisms of Polyaromatic Hydrocarbon and Nitro-Polyaromatic Hydrocarbon Carcinogenesis

Although a great deal is known about the metabolic activation of BaP as a representative PAH, there are many unanswered questions about the process by which exposure to BaP results in tumor induction. It is well established that BaP must undergo enzymatic oxidation to an electrophile that can react with DNA, and that DNA is the key macromolecular target for initiating the tumorigenic process. It is also known that a BaP-7,8-diol-9,10-epoxide is one of the major DNA binding metabolites. However, multiple DNA adducts are formed from BaP and it is not known which of these, or which combination of adducts, is most important in tumor initiation. Other aspects of cellular damage by BaP that may enhance the effects of DNA damage, such as free-radical generation, and their relationship to tumor initiation, remain largely unexplored at present.

The steps by which modified DNA can cause transformation of cells resulting eventually in frank appearance of tumors remain poorly understood, although it is known that BaP-7,8-diol-9,10-epoxide can activate the c-Ha-ras-1 oncogene (Marshall et al. 1984). The role of oncogenes in carcinogenesis is an exciting area of investigation which may provide major leads in understanding the process of tumor development. The effects of BaP metabolites, or of endogenous factors, as cocarcinogens or tumor promoters in BaP carcinogenesis need to be explored.

Whatever gaps exist in our knowledge about BaP carcinogenesis are even greater for most other PAHs. With the possible exception of 7,12-dimethylbenz[a]anthracene (DMBA), no PAH has been so extensively investigated as BaP. Important differences are found in the mechanisms of activation of BaP and DMBA (Dipple et al. 1984). Therefore, further research is needed on the mechanisms of tumor induction by other major carcinogenic PAHs found in mobile source emissions, in particular the benzofluoranthenes.

■ **Recommendation 8.** The mechanisms by which carcinogenic PAHs and nitro-PAHs undergo metabolic activation and detoxification should be determined.

Rational evaluation of the potential carcinogenic effects of PAHs and nitro-PAHs in humans requires a basic understanding of the major pathways of metabolic activation and detoxification of these compounds in laboratory animals. The most important tumorigenic compounds that require further study for elucidation of these basic pathways include the six listed with Recommendation 1 as well as benzo[j]fluoranthene, indeno[1,2,3-cd]pyrene, 1,3-dinitropyrene, and 1,8-dinitropyrene.

These studies should be performed in the species and tissues in which these compounds induce tumors. In general, in vitro experiments are useful for identification of metabolites and evaluation of their role in metabolic activation. However, in vivo studies are essential for relating metabolic pathways to carcinogenesis. In particular, extensive further investigations are necessary on the disposition, metabolism, and DNA binding of PAHs and nitro-PAHs under conditions of inhalation exposure.

Although modifiers of carcinogenesis are important in determining whether or not a PAH or nitro-PAH will induce cancer, research on the mechanisms by which they affect the carcinogenic process is still fairly limited. The studies carried out so far indicate a complex network of effects. It will be important to select one or two environmentally prevalent modifiers of PAH carcinogenesis and to carry out in-depth investigations of their mechanisms of action.

■ **Recommendation 9.** The mechanisms by which environmental compounds modify PAH and nitro-PAH carcinogenicity should be investigated.

Further in vitro and in vivo studies should probe the mechanistic basis for the cocarcinogenic effects of such compounds as fluoranthene, pyrene, and catechol and for the inhibitory effects of certain phenols, isothiocyanates, and indoles. Since these effects can be extraordinarily complex, it is recommended that the studies focus on one or two cocarcinogens, such as fluoranthene and catechol, and one or two appropriate inhibitors such as p-methoxyphenol and indole-3-acetonitrile. These studies should be performed initially with BaP in mice because of the extensive data base that exists on this system.

Summary

Epidemiologic studies have been performed on cohorts exposed to numerous mixtures containing PAHs, including tobacco smoke, coal tars, and soots. Many of these studies have shown that exposure to such mixtures causes cancer at various tissue sites, but individual PAHs have not been specifically incriminated as the causative agents because they occur together with other carcinogens. Nevertheless, these data, taken together with the extensive animal bioassays of PAHs, strongly indicate that PAHs can cause cancer in humans. The role of nitro-PAHs in human cancer is unclear at present.

Studies of PAH metabolism in animal and human tissues have elucidated many of the important pathways of activation and detoxification and have demonstrated that their metabolism is extremely complex. The balance of activation versus detoxification can be influenced by a multitude of genetic and environmental factors. However, it is clear that formation of specific PAH metabolite/DNA adducts is the key step in the initiation of the carcinogenic process. Similar conclusions can be drawn about nitro-PAHs, but their metabolic activation and detoxification pathways are not very well characterized at present.

Summary of Research Recommendations

HIGH PRIORITY

Recommendation 1	The structures of the major DNA adducts and protein adducts formed from representative PAHs and nitro-PAHs should be determined in laboratory animals.
Recommendation 2	Methods should be developed for determining individual uptake and metabolic activation of representative PAHs and nitro-PAHs.
Recommendation 3	Under conditions of chronic administration of PAHs or nitro-PAHs to laboratory animals, the relationship between DNA adduct or protein adduct formation and tumor development should be determined.
Recommendation 4	Pilot studies should be undertaken on individuals potentially exposed to PAHs or nitro-PAHs in order to determine the feasibility of monitoring DNA adducts and protein adducts in humans.
Recommendation 8	The mechanisms by which carcinogenic PAHs and nitro-PAHs undergo metabolic activation and detoxification should be determined.

MEDIUM PRIORITY

Recommendation 5	Inhalation bioassays and intratracheal instillation bioassays of selected PAHs should be performed.
Recommendation 6	The tumorigenicity of environmental nitro-PAHs should be evaluated.

LOW PRIORITY

Recommendation 7	Bioassays should be performed to discover environmental modifiers of PAH and nitro-PAH carcinogenicity.
Recommendation 9	The mechanisms by which environmental compounds modify PAH and nitro-PAH carcinogenicity should be investigated.

Acknowledgments

I thank Ms. Gail Thiede for typing and editing the manuscript. Our studies on PAHs and nitro-PAHs are supported by NCI Grants CA-44377 and CA-35519.

References

Ashurst, S. W., Cohen, G. M., Nesnow, S., DiGiovanni, J., and Slaga, T. J. 1983. Formation of benzo[a]pyrene/DNA adducts and their relationship to tumor initiation in mouse epidermis, Cancer Res. 43:1024–1029.

Autrup, H., Jeffrey, A. M., and Harris, C. C. 1980. Metabolism of benzo[a]pyrene in cultured human bronchus, trachea, colon, and esophagus, In: Polynuclear Aromatic Hydrocarbons: Chemistry and Biological Effects (A. Bjrseth and A. J. Dennis, eds.), pp. 89–105, Battelle Press, Columbus, Ohio.

Becher, G., and Bjrseth, A. 1983. Determination of exposure to polycyclic aromatic hydrocarbons by analysis of human urine, Cancer Lett. 17:301–311.

Beland, F. A., Heflich, R. H., Howard, P. C., and Fu, P. P. 1985. The in vitro metabolic activation of nitro polycyclic aromatic hydrocarbons, In: Polycyclic Hydrocarbons and Carcinogenesis (R. G. Harvey, ed.), pp. 371–396, American Chemical Society, Washington, D.C.

Buening, M. K., Levin, W., Wood, A. W., Chang, R. L., Lehr, R. E., Taylor, C. W., Yagi, H., Jerina, D. M., and Conney, A. H. 1980. Tumorigenic activity of benzo[e]pyrene derivatives on mouse skin and in newborn mice, Cancer Res. 40:203–206.

Busby, W. F., Jr., Garner, R. C., Chow, F. L., Martin, C. N., Stevens, E. K., Newberne, P. M., and Wogan, G. N. 1984. 6-Nitrochrysene is a potent tumorigen in newborn mice, Carcinogenesis 6:801–803.

Calleman, C. J., Ehrenberg, L., Jannsson, B., Osterman-Golkar, S., Segerback, D., Svensson, K., and Wachtmeister, C. A. 1978. Monitoring and risk assessment by means of alkyl groups in hemoglobin in persons occupationally exposed to ethylene oxide, J. Environ. Pathol. Toxicol. 2:427–442.

Carmella, S., LaVoie, E. J., and Hecht, S. S. 1982. Quantitative analysis of catechol and 4-methylcatechol in human urine, Food Chem. Toxicol. 20:587–590.

Conney, A. H. 1982. Induction of microsomal enzymes by foreign chemicals and carcinogenesis by polycyclic aromatic hydrocarbons: GHA Clowes Memorial Lecture, Cancer Res. 42:4875–4917.

Cook, J. W., Hewett, C. L., and Hieger, I. 1933. The isolation of a cancer-producing hydrocarbon from coal tar, J. Chem. Soc. 395–405.

Coon, M. J. 1981. Drug metabolism by cytochrome P-450: progress and perspectives, Drug Metab. Dispos. 9:1–4.

Deutsch-Wenzel, R. P., Brune, H., Grimmer, G., Dettbarn, G., and Misfeld, J. 1983. Experimental studies in rat lungs on the carcinogenicity and dose-response relationships of eight frequently occurring environmental polycyclic aromatic hydrocarbons, J. Nat. Cancer Inst. 71:539–543.

Dipple, A., Moschel, R. C., and Bigger, C. A. H. 1984. Polynuclear aromatic carcinogens, In: Chemical Carcinogens (C. E. Searle, ed.), 2nd ed., pp. 41–174, American Chemical Society, Washington, D. C.

El-Bayoumy, K., and Hecht, S. S. 1982a. Identification of mutagenic metabolites formed by C-hydroxylation and nitroreduction of 5-nitroacenaphthene in rat liver, Cancer Res. 42:1243–1248.

El-Bayoumy, K., and Hecht, S. S. 1982b. Comparative metabolism in vitro of 5-nitroacenaphthene and 1-nitronaphthalene, In: Polynuclear Aromatic Hydrocarbons: Physical and Biological Chemistry (M. Cooke, A. J. Dennis, and G. S. Fisher, eds.), pp. 263–273, Battelle Press, Columbus, Ohio.

El-Bayoumy, K., and Hecht, S. S. 1983. Identification and mutagenicity of metabolites of 1-nitropyrene formed by rat liver, Cancer Res. 43:3132–3137.

El-Bayoumy, K., and Hecht, S. S. 1984a. Metabolism of 1-nitro [U-4,5,9,10-¹⁴C]pyrene in the F344 rat, Cancer Res. 44:4317–4322.

El-Bayoumy, K., and Hecht, S. S. 1984b. Identification of trans-1,2-dihydro-1,2-dihydroxy-6-nitrochrysene, a major mutagenic metabolite of 6-nitrochrysene, Cancer Res. 44:3408–3413.

El-Bayoumy, K., Hecht, S. S., and Hoffmann, D. 1982. Comparative tumor-initiating activity on mouse skin of 6-nitrobenzo[a]pyrene, 6-nitrochrysene, 3-nitroperylene, 1-nitropyrene and their parent hydrocarbons, Cancer Lett. 16:333–337.

El-Bayoumy, K., Hecht, S. S., Sackl, T., and Stoner, G. D. 1984a. Tumorigenicity and metabolism of 1-nitropyrene in A/J mice, Carcinogenesis 5:1449–1452.

El-Bayoumy, K., Reddy, B., and Hecht, S. S. 1984b. Identification of ring oxidized metabolites of 1-nitropyrene in the feces and urine of germfree F344 rats, Carcinogenesis 5:1371–1373.

Fu, P. P., and Yang, S. K. 1983. Stereoselective metabolism of 7-nitrobenz[a]anthracene to 3,4- and 8,9-trans-dihydrodiols, Biochem. Biophys. Res. Commun. 115:123–129.

Fu, P. P., Chou, M. W., Yang, S. K., Beland, F. A., Kadlubar, F. F., Casciano, D. A., Heflich, R. H., and Evans, F. E. 1982. Metabolism of the mutagenic environmental pollutant 6-nitrobenzo[a]pyrene: metabolic activation via ring oxidation, Biochem. Biophys. Res. Commun. 105:1037–1043.

Garner, R. C. 1985. Assessment of carcinogen exposure in man, Carcinogenesis 6:1071–1078.

Geddie, J. E., Amin, S. A., Huie, K., and Hecht, S. S. 1987. Formation and tumorigenicity of benzo[b]-

Correspondence should be addressed to Stephen S. Hecht, Division of Chemical Carcinogenesis, Naylor Dana Institute for Disease Prevention, American Health Foundation, Valhalla, NY 10595.

fluoranthene metabolites in mouse epidermis, *Carcinogenesis* 8:1579–1584.

Gelboin, H. V. 1980. Benzo[*a*]pyrene metabolism, activation, and carcinogenesis: role and regulation of mixed-function oxidases and related enzymes, *Physiol. Rev.* 60:1107–1166.

Gelboin, H. V. 1983. Carcinogens, drugs, and cytochromes P-450, *New Engl. J. Med.* 309:105–107.

Green, L. C., Skipper, P. L., Turesky, R. J., Bryant, M. S., and Tannenbaum, S. R. 1984. In vivo dosimetry of 4-aminobiphenyl in rats via a cysteine adduct in hemoglobin, *Cancer Res.* 44:4254–4259.

Hammond, E. C., and Garfinkel, L. 1980. General air pollution and cancer in the United States, *Prev. Med.* 9:206–211.

Harris, C. C., Vahakangas, K., Newman, M. J., Trivers, G. E., Shamsuddin, A., Sinopoli, N., Mann, D., and Wright, W. E. 1985. Detection of benzo[*a*]pyrene diol epoxide–DNA adducts in peripheral blood lymphocytes and antibodies to the adducts in serum from coke oven workers, *Proc. Nat. Acad. Sci. USA* 82:6672–6676.

Hecht, S. S., Bondinell, W. E., and Hoffmann, D. 1974. Chrysene and methylchrysenes: presence in tobacco smoke and carcinogenicity, *J. Nat. Cancer Inst.* 53:1121–1133.

Hecht, S. S., Carmella, S., Mori, H., and Hoffmann, D. 1981. Role of catechol as a major cocarcinogen in the weakly acidic fraction of smoke condensate, *J. Nat. Cancer Inst.* 66:163–169.

Hecht, S. S., Radok, L., Amin, S., Huie, K., Melikian, A. A., Hoffmann, D., Pataki, J., and Harvey, R. G. 1985. Tumorigenicity of 5-methylchrysene dihydrodiols and dihydrodiol epoxides in newborn mice and on mouse skin. *Cancer Res.* 45:1449–1452.

Hecht, S. S., Melikian, A. A., and Amin, S. 1988. Effects of methyl substitution on the tumorigenicity and metabolic activation of polycyclic aromatic hydrocarbons, In: *Polycyclic Aromatic Hydrocarbons: Structure-Activity Relationships* (B. D. Silverman and S. K. Yang, eds.), in press, CRC Press, Boca Raton, Fla.

Heidelberger, C., and Weiss, S. M. 1951. The distribution of radioactivity in mice following administration of 3,4-benzpyrene-5-C^{14} and 1,2,5,6-dibenzanthracene-9,10-C^{14}, *Cancer Res.* 11:885–891.

Heisig, V., Jeffrey, A. M., McGlade, M. J., and Small, G. J. 1984. Fluorescence-line-narrowed spectra of polycyclic aromatic carcinogen–DNA adducts, *Science* 223:289–291.

Hirao, F., Nishikawa, H., Yoshimoto, T., Sakatani, M., Namba, M., Ogura, T., and Yamamura, Y. 1980. Production of lung cancer and amyloidosis in rabbits by intrabronchial instillation of benzo[*a*]pyrene, *Gann* 71:197–205.

Hirose, M., Lee, M.-S., Wang, C. Y., and King, C. M. 1984. Induction of rat mammary gland tumors by 1-nitropyrene, a recently recognized environmental mutagen, *Cancer Res.* 44:1158–1162.

Hoffmann, D., and Hecht, S. S. 1985. Nicotine-derived *N*-nitrosamines and tobacco related cancer: current status and future directions, *Cancer Res.* 45:935–944.

Hoffmann, D., and Wynder, E. L. 1966. Beitrag zur

carcinogenen Wirkung von Dibenzopyrenen, *Z. Krebsforsch.* 68:137–149.

Hoffmann, D., Rathkamp, G., Newnow, S., and Wynder, E. L. 1972. Fluoranthenes: quantitative determination in cigarette smoke formation by pyrolysis, and tumor-initiating activity, *J. Nat. Cancer Inst.* 49:1165.

Hoffmann, D., Schmeltz, I., Hecht, S. S., and Wynder, E. L. 1978. Tobacco carcinogenesis, In: *Polycyclic Hydrocarbons and Cancer, Vol. 1, Environment, Chemistry and Metabolism* (H. V. Gelboin and P. O. P. Ts'o, eds.), pp. 85–117, Academic Press, New York.

Imaida, K., Hirose, M., Lee, M.-S., Wang, C. Y., and King, C. 1985. Comparative carcinogenicities of 1-, 2-, and 4-nitropyrenes (NP) and structurally related compounds for female CD rats following intraperitoneal injection, *Proc. Am. Assoc. Cancer Res.* 26:93.

International Agency for Research on Cancer. 1983. Chemical, environmental and experimental data, In: *IARC Monographs on the Evaluation of the Carcinogenic Risk of Chemicals to Humans, Vol. 32, Polynuclear Aromatic Compounds*, pt. 1, pp. 13–451, IARC, Lyon, France.

International Agency for Research on Cancer. 1984a. Carbon blacks, mineral oils, and some nitroarenes, In: *IARC Monographs on the Evaluation of the Carcinogenic Risk of Chemicals to Humans, Vol. 33, Polynuclear Aromatic Compounds*, pt. 2, pp. 87–168, IARC, Lyon, France.

International Agency for Research on Cancer. 1984b. Industrial exposures to aluminium production, coal gasification, coke production, and iron and steel founding, In: *IARC Monographs on the Evaluation of the Carcinogenic Risk of Chemicals to Humans, Vol. 34, Polynuclear Aromatic Compounds*, pt. 3, pp. 37–64, 133–190, IARC, Lyon, France.

International Agency for Research on Cancer. 1984c. Industrial exposures to aluminium production, coal gasification, coke production, and iron and steel founding, In: *IARC Monographs on the Evaluation of the Carcinogenic Risk of Chemicals to Humans, Vol. 34, Polynuclear Aromatic Compounds*, pt. 3, pp. 65–131, IARC, Lyon, France.

International Agency for Research on Cancer. 1985a. Bitumens, coal-tars and derived products, shale-oils and soots, In: *IARC Monographs on the Evaluation of the Carcinogenic Risk of Chemicals to Humans, Vol. 35, Polynuclear Aromatic Compounds*, pt. 4, pp. 83–159, IARC, Lyon, France.

International Agency for Research on Cancer. 1985b. Bitumens, coal-tars and derived products, shale-oils and soots, In: *IARC Monographs on the Evaluation of the Carcinogenic Risk of Chemicals to Humans, Vol. 35, Polynuclear Aromatic Compounds*, pt. 4, pp. 161–217, IARC, Lyon, France.

International Agency for Research on Cancer. 1985c. Bitumens, coal-tars and derived products, shale-oils and soots, In: *IARC Monographs on the Evaluation of the Carcinogenic Risk of Chemicals to Humans, Vol. 35, Polynuclear Aromatic Compounds*, pt. 4, pp. 219–241, IARC, Lyon, France.

International Agency for Research on Cancer. 1986.

IARC Monographs on the Evaluation of the Carcinogenic Risk of Chemicals to Humans, Vol. 38, Tobacco Smoking, pp. 199–314, Lyon, France.

Ito, N., Fukushima, S., and Tsuda, H. 1985. Carcinogenicity and modification of the carcinogenic response by BHA, BHT and other antioxidants, *CRC Crit. Rev. Toxicol.* 15(2):109–150.

Jerina, D. M., and Daly, J. W. 1977. Oxidation at carbon, In: *Drug Metabolism—from Microbe to Man* (D. V. Parke and R. L. Smith, eds.), pp. 13–32, Taylor and Francis, London.

Kendrick, J., Nettesheim, P., and Hammons, A. S. 1974. Tumor induction in tracheal grafts: a new experimental model for respiratory carcinogenesis studies, *J. Nat. Cancer Inst.* 52:1317–1325.

Kennaway, E. L., and Hieger, I. 1930. Carcinogenic substances and their fluorescence spectra, *Br. Med. J.* 1:1044–1046.

Ketkar, M., Green, U., Schneider, P., and Mohr, U. 1979. Investigations on the carcinogenic burden by air pollution in man. Intratracheal instillation studies with benzo[a]pyrene in a mixture of Tris buffer and saline in Syrian golden hamsters, *Cancer Lett.* 6:279–284.

Laskin, S., Kuschner, M., and Drew, R. T. 1970. Studies in pulmonary cocarcinogenesis, In: *Inhalation Carcinogenesis* (M. G. Hanna, Jr., P. Nettesheim, and J. R. Gilbert, eds.), pp. 321–351, Division of Technical Information, Oak Ridge, Tenn.

LaVoie, E. J., and Hecht, S. S. 1981. Chemical carcinogens: in vitro metabolism and activation, In: *Hazard Assessment of Chemicals, Current Developments* (J. Saxena and F. Fisher, eds.), Vol. 1, pp. 155–249, Academic Press, New York.

LaVoie, E. J., Bedenko, V., Hirota, N., Hecht, S. S., and Hoffmann, D. 1979. A comparison of the mutagenicity, tumor-initiating activity and complete carcinogenicity of polynuclear aromatic hydrocarbons, In: *Polynuclear Aromatic Hydrocarbons* (P. W. Jones and P. Leber, eds.), pp. 705–721, Ann Arbor Science Publishers, Ann Arbor, Mich.

LaVoie, E. J., Tulley-Freiler, L., Bedenko, V., Girach, Z., and Hoffmann, D. 1981a. Comparative studies on the tumor-initiating activity and metabolism of methylfluorenes and methylbenzofluorenes, In: *Chemical Analysis and Biological Fate: Polynuclear Aromatic Hydrocarbons, Fifth International Symposium* (M. Cooke and A. J. Dennis, eds.), pp. 417–427, Battelle Press, Columbus, Ohio.

LaVoie, E. J., Tulley-Freiler, L., Bedenko, V., and Hoffmann, D. 1981b. Mutagenicity, tumor-initiating activity and metabolism of methylphenanthrenes, *Cancer Res.* 41:3441–3447.

LaVoie, E. J., Bedenko, V., Tulley-Freiler, L., and Hoffmann, D. 1982a. Tumor-initiating activity and metabolism of polymethylated phenanthrenes, *Cancer Res.* 42:4045–4049.

LaVoie, E. J., Amin, S., Hecht, S. S., Furuya, K., and Hoffmann, D. 1982b. Tumor-initiating activity of dihydrodiols of benzo[b]fluoranthene, benzo[j]fluoranthene and benzo[k]fluoranthene, *Carcinogenesis* 3:49–52.

LaVoie, E. J., Hecht, S. S., Bedenko, V., and Hoff-

mann, D. 1982c. Identification of the mutagenic metabolites of fluoranthene, 2-methylfluoranthene and 3-methylfluoranthene, *Carcinogenesis* 3:841–846.

Marshall, C. J., Vousden, K. H., and Phillips, D. H. 1984. Activation of C-Ha-ras-1 proto-oncogene by in vitro modification with a chemical carcinogen, benzo[a]pyrene diol-epoxide, *Nature* 310:586–589.

Melikian, A. A., Leszczynska, J. M., Amin, S., Hecht, S. S., Hoffmann, D., Pataki, J., and Harvey, R. G. 1985. Rates of hydrolysis and extents of DNA binding of 5-methylchrysene dihydrodiol epoxides, *Cancer Res.* 45:1990–1996.

Miller, J. A., Sandin, R. B., Miller, E. C., and Rusch, H. P. 1955. The carcinogenicity of compounds related to 2-acetylaminofluorene. II. Variation in the bridges and the 2-substituent, *Cancer Res.* 15:188–199.

National Cancer Institute. 1978. Bioassay of 5-Nitroacenaphthene for Possible Carcinogenicity, p. 115, Report No. DHEW/PUB/NIH-78-1373, NCI-CG-TR-118, Bethesda, Md.

National Research Council. 1981. *Health Effects of Exposure to Diesel Exhaust: Impacts of Diesel-Powered Light-Duty Vehicles,* pp. 137–138, National Academy Press, Washington, D.C.

National Research Council. 1982. *Diet, Nutrition, and Cancer,* pp. 73–105, National Academy Press, Washington, D.C.

Nebert, D. W. 1981. Genetic differences in susceptibility to chemically induced myelotoxicity and leukemia, *Environ. Health Perspect.* 39:11–22.

Nesnow, S., Triplett, L. L., and Slaga, T. J. 1984. Tumor initiating activities of 1-nitropyrene and its nitrated products in Sencar mice, *Cancer Lett.* 23:1–8.

Oesch, F. 1980. Species differences in activating and inactivating enzymes related to in vitro mutagenicity mediated by tissue preparations from these species, *Arch. Toxicol. Suppl.* 3:179–194.

Ohgaki, H., Negishi, C., Wakabayashi, K., Kusama, K., Sato, S., and Sugimura, T. 1984. Induction of sarcomas in rats by subcutaneous injection of dinitropyrenes, *Carcinogenesis* 5:583–585.

Ohgaki, H., Hasegawa, H., Kato, T., Negishi, C., Sato, S., and Sugimura, T. 1985. Absence of carcinogenicity of 1-nitropyrene, correction of previous results, and new demonstration of carcinogenicity in rats, *Cancer Lett.* 25:239–245.

Phillips, D. H., Hewer, A., and Grover, P. L. 1985. Aberrant activation of benzo[a]pyrene in cultured rat mammary cells in vitro and following direct application to rat mammary gland in vivo, *Cancer Res.* 45:4167–4174.

Pott, P., 1775. Chirurgical observations, reprinted in *Nat. Cancer Inst. Monogr.* 1963(10):7–13.

Randerath, K., Randerath, E., Agrawal, H. P., Gupta, R. C., Schurdak, M. E., and Reddy, M. V. 1985. Postlabelling methods for carcinogen–DNA adducts, *Environ. Health Perspect.* 62:57–66.

Redmond, C. K. 1983. Cancer mortality among coke oven workers, *Environ. Health Perspect.* 52:67–73.

Rice, J. E., Coleman, D. T., Hosted, T. J., Jr., LaVoie, E. J., McCaustland, D. J., and Wiley, J. C.,

Jr. 1985. Identification of mutagenic metabolites of indeno[1,2,3-*cd*]pyrene formed in vitro with rat liver enzymes, *Cancer Res.* 45:5421–5425.

Ryan, D. E., Iida, S., Wood, A. W., Thomas, P. E., Lieber, C. S., and Levin, W. 1984. Characterization of three highly purified cytochromes P-450 from hepatic microsomes of adult male rats, *J. Biol. Chem.* 259:1239–1250.

Saffiotti, U., and Shubik, P. 1956. The effects of low concentrations of carcinogen in epidermal carcinogenesis: a comparison with promoting agents, *J. Nat. Cancer Inst.* 16:961–969.

Saffiotti, U., Montesano, R., Sellakumar, A. R., Cefis, F., and Kaufman, D. G. 1972a. Respiratory tract carcinogenesis in hamsters induced by different numbers of administrations of benzo[*a*]pyrene and ferric oxide, *Cancer Res.* 32:1073–1081.

Saffiotti, U., Montesano, R., Sellakumar, A. R., and Kaufman, D. G. 1972b. Respiratory tract carcinogenesis induced in hamsters by different dose levels of benzo[*a*]pyrene and ferric oxide, *J. Nat. Cancer Inst.* 49:1199–1204.

Santella, R. M., Hsieh, L. L., Lin, C. D., Viet, S., and Weinstein, I. B. 1985. Quantitation of exposure to benzo[*a*]pyrene with monoclonal antibodies, *Environ. Health Perspect.* 62:95–100.

Santella, R. M., Lin, C. D., and Dharmaraja, N. 1986. Monoclonal antibodies to benzo[*a*]pyrene diol-epoxide modified protein, *Carcinogenesis* 7:441–444.

Schuetzle, D. 1983. Sampling of vehicle emissions for chemical analysis and biological testing, *Environ. Health Perspect.* 47:65–80.

Scribner, J. D. 1973. Tumor initiation by apparently noncarcinogenic polycyclic aromatic hydrocarbons, *J. Nat. Cancer Inst.* 50:1717–1719.

Sellakumar, A., and Shubik, P. 1974. Carcinogenicity of different polycyclic hydrocarbons in the respiratory tract of hamsters, *J. Nat. Cancer Inst.* 53:1713–1716.

Shugart, L. 1986. Quantifying adductive modification of hemoglobin from mice exposed to benzo[*a*]pyrene, *Anal. Biochem.* 152:365–369.

Silverman, D. T., Hoover, R. N., Albert, S., and Graff, K. M. 1983. Occupation and cancer of the lower urinary tract in Detroit, *J. Nat. Cancer Inst.* 70:237–254.

Silverman, D. T., Hoover, R. N., Mason, T. J., and Swanson, G. M. 1986. Motor exhaust–related occupations and bladder cancer, *Cancer Res.* 46:2113–2116.

Sims, P., Grover, P. L., Swaisland, A., Pal, K., and Hewer, A. 1974. Metabolic activation of benzo[*a*]pyrene proceeds by a diol-epoxide, *Nature (London)* 252:326–328.

Slaga, T. J., and DiGiovanni, J. 1984. Inhibition of chemical carcinogenesis, In: *Chemical Carcinogens* (C. E. Searle, ed.), 2nd ed., pp. 1279–1321, American Chemical Society, Washington, D.C.

Stanton, C. A., Chow, F. L., Phillips, D. H., Grover, P. L., Garner, R. C., and Martin, C. N. 1985. Evidence for *N*-(deoxyguanosin-8-yl)-1-aminopyrene as a major DNA adduct in female rats treated with 1-nitropyrene, *Carcinogenesis* 6:535–538.

Stanton, M. F., Miller, E., Wrench, C., and Blackwell, K. 1972. Experimental induction of epidermoid carcinoma in the lungs of rats by cigarette smoke condensate, *J. Nat. Cancer Inst.* 49:867–873.

Stenbäck, F., Peto, R., and Shubik, P. 1981. Initiation and promotion at different ages and doses in 2200 mice. III. Linear extrapolation from high doses may underestimate low-dose tumour risks, *Br. J. Cancer* 44:24–31.

Stinson, S. F., and Saffiotti, V. 1983. Experimental respiratory carcinogenesis with polycyclic aromatic hydrocarbons, In: *Comparative Respiratory Tract Carcinogenesis* (H. M. Reznik-Schüller, ed.), pp. 75–93, CRC Press, Boca Raton, Fla.

Stowers, S. J., and Anderson, M. W. 1985. Formation and persistence of benzo[*a*]pyrene metabolite–DNA adducts, *Environ. Health Perspect.* 62:31–40.

Sun, J. D., Wolff, R. K., and Kanapilly, G. M. 1982. Deposition, retention, and biological fate of inhaled benzo[*a*]pyrene absorbed onto ultrafine particles and as a pure aerosol, *Toxicol. Appl. Pharmacol.* 65:231–244.

Sun, J. D., Wolff R. K., Aberman, H. M., and McClellan, R. O. 1983. Inhalation of 1-nitropyrene associated with ultrafine insoluble particles or as a pure aerosol: a comparison of deposition and biological fate, *Toxicol. Appl. Pharmacol.* 69:185–198.

Takayama, S., Ishikawa, T., Nakajima, H., and Sato, S. 1985. Lung carcinoma induction in Syrian golden hamsters by intratracheal instillation of 1,6-dinitropyrene, *Jpn. J. Cancer Res. (Gann)* 76:457–461.

Takemura, N., Hashida, C., and Terasawa, M. 1974. Carcinogenic action of 5-nitroacenaphthene, *Br. J. Cancer* 30:481–483.

Thyssen, J., Althoff, J., Kimmerle, G., and Mohr, U. 1981. Inhalation studies with benzo[*a*]pyrene in Syrian golden hamsters, *J. Nat. Cancer Inst.* 66:575–577.

Tokiwa, H., Otofuji, T., Horikawa, K., Kitamori, S., Otsuka, H., Manabe, Y., Kinouchi, T., and Ohnishi, Y. 1984. 1,6-Dinitropyrene: mutagenicity in *Salmonella* and carcinogenicity in BALB/c mice, *J. Nat. Cancer Inst.* 73:1359–1363.

Topping, D. C., Martin, D. H., and Nettesheim, P. 1981. Determination of cocarcinogenic activity of benzo[*e*]pyrene for respiratory tract mucosa, *Cancer Lett.* 11:315–321.

U.S. Department of Health and Human Services. 1982a. The Health Consequences of Smoking: Cancer, A Report of the Surgeon General, pp. v–viii, U.S. Government Printing Office, Washington D.C.

U.S. Department of Health and Human Services. 1982b. The Health Consequences of Smoking: Cancer, A Report of the Surgeon General, pp. 195–215, U.S. Government Printing Office, Washington D.C.

Van Duuren, B. L., and Goldschmidt, B. M. 1976. Cocarcinogenic and tumor-promoting agents in tobacco carcinogenesis, *J. Nat. Cancer Inst.* 56:1237–1242.

Van Duuren, B. L., Sivak, A., Langseth, L., Goldschmidt, B. M., and Segal, A. 1968. Initiators and promoters in tobacco carcinogenesis, *Nat. Cancer Inst. Monogr.* 28:173–180.

Van Duuren, B. L., Sivak, A., Katz, C., Seidmann, G., and Melchionne, S. 1975. The effect of aging and interval between primary and secondary treatment on two-stage carcinogenesis on mouse skin, *Cancer Res.* 35:502–505.

Wattenberg, L. W. 1985. Chemoprevention of cancer, *Cancer Res.* 45:1–8.

Wislocki, P. G., Bagan, E. S., Lu, A. Y. H., Dooley, K. L., Fu, P. P., Han-Hsu, H., Beland, F. A., and Kadlubar, F. F. 1985. Carcinogenicity of nitro polycyclic aromatic hydrocarbons in the newborn mouse liver and lung, *Proc. Am. Assoc. Cancer Res.* 26:92.

Wood, A. W., Levin, W., Chang, R. L., Lehr, R. E., Schaffer-Ridder, M., Karle, J. M., Jerina, D. M., and Conney, A. H. 1977. Tumorigenicity of five dihydrodiols of benz[*a*]anthracene on mouse skin:

exceptional activity of benz[*a*]anthracene-3,4-dihydrodiol, *Proc. Nat. Acad. Sci. USA* 74:3176–3179.

Wood, A. W., Levin, W., Chang, R. L., Huang, M.-T., Ryan, D. E., Thomas, P. E., Lehr, R. E., Kumar, S., Koreeda, M., Akagi, H., Ittah, Y., Dansette, P., Yagi, H., Jerina, D. M., and Conney, A. H. 1980. Mutagenicity and tumor-initiating activity of cyclopenta[*cd*]pyrene and structurally related compounds, *Cancer Res.* 40:642–649.

Wynder, E. L., and Hoffmann, D. 1959. The carcinogenicity of benzofluoranthenes, *Cancer* 12:1194–1199.

Wynder, E. L., and Hoffmann, D. 1979. Tobacco and health. A societal challenge, *New Engl. J. Med.* 300:894–903.

Yoshimoto, T., Hirao, F., Sakatani, M., Nishikawa, H., Ogura, T., and Yamamura, Y. 1977. Induction of squamous cell carcinoma in the lung of C57BL/6 mice by intratracheal instillations of benzo[*a*]pyrene with charcoal powder, *Gann* 68:343–352.

Health Effects of Aldehydes and Alcohols in Mobile Source Emissions

LAWRENCE J. MARNETT
Wayne State University

Air Pollution, the Automobile, and Public Health. © 1988 by the Health Effects
Institute. National Academy Press, Washington, D.C.

Aldehydes are oxidation products of alcohols; phenols contain an alcohol functionality attached to an aromatic ring. Although they are structurally related, the chemistry and toxicology of the three classes of compounds are different. From a toxicologic standpoint, aldehydes have been more extensively investigated than alcohols and phenols and constitute the most important health hazard. Thus, most of the emphasis of this chapter is placed on them, but the literature on exposure, health effects, metabolism, and chemistry of aldehydes, alcohols, and phenols are also examined. Recommendations are made for research necessary to fill important gaps in our knowledge of these subjects. The literature review highlights key experiments but is not comprehensive; recent review articles and monographs are cited and can be consulted for more complete information.

Ambient Levels and Production by Mobile Sources

Estimates of the atmospheric levels of some common pollutants present in mobile source emissions and the series of compounds that are discussed in this chapter are presented in table 1. Formaldehyde and acetaldehyde have been monitored more extensively than acrolein, alcohols, and phenols. The data for the latter three compounds represent episodic reports rather than averages of extensive compilations. Nevertheless, they are useful for assigning an order of magnitude to the levels likely to be found in urban air. Formaldehyde is usually the most abundant of the compounds of interest and acetaldehyde is next most abundant. Their concentrations are between 100 and 1,000 times lower than carbon monoxide (CO), the principal pollutant in auto exhaust. Although ambient formaldehyde levels range from 4 to 86 parts per billion (ppb), occasional levels in excess of 1,000 ppb—1 part per million (ppm)—have been reported (Goldsmith and Friberg 1977). Acrolein is usually detected at levels below 10 ppb. Most of the aldehydes in urban air are present as gases. Estimates suggest the percentage of aldehydes bound to particles is approximately 1 percent of the concentration in the gas phase (Grosjean 1982). Acrolein is also produced in fires, but by far the highest concentrations are present in cigarette smoke (12 ppm) (Treitman et al. 1980; Carson et al. 1981). Acetaldehyde is also present in very high concentrations in cigarette smoke (1,650–2,500 ppm) (Elmenhorst and Schultz 1968); the formaldehyde concentration in cigarette smoke is approximately equal to that of acrolein (Newsome et al. 1965).

Alcohol concentrations in the atmosphere have only been determined in a cursory fashion. In a recent study in urban and rural sampling stations in Arizona, Snider and Dawson (1985) reported values of 7.9 ppb for methanol in Tucson and 2.6

Table 1. Atmospheric Levels of Compounds Present in Mobile Source Emissions

Compound	Level	% from Mobile Source	Reference
Carbon monoxide	5–13 ppm	70	Graedel, this volume
Nitrogen oxides	20–30 ppb	50	Graedel, this volume
Nonmethane hydrocarbons	1–2 ppmC	37	Graedel, this volume
Sulfur dioxide	9 ppb	<5	Graedel, this volume
Formaldehyde	4–86 ppb[a]	55–75	Grosjean (1982)
Acetaldehyde	2–39 ppb[a]	55–75	Hoshika (1977)
			Grosjean (1982)
Acrolein	2–7 ppb	NA	Beauchamp et al. (1985)
Alcohols	8–100 ppb	NA	Snider and Dawson (1985)
Phenols	NA	NA	

[a] Los Angeles.
NOTE: NA = not available.

Table 2. Concentrations of Formaldehyde and Acetaldehyde in Exhaust Diluted Gases of Internal Combustion Engines Fueled by Gasoline or Ethanol[a]

	Gasoline-Fueled[b]		Ethanol-Fueled	
	Formaldehyde	Acetaldehyde	Formaldehyde	Acetaldehyde
Cold start[c]	540	<8	670	19,800
Hot start[c]	97	<8	100	550
40-mph cruise	48	<8	360	840

[a] Concentrations are given in ppb, and dilution factor is 10:1.
[b] Equipped with a catalyst.
[c] Federal Test Procedure driving cycle.
SOURCE: Based on data from Swarin and Lipari 1983.

ppb at a location 50 km from Tucson. Atmospheric sampling was by a condensation procedure. The concentrations of ethanol at the two stations were reported to be 3.3 and 0.4 ppb, respectively. The origin of the alcohols was not clear. The authors were unable to detect propanol, butanol, or acrolein. Phenol, as well as o- and m-cresol, have been reported in undiluted auto exhaust at levels of 1.4, 0.2, and 0.3 ppm, respectively (Kuwata et al. 1981).

The ambient levels of aldehydes reported to exist in Los Angeles appear to be higher than levels in other cities in the United States and Japan. Therefore, upper limits can be approximated by studies performed in the Los Angeles metropolitan area. Data gathered at a reporting station in Claremont (~50 km east of Los Angeles) indicate that diurnal patterns exist for formaldehyde and acetaldehyde levels (Grosjean 1982). There is a close correlation between variations in the aldehyde levels and the diurnal fluctuations of ozone (O_3) as well as the movement of smog banks. Other reports establish a correlation of fluctuations in aldehyde levels to diurnal variations in CO levels (Cleveland et al. 1977). Diurnal and seasonal fluctuations of formaldehyde and acetaldehyde levels have been observed at a monitoring station on Long Island (Tanner and Meng 1984). These data suggest that mobile sources contribute significantly to ambient concentrations. Estimates of the percentage contribution made by mobile emission sources to the levels of the various compounds are presented in table 1. Such estimates have not been made for alcohols and phenols.

Analysis of the levels of aldehydes in exhaust gases provides evidence for a dependence on the type of engine and fuel (Swarin and Lipari 1983). Table 2 lists the concentrations of formaldehyde and acetaldehyde detected in a 10:1-diluted sample of exhaust gas from an internal combustion engine fueled by gasoline or ethanol. Dramatic increases in acetaldehyde concentrations are detected in ethanol-fueled engine exhaust, as might be expected since acetaldehyde is the two-electron oxidation product of ethanol. Presumably a similar increase in formaldehyde concentrations would be observed in exhaust from engines fueled by methanol.

Diesel engines produce significantly higher amounts of aldehydes than do gasoline engines (table 3). Ratios of acetaldehyde to formaldehyde also increase if catalysts are placed in the exhaust stream (Swarin and Lipari 1983). Taken together, these findings suggest that the character of automotive emissions varies dramatically with the type of engine and fuel. This

Table 3. Concentrations of Selected Aldehydes in Diluted Diesel Exhaust[a]

Aldehyde	Cold Start[b]	Hot Start[b]
Formaldehyde	539	428
Acetaldehyde	115	80
Propionaldehyde	57	24
Acrolein	57	24
Crotonaldehyde	11	6
Benzaldehyde	9	NA

[a] Concentrations are given in ppb, and dilution factor is 10:1.
[b] Federal Test Procedure driving cycle.
NOTE: NA = not available.
SOURCE: Based on data from Swarin and Lipari 1983.

implies that a major shift to, for example, alcohol-containing fuels would have a significant effect on the concentration of certain aldehydes and alcohols in urban air.

The levels of alcohols in urban air are rather low, and it is difficult to imagine that they are high enough to exert any health effects. Levels of aldehydes are normally well below levels at which they induce hazardous effects, although they can occasionally reach high ambient concentrations (Grosjean 1982; Beauchamp et al. 1985). Results of test burns indicate that dramatically increased levels of aldehydes are produced from alcohol-containing fuels (Swarin and Lipari 1983). These higher levels could be well within the range that induces adverse health effects. If methanol- and ethanol-based fuels are widely adapted, it will be important to be able to determine the levels of atmospheric aldehydes and to have a baseline value against which to compare them. Thus, routine monitoring should be initiated now.

■ **Recommendation 1.** Routine monitoring of atmospheric alcohol and aldehyde levels should be performed in regions where alcohol-based fuels are or will be in heavy use.

Reports of the detection of phenols and catechols from mobile source emissions are extremely limited and usually do not provide quantitative information. As a result, knowledge of their levels in urban air and the contribution made by mobile sources is totally inadequate. No realistic risk assessment can be undertaken for this class of compounds without such information.

■ **Recommendation 2.** Methods should be developed to routinely analyze phenols and catechols in urban air.

Health Effects

Aldehydes

The literature on the health effects of aldehydes and alcohols is enormous, but several recent reviews by Consensus Work-

Table 4. Acute Effects of Acrolein on Human Volunteers

Concentration (ppb)	Effects
30–34	Odor threshold for most acrolein-sensitive people.
140–150	Some eye irritation in 2 min; increased annoyance and almost no eye or nose irritation during repeated exposures.
300–500	Slight eye and nose irritation; no effect on respiratory frequency or amplitude; odor perceived.
800–900	Changes in amplitude of respiratory movements; slightly increased respiratory frequency; decreased eye sensitivity to light.
1,200	Extremely irritating to all mucous membranes in 5 min; lacrimation.
1,000–23,000	Medium to severe eye irritation in 5 min.

SOURCE: Based on data from Beauchamp et al. 1985.

shop on Formaldehyde (1984), Beauchamp et al. (1985), and Tephly (1985) are particularly appropriate. The present discussion is restricted primarily to inhalation toxicology.

Acute Effects in Humans. Acute irritant effects of aldehydes on human volunteers have been documented. In general, acrolein is the most potent of the series acrolein, formaldehyde, acetaldehyde, crotonaldehyde. For example, acrolein is approximately two to three times more potent than formaldehyde as an irritant (Beauchamp et al. 1985). Table 4 is a compilation of the doses at which various short-term responses to acrolein have occurred in human volunteers (Carson et al. 1981). Ocular and olfactory irritation is the first detectable response and occurs at doses that are 10–20 times higher than the ambient levels in urban air (table 1). Extreme irritation to mucous membranes and alteration in respiration occur at doses approximately 100 times ambient. At such levels, it is likely that irreversible epithelial damage occurs on chronic exposure (see Acute Effects in Rodents, below). A similar profile of effects is observed for formaldehyde in humans at somewhat higher doses. Irritation occurs at 0.1–3.0 ppm, and respiratory difficulties are evident at 10–20 ppm (Fassett 1963). Acetaldehyde and crotonaldehyde are 10–100 times less active than acrolein and formaldehyde.

Allergic responses to aldehydes have been reported. Hendrick and Lane (1975) documented a case of asthma induced by exposure of a hospital staff member to formalin vapor. A pronounced decrease in respiratory performance was observed after exposure to a 25 percent solution for 15 min; however, the ambient levels of formaldehyde present were not measured. Symptoms were prevented by pretreatment of the patient with betamethasone. Skin allergies have been induced by topical application of solutions of formaldehyde but the dose responses have not been extensively determined (Maibach 1983). Dermal but not respiratory sensitivity has been observed in guinea pigs exposed to 10 ppm formaldehyde for 6 or 8 hr/day for 5 consecutive days. (Lee et al. 1984).

Chronic Effects in Humans. Numerous groups of individuals are occupationally exposed to formaldehyde, acetaldehyde, acrolein, and crotonaldehyde. Epidemiologic studies of the chronic effects of formaldehyde have been conducted with several of these groups, but the results are inconclusive. Despite the ability to identify exposed individuals, there is little information on their smoking and drinking habits, which confounds the interpretation of any detected alterations in disease incidence. The Consensus Workshop on Formaldehyde (1984) evaluated several epidemiologic studies of professional and industrial workers exposed to formaldehyde and concluded that there are insufficient data to establish whether or not it is a human carcinogen. The level of atmospheric exposure in those workers was approximately 0.1–1.0 ppm. In the same groups, there appeared to be no excess mortality associated with formaldehyde exposure. Chronic exposure of humans to high levels of acrolein is considered unlikely because of its extreme irritation. At levels below those that cause olfactory or respiratory damage

(~1 ppm), prolonged exposure to acrolein is intolerable, causing individuals to leave the contaminated environment. Consequently, there is no information on human carcinogenicity or other chronic effects of acrolein, nor are data on potential human carcinogenicity of acetaldehyde or crotonaldehyde available.

Acute and Chronic Effects in Rodents. Pathological changes occur in the upper respiratory, especially nasal, epithelium of rodents exposed to aldehydes. The site and severity are dose dependent (Kutzman et al. 1985). Acute effects have also been noted, and the lesions observed include exfoliation, ciliastasis, cell erosion, ulceration, necrosis, squamous metaplasia, and inflammation (Dalhamn and Rosengren 1971; Buckley et al. 1984). Most of the damage is reversible but some is irreversible. Significantly, these effects are detected when rodents are exposed to the RD_{50}s for formaldehyde and acrolein (Buckley et al. 1984). (The RD_{50} is defined as the level at which a 50 percent reduction in respiratory rate occurs. This level reflects the stimulation of sensory receptors that attempt to limit exposure to irritants.) The RD_{50}s for formaldehyde and acrolein are 3.1 and 1.7 ppm in mice (Buckley et al. 1984); the RD_{50} for acrolein is 6.0 ppm in rats (Babiuk et al. 1985). Since pathological changes occur in experimental animals as a result of exposure to the RD_{50} levels, proposals have been made that the RD_{50} be used to estimate "safe" exposure levels for humans (possibly $0.01–0.1 \times RD_{50}$) (Kane et al. 1979; Alarie 1981). Whether damage occurs in response to exposure to these much lower levels is unknown.

■ **Recommendation 3.** Chronic low-dose inhalation toxicology studies should be undertaken to determine if tissue damage occurs in response to exposure to levels of formaldehyde, acetaldehyde, and acrolein that are 10–100 times lower than their RD_{50}s.

Although pure formaldehyde and acrolein do not cause neutrophil recruitment, an inflammatory response has been observed in response to carbon particles coated with either compound (Kilburn and McKenzie 1978). Paradoxically, exposure of suspensions of neutrophils to formaldehyde and acrolein results in lowered responsiveness to soluble stimuli, such as phorbol esters, and decreased generation of superoxide anion (Witz et al. 1985). This may be responsible for the decreased in vivo killing of bacteria by mice treated with either compound (Jakab 1977).

Carcinogenicity. Exposure of 232 Fischer 344 rats to 14.3 ppm formaldehyde (6 hr/day, 5 days/week for 24 months followed by 6 months of nonexposure) induced squamous cell carcinoma in the nasal epithelium of 103 animals (Kerns et al. 1983). Exposure to 5.6 ppm formaldehyde induced squamous cell carcinoma in only 2 of 235 animals, and at 2.0 ppm no response was observed in 236 animals. In mice (B6C3F$_1$), exposure to 14.3 ppm induced nasal tumors in only 2 of 215 animals (Kerns et al. 1983). This figure did not represent a statistically significant increase but is notable because of the rarity of nasal tumors in mice. Exposure of Syrian golden hamsters to 10 ppm formaldehyde (5 hr/day, 5 days/week for 120 weeks) did not induce any airway tumors (Dalbey 1982). Mixtures of formaldehyde and hydrochloric acid induced nasal cancer in Sprague-Dawley rats that was entirely due to the formaldehyde; no enhancing effect of hydrochloric acid was seen (Albert et al. 1982; Sellakumar et al. 1985). In all of these chronic exposure studies, clearcut evidence was acquired for reversible as well as irreversible damage to respiratory epithelium.

Chronic exposure of groups of Wistar rats (110 animals/group) to acetaldehyde at initial doses of 750, 1,500, or 3,000 ppm (6 hr/day, 5 days/week for 27 months) induced 14, 34, and 38 nasal tumors, respectively, compared to 1 in controls (Woutersen et al. 1984). Severe irreversible degenerative changes of the upper respiratory tract were observed in the high-dose group so the acetaldehyde concentration had to be reduced repeatedly throughout the course of the experiment. The tumors observed at the low and moderate doses of

acetaldehyde occurred in the olfactory epithelium (Woutersen et al. 1984), whereas nasal tumors induced by low levels of formaldehyde occur in the respiratory epithelium (Kerns et al. 1983). Nasal tumors induced by high-level exposure to acetaldehyde and formaldehyde occur in the olfactory and respiratory epithelium. The results at low levels suggest that acetaldehyde is better able to penetrate to remote anatomic locations than formaldehyde. Further evidence for differential effects of aldehydes is provided by the observation that acetaldehyde at levels of 1,650–2,500 ppm (7 hr/day, 5 days/week for 52 weeks) induces tracheal, but not nasal, tumors in Syrian golden hamsters (Feron et al. 1982).

Exposure of Syrian golden hamsters to 4 ppm acrolein (7 hr/day, 5 days/week for 52 weeks) induced a number of pathological changes in the upper respiratory tract, particularly the nasal epithelium, but no tumors were observed in any organs (Feron and Kruysse 1977). Acrolein exhibits potent teratogenic and embryolethal effects when it is administered to rats intraamniotically but not by inhalation (Slott and Hales 1985).

Neither acrolein nor formaldehyde was carcinogenic in Syrian golden hamsters (Feron et al. 1982). Acrolein is similar to formaldehyde in chemical reactivity, irritant activity, and retention in the respiratory tract, so it should be tested in the same species in which formaldehyde has been detected as a carcinogen—the rat (Kerns et al. 1983). Attention should be paid to the development of nasal tumors.

■ **Recommendation 4.** A chronic inhalation toxicology study of acrolein should be undertaken in rats, with emphasis on carcinogenicity.

The species and organ specificities of different aldehydes with respect to their ability to induce respiratory tumors on inhalation exposure is fascinating and has been discussed (Kerns et al. 1983; Swenberg et al. 1983). Stimulation of nasal receptors may play a key role in the difference in the higher sensitivity of rats relative to mice. Rodents attempt to restrict their intake of irritants by reducing their respiratory minute volume (Chang et al. 1981). This response is more pronounced in mice than in rats so, for example, at the same level of exposure to formaldehyde, rats breathe approximately twice as much formaldehyde as mice (Chang et al. 1981). Indeed, the tumorigenic response of mice to the effects of 14.3 ppm formaldehyde is roughly the same as the response of rats to 6 ppm formaldehyde (Kerns et al. 1983). The importance of effective dose on tissue specificity is further emphasized by the observation that no tumors have been detected outside of the respiratory tract with any aldehyde. Studies of the retention (that is, the amount of compound bound to tissue) of various aldehydes in the respiratory tract of dogs indicate that formaldehyde is completely retained and acrolein is nearly completely retained in the upper tract whereas propionaldehyde is much less retained (Egle 1972b). Acetaldehyde is the least retained of all the aldehydes tested in the upper respiratory tract, which is consistent with its ability to induce tumors in hamster trachea (Egle 1972a).

Decreases in minute volume cannot explain the sharpness of the dose response of rats to formaldehyde. However, Swenberg and colleagues (1983) proposed that effects on mucociliary activity may play a role. The nasal respiratory epithelium is normally covered by a dynamic protective layer of mucus. The carbohydrate and protein in this layer may react with molecules such as formaldehyde, preventing their access to epithelial tissue. Interruption of mucous flow might saturate the capacity of these macromolecules to react with formaldehyde over certain anatomic locations, thereby increasing the effective dose. Formaldehyde increases mucous flow at low exposure levels but reduces it at high levels (Swenberg et al. 1983); this may result from the ciliastatic activity exhibited by formaldehyde (Morgan 1983; Morgan et al. 1983). Inhibition of mucociliary clearance introduces an additional step in the carcinogenic process, suggesting that short exposures to high concentrations would be more effective for compound delivery than long exposures to low doses. This is consist-

ent with the nonlinear dose response observed for formaldehyde carcinogenicity in rats. It also suggests that occasional high levels of exposure might exert biological effects not expected from extrapolation of dose responses obtained by chronic low-level exposure (Swenberg et al. 1983). Further support for nonlinearity of formaldehyde action on respiratory epithelium is provided by the dose dependence for induction of squamous metaplasia in the nasal cavity of Fischer 344 rats and B6C3F$_1$ mice (Kerns et al. 1983). Formaldehyde at 2 ppm only induces squamous metaplasia in the anterior-most regions of the nasal cavity in rats. Extensive metaplasia in midlevel and posterior portions of the cavity are observed with 5.6 and 14.3 ppm, respectively. By comparing the extent of squamous metaplasia in mice and rats, it is possible to approximate doses that exert similar pathological effects. Using this criterion, the extent of penetration by formaldehyde appears equivalent in rats and mice at doses of 5.6 and 14.3 ppm, respectively. This result correlates well to the difference in sensitivity of the two species to the carcinogenic action of formaldehyde in the respiratory tract.

Studies indicate that alterations of mucous flow and ciliatoxicity are important components of the nonlinear dose response for the carcinogenic action of formaldehyde (Swenberg et al. 1983). Acrolein exhibits the most potent ciliatoxic activity of any volatile aldehydes (Beauchamp et al. 1985). This may enhance the carcinogenic response to other less ciliatoxic aldehydes such as formaldehyde or acetaldehyde. The most likely combination to test first is acrolein and formaldehyde because they are the most potent ciliatoxins and carcinogens, respectively, in mobile source emissions.

■**Recommendation 5.** A chronic inhalation toxicology study of mixtures of formaldehyde and acrolein should be undertaken in rats and hamsters, with emphasis on carcinogenicity.

How one extrapolates the results of carcinogenicity studies in rodents to human exposure is uncertain. Humans are routinely exposed to atmospheric levels of formaldehyde that are 100–1,000 times lower than the doses that induce nasal tumors in rodents. However, individuals in certain cities are intermittently exposed to much higher levels. Whether long-term damage results from these episodic exposures is uncertain, although there is little doubt that acute effects, such as irritation, occur. High intermittent exposure might serve as an initiating event that provides a focus of transformed cells sensitive to promotion by other pollutants or environmental agents. Epidemiologic data do not provide evidence for a significant contribution of air pollution to human cancer but one might suggest that the combination of exposure to aldehydes in automotive emissions with other environmental agents is important in some individuals such as smokers (Doll and Peto 1981). This seems reasonable enough, but the concentrations of aldehydes in cigarette smoke are several orders of magnitude higher than their concentrations in urban air. Therefore, the significance of the contribution of aldehydes in mobile source emissions to health effects in smokers is uncertain. The other complication of extrapolating results from rodent bioassays to humans is the difference in anatomy and physiology of the two species. Rodents are obligate nose breathers whereas humans are not. This has obvious implications for the amounts of toxic agents that reach respiratory tissues by inhalation.

Cultured Cells. Formaldehyde and other aldehydes exert numerous effects on isolated cells in culture. They are toxic to normal as well as tumor cells and, in fact, certain α,β-unsaturated aldehydes were used in human clinical trials as potential chemotherapeutic agents (Schauenstein et al. 1977; Krokan et al. 1985). The genotoxic effects of formaldehyde have long been recognized (Auerbach et al. 1977). It induces single-strand breaks, DNA-protein cross-links, sister chromatid exchanges, and chromosome aberrations (Ross and Shipley 1980; Bedford and Fox 1981; Ross et al. 1981; Fornace 1982; Fornace et al. 1982; Levy et al. 1983). It is mutagenic in a variety of prokaryotic and eukaryotic cells including human fibroblasts (Chanet and von Borstel 1979; Boreiko et al. 1982; Goldmacher and Thilly 1983; Szabad et al. 1983), transforms rodent cells (Ragan and Boreiko 1981), and enhances viral transformation of Syrian hamster embryo cells (Hatch et al. 1983). Formaldehyde-induced DNA lesions appear to be repaired, but formaldehyde itself inhibits the ability of human bronchial epithelial cells and fibroblasts to repair damage by x rays and methylating agents (Grafstrom et al. 1983, 1984). A similar constellation of events occurs in response to treatment of cells with acrolein (Schauenstein et al. 1977; Beauchamp et al. 1985).

Despite the extensive documentation of the cellular effects of formaldehyde and other aldehydes, the understanding of their actions at the molecular level is incomplete. For example, the critical targets that lead to various cellular pathologies are, for the most part, unknown. Evidence exists linking the toxicity of α,β-unsaturated aldehydes to modification of a critical sulfhydryl protein, but its identity is unspecified (Schauenstein et al. 1977). Certain DNA polymerases contain important sulfhydryl groups that are sensitive to modification, so these are likely candidates (Kornberg 1980). Sulfhydryl reactivity may also contribute to the inhibition of DNA repair by methyl transferases caused by formaldehyde (Krokan et al. 1985).

■ **Recommendation 6.** Experiments should be undertaken in cells cultured from various segments of the upper respiratory tract to determine the mechanisms by which aldehydes exert pathological changes such as toxicity, hyperplasia, ciliatoxicity, and so on.

Such experiments should concentrate on identifying the critical targets for modification by each compound and the extent of modification that triggers the response. For example, despite the extensive literature on killing of prokaryotic and eukaryotic cells by α,β-unsaturated aldehydes, the precise mechanism of killing and the macromolecules involved are uncertain. Does modification of DNA polymerases lead to toxicity or does toxicity result from inhibition of enzymes of ATP generation? At what level of modification does toxicity result? Such knowledge will be important for basic biology as well as for risk assessment based on molecular dosimetry (see below).

A major unresolved question is how one extrapolates the results of experiments demonstrating pathological effects of aldehydes on cultured cells to risk assessment for human exposure. For most of the in vitro experiments, aldehydes are added in solution, whereas in animal exposure experiments they are administered by inhalation. How one relates molar concentrations of liquids to dosages of a gas that may accumulate in a target cell is unknown.

Methanol

Methanol is rapidly absorbed following oral, cutaneous, or respiratory exposure and undergoes general distribution to body water (Yant and Schrenck 1937; Haggard and Greenberg 1939). Its biological half-life is 1.5–2 hr (Sedivec et al. 1981), which means that many of the toxicologic effects triggered by inhalation exposure may be similar to those observed following oral administration. Methanol's oral toxicity to humans has been known for over 100 years (Tephly 1985). Considerable variability is observed in the dose at which toxicity results but best estimates of a dose required for severe intoxication and death are around 1 g/kg (Roe 1982). A lag phase of 12–24 hr is observed before any symptoms of toxicity are seen, which implies that a metabolite is involved in the toxicity (Tephly 1985). Metabolic acidosis occurs followed by visual effects that can lead to blindness. Ocular toxicity is occasionally followed by coma, other central nervous system effects, and death. Ethanol antagonizes the effects of methanol and it may be that varying amounts of ethanol contamination account for the variability in dose at which methanol is toxic to individuals (Roe 1955). Rodents are not susceptible to the toxic effects of methanol but nonhuman primates are; for example, methanol exhibits ocular toxicity in monkeys (Roe 1982)

(see Metabolism, Methanol and Formaldehyde). Exposure of human volunteers to an atmosphere containing 200 ppm methanol results in accumulation of 750 mg of which 50–60 percent is retained in the lung (Sedivec et al. 1981). Considering the dose of methanol estimated to be toxic to humans (1 g/kg), it is unlikely that a normal human being could ever be exposed to enough of it by inhalation to experience acute toxicity.

This author was unable to find carcinogenicity studies of methanol by inhalation exposure. Methanol is metabolized slowly during systemic circulation to formaldehyde, which is quickly metabolized to formic acid (Tephly 1985). A remote possibility is that methanol is oxidized to formaldehyde in the respiratory epithelium which is sensitive to its carcinogenic action. If so, methanol may act as a latent form of formaldehyde leading to accumulation in tissues that formaldehyde is ordinarily inaccessible to. Similar considerations hold for ethanol with respect to acetaldehyde. Taken with the potential importance of methanol and ethanol as alternate fuels, it seems important to test them thoroughly for carcinogenicity via the inhalation route. It is less important to test ethanol in inhalation studies because its oxidation product, acetaldehyde, is 100 times less active as a carcinogen than formaldehyde.

■ **Recommendation 7.** A chronic inhalation toxicology study of methanol should be undertaken in rats and hamsters, with emphasis on carcinogenicity.

Phenols and Catechols

No information is available on the inhalation toxicology of phenols or catechols. Most toxicologic studies have been performed by oral or intravenous administration, so the concentrations used are difficult to relate to inhalation exposure. Phenols are not strongly toxic, and substituted phenols such as butylated hydroxy toluene and butylated hydroxy anisole are used as preservatives in food. There is some speculation that the presence of phenolic antioxidants in food accounts for the steady decrease in stomach cancer in developed countries since 1945 (Doll and Peto 1981). Indeed, phenolic antioxidants such as butylated hydroxy anisole inhibit chemical carcinogenesis and appear to act at the promotion stage (Slaga et al. 1983). However, high doses of phenolic antioxidants actually appear to be tumor promoters themselves (Ito et al. 1982).

Cocarcinogenic Effects of Aldehydes, Alcohols, and Phenols

Cocarcinogenic effects have been reported for formaldehyde and acetaldehyde (Dalbey 1982; Feron et al. 1982). Lifetime exposure of Syrian golden hamsters to 30 ppm formaldehyde concomitant with subcutaneous administration of 0.5 mg diethylnitrosamine resulted in an enhancement of the number of tracheal tumors over treatment with diethylnitrosamine alone (Dalbey 1982). As mentioned above, formaldehyde does not induce tracheal tumors in hamsters. No enhancement of tumorigenesis was seen in the larynx or lung, and the effect on the trachea was only observed when formaldehyde exposure began before diethylnitrosamine administration. Enhancement did not result when formaldehyde exposure began after diethylnitrosamine injections were completed. A similar experiment was performed in Syrian golden hamsters with acetaldehyde (1,650–2,500 ppm) and benzo[a]pyrene administered by intratracheal instillation (Feron et al. 1982). At a dose of 36.4 mg but not 18.2 mg benzo[a]pyrene, enhancement of tracheal and bronchial tumorigenesis was observed after 52 weeks. When a similar experiment was performed with injection of diethylnitrosamine no enhancement of tracheal tumorigenesis was observed. In fact, there appeared to be a decrease over controls but this was considered a casual association. Formaldehyde has been reported to exhibit "initiating" and "promoting" activity in the C3H/10T1/2 in vitro transformation system (Ragan and Boreiko 1981; Frazelle et al. 1983).

Methanol has not been tested for cocarcinogenicity by the inhalation route. An epidemiologic association has been established between consumption of alcoholic beverages and esophageal cancer in smokers, but there is no evidence for direct carcinogenicity of ethanol (Doll and Peto 1981). Its role in enhancing the carcinogenicity of cigarette smoke is uncertain. Catechol has been identified as the major cocarcinogenic component of cigarette smoke (Van Duuren and Goldschmidt 1976; Hecht et al. 1981). However, bioassays were performed using the initiation-promotion model on mouse skin so the importance of catechol as an inhalation cocarcinogen is uncertain.

■ **Recommendation 8.** Attempts should be made to develop an initiation-promotion protocol for carcinogenesis testing of aldehydes and other components of mobile source emissions.

The two-stage mouse skin model has been very useful for detection of potential carcinogens, tumor initiators, and tumor promoters. There is no analogous model that can be used to screen compounds for their effects on respiratory tissues. When aldehydes were administered to rodents simultaneously or after administration of benzo[a]pyrene or diethylnitrosamine (Dalbey 1982; Feron et al. 1982), some stimulatory and inhibitory effects were noted but they were not dramatic, and it was difficult to speculate whether the aldehydes were acting as cocarcinogens or promoters based on the design of the experiments. A reproducible initiation-promotion model would enable rapid testing of mixtures of mobile source emission components by the inhalation route and would provide useful mechanistic information. Considering that formaldehyde and acrolein exert most of their effects on the respiratory epithelium of the nasal tract of rats and that acetaldehyde is a nasal carcinogen in rats and a tracheal carcinogen in hamsters (Feron et al. 1982; Woutersen et al. 1984), efforts should be directed toward developing a model in which the biological effects are monitored in the upper respiratory tract. It appears that most of an inspired dose of these compounds does not reach the bronchi and lungs, so the model should be designed with this in mind. In other words, it would not seem prudent to perform developmental experiments using compounds that exert possible initiating or promoting effects in the lungs.

Metabolism

Although some adverse effects of aldehydes and alcohols have been described in humans, experimental animals, and cell systems, quantification of risk, especially at ambient concentrations, is difficult with the current data base. Additional research is necessary to better estimate the potential toxicity of these compounds. The design and interpretation of experiments will be aided by understanding their metabolism and chemical reactions.

Methanol and Formaldehyde

All of the alcohols and aldehydes considered here are soluble in aqueous and organic solutions, which means they distribute rapidly throughout the body and within cells (Beauchamp et al. 1985). The major pathway of metabolism is oxidative with alcohols oxidized to aldehydes and aldehydes oxidized to acids. For example, methanol is oxidized to formaldehyde, which is oxidized to formic acid:

$$\begin{array}{c} \text{OH} \\ | \\ \text{C} \\ /\,|\,\backslash \\ \text{H} \quad \text{H} \quad \text{H} \end{array} \longrightarrow \begin{array}{c} \text{O} \\ || \\ \text{C} \\ /\ \ \backslash \\ \text{H} \qquad \text{H} \end{array} \longrightarrow \begin{array}{c} \text{O} \\ || \\ \text{C} \\ /\ \ \backslash \\ \text{H} \qquad \text{OH} \end{array}$$

Methanol   Formaldehyde   Formic Acid   (1)

Metabolism of alcohols and aldehydes can result in either detoxification or metabolic activation. The fact that a lag phase is observed before the onset of clinical symptoms of methanol toxicity, coupled with the findings that ethanol and alcohol dehydrogenase inhibitors antagonize methanol toxicity, suggests that a metabolite of methanol is responsible for its observed toxicologic effects. Alcohol dehydrogenase appears to be primarily responsible for the oxidation of methanol (McMartin et al. 1975). Its binding constant for methanol is

approximately six times lower than its binding constant for ethanol, which accounts for the ability of ethanol to antagonize methanol's effects (Makar and Tephly 1975). Catalase is important for methanol metabolism in rats but not in monkeys (Mannering and Parks 1957).

The major metabolite of methanol in monkeys is formic acid (eq. 1). Formate also exhibits ocular toxicity in monkeys (Martin-Amat et al. 1978). It accumulates in monkeys following methanol administration, thereby resulting in metabolic acidosis, but does not accumulate in rats. This is consistent with the differential sensitivity of these species to methanol toxicity and implies a role for formate as a toxic metabolite. Methanol oxidation by alcohol dehydrogenase is the rate-limiting step of metabolism and appears to be equally rapid in rats and monkeys (Watkins et al. 1970; Clay et al. 1975). The accumulation of formate in monkeys relative to rats appears to be due to a decreased rate of its oxidation to carbon dioxide (CO_2) in monkeys (McMartin et al. 1977). Formate metabolism occurs by a folic acid–dependent pathway, so folate deficiency renders monkeys extremely sensitive to methanol toxicity (McMartin et al. 1977). Conversely, folate supplementation lowers their sensitivity (Noker et al. 1980). It appears that folic acid levels are rate-limiting for formate metabolism to CO_2 in monkeys.

The toxic effects of methanol may be enhanced by simultaneous exposure to other compounds. For example, antifolates are used clinically for treatment of psoriasis and cancer, and it is conceivable that individuals undergoing treatment could exhibit enhanced sensitivity to methanol. Acute methotrexate treatment of monkeys does not decrease their rate of formate oxidation, but the effects of chronic treatment are unknown (Noker et al. 1980). Perhaps more important is the observation that nitrous oxide (N_2O) is an inhibitor of an enzyme of folic acid metabolism and leads to folate depletion (Eells et al. 1982). Enhanced sensitivity to methanol toxicity is observed following exposure of monkeys to N_2O, and metabolic acidosis is induced in rats, a species normally resistant to methanol effects (Eells et al. 1981). This raises the possibility that nitrogen oxides (NO_x) in auto exhaust might enhance any toxic effects of methanol and formaldehyde that are mediated by formic acid.

The oxidation of methanol to formic acid most likely involves formaldehyde as an intermediate (eq. 1). In contrast to the slow rate of its formation from methanol, formaldehyde is oxidized quite rapidly. Its half-life is estimated to be 1 min following intravenous infusion (Reitbrock 1969; McMartin et al. 1979). Formaldehyde does not accumulate following methanol administration but is rapidly metabolized to CO_2 (85 percent) and expired (Neely 1964; Mashford and Jones 1982). The initial oxidation of formaldehyde appears to be catalyzed by a formaldehyde dehydrogenase that is glutathione-dependent (Strittmatter and Ball 1955; Goodman and Tephly 1971). The enzyme is quite specific for formaldehyde (Strittmatter and Ball 1955). The remaining 15 percent of formaldehyde that is not oxidized to CO_2 may bind to protein or enter pathways of one-carbon metabolism. Formaldehyde does not appear to contribute significantly to methanol toxicity following oral administration.

Acrolein

The major pathway of acrolein metabolism appears to involve conjugation with glutathione followed by conversion to S-carboxyethyl-mercapturic acid (Draminski et al. 1983).

$$H_2C=CH-C\overset{O}{\underset{H}{\diagup}} + GSH \longrightarrow GS-CH_2CH_2 \cdot C\overset{O}{\underset{H}{\diagup}}$$

$$\longrightarrow Ac-Cys \cdot CH_2 \cdot CH_2 \, C\overset{O}{\underset{H}{\diagup}} \quad (2)$$

Reaction of acrolein with glutathione is a rapid chemical reaction but is also catalyzed by glutathione transferases (Jakoby and Habig 1980). The half-life for conjugation of glutathione with 4-hydroxy-nonenal, a molecule structurally related to acrolein, is

approximately 4 sec in perfused rat heart (Ishikawa et al. 1986). Acrolein is also oxidized to acrylic acid in vivo, but this only accounts for approximately 15 percent of the administered dose (Draminski et al. 1983).

$$H_2C=CH-C\overset{O}{\underset{H}{\diagup}} \xrightarrow{[OX]} H_2C=CH-CO_2H \quad (3)$$

Oxidation of acrolein to acrylic acid by rat liver subcellular fractions is inhibited by diethyldithiocarbamate, an inhibitor of aldehyde dehydrogenase (Patel et al. 1983). Epoxidation of acrolein to glycidaldehyde occurs in rat liver subcellular fractions but it is not known if this transformation takes place in vivo (Patel et al. 1983).

$$H_2C=CH-C\overset{O}{\underset{H}{\diagup}} \longrightarrow H_2C\overset{O}{\diagup}CH-C\overset{O}{\underset{H}{\diagup}} \quad (4)$$

This is a potentially important metabolite of acrolein because it has been classified as an animal carcinogen (International Agency for Research on Cancer 1976). It induced malignant tumors in rats following subcutaneous injection and papillomas in mice following skin painting (Van Duuren et al. 1965, 1966, 1967). Its role in acrolein metabolism and potential carcinogenicity is uncertain.

Measurement of Inspired Methanol and Formaldehyde

Methanol is produced as a result of normal human metabolism; it is detectable in human breath and urine. The best way to assay for inhaled methanol is to quantitate increases in its urinary levels by gas chromatography. Approximately 1 percent of the inspired dose is excreted in human urine. Because there is a significant background level of methanol in human urine due to metabolism or diet, it has been estimated that the lower limit of exposure to methanol that could be detected by an increase in urinary levels would result from inhalation for 8 hr of air containing 100 ppm methanol (Heinrich and Angerer

1982). On this basis, it is unlikely that one could ever detect increases in methanol inhalation resulting from exposure to mobile source emissions. Considering that formaldehyde is an intermediate in methanol and carbohydrate metabolism, one can extend this analysis to reach the conclusion that it is also impossible to detect inhalation exposure to formaldehyde by monitoring urinary levels of it or its initial metabolite formic acid. This conclusion is strengthened by the realization that formaldehyde does not escape the respiratory tissue to which it is administered and, therefore, never achieves appreciable systemic levels.

Chemical Reactions

Aldehydes. *Adduct Formation.* Aldehydes are reactive electrophiles that add reversibly to nucleophiles to form covalent hydroxymethyl and imine adducts.

$$H_2C=O + Nu \longrightarrow H_2C{-}Nu \longrightarrow H_2C=Nu + H_2O \qquad (5)$$

with OH on the hydroxymethyl intermediate.

For simple aliphatic aldehydes such as formaldehyde or acetaldehyde, these adducts are unstable and readily revert to starting materials (Fraenkel-Conrat 1954; Hoard 1960). Thus, despite the fact that formaldehyde is mutagenic, reacts with nucleic acid, and induces major alterations in nucleic acid structure, hydroxymethyl or imine adducts to nucleic acid bases or nucleotides have only recently been isolated and identified (McGhee and von Hippel 1975; Beland et al. 1984). Reaction of hydroxymethyl derivatives with a second nucleotide base produces cross-linked products that are stable to hydrolysis (Feldman 1967).

$$H_2C{-}Nu + Nu \longrightarrow \underset{Nu}{\overset{H}{\diagdown}} C \underset{Nu}{\overset{H}{\diagup}} \qquad (6)$$

Dimeric adducts have been isolated following the reaction of formaldehyde with deoxyguanosine, deoxyadenosine, and deoxycytidine, and mixtures thereof (figure 1) (Feldman 1967; Chaw et al. 1980).

Chemical reduction of imines generates amines, which are stable.

$$H_2C=NH{-}R + NaBH_4 \longrightarrow H_3C{-}NH{-}R \qquad (7)$$

Imine reduction probably does not occur in biological systems in vivo but is often used in in vitro experiments to trap unstable aldehyde/nucleophile addition products (Chio and Tappel 1969). It may be useful as a derivatizing reaction for quantitation of aldehyde/protein or aldehyde/nucleic acid adducts that would otherwise decompose during tissue processing and sample preparation.

Addition of nucleophiles to the β-carbon of α,β-unsaturated aldehydes generates products that are considerably stabler than hydroxymethyl compounds or imines.

$$H_2C=CH{-}\overset{O}{\overset{\|}{C}}{-}H + Nu \longrightarrow Nu{-}H_2C\ CH{-}\overset{O}{\overset{\|}{C}}{-}H \qquad (8)$$

The most reactive nucleophile in proteins toward α,β-unsaturated aldehydes is the sulfhydryl group (Schauenstein et al. 1977). Many proteins and enzymes contain sulfhydryl groups, and a detailed study of the toxicity of α,β-unsaturated aldehydes to virus particles, bacteria, and mammalian cells indicates a correlation of their reactivity toward sulfhydryl groups (Schauenstein et al. 1977). The identity of the critical cellular protein inactivated by α,β-unsaturated aldehydes has not been established, but it is noteworthy that two DNA polymerases are reactive toward sulfhydryl reagents (Kornberg 1980). Another candidate is ribonucleotide reductase, which catalyzes the rate-limiting step in cellular DNA synthesis and contains an unusual dithiol group that is sensitive to α,β-unsaturated aldehydes and sulfhydryl reagents (Thelander and Reichard 1979). Finally, it should be mentioned that the tripeptide glutathione contains a sulfhydryl group that reacts readily with α,β-unsaturated aldehydes. In fact, an important physiological role of glutathione is the scavenging of electrophiles such as α,β-unsaturated aldehydes (Jakoby and Habig 1980). Reaction

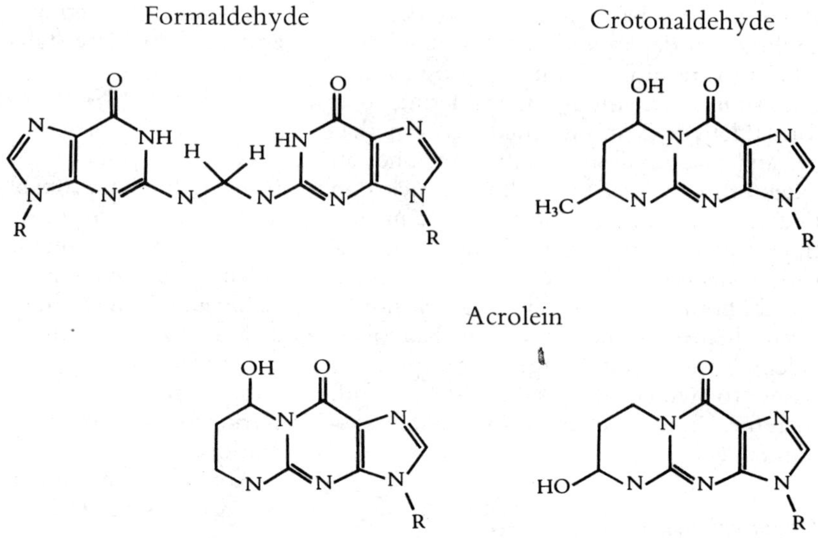

Formaldehyde   Crotonaldehyde

Acrolein

Figure 1. Adducts formed by reaction of guanosine with formaldehyde, croton-aldehyde, and acrolein. The formaldehyde adduct is a dimer of two guanosines with a single molecule of formaldehyde. Analogous dimeric adducts are formed by reaction of formaldehyde with adenosine and cytidine. Unsymmetrical dimeric adducts are also formed when formaldehyde is added to mixtures of guanosine, adenosine, and cytosine. R = deoxyribose.

of glutathione with these compounds occurs spontaneously or is enzyme-catalyzed. Although intracellular glutathione concentrations are high (~6 mM), acute exposure to α,β-unsaturated aldehydes can result in significant glutathione depletion, which lowers the cell's defenses toward electrophilic agents (Jakoby and Habig 1980). This is known to potentiate toxicity of xenobiotics in the short term, but the long-term effects of glutathione depletion are unknown.

Adducts between α,β-unsaturated aldehydes and amine groups also form and are biologically very important (eq. 8). Formation of amine-α,β-unsaturated aldehyde adducts is reversible but sufficiently slow that the adducts can be isolated and identified. The structures of several α,β-unsaturated aldehyde/deoxynucleoside adducts have been elucidated and are listed in figure 1. Adducts to deoxyguanosine have been detected following the reaction of acrolein or crotonaldehyde with DNA (Chung and Hecht 1983; Chung et al. 1984). These adducts survive hydrolysis of DNA to deoxynucleosides, which is a key step in the isolation of any DNA adducts.

They may represent useful potential indicators of DNA damage in cells exposed to high concentrations of α,β-unsaturated aldehydes. It is interesting that cyclic adducts result from the reaction of α,β-unsaturated aldehydes with DNA bases. Cyclization of the aldehyde group of the initial adduct to a suitably disposed amine group of deoxyguanosine is favored by entropy.

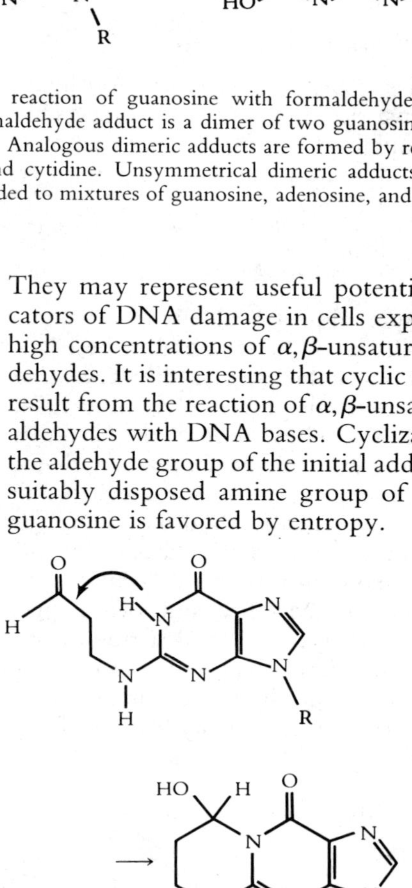

$$\tag{9}$$

A cyclic adduct also forms between deoxyguanosine and glycidaldehyde, an in vitro metabolite of acrolein (Van Duuren and Loewengart 1977).

Reaction of aldehydes with nucleic acids is believed to be responsible for the mutagenic and carcinogenic effects of the compounds. However, saturated and unsaturated aldehydes also inhibit repair of certain adducts formed by methylating agents (Grafstrom et al. 1983, 1984). Inhibition of DNA repair appears to be a result of the covalent reaction of aldehydes with the sulfhydryl group of the methyl acceptor protein that removes the methyl group from O-6-methylguanine residues in DNA exposed to methylating agents. Inactivation of the methyl acceptor protein enhances the mutagenicity of methylating agents. Thus, aldehydes can enhance mutagenesis by a mechanism that does not involve modification of nucleic acid.

Free-Radical Formation. Aldehydes undergo enzyme-catalyzed oxidation and reduction in biological systems. The enzymes involved appear to be dehydrogenases, which implies mechanisms involving hydride transfer—a relatively innocuous transformation. Aldehydes are also prone to autoxidation triggered by one-electron oxidation of the aldehydic carbon-hydrogen bond (Lloyd 1973).

$$\underset{R \quad H}{\overset{O}{\underset{\|}{C}}} \xrightarrow{[OX]} \underset{R}{\overset{O}{\underset{\|}{C\cdot}}} \qquad (10)$$

The resultant radical couples to O_2 under aerobic conditions to form a peroxyl radical.

$$\underset{R}{\overset{O}{\underset{\|}{C\cdot}}} + O_2 \longrightarrow \underset{R \quad O}{\overset{O}{\underset{\|}{C}}}\diagdown O\cdot \qquad (11)$$

The peroxyl radical then carries out one-electron oxidation of another molecule of aldehyde.

$$\underset{R \quad O}{\overset{O}{\underset{\|}{C}}}\diagdown O\cdot + \underset{R \quad H}{\overset{O}{\underset{\|}{C}}}$$

$$\longrightarrow \underset{R \quad O}{\overset{O}{\underset{\|}{C}}}\diagdown O\diagup H + \underset{R}{\overset{O}{\underset{\|}{C\cdot}}} \qquad (12)$$

The latter reactions constitute the propagation steps of a free-radical chain autoxida-tion. Numerous molecules of aldehyde are oxidized and peroxyl radicals produced as a result of a single initiation event. Peroxyl radicals are relatively stable free radicals that are selective in their reaction with cellular constituents (Willson 1985). They have the ability to diffuse far from the site of their generation to react with specific molecules (Pryor 1984). Acyl peroxyl radicals, the type produced by aldehyde autoxidation, are significantly more reactive than alkyl peroxyl radicals and may not be able to diffuse as far intracellularly. The cellular targets for peroxyl radical reactions are unknown, but they epoxidize isolated double bonds and abstract hydrogen atoms from polyunsaturated fatty acid residues in phospholipids (Willson 1985). The latter reaction results in lipid peroxidation, which can disrupt membrane structure, lead to cell death, and release soluble mediators of toxicity and chemotaxis.

Peroxyl radicals also oxidize sulfides to sulfoxides.

$$\underset{R_1 \quad R_2}{\overset{}{S}} \xrightarrow{ROO\cdot} \underset{R_1 \quad R_2}{\overset{O}{\underset{\|}{S}}} \qquad (13)$$

α-1-Proteinase inhibitor contains a critical methionine residue close to the site that combines with a variety of proteinases to inhibit their action (Travis and Salvesen 1983). Oxidation of the methionine residue abolishes the ability of the protein to inhibit catalysis by proteolytic enzymes. This alters the balance of proteolysis and can have dramatic effects on lung function. The physiological target for α-1-proteinase inhibitor appears to be elastase which degrades pulmonary connective tissue by virtue of its action on elastin. Elastase is secreted by neutrophils in response to their activation during inflammation. The activity of elastase is regulated in part by the amount of α-1-proteinase inhibitor that is available to combine with and inactivate it. Certain genetic diseases that are characterized by pulmonary emphysema are associated with decreased amounts of α-1-proteinase inhibitor (see Wright, this volume). The critical methionine residue of α-1-proteinase inhibitor that controls its activity toward proteinases is oxidized by peroxides and hypochlorous acid, both products of activated neutrophils. It is also oxidized by free radicals such as the one present in cigarette smoke (Pryor 1984). The latter reaction may be especially important in the genesis of diseases associated with chronic cigarette smoking such as emphysema. Although it has not been tested, it seems quite likely that acyl peroxyl radicals formed by autoxidation of aldehydes inactivate α-1-proteinase inhibitor by oxidizing its critical sulfide to a sulfoxide. This provides a mechanism by which one-electron oxidation of aldehydes could lead to pulmonary emphysema.

Alcohols.   Alcohols are relatively unreactive chemically with nucleophiles and electrophiles. They can be converted to more reactive derivatives by conjugation with functionalities (for example, glucuronate, sulfate) that render the hydroxyl groups more reactive (Jakoby et al. 1980; Kasper and Henton 1980).

$$H_3C-OH + \cdot OH \longrightarrow H_2\dot{C}-OH + H_2O \quad (14)$$

These conjugates could conceivably act as electrophiles and alkylate nucleophiles, but there is no evidence that the toxicity exhibited by, for example, methanol is a result of such reactions. Furthermore, short-chain alcohols are highly water-soluble, which removes much of the driving force for their conjugation with polar moieties. Alcohols are oxidized to aldehydes and acids by dehydrogenases, which is a reaction of primary importance in the metabolism of alcohols. Alcohols are not oxidized by one electron to free radicals very readily, although they will react with hydroxyl radical to form hydroxylmethyl radicals.

Phenols and Catechols.   Attachment of a hydroxyl group to an aromatic ring greatly enhances the reactivity of the O—H bond. By comparison to aliphatic alcohols, the chemistry of phenols is rich. Phenols are more acidic than alcohols and possess significant nucleophilicity toward reactive electrophiles. For this reason they can serve a protective role as scavengers of metabolically generated electrophiles. They are also

conjugated readily, which effects a marked change in their polarity. The most important reaction of phenols is probably with one-electron oxidants (Simic and Hunter 1983). They readily donate electrons of hydrogen atoms, thereby generating phenoxyl radicals.

$$(15)$$

Phenoxyl radicals are much more stable than aliphatic alkoxyl radicals because of conjugation with the aromatic ring. Phenols with alkyl substitutents *ortho* to the hydroxyl group are widely used as chain-breaking antioxidants (Howard 1973). These compounds donate a hydrogen atom to peroxyl radicals that are formed during the propagation step of autoxidation. The stability and steric hindrance of phenoxyl radicals prevent them from abstracting hydrogen atoms from donors that will react with chain-carrying peroxyl radicals. Phenoxyl radicals couple to second molecules of peroxyl radical to form peroxycyclohexadienones.

$$(16)$$

As a result, every phenol molecule removes two peroxyl radicals from autoxidation mixtures, which interrupts the radical chain and inhibits the autoxidation process.

Phenols that lack alkyl groups *ortho* to the hydroxyl group react with one-electron oxidants to form phenoxyl radicals that are quite reactive. An important reaction of phenoxyl radicals is coupling to O_2, which forms peroxyl radicals.

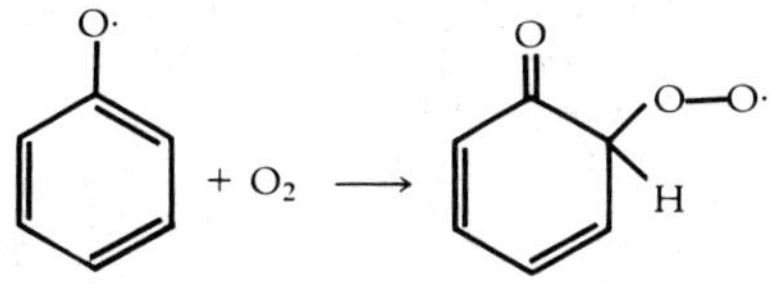

(17)

As discussed above, peroxyl radicals abstract hydrogen atoms from reactive molecules such as unsaturated fatty acids to initiate and propagate radical-chain oxidations. This leads to the paradox that phenols, which are generally thought to be antioxidants, can actually stimulate free-radical autoxidation. The reactions of phenoxyl radicals are especially important because phenols can be oxidized to phenoxyl radicals by high-valence metals as well as alkoxyl and peroxyl free radicals. This provides a mechanism for metal-catalyzed initiation of free-radical oxidations. Free radicals are believed to enhance carcinogenesis, particularly the promotion phase, and free-radical initiators are promoters in the two-stage assay in mouse skin (Slaga et al. 1981). This may contribute to the reported carcinogenicity of certain phenols in mouse forestomach (Ito et al. 1982) and to the cocarcinogenicity of catechols on mouse skin (Van Duuren and Goldschmidt 1976; Hecht et al. 1981).

Unhindered phenols are oxidized to hydroquinones that are extremely air-sensitive and oxidize to quinones (Irons and Sawahata 1985).

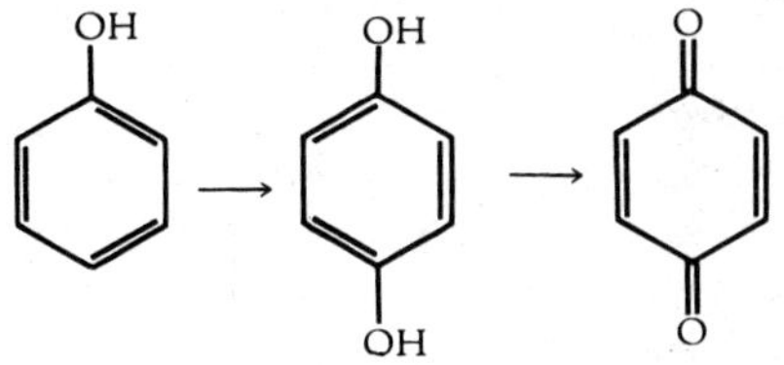

(18)

Quinones are strong electrophiles that undergo addition of nucleophiles to the double bonds of the ring (Irons and Sawahata 1985).

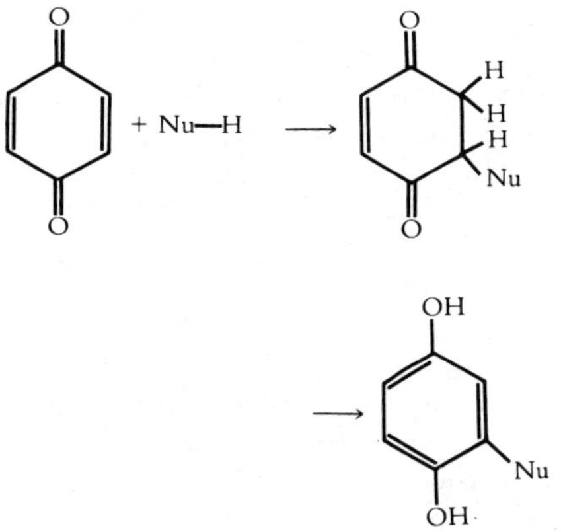

(19)

This can result in the formation of protein and nucleic acid adducts and to depletion of glutathione. Polycyclic quinones are readily reduced to semiquinones that are either further reduced to hydroquinones or are oxidized by O_2 to regenerate the quinone and form O_2^- (Smith et al. 1985). The continuous reduction and oxidation of quinones is called redox cycling and can lead to copious superoxide formation (Smith et al. 1985). This causes DNA strand scission, mutagenicity, and toxicity, all of which probably require metals. Redox cycling may well account for some of the pathophysiological effects of phenols and catechols.

Quantification of Exposure and Estimation of Human Risk

One of the fundamental unsolved problems of toxicology is how to extrapolate dose/response data obtained in animal testing (usually in rodents) to risk assessment in humans. An approach to crossing this species barrier is to quantitate the dose that reaches the target organ at a series of exposure levels. This is called molecular dosimetry. By knowing the amount of compound that must reach a target cell to exert an effect in, for example, rats, it should be possible to more intelligently estimate the risk of a given amount of the same compound reaching the same cell type in humans. The effective dose that reaches target tissues is more relevant to risk assessment than are the atmospheric levels. For example, at an ambient concentration of 10 ppb, the amount of formaldehyde inspired in 24 hr by an average human is 7 μmole. This does not seem a significant amount until one realizes that it is almost exclusively localized in the mucus and epithelial cells lining the upper respiratory tract.

What is the starting point to be if one is to quantitate binding of aldehydes, alcohols, and phenols to critical intracellular targets for their toxic and carcinogenic effects? Most of the toxicity exhibited by these agents is probably due to covalent binding to protein in which the aldehyde or quinone reacts as an electrophile. There is no doubt that protein binding to these compounds or their metabolites occurs and that it can cause toxicity. Saturated and α,β-unsaturated aldehydes as well as quinones bind rapidly to sulfhydryl proteins and inactivate them. Stable adducts also form to lysine residues. Aldehyde/lysine conjugates have been isolated from rat urine that most likely arise from proteolysis of aldehyde/protein conjugates (McGirr et al. 1985). This provides direct evidence for covalent binding of aldehydes to proteins in vivo. A considerable amount of information suggests that covalent binding to sulfhydryl groups of DNA polymerases is responsible for the toxic effects of α,β-unsaturated aldehydes (Schauenstein et al. 1977). Evidence also suggests that the cytostatic effects of quinones derives from their ability to bind specifically and covalently to tubulin (Irons et al. 1981). A similar reaction of tubulin or another component of the flagellar system may also account for the ciliatoxic activity of aldehydes.

Molecular dosimetry offers an approach to the quantitation of physiologically relevant damage resulting from exposure to aldehydes and quinone metabolites of phenols. Methods could be developed for the analysis of covalent adducts to proteins that are involved in pathological responses in target tissues. Of course, this requires that the key protein targets are known. Understanding the role of individual proteins in toxicologic responses represents a major gap in our knowledge. This is why mechanistic toxicology studies in cell culture are so important. Until an adequate knowledge of key protein targets is available, methods should be designed to quantitate adducts to proteins that are not necessarily important in the observed response but are abundant in the cell or tissue in which the response is observed. This approach is similar to the use of hemoglobin for estimation of exposure to methylating agents and carcinogens. Abundant proteins that contain reactive amine groups would be especially useful because amine/aldehyde conjugates are more stable than thiol/aldehyde conjugates. It is likely that the target cells for the effects of airborne aldehydes are in the epithelial tissue of the nasal and respiratory tracts. This is based on the types of effects observed and the extreme reactivity of the molecules. It is very unlikely that aldehydes or quinones escape pulmonary epithelia to reach peripheral tissues. If most of the covalently bound material is localized in the nasal or respiratory tract, it should be possible to sample these matrices in individual human subjects by lavage techniques.

Molecular dosimetry of nucleic acid adducts in target tissue will be important when attempting to relate inhaled dose to carcinogenic response. Thus, method development is recommended. However, quantitation of nucleic acid adducts in vivo is of less value than quantitation of protein adducts for estimation of inhaled dose. The levels of DNA adducts produced in cells are several orders of magnitude lower than protein adducts and are subject to varying degrees of removal by repair enzymes, which further lowers the steady-state concentration of the nucleic acid adducts.

■ **Recommendation 9.** Methods should be developed for quantitation of the amounts of aldehydes that reach target organs or potential target organs in humans and rodents.

Despite the long history of investigation of the reaction of formaldehyde with nucleic acids and the knowledge that formaldehyde is a nasal carcinogen, the identity of the adduct(s) that it forms on reaction with DNA in vivo is unknown. This may be due to the hydrolytic instability of imine and hydroxymethyl derivatives of purines and pyrimidines, and it underscores the need for development of novel methods of isolation and analysis. Less is known of the reaction of acetaldehyde and acrolein with DNA although acrolein/deoxyguanosine adducts have been recently identified and detected following the reaction of acrolein with DNA in vitro (Chung et al. 1984). Acrolein is metabolized by microsomal cytochrome P-450 preparations to glycidaldehyde, which binds to DNA and is a carcinogen (International Agency for Research on Cancer 1976; Patel et al. 1983). The adduct that glycidaldehyde forms on reaction with deoxyguanosine is different than the acrolein/deoxyguanosine adducts, so isolation and quantitation of acrolein/DNA adducts following inhalation of acrolein would determine the extent to which metabolism plays a role in acrolein's genotoxic effects in vivo. The availability of methods for detection and quantitation of aldehyde/DNA adducts formed in vivo might be important as part of a molecular dosimetry approach to risk assessment.

■ **Recommendation 10.** Methods should be developed to detect and quantitate DNA adducts derived from formaldehyde, acetaldehyde, and acrolein. The techniques should then be applied to the detection of DNA adducts formed in target tissues after administration of carcinogenic and subcarcingenic doses of the inhaled compounds. Detection of these adducts in cultured target cells would be a helpful intermediate step in adaption of the analytical methods to detection of in vivo adducts.

There is ample precedent for the importance of electrophilic additions to proteins and nucleic acids in aldehyde and phenol biochemistry and toxicology, but the possibility that free-radical reactions contribute to their health effects has not been rigorously established. Therefore, it would be useful to conduct experiments to probe for the involvement of free radicals as mediators of aldehyde and phenol pathology. This is not a trivial undertaking, because free radicals are species with relatively short half-lives that make them nearly impossible to detect and quantitate directly (usually <1 sec). Nevertheless, it is now possible to trap certain types of free radicals that might be produced from aldehydes and phenols (Packer 1984). In addition, products of in vivo free-radical reactions can be detected and quantitated as indirect evidence for free-radical formation (Packer 1984).

■ **Recommendation 11.** Experiments should be performed to determine if aldehydes exert toxicologic effects by generation of free radicals.

Summary

A review of the literature indicates that aldehydes are the most potent biologically active substances of the compounds under consideration in mobile source emissions. They exert toxicologic effects at concentrations approximately 10–100 times their ambient atmospheric levels. Variations in ambient levels have been reported with occasional toxicologically relevant concentrations reported in heavily polluted metropolitan areas. Inhaled aldehydes exert their toxicologic effects in the upper respiratory tract, and there is no reason to believe that they trigger systemic responses. This may be due to their high reactivity or to the fact that they are rapidly metabolized. Metabolism can result in detoxification or metabolic activation. The formation of substantial amounts of methanol and formaldehyde during normal human metabolism precludes development of analytical methods for their quantitation based on "systemic" approaches, such as plasma or urine analysis. These observations mandate novel approaches to quantitation of exposure and estimation of risk to the human population. Aldehydes and quinones are reactive electrophiles that form adducts with DNA and proteins. The structures of several aldehyde/nucleic acid adducts have been elucidated, but the extent of their formation in vivo is un-

known. Phenols are oxidized to free radicals, which may play a role in tumor promotion. Aldehydes are oxidized to very reactive free radicals in chemical systems but it is uncertain if they form free radicals in biochemical systems in vitro or in vivo.

Summary of Research Recommendations

Exposure
HIGH PRIORITY

Recommendation 1   Routine monitoring of atmospheric alcohol and aldehyde levels should be performed in regions where alcohol-based fuels are or will be in heavy use.

LOW PRIORITY

Recommendation 2   Methods should be developed to routinely analyze phenols and catechols in urban air.

Health Effects
HIGH PRIORITY

Recommendation 4   A chronic inhalation toxicology study of acrolein should be undertaken in rats, with emphasis on carcinogenicity.

Recommendation 5   A chronic inhalation toxicology study of mixtures of formaldehyde and acrolein should be undertaken in rats and hamsters, with emphasis on carcinogenicity.

MEDIUM PRIORITY

Recommendation 3   Chronic low-dose inhalation toxicology studies should be undertaken to determine if tissue damage occurs in response to exposure to levels of formaldehyde, acetaldehyde, and acrolein that are 10–100 times lower than their RD_{50}s.

Recommendation 8   Attempts should be made to develop an initiation-promotion protocol for carcinogenesis testing of aldehydes and other components of mobile source emissions.

LOW PRIORITY

Recommendation 7   A chronic inhalation toxicology study of methanol should be undertaken in rats and hamsters, with emphasis on carcinogenicity.

Cellular Effects
HIGH PRIORITY

Recommendation 6	Experiments should be undertaken in cells cultured from various segments of the upper respiratory tract to determine the mechanisms by which aldehydes exert pathological changes such as toxicity, hyperplasia, ciliatoxicity, and so on.

Molecular Dosimetry
HIGH PRIORITY

Recommendation 9	Methods should be developed for quantitation of the amounts of aldehydes that reach target organs or potential target organs in humans and rodents.
Recommendation 10	Methods should be developed to detect and quantitate DNA adducts derived from formaldehyde, acetaldehyde, and acrolein. The techniques should then be applied to the detection of DNA adducts formed in target tissues after administration of carcinogenic and subcarcingenic doses of the inhaled compounds. Detection of these adducts in cultured target cells would be a helpful intermediate step in adaption of the analytical methods to detection of in vivo adducts.

MEDIUM PRIORITY

Recommendation 11	Experiments should be performed to determine if aldehydes exert toxicologic effects by generation of free radicals.

References

Alarie, Y. 1981. Bioassay for evaluating the potency of airborne sensory irritants and predicting acceptable levels of exposure in man, *Food Cosmet. Toxicol.* 19:623–626.

Albert, R. E., Sellakumar, A. R., Laskin, S., Kuschner, M., Nelson, N., and Snyder, C. A. 1982. Gaseous formaldehyde and hydrogen chloride induction of nasal cancer in the rat, *J. Nat. Cancer Inst.* 68:597–603.

Auerbach, C., Moutschen-Dahman, M., and Moutschen, J. 1977. Genetic and cytogenetical effects of formaldehyde and related compounds, *Mutat. Res.* 39:317–362.

Babiuk, C., Steinhagen, W. H., and Barrow, C. S. 1985. Sensory irritation response to inhaled aldehydes after formaldehyde pretreatment, *Toxicol. Appl. Pharmacol.* 79:143–149.

Beauchamp, R. O., Jr., Andjelkovich, D. A., Kliger-man, A. D., Morgan, K. T., and d'A Heck, H. 1985. A critical review of the literature on acrolein toxicity, *CRC Crit. Rev. Toxicol.* 14:309–380.

Bedford, P., and Fox, B. W. 1981. The role of formaldehyde in methylene dimethylsulphonate-induced DNA cross-links and its relevance to cytotoxicity, *Chem.-Biol. Interact.* 38:119–126.

Beland, F. A., Fullerton, N. F., and Heflich, R. H. 1984. Rapid isolation, hydrolysis, and chromatography of formaldehyde-modified DNA, *J. Chromatogr.* 308:121–131.

Boreiko, C. J., Couch, D. B., and Swenberg, J. A. 1982. Mutagenic and carcinogenic effects of formaldehyde, *Environ. Sci. Res.* 25:353–367.

Buckley, L. A., Jiang, X. Z., James R. A., Morgan, K. T., and Barrow, C. S. 1984. Respiratory tract lesions induced by sensory irritants at the RD_{50} concentration, *Toxicol. Appl. Pharmacol.* 74:417–429.

Carson, B. L., Beall, C. M., Ellis, H. V., Baker, L. H., and Herndon, B. L. 1981. Acrolein Health Effects, NTIS PB82-161282; EPA-68-03-2928; EPA-460/3-81-034, *Gov. Rep. Announce. U.S. Index* 12.

Chanet, R., and von Borstel, R. C. 1979. Genetic

Correspondence should be addressed to Lawrence J. Marnett, 435 Chemistry, Wayne State University, Detroit, MI 48202.

effects of formaldehyde in yeast. III. Nuclear and cytoplasmic mutagenic effects, *Mutat. Res.* 62:239–253.

Chang, J. C. F., Steinhagen, W. H., and Barrow, C. S. 1981. Effects of single or repeated formaldehyde exposure on minute volume of B6C3F1 mice and F-344 rats, *Toxicol. Appl. Pharmacol.* 61:451–459.

Chaw, Y. F. M., Crane, L. E., Lange, P., and Shapiro, R. 1980. Isolation and identification of cross-links from formaldehyde-treated nucleic acids, *Biochemistry* 19:5525–5531.

Chio, K. S., and Tappel, A. L. 1969. Synthesis and characterization of the fluorescent products derived from malonaldehyde and amino acids, *Biochemistry* 8:2821–2826.

Chung, F.-L., and Hecht, S. S. 1983. Formation of cyclic 1,N^2-adducts by reaction of deoxyguanosine with α-acetoxy-*N*-nitrosopyrrolidine, 4-(carbethoxynitrosamino)butanal, or crotonaldehyde, *Cancer Res.* 43:1230–1235.

Chung, F.-L., Young, R., and Hecht, S. S. 1984. Formation of cyclic 1,N^2-propanodeoxyguanosine adducts in DNA upon reaction with acrolein or crotonaldehyde, *Cancer Res.* 44:990–995.

Clay, K. L., Murphy, R. C., and Watkins, W. D. 1975. Experimental methanol toxicity in the primate: analysis of metabolic acidosis in the monkey, *Toxicol. Appl. Pharmacol.* 34:49–61.

Cleveland, W. S., Graedel, T. E., and Kleiner, B. 1977. Urban formaldehyde: observed correlation with source emissions and photochemistry, *Atmos. Environ.* 11:357–360.

Consensus Workshop on Formaldehyde. 1984. Report on the Consensus Workshop on Formaldehyde, *Environ. Health Perspect.* 58:323–381.

Dalbey, W. E. 1982. Formaldehyde and tumors in hamster respiratory tract, *Toxicology* 24:9–14.

Dalhamn, T., and Rosengren, A. 1971. Effect of different aldehydes on tracheal mucosa, *Arch. Otolaryngol.* 93:496–500.

Doll, R., and Peto, R. 1981. *The Causes of Cancer*, pp. 1245–1312, Oxford University Press, Oxford.

Draminski, W., Eder, E., and Henschler, D. 1983. A new pathway of acrolein metabolism in rats, *Arch. Toxicol.* 52:243–247.

Eells, J. T., Makar, A. B., Noker, P. E., and Tephly, T. R. 1981. Methanol poisoning and formate oxidation in nitrous-oxide treated rats, *J. Pharmacol. Exp. Ther.* 217:57–61.

Eells, J. T., Black, K. A., Makar, A. B., Tedford, C. E., and Tephly, T. R. 1982. The regulation of one-carbon oxidation in the rat by nitrous oxide and methionine, *Arch. Biochem. Biophys.* 219:316–326.

Egle, J. L., Jr. 1972a. Retention of inhaled acetaldehyde in the dog, *Arch. Environ. Health* 24:353–357.

Egle, J. L., Jr. 1972b. Retention of inhaled formaldehyde, propionaldehyde, and acrolein in the dog, *Arch. Environ. Health* 25:119–124.

Elmenhorst, H., and Schultz, C. H. 1968. Flüchtige inhaltssotffe des Tabaksrauches. Die chemischen Bestandteile der Gas-Dampf-Phase, *Beitr. Tabakforsch.* 4:90–123.

Fassett, D. W. 1963. Aldehydes and acetals, In: *Industrial Hygiene and Toxicology* (F. A. Patty, ed.), Vol. II, 2nd rev. ed., pp. 1959–1989, Interscience, New York.

Feldman, M. Y. A. 1967. Reaction of formaldehyde with nucleotides and ribonucleic acid, *Biochim. Biophys. Acta* 149:20–34.

Feron, V. J., and Kruysse, A. 1977. Effects of exposure to acrolein vapor in hamsters simultaneously treated with benzo[a]pyrene or dimethylnitrosamine, *J. Toxicol. Environ. Health* 3:379–394.

Feron, V. J., Kruysse, A., and Woutersen, R. A. 1982. Respiratory tract tumors in hamsters exposed to acetaldehyde vapour alone or simultaneously to benzo[a]pyrene or dimethylnitrosamine, *Eur. J. Cancer Clin. Oncol.* 18:13–31.

Fornace, A. J., Jr. 1982. Detection of DNA single-strand breaks produced during the repair of damage by DNA-protein cross-linking agents, *Cancer Res.* 42:145–149.

Fornace, A. J., Jr., Lechner, J. F., Grafstrom, R. C., and Harris, C. C. 1982. DNA repair in human bronchial epithelial cells, *Carcinogenesis* 3:1373–1377.

Fraenkel-Conrat, H. 1954. Reaction of nucleic acid with formaldehyde, *Biochim. Biophys. Acta* 15:307–309.

Frazelle, J. H., Abernethy, D. J., and Boreiko, C. J. 1983. Weak promotion of C3H10T1/2 cell transformation by repeated treatments with formaldehyde, *Cancer Res.* 43:3236–3239.

Goldmacher, V. S., and Thilly, W. G. 1983. Formaldehyde is mutagenic for cultured human cells, *Mutat. Res.* 116:417–422.

Goldsmith, J. R., and Friberg, L. T. 1977. Effects of air pollution on human health. In: *Air Pollution*, Vol. II (A. C. Stern, ed.), pp. 457–610, Academic Press, New York.

Goodman, J. I., and Tephly, T. R. 1971. A comparison of rat and human liver formaldehyde dehydrogenase, *Biochim. Biophys. Acta* 252:489–505.

Grafstrom, R. C., Fornace, A. J., Jr., Autrup, H., Lechner, J. F., and Harris, C. C. 1983. Formaldehyde damage to DNA and inhibition of DNA repair in human cells, *Science* 220:216–218.

Grafstrom, R. C., Fornace, A., Jr., and Harris, C. C. 1984. Repair of DNA damage caused by formaldehyde in human cells, *Cancer Res.* 44:4323–4327.

Grosjean, D. 1982. Formaldehyde and other carbonyls in Los Angeles ambient air, *Environ. Sci. Technol.* 16:254–262.

Haggard, H. W., and Greenberg, L. A. 1939. Studies on the absorption, distribution, and elimination of alcohol. The elimination of methyl alcohol, *J. Pharmacol. Exp. Ther.* 66:479–496.

Hatch, G. G., Conklin, P. M., Chrostensen, C. C., Casto, B. C., and Nesnow, S. 1983. Synergism in the transformation of hamster embryo cells treated with formaldehyde and adenovirus, *Environ. Mutagen.* 5:49–57.

Hecht, S. S., Carmella, S., Mori, H., and Hoffmann, D. J. 1981. A study of tobacco carcinogenesis. 20. Role of catechol as a major cocarcinogen in the weakly acidic fraction of smoke condensate, *J. Nat. Cancer Inst.* 66:163–169.

Heinrich, R., and Angerer, J. 1982. Occupational chronic exposure to organic solvents. X. Biological monitoring parameters for methanol exposure, *Int. Arch. Occup. Environ. Health* 50:341–349.

Hendrick, D. J., and Lane, D. J. 1975. Formalin asthma in hospital staff, *Br. Med. J.* 1:607–608.

Hoard, D. E. 1960. The applicability of formol titration to the problem of end-group determination in polynucleotides. A preliminary investigation, *Biochim. Biophys. Acta* 40:62–70.

Hoshika, Y. 1977. Simple and rapid gas-liquid-solid chromatographic analysis of trace concentrations of acetaldehyde in urban air, *J. Chromatogr.* 137:455–460.

Howard, J. A. 1973. Homogeneous liquid phase autoxidations, In: *Free Radicals* (J. K. Kochi, ed.), Vol. II, pp. 3–62, Wiley-Interscience, New York.

International Agency for Research on Cancer. 1976. Glycidaldehyde, In: *IARC Monographs on the Evaluation of Carcinogenic Risk of Chemicals to Man*, Vol. 11, p. 175, IARC, Lyon, France.

Irons, R. D., and Sawahata, T., 1985. Phenols, catechols, and quinones, In: *Bioactivation of Foreign Compounds* (M. W. Anders, ed.), pp. 259–281, Academic Press, New York.

Irons, R. D., Neptun, D. A., and Pfeifer, R. W. 1981. Inhibition of lymphocyte transformation and microtubule assembly by quinone metabolites of benzene: evidence for a common mechanism, *J. Reticuloendothel. Soc.* 30:359–372.

Ishikawa, T., Esterbauer, H., and Sies, H. 1986. Role of cardiac glutathione transferase and of the glutathione S-conjugate export system in biotransformation of 4-hydroxynonenal in the heart, *J. Biol. Chem.* 261:1576–1581.

Ito, N., Hagiwara, A., Shibata, M., Ogiso, T., and Fukushima, S. 1982. Induction of squamous-cell carcinoma in the forestomach of F344 rats treated with butylated hydroxyanisole, *Gann* 73:332–334.

Jakab, G. J. 1977. Adverse effect of a cigarette smoke component, acrolein, on pulmonary antibacterial defenses and on viral-bacterial interactions in the lung, *Am. Rev. Respir. Dis.* 115:33–38.

Jakoby, W. B., and Habig, W. H. 1980. Glutathione transferases, In: *Enzymatic Basis of Detoxification* (W. B. Jakoby, ed.), Vol. II, pp. 63–94, Academic Press, New York.

Jakoby, W. B., Sekura, R. D., Lyon, E. S., Marcus, C. J., and Wang, J.-L. 1980. Sulfotransferases, In: *Enzymatic Basis of Detoxification* (W. B. Jakoby, ed.), Vol. II, pp. 199–228, Academic Press, New York.

Kane, L. E., Barrow, C. S., and Alarie, Y. 1979. A short-term test to predict acceptable levels of exposure to airborne sensory irritants, *Am. Ind. Hyg. Assoc. J.* 40:207–229.

Kasper, C. B., and Henton, D. 1980. Glucuronidation, In: *Enzymatic Basis of Detoxification* (W. B. Jakoby, ed.), Vol. II, pp. 3–36, Academic Press, New York.

Kerns, W. D., Pavkov, K. L., Donofrio, D. J., Gralla, E. J., and Swedberg, J. A. 1983. Carcinogenicity of formaldehyde in rats and mice after long-term inhalation exposure, *Cancer Res.* 43:4382–4392.

Kilburn, K. H., and McKenzie, W. N. 1978. Leukocyte recruitment to airways by aldehyde-carbon combinations that mimic cigarette smoke, *Lab. Invest.* 38:134–142.

Kornberg, A. 1980. *DNA Replication*, W. H. Freeman, San Francisco.

Krokan, H., Grafstrom, R. C., Sundqvist, K., Esterbauer, H., and Harris, C. C. 1985. Cytotoxicity, thiol depletion, and inhibition of O^6-methyl-guanine-DNA methyltransferase by various aldehydes in cultured human bronchial fibroblasts, *Carcinogenesis* 6:1755–1759.

Kutzman, R. S., Popenoe, E. A., Schmaeler, M., and Drew, R. T. 1985. Changes in rat lung structure and composition as a result of subchronic exposure to acrolein, *Toxicology* 34:139–151.

Kuwata, K., Uebori, M., and Yamazaki, Y. 1981. Reversed-phase liquid chromatographic determination of phenols in auto exhaust and tobacco smoke as *p*-nitrobenzeneazophenol derivatives, *Anal. Chem.* 53:1531–1534.

Lee, H. K., Alarie, Y., and Karol, M. H. 1984. Induction of formaldehyde sensitivity in guinea pigs, *Toxicol. Appl. Pharmacol.* 75:147–155.

Levy, S., Nocentini, S., and Billardon, C. 1983. Induction of cytogenetic effects in human fibroblast cultures after exposure to formaldehyde or X-rays, *Mutat. Res.* 119:309–317.

Lloyd, W. G. 1973. Autoxidations, In: *Methods in Free Radical Chemistry* (E. S. Huyser, ed.), Vol. 4, pp. 1–131, Marcel Dekker, New York.

Maibach, H. 1983. Formaldehyde: effects on animal and human skin, In: *Formaldehyde Toxicity* (J. E. Gibson, ed.), pp. 163–174, Hemisphere Publ. Corp., Washington, D.C.

Makar, A. B., and Tephly, T. R. 1975. Inhibition of monkey liver alcohol dehydrogenase by 4-methylpyrazole, *Biochem. Med.* 13:334–342.

Mannering, G. J., and Parks, R. E., Jr. 1957. Inhibition of methanol metabolism with 3-amino-1,2,4-triazole, *Science* 126:1241–1242.

Martin-Amat, G., McMartin, K. E., Hayreh, S. S., Hayreh, M. S., and Tephly, T. R. 1978. Methanol poisoning: ocular toxicity produced by formate, *Toxicol. Appl. Pharmacol.* 45:201–208.

Mashford, P. M., and Jones, A. R. 1982. Formaldehyde metabolism by the rat: a reappraisal, *Xenobiotica* 12:119–124.

McGhee, J. D., and von Hippel, P. H. 1975. Formaldehyde as a probe of DNA structure. I. Reaction with exocyclic amino groups of DNA bases, *Biochemistry* 14:1281–1296.

McGirr, L. G., Hadley, M., and Draper, H. H. 1985. Identification of N^α-acetyl-ϵ-2-propenal lysine as a urinary metabolite of malondialdehyde, *J. Biol. Chem.* 260:15427–15431.

McMartin, K. E., Makar, A. B., Martin-Amat, G., Palese, G., and Tephly, T. R. 1975. Methanol poisoning. I. The role of formic acid in the development of metabolic acidosis in the monkey and the reversal by 4-methylpyrazole, *Biochem. Med.* 13:319–333.

McMartin, K. E., Martin-Amat, G., Makar, A. B., and Tephly, T. R. 1977. Methanol poisoning. V. Role of formate metabolism in the monkey, *J. Pharmacol. Exp. Ther.* 201:564–572.

McMartin, K. E., Martin-Amat, G., Noker, P. E., and Tephly, T. R. 1979. Lack of a role for formaldehyde in methanol poisoning in the monkey, *Biochem. Pharmacol.* 28:645–649.

Morgan, K. T. 1983. Localization of areas of inhibition of nasal mucociliary function in rats following *in vivo* exposure to formaldehyde, *Am. Rev. Respir. Dis.* 127:166.

Morgan, K. T., Patterson, D. L., and Gross, E. A. 1983. Formaldehyde and the mucociliary apparatus, In: *Formaldehyde: Toxicology, Epidemiology and Mechanisms* (J. J. Clary, J. E. Gibson, and A. S. Waritx, eds.), Marcel Dekker, New York.

Neely, W. B. 1964. The metabolic fate of formaldehyde-^{14}C intraperitoneally administered to the rat, *Biochem. Pharmacol.* 13:1137–1142.

Newsome, J. R., Norman, V., and Keith, C. H. 1965. Vapour phase analysis of tobacco smoke, *Tob. Sci.* 9:102–110.

Noker, P. E., Eells, J. T., and Tephly, T. R. 1980. Methanol toxicity: treatment with folic acid and 5-formyltetrahydrofolic acid, *Alcohol. Clin. Exp. Res.* 4:378–383.

Packer, L. (ed.) 1984. *Methods in Enzymology. Oxygen Radicals in Biological Systems.* Academic Press, New York.

Patel, J. M., Gordon, W. P., Nelson, S. D., and Leibman, K. C. 1983. Comparison of hepatic biotransformation and toxicity of allyl alcohol and [1,1-^{2}H$_2$]allyl alcohol in rats, *Drug Metab. Dispos.* 11:164–166.

Pryor, W. A. 1984. Free radicals in autoxidation and in aging, In: *Free Radicals in Molecular Biology, Aging, and Disease* (D. Armstrong, R. S. Sohal, R. G. Cutler, and T. F. Slater, eds.), pp. 13–41, Raven Press, New York.

Ragan, D. L., and Boreiko, C. J. 1981. Initiation of C3H/10T1/2 cell transformation by formaldehyde, *Cancer Lett.* 13:325–331.

Rietbrock, V. 1969. Kinetik und wege des Methanolumsatzes, *Naunyn-Schmiedebergs Arch. Pharmakol. Exp. Pathol.* 263:88–105.

Roe, O. 1955. The metabolism and toxicity of methanol, *Pharmacol. Rev.* 7:399–412.

Roe, O. 1982. Species differences in methanol poisoning, *CRC Crit. Rev. Toxicol.* 10:275–286.

Ross, W. E., and Shipley, N. 1980. Relationship between DNA damage and survival in formaldehyde-treated mouse cells, *Mutat. Res.* 79:277–283.

Ross, W. E., McMillan, D. R., and Ross, C. F. 1981. Comparison of DNA damage by methylmelamines and formaldehyde, *J. Nat. Cancer Inst.* 67:217–221.

Schauenstein, E., Esterbauer, H., and Zollner, H. 1977. *Aldehydes in Biological Systems*, pp. 35–38, Pion Limited, London.

Sedivec, V., Mráz, M., and Flek, J. 1981. Biological monitoring of persons exposed to methanol vapours, *Int. Arch. Occup. Environ. Health* 48:257–271.

Sellakumar, A. R., Snyder, C. A., Solomon, J. J., and Albert, R. E. 1985. Carcinogenicity of formaldehyde and hydrogen chloride in rats, *Toxicol. Appl. Pharmacol.* 81:401–406.

Simic, M. G., and Hunter, E. P. L. 1983. Interactions of free radicals and antioxidants, In: *Radioprotectors and Anticarcinogens* (O. F. Nygaard and M. G. Simic, eds.), pp. 449–460, Academic Press, New York.

Slaga, T. J., Klein-Szanto, A. J., Triplett, L. L., Yotti, L. P., and Trosko, J. E. 1981. Skin tumor–promoting activity of benzoyl peroxide, a widely used free radical–generating compound, *Science* 213: 1023–1025.

Slaga, T. J., Solanki, V., and Logani, M. 1983. Studies on the mechanism of action of antitumor promoting agents: suggestive evidence for the involvement of free radicals in promotion, In: *Radioprotectors and Anticarcinogens* (O. F. Nygaard and M. G. Simic, eds.), pp. 471–485, Academic Press, New York.

Slott, V. L., and Hales, B. F. 1985. Teratogenicity and embryolethality of acrolein and structurally related compounds in rats, *Teratology* 323:65–72.

Smith, M. T., Evans, C. G., Thor, H., and Orrenius, S. 1985. Quinone-induced oxidative injury to cells and tissues, In: *Oxidative Stress* (H. Sies, ed.), pp. 91–114, Academic Press, New York.

Snider, J. R., and Dawson, G. A. 1985. Tropospheric light alcohols, carbonyls, and acetonitrile: concentrations in the southwestern United States and Henry's law data, *J. Geophys. Res.* 90:3797–3805.

Strittmatter, P., and Ball, E. G. 1955. Formaldehyde dehydrogenase, a glutathione-dependent enzyme system, *J. Biol. Chem.* 213:445–461.

Swarin, S. J., and Lipari, F. 1983. Determination of formaldehyde and other aldehydes by high performance liquid chromatography with fluorescence detection, *J. Liq. Chromatogr.* 6:425–444.

Swenberg, J. A., Barrow, C. S., Boreiko, C. J., d'A Heck, H., Levine, R. J., Morgan, K. T., and Starr, T. B. 1983. Non-linear biological responses to formaldehyde and their implications for carcinogenic risk assessment, *Carcinogenesis* 4:945–952.

Szabad, J., Soos, I., Polgar, G., and Hejja, G. 1983. Testing the mutagenicity of malondialdehyde and formaldehyde by the Drosophila mosaic and the sex-linked recessive lethal tests, *Mutat. Res.* 113: 117–133.

Tanner, R. L., and Meng, Z. 1984. Seasonal variations in ambient atmospheric levels of formaldehyde and acetaldehyde, *Environ. Sci. Technol.* 18: 723–726.

Tephly, T. 1985. The Toxicity of Methanol and Its Metabolites in Biological Systems, Health Effects Institute Report, Cambridge, Mass.

Thelander, L., and Reichard, P. 1979. Reduction of ribonucleotides, *Ann. Rev. Biochem.* 48:133–158.

Travis, J., and Salvesen, G. S. 1983. Human plasma proteinase inhibitors, *Ann. Rev. Biochem.* 52:655–709.

Treitman, R. D., Burgess, W. A., and Gold, A. 1980. Air contaminants encountered by firefighters, *Am. Ind. Hyg. Assoc. J.* 41:796–802.

Van Duuren, B. L., and Goldschmidt, B. M. 1976. Carcinogenic and tumor-promoting agents in tobacco carcinogenesis, *J. Nat. Cancer Inst.* 56:1237–1242.

Van Duuren, B. L., and Loewengart, G. 1977. Reaction of DNA with glycidaldehyde. Isolation and identification of a deoxyguanosine reaction product, *J. Biol. Chem.* 252:5370–5371.

Van Duuren, B. L., Orris, L., and Nelson, N. 1965. Carcinogenicity of epoxides, lactones, and peroxy compounds. II. *J. Nat. Cancer Inst.* 35:707–717.

Van Duuren, B. L., Langseth, L., Orris, L., Teebor, G., Nelson, N., and Kuschner, M. 1966. Carcinogenicity of epoxides, lactones, and peroxy compounds. IV. Tumor response in epithelial and connective tissue in mice and rats, *J. Nat. Cancer Inst.* 37:825–838.

Van Duuren, B. L., Langseth, L., Goldschmidt, B. M., and Orris, L. 1967. Carcinogenicity of epoxides, lactones, and peroxy compounds. VI. Structure and carcinogenic activity, *J. Nat. Cancer Inst.* 39:1217–1228.

Watkins, W. D., Goodman, J. I., and Tephly, T. R. 1970. Inhibition of methanol and ethanol oxidation by pyrazole in the rat and monkey *in vivo*, *Mol. Pharmacol.* 6:567–572.

Willson, R. L. 1985. Organic peroxy free radicals as ultimate agents in oxygen toxicity, In: *Oxidative Stress* (H. Sies, ed.), pp. 41–72, Academic Press, New York.

Witz, G., Lawrie, N. J., Amoruso, M. A., and Goldstein, B. D. 1985. Inhibition by reactive aldehydes of superoxide anion radical production in stimulated human neutrophils, *Chem.-Biol. Interact.* 53:13–23.

Woutersen, R. A., Appleman, L. M., Feron, V. J., and Van Der Heijden, C. A. 1984. Inhalation toxicity of acetaldehyde in rats. II. Carcinogenicity study: interim results after 15 months, *Toxicology* 31:123–133.

Yant, W. P., and Schrenck, H. H. 1937. Distribution of methanol in dogs after inhalation of methanol and administration by stomach tube and subcutaneously, *J. Ind. Hyg. Toxicol.* 19:337–345.

Evaluation of Automotive Emissions as Risk Factors for the Development of Atherosclerosis and Coronary Heart Disease

THOMAS B. CLARKSON
Wake Forest University

Air Pollution, the Automobile, and Public Health. © 1988 by the Health Effects
Institute. National Academy Press, Washington, D.C.

Automotive emissions, particularly carbon monoxide (CO), are thought to be risk factors in the development of atherosclerosis and heart disease. From studies of the components of cigarette smoke, it is known that continued exposure to high levels of CO results in a higher incidence of death and disease among smokers. Thus, it seems reasonable to suppose that significant exposure to automotive emissions could have similar results. Unfortunately, little experimental evidence exists to support that speculation.

In this chapter, the ways in which exposure to automotive emissions might affect atherogenesis are reviewed. The development of atherosclerosis and the susceptibility of various groups are reviewed as well. Although many studies have been done, research to date does not give us a clear picture of the effect of CO on atherogenesis. For that reason, it will be necessary to undertake definitive research on the effects of automotive emissions in general. If air pollution is found to have an effect on atherogenesis, then the components of automotive emissions should be studied separately. Since both types of studies will probably involve animal research, the potential of various animal models for atherosclerosis research is also reviewed.

Atherosclerosis

Natural History

Arteries consist of three layers: the inner layer called the intima, the media, and the outer layer or adventitia. Atherosclerosis is a pathological process causing the intima to be thickened by intra- and extracellular accumulations of lipids, by variable degrees of proliferation of intimal smooth muscle cells, and to some extent, by migration of macrophages into the intima (McGill 1977). The process of atherogenesis occurs in human beings of all races and societies, but differs greatly in degree and extent of clinical complications among various ethnic and geographic groups (Strong et al. 1972). Atherosclerosis in the coronary arteries is the usual cause of coronary heart

disease (a term covering the spectrum of angina pectoris, myocardial infarction, and disturbances of cardiac rhythm). Atherosclerosis of the cerebral arteries is associated with transient ischemic (decreased blood flow) attacks and stroke. Atherosclerosis of leg arteries may result in leg pain (termed "intermittent claudication"), and occasionally gangrene of the extremities.

The first lesions observed in the intima of arteries are fatty streaks; these lesions develop in nearly all children and have no physiological or pathological consequences (Strong et al. 1972). In some, but not all populations, the fatty streaks progress to fibrous plaques, which in turn lead to arterial stenosis and clinical manifestations of atherosclerosis (McGill 1968). Although direct observations of the progression of fatty streaks to fibrous plaques is not possible, the occurrence of such a "progression" is accepted generally (Robertson et al. 1963; National Heart, Lung and Blood Institute 1982).

Depending on the relative exposure to certain risk factors, plaques enlarge to the extent that they reduce blood flow to the point of causing ischemic necrosis (McGill 1968). Arterial stenosis, however, is not a simple matter of continued growth of the plaque. Although much remains to be understood, plaque growth seems to be complicated by events that happen to the plaque itself, and by events that happen to the artery with the plaque (Ross 1981). Plaques may ulcerate and provoke thrombus (blood clot) formation, and the thrombi may accelerate the growth of the plaque. Since the platelets participate in thrombus formation they may liberate physiologically active substances such as thromboxane A_2, which can cause arterial spasm and possibly ischemic necrosis (Neri-Serneri et al. 1981).

Pathogenesis

The cellular and molecular events that lead to the development of fatty streaks have been the focus of intensive research in recent years and have been reviewed by Ross (1986). Most researchers agree that the initial events are injuries to the arterial endothelium (figure 1). Endothelial injury

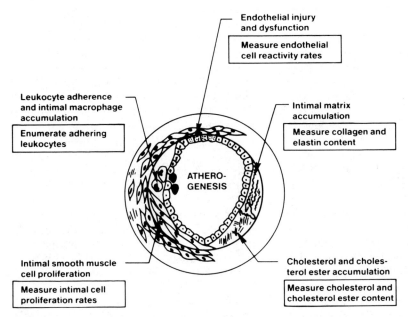

Endothelial injury and dysfunction
Measure endothelial cell reactivity rates

Leukocyte adherence and intimal macrophage accumulation
Enumerate adhering leukocytes

Intimal matrix accumulation
Measure collagen and elastin content

ATHERO-GENESIS

Intimal smooth muscle cell proliferation
Measure intimal cell proliferation rates

Cholesterol and cholesterol ester accumulation
Measure cholesterol and cholesterol ester content

Figure 1. Schematic illustration of the pathogenic components of atherogenesis and methods for the evaluation of these components.

can be diverse in nature and can result from exposure to mechanical forces (Stemerman and Ross 1972; Fry 1973; Moore 1973; Fishman et al. 1975), by exposure to lipoproteins (Faggiotto et al. 1984), by various toxins (Reidy and Bowyer 1978), and by immunologic injury (Minick et al. 1978).

■ Recommendation 1. Research to determine whether exposure of endothelial cells to automotive emissions affects endothelial function should be undertaken with emphasis on endothelial cell replication and prostacyclin production.

■ Recommendation 2. Studies should be done to determine whether chronic exposure to automotive emissions affects the repair of stress-induced endothelial injury and/or the way in which the intima of arteries responds to such injuries.

Intimal smooth muscle cell proliferation is the primary cellular response associated with the progression of atherosclerosis. There is strong evidence that this smooth muscle cell proliferation is the result of a number of growth factors associated with atherogenesis. Particularly relevant is the platelet-derived growth factor (Ross et al.

1974; Ross and Voegl 1978), endothelial-derived growth factor (Fass et al. 1978; Gajdusek et al. 1980), and finally by the monocyte/macrophage-derived growth factor (Liebovich and Ross 1976; Glenn and Ross 1981).

■ Recommendation 3. The effect of automotive emissions on the catabolism of low-density lipoprotein (LDL) particles by smooth muscle cells, the replication rates of these cells, and the kinds of connective tissue proteins they elaborate should be determined.

Many human fibrous plaques have been found to be monotypic for one of the isozymes of glucose-6-phosphate dehydrogenase (G-6-PD) (Benditt and Benditt 1973; Pearson et al. 1978; Thomas et al. 1979). This observation was interpreted as evidence for a monoclonal origin of the smooth muscle cells of the atherosclerotic lesions, leading to the monoclonal theory of atherosclerosis (Benditt and Benditt 1973). Monoclonality suggests that smooth muscle cells of the atherosclerotic plaque arise from a single precursor cell that has superior growth and survival characteristics. Lesions of monoclonal origin are generally considered to be neoplastic. Chemical mutagens and viruses have been suggested as being responsible for this neoplastic change (Benditt and Benditt 1973). There is no doubt that many human fibrous lesions are monotypic with respect to G-6-PD. Whether these monotypic lesions are indeed monoclonal or whether the monotypism results from a selective survival advantage of one or the other cell phenotypes is not clear. Thomas and Kim (1983) have recently reviewed the evidence for and against the "monoclonal hypothesis" versus the "phenotype/selective advantage hypothesis." The results of studies with experimental animals and further examination of human atherosclerotic lesions indicate that monotypism is not a prerequisite for atherosclerosis development. Although the exact roles of chemical mutagens and monotypism are yet to be defined, they are probably important in atherogenesis and provide a rational basis for the assumption that exposure to automotive emissions may be important.

Over the past two decades there has been extensive interest in the role of platelets in atherogenesis. Following injury of the arterial endothelium by such procedures as balloon catheterization, platelets can be seen to adhere to such injuries and presumably release platelet-derived growth factor. It has been shown that in association with the adherence of platelets there is migration of medial smooth muscle cells into the intima (Stemerman and Ross 1972) where they proliferate and form sizable intimal lesions. Intimal smooth muscle cell proliferation following injury can be prevented by treatment of animals with antiplatelet serum (Moore et al. 1976; Friedman et al. 1977).

The most important area of current research in atherogenesis concerns the role of monocyte-derived macrophages. The intimal macrophages are not only important with regard to their phagocytic function, but elaborate important growth factors. Previously, the intimal smooth muscle cell was presumed to be the precursor cell for the predominant numbers of foam cells (lipid-containing cells that are "foamy" in appearance) in lesions. Most workers now agree that monocyte-derived macrophages

represent the majority of the foam cells present in fatty streaks (Gerrity 1981a,b). Macrophages have high-affinity receptors for chemically altered LDLs (Goldstein et al. 1979; Fogelman et al. 1980; Brown et al. 1981; Schechter et al. 1981). Alteration of LDLs by hydrocarbons, aldehydes, or other chemicals in automotive emissions may result in their being internalized by macrophages; thus intimal macrophages may be of particular importance in emissions-exposure atherogenesis.

■ **Recommendation 4.** The effect of automotive emissions on the internalization of β-VLDL (very low–density lipoprotein) and other modified LDL particles by macrophages should be assayed.

Progression of Fatty Streaks to Fibrous Plaques. In humans, fatty streaks appear in the aorta in the first decade, in the coronary and cerebral arteries later, and are not associated with any clinical events. Fibrous plaques with necrotic, lipid–rich cores surrounded by smooth muscle and connective tissue appear during the third decade, and it is these lesions that undergo complications and cause ischemic necrosis in one or another tissue.

Perhaps the most controversial subject in atherosclerosis research is whether fibrous plaques arise from fatty streaks or from some other precursor lesion (McGill 1984). The evidence that fatty streaks are the precursors of fibrous plaques is circumstantial. Fatty streaks occur most commonly in the proximal portion of the left anterior descending coronary artery, and that is also the site in which the most severe fibrous plaques occur (Montenegro and Eggen 1968). Taking advantage of that site's predilection, efforts have been made to trace the pathogenic events that occur in numerous children and young adults (Stary 1983, 1984; Stary and Letson 1983). The general findings from these studies were that the coronary artery intima at the lesion-prone site becomes as thick as the media does in early life, and that the thickened intima is made up of smooth muscle cells, connective tissue fibers, and proteoglycans beneath an intact and normally appearing endothelium. By age 10 there are clusters of monocytes and macrophage foam cells, and adjacent to these clusters are lipid-containing smooth muscle cells. By age 15 these clusters of macrophages and smooth muscle cells begin to undergo necrosis. By age 20 there are necrotic foci associated with foam cells, and the lesions become larger and more frequent. At about this age and as necrosis continues, the plaques appear fibrous, and these temporal relationships lead many to conclude that the progressing fatty streak is indeed the precursor of the fibrous plaque.

Plaque Complications. One of the understudied areas in atherosclerosis research is the pathological processes associated with the progression of fibrous plaques to complicated plaques. The process involves necrosis within the central portions of plaques, mineralization, and ulceration. And, in the process of ulceration, the plaques eventually become covered by thrombi, which can occlude the artery. Although no definitive data exist, the general belief is that necrosis is the primary event in plaque complication and that the process of ulceration and mural thrombosis is associated with necrosis. There are no published accounts of the sequence of pathogenic events in plaque complication in human beings, although the process has been well studied in pigeons (Prichard et al. 1964a).

It seems likely that hypoxia is associated with plaque necrosis and the other events that lead to clinical sequelae of the atherogenic process. Studies of automotive emissions and the relative hypoxia that occurs from the carboxyhemoglobin may be important in the process of plaque complication.

■ **Recommendation 5.** Since automotive emission exposure may result in relative hypoxia, it seems reasonable to suspect that plaque complications could be hastened by that mechanism. Cynomolgus monkeys with diet-induced atherosclerosis of two or three years' duration should be exposed to various concentrations of automotive emissions and their plaque compli-

cations compared with those in appropriate control animals not exposed to automotive emissions.

Coronary Heart Disease. Coronary heart disease is the term used to describe the clinical syndromes and pathological events associated with inadequate or obstructed blood flow to the myocardium. The major clinical manifestation of coronary heart disease is angina pectoris. The pathological event associated with interrupted blood flow to the myocardium is termed a myocardial infarction. In most cases, obstruction of the blood flow in the coronary arteries is caused by stenotic atherosclerosis (Cohen and Braunwald 1980). In some individuals, obstruction of blood flow in the coronary arteries is associated with coronary artery spasm with, or occasionally without, atherosclerotic lesions in the arteries (Dalen et al. 1981). Mural thrombosis associated with coronary artery atherosclerosis is the usual cause of myocardial infarction. Platelet aggregation and fibrin thrombi may be associated with clinical symptoms of coronary heart disease and may also be involved in the progression of coronary artery atherosclerosis (Mustard and Packham 1972).

Research in the area of myocardial ischemia is advancing rapidly. The role of prostaglandins in coronary heart disease has been the subject of a recent monograph by Hegyeli (1981), and the exploration of endothelium-dependent relaxing factors in coronary arteries is in an early stage.

Factors that Influence Atherogenesis

Age

There is emerging evidence that young animals are less susceptible than old animals to the development of diet-induced atherosclerosis. The earliest work was done using *Cebus albifrons* monkeys, and the extent of aortic and coronary artery atherosclerosis was found to be significantly less among juvenile animals compared with adult animals fed the same cholesterol-containing diet for the same lengths of time (Bullock et al. 1969; Clarkson et al. 1976). More recently those observations have been extended to an Old World monkey, *Macaca fascicularis* (Weingand et al. 1986). Although plasma lipid and lipoprotein concentrations were the same, adults had more extensive coronary artery atherosclerosis than juveniles.

Whether such age differences exist in human primates is not known. The possibility that there may be age-related differences in the effect of exposure to CO or to automotive emissions seems possible and should be taken into consideration in the planning of future research.

Gender

The most comprehensive reports comparing coronary artery atherosclerosis of male and female human subjects are those of the International Atherosclerosis Project (Tejada et al. 1968). Males had more coronary artery atherosclerosis than females in all geographic/racial groups except the São Paulo Negroes. The gender-specific differences in extent of coronary artery atherosclerosis are thought to account for the gender-specific differences in coronary heart disease morbidity and mortality, since the differential in clinical events and lesion extent is similar (Kannel et al. 1961; Cassel 1971; Armstrong et al. 1972). Explanations for the gender-specific differences in coronary artery atherosclerosis have included plasma lipoprotein concentrations (particularly higher high-density lipoprotein concentrations among women), a protective effect of estrogens, and the possibility that females do not share the competitive and sometimes hostile behavior of males and are thus spared from the pathophysiological effects of such stresses (Barr 1953; Marmorston et al. 1957; McGill and Stern 1979).

Male and female cynomolgus monkeys appear to be excellent models of the gender-specific difference in coronary artery atherosclerosis seen in some populations, particularly Caucasian North Americans (Kaplan et al. 1982; Hamm et al. 1983; Kaplan et al. 1984a,b). They share with human males and females gender differences in high-density lipoprotein cholesterol (HDLC) concentrations as well as the quotient of total serum cholesterol and HDLC concentrations. There are significant gender-specific differences in the extent of coronary artery atherosclerosis, and the relative degree of female protection seems to be directly related to ovarian function as has been demonstrated by the loss of female protection among chronically stressed submissive females and ovariectomized females.

Studies of humans as well as of nonhuman primates strongly support higher plasma concentrations of the high-density lipoproteins (HDLs) as important in "female protection." As discussed elsewhere in this chapter, there is evidence that cigarette smoking reduces HDL concentrations. By analogy, there is some possibility that exposure to automotive emissions would reduce HDL concentrations.

■ **Recommendation 6.** Studies to determine whether and to what extent automotive emissions exposure may diminish or abolish the gender-specific difference in atherogenesis needs to be explored.

Genetic Susceptibility

At least two genetic mechanisms have been identified that modulate susceptibility to coronary artery atherosclerosis. The first concerns genetic control of the extent to which animals increase their plasma cholesterol concentrations when fed dietary cholesterol (hyper- and hyporesponsiveness). Hyper- and hyporesponsiveness to dietary cholesterol have been studied in squirrel monkeys, rhesus monkeys, and baboons. Although differences in plasma cholesterol concentrations are not apparent among squirrel monkeys fed diets free of cholesterol, some monkeys (hyperresponders) fed cholesterol-containing diets develop considerable hypercholesterolemia, whereas others fed the same diet maintain near normal plasma cholesterol concentrations (Lofland et al. 1970; Clarkson et al. 1971).

Rhesus monkeys have also been studied extensively to understand the mechanisms of hyper- and hyporesponsiveness to dietary cholesterol. Unlike the other species, there are significant differences between hyper- and hyporesponsive rhesus monkeys while they are fed control diets free of cholesterol. Like other species, the trait is exaggerated by the consumption of a high-cholesterol diet (Eggen 1976). Differences in cholesterol absorption appear to be of major importance in hyper- and hyporesponsiveness of rhesus monkeys (Bhattacharyya and Eggen 1977, 1980, 1983).

The strength of the genetic influence on hyper- and hyporesponsiveness to dietary cholesterol among baboons has also been studied (Flow et al. 1981). Estimates of heritability for total serum cholesterol concentration were low in early life, but increased as the animals matured. Studies have also been conducted to determine the relationship between the genetic control of cholesterol metabolism and plasma concentrations of HDLs (Flow and Mott 1984). Strong evidence was provided that the size of the rapidly miscible pool of body cholesterol and the movement of cholesterol in and out of that pool are influenced to a large degree by the same genes that regulate the plasma concentrations of the HDLs.

The second kind of genetic control of atherogenesis has been termed "mesenchymal susceptibility." This phenomenon refers to the reaction of the artery wall to accumulated lipoproteins. Macaques appear particularly useful for research on mesenchymal susceptibility to diet-induced coronary artery atherosclerosis. Using studies of diet-induced atherosclerosis of cynomolgus mokeys, Malinow and coworkers (1976b) were the first to focus attention on this phenomenon. They introduced the terms hyper- and hyporeactivity in coronary artery lesion extent, and the terms hyper- and hyposusceptibility to the trait have subsequently been used. Rhesus monkeys are the macaques that have been studied the most in an attempt to develop colonies of hyper- and hyposusceptible animals. Considerable variation exists among monkeys of this species in their response to an atherogenic diet. The cellular mecha-

nisms that account for these differences have not been explained, although the details of the development of breeding colonies for this trait have been described (Clarkson et al. 1985). It seems useful to determine whether strains of rhesus monkeys that are hyper- and hyposusceptible to the effects of a cholesterol-containing diet are also hyper- and hyposusceptible in response to exposure to CO and automotive emissions.

The work of Watanabe and associates (1985) supports the concept of individual differences in mesenchymal susceptibility to coronary artery atherosclerosis. These investigators developed and characterized a strain of rabbits with a heritable disorder of cholesterol metabolism quite similar to familial hypercholesterolemia of humans (Tanzawa et al. 1980; Kita et al. 1981; Shimada et al. 1981). The more recent work of Watanabe and his colleagues (1985) concerns the selective breeding of animals with increased or decreased coronary artery atherosclerosis while equivalently hyper-β-lipoproteinemic.

■ **Recommendation 7.** Experiments that determine whether chronic exposure to CO affects coronary artery atherosclerosis differently in genetically susceptible and resistant strains are needed.

Psychosocial Phenomena

There is increasing evidence that psychosocial influences might affect the development of coronary artery atherosclerosis and coronary heart disease. The most extensively studied psychosocial variable has been the Type A or "coronary-prone" behavior pattern (Glass 1977). Type A refers to a constellation of overt behaviors and stylistic mannerisms characterized by hard-driving competitiveness, a sense of time urgency and easily evoked hostility. Individuals with a relative absence of these characteristics are termed Type B. Retrospective and prospective studies indicate that individuals with the Type A behavior pattern or elements of this pattern (for example, potential for hostility) develop coronary heart disease more frequently and are more likely to die of myocardial infarction than individuals with the Type B behavior pattern (Rosenman et al. 1975; Zyzanski 1978; Review Panel on Coronary-Prone Behavior and Coronary Heart Disease 1981; Barefoot et al. 1983; Shekelle et al. 1983).

The mechanisms by which psychosocial phenomena influence coronary heart disease and atherosclerosis in humans remain largely unknown. Three findings related to sympathetic arousal are relevant, however. First, individuals vary markedly in their sympathetic responsivity to behavioral stimuli (Manuck and Garland 1980; McCubbin et al. 1983) and these response characteristics are enduring attributes of individuals, reproducible over time under varying stimuli in children as well as adults (Manuck and Garland 1980). Second, when exposed to common laboratory stressors, Type A persons have more appreciable cardiovascular and/or catecholamine responsivity than do their Type B counterparts (Friedman et al. 1975; Dembroski et al. 1978; Glass et al. 1980; Corse et al. 1982; Williams et al. 1982). Third, sympathetic nervous system activity may be more important than usually appreciated since such activity is associated with coronary artery vasospasm and probably with serious problems of cardiac rhythm (Verrier and Lown 1984).

Diet

Dietary influences on atherogenesis and coronary heart disease have been studied extensively. The literature on the subject cannot be reviewed here, but there have been several relatively recent reviews (Gordon et al. 1981; Scott et al. 1981; Grundy et al. 1982; Spector and Johnson 1982; Zilversmit 1982; Samuel et al. 1983).

The majority of attention about diet and coronary heart disease has focused upon dietary fat and cholesterol. Saturated fats have been shown repeatedly to increase plasma cholesterol and LDL concentrations (Ahrens et al. 1957; Keys et al. 1957, 1965a,b; Hegsted et al. 1965). Replacement of saturated fatty acids in the diet with polyunsaturated fatty acids, particularly lin-

oleic acid, has been shown to be effective in the reduction of plasma cholesterol concentration (Ahrens et al. 1957; Hegsted et al. 1965; Keys et al. 1965a,b). The results of carefully controlled metabolic studies with humans have shown that dietary cholesterol increases total plasma cholesterol concentrations (Connor et al. 1961a,b; Connor and Lin 1974). Similarly, for each 100-mg/day decrease in dietary cholesterol, the total plasma cholesterol concentration decreases by an average of about 7 mg/dl (Hegsted et al. 1965; Mattson et al. 1972).

Animal Models

To explore the effects of automotive emissions on atherogenesis the investigator must choose a suitable animal model. The detailed characteristics of all the models cannot be presented here. Rather, the characteristics of the most relevant models have been summarized in tabular form along with key references to use with the models.

Avian Species

Chickens and pigeons are useful for experiments in which large numbers are needed (see table 1). Chickens have been used in experiments designed to explore the possibility that atherosclerosis could be considered a form of benign neoplasm of the arterial intima. Pigeons have been used for that same purpose as well as to study the effects of smoking and CO.

Nonprimates

Several nonprimate models of atherosclerosis are relevant to research on automotive emissions (see table 2). Of these, pigs and hybrid hares have the most relevance. Pigs are useful when large samples of blood or other tissues are required and because of the similarities of their plasma lipoproteins and atherosclerotic lesions to those of humans. Hybrid hares are useful for studies of automotive emission effects on the clonal nature of atherosclerotic lesions.

Table 1. Summary of Atherosclerosis Characteristics of Avian Models Potentially Useful for Studies of Automotive Emissions[a]

Model	Advantages	Disadvantages
Pigeons[b]	Lesions occur naturally and are exacerbated by dietary cholesterol, CO, polycyclic aromatic hydrocarbons (PAHs). Pigeons have a high frequency of plaque complications; breed and strain differences are well established, and they are inexpensive and easy to maintain.	Plasma lipoproteins are different from those in humans, occurring primarily in small intra-myocardial branches; most flocks are infected with pigeon herpesvirus.
Chickens[c]	Lesions occur naturally and are exacerbated by dietary cholesterol, herpesvirus, PAHs; chickens have gender-specific differences in coronary lesions, are readily available, and easy to maintain.	Lesion sites inconsistent, complications uncommon, plasma lipoproteins different from those in humans, unknown infection with herpesvirus can cause unexplained variability, coronary atherosclerosis occurs mostly in intramyocardial branches.

[a] Modified from a previously published summary by Jokinen et al. 1985.
[b] Prichard et al. 1964a,b; Clarkson et al. 1965; Wagner et al. 1973; Wagner and Clarkson 1974; Armitage et al. 1976; Turner et al. 1979; St. Clair 1983; Revis et al. 1984.
[c] Pick and Katz 1965; Simpson and Harms 1969; Albert et al. 1977; Minick et al. 1979; Penn et al. 1980; Bond et al. 1982; Fabricant et al. 1983; Majesky et al. 1985.

Nonhuman Primates

The atherosclerosis of nonhuman primates is more like that of humans than is the case with other models (see table 3). Their primary use for research on automotive emissions will be for experiments on pathogenesis, lipoprotein metabolism, clinical sequela of atherosclerosis, and for situations in which reproductive and/or social function are important.

Effects of Chemicals on Atherogenesis

Past studies on the effect of various chemicals on atherogenesis provide the rationale for speculations about possible automotive emission effects on atherogenesis. To develop working hypotheses about automotive emissions, the mechanisms by which chemical substances could be delivered to the cells of the arterial wall and how these substances may affect the progression of atherosclerosis must be considered. The transport of carcinogens by the plasma lipoproteins is reviewed below, as are the effects of cigarette smoking, CO exposure, and polycyclic aromatic hydrocarbons (PAHs).

Plasma Lipoproteins as Chemical Carriers

Following the intravenous injection of rats with chylomicrons containing benzo[a]pyrene (BaP), the carcinogen was found to be transported primarily by LDLs and VLDLs (Vauhkonen et al. 1980). Subsequently, Shu and Nichols (1981) demonstrated that the in vitro uptake of BaP by LDL, VLDL, and HDL correlated with lipoprotein and total lipid volume.

The plasma lipoproteins have also been shown to play an important role in the removal of carcinogens from cells. Remsen and Shireman (1981a) reported that increasing concentrations of either LDLs, VLDLs, or HDLs resulted in increasing percentages of removal of BaP from cell membranes. Later, Busbee and Benedict (1983) demonstrated that HDL partitioning of lipophilic PAH mutagens from cell culture medium effectively reduces the concentration of carcinogen available for interaction with the cells.

Whether the LDL receptor is necessary for the incorporation of LDLs containing BaP has been investigated by Remsen and Shireman (1981b). Skin fibroblasts derived from a receptor-deficient human with homozygous familial hypercholesterolemia were used for these studies. Benzo[a]py-

Table 2.   Summary of Atherosclerosis Characteristics of Nonprimate Models Potentially Useful for Research on Automotive Emissions[a]

Model	Advantages	Disadvantages
Rabbits[b]	Extensive literature; lesions well character-ized, easily exacerbated by dietary choles-terol, immunologic injury, and perhaps CO; animals reproduce rapidly, are inex-pensive and easy to maintain.	Lesions composed primarily of macro-phages, complications uncommon, lipo-protein metabolism quite different from that of humans, lipid storage occurs in many organs.
Pigs[c]	Naturally occurring lesions common, le-sions exacerbated by dietary cholesterol, lipoprotein metabolism quite similar to that of humans, animals with von Wille-brand's disease useful for studies of plate-let function and atherogenesis, large ar-tery size, miniature breeds available.	Domestic animals are large, difficult to han-dle, expensive to maintain.
Hybrid hares[d] *Lepus timidus* × *Lepus europaeus*	G-6-PD can be used as marker for clonal origins of intimal smooth muscle cells, ef-fects of dietary cholesterol established, good literature base.	Difficult to obtain and maintain, no reports on effects of chemical exposures.
Dogs[e]	Lesions can be induced, complications com-mon, convenient size, easy to handle, experience base in inhalation toxicology.	Antithyroid dogs required for lesion induc-tion, lesions primarily medial, animals expensive to maintain.

[a] Modified from a previously published summary by Jokinen et al. 1985.

[b] Prior et al. 1961; Astrup et al. 1967; Minick and Murphy 1973; Davies et al. 1976; Wilson et al. 1982.

[c] Mahley et al. 1975; Gerrity et al. 1979; Bowie and Fuster 1980; Fritz et al. 1980; Gerrity and Naito 1980; Gerrity 1981a,b; Griggs et al. 1981; Reitman et al. 1982.

[d] Pearson et al. 1979; Lee et al. 1981; Imai et al. 1982; Imai and Lee 1983; Pearson et al. 1983.

[e] Geer and Guidry 1965; Schenk et al. 1965; Robertson et al. 1972; Mahley et al. 1974; Innerarity et al. 1982.

rene from LDLs was found to enter recep-tor-deficient cells, and Remsen and Shire-man suggested that entry was by rapid redistribution between the lipoprotein and cell membrane. Similarly, Plant and co-workers (1985) studied the uptake by cells of BaP from HDLs, LDLs, and VLDLs, as well as from laboratory-prepared vesicles. They concluded that the cellular uptake of BaP from these hydrophobic donors was by transfer through the aqueous phase.

Although all of these reports involve in vitro studies, it seems plausible that the lipoproteins could transport chemicals from the lungs to the cells of the artery wall. In this regard, it is of interest that the urine of cigarette smokers has been shown to contain mutagenic substances (Yamasaki and Ames 1977).

■ **Recommendation 8.** A comparison of LDLs of different size and apoprotein com-position relative to their ability to bind hydrocarbons and deliver them to smooth muscle cells and macrophages in culture is needed.

Cigarette Smoking

There is a vast literature base for the con-clusion that cigarette smoking is a major cause of coronary heart disease in the United States for both men and women. The epidemiologic evidence for that con-clusion, along with supportive pathophys-iological studies, has been the topic of a major recent review monograph (U.S. De-partment of Health and Human Services 1983). In general, the epidemiologic find-ings have established that the risk for de-veloping coronary heart disease increases with increasing exposure to cigarette smoke, that cigarette smokers have a two-fold greater incidence of coronary heart disease, that women who use oral contra-ceptives and who smoke increase the risk of myocardial infarction approximately 10-fold compared with women who neither

Table 3. Summary of Atherosclerosis Characteristics of Nonhuman Primates Potentially Useful for Research on Automotive Emission[a]

Model	Advantages	Disadvantages
Rhesus monkeys[b] (*Macaca mulatta*)	Lesions occur naturally, are exacerbated by dietary cholesterol, similar to those of humans; complications and myocardial infarction relatively common; lipoproteins similar to those in humans; animals are a convenient size.	Not available from countries of origin, must be obtained from domestic breeding colonies, expensive, difficult to handle.
Cynomolgus macaques[c] (*Macaca fascicularis*)	Lesions occur naturally, are exacerbated by dietary cholesterol but not CO, are similar to those of humans; gender differences in coronary artery lesions; have high incidence of myocardial infarction; lipoproteins are well characterized; animals are convenient size, readily available.	Expensive to acquire and maintain, difficult to handle.
African green monkeys[d] (*Cercopithecus aethiops*)	Lipoprotein changes and lesions induced by dietary cholesterol similar to those in humans; relationships between lipoprotein alterations and lesion development documented; convenient size, readily available.	Characterization incomplete, expensive, difficult to handle.

[a] Modified from a previously published summary by Jokinen et al. 1985.
[b] Taylor et al. 1962, 1963; Scott et al. 1967a,b; Manning and Clarkson 1972; Rudel et al. 1979; Bond et al. 1980a; Rudel 1980.
[c] Kramsch and Hollander 1968; Thomsen 1974; Armstrong 1976; Malinow et al. 1976a,b; Bing et al. 1980; Bond et al. 1980b; Armstrong et al. 1985.
[d] Kritchevsky et al. 1977; Trillo and Prichard 1979; Rudel 1980; Rudel et al. 1983.

smoke nor use oral contraceptives, and finally that the cessation of smoking results in a substantial reduction in coronary heart disease events. In this review the pathophysiological basis for the increased coronary heart disease is explored.

Coronary Artery Atherosclerosis. In view of the large effect of cigarette smoking on coronary heart disease morbidity and mortality it is surprising to find that there are few data to suggest that the increased coronary heart disease events are related to more extensive coronary artery atherosclerosis. Viel and coworkers (1968) studied the coronary arteries of 1,150 men and 290 women who died accidentally in Santiago, Chile. They concluded that smoking was not related to the extent of coronary artery atherosclerosis. Earlier, Auerbach and colleagues (1965) concluded that the percentage of men with advanced coronary artery atherosclerosis was higher among cigarette smokers than nonsmokers and increased

with the amount of cigarette smoking. Later, Auerbach and coworkers (1976) reported on detailed studies of the hearts from 2,257 autopsies of male patients at the Veterans Administration Hospital of East Orange, New Jersey. Again they found an association between cigarette smoking and coronary artery atherosclerosis; however, this time, the precise nature of that effect was better defined. They noted that the effect was greater in the intramyocardial arteries than in the larger epicardial arteries, with the most striking effect being found in the myocardial arterioles. As a part of a large international study, Lifsic (1976) studied the relationship between cigarette smoking and coronary artery atherosclerosis as well as its complications. He concluded that there was no clear association between smoking and coronary stenosis, myocardial infarction, or heart weight.

Although the effects of cigarette smoking on coronary artery atherosclerosis may be minimal, it does seem to increase the extent of aortic atherosclerosis. As a part of the Puerto Rican health program, a major study has been conducted to examine the relation of antemortem factors to coronary artery and aortic atherosclerosis at autopsy (Sorlie et al. 1981). In that study, cigarette smoking increased aortic, but not coronary artery, atherosclerosis and was not dose dependent in the case of aortic atherosclerosis. Using the Oslo study as a basis, Holme and coworkers (1981) did not find a significant association between coronary artery–raised lesions and cigarette smoking.

Comparable studies have found a positive association between cigarette smoking and the extent of coronary artery atherosclerosis. Using New Orleans autopsy data, Strong and Richards (1976) concluded that smokers had more raised lesions of the coronary arteries than did nonsmokers, and the differences in lesion extent were more striking among heavy smokers than light smokers. In a study by Vikhert et al. (1976), nutritional status and smoking habit were considered in determining the extent of coronary artery atherosclerosis in subjects from five Russian cities. The investigators found that tobacco smoking in combination with overnutrition had a more positive effect on the development of coronary artery atherosclerotic lesions among white-collar compared with blue-collar workers. In addition, the Honolulu, Oslo, and Puerto Rico heart studies, all using standardized evaluations of atherosclerotic lesions at autopsy, concluded that there was a positive association between cigarette smoking habit and coronary artery atherosclerosis extent (Rhoads et al. 1978; Holme et al. 1981; Sorlie et al. 1981).

Platelets and Platelet Function. The possibility that cigarette smoking may change platelet function and/or prostaglandin metabolism has been of interest to many investigators recently. Hawkins (1972) and Levine (1973) found that smoking enhanced ADP-induced platelet aggregation, and the latter investigator suggested that platelet aggregation might be associated with the increased coronary heart disease morbidity and mortality associated with cigarette smoking. Later, Davis and Davis

(1979) sought to determine whether circulating platelet aggregates were more common among cigarette smokers than among nonsmokers. They found a highly significant increase in circulating platelet aggregates during a 20-min period that followed the smoking of two cigarettes. They speculated that the occurrence of these platelet aggregates may contribute to coronary heart disease morbidity and mortality. Fuster and colleagues (1981) studied platelet survival among cigarette smokers and nonsmokers either with or without a family history of coronary heart disease. A shortened platelet survival half-life was found several times more frequently among individuals who smoked and had a strong family history of coronary heart disease than among normal persons who did not smoke and had no family history of coronary heart disease. The platelet survival could be lengthened either by the administration of dipyridamole plus aspirin or by discontinuing smoking. Siess and coworkers (1982) reported on the effect of smoking and exercise on plasma catecholamine release and platelet aggregation and associated thromboxane formation. They found that physical exercise and smoking markedly increased plasma levels of norepinephrine and epinephrine in healthy men but that these catecholamine changes did not detectably enhance platelet aggregation or thromboxane formation. Similarly, Mehta and Mehta (1982) found no increase in thromboxane A_2 with smoking among habitual smokers and concluded that the lack of such an increase may reflect tolerance to the effects of smoking since increases did occur among nonsmokers made to smoke.

Plasma High-Density Lipoproteins. In recent years there has been considerable interest in determining whether and to what extent cigarette smoking reduces the plasma concentrations of HDLC. This interest stems from the earlier observation that HDLC is independently and inversely correlated with the risk of coronary heart disease. For that reason there has been this interest in the effect of cigarette smoking on this potential protective mechanism. Halfon and coworkers (1982) studied the effect of smoking on HDLC of a group of 17-year-old participants in the Jerusalem Lipid Research Clinic. HDLC was significantly lower in smokers than in nonsmokers (39 versus 42 mg/dl) and in males than females (44 versus 48 mg/dl). Brischetto and colleagues (1983) examined the relationships between smoking and HDLC among the volunteer participants in the Portland heart study. Cigarette smokers, male as well as female, had significantly lower HDLC than former smokers or nonsmokers. The effects seem dose related in that significant differences in HDLC were noted among individuals using more than 25 cigarettes per day, but no effect was seen at 15 or fewer cigarettes per day. Stamford and coworkers (1984a) examined the relationships between cigarette smoking and HDLC among premenopausal females. The HDLC concentrations of smokers were about 10 mg/dl less than those of nonsmokers. Additionally, cigarette smoking was found to attenuate the increasing effect of chronic exercise or alcohol consumption in raising HDLC concentrations. Stamford and colleagues (1984b) studied the effect of cigarette smoking on HDLC of middle-aged males. In that study, cigarette smoking was associated with significant reductions in HDLC, and it was found that high levels of exercise could prevent the HDLC reductions associated with smoking. Later, Haffner and coworkers (1985) studied the effect of cigarette smoking on the HDL subclasses. The studies were based on small groups of men and women from the Northwest Bell Telephone Company health survey. Smoking was not correlated with HDL_2 but was negatively correlated with HDL_3. Since HDL_3, a fraction lowered by cigarette smoking, is not associated with coronary heart disease risk, the authors concluded that the effect of smoking may be mediated through mechanisms other than its effect on HDLC.

In summary, there is a clear effect of cigarette smoking on plasma concentrations of HDLC, but more work needs to be done on its effect on the subfractions of HDL before conclusions can be drawn about the significance of the phenomenon to increase risk for coronary heart disease.

Atherosclerosis of Animal Models.

There have been few relevant studies of the exposure of animal models to cigarette smoke and the resultant atherosclerotic effects. Hojnacki and colleagues (1981) reported on the effect of cigarette smoke exposure and dietary cholesterol on the plasma lipoprotein composition of White Carneau pigeons. They concluded that cigarette smoke mediated alterations and lipoprotein composition that were independent of diet. Sieffert and coworkers (1981) observed endothelial injury and focal platelet aggregation among rats exposed to cigarette smoke. The most comprehensive study of cigarette smoking and experimental atherosclerosis has been done by Rogers and coworkers (1980). In that study, baboons were trained to smoke in a human-like manner. Cigarette smoking caused baboons to have higher LDL/HDL ratios, but there was no effect on the occurrence of atherosclerotic plaques in the coronary arteries. The authors are quoted as indicating that their results do not support the hypothesis that cigarette smoking augments experimental atherosclerosis in the presence of a moderate level of diet-induced hypercholesterolemia (U.S. Department of Health and Human Services 1983).

Carbon Monoxide

Epidemiologic Studies.

Few epidemiologic studies of human exposure to CO have been undertaken. Carboxyhemoglobin concentrations among nonsmokers have been shown to be as high and sometimes higher than those in smokers if there is an occupational exposure to CO (Wald et al. 1973). Carboxyhemoglobin concentrations greater than 5 percent are associated with 20 times the risk of coronary heart disease as compared with individuals with levels below 3 percent.

The prevalence of coronary heart disease among nonsmoking foundry workers (Hernberg et al. 1976) without CO exposure was 2 percent, whereas the prevalence among smokers with job-related CO exposure was 19 percent. A study conducted in England of blast furnace workers found that among individuals with 2.0 to 2.6

percent carboxyhemoglobin concentrations there was no increase in coronary heart disease (Jones and Sinclair 1975); that study supports the hypothesis that 3 percent carboxyhemoglobin may be a critical level. Of particular relevance to concerns about automotive emissions are studies by Decoufle et al. (1977) in which individuals with occupational exposures to CO were found to have an increased frequency of death from coronary heart disease as compared to the general population, and by Edling and Axelson (1984) who reported a fourfold increase in cardiovascular disease among workers exposed to diesel exhaust.

CO exposures may also adversely affect the pathophysiology of coronary heart disease and the expressions of anginal pain and intermittent claudication. Aronow and coworkers (1972) documented that there was aggravation of the symptoms of coronary heart disease among drivers on the Los Angeles freeway. (It should be noted, however, that Aronow's studies have been critically reviewed by the U.S. Environmental Protection Agency [1984]). Anderson and colleagues (1973) found that exercise time was decreased for patients suffering from either angina or intermittent claudication when exposed to CO. Cohen and coworkers (1969) reported that there are increases in fatal myocardial infarctions among individuals in high-pollution areas during periods of relatively increased ambient CO. Taken on balance, the epidemiologic evidence suggests that carboxyhemoglobin concentrations above about 6 percent may exacerbate coronary artery atherosclerosis, and those of around 2 percent may adversely affect the symptoms of individuals with coronary heart disease. On the basis of these data, it would seem that the present industrial standards which allow carboxyhemoglobin levels of 7.36 percent and the newly recommended standard allowing carboxyhemoglobin of 5 percent may be inadequate.

■ **Recommendation 9.** Several types of exploratory investigations could be performed that would take advantage of existing data bases and ongoing population-based cohort studies to evaluate the

association between exposure to automotive exhaust and cardiovascular risk factors, morbidity, and mortality. The validity of responses to simple questionnaires and information on occupation might be evaluated as measures of personal exposure. Noninvasive techniques (β-mode imaging, exercise stress tests, 24-hr ambulatory ECG monitoring) could be used as indices of cardiovascular disease. These preliminary approaches represent cost-effective ways to develop more refined epidemiologic studies.

■ **Recommendation 10.** Epidemiologic studies should be undertaken to detect and measure associations between exposure to automotive emissions and known or suspected risk factors for atherosclerosis or cardiovascular disease, including lipids and lipoproteins, hemostatic factors, platelet function, pulmonary function, exercise capacity, and so on.

■ **Recommendation 11.** Epidemiologic studies should be done to determine whether exposure to automotive emissions is associated with increased risk of atherosclerosis or clinical manifestations of cardiovascular disease in individuals characterized with respect to known risk factors (for example, cigarette smoking, hyperlipidemia, elevated blood pressure, positive family history).

Animal Models. Of all the automotive emissions, CO has been the most extensively studied for its effects on atherogenesis. The published evidence tends to show that CO probably affects atherogenesis only when coadministered with a moderately atherogenic diet, and the coronary arteries are more affected than other arteries. The evidence for the CO effect is summarized below by species.

Pigeons. Exposure to CO has been reported to increase the incidence and severity of coronary artery atherosclerosis in White Carneau pigeons consuming a cholesterol-containing diet (Armitage et al. 1976; Turner et al. 1979). Armitage and coworkers (1976) found that exposure to CO concentrations sufficient to raise their blood carboxyhemoglobin levels to 10 percent had no apparent effect on coronary artery atherosclerosis in birds that were fed a control diet; however, in birds fed a diet containing 1 percent cholesterol, they found a marked increase in the severity of coronary artery disease in birds exposed to CO as compared to nonexposed birds. Later studies by Turner and coworkers (1979) confirmed these findings. Further, they found a greater cholesterol content and a lower content of triglycerides in the aortic tissue of birds exposed to CO than in the aortas of birds that were not exposed.

Rabbits. The earliest work on CO effects on rabbit atherosclerosis was by Astrup et al. (1967). They fed rabbits an atherogenic diet that induced plasma cholesterol concentrations of 1,000 to 2,000 mg/dl and reported that coronary atherosclerosis was exacerbated among rabbits exposed to CO. They suggested that CO produced endothelial injury which then interacted with the hyperlipoproteinemia. Later, this same group of investigators (Wanstrup et al. 1969) reported on the effects of CO on the naturally occurring aortic lesions of rabbits fed normal diets. The authors concluded that the naturally occurring lesions were worsened, although their evidence is not convincing. A later study of cholesterol-fed rabbits exposed to CO (Davies et al. 1976) indicated that coronary but not aortic atherosclerosis was increased by CO exposure. Finally, Hugod and colleagues (1978), working with Astrup, sought to establish that CO exposure of rabbits fed normal diets was capable of inducing endothelial injury. They concluded that there was no effect of CO on the endothelium or any part of the intima. Thus, this group of investigators concluded that CO did not affect atherogenesis.

Cynomolgus Macaques. There have been three reports in the literature concerning the effect of CO exposure on cynomolgus macaque atherosclerosis. Thomsen (1974) reported that CO exposure of young cynomolgus monkeys fed normal diets resulted in an increase in the appearance of monocyte-derived foam cells in the intima. It seems probable that the intimal lesions being studied by Thomsen were naturally occurring lesions of cynomolgus monkeys and probably were not influenced by exposure of the animals to CO, since there was little statistical power in the experiment that only used six animals in a group. Malinow et al. (1976a) studied the effect of 14 months of CO exposure on cynomolgus monkeys fed normal diets and found no effect on total serum cholesterol concentrations, on coronary artery atherosclerosis, or on the occurrence of myocardial infarction. Bing and coworkers (1980) conducted a similar experiment in which cynomolgus monkeys fed normal diets were exposed for 12 months to CO. They concluded that CO had no effect on plasma lipid concentrations or aortic atherosclerosis as determined by measurements of aortic cholesterol.

■ **Recommendation 12.** Research to evaluate components of atherogenesis and extent of coronary artery atherosclerosis of cynomolgus monkeys exposed to varying concentrations of automotive emissions for varying lengths of time should be pursued. Illustrated in figure 1 are methods that could be used for the quantification of the atherogenic components.

Squirrel Monkeys. Webster and colleagues (1970) reported on the effect of CO exposure on squirrel monkeys fed an atherogenic diet, and concluded that CO affected the extent of coronary but not of aortic atherosclerosis. The evidence that coronary artery atherosclerosis had been exacerbated by CO was based upon the authors' estimate of lumen stenosis. The study was conducted before appropriate pressure fixation of coronary arteries had been developed, and thus, their observations on lumen stenosis of nonperfused hearts may be erroneous.

Clonal Character of and Carcinogen Effects on Plaques

As was reviewed under Pathogenesis, cell marker studies have revealed that the cell population of many human atherosclerotic lesions consists of a single phenotype of smooth muscle cells in contrast to normal artery and diffusely thickened intima. These observations suggested the possibility that some atherosclerotic plaques may represent a form of benign neoplasm in the arterial intima. Those innovative studies prompted several investigations of the effect of carcinogens on smooth muscle cell proliferation and arteriosclerotic plaque development in chickens and pigeons.

In 1977, dimethylbenzanthracene and BaP were both shown to increase the frequency and size of atherosclerotic lesions in the abdominal aorta of chickens (Albert et al. 1977). Dimethylbenzanthracene was more potent than BaP. Further studies showed that the chronic administration of dimethylbenzanthracene stimulated the growth of naturally occurring atherosclerotic lesions in the distal aorta of chickens without producing new arterial lesions (Bond et al. 1981; Penn et al. 1981).

A major recent finding has been that focal smooth muscle cell proliferation in the chicken aortic intima can be produced by an initiation/promotion sequence (Majesky et al. 1985). Chickens that were treated with the tumor initiator dimethylbenzanthracene followed by repeated injections of an α-adrenergic agonist, methoxamine, had foci of intimal smooth muscle proliferation in the thoracic aorta that occurred with a greater incidence and severity than elicited by either agent alone. The results were consistent with initiation- and promotion-like stages in the development of intimal lesions, and suggested another way that such proliferation in the arterial intima resembles benign neoplasia in other tissues. Since methoxamine is an α-adrenergic agonist, there may be some relationship between this observation and the effects of psychosocial stress on atherogenesis.

The effect on atherosclerosis of BaP and dimethylbenzanthracene has been studied in another avian model, the White Carneau pigeon (Revis et al. 1984). Dimethylbenzanthracene, but not BaP, was found to increase atherosclerosis.

Extensive studies have been reported to better clarify the cellular metabolic events that may relate to the effect of carcinogens on smooth muscle cell proliferation. Some

of these studies have compared atherosclerosis-susceptible and -resistant pigeons (Hogg and Cryer 1982; Majesky et al. 1983). Others have involved studies of the bioactivation of mutagens in cultured smooth muscle cells (Bond et al. 1979, 1980a).

Summary

Epidemiologic data provide suggestive evidence that automotive emissions may be risk factors for the development of atherosclerosis and coronary heart disease. Further concern about that possibility derives from analogies to cigarette smoking. Cigarette smoking is a major risk factor for coronary heart disease, peripheral vascular disease, and other clinically significant sequelae of atherosclerosis. There is evidence for and controversy about the extent to which atherosclerotic plaque progression among cigarette smokers can lead to the very exaggerated morbidity and mortality in that group. Studies of the atherogenic components of cigarette smoke have tended to implicate CO as the most likely substance promoting atherogenesis. The link between cigarette smoking and exposure to automotive emissions, in the minds of many, relates to the fact that both result in the exposure of human subjects to considerable amounts of CO. In reviewing the rise and fall of ischemic heart disease, Stallones (1980) points out that among the four common risk factors (hypertension, high serum cholesterol, physical activity, and cigarette smoking) only cigarette smoking fits the observed pattern of the rise and fall in mortality from ischemic heart disease. Interestingly, one could speculate that air pollution and its control may be similar.

In this chapter the pathophysiological rationales were provided for ways in which automotive emissions exposure might affect atherogenesis. On the basis of previously reported research, it is not possible to draw a clear conclusion regarding whether and to what extent CO affects the development of atherosclerosis. For that reason, recommendations refer to exposures to automotive emissions. If effects on atherogenesis are found, then components of automotive emissions should be studied separately.

Atherosclerosis and its clinical sequelae were described in the context of the natural history of its development. Emphasis was on the origin of fatty streaks, whether fatty streaks progress to fibrous plaques, and on the development of plaque complications. Age differences in susceptibility to coronary artery atherosclerosis, gender differences in the progression of coronary lesions, genetic susceptibility to coronary atherosclerosis, and psychosocial influences on plaque progression have been reviewed. Animal models for research on atherosclerosis have also been reviewed briefly.

Since little is known about the effect of automotive emissions, other than CO, on atherogenesis, analogies were made to other chemical exposures. Finally, research needs for examining the question of automotive emission effects on atherogenesis were presented, as well as needs for epidemiologic studies of the association between exposure and cardiovascular risk factors, morbidity, and mortality. Those needs are presented in the context of initial research to determine if and to what extent automotive emissions exposure affects atherosclerosis and directions that could be undertaken if initial studies establish an effect of automotive emissions exposure.

Summary of Research Recommendations

HIGH PRIORITY

The studies most likely to yield useful data are listed below.

Recommendation 9 Several types of exploratory investigations could be performed that would take advantage of existing data bases and ongoing population-based cohort studies to evaluate the association between exposure to automotive exhaust and cardiovascular risk factors, morbidity, and mortality. The validity of responses to simple questionnaires and information on occupation might be evaluated as measures of personal exposure. Noninvasive techniques (β-mode imaging, exercise stress tests, 24-hr ambulatory ECG monitoring) could be used as indices of cardiovascular disease. These preliminary approaches represent cost-effective ways to develop more refined epidemiologic studies.

Recommendation 10 Epidemiologic studies should be undertaken to detect and measure associations between exposure to automotive emissions and known or suspected risk factors for atherosclerosis or cardiovascular disease, including lipids and lipoproteins, hemostatic factors, platelet function, pulmonary function, exercise capacity, and so on.

Recommendation 11 Epidemiologic studies should be done to determine whether exposure to automotive emissions is associated with increased risk of atherosclerosis or clinical manifestations of cardiovascular disease in individuals characterized with respect to known risk factors (for example, cigarette smoking, hyperlipidemia, elevated blood pressure, positive family history).

Recommendation 5 Since automotive emission exposure may result in relative hypoxia, it seems reasonable to suspect that plaque complications could be hastened by that mechanism. Cynomolgus monkeys with diet-induced atherosclerosis of two or three years' duration should be exposed to various concentrations of automotive emissions and their plaque complications compared with those in appropriate control animals not exposed to automotive emissions.

Recommendation 12   Research to evaluate components of atherogenesis and extent of coronary artery atherosclerosis of cynomolgus monkeys exposed to varying concentrations of automotive emissions for varying lengths of time should be pursued. Illustrated in figure 1 are methods that could be used for the quantification of the atherogenic components.

MEDIUM PRIORITY

Since there are no definitive data that establish an association between automotive emissions exposure and atherogenesis, the high-priority research should be completed before the research recommended below is considered.

Recommendation 1   Research to determine whether exposure of endothelial cells to automotive emissions affects endothelial function should be undertaken with emphasis on endothelial cell replication and prostacyclin production.

Recommendation 3   The effect of automotive emissions on the catabolism of LDL particles by smooth muscle cells, the replication rates of these cells, and the kinds of connective tissue proteins they elaborate should be determined.

Recommendation 4   The effect of automotive emissions on the internalization of β-VLDL and other modified LDL particles by macrophages should be assayed.

Recommendation 8   A comparison of LDLs of different size and apoprotein composition relative to their ability to bind hydrocarbons and deliver them to smooth muscle cells and macrophages in culture is needed.

LOW PRIORITY

Recommendation 2   Studies should be done to determine whether chronic exposure to automotive emissions affects the repair of stress-induced endothelial injury and/or the way in which the intima of arteries responds to such injuries.

Recommendation 6   Studies to determine whether and to what extent automotive emissions exposure may diminish or abolish the gender-specific difference in atherogenesis need to be explored.

Recommendation 7   Experiments that determine whether chronic exposure to CO affects coronary artery atherosclerosis differently in genetically susceptible and resistant strains are needed.

References

Ahrens, E. H., Jr., Insull, W., Jr., Blomstrand, R., Hirsch, J., Tsaltas, T. T., and Peterson, M. L. 1957. The influence of dietary fats on serum-lipid levels in man, *Lancet* 1:943–953.

Albert, R. E., Vanderlaan, M., Burns, F. J., and Nishizumi, M. 1977. Effect of carcinogens on chicken atherosclerosis, *Cancer Res.* 37:2232–2235.

Anderson, E. W., Andelman, R. J., Strauch, J. M., Fortuin, N. J., and Knelson, J. H. 1973. Effect of low-level carbon monoxide exposure on onset and duration of angina pectoris: a study in ten patients with ischemic heart disease, *Ann. Intern. Med.* 79:46–50.

Armitage, A. K., Davies, R. F., and Turner, D. M. 1976. The effects of carbon monoxide on the development of atherosclerosis in the White Carneau pigeon, *Atherosclerosis* 23:333–344.

Armstrong, A., Duncan, B., Oliver, M. F., Julian, D. G., Donald, K. W., Fulton, M., Lutz, W., and Morrison, S. L. 1972. Natural history of acute coronary heart attacks. A community study, *Br. Heart J.* 34:67–80.

Armstrong, M. L. 1976. Atherosclerosis in rhesus and cynomolgus monkeys, *Primate Med.* 9:16–40.

Armstrong, M. L., Heistad, D. D., Marcus, M. L., Megan, M. B., and Piegors, D. J. 1985. Structural and hemodynamic response of peripheral arteries of macaque monkeys to atherogenic diet, *Arteriosclerosis* 5:336–346.

Aronow, W. S., Harris, C. N., Isbell, M. W., Rokaw, S. N., and Imparato, B. 1972. Effect of freeway travel on angina pectoris, *Ann. Intern. Med.* 77:669–676.

Astrup, P., Kjeldsen, K., and Wanstrup, J. 1967. Enhancing influence of carbon monoxide on the development of atheromatosis in cholesterol-fed rabbits, *J. Atheroscler. Res.* 7:343–354.

Auerbach, O., Hammond, E. C., and Garfinkel, L. 1965. Smoking in relation to atherosclerosis of the coronary arteries, *N. Engl. J. Med.* 273:775–779.

Auerbach, O., Carter, H. W., Garfinkel, L., and Hammond, E. C. 1976. Cigarette smoking and coronary artery disease. A macroscopic and microscopic study, *Chest* 70:697–705.

Barefoot, J. C., Dahlstrom, W. G., and Williams, R. B., Jr. 1983. Hostility, CHD incidence, and total mortality: a 25 year follow-up study of 255 physicians, *Psychosom. Med.* 45:59–63.

Barr, D. P. 1953. Some chemical factors in the pathogenesis of atherosclerosis, *Circulation* 8:641–654.

Benditt, E. P., and Benditt, J. M. 1973. Evidence for a monoclonal origin of human atherosclerotic plaques, *Proc. Natl. Acad. Sci. USA* 70:1753–1756.

Correspondence should be addressed to Thomas B. Clarkson, Arteriosclerosis Research Center, Bowman Gray School of Medicine, Wake Forest University, 300 South Hawthorne Road, Winston-Salem, NC 27103.

Bhattacharyya, A. K., and Eggen, D. A. 1977. Cholesterol metabolism in high- and low-responding rhesus monkeys, In: *Atherosclerosis IV* (G. Schettler, Y. Goto, Y. Hata, and G. Klose, eds.), pp. 293–298, Springer-Verlag, Berlin.

Bhattacharyya, A. K., and Eggen, D. A. 1980. Cholesterol absorption and turnover in rhesus monkeys as measured by two methods, *J. Lipid Res.* 21: 518–524.

Bhattacharyya, A. K., and Eggen, D. A. 1983. Mechanism of the variability in plasma cholesterol response to cholesterol feeding in rhesus monkeys, *Artery* 11(4):306–326.

Bing, R. J., Sarma, J. S. M., Weishaar, R., Rackl, A., and Pawlik, G. 1980. Biochemical and histological effects of intermittent carbon monoxide exposure in cynomolgus monkeys (*Macaca fascicularis*) in relation to atherosclerosis, *J. Clin. Pharmacol.* 20: 487–499.

Bond, J. A., Kocan, R. M., Benditt, E. O., and Juchau, M. R. 1979. Metabolism of benzo[a]pyrene and 7,12-dimethylbenz[a]anthracene in cultured human fetal aortic smooth muscle cells, *Life Sci.* 25:425–430.

Bond, J. A., Yang, H. Y., Majesky, M. W., Benditt, E. P., and Juchau, M. R. 1980a. Metabolism of benzo[a]pyrene and 7,12-dimethylbenz[a]anthracene in chicken aortas: monooxygenation, bioactivation to mutagens, and co-valent binding to DNA in vitro, *Toxicol. Appl. Pharmacol.* 52:323–335.

Bond, J. A., Gown, A. M., Yang, H. L., Benditt, E. P., and Juchau, M. R. 1981. Further investigations of the capacity of polynuclear aromatic hydrocarbons to elicit atherosclerotic lesions, *J. Toxicol. Environ. Health* 7:327–335.

Bond, M. G., Bullock, B. C., Bellinger, D. A., and Hamm, T. E. 1980b. Myocardial infarction in a large colony of nonhuman primates with coronary artery atherosclerosis, *Am. J. Pathol.* 101:675–692.

Bond, M. G., Sawyer, J. K., Bullock, B. C., Barnes, R. W., and Ball, M. R. 1982. Animal studies of atherosclerosis progression and regression, In: *Clinical Diagnosis of Atherosclerosis* (M. G. Bond, W. Insull, Jr., S. Glagov, A. B. Chandler, and J. F. Cornhill, eds.), pp. 435–449, Springer-Verlag, New York.

Bowie, E. J. W., and Fuster, V. 1980. Resistance to atherosclerosis in pigs with von Willebrand's disease, *Acta Med. Scand.* 208(Suppl. 642):121–130.

Brischetto, C. S., Conner, W. E., Connor, S. L., and Matarazzo, J. D. 1983. Plasma lipid and lipoprotein profiles of cigarette smokers from randomly selected families: enhancement of hyperlipidemia and depression of high-density lipoprotein, *Am. J. Cardiol.* 52:675–680.

Brown, M. S., Kovanen, P. T., and Goldstein, J. L. 1981. Regulation of plasma cholesterol by lipoprotein receptors, *Science* 212:628–635.

Bullock, B. C., Clarkson, T. B., Lehner, N. D. M., Lofland, H. B, and St. Clair, R. W. 1969. Atherosclerosis in *Cebus albifrons* monkeys. III. Clinical and pathologic studies, *Exp. Mol. Pathol.* 10:39–62.

Busbee, D. L., and Benedict, W. F. 1983. High-density lipoproteins decrease both binding of a polynuclear aromatic hydrocarbon carcinogen to DNA and carcinogen-initiated cell transformation, *Mutat. Res.* 111:429–439.

Cassel, J. C. 1971. Summary of major findings of the Evans County cardiovascular studies, *Arch. Intern. Med.* 128:887–889.

Clarkson, T. B. 1974. Arteriosclerosis of African green and stump-tailed macaque monkeys, In: *Atherosclerosis III* (G. Schettler and A. Weizel, eds.), pp. 291–294, Springer-Verlag, New York.

Clarkson, T. B., Middleton, C. C., Prichard, R. W., and Lofland, H. B. 1965. Naturally occurring atherosclerosis in birds, *Ann. N.Y. Acad. Sci.* 127: 685–693.

Clarkson, T. B., Lofland, H. B, Bullock, B. C., and Goodman, H. O. 1971. Genetic control of plasma cholesterol. Studies on squirrel monkeys, *Arch. Pathol.* 92:37–45.

Clarkson, T. B., Lehner, N. D. M., Bullock, B. C., Lofland, H. B, and Wagner, W. D. 1976. Atherosclerosis in New World monkeys, *Primate Med.* 9:90–144.

Clarkson, T. B., Kaplan, J. R., and Adams, M. R. 1985. The role of individual differences in lipoprotein, artery wall, gender, and behavioral responses in the development of atherosclerosis, *Ann. N.Y. Acad. Sci.* 454:28–45.

Cohen, P. F., and Braunwald, E. 1980. Chronic coronary artery disease, In: *Heart Disease: A Textbook of Cardiovascular Medicine* (E. Braunwald, ed.), pp. 1387–1436, W.B. Saunders Company, Philadelphia.

Cohen, S. I., Deane, M., and Goldsmith, J. R. 1969. Carbon monoxide and survival from myocardial infarction, *Arch. Environ. Health* 19:510–517.

Connor, W. E., and Lin, D. S. 1974. The intestinal absorption of dietary cholesterol by hypercholesterolemic (type II) and normocholesterolemic humans, *J. Clin. Invest.* 53:1062–1070.

Connor, W. E., Hodges, R. E., and Bleiler, R. E. 1961a. Effect of dietary cholesterol upon serum lipids in man, *J. Lab. Clin. Med.* 57:331–342.

Connor, W. E., Hodges, R. E., and Bleiler, R. E. 1961b. The serum lipids in men receiving high cholesterol and cholesterol-free diets, *J. Clin. Invest.* 40:894–901.

Corse, C. D., Manuck, S. B., Cantwell, J. D., Giordani, B., and Matthews, K. A. 1982. Coronary prone behavior pattern and cardiovascular response in persons with and without coronary heart disease, *Psychosom. Med.* 44:449–459.

Dalen, J. E., Ockene, I. S., and Alpert, J. S. 1981. Coronary spasm, coronary thrombosis and myocardial infarction, *Trans. Am. Clin. Climatol. Assoc.* 93:87–97.

Davies, P. F., Reidy, M. A., Goode, T. B., and Bowyer, D. E. 1976. Scanning electron microscopy in the evaluation of endothelial integrity of the fatty lesion in atherosclerosis, *Atherosclerosis* 25:125–130.

Davis, J. W., and Davis, R. F. 1979. Acute effect of tobacco cigarette smoking on the platelet aggregate ratio, *Am. J. Med. Sci.* 278:139–143.

Decoufle, P., Lloyd, J. W., and Salvin, L. G. 1977.

Mortality by cause among stationary engineers and stationary firemen, *J. Occup. Med.* 19:679–682.

Dembroski, T. M., McDougall, J. M., Shields, J. L., Petitto, J., and Lushene, R. 1978. Components of the Type A coronary-prone behavior pattern and cardiovascular responses to psychomotor performance challenge, *J. Behav. Med.* 1:159–176.

Edling, C., and Axelson, O. 1984. Risk factors of coronary heart disease among personnel in a bus company, *Int. Arch. Occup. Environ. Health* 54:181–183.

Eggen, D. A. 1976. Cholesterol metabolism in groups of rhesus monkeys with high or low response of serum cholesterol to an atherogenic diet, *J. Lipid Res.* 17:663–673.

Fabricant, C. G., Fabricant, J., Minick, C. R., and Litrenta, M. M. 1983. Herpesvirus-induced atherosclerosis in chickens, *Fed. Proc.* 42:2476–2479.

Faggiotto, A., Ross, R., and Harker, L. 1984. Studies of hypercholesterolemia in the nonhuman primate. I. Changes that lead to fatty streak formation, *Arteriosclerosis* 4: 323–340.

Fass, D. N., Downing, M. R., Meyers, P., Bowie, E. J. W., and Witte, L. D. 1978. Cell growth stimulation by normal and von Willebrand porcine platelets and endothelial cells, *Blood* 52(Suppl. 1):181 (abstr.).

Fishman, J. A., Ryan, G. B., and Karnovsky, M. J. 1975. Endothelial regeneration in the rat carotid artery and the significance of endothelial denudation in the pathogenesis of myointimal thickening, *Lab. Invest.* 32:339–351.

Flow, B. L., and Mott, G. E. 1984. Relationship of high density lipoprotein cholesterol to cholesterol metabolism in the baboon (*Papio* sp.), *J. Lipid Res.* 25:469–473.

Flow, B. L., Cartwright, T. C., Kuehl, T. J., Mott, G. E., Kramer, D. C., Kruski, A. W., Williams, J. D., and McGill, H. C., Jr. 1981. Genetic effects on serum cholesterol concentrations in baboons, *J. Hered.* 72:97–103.

Fogelman, A. M., Schechter, I., Seager, J., Hokom, M., Childs, J. S., and Edwards, P. A., 1980. Malondialdehyde alteration of low density lipoproteins leads to cholesteryl ester accumulation in human monocyte-macrophages, *Proc. Natl. Acad. Sci. USA* 77:2214–2218.

Friedman, M., Byers, S. O., Diamant, J., and Rosenman, R. H. 1975. Plasma catecholamine response of coronary-prone subjects (Type A) to a specific challenge, *Metabolism* 24:205–210.

Friedman, R. J., Stemerman, M. B., and Wenz, B. 1977. The effect of thrombocytopenia on experimental atherosclerotic lesion formation in rabbits. Smooth muscle cell proliferation and re-endothelialization, *J. Clin. Invest.* 60:1191.

Fritz, K. E., Daoud, A. S., Augustyn, J. M., and Jarmolych, J. 1980. Morphological and biochemical differences among grossly-defined types of swine aortic atherosclerotic lesions induced by a combination of injury and atherogenic diet, *Exp. Mol. Pathol.* 32:61–72.

Fry, D. L. 1973. Responses of the arterial wall to

certain physical factors, In: *Atherogenesis: Initiating Factors*, pp. 93–125, Elsevier, Amsterdam.

Fuster, V., Chesebro, J. H., Frye, R. L., and Elveback, L. R. 1981. Platelet survival and the development of coronary artery disease in the young adult: effects of cigarette smoking, strong family history and medical therapy, *Circulation* 63:546–551.

Gajdusek, C., DiCorleto, P., Ross, R., and Schwartz, S. M. 1980. An endothelial cell-derived growth factor, *J. Cell Biol.* 85:467–472.

Geer, J. C., and Guidry, M. A. 1965. Experimental canine atherosclerosis, In: *Comparative Atherosclerosis* (J. C. Roberts, Jr., and R. Straus, eds.), pp. 170–185, Harper & Row, New York.

Gerrity, R. G. 1981a. The role of the monocyte in atherogenesis. I. Transition of blood-borne monocytes into foam cells in fatty lesions, *Am. J. Pathol.* 103:181–190.

Gerrity, R. G. 1981b. The role of the monocyte in atherogenesis. II. Migration of foam cells from atherosclerotic lesions, *Am. J. Pathol.* 103:191–200.

Gerrity, R. G., and Naito, H. K. 1980. Ultrastructural identification of monocyte-derived foam cells in fatty streak lesions, *Artery* 8:208–214.

Gerrity, R. G., Naito, H. K., Richardson, M., and Schwartz, C. J. 1979. Dietary induced atherosclerosis in swine. Morphology of the intima in prelesion stages, *Am. J. Pathol.* 95:775–786.

Glass, D. C. 1977. Behavior patterns, stress, and coronary disease, Lawrence Erlbaum Associates, Hillsdale, N.J.

Glass, D. C., Krakoff, L. R., Contrada, R., Hilton, W. F., Kehoe, K., Mannucci, E. G., Collins, C., Snow, B., and Elting, E. 1980. Effect of harassment and competition upon cardiovascular and plasma catecholamine responses in Type A and Type B individuals, *Psychophysiology* 17:453–463.

Glenn, K. C., and Ross, R. 1981. Human monocyte-derived growth factor(s) for mesenchymal cells: activation of secretion by endotoxin and concanavalin A, *Cell* 25:603–615.

Goldstein, J. L., Ho, Y. K., Basu, S. K., and Brown, M. S. 1979. Binding site on macrophages that mediates uptake and degradation of acetylated low density lipoprotein, producing massive cholesterol deposition, *Proc. Natl. Acad. Sci. USA* 76:333–337.

Gordon, T., Kagan, A., Garcia-Palmieri, M., Kannel, W. B., Zukel, W. J., Tillotson, J., Sorlie, P., and Hjortland, M. 1981. Diet and its relation to coronary heart disease and death in three populations, *Circulation* 63:500–515.

Griggs, T. R., Reddick, R. L., Sultzer, D., and Brinkhous, K. M. 1981. Susceptibility to atherosclerosis in aortas and coronary arteries of swine with von Willebrand's disease, *Am. J. Pathol.* 102:137–145.

Grundy, S. M., Bilheimer, D., Blackburn, H., Brown, V., Kwiterovich, P. O., Jr., Mattson, F., Schonfeld G., and Weidman, W. H., 1982. Rationale of the diet-heart statement of the American Heart Association. Report of the nutrition committee, *Circulation* 65:839A–854A.

Haffner, S. M., Applebaum-Bowden, D., Wahl, P. W., Hoover, J. J., Warnick, G. R., Albers, J. J.,

and Hazzard, W. R. 1985. Epidemiological correlates of high density lipoprotein subfractions, apolipoproteins A-I, A-II, and D, and lecithin cholesterol acyltransferase. Effects of smoking, alcohol, and adiposity, *Arteriosclerosis* 5(2):169–177.

Halfon, S.-T., Kark, J. D., Baras, M., Friedlander, Y., and Eisenberg, S. 1982. Smoking, lipids and lipoproteins in Jerusalem 17-year-olds, *Isr. J. Med. Sci.* 18:1150–1157.

Hamm, T. E., Jr., Kaplan, J. R., Clarkson, T. B., and Bullock, B. C. 1983. Effects of gender and social behavior on the development of coronary artery atherosclerosis in cynomolgus macaques, *Atherosclerosis* 48:221–233.

Hawkins, R. I. 1972. Smoking, platelets and thrombosis, *Nature* 236:450–452.

Hegsted, D. M., McGandy, R. B., Myers, M. L., and Stare, F. J. 1965. Quantitative effects of dietary fat on serum cholesterol in man, *Am. J. Clin. Nutr.* 17:281–295.

Hegyeli, R. J. 1981. Prostaglandins and cardiovascular disease, *Atheroscler. Rev.* 8:1–195.

Hernberg, S., Karava, R., Koskela, R. S., and Luoma, K. 1976. Angina pectoris, ECG findings and blood pressure of foundry workers in relation to carbon monoxide exposure, *Scand. J. Work Environ. Health* 2:54–63.

Hogg, S. I., and Cryer, A. 1982. Aortic arylhydrocarbon hydroxylase activity in atherosclerosis-susceptible and atherosclerosis-resistant species, *Comp. Biochem. Physiol.* 73B:669–671.

Hojnacki, J. L., Mulligan, J. J., Cluette, J. E., Kew, R. R., Stack, D. J., and Huber, G. L. 1981. Effect of cigarette smoke and dietary cholesterol on plasma lipoprotein composition, *Artery* 9:285–304.

Holme, I., Enger, S. C., Helgeland, A., Hjermann, I., Leren, P., Lund-Larsen, P. G., Solberg, L. A., and Strong, J. P. 1981. Risk factors and raised atherosclerotic lesions in coronary and cerebral arteries: statistical analysis from the Oslo study, *Arteriosclerosis* 1:250–256.

Hugod, C., Hawkins, L. H., Kjeldsen, K., Thomsen, H. K., and Astrup, P. 1978. Effect of carbon monoxide exposure on aortic and coronary intimal morphology in the rabbit: a revaluation, *Atherosclerosis* 30:333–342.

Imai, H., and Lee, K. T. 1983. Mosaicism in female hybrid hares heterozygous for glucose-6-phosphate dehydrogenase. IV. Aortic atherosclerosis in hybrid hares fed alternating cholesterol-supplemented and nonsupplemented diets, *Exp. Mol. Pathol.* 39:11–23.

Imai, H., Lee, K. T., and Janakidevi, K. 1982. Coronary arterial heart disease in hybrid hares fed alternating cholesterol-enriched and normal diets, *Atherosclerosis* 45:149–160.

Innerarity, T. L., Pitas, R. E., and Mahley, R. W. 1982. Modulating effects of canine high density lipoproteins on cholesteryl ester synthesis induced by beta–very low density lipoproteins in macrophages: possible *in vitro* correlates with atherosclerosis, *Arteriosclerosis* 2:114–124.

Jokinen, M. P., Clarkson, T. B., and Prichard, R. W. 1985. Recent advances in molecular pathology. Animal models in atherosclerosis research, *Exp. Mol. Pathol.* 42:1–28.

Jones, J. G., and Sinclair, A. 1975. Arterial disease amongst blast furnace workers, *Ann. Occup. Hyg.* 18:15–20.

Kannel, W. B., Dawber, T. R., Kagan, A., Revotskie, N., and Stokes, J. 1961. Factors of risk in the development of coronary heart disease—six-year follow-up experience: the Framingham study, *Ann. Intern. Med.* 55:33–50.

Kaplan, J. R., Manuck, S. B., Clarkson, T. B., Lusso, F. M., and Taub, D. M. 1982. Social status, environment and atherosclerosis in cynomolgus monkeys, *Arteriosclerosis* 2:359–368.

Kaplan, J. R., Adams, M. R., Clarkson, T. B., and Koritnik, D. R. 1984a. Psychosocial influences on female "protection" among cynomolgus macaques, *Atherosclerosis* 53:283–295.

Kaplan, J. R., Clarkson, T. B., and Manuck, S. B. 1984b. Pathogenesis of carotid bifurcation atherosclerosis in cynomolgus monkeys, *Stroke* 15:994–1000.

Keys, A., Anderson, J. T., and Grande, F. 1957. Prediction of serum-cholesterol responses of man to changes in fats in the diet, *Lancet* 2:959–966.

Keys, A., Anderson, J. T., and Grande, F. 1965a. Serum cholesterol response to changes in the diet. II. The effect of cholesterol in the diet, *Metabolism* 14:759–765.

Keys, A., Anderson, J. T., and Grande, F. 1965b. Serum cholesterol response to changes in the diet. IV. Particular saturated fatty acids in the diet, *Metabolism* 14:766–787.

Kita, T., Brown, M. S., Watanabe, Y., and Goldstein, J. L. 1981. Deficiency of low density lipoprotein receptor in liver and adrenal gland of the WHHL rabbit, an animal model of familial hypercholesterolemia, *Proc. Natl. Acad. Sci. USA* 78:2268–2272.

Kramsch, D. M., and Hollander, W. 1968. Occlusive atherosclerotic disease of the coronary arteries in monkeys (*Macaca irus*) induced by diet, *Exp. Mol. Pathol.* 9:1–22.

Kritchevsky, D., Davidson, L. M., Kim, H. K., Krendel, D., Malhotra, S., Vander Watt, J. J., Du Plessis, J. P., Winter, P. A. D., Mendelsohn, T. I. D., and Bersohn, I. 1977. Influence of semipurified diets on atherosclerosis in African green monkeys, *Exp. Mol. Pathol.* 26:28–51.

Kritchevsky, D., Davidson, L. M., Weight, M., Kriek, N. P. J., and Du Plessis, J. P. 1982. Influence of native and randomized peanut oil on lipid metabolism and aortic sudanophilia in the vervet monkey, *Atherosclerosis* 43:53–58.

Lee, K. T., Thomas, W. A., Janakidevi, K., Kroms, M., Reiner, J. M., and Borg, K. Y. 1981. Mosaicism in female hybrid hares heterozygous for glucose-6-phosphate dehydrogenase (G-6-PD). II. Changes in the ratios of G-6-PF types in skin fibroblast cultures carried through multiple passages, *Exp. Mol. Pathol.* 34:202–208.

Levine, P. H. 1973. An acute effect of cigarette smoking on platelet function: a possible link between smoking and arterial thrombosis, *Circulation* 48:619–623.

Liebovich, S. J., and Ross, R. 1976. A macrophage-dependent factor that stimulates the proliferation of fibroblasts in vitro, *Am. J. Pathol.* 84:501–513.

Lifsic, A. M. 1976. Atherosclerosis in smokers, *Bull. WHO* 53:631–638.

Lofland, H. B, Jr., Clarkson, T. B., and Bullock, B. C. 1970. Whole body sterol metabolism in squirrel monkeys (*Saimiri sciureus*), *Exp. Mol. Pathol.* 13:1–11.

Mahley, R. W., Weisgraber, K. H., and Innerarity, T. 1974. Canine lipoproteins and atherosclerosis. II. Characterization of the plasma lipoproteins associated with atherogenic and nonatherogenic hyperlipidemia, *Circ. Res.* 35:722–733.

Mahley, R. W., Weisgraber, K. H., Innerarity, T., Brewer, H. B., Jr., and Assmann, G. 1975. Swine lipoproteins and atherosclerosis. Changes in the plasma lipoproteins and apoproteins induced by cholesterol feeding, *Biochemistry* 14:2817–2823.

Majesky, M. W., Yang, H. Y. Benditt, E. P., and Juchau, M. R. 1983. Carcinogenesis and atherogenesis: differences in monooxygenase inducibility and bioactivation of benzo[a]pyrene in aortic and hepatic tissues of atherosclerosis-susceptible versus resistant pigeons, *Carcinogenesis* 4:647–652.

Majesky, M. W., Reidy, M. A., Benditt, E. P., and Juchau, M. R. 1985. Focal smooth muscle proliferation in the aortic intima produced by an initiation-promotion sequence, *Proc. Natl. Acad. Sci. USA* 82:3450–3454.

Malinow, M. R., McLaughlin, P., Dhindsa, D. S., Metacalfe, J., Ochsner, A. J., III, Hill, J., and McNulty, W. P. 1976a. Failure of carbon monoxide to induce myocardial infarction in cholesterol-fed cynomolgus monkeys (*Macaca fascicularis*), *Cardiovasc. Res.* 10:101–108.

Malinow, M. R., McLaughlin, P., Papworth, L., Naito, H. K., Lewis, L., and McNulty, W. P. 1976b. A model for therapeutic interventions on established coronary atherosclerosis in a nonhuman primate, *Adv. Exp. Med. Biol.* 67:3–31.

Manning, P. J., and Clarkson, T. B. 1972. Development, distribution and lipid content of diet-induced atherosclerotic lesions of rhesus monkeys, *Exp. Mol. Pathol.* 17:38–54.

Manuck, S. B., and Garland, F. N. 1980. Stability of individual differences in cardiovascular reactivity: a thirteen month follow-up, *Physiol. Behav.* 24:621–624.

Marmorston, J., Lewis, J. J., Bernstein, J. L., Sobel, H., Kuzma, O., Alexander, R., Magidson, O., and Moore, F. J. 1957. Excretion of urinary steroids by men and women with myocardial infarction, *Geriatrics* 12:297–300.

Mattson, F. H., Erickson, B. A., and Kligman, A. M. 1972. Effect of dietary cholesterol on serum cholesterol in man, *Am. J. Clin. Nutr.* 25:589–594.

McCubbin, J. A., Richardson, J. E., Langer, A. W., Kizer, J. S., and Obrist, P. A. 1983. Sympathetic neuronal function and left ventricular performance during behavioral stress in humans: the relationship between plasma catecholamines and systolic time intervals, *Psychophysiology* 20:102–110.

McGill, H. C., Jr. 1968. *The Geographic Pathology of Atherosclerosis*, Williams & Wilkins, Baltimore, Md.

McGill, H. C., Jr. 1977. Atherosclerosis: problems in pathogenesis, *Atheroscler. Rev.* 2:27–65.

McGill, H. C. 1984. Persistent problems in the pathogenesis of atherosclerosis, *Arteriosclerosis* 4:443–451.

McGill, H. C., Jr., and Stern, M. P. 1979. Sex and atherosclerosis, *Atheroscler. Rev.* 4:157–242.

Mehta, P., and Mehta, J. 1982. Effects of smoking on platelets and on plasma thromboxane-prostacyclin balance in man, *Prost. Leuko. Med.* 9:141–150.

Minick, C. R., and Murphy, G. E. 1973. Experimental induction of athero-arteriosclerosis by the synergy of allergic injury to arteries and lipid rich diet. II. Effect of repeatedly injected foreign protein in rabbits fed a lipid-rich, cholesterol-poor diet, *Am. J. Pathol.* 73:265–300.

Minick, C. R., Alonso, D. R., and Rankin, L. 1978. Role of immunologic arterial injury in atherogenesis, *Thromb. Haemost.* 39:304–311.

Minick, C. R., Fabricant, C. G., Fabricant, J., and Litrenta, M. M. 1979. Atheroarteriosclerosis induced by infection with a herpesvirus, *Am. J. Pathol.* 96:673–706.

Montenegro, M. R., and Eggen, D. A. 1968. Topography of atherosclerosis in the coronary arteries, *Lab. Invest.* 18:586–593.

Moore, A., Friedman, R. J., Singal, D. P., Gauldie, J., and Blajchman, M. 1976. Inhibition of injury induced thromboatherosclerotic lesions by antiplatelet serum in rabbits, *Thromb. Diath. Haemorrh.* 35:70.

Moore, S. 1973. Thromboatherosclerosis in normolipemic rabbits: a result of continued endothelial damage, *Lab. Invest.* 29:478–487.

Mustard, J. F., and Packham, M. A. 1972. Thrombosis and the development of atherosclerosis, In: *The Pathogenesis of Atherosclerosis* (R. W. Wissler and J. C. Geer, eds.), pp. 214–226, Williams & Wilkins Company, Baltimore, Md.

National Heart, Lung and Blood Institute. 1982. Report of the Working Group on Arteriosclerosis, Vol. 2. Publication NIH 82-2035, U.S. Department of Health, Education and Welfare, Washington, D.C.

Neri-Serneri, G. G., Masotti, G., Gensini, G. F., Abbate, R., Poggesi, L., Galanti, G., and Favilla, S. 1981. Prostacyclin thromboxane, and ischemic heart disease, *Atheroscler. Rev.* 8:139–157.

Pearson, T. A., Dillman, J. M., Solez, K., and Heptinstall, R. H. 1978. Clonal characteristics in layers of human atherosclerotic plaques. A study of the selection hypothesis of monoclonality, *Am. J. Pathol.* 93:93–102.

Pearson, T. A., Dillman, J., Williams, K. J., Wolff, J. A., Adams, R., Solez, K., and Heptinstall, R. H. 1979. Clonal characteristics of experimentally induced "atherosclerotic" lesions in the hybrid hare, *Science* 206:1423–1425.

Pearson, T. A., Dillman, J., Malmros, H., Sternby, N., and Heptinstall, R. H. 1983. Cholesterol-induced atherosclerosis. Clonal characteristics of arterial lesions of the hybrid hare, *Arteriosclerosis* 3:574–580.

Penn, A. L., Batastini, G. G., and Albert, R. E. 1980. Age-dependent changes in prevalence, size and proliferation of arterial lesions in cockerels. I. Spontaneous lesions, *Artery* 7:448–463.

Penn, A. L., Batastini, G. G., Soloman, J., Burns, F., and Albert, R. E. 1981. Dose-dependent size increases of aortic lesions following chronic exposures to 7,12-dimethylbenz[a]anthracene, *Cancer Res.* 41:588–592.

Pick, R., and Katz, L. N. 1965. The morphology of experimental, cholesterol- and oil-induced atherosclerosis in the chick, In: *Comparative Atherosclerosis* (J. C. Roberts, Jr., and R. Straus, eds.), pp. 77–84, Harper & Row, New York.

Plant, A. L., Benson, D. M., and Smith, L. C. 1985. Cellular uptake and intracellular localization of benzo[a]pyrene by digital fluorescence imaging microscopy, *J. Cell. Biol.* 100(4):1295–1308.

Prichard, R. W., Clarkson, T. B., Goodman, H. O., and Lofland, H. B, Jr. 1964a. Aortic atherosclerosis in pigeons and its complications, *Arch. Pathol.* 77:244–257.

Prichard, R. W., Clarkson, T. B., Lofland, H. B, and Goodman, H. O. 1964b. Pigeon atherosclerosis, *Am. Heart J.* 67:715–717.

Prior, J. T., Kurtz, D. M., and Ziegler, D. D. 1961. The hypercholesterolemic rabbit. An aid to understanding arteriosclerosis in man? *Arch. Pathol.* 71:672–684.

Reidy, M. A., and Bowyer, D. E. 1978. Distortion of endothelial repair. The effect of hypercholesterolaemia on regeneration of aortic endothelium following injury by endotoxin. A scanning electron microscope study, *Atherosclerosis* 29:459–466.

Reitman, J. S., Mahley, R. W., and Fry, D. L. 1982. Yucatan miniature swine as a model for experimental atherosclerosis, *Atherosclerosis* 43:119–132.

Remsen, J. F., and Shireman, R. B. 1981a. Removal of benzo[a]pyrene from cells by various components of medium, *Cancer Lett.* 14(1):41–46.

Remsen, J. F., and Shireman, R. B. 1981b. Effect of low-density lipoprotein on the incorporation of benzo[a]pyrene by cultured cells. *Cancer Res.* 41:3179–3185.

Review Panel on Coronary-Prone Behavior and Coronary Heart Disease. 1981. Coronary-prone behavior and coronary heart disease: a critical review, *Circulation* 63:1199–1215.

Revis, N. W., Bull, R., Laurie, D., and Schiller, C. A. 1984. The effectiveness of chemical carcinogens to induce atherosclerosis in the White Carneau pigeon, *Toxicology* 32(3):215–227.

Rhoads, G. G., Blackwelder, W. C., Stemmermann, G. N., Hayashi, T., and Kagan, A. 1978. Coronary risk factors and autopsy findings in Japanese-American men, *Lab. Invest.* 38:304–311.

Robertson, W. B., Geer, J. C., Strong, J. P., and McGill, H. C., Jr. 1963. The fat of the fatty streak, *Exp. Mol. Path.* 2(Suppl. 1):28–39.

Robertson, A. L., Jr., Butkus, A., Ehrhart, L. A., and Lewis, L. A. 1972. Experimental arteriosclerosis in dogs. Evaluation of anatomopathological findings, *Atherosclerosis* 15:307–325.

Rogers, W. R., Bass, R. L., III, Johnson, D. E., Kruski, A. W., McMahan, C. A., Montiel, M. M., Mott, G. E., Wilbur, R. L., and McGill, H. C., Jr. 1980. Atherosclerosis-related responses to cigarette smoking in the baboon, *Circulation* 61:1188–1193.

Rosenman, R. H., Brand, R. J., Jenkins, C. D., Friedman, M., Straus, R., and Wurm, M. 1975. Coronary heart disease in the Western Collaborative Group Study: final follow-up experience of 8½ years, *J. Am. Med. Assoc.* 233:872–877.

Ross, R. 1981. Atherosclerosis: a problem of the biology of arterial cell walls and their interaction with blood components, *Arteriosclerosis* 1:293–311.

Ross, R. 1986. The pathogenesis of atherosclerosis—an update, *New Engl. J. Med.* 314:488–500.

Ross, R., and Vogel, A. 1978. The platelet-derived growth factor, *Cell* 14:203–210.

Ross, R., Glomset, J., Kariya, B., and Harker, L. 1974. A platelet-dependent serum factor that stimulates the proliferation of arterial smooth muscle cells *in vitro*, *Proc. Natl. Acad. Sci. USA* 71:1207–1210.

Rudel, L. L. 1980. Plasma lipoproteins in atherogenesis in nonhuman primates, In: *The Use of Nonhuman Primates in Cardiovascular Diseases* (S. S. Kalter, ed.), pp. 37–57, University of Texas Press, Austin.

Rudel, L. L., Shah, R., and Greene, D. G. 1979. Study of the atherogenic dyslipoproteinemia induced by dietary cholesterol in rhesus monkeys (*Macaca mulatta*), *J. Lipid Res.* 20:55–65.

Rudel, L. L., Parks, J. S., and Carroll, R. M. 1983. Effects of polyunsaturated versus saturated dietary fat on nonhuman primate HDL, In: *Dietary Fats and Health* (E. G. Perkins and W. J. Visek, eds.), pp. 649–666, American Oil Chemists Society, Champaign, Ill.

St. Clair, R. W. 1983. Metabolic changes in the arterial wall associated with atherosclerosis in the pigeon, *Fed. Proc.* 42:2480–2485.

Samuel, P., McNamara, D. J., and Shapiro, J. 1983. The role of diet in the etiology and treatment of atherosclerosis, *Annu. Rev. Med.* 34:179–194.

Schechter, I., Fogelman, A. M., Haberland, M. E., Seager, J., Hokom, M., and Edwards, P. A. 1981. The metabolism of native and malondialdehyde-altered low density lipoproteins by human monocyte–macrophages, *J. Lipid Res.* 22:63–71.

Schenk, E. A., Penn, I., and Schwartz, S. 1965. Experimental atherosclerosis in the dog, *Arch. Pathol.* 80:102–109.

Scott, D. W., Gorry, G. A., and Gotto, A. M. 1981. Diet and coronary heart disease: the statistical analysis of risk, *Circulation* 63:516–518.

Scott, R. F., Jones, R., Daoud, A. S., Zumbo, O., Coulston, F., and Thomas, W. A. 1967a. Experimental atherosclerosis in rhesus monkeys. II. Cellular elements of proliferative lesions and possible role of cytoplasmic degeneration in pathogenesis as studied by electron microscopy, *Exp. Mol. Pathol.* 7:34–57.

Scott, R. F., Morrison, E. S., Jarmolych, J., Nam, S. C., Kroms, M., and Coulston, F. 1967b. Exper-

imental atherosclerosis in rhesus monkeys. I. Gross and light microscopy features and lipid values in serum and aorta, *Exp. Mol. Pathol.* 7:11–33.

Shekelle, R. B., Gale, M., Ostfeld, A. M., and Oglesby, P. 1983. Hostility, risk of coronary heart disease and mortality, *Psychosom. Med.* 45:109–114.

Shimada, Y., Tanzawa, K., Kuroda, M., Tsujita, Y., Arai, M., and Watanabe, Y. 1981. Biochemical characterization of skin fibroblasts derived from WHHL rabbit, a notable animal model for familial hypercholesterolemia, *Eur. J. Biochem.* 118:557–564.

Shu, J. P., and Nichols, A. V. 1981. Uptake of lipophilic carcinogens by plasma lipoproteins. Structure-activity studies, *Biochim. Biophys. Acta* 665(3):376–384.

Sieffert, G. F., Keown, K., and Moore, W. S. 1981. Pathologic effect of tobacco smoke inhalation on arterial intima, *Surg. Forum* 32:333–335.

Siess, W., Lorenz, R., Roth, P., and Weber, P. C. 1982. Plasma catecholamines, platelet aggregation and associated thromboxane formation after physical exercise, smoking or norepinephrine infusion, *Circulation* 66:44–48.

Simpson, C. F., and Harms, R. H. 1969. Aortic atherosclerosis of turkeys induced by feeding of cholesterol, *J. Atheroscler. Res.* 10:63–75.

Sorlie, P. D., Garcia-Palmieri, M. R., Castillo-Staab, M. I., Costas, R., Jr., Oalmann, M. C., and Havlik, R. 1981. The relation of antemortem factors to atherosclerosis at autopsy. The Puerto Rico Heart Health Program, *Am. J. Pathol.* 103:345–352.

Spector, A. A., and Johnson, M. R. 1982. Diet, serum lipids, and atherosclerosis, In: *Animal Products in Human Nutrition* (D. C. Beitz and R. G. Hansen, eds.), pp. 501–534, Academic Press, New York.

Stallones, R. A. 1980. The rise and fall of ischemic heart disease, *Sci. Am.* 243(5):53–59.

Stamford, B. A., Matter, S., Fell, R. D., Sady, S., Cresanta, M. K., and Papanek, P. 1984a. Cigarette smoking, physical activity, and alcohol consumption: relationship to blood lipids and lipoproteins in premenopausal females, *Metabolism* 33:585–590.

Stamford, B. A., Matter, S., Fell, R. D., Sady, S., Papanek, P., and Cresanta, M. 1984b. Cigarette smoking, exercise and high density lipoprotein cholesterol, *Atherosclerosis* 52:73–83.

Stary, H. C. 1983. Evolution of atherosclerotic plaques in the coronary arteries of young adults, *Arteriosclerosis* 3:471A.

Stary, H. C. 1984. Comparison of the morphology of atherosclerotic lesions in the coronary arteries of man with the morphology of lesions produced and regressed in experimental primates, In: *Regression of Atherosclerotic Lesions* (M. R. Malinow and V. Blaton, eds.), pp. 235–254, Plenum Press, New York.

Stary, H. C., and Letson, G. D. 1983. Morphometry of coronary artery components in children and young adults, *Arteriosclerosis* 3:485A.

Stemerman, M. B., and Ross, R. 1972. Experimental arteriosclerosis. I. Fibrous plaque formation in primates, an electron microscope study, *J. Exp. Med.* 136:769–789.

Strong, J. P., and Richards, M. L. 1976. Cigarette

smoking and atherosclerosis in autopsied men, *Atherosclerosis* 23:451–476.

Strong, J. P., Eggen, D. A., and Oalmann, N. C. 1972. The natural history, geographic pathology and epidemiology of atherosclerosis, In: *The Pathogenesis of Atherosclerosis* (R. W. Wissler and J. C. Geer, eds.), pp. 20–40, Williams & Wilkins, Baltimore, Md.

Tanzawa, K., Shimada, Y., Kuroda, M., Tsujita, Y., Arai, M., and Watanabe, Y. 1980. WHHL-rabbit: a low density lipoprotein receptor-deficient animal model for familial hypercholesterolemia, *Febs. Lett.* 118:81–84.

Taylor, C. B., Cox, G. E., Manalo-Estrella, P., and Southworth, J. 1962. Atherosclerosis in rhesus monkeys. II. Arterial lesions associated with hypercholesterolemia induced by dietary fat and cholesterol, *Arch. Pathol.* 74:16–34.

Taylor, C. B., Patton, D. E., and Cox, G. E. 1963. Atherosclerosis in rhesus monkeys. VI. Fatal myocardial infarction in a monkey fed fat and cholesterol, *Arch. Pathol.* 76:404–412.

Tejada, C., Strong, J. P., Montenegro, M. R., Restrepo, C., and Solberg, L. A. 1968. Distribution of coronary and aortic atherosclerosis by geographic location, race, and sex, *Lab. Invest.* 18:509–526.

Thomas, W. A., and Kim, D. N. 1983. Biology of disease. Atherosclerosis as a hyperplastic and/or neoplastic process, *Lab. Invest.* 48:245–255.

Thomas, W. A., Reiner, J. M., Janakidevi, K., Florentin, R. A., and Lee, K. T. 1979. Population dynamics of arterial cells during atherogenesis. Study of monotypism in atherosclerotic lesions of black women heterozygous for glucose-6-phosphate dehydrogenase (G-6-PD), *Exp. Mol. Pathol.* 31:367–386.

Thomsen, H. K. 1974. Carbon monoxide-induced atherosclerosis in primates. An electron-microscopic study on the coronary arteries of *Macaca irus* monkeys, *Atherosclerosis* 20:233–240.

Trillo, A. A., and Prichard, R. W. 1979. Early endothelial changes in experimental primate atherosclerosis, *Lab. Invest.* 41:294–302.

Turner, D. M., Lee, P. N., Roe, F. J. C., and Gough, K. J. 1979. Atherogenesis in the White Carneau pigeon. Further studies of the role of carbon monoxide and dietary cholesterol, *Atherosclerosis* 34:407–417.

U.S. Department of Health and Human Services. 1983. The health consequences of smoking. Cardiovascular disease. A Report of the Surgeon General, Public Health Service, Office of Smoking and Health, Rockville, Md.

U.S. Environmental Protection Agency. 1984. Review of the NAAQS for Carbon Monoxide: Reassessment of Scientific and Technical Information, EPA-450/5-84-004, Research Triangle Park, N.C.

Vauhkonen, M., Kuusi, T., and Kinnunen, P. K. 1980. Serum and tissue distribution of benzo[*a*]pyrene from intravenously injected chylomicrons in rat in vivo, *Cancer Lett.* 11(2):113–119.

Verrier, R. L., and Lown, B. 1984. Behavioral stress and cardiac arrhythmias, *Ann. Rev. Physiol.* 46:155–176.

Viel, B., Donoso, S., and Salcedo, D. 1968. Coronary atherosclerosis in persons dying violently, *Arch. Intern. Med.* 122:97–103.

Vikhert, A. M., Zhdanov, V. S., and Lifshits, A. M. 1976. Influence of nutritional status and tobacco smoking on the development of atherosclerosis in male manual and brain workers, *Cor Vasa* 18:202–208.

Wagner, W. D., and Clarkson, T. B. 1974. Mechanisms of the genetic control of plasma cholesterol in selected lines of Show Racer pigeons, *Proc. Soc. Exp. Biol. Med.* 145:1050–1057.

Wagner, W. D., Clarkson, T. B., Feldner, M. A., and Prichard, R. W. 1973. The development of pigeon strains with selected atherosclerosis characteristics, *Exp. Mol. Pathol.* 19:304–319.

Wald, N., Howard, S., Smith, P. G., and Kjeldsen, K. 1973. Association between atherosclerotic diseases and carboxyhaemoglobin levels in tobacco smokers, *Br. Med. J.* 1(856):761–765.

Wanstrup, J., Kjeldsen, K., and Astrup, P. 1969. Acceleration of spontaneous intimal-subintimal changes in rabbit aorta by a prolonged moderate carbon monoxide exposure, *Acta Pathol. Microbiol. Scand.* 75:353–362.

Watanabe, Y., Ito, T., and Shiomi, M. 1985. The effect of selective breeding on the development of coronary atherosclerosis in WHHL rabbits. An animal model for familial hypercholesterolemia, *Atherosclerosis* 56:71–79.

Webster, W. S., Clarkson, T. B., and Lofland, H. B. 1970. Carbon monoxide–aggravated atherosclerosis in the squirrel monkey, *Exp. Mol. Pathol.* 13:36–50.

Weingand, K. W., Clarkson, T. B., Adams, M. R., and Bostrom, A. D. 1986. Effects of age and/or puberty on coronary artery atherosclerosis in cynomolgus monkeys, *Atherosclerosis* 62:137–144.

Williams, R. B., Jr., Lane, J. D., Kuhn, C. M., Melosh, W., White, A. D., and Schanberg, S. M. 1982. Type A behavior and elevated physiological and neuroendocrine responses to cognitive tests, *Science* 218:483–495.

Wilson, R. B., Miller, R. A., Middleton, C. C., and Kinden, D. 1982. Atherosclerosis in rabbits fed a low cholesterol diet for five years, *Arteriosclerosis* 2:228–241.

Yamasaki, E., and Ames, B. N. 1977. Concentration of mutagens from urine by absorption with the nonpolar resin XAD-2: cigarette smokers have mutagenic urine, *Proc. Natl. Acad. Sci. USA* 74:3555–3559.

Zilversmit, D. B. 1982. Diet and heart disease. Prudence, probability, and proof, *Arteriosclerosis* 2:83–84.

Zyzanski, S. J. 1978. Associations of the coronary prone behavior pattern, In: *Proceedings of the Forum on Coronary Prone Behavior* (T. M. Dembroski, S. M. Weiss, J. L. Shields, S. G. Haynes, and M. Feinleib, eds.), Publication number NIH 78-1451, U.S. Department of Health, Education and Welfare, Washington, D.C.

Identifying Neurobehavioral Effects of Automotive Emissions and Fuel Components

New York University Medical Center

Air Pollution, the Automobile, and Public Health. © 1988 by the Health Effects
Institute. National Academy Press, Washington, D.C.

The automobile can be regarded as a mixed blessing. Although it has become a necessary part of daily life, the changes potentially induced by automotive emissions are not necessarily welcomed. Among these are adverse neurobehavioral effects that range in severity from the annoyance provoked by unpleasant odors and the eye irritants to overt behavioral and neurological dysfunction.

Airborne contaminants can alter behavior and the functions of the nervous system in a variety of ways. Chemicals can damage the structure of the nervous system directly, or can alter behavior and nervous system function without pathological changes by affecting neurotransmitter systems, perturbing membranes, or altering cellular metabolism. Since behavior depends upon a wide variety of nervous system functions, behavioral changes can sometimes provide early indications of adverse effects on other organ systems.

Automotive emissions can also alter behavior by stimulating sensory systems. These stimuli may be unpleasant events that alter the conduct of daily life, or may serve as important discriminative or warning stimuli.

The neurobehavioral toxicity of the chemicals involved in automotive technology is not well understood. Emissions may not produce obvious effects at concentrations commonly found in the environment; moreover, some individuals may be exposed occupationally to higher levels. Although people are exposed to the chemicals in automotive emissions environmentally as well as occupationally, and although studies of such people offer unique opportunities, it is inappropriate to rely entirely on these exposures to detect neurotoxicity, especially when suitable techniques exist using animals.

In this chapter, methods are described for the detection of adverse neurobehavioral effects of automotive emissions, and recommendations for research in this area are offered. First, ways are described of identifying, in laboratory animals, the adverse neurobehavioral effects of hazardous substances following acute exposures and repeated exposures, of characterizing subtle

effects, and of determining mechanisms of toxicity.

Second, a review is provided of what is currently known about the neurobehavioral toxicity of automotive emissions and fuel constituents. Third, recommendations are offered on how to proceed with a program to address this class of health effects. Although there is reason to suspect that many emissions and fuel constituents are hazardous, little information is available for most. To address this large group of chemicals, a committee should select the substances to test and the order of testing. For the substances about which some knowledge exists, focused recommendations are offered for detailed evaluation of their hazards. These include whole emissions, carbon monoxide (CO), petroleum hydrocarbons, methanol, and metals and their compounds.

Neurobehavioral Effects and Mechanisms of Toxicity

The behavioral and neurological sciences have made extraordinary progress in the past 25 years. Although further progress will continue to yield new methods, in areas ranging from the subcellular to the behavioral levels of analysis, adequate methods are now available to explore the neurobehavioral effects of automotive emissions and fuel components.

Various approaches to neurobehavioral hazard identification have been recommended over the years by a variety of experts. Although no single comprehensive approach has yet been formalized, a responsible screening effort should include:

• identification of the acute hazards of chemicals—this includes seeking evidence of mortality, morbidity, and morphological changes (as in any acute toxicity evaluation), but with particular emphasis accorded behavioral function, learned as well as unlearned;

• characterization of their toxicity in repeated or continuous exposures—this provides an opportunity to characterize toxicity that is delayed or cumulative, to observe the development of tolerance (or reverse tolerance), and to characterize the reversibility of adverse effects; and

• detailed study of mechanisms of injury and special impairments—this includes initial screening for subtle sensory or perceptual impairments, affective disorders, or cognitive and intellectual dysfunction, and appropriate specialized evaluations using refined neuropathological, neurochemical, and neurophysiological techniques.

The U.S. Environmental Protection Agency (EPA) has issued a series of test guidelines that are appropriate for use in acute and chronic neurobehavioral toxicity evaluation. These include guidelines for the examination of neuropathology (U.S. Environmental Protection Agency 1985c), motor activity (U.S. Environmental Protection Agency 1985b), schedule-controlled behavior (U.S. Environmental Protection Agency 1985f), and a functional observation battery (U.S. Environmental Protection Agency 1985a).

Identifying Acute Hazards

It is important to determine the acute effects of chemicals on behavior and nervous system function. Acute performance impairment can increase accident proneness and lower work efficiency; thus there can be serious consequences of even small lapses of coordination, vigilance, or visual sensitivity in operators of all types of transportation machinery. Irritation and sensory effects—perhaps the most common complaints elicited by automotive emissions—reduce the perceived quality of life, cause people to change their lifestyles (by allocating their time to less distressful activities), and incapacitate sensitive individuals. Hence, even apparently reversible adverse effects are of concern. In fact, acute but reversible effects may well be the ones of most concern (see discussion in Whole Emissions and Their Photochemical Byproducts), and information about them, therefore, is of particular importance when describing acceptable limits of exposure. With care, it is possible to design and

conduct statistically valid experiments to accurately estimate the pollutant concentrations that produce small but consistent adverse effects (Wood and Colotla 1986; Wood and Cox 1986).

Acute toxicity testing requires evaluation of function and morphology, and can have outcomes that are positive or negative with respect to either. Thus the possible outcomes can be expressed in array form as:

		Morphology	
		No Effect	Effect
Function	No Effect	A	B
	Effect	C	D

Morphological changes (outcomes B or D) are clearly of immediate concern. However, many chemicals can produce observable functional changes without any morphological correlates (outcome C). Lead, for example, is an automotive emission that produces functional impairments in humans and animals without marked neuropathological changes. Consequently, a complete safety evaluation requires functional tests at levels of exposure so low that they do not produce detectable morphological changes.

Examinations of conditioned behavior can be constructed to detect impairment of a variety of functions. Evaluation of function may, however, fail to warn of morphological changes in the nervous system (outcome B) until the loss of "functional reserve" is sufficiently great (outcome D). Challenges with pharmacological agents may be useful in such situations to unmask silent damage (outcome A or B).

Behavioral tests detect performance disruptions that are indirect results of effects on other systems, just as a disinclination to dance might precede the onset of diarrhea (Dews 1975). Thus motor activity or food and water consumption can be changed by chemicals that do not enter the central nervous system, and without concomitant changes in body weight (Evans et al. 1986). Irritation is another possible indirect manifestation of toxic impairment, and is discussed further under Whole Emissions and Their Photochemical By-products. There are important gaps in our knowledge of the relationship between exposure and the acute behavioral changes it produces.

Toxicity in Repeated or Continuous Exposures

After acute toxicity has been examined, repeated exposure experiments should be undertaken, typically 28- or 90-day subchronic exposures with routine examination of neuropathology, measurement of motor activity (preferably in the home cage), and routine use of functional observational batteries. With the insights afforded by the acute toxicity experiments, repeated-exposure experiments can be tailored to further characterize those effects and to yield improved sensitivity. For example, special histological studies might be indicated; examining learned behavior in detail can provide important insights into the nature of toxic impairment (Laties and Wood 1986).

Such repeated-exposure experiments become particularly important if acute studies indicate the potential for irreversible toxicity. In such cases, for example, tolerance may play an important role, and the repeated administration of other compounds or reference substances as probes may demonstrate a forward or reverse tolerance. The mechanism of tolerance may then be revealed with an appropriate design, such as taking concurrent blood level and behavioral measurements or using satellite groups of animals for tissue-level determinations of whether tolerance results from an increase in metabolism or elimination of the toxic substance. In some cases, the mechanism of tolerance may be behavioral, and may depend only on an animal's opportunity to respond in the presence of the material. The observations of Kane and Alarie (1977) with formaldehyde and acrolein illustrate how the context of previous experience with the toxicant can affect biological response (see Conditioned Responses to Exposure).

Identifying Subtle Effects

Many sensory and perceptual deficits, affective disorders, and cognitive and intellectual dysfunctions are sufficiently subtle that they can be missed in routine acute and subchronic studies.

Sensory and Perceptual Deficits. There is a recognized need for rapid screening procedures (National Institutes of Health 1977), as well as comprehensive studies of the complex functions of sensory systems.

Vision. Toxic effects on this highly complex sensory system are easily missed, especially when studies are undertaken with rodents. For example, experiments with rodents do not reveal the profound restriction of visual fields produced by methylmercury in humans and nonhuman primates. Carbon disulfide can produce selective impairment of the discrimination of coarser features of visual stimuli, leaving the discrimination of fine features unimpaired (Merigan et al. 1985b). Carbon disulfide also affects color vision and hue discrimination (Raitta et al. 1981), an effect that can only be observed in experimental animals having color vision.

Agents that produce peripheral neuropathies (for example, hexacarbons) are likely to affect vision. Measurement of visual function in primates apparently provides a very sensitive index of central nervous system injury (Eskin et al. 1985; Merigan et al. 1985a). Appropriate studies in primates as well as rodents might support the inference of a defensible safety factor from rodent data on peripheral neuropathy. However, inferences about other compounds (for example, those with a cortical distribution of injury) are much more problematic, and reveal functional deficits only with detailed evaluation.

Audition. In most species, the inner ear is difficult to remove from the surrounding bone, so its pathologies are rarely uncovered. Lesions may be missed if specific functional evaluations are not performed. Toluene, whose toxicity has been the focus of much study for years, offers an excellent illustration. It was only recently that Pryor and coworkers (1984a,b) and Rebert and colleagues (1983) documented a selective high-frequency hearing loss following intense toluene exposure. They discovered this damage using a pole-climb avoidance test that required animals to jump on a pole to avoid shock in response to a warning stimulus; they varied the characteristics of the tone to demonstrate the frequency-specific loss. Techniques have also been developed that do not rely on training, for example, the inhibition of noise-elicited startle by barely detectable sounds (Hoffman and Ison 1980; Young and Fechter 1983).

The loss of sensation can readily be detected using psychophysical procedures with animals, but detecting the loss of "perception" poses a more difficult problem. Some chemicals produce no impairment of hearing for pure tones, but produce profound impairment of "perceptual speed." Thus methylmercury poisoning does not alter "hearing" per se, but does impair language comprehension in such a way that individuals function only when spoken to extremely slowly.

■ **Recommendation 1.** Quick and simple tests of sensory impairment should be developed that can be used for screening in conjunction with functional observation batteries. At the same time, psychophysical procedures that are more comprehensive should be developed; these should be directed at specific sensory and perceptual impairments, for example impairment of complex auditory and visual discriminations.

Olfaction, Gustation, Somesthesis, and Proprioception. Toxic effects on these functions might be detected following high doses of chemicals, albeit with great uncertainty. Experiments could be designed to assess these end points in detail. The technologies are readily available, but knowing when best to use them can be a problem, except for chemicals that affect or react with the olfactory mucosa (and potentially the olfactory nerve and bulb), such as oxidants, aldehydes, and large particles.

Changes in olfactory sensitivity may or may not be reversible, and they can alter behavior and quality of life without the affected individual's being aware of it. People with injuries to their sense of smell complain of burning their food while cooking, or of having bouts of food poisoning from eating undetected spoiled food. These effects, as well as the techniques for studying them, have been reviewed (Wood 1982). Olfactory psychophysics is a highly developed research area in which trained observers (human or animal) are used to establish thresholds for detection of odors or for the detection of differences in intensity of odor (Cain and Moskowitz 1974; Moulton et al. 1975). Pursuing research in this area on human subjects could bear fruit. Neurophysiological techniques can detect acute alteration of nasopalatine nerve function following 1-hr exposures to formaldehyde or ozone (O_3) (Kulle and Cooper 1975).

Odor preference studies with animals are readily done but offer little predictive utility for human preferences. Subjective responses to environmental odors have been studied in humans (see, for example, Turk et al. 1974) including work on diesel odor (Springer 1974).

■ **Recommendation 2.** Studies of olfactory sensitivity following aldehyde and oxidant exposures should be undertaken in humans and rodents.

Affective Disorders. Exposure to some chemicals can produce apparently aberrant "affective" or "emotional" behavior in animals. Normally docile strains of rats exposed to inorganic mercury vapor and housed in groups have been observed to spontaneously assume aggressive postures in the home cage (Beliles et al. 1968). Inorganic mercury vapor is well known for its early production of a "neurotic" syndrome in humans. Other chemicals may exaggerate startle responses to sudden stimulation, or make animals very difficult to handle. Carbon disulfide can make dogs extremely aggressive (Lewey 1941); its effects on humans range from the induction of affective disorders to suicide (Wood 1981b). These effects, some of which can be detected only by a careful observer, have received insufficient attention (National Institutes of Health 1977). Procedures that rely on conditioned behavior to detect affective changes produced by toxic chemicals have yet to be put to use; avoidance or punishment procedures should be well suited for this purpose.

■ **Recommendation 3.** Behavioral models of affective disorders should be developed. The models should be calibrated with reference substances, drugs as well as toxicants, before their application to test chemicals or mixtures.

Cognitive and Intellectual Dysfunction.

Short of epidemiologic investigations, the only practical way to study the impairment of normal cognitive or intellectual functioning resulting from the intake of chemicals is to undertake conditioning experiments with animals (National Institutes of Health 1977; Laties 1982). Studying simple performances in well-trained animals can provide useful information, but the findings may not necessarily be associated with learning impairments. More complex performances conditioned explicitly to examine learning can provide us with insight into the likelihood of injury to this important function.

The repeated-acquisition technique has been used considerably as a model of learning impairment (Thompson and Moerschbaecher 1978). This procedure requires an animal to perform two tasks, both requiring the animal to respond in a particular sequence. For one task, the response sequence is constant from day to day, so that performance of a well-learned task is measured; for the second task, the sequence is changed daily. The animal "learns to learn" a new sequence every day, so that acquisition of a response pattern can be studied repeatedly. Toxicants and psychoactive drugs disrupt this acquisition of new response sequences. A variety of other procedures could also be used to model other aspects of intellectual functioning, including alternation procedures, conditional discriminations, and respondent conditioning procedures, to name a few.

■ **Recommendation 4.** Existing procedures should be used to examine whether toxicants of concern impair learning, memory, cognition, and intellectual functioning.

Pursuing the Mechanisms of Toxicity

The detection of toxicity in either acute or subchronic tests will generate questions that should be pursued, since the conclusive demonstration of physiological or biochemical mechanisms of action can be enormously important.

Neurochemical Evaluation.

Normal behavioral activities of animals (Sparber and Tilson 1972) as well as adverse behavioral effects of exposure result in neurochemical changes. It is therefore difficult to interpret the significance of acute neurochemical changes without correlated functional observations. If large numbers of false positives are acceptable, then routine neurochemical evaluation may be desirable as a screening test; otherwise, such use of neurochemical tests is likely to be a poor allocation of resources.

Neurochemical evaluations can be very useful as adjuncts to other tests and to determine mechanisms of toxicity. Manganese (Mn) is an example where such studies could be especially useful, because Mn is likely to have specific interactions with neurochemical substrates. Furthermore, prolonged or irreversible changes in neurotransmitter (or neuromodulator) levels, turnover rates, or receptor numbers, which might be suspected following the observation of other forms of toxicity, are always important. Several good examples exist in the literature on sympathomimetic amines. Repeated amphetamine administration can produce behavioral changes and death in animals, in the absence of obvious pathology. Amphetamine-like drugs can also produce long-lasting reductions in dopamine and serotonin, and a decreased number of uptake sites in brain (Ricaurte et al. 1985). The changes found would not be obvious in a first-pass neuropathological examination. However, the destruction of some receptor populations could be demonstrated histochemically (using Fink-

Heimer silver stains), after the identification of neurochemical changes.

There probably will not be tests for neurobehavioral toxicity comparable to in vitro tests for mutagenicity (that is, the Ames test). A high rate of false positives could be expected from most such test systems; an effect in the test tube does not mean that an effect will necessarily occur in vivo, because the agent may not reach the site of action. False negatives might occur less frequently, but be of greater concern. However, nerve culture techniques will assist in clarifying the mechanisms of neurotoxicity (Veronesi et al. 1980).

On the other hand, biochemical tests can contribute to identifying particular kinds of neurotoxicity. Assays for neurotoxic esterase induction in the brain and the spinal cord have been useful for the identification of neuropathic organophosphates, and an EPA test guideline has been written for this purpose (U.S. Environmental Protection Agency 1985d). Glial fibrillary acidic protein assays reflect the astrocytic response to central nervous system injury, and may be useful as a screening technique (Brock and O'Callaghan 1987). Furthermore, biochemical assays may be able to help steer the process of chemical synthesis; thus if a test indicates possible toxicity for a chemical being developed, further synthesis work might be directed toward developing less toxic alternatives.

Neuropathological Examinations. Neuropathological examinations should be included in toxicity assessment; frequently, the same animals can be used for behavioral testing and neuropathological examination. Subsequent studies of mechanisms of toxicity, of special systems, or of chronic effects may require the study of satellite groups because immersion tissue fixation and routine staining procedures are inadequate for the description of some types of nervous system injury. For example, histochemical techniques (fluorescence, metal stains, Golgi) and immunohistochemistry can provide insights into the mechanisms underlying the neurotoxic effects of certain chemicals. Such techniques are not routinely used for hazard identification.

Special procedures are sometimes required for some classes of injuries; for example, techniques to examine axoplasmic transport. Most such techniques are incompatible with routine toxicologic evaluation. Similarly, nerve teasing and electron microscopy can demonstrate subtle neuropathies. Both are extremely labor intensive. Neuropathological studies can use special chemicals for studies of the mechanisms of toxic injury. The description of the role of γ diketones in the production of hexacarbon neuropathies offers a good example; subsequent studies of 3,4-dimethyl-2,5-hexanedione have examined accelerated pyrrole formation and its role in potential neurofilament cross-linking (Anthony et al. 1983a,b).

Neurophysiological Examinations. Peripheral nerve conduction velocity is used clinically to assess peripheral nerve function in humans. In animals, hind limb weakness, gait disturbance, and peripheral neuropathology provide adequate sensitivity for hazard identification, given that the exposure concentrations contemplated are high enough to induce frank peripheral neuropathy. Measurement of peripheral nerve conduction velocity may be useful for the detection of demyelination, and the U.S. Environmental Protection Agency (1985e) has promulgated a test standard for this purpose. However, it may not detect axonopathies. Most important, the reversibility of peripheral nerve impairments need not imply the reversibility of central nervous system injury.

Evoked-potential studies might be cost-effective for some classes of sensory effects, especially to determine low-level effects. Brainstem auditory evoked responses are useful in detecting hearing loss; the sensitivity of these procedures is comparable to that for behavioral procedures (Rebert et al. 1983). Flash-evoked responses are less informative about injuries to the visual system than are counterphase spatial-frequency reversal experiments. More advanced procedures are available to characterize the interplay of structural and functional alterations produced by toxicants. By simultaneous measurement throughout a brain structure, it should be possible to describe the pathophysiological progression of injury.

Evaluating Whole Emissions, Fuels, and Their Components

Although it is possible to study whole emissions directly as complex mixtures, this is usually not the most fruitful course to pursue. The difficulties encountered in studying such complex mixtures are profound, because the myriad interactions possible among the reactive components of automotive emissions may result simultaneously in potentiation of, and protection against, adverse effects. The study of mixtures should not preclude the continued evaluation of the toxicity of individual components of mixtures, especially since the reduction or elimination of a single component could have major health impacts. We learn relatively little from the study of a single idiosyncratic mixture or simulated automotive emission; indeed, in such studies, the data collected on the reference substance used to calibrate the sensitivity of the experiment may well constitute a greater scientific contribution than the data from the mixture (Laties 1973; Horvath and Frantik 1974).

More information about the behavioral effects of carcinogenic emissions is needed. When testing for carcinogenesis, pollutant doses and concentrations are usually maximum tolerated doses; those that typically produce behavioral effects are often much lower. Thus, under ambient conditions of exposure, the behavioral, rather than the carcinogenic, effects of carcinogenic components of emissions may be the principal effects of concern; these effects might include malaise or performance degradation.

Whole Emissions and Their Photochemical By-products

Air pollution episodes alter human behavior. Weather reports in certain metropolitan areas regularly include air quality reports, and the elderly, those with respiratory problems, and athletes modify their behavior accordingly. The description of atmospheric conditions (even an erroneous prediction) may change the activities people engage in; for example, in a form of conditioned avoidance behavior, the heightened probability of chest discomfort in a smog alert may lead an athlete to change his or her training regimen. During the 1984 Olympics in Los Angeles, concern about the potential effects of automotive emissions and photochemical products on performance and health prompted recommendations for the siting of athletic competitions and for traffic control, as well as for training and competition schedules (McCafferty 1981).

Even if individuals cannot articulate the association between verbal warnings about air quality and discomfort during exercise, they may nevertheless avoid the circumstances in which unpleasant sensations occur, without directly attributing their avoidance to atmospheric quality. The extent to which unpleasant or uncomfortable sensations alter the behavior patterns of daily life has not been recognized.

Nervous system function can be directly affected by constituents of whole emissions, such as CO. Emissions can also produce effects mediated by less direct mechanisms. Functional disturbances that occur in response to emissions may be conditionable in and of themselves, as one might condition a dog to salivate following the ringing of a bell; the evidence for such conditioning will be discussed later.

Eye Irritation. Eye irritation is the most frequent complaint evoked by emission exposure. Numerous studies have documented the increasing frequency of complaints with increasing contaminant concentrations. In a report on O_3 and other photochemical oxidants, the National Academy of Sciences (1977a, p. 430) asserted "For the two most prevalent symptoms related to photochemical-oxidant exposure—eye irritation and lacrimation—no method of quantification has been developed. Eye irritation, although real, is a subjective response of the subject, and no measurement, other than the complaint itself, has yet been developed. . . ." In another report on pollutants, the National Academy of Sciences (1976) stated that the first uncomfortable reactions are usually felt in the eye tissues. That report examined several studies of eye irritation, including one by Schuck and collaborators (1966). Those investigators conducted eye-only exposures, and permitted subjects to turn a knob that adjusted the position of a pointer on a scale to indicate the eye irritation intensity experienced at any instant during the 5-min exposure. Concurrently, the rate of eye blinking was measured. Orderly functional relationships to exposure emerged from this study, demonstrating the value of such experiments. Comparable studies were undertaken more recently with formaldehyde and sidestream cigarette smoke (Weber-Tschopp et al. 1977).

Studies are most frequently performed by having questionnaires completed by members of exposed populations (for example, Heuss and Glasson 1968; Hagberg et al. 1985), or by exposing panelists to test atmospheres, measuring the elapsed time to a complaint, and eliciting a subjective rating (Bender et al. 1983). Such scaling studies could profit from better utilization of psychophysical scaling procedures developed for the study of odorants (Cain and Moskowitz 1974).

The development of procedures to assay eye irritancy in animals would be desirable. In vitro tests designed to replace the Draize test (instillation of chemicals into the eye of a rabbit) are currently being validated, but neither in vitro tests nor the Draize test can directly address the subjective response to low-level sensory irritation. Blink rate measurements are useful, but with repeated experience in the test situation, animals may learn to close their eyes to avoid exposure. Procedures that permit animals to terminate but not to avoid graded concentrations of irritants are discussed in some detail below. These latter procedures give the animal the opportunity to control the unpleasant stimulation, unlike techniques that require measurement of structural changes following instillation of the material into the eye. Another potential advantage of these procedures is that the cornea is served by the trigeminal nerve, and stimulation of this nerve decreases respiratory and heart rate. These physiological changes are of interest per se and may yield sensitivity comparable to behavioral measurements without training the subject. In any circumstance, the sensitivity of these tissues to irritants may permit the identification of the threshold for aversive stimulation.

■ **Recommendation 5.** Quantitative procedures should be developed using eye-only exposures to provide estimates of aversiveness derived from behavior under the control of irritant stimulation, and from measuring blink, heart, and respiratory rates.

Behavioral Effects. *Wheel Running.* Several experiments describing the effects of whole emissions on animal behavior provide leads for further work. Boche and Quilligan (1960) accustomed mice to running wheels, and then put one in a chamber with filtered air, and another in a smog-like mixture of O_3 and gasoline. The mice were moved from one chamber to the other every day (three 1-day exposures in each chamber), permitted to rest for a week, and then subjected to the same process at a higher smog concentration (figure 1). Although there was no assessment of running in the exposure chamber in the absence of smog, a concentration-related reduction in running was observed. This set the stage for a series of experiments demonstrating the utility of behavior for the study of emissions and photochemical products.

Gage (1979) exposed mice in running wheels to emissions from an automobile engine burning unleaded gasoline. The exhaust from the engine under lean tuning suppressed running in proportion to exposure concentration. The exhaust of an optimally tuned engine had no effect. Ultraviolet irradiation of the exhaust simulated photochemical smog, which suppressed running to a greater extent than nonirradiated exhaust. Activity returned to normal several days after the termination of irradiated exhaust exposure. However, termina-

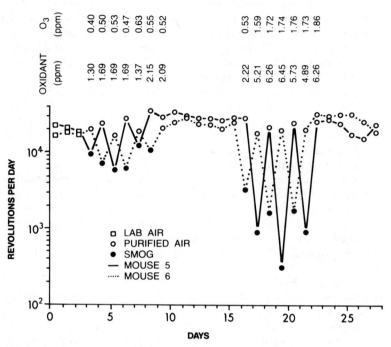

Figure 1. Spontaneous wheel-turning activity of two C57 black male mice in different environments. The total oxidant and O_3 determinations are shown at the top of the graph for each day of exposure to synthetic air pollutant mixture of O_3 and gasoline. (Adapted from Boche and Quilligan 1960.)

tion of nonirradiated exhaust produced rebound hyperactivity (figure 2, upper right panel). This study was not the first report of exhaust-induced hyperactivity (Hueter et al. 1966; Emik and Plata 1969; Stinson and Loosli 1979). These studies could have been improved by more complete reporting of the effects across time, and of their relationship to concentration.

Several investigators have used wheel running to examine biological effects of smog constituents (see, for example, the review by Murphy 1964). Murphy and colleagues (1964) observed a 46 percent reduction in activity following a 6-hr exposure to 0.2 ppm O_3, and a 20 percent reduction following exposure to 7.7 ppm nitrogen dioxide (NO_2). Campbell and coworkers (1970) demonstrated that peroxyacetyl nitrate, a constituent of photochemical air pollution, depressed running in proportion to the exposure concentration. Emik and Plata (1969) and Emik et al. (1971) set mice in wheel-running cages between the lanes of the Hollywood freeway and demonstrated an association between oxidant concentration and depressed running.

Because humans complain following exercise in O_3, Tepper and colleagues (1982) undertook a detailed analysis of the temporal patterning of wheel-running behavior that revealed disruption during 6 hr of exposure to 0.12 ppm O_3, the current ambient air quality standard. Low concentrations increased the duration and number of pauses, but did not change the speed at which the animal ran, or the length of an individual running burst. Unlike the effects on learned behavior described below (Weiss et al. 1981), effects were obvious in the first hour at lower concentrations, and performances declined throughout exposure.

Thus low concentrations of automotive emissions or photochemical reaction products can impair wheel-running performance. The mechanism by which these changes are produced should be a focus of the research agenda because it is of interest per se and because it contributes to risk analysis.

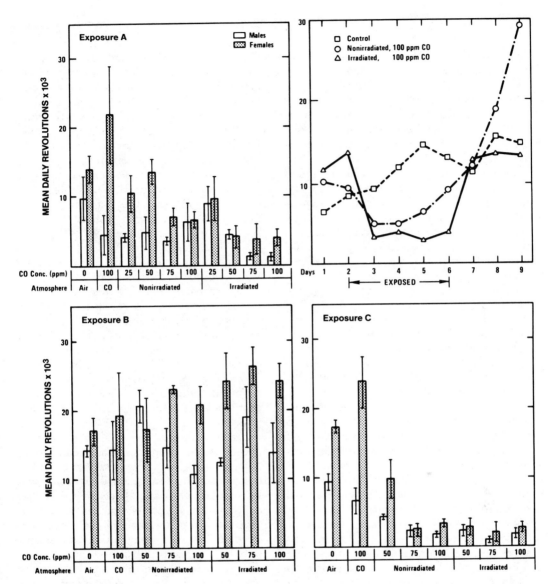

Figure 2. Mean daily activity of mice groups during emission exposure periods of the three exposure tests. Exhaust level is described by the nominal CO concentration. During exposures A and C, engine was tuned to factory specifications. During B, engine was optimally tuned. Upper right panel shows activity over the course of exposure A for male and female mice exposed to concentrations indicated. (Adapted with permission from Gage 1979.)

Mechanistic Experiments. Several mechanistic experiments indicate that the suppression of wheel running performance was probably not due to acute aversiveness of O_3. Wood (1979, 1981a) developed a procedure that permits direct behavioral assessment of the aversive properties of inhaled materials. Mice poked their noses into a conical recess to terminate the delivery of an irritant, and simultaneously produced a facial shower of clean air. Tepper and Wood (1985) demonstrated that O_3 reliably maintained escape behavior, and that its aversive properties were not dependent on previous experience with irritants. Performance was related to concentration, and escape behavior was maintained at concentrations above 0.5 ppm. However, the concentration for the mouse that produces observable reductions in running lies between 0.2 and 0.5 ppm (Murphy et al. 1964; Tepper et al. 1985), and since running was not immediately reduced at 0.2 ppm, acute aversiveness is probably not the principal mechanism responsible for performance degradation.

Furthermore, the effects on running produced by O_3, a typical lower-airway irritant (Alarie 1973), differed from those produced by typical upper-airway irritants. Ammonia immediately reduced running; at the lowest concentration (100 ppm), mice displayed a transient increase in running, followed by a large decrease that was immediately reversible after exposure termination. In contrast, O_3 produced delayed decreases in wheel running that were sustained after the termination of exposure. Although the direction of the effect was the same for the two irritants, the time course of the effects and recovery differed dramatically (Tepper et al. 1985).

Although mice and rats responded similarly and in the same order of potency to O_3 and ammonia, rats were more sensitive, as measured by associated morbidity, to both these irritants. At first glance, this seems to contradict other investigations of sensory irritants: mice display decreased respiratory rate to upper-airway irritants at lower concentrations than rats (Chang et al. 1981). Formaldehyde offers a striking example; since decreased respiratory rate minimizes the delivered dose of formaldehyde, the species difference could account for the selective induction of nasal carcinoma in rats (Swenberg et al. 1980; Albert et al. 1982). Comparable mechanisms may be at work in wheel running in the presence of O_3 as well; rats may display greater reductions in wheel running than mice because they receive a larger effective dose (Tepper et al. 1985).

Exertion appears to be the predominant determinant of sensitivity. Weiss and colleagues (1981), interested in O_3 effects on learned behavior, conditioned rats to press a lever that produced a food pellet; 5 min later, another food pellet was available contingent upon a response. This procedure is called a fixed-interval (FI) schedule of reinforcement (reinforcement is the process by which behavior is maintained in frequency by its consequences). This relatively sedentary schedule generated response rates that gradually increased until food delivery. Rats were exposed to O_3 once a week for 6 hr, thereby decreasing their response rates late in the exposure. Increased concentrations of O_3 caused lever-pressing rates to decrease earlier in the session, resulting in reduced total output. Although effects occurred at 0.71 ppm, this performance was not as sensitive to O_3 concentration as wheel running performance was (Tepper et al. 1982). Tepper and Weiss (1986) also demonstrated comparable insensitivity with another sedentary learned performance, stochastic reinforcement of waiting (SRW), shown in figure 3. But when conditioned wheel running was used, instead of conditioned lever pressing, sensitivity was comparable to that of free-access wheel running. Thus if physical effort was required by the experimental animal, the sensitivity of the assay increased, regardless of whether the performance was conditioned.

Ozone produced specific "motivational" effects. Wheel running maintained by food reward was sensitive; sedentary conditioned behavior maintained by food was insensitive. Thus the effects of O_3 at low concentrations were not dependent on the motivational properties of food. Ozone between 0.2 and 0.5 ppm did not affect lever pressing maintained by food, but did reduce comparable performance maintained by the opportunity to exercise (fixed ratio lever press, figure 3). Compared to lever pressing for food, lever pressing for wheel running was much more sensitive; thus the effects of O_3 cannot be attributed to an effect on lever-pressing behavior per se.

The most important implication of these mechanistic experiments is that O_3 at low concentrations reduces the rewarding properties of exercise. Techniques are available to study this explicitly (for example, Pierce et al. 1986).

Effects of Repeated Exposure. Chronic exposure experiments with animals have raised several concerns. Many of the acute

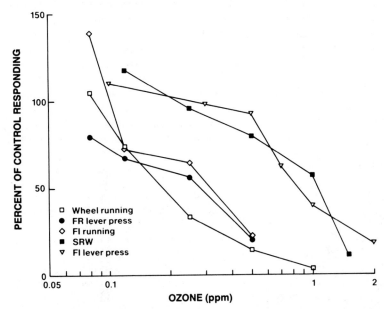

Figure 3. A summary of five behavioral experiments. The fixed-interval lever press (▽) and SRW experiments examined the effects of O_3 on sedentary conditioned behavior maintained by food, and were relatively insensitive. Wheel running was more sensitive (■); wheel running for food reward (FI running, ◇) was comparably sensitive. Pressing a lever to produce the opportunity to run for 15 sec was comparably sensitive (FR lever press, ●). Effects of O_3 at low concentrations were not a function of effects on the motivational properties of food, were exacerbated by exercise, and reduced the rewarding properties of exercise. O_3 between 0.2 and 0.5 ppm did not reduce performance maintained by food, but did reduce comparable performance maintained by the opportunity to exercise. (Adapted with permission from Tepper and Weiss 1986.)

manifestations of exposure described above wane in intensity with repeated exposure. This response could be interpreted as simple adaptation to exposure or habituation to aversive stimulation. However, Hueter and coworkers (1966) demonstrated that tolerance to automotive exhaust was followed by delayed degradation of performance. In a similar experiment (Stupfel et al. 1973), rats exposed to exhaust gases 8 hr/day, 5 days/week for 24 months were slower to learn to run across the box to a shock-free area when a tone warned of impending shock; it has not been determined whether the rats' ability to learn, or their ability to perform the task being used to assess learning, was impaired, although procedures are available to differentiate such effects. These findings suggest that the waning of responsiveness to acutely effective materials might bear some predictive relation to delayed toxicity, perhaps via the loss of protective mechanisms.

■**Recommendation 6.** Several types of behavioral experimentation are needed to characterize the active agents in whole emissions and photochemical by-products. Acute studies should elucidate the mechanisms of the effects by comparing automotive emissions with agents that have well-described effects on the respiratory system. Repeated exposures should reveal the agents responsible for rebound hyperactivity. Mechanistic studies should focus on behavioral determinants of sensitivity and associated changes in irritant receptors in the upper airways, lung or lung innervation, frank lung injury, and alterations in the peripheral or central nervous systems.

■**Recommendation 7.** An animal model of compromised pulmonary function should be developed, because compromised humans are particularly sensitive to oxidant exposure. Experimental models of chronic obstructive pulmonary disease would permit quantitative evaluation of exaggerated oxidant sensitivity and might display behavioral effects that resemble those produced by chronic exposure to oxidants.

Conditioned Responses to Exposure. It is well known that an aversion to a favorite food can develop through association of that food with an illness. There is reason to expect that disruptions of function that occur in response to emissions may be equally conditionable.

Acute exposure to sensory irritants produces a reduction in respiratory rate. Kane and Alarie (1977) studied the effects of single and repeated exposures to formaldehyde and acrolein. In one experiment, a large number of animals were assigned to either an exposure or a control group. The animals in the exposure group underwent repeated irritant exposures in an exposure chamber (lower panels, figure 4), but the animals in the control group were exposed to air. Animals in both groups were then placed in a plethysmograph and exposed to the irritant once only, at a single test concentration. Although a range of test concentrations was used, no animal was subjected to more than one such test exposure. The animals preexposed to formaldehyde did not show altered sensitivity to its acute effects; the animals preexposed to acrolein appeared to exhibit some tolerance—a lessening of response—to acrolein following repeated exposure. But, in the second experiment (top two rows of panels, figure 4), the animals were repeatedly exposed in the plethysmograph; subsequent exposures produced large, prompt respiratory rate reductions that became more pronounced with repeated exposure. The magnitude of the effect was related to concentration, and occurred for both acrolein and formaldehyde. Similarly, in our own work on irritant escape, we have observed progressively greater sensitivity to both O_3 (Tepper and Wood 1985) and formaldehyde (Wood and Coleman 1984). Repeated exposure in the same context may produce conditioned hypersensitivity to airborne contaminants reflected in changes in respiratory function.

There are other examples where airborne irritants function as unconditioned stimuli that can condition responses to occur to previously neutral cues. Alarie (1966) demonstrated the conditionability of respiratory rate decreases in his early work. Hoffman and Fitzgerald (1978) conditioned heart rate and blood pressure using ammonia. Jamison (1951) measured hearing by conditioning ammonia-induced bradycardia to occur following the presentation of a previously neutral tone.

Animal models have also been used to examine the conditioning of allergic responses in animals. Asthmatic responses have been conditioned in guinea pigs (Noelpp and Noelpp-Eschenhagen 1951a,b,c; Ottenberg et al. 1958; Justesen et al. 1970). Russell and colleagues (1984) conditioned histamine release to occur following the presentation of an odor by pairing the odor with a substance to which the guinea pigs had been immunologically sensitized.

These efforts bear broad implications for other potentially conditionable signs and symptoms, ranging from asthmatic attacks to complaints of breathing difficulties. Concentrations that were once ineffective for the elicitation of complaints or behavioral disruption can become effective through respondent (Pavlovian) conditioning mechanisms. If, however, exposure levels are held below those that initially produce the unconditional response, conditioning does not occur.

■ **Recommendation 8.** Conditioning experiments should study the range of effective concentrations and the temporal parameters that are necessary for automotive emissions to cause conditioned hypersensitivity. Threshold concentrations for the appearance of signs and symptoms of toxicity, ranging from asthmatic attacks to complaints of breathing difficulties, could then be determined.

Carbon Monoxide

The neurobehavioral effects of CO have been intensively studied (Laties and Merigan 1979). Most effects on behavior generated by exposure to low concentrations of

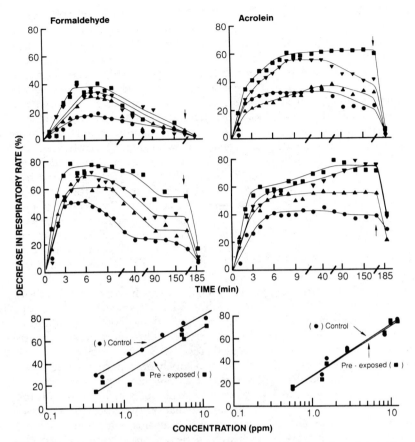

Figure 4. Progressively increasing sensitivity to formaldehyde and acrolein with repeated exposure. The time course of respiratory depression was studied during repeated exposure to two concentrations of formaldehyde (1.0 ppm, top left; 3.1 ppm, middle left) or acrolein (0.5 ppm, top right; 1.7 ppm middle right). The first, second, third, and fourth days of exposure are designated by ●, ▲, ▼, and ■, respectively. Arrows indicate the end of exposure periods. The bottom panels depict concentration/effect relationships for formaldehyde (left) and acrolein (right) for control mice, and for mice previously exposed in a different context to either 0.31 ppm formaldehyde or to 0.17 ppm acrolein for 3 hr/day for 3 consecutive days. No sensitization is apparent for these mice, suggesting that the increased responsiveness observed in the upper two panels was a conditioning phenomenon dependent on the context of exposure. (Adapted with permission from Kane and Alarie 1977.)

CO appear to be marginal; nonetheless, even small effects can have serious consequences for the operators of heavy machinery. Furthermore, the effects of CO depend on the behavior under study: one behavior may be disrupted while another is unimpaired, and the impairment may depend on the assessment procedures. Adequate studies of behavioral impairment need to examine highly reliable effects studied intensively in individual subjects, using multiple exposure concentrations. This is especially important since concentration/effect and time/effect functions may not be monotonic. This is an area that could profit from further experimental work with humans.

Additional effort should be made to examine teratological effects, since prenatal exposure to CO can alter the behavioral performance of offspring (see, for example, Fechter and Annau 1977; Mactutus and Fechter 1984, 1985; Storm and Fechter 1985a,b). This class of effects has received inadequate attention.

The following research recommendations are derived from a comprehensive critical review of the behavioral effects of CO (Laties and Merigan 1979), from the recommendations of the Clean Air Scientific Advisory Committee of the EPA Science Advisory Board (letter from M. Lippmann to W. Ruckelshaus, December 30, 1983), and from the report of a task force on research planning (National Institutes of Health 1977).

■ **Recommendation 9.** Systematic studies should be performed to determine which aspects of performance are most susceptible to disruption by CO.

■ **Recommendation 10.** Studies should be performed to determine which parameters of CO exposure are most important in producing behavioral impairment. Furthermore, since operators of motor vehicles are frequently under the influence of prescription and over-the-counter drugs, as well as drugs of abuse, the interaction of these substances with CO should be described.

■ **Recommendation 11.** Since the elderly and those with cardiovascular or respiratory insufficiency may display exaggerated sensitivity to CO, studies of behavioral effects with these groups should be undertaken.

■ **Recommendation 12.** Experiments utilizing the techniques of modern developmental neurobiology should be undertaken to further elucidate the effects of pre- and perinatal CO exposure.

Petroleum Hydrocarbons

Aromatic and aliphatic hydrocarbons and alcohols all can produce acute performance impairment; the more volatile and lipophilic the fuel, the more likely that such impairments could result following inhalation. Petroleum hydrocarbons display a variety of effects: some may resemble central nervous system depressants in their ability to increase the frequency of behavior suppressed by punishment, in their anti-

convulsant effects (Wood et al. 1984), and in their discriminative stimulus properties (Rees et al. 1986). Others may be frankly convulsant or proconvulsant; alkylcycloparaffins offer a good example (Lazerew 1929; Lazerew and Kramneva 1930; Pryor et al. 1978). Interactions can occur that can mask or potentiate such effects; Pryor and coworkers (1978) demonstrated that the expression of overt methylcyclohexane seizures was blocked when it was administered in a mixture of heptanes. Some hydrocarbons are neuropathic—for example, n-hexane (Spencer et al. 1980). The production of peripheral neuropathies can be potentiated by components of the mixture; in fact, the toxicity of n-hexane is most evident when it is coadministered with other solvents.

The Health Effects Institute (1985) report on Gasoline Vapor Exposure and Human Cancer recognized the potential induction of neurotoxicity. That report recommends the evaluation of neurotoxicity in future chronic and subchronic animal studies. The clinical and animal literature is sparse, except for the large body of literature on the deliberate inhalation of leaded fuels. However, several Scandinavian countries have identified a new disease entity called the solvent syndrome, which has been linked to complex solvent mixtures, components of which are also automotive fuel constituents.

Although controversy surrounds the studies cited to document the syndrome's existence (Grasso et al. 1984; Errebo-Knudsen and Olsen 1986), it is well enough accepted to have become a compensable injury in several of these countries (Flodin et al. 1984). In these studies, painters were exposed for long periods to organic solvent vapors (Arlien-Soborg et al. 1979; Lindstrom et al. 1984). Solvent abusers, on the other hand, expose themselves to extremely high concentrations and experience toxicity more promptly. But both groups display similar impairments: reports of neurological examinations and computerized axial tomography provide evidence of brain atrophy in both groups (Bruhn et al. 1981; Schikler et al. 1982; Fornazzari et al. 1983; Lazar et al. 1983). These changes occur without striking localized lesions, and are best revealed with detailed measurements of brain dimensions (morphometry) (for example, Rodier and Reynolds 1977; Rodier and Gramann 1979; Fornazzari et al. 1983). Recent claims that this syndrome occurs at or below current recommended occupational exposure limits raises concerns about allowable exposure. A recent Consensus Conference recommended directions that could be adopted for human as well as animal research efforts (Baker et al. 1985). A primate model of the solvent syndrome should be developed. Repeated acquisition procedures would be an appropriate baseline for the study of learning impairment (Thompson and Moerschbaecher 1978; Dietz et al. 1979; Howard and Pollard 1983). Repeated acquisition procedures are dependent on intact memory function. Procedures for the explicit evaluation of impairment of remembering are available (Heise 1975; Bartus 1979; Taylor and Evans 1985).

Although it is less likely that profound nervous system injury will occur at environmental levels of exposure, the likelihood of injury for fuel handlers and consumers is unknown. The neurotoxic hazards of gasoline and its constituent fractions are poorly characterized; neurotoxicologic investigations thus far have failed to use the appropriate protocols in examining the hazards. As discussed earlier, selecting a strategy for establishing priorities is a complex issue. Thus, to document the hazard posed by a particular fuel, it is necessary to study a complete complex mixture. However, fractions could perhaps be examined by class to identify hazards, an approach that might be well suited to petroleum-refining technology as well. But examination of representative constituents—that is, single chemical entities—provides the most useful information to the neurotoxicologist. Therefore, a combination of these approaches is undoubtedly best. And since the problem of potential nervous system injuries resulting from exposure to hydrocarbons is not restricted to the motor vehicle and allied industries, it may be possible to examine it in a coordinated interindustry effort.

■ **Recommendation 13.** Quantitative brain morphometry should be performed following prolonged exposure to automotive fuels. This is necessary because chronic solvent exposure does not usually produce gross lesions, but rather a selective wasting syndrome that may not be evident in measurements of gross brain weight.

■ **Recommendation 14.** Acute effect studies would help determine if the prevention of acute effects of petroleum hydrocarbons is sufficient to prevent chronic toxicity. Acute performance impairment should be a major focus of effort.

■ **Recommendation 15.** A primate model of the solvent syndrome should be developed. Repeated acquisition procedures would be an appropriate baseline for the study of learning impairment; procedures for the explicit evaluation of memory impairment would assist in clarifying the nature of any learning impairment observed.

Methanol

Methanol deserves special scrutiny because of the potentially increasing use as a fuel. Methanol produces blindness and life-threatening acidosis in humans following the ingestion of moderately large doses. Although neurotoxicity expressed in the visual system is a hallmark of methanol toxicity, the syndrome remains incompletely characterized. Work with primates has focused on retinal changes; other changes in the visual system or in visual function in nonhuman primates are unknown. Visual evoked potentials and electroretinography are valuable for descriptive purposes; the magnitude of these changes might predict outcome and might also prove to be a useful index of the therapeutic efficacy of methanol antidotes.

■ **Recommendation 16.** Because methanol offers special risks to the visual functions of primates, additional effort should be made to characterize the visual hazards of methanol and their relationship to exposure.

There are a number of clinical reports of damage to the basal ganglia, delayed motor dysfunction resembling parkinsonism, and gross brain injury following methanol exposure (Erlanson et al. 1965; Guggenheim et al. 1971; Aquilonius et al. 1978, 1980; Ley and Gali 1983). This has yet to be studied in primate models of methanol intoxication.

■ **Recommendation 17.** An animal model of methanol-induced motor disorders should be developed. Such a model would assist in determining if acidosis is a necessary precondition for the emergence of methanol-induced motor disorders and if antidotes are effective.

It has been demonstrated that the expression of acidosis is mediated by a folate-dependent metabolic pathway, and that the inhibition of methionine synthetase (produced with nitrous oxide exposure) can provoke methanol-induced acidosis in rats, a normally insensitive species. Furthermore, following a few hours of exposure to nitrous oxide, monkeys treated with 1 g methanol/kg body weight developed acidosis (Eells et al. 1983); this is a much lower dose than is normally required to produce the acute syndrome. This observation might provide a basis for a more sensitive model.

■ **Recommendation 18.** More sensitive animal models of methanol toxicity should be developed, perhaps through the manipulation of folate metabolism. Repeated or continuous exposure to low concentrations of methanol should then be undertaken to determine if acidosis can be produced, and if systemic acidosis is a precondition for the expression of toxicity.

■ **Recommendation 19.** Clinical trials should be undertaken with 4-methylpyrazole, a new drug that is a promising methanol antidote and a candidate for widespread deployment in emergency facilities.

Metals and Inorganic Compounds

Lead (Pb) is the prototypical metallic automotive emission; it is instructive to review our experience with Pb before other metals are added to automotive fuels. The fuel additive tetraethyllead has been responsible for multiple deaths from neurotoxicity and was the focus of a major public health controversy (Rosner and Markowitz 1985). The neurobehavioral toxicity of Pb would most probably have been detected had some systematic approach to testing been in place. Exposure increases the frequency of conditioned behavior of rats at blood levels of Pb below the current clinical definition of elevated Pb burden that requires further diagnostic intervention (Cory-Slechta et al. 1985). Morphological changes in the brain that accompany these functional changes have not been identified.

Another metal, manganese (Mn), however, can produce functional as well as morphological changes. Methylcyclopentadienyl manganese tricarbonyl (MMT) is a potential gasoline additive that has antiknock properties. When used in diesel fuels, Mn additives improve combustion and reduce smoke. The chronic toxicity of the emission product may be subtle, delayed, and readily confused with diseases of other etiology. The syndrome progresses from manic psychosis and disturbances of speech and gait in the early phase, to disturbances of speech, a fixed jovial facial expression, clumsiness and hyperemotionalism in the intermediate stage, and later, muscular hypertonia and tranquil euphoria with memory loss and intact sensory function (Rodier 1955; Mena et al. 1967; Barbeau 1984).

The National Academy of Sciences (1973) reviewed the toxicity of Mn, expressed concern about the Mn emission problem, and made twenty-five research recommendations. Although Mn is not now regulated as a hazardous pollutant, knowledge of the health effects of Mn is limited, especially about the relationship of exposure concentration and duration to effects. The existing data are inadequate to conduct a risk analysis. The risk of adverse health effects from Mn emissions should be characterized as unknown but not necessarily unlikely. The potential injuries are great enough to warrant further study before any significant increase in exposure is contemplated.

Mn is an essential trace element, but at some higher dose it becomes toxic. The relationship between dose and adverse effect remains unclear, despite the fact that the toxicokinetics of Mn have received fairly detailed attention (Dastur et al. 1971; Newland et al. 1987). The literature provides only limited insight into whether increased exposure increases the severity of effect, as well as the number of individuals affected, or merely shortens the time to toxicity. Children, in particular, may be at high risk, because, as with Pb emissions, (1) the amount of Mn in dust and soil should increase with proximity to the emission source, (2) children put their hands and many other things in their mouths, (3) children become more mobile at 6 months of age, and (4) younger organisms absorb a larger proportion of the administered dose (Cahill et al. 1980; Rehnberg et al. 1981).

Most of the clinical literature consists of effects reported following an idiosyncratic exposure, rather than well-controlled experiments. The emphasis has been on identifying toxicity, and not on generating an orderly dose/effect or dose/response function. There are anecdotal reports of delayed toxicity commencing several years after the termination of exposure (Cook et al. 1974). Changes in human behavior from exposure to Mn at levels that are 20 percent of current recommended exposure limits have been reported by Roels et al. (1985).

■ **Recommendation 20.** Should any increase in Mn emissions be contemplated, an expert committee should be formed to review the neurotoxicity of Mn and the adequacy of current exposure estimates, and to consider the benefits of a chronic study in primates. If significant exposure is contemplated, several experiments must be undertaken to provide data for risk analysis that provide close attention to neurobehavioral function, and consequent regional neurochemical assessment. Early signs of altered dopaminergic function should be examined.

■ **Recommendation 21.** Should any increase in Mn emissions be contemplated, kinetic studies of Mn should be undertaken, with the emphasis on brain uptake and the ingestion of accumulated dust by neonates. Estimates of possible intake should be modeled; exposure of neonatal primates and evaluation of neurobehavioral toxicity should be considered.

Studies of the acute toxicity of MMT have identified seizures and Clara cell necrosis as sequelae to exposure. The delayed effects of acute or chronic low doses have not been well characterized, and are of interest per se; the nervous system is of interest because of the greater access an organic metal should have to the central nervous system and because of the evident neurotoxicity of Mn in primates. Organic complexes or salts might facilitate entry and shorten the time to toxicity.

■ **Recommendation 22.** The acute and chronic neurobehavioral toxicity of MMT should be more adequately characterized.

Conclusions

Determining the acute effects of chemicals on behavior and nervous system function is important to the evaluation of neurotoxicity: acute reversible effects may be the effects of major concern. Similarly, repeated or continuous exposures are needed to characterize toxicity that is delayed or cumulative, to observe the development of tolerance (or reverse tolerance), and to characterize the reversibility of adverse effects. Initial screening for subtle sensory or perceptual impairments, affective disorders, or cognitive and intellectual dysfunction needs to be conducted, and, finally, highly focused studies may be needed to fully characterize hazards using methods that are dictated by the nature of the system or function affected, such as specialized evaluations and refined neuropathological, neurochemical, and neurophysiological techniques.

Since the neurobehavioral toxicity of

many compounds present in automotive emissions is unknown, they should be studied systematically, proceeding from acute, then through repeated administration, experiments and subchronic exposures, and finally to detailed characterizations of injuries and mechanisms. The U.S. Environmental Protection Agency has promulgated guidelines for a variety of neurobehavioral studies, including neuropathology (1985c), motor activity—for example, wheel-running studies (1985b), schedule-controlled (learned) behavior (1985f), and a battery of structured functional observations (1985a). Adopting such tests and incorporating them into a more comprehensive testing strategy is a logical step in developing a rational approach to answering the many unanswered questions surrounding safety evaluation of automotive emissions and their fractions.

Such a testing strategy, depending on a network of methods and techniques, offers several advantages. First, it reduces the likelihood of overlooking a hazard or class of hazards, which any single screening method might miss.

Second, it can be structured in "tiers," in such a way that each tier provides essential information for subsequent tiers as well as useful information in its own right. Such a tiered format is likely to speed up the testing process as well as achieve significant economies in laboratory operations that would serve to offset the potential expense of an extensive screening program. For example, acute toxicity evaluations may serve a dose-ranging function for subsequent repeated-exposure experiments; the findings in both may later give direction to detailed mechanistic studies.

■ **Recommendation 23.** The neurobehavioral toxicity of many constituents present in automotive emissions is unknown and should be studied using a tiered-testing strategy. An expert committee should prioritize the selection of compounds for a systematic testing program, and recommend specialized evaluations when appropriate.

Summary

This chapter has described adverse neurobehavioral effects of automotive emissions and offered research recommendations to facilitate their detailed characterization.

● Neurobehavioral toxicity has not been evaluated thoroughly for most of the compounds that are known to produce adverse effects, nor have screening and evaluation procedures for such toxicity been applied systematically to the universe of chemicals that mobile sources produce.

● To some extent, the absence of data is attributable to a lack of process. Those deficits could be remedied by subscribing to a tiered-testing strategy, with its initial agenda of test substances prescribed by an expert committee. The strategy will provide a disciplined approach, relying on animal studies and supplemented by available literature, for dealing with a large collection of chemicals with largely unknown effects. However, many new tests, procedures, and models need to be developed to assess the adverse health effects that can reasonably be expected to be encountered.

● To another extent, the lack of visibility, prominence, and/or appreciation accorded to neurobehavioral toxicity, even with respect to substances where such effects are well-documented and widespread, contribute to the data shortage. In particular, we should be immediately concerned with:

–Sensory irritation and/or repeated and chronic effects that may alter the quality of life;

–Effects of CO on complex human performances, and on special populations with exaggerated sensitivity (for example, the fetus, the infant, and the aged);

–Effects of petroleum hydrocarbons on behavior, and brain structure and function;

–Methanol hazards, should methanol come into wider use; and

–Metallic fuel additives, especially Mn, should that come into wider use.

Summary of Research Recommendations

Although the behavioral and neurosciences have already been used to good effect in toxicity evaluation, some areas need further research and development effort, both to demonstrate feasibility and to improve cost-effectiveness. Such a research effort should include attempts to:

- Undertake testing via a tiered approach for chemicals about which little is known (Recommendation 23) before sophisticated and expensive procedures are used to examine candidate toxicants for their capacity to impair learning, memory, cognition, and intellectual functioning (Recommendation 4).
- Develop rapid tests of sensory impairment that can be used in conjunction with functional observation batteries, and develop comprehensive psychophysical studies directed at specific functional impairments, including impairments of complex auditory and visual discriminations (Recommendation 1).
- Develop animal models of affective disorders. Selected reference substances, drugs as well as toxicants, should be used to validate these models before they are used to test uncharacterized chemicals or mixtures (Recommendation 3).

In addition, more specific recommendations for certain automotive emissions can be made immediately. These are outlined here according to priority and contingent upon increases in the level of exposure to a particular emission or deployment of new technology.

HIGH PRIORITY

Recommendations 6, 23

Whole emissions and photochemical by-products deserve immediate attention. Acute and repeated-exposure wheel-running studies should be used to characterize emissions, to differentiate the behavioral consequences of materials that affect primarily the upper or lower airways, and to determine whether the observed effects are attributable to alterations in irritant receptors in lung or lung innervation, frank lung injury, or actual alterations in the peripheral or central nervous systems. The agents responsible for rebound hyperactivity should be determined.

Recommendation 5

Eye irritation is one of the most frequently complained-about effects of automotive emissions and photochemical products. Quantitative procedures should be developed using eye-only exposures to provide estimates of aversiveness derived from behavior under the control of irritant stimulation, and from measurement of blink, heart, and respiratory rates. The development and validation of animal models would be particularly useful.

Recommendation 8

Signs and symptoms of exposure to automotive emission and photochemical by-products, ranging from asthmatic attacks to complaints of breathing difficulties, may be conditioned to occur after exposure to concentrations that evoke no response in naive subjects. Experiments to define the exposure parameters that produce conditioned alterations in sensitivity to airborne irritants are needed.

Recommendation 12   Prenatal exposure to CO can alter the behavior of offspring. Such effects may be of great importance and should be studied further with the techniques of modern developmental neurobiology.

Recommendation 14   The neurobehavioral toxicity of petroleum hydrocarbons in unleaded automotive fuels has received virtually no serious attention. Acute performance impairment should be a major focus of investigation; acute effect determinations are also necessary to ascertain if preventing acute effects will prevent chronic toxicity.

MEDIUM PRIORITY

Recommendation 2   In humans and rodents, the reduction in olfactory sensitivity produced by oxidants and aldehydes should be described as a function of concentration and duration of exposure.

Recommendation 7   A model of human populations with compromised pulmonary function should be developed, because such populations are more sensitive than others to oxidant exposure. Experimental models of chronic obstructive pulmonary disease should be developed that will permit quantitative estimation of exaggerated oxidant sensitivity and display behavioral effects that resemble those produced by chronic exposure to oxidants.

Recommendations 9, 10, 11   The aspects of performance most susceptible to disruption by CO should be identified. The relative importance of different exposure parameters in determining the extent of behavioral impairment should be described. Exaggerated sensitivity in the elderly and those with cardiovascular or respiratory insufficiency should be examined.

Recommendation 13   Repeated-exposure studies utilizing quantitative morphometric neuropathology should be undertaken in experimental animals exposed to petroleum hydrocarbons. Chronic solvent exposure usually does not produce gross central nervous system lesions, but rather a selective wasting syndrome that may not be evident in measurements of gross brain weight.

Recommendation 15   A primate model of the solvent syndrome should be developed. Repeated-acquisition procedures would be an appropriate baseline for the study of learning impairment; procedures for the explicit evaluation of memory impairment would assist in clarifying the nature of any learning impairment observed.

CONTINGENT PRIORITY

Recommendation 16   Because methanol offers special risks to the visual functions of primates, additional work should be performed to assist in characterizing the hazards and their relationship to exposure.

Recommendation 17   An animal model of methanol-induced motor disorder should be developed. Such a model would assist in determining if acidosis is a necessary precondition for the emergence of methanol-induced motor disorders, and if antidotes are effective.

Recommendation 18   More sensitive animal models of methanol toxicity should be developed, perhaps through the manipulation of folate metabolism. Repeated or continuous exposure of susceptible individuals to low concentrations of methanol should be undertaken to determine if acidosis can be produced, and if systemic acidosis is a precondition for the expression of toxicity.

Recommendation 19   Clinical trials should be undertaken with 4-methylpyrazole, a drug that is a promising methanol antidote, and a new candidate for widespread deployment in emergency facilities.

Recommendation 20   Should any increase in Mn emissions be contemplated, an expert committee should be formed to review the neurotoxicity of Mn and the adequacy of current exposure estimates, and to consider the benefits of a chronic study in primates. If significant exposure is contemplated, several experiments must be undertaken to provide data for risk analysis that provide close attention to neurobehavioral function, and consequent regional neurochemical and neuropathological assessment. Early signs of altered dopaminergic function should be examined.

Recommendation 21   Toxicokinetic studies of Mn should be undertaken, with the emphasis on brain uptake and the ingestion of dust by neonates. Crawling infants ingesting Mn accumulated in dust may be of great concern, as indicated by past experience with Pb. Estimates of possible intake should be modeled; exposure of neonatal primates and evaluation of neurobehavioral toxicity should be considered.

Recommendation 22   The acute neurobehavioral toxicity of any Mn additive should also be evaluated. The delayed effects of single, lower doses have not been well characterized and would be of some interest per se, as might chronic, lower-level exposure. Organic Mn compounds should be very interesting to study in the primate.

References

Alarie, Y. 1966. Irritating properties of airborne materials to the upper respiratory tract, *Arch. Environ. Health* 13:433–449.

Alarie, Y. 1973. Sensory irritation by airborne chemicals, *CRC Crit. Rev. Toxicol.* 2:299–363.

Albert, R. E., Sellakumar, A. R., Laskin, S., Kuschner, M., and Nelson, N. 1982. Gaseous formaldehyde and hydrogen chloride induction of nasal cancer in the rat, *J. Nat. Cancer Inst.* 68: 597–603.

Anthony, D. C., Boekelheide, K., and Graham, D. G. 1983a. The effect of 3,4-dimethyl substitution on the neurotoxicity of 2,5-hexanedione. I. Accelerated clinical neuropathy is accompanied by

Correspondence should be addressed to Ronald W. Wood, Department of Environmental Medicine, New York University Medical Center, 550 First Avenue, New York, NY 10016.

more proximal axonal swellings, *Toxicol. Appl. Pharmacol.* 71:362–371.

Anthony, D. C., Boekelheide, K., Anderson, C. W., and Graham, D. G. 1983b. The effect of 3,4-dimethyl substitution on the neurotoxicity of 2,5-hexanedione. II. Dimethyl substitution accelerates pyrrole formation and protein crosslinking, *Toxicol. Appl. Pharmacol.* 71:372–382.

Aquilonius, S. M., Bergstrom, K., Enoksson, P., Hedstrand, U., Lundberg, P. O., Mostrom, U., and Olsson, Y. 1980. Cerebral computed tomography in methanol intoxication, *J. Comp. Assist. Tomogr.* 4:425–428.

Aquilonius, S. M., Askmark, H., Enoksson, P., Lundberg, P. O., and Mostrom, U. 1978. Computerized tomography in severe methanol intoxication, *Br. Med. J.* ii:929–930.

Arlien-Soborg, P., Bruhn, P., Gyldensted, C., and Melgaard, B. 1979. Chronic painters' syndrome. Chronic toxic encephalopathy in house painters, *Acta Neurol. Scand.* 60:149–156.

Baker, E. L., Bus, J. S., Cranmer, J. M., Curtis, M. F., Golberg, L., Grasso, P., Keller, L. W., Merigan, W. H., Morgan, R. W., Scala, R. A., and Seppalainen, A. M. 1985. Workshop on neurobehavioral effects of solvents. Consensus summary, *Neurotoxicology* 6:99–102.

Barbeau, A. 1984. Manganese and extrapyramidal disorders. A critical review and tribute to Dr. George C. Cotzias, *Neurotoxicology* 5:13–36.

Bartus, R. T. 1979. Physostigmine and recent memory: effects in young and aged nonhuman primates, *Science* 206:1087–1089.

Beliles, R. P., Clark, R. S., and Yuile, C. L. 1968. The effects of exposure to mercury vapor on behavior of rats, *Toxicol. Appl. Pharmacol.* 12:15–21.

Bender, J. R., Mullin, L. S., Graepel, G. J., and Wilson, W. E. 1983. Eye irritation response to humans to formaldehyde, *Am. Ind. Hyg. Assoc. J.* 44:463–465.

Boche, R. D., and Quilligan, J. J. 1960. Effect of synthetic smog on spontaneous activity of mice, *Science* 131:1733–1734.

Brock, T. O., and O'Callaghan, J. 1987. Quantitative changes in the synaptic vesicle proteins synapsin I and p38 and the astrocyte-specific protein glial fibrillary acidic protein are associated with chemical-induced injury to the rat central nervous system, *J. Neurosci.* 7(4):931–942.

Bruhn, P., Arlien-Soborg, P., Gyldensted, C., and Christensen, E. L. 1981. Prognosis in chronic toxic encephalopathy. A two-year follow-up study in 26 house painters with occupational encephalopathy, *Acta Neurol. Scand.* 64:259–272.

Cahill, D. F., Bercegeay, M. S., Haggerty, R. C., Gerding, J. E., and Gray, L. E. 1980. Age-related retention and distribution of ingested Mn_3O_4 in the rat, *Toxicol. Appl. Pharmacol.* 53:83–91.

Cain, W. S., and Moskowitz, H. R. 1974. Psychophysical Scaling of Odor, In: *Human Responses to Environmental Odors* (A. Turk, J. W. Johnston, and D. G. Moulton, eds.), pp. 1–32, Academic Press, New York.

Campbell, K. I., Emik, L. O., Clarke, G. L., and

Plata, R. L. 1970. Inhalation toxicity of peroxyacetyl nitrate, *Arch. Environ. Health* 20:22–27.

Chang, J. C. F., Steinhagen, W. H., and Barrow, C. S. 1981. Effect of single or repeated formaldehyde exposure on minute volume of B6C3F1 mice and F-344 rats, *Toxicol. Appl. Pharmacol.* 61:451–459.

Cook, D. G., Fahn, S., and Brait, K. A. 1974. Chronic manganese intoxication, *Arch. Neurol.* 30:59–64.

Cory-Slechta, D. A., Weiss, B., and Cox, C. 1985. Performance and exposure indices of rats exposed to low concentrations of lead, *Toxicol. Appl. Pharmacol.* 78:291–299.

Dastur, D. K., Manghani, D. K., and Raghavendran, K. V. 1971. Distribution and fate of [54]Mn in the monkey: studies of different parts of the central nervous system and other organs, *J. Clin. Invest.* 50:9–20.

Dews, P. B. 1975. An overview of behavioral toxicology, In: *Behavioral Toxicology* (V. G. Laties and B. Weiss, eds.), pp. 439–445, Plenum Press, New York.

Dietz, D. D., McMillan, D. E., Mushak, P. 1979. Effects of chronic lead administration on acquisition and performance of serial position sequences by pigeons, *Toxicol. Appl. Pharmacol.* 47:377–384.

Eells, J. T., Black, K. A., Tedford, C. E., and Tephly, T. R. 1983. Methanol toxicity in the monkey. Effects of nitrous oxide and methionine, *J. Pharmacol. Exp. Ther.* 227:349–353.

Emik, L. O., and Plata, R. L. 1969. Depression of running activity in mice by exposure to polluted air, *Arch. Environ. Health* 18:574–579.

Emik, L. O., Plata, R. L., Campbell, K. I., and Clarke, G. L. 1971. Biological effects of urban air pollution, *Arch. Environ. Health* 23:335–342.

Erlanson, P., Fritz, H., Hagstam, K. E., Liljenberg, B., Tryding, N., and Voigt, G. 1965. Severe methanol intoxication, *Acta Med. Scand.* 177:393–408.

Errebo-Knudsen, E. O., and Olsen, F. 1986. Organic solvents and presenile dementia (the painters' syndrome). A critical review of the Danish literature, *Sci. Total Environ.* 48:45–67.

Eskin, T. A., Lapham, L. W., Maurissen, J. P., and Merigan, W. H. 1985. Acrylamide effects on the macaque visual system. II. Retinogeniculate morphology, *Invest. Ophthalmol.* 26:317–329.

Evans, H. L., Bushnell, P. J., Taylor, J. D., Monico, A., Teal, J. J., and Pontecorvo, M. J. 1986. A system for assessing toxicity of chemicals by continuous monitoring of homecage behaviors, *Fundam. Appl. Toxicol.* 6:721–732.

Fechter, L., and Annau, Z. 1977. Toxicity of mild prenatal carbon monoxide exposure, *Science* 197: 680–682.

Flodin, U., Edling, C., and Axelson, O. 1984. Clinical studies of psychoorganic syndromes among workers with exposure to solvents, *Am. J. Ind. Med.* 5:287–295.

Fornazzari, L., Wilkinson, D. A., Kapur, B. M., and Carlen, P. L. 1983. Cerebellar, cortical and functional impairment in toluene abusers, *Acta Neurol. Scand.* 67:319–329.

Gage, M. I. 1979. Automotive exhaust and mouse activity: relationships between pollutant concentrations and decreases in wheel running, *Arch. Environ. Health* 34:164–168.

Grasso, P., Sharratt, D., Davies, D. M., and Irvine, D. 1984. Neurophysiological and psychological disorders and occupational exposure to organic solvents, *Food Chem. Toxicol.* 22:819–852.

Guggenheim, M. A., Couch, J. R., and Weinberg, W. 1971. Motor dysfunction as a permanent complication of methanol ingestion, *Arch. Neurol.* 24:550–554.

Hagberg, M., Kolmodin-Hedmin, B., Lindahl, R., Nilsson, C.-A., and Norstrom, A. 1985. Irritative complaints, carboxyhemoglobin increase and minor ventilatory function changes due to exposure to chain-saw exhaust, *Eur. J. Respir. Dis.* 66:240–247.

Health Effects Institute. 1985. *Gasoline Vapor Exposure and Human Cancer: Evaluation of Existing Scientific Information and Recommendations for Future Research*, Health Effects Institute, Cambridge, Mass.

Heise, G. A. 1975. Discrete trial analysis of drug action, *Fed. Proc.* 34:1898–1903.

Heuss, J. M., and Glasson, W. A. 1968. Hydrocarbon reactivity and eye irritation, *Environ. Sci. Technol.* 2:1109–1116.

Hoffman, H. S., and Ison, J. R. 1980. Reflex modification in the domain of startle: some empirical findings and their implications for how the nervous system processes sensory input, *Psychol. Rev.* 87:175–189.

Hoffman, J. W., and Fitzgerald, R. D. 1978. Classically conditioned heart rate and blood pressure in rats based on either electric shock or ammonia fumes reinforcement, *Physiol. Behav.* 21:735–741.

Horvath, M., and Frantik, E. 1974. Quantitative interpretation of experimental toxicological data: the use of reference substances, In: *Adverse Effects of Environmental Chemicals and Psychotropic Drugs: Quantitative Interpretation of Functional Test* (M. Horvath, ed.), Vol. 1, pp. 2–8, Elsevier, New York.

Howard, J. H., and Pollard, G. T. 1983. Effects of *d*-amphetamine, Org-2766, scopolamine, and physostigmine on repeated acquisition of four response chains in rat, *Drug Develop. Res.* 3:37–48.

Hueter, F. G., Contner, G. L., Busch, K. A., and Hinners, R. G. 1966. Biological effects of atmospheres contaminated by auto exhaust, *Arch. Environ. Health* 12:553–560.

Jamison, J. H. 1951. Measurement of auditory intensity thresholds in the rat by conditioning of an autonomic response, *J. Comp. Physiol. Psychol.* 44:118–125.

Justesen, D. R., Braun, E. W., Garrison, R. G., and Pendleton, R. B. 1970. Pharmacological differentiation of allergic and classically conditioned asthma in the guinea pig, *Science* 170:864–866.

Kane, L. E., and Alarie, Y. 1977. Sensory irritation to formaldehyde and acrolein during single and repeated exposures in mice, *Am. Ind. Hyg. Ass. J.* 38:509–522.

Kulle, T. J., and Cooper, G. P. 1975. Effects of formaldehyde and ozone on the trigeminal nasal sensory system, *Arch. Environ. Health* 30:237–243.

Laties, V. G. 1973. On the use of reference substances in behavioral toxicology, In: *Adverse Effects of Environmental Chemicals and Psychotropic Drugs: Quantitative Interpretation of Functional Test* (M. Horvath, ed.), Vol. 1, pp. 83–88, Elsevier, New York.

Laties, V. G. 1982. Contributions of operant conditioning to behavioral toxicology, In: *Nervous System Toxicology* (C. L. Mitchell, ed.), pp. 199–212, Raven Press, New York.

Laties, V. G., and Merigan, W. H. 1979. Behavioral effects of carbon monoxide on animals and man, *Ann. Rev. Pharmacol. Toxicol.* 19:357–392.

Laties, V. G., and Wood, R. W. 1986. Schedule-controlled behavior: its role in behavioral toxicology, In: *Behavioral Toxicology* (Z. Annau, ed.), Johns Hopkins Press, Baltimore.

Lazar, R. B., Ho, S. U., Melen, O., and Daghestani, A. N. 1983. Multifocal central nervous system damage caused by toluene abuse, *Neurology* 33:1337–1340.

Lazerew, N. W. 1929. Toxicity of various hydrocarbon vapors, *Naunyn-Schmiedebergs Arch. Pharmkol. Exp. Pathol.* 143:223–233 (Transl. from the National Translation Center).

Lazerew, N. W., and Kramneva, S. N. 1930. Berkungen uber die giftigkeit der dampfe des zyklopentans und siener homogen, *Naunyn-Schmiedebergs Arch. Pharmkol. Exp. Pathol.* 149: 116–118.

Lewey, F. H. 1941. Experimental chronic carbon disulfide poisoning in dogs, *J. Ind. Hyg. Toxicol.* 23:415–436.

Ley, C. O., and Gali, F. G. 1983. Parkinsonian syndrome after methanol intoxication, *Eur. Neurol.* 22:405–409.

Lindstrom, K., Riihimaki, H., and Hanninen, K. 1984. Occupational solvent exposure and neuropsychiatric disorders, *Scand. J. Work Environ. Health* 10:321–323.

Mactutus, C. F., and Fechter, L. D. 1984. Prenatal exposure to carbon monoxide: learning and memory deficits, *Science* 223:409–411.

Mactutus, C. F., and Fechter, L. D. 1985. Moderate prenatal carbon monoxide exposure produces persistent, and apparently permanent, memory deficits in rats, *Teratology* 31:1–12.

McCafferty, W. B. 1981. *Air Pollution and Athletic Performance*, C. C. Thomas, Springfield Ill.

Mena, I., Marin, O., Fuenzalida, S., and Cotzias, G. C. 1967. Chronic manganese poisoning. Clinical picture and manganese turnover, *Neurology* 17:128–136.

Merigan, W. H., Barkdoll, E., Maurissen, J. P., Eskin, T. A., and Lapham, L. W. 1985a. Acrylamide effects on the macaque visual system. I. Psychophysics and electrophysiology, *Invest. Ophthalmol.* 26:309–316.

Merigan, W. H., Wood, R. W., and Zehl, D. N. 1985b. Recent observations on the neurobehavioral toxicity of carbon disulfide, *Neurotoxicology* 6:81–88.

Moulton, D. G., Turk, A., and Johnston, J. W. 1975. *Methods in Olfactory Research*, Academic Press, London.

Murphy, S. D. 1964. A review of effects on animals of

exposure to auto exhaust and some of its components, *J. Air Pollut. Control Assoc.* 14:303–308.

Murphy, S. D., Ulrich, C. E., Frankowitz, S. H., and Xintaras, C. 1964. Altered function in animals inhaling low concentrations of ozone and nitrogen dioxide, *Am. Ind. Hyg. Assoc. J.* 25:246–253.

National Academy of Sciences. 1973. *Manganese,* National Academy of Sciences, Washington, D.C.

National Academy of Sciences. 1976. *Vapor-Phase Organic Pollutants,* National Academy of Sciences, Washington, D.C.

National Academy of Sciences. 1977. *Ozone and Other Photochemical Oxidants,* National Academy of Sciences, Washington, D.C.

National Institutes of Health. 1977. *Human Health and the Environment—Some Research Needs,* Report of the Second Task Force for Research Planning in Environmental Health Science, DHEW Pub. No. NIH 77-1277, U.S. Government Printing Office, Washington, D.C.

Newland, M. C., Cox, C., Hamada, R., Oberdörster, G., and Weiss, R. 1987. The clearance of manganese chloride in the primate, *Fundam. Appl. Toxicol.* 9:314–328.

Noelpp, B., and Noelpp-Eschenhagen, I. 1951a. Die rolle bedingter beim asthma bronchiale. Ein experimenteller beitrag zur pathogenese des asthma bronchiale, *Helv. Med. Acta* 18:142–158.

Noelpp, B., and Noelpp-Eschenhagen, I. 1951b. Das experimentelle asthma bronchiale des meerschweinchens. II. Mitteilung. Die rolle bedingter reflexe in der pathogenese des asthma bronchiale, *Int. Arch. Allergy* 2:321–329.

Noelpp, B., and Noelpp-Eschenhagen, I. 1951c. Das experimentelle asthma bronchiale des meerschweinchens. III. Mitteilung. Studien zur bedeutung bedingter reflexe. Bahnungsbereitschaft und haftfahigkeit unter stress, *Int. Arch. Allergy* 3:108–136.

Ottenberg, P., Stein, M., Lewis, J., and Hamilton, C. 1958. Learned asthma in the guinea pig, *Psychosom. Med.* 20:395–400.

Pierce, W. D., Epling, W. F., and Boer, D. P. 1986. Deprivation and satiation: the interrelations between food and wheel running, *J. Exp. Anal. Behav.* 46:199–210.

Pryor, G. T., Howd, R. A., Malik, R., Jensen, R. A., and Rebert, C. S. 1978. Biomedical studies on the effects of abused inhalant mixtures. Annual Progress Report No. 2 of NIDA Contract No. 271-77-3402, pp. 62–67, 97–104.

Pryor, G. T., Dickinson, J., Feeney, E., and Rebert, C. S. 1984a. Hearing loss in rats first exposed to toluene as weanlings or as young adults, *Neurobehav. Toxicol. Teratol.* 6:111–119.

Pryor, G. T., Rebert, C. S., Dickinson, J., Feeney, E. 1984b. Factors affecting toluene-induced ototoxicity in rats, *Neurobehav. Toxicol. Teratol.* 6:223–238.

Raitta, C., Teir, H., Tolonen, M., Nurminen, M., Helpio, E., and Malmstrom, S. 1981. Impaired color discrimination among viscose rayon workers exposed to carbon disulfide, *J. Occup. Med.* 23:189–192.

Rebert, C. S., Sorenson, S. S., Howd, R. A., and Pryor, G. T. 1983. Toluene-induced hearing loss in rats evidenced by the brainstem auditory-evoked response, *Neurobehav. Toxicol. Teratol.* 5:59–62.

Rees, D. C., Coggeshall, E., and Balster, R. L. 1986. Inhaled toluene produces pentobarbital-like discriminative stimulus effects in mice, *Life Sci.* 37:1319–1325.

Rehnberg, G. L., Hein, J. F., Carter, S. D., Linko, R. S., and Laskey, J. W. 1981. Chronic ingestion of Mn_3O_4 by young rats: tissue accumulation, distribution, and depletion, *J. Toxicol. Environ. Health* 7:263–272.

Ricaurte, G., Bryan, G., Strauss, L., Seiden, L., and Schuster, C. 1985. Hallucinogenic amphetamine selectively destroys brain serotonin nerve terminals, *Science* 229:986–988.

Rodier, J. 1955. Manganese poisoning in Moroccan miners, *Br. J. Ind. Med.* 12:21–35.

Rodier, P. M., and Gramann, W. J. 1979. Morphologic effects of interference with cell proliferation in the early fetal period, *Neurobehav. Toxicol.* 1:129–135.

Rodier, P. M., and Reynolds, S. S. 1977. Morphological correlates of behavioral abnormalities in experimental congenital brain damage, *Exp. Neurol.* 57:81–93.

Roels, H., Sarhan, M. J., Hanotiau, I., de Fays, M., Genet, P., Bernard, A., Buchet, J. P., and Lauwerys, R. 1985. Preclinical toxic effects of manganese in workers from a Mn salts and oxides producing plant, *Sci. Total Environ.* 42:201–206.

Rosner, D., and Markowitz, G. 1985. A "gift of God"?: the public health controversy over leaded gasoline during the 1920s, *Am. J. Public Health* 75:344–352.

Russell, M., Dark, K. A., Cummins, R. W., Ellman, G., Callaway, E., and Peeke, H. V. S. 1984. Learned histamine release, *Science* 225:733–734.

Schikler, K. N., Seitz, K., Rice, J. F., and Strader, T. 1982. Solvent abuse associated cortical atrophy, *J. Adolesc. Health Care* 3:37–39.

Schuck, E. A., Stephens, E. R., and Middleton, J. T. 1966. Eye irritation response at low concentrations of irritants, *Arch. Environ. Health* 13:570–575.

Sparber, S. B., and Tilson, H. A. 1972. Schedule controlled and drug induced release of norepinephrine-7-^{3}H into the lateral ventricle of rats, *Neuropharmacology* 11:453–464.

Spencer, P. S., Couri, D., and Schaumburg, H. H. 1980. *n*-Hexane and methyl *n*-butyl ketone, In: *Experimental and Clinical Neurotoxicology* (P. S. Spencer and H. H. Schaumburg, eds.), pp. 456–475, Williams & Wilkins, Baltimore.

Springer, K. L. 1974. Combustion odors—a case study, In: *Human Responses to Environmental Odors* (A. Turk, J. W. Johnston, and D. G. Moulton, eds.), pp. 227–262, Academic Press, New York.

Stinson, S. F., and Loosli, C. G. 1979. The effect of synthetic smog on voluntary activity of CD-1 mice, In: *Animals as Monitors of Environmental Pollutants,* pp. 233–239, National Academy of Sciences, Washington, D.C.

Storm, J. E., and Fechter, L. D. 1985a. Alteration in the postnatal ontogeny of cerebellar norepinephrine content following chronic prenatal carbon monoxide, *J. Neurochem.* 45:965–969.

Storm, J. E., and Fechter, L. D. 1985b. Prenatal carbon monoxide exposure differentially affects postnatal weight and monoamine concentration of rat brain regions, *Toxicol. Appl. Pharmacol.* 81:139–146.

Stupfel, M., Magnier, M., Romary, F., Tran, M-H., Moutet, J-P. 1973. Lifelong exposure of SPF rats to automotive exhaust gas, *Arch. Environ. Health* 26:264–269.

Swenberg, J. A., Kerns, W. D., Mitchell, R. E., Gralla, E. J., and Pavkov, K. L. 1980. Induction of squamous cell carcinomas of the rat nasal cavity by inhalation exposure to formaldehyde vapor, *Cancer Res.* 30:3398–3402.

Taylor, J. D., and Evans, H. L. 1985. Effects of toluene inhalation on behavior and expired carbon dioxide in macaque monkeys, *Toxicol. Appl. Pharmacol.* 80:487–495.

Tepper, J. S., and Weiss, B. 1986. Determinants of behavioral response with ozone exposure, *J. Appl. Physiol.* 60:868–875.

Tepper, J. S., and Wood, R. W. 1985. Behavioral evaluation of the irritating properties of ozone, *Toxicol. Appl. Pharmacol.* 78:404–411.

Tepper, J. S., Weiss, B., and Cox, C. 1982. Micro-analysis of ozone depression of motor activity, *Toxicol. Appl. Pharmacol.* 64:317–326.

Tepper, J. S., Weiss, B., and Wood, R. W. 1985. Alterations in behavior produced by inhaled ozone or ammonia, *Fundam. Appl. Toxicol.* 5:1110–1118.

Thompson, D. M., and Moerschbaecher, J. M. 1978. Operant methodology in the study of learning, *Environ. Health Perspect.* 26:77–87.

Turk, A., Johnston, J. W., and Moulton, D. G. 1974. *Human Responses to Environmental Odors*, Academic Press, New York.

U.S. Environmental Protection Agency. 1985a. Functional observational battery, 40 CFR 798.6050 in Toxic Substances Control Act Test Guidelines; Final Rules, *Fed. Reg.* 50:39458–39460.

U.S. Environmental Protection Agency. 1985b. Motor activity, 40 CFR 798.6200 in Toxic Substances Control Act Test Guidelines; Final Rules, *Fed. Reg.* 50:39460–39461.

U.S. Environmental Protection Agency. 1985c. Neuropathology, 40 CFR 798.6400 in Toxic Substances Control Act Test Guidelines; Final Rules, *Fed. Reg.* 50:39461–39463.

U.S. Environmental Protection Agency. 1985d. Neurotoxicity assay, 40 CFR 798.6450 in Toxic Substances Control Act Test Guidelines; Final Rules, *Fed. Reg.* 50:39463–39465.

U.S. Environmental Protection Agency. 1985e. Peripheral nerve function, 40 CFR 798.6850 in Toxic Substances Control Act Test Guidelines; Final Rules, *Fed. Reg.* 50:39468–39470.

U.S. Environmental Protection Agency. 1985f. Schedule-controlled operant behavior, 40 CFR 798.6500 in Toxic Substances Control Act Test Guidelines; Final Rules, *Fed. Reg.* 50:39465–39466.

Veronesi, B., Peterson, E. R., and Spencer, P. S. 1980. Reproduction and analysis of methyl *n*-butyl ketone neuropathy in organotypic tissue culture, In: *Experimental and Clinical Neurotoxicology* (P. S. Spencer and H. H. Schaumburg, eds.), pp. 863–871, Williams and Wilkins, Baltimore.

Weber-Tschopp, A., Fischer, T., and Grandjean, E. 1977. Irritating effects of formaldehyde on men, *Int. Arch. Occup. Environ. Health* 39:207–218.

Weiss, B., Ferin, J., Merigan, W. H., Stern, S., and Cox, C. 1981. Modification of rat operant behavior by ozone exposure, *Toxicol. Appl. Pharmacol.* 58:244–251.

Wood, R. W. 1979. Behavioral evaluation of sensory irritation evoked by ammonia, *Toxicol. Appl. Pharmacol.* 50:157–162.

Wood, R. W. 1981a. Determinants of irritant termination behavior, *Toxicol. Appl. Pharmacol.* 61:260–268.

Wood, R. W. 1981b. Neurobehavioral toxicity of carbon disulfide, *Neurobehav. Toxicol. Teratol.* 3:397–405.

Wood, R. W. 1982. Stimulus properties of inhaled substances: an update, In: *Nervous System Toxicology* (C. L. Mitchell, ed.), pp. 199–212, Raven Press, New York.

Wood, R. W., and Coleman, J. B. 1984. Behavioral evaluation of the irritant properties of formaldehyde, *Toxicologist* 4:119.

Wood, R. W., and Colotla, V. A. 1986. Increased locomotor activity of mice during low-level exposure to toluene, *Toxicologist* 6:220.

Wood, R. W., and Cox, C. C. 1986. A repeated measures approach to the detection of the minimal acute effects of toluene, *Toxicologist* 6:221.

Wood, R. W., Coleman, J. B., Schuler, R., and Cox, C. 1984. Anticonvulsant and antipunishment effects of toluene, *J. Pharmacol. Exp. Ther.* 230:407–412.

Young, J. S., and Fechter, L. D. 1983. Reflex inhibition procedures for animal audiometry: a technique for assessing ototoxicity, *J. Acoust. Soc. Am.* 73:1686–1693.

Index